INFORMATION TECHNOLOGY AND COMPUTER APPLICATION ENGINEERING

PROCEEDINGS OF THE 2013 INTERNATIONAL CONFERENCE ON INFORMATION
TECHNOLOGY AND COMPUTER APPLICATION ENGINEERING (ITCAE 2013),
HONG KONG, P.R. CHINA, AUGUST 27–28, 2013

Information Technology and Computer Application Engineering

Editors

Hsiang-Chuan Liu
Department of Biomedical Informatics, Asia University, Taichung, Taiwan

Wen-Pei Sung
National Chin-Yi University of Technology, Taiping City, Taiwan

Wenli-Yao
*Control Engineering and Information Science Research Association (CEIS),
Chongqing University, Chongqing, P.R. China*

CRC Press
Taylor & Francis Group
Boca Raton   London   New York   Leiden

CRC Press is an imprint of the
Taylor & Francis Group, an **informa** business

A BALKEMA BOOK

Selected, peer-reviewed papers of the 2013 International Conference on
Information Technology and Computer Application Engineering

CRC Press/Balkema is an imprint of the Taylor & Francis Group, an informa business

© 2014 Taylor & Francis Group, London, UK

Typeset by MPS Limited, Chennai, India
Printed and bound in Great Britain by CPI Group (UK) Ltd, Croydon, CR0 4YY.

Published by: CRC Press/Balkema
              P.O. Box 11320, 2301 EH, Leiden, The Netherlands
              e-mail: Pub.NL@taylorandfrancis.com
              www.crcpress.com – www.taylorandfrancis.com

ISBN: 978-1-138-00079-7 (Hardback)
ISBN: 978-1-315-81328-8 (eBook PDF)

Table of contents

Committees

CONFERENCE CHAIRMAN

Prof. Hsiang-Chuan Liu, *University, Taiwan*
Prof. Wen-Pei Sung, *National Chin-Yi University of Technology, Taiwan*

PROGRAM COMMITTEE

Ghamgeen Izat Rashed, *Wuhan University, China*
Andrey Nikolaevich Belousov, *Laboratory of Applied Nanothechnology, Ukraine*
Krupa Ranjeet Rasane, *KLE Society's College of Engineering, India*
Sajjad Jafari, *Semnan University, Iran*
Ahmed N. Abdalla, *Universiti Malaysia Pahang, Malaysia*
BUT ADRIAN, *ELECTROMOTOR company, Timisoara, Bulevardul, Romania*
Yan Wang, *The University of Nottingham, U.K.*
Prof. Yu-Kuang Zhao, *National Chin-Yi University of Technology, Taiwan*
Yi-Ying Chang, *National Chin-Yi University of Technology, Taiwan*
Darius Bacinskas, *Vilnius Gediminas Technical University, Lithuania*
Viranjay M. Srivastava, *Jaypee University of Information Technology, Solan, H.P., India*
Chenggui Zhao, *Yunnan Normal University, China*
Hsiang-Chuan Liu, *Asia University, Taiwan*
Hao-En Chueh, *Yuanpei University, China*
Zhou Liang, *Donghua University, China*
Liu Yunan, *University of Michigan, USA*
Wang Liying, *Institute of Water Conservancy and Hydroelectric Power, China*
Chenggui Zhao, *Yunnan University of Finance and Economics, China*
Rahim Jamian, *Universiti Kuala Lumpur Malaysian Spanish Institute, Malaysia*
Lixin Guo, *Northeastern University, China*
Wen-Sheng Ou, *National Chin-Yi University of Technology, Taiwan*
Mostafa Shokshok, *National University of Malaysia, Malaysia*
Ramezan ali Mahdavinejad, *University of Tehran, Iran*
Wei Fu, *Chongqing University, China*
Anita Kovač Kralj, *University of Maribor, Slovenia*
Tjamme Wiegers, *Delft University of Technology, Netherlands*
Gang Shi, *Inha University, South Korea*
Bhagavathi Tarigoppula, *Bradley University, USA*

CO-SPONSOR

International Frontiers of Science and Technology Research Association
Hong Kong Control Engineering and Information Science Research Association

Information Technology and Computer Application Engineering – Liu, Sung & Yao (Eds)
© 2014 Taylor & Francis Group, London, ISBN 978-1-138-00079-7

Preface

The 2013 International Conference on Information Technology and Computer Application Engineering (ITCAE 2013) will be held in Hong Kong, during August 27–28, 2013. The aim is to provide a platform for researchers, engineers, academics as well as industrial professionals from all over the world to present their research results and development activities in Computer Application Engineering and Information Science.

For this conference, we received more than 400 submissions via email and the electronic submission system, which were reviewed by international experts, and some 189 papers have been selected for presentation, representing 9 national and international organizations. I believe that ITCAE 2013 will be the most comprehensive conference focused on Computer Application Engineering and Information Science. The conference will promote the development of Computer Application Engineering and Information Science, strengthening international academic cooperation and communications, and the exchange of research ideas.

We would like to thank the conference chairs, organization staff, the authors and the members of the International Technological Committees for their hard work. Thanks are also given to Alistair Bright.

We hope that ITCAE 2013 will be successful and enjoyable for all participants. We look forward to seeing all of you next year at ITCAE 2014.

June, 2013

Wen-Pei Sung
National Chin-Yi University of Technology

Information Technology and Computer Application Engineering – Liu, Sung & Yao (Eds)
© 2014 Taylor & Francis Group, London, ISBN 978-1-138-00079-7

A new hybrid architecture framework for system of systems engineering in the net centric environment

P.L. Rui & R. Wang

The 28th Research Institute of China Electronics Technology Group Corporation, Nanjing, China

ABSTRACT: As the emergence of the Net Centric Warfare (NCW), the military information system has been evolved from platform-centric to be net-centric, which brings great challenges for System of Systems (SoS) engineering in the net centric environment. A major task of system engineering is to build system architecture. Although classical system engineering deals very well with architecting problems for a single system, it has no good solutions for SoS architecting problems. In this paper, existing architecture frameworks is evaluated, and a novel architecture framework model for SoS engineering is presented, which combines both advantages of enterprise architecture and system architecture, and enables SoS architecting with the kind of capability based development process.

1 INTRODUCTION

As the military information system moves through the brave new world of Net Centric Warfare (NCW) [1] or Net Enabled Operations (NCO) and the evolution of the U.S. Department of Defense (DoD) Global Information Grid to help implement that vision, the importance of engineering system of system (SoS) in the net centric environment becomes more urgent [2]. The field of system engineering has emerged to address the challenges inherent in these systems, or systems-of-systems. This has necessitated an evolution of the architecting approach, intensified focus on system properties (such as changeability, flexibility, agility, etc.), and recognition of the inseparability of technological system and the enterprise developing and operating such systems.

Architecture frameworks are methods used in system engineering. They provide a structured and systematic approach to designing systems. To date, there are many existing architecture frameworks [3–8] which can be divided into two categories as Enterprise Architecture based Frameworks (EAF) and System Architecture based Frameworks (SAF). These classical architecture frameworks work well with the straightforward requirement and the defined specification for single system design in the stove-piped environment. However, they have no good solutions for SoS design in the net centric environment when optimality and efficiency is not as important as run-time interoperability with services that were not envisioned at design time, and flexibility, compose-ability, and extensibility are now much more important.

The aim of this paper is therefore to develop a new architecture framework to resolve weaknesses in previous frameworks in order to support SoS architecting problems. For this purpose, an overview and evaluation of existing architecture frameworks is given in section 2. Building from here, a novel hybrid architecture framework is presented and analyzed in more details in section 3. The paper concludes with a summary of the proposed method and an outlook of further research in section 4.

2 ARCHITECTURE FRAMEWORKS OVERVIEW

The term "architecture" refers to any kind of socio-technical system, and stands for the fundamental organization of its components and their relationships to each other and the environment as well as the design rules for developing and structuring the system [9]. In order to support architecture descriptions, many architecture frameworks have been developed, which provides directions for developing various architectures and organizing detailed architecture models and architectures that manage tasks inside an enterprise as well as communication to develop the complicated structures of an enterprise [10].

To date, there exists many architecture frameworks, which can be divided into EAF (e.g. Zachman framework [3], FEAF [4], TOGAF [5], etc) and SAF (e.g. C^4ISRAF [6], DoDAF [7], MoDAF [8], etc). The EAF selects a higher level of an enterprise as one scope and uses it as a framework to develop architecture, while the SAF is based on the specific detailed structure of the enterprise, and it selects a sub-enterprise for one scope and applies it to the framework for systematic architecture development.

2.1 Zachman Architecture Framework (ZAF)

The ZAF was proposed by John A. Zachman in 1987. It is described in a matrix with (30 cells) which provides on the vertical axis five perspectives (i.e. planner, owner, designer, builder, and sub-contractor) and on the horizontal axis six classifications of the various stakeholders (i.e. Planner, Owner, Designer, Builder and Subcontractor). The ZAF provides clarity to a complicated enterprise, making it possible to identify models for some projects, and is an important factor in alignment. The ZAF is the de-facto framework to provide a model that describes an enterprise well, but this framework is too idealistic. Furthermore, it is difficult to apply because there is no definition of specific products or templates. An additional disadvantage is that there is no process for application of the architecture, so it is difficult to develop architectures.

2.2 Federal Enterprise Architecture Framework (FEAF)

The FEAF introduced in 1998 by the Chief Information Office consortium provides an enduring standard for developing and documenting architecture descriptions of high-priority areas. It divides a given architecture into business, data, applications and technology architecture descriptions, which are the four levels the FEAF consists of. In Version 1.0 the FEAF includes the first three columns of the Zachman Framework, so that the FEAF is graphically represented as a 3×5 matrix with architecture types (data, application, and technology) on one axis of the matrix and perspectives (planner, owner, designer, builder and subcontractor) on the other. The FEAF defines and clearly explains architecture descriptions for each level to allow better understanding of enterprise architecture concepts. However, even though the framework deals with high-level concepts, it has no template or product for development.

2.3 The Open Group Architecture Framework (TOGAF)

The TOGAF is an industry standard architecture framework that may be used freely by any organization wishing to develop enterprise architecture descriptions for the use within that organization. It is a detailed framework using a set of supporting tools [11]. It enables designing, evaluating, and building the right architecture for any organization. The key to TOGAF is the TOGAF Architecture Development Method (ADM) – a reliable, proven approach for developing enterprise architecture descriptions that meets the needs of the specific business. Even though TOGAF ADM describes the different inputs and outputs for each phase of the architecture development cycle, there are no specification documents that describe the output.

2.4 C^4ISR Architecture Framework (C^4ISRAF)

The Command, Control, Communication, Computer, Intelligence, Surveillance, and Reconnaissance Architecture Framework (C^4ISRAF) was developed by the Architecture Working Group (AWG) of the United States Department of Defense in 1997. It provides 27 concrete templates to facilitate target information system development by using operational view (OV), system view (SV) and technical view (TV). Besides it contains four main types of guidance for architecture development: (1) guidelines, (2) a high level process for using the framework, (3) a discussion of architecture data and tools, and (4) a detailed description of the products. However, it does not provide conceptual perspectives and views as in the ZAF, and there are no specific descriptions about who is responsible or needed in each step of the procedure model to develop architecture descriptions.

2.5 Department of Defense Architecture Framework (DoDAF)

The DoDAF is developed specifically for the US DoD to support its war-fighting operations, business operations and processes. It grew from and replaced the previous architecture framework, C4ISRAF. The DoDAF includes guidelines on determining architecture content based on intended use; focus on using architectures in support of DoD's Programming, Budgeting, and Execution process; Joint Capabilities Integration and Development System; and the Defense Acquisition System; and increasing emphasis on the architecture data elements. Architecture development techniques have been provided in DoDAF to specify processes for scope definition, data requirements definition, data collection, architecture objectives analysis and documentation. However, a role model for the development process is also missing in the DoDAF.

2.6 Ministry of Defense Architecture Framework (MoDAF)

The MoDAF was evolved from U.S. DoDAF with the purpose of facilitating architecture information exchange with U.S. forces. Therefore, the MODAF is consistent with DoDAF in most views, such as OV, SV and TV, and augments it with two new views, i.e. strategy view (StV) and acquisition view (AcV) for analyzing and optimizing ministry capabilities and providing support to associated acquisition plans. Although the MODAF divides architecture users into three kinds and provides guides of architecture development for each kind of users, it also does not provide conceptual perspectives as in the ZAF, and there are no specific descriptions of user role in the architecture development process.

A comprehensive comparison of existing architecture frameworks is shown in Table 1, where "Product/ Template" denotes specification document of the

Table 1. Comparison of Current Architecture Frameworks.

	EAF			SAF		
	ZAF	FEAF	TOGAF	C⁴ISAF	DoDAF	MoDAF
Products/Template	⊙	⊙	○	●	●	●
Architecture role	●	●	○	○	○	⊙
Meta model	○	○	○	⊙	●	⊙
Supporting technique	○	○	⊙	●	●	●
Development process	○	⊙	●	●	●	●

*Legend: ● Fully accomplished; ⊙ Partly accomplished; ○ Not accomplished

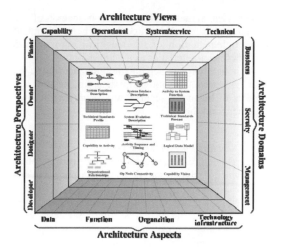

Figure 1. The Overall HAF Model.

architecture, "Architecture role" denotes participating roles for the development and management of the architecture descriptions, "Meta model" denotes how the architecture data normally collected, organized, and maintained, "Supporting technique" denotes the modeling technique for architecting, and "Development process" denotes how the architecture (product or template) is constructed [12].

It can be observed from Table 1 that EAFs usually have strengths in describing architecture roles due to its conceptual perspectives and views. But they have weaknesses in providing specific products and development process, so they are very difficult to apply in reality. In contrast, SAFs have considered no or partly architecture roles, but generally have specification document, supporting technique (e.g. UML), and elaborate development process. Furthermore, it should be mentioned that roles and the procedure are related to each other. If there is no procedure model provided by a method, the definition of roles for the development process would not make any sense.

3 HYBRID ARCHITECTURE FRAMEWORK (HAF)

3.1 The overall architecture

In the following, a new hybrid architecture framework (HAF) resolving the weaknesses mentioned above, is proposed by combing both advantages of EAF and SAF and by introducing a new set of architecture products and its associated development process. The overall HAF is shown in Figure 1 and a description of the framework is given in the subsequent paragraphs.

3.1.1 Architecture views

The architecture is split up into four views: the Capability View (CV), the Operational View (OV), the System/Service View (SV) and the Technical View (TV).

The Capability View (CV) captures the enterprise goals associated with the overall vision for executing a specified course of action, or the ability to achieve a desired effect under specific standards and conditions through combinations of means and ways to perform a set of tasks. It provides a strategic context for the capabilities described in an architectural description, and an accompanying high-level scope, more general than the scenario-based scope defined in an operational concept diagram. The models are high-level and describe capabilities using terminology, which is easily understood by decision makers and used for communicating a strategic vision regarding capability evolution.

The Operational View (OV) helps to give an understanding of the operational environment (the operational scenarios, processes and organization) for which systems will developed to support the operational (command and control) processes. Understanding of the operational processes is a prerequisite for the design and development of flexible solutions in the sense of information and communication systems. The OV describes the operational processes, their relationships, process threads that will be triggered by operational events and the description of the process by operational services.

The System/Service View (SV) captures system, service, and interconnection functionality providing for, or supporting, operational activities. It describes which applications and communication systems will be present, how they will interact and where the operational services will be implemented. Identified applications can be existing legacy applications, can be part of a newly installed package or can be newly built within or outside a program. It also describes the architecture of the individual systems by means of components that deliver services to support operational services for specific operational processes. Over time, the emphasis on service oriented environment and cloud computing may transform system view into service view.

The Technical View (TV) defines the infrastructure (middleware, hardware, network, transmissions media, protocols etc.) required to run systems. The other views mainly trigger the development and change, not only by the functionality but also by the characteristics of those views. Characteristics include performance requirements, volume figures,

3

frequencies, actuality of information, method of use of functionality and resources, etc. The development and implementation of the technical infrastructure take these characteristics as a major input.

Although they are separate architecture views, the four have strong relationships and for the different aspects of business, security and management, they together form the HAF.

3.1.2 Architecture perspectives

The architecture consists of four perspectives: planner, owner, designer, and developer. A perspective is simply a point of view of the EA, and is mapped to a particular set of work products. Perspectives have a specific role in representing the enterprise or examining an organizational entity in the enterprise.

The Planner's Perspective identifies a skeleton of the organization and its function and category, and defines the function, size, and relativity to other systems so that the information system can be finally implemented. The planner is usually the information system project manager.

The Owner's Perspective creates a blueprint for an end-state information system and defines organizational function, the entities included in the process, and the relationship among those. The owner brings forward requirements for the information system.

The Designer's Perspective is a detailed specification for information system at a high level, based on an organization's function model.

The developer's Perspective is redefined at a high level, during which process the developer is constrained by developing tools, IT, and resources. Especially, the technology model specifies the concrete architecture from overall to atomic system scope and a specific part of sub-domains, for example, a programming language, I/O device, etc.

3.1.3 Architecture aspects

The architecture is composed of four aspects: data, function, organization, and technology infrastructure, an aspect means a specific view for observing a related special feature. As a general concept of information technology, applications consist of data and functions. In this case, the sub-hierarchy of an application is the shared data and common functions in the overall enterprise architecture.

The Data Aspect describes the set of data needed to perform enterprise data flow and the relationships in the EA database.

The Function Aspect describes enterprise functions, processes, and activities that act on enterprise information to support enterprise operations.

The Organization Aspect consists of the organizational structure of the enterprise, the major operations performed by organizations, the types of workers, the organization breakdown structure, and the distribution of the organizations to locations.

The Technology Infrastructure Aspect consists of the hardware, software, network, telecommunications, and general services that constitute the operational environment in which business applications operate.

3.1.4 Architecture domains

The architecture covers three main domains: Business, Security and Management.

The Business Architecture is the most important one that describes the core functionality of a business. This functionality deals with the vision, mission and goals of the organization. The Business Architecture is therefore the primary architecture and the others are supporting architectures for other aspects.

The Security Architecture describes the security that must be taken into account for the formulated business functionality. The architecture of the other domains follows the same structure and also covers the same four views, i.e. CV, OV, SV and TV. For example, the Security at the SV level describes the security with respect to the Systems (e.g. information systems and communication systems) in the Business Architecture.

The Management Architecture describes the management domain that is needed for the control and changes of the implemented business functionality, as well as the implemented security. It also encompasses the management of the system operations, the control, administration and management of the objects which will be taken into operation and which are liable to change. This domain also covers the administration and maintenance of the results of the business process modeling activities.

3.2 Architecture products

3.2.1 Product list

The architecture has a total of 33 products [7], which are divided into 5 categories according to architecture views, as shown in Table 2. The first column indicated the view applicable to each product. The second column provides an alphanumeric identifier and the formal name of the product. The fourth column captures the general nature of the product's content.

As shown in Table 2, most of architecture products are obtained from DoDAF. The framework also defines 2 products in the All View (AV) to describe the overview and summary information and the definition of architecture data. Additionally, in order to describe high level concepts of system/service from the Technology infrastructure aspects on the perspective of a planer, a Technical Reference Model (TV-1) used to define the interface within or without systems/services, is introduced. Furthermore, it should be noted that the sequence of products in the table does not imply a sequence for developing the products.

3.2.2 Product Mapping

A mapping of architecture products listed in the above paragraph on the perspectives and aspects of the framework is given in Figure 2. It can be seen that in the framework, Rows 1 and 2 (on the perspective of Planer and Owner) contain the products for the operation,

Table 2. The HAF Products List.

Views	Products	Descriptions
All View	AV-1:Overview and Summary Information	Describes a Project's Visions, Goals, Plans, Activities, Events, Conditions, Effects, and produced objects.
	AV-2:Integrated Dictionary	An architectural data repository with definitions of all terms used.
Capability View	CV-1:Vision	The overall vision for endeavors, which provides a strategic context for the capabilities described.
	CV-2:Capability Taxonomy	A hierarchy of capabilities which specifies all the capabilities throughout Architectural Descriptions.
	CV-3:Capability Phasing	The planned achievement of capability at different points in time or during specific periods of time.
	CV-4:Capability Dependencies	The dependencies between planned capabilities and the definition of logical groupings of capabilities.
	CV-5a:Capability to Systems/Services Mapping	A mapping between the capabilities and the systems/services that these capabilities required.
	CV-5b:Capability to Operational Activities Mapping	A mapping between the capabilities required and the operational activities that those capabilities support.
Operational View	OV-1:High Level Operational Concept Graphic	The high-level graphical/textual description of the operational concept.
	OV-2:Operational Node Connectivity Description	A description of the operational nodes with needlines between those nodes that indicate exchange information.
	OV-3:Operational Information Exchange Matrix	A description of the information flows exchanged between operational activities.
	OV-4:Organizational Relationships Chart	The organizational context, role or other relationships among organizations.
	OV-5:Operational Activity Model	The context of capabilities and activities and their relationships among activities, inputs, and outputs
	OV-6a:Operational Rules Model	It identifies business rules that constrain operations.
	OV-6b:State Transition Description	It identifies business process responses to events.
	OV-6c:Event-Trace Description	It traces actions in a scenario or sequence of events.
	OV-7:Logical Data Model	The documentation of the data requirements and structural business process (activity) rules.
System/ Service View	SV-1:Systems/Services Interface Description	Identification of systems nodes, systems, system items, services, and service items and their interconnections
	SV-2:Systems/Services Communications Description	Systems nodes, systems, and services and their related communications laydowns.
	SV-3:Systems/Services -Systems/services Matrix	Relationships among systems and services; can be designed to show relationships of interest.
	SV-4:Systems/Services Functionality Description	Functions performed by systems/services and the /services data flows among system functions
	SV-5:Operational Activity to Systems/Services Traceability Matrix	Mapping of system, system functions, or service back to operational activities.
	SV-6:Systems/Services Data Exchange Matrix	Provides details of system or service data elements being exchanged and the attributes of that exchange.
	SV-7:Systems/Services Performance Parameters Matrix	Performance characteristics of Systems and Services View elements for the appropriate time frame(s).
	SV-8:Systems/Services Evolution Description	Planned incremental steps toward migrating a suite of systems or services to a more efficient suite.
	SV-9:Systems/Services Technology Forecast	Emerging technologies and software/hardware products that are expected to be available.
	SV-10a:Systems/Services Rules Model	Identification of constraints that are imposed on systems/services functionality
	SV-10b:Systems/Services State Transition Description	Identification of responses of a system/service to events
	SV-10c:Systems/Services Event-Trace Description	Identification of system/service-specific refinements of critical sequences of events in the Operational View.
	SV-11:Physical Data Model	The physical implementation format of the Logical Data Model entities.
Technique View	TV-1:Technical Reference Model	The model that defines the interface with systems/services in the System/Service View.
	TV-2:Technical Standards Profile	Listing of standards that apply to Systems and Services View elements in a given architecture.
	TV-3:Technical Standards Forecast	Description of emerging standards and potential impact on current Systems and Services View elements.

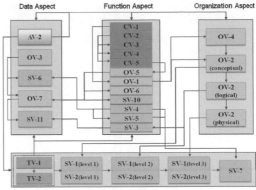

Figure 3. Relationship among Products on the Aspects of HAF.

Aspects Perspectives	Data	Function		Organization	Technology infrastructure	
		AV-1				
Planer	AV-2	CV-1	CV-2	OV-4	TV-1	TV-3
		OV-1	CV-3		TV-2	CV-3
Owner	OV-3	OV-5	CV-4	OV-2 (conceptual)	SV-1(level 1)	
			CV-5b		SV-2(level 1)	
					SV-8	
Designer	OV-7	SV-3	OV-6a	OV-2 (logical)	SV-1(level 2)	
		SV-4			SV-2(level 1,2)	
		SV-5	CV-5a		SV-9	
	SV-6					
Developer	SV-11	SV-10a	OV-6b	OV-2 (physical)	SV-1(level 3)	
		SV-10b	OV-6c		SV-2(level 3)	
		SV-10c			SV-7	

Figure 2. Mapping of architecture products on the perspectives and aspects of the framework.

and Rows 3 and 4 (on the perspective of Designer and Development) contain the products for the system.

Moreover, as the architecture development goes from plan to develop or implementation, the phases or level of associated products will be refined. For example, OV-1 on the perspective of Owner gives conceptual relationships among operational nodes, while on the perspective of Designer, it should describe information exchange (i.e. needlines) of nodes logically in more details.

3.2.3 Relationship among products

All products in HAF have a mutual relationship among themselves from the enterprise point of view. Figure 3 shows the relationship among products according to

each aspect. In each aspect, a sub-component function is inherited using a top-down methodology in stages.

In the aspect of Data, integrated dictionary (AV-1) defines all products and affects the Data, Function, and Technology infrastructure aspects. It must therefore be defined and updated until the product is fully completed. In the aspect of Function, the activity model (OV-5) is related with the operation rule model (OV-6a), operational state diagram (OV-6b), event trace diagram (OV-6c), system/service rule model (SV-10a), system/service state diagram (SV-10b), and system/service event trace (SV-10c), which describe sequence and timing. Moreover, the high level operational concept (OV-1) connects the operation node (OV-2) in the aspect of Organization, which in turn connects with the system node/interface (SV-1) in the aspect of Technology infrastructure at the corresponding level.

3.3 Architecture development process

With respect to a software development lifecycle [13], we propose a 5-step development process for the HAF as shown in Figure 4. The first step is to get organized, which consists of scoping the project, setting up the development team, and defining a target vision. The arrows represent initial relationships, and for implementing the target architecture at least one or two iteration of steps 2 through 5 should be performed. However, this is only the iteration at a high level. Iteration also occurs within steps. Steps 2 through 5 each have their own loops. Within step 3, for example, you may go back and forth between two aspects or loop through all the aspects more than once.

Furthermore, a capability based analysis process for architecture development is also proposed for the architecture development with steps, especially for step 2 and 3, as shown in Figure 5. The main idea is in that architecture development starts from or is based on capability vision, which is used to determine operational concepts and associated activities or tasks. In contrast to activity based method (ABM) that subjects to support specified tasks or requirements, the

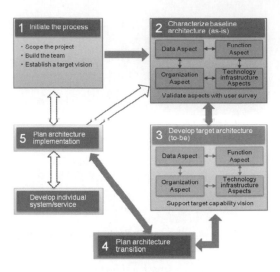

Figure 4.   Development Process for HAF.

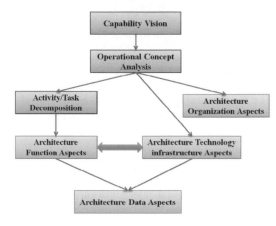

Figure 5.   Capability based Analysis Process for Architecture Development.

capability based method (CBM) [14] is very suitable for building system of systems with various tasks or requirements, because its focus on capability design and implementation.

4   CONCLUSIONS

The implications behind Net Centric Warfare (NCW) or Net Enabled Operations (NCO) bring great challenges for architecting system of system (SoS) in the net centric environment. This has necessitated an evolution of the architecting approach considering SoS properties (such as changeability, flexibility, agility,

etc.). In this paper, a new architecture framework for SoS engineering is proposed, which combines advantages existing frameworks by defining various views, perspectives, aspects and domains of the architecture. Furthermore, a capability based method (CBM) for SoS architecture development is introduced in order to support SoS engineering in the net centric environment. Further researches will be done to validate the effectives of the proposed framework and its associated development process.

REFERENCES

[1] D.S. Alberts, Information Age Transformation: Getting to a 21st Century Military, *Washington, DC, CCRP Publications*. pp. 7–8. 2002

[2] A. Meilich. System of systems (SoS) engineering & architecture challenges in a net centric environment. *IEEE/SMC International Conference on System of Systems Engineering*, April 2006 pp. 5–9

[3] Zachman, John A. A Framework for Information System Architecture. *IBM System Journal,* Vol. 26 No. 3, pp. 276–292, September 1987

[4] CIO Council. Federal Enterprise Architecture Framework. 1999

[5] The Open Group. The Open Group Architecture Framework

[6] Department of Defense. Architecture Framework Working Group, C4ISR Architecture Framework Version 2.0, 18 December 1997

[7] Department of Defense. Architecture Framework Working Group, DoD Architecture Framework Version 2.0 Volume I, II, III. 2009

[8] Ministry of Defense. MoD Architecture Framework Version 1.0, August 2008

[9] IEEE. IEEE Recommended Practice for Architectural Description of Software-Intensive Systems, 2000

[10] Alexander H. Levis, Architecting Information System (lecture notes), George Mason University, 2000

[11] The Open Group, Welcome to TOGAF – The Open Group Architectural Framework, 2002

[12] Antony Tang, Jun Han and Pin Chen, A Comparative Analysis of Architecture Frameworks, *Technical Report: SUTIT-TR2004.01*

[13] Steven H Spewak Enterprise Architecture Planning-Developing a Blueprint for Data, 2001

[14] P.L. Rui, R. Wang, and H. Yu. A Capability-Based Method for System of Systems Architecting in the Net-Centric Environment. *International Journal of Computer and Communication Engineering*, vol. 1, no. 4, 2012

Information Technology and Computer Application Engineering – Liu, Sung & Yao (Eds)
© *2014 Taylor & Francis Group, London, ISBN 978-1-138-00079-7*

Applications of semi-supervised subspace possibilistic fuzzy c-means clustering algorithm in IoT

Y.F. Zhang & Wei Zhang
School of IOT Engineering, Jiangnan University, Wuxi, China

ABSTRACT: For massive and high-dimensional characteristics of IoT data, a novel semi-supervised subspace possibilistic fuzzy c-means clustering algorithm is proposed in this article, sSPCM for short. The algorithm improved clustering accuracy by using a small amount of known data in massive data effectively in semi-supervised fashion. On the other hand, taking into account the characteristics of the high-dimensional data of IoT, we use subspace clustering techniques to excavate the useful information in each space, so as to further improve the clustering performance. The experimental results on simulated data sets and UCI standard datasets show that the algorithm has a better clustering performance compared with traditional clustering algorithm for complex data.

Keywords: IoT; semi-supervised; subspace; PCM

1 INTRODUCTION

Internet of things (IoT) is the Internet connected all objects [1–2]. The birth and development of the IOT will bring the explosive growth of the data. In a meanwhile, the diversification of the sensor makes dimension of data collected by IoT technology generally higher, then how to mine the potential value of the vast amounts of high-dimensional data by analysis is a problem that needs to be solved urgently. Therefore, it is a cornerstone of the stable development of IoT to develop efficient and practical data mining algorithms.

Based on the above reasons, cluster analysis [3] of data mining algorithms is chosen to apply to the IoT data processing. In this article, we choose PCM algorithm [4] as a basis algorithm for its better robustness and the simple mathematical expression [5–7]. On the other hand, for the complex and high-dimensional features of IoT data, the idea of subspace clustering is introduced on the basis of PCM algorithm. Subspace PCM algorithm can detect subspace of high-dimensional data, it has a better adaptability to high-dimensional complex data. Taking into account that a small amount of known information in the real world is easy to obtain, and the small amount of known information has a good guide on clustering algorithm, so the known information can be used effectively in the clustering process. Based on the above analysis, a novel semi-supervised subspace possibilistic fuzzy c-means clustering algorithm is proposed in this article, and the algorithm was successfully applied to the data processing of IoT. The experiments show that the proposed algorithm has better applicability and higher clustering accuracy for the huge and complex IoT data.

2 FUNDAMENTALS

Given a data set $X = \{x_i \mid i = 1, 2, \ldots, N\}$, $x_i \in R^D$, the number of cluster is C, m is the fuzzy weighted index, η_i is a penalty factor, $\widehat{u}_{ij}$ is the typical value of labeled samples, cluster centers $V = \{v_i \mid i = 1, 2, \ldots, C\}$, v_i denotes the i-th cluster center. Let $U = \{u_{ij} \mid i = 1, 2, \ldots, C, \ j = 1, 2, \ldots, N\}$, to be the membership matrix, u_{ij} represents the membership degree of x_j corresponding to the i-th cluster, d_{ij} represents the distance between x_j and v_i, w_{ik}^{τ} presents the fuzzy weighting coefficient of k-th dimensional feature of i-th cluster.

2.1 PCM algorithm

Krishnapuram and Keller proposed PCM algorithm and abandoned the constraint condition of membership, which means the sum of the membership of each sample in each cluster $\sum_{i=1}^{C} u_{ij}$ is no longer subject to a limit of 1, with typical values instead of fuzzy membership in FCM. The noises and outliners for each cluster have smaller memberships, so that the noise and outliners have a smaller impact on the clustering results by PCM algorithm, and PCM also has solved the problem that FCM is sensitive to noise and outliners. PCM algorithm can be expressed in many forms, this article uses the following objective function:

$$J_{PCM}(U,V) = \sum_{i=1}^{C}\sum_{j=1}^{N} u_{ij}^{m} d_{ij}^{2} + \sum_{i=1}^{C} \eta_i \sum_{j=1}^{N}\left(1 - u_{ij}\right)^{m} \quad (1)$$

$$s.t. \ 0 \le u_{ij} \le 1, 0 < \sum_{j=1}^{N} u_{ij} < N$$

where $\eta_i = K\sum_{j=1}^{N} u_{ij}^m d_{ij}^2 / \sum_{j=1}^{N} u_{ij}^m, K=1$

In (1), the first term in FCM item represents the intra-cluster distance and the second item forces the membership to be as large as possible. Thus, it avoids trivial solution. PCM relaxes the column sum constraint of the membership matrix in FCM, so that the sum of each column of PCM partition matrix satisfies the looser constraint. The advantage of PCM compared with FCM is its capability in identifying outliers in dataset and weakening the influence of outliers and noise on clustering results.

2.2 Subspace clustering

Duo to high-dimensional data space usually contains irrelevant attributes, while the target cluster may exist only in some low-dimensional subspace, and the different cluster of its associated sub-space often is not the same [8], which needs to dig out the hidden clusters in different low-dimensional subspace in high-dimensional space. The mining process is called subspace clustering. Subspace clustering can not only find the subspace existed in the cluster, but also find clusters existed in subspace. In 2004, [9] proposed the classic subspace clustering algorithm, the objective function is as follows:

$$J_{AWA} = \sum_{i=1}^{C}\sum_{j=1}^{N} u_{ij} \sum_{k=1}^{D} w_{ik}^{\tau}\left(x_{jk}-v_{ik}\right)^2$$
$$= \sum_{i=1}^{C}\sum_{k=1}^{D} w_{ik}^{\tau} \sum_{j=1}^{N} u_{ij}\left(x_{jk}-v_{ik}\right)^2 \qquad (2)$$

$$s.t.\ 0\le u_{ij}\le 1, 0\le w_{ik}\le 1, \sum_{k=1}^{D} w_{ik}=1$$

3 A NOVEL SEMI-SUPERVISED SUBSPACE POSSIBILISTIC FUZZY C-MEANS CLUSTERING ALGORITHM

The noises and outliners for each cluster have smaller memberships in PCM, so that the noise and outliners have a smaller impact on the clustering results. The algorithm can apply to the collected IoT data sets which contains noises for its robustness. The introduction of the classical subspace clustering has a great significance to high-dimensional complex data. Subspace clustering can not only detect the subspace presence in each cluster of every data, but also detect the cluster in subspace, with the full and efficient use of the data information. The PCM fusion of subspace clustering applied to semi-supervised areas, which is more in line with the objective reality. Because in the actual production, it is usually easy to obtain a small amount of known information, the known information plays an important supervision and guidance role in the clustering process.

Based on the above analysis, a novel semi-supervised subspace possibilistic fuzzy c-means clustering sSPCM algorithm is proposed. The form of

semi-supervised based on the attribute information of the known sample. The objective function of sSPCM algorithm as follows:

$$J_{sSPCM} = \sum_{i=1}^{C}\sum_{k=1}^{D} w_{ik}^{\tau} \sum_{j=1}^{N} u_{ij}^m\left(x_{jk}-v_{ik}\right)^2$$
$$+ \sum_{i=1}^{C} \eta_i \left(\sum_{j=1}^{N}\left(1-u_{ij}\right)^m\right)$$
$$+ \alpha\sum_{i=1}^{C}\sum_{k=1}^{D} w_{ik}^{\tau} \sum_{j=1}^{N}\left(u_{ij}-\hat{u}_{ij}\right)^m\left(x_{jk}-v_{ik}\right)^2$$

$$s.t.\ 0\le u_{ij}\le 1, 0\le w_{ik}\le 1, \sum_{k=1}^{D} w_{ik}=1 \qquad (3)$$

where α denotes a scaling factor used to maintain the balance between supervised and unsupervised component. Set the membership of known sample be 1 and the membership of unknown sample be 0, the above formula is equivalent to:

$$J_{sSPCM} = \sum_{i=1}^{C} \eta_i \left(\sum_{j=1}^{N}\left(1-u_{ij}\right)^m\right)+$$
$$\sum_{i=1}^{C}\sum_{k=1}^{D} w_{ik}^{\tau} \sum_{j=1}^{N}\left(u_{ij}-\hat{u}_{ij}\right)^m\left(x_{jk}-v_{ik}\right)^2 \qquad (4)$$

$$s.t.\ 0\le u_{ij}\le 1, 0\le w_{ik}\le 1, \sum_{k=1}^{D} w_{ik}=1$$

Minimizing the objective function (4) by Lagrangian multipliers, we obtain the updating equation of the membership, the cluster center and the weight:

$$u_{ij} = \frac{1-\hat{u}_{ij}}{\left(\frac{\eta_i}{\sigma_{ik}^2}\right)^{\frac{1}{m-1}}-1}+1 \qquad (5)$$

$$v_{ik} = \frac{\sum_{j=1}^{N} u_{ij} x_{jk}}{\sum_{j=1}^{N} u_{ij}} \qquad (6)$$

$$w_{ik} = \frac{\left(\sum_{j=1}^{N}\left(u_{ij}-\tilde{u}_{ij}\right)^m d_{ik}^2\right)^{\frac{1}{1-\tau}}}{\sum_{k'=1}^{D}\left(\sum_{j=1}^{N}\left(u_{ij}-\tilde{u}_{ij}\right)^m d_{ik'}^2\right)^{\frac{1}{1-\tau}}} \qquad (7)$$

where

$$\sigma_{ik}^2 = \sum_{k=1}^{D} w_{ik}^{\tau}\left(x_{jk}-v_{ik}\right)^2 \quad d_{ijk}^2 = \left(x_{jk}-v_{ik}\right)^2.$$

The algorithm of sSPCM is described as follows.

sSPCM

1) Set the clustering number C, parameter $m>0$, $\eta_i>0$, $\tau>0$, $\alpha>0$, $\beta>0$, threshold $\varepsilon=0.001$,

the maximal number of iterations $t\_max = 100$, randomly initialize cluster centers v_i, the typical values of labeled patterns $U\_label = \{\hat{u}_{ij}\}$ and weight matrices $W(0)$ where $w_{ik}^{\tau} = 1/D$.
2) Compute the partition matrix by (5).
3) Compute cluster center matrix by (6).
4) Compute the weight matrix by (7).
5) Repeat step 2 to step 4, until the termination criterion is satisfied.

4 EXPERIMENTAL RESULTS

In this section, numerical experiments are conducted on artificial and UCI standard data sets to investigate the performance of sSPCM. The comparison algorithms in experiment: classic FCM algorithm, PCM algorithm, SPC algorithm [7], sFCM algorithm [10], and the proposed sSPCM algorithm. In order to reflect the fairness of the comparison, we fixed the parameters used in our experiments as follows: the maximal number of iterations $t\_max = 100$, parameter $m = 2, \tau = 1.1$, the threshold $\varepsilon = 0.001$, the number of labeled patterns is 0.2 of the total number of patterns. The principle of labeled patterns selected as follows: assuming that the category properties of labeled patterns are known in advance, the membership of labeled pattern x_j is defined as $u_{ij} = 1$, and the membership of unlabeled pattern x_j is defined as $u_{ij} = 0$.

The rand index (RI) and the normalized mutual information (NMI) are used for revaluating the performance of the proposed sPCM algorithm. Both RI and NMI take the value within the interval between 0 and 1. The higher the values are, the better the clustering performance is.

4.1 A synthetic dataset

In this subsection, a synthetic dataset with controlled cluster structures is used to investigate the performance of the proposed sSPCM algorithm. The features of the synthetic dataset are as follows: 1) it contains three clusters with 900 samples and dimension of 200; 2) each cluster of data is located in a different subspace; 3) the size of each clusters are different. Figure 1 shows the distribution of the data in the different subspace. Table 1 shows the performance comparison of each algorithm on synthetic dataset.

It can be found from Table 1 that the clustering effect of traditional unsupervised clustering algorithm such as FCM, PCM algorithm is not ideal for the data set with subspace characteristics. The SPC algorithm which is introduction of subspace clustering has improved the clustering accuracy than PCM algorithm, because subspace clustering can effectively detect the fuzzy subspace in each cluster, and improve the clustering accuracy and adaptability of the algorithm. With the introduction of the small amount of semi-supervised information to SPC algorithm, it effectively guides the clustering process and makes the clustering

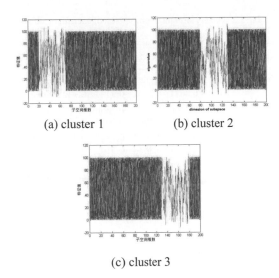

(a) cluster 1   (b) cluster 2

(c) cluster 3

Figure 1. The distribution of the synthetic dataset in different sub-space.

Table 1. The performance comparison of each algorithm on synthetic dataset.

Index	FCM	PCM	SPC	sFCM	sSPCM
NMI	0.7140	0.7211	0.8912	0.9013	**0.9567**
RI	0.8166	0.8879	0.9596	0.9691	**0.9893**

Table 2. The description of experimental datasets.

Dataset	N	D	C
Iris	150	4	3
Wine	178	13	3
Zoo	101	16	7
IS	2310	18	7
MF	2000	649	10

performance of proposed sSPCM algorithm much better than SPC algorithm.

4.2 UCI datasets

The performance of the proposed sSPCM algorithm has been evaluated and compared with four clustering algorithms using five UCI datasets. Table 2 shows the description of experimental datasets Table 3 shows the performance comparison of each algorithm on UCI datasets.

It can be found from Table 3 that the superiority of sSPCM algorithm is not particularly evident for the conventional small data sets, but the clustering accuracy is still a slight rise, it can be seen from experimental results of Iris, Wine and Zoo. For the scale of experimental data set is small and the structure is simple, it fails to reflect the advantage

Table 3. The performance comparison of each algorithm on UCI datasets.

Dataset	Index	FCM	PCM	SPC	sFCM	sSPCM
Iris	NMI	0.7957	0.8046	0.8332	0.8545	**0.8625**
	RI	0.8611	0.8742	0.9117	0.9427	**0.9601**
Wine	NMI	0.4168	0.7570	0.7660	0.7882	**0.8011**
	RI	0.7105	0.8969	0.9068	0.9117	**0.9257**
Zoo	NMI	0.6733	0.6902	0.7882	0.7984	**0.8110**
	RI	0.8331	0.8634	0.9103	0.9192	**0.9379**
IS	RI	0.4854	0.4993	0.5096	0.5349	**0.6125**
	RI	0.8305	0.8537	0.8622	0.9013	**0.9211**
MF	NMI	0.3575	0.3596	0.4121	0.4482	**0.5623**
	RI	0.5746	0.5900	0.6659	0.7256	**0.8427**

of proposed algorithm. For the large amount and complex structure of data sets, such as IS, MF, the advantages of the proposed algorithm becomes very significant.

5 CONCLUSION

The collected data's quantity is growing with a vigorous development of the IoT, and the structure of the data is more and more complex. Some existing algorithms can not satisfy the demand of the data processing. In this context, a possibilistic clustering algorithm combined with the ideal of subspace clustering, supervised by a small amount of known information, a semi-supervised subspace possibilistic fuzzy c-means clustering algorithm is proposed. Potential structure of the subspace in the complex data is considered in the algorithm. In addition, using a few supervised information in the algorithm is more realistic. The experimental results on simulated data sets and UCI standard datasets show that the algorithm has a better clustering performance and better adaptability compared with traditional unsupervised clustering algorithm and the normal semi-supervised clustering algorithm.

ACKNOWLEDGEMENT

The authors would like to thank the reviewers for their valuable comments that have greatly improved the quality of our manuscript in many ways.

REFERENCES

[1] He Qing. Internet of Things and data mining cloud services. Intelligent Systems, 2012, 7(3):1–5.

[2] Zhang Wei, Li Liang. Applications of Multi-sensor Data Acquisition Technology in the Internet of Things. Journal of GuangZhou University, 2012, 11(3):75–80.

[3] Zhang Min, Yu Jian. Fuzzy clustering algorithm Based on Partitioning. Journal of Software, 2004, 15(6):859–868.

[4] Krishnapuram R, Keller J. A Possibilistic Approach to Clustering [J]. IEEE Transactions on Fuzzy Systems, 1993, 1(2):98–110.

[5] Miin S Y, Kuo L W. Unsupervised possibilistic clustering. Pattern Recognition, 2006, 39:5–21

[6] Han X D, Xia S X, Liu Bin. A Fast Possibilitic Clustering Algorithms Based Nuclear. Computer Engineering and Applications. 2011, 47(6): 176–180.

[7] Guan Qing, Deng Z H, Wang S T. Research on Subspace Possibilistic Clustering Mechanism. Computer Engineering, 2011, 37(5):224–226.

[8] Chen L F, Guo G D, Jiang Q S. Adaptive soft subspace clustering algorithm. Journal of Software, 2010, 21(10):2513–2523.

[9] Elaine Y C, Ching Waiki, Michael K N, et al. An Optimization Algorithm for Clustering Using Weighted Dissimilarity Measures[J]. Pattern Recognition, 2004, 37(5):943–952.

[10] Endo Y, Hamasuna Y, Yamashiro M and Miyamoto S. On semisupervised fuzzy c-means clustering [C], IEEE International Conference on Fuzzy Systems, 2009.

[11] Deng Z H, Choi K S, Chung F L, Wang S T. Enhanced soft subspace clustering intergrating within-cluster and between-cluster information[J]. Pattern Recognition. 2010, 43(3):767–781.

Information Technology and Computer Application Engineering – Liu, Sung & Yao (Eds)
© 2014 Taylor & Francis Group, London, ISBN 978-1-138-00079-7

Low speed operation analysis of PMSM DTC

Zikuan Zhang & Lin He
School of Mechanical Engineering, Guizhou University, Guiyang City, Guizhou Province, China

ABSTRACT: To study low speed operation of permanent magnet synchronous motor of direct torque control, it was used to model and simulate based on Matlab/Simulink software, reached different simulation results through set up different speed in low speed range. The results indicate that permanent magnet synchronous motor can operate smoothly under low speed condition and transit smoothly with different torque. The proposed simulation system can achieve stable control, and its effectiveness is confirmed experimentally. Results of simulation analysis have some certain practical value for permanent magnet synchronous motor of direct torque control.

Keywords: PMSM; DTC; low speed; Simulink

1 INTRODUCTION

The technology of direct torque control (DTC) is developing technique of asynchronous motor frequency conversion after technique of vector conversion technique. To a large extent, DTC solves the problem that calculation complicated, character easy to influence by motor parameter. Under low speed condition, influence of voltage drop of stator resistance lead to that flux linkage track happened distortion. Consequently, the track is circle approximately with voltage vector to control. DTC is useful for permanent magnet synchronous motor (PMSM) and can improve rapid torque respond. In order to get great control effect of motor torque and flux linkage, this paper uses Matlab/Simulink to model and simulate, and analyze control performance with simulation results. The results indicate that PMSM can operate smoothly under low speed condition.

2 MATHEMATICAL MODEL OF PMSM DTC

DTC uses space vector analysis method to calculate and control torque at stator coordinate system directly, also uses stator field orientation, process optimum control to switch status of inverter that rely on discrete method of two point to adjust PWM signal, and obtain high dynamic performance of torque. DTC is regulate speed of stator flux linkage through space voltage vector with maintain flux linkage amplitude constant to control torque and speed.

PMSM DTC is based on coordinate system α-β.

Transformation of α-β coordinate system into d-q coordinate system:

$$\begin{bmatrix} V_d \\ V_q \end{bmatrix} = \begin{bmatrix} \cos\delta & -\sin\delta \\ \sin\delta & \cos\delta \end{bmatrix} \begin{bmatrix} V_\alpha \\ V_\beta \end{bmatrix} \tag{1}$$

where V is any vector.

The stator flux linkage that can be expressed in the stationary reference frame is

$$\psi_s = \int (u_s - i_s R)\, dt \tag{2}$$

Transform three phase variables into two phase variables in α-β reference frame

$$\begin{bmatrix} f_\alpha \\ f_\beta \end{bmatrix} = \begin{bmatrix} 1 & -1/2 & -1/2 \\ 0 & \sqrt{3}/2 & -\sqrt{3}/2 \end{bmatrix} \begin{bmatrix} f_a \\ f_b \\ f_c \end{bmatrix} \tag{3}$$

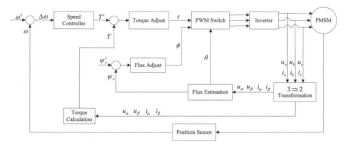

Figure 1.   Block diagram of direct torque control system.

where f represents voltage, current, and flux linkage.

The magnitude of the stator flux linkage vector can be derived from ψ_α and ψ_β as

$$\psi_\alpha = \int (u_\alpha - i_\alpha R)dt$$
$$\psi_\beta = \int (u_\beta - i_\beta R)dt \tag{4}$$

$$\psi'_s = \sqrt{\psi_\alpha^2 + \psi_\beta^2} \tag{5}$$

The angular position of the stator flux linkage vector can be calculated as following:

$$\theta = \arctan(\psi_\alpha / \psi_\beta) \tag{6}$$

Electromagnetic torque equation can be expressed as follows:

$$T = 1.5p(\psi_\alpha i_\beta - \psi_\beta i_\alpha) \tag{7}$$

Here u_α and u_β are the armature voltages, i_α and i_β are the armature currents, R is the armature resistance, ψ_α and ψ_β are respectively the estimated stator flux linkage and θ is the estimated position of the stator flux linkage, p is the number of pole pairs.

In order to select voltage vector to control amplitude of the stator flux linkage, voltage vector is divided into 6 sections. In every section, selects two adjacent vectors to control the value of flux linkage.

Output voltage of inverter can be calculated as

$$\begin{vmatrix} u_a \\ u_b \\ u_c \end{vmatrix} = \frac{1}{3} \begin{vmatrix} 2 & -1 & -1 \\ -1 & 2 & -1 \\ -1 & -1 & 2 \end{vmatrix} \begin{vmatrix} S_a \\ S_b \\ S_c \end{vmatrix} U_{dc} \tag{8}$$

where S_a, S_b, S_c represent three on-off state.

3 ESTABLISH SIMULATION MODELING OF CONTROL SYSTEM

Through measure three phase current of motor stator, according to equation (3), transformed into electric power by Clark transformation in two phase stationary reference frame. The three phase voltage transformation has the same theory.

By equation (4), establish the module of flux linkage calculation

The angular position of the stator flux linkage vector can obtain from equation (6).

Compared with the reference value of torque and actual calculation of torque, change torque value relies on error.

Build torque module by equation (7).

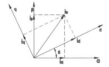

Figure 2.   Vector diagram of different reference frames.

Table 1.   Switching table for inverter.

| ϕ | τ | θ | | | | | |
		θ_1	θ_2	θ_3	θ_4	θ_5	θ_6
$\phi=0$	$\tau=1$	$U_2(110)$	$U_3(010)$	$U_4(011)$	$U_5(001)$	$U_6(101)$	$U_1(100)$
	$\tau=0$	$U_6(101)$	$U_1(100)$	$U_2(110)$	$U_3(010)$	$U_4(011)$	$U_5(001)$
	$\tau=-1$	$U_5(001)$	$U_6(101)$	$U_1(100)$	$U_2(110)$	$U_3(010)$	$U_4(011)$
$\phi=1$	$\tau=1$	$U_3(010)$	$U_4(011)$	$U_5(001)$	$U_6(101)$	$U_1(100)$	$U_2(110)$
	$\tau=0$	$U_5(001)$	$U_6(101)$	$U_1(100)$	$U_2(110)$	$U_3(010)$	$U_4(011)$
	$\tau=-1$	$U_6(101)$	$U_1(100)$	$U_2(110)$	$U_3(010)$	$U_4(011)$	$U_5(001)$

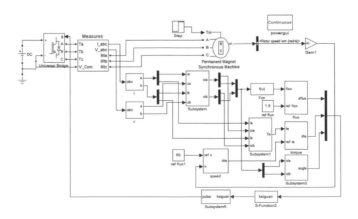

Figure 3.   Simulation model of PMSM DTC.

Module of speed regulation is use proportion and integration coefficient in PID control. Proportion and integration coefficient is obtain by adjust and compare.

Module of voltage vector switch signal selection is input from result of flux linkage hysteresis comparator and torque hysteresis comparator and the angular position of flux linkage, so that selects relevant space voltage vector. According to Table 1, builds s-function in Simulink to achieve this function.

4 SIMULATION RESULTS AND ANALYSIS

(1) Rated speed is 50 rpm, torque take step-input that 5 N·m from 0 sec to 1 sec and 10 N·m from 1 sec to end, which have 2 seconds totally.
(2) Rated speed is 100 rpm, others are fixed.
(3) Rated speed is 150 rpm, others are fixed.

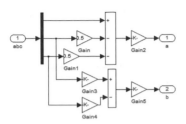

Figure 4. Module of stator three phase current Clark transformation.

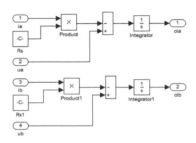

Figure 5. Module of stator flux linkage in stationary reference frame.

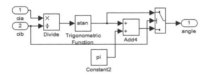

Figure 6. Module of the angular position of the stator flux linkage vector in stationary reference frame.

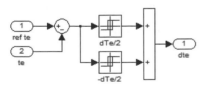

Figure 7. Torque error signal.

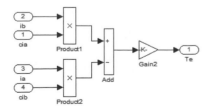

Figure 8. Electromagnetic torque module.

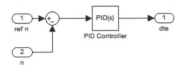

Figure 9. Module of speed control.

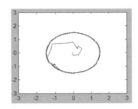

Figure 10. Stator flux linkage track simulation.

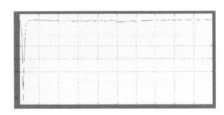

Figure 11. Speed simulation response.

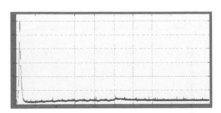

Figure 12. Torque simulation response during load change.

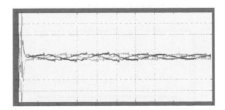

Figure 13. Three phase current of PMSM.

Figure 14. Stator flux linkage track simulation.

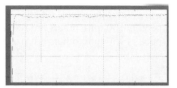

Figure 15. Speed simulation response.

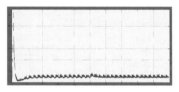

Figure 16. Torque simulation response during load change.

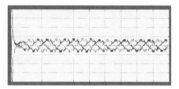

Figure 17. Three phase current of PMSM.

Figure 18. Stator flux linkage track simulation.

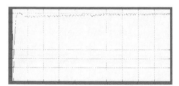

Figure 19. Speed simulation response.

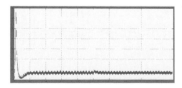

Figure 20. Torque simulation response during load change.

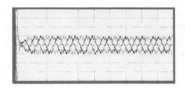

Figure 21. Three phase current of PMSM.

Table 2. Experimental system parameters.

Pole Pairs	2
Flux Linkage (V · s)	0.08
Stator Phase Resistance (Ω)	1.9
Moment of Inertia (kg · m^2)	0.08
Inductances (H)	0.0072
Proportional	2.5
Integral	3
Voltage (V)	380

From above waveforms, we can see that simulation fluctuates at start phase, nevertheless adjusts to normal state immediately at three low speed range. Therefore, as a whole, the simulation reaches expected effect.

5 CONCLUSIONS

According to above results of simulation, the control of PMSM DTC can be estimated relative stable totally. Although have a littleinstability at launch phase, it can meet apply requirement of motor during load torque. This time of simulation PMSM DTC can provide the basis of PMSM actual control means, and lay a solid foundation for the next step research.

REFERENCES

C. French & P. Acarnley. 1995. Direct Torque Control of Permanent Magnet Drive. *Proc. of IEEE Industry Application Society Annual Meeting*, Vol. 1, pp. 199–206, Florida, USA.

Luukko J & Pyrhonen J. 1998. Selection of the flux linkage reference in direct torque controlled permanent magnet synchronous motor drive. *Adv Motion Control 5th Int Workshop*: 198–203.

Matins C A et al. 2002. Switching Frequency Imposition and Ripple Reduction in DTC Drives by Using a Multilevel Convert [J]. *IEEE Transactions on Power Electronics*, vol. 17, pp. 286–297.

Rahman MF, Zhong L & Lim KW. 1997. Analysis of direct torque control in permanent magnet synchronous motor drive. *IEEE Trans Power Electron*, 12(3): 526–8.

S. Morimoto, M. Sanada & Y. Takeda. 1994. Wide-speed operation of interior permanent magnet synchronous motors with high-performance current regulator. *IEEE Transactions on Industry Applications*, vol. 30, no. 4, pp. 920–926, Jul./Aug.

Zhong L & Rahman M F. 1997. Analysis of direct torque control in permanent magnet drives [J]. *IEEE Transactions on Power Electronics*, 12(3): 528–535.

Information Technology and Computer Application Engineering – Liu, Sung & Yao (Eds)
© 2014 Taylor & Francis Group, London, ISBN 978-1-138-00079-7

Finite-time receding horizon control for Markovian jump linear systems with partly unknown transition probabilities

Ji Wei Wen

Key Laboratory of Advanced Process Control for Light Industry, (Ministry of Education),
School of Internet of Things Engineering, Jiangnan University, Wuxi, P.R. China

Jing Liu

The Chinese People's Liberation Army 61769, Lvliang, P.R. China

ABSTRACT: This paper solves finite-time receding horizon control problems for discrete-time Markovian Jump Linear Systems (MJLSs) subject to partly unknown Transition Probabilities (TPs) by minimizing targeted quadratic cost on-line. The motivation of the proposed control strategy is to pay more attention on the transient response rather than the stochastic stabilizability. The Finite-Time Stability (FTS) can be achieved by properly employing stochastic Lyapunov functional approach. A stabilizing receding horizon controller for the underlying system is obtained via Semi-Definite Programming (SDP), which can be solved efficiently by calculating the Linear Matrix Inequalities (LMIs).

Keywords: Markovian jump linear systems; Receding horizon control; Finite-tine stability; Partly unknown transition probabilities

1 INTRODUCTION

Technical and economical reasons motivate the development of Markovian jump linear systems (MJLSs) with an ever-increasing complexity [1]. Some basic issues of MJLSs, such as stochastic stabilizability, finite-/infinite-horizon filtering, quadratic optimal control and H_2/H_∞ performance, etc, have been extensively studied over the past two decades. It is worth mentioning that the transition probabilities (TPs) or jump rates (JRs) play important roles in system behavior, thus rich references appear with burgeoning research interest (see [2] and references therein).

On another research front line, receding horizon control (RHC), which is often known as model predictive control (MPC), has become a popular strategy to handle hard/soft constraint, guaranteed cost and stability for MJLSs. The RHC has been developed for classical discrete-time MJLSs [3] and it has been extended to NCSs with time delay or random data packet loss, which is modeled by Markov chain [4]. Among these references, the feedback RHC approach has the advantages on fast computation, quick deployment and the ability to consider both control performance and feasible solution space.

With the references review, the rationale of stochastic stability based on the stochastic Lyapunov function (SLF) has been extensively applied. However, it is worth mentioning that the Lyapunov stochastically

stable systems may not possess good or expected transient characteristics over a finite horizon. In terms of engineering application, such as communication network system, biochemistry reaction system and robot control system (see [5] and references there in), more attention must be paid on their behaviors over a fixed finite-time interval. Therefore, it is necessary to limit the state in an acceptable region that is to consider the finite-time stability (FTS) of the system [6, 7].

Inspired by the stability criterion based on the FTS, the aim of this paper is to deal with the FTS for MJLSs under the feedback RHC framework because it is rarely addressed how to guarantee a quadratic performance index over a finite-time interval with a relatively good transient response. The main procedure of this paper is focused on the design of a feedback receding horizon controller so that the given index can be minimized when the closed-loop MJLS is stable in the FTS sense. First, a standard SLF is constructed to obtain the minimum value of the performance index and analyze the FTS of the controlled system. Then, based on the SLF, a feedback receding horizon controller is developed to reduce the minimum cost and achieve better dynamic character. The addressed optimization problem is solved in terms of semi-definite programming (SDP) which can be efficiently calculated by some available numerical software such as LMI toolbox of Matlab.

2 SYSTEM DESCRIPTIONS AND PROBLEM FORMULATION

We consider a discrete-time MJLS which can be described by the following mathematical models:

$$x_{k+1} = (A(r_k) + \Delta A(r_k))x_k + (B(r_k) + \Delta B(r_k))u_k, \qquad (1)$$

where $k \in \{1, \ldots, N\}, N \in \mathbb{N}$, and $\mathbb{N}$ is the set of positive integers. $x_k \in \mathbb{R}^n$ is the state vector, $u_k \in \mathbb{R}^m$ is the control input vector. For each possible value of $r_k = i$, we denote $A(r_k) = A_i, B(r_k) = B_i, \Delta A(r_k) = \Delta A_i, \Delta B(r_k) = \Delta B_i$, for simplicity. A_i and B_i are constant matrices with appropriate dimensions. ΔA_i and ΔB_i represent time varying parameter uncertainties, which are assumed to be norm bounded and can be given as

$$\begin{bmatrix} \Delta A_i & \Delta B_i \end{bmatrix} = E_i \Delta_i \begin{bmatrix} H_{1i} & H_{2i} \end{bmatrix}, \qquad (2)$$

where E_i, H_{1i} and H_{2i} are known constant matrices which characterize the structure of the uncertainties. Δ_i are unknown time-varying matrix functions with Lebesgue measurable elements satisfying $\Delta_i^T \Delta_i \leq I$. For notational ease, we also denote

$$A_i(k) = A_i + E_i \Delta_i H_{1i}, B_i(k) = B_i + E_i \Delta_i H_{2i}. \qquad (3)$$

$r(k)$ is a discrete-time, discrete-state Markov chain taking values in $S = \{1, 2, \ldots, s\}$ with transition probabilities $\mathrm{Prob}\{r_{k+1} = j \mid r_k = i\} = \pi_{ij}$, where π_{ij} is the TPs from mode i to mode j that satisfies $\pi_{ij} \geq 0, \sum_{j=1}^m \pi_{ij} = 1, \forall i, j \in S$.

In addition, the TPs of Markov process are assumed to be partly unknown and partly accessed. For example, for system (1) with four operation modes, the TPs matrix [2] can be viewed as

$$\Pi = \begin{bmatrix} \pi_{11} & \pi_{12} & ? & ? \\ ? & ? & \pi_{23} & \pi_{24} \\ \pi_{31} & ? & ? & ? \\ ? & \pi_{42} & \pi_{43} & \pi_{44} \end{bmatrix}, \qquad (4)$$

where ? represents the unknown element. For notation clarity, $\forall i \in S$, we denote that

$$S = S_k^i + S_{uk}^i, \qquad (5)$$

where

$$S_k^i = \{j : \pi_{ij} \text{ is known}\}, S_{uk}^i = \{j : \pi_{ij} \text{ is unknown}\}. \qquad (6)$$

If $S_k^i \neq \phi$, it can be described as

$$S_k^i = \{k_1^i, k_2^i, \cdots, k_q^i\}, \quad \forall 1 \leq q \leq s, \qquad (7)$$

with k_q^i represents the jump mode j corresponding to known element located in the ith row, qth element of matrix Π. Also, we denote

$$P^i = \sum_{j \in S} \pi_{ij} P_j, P_k^i = \sum_{j \in S_k^i} \pi_{ij} P_j, \pi_k^i = \sum_{j \in S_k^i} \pi_{ij}, \qquad (8)$$

throughout this paper.

The general idea of FTS puts a restriction on the state and it can be viewed as the quadratic hard time-domain constraint in a period of time. This concept is formalized through the following definitions, which is an extension of discrete-time linear systems given in [6].

Definition 1. (finite-time stability): The MJLS

$$x_{k+1} = A_i(k)x_k, \qquad k \in N$$

is said to be FTS with respect to (c_1, c_2, G_i, N), if

$$x_0^T G_i x_0 \leq c_1^2 \Rightarrow x_k^T G_i x_k \leq c_2^2, \qquad \forall k \in \{1, \cdots, N\} \qquad (9)$$

where G_i is a positive-definite matrix, $0 < c_1 < c_2$.

Definition 2. (finite-time stabilizability via state feedback): The MJLS is said to be finite-time stabilizable with respect to (c_1, c_2, G_i, N), if there exist a mode-dependent control law (constant for each value of i),

$$u_k = K_i x_k \quad when \quad r_k = i, \qquad (10)$$

such that the closed-loop system

$$x_{k+1} = (A_i(k) + B_i(k)K_i)x_k \qquad (11)$$

is FTS.

In general, a feedback receding horizon controller should be written as

$$u_{k+f|k} = K(r_{k+f|k})x_{k+f|k}$$

where f and k represents predictive step and current time index, respectively. However, such a predicted controller is very difficult to be calculated because there is no exact mode information at future time instant. Therefore, the predictive step f is set as zero in this paper to obtain a feasible feedback receding horizon controller (3) and $x_{k|k}$ is always denoted as x_k.

Definition 3. A finite-time performance index is given by the quadratic cost

$$J_N(k) = E \sum_{k=0}^N \{x_k^T Q(r_k)x_k + u_k^T R(r_k)u_k \mid \xi_0\}, \qquad (12)$$

where ξ_0 represents the σ-algebra generated by x_0 and r_0. For mode $r_k = i$, we have $Q(r_k) = Q_i > 0$, $R(r_k) = R_i > 0$.

Lemma 1. Let Y, E, H be given matrices with appropriate dimensions. For matrix F satisfying $F^T F \leq I$, we have $Y + EFH + H^T F^T E^T \geq 0$, if and only if there exists a constant $\delta > 0$ satisfying $Y - \delta EE^T - \delta^{-1} H^T H \geq 0$.

This paper is concerned with the design of the controller (10) via receding horizon approach, such that the closed-loop system (11) is FTS with guaranteed cost (12). In the development, we always assume the full access of the current time state x_k and jump mode r_k.

16

3 FINITE TIME RECEDING HORIZON CONTROL

In this section, we seek to obtain a feedback receding horizon control move through minimizing finite-time quadratic performance index (12) for MJLS (1). First, the optimization problem is transferred into a trackable SDP and is solved on-line to reduce the minimum value of the cost. Then, the feasibility of SDP at every sampling time and FTS of the closed-loop system is discussed.

Theorem 1. Given a scalar $\alpha \geq 1$. The sufficient condition for the existence of the finite-time receding horizon controller for disturbance-free MJLS (1) can be transformed into the following SDP.

$$\min_{\gamma, \lambda, X_i, Y_i} \gamma \tag{13}$$

subject to

$$\begin{bmatrix} 1 & \alpha^{\frac{1}{2}} x_k^{\mathrm{T}} & c_1 \\ * & X_i & 0 \\ * & * & \gamma \end{bmatrix} \geq 0, \tag{14}$$

$$\begin{bmatrix} \alpha X_i & \overline{\phi}_i^{\mathrm{T}} & \Xi_i^{\mathrm{T}} \\ * & X_j - \delta E_i E_i^{\mathrm{T}} & 0 \\ * & * & \delta I \end{bmatrix} \geq 0, \forall j \in S_{uk}^i, \tag{15}$$

$$\begin{bmatrix} \alpha \pi_k^i X_i & V_{1i}^{\mathrm{T}} & X Q_i^{\frac{1}{2}} & Y_i R_i^{\frac{1}{2}} & V_{3i}^{\mathrm{T}} & 0 \\ * & W_j & 0 & 0 & 0 & \delta V_{2i} \\ * & * & I & 0 & 0 & 0 \\ * & * & * & I & 0 & 0 \\ * & * & * & * & \delta I & 0 \\ * & * & * & * & * & \delta I \end{bmatrix} \geq 0, \tag{16}$$

$$X_i < G_i^{-1}, \tag{17}$$

$$\lambda G_i^{-1} < X(r_0), \tag{18}$$

$$\begin{bmatrix} \dfrac{c_2^2}{\alpha^N} & c_1 \\ * & \lambda \end{bmatrix} \geq 0, \tag{19}$$

where

$$W_j = \mathrm{diag}\left\{ X_{k_1^i}, X_{k_2^i}, \cdots, X_{k_q^i} \right\},$$

$$\overline{\phi}_i = A_i X_i + B_i Y_i, \Xi_i = H_{1i} X_i + H_{2i} Y_i,$$

$$V_{1i}^{\mathrm{T}} = \left[(\pi_{ik_1^i})^{\frac{1}{2}} \overline{\phi}_i^{\mathrm{T}} \quad (\pi_{ik_2^i})^{\frac{1}{2}} \overline{\phi}_i^{\mathrm{T}} \quad \cdots \quad (\pi_{ik_q^i})^{\frac{1}{2}} \overline{\phi}_i^{\mathrm{T}} \right],$$

$$V_{2i}^{\mathrm{T}} = \left[(\pi_{ik_1^i})^{\frac{1}{2}} E_i^{\mathrm{T}} \quad (\pi_{ik_2^i})^{\frac{1}{2}} E_i^{\mathrm{T}} \quad \cdots \quad (\pi_{ik_q^i})^{\frac{1}{2}} E_i^{\mathrm{T}} \right],$$

$$V_{3i}^{\mathrm{T}} = \left[(\pi_{ik_1^i})^{\frac{1}{2}} \Xi_i^{\mathrm{T}} \quad (\pi_{ik_2^i})^{\frac{1}{2}} \Xi_i^{\mathrm{T}} \quad \cdots \quad (\pi_{ik_q^i})^{\frac{1}{2}} \Xi_i^{\mathrm{T}} \right].$$

The receding horizon control can be obtained by $u_k = K_i x_k = Y_i X_i^{-1} x_k$, if there exist scalars $\gamma, \lambda > 0$, matrices $X_i = X_i^{\mathrm{T}} > 0$ and Y_i satisfying LMIs (14)~(19). If SDP (13) has a solution at every sampling time k, then the RHC law u_k stabilizes the MJLS (1) in the FTS sense with respect to given α and (c_1, c_2, G_i, N) over the finite-time interval $[0, N]$.

Proof The stochastic Lyapunov function is taken as $V(x_k) = x_k^{\mathrm{T}} P_i x_k$. First, the upper bound of the optimized index (12) must be found to make the minimization of $J_N(k)$ computable. Assume an additional constraint should be satisfied, that is

$$\mathrm{E}\{V(x_{k+1}) \mid \xi_0\} - \alpha \mathrm{E}\{V(x_k) \mid \xi_0\}$$
$$\leq -\mathrm{E}\{x_k^{\mathrm{T}} Q(r_k) x_k + u_k^{\mathrm{T}} R(r_k) u_k \mid \xi_0\}. \tag{20}$$

Summing both sides of (20) from $k = 0$ to $N - 1$, we have

$$J_N(k) \leq \alpha V(x_0) + (\alpha - 1)(V(x_1) + V(x_2) + \cdots$$
$$+ V(x_{N-1})) - V(x_N) \tag{21}$$

Because $\alpha \geq 1$, (21) holds only if

$$J_N(k) < \alpha V(x_0) - V(x_N). \tag{22}$$

Putting an upper bound γ to $J_N(k)$ and considering $c_1^2 \leq V(x_N) \leq c_2^2$, we have (22) hold only if

$$\alpha V(x_0) + c_1^2 \leq \gamma. \tag{23}$$

The condition (23) is strongly dependent on the initial knowledge. In such a deterministic case, a RHC strategy shows significant reduction on the cost as opposed to using a linear state feedback gain which only depends on the x_0 and r_0. Combining with a feedback RHC strategy, the control move is recomputed at each sampling time k with measured mode and state. Thus we take the following condition instead of (23):

$$\alpha V(x_k) + c_1^2 \leq \gamma. \tag{24}$$

Denoting $X_i = \gamma P_i^{-1}$ and applying *Schur* complement to (24), we have optimization objective (13) and LMI constraint (14).

Next, we consider the additional constraint (20). It gives a feasible upper bound of index (12) and also has immediate impact on the FTS of the closed-loop system. Denoting

$$\phi_i(k) = A_i(k) + B_i(k) K_i, \phi_i = A_i + B_i K_i,$$
$$\Xi_i(k) = E_i \Delta_i (H_{1i} + H_{2i}) K_i,$$
$$\Theta_i = \alpha P_i - \phi_i^{\mathrm{T}}(k) P^i \phi_i(k) - Q_i^{\mathrm{T}} - K_i^{\mathrm{T}} R_i K_i.$$

It can be inferred from (20), that

$$\alpha x_k^{\mathrm{T}} P_i x_k - x_k^{\mathrm{T}} \phi_i^{\mathrm{T}}(k) P^i \phi_i(k) x_k$$
$$- x_k^{\mathrm{T}} Q_i x_k - x_k^{\mathrm{T}} K_i^{\mathrm{T}} R_i K_i x_k \geq 0.$$

The above inequality holds only if $\Theta_i \geq 0$ holds. Because $\sum_{j=1}^m \pi_{ij} = 1$, we have

$$\Theta_i = \alpha\left(\sum_{j\in S}\pi_{ij}\right)P_i - \phi_i^{\mathrm{T}}(k)\left(\sum_{j\in S}\pi_{ij}P_j\right)\phi_i(k)$$
$$-Q_i^{\mathrm{T}} - K_i^{\mathrm{T}}R_iK_i$$
$$= \alpha\left(\sum_{j\in S_k^i}\pi_{ij}\right)P_i - \phi_i^{\mathrm{T}}(k)\left(\sum_{j\in S_k^i}\pi_{ij}P_j\right)\phi_i(k)$$
$$+\alpha\left(\sum_{j\in S_{uk}^i}\pi_{ij}\right)P_i - \phi_i^{\mathrm{T}}(k)\left(\sum_{j\in S_{uk}^i}\pi_{ij}P_j\right)\phi_i(k)$$
$$-Q_i^{\mathrm{T}} - K_i^{\mathrm{T}}R_iK_i$$
$$= \alpha\pi_k^iP_i - \phi_i^{\mathrm{T}}(k)P_k^i\phi_i(k) - Q_i^{\mathrm{T}} - K_i^{\mathrm{T}}R_iK_i$$
$$+\sum_{j\in S_{uk}^i}\pi_{ij}P_j(\alpha P_i - \phi_i^{\mathrm{T}}(k)P_j\phi_i(k)).$$

Thus $\Theta_i \geq 0$ holds only if

$$\alpha P_i - \phi_i^{\mathrm{T}}(k)P_j\phi_i(k) \geq 0, \forall j \in S_{uk}^i \qquad (25)$$

and

$$\alpha\pi_k^iP_i - \phi_i^{\mathrm{T}}(k)P_k^i\phi_i(k) - Q_i^{\mathrm{T}} - K_i^{\mathrm{T}}R_iK_i \geq 0 \qquad (26)$$

hold, respectively.

Considering (25), we have

$$(25) \Leftrightarrow \alpha P_i - (\phi_i + \Xi_i(k))^{\mathrm{T}}P_j(\phi_i + \Xi_i(k)) \geq 0$$
$$\Leftrightarrow \begin{bmatrix} \alpha P_i & (\phi_i + \Xi_i(k))^{\mathrm{T}} \\ * & P_j^{-1} \end{bmatrix} \geq 0.$$

By denoting $X_i = \gamma P_i^{-1}$ and performing a congruence to the above by $\mathrm{diag}\{\gamma^{\frac{1}{2}}P_i^{-1}, \gamma^{\frac{1}{2}}I\}$, we know (25) is equivalent to

$$\begin{bmatrix} \alpha X_i & (\bar{\phi}_i + \Xi_i(k)X_i)^{\mathrm{T}} \\ * & X_j \end{bmatrix} \geq 0. \qquad (27)$$

By denoting

$$Z_{1i} = \begin{bmatrix} \alpha X_i & \bar{\phi}_i^{\mathrm{T}} \\ * & X_j \end{bmatrix},$$

we know (27) can be written as

$$Z_{1i} + \begin{bmatrix} 0 \\ E_i \end{bmatrix}\Delta_i[\Xi_i \quad 0] + \begin{bmatrix} \Xi_i^{\mathrm{T}} \\ 0 \end{bmatrix}\Delta_i^{\mathrm{T}}[0 \quad E_i^{\mathrm{T}}] \geq 0.$$

According to lemma 1, we know that (27) holds, only if

$$Z_{1i} - \delta\begin{bmatrix} 0 \\ E_i \end{bmatrix}[0 \quad E_i^{\mathrm{T}}] - \delta^{-1}\begin{bmatrix} \Xi_i^{\mathrm{T}} \\ 0 \end{bmatrix}[\Xi_i \quad 0] \geq 0$$
$$\Leftrightarrow \begin{bmatrix} \alpha X_i - \delta^{-1}\Xi_i^{\mathrm{T}}\Xi_i & \bar{\phi}_i^{\mathrm{T}} \\ * & X_j - \delta E_iE_i^{\mathrm{T}} \end{bmatrix} \geq 0$$

holds. Applying Schur complement to the above inequality, we obtain LMI constraint (15).

Considering (26), we have

$$(26) \Leftrightarrow \alpha\pi_k^iP_i - (\phi_i + \Xi_i(k))^{\mathrm{T}}P_k^i(\phi_i + \Xi_i(k))$$
$$-Q_i - K_i^{\mathrm{T}}R_iK_i \geq 0$$
$$\Leftrightarrow \begin{bmatrix} \alpha\pi_k^iP_i & (\phi_i + \Xi_i(k))^{\mathrm{T}} & Q_i^{\frac{1}{2}} & R_i^{\frac{1}{2}} \\ * & (P_k^i)^{-1} & 0 & 0 \\ * & * & I & 0 \\ * & * & * & I \end{bmatrix} \geq 0.$$

By denoting

$$X_i = \gamma P_i^{-1},$$
$$U_i^{\mathrm{T}} = \left[(\pi_{ik_1^i})^{\frac{1}{2}}(\bar{\phi}_i + \Xi_i(k)X_i)^{\mathrm{T}} \quad (\pi_{ik_2^i})^{\frac{1}{2}}(\bar{\phi}_i + \Xi_i(k)X_i)^{\mathrm{T}}\right.$$
$$\left.\cdots \quad (\pi_{ik_q^i})^{\frac{1}{2}}(\bar{\phi}_i + \Xi_i(k)X_i)^{\mathrm{T}}\right]$$

and performing a congruence to the above by $\mathrm{diag}\{\gamma^{\frac{1}{2}}P_i^{-1}, \gamma^{\frac{1}{2}}I, I, I\}$, we know (26) is equivalent to

$$\begin{bmatrix} \alpha\pi_k^iX_i & U_i^{\mathrm{T}} & X_iQ_i^{\frac{1}{2}} & Y_i^{\mathrm{T}}R_i^{\frac{1}{2}} \\ * & W_j & 0 & 0 \\ * & * & I & 0 \\ * & * & * & I \end{bmatrix} \geq 0. \qquad (28)$$

By putting some simple matrix arrangements to the above inequality, we obtain LMI constraint (16).

LMI constraints $(17) \sim (19)$ can be obtained following the same lines of the proof of theorem 1 of [6] and we omitted the proof details here. In summary, (14) gives upper bound of index (12), and $(15) \sim (19)$ guarantee the FTS of MJLS (11). This completes the proof.

4 NUMERICAL EXAMPLE

In this section, a numerical example is given to show the potential of the finite-time strategy. We borrowed MJLS (1) with three operation modes from [2].

$$A_1 = \begin{bmatrix} 0.88 & -0.05 \\ 0.4 & -0.72 \end{bmatrix}, B_1 = \begin{bmatrix} 2 \\ 1 \end{bmatrix}, E_1 = \begin{bmatrix} 0.2 & 0 \\ 0 & 0.2 \end{bmatrix},$$
$$H_{11} = \begin{bmatrix} 0.1 & 0 \\ 0 & 0.1 \end{bmatrix}, H_{21} = \begin{bmatrix} 0.1 \\ 0 \end{bmatrix}, \Delta_1 = \begin{bmatrix} \sin(k) & 0 \\ 0 & \cos(k) \end{bmatrix},$$
$$A_2 = \begin{bmatrix} 2 & 0.24 \\ 0.8 & 0.32 \end{bmatrix}, B_2 = \begin{bmatrix} 1 \\ -1 \end{bmatrix}, E_2 = \begin{bmatrix} 0.1 & 0 \\ 0 & 0.2 \end{bmatrix},$$
$$H_{12} = \begin{bmatrix} 0.1 & 0 \\ 0 & 0.2 \end{bmatrix}, H_{22} = \begin{bmatrix} 0.2 \\ 0 \end{bmatrix}, \Delta_2 = \begin{bmatrix} \cos(k) & 0 \\ 0 & \sin(k) \end{bmatrix},$$
$$A_3 = \begin{bmatrix} -0.8 & 0.16 \\ 0.8 & 0.64 \end{bmatrix}, B_3 = \begin{bmatrix} 1 \\ 1 \end{bmatrix}, E_3 = \begin{bmatrix} 0.1 & 0 \\ 0 & 0 \end{bmatrix},$$
$$H_{13} = \begin{bmatrix} 0 & 0 \\ 0 & 0.2 \end{bmatrix}, H_{23} = \begin{bmatrix} 0 \\ 0.1 \end{bmatrix}, \Delta_3 = \begin{bmatrix} \sin(k) & 0 \\ 0 & \sin(k) \end{bmatrix},$$
$$\Pi = \begin{bmatrix} 0.2 & ? & ? \\ ? & ? & 0.4 \\ 0.2 & ? & ? \end{bmatrix}.$$

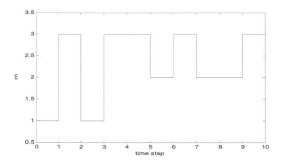

Figure 1. Jump Modes.

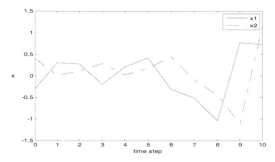

Figure 2. State response of free MJLS.

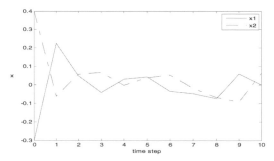

Figure 3. State response under finite-time RHC.

where Π is a partly unknown one-step transition probability matrix.

The weighting matrix is taken as $Q_i = R_i = G_i =$ diag$\{1, 1\}$. The boundary of the ellipsoids are taken as $c_1 = 0.5$ and $c_2 = 1.0$, respectively. α is chosen as 1.65. The initial state is set to be $x_0 = [-0.3 \quad 0.4]^T$ and the initial mode is $r_0 = 1$. Simulation time is chosen as 10 time units and each unit is taken as Ts $= 1$. The mode path from time step 0 to the time step 10 is generated randomly, 10 times. The cumulated cost is taken as $\sum_{k=0}^{10} x_k^T Q_i x_k + u_k^T R_i u_k$.

We solve SDP (13) subject to LMIs (14) $\sim$ (19) at every sampling time (on-line) to regulate the system into the mean square stable sense while optimizing quadratic performance index and satisfying the ultimate state constraint under a given mode evolution. The simulation results are shown in the following figures.

From the simulated graphs, one can observe that the proposed finite-time RHC strategy for MJLS (1)

is effectively justified. In Fig. 2, it is obviously to see that the MJLS, under a given mode evolution shown in Fig. 1, is unstable and the overshot of the state response is quite large, which violates the FTS requirement. However, as shown in Fig. 3, after applying the finite-time RHC strategy, the state responses are limited in a small region. Note that the state may not converge to zero on the horizon [0, 10], actually its ultimate value is $[0.0602 \quad -0.0924]^T$. That is to say, our proposed strategy guarantees FTS for the controlled system while optimizing the performance index.

5 CONCLUSIONS

In this paper, the LMI approach is utilized to study the finite-time RHC problem for MJLS. The resulting receding horizon controller guarantees the FTS of the closed-loop system and provides a guaranteed cost index over a finite-time interval. It is noted that the concept of FTS is not the same as LSS. We pay more attention on the transient response of MJLS by relaxing the dissipation constraints. In order to reveal the advantages of the proposed method, a numerical result is shown in graphs. Moreover, some comparisons with previous reports are also discussed.

FUNDING ACKNOWLEDGEMENTS

This project was jointly supported by NSFC (60973095), self-determined research program of Jiangnan University (JUSRP11233), start-up fund of scientific research of Jiangnan University (20122837).

REFERENCES

[1] Costa O L V, Fragoso M D, Marques R P. Discrete time Markovian jump linear systems, *London: Springer-Verlag*, 2005.

[2] Zhang L X, Boukas E K. Stability and stabilization of Markovian jump linear systems with partly unknown transition probabilities, *Automatica*, 2009, 45(2): 463–468.

[3] Vargas A N, Furloni W, do Val J B R. Constrained model predictive control of jump linear systems with noise and non-observed Markov state, *American Control Conference*, Minneapolis, 2006, 929–934.

[4] Liu A D, Yu L, Zhang W A. One-step receding horizon H_∞ control for networked control systems with random delay and packet disordering control, *ISA Transactions*, 2011, 50(1): 44–52.

[5] Weiss L, Infante E. Finite time stability under perturbing forces and on product spaces, *IEEE Transactions on Automatic Control*, 1967, 12(1): 54–59.

[6] Amato F, Ariola M. Finite-time control of discrete-time linear systems, *IEEE Transactions on Automatic control*, 2005, 50(5): 724–729.

[7] Amato F, Ariola M, Cosentino C. Finite-time control of discrete-time linear systems: Analysis and design conditions, *Automatica*, 2010, 46(5): 919–924.

Information Technology and Computer Application Engineering – Liu, Sung & Yao (Eds)
© *2014 Taylor & Francis Group, London, ISBN 978-1-138-00079-7*

Equivalent interpolation algorithm for NIRS data

Yan Jiang
Department of Software, Shenyang University of Technology, China

Xin Li
Department of information science and technology, Shenyang University of Technology, Shenyang, China

Xin Jin
Sheng yang Testing Equipment Co., Ltd., China

ABSTRACT: When the NIRS data are handled smoothly, we extract the characteristic points, and interpolation calculation will be executed based on these characteristic points, obtain a more accurate and smooth curve at last. The aim of research presents a way is how to avoid the emergence of the "Runge phenomenon" and implement equal value interpolation that the smoothed curve passes through the characteristic pints of original data. This paper presents new algorithm of equal value interpolation, and the process of data recovery with the characteristic points. The auxiliary points could induce interpolation algorithm to realize computation of the interpolation, and will meet the requirements of equal value interpolation. In this paper the different methods of computation auxiliary points and the analysis of the similarity between interpolation curve and original data curve with different auxiliary points are also provided. With the different auxiliary points we implement symmetric and asymmetric interpolation curve. The curve with equal value interpolation algorithm based on auxiliary points can preferably describes the variation tendency of original data.

Keywords: Interpolation algorithm; Time series analysis; Near-infrared spectroscopy (NIRS); Oxygenated hemoglobin

1 INTRODUCTION

The main aim of Brain Computer Interface (BCI) builds a communicating bridge between brain and peripheral devices [1]–[3]. One of the essential conditions to want better development and popularizing application of the BCI system is finding a kind of signal which could reflect different mental state of brain and could be extracted and classified in real time or short term. Electroencephalogram (EEG) is a non-invasive technology of brain activity, with high resolution, reliability, the amount of information, visual images of features, so it becomes one of the best choices for BCI [4–5].

In the field of signal processing, the empirical mode decomposition (EMD)[6]–[9] has been recognized as the driving signal decomposition method of effective data, and has been widely applied to multiscale signal analysis. EMD method is to remove the average of superior envelope and inferior envelope in the source data, and it will inevitably affect the true value to the original data.In the new algorithm, we ensure the effective characteristics of the data at the same time, and use a smaller number of feature points to complete the description of the source data. Through calculation we can get the new data replacing feature points. These data not only describe the characteristics of the source data, but also facilitate interpolation algorithm to interpolate. The source data is described as a smooth curve by interpolated data we need. The curves can describe the basic characteristics of the source data, and output stable data.

2 METHOD

2.1 Improved algorithm

After we get the characteristic points, we hope to restore the smooth curve from the characteristic points by interpolation algorithm. Ordinary interpolation algorithm could form interpolation curve with the characteristic points, but this curve could not meet the requirements. Because interpolation curve could not be formed in accordance with the trend of the original data, and could not guarantee that the characteristic points appear in the peaks and troughs of the curve. Therefore, we need improve the interpolation algorithm.

In the process of traditional interpolation algorithm, when the number of interpolation points and operations are too many, so the insertion value is uncertainty.it is said that although we can get the given value, there will be a great deviation between "fact" and the value in the vicinity, this kind of phenomenon is also called "the Runge phenomenon". The solution is piecewise interpolation polynomial in low degree, to reduce the interpolation point. According to the features of the data we need interpolation, we need to be closed interval segmentation, to make the relatively small number of interpolation points in each cell and then to interpolate between each cell.

After determining the number of interpolation points, according to the experiment data, the effect between the two interpolation points in about 5 is relatively satisfactory, but more than 10 will be more serious, so if the interpolation points are not too many, every interval with 5 interpolations can get good effect. But to ensure that each contact more closely, make the interpolated curve more smooth, we should let the interval contains several identical interpolation point in the interval distribution. The number of selected the interpolation points for N, The interpolation points for each interval contains m. Q[i] defines a set of interpolation points every time.

Our approach is in accordance with the sliding window, the window contains the M points, each window sliding down the instrument point. This two time interpolation calculation of a point M-1 is repeated last used interpolation point. The algorithm only needs to compute at point N-m +1, the end of the computation. The smoothness of the m value determination will affect the curve.

This interpolation method avoids the "Runge phenomenon", but there is phenomenon of deviation from the original data in interpolation curve. Although with three points we can describe trough and wave crest of the curve, the curve is not necessarily expected curve. The curve passes through the interpolation point, but the different coordinates of the interpolation points affects the accuracy of the calculated result greatly. With the aid of the auxiliary points, interpolation algorithm can be guided to meet the requirements during the interpolation process.

2.2 The Generation of auxiliary points

The auxiliary point is generated by the characteristic points. We can calculate the auxiliary points with characteristic points, and the auxiliary points could induce interpolation algorithm to realize computation of the interpolation, and will meet the requirements of equal value interpolation. Now we will introduce some methods to generate the auxiliary points.

Firstly, we introduce the principle of calculating the auxiliary point.

To obtain a piecewise smooth curve, we can define every segment as a parabolic curve. This method cannot guarantee that three points will be able to determine a parabola, so according to the two points outside the peak value and y_2 can calculate the equation of the curve. We use (x_2',y_2) to replace (x_2,y_2). Based on coordinates of two points and ordinate of peak point, we determine curve equation (Fig. 1).

Because there is square root in the operation, the calculation result of M is two values. We can

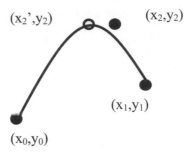

Figure 1. Position of three points (x_0,y_0), (x_1,y_1) and $(x_2,y_2) \cdot (x_2',y_2)$ is the position of new point. Using (x_2',y_2) to replace (x_2,y_2). Calculate new points with (x_0,y_0), (x_1,y_1) and (x_2',y_2).

determine which one of the two values is consistent with the curve we design. The result of data analysis shows that m is the abscissa of parabola, and the value of m is between x_0 and x_1.

$$\begin{cases} y_0 = a*(x_0-m)^2 + y_2 \\ y_1 = a*(x_1-m)^2 + y_2 \end{cases} \tag{1}$$

$$m = \frac{x_1-x_0}{\pm 2\sqrt{\frac{y_0-y_2}{y_1-y_2}-1}} + x_1 \tag{2}$$

$$a = \frac{y_0-y_2}{\sqrt[2]{x_0-a_0}} \tag{3}$$

With the formula, we can calculate the auxiliary points to interpolate. First of all, we analyze how to select the two points of (x_0, y_0) and (x_1, y_1) with the regular characteristic points. We analyze that the auxiliary points have the effects on the procession of interpolation S. The five points we selected are assumed as the feature points. If we want the interpolated curve to pass through these five points, and these five points are continuous peaks of the curve, the curve will pass through the intermediate points of the connection of each two adjacent points. The intermediate points will be added into original peaks to calculate auxiliary points' position. We use the auxiliary points to replace original the characteristic points to realize interpolation calculation. We find six intermediate points as characteristic points in the connectivity, and figure out 12 new points as the auxiliary points to complete interpolation calculation (Fig. 2). The curve after interpolation passes through five characteristic points with the auxiliary points (Fig. 3).

In the practical application, the distribution of feature points is usually asymmetric triangular arrangement, so we change the coordinate of the point (x_2, y_2) from (400,450) to (450,450). There is an asymmetric curve in the second crest. In this situation, we have two methods to calculate auxiliary points one is that the abscissa of peak is changed, the other is unchanged.

$$m = \begin{cases} \frac{x_1-x_0}{-2\sqrt{\frac{y_0-y_2}{y_1-y_2}-1}} + x_1 & x_1 > x_0 \\ \frac{x_1-x_0}{2\sqrt{\frac{y_0-y_2}{y_1-y_2}-1}} + x_1 & x_1 < x_0 \end{cases} \tag{4}$$

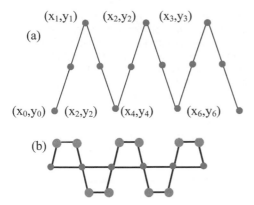

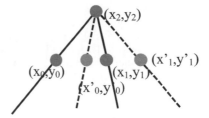

Figure 4. Position of 5 points. Red points are auxiliary points. Green points are new auxiliary points.

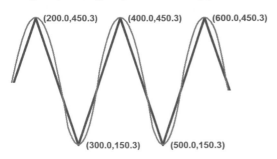

Figure 2. For both (a) and (b), the red points are characteristic points, and the green points are auxiliary points.

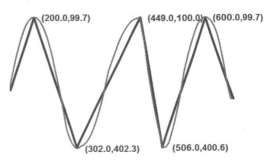

Figure 5. The processing of the second group of data. Interpolation curve is depicted as red curve. Characteristic curve is depicted as blue curve.

the reason and propose some advanced strategies and suggestions.

$$\begin{cases} x_1' = 2 \times x_2 - x_0 \\ y_1' = y_0 \end{cases} \tag{6}$$

Figure 3. The processing of the first group of data. Interpolation curve is depicted as red-curve. Characteristic curve is depicted as blue curve.

2.3 The interpolation of auxiliary points

In the first case, the method of calculating the auxiliary points is the same as the method of the above introduction. There is deviation between the new peak abscissa and original peak abscissa.

In the second case, in order to ensure the abscissa of peak is unchanged, we choose different auxiliary points on each side of peak.

$$\begin{cases} x_0' = 2 \times x_2 - x_1 \\ y_0' = y_0 \end{cases} \tag{5}$$

When we calculate the auxiliary points of the left side of the peak, we use the point of (x_1',y_1') to replace the point of (x_1,y_1), and use the points of (x_0,y_0), (x_2,y_2) and (x_1',y_1') to calculate auxiliary points with formula. Because the point of (x_2,y_2) is equidistant from the point of (x_0,y_0) and the point of (x_1',y_1'), according to this auxiliary points interpolation curve must go through the extreme point. The same method can be used to replace (x_0,y_0) with (x_0',y_0'). use the points of (x_0,y_0), (x_2,y_2) and (x_0',y_0') to calculate auxiliary points in right of the extreme points (Fig. 4). We use both sides of the extreme points to complete the interpolation calculation. But in the interpolation process, we found the deviation between the new peak point and original peak point, and two points cannot be better coincidence. Based on this, we explore

To avoid interpolation algorithm "Runge phenomenon" and according to the characteristics of the data we need to interpolation, interpolation interval is divided. There are small numbers of interpolation points in every interval, and we realize interpolation computation respectively in each interval. There are 3 auxiliary points in every interval in new algorithm. When interpolation algorithm is in the process of the realization of the interpolation calculation, three auxiliary points appear on either side of extreme point, and these three points are calculated by the same formula, so the interpolation data meet the requirements. But when the three auxiliary points appears on both sides of the extreme point, the formula on both sides of the extreme points is different, and the data after interpolation will deviate from the original extreme points (Fig. 5). The interpolation curve peak will be less than the extreme point.

In this case, there are two methods to solve this problem. The first method is when the auxiliary point appears on both sides of the extreme point, and we use two sets of auxiliary points. When two points is on the left side of the extreme point, the auxiliary points on the right side need to be changed. When two points is on the tight side of the extreme point, the auxiliary points on the left side need to be changed. That is to say that there are two groups auxiliary points, and will be used to be interpolated on each side of the extreme

23

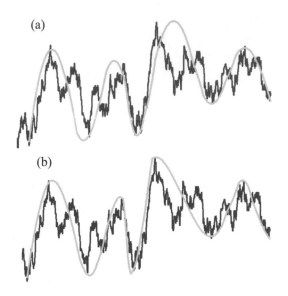

(a)

(b)

Figure 6. (a) the blue curve is original time series. (b) the superior and inferior envelopes are depicted as red and green dashed curves.

point. The second method is that we can select the auxiliary point near one side of the extreme point, so we can ensure the curve pass through extreme point.

3 RESULT

Here we use a set of actual data to achieve the extraction of the characteristic curve. We use two methods to design the auxiliary point for interpolation. There are the results of interpolation calculation with a group of auxiliary points in Fig. 6(a). The interpolation curve with this method is smooth, but the abscissa of peak will deviate from the original peak point. There are the results of interpolation calculation with two groups of auxiliary points in Fig. 6(b). The interpolation curve with this method passes through original characteristic points, but it is not as good as the first one. The results showed that when the position of characteristic point is center of every interval, the peak of interpolation curve is close to characteristic points. With the induction of auxiliary point, the interpolated curve may reflect trends of original data and ensure that the peak of interpolation curve is the same as the value of characteristic points. We choose the generation method of auxiliary points based on the requirements of curve.

4 CONCLUSION

According to the characteristics of the data points, we use the interpolation algorithm to achieve the restoration of the characteristics of the original data. This paper presents new algorithm of equal value interpolation, and the new algorithm can solve the

"Runge phenomenon" that possibly happens in the conventional interpolation algorithm. The auxiliary points could induce interpolation algorithm to realize computation of the interpolation, and will meet the requirements of equal value interpolation. The analysis of the similarity between interpolation curve and original data curve with different auxiliary points are also provided. Experimental result show that the curve with equal value interpolation algorithm based on auxiliary points can preferably describes the variation tendency of original data, and can be used as output signal.

ACKNOWLEDGEMENT

The author finished this paper at the Intelligent Robotics Laboratory of the Kochi University of Technologyas a visiting scholar. The authors would like to express our hearty thanks for their support to conduct NIRS experiments.

REFERENCES

[1] Flandrin, Patrick, and Paulo Goncalves. 2004. Empirical mode decompositions as data-driven wavelet-like expansions. *International Journal of Wavelets*, 2(4):477–496.
[2] M. Alamgir, M. Grosse-Wentrup, and Y. Altun. Multitask learning for Brain-Computer Interfaces. *International Conference on Artificial Intelligence and Statistics (AISTATS2010)*, *May 13–15, 2010, pp. 17–24*.
[3] B. Blankertz, C. Sannelli, S. Halder, E.M. Hammer, A. Kübler, K.R. Müller,G. Curio, and T. Dickhaus. 2010. Neurophysiological predictor of SMR-based BCI Performance. *NeuroImage*, 51(4): 1303–1309.
[4] S. Fazli, J. Mehnert, G. Curio, A. Villringer, K.-R. Müller, J. Steinbrink, and B. Blankertz. 2012. Enhanced performance by a hybrid NIRS-EEG Brain Computer Interface. *NeuroImage*, 59(1): 519–529.
[5] Daly, Janis J., and Jonathan R. Wolpaw. 2008. Brain–computer interfaces in neurological rehabilitation. *Lancet neurology*, 7(11):1032–1043.
[6] Fan Deng, Dajiang Zhu, Jinglei Lv, Lei Guo, and Tianming Liu. 2013. FMRI signal analysis using empirical mean curve decomposition. *IEEE transactions on biomedical engineering*, 60(1):42–54.
[7] M. De Luca, C.F. Beckmann, N. De Stefano, P.M. Matthews, and S.M. Smith. 2006. fMRI resting state networks define distinct modes of long-distance interactions in the human brain, *Neuroimage*, 29(4):1359–1367.
[8] D.J. Heeger and D. Ress. 2002. What does fmri tell us about neuronal activity? *Nature*, 3(2):142–151.
[9] N.K. Logothetis. 2008. What we can do and what we cannot do with fmri. *Nature*, 453(7197):869–878.

Information Technology and Computer Application Engineering – Liu, Sung & Yao (Eds)
© 2014 Taylor & Francis Group, London, ISBN 978-1-138-00079-7

Improvement and implementation of the MD5 algorithm based on user security

Guozhu Liu & Huaxin Qi

Information College of Science and Technology, Qingdao University of Science & Technology, Qingdao, China

ABSTRACT: The MD5 algorithm is an algorithm that is able to turn arbitrarily different lengths of messages into a 128-bit digest. Each digest of corresponding to the message is unique, and it is difficult to calculate the message itself by digest. It specifically elaborates the realization of the principle and the security of the MD5 algorithm. For the brute-force method and collision attack mean, improvement and implementation of the MD5 algorithm in user security application is proposed. Changing the structure of the plaintext and ciphertext can ensure the security of the user security.

Keywords: MD5 algorithm; abstract; message; User Security

1 INTRODUCTION

With the rapid development of network technology, information security has been a great threat; the security of network information becomes particularly important. User login password verifies that user is legitimate or not, which is an important basis. How to ensure user login password security becomes the focus of user security. Once the user login information was leaked, intercept or tamper, it will bring enormous threat to users, the government and enterprises. Traditionally, the user login username and password are encrypted using a simple encryption technology directly in the network remote transmission, so it is easy to be intercepted and listened, and obtain the user's important information. Database stores the user login user name and password, which is very easy for illegal personnel to be attacks on[1], user login information are large area leaked. Therefore, the MD5 algorithm is proposed to encrypt the user login password, it can protect accuracy and integrity of the user login information. However, with the progress of science and technology, MD5 algorithm has been cracked constantly. The MD5 algorithm has been cracked constantly, for the exhaustive attack and collision attack. A new MD5 algorithm improvement scheme is put forward. The improved MD5 algorithm changes and processes the user login password itself and messages through the MD5 algorithms encrypt. The user login password encryption adjustment effectively increase difficulty of the exhaustive attack, the changes of ciphertext can effectively prevent the collision attack.

2 MD5 ALGORITHM

MD5 algorithm[2], called Message Digest algorithm 5 algorithm, also known as Message-Digest algorithm. The algorithm inputs an arbitrary length message, and the output is a compressed 128 digest. Arbitrary message does not produce two different digests, at the same time two arbitrary different messages are also very difficult to calculate the same digest, so as long as the message has been any tiny changed, it will affect the final digest, and lead to great changes itself. The MD5 algorithm is an one-way and irreversible algorithm, it is widely used for user authentication, digital signature and data integrity verification.

3 MD5 ALGORITHM

MD5 algorithm is based on the 512 bit as a data packet, after processing algorithm encryption arithmetic, output is a 128 big integer that consists of four 32-bit packets. There are 5 step processes that mainly include complement, adding data length, variable initialization, packet transform operations, the final output [3]. The details are as follows:

(1) Complement. The input data is carried on unified complement, after unified the data into a standard length, the next step of operation is effectively continued. Complement method is that the length of complement data take more than 448,512. that is L% $512 = 512$ (L stand for the length of complement data), the first bit is 1, all the other data bits is 0. If the input data is 120, 512 $(120 + 328)\% = 448$, so adding

the length of the data is 328, one of the "1" and "0" of 327.

(2) Adding the data length. The length of the input data expresses as a 64 bits binary. it is appended to the complement data, if the length of input data is over long, 64 bits binary cannot complete express. So it is retained after 64 bits, remove the previous binary digits. The final length of such data is added to an integer multiple of 512 bits.

(3) Variable initialization. Four 32-bit registers store variable initialization A, B, C and D that respectively expresses hexadecimal, it uses to calculate the information digest, they respectively are:

A: 0x01234567, B: 0x89abcdef, C: 0xfedcba98, D: 0x76543210

(4) Main operation.

1) Processing blocks of 512 bits is divided into 16 small blocks, each small block is 32 bits.

2) The (3) of four variable are assigned to the variable a, b, c, d, $a = A$. $B = B$, $c = C$, $d = D$.

3) Each processing block is the 4 rounds main loop. Each round has 16 operations, so it needs a total of 64 arithmetic operations. Every operation is very similar. The specific operation is: three of the four variables such as a, b, c, d have a nonlinear function, then the results are combined with fourth variables, a grouping of the text and a constant, the result ring right shift an indefinite number, and plus a variable of a, b, c or d. The final result is assigned to a variable of a, b, c or d. Operation [4] as follows:

FF (a, B, C, D, Mi, s, Ti) a=b+ ((a+ (F (B, C, d) +Mi+ti) <<<s)

GG (a, B, C, D, Mi, s, Ti) a=b+ ((a+ (G (B, C, d) +Mi+ti) <<<s)

HH (a, B, C, D, Mi, s, Ti) a=b+ ((a+ (H (B, C, d) +Mi+ti) <<<s)

II (a, B, C, D, Mi, s, Ti) a=b+ ((a+ (I (B, C, d) +Mi+ti) <<<s)

Among them, the following definition of the auxiliary function[5]:

$F(x, y, z) = (x \& y) | ((\sim x) \& z)$ $G(x, y, z) = (x \& z) | (y \& (\sim z))$

$H(x, y, z) = x^\wedge y ^\wedge z$ $I(x, y, z) = y^\wedge (x | (\sim z))$

(X, Y and Z are 32 bit integer; & expressed as AND, | expressed as OR, $\sim$ expressed as NO, $^\wedge$ expressed as XOR)

The above four functions: if the corresponding of x, y and z are independent and uniform, then each one of the results is independent and uniform, the function F operates by bitwise; if X, then Y, or Z, function H operates by bit-by-bit parity

(5) Output. After all the packet are operated, finally, the MD5 algorithm generates a 128 bits output.

4 SECURITY ANALYSIS OF MD5 ALGORITHM

According to the characteristics of MD5 algorithm, it is applied to the user login password, hacker attack user login password, and their main methods are:

(1) The exhaustive method attack. This method firstly knows the user login password MD5, then it can decode MD5 algorithm. The common characters and numbers are combined into a password, and then the corresponding digest is calculated with the MD5 algorithm, according to plaintext – ciphertext table, user login password is searched. We assume that the maximum length of the password cannot be more than 12 bytes, while assuming that the password is only uppercase letters, lowercase letters and numbers, not the other characters, a total of $26 + 26 + 10 = 62$ characters, this combination has $P(62, 1) + p(62, 2) + \cdots + p(62, 12)$, there are about Billions of combinations. If the number of characters increases or user login password Increases the length, then the combined number will be geometrically increase.

(2) Collision attack. Collision is two different messages to get the same digest. Hackers try to find two different messages to generate the same digest through the MD5 algorithm, and if you find such digest, then it can produce a collision. Therefore, calculation of collision is not the specified legitimate user login password, and need to spend a lot of time to calculate collision.

5 IMPROVED OF MD5 ALGORITHM IN THE APPLICATION OF USER SECURITY

5.1 Improved scheme of the MD5 algorithm

In order to increase the difficulty of cracking MD5 algorithm, an encryption change scheme of plaintext and ciphertext is proposed. Specific improving plan follows: (1) the user login password makes a simple change before using MD5 encryption algorithm. First of all, additional strings are added with digitals and characters in order to lengthen the length of password and increase the difficulty of break. Then, of the lengthened user login password make a simple replacement, the formation of lengthened password is written using diagonal, then it is written in the order. This replacement is made two times, so the disruption of the additional characters and user password in the order to increase use security. (2) The MD5 algorithm encrypted ciphertext make a change. A random intercept a data of the ciphertext, and then the equal length data is obtained using random function, it instead of the intercepted data. So the data length is unchanged, and increases the ability to fight against collision. The concrete encryption method follows:

The process of plaintext

1) In the back of the user input password joining randomly a character receives password1. Character table that consists of the combination of tens of thousands of digitals and characters is stored in the server database.

2) Password 1 makes a replacement. The specific method is: the lengthening password is written by the diagonal form, password2 is written in the order.

3) Repeating operation 2), and finally password3 is received.

(2) The process of ciphertext.

1) The final plaintext password3 is encrypted by using the MD5 algorithm, the ciphertext password4 = md5 (password 3).

2) The ciphertext password4 is intercepted by using of interception function. The data is intercepted num bits from the position of begin, so password4 is intercepted into three sections. L = left (password4, begin-1), M = (begin, begin + num-1), R = right (begin+num, password4). password5 = L & M & R.

3) Using a random function random (Num) replaces the intercepted num bits M.

4) Finally, the transformed code value is newpassword = L & random (Num) & R. The user login password MD5 value is newpassword.

Algorithm flow chart is the following figure 1:

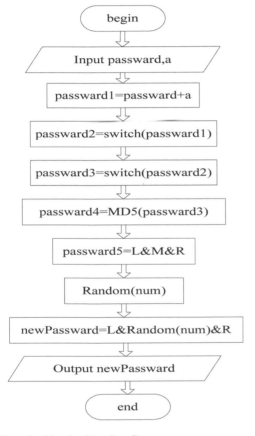

Figure 1. The algorithm flow figure.

5.2 *Implementation of the improved MD5 algorithm*

When users enter the login interface, they send the login request to the server, after the server get the user's request, it will randomly choose a characters from the character table to the user, and record extraction character. On the client the user make a simple replacement of password two times, break into password sequence. Then it is encrypted using the MD5 algorithm, the encrypted ciphertext make an interception and replacement, finally the user login password corresponds to the MD5. The user will send the encrypted login password, the L and R of ciphertext to the server. The server will operate the login password and the recorded characters like the client, finally the server compares to L and R came from the client. If the two values are the same, then the user normally lands. So the safety and reliability of user login password are higher. The test result follows Figure 2:

Please input the text: abd345

Operate the plaintext for the first time: abd345abcdef

The first diagonal replacement operation: ad4aceb35bdf

The second diagonal replacement operation: a4cb5ddae3bf

MD5 encryption operation:
1db97f478684226e55f107597f9aflfe

Replacement operation:
1db97BTsu684226e55f107597f9aflfe

Figure 2. Test results figure.

6 CONCLUSION

The improved MD5 algorithm enhances the confidentiality and security of user login password. Added user login password length and disrupted its own order will hide characteristics of the user login password and increase the difficulty of exhaustive attack. At the same time, after the random function replaces the MD5 value, which will hide characteristics of the MD5 value and increase the difficulty of collision attack. Therefore, plaintext and ciphertext make structural change to ensure the security of user information.

ACKNOWLEDGEMENT

The work described in this paper was supported by a grant from Project of Shandong Province Higher Educational Science and Technology Program of China (Project No. J08LJ21), and Shandong Province Natural Science Foundation of China (Project No. ZR2009GL006).

REFERENCES

[1] Luo Jianghua. MD5-Base 64 based hybrid encryption algorithm [J]. Computer Application. 2012, 32 (A01):47–49.

[2] Zhang Yizhi, Tang Xiao bin, Zhao Yi. MD5 algorithm [J]. Computer Science. 2008, 35 (7):295–297.

[3] Wang Jinzhu, Li Yuancheng. Application of MD5 algorithm based on J2EE in user management system [J]. Computer Engineering and Design. 2008, 29 (18):4728–4730.

[4] Zhang Xinyan, Run Deqin. Generating stronger avalanche effect in MD5 algorithm [J]. Computer Application. 2010, 30 (5):1026–1029.

[5] Zhang Runmei, Wang Xiao. MD5 Crack Method Based on Compute Unified Device Architecture [J]. Computer Science. 2011, 38 (2):302–305.

Information Technology and Computer Application Engineering – Liu, Sung & Yao (Eds)
© *2014 Taylor & Francis Group, London, ISBN 978-1-138-00079-7*

Worm detection research based on biological immune principle and FCM algorithm

Guozhu Liu, Huimin Kang & Huaxin Qi
Information College of Science and Technology, Qingdao University of Science & Technology, Qingdao, China

ABSTRACT: Today, the frequency of worm outbreak is faster and faster. Especially, more and more worms emerged in the latest two years such as blaster worm, sassier worm, SQL worm, W32/Netsky2P worm and W32/Zafi2B etc... So, the detection and defense to worm has become one of the important research topics in network security. Worm detection technology is fully analyzed in this paper. Some weak points in worm detection are found. One method about worm detection based on biological immune principle and FCM fuzzy clustering algorithm is proposed. The method can detect unknown worms better through simulation experiment. It can be concluded that the method can prevent the outbreak of unknown worms effectively...

Keywords: FCM; fuzzy clustering; network security

1 INTRODUCTION

With the growing popularity of the network in modern life, computer worms and the increasingly rapid spread of the resulting harm is also growing. The detection and prevention technology to the computer worm has become an important research topic in network security[1].

Currently, the research to the worm has also focused on the detection and prevention. The worm control measures are in a very passive state[2]. Therefore, early detection and early warning research to new worm active defense technology are important to reduction the worm outbreak and harm. Because the unknown worm can be early detected and warned.

Biological immune system is a highly distributed, parallel and adaptive system. The system has many fine features in maintaining health of the body and eliminating invasive pathogens[3]. So, we are inspired by the idea of biological immune system to solve the problem of the computer worm detection.

Through the introduction of the method that constructs self and non-self from biological immune principle, the use of fuzzy clustering algorithm FCM computer system to optimize the clustering process, abnormal behavior by monitoring the process to monitor possible worm attacks, because the prior does not need specific information about any worms, so can the unknown worms and worm variants were detected, and finally proved by experiments that the detection method is reasonable.

2 THE CURRENT WORM DETECTION TECHNOLOGY

Worm detection methods currently used are: checksum, feature code, behavior detection, and some other advanced detection techniques[4]. These methods are based on different principles. When implementing with deferent methods, these methods need deferent cost and have different detection range. Each method has itself merit.

2.1 Checksum

The detection method of checksum is to calculate the checksum of the file, and save the checksum. After that, periodically check the checksum calculated by contents of the file with the checksum saved originally. We can find the two checksums are the same or not[5]. So, we can know the file infected by virus or not. By the method, we can find the virus known or unknown before. But we cannot realize the kind of the virus. Furthermore, the operation to file has other methods. Therefore the method of checksum used to virus detection has high check error rate.

2.2 Feature code

Feature code is the simplest, the minimum cost method to detect known viruses. The feature code is collected by taking samples of known virus, and the virus feature code library is constructed[6]. When we detect the virus, we can compare the file with feature code in the library directly.

2.3 Behavior detection

The method of behavior detection is to use virus-specific behavioral characteristics to check the virus[7]. Through research and summarizing on the virus in many years, some of the special behavior of the virus is summarized. When the program is running, the virus behavior is checked. When the behavior of known virus is fined, the virus warning is issued immediately.

Existing computer virus detection technology is difficult to detect unknown virus, and is in a passive defense to the virus. The severe form of the current virus detection, the current virus detection need a method urgently which have a self-learning ability, can classify the virus actively, identify and detect unknown computer virus. We build a new virus detection system by evolving semi-supervised fuzzy clustering algorithm.

3 PRINCIPLE OF BIOLOGICAL IMMUNE SYSTEM

When foreign bacteria invade the body, the biological immune system organize the lymphocyte cells to recognize "self" and "nonego" antigens, and to eliminate and remove foreign bacteria[8]. From computer security and virus detection, the biological immune system shows great potential. Its operating mechanism, system structure, level models etc. are very similar with computer security systems. And it has the following characteristics that the computer security system is required:

Distributed: Biological immune system is a highly distributed computing system.

System reliability: The immune system detects bacteria by lymphocyte. The cell is considered "negative". Thus, the autoimmune response is avoided. The safety and reliability of the system are ensured.

Diversity: Organism can produce large amounts of a variety of immune antibodies.

Fuzzy matching detection: The immune system to produce antibodies not only reacts to the corresponding antigen, but also reacts to a similar antigen by fuzzy matching.

Self-learning and memory: In the immune system, the "learning" and "memory" to the unknown antigen are completed by lymphocyte. So, the immune system responds quickly to the similar antigen invasion.

The core to the immune computation is how to define self body and non self body. The self body must reflect the characteristics of the protected system completely. The characteristics are relatively stable. But non self body is the unusual features of the system (infected by virus)[9].

4 THE BASIC PRINCIPLES OF FUZZY CLUSTERING ALGORITHM (FCM)

Suppose every element in the set $X = \{x_1, x_2, \ldots, x_n\}$ have m characteristics, i.e. $x_i = (x_{i1}, \ldots x_{im})$. The set X should be divided into c classes ($2 \leq c \leq$ n). Suppose X have c cluster centers, these centers are $V - \{v_1, v_2 \ldots v_c\}$, $v_i \in \{v \mid v = \sum_{i=1}^{n} a_i x_i, a_i \in R, x_i \in X\}$.

Set $d_{ik} = \|x_k - v_i\| = \left[\sum_{j=1}^{m}\left(x_{kj} - v_{ij}\right)^2\right]^{\frac{1}{2}}$ d_{ik} is the Euclidean distance between the Samples x_k and cluster centers v_i. The ideal objective is to minimize the objective function $J\left(U, V\right) = \sum_{k=1}^{n} \sum_{i=1}^{c} u_{ik}\left(d_{ik}\right)^2$, so we can get the value U. u_{ik} is the membership degree of the samples x_k for the cluster centers v_i.

5 THE OVERALL PROCESS OF VIRUS DETECTION

Using cluster analysis to detect the virus include many processes mainly as follows: data collection, standardization of data, cluster analysis, tag instance data etc. . . Virus detection process is described as follows:

5.1 Data collection (collected from the self body attribute)

Randomly collected m processes (denoted as sample G), each process has k types of properties (relatively stable process property of the system). On the attribute vector $P_x = (p_1, p_2, \ldots, p_k)(x = 1, 2, \ldots, m)$, each attribute value is respectively converted into two bits string. The string is composed of two numbers 0 and 1. Conversion function g, see Equation (1), where min and max are the lower and upper of its property under normal circumstances.

$$g(p_i) = \begin{cases} 00, p_i < \min \\ 01, \min < p_i \leq (\max+\min)/2 \\ 10, (\max+\min)/2 < p_i \leq \max \\ 11, p_i > \max \end{cases} \quad (i=1,2,\ldots k) \quad (1)$$

In the Equation (1), $g(p_i)$ is the instance data of the process attribute. Denote $P_x' = (g(p_i), g(p_2) \ldots g(p_k))$, then $G = P_1' P_2' \ldots P_m'$.

5.2 Standardization of instance data

In order to eliminate the impact on similarity because of different measurement, the measurements data to the instance should be standardized before the use of clustering algorithms. Standardized method is as follows:

If there are n samples and each sample have m data, then each variable can be written as $X_{i,j}$ (i = 1, 2, . . . , n;

$j = 1, 2, \ldots, m$). After standardized, the variable are $X'_{i,j}$, then:

$$X_{i,j}{}' = \frac{X_{i,j} - \overline{X_j}}{S_j}\begin{pmatrix} i = 1,2,\ldots n \\ j = 1,2,\ldots m \end{pmatrix}$$

$\overline{X_j}$ and S_j are expressed as follows:

$$\overline{X_j} = \frac{1}{n}\sum_{i=1}^{n} X_{i,j}$$

$$S_j = \sqrt{\frac{1}{n-1}\sum_{i=1}^{n}\left(X_{i,j} - \overline{X_j}\right)^2}$$

In the two formulas, $\overline{X_j}$ is the average of jth variable, S_j is the standard deviation of the jth variable. $X'_{i,j}$ is the standardized instance data.

5.3 Implementation of clustering

In order to change the relative membership degree of elements more flexible, the objective function can be generalized as:

$$J\left(U, V\right) = \sum_{k=1}^{n}\sum_{i=1}^{c}\left(u_{ik}\right)^r \left\| x_k - v_i \right\|^2 \qquad (2)$$

In the formula (2), r is a parameter waiting for decision and $r \geq 1$, $\| \cdot \|$ is a norm in space R.

The calculation steps are as below:

1) Set condition and initial value.

Get c and set $2 \leq c \leq n$, set the stop condition ε, initialize the cluster center $V^{(0)}$, iterate step by step ($l = 1, 2, \ldots$).

2) Use $V^{(l)}$, modify $U^{(l)}$

$$U^{(l)} = \frac{1}{\sum_{j=1}^{c}\left(\frac{\left\|x_k - v_i\right\|}{\left\|x_k - v_j\right\|}\right)^{\frac{1}{r-1}}}, \forall i, \forall k \left\|x_k - v_i\right\| \neq 0 \; and \; \left\|x_k - v_j\right\| \neq 0 \qquad (3)$$

$$u_{ik}^{(l+1)} = \begin{cases} 1 & when \; \left\|x_k - v_i\right\| = 0 \\ 0 & when \; \left\|x_k - v_j\right\| = 0 \end{cases} \qquad (4)$$

3) Calculate the cluster center

$$v_i^{(l)} = \frac{\sum_{k=1}^{n}\left(u_{ik}^{(l)}\right)^r x_k}{\sum_{k=1}^{n}\left(u_{ik}^{(l)}\right)^r x_k} \qquad (5)$$

4) Stop iteration

Compare $V^{(l)}$ and $V^{(l+1)}$, for the stop condition $\varepsilon > 0$ (ε value is between 0.001 and 0.01), if $\| V^{(l+1)} - V^{(l)} \| \leq \varepsilon$, then stop, else $l = l + 1$ go to (b).

5) Clustering

U is fuzzy dividing matrix. The matrix U is corresponding to fuzzy dividing X. We can get common cluster to X through two method as below.

a) Method one: $\forall x_k \in X$ if $\|x_k - v_{i0}\| = \min \|x_k - v_i\|$, then x_k belong to i_0 the cluster. v_{i0} is the cluster center in i_0 the cluster. That is if x_k nearest to a cluster center, then x_k belong to the cluster. Highlight all author and affiliation lines.

b) Method two: For the column in matrix U, if $u_{i_0 k} = \max_{1 \leq i \leq c}(u_{ik})$, then x_k belong to i_0 the cluster. That is if membership degree of x_k to a cluster is greatest, then x_k belong to the cluster. This method is the greatest principle method.

6 EXPERIMENT RESULTS

Test experiment is made to unknown worm virus in Windows XP system. Five protected programs are selected. The five protected programs are svchost, winlogon, system, nvsvc32, taskmgr. The process properties have twelve kinds of user state execution time, system state execution time, authority value, resident pages, the cumulative number of pages, the virtual memory page, number of text pages, number of data pages, number of the stack pages, the files number opened by process, the number of TCP ports and UDP ports opened by process and its Subroutine.

The non-self is the file infected by worm virus in the experiment. Forty eight family viruses (total is 148 including deformation of the viruses) infect the system file. Worm detection module identifies the worm by above algorithm (FCM algorithm). At the same time, we use the traditional character code method, check sum method and behavior detection method to detect the worm. The detection results by the four methods are in figure 1.

From figure 1, we know that detection rate by character code method is the lowest to the same unknown worm. When detecting worm with character code method, the method must use worm character code file. If there is no pre-defined character code file, then detection rate will be very low and false alarm rate will be high. For checksum method, according to current file contents to calculate the checksum regularly or before using file, afterwards, we can compare

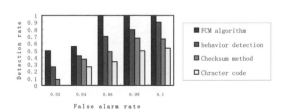

Figure 1. Detection effect.

the calculated checksum with saved checksum. Normal access to a file would be regarded as illegal. So, false alarm rate will be high by this method. For behavior detection method, we summarize same behavior about some worms by long term observing and researching. When running program, the behavior of the program is monitored. If the attack behaviors are found, the alarm is given at once. This method is brought into effects by worm's commonness. The commonness is accumulated by "experience". But, when new worm appeared, the "experience" showed to be inadequate. So, the effect on the detection of unknown worms is greatly reduced. By compared, FCM algorithm have better check ability to worm.

7 CONCLUSION

Check worm method proposed by this paper (FCM algorithm) focus on process dynamic behavior attributes to define self and to cluster the progress attributes. This method has a greater advantage over conventional worm detection technology. Test results also show that this method of detection of unknown worms with good effects. So, the method to worm detection technology for the future research has great application value.

ACKNOWLEDGEMENT

The work described in this paper was supported by a grant from Project of Shandong Province Higher Educational Science and Technology Program of China (Project No. J08LJ21), and Shandong Province Natural Science Foundation of China (Project No. ZR2009GL006).

REFERENCES

[1] MOORED, SHANNON C, CLAFYK. Code-red: a case study on the spread and victims of an Internet Worm [C]Proc of Internet Measurement Workshop. New York: ACM Press, 2002:273–284.

[2] Wen Wei ping, Qing Si han, Jiang Jian chun, et al. Research and development of internet worms[J]. Journal of Software, 2004,15(8):1028–1032.

[3] Dennis L Chao, Stephanie Forrest. Information immune systems [C]. Proceedings of the first international conference on artificial immune systems, England: University of Kent at Canterbury Printing Unit, 2002. 132–140.

[4] Zheng Hui Research of internet worms[D]. Nankai University, 2003.

[5] Cliff Changchun Zou, Weibo Gong, Don Towsley. Code red worm propagation modeling and analysis [C]. ACM Symp on Computer and Communication Security, 2002: 138–147.

[6] Carey Nachenberg. The evolving virus threat[C]. 23rd NISSC Proceedings, Baltimore, Maryland, 2000:201–206.

[7] Zesheng Chen, Lixin Gao, Kevin Kwiat. Modeling the spread of active Worms [C]. INFOCOM 2003. Twenty-Second Annual Joint Conference of the IEEE Computer and Communications Societies. IEEE. Vol 3: 1890–1900.

[8] Jeffrey O Kephart. A biologically inspired immune system for computers[C]. Integrity computing laboratory. Artificial life IV: Proceedings of the fourth international work-shop on the synthesis and simulation of lving systems, US:MIT Press, 1994:130–139

[9] Stephanie Forrest, Alan S Perelson, Lawrence Allen, et al. Selfnonself discrimination in a computer [C]. Proceedings of the 1994 IEEE symposium on research in security and privacy, Los Alamitos, CA: IEEE Computer Society Press, 1994:372–377.

Information Technology and Computer Application Engineering – Liu, Sung & Yao (Eds)
© 2014 Taylor & Francis Group, London, ISBN 978-1-138-00079-7

Fusion of multifocus based on NSCT and NMF

ZiJuan Luo
Science and Technology on Information Systems Engineering Laboratory, Nanjing, China

Shuai Ding
The 28th Research Institute of China Electronics Technology Group Corporation, Nanjing, China

ABSTRACT: It is mostly difficult to get an image that contains all relevant objects in focus. There is multi-focus image fusion method which can solve the problem effectively. A new method of multi-focus image fusion based on Nonsubsampled contourlet transform (NSCT) and Nonnegative Matrix Factorization (NMF) is proposed. After two multi-focus images are decomposed by NSCT, NMF is applied to their high and low-frequency components, respectively, and finally an image is synthesized. Subjective-visual-quality of the image fusion result is compared with those image fusion methods which are based on NSCT and the combination of wavelet/ contourlet with NMF. After evaluated the experimental results, and contrasted the running time, we found that the new method can get larger information entropy, standard deviation and mean gradient. It also means integrate featured information from all source images effectively avoid background noise and promote space clearness in the fusion image.

Keywords: Image fusion; Nonsubsampled contourlet transform; Nonnegative matrix factorization

1 INTRODUCTION

The imperfection of optical lenses system is that it can only focus on some specific object. That is to say we can make all the objects clearly in one image capture by this system. Image fusion (Ying, L. et al. 2007) is an effective way to solve such issue. Using the same system, we can capture several images focus on different objects. After that a new image could be generated by assembling the clear objects of different focus. Then it will be easy for human visual perception and the compute-processing.

Wavelet transform (Shutao, L. & Bin, Y. 2008), as a way to analyze Multiscale Geometric, has been widely used in image fusion fields. Firstly, the original images are decomposed into a series of frequency channels. Then, the different features and details are combined at multiple decomposition levels and in multi-frequency bands. Based on WT, the fusion image generated by Multifocus Image fusion system is suitable for multi-scale properties of the human vision. Meanwhile, the defect is that only limited directional information will be captured. The contourlet transform (CT) can solve this issue (Do, M.N. & Vetterli, M. 2005). Based on efficient two-dimensional multi-scale and directional filter bank (DFB), the contourlet transform can process images with smooth contours effectively. Both WT and CT lack the translation invariant property, leading to pseudo-Gibb's phenomenon with visually disturbing artifacts in fused images. The Nonsubsampled Contourlet transform combining dualtree Nonsubsampled Pyramid (NSP) and Nonsubsampled Directional Filter Bank (NSDFB) effectively solved this problem. Recently Lee and Seung proposed a nonnegative matrix factorization (NMF) to obtain a reduced representation of data (Lee, D.D. & Seung, H.S. 1999), which accesses data representation by non-negativity constraints and makes the representation purely additive (allowing no subtractions). NMF can obtain a basis of localized features in an unsupervised way and retain many detailed edge features of the original image with strong robustness.

In view of the above-mentioned methods, a novel multi-focus image fusion algorithm based on NSCT and NMF is proposed in this paper. This paper is organized as follows. In Section 2, we introduce NSCT and NMF model as a preliminary work. In Section 3, the detailed fusion algorithm is described. Experimental results are presented in Section 4 and conclusions are drawn in Section 5.

2 NONSUBSAMPLED CONTOURLET TRANSFORM AND NONNEGATIVE MATRIX FACTORIZATION

2.1 *Nonsubsampled contourlet transform*

The NSCT is an over-complete transform that uses a flexible multi-scale, multi-direction and shift-invariant image decomposition that can be efficiently implemented viatrous algorithm. The NSCT is the shift-invariant version of the CT and is built on iterated non-separable two channel two channel nonsubsampled filter banks to obtain the shift in variance. The NSCT can be divided into two shift-invariant parts: 1) a nonsubsampled pyramid structure that ensures the multi-scale property and 2) a non-subsampled DFB structure that gives directionality.

Nonsubsampled Pyramid (NSP): The multi-scale property of the NSCT is obtained from a shift invariant filtering structure that achieves a subband decomposition similar to Laplacian pyramid. This is achieved by using two-channel nonsubsampled 2-D filter banks. The filters of subsequent scales can be acquired through upsamping that of the first stage, which gives the multiscale property without the need of additional filters design.

Nonsubsampled Directional Filter Bank (NSDFB): the DFB is constructed by combining critically sampled two-channel fan filter banks and resampling operations. The result is a tree-structured filter bank that splits the 2-D frequency plane into directional wedges. The NSDFB is constructed by eliminating the downsamplers and upsamplers in the DFB. This is done by switching off the downsamplers/upsamplers in each two-channel filter bank in the DFB tree structure and upsampling the filters accordingly.

Combining the NSP and NSDFB in the NSCT: The NSCT is constructed by combining the NSP and the NSDFB. Because of no downsampled in pyramid decomposition, the lowpass subband has no frequency aliasing, even the band width of low-pass filter is larger than $\tau/2$. Hence, the NSCT have better frequency characteristics than CT. property of the NSCT is obtained from a shift invariant filtering structure that achieves a sub-band decomposition similar to Laplacian pyramid. This is achieved by using two-channel nonsubsampled 2-D filter banks. The filters of subsequent scales can be acquired through upsamping that of the first stage, which gives the multiscale property without the need of additional filters design.

Nonsubsampled Directional Filter Bank (NSDFB): the DFB is constructed by combining criticallysampled two-channel fan filter banks and resampling operations. The result is a tree-structured filter bank that splits the 2-D frequency plane into directional wedges. The NSDFB is constructed by eliminating the downsamplers and upsamplers in the DFB. This is done by switching off the downsamplers/upsamplers in each two-channel filter bank in the DFB tree structure and upsampling the filters accordingly.

Combining the NSP and NSDFB in the NSCT: The NSCT is constructed by combining the NSP and the NSDFB. Because of no downsampled in pyramid decomposition, the lowpass subband has no frequency aliasing, even the band width of low-pass filter is larger than $\tau/2$. Hence, the NSCT have better frequency characteristics than CT.

2.2 Nonnegative matrix factorization

Given a nonnegative matrix $V_{n \times m}$, NMF finds nonnegative matrices $W_{n \times r}$ and $H_{r \times m}$ such that:

$$V \approx WH \tag{1}$$

Usually r is chosen to be smaller than n or m, so that the size of W and H is smaller than that of the original matrix V, which results in a compressed version of the original data matrix. NMF produces a sparse representation of V, which depicts much of the data by using few "active" components of W. This means that NMF (Dipeng, C. & Qi, L. 2005) can express a kind of physical significance. The objective function is defined by:

$$\min F = \sum_{i=1}^{n}\sum_{j=1}^{m}[V_{ij} - (WH)_{ij}] \quad s.t. W_{ij} \geq 0, H_{ij} \geq 0$$
$$i = 1, 2, .., n, j = 1, 2, \cdots, m \tag{2}$$

$$W_{i\alpha} = W_{i\alpha}\sum_{i}\frac{V_i}{(WH)_i}H_\alpha \tag{3}$$

Iteration steps are adopted here, and the updating rules are defined by:

$$W_{i\alpha} = \frac{W_{i\alpha}}{\sum_{j}W_{i\alpha}} \tag{4}$$

$$H_\alpha = H_\alpha\sum_{i}W_{i\alpha}\frac{V_i}{(WH)_i} \tag{5}$$

It is proved that the above updating rules are guaranteed to converge to a locally optimal matrix factorization.

3 FUSION ALGORITHM BASED ON NSCT-NMF

NMF is to conduct the decomposition of a matrix under the constraint that all values should be non-negative. The local characteristic of the original data can be incorporated effectively when the dimension of characteristic space is chosen properly. With good directionality and anisotropy, NSCT can well describe the edge and detailed information of images. After the acquirement of sub-band images containing specific directional information, NMF can be applied here to conduct fusion. Besides, NSCT has the property of translation invariant, which will suppress the pseudo-Gibb's phenomenon.

3.1 Decomposition

NSP is applied to the two source images by using the filters of "antonini" and "qshift_a", respectively. On each scale, DFB is applied to every obtained sub-band, with the filter of "pkva". Because NMF is an algorithm for nonnegative data, we have to normalize each sub-band image. The operator of normalization is:

$$N(I) = \frac{I - \min(I)}{\max(I) - \min(I)} \qquad (6)$$

In this way, every pixel of sub-band images can be adjusted to the values ranging between 0 and 1.

3.2 Fusion

Step 1. The pixels of every two corresponding sub-band images are saved into matrix V in terms of row priority.

$$v = \begin{bmatrix} x_{11} \cdots x_{1m} \cdots x_{n1} \cdots x_{nm} \\ y_{11} \cdots y_{1m} \cdots y_{n1} \cdots y_{nm} \end{bmatrix}^T \qquad (7)$$

Then, The image quality degradation model can be expressed as:

$$V = WH + \varepsilon \qquad (8)$$

where ε is the noise. The image fusion based on NMF uses iterative computing to make the noise ε tend to converge. NMF can extract the basis matrix W constructed by global eigenvectors from matrix V, and the fused image can be obtained from the basis matrix W. In particular, the column number r of W determines the dimension of characteristic sub-space. Assuming $r = 1$, we can get the eigenvector W with only one column which contains all the characteristics of the sub-band images. That means NMF can integrate featured information from all source images and avoid background noise effectively.

Step 2. The basis matrix W and the coefficient matrix H are initialized to be two random nonnegative matrices.

Step 3. W and H are kept updated according to Eqs. (3)–(5) until the objective function is convergent.

Step 4. When the calculation is completed, the basis matrix W is obtained. By transforming W using the inverse approach of Step 1, we can get the fused sub-band images.

3.3 Reconstruction

Firstly, using the inverse approach of normalization, the coefficients suitable for the inverse NSCT are obtained. The inverse normalization is achieved by:

$$N^{-1}(I) = I^*[\max(I) - \min(I)] + \min(I) \qquad (9)$$

Then, the output image is reconstructed by inverse DFB and inverse NSCT.

4 EXPERIMENTAL RESULTS AND DISCUSSION

In this section, we apply the proposed image fusion method based on NSCT and NMF. We compare our NSCT-NMF model with three other methods: NSCT (Junying, Z. et al. 2004), CT-NMF (Meili, L. et al. 2010), WT-NMF (Zhijun, W. et al. 2005). Note that CT-NMF is to replace WT in WT-NMF with CT. (Fig. 1) shows the fusion results of image 1 and image 2, and Fig. 4 is the fusion results of image 3 and image 4. We can see that (Fig. 1e), (Fig. 1d), (Fig. 2e) and (Fig. 2e) have obvious disturbing artifacts, which are caused by the lack of translation invariant property. Hence, WT-NMF and CT-NMF are not good enough for image fusion. It is shown in (Fig. 1c) and (Fig. 2c)

     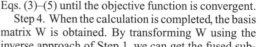

| (a) image1 | (b) image2 | (c) NSCT | (d) WT-NMF | (e) CT-NMF | (f) this method |

Figure 1. The fusion results of image 1 and image 2.

    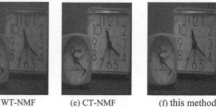

| (a) image3 | (b) image4 | (c) NSCT | (d) WT-NMF | (e) CT-NMF | (f) this method |

Figure 2. The fusion results of image 3 and image 4.

35

Table 1. Comparisons of image fusion results.

	IE	STD	MG
NSCT	6.817	42.151	44.073
WT-NMF	6.819	43.652	4.122
CT-NMF	6.820	44.230	4.163
NSCT-NMF	6.836	46.221	4.664

that the results of NSCT are not so clear, because the ordinary fusion methods cannot suppress noise as NMF does. In general, NSCT, WT-NMF and CT-NMF can get good results, but the fused images of NSCT-NMF are brighter and with less noise, and they contain clearer structure and texture information of the original images.

In order to validate the accuracy of our NSCT-NMF model, we evaluate the fusion results of infrared and visual images through three items: information entropy IE), standard deviation (Std) and mean gradient (MG).

$$IE = -\sum_{i=0}^{L-1} p_i \lg p_i \qquad (10)$$

$$Std = \sqrt{\frac{\sum_{x=1}^{M}\sum_{y=1}^{N}(f(x,y)-u)^2}{MN}} \qquad (11)$$

$$MG = \frac{1}{MN}\sum_{x=1}^{M}\sum_{y=1}^{N}\sqrt{\frac{\partial f_x^2 + \partial f_y^2}{2}} \qquad (12)$$

where $f(x,y)$ is the fused image; $M \times N$ is the image size; μ is the average value of the image; L is the gradient level; and pi is the probability of the *ith* gradient. IE shows the quantity of information that the image contains. With larger IE, the image can convey more information. MG reflects the ability of the image to show fine details, and it can be used to evaluate the clearness of the image. Std is a measure of the extent to which the image pixels and the average value vary. The larger the std, the bigger the contrast of the image. From Table 1, we can see that the fused image of the proposed algorithm has larger IE, Std and MG, which means that the fused image contains more information, and expresses the variation of texture features and fine details better.

5 CONCLUSIONS

An effective fusion method based on NSCT and NMF is proposed by considering the different properties of the two methods. Compared with NSCT and the methods based on the combination of wavelet/contourlet with NMF, the proposed algorithm can get more characteristic information of the source images, reduce the noise to a certain extent and be more faithful to the real situation.

REFERENCES

Dipeng, C. & Qi, L. 2005. *The use of complex contourlet transform on fusion scheme.* France: Balkema.
Do, M.N. & Vetterli, M. 2005. The contourlet transform: An efficient directional multiresolution image representation. *IEEE Transactions on Image Processing*, 14(12):2091–2106.
Shutao, L. & Bin, Y. 2008. Multifocus image fusion by combining curvelet and wavelet transform. *Pattern Recognition Letters*, 29(9):1295–1301.
Lee, D.D. & Seung, H.S. 1999. Learning the parts of object by nonnegative matrix factorization. *Nature*, 401(6755):788–791.
Ying, L. et al. 2007. Application of image fusion in urban expanding detection. *Journal of Geomatics*, 32(3):4–5.
Zhijun, W. et al. 2005. A comparative analysis of image fusion methods. *IEEE Transactions on Geoscience and Remote Sensing*, 43(6):1391–1402.
Junying, Z. et al. 2004. Image fusion based on Nonsubsampled contourlet transform. *Image Processing*, (6):73–76.
Meili, L. et al. 2010. Image fusion algorithm based on nonnegative matrix factorization. *Computer Engineering and Applications*, 46(8):21–24.

Information Technology and Computer Application Engineering – Liu, Sung & Yao (Eds)
© 2014 Taylor & Francis Group, London, ISBN 978-1-138-00079-7

System design of detection device for special cigarette filter

Tiejun Li, Deqiang Yang, Qiang Li & Guangxiong Liao
China Tobacco Jiangxi Industrial LLC Nanchang Cigarette Factory, Nanchang, China

ABSTRACT: Composition and characteristics of detection system for cigarettes based on machine vision technology was described in this paper, the system scheme of hardware and software were designed, a method was proposed that Integrity of the depression pattern in the cigarette filter, the system applied in some cigarette enterprises, achieved good effect.

Keywords: Depression pattern; Machine vision; Industrial control panel; image capture; Interface board of control system

1 GENERAL INSTRUCTIONS

With the development of tobacco technology, application of machine vision technology is used more and more, which including the appearance quality of packet detection, case detection and spinning process of foreign body detection, but a few application of cigarette detection. In the production process of special filter cigarettes, the devices without detection function of integrity in depression pattern which leads unqualified cigarette packs, workload was increased and poor quality by traditional manual sampling, so the design of detection device of special cigarette filter is particularly important for us.

2 INTRODUCTION TO THE SYSTEM

The detector of special filter cigarettes is used to detect integrity of depression pattern on cigarette filter tips in the cigarette packets. The detector sent the removal signal to the packaging unit control system when the flawed package is detected, than the packaging unit control system records the package's location, and remove it when it arrives in removal place[1]. The detector of special filter cigarettes achieves synchronization with the complete machine through the 1800-step incremental encoder. The incremental encoder runs a cycle when a pack of cigarettes is passed. The system can determine the accurate location of the cigarette package through the encoder, thus ensuring each cigarette in the package being tested when the cigarette package is passing the image acquisition component. The detector of special filter cigarettes collects images of the filter tip's end face of cigarette package which wait for inspecting by the image acquisition component, and process the image data by the controller and judge integrity of depression pattern of all cigarettes in the cigarette packets, and according to the result of the judgment, it sends the appropriate removal signal to the packaging unit control system[2]. The packaging unit control system remove the corresponding cigarette package after receiving the removal signal sent by the detector of special filter cigarettes.

The main features and indicators of the detector of special filter cigarettes include:

1) Non-contact detecting, suitable for high speed packaging machines, up to 600 packs per minute.
2) Dynamic automatic center positioning system, the jitter of cigarettes in the package will not affect the final result.
3) Efficient self-diagnostic capabilities, can do the alerts notification when the trouble happen by industrial cameras, encoders and so on.
4) Applicable to different specifications of cigarette packages, just need to do the simple operation via the touch screen (including specification such as 767, 776, etc.).
5) 12.1-inch color LCD with the touch screen, easy operation.
6) Have powerful statistical functions, can count the situations of missing production, depression patterns and filter at length.

3 OVERALL HARDWARE PLAN

The detector of special filter cigarettes mainly consists of peripheral devices such as the image acquisition component, the controller, and the encoder, and its hardware schematic diagram shown in Figure 1.

3.1 *The image acquisition component*

The image acquisition component is responsible for collecting images of the filter tip's end face of

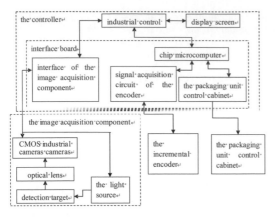

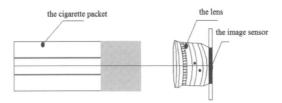

Figure 1. The schematic diagram of the detector of special filter cigarettes.

If there is a cigarette that is reversed in the package, the image captured by the system is the end face of tobacco, and it draws sharp contrasts with the white image of the normal filter tip's end face, so the system can detect the cigarette that is reversed in the package[6].

3.2 *The controller*

The controller process the image data collected by the image acquisition component through image processing techniques, and determine whether the cigarette packs have images or filter tips that are missing, then according to the result of determination, the controller sends the appropriate removal of signal to the packaging unit control system. The controller mainly consists of industrial control panels, interface boards and color LCD with the touch screen.

The industrial control panel used for running the operating system and the upper machine software, and this system uses a COM Express module. The COM Express module is a highly integrated computer module based on the PCI Express bus, and it is installed on a substrate designed by the customers for a particular application[7]. The smallest size of the COM Express module is only 84 mm × 55 mm, and it can provide application interfaces needed for most functionality, including video, audio, Ethernet, storage and USB interfaces, and so on. Customers only need to focus on the special applications to fill the functionality that is not offered by the COM Express.

The interface board, which mainly consists of the interface board's chip microcomputer, the encoder's signal acquisition circuit, the image acquisition component's communication interface and the packaging unit control system's communication interface, and so on, used for the perimeter expansion of the industrial control panel. The interface board's chip microcomputer use the current popular TI company's 32 bit chip microcomputer that based on Cortex-M3 called LM3S9B96, and it is responsible for executing the order of upper machine software, including reading the encoder's phase, and communicating with the packaging unit control system.

The encoder's signal acquisition circuit used for reading the encoder's phase by the interface board's chip microcomputer and transmit to the industrial control panel, to achieve the phase's synchronization of the system and the whole packaging unit control system. The packing unit control system's communication interface used for sending removal signals to the control system by the interface board's chip microcomputer.

As the human-computer interaction interface, the display screen's design principles are: the interface is organized, clear, comprehensive, coherent, and simple operation.

In order to meet these require-ments, the system uses a 12.1-inch color LCD screen and uses the touch screen with the resolution of 1024*768, and users can view the information about running the system

Figure 2. The installation diagram of the lens.

cigarette package which wait for inspecting through the image sensor, and transmit to the controller after initial processing[3]. The image acquisition component mainly consists of the light source, optical lens and CMOS industrial cameras[4].

The light source is the important guarantee for stable performance of the image acquisition component, and the LED light source has advantages of low cost, high stability, long service life and maintenance-free, so this system use the white bar LED light source as the light source of the image acquisition component.

The quality of the lens directly affects the overall performance of the system, so reasonable selection and installation of optical lenses are basis to ensure clear imaging and obtain the normal image signals. The installation diagram of optical lens in image acquisition component is shown in Figure 2, and optical lens and filter tip's end face are installed vertically. The captured image by the system is the positive image of the filter tip's end face.

The brightness of the part of the depression pattern on cigarette filter tip is significantly lower than the surrounding white filter tip's end face in the image, so computers can recognize the depression pattern and figure out whether the depression pattern of cigarettes is missing. When a cigarette or a filter tip is missing in the mold box, it occurs a significantly darken region in the image that the system captured, and it draws sharp contrasts with the white image of the normal filter tip's end face, so the system can determine that the cigarette packs have cigarettes or filter tips that are missing[5].

through the display screen, as well as can set the related parameters via the touch screen system.

4 PHOTOGRAPHS AND FIGURES

The software of the detector of special filter cigarettes mainly includes the controller's upper machine software, and the control software of the controller's interface board.

4.1 *The controller's upper machine software*

The controller's upper machine software is mainly responsible for image processing and the system's overall control, and its basic features include the follows.

Firstly, it communicates with the image acquisition component through the USB interface on the controller's the interface board, and receive the image data from the image acquisition component.

Secondly, through the image processing technology, the collected image was processed, such as gray scale, binary image, corrosion, and expansion processing, then find and draw image outlines, and then judge whether it exists images or filter tips that are missing (deficiency or reverse) according to total area of the outlines.

Besides, if the current cigarette packet exists deficiency, such as there are images or filter tips that are missing (deficiency or reverse), then it will send the order to the interface board's chip microcomputer, to inform the chip microcomputer send the removal signal to the packing unit control system on the specified phase. What's more, it shows the relevance information about the system's operation through the LCD display screen.

Lastly, it responds to the user parameter settings via the touch screen.

The control component divides up the images of filter tips' end face in the cigarette package photographed by the image sensor, so that the location of each cigarette corresponds to a small image. And then do the grayscale, binary image, corrosion, and expansion processing to image corresponding to the location of each cigarette via image processing technique, then find and draw outlines of depression images, and finally calculate outlines' area, the same as the number of pixels of depression pattern[8].

Grayscale is intended for transforming the color images of filter tips end face photographed by the image sensor to the grayscale images, and it is not necessary if the black and white image sensor is used. Binary image is intended for making a distinction between the depression pattern on cigarette filter tips and the surrounding white end face of filter tips, so the grayscale value of the depression pattern's pixel is set for 255, and the white end face of filter tips is set for 0. Corrosion and expansion are intended for eliminating some small noise points in images after binary image processing, to increase the speed of subsequent finding outlines of depression pattern.

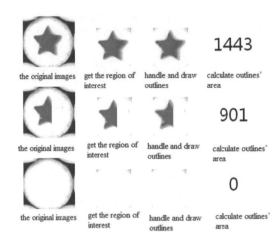

the original images   get the region of interest   handle and draw outlines   calculate outlines' area

Figure 3. Original images and ultimate results of cigarette's location in various situations.

Finding outlines of depression pattern is intended for drawing the outlines of depression pattern, and it is the premise of calculating the number of pixels of depression pattern. Calculating outlines' area of depression pattern is intended for calculating the number of pixels of depression pattern surrounded by outlines of depression pattern[9-11].

Figure 3 is a collection of original images and ultimate results of a cigarette in a cigarette package that are photographed by the image sensor in situations of the normal situation, the depression patterns being partially missing and all missing. It can be seen that the number of pixels of depression patterns is significantly lower than the normal situation when the depression patterns are partially missing or all missing, and the computer judge that the cigarette's depression pattern is missing when the number of pixels of depression patterns is lower than sensitivity value set by users, finally, we can judge whether the cigarette package as a whole is flawed according to the judging results of all cigarettes.

System sensitivity setting method: assuming that the sensitivity the setting range is 0 to 100 and adjustment is 1, lower sensitivity value indicates a more sensitive detection.

When the depression pattern is completed, the captured number of pixels of depression patterns is S, and the proportion between the sensitivity and pixel can be given: $K = S/100$.

So when the user sets a sensitivity of 50, it will be ineligible cigarette when the number of pixels of depression pattern is lower than $S-50*K$.

4.2 *The control software of controller's interface board*

The control software of the controller's interface board is primarily responsible for implementing orders of the controller's upper machine software, and does the related operations on peripherals. Complete features mainly include the follows.

Firstly, it read the current phase of the encoder's signal acquisition circuit, and delivery to the controller's upper machine software.

Secondly, it controls the image acquisition component to do the image acquisition in the specified phase.

Lastly, it sends the removal signals to the packaging unit control system.

5 CONCLUSION

The special filter cigarette detector with reliable performance and simple operation, which suitable for all kinds of cigarette packaging on the market, it have been used in Nanchang, Guangshui cigarette factories and unanimously affirmed by the users.

REFERENCES

[1] Fan Lu. 2010. Research on Detection Technology of the Packaging Machines Loose Ends Based on Machine Vision. *Light Industry Machinery*, 28(02):65–67, 71.

[2] Haikuan Zhu. 2009. Design of Image Acquisition and Disposal System Based on Machine Vision. *Electronic Test*, (01):53–56, 89.

[3] Shuren Tan, MaoJun Zhang, Wei Xu. 2006. Design of Multi-sensor Synchronization Image Capture Sytem. *Television Technology*, (9):84–87.

[4] JianLu Jia, JianLi Wang, Shuang Guo, etc. 2010. High Speed Image Grabber and Processor Based on Camera Link. *Chinese Journal of Liquid Crystals and Displays*, 25(06):914–918.

[5] N.M. Desai, Rinku Agrawal, J.G. Vachhani, V.R. Gujraty, S.S. Rana. 2003. High speed data acquisition systems for ISRO's airborne and spaceborne radars. *8th International Conference on Electromagnetic Interference and Compatibility*, 1(1):29–36.

[6] Andrei C. Jalba, Michanel H.F. Wilkinson, Jos B.T.M. ROERDINK. 2004. Automatic Segmentation of Diatom Images for Classification. *Microscopy Research and Technique*, (65):72–85.

[7] Wei Wang, JingHong Jiang, Yao Liu, etc. 2011. The Design and Implementation of a Data Recorder Based on COM Express Architecture. *Embedded Technology*, 37(12):29–32.

[8] HuoRong Ren. 2004. *Application of Mathematical Morphology*. Xidian University.

[9] Kenneth. R. Cast Leman. 2002. *Digital Image Processing*. Beijing: Publishing House of Electronics Industry.

[10] HoLung Yun, Jae Seok Kim, Bong Soo Hur. 2000. Design of real-time image enhancement preprocessor for CMOS image sensor. *Consumer Electronics*, 46(1):68–75.

[11] M.Y. Jiang, D.F. Yuan. 2002. A multi-grade mean morphologic edge detection. *6th International Conference on Signal Processing*, 1079–1082.

Information Technology and Computer Application Engineering – Liu, Sung & Yao (Eds)
© 2014 Taylor & Francis Group, London, ISBN 978-1-138-00079-7

The ant colony optimal algorithm based on neighborhood search to solve the traveling salesman problem

Huanli Pang, Yonghe Li & Xianmin Song
Department of Computer Science and Technology, Changchun Univercity of Technology, Changchun, Jilin

ABSTRACT: The ant colony optimal algorithm is a new modern intelligent heuristic search algorithm can effectively solve combinatorial optimal problems. This paper gives an improved ant colony algorithm, combined with the idea of iterative local search algorithm, strengthen the local optimal solution and improve the accuracy of the ant colony algorithm. Each ant will select the next node as short as possible by the 2-opt neighborhood search algorithm. To avoid algorithm premature convergence trap, the improved algorithm limits the value of the pheromone on the path. The paper gives the results of the improved algorithm by testing the data of TSPLIB, the results show that the improved ant colony algorithm has good characteristics in convergence speed and accuracy.

Keywords: Ant Colony Algorithm; Neighborhood Search; Path optimization; 2-opt

1 INTRODUCTION

Ant colony optimal algorithm (ACO) is a modern intelligent and bionic algorithm, as the behavior of current analog nature ants looking for food, first proposed by Italy scholar M. Dorigo in 1991. In reality the ant colony pass information to each other and to find the shortest path, especially, the algorithm has the characteristics of positive feedback and distributed computing, which can effectively solve many complicated combinatorial optimization problems. ACO is a simulated evolutionary algorithm of simulating to seek food of ants in nature.

Traveling salesman problem (TSP) is one of the classic problems of combinatorial optimization, which is a typical NP complete problem and generally be described as: Given a collection of cities and the length of travel between each pair of them, then TSP is to find the shortest route of visiting all of the cities and return the started point. At present, for solving the TSP problem commonly used algorithm with heuristic t method, genetic algorithm, ant colony algorithm, immune algorithm and simulated annealing algorithm. But these algorithms have their own advantages and disadvantages, such as some of them have high time efficiency or the better precision.

Traveling salesman problem is to solve the global shortest path problem. In this paper the ant colony algorithm based on neighborhood search (ACAN) to solve the path optimization problem, and test the effectiveness of the ACAN algorithm by solving the TSP problem. ACAN algorithm is based on the traditional ant colony algorithm in combination with the iterated local search algorithm, using 2-opt algorithm and some accelerated techniques to enhance the local search ability. Finally, the research indicates that the ACAN algorithm is robust and adaptable as novel optimization methods.

2 THE BASIC PRINCIPLE AND CHARACTERISTICS OF ACO

2.1 *The basic principle of ACO*

ACO is inspired by the scholars observed carefully about the method of the real ants looking for food, and then sholars prove that it is a modern intelligent optimal algorithm by experiments. In nature the ant individuals through cooperation can always find the approximate shortest path from the nest to the food, and also can change with the natural environment changes. Zoologists through long-term observation showed that when the ants are foraging, they will release a chemical called pheromone chemicals and tend to choose the direction of more pheromone concentrations. As the time goes, the pheromone concentrations on the short path will be higher, so the ants on the short path will be more and more. This is a positive feedback process of ACO, all the ants will be on the approximate shortest path.

2.2 *The characteristics of ACO*

The main advantage of ACO is the mechanism of positive feedback, distributed, computation, and the strong global search better solution ability. In addition, it has strong robust and good expansibility, which is easy to combine with other heuristic algorithm. However, the ACO has some shortcomings. When the scale of the problem is very large, its efficiency decreased rapidly and the search process will need more time. Ant colony algorithm is prone to be stagnation behaviour, so when to be a certain level, it cann't find the better solutions. Many scholars has conducted in-depth research on the algorithm and proposes some improved method. The [4] is proposed two kinds of methods to update the pheromone, one is to update the pheromone each loop, the other is to update the pheromone in global process. Put forward in [5], using 2-opt and 3-opt local iterative idea for the approximate solution of restructuring, so as to get the better solutions. Presented in [7] to set the value of upper and lower limit about pheromone, so as to avoid the algorithm premature convergence.

3 ANT COLONY ALGORITHM BASED ON NEIGHBORHOOD SEARCH

3.1 *The basic model of ACO*

We explain the basic process of ACO by solving the TSP problem. Introducing the following sign to simulate behavior about the natural ant: "m" represents the total count of ants, "n" represents the total number of city, "$d_{ij}(i,j = 1, 2, 3, \ldots n)$" represents the distance between city nodes 'i' and 'j', "$\tau_{ij}(t)$" represents the total remaining pheromone on the path between city nodes 'i' and 'j' in the 't' time. Every ant is randomly initialized on one city and the total pheromone on the path between the citys is 'q'. Using the basic Equation1 of ant colony algorithm in the literature [4] and $p_{ij}^k(t)$ represents the probalility about the ant (k) from city(i) to city(j) at the 't' time.

$$
p_{ij}^k(t) = \begin{cases} \dfrac{\tau_{ij}^\alpha(t)\eta_{ij}^\beta(t)}{\sum\limits_{k \in allowed_k} \tau_{ij}^\alpha(t)\eta_{ij}^\beta(t)} & if \ j \in allowed_k \\ \\ 0 \end{cases} \quad (1)
$$

$allowed_k = \{0, 1, \ldots, n-1\}$ -$tabu_k$ resprents the set of city nodes which the ant 'k' can select. The pheromone 'q' will decay as the time go on. The list $tabu_k$ records city node which the ant 'k' have already arrived. When the all city nodes are added into the $tabu_k$, the ACO complete a cycle and it represents a candidate solution. Expectation factor α represents the dependence of the pheromone on the path and heuristic factor β represents the dependence of the ants select the probable short path. The $\eta_{ij}(t)$ represents the expectation about city(i) transfers to city(j), it is up to the actual situation of the heuristic algorithm. When the all ants complete one loop, the pheromone on the path between the nodes will be updated according to the below Equation 2.

$$
\begin{cases} \tau_{ij}(t+n) = (1-\rho)^* \tau_{ij}(t) + \Box \tau_{ij}(t) \\ \\ \Box \tau_{ij}(t) = \sum\limits_{k=1}^{m} \Delta \tau_{ij}^k(t) \end{cases} \quad (2)
$$

In the Equation 2, the ρ represents the evaporation parameters and the $\Delta\tau_{ij}(t)$ represents the remaining pheromone in this cycle. The $\Delta\tau_{ij}^k(t)$ represents that the ant(k) leave the pheromone between node(ij). In the most comomnly used ACS(ant cycle system) model, the value of $\Delta\tau_{ij}^k(t)$ is calculated by the below Equation 3.

$$
\Delta\tau_{ij}^k(t) = \begin{cases} \dfrac{Q}{L_K} & if \ the \ ant(k) \ pass \ the \ path(ij) \ on \ the \ current \ loop \\ \\ 0 & else \end{cases} \quad (3)
$$

In the Equation 3, the 'Q' is a constant and it represents the total initialized phereomone. The L_k means the total length of the path which the ant(k) have already arrived.

3.2 *The ant colony algorithm based on local iterated search*

When the scale of the TSP problem is very large, the ant colony algorithm need the longer time to run and get the result. We think that is because that the candidate solution maybe have cross path, so the pheromone on the short path can not increase quickly. In this paper, combined with the local iterated search algorithm, the ant will select the next node as short as possible. At first, we create the two-dimensional array to store the distance about the current node to the others node. In addition, wo sort the array order by the distance from small to large. According to the neighborhood array, it is easy to find the next nearest node. Consequencely, the ant will seletct the probale near node. If the ant has not selected one node, we will use the above Equation 1 to choose one node. This method not only can reduce the algorithm complexity, but also can improve the accuracy of the solution.

The 2-opt algorithm means that it is to alternative the any two nodes among the nodes in the candidate solution until the total length is not better than the current value. Using the 2-opt algorithm, each ant can get the current best optimal solution and it avoid the problem in the Picture 1. As shown in Picture 1, the original path is "a-e-d-c-b-a", through the 2-opt algorithm it turns "a-c-d-e-b-a", thus strengthening the local better optimal solution.

In the improved algorithm (ACAN), we also use the three standard accelerated technique to help the ant as soon as possible to find the optimal solution. First, the ants will be prioritized to select the short nodes. Second, the algorithm only attempt to exchange with the close nodes on the iterated step. Third, if the

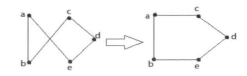

Figure 1.   Eliminate the cross paths in the search process.

current node has no better results to exchange the position with its all adjacent nodes, we will make the sign that "don't look" on the current node. Lastly, we change the rule about how to update the pheromone and limit the value of the pheromone on the path, because this can avoid the state that the algorithm get the local optimal solution.

The steps of the ACAN algorithm are as follows:

Step 1: To declare and initialize the parameter, for example, we declare the 'total_cycle' as the loop times, and the 'demandcycle' means the maximum cycle time, then the 'citycount' means the total number of the city. we also define that the number of the ants, expectation factor, heuristic factor, evaporation coefficient, and so on. Lastly we define the 'Nearcity_of_next' as the array means the mutual sequenced distance between each city.

Step 2: totalcycle = totalcycle + 1;

Step 3: each ant has a random city node at the beginning and modify the ant's tabu list index about 'k = 1';

Step 4: antcount = antcount + 1;

Step 5: each ant will try to select the near node one by one, but if the time of selecting is more than we limit its length, the algorithm will use the Equation 1 to select the next city(j) and add the city(j) into the list {C-tabuk}, then the ants move to the new node.

Step 6: if city nodes are not completely ergodic and it means 'K < citycount', then goto the Step 5.

Step 7: if the nodes are all arrived, the algorithm will call the function about 2-opt. Then the program will use the Equation 2 and Equation 3 to update the pheromone on the path, meanwhile, the algorithm will judge the value of pheromone $\tau_{ij}(t)$ and limit its value between τ_{min} and τ_{max}.

Step 8: if the ants have not been cruising the all nodes and it means 'antcount > anttotal', then goto Step 4.

Step 9: if this algorithm meet the end conditions, that means the value of 'totalcycle is greater than the value of 'demandcycle', so the program will output the solution and end, otherwise the algorithm will empty the candidate solutions of the list {C-tabuk} and goto step 2.

4 EXPERIMENTAL RESULT

This paper selects data from TSPLIB to simulate. The parameter is $\alpha = 1$, $\beta = 2$, $\rho = 0.6$. The total number of ants is 20 and the circle of program is 400 times. When the ants select the next node, we consider only twenty nodes which are more close than others nodes. Experiments are carried out for 50 times, then we calculate the average results of ACO algorithm and the ACAN algorithm. It compares with the experiments in [8] and its experimental results are shown in table 1.

In the table 1 there are three experimental results about the different ant colony algorithm. we can conclude that the ACAN algorithm has good global searching better solution ability by comparing the data in the table. It enhance the efficiency and accuracy of

Table 1. The experimental results of the algorithms.

| TSP | Experimental results | | | |
	ACO	*ACAN*	*The [8]*	*Public solution*
Olive30	422.5	420	427.611	420
Eil51	447.3	426	436.007	426
Eil76	568.6	538	562.508	538
Ch150	6827.8	6564.8		6528
Pr299	59587	49140		48191

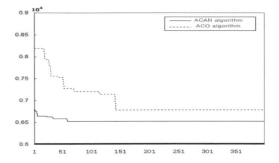

Figure 2. Two different convergence of the curve to show the length in each circle.

the ant colony algorithm. What's more important, the improved ant colony algorithm can always find the global optimal solution about the nodes of the city which amount below one hundred nodes. As shown in Figure 2, it draw two different convergence of the curve to show the efficiency of the algorithm to solve the problem of TSP about the Ch150 City. The X-axis represent the current shortest length of the path and Y-axis represent the loop count of the algorithm, so we can see clearly that the ACAN algorithm is better than ACO from the Figure 2.

5 CONCLUSION

According to the above experimental results, we can draw a conclusion that the ACO algorithm is a new intelligent heuristic search algorithm, which can quickly solve large-scale combinatorial optimization problems. In this paper, the ACAN algorithm is applied to the ant colony optimization algorithm. Using 2-opt algorithm to avoid cross paths in this paper can improve local search ability. Besides it can effectively improve ant colony algorithm accuracy and shorten the convergence time by using three standard accelerated techniques.

REFERENCES

[1] Guohui Lin, Zhengxin Ma, Yongqian Wang, Zhigang Cao, Zhongzhi Shi. The congestion avoidance algorithm based on ant colony algorithm[J]. Journal of Tsinghua University (natural science edition), 2003, 43(1):1–4.

[2] Zhan Fan, Guolong Liang, Wangsheng Lin, Kai Liu. The adaptive neighborhood search to solve TSP problem and its extension[J]. Computer Engineering and Applications, 2008, 44(12):71–74.

[3] Yezheng Liu, Haifeng Ling, Shanlin Yang. The research and application of ant colony optimization[J]. Journal of Hefei University of Technology (natual science edition), 2006, 7(10):11 11.

[4] Gambardella, L.M., Dorigo, M. Ant-Q: a reinforcement learning approach to the traveling salesman problem[A]. Proceedings of the 12th International Conference on Machine Learning[C]. Tahoe City, CA: Morgan Kaufmann, 1995, 252–260.

[5] Qunxing Le, Fajie Wei. The basic principles of ant colony algorithm and its development[J]. Journal of Beihang University (social science edition), 2005, 12(18):5–8.

[6] Dorigo, M., Gambardella, L.M. Ant colonysystem: a cooperative learning approach to the traveling salesman problem[J]. IEEE Transactions on Evolutionary Computation, 1997, 1(1):53–56.

[7] Bullnheimer, B., Hartl, R.F., Strauss, C. A new rank-based version of the ant system: a computational study, Tech Rept POM-03/97[R]. Viena: Institute of Management Science, University of Vienna, Austria, 1997.

[8] Guoqiang Ding, Zeyu Sun, Chuanfeng Li. Design and implementation of genetic ant colony algorithm for solving optimization problems[J]. Computer Measurement & Control, 2011, 19(10):2558–2561.

Information Technology and Computer Application Engineering – Liu, Sung & Yao (Eds)
© *2014 Taylor & Francis Group, London, ISBN 978-1-138-00079-7*

Design and implementation of FORTRAN to C Translator based on Perl scripts

G.M. Peng, Y.J. Shen, Z.M. Yi & G.D. Zhang
College of Information Science and Engineering, Lanzhou University, Lanzhou, Gansu Province, China

ABSTRACT: According to the specific requirements of the translation tools from FORTRAN to C Translator, this paper proposes implementing the functions of the translator using the Perl scripting language, introduces in detail the formulation of translation rules, the implements of the Perl script and the design and implementation of the translator, and summarizes the experimental results and the advantages and disadvantages of the translator. The experimental results demonstrate that the translator can perform translation from FORTRAN to C Translator.

Keywords: FORTRAN; Perl script; translation rule

1 INTRODUCTION

The FORTRAN language is the first officially promoted high-level language in the world. It is designed for science and engineering problems and the problems which can be expressed in a mathematical formula in enterprise management. Numerical computation has been ruled by FORTRAN for decades. The predecessors have written many procedures with high quality when software and hardware are confined, now the performance of these procedures has become very stable after modifications.

Now, in order to facilitate secondary development and application of the original program, some research institutes hope to transform FORTRAN source code to the C language application with the same functions, so that it continues to play a strong role. Therefore, it is valuable to design a simple converter which translates FORTRAN language to C language. The existing f2c automatic translation tool can translate FORTRAN 77 source code into the corresponding C code. The advantage of f2c translation tool is in that it is more convenient than manual implementation in the link of the run-time library function, But there are many disadvantages as follows: firstly, the code assumes that the array elements are arranged in columns, instead of in rows like the C language; secondly, the readability of the conversed C code is poor; thirdly, parameters are passed through transferring address; fourthly, temporary variables are introduced for the exchange of array elements; finally, f2c tools can only translate the FORTRAN 77 code. At present, few studies have been made on the translation of the FORTRAN 90/95 program. Individual researchers analyze in-depth the design of translators which convert FORTRAN 90/95 language to the C language from the perspective of

compilation principle. But the translators are still not for public use. A large amount of programs still need programmers who are familiar with both languages to translate manually, which is time-consuming, labor-intensive and error-prone. So it is necessary to design a translator which can convert all versions of FORTRAN source code to C.

Although the full functionality of translators can be achieved through designing from the perspective of compilation principle, it must go through lexical analysis, syntax analysis and semantic analysis, which is obviously time-consuming and labor-intensive. Perl is a free and powerful text processing language. Its perfect regular expression functions can't be achieved by other programming languages, and the best tools to lexical analysis are regular expressions which can perform exact search and match. Therefore, it can greatly improve the efficiency of the translators to design a simple translator using Perl script.

2 REPLACEMENT RULE AND PERL SCRIPT GENERATION

We can know, from a large number of FORTRAN source code, that FORTRAN statement is divided into data definitions, process control and output control. 95% of the FORTRAN statements can be replaced with the corresponding C statements, which simplifies rule formulation and script generation. However, the differences between FORTRAN language and C language array are as follows:

1) *FORTRAN array subscript default starts from 1, while the subscript C array indexes starts from 0;*

Table 1. translate_array.pl transformation rules.

FORTRAN	C	Remark
'('	'['	Mark is 1 replacement
')'	'−1]'	Mark is 1 replacement
','	'−1]['	Mark is 1 replacement

2) *FORTRAN array subscript is in parentheses, while the C array subscript is in square brackets;*
3) *FORTRAN array is stored in column, while C array is in row.*

The third case is not considered in translation since it is processed in the initial array. The wide use of parentheses in FORTRAN makes the translation of the array troublesome. In view of this, Perl statement of translation array is written in a separate file of translate_array.pl and the other two scripts translate_if.pl and tanslate_write.pl, with independent functions are responsible for translate process control and output control. The following describes the replacement rules and general process of three kinds of Perl scripts.

2.1 The translate_array.pl script

A large number of arrays are used in every FORTRAN project. If these arrays are translated manually, it not only needs to translate parentheses into the brackets, but also make the array subscript minus 1, which is prone to occur unpaired or error-paired brackets. translate_array.pl script can realize the conversion between FORTRAN array and C array perfectly. Parentheses and commas are used widely in FORTRAN; we must pinpoint the parentheses and commas when dealing with them. The key words and function names which are with parentheses and do not need processing are written into the file keywords.txt. We compare scanned words and words in the file keywords.txt and mark them when translating documents. Then we use stack to solve the problem of matching brackets level. Table 1 lists the conversion rules. The process of translate_array.pl script is shown in Figure 1.

2.2 The translate_if.pl script

Data definition and flow control statements in FORTRAN source code is translated by translate_if.pl script. The general process is as follow: Scanning Fortran source code, matching and replacing words and phrases in the FORTRAN code exactly using the matching function of Perl's regular expression. Table 2 lists some of the conversion rules.

2.3 The translate_write.pl script

FORMAT is the output control command of FORTRAN language. The FORMAT command includes

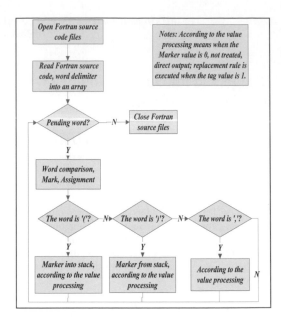

Figure 1. translate_array.pl the process.

Table 2. Some of translate_if.pl transformation rules.

FORTRAN	C	Remark
use module filename	# include "filename.h"	Module file processing
character	char	Date Definition
real	double	Date Definition
integer	int	Date Definition
!	//	Line comment
cycle	continue	Loop control
exit	break	Loop control
/=	!=	Not Equivalent
.AND.	&&	Logical operators 'And'
.NOT.	!	Logical operators 'Not'
.OR.	||	Logical operators 'Or'
str (1)= '......'	strcpy (str, "...")	String assignment
then	{	Multi-branch statement
else	} else {	Multi-branch statement
elseif	} elseif	Multi-branch statement
endif/enddo	}	Loop control
do s=n1,e,n2	for(s=n1;s<=e;s+=n2){	Loop control
go to + number	goto l + number	goto statement
number + continue	l+number	Loop control

rich formatting control, and some formats that don't have the same meaning in C language is marked. Output control of FORTRAN language is rather complex; however, large-scale projects is still based on more basic output control and the output is written into a file. Therefore, write statement is directly translated into FPRINTF statement. SPRINTF statement can be used in the case where write statement is assigned to a string. Table 3 lists some of the conversion rules.

Table 3. Some of translate write.pl transformation rules.

FORTRAN	C	Meaning
Aw	%ws	w characters wide output string
Dw.d	%w.df	w characters width output index type of floating-point number, Fractions width d characters.
Ew.d[Ee]	%w.de	w characters width output index type of floating-point number, fraction width d and exponent e characters.
ENw.d[Ee]	%w.de	Index type to output floats number.
ESw.d[Ee]	%w.de	Index type to output floats number.
Fw.d	%w.df	w characters width output floating-point number, fractions width d characters.
Iw[.m]	%wd	w character width output integer number, output m numbers at least.
Lw	%wd	w characters width output T or F truth value.
/	\ n	Wrap.
Ow[.m]	%o	Integer conversion into octal, w characters width output, fixed output m values. m value can't be given.
Zw[.m]	%x	Integer conversion into hexadecimal, w characters width output, fixed output m values. m value can't be given.

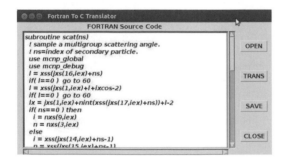

Figure 2. The interface after opening file.

3 DESIGN AND IMPLEMENTATION OF TRANSLATORS

3.1 Design and implementation of interface

T The interface of translation use Java language to design, and its development environment is Eclipse + Linux. In order to shorten the development cycle, the interface design is relatively simple, including two text boxes (one display source FORTRAN code and the other one display the translated C code), four buttons (OPEN, TRANS, SAVE, CLOSE), and two labels. The open interface is shown in Figure 2.

3.2 The process of calling script

After pressing the translation button, Java program, Java program calls the tranlate_array.pl script and then

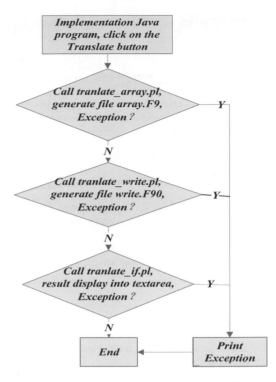

Figure 3. The processing after clicking the TRANS button.

the temporary file array.F90 is generated; then, Java program calls the tranlate_write.pl script to handle the temporary files array.F90 and the temporary files write.F90 is generated; next, Java program calls tranlate_if.pl script to process the temporary file write.F90 and the results are displayed in the text box. You can save the result by clicking the save button.

The key codes of calling the script and saving the processed results to a temporary file as follows:

```
String cmd = "/home/pgm/workspace/./translate_array.pl "+op.getDirectory()+op.getFile();
String lineToRead = "";
File f1 = new File("/home/pgm/workspace/array.F90");
FileWriter fw = new FileWriter(f1);
BufferedWriter bw = new BufferedWriter(fw);
Process proc = Runtime.getRuntime().exec(cmd);
InputStream inputStream = proc.getInputStream();
BufferedReader bRreader = new BufferedReader(new InputStreamReader(inputStream));
if ((lineToRead = bRreader.readLine()) != null){
bw.write(lineToRead);
bw.write("\n");
}
```

The process after clicking the Translate button is shown in Figure 3.

The result after the processing of the program is shown in Figure 4.

4 EXPERIMENTAL RESULTS AND ANALYSIS OF THE ADVANTAGES AND DISADVANTAGES OF TRANSLATOR

Comparing a large number of C language files processed by the translators with complete manual

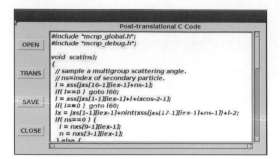

Figure 4. The results after the procedure.

translation, we can see that the logical structure and data processing part are completely consistent and there is some markers in output control statements. The translator has some advantages compared with f 2 translation tool and human translation.

1) *Solving the problem that f2c doesn't translate FORTRAN 90/95;*
2) *Greatly improving the efficiency and accuracy of manual translation, in particular with many arrays or output statements;*
3) *Perl script can be dynamically modified and it is easy to add new rules;*
4) *Perl script processing is faster and the efficiency and accuracy of translation are higher;*
5) *The translator has a shorter development cycle and with perfect function.*

The translator also has some imperfections.

1) *The uncertain format of post-translational target statements is very difficult to achieve by Perl script.*
2) *The translated output format is not very beautiful.*
3) *The translator can't dynamically define the implicit local variables in the FORTRAN file.*
4) *Address parameters when the function is called are not processed.*

5 CONCLUSION

This paper focuses on the translation of FORTRAN to C language by Perl script, including formation of the rules, generation of the Perl script and the process of developing the translator. About 95% FORTRAN source code can be processed through existing translation rules and script processing which achieves a better translation of the FORTRAN source code to the C language code and saves much time for the FORTRAN language translation workers, thereby greatly improving the efficiency. Of course, the translator has some imperfections; we will continue to improve the rules of FORTRAN to C translation and the performance of the translator.

REFERENCES

[1] Knudsen J., Niemeyer P. 2005. Learning Java 2005. Inc, USA: O'Reilly Media.
[2] Junfeng ZHAO, Jiantao ZHOU, Jing LIU. 2012. "Translation Rules and a Supporting Tool for Model-Based Reuse," Izmir, Turkey COMPSACW: 310–315.
[3] Muyun Yang, Zhanyi Liu, Tiejun Zhao. 2003. "TBED based Chinese-English translation rule acquisition," Beijing, China NLPKE: 158–162.
[4] R.L. Schwartz. B.D. Foy, T. Phoenix. 2006. Intermediate Perl. Inc, USA: O'Reilly Media.
[5] R.L. Schwartz. B.D. Foy, T. Phoenix. 2011. Learning Perl. Inc, USA: O'Reilly Media.
[6] SHU-JIE LIU, MU-YUN YANG,TIE-JUN ZHAO. 2006. "A Cascaded Approach to the Optimization of Translation Rules," Dalian, China MLC: 4089–4092.
[7] S'ebastien Darfeuille. 2012. "PERL scripts-based netlist analysis tool for the detection of ESD "big buffer" configurations," Seville, USA SMACD: 269–272.

Information Technology and Computer Application Engineering – Liu, Sung & Yao (Eds)
© 2014 Taylor & Francis Group, London, ISBN 978-1-138-00079-7

Application of computer simulation technique in field waterflood system

Buxiao Fan, Gang Wu, Shuo Zhang, Guojun Zhang, Hao Sun, Xinde Han & Weiyi Xie
Huabei Oilfield Company, Petrochina, Renqiu, China

ABSTRACT: On the basis of system engineering theory, topology method, digital simulation and optimization technology, taking least unit power consumption for waterflood as objective function, computer simulation technique used in oilfield waterflood system adopts such means as pipe unit calculation, pipe network node point optimization and constraint conditions setting to build up operation parameter optimization mathematics model for waterflood system. Based on actual operating data of waterflood, simulation computation is performed to determine pressure loss of each node point in whole pipe network and system operating efficiency, providing important data for pipe network upgrading and motor/pump parameter optimization.

Keywords: Waterflood; Surface pipe network; Simulation computation; Software design

1 INSTRUCTION

Field waterflood system, comprised of water injection station, pipe network, water distribution station and injectors, is in charge of injected water boosting, delivery, flow restriction and distribution. Abundant electric power is consumed for system running. According to block diagram of motor/pump units, pipeline, valves, water distribution stations and injectors in waterflood pipe network, simulation technique for waterflood establishes pipe network tree and topology relationship to build up hydraulics computation mathematics model in computer. Based on actual operating data of waterflood system, simulation computation is performed to determine pressure loss of each node point in whole pipe network and system operating efficiency, providing important data for pipe network upgrading and motor/pump parameter optimization.

2 HYDRAULICS COMPUTATION MATHEMATICS MODEL FOR WATERFLOOD PIPE NETWORK

Finite element method is used to build up hydraulics computation mathematics model for water flood pipe network. In this model, waterflood pipe network is regarded as make-up of units (pipe units) jointed by node points. Due to slight impact of water resource on waterflood system, this network mathematics model is conditioned at constant flow status.

2.1 Pipe units mathematics model

For constant flow, both single pipe and pipe units are quantitatively described by algebraic equations derived from mass conservation and energy

conservation laws. Optional pipe unit is i (i.e. unit) and node points linked with this unit are k and j.

Energy equation for pipe unit i is as follow:

$$\Delta H^i = H_k - H_j \tag{1}$$

In which, ΔH^i = pressure loss of pipe unit i; H_k = pressure of node point k; H_j = pressure of node point j.

It is assumed that water flows from k to j, $H_k > H_j$. Formula for ΔH pressure loss is as folow:

$$\Delta H = \frac{Q^\alpha L}{\bar{K}^2} \tag{2}$$

In which, $\bar{K}$ = flow coefficient; L = pipe unit length; Q = pipe unit flow rate; α = coefficient.

Thus, energy equation of pipe unit i is:

$$Q^i = K^i \Delta H^i \tag{3}$$

In which, $K^i = \sqrt[\alpha]{\frac{(\bar{K}^i)^2}{L^i \cdot (\Delta H^i)^{\alpha - 1}}}$

2.2 Pipe network mathematics model

Currently, water flood activity primarily adopts branched and circular pipe networks. Connection form is shown in Fig. 1 and mathematics model building is stated below.

2.2.1 Branched pipe network
Branched pipe network is made up of in-line pipe units (shown in Fig. 1(a)). The energy equation is shown as follow:

$$\Delta H_{1 \to 5} = \sum_{i=1}^{5} \frac{Q^i}{K^i} \tag{4}$$

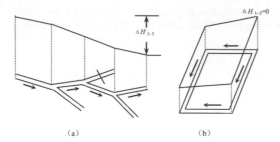

Figure 1.

In which, the sign can be determined by positive flow direction of pipeline one to five.

Equation of continuity for each node point is expressed as:

$$\sum_{i=1}^{m} Q^i = 0 \qquad (5)$$

In which, m represents number of pipe units linked with node points. It is assumed that flow rate from node point is positive.

2.2.2 Circular pipe network

Circular pipe network is one particular closed loop as shown in Fig. 1(b). The energy equation is as follow:

$$\sum_{i=1}^{m} \Delta H^i = 0 \qquad (6)$$

In which, m represents pipe units number in closed loop. Each node point in loop should meet equation of continuity (5).

3 FIELD WATERFLOOD SYSTEM OPERATION PARAMETER OPTIMIZATION

With known pipe network architecture and system load, ensuring total content of water flood, waterflood system operation parameter optimization determines proper pump employment scheme (pump employment numbers and arrangement) at each water injection station and water injection pump operation parameters, minimizing power consumption. The algorithm is one nonlinearity mathematics model containing continuous variables and whole number variables. Based on analysis of waterflood system features and study on waterflood system optimization from both theory and method, the solution can be found.

3.1 Waterflood system parameter optimization mathematics model building-up

Mathematics model building-up includes objective function establishment and constraint condition definition. The objective function should be minimal system unit power consumption. The expression is as follow:

$$\min f(u) = \alpha \frac{\sum_{i=1}^{m} \sum_{j=1}^{n_i} \frac{u_{ij} p_{ij}}{\eta_{pij} \eta_{eij}}}{\sum_{i=1}^{m} \sum_{j=1}^{n_i} u_{ij}} \qquad (7)$$

In which, u_{ij} = flow rate of injection pump j at water injection station i; p_{ij} = head of injection pump j at water injection station i; η_{pij} = efficiency of injection pump j at water injection station i; η_{eij} = motor efficiency driving injection pump j at water injection station i; α = units conversion coefficient; m = numbers of water injection stations; n_i = numbers of injection pumps at water injection station i.

First of all, waterflood system optimization scheme must fulfill fundamental constraint conditions of system. For waterflood pipe network system with n node points, all parameters related with optimal scheme should meet the equation:

$$\boldsymbol{KH} - \boldsymbol{C} = 0 \qquad (8)$$

Secondly, pressure at all injector node points must be greater than or equal to lower limit of requested injection pressure for oilfield development. The expression is as follow:

$$H_i \geq H_i^{\min} \qquad\qquad i = 1, 2, \cdots, t \qquad (9)$$

In which, t = injector numbers; $H_i^{\min}$ = lower limit of requested injection pressure at node point i.

Due to various waterflood pump models used at different water injection station, the waterflood pumps should meet the demand of maximum discharge capacity (or peak power) and minimal discharge capacity (for pumps on duty). The expression is as follow:

$$u_{ij} \leq u_{ij}^{\max} \qquad\qquad i = 1, 2, \cdots, m \qquad (10)$$

$$u_{ij} \geq u_{ij}^{\min} \qquad\qquad j = 1, 2, \cdots, n_m \qquad (11)$$

Consideration of water supply, waste water reinjection and electricity quantity, injection flow rate at each water injection station should fulfill the following expression:

$$\sum_{j=1}^{n_i} u_{ij} \leq u_i^{\max} \qquad\qquad i = 1, 2, \cdots, m \qquad (12)$$

$$\sum_{j=1}^{n_i} u_{ij} \geq u_i^{\min} \qquad\qquad i = 1, 2, \cdots, m \qquad (13)$$

In which, $u_i^{\max}$ and $u_i^{\min}$ are respectively upper limit and lower limit of injection flow rate at water injection station i.

In addition, according to waterflood pipe network characteristics, total injection flow rate should be equal to the sum of total injection rate of injector and flow rate loss of pipe network. Thus,

$$\sum_{i=1}^{m} \sum_{j=1}^{n_i} u_{ij} = \sum_{i=1}^{n} Q_i + \Delta Q \qquad (14)$$

In which, $\Delta Q =$ water loss; $n =$ injector numbers.

Expressions from (7) to (14) constitute mathematics model for waterflood system operation parameter optimization.

4 WATERFLOOD SYSTEM SIMULATION AND OPTIMIZATION SOFTWARE DESIGN

4.1 Waterflood system operation parameter optimization model building-up

4.1.1 Waterflood system parameter control design

Oilfield waterflood pipe network is a large scale network system connected by pipings with varying diameters. Pipings can be classified as water transmission line from water injection station to water distribution station and water distribution line from water distribution station to injectors. Pipe network connection is complicated, some injectors are connected to water distribution station while some are directly connected with water transmission line, bringing difficulties in parameter computation for pipe network. Therefore, injection allocation of each injector in pipe network can be simplified as node flow rate of water distribution station, and injection pressure allocation of injector is successively accumulative against water flow direction till host node. In this way, pressure and flow rate control parameters of injectors can be transformed to host node of pipe network.

4.1.2 Simulation and optimization computation method

Waterflood system pipe network simulation and optimization adopts bi-level hierarchical iteration method. The process is as follow: pump employment scheme and initial pressure at water injection station as well as initial flow u of water injection pump are given in advance. And then, flow rate and other operation parameters of each water injection pump are optimized by pipe network optimization method. Pressures of node points $H_i(i = 1, 2, \ldots, n)$ can be computed and system pressures at all water injection station can be finalized. After many times of computation, the proper pump employment scheme and optimum operation parameters will be derived.

4.2 Pipe network simulation and optimization program design

Optimization program adopts nonlinear programming algorithm. The basic clue is: optimization computation creates Lagrange function, which is used to builds up one parameter control range at each iteration point. Using numerical method derives extremal solution. Secondary extremal solution of each iteration acts as its search direction. Adoption of a series of such iteration algorithm enables to finalize pipe network optimization result. The program design block diagram is shown in Fig. 3-1.

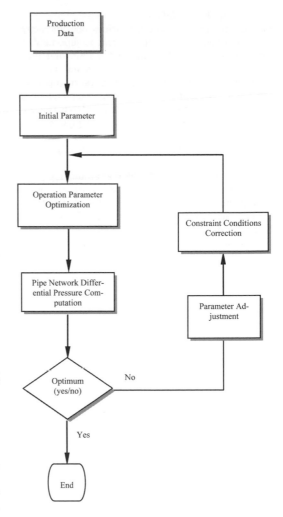

Figure 2. Computer block diagram for pipe network simulation and optimization.

4.3 Simulation and optimization software characteristics

4.3.1 Compatibility

Software database is compatible with Database A1 and A2 used by Petrochina.

4.3.2 Adaptability

This system adopts modularized composition. Universality of software configuration is united with flexibility. Key processing model is isolated from application software body. Multilayer logic analysis is dynamically linked with software system.

4.4 Simulation and optimization software functions

Software functions primarily include graphic modeling, system simulation, system optimization, pipe network optimization, production database and injector data inquiry.

4.4.1 *Graphic modeling*

Graphic modeling function provides one interactive mode interface for operator to amend pipe network graph at all times. It includes the following functions: adding injectors, water injection station, water distribution station, intermediate nodes and pipelines; amendment of unit parameters; searching wells, stations and water distribution stations; DXF file output; various parameters display; and pipe network check.

4.4.2 *System simulation*

This function can provide the day production simulation for entire waterflood system. Base on the day production data, such parameters as flow rate and pressure at each point of pipe network can be found out. Dynamic display of flow state and floating display of unit parameters (including flow rate, pressure and differential pressure) can be realized.

4.4.3 *System optimization*

According to the current production status, system optimization can achieve the goal of least unit power consumption. Computation can give optimum pump employment scheme to operating personnel.

4.4.4 *Pipe network upgrading*

Through current production data simulation of waterflood system, pipe network efficiency within injected zone and maximum pressure loss between water injection station and end point wells can be determined, and pipe network operation parameters can be optimized

as well, achieving the goal of upgrading pipe network configuration.

5 CONCLUSION

On the basis of system engineering theory, topology method, digital simulation and optimization technology, through pipe unit computation, pipe network node points optimization and constraint condition setting, taking least unit consumption as objective function, computer simulation technology for oilfield waterflood is used to build up mathematical model for waterflood system operation parameter optimization. By using pipe network simulation and optimization method, optimal running scheme for waterflood activity is designed, realizing proper allocation of power consumption for waterflood system and achieving the goal of improving field waterflood system efficiency.

REFERENCES

[1] ZHONG Weijun. *Municipal Water Supply System Simulation and Optimization*, Southeastern University Press, 1995.
[2] LIU Dongsheng, YUAN Guoying, HAN Zhiguo. *Power Saving Technology for Oilfield Waterflood System*, Petroleum Industry Press, 2003.
[3] ZHANG Qunhui, et al. *Computer Optimization Design for Oilfield Waterflood Pipe Network*, Xi'an Petroleum Institute Journal, 1995.

Information Technology and Computer Application Engineering – Liu, Sung & Yao (Eds)
© 2014 Taylor & Francis Group, London, ISBN 978-1-138-00079-7

The thinking of the introduction of computer technology course

Lei Deng, Xuefeng Jiang & Junrui Liu
School of Computer, Northwestern Polytechnical University, Xi'an, China

ABSTRACT: According to the characteristics of the introduction of computer technology course and the existing problems in course teaching, the author suggests that the teaching goal of the course, the teaching method and the teaching method should be reformed, so that the teaching level is boost and the teaching effect is improved, and the course has been will keep exuberant vitality and competitiveness.

Keywords: introduction of computer technology; computation ability; innovative thinking; teaching methods; the study contents average residual rate

1 INSTRUCTIONS

The introduction of computer technology course is called as the computer cultural foundation in some schools, and it's a curriculum in the first semester in our country higher education. Its goal is to enable the students to master the basic knowledge of the computer software and the computer hardware, to train the students to have the computer application ability, and to train the students to analyze and solve the problem with the computer, and to improve the students' computer culture. The course plays a important role in the cultivation of students' basic skill of operating computer, information literacy, autonomous learning and the ability of sustainable development. However, with the enhancing of the domestic education, some introductory courses of the computer application have been beginning in some middle school, even primary school. This situation caused a huge impact to the introduction of computer technology course, some experts even have begun to discuss whether this course is need. Therefore, analyzing the existing problems in this course and thinking the corresponding countermeasure are urgent issues.

2 COURSE'S CHARACTERISTICS AND EXISTING PROBLEMS

The introduction of computer technology course is the first public computer course in university, and is a compulsory course to the students of most majors. This course mainly introduces the basic theory and the basic application knowledge of the computer, and trains the students' basic skill of operating computer. Most of the course contents need to verify and improve through the practice, and because of the computer industry's rapid development, the course content is updated faster and it has a strong frontier. The main problems existing in the course teaching is shown as following[1].

(1) The course content is much and miscellaneous, lack of penetration, and is difficult to become a system.

The course contents include the computer architecture, the computer fundamentals, the office software (such as word, Excel, power, Point), the image processing software (flash, Photoshop) and the network foundation, is much and very complex, and lacks strong connection between these contents, so the knowledge system can't be built.

(2) The level of the course's students is uneven.

Because the computer science course has been opened in part of primary and middle schools, the level of the course's students is uneven. The students who have study this course can use the computer skillfully, but the others may never use the computer, and even can't boot and shutdown.

(3) The course has lost contact with the application and is far away from the goal of cultivating talents.

In the process of teaching, the students' learning content is unified, and is not contact with the actual application and professional needs.

3 SUGGESTIONS ON TEACHING

Aimed to the curriculum characteristics and the problems, the author combines the actual situation of our college, and gives some teaching suggestions.

(1) Suggesting the curriculum aims to cultivate students' calculating ability and innovative thinking, and separates from the teaching of the primary tools.

With the enhance of the domestic education resources and the education level, many primary and middle schools in towns have opened computer basic course, such as the information technology, the students have master the basic knowledge and basic operational skills before entering university. In this situation, the basic computer courses need find only another way to keep its vitality. Therefore, the author suggests its curriculum objectives to adjust. Because the teaching object already has a primary computer operating ability, this course should pay attention to the calculating ability cultivation of the students to use the computers to solve practical problems, and pay attention to the cultivation of the students' innovative thinking, and not to stay in teaching the students to use several commonly computer software.

To cultivate the students' calculating ability and innovative thinking, we should change firstly the teachers' and students' curriculum concept in the teaching process, and at the beginning let the students have a thinking that the computer is merely a tool and their ultimate learning goal is that they can use the computer to solve the application problems and they obtain the thinking ability. Therefore, the author suggests using two kinds of means to cultivate the students' innovation ability and computational thinking ability in the teaching process.

Firstly, we should learn the advanced thinking of the case teaching method[2], and should not take the teaching of software usage as the teaching goal, but should let the students not only know how to use the software, but also what is the software's function through a case demonstration and explanation. The teachers should teach the students to use the software to solve the application and think the application method in their own major field, and then the students will fully understand the learning purpose and significance of every knowledge unit. Through this teaching method closely combined with the practical application, the students can get a deeper understanding of the computer tool characteristics, are exposed to a large number of application examples, and have the experience of the method and means that other people use the computer to solve the practical problems, and then lay a solid foundation for their practical application in problem solving, so as to the students can obtain good calculation ability.

Secondly, the teachers should abandon the homework method in the traditional teaching process, and should no longer arrange the homework alone for a course unit. They should learn the project design in the project teaching method[3], and encourage the students to choose a real application as the curriculum design to practice, such as a full resume or the school supermarket web, finally the students can obtain the innovation thinking ability from the practice process.

(2) Suggesting the teaching methods must be realistic, do not copy.

As a first basic computer popularization course in domestic most colleges, the basic computer courses' teaching method is the subject studied by a lot of experts. Especially under the new education form, the introduction of computer technology course is impacted by the information technology curriculum opened in the middle and primary school, so a variety of coping strategies and advanced teaching method is more emerge in an endless stream, such as the case teaching, the project teaching method, the task-based teaching method[4], the happy teaching method[5], the interest teaching method[6], and so on. However, the introduction of computer technology course in each school status are different, many advanced teaching methods can't obtain good teaching effects, even can't be carried out smoothly. Therefore, the author suggests the teaching method must be combined with the school own actual situation, combined with the teachers own actual situation, and do not copy the advanced teaching methods. The following is the author's teaching strategy according to own actual situation.

First, due to the current form that the course's teaching objects' uneven levels, the author uses an improved hierarchical teaching method to take into account the overall course teaching effect.

The stratified teaching method which is described in the reference[7–9] divides the introduction of computer technology course into several sub curriculums according to the students' level, and the students can choose some a course to learn according to their actual situation, so all students can obtain good teaching effect. But the author thinks that the teaching method has some disadvantages: (a) because the curriculum is divided into several sub curriculum, the management, the implementation and the teaching resources all faces a severe test; (b) because of the students' energy and time constraints, they can choose only one branch, the results is still a batch of uneven students after this course, and the follow-up course still need to solve the problem. Therefore, in the implementation of teaching process, the author draws the merits of the stratified teaching method and overcome the above problem.

Specific approach is as follows: (a) according to the students' level, the courses are divided into three levels: the entry knowledge for the students who has no computer ability; the middle knowledge for skills training; the difficult knowledge for improving the students' accomplishment; (b) making full use of the multimedia and network teaching means, all contents of the above three levels is made into the teaching videos which are placed in the courses teaching web.

(c) In class, the middle knowledge is described carefully, and the other contents will be learned by themselves through the network teaching resources, so the students at different levels will be meeting the personalized requirement.

Because the author improved the lamination teaching method, the students' levels after this course are not differ too much, and their ability have been greatly improved while the implementation of teaching process is very smoothly.

Second, because of the wide cover range of this course content, each knowledge unit is not very good for the series, and easy to make students produce boredom in the learning process. Aimed to this, the author uses the project teaching method and the interest teaching method in the teaching process, and shows the course content as an organic whole through the demonstration of an actual project from scratch. At the same time, in the course teaching, the author emphasizes to stimulate the students' interest in learning through a variety of means, for example, carrying out a variety of small course knowledge contest, giving every student an opportunity to show himself, suggesting the students to put individual elements into the practice works, and learning Photoshop to make a picture of himself and a own advice and a celebrity. Various teaching means and practical way has fully mobilized the students' interest in learning and can greatly improve the effect of class teaching.

(3) Suggesting that the teaching methods don't stick to one pattern and take improving the teaching effect as the ultimate goal.

In the nineteen seventies, the United States National Training laboratories had made investigation on various learning style learning effect. The results showed that, all the way of study, the classroom lectures link which has been considered by the domestic widely the most important part has only 5% learning content average retention rate, the reading, the listening, and the showing has 10%, 20% and 30% learning content retention rates. The discussion, the practice and the teaching others, which are excellent methods but is weak in the domestic, especially the practice and the teaching others has as high as 75% and 90% learning content retention rates. Therefore, the author has changed the traditional single teaching mode which is mainly to the teachers in classroom, the introduction of a variety of means makes the teaching effect improved. Specific practices are as follows.

First, the guide teaching is important in the classroom, and the teachers teach everything no longer like the primary and secondary school, the reasons are: (a) Now, the computer industry has the rapid development, our teaching can't keep up with the scale of software, the updating and upgrading speed of software. If we introduce everything in course design, the teaching resource is not allowed. (b) Very detailed and comprehensive teaching will confine the students' thinking and put an end to the students' innovation and the desire of the autonomic learning ability, the students still have serious dependence to the teachers. Therefore, in classroom teaching, the author mainly focuses on teaching the students how to learn the method, lets the students gain the independent learning and lifelong learning ability. When the teachers have no detailed introduction of the classroom content, the students can meet their own learning requirements through self study. Specific practices include that the teachers should introduce more software functions, how to use the software help manuals and how to solve the problem with the software when the teachers introduce the software, all these is more important than the usage of a function menu.

Second, at the premise of leading teaching, class discussion should be frequently carrying on. The teachers should urge the students to think through classroom discussion form and change the students from the passive learning to the active learning. When the students who have close discuss, they will find the communication more easy, and the general problems existing in the teaching process can be more easily found, so that teachers can solve timely.

Third, the teachers should arrange the student assistants. According to students' learning situation, the teachers can pick out good students as student assistants and they can assist the teacher to solve the students' learning problems encountered in the learning process timely; and a student assistant is not immutable and frozen, a student who only master a part of the content well, then he can act as the student assistant for this knowledge unit.

Fourth, the teachers should arrange a certain period as a student presentation period, at the same time through the network platform to build the students' works and the exchange zone. The students can display their works in class or through network, then find the shortcoming of own works, and all students can find the advantages and disadvantages of other work and this can be more conducive to raising the level of self.

Fifth, the teachers should provide the students with more opportunities for practice, and give the students work with timely review and guidance by the aid of the network platform.

4 CONCLUSIONS

The teaching of the introduction of computer technology course is the difficult problem in basic computer course. In this article, the author analyzes the characteristics of the course and the problems existing in the teaching process, suggests a change of teaching objectives, and enhances the teaching level, so as to maintain the vitality, also suggests to take a variety of teaching methods in the teaching process according to their own situation to guarantee the course's teaching effect.

REFERENCES

[1] Miao Feng. The teaching and learning of the fundamentals of computer culture, China modern education equipment. 2011.1:147–149

[2] Wu Chunying. The case teaching method in the Visual Basic program design courses. Education Theory and Practice 2009.9:57–58

[3] Gao zhongyi. The project teaching method in "computer culture basis". The teaching education front 2011:90

[4] Zhang Zhifen. The "task driven" in "computer culture basis" teaching. Fujian computer 2010.10:52

[5] Zhu Rui. The happy teaching method in C language teaching. Union Forum 2010.11:114–115

[6] Chen Yijun. The interest teaching of "university computer cultural foundation". Information and Computer 2010.12:160

[7] Dan Haomin. The factors analysis and Countermeasures of "computer culture basis" in higher vocational colleges teaching quality. Computer Education 2010.12:112–115

[8] Wang Qi. The reform of college "computer culture basis". The science and technology innovation 2010.31:175

[9] Qian Fang. The practice and thinking of the open education "computer culture basis" asynchronous layered teaching. Neijiang science and technology 2011.1:73

Information Technology and Computer Application Engineering – Liu, Sung & Yao (Eds)
© 2014 Taylor & Francis Group, London, ISBN 978-1-138-00079-7

Discussion on the method of Matlab Web application development

Song Gao
College of Machinery Manufacturing and Automation, Shandong Yingcai College, Jinan Shandong, China

ABSTRACT: This paper introduces the principle of the Matlab web application, the flow of Matlab Web application is given. The key problems such as how to receive parameters from web pages, how to create and call M-files, how to call Simulink model and how to make the output HTML files including data table and images are mainly discussed.

Keywords: Matlab; Web application; M-files

1 GENERAL INSTRUCTIONS

MATLAB Web Server is a Matlab Release11 (Matlab 5.X) and later published components, by using standard HTML documents and forms, developers can Matlab/Simulink-based applications over the Internet to publish [1]. To using Matlab powerful computing and simulation Matlab Web applications to solve various Web numerical poor analysis capabilities, drawing less powerful, has a friendly interface, simple operation, strong distribution, expansion and easy maintenance. It can be widely used in scientific computing, control systems, information processing and other areas of research and teaching [2].

2 FUNDAMENTAL

MATLAB Web Server is one of the core of the network computing functions Kit, which contains Matlabserver, Matweb.exe components. Matlabserver manage the communication between the Web application and Matlab is a multi-threaded TCP/IP server, create a Matlab process and calls the corresponding M-file in response to the input page calculation request. Matweb.exe is the Matlabserver a TCP/IP client, CGI program HTML form data is translated into a Matlab object, pass it to Matlabserver calculated from Matlabserver accept the calculated results will eventually return to the client browser displays [3].

The Matlab Web application development process on the Web server and Matlab Web Server configuration issues, related documents have been described in detail, not repeat them here. Each of the functions of the system consists of a Web-input or output page and the background of the M-file to achieve the client simply enter the appropriate control parameters can be submitted through the Web page to call Matlab process server-side and make the appropriate calculation and simulation, the user can view the results of the run from the returned page [4]. On the Web server Apache2.2 and Windows platforms, introduced the common development of methods and techniques to enter the page, M file, the output page in the the Matlab Web application development process.

3 THE KEY ISSUES OF THE WEB APPLICATION IN THE PREPARATION OF THE MATLAB

3.1 The input page

The common way to enter the page text box to enter the main input pull-down menu.

3.1.1 Text box to enter

The basic framework of the text box to enter the page as follows:

```
<formaction="/cgi-bin/matweb.exe" method="POST">
<input type="hidden" name="mlmfile" value="in2_1">[4]
My input variable 1:
<inputtype="text" name="my_input_variable_1">
<input type="submit" name="Submit" value="Submit">
...
```

matweb.exe is the Matlab Web Server's entrance, `<form action="/cgi-bin/matweb.exe" method="POST">` the text box to send data to the virtual directory/the CGI program matweb CGI-bin under an exe. Note that this one in actual use may occur on the executable file matweb.exe download phenomenon, the solution is to add a question mark after the matweb.exe that matweb.exe? So that it no longer is an executable file available for download. The code `<input type="hidden"name="mlmfile"value="in2_1">`, input form a hidden variable called: mlmfile, attribute value in2_1, the value is the name of the file called M-file is to be placed in the MATLAB default storage

M file directory, the file name must be unified with the corresponding M-file will not be found. <input type="text" name="my_input_variable_1"> page text box to enter input parameters the text box identifier my_input_variable_1 submitted to the M-file to calculate [5]. <input type="submit" name="Submit" value="Submit"> placed a submit button. The input page enrich and improve as long as follow the HTML provisions can.

3.1.2 Drop-down menu input

The drop-down menu, input the file format of the page and the text box to enter is basically the same, but the corresponding HTML vary.

3.2 M-files

There are two types of M-files: script files and function files, MATLAB Web Server using the M function file.

3.2.1 M-file format

The basic framework of the M-file as follows, the M file storage must be the same asthe function name mfile_template.

```
function retstr = mfile_template(instruct)
retstr = char('');   % To retstr initial value is empty.
mlid = getfield(h, 'mlid');   % Get Matlab process ID
cd(instruct.mldir);   % The establishment of the path
of the workspace
my_input_variable_1 = instruct.my_input_variable_1;
...
```

Matlab calculation and simulation code; ...

```
outstruct.GraphFileName = sprintf('figure_name');
% The file name of the picture as the output variable
stored in outstruct
wsprintjpeg(f, outstruct.GraphFileName);
% The images generated image file named out-
struct.GraphFileName in the handle f
templatefile = which('output_template.html');%
% Specify the output HTML template file
retstr = htmlrep(outstruct, templatefile);
```

Submitted by matweb.exe all over the M-file argument, all types of domains is stored as a string in the input parameters structure, even if the mass of numbers, and each submission will be by the system automatically generates a unique process ID of the mlid structure body area. For example, the text box to enter the page the identifier my_input_variable_1 the parameter assignment the local variables my_input_variable_1 to M-file structure assignment method requires the use of my_input_variable_1 = instruct.my_input_variable_1. Stored in the local variables my_input_variable_1, string '8' For numeric types, need to use str2double or str2num string type data into a numerical value.

3.2.2 Call other M-files

M-file can also call other M-file, the same rules and Matlab main functions and subroutines call the main function that is a function, the function name is the

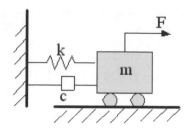

Figure 1. Example of dynamical systems.

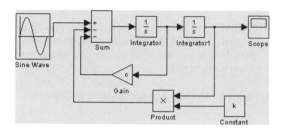

Figure 2. Simulink model.

same as the file name, the other function is called Functions they can only be referenced by the main function of the M-function files or other functions. Each subroutine is defined by the function, followed by a subsequent statement to the body of the function. Various subroutines in any order, but the main function must be the first function. Examples are as follows:

```
function PageString = webstockrnd(InputSet)
spot = str2double(InputSet.spot);
r = str2double(InputSet.r);
sigma = str2double(InputSet.sigma);
numsim = str2double(InputSet.numsim);
t = (0:365)'/365;
S = localstockrnd(spot, r/100, t, sigma/100, numsim);
...
retstr = htmlrep(s, templatefile);      % The main
function ends;
% The following Functions;
function S = localstockrnd(s0, r, t, sig, NUMRND)
...
a = r+sig;
...      % Functions body;
S = s0*exp(a);
```

3.2.3 Simulink model is called

According to the system principle in Simulink model, and then write the corresponding command in the M-file. Specific Examples are as follows, as shown in Figure 1 shows a single-degree-of-freedom dynamic system.

The kinetic equation: $m\frac{d^2y}{dt^2} + c\frac{dy}{dt} + ky = F$

For simplicity, $m = 1$, $F = 2\sin\left(2t + \frac{\pi}{3}\right)$

c is the damp, k is the Stiffness, parameters into and convert the form of $\frac{d^2y}{dt^2} = 2\sin\left(2t + \frac{\pi}{3}\right) - c\frac{dy}{dt} - ky$. Created in Simulink the corresponding model named dynamic.mdl, as shown in Figure 2.

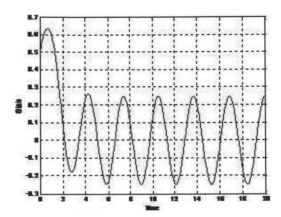

Figure 3.   Output page Image.

Reserved in the input page identifier and a text box, and then write the corresponding M-file as follows:

function responce = single_degree(h)
...
set_param('dynamicr/Gain','Gain',h.c);
set_param('dynamic/Constant','Value',h.k);
[t,x,y]=sim('dynamic.mdl',[0,20]);
f = figure('visible','on');
plot(t,y);
grid on;
...
responce = htmlrep(outstruct, templatefile);

Gain parameter value hc, parameter values modified by entering the page pass parameter values to Simulink model with command set_param set dynamic system Gain module is similar. Modify the parameters required because set_param command value is of type String, h.c fits. The The key is to find the corresponding parameters of the module to be modified, such as 'Gain' and 'Value'. Module parameters can be through callback function get_param ('dynamic/Gain', 'Object Parameters'). Then command simulation, and simulation data stored in the [t, x, y], in order to function for drawing the workspace. Shown in Figure 3 simulation pictures, the htmlrep command generated when the input parameters for the output page, explained in detail in the next section.

Therefore, in the M-file calls other M-file, M functions and Simulink models can utilize, you can take full advantage of the Matlab programming functions to develop a more powerful application.

3.3 Output page

Output page of the form of variable output, image output table output. Regardless of what form of output, must be output variable values or pictures to structure outstruct stored in the form, because the htmlrep a direct replacement for use in a variety of field values in the outstruct variable output page HTML file with the same domain name.

Figure 4.   The example of table output.

3.3.1 Variable output
The basic framework of the output page as follows:
...
My output variable 1 has been computed to be: $my_output_variable_1$
...
Calculated in the M-file variable my_output_variable_1's replace the value of the same variable name, the location of the output page.

3.3.2 Image output
The basic framework of the picture output is as follows:
...
<img src="$ GraphFileName$ ">
...
The M-file via the command outstruct. GraphFileName = sprintf ('figure_name') the file name of the picture as the output variables stored in outstruct. In addition, the generation of the picture, the adjustment of the image size, image format conversion in the M-file with the appropriate command to complete [3]. Last, domain refers outstruct GraphFileName substituting the pictures will be directly replaced by having same variable name GraphFileName the place on the output page display shown in Figure 3.

3.3.3 Table output
Page to the output of the output variable output and pictures output. For example, in the M-file to generate a magic matrix code outstruct.msquare = the magic (msize).
On the output page write code as follows:
<table border="1" cellspacing="1" autogenerate="$ msquare$">
  <tr>
    <td align="right">
    </td>
  </tr>
</table>
Showing results are shown in Figure 4:

4 CONCLUSION

This article explores some of the key technologies for the development of Web applications based on Matlab. With Net, Java AcitveX, technology development, it will render the Matlab Web application function more powerful and rich, and further play Matlab strong engineering computing power to solve the problem of Web application in numerical computation, image processing cumbersome, which is bound to distance education, virtual laboratory has a profound impact [6].

REFERENCES

[1] Lou Shuntian, Shentan Chen, Humin Lei. MATLAB 5.x Programming Language. Xi Dian University Press, Xi'an, 2000.

[2] Yong-an Huang, Lu Ma, Huimin Liu. Matlab 7.0/Simulink to 6.0 modeling and simulation development and process applications. Tsinghua University Press, Beijing, 2005.

[3] Chunxia Tang, Xiaobei Wu, Zhiliang Xu. Development of Web Application. Control Engineering of China. 2005. 12(2), pp. 159–161.

[4] Song Gao, Na Yao. The Design for Web and Matlab-based Virtual Experimental Platform of Automatic Control Theory. Laboratory Science, 2010. 13(4), pp. 67–69.

[5] Min Gao, Yi Zeng, Zhengguang Tu, Weiping Song, Ying Li. Web Application and Method Discussion of Matlab. Computer Application. 2004. 24(6), pp. 188–190.

[6] Zhaoxia Guo, Jian-an Fang. Web-based System Simulation Technology. Computer Engineering. 2005. 31(10), pp. 228–230.

Information Technology and Computer Application Engineering – Liu, Sung & Yao (Eds)
© 2014 Taylor & Francis Group, London, ISBN 978-1-138-00079-7

Adaptive terminal sliding mode control for AC servo system based on neural network

Yue-Fei Wu, Da-Wei Ma & Gui-Gao Lei
School of Mechanical Engineering, Nanjing University of Science and Technology, Nanjing, Jiangsu, China

ABSTRACT: Terminal sliding mode variable structure control (TSMC) is a special nonlinear control strategy, which has strong robustness against parameter variations, load disturbances and uncertainty of system. Nevertheless, the control precision and stability of system will be affected by the chattering, caused by sliding mode switch control. This control method can approach the nonlinear model through RBF neural network when the accurate model of the nonlinear system is unknown. The parameter adaptive law is designed according to Lyapunov theory, without sacrificing the strong robustness of TSMC. Simulation results show the control input is chattering-free, and both load disturbance and parameter perturbation can be compensated.

Keywords: Automatic control technology; Terminal sliding mode; neural network

1 GENERAL INSTRUCTIONS

It is well known that servo system have nonlinear and time-varying behavior with various uncertainties and disturbances. In recent years, control of such systems has attracted great research interest[1][2]. Since the traditional control method cannot meet the requirements of its servo system any longer, some advanced control strategy has been proposed for position control[3][4], such as neural network control and terminal sliding mode control (TSMC).

TSMC has been attracted lots of attention over the past decades for its simple structure, strong robustness to model uncertainty and to parameter variations, and its good disturbance rejection properties. However, its major drawback in practical applications is the chattering problem, which has been well researched by experts[5][6], and lots of solutions to eliminate chattering like, adaptive neural network TSMC have been introduced.

Therefore, in this paper, an improvement of TSMC sliding surface has been made, and RBF network approach the nonlinear model, based on which self-tuning law is derived. Simulation results show the dynamic performance of system is well guaranteed with strong robustness, besides, the control input is smooth.

2 GETTING STARTED

2.1 *Model of PMSM*

Assumption[1]: (1) ignoring the saturation effect; (2) uniform distribution of the motor air-gap field and induced EMF is sinusoidal waveform; (3) no magnetic hysteresis and eddy current loss; (4) no dynamic response of excitation current; (5) no field winding in rotor; (6) under vector control of orientation for the magnetic pole position, the component of the stator current excitation $i_d = 0$.

$$u_q = Ri_q + Lpt + \omega_r \psi_f \qquad (1)$$

$$u_d = -\omega_r Li_q \qquad (2)$$

$$T_{em} = 1.5 p_n \psi_f i_q = K_t i_q \qquad (3)$$

$$T_{em} = T_L + B(\psi_r/p_n) + (J/p_n)p\omega_r \qquad (4)$$

where u_d, u_q are component of armature voltage in d-q coordinate respectively; i_d, i_q is component of armature current in d-q coordinate respectively; L is equivalent armature inductance; R is armature winding resistance; ω_r is electrical angular velocity; Ψ_f is rotor flux; p_n is motor pole pairs; T_L is load moment; B is damping coefficient; K_t is electromagnetic torque coefficient; T_{em} is output moment of the motor, and p is differential operator.

Assuming current loop of PMSM is ideal[1][2], i.e. $G_c(s) = 1$, meanwhile the reducer can be simplified as $G_j(s) = K_j/s$, where K_j is reducer coefficient. With Eq. (1)–(4), the decoupled servo system is shown in Fig. 1, where K_v is velocity feedback coefficient, K_c is angular converting coefficient.

θ_m is a position feedback signal, u is control input, and ω_m is angular velocity of PMSM. Define $x_1 = \theta_m, x_2 = \dot{\theta}_m, x = [x_1 \quad x_2]^T$, the system matrix can be expressed as:

$$\dot{\mathbf{x}} = \begin{bmatrix} 0 & 1 \\ 0 & a \end{bmatrix} \mathbf{x} + \begin{bmatrix} 0 \\ b \end{bmatrix} u + \begin{bmatrix} 0 \\ d \end{bmatrix} \qquad (5)$$

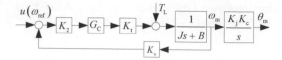

Figure 1. Block diagram of decoupled system.

where $a = -(B + K_2K_tK_v)/J$, $b = K_2K_tK_jK_c/J$, $d = K_jK_cT_L/J + \Delta ax_2 + \Delta bu$, d is uncertainty term, Δa and Δb are uncertainties of a and b respectively. The relative degree of the system is 2.

2.2 Adapative terminal sliding mode control

The TSMC based on it can achieve favorable tracking performance, and it's robust with regard to uncertainties and disturbance. The algorithm is optimized in this section, and sliding surface s is of the following form:

$$s = \dot{e} + c_1 e + c_2 e^{q/p} \tag{6}$$

where $e = x_1 - \theta_{ref}$, θ_{ref} is a reference signal, c_1, c_2, are design parameters respectively.

A general nonlinear function $f(Z)$ can be approximated by an RBFNN as

$$f(\mathbf{Z}) = \mathbf{W}^{*T}\mathbf{\Phi}(\mathbf{Z}) + \varepsilon(x),$$
$$\Phi_i = \exp(-\frac{\|\mathbf{Z} - \xi_i\|^2}{\sigma_i^2}), i = 1, \cdots, l \tag{7}$$

Where Z is the input vector, $\mathbf{\Phi} = [\Phi_1, \ldots, \Phi_1]^T$ is the bias function, $W = [w_1, \ldots, w_1]$ and $\xi_i = [\xi_{i1}, \ldots, \xi_{in}]^T$ are dilation and translation parameters respectively, with gaussian function as the bias function each node, $\varepsilon(x)$ is a RBFNN functional reconstruction error vector, and W* are the optimal parameter vectors of w. Let d* be the optimal function approximation using an ideal approximator then

$$d(x) = \omega^* \Phi(x, \xi^*) + \varepsilon(x) \tag{8}$$

Where ω^*, ξ^* are the optimal parameter vectors of w and ξ, respectively, and $\varepsilon(x)$ denotes the approximation error and $\varepsilon(x)$ is assumed to be bounded by $\|\varepsilon(x)\| \leq \varepsilon$, in which ε is positive constant. Optimal parameter vector needed for best approximation of the function are difficult to determine so defining an estimate function as

$$\hat{d}(x) = \hat{\omega}^T \Phi(x, \hat{\xi}) \tag{9}$$

Where $\hat{\omega}, \hat{\xi}, \hat{d}$ are the estimates of ω^*, ξ^*, d^*, respectively. Define the estimation error as

$$\tilde{\xi} = \xi^* - \hat{\xi}, \tilde{\omega} = \omega^* - \hat{\omega}, \tilde{d} = d^* - \hat{d} \tag{10}$$

The closed loop system with proposed controller is chosen as

$$u = -b^{-1}(t)\left(ax_2 - \ddot{r} + c_1\dot{e} + c_2\frac{d}{dt}e^{\frac{q}{p}} + y + v + cs + \eta\,\mathrm{sgn}(s)\right) \tag{11}$$

Where $y = \hat{\omega}^T\Phi(x, \hat{\xi})$ – the disturbance compensation of the RBF networks, $v = \hat{\lambda}\mathrm{sgn}(s)$ – the approximation error of the RBF networks.

Adaptation law for the RBF networks used to approximate are:

$$\dot{\hat{\omega}} = \Gamma_\omega\left(\Phi(x, \hat{c})s^T - \lambda_\omega \hat{\omega}\right)$$
$$\dot{\hat{\xi}} = \Gamma_c'\left((\Phi_c')^T \hat{\omega}s - \lambda_c \hat{\xi}\right) \tag{12}$$

Adaptation law used to approximation error are:

$$\dot{\hat{\lambda}} = \tau_v\left(\|s\| - \sigma_v \hat{\lambda}\right) \tag{13}$$

Where Φ_c' is the partial derivative of $\Phi(x, \hat{\xi})$, and $\Gamma_\omega, \Gamma_c, \lambda_\omega, \lambda_c, \tau_v, \sigma_v$ are the learning rates with positive constants.

Using Taylor expansion linearization technique to transform the nonlinear function into a partically linear form:

$$\Phi(x, \xi^*) = \Phi(x, \hat{\xi}) + \Phi_c'\tilde{\xi} + O(x, \tilde{\xi}) \tag{14}$$

With $O(x, \tilde{c})$ are the vector of higher order terms and

$$\|O(x, \tilde{\xi})\| \leq \|\Phi(x, \hat{\xi})\| + \|\Phi_c'\|\|\tilde{\xi}\| \tag{15}$$

The approximation error of the nonlinear function $d(x)$ through RBFNN can be presented as:

$$d(x) - y - v = \tilde{\omega}^T\Phi(x, \hat{\xi}) + \hat{\omega}^T\Phi_c'\tilde{\xi} +$$
$$\tilde{\omega}^T\Phi_c'\tilde{\xi} + \omega^T O(x, \tilde{\xi}) + \varepsilon_f(x) - v \tag{16}$$

With $\tilde{\omega} = \omega^* - \hat{\omega}$, and

$$E = \tilde{\omega}^T\Phi_c'\tilde{\xi} + \omega^T O(x, \tilde{\xi}) + \varepsilon_f(x) \tag{17}$$

According to (14) and (17), it can be derived that E is bounded, and E is assumed to be bounded by $E \leq \|\lambda\|$, in which ε is positive constant, it result in:

$$d(x) - y - v = \tilde{\omega}^T\Phi(x, \hat{\xi}) +$$
$$\hat{\omega}^T\Phi_c'\tilde{\xi} + E - v \tag{18}$$

Substituting (17), (23) into (6), it result in:

$$\dot{s} = \tilde{\omega}^T\Phi(x, \hat{\xi}) + \hat{\omega}^T\Phi_c'\tilde{\xi} + E -$$
$$\hat{\lambda}\mathrm{sgn}(s) - \eta\,\mathrm{sgn}(s) - cs \tag{19}$$

To examine the control scheme consider the Lyapunov function:

$$V = \frac{1}{2}s^T s + \frac{1}{2}tr\left(\tilde{\omega}^T\Gamma_\omega^{-1}\tilde{\omega}\right) +$$
$$\frac{1}{2}tr\left(\tilde{\xi}^T\Gamma_c^{-1}\tilde{\xi}\right) + \frac{1}{2\tau_v}\tilde{\lambda}^2 \tag{20}$$

Its derivative is:

$$\dot{V} = s^T \dot{s} + tr\left(\tilde{\omega}^T \Gamma_\omega^{-1} \dot{\tilde{\omega}}\right) + tr\left(\tilde{\xi}^T \Gamma_c^{-1} \dot{\tilde{\xi}}\right) + \frac{1}{\tau_v} \tilde{\lambda}\dot{\tilde{\lambda}} \tag{21}$$

Substituting (17), (18), (19) into the above equation:

$$\dot{V} = s^T\left(\tilde{\omega}^T \Phi(x,\hat{\xi}) + \hat{\omega}^T \Phi_c' \tilde{\xi} + E - \hat{\lambda}\,\mathrm{sgn}(s) - \right.$$
$$\mu\,\mathrm{sgn}(s) - cs) - tr\left(\tilde{\omega}^T\left(\Phi(x,\hat{c})s^T - \lambda_\omega \hat{\omega}\right)\right) - \tag{22}$$
$$tr\left(\tilde{\xi}^T\left((\Phi_c')^T \hat{\omega}s - \lambda_c \hat{\xi}\right)\right) - \tilde{\lambda}\left(\|s^T\| - \sigma_v \hat{\lambda}\right)$$

Because

$$tr\left(\tilde{\omega}^T\left(\Phi(x,\hat{\xi})s^T\right)\right) = s^T \tilde{\omega}^T \Phi(x,\hat{\xi}) \tag{23}$$

$$tr\left(\tilde{\xi}^T\left(\Phi_c'\right)^T \hat{\omega}s\right) = s^T \hat{\omega}^T \Phi_c' \tilde{\xi} \tag{24}$$

Substituting (23), (24) into (22), it result in

$$\dot{V} = s^T\left(E - \hat{\lambda}\,\mathrm{sgn}(s) - \mu\,\mathrm{sgn}(s) - cs\right) + \lambda_\omega tr\left(\tilde{\omega}^T \hat{\omega}\right) +$$
$$\lambda_c tr\left(\tilde{\xi}^T \hat{\xi}\right) - \tilde{\lambda}\left(\|s^T\| - \sigma_v \hat{\lambda}\right) \tag{25}$$

Considering ω^*, c^*, λ^* is assumed to be bounded by $\|\omega^*\| \le \bar{\omega}$, $\|\xi^*\| \le \bar{\xi}$, $\|\lambda^*\| \le \bar{\lambda}$, it result in

$$tr\left(\tilde{\omega}^T \hat{\omega}\right) = tr\left(\tilde{\omega}^T\left(\omega^* - \tilde{\omega}\right)\right) \le \frac{1}{2}\bar{\omega}^2 - \frac{1}{2}\|\tilde{\omega}\|^2 \tag{26}$$

$$tr\left(\tilde{\xi}^T \hat{\xi}\right) \le \frac{1}{2}\bar{\xi}^2 - \frac{1}{2}\|\tilde{\xi}\|^2, \tag{27}$$

$$\tilde{\lambda}\hat{\lambda} = \tilde{\lambda}\left(\lambda^* - \tilde{\lambda}\right) \le \frac{1}{2}\bar{\lambda}^2 - \frac{1}{2}\tilde{\lambda}^2 \tag{28}$$

From (27) and (28) the dynamic of V become:

$$\dot{V} \le \lambda\|s\| - \hat{\lambda}\|s\| - \mu\|s\| - c\|s\|^2 - \tilde{\lambda}\|s^T\| + \frac{\lambda_\omega}{2}\bar{\omega}^2 +$$
$$\frac{\lambda_c}{2}\bar{\xi}^2 + \frac{\sigma_v}{2}\bar{\lambda}^2 - \frac{\lambda_\omega}{2}\|\tilde{\omega}\|^2 - \frac{\lambda_c}{2}\|\tilde{\xi}\|^2 - \frac{\sigma_v}{2}\tilde{\lambda}^2 \tag{29}$$
$$\le -\mu\|s\| - c\|s\|^2 + \frac{\lambda_\omega}{2}\bar{\omega}^2 + \frac{\lambda_c}{2}\bar{\xi}^2 +$$
$$\frac{\sigma_v}{2}\bar{\lambda}^2 - \frac{\lambda_\omega}{2}\|\tilde{\omega}\|^2 - \frac{\lambda_c}{2}\|\tilde{\xi}\|^2 - \frac{\sigma_v}{2}\tilde{\lambda}^2$$

From above the dynamic of V become:

$$\dot{V} \le -kV + V_0 \tag{30}$$

where

$$V_0 = \frac{\lambda_\omega}{2}\bar{\omega}^2 + \frac{\lambda_c}{2}\bar{c}^2 + \frac{\sigma_v}{2}\bar{\lambda}^2 \tag{31}$$

$$k = \min\left\{\frac{\lambda_\omega}{\lambda_{\max}\left(\Gamma_\omega^{-1}\right)}, \frac{\lambda_c}{\lambda_{\max}\left(\Gamma_c^{-1}\right)}, \ \sigma_v \tau_v\right\} \tag{32}$$

is bounded above by

$$V(t) \le V(t_0)e^{-k(t-t_0)} + \int_0^t V_0 e^{-k(t-\rho)}d\rho =$$
$$V(t_0)e^{-k(t-t_0)} + \frac{V_0}{k}\left(1 - e^{-k(t-t_0)}\right) \tag{33}$$

If after a finite time t_0, there exist parametric uncertainties only, then, in addition to result, zero final tracking error is achieved, ie, $e \to 0$ as $t \to \infty$.

3 SIMULATION

Simulation is performed by Matlab/Simulink solver chosen as ODE45. Parameters of pitching subsystem are listed below: $K_2 = 10$, $K_t = 1.11$, $K_j = 1:231$, $K_c = 57.2957$, $K_v = 0.0318$, $B = 0.000143\,\mathrm{N\cdot m\cdot s}$, $J = 0.002627\,\mathrm{kg\cdot m^2}$. Parameters of the controller are listed below: $\eta = 2$, $c = 5$, $c_1 = 45$, $c_2 = 1$, $\lambda_c = 0.1$, $\lambda_\omega = 2.5$, $\tau_v = 0.5$, $\sigma_v = 0.15$. $\Gamma_\omega = 0.5 \times I_{10\times10}$, $\Gamma_c = 3 \times I_{10\times10}$.

Scenario 1: To test response speed and robustness to firing impact, step signal $\theta_{\mathrm{ref}} = 15°$ at 0 s is selected as excited signal. Impulse disturbance is added to motor output axis at 1 second, whose duration is 0.01 s and amplitude is $15\,\mathrm{N\cdot m}$. The simulation results are shown in Fig. 2.

Scenario 2: To test tracking performance and robustness to parameter perturbation, $\theta_{\mathrm{ref}} = 15° \sin(0.5\pi t)$ is selected as the excited signal. Moment of inertia is increasing till 4 times as its initial value during the whole process. The simulation results are shown in Fig. 3 and Fig. 4.

Fig. 2 to Fig. 4 show the proposed control method is robust to disturbance impact, and the control input is smooth, it show smooth control activity and excellent tracking performance, moreover, system shows better robustness to changes of inertial.

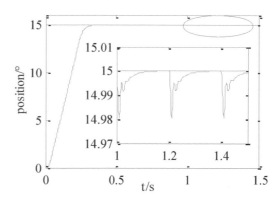

Figure 2. Response to a step command position change with impulse disturbance.

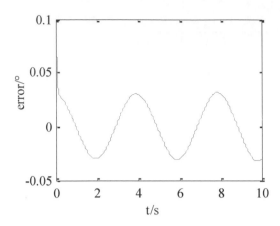

Figure 3. Tracking error with inertial perturbation.

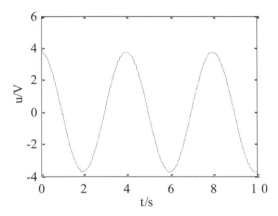

Figure 4. Control input.

4 CONCLUSIONS

An adaptive TSMC based on RBF neural network identification is introduced in this paper. An improvement of TSMC sliding surface is designed in proposed controller to smooth control input. Simulation results show the proposed scheme can guarantee the response speed and tracking precision, furthermore, it isn't sensitive to either external load disturbance or internal parameters perturbation, which indicates its potential of being tailored for the AC servo system.

ACKNOWLEDGEMENT

It is a project supported by the ministry of industry and information technology based research (B2620110005).

REFERENCES

[1] GUO Ya-jun, WANG Xiao-feng, MA Da-wei, 2011. Application of Adaptive Backstepping Sliding Mode Controlin in Alternative Current Servo System of Rocket Gun. Acta Armamentarii. 28(1):1272–1275.
[2] CHAI Hua-wei. 2008. Robust Control and Application Research on Position Servo System of Cluster Antiaircraft Rocket Launcher. PhD dissertation, Nanjing University of Science and Technology.
[3] CHUNG S C Y, LIN C L. 1999. A transformed Lure Problem for Sliding Mode Control and Chattering Reduction. IEEE Trans on Automatic Control. 3(1): 563–568.
[4] WANG Yanmin, FENG yong, HAN Xiangwei. 2008. High-order sliding mode control of uncertain multi-variable system. Control and Decision. 455–459.
[5] Oscar Barambones, Patxi Alkorta. 2011. A robust vector control for induction motor drives with an adaptive sliding-mode control law. Journal of the Franklin Institute. 2: 300–314.
[6] Tzu-Chun Kuo, Ying-Jeh Huang2008, Shin-Hung Chang. Sliding Mode Control With Self-tuning Law for Uncertain Nonlinear Systems. ISA Transaction. 171–178.
[7] İlyas Eker. Sliding Mode Control with PID Sliding Surface and Experimental, 2006. Application to an Electromechanical Plant. ISA Transactions. 6: 109–118.

Research and application of information security evaluation in website

Chunjing Si
School of Information Engineering, Tarim University, Alaer, Xinjiang, China

ABSTRACT: According to the safety criteria of international information system and regarding the website characteristics, at first, this paper proposes an evaluation system for website safety. And then used Analytic Hierarchy Process (AHP) to calculate index weight. Finally, an application example is given to show rationality and effect of the security assessment system of information and a realistic technical frame of group decision support system can been built.

Keywords: Information safety evaluation system; Analytic hierarchy process; Index weight

1 INTRODUCTION

The technical infrastructures that enable them are showing a strong trend towards convergence and net-working. The information obtaining, processing and security guarantee capability are playing critical roles in comprehensive national power, and information security is related to the national security and social stability. At the same time, the threat events of information security often happen, having severe security problems. Information policy of terror is a new way to threaten society development. Recently, the inter-net is used to publicize and contact tool by terrorist in Xizang and Xinjiang. These problems, together with pressure from information security and privacy legislation, are increasing the need for adequately information security evaluation system in website[1].

2 EVALUATION CRITERIA OF INFORMATION SECURITY

2.1 *Trusted Computer System Evaluation Criteria (TCSEC)*

TCSEC is a United States Government Department of Defense (DOD) standard that sets basic require-ments for assessing the effectiveness of computer security controls built into a computer system. The TCSEC was used to evaluate, classify and select computer systems being considered for the process-ing, storage and retrieval of sensitive or classified information.

2.2 *Information Technology Security Evaluation Criteria (ITSEC)*

ITSEC is a structured set of criteria for evaluat-ing computer security within products and systems. Since the launch of the ITSEC in 1990, a number of other European countries have agreed to recognize the validity of ITSEC evaluations. The ITSEC has been largely replaced by Common Criteria which-provides similarly-defined evaluation levels and implements the target of evaluation concept and the Security Target document.

2.3 *Common Criteria (CC)*

CC is a framework in which computer system users can specify their security functional and assurance requirements, vendors can then implement and/or make claims about the security attributes of their products, and testing laboratories can evaluate the products to determine if they actually meet the claims. In other words, Common Criteria provides assurance that the process of specification, implementation and evaluation of a computer security product has been conducted in a rigorous and standard manner[2–4].

3 EVALUATION ITEMS OF INFORMATION SECURITY IN WEBSITE

Evaluation items are gathered from reference [5–11] and website safe factors as input to the decision pro-cess of security evaluation. The items collection should be arranged in a way that supports evaluation of secu-rity behavior and security actions. The process of gathering measured or assessed information uses secu-rity metrics as its basis. Some examples of measured security evidence are listed as follow, including first level assessment factor set and second level assess-ment factor set. The first level assessment factor set is constructed by (1). The second level assessment factor set is constructed by (2).

$$B = \{B_1, B_2, B_3, B_4, B_5\} \tag{1}$$

$$C = \{C_1, C_2, \cdots, C_j\}, 1 \le j \le 27 \tag{2}$$

Table 1. Assessment Items of First Level.

Symbol	Assessment Items
B1	entity and environment safety
B2	organize management and safe system
B3	safe technique measure
B4	network and correspondence safety
B5	software and information safety

Table 2. Assessment Items of Second Level.

	Symbol	Assessment Items
B$_1$	C1	Assess the influence of environment of computer room
	C2	Assess control and supervisory system
	C3	Made waterproof and fireproof
	C4	Environment Monitoring System
	C5	Avoid power fails
	C6	Against burglary
B$_2$	C7	Set up organization and people of information security
	C8	Institute rules and regulations
	C9	Made registration and storage documents
	C10	Establish contingency plan for accidents
	C11	Made system of information security training
	C12	Make sure people is competent at a job
B$_3$	C13	Made Countermeasure for restoring computer function
	C14	Made network security auditing system
	C15	Made system operation log
	C16	Made security and data backup of server
	C17	Avoid hacker attack
	C18	Avoid computer viruses
B$_4$	C19	Take measures to encrypt
	C20	Lead safety audit and follow-up correction activities for system
	C21	Made access control for information system
	C22	Perform a backup of communications lines or equipment
B$_5$	C23	Made access control for OS and data
	C24	Protect applications software
	C25	Made control system for data state
	C26	Made user authentication
	C27	Perform a backup of user information

3.1 First level assessment factor set

Table 1 gives assessment items of first level.

3.2 First level assessment factor set

Table 2 gives assessment items of second level.

4 AHP APPLIED IN WEBSITE EVALUATION

4.1 Constructed hierarchy structure

Evaluation index system show step-by-step hierarchy structure, and divides 5 first level index and 27 second level index.

4.2 Constructed judgment matrix

Judgment matrix is matrix whose includes relatively important judgment value of index in same level. After

Table 3. Important Scale of Judgment Matrix.

Important scale	Meaning
1	The two indexes are the same importance
3	One index more important than the other
5	The former is more obviously important than the latter
7	The former is more strongly important than the latter
9	The former is more extremely important than the latter
2,4,6,8	The center value
Reciprocal	Comparing index i to index j is a_{ij}, so the reciprocal is $a_{ji} = 1/a_{ij}$

comparing one index to another index, a judgment matrix is set up. Experts will give remarks of the satisfied degree to each assessment factor according to the judge rules. The remarks are divided into several levels to measure the behavior on the rule and the value of the assessment factors.

$$B = (a_{ij})_{n \times n} \qquad (3)$$

a_{ij} is multiple which is incremental partand i part j Judgment matrix is single objective decision which can discriminate well from bad easily. Table 3 give the value of judgment matrix, according to 1–9 scale of T.L. Saaty[12].

4.3 Calculate index weight

Judgment matrix is AHP foundation. We can obtain eigenvectors of λ_{max} basing on $BW = \lambda_{max} W$, and normalize index weight which can obtain relative importance coefficient form two different hierarchy structure. The main steps are as follows:

Calculate each line product of judgment matrix B.

$$M_i = \prod_{j=1}^{n} a_{ij} (i, j = 1,2,3,\cdots,n) \qquad (4)$$

Calculate each line Mi to the power of $1/n$

$$\overline{W_i} = \sqrt[n]{M_i} \qquad (5)$$

Normalize vector: $\overline{W} = (\overline{W_1}, \overline{W_2} \cdots \overline{W_n})^T$

$$W_i = \overline{W_i} \Big/ \sum_{i=1}^{n} \overline{W_i} \quad W_i \text{ is index weight.} \qquad (6)$$

4.4 Test data compatibility

Because of complexity of evaluation object and difference of people view for the same object, so the judgment result may deviate form its actual value. We test data compatibility. The main steps are as follows.

Calculate the largest eigenvectors of judgment matrix

$$\lambda_{max} = \frac{1}{n} \sum_{i=1}^{n} \frac{(BW)_i}{(W)_i} \qquad (7)$$

Table 4. Table of 2–9 Order of the Matrix.

n	2	3	4	5
RI	0	0.52	0.9	0.12
n	6	7	8	9
RI	1.24	1.32	1.41	1.45

Calculate compatibility of evaluation index

$$CI = \frac{\lambda_{max} - n}{n-1} \qquad (8)$$

n is order of judgment matrix, CI is index that judgment matrix deviate form its actual value.

Calculate conformance ratio CR

According to average random consistency table, we obtain random conformance ratio. Table 4 gives RI of 2–9 order of judgment matrix.

$$CR = CI/RI \qquad (9)$$

If $CR < 0.1$, we think the judgment matrix have good conformance; If $CI > 0.1$, we think the judgment matrix do not have conformance. So we must adjust value until the matrix have good conformance.

4.5 Calculate relative weight

Weight ratio of C to A = weight of C × weight of B to A.

4.6 Confirm evaluation result

If website evaluation use hundred-mark, and experts give remarks for result. Table 5 gives the remarks and security analysis.

5 APPLICATION EXAMPLE

We give an example of E-government system for explaining the application of website evaluation.

Construct judgment matrix: The value of judgment matrix of $A \sim B$ is constructed as follow in this paper.

$$A \sim B = \begin{vmatrix} 1 & 0.5 & 0.3 & 0.25 & 0.2 \\ 2 & 1 & 0.67 & 0.5 & 0.4 \\ 3 & 1.5 & 1 & 0.75 & 0.6 \\ 4 & 2 & 1.3 & 1 & 0.8 \\ 5 & 2.5 & 1.67 & 1.25 & 1 \end{vmatrix} \qquad (10)$$

Calculate index weight: The index weight of judgment matrix of $A \sim B$ is calculated as follow, using above method.

$$W_i = \{0.667, 0.1333, 0.2000, 0.2667, 0.3333\}(i = 1,2,3,4,5)$$

Table 5. Remarks and Security Analysis.

remarks	conclusion	interpret
Above 90	excellent	Need to carry on an improvement in the small place, try hard for perfect
80 ∼ 89	good	The network and information is safe, need to carry on to the insecurity factor perfect
70 ∼ 79	pass	Network and information safety exist a certain problem, need to strengthen measure to carry on an improvement
Below 70	No pass	Network and information safety exist greater problem, probably will appear a safe problem at any time, need to enlarge measure to carry on an improvement in time

Test data compatibility:
The largest eigenvectors of judgment matrix:
$\lambda_{max} = 4.9693$
Compatibility of evaluation index: $CI = -0.0076$
Conformance ratio: $CR = 0.0067$, the result is good.
Calculate the second level index weight using the same method as follow.

$$W_{B_1-C_i} = \{0.1298, 0.4298, 0.1205, 0.0696, 0.1298, 0.1298\}$$
$$(i = 1,2,3,4,5,6)$$
$$W_{B_2-C_i} = \{0.1205, 0.1298, 0.1298, 0.3696, 0.1298, 0.1205\}$$
$$(i = 7,8,9,10,11,12)$$
$$W_{B_2-C_i} = \{0.1205, 0.1298, 0.1298, 0.1696, 0.2298, 0.2205\}$$
$$(i = 13,14,15,16,17,18)$$
$$W_{B_2-C_i} = \{0.2601, 0.2601, 0.2269, 0.2601\}(i = 19,20,21,22)$$
$$W_{B_2-C_i} = \{0.1429, 0.2857, 0.1429, 0.1429, 0.2857\}$$
$$(i = 23,24,25,26,27)$$

Calculate relative weight
Table 6 gives the final result.

Table 6. Index Weight.

First level index weight	Second level index weight
$B_1 = 0.0667$	$C_1 = 0.0087; C_2 = 0.028; C_3 = 0.008;$ $C_4 = 0.0046; C_5 = 0.0087; C_6 = 0.0087$
$B_2 = 0.1333$	$C_7 = 0.0161; C_8 = 0.0173; C_9 = 0.0173;$ $C_{10} = 0.0493; C_{11} = 0.0173; C_{12} = 0.0161$
$B_3 = 0.2$	$C_{13} = 0.0241; C_{14} = 0.026; C_{15} = 0.026;$ $C_{16} = 0.0339; C_{17} = 0.046; C_{18} = 0.0441$
$B_4 = 0.2667$	$C_{19} = 0.0694; C_{20} = 0.0694; C_{21} = 0.060;$ $C_{22} = 0.0694$
$B_5 = 0.3333$	$C_{23} = 0.0476; C_{24} = 0.095; C_{25} = 0.0476;$ $C_{26} = 0.0476; C_{27} = 0.0952$

Resulting score
Table 7 gives the resulting score.

Table 7. Resulting Score.

First level index score	Second level index score
$B_1 = 6.7$	$C_1 = 1; C_2 = 2.8; C_3 = 0.8; C_4 = 0.4;$ $C_5 = 0.9; C_6 = 0.8$
$B_2 - 13$	$C_7 = 1.6; C_8 = 1.7; C_9 = 1.7; C_{10} = 4.7;$ $C_{11} = 1.7; C_{12} = 1.6$
$B_3 = 20$	$C_{13} = 2.4; C_{14} = 2.6; C_{15} = 2.6; C_{16} = 3.4;$ $C_{17} = 4.6; C_{18} = 4.4$
$B_4 = 26.7$	$C_{19} = 6.9; C_{20} = 6.9; C_{21} = 6.0; C_{22} = 6.9;$
$B_5 = 33.6$	$C_{23} = 4.8; C_{24} = 9.6; C_{25} = 4.8; C_{26} = 4.8;$ $C_{27} = 9.6$

6 CONCLUSIONS

This paper proposes an evaluation system of information security for website based on AHP which make an objective and scientifically appraisal for website. The resulting score table provides reference and support for evaluating and designing website.

REFERENCES

[1] Mikolajczak B, Joshi S. "Modeling of information systems security features with colored Petri nets," J. Systems, Man and Cybernetics, vol. 2, 2004, pp. 4879–4884.

[2] National Technical Committee on Information Technology Security of Standardization Administration of China. "GB/T 20984-2007 Information security technology-Risk assessment specification for information security," S. Beijing: China standards press, 2007.

[3] Dai Z., Luo W., "Security of information system," M. Beijing: Publishing house of electronic industry, 2002.

[4] Lu L., "Research on the evaluation criteria for IT security and Design of information security system," D. Henan: University of Information Engineering, 2001.

[5] Huang L., "Research on information evaluation of network," D. Shandong: Shandong University, 2004.

[6] Wu X., Li S., "Establishment of website' evaluation index from the types of index," J. Journal of the China Society for Scientific and Technical Information, vol. 3, 2005, pp. 352–356.

[7] Cheng W., Gong J., "Network security evaluation," J. Computer engineering, vol. 29, 2003, pp. 182–184.

[8] Wang Z., Bai R., Bao C., "Research on information evaluation system and library network," J. Document, information and knowledge, vol. 3, 2004, pp. 57–59.

[9] Wu G., Hong G., "Application of information security evaluation system in network security evaluation," J. Agriculture Network information, vol. 4, 2007, pp. 97–100.

[10] Shi D., Liu Y., "Research on information evaluation index weight of E-government," J. Coop ergative economy and science, vol. 2, 2008, pp. 44–46.

[11] Chen J., Chen J., "Interactive group decision model and its realization for security investment of information system," J. Application research of computers, vol. 3, pp. 1124.

[12] T.L. Saaty. The Analytic Hierarchy and Analytic Network Processes for the Measurement of Intangible Criteria and for Decision-Making[C]//*Multiple Criteria Decision Analysis: State of the Art Surveys*. Springer Verlag, Boston, Dordrecht, London, 2005:345–408.

Knowledge management approach for serious game development based on user experience

Qingqiang Wu, Xiaoxia Zhang, Yingying She, Yingyao Bi & Baorong Yang
Software School of Xiamen University, Xiamen, Fujian Province, China

ABSTRACT: With the innovation and development of information technology, the new media has been acquiring an increasingly important position in our lives. Serious game compromises the educational element rather than just being a pure entertaining process, and their application domain is more and more widely. This paper aimed to be user-centered, and designed a serious game model combined with knowledge management method on the basis of user experience. The model provides a framework for serious game design process to enhance entertainments in study process, and analyzed the level difficulty with the difficulty graphs algorithm, the study aims to enrich design forms of serious game and related theories.

Keywords: User Experience; Knowledge Management; Serious Game; Game design

1 INTRODUCTION

In recent years, electronic game has been very popular with teenagers and young adults and playing computer games has become a common social phnomenon, so more and more companies would like to use electronic game to train employees. The growing te-nydency of demand for serious games calls for a standard method to guide the game designers to produce more favorite works. Serious game design is a set of formal methods, making the rules and planning for the game. Serious games' goal is both usability and playability, which makes the game a new educational experience integrated with characteristics, interaction and entertainments (Alyson & Selan. 2012), so it must possess educational and entertaining features. The designer of serious game must consider adding elements that make the game fun and attractive. At the same time, they should also think about how to add usability elements, making the game not only be more appealing and fun but educational.

2 RELATED WORK

2.1 *User experience*

Experience is the result of the interaction between the individuals' own mental state and planned scenario, and their physiological and psychological feelings and emotional sublimation after the experience of external things and environments (B. Joseph & James. 2011). The essence of the game lies in the interaction, with the user experience in games constructed on the components of game mechanics, user interface, narrative (storytelling) and also evolves from the game play (Zoe et al. 2012). The user experience this paper referring to is the whole feelings and reflections in the entire game playing process.

Jesse James Garrett divided user experience into five levels, which are Strategy Plane, Scope Plane, Structure Plane, Skeleton Plane and Surface Plane (Jesse. 2002). The user experience and design process of game are presented according with the divinations. Strategy plane solves the problems that what game designers and players want to get. Scope plane needs to make sure the content and function of game. Structure plane connects all the game elements through action series together to form a storyboard. Skeleton plane determines how game interfaces layout game elements. Surface plane is responsible for presenting the goals of the above four planes to the players.

2.2 *Knowledge management approach-knowledge-centered*

At present, the situation that the development of the game presenting on the market is based on the premise of technical conditions, and the creativeness of game content is still in the emotional needs in the human initial period, is mainly due to the lack of developers' knowledge and the restrictions of market demand. However, serious games' development objectives are very clear and the develop contents which are converted to the game in terms of the process which human explore the world. But many serious game designers are always just considering users' entertainments, ignoring the study function of the game which results in the games bringing nothing beneficial to the players. For example, many games for the purpose of education were not built on solid educational principles and theories. They are likely to lose strength as an educational tool. Meanwhile, some serious games just put the application scenarios as the cloak of the game and are lack of entertainment (Glenda et al. 2006).

This is because the game designers do not understand professional knowledge applied in game clearly, and cannot put the required knowledge or rules into the game. Knowledge management can be adopted to help experts and game designers to discuss game requirement with knowledge-centered. Firstly, knowledge experts will divide the goal into a series of knowledge, making them as a knowledge base for the game, then the game designers shall put them into the virtual game environment and give them different attributes or functions to encourage users to refer to knowledge base with the aid of the related game elements, so that players can naturally role and acquire game experience, making the usability and entertainment of the serious game supplement each other (Bruno et al. 2009).

2.3 Serious game and knowledge management approach

Serious game is the branch of electricity game, its main purpose is not for entertainment, but takes the game form of teaching through active learners' involvement through exploration, experimentation, competition and cooperation (Johann et al. 2011), let the users to accept some information through the games, so that they can be trained or cured. Vocational skills are needed for vocational careers, or those that entails using manual or training abilities in a variety of jobs. Having one set of vocational skills sometimes can be applicable to a number of different careers. For example, Time Management is a very important skill for many types of jobs (Ling He et al. 2011).

So, in serious games, as the project unfolded, integrating different kinds of knowledge into the game became the critical issue. Knowledge management is best choice that helps us to extract knowledge from applied scenes. Such an approach is motivated by the perceived lack of clear guidelines in the literature on how to take the best concepts and research of the applied field, and apply them in a game development process (Ben et al. 2011).

3 USER EXPERIENCE BASED KNOWLEDGE MANAGEMENT APPROACH TO SUPPORT A SERIOUS GAME DEVELOPMENT

In the process of game design, the five planes of user experience are considered as framework and the requirement knowledge and game elements are put into framework based on knowledge management approach.

3.1 Strategy plane

A successful user experience bases on a clear expression of the "strategy". In order to design a good game, the position of the game should be determined at the beginning. For serious game, requirement is still indispensable, and its educational symbols and situations rather than just "pure game" symbols and situations should be taken into account. Then game designers must determine what they want to get through this game, how to inject the company's philosophy into the game, and what are the success criteria of the game, etc. Finally, they need to identify the aimed users of the game, including their genders, ages, educational level, proficiency level to the applied profession and the goal to be achieved. All of these constitute the concept of the game and will be transferred to the players in the process of interacting with game.

The "strategy" of serious game is more single from which players usually hope to learn some professional knowledge in a particular area and the user group is relative single. Game designers should develop the game simulating the real event or process, while taking care of the games' appearance and sensory experience. For example, <America's Army> is used to attract young to join the army by America government; <SimCity> is used to facilitate respective history and social studies (Zhaoyan. 2011).

3.2 Scope plane

When the users' requirements and product objectives are converted into the products' content and functionality, a strategy becomes a scope. When the strategy document is finished, based on the game background story and game types, we can extract the relevant elements of the game. The game elements just stay on the conceptual document format right now.

3.3 Knowledge extraction

The nature of the knowledge deals with its proper characteristics according to objective criteria. According to Chi's approach (Chi. 1988), we have decided to classify knowledge in factual, declaratory and procedural knowledge (Bruno. 2009).

Once all the knowledge has been characterized, we can refine the teaching goal of the knowledge. In the early stage of development, the game teaching goal has been defined. In this stage, we will assign these teaching objectives to different types of knowledge. Finally, we get a hierarchical structure view of knowledge.

3.3.1 Game cognitive model

Game designers design game background story and produce the game cognitive model with the right game symbol representing for the background knowledge. The development of cognitive model should also result from a meta-knowledge model. Then designers divide the cognitive model goal into a series of tasks that can be done by a character or a group of characters.

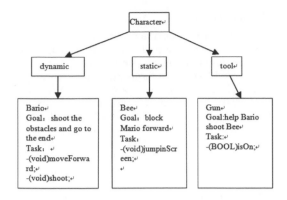

Figure 1. Characters Tree Structure.

Figure 2. Specifying the Task Ask Information.

A tree structure can used to classify character which is the root node into several categories by their goals which are the child nodes, then add all characters as child nodes to corresponding parent node. For example, as shown in Figure 1: we can see that there are three subgroups: Dynamic Players, Static Players and Tools. This division has been made on the basis of the characters that are controlled by the users (Dynamic), those who are autonomous (Static), and those who are responsible for ensuring the integrity of transactions between characters and the environment (Tool) (Alma et al. 2011).

3.3.2 Knowledge conversion

Now game designers and experts put the extracted knowledge into game scenes, and correspond hierarchical structure view of knowledge with tree structure, make sure which variables are player-controlled, which variables are NPC-controlled or as props. This process is a pre-process, allowing us to connect the teaching model and the game. According to nature of knowledge, teaching target is converted into the description of game design, which determines that the teaching target is expressed as a game element or the game background.

3.3.3 Goal decomposition

The game goal has already been divided into many tasks in terms of characters' goal according to the goals and rules of each character in different scene, tasks are specific to a series of executable behaviors. From this model, we can deduce the learning activities (M. Prensky. 2001) and mental processes that learners/players will use in order to make the knowledge transfer and learning work. The learners learn about their own learning experience as long as they know their actions in the game as well as the consequences. As shown in Figure 1, the goal that Bario reaches the end is decomposed into behaviors of move forward and shoot, etc.

3.3.4 Other work

In this phase, the relevant game elements should also be extracted, including the characters, the props, the

main scenes, the length of the text, the picture pixels and other basic elements. The interactive methods between players and NPCs or props, and the actions of props and NPCs will also affect the players experience during the interactive process. Serious games have more rich interaction modes, for example, we can interact with screen by using external equipment or controller, which can enhance the playability of the game (Ben. 2011). The sports competition game can increase the reality sense with kinects' motion capture technology.

3.4 Structure plane

After collecting users' requirements, the scope plane does not specify how to integrate the decentralized design fragments extracted as a whole; structure plane is to create an overall structure for the game. Since being the most important part of the game design, the structure plane determines the mode and order of option presented to the players and organizes the story board with the extracted elements and interactive mode.

In the process of designing serious game, designers specify every character's tasks and behaviors in scene and the system response when goal is not finished or achieved according to the game background, then order them on the basis of hierarchical structure view of teaching goal. In this sense, the Figure 2 shows an example of this kind of diagram in which we can find out how the task is initiated when the initial event takes place. This event may be invoked by a mouse click or by an action from the keyboard. It is created, at that point, the conversation Information Dialogue, producing the Required Information, and completing the goal pursued Get Clues. Finally, it updates the agent's mental state as a result of the new knowledge.

Then, difficulty graphs (Rafael. 2011) can be as graphical representations of how difficulty changes throughout the game and connect all levels. In general, they need two parameters, threat level and situation multiplier, and the player power also has an influence on them.

So the base formula to find a wave's difficulty would be:

$$\Sigma_{n=0} = (ET_n)(ES_n) \tag{1}$$

Where n is a specific enemy in a wave; ET is the enemy threat level; ES is the situation multiplier.

And we take <Super Mario> for example, the steps as follows (Rafael. 2011):

- First, what the game's standard conditions are determined. This is the minimum power level the player has during a determined segment of the game.

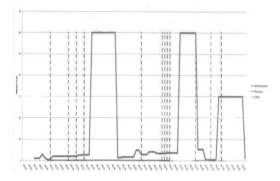

Figure 3.   Difficulty Graph of Super Mario.

The standard condition of <Super Mario> is Mario runs and escapes or kills all enemies and gets to the next level.

- Then, set the threat level and the situation multiplier: the little monster: $ET = 5$; the tortoise: $ET = 5$; the gap: $ET = 5$ or $ET = 10$ (according to the width of the gap);

  If the enemy is on a flat space on the ground: $ES = 1$; if it is placed within a close distance to the edge of a gap: $ES = 6$; if it is on o rudder a set of horizontal blocks: $ES = 3$;

- Playtest. Have different players with different play styles move through the level so that the average time or average distance of each enemy encounter can be gotten.
- Assign values to those situations depending on how the mechanics and enemy behaviors allow fighting them.
- Determine how the encounters work. Figure out how many waves there are in each encounter and what enemies they are composed of.
- $\Sigma_{n=0} = (Etn)(ESn)$ Multiply each enemy's threat level by the situation multiplier and add them all up for each wave.
- Plot them in a graph according to the time when they appeared or the distance from the starting point where players meet them.

As shown in Figure 3, Super Mario, the shortest of the bunch (clocking at around 1:20). We can see that it presents enemy encounters every two to three seconds.

In the process of specification, designers also have to deal with details of the interactive process, such as information feedback suits to the scene, the balance of game narrative and interaction, players' controlling the game, which can enrich the content of interaction design. Besides, it is important to deal with players' errors friendly.

At the end of this phase, a series of smooth and moderately action lines can be formed. It guides players to enter the game, interact with the game and provide feedback immediately after players' every performance and then go to the next scene at an appropriate time.

3.5   Skeleton plane

Lots of demands, which coming into being in the structure plane, need to be further refined, including the interface design, navigation design and information design. Skeleton design forms the foundation of the surface plane, and its information design, that is the functional visual design, focuses on how to display the interface design and navigation design well. Here, using user-centric terminology is important, which gives players information to help them interact with the game. Designers convey the ideas to the players, helping the players be able to clearly know function of the elements once they see.

Successful interface design stress its core part, which makes the user concentrate on it and fall in love with it at first sight. In addition, it is always important to make the components accessible in an easy way and in case that players give up the game for its complex interface design. Designers should also follow the principle of top-down thinking, enhance weight of users concerned contents, principle of being easily used and etc. (Apple Inc. 2012).

3.6   Surface plane

The surface plane is the visual design, which is the collection of contents, functions and aesthetic. The structure plane presents the final game interface to users and transmits the game concept to players under the guidance of the game mechanics. It is in the surface plane where players can interact directly with the game.

And the beauty does not just refer to the interfaces' appearance but also including whether the appearrance and content is match or not. In serious game, the decorative elements are always low-key, it highlights its tasks by using standard controllers and actiions unifies interface layout with scenarios, helping players get information from the game. And it must give consideration to both playability and usability, so it displays entertainment of the game with auxiliary method in interface in the boring long-term and efficient gaming process. For example, we can use animation to provide perceptible feedback information instead of text, slightly exaggerate pictures to en-hance the realism and please players with perfect pictures (Apple Inc. 2012).

4   APPLICATION OF THE MODEL

<Concentration Train> is an educational game, and aims to enhance children's attention through a series of games. Its users are children that lack concentration, doing it under supervision of their parents.

For example, as shown in Figure 4, <Concentration Train> is to improve the ability of thinking distinguish and reduce the wrong because of be carless. Player need to find the number which the picture lacks and drag the digital into corresponding position. The less

Figure 4. Concentration Train.

time they spend, the better performance they get. Players can also set the level difficulty in "Menu": Easy, Difficulty, Intermediate, and parents can check out grades clicking the button "Grade".

5 CONCLUSIONS

Serious game is a new field of the game industry, with its business pattern, profit formula and development process, is different from the other types of game, and becomes the potential stock of game industry. The model the paper presented showed how to take the advantage of serious game as teaching method to make players learn necessary knowledge, especially how to compromise game scene and knowledge background perfectly to add interest to the teaching and also provided an insight into the level design with the difficulty graphs algorithm.

ACKNOWLEDGEMENTS

This work was financially supported by the project Undergraduate Courses Platform Construction of Top-notch Innovative Talents Training Projects of Xiamen University (Z02101) and corresponding author Baorong Yang.

REFERENCES

Alma Juan Carlos, David Ramos. 2011. Modeling Serious Games using AOSE methodologies. *11th International Conference on Intelligent Systems Design and Applications:* 53–58.

Alyson Matheus de Carvalho Souza, Selan Rodrigues dos Santos. 2012. Handcopter Game: A Video-Tracking Based Serious Game for the Treatment of Patients Suffering from Body Paralysis Caused by a Stroke *14th Symposium on Virtual and Augmented Reality*: 201.

B. Joseph Pine II, James H. Gilmore. 2011. *The Experience Economy*. USA. Perseus.

Ben Cowley, Jose Luiz Moutinho, Chris Bateman, Alvaro Oliveira. 2011. *Serious Games Development and Applications. Entertainment Computing: 103*.

Bruno Capdevila Ibanez, Valerie Boudier, Jean-Marc Labat. 2009. Knowledge Management Approach to Support a Serious Game Development. *2009 Ninth IEEE International Conference on Advanced Learning Technologies:* 420–422.

Chi, M. T. H. 1988. *The nature of expertise*. USA. Glaser, R., & Farr, M. (Eds.).

Glenda A Gunter, Ph. D. Robert F. Kenny, Ph. D. Erik Henry Vick, Ph. D. 2006. A Case for a Formal Design Paradigm for Serious Games. *The Journal of the International Digital Media and Arts Association: 93*.

IOS Human Interface Guide. *2012 Apple Inc.*

Jesse James Garrett. 2002. *The Elements of User Experience*. USA. Peachpit Press.

Johann c.k.h. Riedell, Jannicke Baalsrud Hauge. 2011. State of the Art of Serious Games for Business and Industry. *17th International Conference on Concurrent Enterprising.*

Ling He, Xiaoqiang Hu, Dandan Wei. 2011. The Case Analysis of Serious Game in Community Vocational Educationl. *2011 International Conference on Computer Science and Network Technology:* 1863–1866.

M. Prensky. 2001. *Digital game-based learning*. Paragon House Edition. USA: 105–200.

Rafael Vazquez. 2011. *Information on http://www.gamasutra.com/view/feature/134917/how_tough_is_your_game_creating_.php.*

Ying Liu, Zoe Kosmadoudi, Raymond C.W Sung, Theodore Lim, Sandy Louchart, James Ritchie. 2010. Capture User Emotions during Computer-Aided Design. *IDMME – Virtual Concept 2010:23.*

Zhaoyan HU. 2011. Developing Serious Game does not Mean Reject other Pure Entertainment Game-Whether "Serious" and "Game" Can Be Compatible. *China Finance newspaper.*

Information Technology and Computer Application Engineering – Liu, Sung & Yao (Eds)
© 2014 Taylor & Francis Group, London, ISBN 978-1-138-00079-7

The non-fragile controller design based on Lyapunov theory

Chang-Wei Yang
The 365 institute, Northwestern Polytechnical University, People's Republic of China

Jie Chen
School of Aeronautics, Northwestern Polytechnical University, People's Republic of China

Xin-Feng Zhang
XI YI Group CO. LTD, People's Republic of China

ABSTRACT: A non-fragile controller design method using Lyapunov theory is presented in this paper. According to the additive controller gain perturbation, the necessary and sufficient conditions for the existence of non-fragile state feedback controller are given and transformed to the LMI problems, which approach simplifies the solutions to obtain non-fragile state feedback controllers. Simulation results on the UAVs (Unmanned Aerial Vehicles) flight control prove the reliability and validity of the method.

1 GENERAL INSTRUCTIONS

System modeling always can only describe the engineering plant approximately (e.g. there exists the error between them), so there have been great deals of literatures which show interesting in the study of system robust control problems which subject to system and external uncertainty [1, 2]. But such robust controller which tackles the uncertainty of controlled plant is assumed to be implemented exactly. This demand seems validity, but actually, as the inevitable uncertainty in advanced controller operation and engineering implementation, such as environment temperature variation, the digital device's word capacity, and data transform error, etc. controller implementation exactly is impossible in the actual engineering, e.g. the closed system is fragile to the uncertainty in controllers itself.

Based on the analysis of closed system which is composed of plant and controller, the robust control theory has already considered the controller fragile problem, and the fruitful results of robust control theory are helpful to tackle the non-fragile problem. Theoretical research and engineering practice prove that, if we split the uncertainty of plant and controller, the robust and non-fragile is interactive and conflict. As the robust to the uncertainty of the plant is stronger, then the closed system is fragile or sensitive to the uncertainty of controller. Keel [3] concludes that the fragile of controller will be worse as the order of controller become higher. So it's necessary to consider the controller's uncertainty during the robust controller design and synthesis, or non-fragile control design. That is, we should consider both the uncertainty of plant and controller in the same time.

To tackle these problem, many remarkable works have been achieved in developing advanced non-fragile controller [4–6], but the general methods is given by the operator theory or Riccati inequality, which is complex and hard to practice in engineering. It is well known that the stability analysis and stabilization using Lyapunov theory has gained more attention in control field as it can unify the complex stability problem to the standard linear matrix inequality, and simplify the problem solution procedure.

Based on the additional controller gain uncertainty in system description, this paper presents the non-fragile controller synthesis problem based on the Lyapunov theory and this controller can be solved by the LMI toolbox in Matlab® software. A numerical example of UAVs flight control is provided to show the validity of the proposed method.

2 PROBLEM STATEMENTS

For the linear system

$$\begin{cases} \dot{x}(t) = Ax(t) + Bu(t) \\ y(t) = Cx(t) \end{cases} \tag{1}$$

where $x(t) \in \mathbb{R}^n$ is the state vector, $u(t) \in \mathbb{R}^m$ is the control input vector, $y(t) \in \mathbb{R}^r$ is the system output vector, A, B and C is the matrices with proper dimension.

Definition 1 The system (1) is system stabilizable, if there has the feedback controller $u(t) = Kx(t)$ which makes the closed system stable.

Lemma 1 [7, 8] For the given symmetric matrix

$$S = \begin{bmatrix} S_{11} & S_{12} \\ S_{12}^T & S_{22} \end{bmatrix}$$

where the S_{11}, S_{22} is nonsingular invertible matrix, then the three conditions is equivalent:

(1) $S < 0$;

(2) $S_{11} < 0$, $S_{22} - S_{12}^T S_{11}^{-1} S_{12} < 0$;

(3) $S_{22} < 0$, $S_{11} - S_{12}^T S_{22}^{-1} S_{12} < 0$;

Lemma 2 [7, 8] For random given proper symmetric matrix X, Y, the following inequality is existed for random $\beta > 0$

$$X^T Y + Y^T X \leq \beta X^T X + \frac{1}{\beta} Y^T Y$$

Lemma 3 [7, 8] For given proper dimension matrix Y, D and E, in which the Y is symmetric, then

$$Y + DFE + E^T F^T D^T < 0$$

is existed for all the matrix F which satisfy $F^T F < I$, if and only if there exists the constant $\varepsilon > 0$ and

$$Y + \varepsilon DD^T + \varepsilon^{-1} E^T E < 0$$

3 MAIN RESULTS

3.1 Stable controller design based on LMI approach

Theorem 1 The system (1) is stabilizable if there exists positive define matrix Q, Y, and real constant ε that satisfy the following LMIs

$$\begin{bmatrix} AQ + QA^T & \varepsilon B & Y^T \\ \varepsilon B^T & -\varepsilon I & 0 \\ Y & 0 & -\varepsilon I \end{bmatrix} < 0 \qquad (2)$$

Moreover, if there has a feasible solution to the inequality (2), then the controller $u(t) = YQ^{-1}x(t)$ is the state feedback stability solution for this problem.

Proof. Define $u(t) = Kx(t)$ in (1), and then substitute it into the equation (1)

$$\begin{cases} \dot{x}(t) = (A + BK)x(t) \\ y(t) = Cx(t) \end{cases} \qquad (3)$$

According to the Lyapunov stability theory, choose the positive Lyapunov function for the Eq. (3)

$$V(x(t)) = x^T(t) Px(t) > 0$$

If the $\dot{V}(x(t))$ is negative definite

$$\frac{dV(x(t))}{dt} < 0$$

Then the closed system (3) is stable. So

$$\frac{dV(x(t))}{dt} = \dot{x}^T(t) Px(t) + x^T(t) P\dot{x}(t)$$
$$= x(t)\left[(A + BK)^T P + P(A + BK) \right] x(t) < 0 \quad (4)$$

The closed system (3) is stable if there exists positive define matrix $Q = P^{-1}$ which makes

$$Z = \left[(A + BK)Q + Q(A + BK)^T \right] < 0$$

Define the variable

$$KQ = Y \qquad (5)$$

According to the Lemma 2

$$Z = (A + BK)Q + Q(A + BK)^T$$
$$= AQ + QA^T + BY + Y^T B^T$$
$$\leq AQ + QA^T + \varepsilon BB^T + \varepsilon^{-1} Y^T Y < 0 \quad (6)$$

According to the Lemma 1, $Z < 0$ is equal to the LMIs (2). $\qquad \square$

3.2 Non-fragile controller design based on LMI approach

Consider gain perturbation in feedback controller for system (1)

$$u(t) = (K + \Delta K)x(t) \qquad (7)$$

ΔK is the perturbation in controller matrix

$$\Delta K = E\Delta(t)F \qquad (8)$$

where E, F are real constant matrix with proper dimension. $\Delta(t)$ satisfies

$$\Delta^T(t)\Delta(t) \leq I$$

Theorem 2 The controller (6) is non-fragile state feedback controller for system (1), then for given positive constant λ, there exists constant $\varepsilon_1 > 0$, $\varepsilon_2 > 0$ and positive definitive symmetric matrix $Q > 0$, $Y > 0$ which satisfy the LMIs

$$\begin{bmatrix} AQ + QA^T + \lambda I & \varepsilon_1 B & Y^T & \varepsilon_2 BE & (FQ)^T \\ \varepsilon_1 B^T & -\varepsilon_1 I & 0 & 0 & 0 \\ Y & 0 & -\varepsilon_1 I & 0 & 0 \\ \varepsilon_2 (BE)^T & 0 & 0 & -\varepsilon_2 I & 0 \\ FQ & 0 & 0 & 0 & -\varepsilon_2 I \end{bmatrix} < 0 \quad (9)$$

then $K = YQ^{-1}$ is the state feedback control law for the system (1).

Proof: Consider the closed system

$$\begin{cases} \dot{x}(t) = \left[A + B(K + \Delta K)\right]x(t) \\ y(t) = Cx(t) \end{cases}$$

Choose the positive Lyapunov function for the Equation above

$$V(x(t)) = x^T(t)Px(t) > 0$$

If the $\dot{V}(x(t))$ is negative definite

$$\frac{dV(x(t))}{dt} < 0$$

Then the closed system is stable. So

$$\frac{dV(x(t))}{dt} = \dot{x}^T(t)Px(t) + x^T(t)P\dot{x}(t)$$

$$= x(t)\left[(A + B(K + \Delta K))^T P + P(A + B(K + \Delta K))\right]x(t) < 0$$

The closed system is stable if there exists positive define matrix $Q = P^{-1}$ which makes

$$Z = \left[(A + B(K + \Delta K))Q + Q(A + B(K + \Delta K))^T\right] < 0$$

Ensuring the dynamical performance, for given constant $\lambda > 0$

$$Z + \lambda I = \left[(A + B(K + \Delta K))Q + Q(A + B(K + \Delta K))^T\right] + \lambda I < 0$$

Then the closed system is stable.

$$Z + \lambda I = \left[(A + B(K + \Delta K))Q + Q(A + B(K + \Delta K))^T\right]$$
$$+ \lambda I$$
$$= AQ + QA^T + \lambda I + BKQ + QK^T B^T + B\Delta KQ$$
$$+ Q(\Delta K)^T B^T$$
$$= AQ + QA^T + \lambda I + BKQ + QK^T B^T$$
$$+ BE\Delta(t)FQ + QF^T\Delta^T(t)E^T B^T$$

Let $Y = KQ$, according to the Lemma 2 and 3, if there exists constant $\varepsilon_1 > 0$, $\varepsilon_2 > 0$, then

$$BKQ + QK^T B^T = B(KQ) + (KQ)^T B^T$$
$$= BY + Y^T B^T \le \varepsilon_1 BB^T + \varepsilon_1^{-1}Y^T Y$$
$$BE\Delta(t)FQ + QF^T\Delta^T(t)E^T B^T$$
$$= (BE)\Delta(t)(FQ) + (FQ)^T\Delta^T(t)(BE)^T$$
$$\le \varepsilon_2(BE)(BE)^T + \varepsilon_2^{-1}(FQ)^T(FQ)$$

So

$$AQ + QA^T + \lambda I + \varepsilon_1 BB^T + \varepsilon_1^{-1}Y^T Y + \varepsilon_2(BE)(BE)^T$$
$$+ \varepsilon_2^{-1}(FQ)^T(FQ) < 0 \Rightarrow$$
$$(AQ + QA^T + \lambda I) - \left[\varepsilon_1 B \quad Y^T \quad \varepsilon_2(BE) \quad (FQ)^T\right]$$
$$R^{-1}\left[\varepsilon_1 B \quad Y^T \quad \varepsilon_2(BE) \quad (FQ)^T\right]^T < 0$$

where $R = diag(-\varepsilon_1 I, -\varepsilon_1 I, -\varepsilon_2 I, -\varepsilon_2 I)$.

Using the Lemma 1, the inequality above is equality to the LMI (8). And then $K = Y(Q)^{-1}$ is the state feedback control law for the system (1).

4 SIMULATION EXAMPLES

For the following UAV's flight model

$$\dot{x}(t) = Ax(t) + Bu(t) \tag{10}$$

$$A = \begin{bmatrix} -0.2029 & 0.6110 & -1.7044 \\ -0.5670 & -3.8086 & 22.4291 \\ 0.4906 & -4.2213 & -4.3901 \\ 0 & 0 & 1 \\ 0.0753 & -0.9972 & 0 \end{bmatrix} \rightarrow$$

$$\leftarrow \begin{bmatrix} -9.7991 & 22.9997 \\ -0.7397 & 0.0010 \\ 0 & 0 \\ 0 & 0 \\ 22.9997 & 0 \end{bmatrix},$$

$$B = \begin{bmatrix} 0.3132 & 0 \\ -1.9847 & 0 \\ -27.5485 & 0 \\ 0 & 0 \\ 0 & 4.4584 \end{bmatrix}.$$

where the state variables are: flight velocity V, angle of attack α, pitch rate q, pitch angle θ, and flight height h, control input is elevator deflection δ_e and engine thrust δ_T.

The controller addictive gain perturbation can be written as the Eq. (7)

$$E_k = \begin{bmatrix} 1 \\ 1 \end{bmatrix}, \quad F_k = \begin{bmatrix} 0.1 & 0.1 & 0.1 & 0.1 & 0.1 \end{bmatrix},$$

$$|\Delta_k(t)| \le 1$$

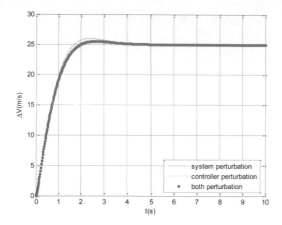

Figure 1.   Flight velocity ΔV response curves.

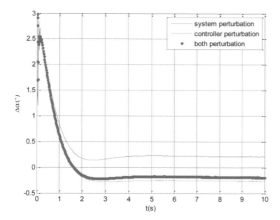

Figure 2.   Angle of attack $\Delta \alpha$ response curves.

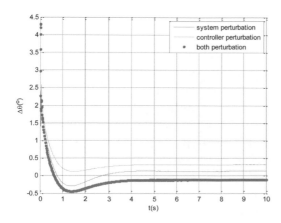

Figure 3.   Pitch angle $\Delta \theta$ response curves.

According to the LMIs (8), the state feedback controller is designed as

$$K = \begin{bmatrix} 4.6500 & -0.4028 & 3.4010 & -4.4320 & 0.9117 \\ 0.3322 & -0.1573 & -0.2933 & 5.2041 & 0.3239 \end{bmatrix}$$

For validating this controller, the closed system response is presented. Fig. 1 to 3 gives the flight velocity, angle of attack and pitch angle response curves respectively with perturbation. These results show that, the system can stable very soon, and the perturbation from the system itself and the controller have just little impact on the system response.

5   CONCLUSIONS

Based on the Lyapunov function, the system stability problem with controller additive gain perturbation is presented in this paper, and then the state feedback controller design method is derived. Moreover, a numerical example has been provided by the LMI toolbox. The desired performance of closed-loop aircraft has been investigated via computer simulation in Section 4 to validate the proposed control synthesis approach.

REFERENCES

[1]   Zhou K, Doyle J C, Glover K. Robust and optimal control [M] . New Jersey: Prentice Hall Englewood Cliffs, 1996.
[2]   Fang Qiu. Further results on robust stability of neutral system with mixed time-varying delays and nonlinear perturbations [J]. Nonlinear Analysis: Real World Applications, 2010, 11:895–906.
[3]   L.H. Keel, S .P. Bhattacharyya, Robust, Fragile or Optimal? [C], Proceedings of the American Control Conference Albuquerque, New Meixco, June 1997:1307–1313.
[4]   An Shoumin, Huang Lei, Gu Shusheng, Wang Jianhui. Robust Non-fragile State Feedback Control of Discrete Time-Delay Systems, control and automation, 2005.52(2):794–799.
[5]   S. Lakshmanan. Improved results on robust stability of neutral systems with mixed time-varying delays and nonlinear perturbations [J]. Applied Mathematical Modeling, 2011.35:5355–5368.
[6]   G.H. Yang, J.L. Wang, C Lin. H_∞ control for linear systems with additive controller gain variations [J]. International Journal of control, 2000, 73(16):1500–1506.
[7]   YU Li. Robust control [M]. Beijing: Tsinghua University Press, 2002.
[8]   WU Ming, GUI Wei-hua, HE Yong. Modern robust control [M]. Changsha: Central South University Press, 2006.

Information Technology and Computer Application Engineering – Liu, Sung & Yao (Eds)
© 2014 Taylor & Francis Group, London, ISBN 978-1-138-00079-7

Exploration and development of text knowledge extraction

Liping Zhu, Hongqi Li, Siyao Wang & Chuan Li
College of Geophysics and Information Engineering, China University of Petroleum, Beijing, China

ABSTRACT: Text knowledge extraction technology has been applied in many fields, but few practices in the field of petroleum exploration and development. In this paper, we comprehensively utilize a statistical and natural language understanding technology to extract knowledge from the articles in the field of petroleum exploration and development. First of all, we get the key words and the core sentences containing article process model. Then after the word segmentation and phrase recognition, we use semantic templates to match and extract semantic information from the key sentences. The experimental results show that this method achieves 70% accuracy rate in keywords spotting and process model extraction.

Keywords: Knowledge extraction; Text mining; Phrase recognition; Exploration and development; Rules match

1 INTRODUCTION

After decades of research, it has accumulated a large number of research results in the field of oil exploration and development. These results are mostly in the form of words reports and a series of unorganized documentations. Most of the useful information is scattered in the engineering investigation reports in the form of words which haven't be fully utilized and are difficult to be shared. Text mining technology can be used as an excellent knowledge extraction method for these documents. What is more important, we could take a good control of these documents from knowledge management angle which makes knowledge sharing more convenient.

Text knowledge extraction is to address text data containing useful information through a structured process means. And in this way, we convert textual data to the knowledge with a certain form of organization, and also make it more convenient for storage.

The input of text knowledge extraction system is the original literature, and the output is information in a certain format. This information extracted from a large number of professional literatures stored in a unified form, which bring many advantages. It is more convenient to manage as well as further processing theses structural information.

Most of the oil professional literatures are about purpose, method, process, conclusions. That is, using a certain kind of mean to process textual data and get results within a few steps. It concluded that catching a few core sentences in the article will be able to grasp this article from a technical level. Therefore, article knowledge extraction is to process these core sentences and stored them.

This article uses a method combination of statistical methods based on natural language understanding to extract knowledge from the Chinese oil exploration and development literatures. The process relates to the petroleum professional lexicon and predicate vocabulary building, text pre-processing and Chinese word segmentation, keywords and procedure sentence extraction, phrase recognition, and storage. The framework of this system is shown in Figure 1 below.

2 DICTIONARY CONSTRUCTING

Faced with the knowledge of extraction of oil and professional literature, we must first establish a professional vocabulary. In this paper, we mainly deal with the articles in the field of petroleum exploration and development, the first step is to establish a professional term library. At the same time, giving an important role of the verb in language, building a suitable verb lexicon in the oil fields of expertise has a significant impact on the effect of knowledge extraction. So, professional thesaurus including: the professional concepts thesaurus in the field of petroleum exploration and development and predicate vocabulary.

2.1 Build professional thesaurus in the field of exploration and development

As segmentation tool does not recognize the technical terms of the oil fields, in order to get a better effect of segmentation, we need to establish professional vocabulary for the field of petroleum exploration and development. Our processing platform mainly addresses two areas of literature: petroleum

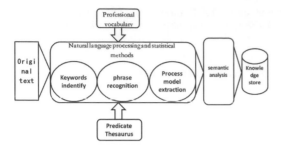

Figure 1. System framework.

Relationship words	根据、取得、使用、利用、属于、造成 etc.
Default word	本文
DE word	的

Figure 2. Special words.

exploration wells in the deployment and development programs. Therefore, in addition to the oil common vocabulary, we collect explore well deployment and development plan vocabularies. The vocabulary relevant with exploration wells deployments are mainly about source, reservoir, cap, ring, transportation and security. The collection and sort of development vocabulary are mainly related to the reservoir type, oil extraction technology, monitoring and observation technologies, reservoir modification technologies and management methods. The collection of the jargon dictionary is a comprehensive, standard and professional process. It ensures that the technical terms can be correctly identified. We collect more than 20 million entries covering almost all the professional vocabularies in the field of exploration and development based on China Petroleum Exploration and Development Encyclopedia, oil ceremony.

2.2 Construction of the predicate lexicon

In Chinese, the predicate is extremely important in a sentence; it connects the subject and object, and plays an irreplaceable role for understanding the meaning of whole sentence. Analyzing the role of the predicate in different sentences is the first step to extract the semantic information of a sentence. So it is crucial to establish a thesaurus of the predicate relationship. In Chinese sentences, the association relationship always said with the verb, and some common nominal (as a subject, object and that some other words) and descriptive words in the field of non-engineering rarely used in the documents about oil exploration and development, so using the verb as the relationship word, such as the words shown in Figure 2.

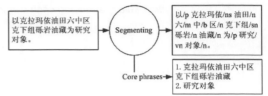

Figure 3. Segmenting process.

3 TEXT KNOWLEDGE EXTRACTION

3.1 Abbreviations and acronyms

The type of professional documents in field of petroleum exploration and development are mainly PDF, DOC, DOCX, PPT, TXT, etc. As it is more convenient to operate TXT files, so before processing these documents we have to change them into TXT forms.

When we convert all the documents into a standard TXT form, the first step is to read the content and store it as a string. And then we parse the string to get titles and their content. Finally we store them into a Hashmap according to the structure of the article in a standard form as "title" – "content".

To the next semantic processing, we must first do word segmenting on the content of the document, the result is the form of: word/speech tag. Here we use the Chinese Academy of Sciences word segmenting tool ICTCLAS to do the job. One of the experiments is shown in Figure 3. The speech tag "/sn" is exploration and development of professional entry.

3.2 Keywords and the process of sentence extraction

Nowadays, keyword has become quite an important judgment index of literatures like oil exploration and exploitation engineering field and so on. At same time, literature's contents are mostly related to Research Purpose, Research Background, Research Method, Research Process, Research Result and Research Conclusion. For Engineers, it is enough to capture few core processes. So that, the first step is extracting the keywords of articles and some sentences, which can reflected the core process. Process sentences and abstract extraction process is very similar, both are aims at extracting the core of the article. But Process sentence paid more attention on research process.

Assume that document D consists of m sentences, they are S1, S2 . . . Sm. And the definition of sentence Si is as follows:

$$S_i = (w_{i1}, w_{i2}, \ldots w_{in})$$

Of which: Win is the word in the sentence;

Sentence weight is calculated as follows:

$$score(S_i) = length(S_i) * position(S_i) * \sum word(w_{ij}) \qquad (1)$$

The importance of a word to the article differs. Here, we handle only three kinds of words, the term noun phrases (including pronouns, location words), verb in the verb phrase and adjective phrase.

The length of sentence plays a very important role for the core semantics of the article. In the engineering literature, the short sentences often cannot be view as core sentence of the article, so in (1), the length(S_i) is the weight of the length of sentence S_i. Here the length of the sentence is the sum of above three kinds of words length, while the other words arc negligible.

Among all sentences of the article, the first and the last sentence of the paragraph is often the kernel sentence. What is more, the sentence in the summary will always be the core content of the article. So, when weight sentence, the position of the sentence is especially important. Position(S_i) represents the position of the weight of the sentence.

In (1), word(w_{ij}) represents weight of each word in the sentence. The weight of the word is related to its speech tag, its frequency in the article, keyword flag and ontology concept word flag. The formula is as follows:

$$word(w_{ij}) = freq(w_{ij}) * POS(w_{ij}) * key(w_{ij}) * title(w_{ij}) \qquad (2)$$

In (2), freq(w_{ij}) is the frequency of w_{ij} in the article, POS(w_{ij}) is the weight of the word's speech tag. Here, the word with a verb speech tag weights most among all the speech tags. We give verb word with a weight of 0.45, noun word 0.4 and adjective word 0.15. The title(w_{ij}) means the word's weight for its position in the title, first title word weight for 5, followed by two, three, four title weight; their weights are 3, 2, and 1.5. Other sentences get a title(w_{ij}) 1.

According to (1) and (2), the key words and process sentences can be drawn from the article.

3.3 Phrase recognition

After segmenting of the sentence, the outcome becomes relatively complex. Words in the sentence cannot convey the meaning of it. There is a sentence as an example shown in the Figure 2. In this example, it is the core phrases shown in the Figure 2, which can convey the meaning of this sentence accurately. Analysis and identification the Noun phrases is one of the important missions of analysis the natural language syntax. As a result, identification the Noun phrases in sentence of segmentation is necessary for the next mission i.e. semantic analysis.

After using the ICTCLAS word segmentation tool, there are a total number of 21 classes' speeches, including nouns, verbs, prepositions, time words, adjectives, particle, interjection, quantifiers, and numerals and so on, not including the punctuation. In the literature, there are a large number of stop words that have the function of initiating, bearing, transferring, these words almost have no effect for the expression of semantics which including the interjection,

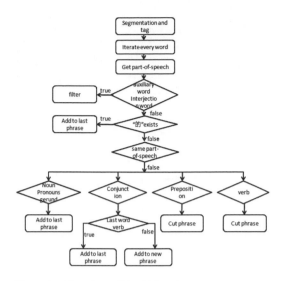

Figure 4.   Flow chart of identifying phrase.

modal, auxiliary, etc. excepting the DE word shown in Figure 2. The DE word have special function in the Chinese phrase, when the adjectives, nouns, or auxiliary be joined to the DE word face or back, they all can form one phrase. For example, research DE object (this is in Chinese, which means the research object), run into a probable DE problem (this is in Chinese, which means may encounter a problem). The phrase recognition flow chart is shown in Figure 4.

Rules show as follows:

Rule No. 1: filter the interjection, modal particle and auxiliary word and so on.

Rule No. 2: if the last word is the DE word or the next one is the DE word, it will be the part of a noun phrase.

Rule No. 3: it will be added to the last phrase, if the word of the speech is same as the previous one.

Rule No. 4: if word is n. speech as noun, adj., numeral, measure word and pronoun, it will be added to previous phrase.

Rule No. 5: delete the conjunction or it will be the part of a phrase, if the word is coordinating conjunction, and the next word is verb.

Rule No. 6: split the phrase as two parts, if the word is verb. Or prep.

The table shown in Figure 5 is the instance after classifying words and conducting phrase identification follow the rules above.

3.4 Extracting procedural semantic

Usually, the content of extracting base on the rules has a severe requirement, but it has good results for solving the special problems. The extracting text semantic of this paper base on rule base, so, the quality of rule base will determine the precision of the results. After identifying phrase, the sentence, only, has noun, verb and prep. Under the condition of a sentence only has 3 part of speech, we can fully

The original sentence	Classify the words	After phrase identification	Memo
以拉克拉依油田六中区克下组砾岩油藏为研究对象	以/p 克拉玛依油田/sn 六/m 中/b 区/n 克/sn 下组/n 砾岩/sn 油藏/sn 为/p 研究/vn 对象/n	以/p克拉依油田六中区克下组砾岩油藏 为/p 研究对象	Rule No.4 and rule NO.6, Nominal words synthesis of a phrase, preposition play as a separate phrase.
利用所建立的凝析气藏数值模拟模型	利用/v 所/usuo 建立/v 的/ude1 凝析气藏/sn 数值/n 模拟/vn 模型/n	利用/v建立的凝析气藏数值模拟模型	Rule No.1 filter the interjection. As the Rule No.2 because the next word is DE, the verb. is token as a part of noun phrase.
该区块地质模型的建立是以完备数据为基础的	该/rz 区块/n 地质/n 模型/n 的/ude1 建立/v 是/vshi 以/p 完备/a 数据/n 为/p 基础/n	该区块地质模型的建立 是以/p 完备数据 为/p 基础	The verb in front of the DE word is token as a part of the noun phrase.
提高采收率和降低成本	提高/v 采收率 和/c 降低/v 成本	提高/v 采收率 降低/v 成本	Rule No.5 if the word after the conjunction is verb., delete the conjunction.
东营期末、馆陶镇末以及明化镇末	东营/ns 期末/t 、/wn 馆陶/ns 期末/t 以及/cc 明化镇/sn 期末/t	东营期末、馆陶期末以及明化镇期末	Rule No.5 if the word after the conjunction is not the verb., front and back sentence make up one sentence.

Figure 5.   The instance of phrase identification.

Model	Phrase	Extracting results
n.1+verb.+n.2	济阳坳陷异常高压具有/v 晚期形成的特征	济阳坳陷异常高压; 具有: 晚期形成的特征.
Prep.1+n.1+prep.2+n.2	以/p 克拉玛依油田六中区克下组砾岩油藏 为/p 研究对象	研究对象: 为: 克拉玛依油田六中区克下组砾岩油藏.
N.1+verb.1+n.2+verb.2+n.3	地球物理学家 利用/v 其背景知识 选择/v 一些重要的信息 构造/v 一些新的参数	1.地球物理学家: 利用: 其背景知识. 2.地球物理学家: 选择: 一些重要的信息. 3.地球物理学家: 构造: 一些新的参数.
Verb.+n.	利用/v 决策树的分析方法	本文: 利用: 决策树的分析方法
N.1+verb.1+n.2+prep.2+n.3	本文 以/p 密闭取心资料和测井曲线值 为/p 数据基础	本文: 数据基础: 密闭取心资料和测井曲线值

Figure 7.   Matching model of the semantics and instances.

Table 1.   The accuracy of extracting.

	Identify keywords	Extract sentence of procedural semantics	Extract semantics
Precision	82.6%	72.3%	70.5%
Recall	84.2%	74.8%	71.8%

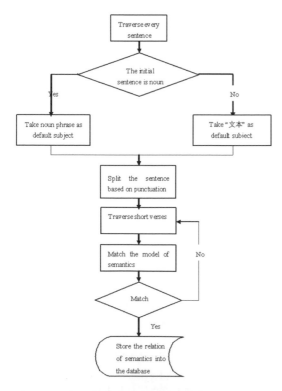

Figure 6.   Flow chart of extracting semantics.

establish the model of Semantic Matching to make syntactic analysis of sentence and extract the semantics according to the method of "subject : predicate : object" base on the rules. Then, store the results to the database. Figure 6 is the flow chart of extracting semantics:

The concrete steps as follows:

Step No. 1: judge whether the sentence-initial is noun phrase. If yes it will be the default subject of this sentence, else use default word shown in Figure 2 as the default subject.

Step No. 2: divide the sentence into several sentences base on dot and semicolon.

Step No. 3: match every phrase one by one, if it is success, store the results into the database, else match next one.

Table shown in Figure 7 shows some model and instances about it.

4   EXPERIMENT RESULT

This paper selects 500 literatures of Petroleum exploration and development class o test the extracting of keywords and procedural semantics, results show as Table 1.

In the table shown in Figure 8, there is the information of the article which is used in the experiment. And it also shows the processing results in the form of "subject: predicate: object".

Title	基于决策树方法的砾岩油藏岩性识别
Keywords	测井解释；决策树；数据挖掘；砾岩油藏；岩性识别；克拉玛依油田。
Input Sentences	Results in the form of "Subject: predicate: object"
以克拉玛依油田六中区克下组砾岩油藏为研究对象，利用决策树的分析方法，结合密闭取心资料对砾岩油藏8种岩性的测井曲线敏感性进行研究，优选出原状地层电阻率、声波时差和自然伽马3个测井响应值作为砾岩油藏岩性识别的特征参数，建立决策树岩性识别模型，模型的综合判断准确率达到96.197%。	1)研究对象：为：克拉玛依油田六中区克下组砾岩油藏 2)本文：利用：决策树的分析方法 3)本文：结合：密闭取心资料 4)本文：研究：砾岩油藏8种岩性的测井曲线敏感性 5)本文：建立：决策树岩性识别模型
本文以密闭取心资料和测井曲线值为数据基础，利用决策树方法分析了各种测井井参数对砾岩岩性判断的权重，建立决策树岩性识别模型并且取得了很好的应用效果。	1)本文：数据基础：密闭取心资料和测井曲线值
因此，克拉玛依油田六中区克下组属于典型的砾岩油藏，储层类型多样、岩性变化复杂，孔喉结构多峰分布、非均质性强等是该油藏最大的特点。	1)克拉玛依油田六中区克下组：属于：典型的砾岩油藏 2)储层类型多样、岩性变化复杂，孔喉结构多峰分布、非均质性强：是：该油藏最大的特点
六中区克下组砾岩岩性按其粒度可以分为砾岩、砂质砾岩、砂砾岩、含砾粗砂岩、细砂岩，含砾泥岩、粉砂质泥岩和泥岩等8种岩性。	1)六中区克下组砾岩岩性按其粒度：分为：砾岩、砂质砾岩、砂砾岩、含砾粗砂岩、细砂岩、含砾泥岩、粉砂质泥岩和泥岩等8种岩性
决策树算法采用C510进行砾岩岩性的数据挖掘，算法利用信息增益率作为划分度量，选择信息增益率值最大的属性作为分裂节点。	1)决策树算法：采用：C510 2)决策树算法：进行：砾岩岩性的数据挖掘 3)算法：选择：信息增益率值最大的属性作为分裂节点
为了综合3个岩性敏感参数的信息，采用原状地层电阻率和声波时差与自然伽马的乘积制作砾岩岩性交会图版。	1)本文：综合：3个岩性敏感参数的信息 2)本文：采用：原状地层电阻率和声波时差与自然伽马的乘积 3)本文：制作：砾岩岩性交会图版

Figure 8. The experimental results.

5 CONCLUSION

(1) The integrity of constructing Professional vocabulary and Predicate Thesaurus play a fundamental role in the precision of Text-Knowledge Extracting, so it is a cornerstone to construct a scientific and covering a wide range of professional vocabulary and a Predicate Thesaurus conforming to Petroleum exploration and development literature in the Knowledge extraction of the Petroleum disciplines.

(2) Chinese word segmentation and Part of Speech Tagging is the most central and hardest step of the Natural language understanding, this paper use the ICTCLAS of Chinese Academy of Sciences to establish a complete thesaurus of petroleum field, which can reach 99% precision of Chinese word segmentation.

(3) Compared with the widely used summary extract, like petroleum engineering literature, the performance of the method's results can be better said that the core of the literature, and it has greater value to reference staff.

(4) Be an important part of natural language analysis, indentify phrase can reach an high level of identifying especially in the field of Petroleum Engineering which has rigorous writing style through judging by the simple part of specch in practical.

(5) Through the identifying phrase, the sentence only has several parts of speech, so it can reach a good result through establish database of rules to make syntactic analysis.

Compare with Botany and Medicine which has more researches in extracting Text-Knowledge, more complicated operation, and has a harder way to extract than the field of exploring petroleum exploration. Through the results we can see that this paper's results of extracting can show the article's center content and procedural semantics, and can reach the goal of knowledge management and sharing.

ACKNOWLEDGEMENT

This paper was supported by Science Foundation of China University of Petroleum, Beijing (KYJJ2012-05-36). The serial number of the major project which this work belongs to is 2011ZX05020-009.

REFERENCES

Wang, Z. 2008. *The acquisition of knowledge in the text*, Beijing: Beijing university of Posts and Telecommunications.
Raman, M. & Besancon, R. 2010. *Text Mining-Knowledge Extraction from Unstructured Textual Data*, http://www.epfl.ch/Publications/Archive/RajmanBesancom98apdf
Hahn, U. & Schnattinger, K. 1997. *Knowledge Mining from Textual Sources*, Las Vegas, Nevada: Proceedings of the Sixth International Conference on Information and Knowledge Management (CIKM'97): 83–90.
Dorre, J., Gerstl, P. & Seiffert, R. 1999. *Text Mining: Finding Nuggets in Mountains of Texts,* San Diego: Fifth ACM SIGKDD International Conference on Knowledge Discovery & Data Mining: 398–401.

Information Technology and Computer Application Engineering – Liu, Sung & Yao (Eds)
© 2014 Taylor & Francis Group, London, ISBN 978-1-138-00079-7

Design of interactive spoken English teaching platform in MCALL model

Qian Zhao
Harbin Normal University, Harbin, China

Di Chen
Harbin Institute of Technology, Harbin, China

ABSTRACT: With the rapid development of information technology, especially the wide application of network technology, the integration of multimedia techniques and second language teaching has become a focus in the field of foreign language teaching and research. At present, the application of multimedia and computer technology is playing more and more significant roles in spoken English teaching and learning in China, among which MCALL (Multimedia Computer Assisted Language Learning) has shown prominent influence. This paper lists some main characteristics of spoken English teaching in MCALL model such as: increasing understandable language input, modification to language input and improving learner's interest from perspective of emotion. Then the paper explores the design of MCALL teaching platform based on interaction from the angles of the design of functional modules and database, which may contribute to the second language teaching.

Keywords: MCALL; Spoken English teaching; Interactive teaching platform

1 INTRODUCTION

The rapid development of science and technology in the 21st century has laid a solid technical foundation for the development of knowledge economy, hence provides broad space for the revolution of education. The integration of multimedia techniques and second language teaching has become a focus in the field of foreign language teaching and research. Nowadays, the application of multimedia and computer technology is playing more and more significant roles in college English teaching and learning in China, among which MCALL (Multimedia Computer Assisted Language Learning) has shown prominent influence.

Levy (1997) proposes that Computer Assisted Language Learning (CALL) is a comparatively new and rapidly changing academic field that explores the role of information and communication technology in language learning and teaching. However, multimedia is far from a new term in language teaching. Vocal materials such as cassettes and CDs, and visual materials such as pictures, slides and videos, can all provide vivid scenarios and be widely used in teaching. These different conventional media require different equipments to record, transfer and replay, which may be quite inconvenient. If we can change all the signals no matter literal, graphical, audio or video into digital ones, then transfer them via the same channel, and replay them at the same terminal, and then the limits will be broken through. This is called multimedia computer technology. MCALL, in a broad sense,

is a teaching method covering various media that is on based and dominated by the computer. According to the teaching aims and requirements, the multimedia information combines to form reasonable teaching structures, which enables a series of man-machine alternations in order to ensure that students are learning in best surroundings. Its characteristics are to give a three-dimensional display of teaching contents and to integrate sound, text, graph, chart, image and animation through information technology. Studies show that 80% of the information received by humans is obtained from images through visual sense; and psychologist Treicher has also proved that 94% of the information obtained by humans is by sense of seeing and hearing. Therefore, for language teaching and learning, MCALL is a must and will deeply influence its method and development. Most papers on MCALL are about the importance, the development and the features of it. This paper intends to explore the design of interactive spoken language teaching platform on the basis of database.

2 CHARACTERISTICS OF SPOKEN ENGLISH TEACHING IN MCALL

2.1 *Increasing understandable language input*

It is well-known that the constantly developing technical function of multimedia offers infinite possibilities to second language acquisition especially spoken English teaching. American linguist, Stenphen

D. Krashen, put forward a series of significant theories on second language acquisition in the 1980s, which proposed that the language acquisition is achieved through language input; hence, we teachers should lay emphasis on providing students with best input. He further listed the four necessities of best language input, i.e. it is understandable, interesting, and closely-related, not based on grammar, and abundant. According to this theoretical model, the using of multimedia in teaching, which conveys knowledge by network, has totally changed the traditional form of teacher's dictation and blackboard writing and has greatly increased classroom knowledge capacity. Furthermore, the abundant words, pictures, video and audio materials may stimulate students' senses properly to arouse their learning interest, which will improve the classroom instruction efficiency to a great extent.

2.2 Modification to language input

In concrete teaching activities, language input can be presented to learners in the form of written or spoken language. And the forms of modification are also various. They express repetition, simplification, enunciation, non-language hints, providing references and so on. The request to modify audio materials can be designed like the mode of "help" button whose function is to help to correct language input. For instance, when a learner can't understand the sentence he heard, he may press the modification button once to get a second language input; if he still can't grasp the meaning, pressing the same button again will enable the sentence to be read slowly. The design of the button entitles learners to choose the way of modification. As for text materials, the pattern of modification can adopt the way of word transition, referring to other materials and so on. The main point of the design is to supplement and modify original materials by hypertext links. When one or two words or cultural background is beyond learners' understanding, they can point the mouse to the highlighted words, and then the hypertext links may be activated to show the substituted words or explanations. Modification may well interpret written contents by spoken language. And then we can observe the learner's need of modification by recording their interactions and analyzing detailed contents.

2.3 Improving learner's interest from perspective of emotion

American linguist Stenphen.D.Krashen in the 1980's proposed a series of second language acquisition theories, among which the Hypothesis caused great repercussions. In addition to proper understandable language input, language acquisition should take the learner's psychological factors into account, such as learning motivation, attitude to second language acquisition, and the learner's self-confidence and emotion, which may filter the language input and restrict the amount of information the learner accepted. Krashen uses the following mode to interpret the affective filter in the process of second language acquisition.

language input → | ... language acquisition device → ability of acquisition

From the above figure, we can know that the filter net lies between the input and acquisition device, thus the thickness of the net limits the amount of input, and therefore influences the ability of acquisition. If a learner is confident enough and has large learning motivation, he will hold active and positive attitude to second language acquisition. Hence the filter function to the language input must be minor, and get much input. According to this hypothesis, MCALL is helpful in the following aspects. First, interesting and related language input information can arouse the interest of learners to enable them to grasp the language easily. For this learning interest is the most powerful motive force. Modern techniques may combine the traditional teaching media (i.e. text book, slide, projection and so on) with modern teaching media (i.e. computer and Internet). MCALL can stimulate nearly all the senses of the learner to strengthen their listening and speaking abilities. Teachers should take full advantage of these features to inspire them to learn and involve in classroom activities actively.

2.4 Instant information feedback

During different processes of MCALL, the computer will evaluate the learner's parameter of learning and will generate diagnostic reports in the forms of text description, chart and histogram. Thus learners may get to know their own spoken abilities evidently. Furthermore, this will also make sure their positions in a certain group. Without emotional interference, the learner may concentrate more on what they are learning avoiding exterior factors. The information feedback can also be achieved through Internet consultation to teachers. And a dialogue space may be built. Students can exchange their questions, opinions and experience by e-mail, BBS, blog and MCALL platform. Also, they can directly consult their teachers when necessary, or questions can not be solved.

3 THE DESIGN OF MCALL TEACHING PLATFORM BASED ON INTERACTION

3.1 MCALL teaching platform

With the rapid development of information technology, especially the wide application of network technology, online learning has become an important way of teaching and learning. Compared to traditional classroom teaching, online teaching is of high efficiency in information processing, and is quick in information feedback with low running cost. Under online learning system, students can choose to surf the Internet according to their own convenience which greatly enhances the flexibility of student learning and the effect of

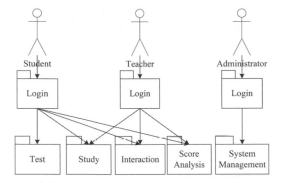

Figure 1. Relationship of the six functional modules.

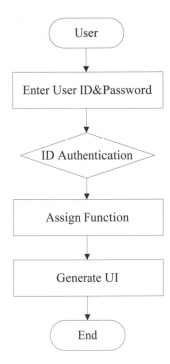

Figure 2. Flowchart of login module.

school education. Teachers can have immediate access to students' feedback on the effectiveness of teaching through an online teaching system, and proper adjustments may be made to teaching and learning activities.

The online teaching platform based on interaction is developed on the basis of database, which enables students to log on to the system through the Internet anytime and anywhere. They can learn the spoken materials related to the second language acquisition and exchange their feelings and opinions to ensure effective and timely teaching. The teaching platform mainly uses the technology of the Internet, databases or others, whose subjects are mostly students, teachers, and administrators. Users enter the teaching platform through the system login screen, and then get corresponding system functions based on user permissions. After that, students will be able to carry out learning, testing, interaction and system management of second language learning according to the obtained functions.

3.2 *Design of functional modules*

According to the above demand, interactive online learning platform consists of six functional modules: login module, learning materials query and browsing module, test module, test scores query and analysis module, interaction module, and system management module. The relationship of these functional modules is shown in Figure 1.

a) Login module: The function of login module is to determine the identity of a user based on the user name. The login module is required to ensure the uniqueness and integrity of the identity of the user as well as to ensure the security of the system. It must also be able to grant the appropriate operating authority based on the user's identity. By entering the user ID and password, in the login process, the system will load the operating interface based on the user's permissions. Figure 2 shows the flowchart of this module.

b) Learning materials query and browsing module: The query and browsing module is to provide students with query and browsing functions of spoken English learning materials. Students can log on to the query interface to select the appropriate texts, audio or video

materials for learning according to their individual needs.

c) Test module: It generates automatic test questions according to the difficulty set by students. And after students have been tested, the test results will be archived.

d) Test scores query and analysis module: Teachers and students can query the test results, and analyze the results within a certain time. This is to facilitate the teachers, or students, to understand the effect of spoken English learning and modify the learning program in a timely manner.

e) Interaction module: Interaction module is designed based on Internet technology. On the one hand, teachers can explore on teaching content with students and answer their questions, therefore promptly get access to student feedback on the effectiveness of teaching to improve teaching activities. On the other hand, students may communicate with each other as well as correct each other to improve their spoken English. In order to encourage students to take the initiative to exchange, teachers and students within the platform ought to communicate anonymously. And depending on student activity within the platform, teachers can add corresponding points in their final examinations.

f) System management module: The module consists of two parts: user management and database maintenance. User management mainly includes personal information settings and permission settings, through which teachers manage students' information and permissions. Database maintenance mainly refers to the management of learning database, management of test question bank, and data backup and restoration.

3.3 *Design of database*

Database is the main part of the online teaching platform, which is based on interaction. Whether the design of database is reasonable and easy to maintain is the key to the success of database design.

The teaching platform normalized database is designed to avoid complex data structures and improve the efficiency of the application. And the databases include spoken English teaching database, multimedia database, and test questions database. The spoken English teaching database is composed of learning methods, knowledge and skills used for self-access language learning. Multimedia databases store large amounts of text, audio, video and spoken English learning materials, such as the listening and speaking materials of New Start College English and New Horizon College English to assist students' learning. The test questions database is used for storing large number of different levels of speaking test questions. This can form test questionnaires according to the set level automatically to enable the students to detect and understand their levels of spoken English.

4 CONCLUSION

The MCALL spoken English teaching platform based on interaction will definitely bring an impact to the method of second language teaching. At the meantime, there are many shortcomings in the design of the platform because of our lacking of theories and techniques. However, if we continue to explore this matter, MCALL is bound to adapt to the rapidly changing need of teaching to make it fully play the role of assisted language teaching and learning.

ACKNOWLEDGEMENTS

This paper is one of the outcomes of the 2012 Heilongjiang Philosophical and Social Science Project: Localized Empirical Study of the Motivation, Strategy and Evaluation of Autonomous Learning of Oral L2 in MCALL Model. It is also the outcome of the 2011 Harbin Normal University New Century Teaching and Reformation Project: The Quantitative Study of the Feasibility on Evaluating the Oral English Test by Spectrum Analysis.

REFERENCES

[1] Levy, M. Computer-Assisted Language Learning: Context and Conceptualization [M]. Oxford University Press, 1997.
[2] Rebecca Oxford, Language Learning Motivation: Expanding the Theoretical Framework, The Modern Language Journal, 1994, 78(1).
[3] Ellis, R. The Study of Second Language Acquisition [M]. Shanghai:Shanghai Foreign Language Education Press, 1999.
[4] Krashen. S. Prirwiples and Practice in Second Language Acquisition[M]. Oxford:Pergamon, 1982.

[Introduction of the Author]
Qian Zhao (1984–), Lecturer, College English Teaching and Research Department, Harbin Normal University, Harbin, China; 150080;
Di Chen (1979–), Lecturer, School of Electronics and Information Engineering, Harbin Institute of Technology, Harbin, China; 150001; Email: dchen@hit.edu.cn

Information Technology and Computer Application Engineering – Liu, Sung & Yao (Eds)
© 2014 Taylor & Francis Group, London, ISBN 978-1-138-00079-7

Suboptimal controller design for aircraft based on genetic algorithm

Y.C. Li
School of Aeronautic Science and Engineering, Beihang University, Beijing, China

ABSTRACT: In this paper, a suboptimal controller for the longitudinal attitude of an aircraft is presented. Linear quadratic form control method is used to design a LQR controller for the system to improve its longitude short-period dynamic performance. Genetic algorithm is applied to select two weighting matrixes, and the optimal selection is obtained by time-domain performance index which corresponds to the fitness function in the algorithm. The simulation results indicate that using genetic algorithm for the selection of weighting matrixes is a feasible method. The proposed controller's performance is better than another LQR controller with manual selection of weighting matrixes.

Keywords: Suboptimal Control; Linear Quadratic Regulator; Genetic Algorithm; Weighting Matrix

1 INTRODUCTION

With the development of modern aircraft design, aircraft's aerodynamic layout often cannot meet the requirements of stability, and the designers sometimes even intend to make it an unstable system. In this situation, the flight stability augmentation control system with good performance is necessary to realize ideal flight qualities. Various control methods are applied to the attitude controller for aircraft, such as adaptive control, robust control and optimal control (Ali 2006, Hwang 2006).

Among these methods, the optimal control method is widely used in the design of control system due to its ease of designing. However, the number of measurements is required to be equal to that of state variables when optimal control method is used. Therefore, suboptimal control, which needs fewer measurements and is easy to be implemented by physical hardware, is an appropriate design method for control system.

In this paper, a LQR (i.e., Linear Quadratic Regulator) controller is designed as stability augmentation control system to improve an aircraft's longitude short-period dynamic performance, and enhance its flying qualities. Considering from the perspective of manipulation, it is designed to make the aircraft react faster, decrease the oscillation amplitude and shorten the time of reaching steady state. According to the analysis above, the requirements of performance indicators and parameters are as follows:

1) Decrease rise time and adjustment time of dynamic response of kinematic parameters (e.g., the angle of attack);
2) Reduce the maximum overshoot;
3) Appropriately increase the longitudinal short-period damping ratio.

To meet these demands, the designing of controller is transfer to the task of finding an optimal selection of the weighting matrices with genetic algorithm.

The remainder of the paper is organized as follows. Section 2 introduces the aircraft dynamic model. Section 3 establishes a suboptimal controller and presents the genetic algorithm for searching an optimal selection of the weighting matrices. In section 4, the proposed controller is verified on the dynamic model of pitch motion. Finally, the conclusion is made in Section 5.

2 SYSTEM DYNAMICS

In this paper, the stability augmentation control system is designed in the condition that the aircraft is flying in constant speed and height. The flight dynamic model and assumptions adopted are as follows.

1) The second-order state-space model, which presents the longitudinal short-period motion and adopts linear small perturbation assumption, is given by

$$\dot{x} = Ax + Bu \tag{1}$$

$$A = \begin{pmatrix} -0.6981 & 1 \\ -1.6991 & -0.5057 \end{pmatrix}$$

$$B = \begin{pmatrix} -0.0571 \\ -0.9828 \end{pmatrix} \tag{2}$$

where the state $x = (\alpha, q)^T$ and u presents the elevator deflection angle, α is the angle of attack, and q denote the pitch angle rate.

2) The control problem is simplified to the regulator design, which means just considering the control

system's inhibiting ability against the disturbance input, and ignoring the system input.

3) The disturbance input is vertical wind, and the initial angle of attack is 5°.

3 SUBOPTIMAL CONTROL AND GENETIC SELECTION

3.1 Linear quadratic form control method

For linear system, if the integration of quadratic function with state variables and control variables are set as the performance index, the dynamic system optimization problem is called as optimal control problem based on linear system's quadratic form performance index, referred to as linear quadratic form problem. The optimal solution can be written as a unified analytic expression, and can elicit a simple linear state feedback control law (Ouyang 2001). In the specific situation, if the system is linear and the index function is quadratic form (i.e., the index functions $X(t)$ and $U(t)$ are quadratic functions), we can obtain the linear optimal feedback control law $U(t) = -G(t)X(t)$. The function $G(t)$ can be obtained by solving a nonlinear Riccati matrix differential equation or algebraic equation (Zhang & Wang 2006).

General linear quadratic form problem can be expressed as follows.

The equation of a linear time-invariant system is:

$$\dot{X}(t) = AX(t) + BU(t)$$

$$Y(t) = CX(t) \tag{3}$$

where $X(t)$ is n-dimensional state vector, $U(t)$ is m-dimensional control vector, and $Y(t)$ is output vector. Assume that $U(t)$ is unlimited, and error vector is:

$$e(t) = Z(t) - Y(t) \tag{4}$$

where $Z(t)$ is one-dimensional ideal output vector. It is desired to find the optimal controller, so that the following quadratic performance index is minimized,

$$J = \frac{1}{2}\int_{t_0}^{t_f} \left[e^T(t)Q(t)e(t) + U^T(t)R(t)U(t) \right] dt \tag{5}$$

where Q is symmetric positive semi-definite matrix, R is symmetric positive definite matrix. Generally, the Q and R are set as diagonal matrix.

In this situation, by solving the Riccati algebraic equation:

$$PA + A^T P - PBR^{-1}B^T P + Q = 0 \tag{6}$$

we can obtain the solution of Riccati equation P and the optimal feedback gain matrix K, such that the optimal control law that is described as formula (7) can minimize the performance index (6).

$$U(t) = R^{-1}B^T PX(t) = -KX(t) \tag{7}$$

It's unnecessary to determine the location of the closed-loop pole according to the requirement of

performance index, but just find out appropriate weighting matrix of state variables and controlled quantity, which is the most advantage of using LQR controller as the method of control system design and correction.

Using the performance index J as the constraint condition of controller design, we can obtain the optimal system performance with the standard of that index. But it doesn't mean the satisfactory control system can be designed. The key of LQR controller design is selecting appropriate state variable weighting matrix Q and control variable weighting matrix R.

In general, the optimal linear quadratic form controller can be obtained in the following steps:

Step 1: Describe the system's dynamic model in the state space

Step 2: Set the weighting matrix Q and R (usually take constant diagonal matrix)

Step 3: Solve the Riccati equation

Step 4: Calculate the system's time-domain response of state variables. If the performance index cannot reach the standard, return to the Step 2 and change the weighting matrixes until the satisfactory performance is obtained.

From the steps above, we can see the controller design as a process of trial. Appropriate selection of the weighting matrix can decrease the number of iterations. In general, increasing a weighting coefficient in matrix Q can make the corresponding state variable convergence faster, and a larger coefficient in matrix R can decrease the control values.

3.2 Genetic selection for the weighting matrices

In this section, the genetic algorithm is used to search the optimal selection of the weighting matrixes. The suboptimal controller will be tested by an index function which represents the time-domain requirements of the system's performance. During the process of genetic search, every individual corresponds to a selection of the weighting matrix, and the fitness function consists of some selected time-domain indexes. The selection settings and manners based on genetic algorithm are illustrated in the following.

As the Q is positive semi-definite matrix and R is positive definite matrix, they are set as diagonal matrix. In order to apply genetic algorithm to select the optimal controller, the weighting matrixes Q and R should be encoded as the following form:

$$\Lambda = \{q_1, \ldots, q_n, r_1, \ldots, r_m\} \tag{8}$$

where $q_i > 0$ ($i = 1, \ldots, n$) are diagonal elements of Q, and $r_i \geq 0$ ($i = 1, \ldots, m$) are diagonal elements of R. In this study, the weighting matrixes Q and R is given as formula (9), and the individual of genetic algorithm can be encoded as $\Lambda = \{q_1, q_2, r\}$.

$$Q = \begin{pmatrix} q_1 & 0 \\ 0 & q_2 \end{pmatrix}$$

$$R = r \tag{9}$$

Set the individual of initial population as

$$Q = C^T C$$

$$R = B^T B \qquad (10)$$

Furthermore, some parameters should be set, such as the size of population, crossover rate and mutation rate (Lee 2005).

The size of population can influence the effectiveness of genetic algorithm. Small number of population is unable to provide enough sample points and leads to the optimal solution to be uncertain; too many number of population can increase the amount of calculation and extend the time of convergence. The size of population is equal 80.

The crossover rate in the algorithm controls the frequency of the crossover operation. Little crossover rate will kill the solution with high fitness, while too high rate will reduce the search efficiency. Set the crossover rate as 0.25 in this example.

The mutation rate is a key factor that can increase the diversity of population. New gene will not born, if this rate is too small; too high rate will make the optimal algorithm degenerate as random search. The rate of 0.03 will be appropriate.

The fitness function is set to be:

$$J_{fit} = \frac{1}{\max(u) + \sigma(\alpha) + t_p(\alpha)} \qquad (11)$$

where $\sigma(\alpha)$ is the overshoot of angle of attack output, and $t_p(\alpha)$ is the peek time, $\max(u)$ is the maximum value of elevator control.

By giving a predetermined size of population, and set proper genetic algorithm parameters (such as crossover rate, mutation rate and selection rate), we use Matlab 7.1 Genetic Algorithm Toolbox to calculate the evolution process with the initial generation. Then the optimal selection of weighting matrix which makes the value of fitness function minimum will be achieved.

It should be emphasized that for each individual, the elements of Q and R must be positive, or the value of fitness function would be infinite during the genetic search (Wang & Wang 2009).

4 NUMERICAL RESULTS

The suboptimal attitude controller is validated by the numerical simulation of genetic algorithm, and its performance will be compared with another controller without optimal design.

By genetic search, the weighting matrixes are achieved that:

$$Q = \begin{bmatrix} 165.98 & 0 \\ 0 & 9.89 \end{bmatrix}, \quad R = 5.61 \qquad (12)$$

The corresponding suboptimal controller is:

$$U(t) = -KX(t)$$

$$K = [-2.677, -1.99] \qquad (13)$$

In order to illustrate the effectiveness of optimal design based on genetic algorithm, we select another controller as a comparison whose weighting matrixes are as follows:

$$Q' = \begin{bmatrix} 1 & 0 \\ 0 & 1 \end{bmatrix}, \quad R' = 1 \qquad (14)$$

The corresponding feedback gain matrix is:

$$K' = [0.013, -0.55] \qquad (15)$$

The evolution process of genetic search is depicted in Figure 1. The attitude response as output and the elevator control as input are compared between two controllers in Figure 2 and Figure 3.

Figure 1 indicates that fitness value of the best individual in every generation is decreasing until a stable value, which means the genetic algorithm is effective in selecting suboptimal controller with better performance.

By analyzing Figure 2, it shows that the performance of suboptimal controller is better than that of the controller which is obtained by manual selection of weighting matrixes. Both the index of overshoot and peek time of suboptimal control system are smaller than that of manual selected controller. From Figure 3, the maximum value of control input of suboptimal control system is larger.

From the result of the simulation, we can conclude that a better design effect can be achieved by optimal selection of weighting matrixes based on genetic algorithm.

By using genetic algorithm to search optimal weighting matrixes and design suboptimal controller, the heavy and complicated work of trial can be avoided.

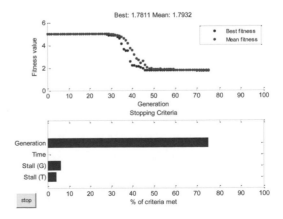

Figure 1. Evolution of the fitness value.

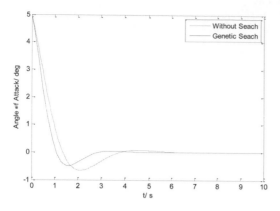

Figure 2. Output of angle of attack.

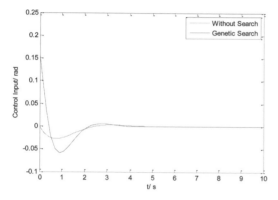

Figure 3. Input of elevator control.

The model adopted in this paper is relatively simple. It should be emphasized that the global optimum weighting matrixes and control law would be harder to be obtained by using genetic algorithm.

5 CONCLUSIONS

A suboptimal controller for the longitudinal attitude of an aircraft is designed in this paper. The dynamic model of the aircraft is built with linear assumption, such that the linear quadratic form control method can be applied to design a LQR controller for the system. In this paper, we use genetic algorithm to select the two weighting matrixes, and the optimal selection is obtained by time-domain performance index which corresponds to the fitness function in the algorithm. The simulation results indicate that using genetic algorithm for the selection of weighting matrixes is a feasible method. Comparing with another LQR controller with manual selection of weighting matrixes, the proposed controller's performance is relatively satisfactory. Future works may study appropriate principle of selection of genetic algorithm parameters and the fitness function to make the optimal method practical.

REFERENCES

Ali, R. Mehrabian, C. & Jafar, R. 2006. Aerospace launch vehicle control: an intelligent adaptive approach, *Aerospace Science and Technology* 10: 149–155.

Hwang, S.F. & He, R.S. 2006. Improving real-parameter genetic algorithm with simulated annealing for engineer problems. *Advance in Engineering Software* 37: 406–418.

Lee, Y.J. 2005. *MATLAB Genetic Algorithm Toolbox and its Application*. Xi'an: Xi 'an University of Electronic Science and Technology Press.

Ouyang, L.M. 2001. *MATLAB Control System Design*. Beijing: National Defence Industry Press.

Wang, T. & Wang, Q. 2009. Suboptimal controller design for flexible launch vehicle based on genetic algorithm: selection of the weighting matrices Q and R. *IEEE International Conference on Intelligent Computing and Intelligent Systems*: 720–724.

Zhang, H.Y. & Wang, Q. 2006. *The Optimal Ccontrol Theory and Application*. Beijing: Higher Education Press.

Information Technology and Computer Application Engineering – Liu, Sung & Yao (Eds)
© *2014 Taylor & Francis Group, London, ISBN 978-1-138-00079-7*

On the dilemma and way out for the cultivation of the cultural consciousness in contemporary China

Yongfang Deng & Wenjuan Hu
Jiangxi University of Science and Technology, Ganzhou, Jiangxi, China

ABSTRACT: In contemporary China, cultural consciousness has already become one requirement of modern transformation, however, it is faced with multiple difficulties during cultivating cultural consciousness, so, we need to seek the path to cultivate cultural consciousness.

Keywords: Cultural consciousness; Dilemma; Path; Modernization

In November 2012, the eighteenth congress of the communist party of China were held smoothly in Beijing, president Jintao Hu said: "We must keep advanced socialist culture as our orientation, setting up a high level of the cultural consciousness and cultural confidence." This is our party once again stressed the importance of cultural consciousness, it is one of necessary conditions to construction of socialist culture power, so, we need to cultivate Chinese cultural consciousness.

1 THE PURSUIT OF CULTURAL CONSCIOUSNESS

The research of contemporary cultural consciousness can be traced back to the 80s. we can divided it into three periods: the earlier stage is the cultural consciousness research by Sumin Xu, the representative of the medium-term (1997 years later) is the Xiaotong Fei's cultural consciousness theory. In recent years, cultural consciousness research has entered a new stage. In October 2011, the communist party of China held the sixteenth plenary session of the 17th, which Emphasize to develop high cultural consciousness from the academic level to policy level, the word of Cultural consciousness has become the official willing to advocate cultural discourse.

1.1 *The connotation of cultural consciousness*

What is the cultural consciousness? Sumin Xu made clear definition of cultural consciousness: "The so-called cultural consciousness, but refers to a kind of cultural mentality, it is through the introspection way to know the inevitable trend of the old culture demise and new culture produce, which is clearly aware of their own historical mission, and put into practice."[1]

Xiaotong Fei think that cultural consciousness is a kind of method to deal with interpersonal relationship, that "people who live in a certain culture have self-knowledge, understand the origin of its forming process, its characteristics and its development trend."[2] Yunshan Liu point out: "cultural consciousness, mainly refers to a national or a party have the consciousness and awareness in the culture, including the understanding of the role of culture in historical progress, the mastering of the law of culture development, and the assuming history responsibility for development culture."[3] itrequires the subject not only to grasp the history and face the realization, but also to take the historical responsibility for development cultural. in fact, cultural consciousness is the prelude for a national to rise.

1.2 *The cultural consciousness in contemporary Chinese*

From the perspective of philosophy, the essence of cultural consciousness is culture subject's spiritual consciousness. From the perspective of subject, cultural consciousness is the cultural subject's spiritual consciousness; From the perspective of object, cultural consciousness is a mark of high level understanding of the connotation, the function, the law of development, the construction path of a certain times culture, it is a kind of profound cultural reflection. However, cultural consciousness is not a purely subjective mental activities, but a kind of practical activities. It reflects culture subject's critical, historical, profundity, and practical and so on. since modern times, in essence, cultural consciousness is a modernity topic, that is a kind of cultural consciousness of modernity. So to speak, the modernity situation constitutes the grand background of cultural consciousness.[4]

2 THE DIFFICULTIES OF CULTIVATE CULTURAL CONSCIOUSNESS IN CONTEMPORARY CHINA

"Culture is the soul of the nation, cultural consciousness is the key of promote Chinese nation culture spirit character."[5] Cultivate cultural consciousness, is the need for building harmonious culture. However, in contemporary China, to cultivate cultural consciousness is faced many difficulties.

2.1 The cultural dilemma in the process of modernization

In comparison with western countries, China's modernization is special, China faced with double modernization task: complete social thought enlightenment and establish modern nation-state, and the historical development inevitable to prefer the latter. Since then, Chinese are too busy to spread and the development of modern thought, also have no energy to inheritance and absorption traditional culture. However, in the process of developing socialism with Chinese characteristics, we encourage to develop the social productivity, and relatively ignored the development of cultural, especially to shape the social values. Under the impact of market economy, more and more things to be labeled as price tag, people are more keen to pursue wealth and honor, ignoring all kinds of social due ethics and values. In this atmosphere, people only care about their economic interests, pay little attention to national culture, there would be no cultural consciousness.

2.2 The nonconfidence be brought by cultural globalization

Globalization is one of the most striking features of the world. Globalization is not only refers to the various countries and regional economic integration, it is also refers to the cultural values and life-style spread around the world. In the cultural globalization background, the strong culture often occupy propagation advantage, forming a kind of cultural information with no peer communication. Since the reform and opening up, all kinds of foreign cultural concepts and cultural products will poured into China. In today's situation, the western capitalist culture with its economy, science and technology advantages, gain a powerful position in the world cultural competition, it has great attraction and impact force (including new traditional culture, such as Marxism) for developing countries. The outside environment formed a huge cultural pressure, and even lead to internal low self esteem and quitting, in the circumstances, to cultivate people's cultural consciousness will have to face powerful challenges.

2.3 Western capitalist countries' cultural infiltration

Since China took the way to socialism, the western capitalist countries never give up in sabotaging China. "Subversion socialist countries is their basic strategic target, they are absolutely not allow socialist country vigorous development, when the force can't achieve this goal, they will use peaceful evolution strategy."[6] They subversion China at the use of various means, including strong cultural offensive, of which the most typical is the United States. Frank's A NingKe, The United States diplomatic historian, he said: "Culture means, the same as political, economic and military means, is an important part of the United States foreign policy, under the limited military action condition, especially can't protect homeland security during modern war, cultural means become a more important strong penetration tool for the United States through obstacles"[7] Western capitalist countries through the various cultural education exchange program and news media to spread their ideology and values, people will easily fall into their trap: accept their ideology and values, with lose our own national culture. To cultivate cultural consciousness, this is highly destructive.

2.4 The limitations of the low education level of culture subject

Education is a product by the historical development, it is a main way to promote culture subject's all-round development, and is also an important way to make culture subject to understand national culture. A culture subject that do not study national culture, do not understand the national culture's status, impact or law of development, won't produce cultural consciousness. cultivating cultural consciousness depends on the development of education, From the sixth national census data, we can see that the education level of China's population is low. "In Mainland China, the population of people with college degrees is 120 million, the population of people with high school-level education is 188 million; the population of people with junior middle school education is 520 million; the population of people with primary education is 359 million."[8] Low level education will limit the scope of culture subject to understand and comprehend cultural. only a high level of education can foster a high level of cultural consciousness.

3 THE PATH OF CULTIVATE THE CULTURAL CONSCIOUSNESS IN CONTEMPORARY CHINA

To cultivate cultural consciousness, promoting great cultural development and prosperity, we need to explore the path to cultivate Cultural Consciousness.

3.1 Promoting traditional culture transform to the contemporary culture to get out of cultural dilemmag

To talk about Chinese culture is inevitably involve China's traditional culture, there is no Chinese traditional culture, there is no socialist culture with Chinese characteristics. Mao tse-tung in the "New Democratic Theory" so said: "China's current culture is development from ancient old culture, therefore, we must respect ourselves history, never partition history. But this respect, is to give a scientific position to history, and to respect history's dialectical development." To get out of cultural dilemma, we should be based on the principle of absorb the essence, to its dregs. on the one hand, we should strengthen the work to protect traditional culture (including physical and nonphysical two forms), protect the traditional culture is the same as protecting the national characteristics. On the other hand, we should fully dig the modern sense in traditional culture, traditional culture has a lot of regulating social relations point of view, which is suitable for modern society, such as "Da tong", "oneness of man and nature" and so on. At the same time, we can absorb foreign advanced culture to inject fresh blood to traditional culture, make it walk into people's vision and improve national culture's sense of self-identity.

3.2 Developing cultural industry to enhance cultural confidence

Speed up developing cultural industry can make full use of cultural resources, rich cultural products, and reflect national characteristics. "We can show broad and profound of the Chinese nation culture through the physical form of cultural products, set up the good international image, and reflect the charm of Chinese culture."[9] The sixteenth plenary session of the 17th pointed out: "to speed up the development of cultural industry, we must construct a modern cultural industry system with reasonable structure, complete range, tech-heavy, be creative and competitive." Therefore, we must be guided by the market, optimizing the culture industry structure, strengthening the ability of cultural independent innovation, expanding the scale of the cultural industry. So, in the process of promoting the development of cultural industry, can improve the charm of Chinese culture, strengthening cultural confidence, which help can to cultivate people's cultural consciousness.

3.3 Shaping socialist core values to resist cultural infiltration

Jintao Hu in the party's 17th points out: "Culture becomes increasingly a source of national unity and creativity", and "what is culture's sou, it is condensation in the culture, decision rules and the direction of the deep elements, that is the core values. We have what kind of values, there is what kind of cultural position, culture orientation and culture choice."[10] Today, the world is in the midst of major developments, major changes and major adjustments, along with the further development of globalization, all kinds of ideological and cultural communication more frequently, to cultivate people's cultural

consciousness, it is necessary to improve Chinese culture's competitiveness, we must speed up the shape of socialist core values, and enhance national cohesion. Shaping the socialist core values need to combination the socialist core value system and Chinese traditional culture, which should not only reflect changes of the times, but also reflect national character, with whole force and leading force. Only the socialist core values can over social, enhance national cohesion and resist consciously western cultural infiltration.

3.4 Developing education to improve culture subject's education level

Developing education is the necessary way to rejuvenate the nation strategy, also the only way to improve the cultural subject's cognitive level. At present, Chinese education level is far below developed countries, only by vigorously developing education, can we improve people's the level of education and scientific, promote the cultural subject understand the national culture. Vigorously developing education can promote people to understand and love their own cultural by consciously comparative, analysis and identification. It is need to do the following: the first, strengthen the investment of education and constantly broaden financing channel to make education enterprise vigorous development; Second, we should continuously push forward the reform of education, implementation of quality education, and further perfect the education management system; The third, we need to provide a good environment for the development of education, to mobilize social people to support and participate the reform and development of education.

REFERENCES

[1] Su-Min Xu, "Philosophy of Culture," Shanghai people's publishing house. p. 299, 2003.
[2] Xiao-Tong Fei, "Reflection Dialogue Cultural Consciousness," Journal of Peking University: Philosophy and Social Sciences Edition. 1997(3).
[3] Spruce, "Cultural Consciousness Cultural Confidence Culture Self-improvement – The Thinking about The Prosperity and Development of Socialist Culture With Chinese Characteristics (Part one)," Red Flag Manuscript. pp. 4–5, 2010.
[4] Yong-Fang Deng, "Cultural Modernity in The Perspective of Philosophy," Jiangxi people's publishing house. pp. 273–279, 2009.
[5] Jian-Bo Ouyang, "The Thinking of The Relationship Between The development of Harmonious Society and Cultural Consciousness," Huxiang Forum, pp. 16–17, 2008.
[6] Cong-Cong Wang, Cheng-wen Yan, "Analyses China's Survival Strategy From The Collapse of The Soviet Union Cause," Journal of Chongqing

University of Science and Technology: Social Science edition, pp. 21–23, 2011(22).

[7] Xiu-Min Jiang, Hong-Xia Lin, "Analysis The Cultural Security Problem Under The Background of Globalization," Journal of Dalian Maritime University: Social Science Edition, pp. 42–46, 2004(4).

[8] Xin Yuan, "Six General Population Quality Big Bottom Up China's Population is Jumped From The Mouth to The Hand," Social Watch, pp. 58–59, 2011(6).

[9] Lei Wang, "The Way to Promote China's Cultural Soft power,"

[10] Spruce, "Cultural Consciousness Cultural Confidence Culture Self-improvement – The Thinking about The Prosperity and Development of Socialist Culture With Chinese Characteristics (Part one)," Red Flag Manuscript, pp. 4–5, 2010.

Information Technology and Computer Application Engineering – Liu, Sung & Yao (Eds)
© 2014 Taylor & Francis Group, London, ISBN 978-1-138-00079-7

The analysis of the development of China news media's soft power

Xianchao Deng & Xiangyu Yao
Jiangxi University of Science and Technology, Ganzhou, Jiangxi, China

ABSTRACT: The China Media, as an effective way to disseminate information, is an important carrier of cultural soft power. It plays important roles on instructing the right values, inheriting excellent culture, improving the national image and transmitting the culture and so on. Facing the emerging challenges and problems such as the voice of small and vulnerable in the international communication, moral loss in the medias and the concussion of the new media, we need to increase the efforts to reform and develop China media, accelerate the news media soft power of promotion our country, build the large media group with great international influence, firmly grasp the correct direction and actively use, guide the new media and occupy the commanding heights of distributing the information of China media.

Keywords: News Media; Soft Power; Analysis

The world today is undergoing great changes and major adjustments. With the globalization of world economy and revolution in science and technology quicken the human civilization. In addition, the world competition of synthetically national power becomes more intense because of that. The cultural soft power, as a vital factor that weighs comprehensive national power of a state to lose by force, plays a growing role in the development of economy and society. If China wants to be in intense competition remain invincible, achieve competitive advantage, just constantly promoting the construction of Hard power which included economy power, technology power and country defense power does not go far enough. We should also focus on developing the construction of national culture and promoting the national cultural soft power. Science news media is an important carrier of cultural soft power, intensifying the development and reforms of news media, increasing the communication capacity and influence power of news media are of great importance.

1 THE FUNCTIONS OF NEWS MEDIA ON ENHANCING THE NATIONAL CULTURAL SOFT POWER

In the early 1990s, Science Joseph Nye, a Harvard professor and doyen of US foreign policy analysts, proposed the theory of soft power, the soft power theory has become a hot area of research in the world. A country with soft power, Mr. Nye contended, could bend others to its will without resorting to force or payment. In his view, soft power is an influence power of a country's culture and ideology. It includes information dissemination force, cultural influence force, and public opinion guiding force and so on.[1] News Media as the most effective vehicle of propagating information has many effects on instructing the right values, inheriting excellent culture, improving the national image and transmitting the culture and so on. The detail functions are in the follows.

1.1 News media affects the construction of the system of socialist core values

Building up the system of socialist core values is the core section of national cultural soft power construction. With the development of society and science technology, the human society entered the informationization times. People get the numeral messages through TV, network, newspaper and so on. The message disseminated by those news media mentioned in the above can easily effects people moral concept and values. It will enhance national cohesion and activate the national spirit if people accept the correct positive public opinion. Otherwise, it will disturb audio-vision if people accept the negative public opinion. Therefore, what the messages the news media disseminated will affect the construction of the system of socialist core values.

1.2 News media affects the heritage of the China outstanding culture and transmission

On one side, the news media disseminates China outstanding culture. Also may say that, is the media discover and promote many aspects of the content, form and value of Chinese culture through book publishing, commonweal advertisement, network media, television broadcast and many other reporting form, let people get more information about Chinese culture. On the other side, the news media protects China outstanding culture. It can use its social resource and information transmissibility to universe the knowledge of the law concerns cultural heritage protection. Further more, the information related Chinese culture can also let more and more people know the diversity of social culture, and participate into the activities of guardian of social culture.

1.3 News media affects molding national image

With increasing international exchanges, national image plays more and more important roles on international cooperation and international affairs handling. To establish a country's national image need effective international propaganda of media, building largely to the effective promotion of media. As an authoritative source of information and an important channel of information dissemination, the news media is an important participant and performer in international communication. When the news media develop international reports and analyze international events, full consideration should be given to reports focus, attitude and style, because these factors will affect public's opinion on assessing national image. Therefore, a powerful influential mainstream news media has an important role in shaping Chinese international image, strive for a greater voice and enhance China's cultural soft power.

1.4 News media affects Chinese culture "going global"

To improve the country's soft power, promote Chinese culture "going global" is an important development strategy in China. The dissemination capacity of news media of China is gradually increasing and affects Chinese culture "going global". Many examples of this could be given. China Daily and Global Times are the best examples of that. It is attracting increasing attention for China Daily, Global Times and many other types of these kinds of English newspaper publish. CCTV 4 held a program called Meet China, which aims to spread Chinese culture to the world through high technology [2]. Therefore, news media can really do many things on Chinese culture "going global".

2 THE PROBLEMS IN NEWS MEDIA DEVELOPMENT ON IMPROVING THE NATIONAL CULTURAL SOFT POWER

Since China's reform and opening up, China news media development has entered a steadily and rapidly developing stage in history, and China news media total strength rises greatly. However, compared with foreign countries, China's news media development started late, is still in the exploratory stage, and foreign have been relatively mature. Besides, the construction of news media soft power is facing many problems

such as weaker say in the international communication, moral loss in the media and the concussion of the new media.

2.1 The voice of small and vulnerable in the international communication

Three decades of reform and opening up, although the number and strength of China media has increased dramatically, the western media such as Times still seriously impact the world opinion and China medias lack influence on discourse. Because of the voice of small and vulnerable in the international communication, most of western countries still hold many negative impressions on China, such as China is a country without human rights, a country produces poor-quality products and a country steals core technology. The biased media hyped the negative message of China, causing serious damage to China's national image. In addition, many western countries make full use of its powerful media to hype their western countries culture and values, threaten native culture. These kinds of action finally lead to many people idolize the western culture and ignore native culture.

2.2 Moral loss in the media

Since China enters transition period of the society, all kinds of ideas and cultures are colliding. All kinds of news papers, magazines, and networks and many other kinds of media have rushed into our life, many media begin to report many false information in order to get more attentions and profits. For example, the news papers and TV always focus on the profits from the disseminating advertisement and ignore whether this advertisement is truth or false; some websites, news papers and magazines even spread pornographic messages in order to get a higher attention. All these phenomena disobey the social code of ethics and create many serious problems, such as affecting juvenile's sound growth, hindering our country socialism core value system construction.

2.3 The concussion of the new media

The new media refers to a new media form that uses the new technology, the new propagation mode, the new terminal platform and new applying means. Because this kind of new media include with high technology, it has many new characteristics, such as spread fast, wide coverage, without borders limit, open access and larger number of users and so on. The route transmission of new media is showing in the below drawing 1. The new media, to some extent, has many special functions to compensate the lacks of traditional media. However, there are many problems in the process of the new media development. The problems details are showing in the follows.

- The diversity of disseminators and contents lead the users of new media to a stereotype of the information.

- Lack the gatekeeper of the news and increase the difficulties of public opinion guidance.
- The users of new media lack the professional ability to express the viewpoint.
- And other problems.

3 THE PATH CHOICE OF THE NEWS MEDIA SOFT POWER CONSTRUCTION

The 6th Plenary Session of the 17th Central Committee of the Communist Party of China announced two decisions of immense importance which are strengthening the work of public opinion of news media and developing the modern transmission system [3]. Building the news media soft power not only is a symbol of enhancing national cultural soft power, but also is a necessity of enlarging the news media's discourse in the international situations.

3.1 Develop a huge news media with international influence

According to statistics, nine of the world's biggest media groups – Time Warner, Disney, Bertelsmann, Viacom, News Corp, Sony, TCL, Universal, Japan Broadcasting Corp – control 50 media companies and 95% of the world's media market [4]. These large media group not only has a large scale, but also has advanced technology and management concepts. There is a big gap between China's news media and international first-class media in the dissemination of means, management, technical equipment and content creation. Therefore, to accelerate the development of core news media in software and hardware construction has become an important choice.

- The government should increase the investment of the news media, absorb the civil and social capital extensively, widen financing source channel, and realize the diversification of fund sources. The core news media should make effective use of existing resources, improve the means of information dissemination, and innovate management concepts, grasp the usage of advanced equipment, enlarge international communication and influence.
- The government should strengthen the media policy service, provide policy support to the core news media in good fiscal, taxation, financial, and land and so on. At the same time, the government should introduce policy measure to break the news media "compartmentalization, low-level repeat" and "act of one's own free will, non-contacts", realize media collaborates and complementary advantages, promote the leaping development of news media.
- The domestic media should strengthen cooperation and exchanges with foreign media. By establishing a platform for international cooperation mechanisms, on the one hand, make foreign media know about China, understand China, evaluate China more objectively; on the other hand, study the abroad audience's cultural habits and consumer psychology, strengthen propaganda, enhance the attraction and influence of international communication, create the condition for better implementation of "going out" culture strategy.

3.2 Firmly grasp the correct direction

The news media plays important roles not only on disseminating the socialism core value system, but also on the propagating integrity, guiding and supervising the social opinion. In the 6th Plenary Session of the 17th Central Committee of the Communist Party of China, President Hu pointed out that the news media and the journalists should take social responsibilities and abide the professional morality, spread the information correctly, resist consciously the mistaken views, resolutely put an end to the false news.[5] The detail countermeasures are in the follows.

- The government should take the public interest especially teenagers' benefit as the standard, strengthen supervision of news media's content, put an end to information on flattery, pornography, violence, money worship and prevent the spread of information threatening citizens and society. The news media should strengthen analysis on public opinion and provide a platform for people to discuss and make people can express their views and opinions seriously.
- The news media should strengthen the guide of hot and difficult social problems, eliminate doubt and confusion and make effective consensus. Secondly, we should strengthen the self-discipline of news media industry. In order to guarantee the normal operation and benign competition, the news media industry association should formulate special rules and regulations for the industry to ensure that the industry's words and deeds accord with national laws and moral requirements. Industry associations should exert pressure and measures to these phenomenon that is harmful to the social core value system, for example, enhancing ratings, exposure rate and click rate. They should change their words and deeds, adjust the communication orientation.
- The news media worker should correct their attitude and base on objective facts. They should not add personal bias in the dissemination of information and should restore the truth objectively and fairly.

3.3 Actively use, guide the new media and occupy the commanding heights of distributing the information of China media

With the rapid development of modern technology, the new media has become a necessary part of our daily life and work. However, the using of new media is a double-edged sword for constructing the national cultural soft power. On the one hand, the new media could supply a platform for disseminating social opinion and culture. On the other hand, the new media may also bring negative effects for social stability and cultural harmony. In order to take a full part of new media's advantages and reduce the disadvantages of new media, the ways to standard the media are proposed in the followings.

- The government should put a high value on the new media's functions of information disseminating, establish and improve the regulations and policies concerned about new media, enhance the ability of guiding, supervising and managing of the disseminating the information through the Internet.
- The new media companies should increase the media scientific and technological capacity to communicate pay attention on technology development and utilization. In addition, the companies should also encourage their staffs to try their best to do the excellent culture works by using the high technology.

REFERENCES

[1] Joseph S. Nye, The Paradox of American Power, Page 13 Beijing: World Knowledge Press, 2002

[2] Hongyan Zhao, Yang Xu. "Televison Diplomacy: A kind of soft power construction", China's cultural soft power research (2010), Page 361, Beijing: Social Sciences Academic Press, 2011

[3] http://news.xinhuanet.com/politics/2011-10/25/c_122197737.htm

[4] Min Wu, "The Analysis of Increasing the Public Opinion Guidance of the New Media", Southeast Academic Research, 2011(5):233–239.

[5] http://news.xinhuanet.com/politics/2011-10/25/c_122197737.htm

Information Technology and Computer Application Engineering – Liu, Sung & Yao (Eds)
© 2014 Taylor & Francis Group, London, ISBN 978-1-138-00079-7

Modeling for joint operation information cognitive chain based on information life cycle

M. Yang, M.L. Wang & M.Y. Fang
The Second Artillery Engineering University, Xi'an, Shanxi, China

ABSTRACT: Aiming at the requirement of information modeling for the research of joint operation system of systems, a Joint Operation Information Cognitive Chain (JOICC) model was proposed based on Information Life Cycle (ILC). From the new angle of life evolution, the new comprehension and concept for ILC were presented. Combined with cognitive process, the Information Cognitive Chain (ICC) model was built and the JOICC model was proposed with three domains. The math's describing models for situation knowledge assessment and mission decision generation were built separately in cognitive domain.

Keywords: joint operation; information life cycle; information cognitive chain; situation; decision; model

1 INTRODUCTION

Joint operation is becoming the important mode of information war step by step. This operation mode is the joint of multi-army units and the antagonism between systems and systems. As one of the important work for operation system of systems modeling, the building-up of information model is always the hotspot and emphasis.

The main methods for information modeling are the ontology model (Zhang 2002, Du 2006, Tang et al. 2008, Gao 2007), meta-data model (Yuan et al. 2008, Xie 2007, Bu 2011, Xu 2008), agent model (Chi 2007, Liao 2005, Tang 2010, Huang 2011), and etc. These methods emphasize particularly on many aspects, but they don't make any researches on the angle of information cognition. As a new exploration for information modeling, this paper will make the pertinent research based on the theory of Information Life Cycle.

Information Life Cycle (ILC) (Ma & Wang 2010, Wan 2009, Suo 2010) which is the science concept for information research was proposed by Levitan in the field of information resource management in 1981. With the help of Taylor, Horton and other scholars for defining ILC, the academe has defined much concept. The traditional concept is that the ILC is a process for confirming requirement, collection, machining, storage, searching, transferring, using and treatment with the management angle. Fewer models analyze the hypostasis of information. It is a new approach to discuss the joint operation information modeling based on the new understanding and definition of ILC with the process of life evolution.

2 INFORMATION LIFE CYCLE

Information is the state and method of matter interior structure and exterior movement. The information movement is not only external being but also complexness. It is decided with the interior value and exterior environment, and it has much character such as abstract, multiplicity, periodicity and phase.

Life Cycle is a series of period for an individual form birth to death. For example, a simple plant growth is a natural evolution process of seed, pullulating, efflorescence, fructification, and become the seed lastly. This process is a complete plant life cycle. Then we can use this cognition to make a new decision that the ILC is an impersonal rule to describe the whole information life process from generation to disappearance. In other words, ILC can be regarded as a disciplinarian cycle and movement of the information self-interior movement.

Then, the definition of ILC is not only to reflect the character of information but also to advise the five factors: 1) ILC must be the organic combine with the human cognition and information existing mode. 2) All phases of ILC must be ordered and related to each other. 3) The every phase transformation of ILC must be corresponding to the transformation of value. 4) ILC is the information action which is human participated in. Here it is a broad concept, not only comprising the nature information but also containing the information resource developed and organized by human. 5) ILC can be described, simulated, and measured with the quantitative method.

Based on the above analysis and referencing the Taylor's ILC model, the new concept of ILC is proposed with the information evolvement process from generation, integration, and organization to application. The related terminology is defined:

Definition 1 Data: It is the single fact, concept or dictate which is obtained by observation and survey separately. It's existing with the mode of human or auto-machine communication, explanation, and disposal.

Definition 2 Information: It is the data gather that used some methods to dispose, or it is the result of some significant relationship with data element.

Definition 3 Knowledge: It is the accumulation of the kindred information based on some kind of mode, and it is the abstract and general information summarization which is to carry out some special purpose.

Definition 4 Decision: It is the selected action based on the knowledge which is used some scientific methods to collect and appraise knowledge for completing some objectives.

After these, the concept of ILC can be defined:

Definition 5 Information Life Cycle (ILC): It is the whole process which is comprised of the generation, collection, organization, and application of information based on data, information, knowledge, and decision. ILC can be called the process of DIKD. The core essentials are data, information, knowledge, and decision.

Being compared and contrasted with plant life cycle, it is concluded that the data is seed, information is pullulating, knowledge is efflorescence, and decision is fructification. With this angle, the connotation of ILC is the transformation between the existing form and the value of application.

3 INFORMATION COGNITIVE CHAIN

Definition 6 Cognitive: It is the process of the cognitive action usually. In this paper, the cognition is the action including information obtaining, understanding, and decision; it is the series of cognitive action that is accepting and evaluating the data, generating the knowledge and methods for dealing problems, forecasting and estimating the results, and making some decisions.

The four essentials of information cognition include the data, information, knowledge, and decision in definition 6. Combined with the ILC, the abstract description of information cognitive process can be obtained as Information Cognitive Chain (ICC):

The first is the primary cognition for receiving the objectives or the data elements in realistic environment. The second is the more cognition for the information based on the data integration. The third is the apprehending cognition for the knowledge based on the information integration. The last is the decision. That is the whole ICC.

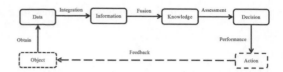

Figure 1. Information cognitive chain model.

After that, the ICC can be described as the other cycle.

When the decision is made, the action of execution must be occurred. It will act on the real environment and make some changes in reality. The new obtained information data will react on the cognition and will build a new information cognitive process, as depicted in Fig. 1.

4 THE JOINT OPERATION INFORMATION COGNITIVE CHAIN

The existing space and active category of joint operation have happened in three domains: physical domain, information domain, and cognitive domain. The physical domain is the existing space of combat entity, and has the battlefield facts of battlefield environment, combat operation, and etc. Information domain is the cyberspace for data integration, information fusion, and information transmission. Cognitive domain is the cognitive space for situation representation, thread evaluation, and mission decision.

With the angle of information, the main function of joint operation is bringing out the transformation from the information to decision. It should be transforming the battlefield external data to situation apperceiving with high quality and real-time sharing, and making the high-efficient and accurate decision.

With the angle of operation, the data which is apperceived with the sensor system in physical domain is endowed with definite context such as environment, goal, mission, state, and etc. Then, the data integrates the information with the series of disposal such as filtration, classification, analysis, and etc. The information is fused and transforms the situation knowledge from the information network system to command and control (C2) system in cognitive domain. The comperes comprehend and analyze the situation, and make the mission decision. At last, the decision is implemented with firepower system, combat troops, and combat operations. The effect will act on and make some changes of the battlefield entity, and then it will generate the new information data of battlefield.

Combining with the ICC, the Joint Operation Information Cognitive Chain (JOICC) can be obtained, as depicted in Fig. 2.

The JOICC has four steps in Fig. 2:

1) Apperceiving Data: Using the sensor system to detect the battle-space entity and collecting the battlefield data.

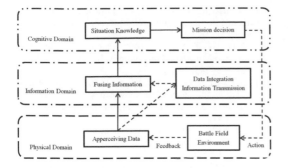

Figure 2. Joint operation information cognitive chain model.

2) Fusing Information: Using the sensor system to integrate the apperceiving data, and fusing information in C2 system based on the information network system.
3) Situation Knowledge: Using the distribution of information network system, the fusing information which is shown by the situation knowledge style should be transmitted to different decision-makers and executants.
4) Mission decision: The decision-maker and executant are cognizing the situation information together, making the uniform estimation and judgment, and making the best advantaged acting project.

5 THE MATH'S DESCRIPTION OF COGNITIVE DOMAIN

5.1 The math's description of situation knowledge assessment

Supposing: the known experience knowledge of operation is $K = (K_1, K_2, \ldots, K_M)$, the real-time data is $S = (S_1, S_2, \ldots, S_N)$, the estimation of situation can be described as the confidence $P = (H/K, S)$ for the uncertain situation $H = (H_1, H_2, \ldots, H_t)$. Now the operation knowledge has the decisive function in situation estimating, the current situation can be classified and identified based on the opposite relationship of situation character and situation identification with knowledge.

Supposing the frame of situation space is $\theta = (\theta_1, \theta_2, \ldots, \theta_k)$, the element shows the whole possible of situation sorts. The aggregate of situation character is $M = (M_1, M_2, \ldots, M_s)$ and it shows the occurrence in battle space. So the situation assessment is the mapping from the aggregate of situation character to the frame of situation space:

$$f : M \rightarrow \theta \tag{1}$$

Classifying the situation:

$$M / f = \left\{ \tilde{M} \middle| \tilde{M} = f^{-1}(\theta) \right\} \tag{2}$$

This process is multi-level identified. There has the aggregate of mapping function $F = (f_1, f_2, \ldots, f_n)$. In this function the f_i describes the opposite relationship of situation identification and sort character which belongs to the element of M in layer i. The cognition of whole situation can be achieved based on the multi-level identification with F to M.

5.2 The math's description of mission decision generating

The situation of battle space can be supposed as situation space S, and the mission decision can be supposed as decision space D. The generating process of mission decision is a mapping function form situation space to decision space $F : S \rightarrow D$, it reflects the decision-maker's ability of comprehension and decision for the current battle space situation.

In the situation space S, the increment of situation $\delta s_{1 \rightarrow 2}$ can be used to show the difference of two discretional situations S_1 and S_2, and it shows the decision information which is needed from S_1 to S_2:

$$S_2 = S_1 + \delta s_{1 \rightarrow 2} \tag{3}$$

The actualizing function of decision $\delta(d, s_C(t))$ can be defined a function from $D \otimes S$ to S:

$$S_n = S_i + \delta \left(d, s_c(t) \right) \tag{4}$$

This function shows the new battle space situation which is generated with adding by the situation S_i and the new operational situation based on the decision d used in $s_c(t)$. A feedback mechanism is provided from decision space to situation space by $\delta(d, s_C(t))$, and it makes the two space consist a closed system of feedback controlling. The objective of decision is to find the opposite decision action with current situation, and changes the expectation of situation step by step based on the implement of decision action.

6 CONCLUSIONS

Firstly, the concept of IIC was researched and the new definition was proposed based on the angle of life evolution. Combining with the cognitive process, the model of ICC was presented. Secondly, the model of JOICC was built up on the three domains: physical domain, information domain, and cognitive domain. Lastly, the math's descriptions of situation knowledge and mission decision were built in cognitive domain. It's a new attempt for researching the modeling of operation information based on the theory of IIC, and it provides a new method to research information modeling and information superiority in operation system of systems.

REFERENCES

Bu, H. 2011. Research on sharing command information management technology based on ontology. Network Security Technology & Application (9): 11–13.

Chi, Y., Deng, H.Z. & Tan, Y.J. 2007. Research on the attacking behavior model of combat agent. Systems Engineering and Electronics 29(11): 1897–1899.

Du, W.H. 2006. Research of the information architecture model based on ontology. Information Research (9): 66–68.

Gao, Z.P. 2007. Ontology-based Modeling Methods, Model and Application of Shared Management Information. China: Beijing University of Posts and Telecommunications.

Huang, J.X. Li, Q., Jia, Q., et al. 2011. Research on composable agent model framework for SoS. Systems Engineering and Electronics 33(7): 1553–1557.

Liao, S.Y. 2005. Research on Methodology of Agent Based Modeling and Simulation for Complex Systems and Application. China: National University of Defense Technology.

Ma, F.C. & Wang, J.C. 2010. A literature review of studies on information life cycle (I) – the perspective of value. Journal of the China Society for Scientific and Technical Information 29(5): 939–947.

Ma, F.C. & Wang, J.C. 2010. A literature review of studies on information life cycle (II) – the perspective of management. Journal of the China Society for Scientific and Technical Information 29(6): 1080–1086.

Suo, C.J. 2010. Remarks on the conception and research contents of information life cycle. Library and Information Service 54(7): 5–9.

Tang, X.B., Wei Z. & Xu L. 2008. The approach to information modeling based on ontology. Information Science 26(3): 391–395.

Tang, S.Y., Yu, W.G., Zhu, Y.F., et al. 2010. Agent-based simulation and analysis of networking air defense missile systems. Systems Engineering and Electronics 32(12): 2632–2637.

Wan, L.P. 2009. The paradigm and theoretical defectiveness in the information life cycle research. Journal of Library Science in China 35(9): 36–41.

Xie, Z.H. 2007. Information Integration and Modeling Research Based on Metadata to the CIS of CAPF. China: National University of Defense Technology.

Xu, Y.T. 2008. Research on Models and Application of E-Government Information Resources Metadata Based on the E-R-P Modeling System. China: Dalian University of Technology.

Yuan, Y.N., Wang, B., Zhang, L., et al. 2008. Metadata-based MIG information integration and service framework. Journal of Beijing University of Posts and Telecommunications 31(2): 120–124.

Zhang, W.M. 2002. *Information System Modeling*. China: Publishing House of Electronic Industry.

Information Technology and Computer Application Engineering – Liu, Sung & Yao (Eds)
© *2014 Taylor & Francis Group, London, ISBN 978-1-138-00079-7*

Design of remote education system based on agent

Lifen Wang
Jilin Agricultural Science and Technology College, Jilin

ABSTRACT: This paper mainly discusses the design of remote education system based on Agent. The implementation introduces the theory of Mobile Agent Cooperative Information System Middleware (MACISM) and gives specific ideas and solutions on which the remote learning system is based. Therefore the cooperative and mutual remote learning and the intelligence and characteristics in it are preferably solved. This can meet the requirements of the current web-based distance education system well.

Keywords: Distance Education; System; Agent

1 INTRODUCTION

Traditional distance education system focuses on itself, does not fully consider the needs and habits of the students and the law of human learning, and requires the user to adapt to the system rather than being adapted, which result in poor learning effect. For these reasons, we apply the agent to the distance education system, which is an extension and supplement to traditional distance education.

2 THE DESIGN OF DISTANCE EDUCATION SYSTEM BASED ON AGENT

Agent is a basic computational unit of agent computing model, a software entity representing interests of users and of independence, which is always in a certain environment and accomplish the tasks assigned by user through migration of operating environment and interaction with the environment.

On the basis of in-depth study on distance education and agent computing model, we conducted design and implementation of agent collaboration information middleware, which is abbreviated, as MACISM (Mobile-Agent-Cooperative information System Middleware) in the paper; then we developed a distance education model system[1] by it, and achieved it on jsp platform.

Distance learning system is a complex system. The solution is to divide the problems, construct some inspiriting and state recording agents which deal with the sub-problems; Agent in the system refers each other through collaboration when interdependence among sub-problems exist; MACISM middleware based on the Agent technology extends it into distance learning, that is Agent-based distance learning system.

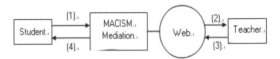

Figure 1. Distance education process diagram.

The process of collaborative distance learning, as shown in figure 1, its basic functions are as follows:

2.1 Students access the distance learning system and register on it through browser. For legitimate users, the student information agent accepts user requests in the friendly way and records the student's specific learning requirements; for first-time users, after initial registration, the system detailed records the student learning levels, objectives and preferences. Student information agent could convert student requests to the command that could be identified, and then produces a corresponding remote student agent through intelligent negotiation in the context of the local host; at last, students agent send a request to MACISM Mediation Server for query request of distance learning server.

2.2 Mediation Server delivers the address and the port of Information Server to student agent, and start MACISM Agent subsystem proxy dispatch it to teacher;

2.3 On the teacher's side, teachers would recommend or specify what to learn to student according to learning of each student, after the interaction, agent unload the request information and return the result to the front-end server;

2.4 The front-end server gets the result and returns it to users through web server;

2.5 Logout, when student completes the job or interrupt the job, student information agent would record the situation of learning and update relative information of student information database in teacher module.

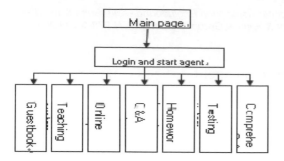

Figure 2.   System functional configuration of FIG.

3   THE REALIZATION OF THE DISTANCE EDUCATION SYSTEM

3.1   *System performance analysis*

The system must meet the basic conditions for running on the Internet, and should also have a safe, stable, reliable, and fast download character; the function of the system is shown in Figure 2: The system consists of seven sub-modules. Before login, each module should start learning agent[2], and then enter them.

3.2   *Homepage*

The composition of the system home page includes registration, login, and distance learning subsystem login and other information, and a brief introduction of the distance learning system. User type includes teacher, student, and the system administrator. If this is the first time for the user to access the system, the user should be registered, enter personal information of the user, if already registered, the user could log in directly. User need to enter a user name and password on the sign-in page and then into the system after verification, and the system will grant users different access rights based on the logged-on user type. If personal information of users changed, the user can modify personal data anytime.

When the authorized user logs in, the user could enjoy various services provided by the system, such as online lectures, do some function in the teaching system, send a message in the guestbook, see the history of message, and enter the examination system as well as management system.

3.3   *The realization of user login subsystem*

The source of user login subsystem contains the following files: trylogin.jsp, login.jsp, register.jsp, loginout.jsp. Trylogin.jsp is a user login interface login.jsp is for user authentication, if the user log in successfully, then the page transfers into the system, or else to the user login interface. Register.jsp used to complete user registration, then keep the user information into the database, when users modify their information, the change would be saved into the database. Loginout.jsp used to exit the system.

3.4   *The realization of Guestbook subsystem*

Guestbook is essential to promote the exchange of students and teachers. Guestbook will save the message sent by students, teachers as well as potential users in the message table. Message is not only used to communicate, there is another thread in the server to analyze the message in order to get useful information.

3.4.1 Main interface. In this interface, the required information must be filled before submission; There is another function that if the message is too long, a message to warn would appear; Which is useful to manage the guestbook system, and easy to maintain;

3.4.2 View interface, viewing message usually retrieve relative information from message database, which will involves the interaction with database; substantially the part of the information is completed by server, the interface only show the satisfied message.

3.4.3 Message table, the table, used to keep messages, which have five fields: the total length, name, time, content and e-mail. Data type and length are also defined respectively.

3.5   *Teaching systems*

Teaching systems using web-based asynchronous teaching, it is mainly achieved by Web browser. Teaching content is kept independently on the web server in the form of HTML; User could access the server through browser, choose their own time and content of learning; The name of page is the course, chapter, number of knowledge point; learning progress and current status are managed by Agent.

3.6   *Online practice and testing systems*

3.6.1   *The architecture of online practice and testing system*

In order to combine interactive practice and tests with the presentation of the teaching content closely, and make users examine how they learn in the process of learning, make practice and testing and teaching content divided based on function and easy to maintain, we adopt the following resolutions:

3.6.1.1 Exercise and test questions are stored in a dedicated table, which includes the content, answers, pictures, comments. At the same time there is a Web page index field is used to associate the teaching content.

3.6.1.2 Separate the function of interactive practice ,test and evaluation from teaching content display and make it to be a independent practice and testing subsystems, which have three functions: First, the maintenance of practice and the test questions, including Add, delete, modify, etc; Second, interactive practice and testing, show practice and test questions to the learner, accept information the learner enter and give the response; Third results of practice and test, including analysis and assess on what users learn.

3.6.1.3 There is a button on the Web page to enter into the practice and test subsystems from teaching

Figure 3.   Three-tier architecture of online practice and test system.

content sub-system; learners can enter the practice and test subsystems, practice and test what they learn on the Web page. According to the results of the feedback, learners can decide the next Web page to learn or continue to review the Web page.

According to the J2EE specification, online practice and test system is realized by the three-tier architecture shown in Figure 3.

The second layer is the application server, which complete all logic relevant of practice and test; The Web server is an important part of the layer, which starts the process to complete the transaction according to the client's request, and sent the results to the client browser in the form of HTML and show to users; the system uses Tomcat as the Web server. EJB Container is EJB component container, used to complete business logic and rules and the connection to the database, interactive practice and test is managed by learning Agent currently.

3.6.2   *Realization of functional modules*

Online practice test system consists of three functional modules, the practice and test module for online interactive practice and test, questions database module used to do insert, delete, update operation on questions database, statistical module for in summary of the results of learning and test. Practice and test question are kept in the SQL server, different types of practice and test questions are stored in different tables.

3.7   *Data security and maintenance of the system*

The system maintains the data security through account and role; only user logins the system successfully and has a corresponding role, then user have right to accomplish the task; Meanwhile, both account and password information through data encryption are stored in database according to different role. When user have data update permission, the data related to the user will be stored in database; the user could do work through the user login authentication before login.

4   CONCLUSIONS

This paper introduces the advantage of Mobile Agent technology, provides a framework of a Mobile Agent-based distance education system. Compared with other distance education mode, the system highlights the interaction in the process of distance learning; it is more flexible, accurate, timely, targeted to achieve the most urgent needs of the distance learners – interactive learning mode. However, in order to achieve better collaboration and interoperability between systems, there is some work to further improve the system security, system architecture standards and how to use Java-related technologies to achieve and improve the function of the platform and other issues.

REFERENCES

Chaoyang Qu etc. The accomplishment of remote education system based on mobile Agent. *Computer Engineering and Science* 2004, 26(12):26–29.
Wenlong Zhao etc. The conceptual model and application technology of Agent. *Computer Engineering and Science* 2000, 22(6):75–79.

Information Technology and Computer Application Engineering – Liu, Sung & Yao (Eds)
© 2014 Taylor & Francis Group, London, ISBN 978-1-138-00079-7

Crank train multi-body dynamic and finite element analyses for one-cylinder engine

Yong Wang
Chong Qing College of Electronic Engineering, Department of Automotive Engineering, Chong Qing, China

Jing Wang
Chong Qing Technology and Business Institute, Department of Automotive Engineering, Chong Qing, China

ABSTRACT: Multi-Body Dynamic (MBD) and Finite Element Analysis (FEA) of One-Cylinder Engine Crank Trains have been completed. MBD analysis of it provided the loads for subsequent FEA. The purpose of One-Cylinder Engine Crank Trains FEA study has been to establish stress upper limit targets for new engine crankshaft and connecting rod design.

Keywords: Engine; Crank Train; Multi-Body Dynamic; Finite Element

1 INTRODUCTION

Crank and connecting rod mechanism is an important part of an internal combustion engine, that it to do high-speed reciprocates under the action of high temperature and high pressure, the force is very complicated, In addition, its geometry, material properties, weight are different, therefore, the kinetic analysis has always been the important and difficult for engine design [1,2]. Study on the kinematics and dynamics of the crank trains, it can built a clear understanding of motion, in order to better product design and performance analysis.

In this Paper, the multi-body dynamic system is taken as the theoretical basis. Based on dynamics simulation analysis software, the process of engine crank trains dynamics simulation analysis is built. And the commonly procedure and theory of dynamics analysis was also described. The dynamical results of the crank and connection mechanism of a practical engine are obtained by means of multi -body system simulation. The results provide the optimized design of the mechanism itself and as a boundary condition of the connecting rod finite element analysis, to provide a basis for the follow design and development.

2 CRANK TRAIN MULTI BODY DYNAMIC (MBD) ANALYSIS

In this Paper, the multi-body dynamic system is taken as the theoretical basis. In order to obtain dynamic characteristics of engine, both rigid and flexible body dynamics simulation is done using software ADAMS, UG and ANSYS [3].

2.1 Crank train CAD modeling

It was built in UG CAD software package; crankshaft and connecting rod (Con-Rod) have been modeled with attention to detail for reliable prediction of stress. Piston and Flywheel have been modeled for accurate representation of inertia properties. Main bearings internal races and gears were modeled with simplified shapes. The crank-connecting rod mechanism of engine is shown in Figure 1.

2.2 Crank train MBD modeling

The dynamics analysis of crank train is using the mechanical system simulation software ADAMS, which uses Lagrange equation to establish a system

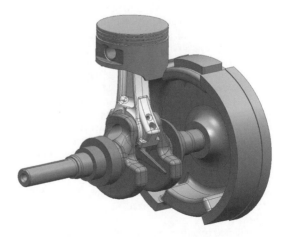

Figure 1. One-Cylinder Engine Crank Train Model.

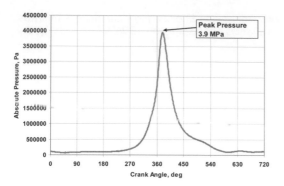

Figure 2. Engine Cylinder Pressure.

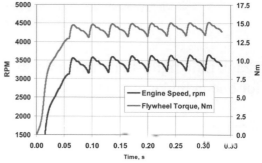

Figure 3. Predicted Rotational Speed and Flywheel Torque.

of differential equations of motion. Firstly, it selects the appropriate generalized coordinates to describe the object: For rigid body, using Cartesian coordinates where the center of mass in the inertial frame of reference and Euler angles which reflects the rigid body orientation as generalized coordinates [4].

$$\xi = \{x, y, z, \psi, \phi, \theta\}^T \qquad (1)$$

Where, translational coordinates (x, y, z) represents the center of mass, Euler angles (ψ, ϕ, θ) used to determine its orientation.

Body equations of motion based on Lagrangian differential equations is this form as follows:

$$Q = M\ddot{\xi} + \dot{M}\dot{\xi} - \frac{1}{2}[\frac{\partial M}{\partial \xi}\dot{\xi}]^T\dot{\xi} + K\xi + f_g + D\dot{\xi} + [\frac{\partial \psi}{\partial \xi}]^T\lambda \ (1)$$

Where, $\xi, \dot{\xi}$ is the generalized coordinates for the flexible body and its time derivative, M, $\dot{M}$ is the mass matrix of flexible body and its time derivative, $\frac{\partial M}{\partial \xi}$ is the mass matrix of flexible body generalized coordinates of the partial derivatives.

2.3 Dynamic simulation and result

Simulated is using ADAMS MBD software package. Fully transient dynamic excitation to capture variations in rotational speed. Loads include: Cylinder pressure force as a function of crank angle (from measured pressure trace on a 4 cylinder 4 stroke engine at 3500 rpm and 106 Pa BMEP, see Figure 2); Gauge pressure data was used to calculate the force applied to the specific piston; Flywheel starting torque (non zero for the first two 360° cycles only) to set the model into motion; Flywheel resistance torque proportional to the crankshaft rotational speed to emulate the brake load; The resistance torque was adjusted so that the crankshaft's rotational speed settles down to about 3400 rpm in mean value; The corresponding flywheel torque mean value is about 15 Nm. MBD Model results are summarized figure 3, figure 4 and figure 5.

Predicted rotational Speed and Flywheel Torque by crank train MBD model is shown in figure 3. Using

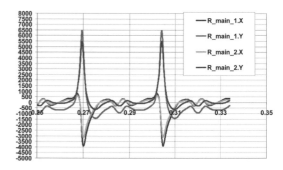

Figure 4. Predicted Con-Rod Big End and Small End Reaction Forces in Con-Rod coordinate system.

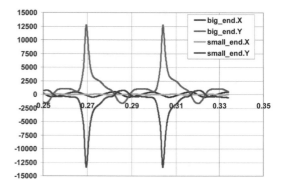

Figure 5. Predicted Main Bearing Reaction Forces in Crankshaft coordinate system.

the crank train dynamic simulation model, the author sets the crankshaft speed as 3500 rpm, and steady-state simulate it with 0.35 seconds. Predicted variation in speed is about 380 rpm peak-to-peak. The speed fluctuation of crankshaft is normal, because of the inertance in this system.

The Con-Rod big end and small end loads of one-cylinder engine are shown in Fig. 4. When the cylinder is working under compression and expansion stroke, the piston pin load is downward, which is described negative in the figure; when the piston is near TDC of the exhaust stroke, the gas force is smaller and the piston reciprocating inertia force became larger, so the piston pin load is upward, which is described positive

Table 1. Con-Rod FEA load cases.

	Big end FX(N)	Big end FY(N)	Small end FX(N)	Small end FY(N)
CASE 1: Max Compression	−137	12,305	−31	−13,633
CASE 2: Max Tension	−55	−1355	−21	675

in the figure. The peak load of the piston pin (Con-Rod small end) is found at the time of the outbreak of the cylinder, the maximum load is X: 13,633 N, Y: 31 N, less than the maximum breakout gas force 174010 N. The maximum load for Con-Rod big end is X: 12,305 N, Y: 137 N. The normal force of piston is X and the side thrust of piston is Y.

Predicted main bearing reaction forces in crankshaft coordinate system are shown in Fig. 5. Through the combined influence of piston reciprocating inertia force and crankshaft rotary centrifugal force, the maximum load of crank pin is 5500 N, which is not only less than the maximum breakout gas force but also less than the maximum load of piston pin. Therefore, the result shows that applied the maximum breakout gas force to the crank pin during strength calculation of the crankshaft are reasonable.

3 CRANK TRAIN FINITE ELEMENT ANALYSIS

3.1 Introduction

ANSYS FEA software was used to complete the analyses. Using the three-dimensional modeling software to establish the full-scale model of the crankshaft, and then output the model to the finite element analysis software. Defining the material properties, meshing and doing other pre-treatment work, the model using tetrahedron element, is divided into 26530 nodes, 19397 elements.

3.2 Connecting rod

The load cases summarized in Table 1 were derived from the data shown in Figure 4. The loads are shown in the local coordinate system moving together with the Con-Rod.

The external loads were applied to the connecting rod model as bearing loads on the piston pin (Small End) and Big End cylindrical surfaces. Inertia Relief effects were turned on to balance the external forces. Weak spring models were used for additional support of the Con-Rod model for numerical stability. The resulting equivalent stress (Von Misses) contour plots are shown in figure 6 and figure 7 and summarized in table 2.

Figure 6 shows the maximum predicted alternating stress of connecting rod is 92 MPa for the compression load case. This is higher than typical conservative

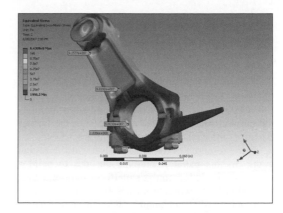

Figure 6. CASE 1: Predicted Von Misses Stress.

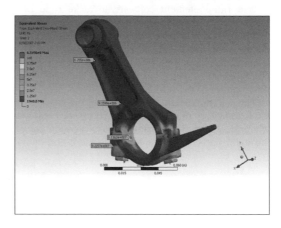

Figure 7. CASE 2: Predicted Von Misses Stress.

Table 2. Max Predicted Equivalent (Von Mises) Stress Summary.

	Assembly stress, MPa	Alternating Stress, MPa
Case 1, Compression	92	91
Case 2, Tension	87	9

target 75 MPa for cast Aluminium material. Therefore, the result applied the maximum breakout gas force to the crank pin during strength calculation of the connecting rod is unreasonable. It need to improve its design.

From the data of the simulation results, the stress of the crank and connecting rod mechanism in a cycle is complex. The size and the direction of each component force changes frequently, therefore, when calculating the strength and analysing the fatigue of the main components of the mechanism such as the crankshaft, connecting rod, piston, we need to take into account the cylinder gas force, piston reciprocating inertia force, crankshaft rotation inertia force, side thrust of piston and other alternating loads [5].

4 CONCLUSIONS

The virtual prototyping parametric model is imported into ADAMS software by UG module interface, and the characteristic of movements and force of the main parts such as piston, crank, connecting rod are obtained. The results show that the simulation results are accordant with the true working state of internal combustion engine, the references for the further design of internal combustion engine are offered.

To the research of dynamic analysis for the engine crank, in this paper it presents the idea to build a dynamic simulation analysis platform and build platform internal common processes. It provides a unified approach for different engine crank train, and also provides a reference to develop the modern engine digital simulation platform.

Use finite element software to make rigid discrete entities and the rigid body dynamics models including the crankshaft, connecting rod group, the piston group and the flywheel, dynamic model of the crankshaft, connecting rod set, piston set, flywheel and other parts are analyzed to get the loads of crank train. it is superior to the traditional method.

Through the dynamic simulation of the crank train the load history of all the constraints can be obtained, which describes the dynamic boundary conditions of the main parts in the engine cycle more accurately.

The result can be used to predicte strength and the fatigue life.

REFERENCES

[1] Wu xiugeng, Zhen Bailin. Multi-flexible body dynamics simulation and fatigue analysis on diesel engine crankshaft [J]. Computer aided engineering, Vol. 16(02), p. 1.

[2] Zhang Xiaoming, Wei Dechao, Guo Xiaojie. Dynamics Simulation Analysis of Crank Connecting Rod Mechanism of Multi-cylinder Internal Combustion Engine based on ADAMS [J]. Journal of Mechanical Transmission, Vol. 36(12), p. 95.

[3] Yimin shao, Lianghua wan. Simulation on Vibration of Engine Crank-Connecting Rod Mechanism with Manufacturing Errors Using ADAMS. Applied Mechanics and Materials Vols. 34–35 (2010) pp. 1088–1091.

[4] Wu Nan, Liaori dong. Multi Body System Dynamics Analysis of the Crank and Connecting Rod Mechanism in Diesel Engines. Chinese Internal Combustion Engine Engineering Vol. 26 (5), p. 69.

[5] Qingguo Luo, Xudong Wang. Dynamics Simulation of the Crank and Connecting Rod Mechanism of Diesel Engine. Advanced Materials Research Vols. 354–355 (2012) pp. 438–441.

Information Technology and Computer Application Engineering – Liu, Sung & Yao (Eds)
© 2014 Taylor & Francis Group, London, ISBN 978-1-138-00079-7

An invariance principle of SLLN for G-quadratic variational process under capacities

X.Y. Chen

Graduate Department of Financial Engineering, Ajou University, Suwon, Gyeonggi-do, South Korea

ABSTRACT: In this paper, under a pair of capacities (V, v), we will investigate an invariance principle of strong law of large numbers for G-quadratic variational process $\{\langle B \rangle_t, t \geq 0\}$, which was introduced by Peng in 2005 (see Peng 2007a) in a framework of G-expectation. We prove the cluster set of limit points of the sequence of stochastic processes $\eta_n(t) := \frac{\langle B \rangle_{nt}}{n}, t \in [0, 1]$, denoted by $C(\eta_n)$, is a subset of a compact space K of absolutely continuous functions on $[0, 1]$ with initial value 0 and first derivatives between sub-mean $-E_G[-\langle B \rangle_1]$ and super-mean $E_G[\langle B \rangle_1]$ with probability 1 under the lower probability v. On the other hand, each element of K lies in the set $C(\eta_n)$ with probability 1 under the upper probability V. Finally, we also extend our results to more general processes.

Keywords: Invariance principle; capacity; G-quadratic variational process; stationary and independent increments

1 INTRODUCTION

Recently, the law of large numbers (SLLN for short) and the central limit theorem (CLT for short) were given under sublinear expectations by Peng in 2007 (see Peng 2007a,b, 2008–2010). Later, Chen (2010) investigated a strong law of large numbers (SLLN for short) under capacities generated by a sublinear expectation. He showed that for a sequence of independent and identically distributed (IID for short) random variables $\{X_n\}_{n=1}^\infty$ under a sublinear expectation $\tilde{E}$, if $\tilde{E}[|X_1|^{1+\alpha}] < \infty$ for some $\alpha \in (0, 1)$, and there exist two real constants $\mu_1 \leq \mu_2$ such that $-\tilde{E}[-X_1] = \mu_1$ and $\tilde{E}[X_1] = \mu_2$, then we have

$$v(\mu_1 \leq \liminf \tfrac{S_n}{n} \leq \limsup \tfrac{S_n}{n} \leq \mu_2) = 1,$$

where $v(A) = -\tilde{E}[-I_A], \forall A \in F$, is a lower probability on a measurable space (Ω, F), $S_n = \sum_{i=1}^n X_i$, and if we further assume the upper probability $V(A) = \tilde{E}[I_A]$, $\forall A \in F$, is continuous, then for any $x \in [\mu_1, \mu_2]$, we have $V(x \in C(\frac{S_n}{n})) = 1$, where $C(x_n)$ is the cluster set of limit points of a real sequence $\{x_n\}_{n=1}^\infty$. For more results of strong law of large numbers under capacities one can refer to Marinacci (1999), Maccheroni & Marinacci (2005), Chen & Wu (2011), Hu (2012), Chen (2012) and the references therein.

In 2010 Zengjing Chen conjectured that when $\mu_1 = -1$ and $\mu_2 = 1$, for a sequence of stochastic processes

$$\xi_n(t) = \frac{1}{n}[(1 + nt - [nt])S_{[nt]} + (nt - [nt])S_{[nt]+1}], t \in [0,1], \quad (1)$$

which is generated by linearly interpolating $\frac{S_i}{n}$ at $\frac{i}{n}$ for $1 \leq i \leq n, n \geq 1$, the cluster set of $C(\xi_n)$ of limit points of this sequence of stochastic processes would be a subset of some compact space D of absolutely continuous functions f on $[0,1]$ satisfying $f(0) = 0$ and

$$\int_0^1 |\tfrac{df}{dt}|^{1+\alpha}\, dt \leq M_\alpha$$

with probability 1 under v, that is, $v(C(\xi_n) \subseteq D) = 1$, where $M_\alpha = \tilde{E}[|X_1|^{1+\alpha}]$. Conversely, any $x \in D$ would belong to the set $C(\xi_n)$ with probability 1 under V, that is, $V(x \in C(\xi_n)) = 1$.

Motivated by uncertainty in finance and statistics, Peng (2007a) introduced a new kind of normal distribution and Brownian motion $(B_t)_{t \geq 0}$ under a sublinear expectation E_G in 2005 (see Peng 2007a, 2009–2010). They are called G-normal distribution and G-Brownian motion, respectively. The expectation E_G is called G-expectation. He also introduced a G-quadratic variational process $(\langle B \rangle_t)_{t \geq 0}$ of the G-Brownian motion. If the G-Brownian motion is nontrivial (it has variance uncertainty, $E_G[B_1^2] \neq -E_G[-B_1^2]$), then $(\langle B \rangle_t)_{t \geq 0}$ is no longer a deterministic function. This is different from probability framework. But it is still increasing with initial value 0 and with stationary and independent increments.

In this paper, we consider the following sequence of stochastic processes

$$\eta_n(t) := \frac{\langle B \rangle_{nt}}{n} = \frac{\sum_{i=1}^n (\langle B \rangle_{it} - \langle B \rangle_{(i-1)t})}{n}, \forall n \geq 1, t \in [0,1], \quad (2)$$

where $(\langle B \rangle_t)_{t \geq 0}$ is nontrivial. η_n has a little different form from ξ_n given in Equation (1). We will prove that the cluster set $C(\eta_n)$ of limit points of η_n will be a subset of K (which is different from D, see Section 3) with probability 1 under ν, a lower probability generated by $\tilde{E}_G$ which is an upper expectation with the same expectation as E_G for each random variable $X \in L_G^1(\Omega)$ (see Section 2.3). On the other hand, each element of K will lie in the set $C(\eta_n)$ with probability 1 under a conjugate capacity V (upper probability) of ν. We also extend these results to more general processes with independent but not necessarily identically distributed increments, in these cases, the corresponding function set $\tilde{K}$ (see Section 4) is strictly contained in D if $\mu_1 = -1$ and $\mu_2 = 1$.

Finally, this paper is organized as follows. Section 2 recalls some basic concepts and useful lemmas. Section 3 states the main results and gives their proofs. In Section 4 we extend our results to more general cases.

2 PRELIMINARIES

In this section we will recall some basic concepts and useful results (see Peng 2007a, 2009–2010, Chen & Wu 2011 for more details).

2.1 Sublinear expectation space

Let Ω be a given sample space. F denotes a Borel σ-algebra of Ω. H denotes a linear space of random variables on (Ω, F) such that for any $n \in N$, if $X_i \in H, 1 \leq i \leq n$, then for all $\varphi \in C_{l,Lip}(R^n)$, $\varphi(X_1, \cdots, X_n) \in H$, where $C_{l,Lip}(R^n)$ is a linear space of all real-valued functions φ on R^n satisfying

$$|\varphi(x) - \varphi(y)| \leq C_\varphi(1 + |x|^m + |y|^m)|x - y|, \forall x, y \in R^n,$$

where $C_\varphi \geq 0$ and $m \in N$ are two constants only depending on φ, $|\cdot|$ denotes the Euclidean norm.

We call a functional $\tilde{E}: H \to R$ sublinear expectation, if it satisfies for all $X, Y \in H$,

(i) monotonicity: $\tilde{E}[X] \leq \tilde{E}[Y]$, if $X \leq Y$.
(ii) constant preserving: $\tilde{E}[c] = c, \forall c \in R$.
(iii) sub-additivity: $\tilde{E}[X + Y] \leq \tilde{E}[X] + \tilde{E}[Y]$.
(iv) positive homogeneity: $\tilde{E}[\lambda X] = \lambda \tilde{E}[X], \forall \lambda \geq 0$.

The triple $(\Omega, H, \tilde{E})$ is called a sublinear expectation space. For an n-dimensional random vector $X = (X_1, X_2, \ldots, X_n)$ on this space where $X_i \in H$, for each $1 \leq i \leq n$, we denote $X \in H^n$. For $X, Y \in H^n$, if for all $\varphi \in C_{l,Lip}(R^n)$, $\tilde{E}[\varphi(X)] = \tilde{E}[\varphi(Y)]$, then we call X and Y are identically distributed, denoted by $X \cong Y$. For $Y \in H^m$ and $X \in H^n$, if

$$\tilde{E}[\varphi(X, Y)] = \tilde{E}[\tilde{E}[\varphi(x, Y)]_{x=X}], \forall \varphi \in C_{l,Lip}(R^{n+m}),$$

where $\tilde{E}[\varphi(x, Y)]_{x=X} \in H$, we call Y is independent from X.

For $X \in H$, if there exist real constants μ_1 and μ_2 such that for any real continuous function φ on R we have

$$\tilde{E}[\varphi(X)] = \sup_{\mu_1 \leq \mu \leq \mu_2} \varphi(\mu)$$

then we call X is maximally distributed with real parameters μ_1 and μ_2.

If $X \in H$ satisfies the following conditions

(i) $aX + b\bar{X} \cong \sqrt{a^2 + b^2} X, \forall a, b \geq 0,$

(ii) $G(a) = \tilde{E}[aX^2], a \in R,$

where $\bar{X}$ is an independent copy of X, $G(a) := 1/2(a^+ - \sigma^2 a^-), \forall a \in R, \sigma \in [0, 1]$ is a constant, then we call X is G-normally distributed.

2.2 G-expectation and G-quadratic variational process

Let $\Omega = C_0(R_+)$ denote the space of all real-valued continuous paths $\{\omega_t, t \in [0, \infty)\}$ with $\omega_0 = 0$, equipped with the distance

$$\rho(\omega^1, \omega^2) := \sum_{i=1}^{\infty} 2^{-i}[(\max_{r \in [0, i]} |\omega_r^1 - \omega_r^2|) \wedge 1], \forall \omega^1, \omega^2 \in \Omega.$$

In the sequel all the arguments are based on this sample space.

For each fixed $T \geq 0$, we set

$$H_T^0 = \{\varphi(B_{t_1}, \cdots, B_{t_n}) : 0 \leq t_1, \cdots, t_n \leq T, \varphi \in C_{l,Lip}(R^n),$$
$$n \in N\},$$

where $B_t(\omega) = \omega_t, \forall \omega \in \Omega$, is the canonical process. It is obvious that for any $0 \leq s \leq t < \infty$, $H_s^0 \subseteq H_t^0$. We denote $H^0 := \bigcup_{n=1}^{\infty} H_n^0$.

Then an expectation $E_G: H^0 \to R$ is called a G-expectation if for any $X \in H^0$ with the form $X = \varphi(Y)$ for some $n \in N$, $\varphi \in C_{l,Lip}(R^n)$ and $0 = t_0 < t_1 < \cdots < t_n < \infty$, where

$$Y = (B_{t_1} - B_{t_0}, B_{t_2} - B_{t_1}, \cdots, B_{t_n} - B_{t_{n-1}}),$$

we have

$$E_G[X] = \tilde{E}[\varphi(\sqrt{t_1 - t_0} Z_1, \sqrt{t_2 - t_1} Z_2, \cdots, \sqrt{t_n - t_{n-1}} Z_n)],$$

where for $1 \leq i < n$, Z_{i+1} is independent from $(Z_1, \ldots, Z_i)$ and identically distributed with Z_1 on a sublinear expectation space $(\tilde{\Omega}, \tilde{H}, \tilde{E})$, Z_1 is G-normally distributed.

The canonical process $\{B_t, t \geq 0\}$ is called a G-Brownian motion on (Ω, H^0, E_G).

For any $p \geq 1$, we can construct the complete space $L_G^p(\Omega)$ (resp. $L_G^p(\Omega_t)$) of H^0 (resp. H_t^0) under the norm

$$\|X\|_p := E_G[|X|^p]^{1/p}.$$

G-expectation E_G can be continuously extended into the complete spaces.

114

We say $\{X_t, t \geq 0\}$ is a stochastic process on $(\Omega, L_G^1(\Omega), E_G)$ if for any fixed $t \geq 0$, $X_t \in L_G^1(\Omega_t)$. If for any $t_n > t_{n-1} > \cdots > t \geq 0$, and $n \in \mathbb{N}$ we have

$$(X_{t_n} - X_{t_{n-1}}, \cdots, X_{t_1} - X_t) \cong (X_{t_n - t_{n-1}}, \cdots, X_{t_1 - t})$$

under E_G, then the stochastic process $\{X_t, t \geq 0\}$ is said to have stationary increments. If for any $n, m \in \mathbb{N}$ and $t_n > t_{n-1} > \cdots > t \geq s > s_1 > \cdots > s_m \geq 0$, we have $(X(t_n) - X(t_{n-1}), \ldots, X(t_1) - X(t))$ is independent from $(X(s), X(s_1), \ldots, X(s_m))$, then the stochastic process $\{X_t, t \geq 0\}$ is said to have independent increments.

A quadratic variational process $\langle B \rangle_t$ of G-Brownian motion B_t is a stochastic process which is continuous in (t, ω) and increasing in t for every $\omega \in \Omega$ and satisfies

$$\langle B \rangle_t = B_t^2 - \int_0^t 2B_s dB_s, \quad \text{in } L_G^1(\Omega_t), t \geq 0,$$

where the above integral w.r.t. B_t is G-Ito integral (see Peng 2010 for details). Here $\langle B \rangle_t$ has finite variation, independent and stationary increments. Thus $d\langle B \rangle_t / dt$ exists a.e. t for every $\omega \in \Omega$. Furthermore, from Xu (2010) $\langle B \rangle_t$ and $d\langle B \rangle_t / dt$ are both maximally distributed under E_G with parameters $\sigma^2 t$ and t and parameters σ^2 and 1, respectively, for a.e. $t > 0$.

2.3 Capacities generated by G-expectation

From Denis et al. (2011) there exists a weakly compact set of countable additive probabilities Θ such that $\tilde{E}_G[X] = \tilde{E}_G[X], \forall X \in L_G^1(\Omega)$, where $\tilde{E}_G[X] := \sup_{P \in \Theta} E_P[X]$, $\forall X \in \sigma(\Omega)$, is a sublinear expectation. Then we can introduce a pair of conjugate capacities (V, v) by $V(A) = \sup_{P \in \Theta} P(A)$ and $v(A) = \inf_{P \in \Theta} P(A)$ for any $A \in \sigma(\Omega)$. Since Ω is a complete separable metric space and Θ is weakly compact, (V, v) also has the following useful properties (see Chen & Wu 2011 for details).

PROPOSITION 2.1 For any events A_n and A in $\sigma(\Omega)$, $n \in \mathbb{N}$,

(1) $V(\sum_{n=1}^{\infty} A_n) \leq \sum_{n=1}^{\infty} V(A_n)$.
(2) V is left-continuous, i.e. if $A_n \uparrow A$, then $V(A_n) \uparrow V(A)$.
(3) V is right-continuous for closed sets, i.e. if A_n are all closed and $A_n \downarrow A$, then $V(A_n) \downarrow V(A)$.
(4) v is right-continuous, i.e. if $A_n \downarrow A$, then $v(A_n) \downarrow v(A)$.
(5) v is left-continuous for open sets, i.e. if A_n are all open and $A_n \uparrow A$, then $v(A_n) \uparrow v(A)$.

V (resp. v) is not always right-continuous (resp. left-continuous) for open (resp. closed) sets on $\sigma(\Omega)$ (see Xu & Zhang 2010).

In this paper, for convenience we say a property holds quasi-surely (q.s. for short) if the event A that the property does not hold satisfies $V(A) = 0$ (resp. $V(A^c) = 1$, A^c denotes the complementary set of A).

3 INVARIANCE PRINCIPLE FOR G-QUADRATIC VARIATIONAL PROCESS

Let $C[0,1]$ be the linear space of all real-valued continuous functions on $[0,1]$ with the supremum norm $\|\cdot\|$. $K \subseteq C[0,1]$ is a linear space of absolutely continuous functions such that for any $x \in K$, $x(0) = 0$ and $\sigma^2 \leq x'(t) \leq 1$, a.e. $t \in [0,1]$, where x' is the first derivative of x.

PROPOSITION 3.1 K is compact.

Proof. Obviously K is bounded. Noticing that for any $0 \leq a \leq b \leq 1$ and any $x \in K$, we have $|x(a) - x(b)| = |\int_a^b x'(t)dt| \leq b - a$, thus K is equi-continuous. By Arzela-Ascoli theorem K is relatively compact. We only need to prove K is also closed. For any sequence $\{x_n\}_{n=1}^{\infty}$ in K, if it is convergent, then it has a limit $x \in C[0,1]$. Since x_n' is bounded a.e. on $[0,1]$, by Lebesgue dominated convergence theorem it follows that

$$x(t) = \lim_{n \to \infty} \int_0^t x_n'(s)ds = \int_0^t \lim_{n \to \infty} x_n'(s)ds, \forall t \in [0,1].$$

Thus x' a.e. exists and $x'(t) = \lim_{n \to \infty} x_n'(t)$, a.e. $t \in [0,1]$. That is, x is absolutely continuous. It is obvious that $x(0) = 0$ and $\sigma^2 < x'(t) \leq 1$, a.e. $t \in [0,1]$. Hence $x \in K$. The proof is complete. □

In the following we will give some useful lemmas.

LEMMA 3.2 For any $T > 0, X_t \leq 0$ (resp. $X_t \geq 0$) a.e. $t \in [0,T]$ quasi-surely is equivalent to $\tilde{E}_G[\int_0^T X_t^+ dt] = 0$ (resp. $\tilde{E}_G[\int_0^T X_t^- dt] = 0$). (The proof is obvious.)

LEMMA 3.3 For any $T > 0, \langle B \rangle_t \in [\sigma^2 t, t]$ and $d\langle B \rangle_t / dt \in [\sigma^2, 1]$ a.e. $t \in [0,T]$ quasi-surely.

Proof. Since

$$\tilde{E}_G[\int_0^T (\langle B \rangle_t - t)^+ dt] \leq \int_0^T \tilde{E}_G[(\langle B \rangle_t - t)^+] dt = 0,$$

then by LEMMA 3.2 we have $\langle B \rangle_t \leq t$ a.e. $t \in [0,T]$ quasi-surely. $\langle B \rangle_t \geq \sigma^2 t$ a.e. $t \in [0,T]$ quasi-surely can be similarly proved. Since $d\langle B \rangle_t / dt$ is maximally distributed with parameters σ^2 and 1, thus its desired result can be similarly proved as $\langle B \rangle_t$.

LEMMA 3.4 For any $\sigma^2 \leq a < b \leq 1$, we have for any $m \in \mathbb{N}$ and $0 \leq t_1 < t_2 < \cdots < t_m < \infty$,

$$V(\langle B \rangle_{t_{i+1}} - \langle B \rangle_{t_i} \in [a(t_{i+1} - t_i), b(t_{i+1} - t_i)], 1 \leq i \leq m) = 1.$$

Proof. Let $X_i(t_{i+1} - t_i) = \langle B \rangle(t_{i+1}) - \langle B \rangle(t_i)$ for $1 \leq i \leq m$, then it is easy to see that X_{i+1} is independent from $(X_1, X_2, \ldots, X_i)$ under E_G. It follows that

$$\prod_{i=1}^m E_G[g_i(X_i)] \leq V(a \leq X_i \leq b, 1 \leq i \leq m)$$
$$\leq \prod_{i=1}^m E_G[f_i(X_i)],$$

where $f_i(x)$ is a bounded and continuous function on R taking value 1 on interval $[a,b]$ and value 0 on intervals $(-\infty, a-\delta] \cup [b+\delta, +\infty)$, $g_i(x)$ is also a bounded continuous function on R taking value 1 on interval $[a+\delta, b-\delta]$ and value 0 on intervals $(-\infty, a] \cup [b, +\infty)$, $0 < \delta \leq b - a/2$, $1 \leq i \leq m$. By the distribution of X_i, we can obtain

$$E_G[\phi(X_i)] = \sup_{\sigma^2 \leq y \leq 1} \phi(y) = 1,$$

for all $\phi = f_i, g_i, 1 \leq i \leq m$, since $[a,b]$ and $[a+\delta, b-\delta]$ are both nonempty sets and subsets of $[\sigma^2, 1]$. The proof is complete. $\square$

Now for each $n \in \mathrm{N}$ and $t \in [0,1]$, let $\eta_n(t) = \langle B \rangle_{nt}/n$. Then we have the following theorem.

THEOREM 3.5 $\{\eta_n, n \geq 1\}$ is relatively compact in $C[0,1]$ quasi-surely. Furthermore,

$$\nu(C(\eta_n) \subseteq K) = 1 \quad \text{and} \quad V(x \in C(\eta_n)) = 1, x \in K,$$

where $C(x_n)$ is the cluster set of limit points of a real sequence $\{x_n\}_{n=1}^{\infty}$.

Proof. By LEMMA 3.2 it is easy to see $\{\eta_n, n \geq 1\}$ is relatively compact in $C[0,1]$ quasi-surely and $\eta_n \in K$ quasi-surely. Since K is compact, thus $C(\eta_n) \subseteq K$ quasi-surely. The first equality is obtained. Now we prove another equality. For any $x \in K$ and $\varepsilon > 0$, noticing that

$$V(\|\eta_n - x\| \leq \varepsilon)$$
$$\geq V(\sup_{t \in [\frac{k-1}{l}, \frac{k}{l}], 1 \leq k \leq l} |\eta_n(t) - \eta_n(\tfrac{k-1}{l})| + |\eta_n(\tfrac{k-1}{l}) - x(\tfrac{k-1}{l})|$$
$$+ |x(\tfrac{k-1}{l}) - x(t)| \leq \varepsilon)$$
$$\geq V(|[\eta_n(\tfrac{k-1}{l}) - \eta_n(\tfrac{k-2}{l})] - [x(\tfrac{k-1}{l}) - x(\tfrac{k-2}{l})]| \leq \tfrac{\varepsilon}{3l},$$
$$2 \leq k \leq l),$$

where $l \geq 3/\varepsilon$ is an integer. For $n_m = l^m, m \in \mathrm{N}$, set

$$A_m = \{\|\eta_{n_m} - x\| \leq \varepsilon\}.$$

Similarly we can prove for any fixed $j \in \mathrm{N}$,

$$V(\bigcap_{m=j}^{\infty} A_m) \geq V(\bigcap_{m=j}^{\infty} \bigcap_{k=2}^{l} \{|[\langle B \rangle_{l^{m-1}(k-1)} - \langle B \rangle_{l^{m-1}(k-2)}]$$
$$- l^m [x(\tfrac{k-1}{l}) - x(\tfrac{k-2}{l})]| \leq \tfrac{l^{m-1}\varepsilon}{3}\}),$$

since V is right-continuous for closed sets (by PROPOSITION 2.1). Note that for $2 \leq k \leq l$, $x(\tfrac{k-1}{l}) - x(\tfrac{k-2}{l}) \in [\tfrac{\sigma^2}{l}, \tfrac{1}{l}]$, then by using the right-continuity for closed sets of V again, from LEMMA 3 4 we can obtain $V(\bigcap_{m=k}^{\infty} A_m) = 1, \forall k \in \mathrm{N}$.

And then by the right-continuity of ν we have

$$\nu(\bigcap_{k=1}^{\infty} \bigcup_{m=k}^{\infty} A_m^c) = \lim_{k \to \infty} \nu(\bigcup_{m=k}^{\infty} A_m^c) = 1 - \lim_{k \to \infty} V(\bigcap_{m=k}^{\infty} A_m) = 0.$$

This implies

$$\nu(\limsup_{m \to \infty} \|\eta_{n_m} - x\| > \varepsilon) = 0.$$

Then

$$V(\liminf_{m \to \infty} \|\eta_{n_m} - x\| \leq \varepsilon) \geq V(\limsup_{m \to \infty} \|\eta_{n_m} - x\| \leq \varepsilon) = 1.$$

Thus $V(\lim_{n \to \infty} \|\eta_n - x\| \leq \varepsilon) = 1$. Letting ε tend to 0, we have $V(\liminf_{n \to \infty} \|\eta_n - x\| = 0) = 1$, i.e. $V(x \in C(\eta_n)) = 1$. The whole proof is complete. $\square$

COROLLARY 3.6 For any real-valued functional φ on $C[0,1]$, the sequence of random variables $\{\varphi(\eta_n)\}_{n=1}^{\infty}$ is relatively compact quasi-surely and the set of its limit points is a subset of $\varphi(K)$ quasi-surely, i.e. $\nu(C\{\varphi(\eta_n)\} \subseteq \varphi(K)) = 1$. On the other hand, for any $y \in \varphi(K)$, $V(y \in C\{\varphi(\eta_n)\}) = 1$.

Since the discreteness of n is inessential for the above arguments, if $u > 0$, let $\eta_u(t) = \frac{\langle B \rangle_{ut}}{u}$ for any $t \in [0,1]$, then the following theorem can be similarly proved.

THEOREM 3.7 $\{\eta_u, u > 0\}$ is relatively compact quasi-surely in $C[0,1]$ and the set of its limit points $C(\eta_u)$ is a subset of K quasi-surely. Conversely, any $x \in K$ is an element of $C(\eta_u)$ with probability 1 under V.

4 EXTENSIONS

Let $\tilde{K}$ be a subset of $C[0,1]$ and satisfy that for any $x \in \tilde{K}$, x is absolutely continuous, $x(0) = 0$ and $x'(t) \in [\mu_1, \mu_2]$ a.e. $t \in [0,1]$, where $\mu_1 \leq \mu_2$ are two given real constants.

THEOREM 4.1 Suppose $\{X_t, t \geq 0\}$ is a stochastic process on $(\Omega, L_G^1(\Omega), E_G)$ with independent and stationary increments and $X_0 = 0$. We also assume $E_G[X_1] = \mu_2$, $-E_G[-X_1] = \mu_1$ and $\lim_{t \to 0} E_G[X_t^2 t^{-1}] = 0$. Set $\eta_n(t) = \frac{X_{nt}}{n}$ for all $t \in [0,1]$ and $n \in \mathrm{N}$. Then $\{\eta_n, n \geq 1\}$ is relatively compact in $C[0,1]$ quasi-surely. Furthermore, $\nu(C(\eta_n) \subseteq \tilde{K}) = 1$ and $V(x \in C(\eta_n)) = 1$, $\forall x \in \tilde{K}$.

Proof. By Proposition 5.2 of Chapter III in Peng (2010) we can easily get the proof by using the arguments of THEOREM 3.5. $\square$

Moreover, the following lemma can be easily obtained from the proof of Proposition 5.2 of Chapter III of Peng (2010).

LEMMA 4.2 Suppose $\{X_t, t \geq 0\}$ is a stochastic process on $(\Omega, L_G^1(\Omega), E_G)$ with $X_0 = 0$ and satisfies $E_G[X_t] = \mu_2 t$, $-E_G[-X_t] = \mu_1 t$ and $\lim_{t \to 0} E_G[X_t^2 t^{-1}] = 0$. Then X_t is maximally distributed with parameters $\mu_1 t$ and $\mu_2 t$ for any $t \geq 0$.

Then by this lemma we can similarly prove the following theorem as THEOREM 3.5.

THEOREM 4.3 Suppose $\{X_t, t \geq 0\}$ satisfies the conditions of LEMMA 4.2 and $\{\eta_n(t)\}$ is defined as in THEPREM 4.1. Then the results of THEOREM 4.1 still hold.

REMARK 4.4 Condition $\lim_{t\to 0} E_G[X_t^2 t^{-1}] = 0$ in THEOREM 4.3 can be reduced to $\lim_{t\to 0} E_G[X_t^{1+\alpha} t^{-1}] = 0$ for some $\alpha \in (0, 1)$. And under the conditions of THEOREM 4.3, COROLLARY 3.6 still holds.

ACKNOWLEDGEMENTS

This research was supported by WCU (World Class University) program through the National Research Foundation of Korea funded by the Ministry of Education, Science and Technology (R31-20007). We thank Zengjing Chen and Wenjuan Li for their helpful comments.

REFERENCES

Chen, X.Y. 2012. Strong law of large numbers under an upper probability. *Applied Mathematics, 3: 2056–2062.*

Chen, Z.J. 2010. Strong law of large numbers for capacities. *arXiv:1006.0749v1 [math. PR] 3 Jul 2010.*

Chen, Z.J. & Wu, P.Y. 2011. Strong law of large numbers for Bernoulli experiments under ambiguity. *Nonlinear Mathematics for Uncertainty and its Applications. Advances in Intelligent and Soft Computing. Vol. 100/2011, 19–30.*

Denis, L., Hu, M.S. & Peng, S.G. 2011. Function spaces and capacities related to a sublinear expectation: applications to G-Brownian motion paths. *Potential Anal (2011) 34: 139–161.*

Hu. F. 2012. General law of large numbers under sublinear expectations. *arXiv:1104.5296v2 [math.PR] 9 Feb 2012.*

Maccheroni, F. & Marinacci, M. 2005. A strong law of large numbers for capacities. *Annals of Probability 33 (3): 1171–1178.*

Marinacci, M. 1999. Limit law of non-additive probabilities and their frequentist interpretation. *Journal of Economic Theory 84 (2): 145–195.*

Peng, S.G. 2007a. G-expectation, G-Brownian motion and related calculus of Ito's type. *In: F.E. Benth et al. (Eds.), Stochastic Analysis and Applications: The Abel Symposium 2005,* Berlin, Springer, 2007: 541–567.

Peng, S.G. 2007b. Law of large numbers and central limit theorems under nonlinear expectations. *arXiv:mathPR/0702358v1 13 Feb 2007.*

Peng, S.G. 2008. A new central limit theorem under sublinear expectations. *arXiv:0803.2656v1 [math.PR] 18 Mar 2008.*

Peng, S.G. 2009. Survey on normal distributions, central limit theorem, Brownian motion and the related stochastic calculus under sublinear expectations. *Science in China Series A: Mathematics 52 (7): 1391–1411.*

Peng, S.G. 2010. Nonlinear expectations and stochastic calculus under uncertainty-with robust central limit theorem and G-Brownian motion. *arXiv:1002.4546v1 [math.PR] 24 Feb 2010.*

Xu, J. & Zhang, B. 2010. Martingale property and capacity under G-framework. *Electronic Journal of Probability 15 (67): 2041–2068.*

Xu, Y.H. 2012. Backward stochastic differential equations under super-linear G-expectation and associated Hamilton-Jacobi-Bellman equations. *arXiv:1009.1042v1 [math.PR] 6 Sep 2010.*

Information Technology and Computer Application Engineering – Liu, Sung & Yao (Eds)
© *2014 Taylor & Francis Group, London, ISBN 978-1-138-00079-7*

The superiority of adaptive fuzzy PID control algorithm in sintering furnace temperature control

Ya-Juan Ji & Hui-Ran Jia
Institute of Electrical Engineering, Hebei University of Science and Technology, Shijiazhuang, China

ABSTRACT: The temperature control of sintering furnace is an important part of powder metallurgy. Due to the complicated working conditions, large inertia, time-varying system and obvious lagging phenomenon, it is difficult to establish such precise mathematical model and the traditional PID control method can't meet the system requirements any more. Orienting the complexity of the temperature control of sintering furnace processes, this paper presents an adaptive fuzzy PID control system for sintering furnace temperature, which is proved superior through simulation compared with the traditional PID control.

Keywords: Sintering furnace; The PID controller; Adaptive fuzzy PID controller; The simulation

1 INTRODUCTION

Powder metallurgy, vacuum sintering furnace is a time varying, big lag and nonlinear heating system. Due to the complexity of its physical and chemical mechanism and affected by many factors, it is difficult to obtain accurate mathematical model. And it is difficult for the traditional PID control effect to be satisfactory when parameters change greatly and the control precision is highly demanded. The adaptive Fuzzy PID control does not require the system to give a precise mathematical model. The control rules are set up according to practical experience of people, thus the control decision table is made. According to the decision table to control, the system works especially well for sintering furnace which is a complex object of industrial production. The main factors of affecting product quality in the sintering process are temperature and time. Particularly, the temperature is required strictly. In this paper, adaptive fuzzy PID algorithm is adopted and simulation was carried out for the temperature.

2 THE PID CONTROLLER

PID is the general term for the proportion the integration and the differential of a deviation between the control parameters and set value.

The PID controller system schematic diagram is as follows:

According to the given value $y_d(t)$ and the actual output value $y(t)$, the control deviation is made up by PID controller which is a linear controller,

$$error(t) = y_d(t) - y(t)$$

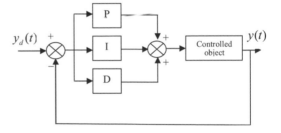

Figure 1. The PID controller system schematic diagram.

The PID control law is:

$$u(t) = k_p[error(t) + \frac{1}{T_1}\int_0^t error(t)dt + \frac{T_D derror(t)}{dt}] \quad (1)$$

Type: k_p – Scale factor T_I – Integral time constant
T_D – Differential time constant
After discretization, following is :

$$u(k) = k_p e(k) + k_i \sum_{j=0}^{k} e(j) + k_d[e(k) - e(k-1)] \quad (2)$$

Type: k – Sampling sequence number $k = 1, 2, \ldots,$
$u(k)$ – The first k times of sampling time controller output value;
$e(k)$ – The first input k times of sampling time deviation;
$e(k - 1)$ – The first input k - 1 times of sampling time deviation
k_i – Integral coefficient, $k_i = k_p \frac{T}{T_i}$
k_d – Differential coefficient $k_d = k_p \frac{T_d}{T}$

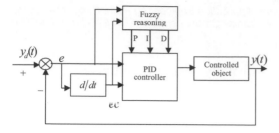

Figure 2. The adaptive fuzzy PID controller system schematic diagram.

By incremental PID control algorithm and according to the recursive principle, the following should be :

$$u(k-1) = k_p e(k-1) + k_i \sum_{j=0}^{k-1} e(j) + k_d[e(k-1)-e(k-2)] \quad (3)$$

With the Eq. (2) minus the Eq. (3), the following should be:

$$\triangle u(k) = k_p \triangle e(k) + k_i e(k) + k_d(\triangle e_k - \triangle e_{k-1}) \quad (4)$$

With the value $\triangle e_k = e_k - e_{k-1}$, Eq. (4) becomes:

$$\triangle u_k = A e_k + B e_{k-1} + C e_{k-2}$$

Among them:

$$A = k_p (1 + \frac{T}{T_i} + \frac{T_d}{T})$$

$$B = -k_p (1 + \frac{2T_d}{T})$$

$$C = k_p \frac{T_d}{T}$$

All of them are constant coefficient concerning with the sampling period, the proportion coefficient, integral time constant, and derivative time constant.

3 THE ADAPTIVE FUZZY PID CONTROLLER

The adaptive Fuzzy PID parameters self-tuning realization thought is to find out the fuzzy relationship between three parameters of PID and deviation e and the deviation change rate ec, in progress of which, the values e and ec continuously tested. According to the principle of fuzzy control, the three parameters are modified on-line then so as to meet different requirements of the values e and ec concerning to the control parameters. As a result, the controlled object has a good dynamic and static performance [1].

Adaptive fuzzy PID control structure diagram is as shown in Figure 2.

3.1 The Adaptive fuzzy PID algorithm

Based on the PID algorithm, through calculating the error e and error change rate ec in the current system,

Table 1. Fuzzy control rules table of k_p.

$e/\Delta k_p/ec$	NB	NM	NS	NO	PS	PM	PB
NB	PB	PB	PM	PM	PS	ZO	ZO
NM	PB	PB	PM	PS	PS	ZO	NS
ZS	PM	PM	PM	PS	ZO	NS	NS
ZO	PM	PM	PS	ZO	NS	NM	NM
PS	PS	PS	ZO	NS	NS	NM	NM
PM	PS	ZO	NS	NM	NM	NM	NB
PB	ZO	ZO	NM	NM	NM	NM	NB

Table 2. Fuzzy control rules table of k_i.

$e/\Delta k_i/ec$	NB	NM	NS	NO	PS	PM	PB
NB	NB	NB	NM	NM	NS	ZO	ZO
NM	NB	NB	NM	NM	NS	ZO	ZO
NS	NB	NM	NS	NS	ZO	PS	PS
ZO	NM	NM	NS	ZO	PS	PM	PM
PS	NM	NS	ZO	PS	PS	PM	PM
PM	ZO	ZO	PS	PS	PM	PB	PB
PB	ZO	ZO	PS	PM	PM	PB	PB

Table 3. Fuzzy control rules table of k_d.

$e/\Delta k_d/ec$	NB	NM	NS	NO	PS	PM	PB
NB	PS	NS	NB	NB	NB	NM	PS
NM	PS	NS	NB	NM	NM	NS	ZO
NS	ZO	NS	NM	NM	NS	NS	ZO
ZO	ZO	NS	NS	NS	NS	NS	ZO
PS	ZO	ZO	ZO	ZO	ZO	ZO	ZO
PM	PB	NS	PS	PS	PS	PS	PB
PB	PB	PM	PM	PM	PS	PS	PB

Fuzzy self-tuning PID applies fuzzy rules to conduct fuzzy reasoning, and refer to fuzzy matrix table to fulfill the parameter adjustment. The core of fuzzy controller design is summarizing engineering design technician's knowledge and practical operation experience, and establishing proper fuzzy rule table, so as to get the fuzzy control table concerning to three parameters k_p, k_i, k_d (refer to table 1~3).

The deviation and deviation vary and control the amount of k_p, k_i, k_d range blur into the theory of fuzzy sets on the domain

$E, E_C = \{-3,-2,-1,0,+1,+2,+3\}$;

$k_p = \{-0.3,-0.2,-0.1,0,+0.1,+0.2,+0.3\}$;

$k_i = \{-0,06,-0.04,-0.02,0,+0.02,+0.04,+0.06\}$;

$k_d = \{-3,-2,-1,0,+1,+2,+3\}$;

And fuzzy word set shall be defined as {NB,NM,NS,O,PS,PM,PB}. With the Fuzzy variable deviation E and deviation vary E_C and control quantity of fuzzy sets and theory of domain being determined, the membership function of fuzzy language variables shall be defined then, namely the fuzzy

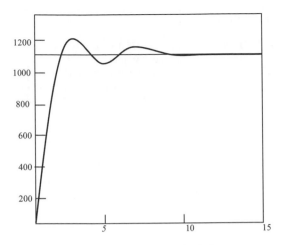

Figure 3.　The PID controller system simulation.

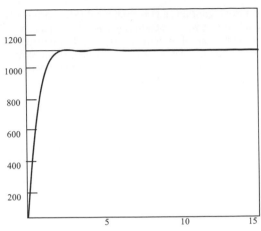

Figure 4.　The adaptive fuzzy PID controller system simulation.

variable assignment, and it is to determine element theory domain membership degree of fuzzy language variables. Membership function curve is defined as sigmoidal function and triangle functions and z function [2]. Established fuzzy control rule which is shown in the above table fig. 1–3, a total of 49 fuzzy rules:

if (E is NB) and (EC is NB) then (k_p is PB) (k_i is NB) (k_d is PS)

if (E is NM) and (EC is NM) then (k_p is PB) (k_i is NB) (k_d is NS)

$$\vdots \qquad\qquad \vdots$$

if (E is PB) and (EC is PB) then (k_p is NB) (k_i is PB) (k_d is PB)

4　THE EXPERIMENTAI SIMULATION

Temperature is defined as the controlled object, the stability of control determines the quality of the products directly. For sintering furnace temperature control system is endowed with such characteristics as nonlinear, time lag and parameter time-varying, the transfer function is:

$$G(s) = \frac{ke^{-\tau s}}{Ts + 1}$$

Hereby, it is set $k = 1.12$　$\tau = 30$　$T = 25$, the given temperature is 1100°C, the simulation result is obtained by MATLAB simulation [3] which is shown in the figure 3.

For the PID parameters $P = 18$　$I = 1.6$　$D = -10$, As is shown in the chart, system overshoot volume has reached 100°C. It is taken too long to adjust time, it is prone to shocks and cannot meet the requirement of system.

It is presented the MATLAB simulation figure of adaptive fuzzy PID control algorithm below.

As is seen from the curve diagram above, the system steady-state error is small, the adjusting time is

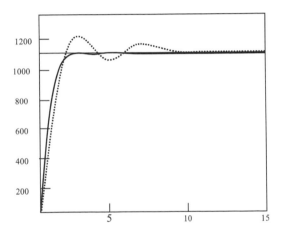

Figure 5.　The comparison.
Notes: . . . stands for the traditional PID;
— stands for the adaptive fuzzy PID.

short, overshoot volume is less than 0.5% and the steady-state error is nearly 0. Thus the requirements of performance indicators can be achieved.

A comparison is made between the traditional PID and adaptive fuzzy PID control effect.

From the simulation curve, it can be seen that compared with the traditional PID control, the fuzzy control enjoys such qualities as less overshoot, faster adjustment and more time-saving, which proves that the fuzzy PID control has better control characteristics [4].

5　CONCLUSION

In response to the nonlinear timely changing characteristics of the temperature of vacuum sintering furnace, the adaptive fuzzy PID control algorithm is put forward in this thesis. The combination of fuzzy control algorithm and the traditional PID controller is applied to powder metallurgy sintering furnace temperature control system, which also has carried on MATLAB simulation. As a consequence, the control effect is better achieved, which proves the superiority of adaptive fuzzy PID control algorithm.

ACKNOWLEDGEMENT

This paper is supported by School of electrical engineering, Hebei University of Science and Technology. Corresponding author: Hui-ran Jia Email: jiahuiran@263.net

REFERENCES

[1] Defeng Zhang. *The MATLAB fuzzy system design.* National defence industry press.

[2] Dongjin Chen. Xinhua Jiang. *Carbide sintering furnace temperature controller based on fuzzy control design Journal of engineering in fujian.* In December 2005.

[3] Baoxiong Chen. *Vacuum furnace of PID parameter adjustment method Metal heat treatment.* 2004, 29(10).

[4] Xiangqian Chu. Wu Zhu. Shengfeng. Huang *Vacuum sintering furnace fuzzy control system simulation* 2011, 48(3).

Information Technology and Computer Application Engineering – Liu, Sung & Yao (Eds)
© *2014 Taylor & Francis Group, London, ISBN 978-1-138-00079-7*

Design and implementation of the control system to ovenware furnace based on MOCVD equipment

Cong Li, Chunmei Li & Yaning Zhang
School of Electrical Engineering, University of Jinan, China

ABSTRACT: This paper put forward a design scheme of the control system for ovenware furnace based on MOCVD equipment. The control system composed field bus control system (FCS) based on industrial computer, PLC, intelligent PID instrument, digital mass flow meter etc. The whole system realize the controlling unit including temperature control of the furnace, tracking control of gas flow, cooling control, control system of recirculating air, PI control of the vacuum in the furnace. It has been found that the system operation is reliable and meets the demand of precision, and can completely replace the imported products.

Keywords: ovenware furnace; PLC; VB.NET; Temperature Instrument; recirculating air control system; vacuum system; cooling system

1 INTRODUCTION

Vacuum ovenware furnace equipment is new vacuum equipment in LED manufacturing industry which is supporting MOCCD equipment. Furnace is pumped to vacuum, processed work pieces are placed in the furnace, take dry cleaning by heating the cleaning gas (cleaning gas: N_2, H_2), which mainly used to clean the ammoniation gallium and nitrogen aluminum etc. on the MOCVD susceptor (sic coating Graphite disc or quartz disc) effectively, so we can achieve effective cleaning and improve the quality. We need equip one vacuum ovenware furnace for two or three MOCVD equipments. Our country has begun to independent research and development MOCVD equipment, and then developed ovenware furnace. Each year the demanded quantity is about 500 sets which mainly rely on the import of Taiwan (Han T. & Ge Y. 2012). Based on this situation, we design and implement a control system to ovenware furnace based on one kind of MOCVD equipment.

2 CONTROL SYSTEM OVERALL DESIGN

2.1 The working principle of ovenware furnace

The working principle of Vacuum ovenware furnace is simple; firstly open the furnace door and put the graphite covered with depositional impurity into the furnace in the studio, and then closed the furnace door, restart the vacuum system and heating system; After reach certain vacuum target value and temperature target value, start air source system, filling the gas mixed by N_2, N_2 and H_2 to the furnace according

to the corresponding requirements of this process. In the high temperature and low pressure furnace environment, after a period of time, sediment will volatile into gas totally, and pumped away through the vacuum system, and then achieves the purpose of purpose of cleaning graphite. After cleaning fished, the cooling system begins to work. The pressure in the furnace will slowly down below the set value in the cooling process, according to the principle of the expansion and contraction of gas. In order to recover pressure, the air source system will automatically supply N_2 gas into furnace until reach normal temperature and pressure. The whole process is finished until this time, open the furnace door and take out the cleaned graphite disc.

2.2 Ovenware furnace control system principle

According to the demons of semiconductor technological process, the control system composed FCS through taking PLC as the control core and industrial personal computer as human-computer interface, the communication between master control unit (PLC) and the upper computer adopt RS232 / RS485 communication. The communication between PLC and temperature instrument, vacuum gauge adopt RS485 bus (Jin Yihui et al. 1997). The overall structure of the ovenware furnace control system as shown in Figure 1.

The system includes four closed loop control as shown in Figure 1:

1. Temperature control system: The temperature in the ovenware furnace can reach several hundreds even one thousand Celsius degrees; we need multiple setting value control in order to realize heating, cooling and constant temperature control. Temperature control

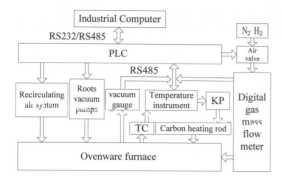

Figure 1. Ovenware furnace control system overall structure.

is the most important link of the technical control system; the cleaning directly determined by control effect. In order to meet the demand, we take closed loop control to temperature by adopting self-tuning PID high precision instrument (Bai Ruilin et al. 2005). This system is a single point temperature control whose control theme is the temperature in the ovenware furnace. The control system adopts different grade temperature instruments according to the industrial requirements; so we can choose high precision temperature instrument or adopt cascade control system if require very high control precision (Chen Zengqiang et al. 1994); We choose the thermocouple in the furnace as the input parameter of the main control loop when apply the cascade control, and add a thermocouple on the furnace wall as the input parameter of subsidiary control loop. Furnace temperature control adopting phase shift trigger, heating through voltage controlled by Silicon Controlled Rectifier (SCR). This heating form adopt graphite heating rod that is different from diffusion furnace control. The principle is: three-phase sources AC380V step down to about 90V through the transformer, and then connect with many graphite heating rods by delta connection for heating. Temperature instrument make the corresponding heating voltage change between 0 to 90V through the silicon controlled rectifier PWM template, so as to realize the temperature control of ascending, descending, thermostatic.

2. Air source control system: PLC load the actual gas flow value directly from the digital interface of the mass-flow gas meter, and take tracking control for flow output through the digital interface according to the technological requirements for gas flow, and then realize closed loop control. The gas flow control for the ovenware furnace is a very important technological requirement in production. In production, we complete aeration process mainly through taking mass flow controller (MFC) as monitoring equipment of flow. Digital mass-flow gas meter can directly communicate with PLC and industrial computer by RS485 communication protocol.

3. Cooling control system: The cooling system is consisted of gas pipeline system and recirculating air system. The pipeline system automatically supply N_2,

N_2 and H_2 for ovenware furnace under the control of PLC according to the technological requirements, then recirculating air system start, make the air circulate in the work area of the furnace and take away the heat of the furnace through the furnace wall, accordingly achieve the cooling effect.

4. Vacuum control: Vacuum control mainly adjust roots vacuum pump through vacuum gauge in order to achieve corresponding vacuum control. Achieve corresponding PI control by vacuum gauge according to the requirements (Gao Aihua et al. 2008). As a general rule, roots always select water ring pump as the backing pump which is more favorable than the others, which mainly because the water ring pump can pump out a large number of condensable steam, especially when the pump ability of the gas ballast oil seal mechanical vacuum pump for condensable steam is not enough, it is also more obvious when the vacuum system oil pollution is forbidden. The equipment is placed in the purification workshop because the solar energy equipment demands a clean environment; therefore it is suitable for using roots vacuum unit.

3 HARDWARE DESIGN

The hardware of ovenware furnace control system mainly consisted of computer control cabinet and electrical control cabinet.

Main hardware of computer control cabinet:

- Advantech IPC: CPU GHz: 2.0 GHz; memory: 500 M and above; hard disk: 80 GB and above; serial port: 2.
- PLC adopts Mitsubishi PLCFX2N-48 module.
- RKC FB400 temperature controller.
- Oerlikon leybold vacuum gauge.
- PWM heating driver module.
- The other components mainly realize the start-up and shut down, various kinds. of alarm, start-up and shut down of heating of equipment.

Main hardware of electrical control cabinet:

- Digital mass-flow gas meter CS200.
- Gas solenoid valve.
- Three-phase step-down transformer.

4 SOFTWARE DESIGN

Upper computer software written by VB.NET and Access database, adopt Combobox control to replace the Textbox control, adopt MSComm to program upper computer and PLC in order to realize Modbus communication and CRC checking routine (Chen Bin et al. 2006). Adopt the RTU verify mode of the Modbus to realize the communication between PLC and instrument. The fundamental function of the upper computer control system is achieve collection, display, control, alarm, storage, query of the data of gas flow, vacuum, temperature, the valve of the furnace tube and so on. Keep the lag time of acquisition

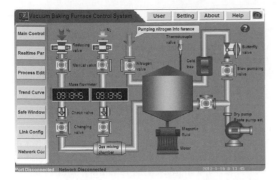

Figure 2. The upper computer main interface of the oven-ware furnace.

and control as short as possible, the history data within ten years can be queried; the curve of the historical data can be obtained. The main idea of the programming design of the control system is accomplish the data collection of multiple furnace tubes at the same time through taking control operation to multi-thread in major cycle. Create user to manage database and save data to database automatically in the initialization, achieve the initialization of various parameters and the creation of the thread. The upper computer main interface of the ovenware furnace as shown in Figure 2. The main function of the system software:

- Set the process temperature curve function.
- Data real-time display function.
- All kinds of alarming display function.
- System time sampling record function.
- Temperature curve real-time display function/.
- The switch of system control function.
- Online operation of various parameters.
- Classification user purview control system.

4.1 Initialization of parameter

The software of ovenware furnace control system not only shows the current state of the PLC, the current operation information of craft curve in the PLC and the status of each valve switch, but also includes the craft curve (Michael Quirk & Julian Serda 2009) loading and craft curve downloading to the PLC. All parameters that appeared in the ovenware furnace control system software require a certain parameter to save. When starting the software, must have the parameters initialization process. In fact, the process of parameter initialization is that when initiating the software, loading the default data in the software system. It's mainly includes the following points:

- The name of the last craft downloading to PLC.
- The lot number of the last product downloading to PLC.
- The data sampling cycle of software.
- The data preservation cycle of software.

Setting process is one of the biggest functions of the ovenware furnace. Under the control of PLC, it is necessary to complete a certain process steps. The inland

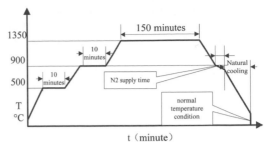

Figure 3. Temperature control process of ovenware furnace.

are doing research about ovenware furnace process. The technological process includes temperature curve, gas flow, ventilation cooling period, vacuum degree, etc. The temperature curve is shown in Figure 3.

With the reference of the imported and domestic developed ovenware furnace characteristics, the technological process is given as follows:

Firstly, understanding the function of partial pressure valve, preparation pressure valve and charging valve.

1. The function of partial pressure valve: In the process of operation, the target value set by vacuometer determines the valve's open and close. when partial pressure to inflate N_2 or the mixture of N_2, H_2 to furnace. Partial pressure condition is the process of heating and temperature is maintained at 1350 degrees, and temperature reduces from the highest temperature of 1350 degrees to 900 degrees, in the absence of up to 1350 degrees for maintaining the temperature and 900 degrees below not in use.

2. The function of preparation pressure valve: In order to open the fan, the temperature inside the furnace needs to reduce to the right temperature that can open the furnace door. The condition of preparation pressure is when the temperature downs to 900 Celsius degrees, charging from -100 to -200 mmHg.

3. The function of charging valve: Fill gas, open the door.

- Open the vacuum pump heat to 500 degrees. During this process, when the vacuum target reaches to $5*10^{-2}$ torr, open the partial pressure valve, and charge in N_2 or mixture of N_2 and H_2 for partial pressure, and when the vacuum target reaches to 3 torr, close the partial pressure valve.
- Temperature is maintained at 500 degrees for 10 minutes, the vacuum pump is open and the partial pressure valve is closed.
- When the temperature reaches to 900 degrees, Open the vacuum pump, During this process, when the vacuum target reaches to $5*10^{-2}$ torr, open the partial pressure valve, and charge in N_2 or mixture of N_2 and H_2 for partial pressure, and when the vacuum target reaches to 3 torr, close the partial pressure valve.
- Temperature is maintained at 500 degrees for 10 minutes, the vacuum pump is open and the partial pressure valve is closed.

- When the temperature reaches to 1350 degrees, Open the vacuum pump, During this process, when the vacuum target reach to $5*10^{-2}$ torr, open the partial pressure valve, charge in N_2 or mixture of N_2 and H_2 for partial pressure, when the vacuum target reaches to 3 torr, close the partial pressure valve.
- Temperature is maintained at 1350 degrees for 2 hours and 30 minutes, the vacuum pump is opening and the partial pressure valve is closed. During this process, when the vacuum target reach to $5*10^{-2}$ torr, open the partial pressure valve, charge in N_2 or mixture of N_2 and H_2 for partial pressure, when the vacuum target reaches to 3 torr, close the partial pressure valve.
- Cool to 900 degrees, close vacuum pump, open the preparation pressure valve and charge in N_2.when the Pressure value reaches to $-100 \sim -200$ mmHg, close the preparation pressure valve, Open the fan and cool down to the appropriate temperature for opening the furnace door.
- Open the charging valve to inflate N_2 until the pressure to normal, open the door.

4.2 PLC communication

The system use PC computer as the host computer and PLC as centrally-controlling machine, PLC collects data and controls equipments, and the host computer is used to complete various complicated data processing and control of PLC. Further analysis and processing of the collected data are finished by the powerful industrial PC computer. Therefore, the PC computer and the PLC have a great deal of data interchange. According to the process curve, PLC controls the temperature rising, cooling and constant process of the instrument time sharing. The PC computer communicate with PLC with ModBus and CRC mode; PLC and temperature instrument follow instrument manufacturer agreement, and commonly adopt ModBus and RTU mode.

4.3 Data processing

After getting the present temperature value from the specific address of temperature control table, PLC puts it at the designed address. The host computer control system gets this temperature value from the agreed address. After scale transformation, the collected data are showed on the corresponding interface.

4.4 Storage and display

The host computer software of ovenware furnace control system requires show the data to the user real-time and intuitively, and save the historical data in database. This control system uses ACESS database.

5 CONCLUSIONS

The ovenware furnace control system designed in full accordance with the field control bus. The fieldbus control scheme increases the reliability of this control system greatly, and reduces the system connection. The system still works normally when the industrial control machine meets fault. Reliable PLC temperature instrument and digital mass flowmeter ensure the reliable operation and the control accuracy of the system. During software design, we refine statement and do repeated tests to ensure no crash phenomenon. Practice proves that this control system runs very reliably and can replace the imported products completely, which broadens the channels for our ovenware control system localization.

REFERENCES

Bai R. L., Jiang L. F., & Wang J. 2005. The study of fuzzy self-tuning PID controller based on FPGA. *Chinese Journal of Scientific Instrument* 26(8):833–837.

Chen B., Zhang B., An C. S., & Qiu D. Y. 2006. Electric device network monitoring system based on ModBus / TCP protocol and web. *Chinese Journal of Scientific Instrument* 27(9):1062–1066.

Chen Z. Q., Che H. P., Yuan Z. Z., & Cui B. M. 1994. Predietive-PID caseade control and its applications on boiler water height system. *Control and Decision* 9(5):379–382.

Gao A. H., Liu W. G., Zhou S., & Zhang W. 2008. Design of vacuum measurement and autocontrol for LS-MOCVD. *Journal of Xi'an Technological University* 28(2): 103–106.

Han T., & Ge Y. 1980. Design and Implementation of Vacuum ovenware furnace control system. *Sciences and Wealth* (5):178–179.

Jin Y. H., Wang S. F., & Wang G. Z. 1997. Developments and prospects of process control. *Control Theory and Applications* 14(2):145–148.

Michael Q., & Julian S. 2009. *Semiconductor manufacturing technology*. Beijing: Electronic Industry Press.

Application of intellectual control treatment for grinding process based on expert knowledge base

Cong Li, Yaning Zhang & Houzhao Dai
School of Electrical Engineering, University of Jinan, China

ABSTRACT: This paper introduces a control scheme for the load in the grinding process of the cement raw mill. An expert control algorithm is proposed for the control of the grinding process of the cement raw mill. According to the three parameters including mill sound, the back powder of the separator, the power load of the bucket elevator, this algorithm has nine rules and divides the grinding process into three stages: "empty mill", "normal" and "blocked mill". Practice shows that the algorithm can increase the efficiency up to about 82%. Based on the above algorithm, turn the unilateral control algorithm of the knowledge base into fusion intelligent control, which can increased the efficiency to about 86%. Considering the lag time and the error value of the three parameters in grinding process are different, we take variable period sampling, which can adjust the mill load more quickly, and further enhance the efficiency to around 90%. The MATLAB simulation results proves that.

Keywords: grinding process; expert knowledge base; intelligent control; variable period; fusion intelligent

1 INTRODUCTION

Grinding process is a key link in modern industrial production, which directly affects the quality and various performance indicators of the products (LI Peiran et al. 2008).Taking automatic control for cement raw mill load not only can save large amount of electricity, also can increae the production. However, the grinding of the cement raw meal is a complex production process, which is nonlinear, big lag and difficult to establish mathematical model. Using some advanced or routine algorithm such as PID control algorithm can't solve above problem completely.We summarized an expert control system according to experience through long-term practice and theoretical analysis, which is based on three parameters including mill sound, the back powder of the separator and the power load of the bucket elevator. This system has nine rules and divides the grinding process into three stages: "empty mill", "normal" and "blocked mill". Different control scheme will be implemented according to different stage and different control rule. Practice shows that the overall efficiency of this control algorithm can reach about 80% and it can avoid appearing "empty mill" and "blocked mill" phenomenon. In order to increase the efficiency, turn the unilateral control algorithm for each section into fusion intelligent control,the efficiency can be improved to about 85%. According to the different error value of the three parameters in grinding process, we take variable period sampling on the basis of the intelligent control, the operation

efficiency of the mill can be further improved to about 90%.

2 GRINDING PRODUCTION PROCESS AND CONTROL PRINCIPLE

The product system of the cement mill is showed in Figure 1. The materials are sent into the ball mill for grinding by the conveyor belt, and are unloaded through the mill tail after ground by mill, and then delivered to the separator through the bucket elevator. The materials are divided into fine grinding material flow and coarse grinding material flow after classification. The fine grinding material flow directly output as a finished product, the coarse grinding material flow returns to the feed inlet of the mill for grinding again. Dust collector is for energy conservation and environmental protection. The grinding condition can be divided into three stages: "Empty mill"; "Normal"; "Blocked mill".

Figure 1 shows the control system structure of the cement mill which consisted of industrial computer, measurement circuit, control circuit. The mill parameter measurement part: mill sound, the back powder of the separator, the power load of the bucket elevator. The measuring means for the three parameters: mill sound ($F(k)$) measured by electric ear. the back powder of the separator ($I(k)$) measured by impact flow meter. the power load of the bucket elevator ($P(k)$) obtained through measuring the motor power. Control

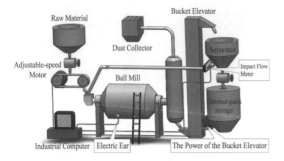

Figure 1. The principle diagram of the grinding production process and control system.

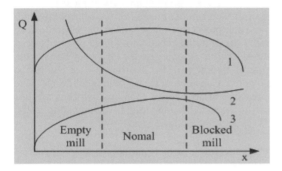

Figure 2. The curve graph of the production parameters in grinding process and the production efficiency.

circuit: control feed quantity through combining the system output signal with actuator (control the motor for material conveying through frequency converter as shown in Figure 1).

3 THE ESTABLISHMENT OF THE EXPERT KNOWLEDGE BASE

There are many factors affect the cement grinding process, and the three parameters including mill sound, the back powder of the separator, the power load of the bucket elevator can effectively reflect the motion characteristics. We obtain the operating characteristic of the three parameters as shown in Figure 2. It can roughly reflect the moving trajectory of the mill operation (ZHOU Rongliang et al. 2003) by superposing the three curves together.

As shown, 1: The curve of the power of the bucket elevator and the grinding efficiency; 2: The curve of the mill sound and the grinding efficiency; 3: The curve of the current of the impact flow meter and the grinding efficiency,there are totally nine rules, we give different algorithm for different rule as shown in Table 1.

In table 1, SP: sound pressure, BP: the back powder, LE: the load of bucket elevator, uv: upper limit value, lv: lower limit value. This algorithm has good robustness and can increase the efficiency up to about 82%. The abnormal phenomenon of "empty mill"

Table 1. The rule table of the expert knowledge base.

Pheo menn	SP	BP	LE	The new Additive amount	waiting time	rule
Empty mill SP > uv		BP > uv nomal BP < lv		Increase a certain value	τ_1 τ_2 τ_3	1 2 3
Nomal nomal		BP>uv nomal BP < lv		Slowly reduced keeping Slowly increased	τ_4 τ_5 τ_6	4 5 6
Stifled mill SP < lv			LE > uv nomal LE < lv	reduce a certain value	τ_7 τ_8 τ_9	7 8 9

and "blocked mill" can also be basically avoided which has been proved in industry production. In order to further improve the production efficiency, on the base of the rules of above expert system which has been stably running for many years, we make an improvement and propose the expert intelligent control system. The stability of the mill load can be effectively controlled and the grinding efficiency get further improvement.

4 EXPERT INTELLIGENT CONTROL SYSTEM

According to different conditions and characteristics of the system operation, the expert intelligent control takes different control modes to obtain the best control effect, and this is expert intelligent control's general features (PANG Quan & YANG Cuirong 1998), the principle is as follows:

According to the influence degree of the three parameters on the mill load in different stages, The article gives nine different rules.

In "empty mill" stage and "blocked mill" stage, when the error between the measured value and the given value is big (exceed threshold), using proportion algorithm can make the actual load value approximate the given value quickly, and we call this "coarse adjustment", When deviation is within the setting range, we use the incomplete differential PID control algorithm (ZHU Zhong sui 2005), According to parameter values collected and setting values, we obtain new feeding after doing the computations on the deviation between them, Because of the oversize amplitude, it's easy to cause overflow of computer data, and this is the defect of the ideal PID control algorithm. This algorithm can avoid this, so we call this "fine-tuning",

(1) "Empty mill" stage

In "empty mill" stage, we can see from table 1 that mill load is relevant to F, P.

Rule 1–3:

IF $(F(k) > F_{\max})$

THEN $U(k) = \{l_0 F_s - l_1 F(k)\} \cdot \dfrac{P_s}{P(k)}$ (1)

Coefficient l_0, l_1 may not be constant coefficients, and they can also be concerned with expressions. They are based on a lot of field experience. Wait time τ is based on the scene's concrete application example.

F_s, F_{min}, F_{max} stand for the set value, the minimum value, the maximum value of SP; P_s, P_{min}, P_{max} stand for the set value, the minimum value, the maximum value of BP; I_s stand for the set value.

(2) Nomal stage

Raw mill is a pure delay control object , and this article uses the incomplete differential PID control algorithm to control the mill load in this stage.

Control output $U(k)$ uses the following algorithm:

$$U(k) = U_p(k) + U_i(k) + U_d(k) \qquad (2)$$

when the error $\ell(k)$ is big, we use the incomplete differential PID control algorithm whos Integral action is weak and proportional action is strong, when the error $\ell(k)$ is small, we use the incomplete differential PID control algorithm whos Integral action is strong and proportional action is weak. These two algorithms are defined by the specific system.

In addition, in order to make the system keep stable, we let $\Delta u(k)$ equals 0 when the error $\ell(k)$ is smaller, that means we do nothing about the system.

Rule 4:

IF $(F(k) = F_s)$ AND $P(k) > P_{max}$
THEN $U(k) = U_p(k) + U_i(k) + U_d(k) \qquad (3)$

$(\text{IF } |\ell(\text{k})| > E_1)$

THEN $\quad U(k) = U_p(k) + U_i(k) + U_d(k)$

$$= K_{p1}\ell(k) + K_{i1}\sum_{j=0}^{k} e(j) + U_d(k)_1 \qquad (4)$$

$(\text{IF } E_2 < |\ell(\text{k})| < E_1)$

THEN $U(k) = U_p(k) + U_i(k) + U_d(k)$

$$= K_{p2}\ell(k) + K_{i2}\sum_{j=0}^{k} e(j) + U_d(k)_2 \qquad (5)$$

$$U_d(k)_i = \frac{K_{pi}T_{di}}{T_{fi}+T}[\ell(k)-\ell(k-1)] + \frac{T_{fi}}{T_{fi}+T}U(k-1)$$

Rule 5 : IF $(F(k) = F_s)$ AND $P(k) = P_s$

THEN $U(k) = U(k-1) \qquad (6)$

In this stage, the mill load will remain stable if keeping the last feeding.

Rule 6:

IF $(F(k) = F_s)$ AND $P(k) < P_{min}$
THEN $U(k) = U_p(k) + U_i(k) + U_d(k) \qquad (7)$

$(\text{IF } |\ell(\text{k})| > E_3)$

THEN $\quad U(k) = U_p(k) + U_i(k) + U_d(k)$

$$= K_{p3}\ell(k) + K_{i3}\sum_{j=0}^{k} e(j) + U_d(k)_4 \qquad (8)$$

$(\text{IF } E_4 < |\ell(\text{k})| < E_3$

THEN $U(k) = U_p(k) + U_i(k) + U_d(k)$

$$= K_{p4}\ell(k) + K_{i4}\sum_{j=0}^{k} e(j) + U_d(k)_4 \qquad (9)$$

The sampling period adopts the smaller value like the first case, and it can realize the slow increase of feeding. The mill load can be kept on the ideal state.

In "Normal" stage, the waiting time τ is defined by the particular circumstance.

Note: $\Delta u(k)$ express the computer output increment value of the Kth sampling time, $\ell(k-1)$ expresses the input deviation value of the (K-1) sampling time, $E_1 \sim E_4$ express thresholds set by the system, $T_{f1} \sim T_{f4}$ express filter coefficients, $T_{d1} \sim T_{d4}$ express derivative time constants, $K_{p1} \sim K_{p4}$ express proportion coefficients of controller, $K_{i1} \sim K_{i4}$ express integral coefficients, and T expresses time constant.

(3) "Blocked mill" stage

In "blocked mill" stage, we can see from table 3 that mill load is relevant to F, I.

Rule 7 - 9 :

IF $(F(k) < F_{min})$

THEN $U(k) = \{l_0 F_s - l_1 F(k)\} \cdot \dfrac{I_s}{I(k)} \qquad (10)$

Coefficient l_0, l_1 may not be constant coefficients, and they can also be concerned with expressions. They are based on a lot of field experience. Wait time τ is based on the scene's concrete application example. Practice and MATLAB simulation results show that this rule can increase the efficiency up to about 86%.

Sometime, if the last input makes the actual value of mill load close to the setting value and e(k) is small, the mill will go to the dead area. To change this phenomenon, we can adopt the method of changing sampling periods to avoid unnecessary adjustment (YANG Xiaofei et al. 2001, ZHANG Zengmin & XIE Jia 2009)

5 EXPERT INTELLIGENT CONTROL SYSTEM WITH VARIABLE PERIOD SAMPLING

According to different error ,expert intelligent control system with variable period sampling uses different sampling period, The article give three rules that shown in Table 2.

A, B express thresholds set by the system.When the parameters of expert intelligent control system are constant, using intelligent control algorithm with variable period sampling can further enhance the efficiency, Practice and MATLAB simulation results show that this rule can increase the efficiency up to about 90%.

Table 2. The rule of expert intelligent control system with variable period sampling.

Rule	condition	waiting time
1	$|\ell(k)| > A$	3min
2	$B < |\ell(k)| < A$	5min
3	$|\ell(k)| < B$	7min

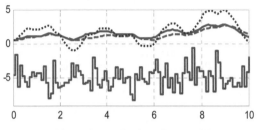

(a)white noise as the interference signal

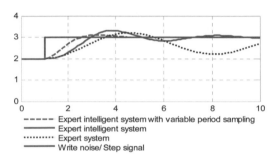

----- Expert intelligent system with variable period sampling
——— Expert intelligent system
·········· Expert system
——— Write noise/ Step signal

(b) step signal as the interference signal

Figure 3. The Matlab curve diagram.

6 SIMULATIONS AND CONCLUSIONS

The above introduction about expert control system, expert intelligent control system and expert intelligent control system of changeable periods. The article makes simulations of mill's normal stages respectively, and regulated by those three different schemes with the same interference signal, we get the following simulation curve.

With two different input signals, we can obviously see from these three simulation curves that the simulation curve of mill load using expert intelligent control system of changeable periods is of the best stability and minimum fluctuation, expert intelligent control system takes the second place, and expert control system is the last one. other stages are same as the nomal stage.

The intelligent control technology based on expert system discussed in this paper has certain reference value for the general industrial powder system and blast furnace feeding system. If only we build the control rule accords practical need and use related algorithm in this paper, the control system can achieve the optimal control.

REFERENCES

LI Pei Ran, SHEN, Tao & WANG Xiao hong. 2008, The optimization control system of grinding process load. *The University of Jinan journals: Natural science edition* 22(2): 116–123.

PANG Quan & YANG Cui Rong. 1998, Expert intelligent pH measurement instrument and its application. *Instrument journal* 19(5): 471–475.

ZHOU Rongliang, ZHANG Yanbin, CUI Donggang & LI Chang. 2003, The study and application of the cement mill's load control system, *Control Engineering of China* 10(6): 518–520.

YANG Xiaofei, CHEN Tiejun & SHANG Haitao. 2001, Humanoid variable cycle controller based on the pattern recognition and its application, *Journal of Zhengzhou University* 22(4): 99–102.

ZHANG Zengmin & XIE Jia. 2009, The application of fuzzy control strategy in cement mill load control system, *Journal of Shandong Agricultural University: Natural Science Edition* 4(4): 572–576.

ZHU Zhongsui. 2005, The application of incomplete differential PID algorithm in the pure lag system, *Control system* 21(9-1): 27–28.

Information Technology and Computer Application Engineering – Liu, Sung & Yao (Eds)
© *2014 Taylor & Francis Group, London, ISBN 978-1-138-00079-7*

Algorithm and implementation of digital PID based on MCGS-DDC

Yaping Lu & Tianlin Song
Applied Technology College, Soochow University, Suzhou, China

ABSTRACT: At present, with the development of control theory and electronic technology, digital PID control is gradually replacing analog PID control, and gradually become the core of modern industrial controller. And with the development of electronic technology, especially the Direct Digital Control module, digital PID controller has been rapid development. This paper is based on the MCGS-DDC single loop controller, using Basic language programming, a detailed description of the PID programming ideas, especially for sensor and actuator specifications (1–5 V or 4–20 mA) signal conversion algorithm, as compared with the analog PID controller has strong flexibility, can according to the test and experience online adjustment parameter, can better improve the performance of control, can easily be implemented control rules of various complex and special control algorithms.
Keywords: Digital PID; MCGS-DDC; Single Loop

1 INTRODUCTION

PID controller is the core component of common industrial control loop applications. There are generally two kinds, in the nineteen eighties, PID controller is simulated by the PID controller, the hardware (mainly electronic components, including resistors, capacitors, inductors) to realize its function, but there are parts of complex, easy to damage the aging problem, not flexible collocation. With the development of electronic technology, especially the processor CPU, digital PID controller has been rapid development, compared with the analog PID controller advantages are: flexibility, can according to the test and experience online adjustment parameter, control performance can be better. Of course, extensive application of digital PID controller is also inseparable from the sensor, actuator and the rapid development of A/D, D/A conversion module.

MCGS (Monitor and Control Generated System) is a set of configuration software for monitoring system for rapid construction and generation of computer, it can run on the Windows platform, based on field data collection and processing, with animation display, alarm processing, process control, real-time curve, historical curve and report output and so on many kinds of ways to solve practical engineering solution to the user. MCGS established by the main window works, equipment window, user window, real-time database and operating strategy is composed of five parts, equipment window established the connection relation between system and external hardware devices, which are connected by serial port and R7017, R7024, the system can read data from an external device and control external equipment status, implementation monitoring of industrial processes. Important user window is vivid graphics interface, realistic animation effect to describe the practical engineering problems. Operating strategy of window are a common Basic language as the foundation, according to predetermined conditions, such as through the digital PID programming, running state of the running process and equipment system for targeted selection and precise control.

DDC Direct Digital Control, has high flexibility, easy maintenance, high openness of the system and the advanced PID algorithm. Here is the use of remote data acquisition of analog input module R7017 and remote data acquisition of analog output module R7024, the main function of R7017 is the external analog signal into digital signal, through the communication is connected with the computer interface, 8 channel analog input. The R7024 is connected with the computer through the communication interface, the digital quantity received signal is converted into analog signal output, 4 channels of analog output. MCGS-DDC control system as shown in figure 1. The DDC module is connected with the computer through the RS232/485 converter, realizes the data communication.

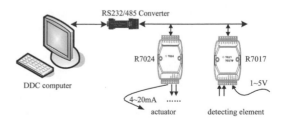

Figure 1.   MCGS-DDC control system.

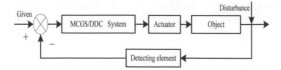

Figure 2. MCGS-DDC single circuit diagram.

Control method using DDC and MCGS software combined with single loop DDC design, for example, as shown in Figure 2, the Basic language as the programming language, algorithm design to specify digital PID and programming method.

2 ALGORITHM DESIGN IDEAS

The control algorithm of PID typical as shown in equation: where KP is the proportional coefficient; TI is the integral time constant; TD differential time constant; E (T) for a given value and the measured value of the deviation value.

$$u(t) = K_p[e(t) + \frac{1}{T_1}\int_0^t e(t)dt + T_D \frac{de(t)}{dt}] \qquad (1\text{-}1)$$

Because the computer control is a kind of sampling control system, it can only be based on the deviation sampling time value to calculate the control quantity. Therefore, type of integral and differential item cannot be used directly, by discretization. We make T as the sampling period, with a series of sampling time point kT represents the continuous time t, the cumulative sum approximation instead of integration, in order to replace the differential difference approximation, do the following approximate transformation:

$$t=kT \qquad (1\text{-}2)$$

$$\int_0^t e(t)dt \approx T\sum_{j=0}^i e(jT) = T\sum_{j=0}^i e(j) \qquad (1\text{-}3)$$

$$\frac{de(t)}{d(t)} \approx \frac{\{e(kT) - e[(k\text{-}1)T]\}}{T} = \frac{e(k)-e(k-1)}{T} \qquad (1\text{-}4)$$

Among them, T is the sampling period, usually 200 ms, e(k) as the system error K sampling time value, e(k-1) for the system (k-1) sampling time offset value, i sampling serial number, i = 0, 1, 2,. . .. The (1-3) and (1-4) type substitution (1-1) type, you can get the PID expression of discrete:

$$u(i) = K_p \{e(i) + \frac{T}{T_1}\sum_{j=0}^i e(j) + \frac{T_D}{T}[e(k)-e(k-1)]\} \qquad (1\text{-}5)$$

If the sampling period T is small enough (200 ms), the formula can be a good approximation analog PID formula, and thereby controlled process with continuous control process is very close. (1-5), usually known

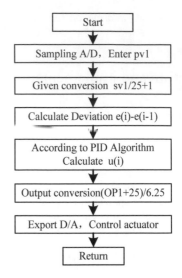

Figure 3. Digital PID control algorithm flowchart.

as the location of the PID control algorithm. If the formula (1-5) is transformed, then:

$$u(i) = k*ei + k*\frac{0.2}{ti}ei + k*\frac{td}{0.2}(pvx\text{-}pv1) \qquad (1\text{-}6)$$

1-6 is a programming expressions of digital PID control algorithm, for PVX (i-1) measurements, the PV1 for the I measurements.

The program flow diagram is shown in figure 3. The conversion of sv1/25+1 is given to set the value of digital-to-analog conversion, such as the requirements of a given level of 100CM, through the 100/25+1=5 conversion, represents the highest level of the measured voltage is 5 V. But for digital-to-analog conversion (OP1+25) /6.25 is to make the output into the current size, if the output of OP1 is 100%, then (100 + 25)/6.25 = 20, representing the output to the maximum current 20 MA.

For input and output, attention should be paid to the conversion between the mathematical model/modulus interior: through the transmitter output (1–5 V) analog, into the analog input module ICP-R7017 into digital, digital quantity to compare to give quantitative, give quantitative must digital-to-analog conversion, the conversion of liquid level, temperature and flow rate of the decimal number given 1.00–5.00 V, data processing formula refer to the following procedures. The digital PID calculated OP1 values for analog-to-digital conversion (4–20 mA), obtained through the analog output module ICP-R7024, to control the actuator.

3 ALGORITHM IMPLEMENTATION

The following procedure is based on the Basic language, realize the core program of digital PID control algorithm. In this process, the process cycle time is 200 ms, set is the start button, EI deviation I sampling time value, for a given value of sv1, PV1 measurement, OP2 for computing the results output. Through the K, Ti, TD three-phase coefficient selection, three kinds of control mode to select the P regulation, PI regulation and PID regulation.

```
if set=1 then
if k=0 and ti=0 and td=0 then
q0=0
q1=0
q2=0
endif
ei=(sv1/250+1)-pv1
if k<>0 and ti<>0 then
q0=k*ei
mx=k*0.2*ei/ti
q2=k*td*(pvx-pv1)/0.2
endif
if k=0 then
q0=0
cb=1
mx=cb*0.2*ei/ti
q2=cb*td*(pvx-pv1)/0.2
endif
if ti=0 then
q0=K*ei
q1=0
mx=0
q2=k*td*(pvx-pv1)/0.2
endif
IF MX>5 THEN
MX=5
ENDIF
IF MX<-5 THEN
MX=-5
ENDIF
if (sv1/250+1)>=pv1 then
if op1>=100 then
q1=q1
else
q1=q1+mx
endif
else
if op1<=0 then
q1=q1
else
q1=q1+mx
endif
endif
if pv1=0 then
pv1=pvx
```

```
endif
pvx=pv1
op1=Q0+Q1+Q2
OP2=(OP1+25)/6.25
IF OP2<4 THEN
OP2=4
endif
if op2>20 then
op2=20
ENDIF
ppv1 = (pv1-1)*25
if ppv1>100 then
ppv1=100
endif
psv1 =sv1
else
q0=0
q1=0
q2=0
op2=4
endif
else
endif
```

4   SUMMARY

Digital PID control based on MCGS-DDC, it has a better control effect, have certain engineering application value. Can be widely used in the middle level, flow and temperature sensor signal, not only for single loop control, can also be used in a double loop control, cascade control and ratio control, such as the complicated control.

REFERENCES

[1] Jianqiang Ding, Xiao Ren, Yaping Lu: "the computer control technology and its application" Qinghua University Press 2012

[2] Tianlin Song, Yaping Lu, Lei Shi: bio-design and develop of robot fish, submitted to key Engineering Materials Vol. 464, (2011), Pages 225–228

[3] Tianlin Song, Yaping Lu, Zhiyao Li: Structural Design and Research of the Bionic Snake-Like Robot, submitted to Advanced Materials Research Vol. 538, (2012), Pages 3034–3037

[4] Liuli Ji: the design based on the DDC boiler temperature control system, the scientific and technological square, 2011 (10)

Information Technology and Computer Application Engineering – Liu, Sung & Yao (Eds)
© 2014 Taylor & Francis Group, London, ISBN 978-1-138-00079-7

Modeling and control of solar powered HALE UAV

W.N. Zhao, P.F. Zhou & D.P. Duan
School of Aeronautics & Astronautics, Shanghai Jiao Tong University, Shanghai, China

ABSTRACT: This paper focuses on the lateral characteristics and control of High-Altitude-Long-Endurance (HALE) solar powered Unmanned Air Vehicle (UAV). The rudder and propeller differential as two kinds of actuators are selected to achieve the yaw control. In order to minimize the consumption of energy, pseudo-inverse control allocation method has been applied. And daisy-chain control allocation method has also been used to improve the cornering ability of the vehicle. The effectiveness of both methods has been proved.

Keywords: solar powered; pseudo-inverse; daisy-chain

1   INTRODUCTION

Solar powered HALE UAV is a kind of long endurance flight vehicle powered by solar energy and flied in the stratosphcre at an altitude of 17–20 km for more than six months (Cestino 2006). The flying height makes the UAV capable of advantages that satellites and conventional vehicles are not equipped. And as solar energy is characterized by green environmental-protective quality and infinity, Solar Powered HALE UAV has been shown to have promising application in both military and civil fields such as forest fire mapping, flood control, hurricane tracking, agriculture remote sensing (Herwitz, Johnson et al. 2002) and surveillance mission, relay communication. Many countries are focused on the development of this kind of UAV and its related key techniques.

In this research the designed UAV was equipped with two propellers on the wings, the wings were covered with solar cells, and there were no ailerons on the aircraft. Dihedral angles were designed to strengthen the roll stabilization. In order to fulfill the demands of high altitude and long endurance flight, high lift to drag ratio and large aspect ratio wing was deployed on the UAV. The layout of UAV was illustrated in Fig. 1. Two control devices, rudder and propeller differential, were configured to establish the yaw control. As these two kinds of actuators were associated different effectiveness, meanwhile the advantages of propellers on large aspect ratio wings could provide effective moment by propeller differential, the control allocation were used to improve the effects of control.

The reminder of this article is organized as follows. The modeling procedures and dynamic characteristics of UAV were presented in Sec. 2. In Sec. 3, several flight controllers were designed. The simulation results

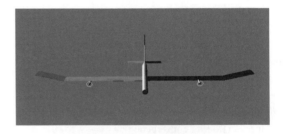

Figure 1.   Layout of Solar Powered HALE UAV.

were discussed in Sec. 4. Finally the conclusion was given in Sec. 5.

2   LATERAL MOTION MODEL OF THE UAV

As static, dynamic and control properties can be analyzed in quick and better ways by DATCOM during the general design stage. The Solar powered HALE UAV researched in this paper was at the initial time, the desired aerodynamics coefficients acquired by DATCOM have a certain reference value.

Because it is convenient to analyze the properties of the vehicle and obtain control law by designing longitudinal and lateral decoupled model, thus the decoupled motion equations are needed to be obtained firstly.

2.1   *Lateral equation*

The lateral equations of the vehicle (Etkin and Reid 1996), are governed as bellows:

$$\dot{x} = Ax + Bu \tag{1}$$

where state variables

$$x = \begin{bmatrix} v & p & r & \phi \end{bmatrix}^T$$

System matrix $A =$

$$\begin{bmatrix} \dfrac{Y_v}{m} & \dfrac{Y_p}{m} & \dfrac{Y_r}{m} - u_0 & g\cos\theta_0 \\[2mm] \dfrac{L_v}{I_x'} + I_{zx}'N_v & \dfrac{L_p}{I_x'} + I_{zx}'N_p & \dfrac{L_r}{I_x'} + I_{zx}'N_r & 0 \\[2mm] I_{zx}'L_v + \dfrac{N_v}{I_z'} & I_{zx}'L_p + \dfrac{N_p}{I_z'} & I_{zx}'L_r + \dfrac{N_r}{I_z'} & 0 \\[2mm] 0 & 1 & \tan\theta_0 & 0 \end{bmatrix}$$

And control matrix B can be simplified as

$$\begin{bmatrix} \dfrac{Y_{\delta r}}{m} & 0 & 0 \\[2mm] \dfrac{L_{\delta r}}{m} & 0 & 0 \\[2mm] \dfrac{\Delta N_{\delta r}}{I_z'} & \dfrac{\Delta N_{\delta r}}{I_z} & \dfrac{\Delta N_{\delta r}}{I_z} \\[2mm] 0 & 0 & 0 \end{bmatrix}$$

where δ_{pL} and δ_{pR} are the percentage of the full power of the left propeller and right propeller, and $N_{\delta pl} = -N_{\delta pr}$.

2.2 Analysis of the lateral mode

The aerodynamics coefficients were obtained from DATCOM, and the data needed in the equations above were calculated by the equations via the classical flight dynamic methods (Etkin and Reid 1996) thus the stabilization of the vehicle can be analyzed.

For the lateral modes on altitude 20 km and speed 30 m/s, the eigenvalues of the system matrix A are:

Mode 1 (Rolling convergence): $\lambda_1 = -2.5380$
Mode 2 (Spiral mode): $\lambda_{2,3} = 0.1971 \pm 1.00117i$
Mode 3 (Dutch Roll): $\lambda_4 = 0.0820$

We see that in the lateral modes rolling convergence and spiral mode are stable but Dutch roll is unstable due to the large dihedral angle for the static stability of roll. This can be inhibited by yaw damper.

3 YAW CONTROLLER DESIGN

The airplane was not capable of roll control as there was no aileron control surface, thus coordinate turning was impossible. On the UAV, rudder and propeller differential were selected to perform turning activities. The yaw control was achieved by feedback of yaw angle rate.

3.1 Yaw control with rudder and propeller differential

Fig. 2 shows the structure diagram of yaw controller with rudder. The control law was $K_p = 1$, $K_i = 0.03$.

Fig. 3 shows the structure diagram of yaw controller with propeller differential. Yaw direction was controlled by propeller differential which producing yaw moment by the different thrust forces on both sides of the wings. The designed control law was $K_p = 12$, $K_i = 0.5$.

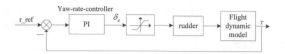

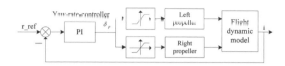

Figure 2. Structure diagram of yaw controller with rudder.

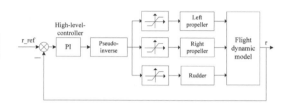

Figure 3. Structure diagram of yaw controller with propeller differential.

Figure 4. Structure diagram of pseudo-inverse control allocation method.

3.2 Control allocation

Control allocation is defined that expected control inputs are provided by control algorithm and dynamically allocated to the actuators in an optimal manner (Enns 1998). Control allocation is widely used in advanced fighter, civil airline, and missile and so on. There are many advantages to divide the control allocation algorithm and control allocation into two independent units such as simplifying the design of control system, fault reconfiguration without redesigning the control algorithm (Johansen 2004).

3.2.1 Pseudo-inverse method

Energy system including solar energy and energy storage cells urgently needs to develop to meet the demands of the long term flight in present technical conditions. As the shortage of power, minimizing the control energy has a significant value. As mentioned above the pseudo inverse control allocation method was taken into consideration. The rudder and the propeller differential were chosen to realize combination control.

Pseudo-inverse method has advantages of minimizing the control quantities of both kinds of actuators and being applied easily in the practical engineering.

Fig. 4 shows the structure diagram of pseudo-inverse control allocation method.

This paper introduced pseudo-inverse with weights which was based on the optimization of relative controlled variable quadratic sum to make the control allocation with limitation of the operational capacity of actuators. This means that the more effective actuators were used more frequently. So the cost function is

$$J = \frac{1}{2}\left[\left(\frac{\delta_e}{|\delta_e|_{max}}\right)^2 + \left(\frac{\delta_p}{|\delta_p|_{max}}\right)^2\right] \quad (2)$$

where δ_e is rudder deviation, $|\delta_p|_{max}$ and $|\delta_e|_{max}$ is the maximum deviation of rudder, δ_p is the percentage of the maximum thrust, and $|\delta_p|_{max}$ is 1 here.

According to Min and Kim (2005), if the optimum index $\min J = 1/2u^T Wu$ is selected, the solution of the question of pseudo-inverse with variable weights is

$$u = W^{-1}B^T\left(BW^{-1}B^T\right)^{-1}v = B^+v \quad (3)$$

where W is weight matrix, B^T is the transpose of the control matrix $B = [b_1\ b_2]^T$, B^+ is the pseudo-inverse of the control matrix B. In general, weight matrix W is selected as a positive diagonal matrix, so

$$W = \begin{bmatrix} \dfrac{1}{|\delta_e|_{max}^2} & 0 \\ 0 & \dfrac{1}{|\delta_p|_{max}^2} \end{bmatrix} \quad (4)$$

According to (3)

$$u = B^+v = \begin{bmatrix} \dfrac{|\delta_e|_{max}\,b_1}{|\delta_e|_{max}\,b_1^2 + |\delta_p|_{max}\,b_2^2} \\ \dfrac{|\delta_p|_{max}\,b_2}{|\delta_e|_{max}\,b_1^2 + |\delta_p|_{max}\,b_2^2} \end{bmatrix} v \quad (5)$$

It can be concluded that the introduction of weight matrix changes the authority of actuators, thus will have effects on the allocation results. The control law was $K_p = 2$, $K_i = 0.03$.

3.2.2 Daisy-chain method

Considering the low speed of solar aircraft and rudder effect will decline as the decline of speed, daisy-chain method was introduced to improve the cornering ability of the vehicle.

In daisy-chain method, rudder as the main executing actuator works all time while the thrust vector as the auxiliary actuator only works when needed. Each actuator work successive increments the input in order.

When the rate and position signals of rudder are in the limited range, the thrust vector does not work. The thrust vector begins to work to compensate the insufficient moment when the detected signals are out of limited range. As a result, this method can be used to compensate the deficiencies carried by the limitation of rate and location of rudder.

Daisy-chain control allocation method is simple and easy to be used in the practical engineering and has the advantage of avoiding full-time work of differential thrust.

Fig. 5 shows the structure diagram of daisy-chain control allocation method.

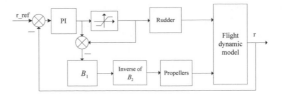

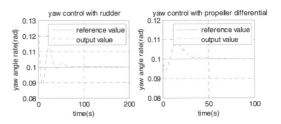

Figure 5.   Structure diagram of daisy-chain control allocation method.

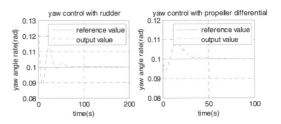

Figure 6.   Control effects of yaw controller with rudder and propeller differential.

The desired moment can be formulated as follows.

$$B_1u_1 + B_2u_2 = v \quad (6)$$

where $B = [B_1\ B_2]$ is the control matrix and $u = [u_1, u_2]^T$ is the control quantity. The control law was $K_p = 2.7$, $K_i = 0.05$.

4   SIMULATION RESULTS

Simulation is achieved by Matlab/Simulink at the altitude of 20 km with the speed of 30 m/s. The setting value of yaw angle rate is 0.1 rad/s.

In the Fig. 6 (yaw control with rudder), the output of yaw angle rate was slightly oscillate while the setting value can be tracked accurately. From the Fig. 6 (yaw control with propeller differential), we can make a conclusion that the yaw angle rate was almost no oscillate with respect to propeller differential control and tracking effect was good.

The varied value of thrust was same in opposite direction and the control allocation method was very good in tracking the setting value from Fig. 7. Yaw moment was mostly provided by rudder with higher effective while the propeller produced small residual moment. The deflection of rudder and value of propellers in relatively high speed were demonstrated that the more effective devices were used with more frequencies. The usage frequency of propeller will increase with the decrease of flight speed and not be discussed in this paper.

In order to validate that the daisy-chain method improves corning ability, the yaw rate of 0.2 rad/s which rudder and propeller differential cannot be achieved singly is selected. From Fig. 8, the method can realize the corning ability and when the rudder is saturated, the propeller differential is used. It is evident that the propeller differential is begin to work

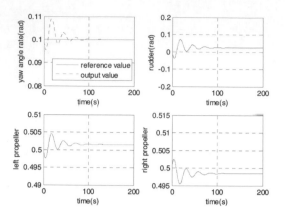

Figure 7. Control effects of yaw controller with pseudo-inverse method and the variation of actuators.

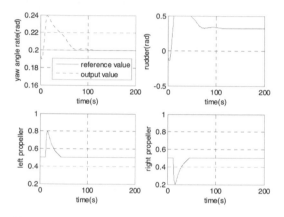

Figure 8. Control effects of yaw controller with daisy-chain method and the variation of actuators.

when needed and then back to initial value when the rudder can achieve the desired moment.

5 CONCLUSION

In this paper, a HALE solar UAV model was established and the lateral dynamic characteristics were analyzed. In order to minimize the consumption of energy, a pseudo-inverse with weights was introduced to control allocation of yaw moment and daisy-chain was introduced to improve the cornering ability. The simulation results reveal the feasibility and effectiveness of UAV model.

REFERENCES

Cestino, E. 2006. Design of solar high altitude long endurance aircraft for multi payload & operations. Aerospace science and technology 10(6): 541–550.

Enns, D. 1998. Control allocation approaches. *AIAA Guidance, Navigation and Control Conference and Exhibit.*

Etkin, B. and L.D. Reid 1996. *Dynamics of flight-Stability and control.* New York: John Wiley & Sons.

Herwitz, S.R., L.F. Johnson, J. Arvesen, R. Higgins, J. Leung and S. Dunagan 2002. Precision agriculture as a commercial application for solar-powered unmanned aerial vehicles. *1st American Institute of Aeronautics and Astronautics UAV Conference.*

Johansen, T.A. 2004. Optimizing nonlinear control allocation. *Decision and Control, 2004. CDC. 43rd IEEE Conference on, IEEE.*

Liu, H.H. 2001. Design combination in integrated flight control. *American Control Conference, 2001. Proceedings of the 2001, IEEE.*

Min, B.-M., E.-T. Kim and M.-J. Tahk 2005. Application of control allocation methods to SAT-II UAV. *Proceedings of the AIAA Guidance, Navigation, and Control Conference.*

Information Technology and Computer Application Engineering – Liu, Sung & Yao (Eds)
© 2014 Taylor & Francis Group, London, ISBN 978-1-138-00079-7

The greatest common factor about bivariate polynomials

Yueliang Xu
School of Mathematic, Southwest Jiaotong University, China

Jian Xiao
School of Electrical Engineering, Southwest Jiaotong University, China

ABSTRACT: In this paper, a counterexample is provided that explains the algorithm's faultiness for computing the H_∞ norm by directly computing the isolated common zero of two special bivariate polynomials (by Madhu N. Belur and C. Praagman). In order to find the isolated common zero, the greatest common factor of two special bivariate polynomials is important. The greatest common factor should meet the conditions discussed. Property and the algorithm of calculating method are offered.

Keywords: Euclidean algorithm; Isolated Common Zero; Greatest Common Factor of bivariate polynomial; H_∞-norm

1 INTRODUCTION

For computing the H_∞ norm of transfer function matrix, various method has been presented, for example, the method of finding large singular value of transfer function matrix [1], and the iterative algorithm under certain conditions [2]. In [3] [4] [5], different methods of computation the H_∞ norm are offered by their authors. Recently, Belur and Praagman offer a new method (2011), the main idea is: by directly computing the isolated common zero of two special bivariate polynomials. Unfortunately, the arguments may have defects. Belur and Praagman's method is based on theorem 5 (2011). To prove theorem 5, it implies using the following proposition.

Proposition: Let $p(x,y) = a_0(x) + a_2(x)y^2 + \cdots + a_{2n}(x)y^{2n}$ to be a bivariate polynomial on real number field $\Re$ ($a_i(x)$ is a polynomial on real number field $\Re$), $q(x,y) = \partial p/\partial y$. Suppose $d(x,y)$ to be the greatest common factor of $p(x,y)$ and $q(x,y)$ in the real domain $\Re$. If $d(x^*,y^*) = 0$, then $d(x^*,y) = 0$ has roots of even multiplicity at y^*.

This proposition is not credible. When $a_0(x) \equiv 0$, the result generally isn't true. For example:

$$p(x,y) = x^3y^2 - x^2y^4 - xy^6 + y^8 = (xy^2 + y^4)(x - y^2)^2$$

then

$$q(x,y) = 2x^3y - 4x^2y^3 - 6xy^5 + 8y^7 = 2y(x-y^2)(x^2 - xy^2 - 4y^4)$$

The greatest common factor of $p(x,y), q(x,y)$ in the real domain $\Re$ is $d(x,y) = (x - y^2)y$, $(x^*,y^*) = (1,1)$ is zero point of $d(x,y)$, but $y^* = 1$ isn't roots of even multiplicity about $d(1,y) = 0$ viz. $(1 - y^2)y = 0$.

In fact, the above proposition is not necessarily true even in $a_0(x) \neq 0$. This shows that the algorithm in paper [6] provided by Madhu N. Belur and C. Praagman has limitations. The main reason for this defect is that the authors believe that the greatest common factor of even power variate polynomial and odd power variate polynomial is even power. In this paper, the properties of the greatest common factor about two bivariate polynomials not only is studied in detail, the method of calculation also is presented. At the end, the greatest factor of $p(x,y), q(x,y)$ like as proposition is studied, and the method of calculation is offered. This paper is organized as follows. The next section presents the properties of one-variate or bivariate polynomial, expounds why the Euclidean algorithm can be used for finding the greatest common factor of two bivariate polynomials. Section 3 states the methods of calculating the greatest common factor and corresponding property about two special bivariate polynomials. One of the two special bivariate polynomials is even power variate polynomial, another is odd power variate polynomial. In this section, the relationship of the greatest factor between two bivariate polynomials $p(x,y), q(x,y)$ and two bivariate polynomials $p(x,y^2), q(x,y^2)$ is studied.

2 PROPERTIES ABOUT TWO BIVARIATE POLYNOMIAL GREATEST COMMON FACTOR

Definition 1: Said about the letters x, y expression $\sum_{i,j} a_{ij}x^iy^j$ for a bivariate polynomial in the field $\mathfrak{F}$, $a_{ij} \in \mathfrak{F}$. In this paper, $\mathfrak{F}$ is a real domain. via. $\mathfrak{F} = \Re$.

Lemma 1: For any bivariate polynomial, it can be written as:

$$a_0(x)y^n + a_1(x)y^{n-1} + \cdots + a_{n-1}(x)y + a_n(x) \tag{1}$$

where $a_k(x)$ is a one-variate polynomial.

Definition 2: The bivariate polynomial $d(x,y)$ is said to be a factor of bivariate polynomial $f(x,y)$, if there is a bivariate polynomial $g(x,y)$, they satisfy

$$f(x,y) = d(x,y)g(x,y) .$$

Definition 3: Suppose $f(x,y), g(x,y)$ are two bivariate polynomials, the bivariate polynomial $d(x,y)$ is said to be a common factor of $f(x,y)$ and $g(x,y)$, if $d(x,y)$ is not only a factor of $f(x,y)$, but also a factor of $g(x,y)$.

Definition 4: The bivariate polynomial $d(x,y)$ is said to be the greatest common factor of two bivariate polynomials $f(x,y), g(x,y)$, if any common factor of $f(x,y)$ and $g(x,y)$ is a factor of $d(x,y)$, written $d(x,y) = (f(x,y), g(x,y))$; if $d(x,y)$ is a one variate polynomial or nonzero constant, then $f(x,y), g(x,y)$ are said to be coprime.

The method of calculation the greatest common factor about one-variate polynomials can be used for finding the greatest common factor of two bivariate polynomials. Firstly, by lemma 1, two bivariate polynomials are written as two one-variate polynomials, and then the Euclidean algorithm can be used for finding the greatest common factor of two one-variate polynomials. In the process of calculation, quotient and remainder may be one-variate polynomial with rational polynomial of another variate as the coefficient. But in the end the common factor will be written in bivariate polynomial form. The reasons are the following conclusion guarntee.

Lemma 2: Suppose one-variate polynomials $u(x)$ and $v(x)$ are coprime, $h(x)$ is another one-variate polynomial that satisfies $u(x)|v(x)h(x)$, then $u(x)|h(x)$ ($u(x)|h(x)$ means that $h(x)$ is divisible by $u(x)$).

Corollary 1: Suppose one-variate polynomials $u(x)$ and $v(x)$ are coprime, and one-variate polynomial $p(x)$ and $q(x)$ satisfy

$$u(x)p(x) = v(x)q(x)$$

then $u(x)|q(x)$.

By corollary 1, there is the following conclusion to the decomposition of bivariate polynomials:

Corollary 2: Suppose

$$a_0(x)y^n + a_1(x)y^{n-1} + \cdots + a_{n-1}(x)y + a_n(x) =$$
$$v(x)/u(x)\left(\bar{b}_0(x)y^k + \bar{b}_1(x)y^{k-1} + \cdots + \bar{b}_{k-1}(x)y + \bar{b}_k(x)\right) \times$$
$$\left(\bar{c}_0(x)y^{n-k} + \bar{c}_1(x)y^{n-k-1} + \cdots + \bar{c}_{n-k-1}(x)y + \bar{c}_{n-k}(x)\right)$$

and $(a_0(x), a_1(x), \cdots, a_n(x)) = 1$, $(u(x), v(x)) = 1$

then $v(x) \equiv \bar{c}$ ($a_i(x), \bar{b}_j(x), \bar{c}_l(x), u(x), v(x)$ are one-variate polynomial, $(a_0(x), a_1(x), \ldots, a_n(x)) = 1$ means that $a_0(x), a_1(x), \ldots, a_n(x)$ are coprime)

Lemma 3: Suppose

$$\begin{cases} \left(\bar{b}_0(x), \bar{b}_1(x), \cdots, \bar{b}_k(x)\right) = 1 \\ \left(\bar{c}_0(x), \bar{c}_1(x), \cdots, \bar{c}_{n-k}(x)\right) = 1 \text{ and} \end{cases}$$
$$a_0(x)y^n + a_1(x)y^{n-1} + \cdots + a_{n-1}(x)y + a_n(x) =$$

then
$$\left(\bar{c}/u(x)\right)\left(\bar{b}_0(x)y^k + \bar{b}_1(x)y^{k-1} + \cdots + \bar{b}_{k-1}(x)y + \bar{b}_k(x)\right)$$
$$\left(\bar{c}_0(x)y^{n-k} + \bar{c}_1(x)y^{n-k-1} + \cdots + \bar{c}_{n-k-1}(x)y + \bar{c}_{n-k}(x)\right)$$
$$u(x) \equiv c$$

Proof: If $u(x) \neq c$, suppose x_0 is a zero point of $u(x)$, there must be minimum i and j, satisfy $\bar{b}_i(x_0) \neq 0$, $\bar{c}_j(x_0) \neq 0$ because $(\bar{b}_0(x), \bar{b}_1(x), \ldots, \bar{b}_k(x)) = 1$ and

$$\left(\bar{c}_0(x), \bar{c}_1(x), \cdots, \bar{c}_{n-k}(x)\right) = 1 . \text{ Let } i+j = r, \text{ then}$$
$$u(x)a_r(x) = \bar{b}_0(x)\bar{c}_r(x) + \bar{b}_1(x)\bar{c}_{r-1}(x) + \cdots + \bar{b}_r(x)\bar{c}_0(x)$$
$$= \sum_{\substack{s+t=r \\ s \geq 0, t \geq 0}} \bar{b}_s(x)\bar{c}_t(x)$$

(here if $s > k$ or $t > n - k$, let $\bar{b}_s(x) = 0$, $\bar{c}_t(x) = 0$).

Due to $u(x_0)a_r(x_0) = 0 \Rightarrow \bar{b}_i(x_0)\bar{c}_j(x_0) = 0$. This is a contradiction. So Lemma 3 is proven.

Corollary 3: Suppose bivariate polynomial $a_0(x)y^n + a_1(x)y^{n-1} + \cdots + a_{n-1}(x)y + a_n(x)$ to satisfy:

(1) $\left(a_0(x), a_1(x), \cdots, a_n(x)\right) = 1$
(2) $a_0(x)y^n + a_1(x)y^{n-1} + \cdots + a_{n-1}(x)y + a_n(x) =$
$$\left(\frac{\bar{b}_0(x)}{b_0(x)}y^k + \frac{\bar{b}_1(x)}{b_1(x)}y^{k-1} + \cdots + \frac{\bar{b}_{k-1}(x)}{b_{k-1}(x)}y + \frac{\bar{b}_k(x)}{b_k(x)}\right)$$
$$\times \left(\frac{\bar{c}_0(x)}{c_0(x)}y^{n-k} + \frac{\bar{c}_1(x)}{c_1(x)}y^{n-k-1} + \cdots + \frac{\bar{c}_{n-k-1}(x)}{c_{n-k-1}(x)}y + \frac{\bar{c}_{n-k}(x)}{c_{n-k}(x)}\right)$$

then we can conclude that there are two bivariate polynomials $\varphi(x,y)$, $\psi(x,y)$, and satisfy

$$a_0(x)y^n + a_1(x)y^{n-1} + \cdots + a_{n-1}(x)y + a_n(x) = \varphi(x,y)\psi(x,y)$$

here

$$\varphi(x,y) = \tilde{b}_0(x)y^k + \tilde{b}_1(x)y^{k-1} + \cdots + \tilde{b}_{k-1}(x)y + \tilde{b}_k(x)$$
$$\psi(x,y) = \tilde{c}_0(x)y^{n-k} + \tilde{c}_1(x)y^{n-k-1} + \cdots + \tilde{c}_{n-k-1}(x)y + \tilde{c}_{n-k}(x)$$

($b_i(x), \bar{b}_i(x), \tilde{b}_i(x), c_j(x), \bar{c}_j(x), \tilde{c}_j(x)$ are one-variate polynomial)

Proof: By corollary 2 and Lemma 3, we can easily proof corollary 3.

Corollary 3 shows: when a bivariate polynomial is factorized, if the factor is a polynomial of vriate y with coefficient of rational polynomial about variate x, then the factors can be written as two bivriate polynomials and the two bivriate polynomials have same degree of polynomial about variate y like before. The greatest common factor's properties about bivariate polynomial is similar as one-variate polynomial:

Lemma 4: Suppose the bivariate polynomial $d(x,y)$ is the greatest common factor of two bivariate polynomials $f(x,y)$ and $g(x,y)$, then there are two bivariate

140

polynomials $u(x,y), v(x,y)$ and obe-variate polyno-mial $h(x)$ to satisfy

$$f(x,y)u(x,y)+g(x,y)v(x,y)=d(x,y)h(x) \qquad (2)$$

and degree of $d(x,y)$ about variate y is fixed.

Proof: Firstly write $f(x,y), g(x,y)$ as (1), then use euclidean algorithm to find the greatest factor $d_1(x,y)$ and two corresponding polynomials $u_1(x,y), v_1(x,y)$ to satisfy:

$$f(x,y)u_1(x,y)+g(x,y)v_1(x,y)=d_1(x,y)$$

Here, $d_1(x,y), u_1(x,y), v_1(x,y)$ are polynomials about variate y, their term coefficients may be ratio-nal polynomials with variate x. $d_1(x,y)$'s degree about variate y is fixed, we can find the least common denom-inator polynomial about $d_1(x,y), u_1(x,y), v_1(x,y)$'s rational polynomials with variate x, let the least common denominator polynomial to be $r(x)$, set

$$u(x,y)=u_1(x,y)r(x), v(x,y)=v_1(x,y)r(x),$$
$$d(x,y)h(x)=d_1(x,y)r(x)$$

So (2) is proven.

Corollary 4: Suppose

$$f(x,y)=a_0(x)y^n+a_1(x)y^{n-1}+\cdots+a_{n-1}(x)y+a_n(x)$$
$$g(x,y)=b_0(x)y^m+b_1(x)y^{m-1}+\cdots+b_{m-1}(x)y+b_m(x)$$

and $(a_0(x),a_1(x),\cdots,a_n(x))=1,(b_0(x),b_1(x),\cdots,b_m(x))=1$

Then there is $k, 0 \leq k \leq \min\{n,m\}$, and the great-est factor $d(x,y)$ of $f(x,y), g(x,y)$ can be written as

$$d(x,y) = c_0(x)y^k + c_1(x)y^{k-1} + \cdots + c_{k-1}(x)y + c_k(x)$$

Here $(c_0(x), c_1(x), \ldots, c_k(x)) = 1, c_0(x) \neq 0$.

Proof: By corollary 3, lemma 3 and lemma 4, we can easily proof corollary 4.

Lemma 5: Two bivariate polynomials $f(x,y), g(x,y)$ are coprime if and only if there be two bivariate polynomials $u(x,y), v(x,y)$ and one-variate polyno-mial $h(x)$ or $h(y)$ satisfy

$$f(x,y)u(x,y)+g(x,y)v(x,y)=h(x)$$

or $h(x)$ satisfy

$$f(x,y)u(x,y)+g(x,y)v(x,y)=h(y)$$

From the above properties, the greatest common factor of $f(x,y), g(x,y)$ can be carried out according to the following steps:

(1) write $f(x,y), g(x,y)$ as follewe (two polynomi-als with variate y, $a_i(x), b_j(x)$ are one-variate polynomials with variate x):

$$f(x,y)=a_0(x)y^n+a_1(x)y^{n-1}+\cdots+a_{n-1}(x)y+a_n(x)$$
$$g(x,y)=b_0(x)y^m+b_1(x)y^{m-1}+\cdots+b_{m-1}(x)y+b_m(x)$$

(2) calculate the greatest common factor $\alpha(x)$ of $a_0(x), a_1(x), \ldots, a_n(x)$: the greatest common fac-tor $\beta(x)$ of $b_0(x), b_1(x), \ldots, b_m(x)$

(3) calculate $f_1(x,y), g_1(x,y)$ that satisty

$$f(x,y)=\alpha(x)f_1(x,y), g(x,y)=\beta(x)g_1(x,y)$$

(4) calculate the greatest common factor $d_1(x,y)$ of $f_1(x,y), g_1(x,y)$ by Euclidean algorithm. (here, if write $d_1(x,y)$ as a polynomial with variate y, it's coefficient polynomials with variate x are coprime)

(5) calculate the greatest common factor

$$d(x,y) \text{ of } f(x,y), g(x,y)$$
$$d(x,y)=(\alpha(x),\beta(x))d_1(x,y)$$

3 MAIN CONCLUSION

The greatest common factor of two bivariate polyno-mials problem involved in [6] is very special, one of two bivariate polynomials is even degree about vari-ate y, another is odd degree about variate y. What are the greatest common factor characteristics of such two bivariate polynomials? The odd degree polyno-mial about variate y can written as product of a even degree polynomial about variate y and variate y. So characters of the greatest common factor like these two special bivariate polynomials must be studied. The following Theorem 1 shows how to find the greatest common factor of one-variate polynomial:

Theorem 1: Suppose $d(x)$ is the greatest common fac-tor of two one-variate polynomials $f(x), g(x)$ then, $d(x^2)$ is the greatest common factor of two one-variate polynomials $f(x^2), g(x^2)$.

Proof: For $(f(x), g(x)) = d(x)$, then there are two one-variate polynomials $u(x,y), v(x,y)$ to satisfy

$$u(x)f(x)+v(x)g(x)=d(x)$$

So $\quad u(x^2)f(x^2)+v(x^2)g(x^2)=d(x^2)$

and $d(x^2)|f(x^2), d(x^2)|g(x^2)$ as a result of

$$d(x)|f(x), d(x)|g(x).$$

Therefore, $d(x^2)$ is the greatest common factor of $f(x^2), g(x^2)$.

Theorem 1 shows that the greatest common factor of $f(x^2), g(x^2)$ can be found by calculating the greatest common factor of $f(x), g(x)$. And one of the polynomi-als is even degree, another polynomial is odd degree, their greatest common also is similar to Theorem 1.

Theorem 2: Suppose $d(x)$ is a nonzero degree polyno-mial to satisfy

$$(f(x), g(x))=d(x) \text{ ,and } f(0) \neq 0$$

Then $d(x^2)$ is the greatest common factor of $f(x^2), xg(x^2)$.

Proof: Firstly, $d(x^2)$ must be a common factor of $f(x^2), xg(x^2)$, therefore the greatest common factor $d_1(x)$ of $f(x^2), xg(x^2)$ must be devided exactly by $d(x^2)$. From $d_1(x)|f(x^2) \Rightarrow f(x^2) = d_1(x)h(x) \Rightarrow d_1(0) \neq 0$ so $(x, d_1(x)) = 1 \Rightarrow d_1(x)|g(x^2)$ (by Lemma 2), this means that $d_1(x)$ also is a greatest common factor of $f(x^2), g(x^2)$. Hence $d_1(x)|d(x^2) \Rightarrow d_1(x) = cd(x^2)$. (c is a nonzero constant in here). So $(f(x^2), xg(x^2)) = d(x^2)$.

There are some similar conclusions about bivariate polynomial:

Theorem 3: Suppose two bivariate polynomials $f(x,y), h(x,y)$ are coprime, $g(x,y)$ is a bivariate polynomial, satisfies $h(x,y)|f(x,y)g(x,y)$.

Then $h(x,y)|g(x,y)$ can be concluded.

Proof: Because $f(x,y), h(x,y)$ are coprime, there are two bivariate polynomials $u(x,y), v(x,y)$ and a one-variate polynomial $\bar{h}(x)$ that satisfy

$$u(x,y)f(x,y) + v(x,y)h(x,y) = \bar{h}(x)$$

$$\Rightarrow u(x,y)f(x,y)g(x,y) + v(x,y)h(x,y)g(x,y) = g(x,y)\bar{h}(x)$$

so $\quad h(x,y)|(u(x,y)f(x,y)g(x,y) + v(x,y)h(x,y)g(x,y))$,

viz $\quad h(x,y)|g(x,y)\bar{h}(x)$, therefore $h(x,y)|g(x,y)$

Similarly, if $u(x,y)f(x,y) + v(x,y)h(x,y) = \bar{h}(y)$, $h(x,y)|g(x,y)$ can also be deduced.

Theorem 4: Suppose bivariate polynomial $d(x,y)$ is the greatest common factor of two bivariate polynomials $f(x,y), g(x,y)$, then $d(x,y^2)$ is the greatest common factor of two bivariate polynomials $f(x,y^2)$, $g(x,y^2)$.

Proof: Because $(f(x,y), g(x,y)) = d(x,y)$, there are two bivariate polynomials $u(x,y), v(x,y)$ and a one-variate polynomial $h(x)$ that satisfy:

$$u(x,y)f(x,y) + v(x,y)g(x,y) = d(x,y)h(x)$$

and $d(x,y)|f(x,y), d(x,y)|g(x,y)$
so $u(x,y^2)f(x,y^2) + v(x,y^2)g(x,y^2) = d(x,y^2)h(x)$
and $d(x,y^2)|f(x,y^2), d(x,y^2)|g(x,y^2)$ also are true
so $d(x,y^2)$ is the greatest common factor of $f(x,y^2)$, $g(x,y^2)$.

Theorem 5: Let the greatest common factor of two bivariate polynomials $f(x,y), g(x,y)$ is $d(x,y)$ and the bivariate polynomial $d(x,y)$ is indeed contain the variate y. $f(x,0) \neq 0$ is satisfied also. Then the greatest common factor of $f(x,y^2), yg(x,y^2)$ is $d(x,y^2)$.

Proof: Firstly $d(x,y^2)$ must be a common factor of $f(x,y^2), yg(x,y^2)$, therefore the greatest common factor $d_1(x,y)$ of two bivariate polynomials $f(x,y^2)$, $yg(x,y^2)$ is divided exactly by $d(x,y^2)$.

Secondly $d_1(x,y)|f(x,y^2) \Rightarrow f(x,y^2) = d_1(x,y)$ $h(x,y) \Rightarrow d_1(x,0) \neq 0$, therefore the degree about variate y of $d_1(x,y)$ at least is 2. It can be concluded that $y, d_1(x,y)$ are coprime. From the Theorem 3 to known $d_1(x,y)|g(x,y^2)$, so $d_1(x,y)$ is a common factor of $f(x,y^2), g(x,y^2)$. Therefore $d_1(x,y)|d(x,y^2)$. Comprehensive above $d(x,y^2)|d_1(x,y)$, we can conclude

$$d_1(x,y) = cd(x,y^2) \quad \text{(c is nonzero constant in here).}$$

This means $(f(x,y^2), yg(x,y^2)) = d(x,y^2)$.

Theorem 5 tell us that, the greatest common factor of $f(x,y^2), yg(x,y^2)$ can be found by calculating the greatest common factor of $f(x,y), g(x,y)$ under certain conditions.

ACKNOWLEDGEMENT

This work supported in part by "the Fundamental Research Funds for the Central Universities-SWJTU.11ZT29".

REFERENCES

[1] Boyd S, Balakrishnan V, Kabamba P. 1989. A bisection method for computing the H∞ norm of a transfer matrix and related problems. *Math. of Control, Signals, and Systems, vol. 2: 207–220.*

[2] N. Bruinsma and M. Steinbuch. 1990 A fast algorithm to compute the H∞-norm of a transfer function matrix. *Syst. Control Lett., vol. 14, pp. 287–293.*

[3] R. Caponetto, L. Fortuna, G. Muscato, and G. Nunnari, A direct method for computing the L-infinity norm of a transfer matrix, *J. Franklin Inst., vol. 329, pp. 591–604, 1992.*

[4] M. Belur and C. Praagman, 2004, Computation of the H∞-norm: An Efficient Method. *Univ. of Groningen, The Netherlands, IWI Tech. Rep., vol. 2004-4-4-01.*

[5] M. Belur, Scilab codes for proposed method of H∞ norm computation. *Dept. Elect. Eng., IIT, Bombay, India, Tech. Rep.,* 2011 [online]. Available: http://www.ee.iitb.ac.in/belur/hnorm/

[6] Madhu N. Belur and C. Praagman, 2011. An Efficient Algorithm for Computing the H-inf Norm. *IEEE Transactions on automatic control vol. 56. no. 7. July pp. 1656–1660.*

Information Technology and Computer Application Engineering – Liu, Sung & Yao (Eds)
© *2014 Taylor & Francis Group, London, ISBN 978-1-138-00079-7*

Digital video communication platform based on cloud computing technology

Han Liu

Wuhan Donghu University, Hubei Province, P.R. China

ABSTRACT: The concept of cloud computing trends will certainly become a part of the triple play, cloud computing and media content resources management and use of the future, can be calculated by the cloud of the contents of the resource management system transformation, building regional video content resources that cloud video which to build a cloud digital video communications platform.

Keywords: cloud computing; cloud video; communication

1 CLOUD COMPUTING AND ITS CHARACTERISTICS

1.1 *What is cloud computing*

Cloud computing is a new information age change, IT is the first time following the PC as the core of IT technology revolution, and with the Internet as the core of the second IT technology revolution after the third technical revolution. Cloud computing is a delivery resources use pattern, namely through resources required for network application. Cloud computing equipment, software and service package, changed people to obtain, process and store information.

In fact, cloud computing is the integration of the product of the traditional computer technology and network technology development. Narrowly cloud computing refers to the delivery of the IT infrastructure and usage patterns through the network to on-demand, scalable way to obtain the necessary resources; generalized cloud computing refers to the delivery of services and usage patterns that the network in an on-demand, easy to expand The way to get the services they need. This service can be IT and software, Internet-related, but also other services. Cloud computing is grid computing, distributed computing, parallel computing, utility computing, network storage, virtualization, load balancing, and other traditional computer and network technology development integration of the product. Applications talking about cloud computing is a broad concept.

1.2 *The main characteristics of cloud computing*

Application of cloud computing in China started relatively late but develops rapidly, and take the lead in put forward the concept of "cloud security", created a cloud security solutions. In recent years, the domestic many powerful companies began to build their own cloud platform, such as alibaba group, China mobile research institute, etc. For vlsi, virtualization, high reliability, versatility, high scalability and on-demand service, very cheap. Cloud computing Hugh showed four characteristics for applications in the field of electronic media:

1.2.1 provides the most reliable, safe data storage center, users don't have to worry about data loss, viruses, intrusion, etc. Cloud computing applications, all user data is stored in highly fortified server group, and there are a large group of professional server in maintenance updates and protection, users don't have to worry about data loss caused by hackers breached and information theft.

1.2.2 the minimum requirement to the client's equipment, is the most convenient for you to use. Even now, the Internet business applications, the cheap netbook, installed after the most common browsers, can meet most of the cloud computing client demand, without the. need for additional equipment and upgrade. For radio, film and digital interactive TV IPTV set-top boxes and even the telecommunications industry, this function has a decisive significance. Cloud computing to the client's hardware requirements low, can greatly ease the development of interactive television early caused by STB and middleware performance is not high value-added business and profits decline.

1.2.3 can easily share data between different devices and applications. In the field of radio and television, as a result of has a threshold policy, in the video always enjoy an advantage in the application of class status. And see the operating in the evolution of radio, film and television, has formed the comprehensive development of land, sea and air. This involves between different user terminal, operational data and application service of sharing same questions. Such as we are on the bus, by moving the CMMB mobile phone to watch TV, in the home can switch to the family of the TV, look more comfortable.

1.2.4 to use for network provides almost unlimited possibilities. Cloud computing shows three nets fusion huge advantage on the business. For example we uploaded via the Internet on PC take tourism video, uploaded to our first digital home sharing service in the "cloud" server group, you can choose by way of P2P information, finally through a TV set-top boxes and mobile phones to play, let friends and family to share your happiness and experience. And "cloud computing" data is stored in the "cloud" of servers, you don't have to carry a dedicated device, at any a client device that connect the "cloud computing" services, you can through the browser to log in and continue to see a part of the movie or wrote half of the article, and talked for about half of the video call.

1.3 *"Cloud computing" in the media industry application prospect*

In digital era, three nets fusion of computer technology and the media industry increasingly development, digital video has increasingly become the main content and form of media. And the cloud to build a cross-media digital video transmission platform provides the opportunity.

Cloud computing technology and cloud-based services has become the focus of the television industry, in the era of cloud computing, data processing, especially for large data processing, audio and video data format conversion, extended to the Internet from a stand-alone server computer cluster, cloud computing is seen as a revolution in the technology industry, it will bring a fundamental change in working methods and business models. By then, TV users can enjoy unlimited Internet era, all comprehensive information on business will be like water, electricity, heating and other infrastructure throughout the tens of thousands of households.

2 MEDIA USE OF CLOUD COMPUTING

As cloud computing technology popularization and application of in the field of mass media dissemination, development trend of the concept of cloud computing will become a part of the three nets fusion, cloud computing and media resources management and use in the future, through cloud transformation content resource management system, build regional video content resource that cloud video, gradually formed multiple independent media cloud "content", at the same time, through the corresponding standard protocols to make different "cloud" media connectivity, gradually formed the media "the media big cloud".

Spread by the mass media characteristics, application requirements scientific classification, summed up the business applications that take full advantage of the benefits of cloud computing, focused, targeted to carry out applied research.

2.1 *The media cloud*

Newspapers, magazines, radio, television, Internet, new media, cable operators, etc. With rich media resources. But at the moment due to the limitation of technology system capability and non-isolating media resources management system, caused a large number of media resources use efficiency is low. Through cloud computing to existing content resource management system reform, establishment of radio, film and television program content resource pool build a radio and television media cloud, digital content resources, mass storage, mining, and data sharing and super retrieval content production capacity, improve the efficiency of resource use, rich content, improve the service ability, reduce operating costs.

2.2 *Service cloud*

With the development of the three nets fusion, not only the traditional newspapers, radio, television and other media to digital, networked two-way transformation, at the same time also will produce a large number of new services and formats, adapt to the need of the users of information comprehensive service ability is the important embodiment of core competence of radio, film and television in the future. With cloud computing technology to build the media cloud service, will help to comprehensive information service, promote the integration of traditional media and new media. At the same time, cloud computing will make a wide range of services computing is accomplished through the cloud of high-performance server group, can reduce the hardware configuration of user set-top boxes, terminal equipment, speed up the deployment of all kinds of new business and promote at a lower cost for more users to provide innovative business experience.

2.3 *Cloud video*

As a primary component of the Internet application, the development of network video, is closely related to "the cloud". Cloud the development of video technology is more and more widely at present, cloud video concept is mainly used in many fields such as video conference, network video, IPTV, video cloud of practical application, to provide users with a large content and audience, provide more choices, related video technology development of enterprises, is planning to cloud video as a strategic development direction, will be in the same cloud video platform as the center, for different types of Internet end users with genuine, vast, clear, convenient video content, so that consumers enjoy the high quality video real life. As cloud computing applications will continue to gather rich video content, gathered a large number of user groups; According to user needs, provide personalized video content, and according to the different requirements of advertisers achieve accurate coverage, coverage involves multiple platforms, and present a massive amounts of content.

3 THE CORE CONCEPT AND APPLICATION OF CLOUD VIDEO

3.1 The core concept of cloud video

Cloud video is through the large scale storage, and distribution of intelligent distributed parallel processing technology, dynamic distributed real-time transcoding technology, intelligent terminal rate adjustment video transmission technology, choice of users to provide services and other advanced processing technology, computer software, and organically blend together to form clouds of highly intelligent video network service group.

Cloud video concepts need to constantly be perfect and abundant, it is a gradual process. Current form has had the user terminal compatibility, completely automatic perception and push the user terminal equipment adaptation ability to the best data flow. The final form to the user terminal equipment greatly simplified zero maintenance, user terminal software.

3.2 Various cloud video application system development

At present, cloud video application system development has been put into practice, and good results, more representative of the cloud video application has the following several developers.

3.2.1 CDV full telecommunications, media and technology, cloud-based architecture

CDV C-MAP-based video cloud computing technologies and new architecture, a unified data center, to build four core platform of the operation and management platform, production produced and broadcast platform, integrated distribution platform and operational platform. Operation and management platform services include production management services, resource management services, editing and broadcasting management services, advertising management services, copyright management services and operations management services. Production produced and broadcast platforms include acquisition included service, program production services, the presentation business services, content management services and Cluster Broadcasting Service. Integrated distribution platform covering the operation and management services, content aggregation services, content processing services, content management services, processing services and content distribution services. Operations publishing platform services have interactive TV platform, network TV platform, IPTV centralized control platform, centralized control of mobile TV, mobile TV platform and Internet TV.

The main advantage is reflected in the platform has a dynamic, efficient, on-demand, flexible, green and so on, the use of resources can be achieved more fully, to adjust more flexible, more environmentally friendly consumption, media processing more efficient.

3.2.2 Sobey network television cloud services solution

Sobey the timely introduction eWebTV network television cloud service solutions. This is based on the cloud platform new media related services, can provide "rapid deployment", "fast operation" high-end content integration and value-added technology platform programs. Users can not only self-built network television operators the ability of the operator network television, you can quickly get through the procurement of services.

In addition, for fine upload workstation integrates the unique advantages of the Sobey non-linear editing systems to edit the material while adding a variety of new media format editing, compositing support. Upload material support the EX/P2/XDCAM/ VTR/DVD and other types of media. System and network television content to be seamlessly integrated, upload, edit the synthetic material sent directly to the network television release.

3.2.3 Tvmining launched media cloud computing platform

Tvmining construction of a large-scale movement of the "Mobile TV" media cloud services platform, the platform is divided into two parts: First CNTV hundreds Road distributed TV signal integrated coding acquisition, while achieving more for mobile Internet terminal large-scale live streaming push service, capable of supporting 500 million users on-line high-performance CDN distribution network to cover the whole process from the signal processing to real-time encoding to multi-protocol, multi-terminal release of live streaming. Second, the rapid release of the terminal software, the platform relies on the the TMFC day clock basis Tianmai unique application development library, able to quickly develop and deploy client applications that support Android, iOS and other intelligent terminal to ensure a first-class user experience.

The project uses the H.264 the commercial coding techniques, 512 Kbps bit rate can be achieved SD quality. The rapid construction of a see 7 days EPG, 10 seconds screenshots enhanced flow of application services. Fragmentation streaming technology based on HTTP protocol, to ensure complete stability in the public network environment lossless transmission, much better than the existing RTMP protocol unsteady flow push mode.

4 DIGITAL VIDEO COMMUNICATION PLATFORM BASED ON CLOUD COMPUTING TECHNOLOGY

With the continuous development of cloud computing technology and innovation, digital video, as a kind of mature technology, widely accepted by people, has the necessity and feasibility of combined with cloud computing technology. But due to the digital video service has the characteristics of large-scale, high interaction, real-time strong, how to use cloud computing technology to build a flexible, reliable, energy saving, security service platform still faces many challenges.

4.1 Building cloud digital video communication platform

In the process of radio and television the IT construction, the introduction of "cloud" computing model and architecture, to reform the existing equipment and network structure, is a kind of solving the problem of rapid growth of multimedia data, improve the efficiency of complicated video processing effective feasible method.

The cloud digital video communication platform by the client, remote client center and close to the client's agent of cloud.

4.1.1 client types varied, can be a computer, television, mobile phones, digital terminals, the client through their network (wireless network, cable TV network and Internet, etc.) enjoy cloud network video services.

4.1.2 far customer center cloud is composed of the directory server and the content server, a directory server lookup service as agent for the cloud to provide video object, content server exists a full backup of all of the video object.

4.1.3 near the client proxy cloud by access to the server, the Tracker code with server, proxy server and server, access to the server as a client request cloud to the agent service interface that receives the client service request, to the Tracker to obtain a list can provide service proxy server and their running condition, choose the appropriate proxy server and the client establish a connection and provide services.

In this communication platform, a proxy server as a provider of video objects, its run on multiple types of service instance, multiple proxy server location near may even Shared storage. Coding is the role of the conversion server providers or users to upload different code of video object converted to conform to the cloud digital visual communication platform code of the video object, cloud digital visual communication platform frequently must be completed using high processing power of CPU server code system conversion.

4.2 The characteristics and advantage of cloud digital video transmission platform

Because of the network video service platform of cloud client variety and quantity is huge, so the cloud network video technology must meet the following large capacity, high throughput, distributed storage, high concurrency, low latency, manageability, scalability, etc.

Compare the network video service platform of cloud and the traditional network video systems, cloud network video technology has the following advantages:

4.2.1 cloud service platform of network video server can run jointly with other services in the form of a virtual server on the same physical resources, isolation is better, safety is higher.

4.2.2 each virtual server can dynamically create, delete or live migration, its possession of resources can be changed along with the demand of dynamic, flexible configuration, good scalability.

4.2.3 Holdings multiple proxy server to access the same video in the form of Shared storage object, saves the storage resources, servers, flexibility is good, has nothing to do with the data source of the scheduling and migration can be implemented successfully.

4.2.4 run when a server fails, the network video service platform of cloud can provide better fault tolerance and data recovery mechanism and higher reliability.

4.2.5 cloud network video services running in a common platform, do not need a dedicated server, can make full use of resources, saving hardware and operating costs.

5 CONCLUSION

As cloud computing technology and the further development of network video technology, the development and application of digital video transmission platform for cloud provides a rare opportunity. Cloud digital video transmission platform based on cloud computing will be more new features. Practical application of cloud video, provides users with more choices. Although there are still some safety problems not radically solve, but believe in the near future of cloud computing in the information security will be mature, its application will also be more and more widely in media communication.

Information Technology and Computer Application Engineering – Liu, Sung & Yao (Eds)
© *2014 Taylor & Francis Group, London, ISBN 978-1-138-00079-7*

The computational model of the airfield capability based on system dynamics

Yaxiong Li, Xinxue Liu, Yue Guo, Hui Hao & Yanping He
Xi'an Research Institute of Hi-Tech, Hongqing Tow, Xi'an, P.R. China

ABSTRACT: The operational capability model is a basis to assess the damage effect of the Airfield or other such similar targets, on which most researches in this area are using static system dynamic methods. As a dynamic system, there are dynamic behaviors and characteristics in the Airfield System. To calculate the operational capability of the Airfield a system dynamic method is proposed, which can properly describe the relationship between the inner structure and dynamic behavior.

Keywords: system dynamics; airfield; capability

1 INTRODUCTION

To estimate the influence of sub-target change to the whole airfield function is the foundation research of the airfield capability. There are many factors caused the change of the airfield sub-target capability. Such as the sub-target may been devastated, replaced or repaired. If we used the traditional simulation which is simulated by time step or triggered by occurrence will caused so many variables, complex interactions and situations. At last, it is hard to achieve the simulation. SD (system dynamics) which research system dynamics' simulation is adapted to calculate complex systematic variables in feedback system and it is can been used to estimate the airfield capability.

2 ANALYSIS OF THE AIRFIELD CAPABILITY

There are two important factors to been restricted the airplane's capability. The first factor is the plane must have enough oil and ammunition and it is influenced by airfield supply. The second factor is the plane must flying-off and it is influenced by flying filed, parked establishment and conning tower. Therefore, the airfield capabilities can analyze as four main capabilities, such as airfield service ability, supply ability, the amount of usable plane, the command and control ability and so on.

Airfield capability: Airfield's sorties in limited time which based on airfield capability at confirmed time-interval.

Airfield capability at confirmed time-interval: Airfield's sorties at confirmed time-interval based on airfield service ability, supply ability, the amount of usable plane, the command and control ability at confirmed time-interval.

Airfield service ability is based on the runway passing capability, contact way passing capability and taxiway passing capability. Runway passing capability is the max sorties in limited time in different timetable; Contact way passing capability is the max sorties flying-off from contact way in timetable; Taxiway passing capability is the max sorties flying-off from taxiway in timetable.

Aviation supply capability: The max sorties can been supplied with oil and ammunition in timetable.

The number of Available plane: The number of Available plane is depend on the original number of available plane, the number of the damaged plane and the number of the complementary plane.

Command and control capability: The max sorties passed runway, contact way and taxiway and been supplied.

3 THE COMPUTATIONAL MODEL OF THE AIRFIELD BASED ON SD

3.1 *The diagram of the cause and effect*

To analyze the cause and effect is used the diagram of the cause and effect to dynamics feedback by qualitative analysis method. The cause and effect of airfield analysis contain one main loop and four sub loop. The main loop is the loop of the airfield capability. As shown in Figure 1.

Airfield ability is depended on airfield service ability, aviation supply capability, the number of Available plane and command and control capability. When the airport been attacked, the various establishment will been damaged and the capability will been descended.

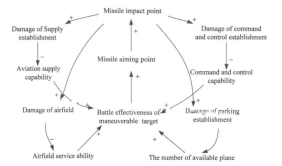

Figure 1. Airfield capability's diagram of the cause and effect.

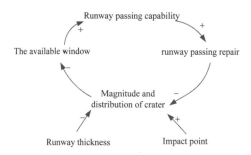

Figure 2. Diagram of the cause and effect of runway passing capability.

a) Diagram of the cause and effect of airfield service capability. Airfield service ability is based on the runway passing capability, contact way passing capability and taxiway passing capability. The diagram of runway passing capability is shown in Figure 2.

b) Diagram of the cause and effect of aviation supply capability. The sub loop of the aviation supply capability is shown in Figure 3.

c) The number of available plane's diagram of the cause and effect. The sub loop of the number of available plane's diagram is shown in Figure 4.

d) Diagram of the cause and effect of the command and control capability. The sub loop of the command and control capability is shown Figure 5.

3.2 SD flow chart

By the relation of variables and the construction of airfield capability, the system dynamics' flow chart can be plotted as Figure 6.

3.3 The variable and equation of the SD model

a) The type and signification of variable
The computational model of the airfield capability has different type variables and can been classified fixed variable and time varying variable. The fixed variable is not changed by time such as damage efficiency of missile to sub-target which is depend on the intrinsic attribute of the weapon and airfield. The time varying variable is changed by time such as the runway

Figure 3. Diagram of the cause and effect of aviation supply capability.

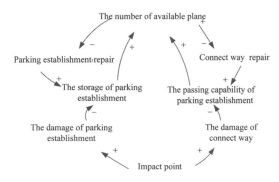

Figure 4. The number of available plane's diagram of the cause and effect.

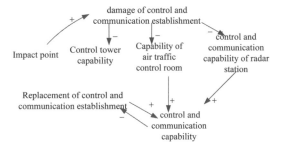

Figure 5. Diagram of the cause and effect of the command and control capability.

passing capability. These time varying variables is also changed by the increasing warhead impact point and the repaired of sub-targets. The time varying variable also includes exogenous and endogenous variables. From the system dynamics' flow chart, the type, name and significance of variable are shown in followed table.

b) The SD equation of airfield capability

$$\text{L} \quad N_{YJQF}^{FJ}(t) = N_{YJQF}^{FJ}(t-1) + A_{PD}^{J \; J \; T}(t^P-1) \quad (1)$$

$$\text{L} \quad N_{JJTJP}^{FJ}(t) = N_{JJTJP}^{FJ}(t-1) - A_{PD}^{J \; J \; T}(t^P-1) \quad (1)$$

$$\text{R} \quad A_{PD}^{JJTJP}(t) = f_{APD}^{JJTJP}(MBSX, DTLD(t), \ldots) \quad (2)$$

$$\text{L} \quad N_{YJQF}^{FJ}(t) = N_{YJQF}^{FJ}(t-1) + A_{PD}^{LLD}(t-1) \quad (3)$$

$$\text{L} \quad N_{DDQF}^{FJ}(t) = N_{DDQF}^{FJ}(t-1) - A_{PD}^{LLD}(t-1) \quad (4)$$

$$\text{R} \quad A_{PD}^{LLD}(t) = f_{APD}^{LLD}(MBSX, DTLD(t), \ldots) \quad (5)$$

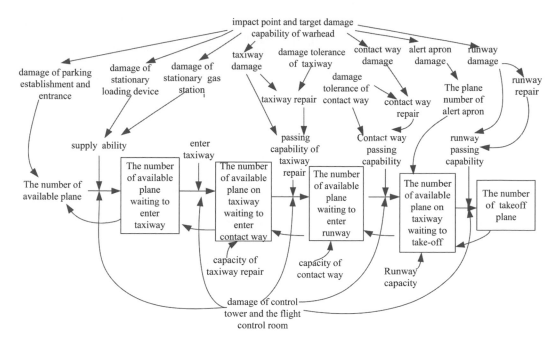

Figure 6.   The flow chart of airfield capability.

Table 1.   The variable table of airfield capability.

Serial number	Variable name	Variable type	Variable significance	Remark
1	$MBSX$	Constant variable(C)	Airfield attribute	Intrinsic attribute of airfield
2	$DTLD(t)$	Assistant variable(A)	Impact point	Time varying exogenous variable
3	$A_{NFJ}(t)$	Level variable (L)	The number of available plane	Time varying endogenous variable
4	$A_{BJ}(t)$	Rate variable(L)	Aviation supply capability	Time varying endogenous variable
5	$N_{DDHXD}^{FJ}(t)$	Level variable (L)	The number of available plane waiting for taxiway	Time varying endogenous variable
6	N_{HXDRR}^{MAX}	Constant variable(C)	Taxiway capability	Intrinsic attribute of airfield
7	$N_{DDLLD}^{FJ}(t)$	Level variable (L)	The number of available plane waiting for contact way	Time varying endogenous variable
8	$A_{HXD}(t)$	Rate variable(R)	taxiway passing capability	Time varying endogenous variable
9	N_{LLDRR}^{MAX}	Constant variable(C)	Contact way capability	Intrinsic attribute of airfield
10	$N_{DDPD}^{FJ}(t)$	Level variable (L)	The number of available plane waiting for runway	Time varying endogenous variable
11	$A_{LLD}(t)$	Rate variable(L)	Contact way passing capability	Time varying endogenous variable
12	N_{PDDRR}^{MAX}	Constant variable(C)	Runway capability	Intrinsic attribute of airfield
13	$N_{DDQF}^{FJ}(t)$	Level variable (L)	The number of available plane waiting flying-off on runway	Time varying endogenous variable
14	$N_{JJTJP}^{FJ}(t)$	Level variable (L)	The number of available plane on watchful parking apron	Time varying endogenous variable
15	$A_{PD}^{LLD}(t)$	Rate variable(R)	Runway passing capability (from the contact way)	Time varying endogenous variable
16	$A_{PD}^{JJTJP}(t)$	Rate variable(R)	Runway passing capability (from the watchful parking apron)	Time varying endogenous variable
17	$N_{YJQF}^{FJ}(t)$	Level variable (L)	The number of flying-off plane	Time varying endogenous variable

L $N_{DDQF}^{FJ}(t) = N_{DDQF}^{FJ}(t-1) + A_{LLD}(t-1)$ (6)

R $A_{LLD}(t) = f_{ALLD}(MBSX, DTLD(t), \ldots\ldots)$ (7)

L $N_{DDPD}^{FJ}(t) = N_{DDPD}^{FJ}(t-1) + A_{HXD}(t-1) - A_{LLD}(t-1)$ (8)

L $N_{DDPD}^{FJ}(t) = N_{DDPD}^{FJ}(t-1) - A_{LLD}(t-1)$ (9)

L $N_{DDPD}^{FJ}(t) = N_{DDPD}^{FJ}(t-1) + A_{HXD}(t-1)$ (10)

L $N_{DDPD}^{FJ}(t) = N_{DDPD}^{FJ}(t-1)$ (11)

R $A_{HXD}(t) = f_{HXD}(MBSX, DTLD(t), \ldots\ldots)$ (12)

L $N_{DDLLD}^{FJ}(t) = N_{DDLLD}^{FJ}(t-1) + 1 - A_{HXD}(t-1)$ (13)

L $N_{DDLLD}^{FJ}(t) = N_{DDLLD}^{FJ}(t-1) - A_{HXD}(t-1)$ (14)

L $N_{DDLLD}^{FJ}(t) = N_{DDLLD}^{FJ}(t-1) + 1$ (15)

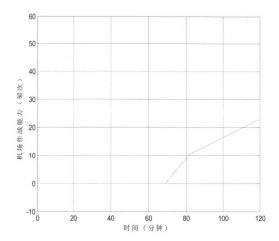

Figure 7.   Airfield capability change with time.

$$\text{L} \quad N_{DDLLD}^{FJ}(t) = N_{DDLLD}^{FJ}(t-1) \quad (16)$$

$$\text{L} \quad N_{DDHXD}^{FJ}(t) = N_{DDHXD}^{FJ}(t-1) + A_{BJ}(t) - 1 \quad (17)$$

$$\text{L} \quad N_{DDHXD}^{FJ}(t) = N_{DDHXD}^{FJ}(t-1) + A_{BJ}(t) \quad (18)$$

$$\text{L} \quad N_{DDHXD}^{FJ}(t) = N_{DDHXD}^{FJ}(t-1) - 1 \quad (19)$$

$$\text{L} \quad N_{DDHXD}^{FJ}(t) = N_{DDHXD}^{FJ}(t-1) \quad (20)$$

$$\text{L} \quad A_{NFJ}(t) = A_{NFJ}(t-1) - A_{BJ}(t-1) \quad (21)$$

$$\text{R} \quad A_{BJ}(t) = f_{BJ}(MBSX, DTLD(t), \ldots\ldots) \quad (22)$$

$$\text{L} \quad A_{NF}(t) = f_{NF}(MBSX, DTLD(t), \ldots\ldots) \quad (23)$$

In these equations, 'L' means level variable equation and 'R' means rate variable equation, while 'C' means constant variable equation and 'A' means assistant variable equation.

The first equation means the increasing number of flying-off plane which is from parking apron to runway, while the second equation means the decreasing number of available plane on parking apron. The third equation means the computational methods of runway passing capability. The fourth equation means the increasing number of flying-off plane which is from contact way to runway. While all planes in watchful parking apron has been flying-off, the available plane will take off from the contact way. The significances of the equation from the fifth to the twenty-fourth are very similar to the first equation, so no more tautology here.

4   SIMULATION ANALYSIS

A typical airfield has 55 planes and 15 of them which take on the combat readiness stop on the watchful parking apron. The other 40 planes are putted on central parking apron, garage, blindage and covering garage. If we use missile to attack the runway, taxi-way and garage, the capability of subsystem will be calculated and the time varying capability of airfield also can be attained by SD model.

The runway ability is zero while the airfield is attacked, so airfield capability is zero between 0 min and 7 min. The sortie is 0.8 per min between 70 min and 82 min and the sortie is 0.3 per min between 82 min and 120 min. The sortie decrement mainly because the available plane need take off from contact way while the all the plane on watchful parking apron has taken off.

From the results of the simulation, the computational model of the airfield capability based on system dynamics is correct and effective.

REFERENCES

[1]   Jiang Hao, Cheng Haoguang. The evaluation of airfield damage effect based on sub-fuzzy criterion [J]. Automation of Armament, vol. 26, 2007.
[2]   Fan Yangtao,Yang ping. The evaluation of airfield damage effect [J]. Sichuan ordnance journal, vol. 30, 2009, pp. 93–95.

Information Technology and Computer Application Engineering – Liu, Sung & Yao (Eds)
© 2014 Taylor & Francis Group, London, ISBN 978-1-138-00079-7

Business strategy for Shaw Communications Inc

Yi Li
School of Economics and Business Administration, Yangtze Normal University, Chongqing Province, China

ABSTRACT: After analyzing the internal and external environments through the tools, such as Porter's Value Chain, VRINE, and Five Forces, this report will provide in-depth strategic analysis about Shaw Communications Inc. Furthermore, the company's corporate level and business level strategies, and its recent history will be discussed step by step. For operating and managing well in Shaw Communications, the details will be elaborated in the report, which are based on the research.

Keywords: Internal environment; strategic analysis; operating managing

1 INTRODUCTION

Demand of telecommunication is being driven by customers and businesses during these centuries. A variety of competitors target in this industry and strive for the market shares. Many companies, such as Rogers Communications Inc., BCE Inc., Telus Corporation, Cogeco Inc., Bell Aliant Inc., and MTS Allstream Inc., are competing with Shaw.

2 SHAW'S CORE COMPETENCY

Shaw's core competency is running a network, and integrating all those telecommunication services through their technology. It's not in producing content, but providing and focusing on diversified deployment and high-quality services for customers. Based on the research in the previous report, Shaw is developing its diversification in the fields of telecommunication. Shaw's leverage its network by providing additional services beyond traditional cable. It enhanced the quality, depth and capacity of its plant and network infrastructure through significant capital investments, and the plant and network is essentially fully digital and two-way capable. Furthermore, as a leading communications and entertainment provider, the company keeps exploiting the arenas: 1) Shaw Media develops some sections such as conventional broadcasting (Global, CJBM), specialty channels (DIY network, HDTV, TVTROPOLIS, SHOW CASE, etc.); 2) Shaw Direct focuses on the two parts of consumers and businesses, like cable, Internet, 100% digital satellite TV, and home phone; 3) Last but not least, Shaw Wireless will be launched in 2012. It holds approximately 20MHz of AWS spectrum across Western Canada. For instance, Shaw is operating in many unrelated businesses: business solutions, broadcasting services, SOHO, and tracking. All these kinds of services are managed as a portfolio of businesses relied on the diversified technical distributions when most competitors have difficulty managing unrelated business simultaneously.

3 SHAW'S DISTINCTIVE COMPETENCY

Comparing with the competitors (Cogeco, Telus, MTS, Bell, etc.), Shaw's distinctive competency is the feature of customization which the consumers could design their own package services in the part of TV channels, Internet, and so on. For example, in the respect of TV channels, the customers can select the different channels and combine them as a package. It's effective and efficient to satisfy their needs and save the money; in the respect of Internet, customization design also provides various options to the customers to set the level of data usage they actually want to utilize. However, other competitors are not able to achieve this situation temporarily because they think that the profit of customization design is low. Even though this way is easy to imitate and copy, Shaw still believes it's important to focus on the high-quality customer service. This point shows Shaw always tries to defend and add the value to their customers, and forms its one of the competitive advantage.

4 SHAW'S FUNDAMENTAL COMPETITIVE

As an integrated distributor, Shaw Communication Inc. which bought CanWest TV assets last year has a competitive advantage when negotiating with each other on the fees it pays to carry specialty stations. This benefit helps Shaw lower the costs on the platform of cable and satellite TV services. In addition, Cancom

Tracking, a Shaw Communications company, provides tracking, two-way messaging, and integrated transportation and logistics solutions to the trucking industry. Cancom's products and services have significant competitive advantages to transportation companies requiring ongoing fleet coordination. ("Improve Route Planning", 2002, ¶10). Although Shaw is a smaller telecommunication company than its competitors, such as Rogers, Bell, and etc., it's still attempting to expand the scale of products and services in order to establish more competitive advantages in the industry. The strategy of Shaw is to deliver quality products and services at competitive prices, to stay focus on their customers, and to manage the capital structure in an efficient and responsible manner. (Shaw Communications, 2009, ¶6). In a word, Shaw's competitive advantages mainly base on the diversified businesses.

5   SHAW'S CORPORATE LEVEL STRATEGY

Corporate strategy is the strategic management of companies with multiple business units. Each business unit is a separate entity that has its own customers and control over the activities needed to satisfy those customers (Carpenter & Sanders, 2012). Shaw communications Inc. has five business units which service different type of customers as follows:

- Shaw Business owns and operates a national fiber-optic backbone network, providing data networking, video, and voice and Internet services to companies of all sizes.
- Shaw Tracking offers integrated on-board computing technology and wireless data solutions for the Canadian transportation, mobile workforce and logistics industries.
- Shaw Media is home to many of Canada's most-loved television brands which have a lot of channels (Shaw.ca, 2011).

The company is horizontally integrated. These business units are related in the telecommunications industry.

6   SHAW'S CORPORATE LEVEL STRATEGY

6.1   *Arenas*

Shaw Communications is a diversified communications company whose core business is providing broadband cable television, High-Speed Internet, Digital Phone, telecommunications services (through Shaw Business Solutions) and satellite direct-to-home services (through Shaw Direct). Shaw serves 3.4 million customers, including 1.8 million Internet subscribers and over 1.0 million Digital Phone customers, through a reliable and extensive network, which comprises 625,000 kilometers of fiber (Marketwire, 2011). To continue growing, Shaw always keeps improving its current product.

6.2   *Vehicle*

To enlarge the company, Shaw provided more products. It launched home phone service in 2005. In 2009, it acquired Mountain Cablevision Ltd. in Hamilton, Ont. and becomes the largest cable company in Canada (shaw.ca, n.d.). In 2010, an opportunity presented itself when CanWest filed for bankruptcy protection. Shaw was successful in acquiring CanWest's broadcast TV network, content ownership and specialty services ("Media"). After receiving such resources, Shaw creates newest business unit to leverage content with current and future distribution systems, opening up new opportunities for growth (Marketwire, 2011).

6.3   *Economic logic*

Shaw's different product share common resources such as equipment, installation staff and salespersons, so low costs can be achieved through scope and replication advantages. Meanwhile, in the western part of Canada, Shaw is a major leader and few companies can compete with it. Shaw's product has a higher quality, thus it can charge premium prices to the customers. Compare with other internet service providers, Shaw has a better internet signal and higher internet speed because of its good network. Furthermore, the acquisition of CanWest offers opportunities for Shaw Communications to leverage CanWest's content with its existing distribution systems across many platforms of cable, DTH, Digital and internet to expand marketability and increase revenue. With 100% ownership, Shaw will be able to integrate all CanWest's media operations seamlessly into its existing business at a low operational cost that provides maximum freedom to create more value with those new assets (Marketwire, 2011).

7   SHAW'S BUSINESS LEVEL STRATEGY

Business level strategy refers to the ways in which in a company goes about achieving its objectives within a particular industry or industry segment. Generally, business level strategy proposed to address two critical questions (a) how the company will achieve its objectives today, when other companies may be competing to satisfy the same customers' needs, and (b) how the company plans to compete in the future (Carpenters, M.A., Sanders, W.G., (2012)).

According to the study, Shaw was locked in the Focused Differentiation block, which is composed with narrow scope of arenas and differentiation economic logic. However, several reasons are given to box Shaw up in this category. Firstly, the marketing geography of Shaw is western of Canada. Secondly, the products and services supported by Shaw do not include personal cell phone, while competitors like Rogers and Telus do. Thirdly, the Shaw's products are expensive R&D cost and high added value, in order to match with the specialized target customers. Fourthly, the products and service of Shaw are high quality (professional skills), reliability and prestige (family brand name), and customization. In a word, the number and breadth of arenas of Shaw are narrow, while the economic logic of Shaw is differentiation.

7.1 Economic drivers of strategic positioning

Shaw has two main differentiators under focused differentiation strategic positioning, which are premium brand image, and convenience and customization. As mentioned in assignment one, most senior customers are familiar with the brand and easy to accept the derivative from Shaw, and keep a deep sense of loyalty to Shaw. Then, convenience and customization is the most differentiator of Shaw to compete with rivalry. Compared the products offered by Shaw with Rogers and Telus, consumers can enjoy more individual choices in Shaw. No matter bundles, television, Internet, or home phone, customers can design their personal style product and service on line immediately. Even trapping into the intractable problems, on-line IT support would be available 24 hours. Anyway, Shaw utilizes the premium brand image and convenience and customization to make itself more competitive advantage than others.

7.2 Threats to successful competitive positioning

In theory, failing to increase buyers' willingness to pay higher prices, underestimating costs of differentiation, over-fulfilling buyers' needs and lower-cost imitation are the four threats to focused differentiation position (Carpenters, M.A., Sanders, W.G., (2012)). However, the most powerful threat to Shaw is the low-cost imitation. Although the entry barrier is slightly high, the substitutions can appear as well as anytime anywhere. Addition to this, Shaw had to face one specified threat because it is focused. Competitors can come along that take away the customers by being more focused than Shaw which the focused strategy. So how to prevent the core resources and capabilities from competitors' imitation and keep the competitive advantage all the time is the most important thing Shaw need to deal with.

7.3 Evaluation

The macroeconomic environment of North America began recession from the first technology bubble burst that the stock prices of all technology firms dramatically increased and were enormously popular with investors in the mid-to late 1990s. However, the capital market of North America was hard beat when it went through the September 11th event. However, the subprime crisis made the whole financial market dropped into the endless bottom. Although the economic environment starts to revive, the transactions are still not active enough. Under kind of this situation, Shaw formulated its strategy on focused differentiation strategic positioning to concentrate on the

wireless infrastructure. Firstly, the demand of wireless service was affected by the financial crisis, and adequate fund cannot be raised from the capital market to support aggressive acquisition strategy directly. Shaw has not additional money to involve into other new and unrelieved products. Only to tighten up pocket and strengthen the core part of company is a judicious decision. Secondly, the most threat in wireless service industry is low cost imitation and substitutions. Shaw should work out how to keep its competitive advantage sustainable, through focus on developing and researching the core resources and capabilities of the company, and makes distinctive from rivals finally.

7.4 Challenges

On one side, Canadian government is eager to attract investment in the information and communications technology (ICT) sector, and numerous incentives are available (ViewsWire-Economist Intelligence Unit, n. On the other side, Canada's "big three" telecommunication operators which are Bell, Rogers and Telus, lobby to squash upstart rivals (ViewsWire-Economist Intelligence Unit, n.d.).

7.5 Opportunities

a) The technology risk rating of Canada is low and stable (ViewsWire-Economist Intelligence Unit, n.d.). b) The demand of wireless service began to heat up, which includes the hardware and software. For example, the sales for semiconductors recovered faster than expected through the second half of 2009 and all of 2010. The sales for the first eight months of 2010 increased 44% over the same period of the previous year, to US$194.6bn (ViewsWire-Economist Intelligence Unit, n.d.). This rising trend will continue in next five years by expert forecast. c) the Economist Intelligence Unit estimates that Canada spent 2.5% of GDP on information technology (IT) alone in 2010, and the amount will keep to increase in next recent years (ViewsWire-Economist Intelligence Unit, n.d.).

Focused differentiation is suitable to Shaw now, but is not sustainable in the future. It is learnt that the business strategy should change timely to catch the step of dynamic environment, the same to Shaw. As an emerging industry, more and more rivals are taking part in to share this cake. The "big three" removed their traditional business from telephone to cable service; the new entrants such like Public Mobile, Globalive Wireless and Mobilicity (formerly DAVE Wireless) devote to the wireless products as well.

8 SHAW AS A "FIRST MOVER" IN THE INDUSTRY

The communications service industries in Canada include wire line, wireless, resellers, satellite and other telecommunications services and broadcast distribution. Shaw contributes about 40 years for the entertainment, broadcasting and communications industry and it starts its business in 1966. When it compares with Rogers Communications, Videtron Ltee, Cogeco communication which enters into the market in 1972 and Perora cable which starts up in 1986, Shaw chose to initiate a strategic action at a early time. It is the one of introductions of a new product such as Capital Cable TV into Canada and service such as internet service as well as the development of a new process that improves quality, lowers price or both so the Shaw is the first mover in Canadian market. Furthermore, the company own first-mover advantages. First advantage is that the firm's image and reputation are hard to imitate at a later date. The Shaw brand is synonymous with diverse product offerings and exceptional customer service. When customers want to get related products and services, they prefer the Shaw which is give them a good image than a new brand although the customer never use the products before. The second advantage is that the firm locks customers' preferences. The last advantage is that the company achieves a scale. Through expanded geographic footprint that stretches across Canada, the company has the ability to connect Canadians allowing them to share, inspire and create. The company has a large network of factories and customer support offices. For clients, this local Shaw's operation area translates to ease of doing business.

9 DYNAMIC APPROACH

Business world is a wide fighting arena where competitors are constantly moving and changing positions in order to get ahead in the business track. Markets are dynamically marked by both opportunities to attract new customers and the threats from competition trying to lure the firm's customers away. When you are the leading company of the industry you belong you need a defensive strategy to maintain your status quo or level in the market field but if you're in the bottom line or the 2nd place then you need to have an offensive strategy to get ahead of the leading company. For the Shaw to deal with the opportunities and threats in the market, it needs to have the ability to engage in both offensive and defensive method in a dynamic sense.

The Canadian cable television industry has moved from a highly price regulated environment to one based on fair and sustainable competition. In such a competitive environment, cable companies have adopted defensive strategies, consolidating and realigning geographically to take advantage of potential administrative, operating and marketing synergies that arise from larger, focused operations. A first defensive

move is to counter the entrant's offer by improving its own offer. The company continued to invest in technology initiatives to recapture bandwidth and optimize their network, including increasing the number of nodes on the network and using advanced encoding and digital compression technologies such as MPEG4. They also continued to enhance their video offerings; introduced the Shaw Wireless Gateway and commenced trials of a 1 Gigabit Internet service. Second direct way to deal with the new competitor is to buy it. During 2010 the company also completed the purchase of Mountain Cablevision Ltd. ("Mountain Cable"). Mountain Cable is based in Hamilton, Ontario and was one of the larger remaining independent cable companies in Canada. The outlook for the Canadian cable industry is attractive and acquiring Mountain Cable represented a unique opportunity to grow their core business. The third way to deal with a competitive entrant is to avoid it by moving on. In April 2010 Shaw announced its intention to move forward on the rollout of its Wireless strategy with planned launches anticipated to commence late in calendar 2011. They commenced their strategic wireless infrastructure build in 2010. In 2009, they acquired approximately 20 megahertz of spectrum across most of their cable footprint and received their ownership compliance decision from Industry Canada. This year they started to expand their infrastructure to include Wireless which is a new business to the company that will open up new market to avoid face to face fight with their competitors. Finally, the company keeps potential entrants out through behavior channel control. Through expanded geographic footprint that stretches across Canada and owns the specialty channels acquired from Alliance Atlantis Communications Inc. in 2007, the company has a close connect with customers. They know what the customers want and their channel is established to satisfy their customers. As results, the company controlled the behaviors of customers.

REFERENCES

[1] Bell Canada Annual Report 2010
[2] Carpenters, M.A., Sanders, W.G., (2012). Strategic Management: A Dynamic Perspective Concepts, First Canadian Edition, Pearson Education Canada.
[3] ICT Sector Statistical Report, (n.d.) Retrieved June 08, 2011 from http://www.ic.gc.ca/eic/site/ict-tic.nsf/eng/h_it06143.html
[4] KPI Telecommunication, (n.d.) Retrieved from http://kpilibrary.com/search?s=&category%5B%5D=607
[5] Maketwire.com (n.d.) Shaw Communications Inc. Announces Executive Leadership Appointments. Retrieved June 07, 2011

Information Technology and Computer Application Engineering – Liu, Sung & Yao (Eds)
© 2014 Taylor & Francis Group, London, ISBN 978-1-138-00079-7

Research on training modes of vocational values for maritime students based on online web technology

Zhongxia Cai & Baohua Zhu
Maritime College, Shandong Jiaotong University, Weihai, ShanDong, China

ABSTRACT: Online survey technology is employed to collect students' thoughts on vocational values. Based on the survey results, problems existing in current training mode to vocational values are found. Through analyzing reasons resulting in these problems and combining with the characteristics of marine education, some advices, like Integrating theory education with practice, applying example education scientifically, improving educator's quality and utilizing internet to help with campus culture building, are applied to modify the training mode. Online survey results show that these measures get noticeable effect.

Keywords: Online web technology; Orientation; Professional ethics; Practices

1 GENERAL INSTRUCTIONS

Most graduators from maritime colleges devote themselves to ocean shipping industry. Compared to other graduators, they are required to have better vocational values and habit because of hard and dull working environment.

Based on marine incident analysis, it is found that 38% marine incidents result from deck officers' bad mood and poor sense of responsibility. At present, maritime graduates generally lack right vocational values. The lack not only cannot guarantee the security of vessels, but also leads to the flight of marine talents. According to the latest research, the wastage rate of graduates employed by shipping industry is 30% in 2 years, and rises to 50% in 5 years.

An excellent maritime graduate should not only be on top of professional knowledge and skill, but also has good vocational values. Maritime colleges shoulder the responsibility to help students to develop right vocational values. Current maritime colleges lay emphasis on curricula education and Competence Test of oceangoing seafarers, but pay less attention on cultivation of vocational values. The core journals in this industry, such as '*Journal of Dalian Maritime University*' and '*Journal of Shanghai Maritime University*', publish scarcely any article about vocational values cultivation. Therefore, it is essential to research the new cultivating mode of vocational values and bring this into maritime talent training plan. Because marine colleges distribute in wide ranges, online survey technology is applied. Based on sufficient practical investigation and discuss, some optimized cultivating methods are put forward.

2 THE CURRENT SITUATION

The current college students' ideology shows diversity, otherness and independence. Besides the main vocational values of marine college students, some misplaced values, such as utilitarianism and hedonism, turn up.

2.1 Weakened socialist core values

The statistics, based on the questionnaires for hundreds of students from several marine colleges, also prove this point. The result is shown in Figure 1 and Figure 2.

The result shows that marine college students take individual vocation and livelihood into account more with the escalation of various social disputes in the period of social transition. They cannot emotionally and rationally realize the historic and social

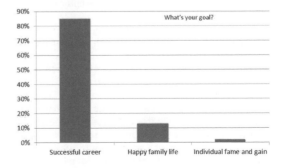

Figure 1. Survey result for life goal.

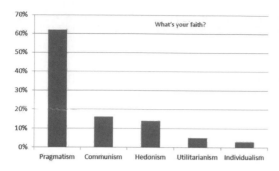

Figure 2.   Survey result for faith.

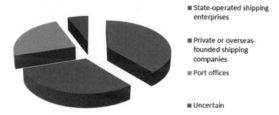

Figure 3.   The survey result of career orientation for 463 maritime graduates.

responsibility on their shoulders. These weaken their socialist core value in varying degrees.

2.2   *Inconstant career orientation*

The survey shows that most maritime students only concern the income and environment when applying for a job with the marketization of employment. Their career ideal is dim. They often change the original intention to engage in other industry. The survey result of career orientation for 463 maritime graduates is shown in Figure 3.

Figure 3 shows that the rank is state-operated shipping enterprises, private or overseas-founded shipping companies, and port offices in descending order. Most graduates take state-operated shipping enterprises as the first choice because of better welfare and abundant fund. Private or overseas-founded shipping companies also attract some graduates by good wages. Though the salary is small, a part of graduates enter into port offices in view of family factors.

The survey of service years engaging in ocean shipping is shown in Figure 4. It shows that majority graduates do not want to devote themselves to ocean shipping for long. They will switch into other industries when their position and qualification reach higher level. Influenced by adverse social impact, some students do not want to engage in the hard dull ocean shipping industry. Though signing employment agreements with ocean shipping enterprises, some of them break the contract and move into other industries in a short period. This leads to talent flight and short of high-quality officers which annoys the employers most.

Figure 4.   The survey of service years engaging in ocean shipping.

2.3   *Missing professional ethics*

Chasing interests is the marked feature of market economy society. Under this social situation, the income gaps are growing. Money worship becomes increasingly widespread. Interests become an important chip in job searching. The value of knowledge and morality is underestimated day by day. For students from marine colleges, professional skills are the key to long-term development. Although all marine colleges practice semi-militarization management to strengthen their professional education, some appearances of professional ethics missing, such as bad sense of organizational discipline, poor team working spirit, and short dedication spirit, still exist.

3   DEFICIENCIES IN CURRENT VOCATIONAL VALUES EDUCATION

3.1   *Contents and methods of vocational values education remain to be improved*

The current contents of vocational values education deviate from social reality. Education contents, such as basic principle of Marxism, Mao Zedong thought, Deng Xiaoping theory, communism, patriotism, and collectivism, still rest on formalized theory study. These contents without examples are unappealing to maritime students. And at the same time, the traditional dull education methods ignore the feeling of students. Therefore, good education result cannot be achieved.

3.2   *Less values education practices*

The knowledge of marine major covers widely. The professional knowledge involves engineering science, natural science, legal science, management science, and so on. Besides all kinds of exams organized by college, Maritime students still have to cope with various tests organized by national maritime safety administration. To satisfy the requirements of various international conventions and domestic regulations, maritime students must receive some special training programs like emergency treatment, advanced firefighting. All above make both college authorities and students have insufficient time and space to carry out effective values education practices.

3.3 Values education cannot integrate with professional education

The vocational values education requires wide participation of educator. Now, teachers to professional courses in marine colleges only assume teaching duty, and yet they shoulder educating duty less. For instance, during professional teaching process, they pay attention to theoretical knowledge teaching, some contents, like cultural background vocational ethics education playing important role in values development, are ignored. The lack of values education results from many reasons, including not only educators' deficiencies in comprehensive quality and capacity but also more is administers' neglect to values education. This education behavior without regulation and guidance develops the vacuum of values education.

4 MODIFIED TRAINING MODES OF VOCATIONAL VALUES

Above survey and analysis illustrate the importance of values education. Higher marine education contains both curricula education and vocational education. Vocational education is close to the requirements of social development. Therefore, based on current training mode, there are various methods can be utilized to modify the current training mode.

4.1 To innovate the carrier of values education

Integrate theory education with practice. Social practice is a key part in nautical education. It is the best way to carry out theory education through social practice. What does the marine colleges train are applied shipping talents. Centralized military training, labor education, social survey, and intensive professional practice should bring into theory education project. The industrial advantages should be brought into full play. Contact with shipping enterprises to carry out various practical activities like visiting new built vessels and navigation practice. Through these social practices, students can know the developing trend of shipping industry and control more professional skill. Hence, right values and thoughts can be internalized which can lay solid foundation for career development in future.

4.2 Apply example education scientifically

A fine example has boundless power. Apply example education scientifically can get twice the result with half the effort. All the maritime students are boys. They have similar personality characteristics. In this group, fine examples can drive others. Many outstanding maritime graduates devote themselves to marine adventure, and have made remarkable achievements. These elites in shipping industry can be set as fine examples. Educators can use various propaganda methods like speeches, reports and posters to encourage students. The vivid experiences from these elites are close to students' thought, are persuasive, and can drive students to build right vocational values.

4.3 Utilize internet to help with campus culture building

One of the most important factors affecting values education is campus culture environment. Maritime students' spare time is limited due to heavy learning tasks. And with the rapid development of internet technology, students spend most arrangements after school on internet. Now, web culture is deeply affecting students' values. Therefore, web culture building has become an important way to perform values education. Website close to maritime students' thoughts should be built to grab the internet position. Well-designed contents on values education in the website can play subtle role. Combine class education with internet education can expand channels and spaces to values education greatly.

4.4 Improve values educator's quality

Administrators from maritime colleges should build up the values educators' ranks. On one hand we should improve welfares to allow more educators take delight in values education, on the other hand administrators should create conditions to enhance the educators' training, for examples, on-the-job training, speeches, guiding activities, visits, and exchanges. Improve values educators' quality can play the guiding role on values education better.

4.5 Improve students' psychological quality

As a special group, seafarers have to resist various working stress and harsh conditions. Long-term work at sea, less communication with family members, and walled social network result in autism, depression, and false vocational values. Therefore, seafarers should have strong psychological enduring and self-adjustment capacity. In order to improve students' psychological quality, marine colleges should set psychological health and education office and staff it. Through grasping students' psychological condition by investigating, this office can carry out various targeted psychological health courses to guide students. Different materials should be provided to students in different grades. These measures can be helpful for the development of students' right vocational values.

5 CONCLUSIONS

Shipping industry is an important basic industry in national economy. The market focus is transferring towards Asia. Our shipping industries should catch the strategic opportunity, and progress towards powerful shipping country. The rapid expansion of shipping industry needs the strong support from. Right vocational values are a basic requirement to high-quality shipping talents. Marine colleges are the vital link. It is essential to modify the current training mode of vocational values. It is hoped that these advices can help with the maritime students' vocational values education.

This research was financially supported by Shandong Jiaotong University Scientific Research Fund Project (R201240). We thank Pro. Ren Wei, who provided constructive criticism and helped improve the manuscript.

REFERENCES

Guan Z.J. & Lv H.G. 2008. Research on marine talents drain and countermeasures. *Maritime Education Research* (2):4–8.
Liu Z. & Tian R. 2010. Research on Strengthening and Improving Ideological and Political Education Work for Maritime Professional Students. *Logistics Engineering and Management* 32(12): 166–168.
Wang C.Q. 2002. New Concepts for China Maritime Education in the New Century. *Journal of Nantong Vocational & Technical Shipping College* (01):1–5.
Liu Z.J. & Wu Z.L. 2008. China's Measures for Sustainable Development of High Quality Seafarers. *China Maritime Safety* (05):28–31.

Information Technology and Computer Application Engineering – Liu, Sung & Yao (Eds)
© *2014 Taylor & Francis Group, London, ISBN 978-1-138-00079-7*

An adaptive context management framework for supporting context-aware applications with QoC guarantee

N. Xu, W.S. Zhang, H.D. Yang, X.G. Zhang & X. Xing
College of Information Science and Technology, Dalian Maritime University, Dalian, China

ABSTRACT: Context awareness has become a necessity for intelligent applications that have the ability to adapt their behaviors to the changing context. However, it is still a great challenge to enable context awareness due to lack of adequate infrastructure to support context-aware applications. In this paper, we present a modular context management framework in order to afford effective supports for context acquisition, preprocessing, representation, interpretation and utilization to applications. It supports not only context-aware applications adaption, but also adaption of the context management framework itself. In addition, a formal context model based on OWL ontology has been investigated to facilitate context modeling and reasoning, and to enable knowledge sharing and reuse. Moreover, we extend the context model with meta-context as the additional characteristics to improve quality of context information which can help taking appropriate adaptation decisions. Finally, we present a performance study for our prototype.

Keywords: Context management; Context awareness; Quality of context (QoC); Context acquisition; Automatic rule generation; Context reasoning

1 INSTRUCTIONS

Context awareness, as one of the key technologies applied in context-aware computing, makes applications aware of the situation of their users and their environments so as to follow changing situations, and offer assistance by adapting their intended functionalities to the current context. According to Dey & Abowd (2000): "Context is any information that can be used to characterize the situation of an entity". This context information includes location, time, user's activities, device's capabilities etc. Additionally, we add the concept of "quality of context" (QoC) (Buchholz et al. 2003) to express the quality characteristics of the context information.

Context-aware computing has been drawing much attention since it was proposed about a decade ago. However, context-aware applications have never been widely available to everyday users due to lack of adequate infrastructure to support context awareness (Gu et al. 2005). Therefore, we argue that a generic context management framework is required to enable effective management of context for supporting context-aware applications. It is responsible for context information acquisition, transformation, retrieval, structured storage and interpretation, and the decision making to initiate certain actions. The benefits for such an infrastructure detail as follows. Firstly, it can reduce the cost by sharing context sources and context management functionality. In addition, it can handle the complexity by taking care of context processing tasks that are too "heavy" for lightweight application devices. Moreover, through providing dynamic discovery of (new) context sources, and supporting information exchange between such sources and applications, it can offer richer functionality.

In this paper, we present a framework for managing context that supports context-aware applications. The proposed context management framework consists of several modules that can acquire, preprocess, fuse and use context information. Using this framework, both the applications and the context management framework are subject to adaptation triggered by a changing context. The proposed context management framework initiates both deployment time and runtime adaptations by providing all the necessary information to activate the adaptations. On the other hand, specific components of the context management framework can be eliminated or replaced if they are of no or improper use to any applications. In addition, an OWL ontology-based context model is described, which is shared by these components so that they can process their tasks. Moreover, in order to improve quality of context information which can help taking appropriate adaptation decisions, we extend the context model with meta-context as the additional characteristics of context information for describing quality of context.

The remainder of this paper is organized as follows. Section 2 describes the works related to this study. In section 3, both context information and quality of context are modeled to support effective and better quality context management. Section 4 is devoted to the architecture of the proposed context management framework. The performance evaluations are discussed in section 5. Finally, section 6 draws conclusions and future work.

2 RELATED WORKS

Over the last decade, many researchers have been working on development of context-aware applications. Numerous context frameworks including toolkits and middlewares have been developed.

Henricksen et al. (2003) suggest an outstanding context model using graphical notations and propose a context management infrastructure based on this model. They model various context characteristics and quality of context. However, their model lacks formality and can't provide context management in interoperable manners.

Chen (2004) proposed an agent-oriented infrastructure CoBrA for context representation, sharing knowledge and user's privacy control, which includes a context broker in order to maintain a shared context model. However, the demonstration domains are limited and the rules are tightly defined the specific relationship among user's intentions, activities and services. Moreover, their approaches did not address the issue of quality of context.

Gu et al. (2005) propose a Service-Oriented Context-Aware Middleware (SOCAM), based on an OWL encoded context ontology called CONON which models context information only.

COIVA (Hervás & Bravo 2011) is a context-aware and ontology-powered information visualization architecture. Its reasoning engine combines ontology reasoning and rule-based reasoning, enabling the generation of high-level context at runtime. However, it is designed with the particular application in mind and without considering the aspect of quality of context. In addition, rules in its reasoning engine are defined manually by developers or users.

In comparison with these approaches mentioned above, we identify three main advantages of our framework. Firstly, not only the context information, but also the quality of context is modeled based on OWL ontology technology. In addition, both applications and the context management framework itself are subject to adaptation triggered by a changing context. Moreover, the high-level and better quality context information can be inferred by the proposed framework. Particularly, rules in our framework can be generated automatically without interacting with users or developers.

3 THE FORMAL CONTEXT MODEL

3.1 Modeling context information

Context-aware applications need a unified context model which is flexible, extendible and expressive to adapt the variety of context attributes and their relations. The ontology technology can fulfill these requirements (Strang & Linnhoff-Popien 2004). Therefore, we deploy an ontological context model called CACOnt to represent context information. Our context model is distributed into the upper general context ontology for the common concepts and relations, and the domain context ontology for domain-specific environment. The ontology for each sub-domain may be dynamically 'plug in' the upper general ontology when the environment changed. Through our research in context-aware computing, we observe that user, device, service, location, time and environment are the most fundamental elements for characterizing the executing situation of general context-aware systems. Details of this context model are illustrated in our previous work (Xu et al. 2013).

3.2 Modeling meta-context

A large number of sensors are included in the context-aware computing environment, and all the sensors may produce errors. Accordingly, the high-level context which is inferred from the sensor data may also contains errors. Especially, context-aware computing environment is dynamic and heterogeneous, so the context information is changed rapidly, and the updating intervals also vary with context types. Taking into account the quality of these data becomes a corner stone of an efficient context management solution (Eunhoe & Jaeyoung 2008). Therefore, the context model needs additional elements to describe quality of context, such as the certainty, accuracy and freshness etc. Accordingly, we extend the context model with meta-context to describe quality of context using "owl:AnnotationProperty" which allow annotations on classes, properties, individuals and ontology headers. This information is generally used to maintain quality of context, rather than interpreted or used in OWL DL reasoning process so as to not influence the performance of reasoning.

We explain mate-context in details as follows. The annotation properties "hasAccuracy" and "hasPrecision" are defined for describing accuracy of context information with different context sources. In addition, we define annotation property "hasCertainty" for certainty of context information whose value measures with some method such as probability or fuzzy logic and is usually between 0 and 1. Moreover, for the freshness of context information, we define "productionTime" and "lifeTime" for describing the creation time of context information and average interval of valid information updating respectively. Figure 1 shows an example description of these annotation properties.

4 THE FRAMEWORK ARCHITECTURE

In this paper, the job of context management is performed by the proposed framework. It consists of

```
<owl:ObjectProperty rdf:ID="locatedIn">
 <hasSource>
  <Source rdf:ID="WirelessNetwork"/>
 </hasSource>
 <hasAccuracy>
  <Accuracy rdf:ID="locatedInAccuracy"/>
   <unit rdf:datatype="http://www.w3.org/2001/XMLSchema#string">centimeter</unit>
   <value rdf:datatype="http://www.w3.org/2001/XMLSchema#float">7</value>
  </Accuracy>
 </hasAccuracy>
 <hasPrecision>
  <Precision rdf:ID="locatedInPrecision"/>
   <unit rdf:datatype="http://www.w3.org/2001/XMLSchema#string">percent</unit>
   <value rdf:datatype="http://www.w3.org/2001/XMLSchema#float">75</value>
  </Precision>
 </hasPrecision>
 ......
</owl:ObjectProperty>
 .....
<Library rdf:ID="library">
 <noise rdf:resource="#SoundMeter">
 <productionTime>
  <Time rdf:ID="library.noise.productionTime">
   <hour rdf:datatype="http://www.w3.org/2001/XMLSchema#int">10</unit>
   <minute rdf:datatype="http://www.w3.org/2001/XMLSchema#int">1</value>
  </Time>
 </productionTime>
 <lifeTime>
  <Time rdf:ID="library.noise.lifeTime">
   <minute rdf:datatype="http://www.w3.org/2001/XMLSchema#int">1</value>
  </Time>
 </productionTime>
 <hasCertainty>
  <Certainty rdf:ID="locatedInCertainty"/>
   <measure rdf:datatype="http://www.w3.org/2001/XMLSchema#string">membership</unit>
   <value rdf:datatype="http://www.w3.org/2001/XMLSchema#float">0.9</value>
  </Certainty>
 </hasCertainty>
</Library>
```

Figure 1.   The example description of meta-context.

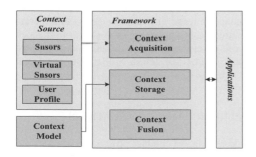

Figure 2.   Overview of the context management framework architecture.

three modules, each with a specific duty: *Context Acquisition*, *Context Storage* and *Context Fusion*. See Figure 2 for a general overview. The following subsections treat each of them respectively for discussing how the context infrastructure is able to manage context information with QoC guarantee and adaptation support.

4.1   Context acquisition

This module gathers information from sensors or other providers on the framework itself or in the neighborhood. We distinguish three sources of context information. Firstly, Information can be acquired by sensors or virtual sensor services. This raw information is prone to measurement errors and needs preprocessing before being usable. Another source of information is acquired through user profiling. Information can also be exchanged with other parties included in the framework. This information can be raw or be derived from some other kinds of context information.

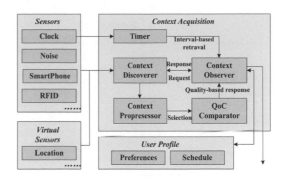

Figure 3.   Overview of the context acquisition module.

An overview of all components involved in context acquisition is given in Figure 3. We have several components acting as information providers: *Sensors*, *Virtual Sensors* and *User Profile*. The *Context Observer* is the initiator of all information requests and monitors information that triggers application adaptations, such as changes in current location.

These requests are sent to the *Context Discoverer*, which forwards them to the information providers and responds the requester with better quality information. In order to improve the efficiency and decrease packet drop rate, we design the Semantic Discovery Protocol (SDP) which is taken by *Context Discoverer*. This protocol cleaves the response packet into meaningful fragments according to whether the fields are required. Each fragment is transmitted after a random time interval. Since each packet of SDP is comprehensive, a more accurate list of context providers, which are more capable of providing the required context information, can be obtained.

The *Context Preprocessor* offers the transformation, normalization and filtering function for the acquired raw context. When several context sources provide similar information in response to a request, the *QoC Comparator* selects "the most reliable, accurate and fresh information" and forwards it back to the requester.

Moreover, a *Clock* also periodically sends a time signal and pushes this information to *Timer* component. Then, *Context Observer* can send a request to *Timer* to be periodically notified to allow interval-based information monitoring. Additionally, in terms of different devices' processing capabilities, components can be replaced or reduced in complexity or even eliminated for supporting self-adaption.

4.2   Context storage

A context repository ensures persistency of context information. This repository is implemented as three different containers and provides a set of API's for other components to query, add, delete or modify context knowledge. See Figure 4 for an overview.

In this module, *Fact Container* which holds instances of concepts from context ontologies, maintains not only the most recent value of a certain context

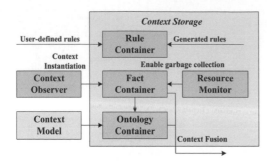

Figure 4. Overview of the context storage module.

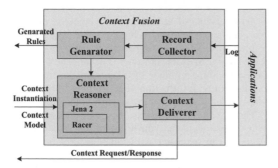

Figure 5. Overview of the context fusion module.

attribute received from *Context Observer*, but the history context information as well. The ontological context model is pre-loaded into the *Ontology Container* at the initiation. And the *Rule Container* is in charge of the maintenance of rules for context reasoning and adaptation decision making. The *Resource Monitor* is responsible for enabling garbage collection on the *Fact Container* when running low on storage capacity for supporting self-adaption. If one of the supplemental ontologies (i.e. not our upper general context ontology, which serves as a common ground), is no longer referred to the current facts, then it can be removed from the *Ontology Container* as well. Storage can be unnecessary or outsourced to a device with more storage capacity.

4.3 Context fusion

This part reasons on context information to provide suitable information for initiating the context-aware applications adaptation.

A general overview of *Context Fusion* is given in Figure 5. The *Context Reasoner* has the functionality of deriving new facts by combining information from the *Fact Container* and the *Ontology Container*. These new facts are stored again in the *Fact Container*. Another functionality of this component is to detect inconsistency and conflict in the *Ontology Container*. In this reasoner, two kinds of reasoning are currently supported: ontology reasoning and rule-based reasoning. In order to solve that rules in most of the existing context-aware systems are defined manually by users

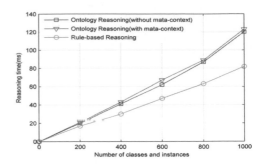

Figure 6. The performance of context reasoning.

or developers, we propose a new method for generating rules automatically which is applied to *Rule Generator* component in terms of the records collected by *Records Collector*. Lack of space forbids further treatment of this method here. The *Context Deliverer* component is responsible for providing the necessary information to the triggers (not shown in the figure) that activate context-aware applications adaptation. This adaptation is initiated by several triggers that fire due to a change in context. The necessary information is delivered from the context repository, or deduced after several reasoning steps. Another functionality of this component is to forward information requests or responses. In the implementation, we investigated Racer (Haarslev & Moller 2003) and Jena 2 (HP Labs 2004) as the reasoning tools.

5 PERFORMANCE EVALUATIONS

In this section, we conduct two groups of experiments over our prototype system to evaluate its performance in terms of context reasoning and automatic rule generation. The results are shown in the following subsections.

Experiment 1: In this experiment, we study the performance of context reasoning. In our framework, not only context information, but also meta-context information is used. However, ontology processing needs a lot of computing resources, even performs poorly according to the weight of the ontology. Therefore we measured the context ontology reasoning performance as the number of triples of context ontology increased for each case of context information only and with meta-context (containing 200 triples) included respectively.

Figure 6 shows the results of context ontology reasoning and rule-based reasoning. It clearly shows that although meta-context increased the size of ontology, it didn't seriously increase the reasoning time for that the meta-context is described using "owl: AnnotationProperty" which is neither interpreted nor used in OWL DL reasoning process. This information is used in the proposed framework for effective context management with QoC guarantee.

Experiment 2: This group of experiment aims to test the effectiveness and efficiency of the automatic

162

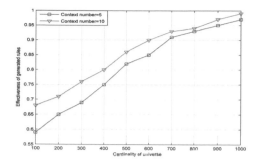

Figure 7. Effectiveness of the generated rules.

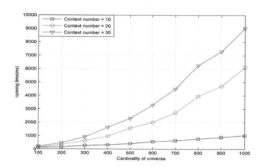

Figure 8. Efficiency of automatic rule generation.

rule generation method in terms of user operation records (i.e. cardinality of universe) increased with different number of involved context attribute. The relationships between the cardinality of universe and the effectiveness of generated rules, with the context number is 5 and 10 respectively, are represented in Figure 7. It clearly shows that effectiveness is increased with cardinality of universe, and approaches the maximum gradually. In other words, with the increasing cardinality of universe, the rules generated through the proposed method approach the rules defined manually.

However, running time is also increased with cardinality of universe. Figure 8 illustrates the relationships between running time and cardinality of universe, when the context number is 10, 20 and 30 respectively. We notice that the running time increases exponentially as a function of the cardinality of universe. Particularly, the more context attributes are, the faster growth is. Therefore, it is necessary to choose suitable cardinality of universe which is the key to guarantee both of effectiveness and efficiency.

6 CONCLUSIONS

This paper has described a context management framework which provided effective management of the context information for satisfying qualities of context information required by context-aware applications.

In summary, our research presents several contributions that set the proposed framework apart from previous works. Firstly, we defined the meta-context including context accuracy, certainty and freshness of the context information using OWL annotation properties for quality of context modeling. In addition, besides the applications, the framework itself is subject to adaptation triggered by a changing context. Furthermore, rules in our framework are generated automatically, rather than defined manually by users or developers. Future work will focus on uncertain context modeling and reasoning. In addition, security and privacy are the issues important to the context-aware computing environment, which must be addressed in the future framework.

ACKNOWLEDGEMENTS

This research is supported by the National Science Foundation of China (Grant No. 61272172) and the Fundamental Research Funds for the Central Universities in China (Grant No. 2011QN027).

REFERENCES

Buchholz, T., Küpper, A. & Schiffers, M. 2003. Quality of Context: What It Is and Why We Need It. *HP OpenView Univ. Assn.; Proc. conf., Geneva, Switzerland, July 2003.*

Chen, H. 2004. An Intelligent Broker Architecture for Pervasive Context-Aware Systems. *Ph.D. thesis,* University of Maryland, Baltimore County, Maryland, 2004.

Dey, A.K. & Abowd, G.D. 2000. Towards a Better Understanding of Context and Context Awareness. *Human Factors in Comp. Sys.; Proc. conf., The Hague, Apr. 2000.*

Eunhoe, K. & Jaeyoung, C. 2008. A context management system for supporting context-aware applications. *Embedded and Ubiquitous Computing.; Proc. IEEE/IFIP intern. conf., New York, 17–20 December 2008.*

Gu, T., Pung, H.K. & Zhang, D.Q. 2005. A Service-Oriented Middleware for Building Context-Aware Services. *Elsevier Journal of Network and Computer Applications* 28(1):1–18.

Haarslev, V. & Moller, R. 2003. Racer: A Core Inference Engine for the Semantic Web. *Evaluation of Ontology-based Tools.; Proc. intern. wkshp., Florida, USA, 2003.*

Henricksen, K. et al. 2003. Generating Context Management Infrastructure from High-Level Context Models. *Mobile Data Management.; Proc. 4th intern. conf., January, 2003.*

Hervás, R. & Bravo, J. 2011. COIVA: context – aware and ontology – powered information visualization architecture. *Software: Practice and Experience* 41(4): 403–426.

HP Labs. 2004. Jena 2 – A Semantic Web Framework. http://www.hpl.hp.com/semweb/jena2.htm.

Strang, T. & Linnhoff-Popien, C. 2004. A Context Modeling Survey. *Advanced Context Modelling, Reasoning And Management at UbiComp.; Proc. intern. wkshp., Nottingham, England, 7 September 2004.*

Xu, N. et al. 2013. CACOnt: An Ontology-Based Model for Context Modeling and Reasoning. *Computer Science, Electronic Technology and Intelligent System; Proc. intern. conf., Hangzhou, 22–23 March 2013.*

Information Technology and Computer Application Engineering – Liu, Sung & Yao (Eds)
© 2014 Taylor & Francis Group, London, ISBN 978-1-138-00079-7

Performance analysis of missile-borne SAR moving target parameter estimation by using Wigner-Ville Hough transform

Ying Liu, Dian-Ren Chen & Lei Chen
College of Electronic Information and Engineering, Changchun University of Science and Technology, Changchun, China

ABSTRACT: In missile-borne SAR imaging system, moving targets in the observation area will produce echo phase errors, resulting in moving target echo wave does not match with matched filter of stationary target, the image of the moving parts is fuzzy. This paper provides the Wigner-Ville Hough Transform (WVHT) formula, analyzes echo characteristics of the missile-borne SAR and deduces the chirp signal's WVHT. At last, the MATLAB simulation results of missile-borne sar moving target parameter estimation demonstrate that the WVHT is effective.

Keywords: SAR; moving target; WVHT; chirp signal

1 INSTRUCTION

In spaceborne and airborne synthetic aperture radar (SAR) system studying, moving target detection (MTI) technology is an important part. Missile-borne SAR/MTI technology since the 1980s continue to develop and be perfect. SAR technology of the high range resolution obtained through the LFMP compression. High azimuth resolution is achieved by the coherent processing of the relative motion that is between SAR and object with the echo signal. If the imaging scene exist unknown way of movement of the target, conventional SAR imaging methods can not work properly, resulting in the fuzzy image and orientation offset of the moving objects. Moving target detection and realization need to achieve Doppler frequency characteristics and try to re-focus imaging[1]. Using time-frequency analysis method to estimate the Doppler frequency and Doppler center frequency of the moving target is currently one of the hot spots. This paper analyzes detection effects of missile-borne SAR moving target with WVHT and detection simulation of point moving target with WVHT by using MATLAB.

2 ECHO MODEL OF MISSILE-BORNE SAR MOVING TARGET

This paper discusses the missile-borne SAR moving target detection, missile flight with high speed and a relatively short period of time, non-linear movement of the missile is approximately regarded as linear motion[2]. For easy analysis, the approximate

geometric relationships based on the the equivalent model of strabismus distance. Ground coordinate system (x-y-z) is regarded as coordinate system without considering the Earth's rotation, based on the geometric equivalent model diagram of strabismus distance shown in Figure 1.

Aircraft flight speed is v_a, Missile coordinates is $(0,0,h)$ at $t = 0$, ground moving target is $P(x_0, y_0, 0)$, the distance between missile and fight path is $R_c = \sqrt{y_0^2 + h^2}$, slant distance is $R_0 = \sqrt{x_0^2 + R_C^2}$, the speed of azimuth and distance are v_x and v_y, in the slant plane distance to speed is v_r, $\dfrac{v_r}{v_y} = \dfrac{y_0}{R_c}$ is achieved from figure 1, the slant distance between moving target and

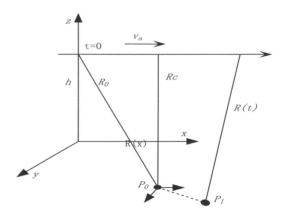

Figure 1. The equivalent geometric model diagram of strabismus distance.

$P_t(x_t, y_t, 0)$ is $R(t) = R(nT)$, we can based on Fresnel approximation

$$R(nT) \approx R_0 + \frac{1}{2R_0}(-2x_0v_a nT + v_a^2 n^2 T^2 + (2x_0v_x + 2y_0v_y)nT$$
$$+ (v_x^2 - 2v_a v_x + v_y^2)n^2 T^2) \tag{1}$$

The compressed echo signal that the moment $t = \tau$ trough the distance can be expressed as:

$$\sigma(x_0) = G(nT)\sqrt{\sigma_0(x_0, R_c)} \tag{2}$$

So visible, the echo signal with radar movement of the doppler frequency is:

$$f_r = \frac{-1}{\lambda R_0}(2x_0v_a - 2v_a^2 nT) \tag{3}$$

The doppler frequency from the target motion is:

$$f_i = \frac{-1}{\lambda R_0}(2x_0v_x + 2y_0v_y + 2(v_x^2 + v_y^2 - 2v_a v_x)nT) \tag{4}$$

So, the doppler center frequency and adjustable frequency generated by the radar's movement respectively are:

$$f_{rc} = \frac{1}{\lambda R_0}2x_0v_a \tag{5}$$

$$K_r = \frac{-2}{\lambda R_0}2v_a^2 \tag{6}$$

The doppler center frequency and adjustable frequency generated by the target motion from respectively are:

$$f_{ic} = \frac{-1}{\lambda R_0}(2x_0v_x + 2y_0v_y) \tag{7}$$

$$K_i = \frac{-2}{\lambda R_0}(v_x^2 + v_y^2 - 2v_a v_x) \tag{8}$$

Therefore, the target motion unpredictability will affect bearing to compression performance. Moving target's doppler center frequency and adjustable frequency is different from static target's, so in the static target radar image, moving target will not only bearing shif, and the image will also defocusing, amplitude decline[3].

3 CHIRP SIGNAL'S WVHT

WVHT algorithm is usually used for LFM signal parameter estimation occasions, its role is in time-frequency image after WVD transformation searching of straight line, the literature [11] giving the LFM signal's WVHT transformation can be expressed as[5]:

$$WH_x(f, g) = \sum_{n=0}^{N/2-1}\sum_{k=-n}^{n} R_{xx}[n, k]e^{-j(4\pi fk\Delta + gn\Delta)}$$
$$+ \sum_{n=N/2}^{N-1}\sum_{k=-(N-1-n)}^{N-1-n} R_{xx}[n, k]e^{-j(4\pi fk\Delta + gn\Delta)} \tag{9}$$

Among $R_{xx}(n, k) = x[n + k]x * [n - k]$, is $x(n)$'s autocorrelation function, $n = 0, 1, 2 \ldots N - 1$, Δ signal sampling period. Show that the type, if $x[n]$ is linear frequency modulation signal, on $x[n]$'s parameters point $WH_x(f, g)$ get maximum, when deviat (f_0, g_0), $WH_x(f, g)$ duration reduced quickly.

In the noise environment, hypothesis sample signal $x(n) = s(n) + v(n)$'s SNR is SNR_i, among $v(n)$'s mean is 0, power spectral density is $S(\omega) = N_0/2$ gaussian noise, $n(t)$'s mean is 0, power spectral density is $S(\omega) = N_0/2$ gaussian noise, Sample signal is $x(n)$, the output signal after the WVHT $P_x(\Omega)$'S SNR can be expressed as:

$$SNR_x = \frac{|P_x(\Omega_0)|^2}{\text{var}\{P_x(\Omega_0)\}} \tag{10}$$

Noise signal $v(n)$'s autocorrelation function:

$$r_{vv}(k) = E(v(n)v^*(n+k)) = \delta_v^2\delta[k] \tag{11}$$

The denominator of type (12) is

$$\text{var}\{P_x(\Omega_0)\} = E(|P_x(\Omega_0)|^2) - E^2(P_x(\Omega_0)) \tag{12}$$

among:

$$E(|P_x(\Omega_0)|^2) = \sum_n\sum_k [s(n) + v(n)]\cdot[s(n) + v(n)]^* F[\Omega_i]$$
$$= \frac{N^2 A^2}{2} + N\delta_v^2 \tag{13}$$

$$E(|P_x(\Omega_0)|^2) = \frac{N^4 A^4}{4} + \frac{3N^3 A^2\delta_v^2}{2} + \frac{3N^2\delta_v^4}{2} \tag{14}$$

So, $$SNR_x = \frac{SNR_i \cdot N^2/2}{SNR_i \cdot N + 1} \tag{15}$$

4 PERFORMANCE ANALYSIS

One of the key of Bomb spaceborne SAR moving target detection is estimating motion target doppler modulation frequency and the doppler center frequency, focusing the moving target on the image. Detaction missile SAR moving target with WVHT block diagram showing in figure 2[4].

When there is a moving target in the local surface observation area, according to the moving of the target state, In the signal of the Wigner time-frequency surface corresponds to appear the corresponding intermediate frequency signal energy distribution line.

Figure 2. Missile-born SAR moving target detection block diagram.

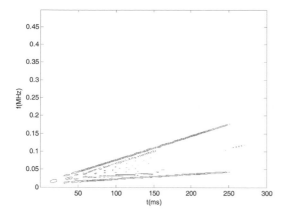

Figure 3. WVD of a moving point target and a stationary target echo.

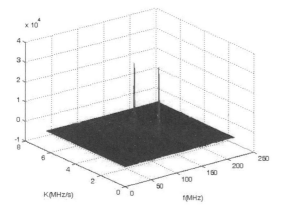

Figure 4. WVHT of a moving point target and a stationary target echo.

But these linear slope and starting point and stationary target echo in the Wigner time-frequency energy distribution on the surface of the straight slope and starting point is not the same. At the same time, In the Wigner time-frequency surface will appear still targets and moving target energy distribution cross-term, noise and interference signal energy distribution. The signal WVHT echo signal directly, in the transformation of the image will appear the corresponding very sharp peak point, According to the characteristics of SAR emission signal, Can eliminate WVHT graph of the stationary target the corresponding peak point, Leave the number of the peak point on behalf of the number of moving target, Then through the peak point coordinates of target motion parameters can be obtained.

Below are a moving target and a stationary targets, and simulation diagram 3 for Wigner time-frequency diagram 4 for WVHT diagram.

As seen in figure 3 and figure 4, WVHT can effectively inhibit the cross-term appeared in WVD figure. The two obvious peaks are the time-frequency fenergy linear mapping of the two target. Thus, it is easy to achive the target's motion parameters by echo signal WVHT.

5 THE CONCLUSION

This paper analyzed the missile-borne SAR echo characteristics, studied the missile-borne SAR moving target detection method based on WVHT, simulated the WVHT of the radar echo when the observations ground have moving target and stationary targets at the same time using MATLAB. The simulation results show that the missile-borne SAR moving target detection using WVHT can effectively inhibit the cross-term, It made the detection result have a batter time and frequency resolution, improved the detection and estimation accuracy of the moving target parameter.

REFERENCES

[1] Wen Jing-yang, Zhang Huan-yu,Wang Yue. Parameters Estimation Algorithm of LFM Puls Compression Radar Signal[J]. *Transactions of Beijing Institute of Technology*, 2012, 32(7): 746–750.

[2] Geroleo, F.G.; Brandt-Pearce, M. Detection and Estimation of Multi-Pulse LFMCW Radar Signals [C]. *Radar Conference, 2010 IEEE, Dept. of the Army, Charlottesville, VA,* USA, 2010: 1009–1013.

[3] Liu Feng Xu Hui-fa, Tao Ran, Detection and Parameter Estimation of Symmetrical Triangular LFMCW Signal Based on Fractional Fourier Transform[J]. *Jounal of Electronics & Information Technology,* 2011, 33(8):1865–1870.

[4] Qu Qiang, Jin Ming-lu, Adaptive Fractional Fourier Transform Based Chirp Signal Detection and Parameter Estimation [J]. *Jounal of Electronics & Information Technology,* 2009, 31(12):2930–2947.

[5] Li Ya-Chao, Wang Hong-xian, Xing Meng-dao, Bao Zheng. The Analysis of the Characteristic and the Method of the Parameters Estimation of LFM Signal Based on STTFD[J]. *Acta Electronica Sinica.* 2009, 37(9):2102–2108.

[6] Geroleo, F.G.; Brandt-Pearce, M. Detection and Estimation of LFMCW Radar Signals [J]. *IEEE Transactions on Aerospace and Electronic Systems,* 2012, 48(1):405–418.

Information Technology and Computer Application Engineering – Liu, Sung & Yao (Eds)
© 2014 Taylor & Francis Group, London, ISBN 978-1-138-00079-7

A matching model of cloud service between information resources supply and demand for networked manufacturing

Xiuzhen Feng & Bin Wang
Economics & Management School, Beijing University of Technology, Beijing, China

ABSTRACT: Networked manufacturing has been greatly promoting the manufacture business, and creating a significant opportunity for developing cloud service of information resource management. Based on the concept of cloud service of information resources, a two-stage matching model in this paper is proposed to deal with the matching problem between supply and demand, which could match possible partners for networked manufacturing enterprises. Firstly, the model was constructed for clustering the characteristics from both sides of supply and demand in order to make matching groups. Then, we adopted a weighted Euclidean distance calculation to calculate the similarity between demand side and supply side, which could find more suitable partners automatically. Finally, a numerical example was presented to test effectiveness and feasibility of the model.

Keywords: matching model; cloud service; smart match; information services

1 INTRODUCTION

Networked manufacturing[1–2] is a dynamic alliance among enterprises. When enterprises have been joining in a cooperative network for their production, they have to face the dynamic relationship in the cooperative network because the relationship may be dissolved or rapidly restructured to form new connections. Accordingly, networked manufacture environment makes the cooperative relationships among the manufacture enterprises are more flexible, more varied, and more complicated.

Taking advantage of the rapid development of information technology and network technology, networked manufacture enterprises get promotion with cooperative relationships from many perspectives, such as design, production, operation, service and so on. For example, manufacture enterprises in the network could be able to access more qualified manufacture resources, coordinate effectively and efficiently, complement advantages mutually, and benefit each other for further development. In the networked manufacture environment, all enterprises have been playing the role as the nodes of manufacture network. The business relationships among enterprises construct the flexible, dynamic, complicated manufacturing network.

In practice, enterprises in the networked manufacture environment have been challenging from two perspectives. One is quickly find potential and qualified suppliers in the network. The other one is efficiently determine the proper suppliers from the queue of many enterprises. According to literature study, many research efforts have been concentrating

on matching study between supply and demand. However, the research contributions on matching show a trend of the diversity. Previously, the matching study interests were focused on the following areas, such as human resources management[3–4], the e-commerce management[5], and the matching in knowledge service[7–8]. Modeling study has been more attractive. For example, a matching model was constructed between supply and demand, which is based on neural network and deals with the matching issues on an integrated platform for third-party logistics[6]. In order to solve the two-side matching on technological knowledge supply and demand, a two-phase decision analysis was proposed[7]. Lu Jia, etc.[8] created a bilateral matching optimization model with multi objectives. Research contributions above paved a consolidate foundation for matching business relationship between supply and demand. However, manufacture enterprises must actively participate in the matching procedure, which means that both supply and demand side have to be effectively interacted repeatedly for matching. To quickly obtain the business relationship between supply and demand, and automatically determine the proper matching groups, our research efforts will concentrate on matching business relationships between supply and demand in the networked manufacture enterprise automatically and intelligently.

Cloud service of information resources utilize cloud computing technology[9] and cloud service mode[10–11] to reintegrate and redeploy information resources distributed in different regions, providing high-quality services which are diversification, automation, intelligent to the users. In the networked manufacturing environment, cloud service of information resources could encapsulate information resources of manufacture enterprises with a variety of service modes. Information resources include all kind of enterprise information, such as research and development information resources, design information resources, production assembly information resources, product device information resources, transportation information and warehouse information resources. It is significantly important to provide networked manufacture enterprises with information matching service automatically, rapidly, and conveniently.

We would adopt the concept of cloud service with information resources, and focus on the automatic matching between supply and demand for networked manufacture emprises, which expected to establish a new business relationship or optimize the existing business relationships.

2 INTELLIGENT MATCHING MODEL BASED ON CLOUD SERVICE WITH INFORMATION RESOURCES

To obtain demand/supply information from manufacture enterprises, we employed a web crawler approach[12] that combined with natural language processing technology and regular expression technology, which could extract information from the relevant manufacturing enterprises websites.

For example, Zhuhai Benxiang tires wholesale Co., Ltd. has the supply information, such as supply continental tires 225/55R16 94V, unit price RMB570, supply 650, valid until March 23, 2013. Some useful information could be further processed and abstracted like tire type, brand, load index, speed level, unit price, quantity and the validity date. To describe the characteristics of the tire, we pick up the following items including load index, speed level, unit price and quantity, which could be represented by y. In this way, the characteristics of products can be expressed as $Y = (y_1 \ldots y_p \ldots y_q)$.

Assuming the products supply information could be presented as $S = \{S_0 \ldots S_i \ldots S_n\}$. Si means products supply information. Accordingly, the features for S_i could be described as $YS_i = (y_{i1} \ldots y_{ip} \ldots y_{iq})$. Moreover, products demand information of demand enterprises could be presented as $D = \{D_0 \ldots D_j \ldots D_m\}$. D_j indicates product demand information, so the features for D_j can be described as $YD_j = (y_{j1} \ldots y_{jp} \ldots y_{jq})$, $j \in 0, 1 \ldots m, i \in 0 \ldots n, p \in 1 \ldots q$.

In order to match the supply and demand among supplying enterprises and multiple demanding enterprises efficiently, we constructed a supply and demand matching model which can be conducted as two stages: the first stage is merging and clustering both supply and demand based on supply and demand characteristics[13]. The second stage is calculating the matching degree between the supply and demand within each class.

2.1 Supply and demand feature clustering

Assuming a supply enterprise and a demand enterprise has business relationship. In practice, this partnership might not be the best cooperation. Networked Manufacturing environment could build up a new dynamic alliance for manufacture enterprises. All of enterprises involved in the networked manufacturing could organize or construct new cooperation. From time to time, those manufacture enterprises could adjust their processes, develop better cooperative relationships, even create new cooperative relationships.

Using K-means clustering algorithm to merge and cluster the supply characteristics and the demand characteristics together, we could get the similar groups based on supply and demand characteristics.

The index set for K-means clustering algorithm are $Y = (y_1 \ldots y_p \ldots y_q)$. The sample set are $\{S_0 \ldots S_n, D_0 \ldots D_m\}$, representing products supply and demand information. $\{S_0 \ldots S_n\}$ are products supply information set, $\{D_0 \ldots D_m\}$ represent the products demand information set. The sample No. i for index y_p can be expressed as $y_{ip}, p \in 1 \ldots q$.

STEP 1: Standardize index value. The indexes of supply and demand characteristics can be divided into benefit index and cost index. If the value of benefit index is larger, the comprehensive performance will be better. However, if the value of cost index is smaller, the comprehensive performance will be better. In order to eliminate the impact of the dimensional differences, we use Equation 1 and 2 to standardize the supply and demand characteristics.

Benefit index standardizing formula:

$$x_{ip} = y_{ip}/\max(y_{ip}) \tag{1}$$

Cost index standardizing formula:

$$x_{ip} = \min(y_{ip})/y_{ip} \tag{2}$$

STEP 2: Determine initial cluster centers. K-means algorithm divides sample into k^* divisions: $C = \{c_k, k = 1, \ldots k^*\}$, each division represents a class c_k, each class c_k has a cluster center v_k. Randomly select k^* objects from data objects as initial cluster centers, each object represents a class cluster center. Set the initial number of iterations b is 0.

STEP 3: Determine the initial classification. Select Euclidean distance as the similarity criterion for the data objects in the sample. If the distance of two data objects is smaller, then similarity is higher, the difference is smaller; while if the distance is greater, then the similarity is lower, the difference is greater. According to the distance between data objects and their cluster center $v_k^{(b)}$, we can assign data objects to the class in accordance with the nearest criteria.

STEP 4: Calculate standard measure function. First, we calculate the sum of squares of distance from points in class c_k to cluster center $v_k^{(b)}$ (Equation 3), then calculate the total sum of squares of distance (Equation 4).

$$J(c_k)^{(b)} = \sum_{x_i \in c_k} \left\| x_i - v_k^{(b)} \right\|^2 \quad (3)$$

$$J(C)^{(b)} = \sum_{k=1}^{k^*} J(c_k)^{(b)} \quad (4)$$

STEP 5: Updating cluster centers and standard measure function. We calculate the mean value of all objects in each category as the new cluster center of the class (Equation 5). $|c_k|$ indicates the number of data objects in the class c_k. New cluster centers for all data objects are divided again. Then, to calculate the sum of squares of distance from all the samples to their class cluster centers, that is $J(C)^{(b+1)}$.

$$v_k^{(b+1)} = \frac{1}{|c_k|} \sum_{x_i \in c_k} x_i \quad (5)$$

STEP 6: Repeat step 3 to 5 until cluster centers and standard measure function does not change, then we could obtain the optimal cluster centers and the convergence standard measure function.

2.2 Calculate the matching degree between supply and demand

Based on the clustering previously, cloud service of information resources could calculate the matching degree between supply and demand. According to the matching degree, we expect automatically to find out the best partnerships for the supply enterprise S_0.

The matching degree between supply and demand could be calculated with the weighted Euclidean distance formula (Equation 6). The procedure is to calculate the distance between supply and demand first, and then to use the distance and similarity conversion formula[14] (Equation 7) to convert the Euclidean distance into matching degree between supply and demand. According to the calculated matching degree, we could sort out the demand enterprises. If the value of matching degree is greater, it means that both products supply and demand characteristics are more similar. In this case, we think that the possibility of cooperation is higher, cooperative relations of enterprises will be more stable, which specified that selecting the best partner for the supply enterprise can make use of the value of the matching degree.

$$Dis(S_i, D_j) = \sqrt[2]{\sum_{p=1}^{k} \omega_p (x_{ip} - x_{jp})^2} \quad (6)$$

$$Sim(S_i, D_j) = \frac{1}{1 + Dis(S_i, D_j)} \quad (7)$$

ω_p is the weight of each indicator, $i \in 0, 1, \dots, n$, $j \in 0, 1, \dots, m, p \in 1, 2, \dots, q$.

3 THE CASE ANALYSIS

The case is to verify the matching model between supply and demand. Cluster sample are 10 supply information of products and 9 demand information of

Table 1. Products supply and demand information table.

No. \ Y	y_1 Unit price	y_2 Quantity	y_3 Load	y_4 Speed
S_0	1590	660	775(99)	240(V)
S_1	1200	1000	690(95)	300(Y)
S_2	1160	1100	690(95)	270(Y)
S_3	1020	900	670(94)	240(V)
S_4	1800	720	775(99)	270(W)
S_5	980	980	690(95)	240(V)
S_6	1020	1000	690(95)	300(Y)
S_7	1700	690	775(99)	300(Y)
S_8	1220	720	710(96)	270(W)
S_9	1200	790	710(96)	270(W)
D_0	1560	650	775(99)	270(W)
D_1	1000	800	670(94)	240(V)
D_2	1160	1050	690(95)	240(V)
D_3	1000	990	690(95)	300(Y)
D_4	1180	1000	670(94)	240(V)
D_5	1680	670	775(99)	240(V)
D_6	1560	920	775(99)	210(H)
D_7	1200	680	710(96)	270(W)
D_8	980	900	670(94)	240(V)

Table 2. Load index conversion table.

Load index	94	95	96	99
Actual load (kg)	670	690	710	775

Table 3. Speed level conversion table.

Speed level	Q	R	T	H	V	W	Y
Speed (km/h)	160	170	190	210	240	270	300

products, as shown in Table 1. ($S_0 \dots S_9$) represent supply information of tires 225/55R16; ($D_0 \dots D_8$) are demand information for tires 225/55R16. Samples include four indicators: unit price, quantity, load, speed. Unit price is cost index, quantity, load and speed is benefit index. The weight of the four indexes are set to $\omega_1 = 0.35$, $\omega_2 = 0.3$, $\omega_3 = 0.2$, $\omega_4 = 0.15$.

The changes of load index and speed level for tire 225/55R16, mainly has the following varieties: {99Q,99T,99H,99V,99W,99Y,96W,95Q,95R,95H,95V, 95W,95Y,94V}. In order to facilitate the calculation, load index and speed level are converted into the actual load and speed. The conversion rules are showing in Table 2 and 3.

3.1 Cluster supply and demand characteristics

We standardized the data presented in Table 1 by Equation 1 and 2. Merging and clustering the standardized supply and demand characteristics. Section 2.1 has introduced relevant steps.

Suppose $k = 3$, SPSS16.0 selects S_1, D_1, D_5 as initial cluster centers. We will obtain the optimal cluster centers $V^{(3)}$ (Table 4) and the convergence standard

Table 4. The optimal cluster centers $V^{(3)}$.

	Cluster centers		
	1	2	3
Unit price	0.5962	0.8968	0.9862
Quantity	0.6530	0.9247	0.7258
Load	1.0000	0.8866	0.8903
Speed	0.8500	0.9000	0.8500

measure function $J(C)^{(3)} = 0.3072$ after three iterations using SPSS16.0. Finally, we get 3 classes: class 1 includes six products information, ie $S_0, S_4, S_7, D_0, D_5, D_6$; class 2 includes seven products information, ie $S_1, S_2, S_5, S_6, D_2, D_3, D_4$; class 3 includes six products information, that is $S_3, S_8, S_9, D_1, D_7, D_8$.

3.2 Calculate the matching degree between supply and demand

Based on clustering results calculate the matching similarity between supply and demand, then sort them by the matching degree, in order to automatically find a more suitable partner for the supply side.

Take S_0 for example. First, S_0 is in class 1. Class 1 has three supply information and three demand information. Calculate the distance between demand information and S_0. Secondly, calculate the matching degree between supply and demand by Equation 7: $Sim(S_0, D_0) = 0.9618$, $Sim(S_0, D_5) = 0.9802$, $Sim(S_0, D_6) = 0.8808$. Finally, we get the sort by matching degree: $Sim(S_0, D_5) > Sim(S_0, D_0) > Sim(S_0, D_6)$, so the matching degree between D_5 and S_0 is the highest, then it is more suitable for their enterprises to establish business relation.

Likewise, the more appropriate supply and demand matching pairs are: $\{S_0, D_5\}$, $\{S_1, D_4\}$, $\{S_2, D_2\}$, $\{S_3, D_8\}$, $\{S_4, D_6\}$, $\{S_6, D_3\}$, $\{S_7, D_0\}$, $\{S_8, D_7\}$, $\{S_9, D_1\}$.

The matching model in the paper has some advantages: first, do not require enterprises' active participation, but provide services actively to enterprises; secondly, combine supply and demand to cluster, do not need to match supply and demand one by one, so the computation is reduced; thirdly, the model just need clustering and matching, reducing matching stage.

4 CONCLUSION

It is very important for manufacturing enterprises to find the suitable partner quickly and accurately among the many-to-many relationships. This will improve manufacturing efficiency and make efficient use of manufacturing resources. In the paper, the supply and demand matching is achieved quickly by clustering and matching. First, gathering complex many-to-many relationship to class, then calculate matching degree between supply and demand in the class. The model provides a new idea for better use of enterprise information resources. Enterprises will find the suitable partners automatically. The model will optimize the business relations between supply and demand for manufacturing enterprises.

ACKNOWLEDGEMENT

The work was supported by Beijing Planning Office of Philosophical and Social Science "Eleventh Five-Project" Fund (10AbJG389).

REFERENCES

[1] Li Bohu, Zhang Lin, Wang Shilong, et al 2010. Cloud Manufacturing: A New Service-oriented Networked Manufacturing Model. Computer Integrated Manufacturing Systems 16(1):1–10.

[2] Liu Fei, Yin Chao, Liu Sheng 2000. Regional Networked Manufacturing System. Chinese Journal of Mechanical Engineering (English Edition) 13: 97–103.

[3] Lin H. T. 2009. A Job Placement Intervention Using Fuzzy Approach for Two-way Choice. Expert Systems with Applications 36(2):2543–2553.

[4] Huang D. K., Chiu H. N., Yeh R. H., Chang J. H. 2009. A Fuzzy Multi-criteria Decision Making Approach for Solving a Bi-objective Personnel Assignment Problem. Computers & Industrial Engineering 56(1):1–10.

[5] Sarne D., Kraus S. 2008. Managing Parallel Inquiries in Agents' Two-sided Search. Artificial Intelligence 172(4-5):541–569.

[6] Ju Chunhua, Sun Bin 2009. Supply and Demand Matching Model for Third Party Logistics Integrated Platform. 2009 International Joint Conference on Artificial Intelligence:798–801.

[7] Chen Xi, Fan Zhiping, Li Yuhua 2010. A Two-phase Decision Analysis Method for Two-sided Matching of Technological Knowledge Supply and Demand. Industrial Engineering and Management 15(6):90–94.

[8] Jia Lu, Fan Zhiping, Shen Kai, Xu Baofu 2011. Bilateral Matching Model of Demand and Supply in Knowledge Service. Journal of Northeastern University (Natural Science) 32(2):297–301.

[9] Yashpalsinh Jadeja, Kirit Modi 2012. Cloud Computing-Concepts, Architecture and Challenges. 2012 International Conference on Computing, Electronics and Electrical Technologies: 877–880.

[10] Tang Longji, Dong Jing, Zhao Yajing, Zhang Liangjie 2010. Enterprise Cloud Service Architecture. IEEE 3rd International Conference on Cloud Computing:27–34.

[11] Joachim Schaper 2010. Cloud Services. 4th IEEE International Conference on Digital Ecosystems and Technologies (IEEE DEST 2010):91.

[12] Feng Xiuzhen, Hao Peng 2011. The Study of Information Mining for Requirements Based on News of Enterprise Website. Computer Science 38(10): 156–159.

[13] Feng Xiuzhen, Wu Gaofeng 2012. Research of Service-on-demand Discovery Process Based on QoS. Journal of Wuhan University of Technology (Information & Management Engineering) 34(3):327–330.

[14] Hans-Dieter Burkhard 2001. Similarity and Distance in Case Based Reasoning. Fundamental Informaticae 47:201–215.

Information Technology and Computer Application Engineering – Liu, Sung & Yao (Eds)
© 2014 Taylor & Francis Group, London, ISBN 978-1-138-00079-7

A review of the agile and geographically distributed software development

Ming Yin & Jing Ma
School of Software and Microelectronics, Northwestern Polytechnic University, China

ABSTRACT: With the development of globalization and widely communication of people all around the world, more and more people are collaborating together in geographically distributed teams. At the same time, ever since the agile manifesto was first put forward, the researchers have paid a great deal of attention on agile software development. As we all know, trust is an indispensable ingredient in the collaborations, while many people seem to overlook the challenges creating by works and social environments. Here we discuss the agile and geographically distributed software development separately to distinguish the advantages and limitations, in order to avoid the disadvantages and make full use of the advantages. After examining publications and citations, this article aims to illustrate: firstly, review the predecessors' work and studies, then discuss their discovering and find out the limitations, at last suggest some future research directions.

Keywords: agile software development; geographically distributed software development; agility; trust

1 INTRODUCTION

Nowadays the businesses and corporations are restructuring and re-engineering themselves in response to the challenges and demands. The 21st businesses and corporations will also have to face the realistic challenges that the developers are distributed in different areas. In order to meet the needs and solve the potential problems, the concept of agile and geographically distributed development was put forward. Agility addresses new ways of operating companies to meet the challenges and cut off the old ways of doing things that are on longer appropriate. In a changing and competitive environment, there is a need to make organizations and corporations more flexible and responsive. Geographically distributed systems have traditionally tackle problems associated with the interconnection of a number of computer systems. It provides a way to organize complex multi-participant software development with the participators is distributed among different geographical boundaries. The agile and geographically distributed development methods were particularly applied to deal with change and uncertainty in different places. Although trust in geographically distributed teams is widely acknowledged, only a few research workers have addressed it with appropriate level of importance. So there is a need to discuss the problem and raise the awareness of researchers. Cognitive cooperation is often neglected in current team software development process. And the issue becomes more important than ever when team members are geographically distributed. Common misunderstandings in development process should also be avoided. The article is organized as follows. Firstly a brief review of the previous work on agile and geographically distributed software development provides a background of the study. Following these is a deeper discussion on the agile and geographically distributed development to find out the discovering and limitations. A description of the study including its development methodologies and problems is next. The strengths and limitations of the research are examined and the article concludes by considering the ways in which further research in this area can make use of the theory.

2 LITERATURE REVIEW

2.1 Agile principles and agility

While agile methods are in uses in industry, little research has been undertaken into what is meant by agility and how a supposed agile method can be evaluated in software development methodological approaches. Recently, many practitioners have adopted the ideas of agile software development. Despite the variety of methods, all of them share the core principles of agile manifesto.

Agility is perceived as the dominant competitive media for all organizations especially the software development teams in an uncertain and ever-changing business environment. Many formal definitions of agility have begun to appear in recent years, mainly in manufacture and management domains, while the appearance of agility has its roots. Many scholars have carried out relevant researches. In Henderson Sellers and Serour's opinion, agility involves both the ability to adapt to different changes and to refine and adjust development process as need. While Lee and

Xia define agile software development as the software team's capability to efficiently and effectively respond and incorporate to users' requirement changes during the project life cycle. Conboy defines software development agility as the continued readiness "to rapidly or inherently create change, proactively or reactively embrace change, and learn from change while contributing to perceived customer value, through its collective components and relationships with its environment. Early experience report using Agile practices some success in the treatment of the problem when the software crisis and the proposed program based Agile practices are not mutually exclusive.

2.2 Geographically distributed development

Geographically distributed software development is a kind of software development management paradigm that focuses on work cooperation and resource sharing among the geographically distributed team members during the development process. So it needs the support of a distributed and across-platform computing environment.

The geographically distributed team is described as the core building block of distributed organization. A traditional team is defined as a social group of individuals who are responsible for the outcomes. Although the geographically distributed team shares the same goal and objectives as traditional teams and interacts through independent tasks, operates across time, geographical location and organizational limitations linked by communication technologies, they often have to cooperate despite multi-culture and multilingual environment. The communication between the members is always e-mails or lively online interactions with limited opportunities for face to face contact. The team's overall goal is to function as a single team, with the same destinations as if they were working together. The advantages of implementing a geographically distributed team are also apparent. What stands in the breach is the cost. For example, the participants distributed in different areas or even different countries would no longer to get together to work or resolve a problem. Thus it saves time and transportation fees. In return, it increases developing efficiency and obtains more rewards.

As the trust is critical to the geographically distributed software development, predecessors have done some researches. And with the recent growth in communication media, distributed software development has become the norm for software developers to jointly develop software artifacts. Distributed software development methods also raise new risks for the partners having different organizational allegiance. Some identified risks in distributed software development teams relate to intellectual property, opportunistic behavior and public relations mishaps. So trust research is necessary and important. In general, trust contains the following aspects. A trust model-based qualitative risk management approach for distributed

software development security is proposed by Lin and Varadharajan. It uses trust assessment results to prevent the access of unwanted users in the model. Meanwhile, it also provides access to the trusted users. From the above we know that the method can be employed to maximize the agility of the geographically distributed system.

2.3 The agile and geographically distributed software development

Here we discuss two popular development approaches: agility and geographically distributed software development. Ever since the agile manifesto was put forward in 2001-a little over a decade ago, the research community has devoted a great deal of attention to agile software development. At present, the following methods are enrolled to agile software development methods. These include Software Development Rhythms, Agile Database Techniques, Agile Modeling, Adaptive Software Development Scrum, and Test-Driven Development Breed. In addition, the book named "Agile Software Development: Current Research and Future Directions" explains foundations and background of agile software development. Furthermore, seven papers were published in the Information System Research attempting to explore or define the central concept of agility and under which circumstance the agile methods are most effective as well as the challenges of agile in distributed projects. Even so, the introduction of the agile methods didn't change the fundamental requirements of the software development.

The roadmap of the agile software development is as follows: Firstly is the Test-Driven Development. It is the most important part in agile development. At Thought Works, we realize that any function is from the start of the test. Secondly is the continuous integration. There is usually a long time before integration, thus will lead a lot of problems. Agile software development advocates continuous integration, integrated a dozen times or even dozens times of a day, to minimize conflict. Thirdly is the reconstruct. It aims to make the code as simple and beautiful as possible. In agile software development, reconstruct is running throughout the entire development process to achieve "clean code that works". The fourth is pair-programming. In development, many things would better be done in pairs, including analysis, writing test, writing to achieve code or reconstructing to generate the spark of ideas. The fifth one is standing up for a meeting. All members of the project team will stand for a short meeting with the aim to answer questions about what they have done. In the standing Conference, team members exchange ideas with the content they are familiar. The sixth is frequent releases, which collecting feedbacks to improve the products. In such a case, the documentation and design are largely simplified. All these ways help increasing the efficiency of the software development.

The agile software development methods strive to reach a balance between three basic goals: satisfactory software quality, the proper cost and timely delivery. For any design of software in all kind of fields, the vital consideration of the products is the software quality, and most of the developments can reach the requirement. Then is the cost of the development. Agile methods have exceedingly advantages. In the distributed software development systems, the cost can be reduced substantially. And with the fast development of communication methods, the timely delivery is no longer a problem. From the above discussion we know that the advantages of the agile and distributed software development are apparent. Because of the simple design and no complex architecture, it can adapt to customers' new requirements and needs quickly. The agile software development method sometimes is mistaken as unplanned and undisciplined; while the more definite statement would be that it emphasizes on adaptability rather than foreseeable ability. The central idea of the adaptability is to accommodate to the changing reality quickly. When the requirements of the project change, the team needs to acclimatize itself to the variation swiftly. It's difficult for the team to describe what the future would be like precisely. While in contrast to the iteration, both of them emphasize on short development cycle, and agile software development cycle may be even shorter, the teams are more highly collaborative. So when refers to the agile and geographically distributed software development, it not only contains the short development cycle but also includes the developers coordinating their activities across time and space. Apparently, the combination of the two aspects increases burdens for the development teams. Although considerable research effort has been made in this area, as yet, no agreement has been reached as an appropriate process for its special reason, which will be discussed later in the context.

3 DISCUSSION OF DISCOVERINGS AND LIMITATIONS ON CURRENT RESEARCH

The papers were all experience reports. Consequently, the evidence presented in the papers was personal experience typically gathered by deliberating a large quantity of related materials and literatures. The obstacles in the agile and the geographically distributed software development are critical but poorly understood. The common misunderstandings are as follows: First of all, some people think that agility requires developers with high quality, so they shrink back with the agile development methods. All in all, software development is creative activities. The crucial deciding factor for the quality of software is the developers' technological level and personal ability and capability. So the requirements of agile for developers are not overly, meanwhile, agile software development methods can also help to train all kinds of abilities as you need, while the precondition is that you are in the real agile development environment. Another blind spot about agile software development is that it doesn't need documents and design. The misunderstanding of software development methods has begun from XP and seems to never stop. XP encourages that "it doesn't need to have documents until the necessary and significant circumstances". The above circumstances is a criteria, it doesn't mean that all the documents aren't need. So does the agile software development methods. Sometimes, the appropriate documents can help increasing the agility of the distributed software development rather than adding its complexity. When refers to agile software development methods, some people cheer for them and think all the agile practices are perfect. As to other methods such as CMMI, they oppose the methods immediately. And anything related to the methods is bad, either. In their opinion, agile software development methods and other development methods are polar opposites. In fact, agile software development methods absorb other's merits and improving efficiency. At the other extreme, the method itself doesn't have the differences from good to bad. It just depends on whether the method is suitable for you to resolve your problems. So it is easy to find the development methods don't share the same criteria, because no two projects are the same.

Limitations to the reviewed papers also limit the degree to which we can draw conclusions in our literature review. We have made efforts to take this into account when reporting our findings. Additional limitations to our review include trust. Trust is an indispensable component in the agile and geographically distributed software development. Network procedures involve a large number of malicious devices which create problems in the framework, while the trust calculation relies heavily upon recommendations. So each framework must have some mechanisms to cope with the false recommendations, while the fact is not as we wish.

Geographically distributed software development teams situated at different locations, collaboration using communication technologies in shared project workspaces. Many iterative cycles are contained in the software development process. Any abnormal condition in a module can impact the overall software design and cost highly in the project development schedule. Thus, there is great need to take a test before the software products are released. The test strategies have been put forward, as geographically distributed software development is associated with risks and challenges due to imbalance in the knowledge base and unfamiliarity between team members belonging to diverse social, cultural and organizational locations. Previously, in order to reduce costs, the distributions of the software development are always in low wage countries like China and India. While the situation will be changed and the phenomenon can no longer be ignored in the near future with the rapid development of this countries and the raising of the price of commodities.

4 SUMMARY AND CONCLUSION

Agile software development has been the tendency of software development. And it is apparent from the article to see that research community has paid attention on it. This is evident from the number of scientific publications, the widespread interest in the topic of agile research. Although the direct research about the agile and geographically distributed software development is insufficiency, the number of studies will increase significantly as well as journal articles. However, our limited analysis of the theoretical basis should be paid more attention. Despite the agile and geographically distributed software development has been used in many fields such as academia, industry, ICT sector, government institutions and so on, a set of integrated and systematic theory has not been established. As the theory helps us to adapt innovations at a faster rate, we thus urge researchers to embrace a more theory-based approach in the future research.

Despite the unique feature of agile methods, it still has many common grounds with other development methods, such as iteration, focus on iterative communication, and reducing resource consumption in the intermediary process and so on. Generally speaking, we can measure the feasibility of agile methods with the following aspects: firstly, from the visual angle of the product, agile software development method is adapted to the situation of changing rapidly, if the highly critical, reliable and security are required, it may not be completely suitable; secondly, from the angle of the organization structure, its culture, membership and communication determine whether the method is suitable.

The rational agile software development scheme has been put forward by IBM; it can be integrated almost in any developing circumstances. Thus for the agile and geographically distributed software development, it would like to be a good solution. As of now, we have studied the agile and geographically distributed software development only for small networks. Our immediate focus is to investigate the effectiveness for various scales of enterprises from mid-size to large-size. The study can also be extended with the functionality that will allow security analysis. Meanwhile, we also aim to incorporate in the study another very interrelated issue called "privacy", which plays a major role in the software development. From above we know that agile software development process is very different from the traditional one. In the process, the developers are passionate and creative, they can be able to adapt to more changes and make higher quality software.

ACKNOWLEDGEMENT

The author thanks all persons involved in the research. The researchers who are listed in the references are also need to be thanked greatly. We also thank for the support of National Natural Science Foundation of China (71201124, 71172124, 71102087), Humanities and Social Science Foundation of Ministry of Education of China (09YJC630188), National Scholarship Fund (201203070070), Research Fund for the Doctoral Program of Higher Education of China (20116102110036), Science and Technology Project of Shaanxi Province (2012k06-39), Humanities and Management Science Foundation of NWPU(RW201111). We appreciate the work of the editors on the paper.

REFERENCES

Conboy, K, 2009. "Agility from first principles: reconstructing the concept of agility in informatin systems development," Information System Research, vol. 20, pp. 329–354.

Lee. G & Xia. W, 2010. "Toward agile: an integrated analysis of quantitative and qualitative field data on software development agility," MIS Quarterly, vol. 34, pp. 87–114.

Manifesto for agile software development. <http://www.agilemanifesto.org> (accessed on 27.02.10).

Mathrani, A & Mathrani, S, 2013. "Test strategies in distributed software development," Computer in Industry, vol. 64, iss. 1, pp. 1–9.

Richardson, I, Casey, V, McCaffery, F, Barton, J & Beecham, S, 2012. "A process framework for global software engineering teams," Information and Software Technology, vol. 54, iss. 11, pp. 1175–1191.

Rodden, T & Blair, G, 1992. "Distributed systems support for computer supported cooperative work," Computer Communications, vol. 15, iss 8, pp. 527–538.

Sellers, H & Serour, B, 2005. "Creating a dual-agility method: the value of method engineering," Journal of Database Management, M.K. vol. 16, pp. 1–23.

Zhage, H, 2002. "Knowledge flow management for distributed team software development," Knowledge-Based Systems, vol. 15, iss. 8, pp. 465–471.

Information Technology and Computer Application Engineering – Liu, Sung & Yao (Eds)
© 2014 Taylor & Francis Group, London, ISBN 978-1-138-00079-7

Developing an application model based on business-oriented

Lian Wei Li & Zhan Liu
College of Geoscience and Technology, China University of Petroleum, Qingdao, China

He Long Wei
Qingdao Institute of Marine Geology, Qingdao, China

Chen Yang
College of Geoscience and Technology, China University of Petroleum, Qingdao, China

Ji Hong Sun
Qingdao Institute of Marine Geology, Qingdao, China

Jie Sheng
College of Geoscience and Technology, China University of Petroleum, Qingdao, China

ABSTRACT: Traditional database was mainly used for the organization and management of data, and the ultimate goal of information technology was the business application of the data. However, the traditional business analysis has many deficiencies. In order to solve the lack of the traditional database and the traditional business analysis, this study adopt from global to local and from macro to micro research methods. This study design and build the architecture of an application model based on the business and the flow of implementation technology. Business model built according to the idea of division of the business domain-division of business-division of business process-division of business activities-structural description of business activities. Logical model of Object-Oriented built according to the building of business model-the analysis of business activities-the abstract of business-the integration of model. According to the application model, the defect of the traditional database and the traditional business analysis can be solved, and data can be applied by Object-Oriented data model.

Keywords: Information Technology; Data Model; Object-Oriented; Business Model; Logical Model

1 INTRODUCTION

Database was initially used to support the processing of applications of the large amounts of structured data[1]. Traditional database was mainly used to organize and manage the data. Nevertheless, with the development of technology, traditional database start to show some limitations: 1) cannot be able to express complex objects of the objective word and is lack of support of the complex data types; 2) impedance mismatch and semantic fault; 3) lack of the ability of managing knowledge and object; 4) cannot be able to express the relationship between object and natural watwork; 5) cannot meet the need of application because of the data type[2][3]. In order to overcome the limitation of the traditional database described above, fortunately, with the development of the Object-Oriented technology, the Object-Oriented database finally generates. There are two approaches to combine the Object-Oriented technology and the database technology: one approach is to build the pure Object-Oriented Database Management System (OODBMS) by using the Object-Oriented language and increasing the functionality of the database. This approach can

support the persistent objects and can realize the sharing of the data, but it is difficult for developing and achieving the data conversion[4]. Another approach is to build an intermediate layer on the currently-mature relational database system, and this layer realizes the conversion from the relational data model to the Object-Oriented data model. This approach is considered as the best scheme between object technology and the traditional relational database.

The business application of the data is the ultimate goal of the data management and the core work in the information work[5]. For traditional method of business analysis, the process were mainly considered as follows: Firstly, the business research in accordance with the organization was carried out and the relevant information was collected; Secondly, diagrams involving the business process and the data flow were designed; Finally, logical design of the data was conducted based on the information[6]. The defects of the traditional business analysis method mainly include: 1) Business may response to crosses and omissions since that it was divided by the organization; 2) It is difficult to carry out the cross-validation between businesses because that the business process is divided

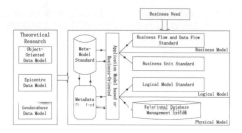

Figure 1.   Model Architecture.

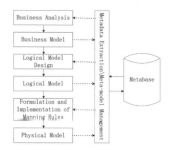

Figure 2.   Technique Flow Diagram.

in accordance with the business management practices of each business domain and the methods are not uniform, 3) Template of the business description is inconsistent and the description is not standardized.

In recent years, lots of research on Object-Oriented data model has been conducted[7][8][9][10]. However, the Object-Oriented data models oriented toward business applications need to be improved and an improved business model still need to be developed. This paper firstly design the architecture and technique flow of the business-oriented application model, and then design the business model and the logical model in order to convert the relational data model stored in relational database into Object-Oriented data model. The defects of the traditional database and the business analysis can be solved by using object in business application.

2   IMPLEMENT SCHEME OF THE APPLICATION MODEL

2.1   *Model architecture*

According to the research of the theory of the Object-Oriented data model, the data model of Epicentre and the spatial data model about Geodatabase, the study designs the model architecture of the business-oriented application (see Fig. 1).

Firstly, through the existing the research results, the data business flow and the data stream standard and the business model standard and description method of the data can be constructed by the analysis of the specific business requirement. Secondly, the business unit standard is constructed and the logical model standard is established by the construction of the business model. The logical model standard is the Object-Oriented and the relatively stable and the scalable. Finally, the relational physical model stored in the traditional relational database management system is mapped to the logical model that has been built, subsequently, the different application models are produced according to the different application purposes. In addition, the standard of the metadata and the metadata model need to be constructed when constructing the application model. The reasons for this are, on one hand, the business model and the logical model need to be managed, on the other hand, the conversion relationship of the business model and the logical model also need to be managed.

2.2   *Technical process*

The technical process of the application model is described as: business analysis – business model – logical model design – logical model – mapping of the physical model by investigating the business model, logical model and physical model (see Fig. 2).

(1) Business analysis and modeling

The standard of the business flow and the data flow are formed through analyzing the business needs. The business model is built according to the construction method and the norm is constructed subsequently.

(2) Design and construction of the logical model

Combing with the business characteristics, the logical model of the business is created in line with the business model that has been built previously. The standard of the business unit is built through analyzing the business unit of the business activity. Subsequently, the relationship between the business model and the logical model is established.

(3) Mapping of the logical model and the physical model

Data stored in the traditional relational database cannot be directly applied to the Object-Oriented logical model since that the physical of the data is the relational data model. An appropriate mapping rule need to be developed in order to achieve the goal of mapping of the logical model object and the relational database of the physical layer.

3   MODELING OF THE BUSINESS MODEL

The construction process of the information system involves in the wide range of the business and the complex business process. In order to carry out the information planning and the design of the overall business, it is necessary to regulate the business process analysis and the construction of the business model. The business process and the business model will change along with the development of the business and the information, since that the construction of the system is an iterative process. However, in previous enactment of the data standard and the business process analysis of the large-scale planning project, there was no design of the business model standard. Consequently, the business requirements are mostly described using flow charts or word documents, which thus lead to difficulties of continued updating of the business process and model. Therefore, the standard of the business process analysis and the business modeling need to be established, and the standard operation procedure of the business requirement analysis and modeling should be developed. The maintenance process and the mechanism of the business model should be established in order to realize the sustainable development of the business model.

The construction of the business model contains three parts: 1) the division of the business domain and the business; 2) the division of the business process; 3) the division of the business activity and the description of the structured data[11][12].

3.1 Business domain and business division

Business domain should be divided according to according to the form of business management and the industry lifecycle in the field or industry management functions and make sure that different business domains are not repeatable and they are able to cover all the business.

The business is divided by the characteristics of different business domains, which generally abides by the following principles: 1) be divided in accordance with the business types; 2) be divided in accordance with business functions; 3) be divided in accordance with work objectives; 4) be divided in accordance with business phases. The business division should cover all the business of the business domain and each business contains a complete business process.

3.2 Business process division

Business process division should be based on the orders of the business, so each business process contains at least a starting business and an ending business and two business processes cannot be combined. Business process division should follow specific methods: 1) business process division should be conducted from the perspective of core business; 2) there is a correlation between business processes, as links between a group of business activities; 3) divided by business functions, different business processes cannot contain the same kind of business activities; 4) Merge them reasonably and keep the consistency of the key results.

3.3 Division of business activity and description of structured data

Business activity is the smallest functional unit, the most basic and disabled to discompose any further. A business process contains a number of business activities. The division of business activity should abide by the following principals: 1) Operators of business activities in the event bear only a single function; 2) There must be objects that activities effect and it can produce clear results of the activities; 3) Associated with the activities.

The description of structured data is the description of all relevant data items of the specific business activities, we analyze the role in the business activities firstly, and then extract the physical object, and finally sort out the roles and all the attributes of the entity object.

4 CONSTRUCTION OF LOGICAL MODEL

The logical model is an object-oriented, relatively stable model, and a stable logical model is a unified interaction language between applications and database. Logical model is designed to provide reference for description and presentation of mutative and progressive businesses by object-oriented description and expression of business activities of the business domain.

4.1 Construction principle

(1) Adoption of object-oriented approach

Adopting object-oriented modeling techniques, entity as a basis for data organization, we can get rid of the dependence on subjective factors such as specific business needs, specific application software.

(2) Systemic and scientific data management system

Design of the model should fully reflect the data management system, to provide the foundation for a smooth transition of existing application software systems and development of application software in the future.

(3) A high level of integration

The model must cover the objects of business domain and these objects are split, not according to their professions, but links between organizations of objects in the objective world. Every professional application software can operate this model, and their datum should be consistent with each other, to achieve the data integration of different application software on the same level.

4.2 Construction method

Logical model is abstract and general expressions of business starting from analysis of the business, that not only be reflected in data management and description, and also in the processing of data and behavior. The construction of logical model should start with business analysis and should be built from bottom to top iteratively, firstly logical model of business unit activities, then the integration of the logical model. The specific modeling process is shown in Fig. 3.

(1) Construction of business model

Take the business model construction method to build business model and measure off specific business activities and carry out standardized description of business activities.

(2) Analysis of business unit activities

Choose business unit activities from the business model which has been built. Business unit is the basic unit of business definition and division which is divided according to business analysis, describes the status of the business, whose description granularity is

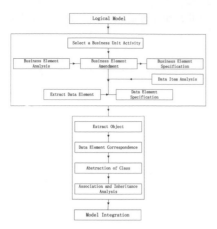

Figure 3. Flow Diagram of Logical Model Building.

relatively coarse, and cannot express the relationship between data items and the actual principal objects and this relationship is exactly the basis of further definition of data elements and data element models. The basic principles of the business unit analysis are consistent with the standardized definitions of business models and data elements.

1) Business element analysis: Extract business elements of the business units from the business model, and amend the elements which cannot fulfill the current business and propose the standardized definition of various business elements and their associated data items.

2) Data element analysis: Analyze data items related to the business elements and extract data elements from business elements and carry out normalization amendment of data elements.

(3) Model abstraction

1) Object extraction: Extract objects involved in the business based on the analysis of business unit activities and these objects contain producer objects and action objects; Extract all objects in the business activities according to the analysis of specific business activity; Establish the relationship between data elements in the analysis of business unit activities and object entity.

2) Classify abstraction: Abstract objects have the same or similar nature and a class forms.

3) associate and inherit analysis of class: Study the relationship between various classes and carry out inheritance and association analysis of the classes.

(4) Model integration

Integrate each logical model of business unit activities to achieve the ultimate logical model.

5 CONCLUSION

We designed architecture and technical flow of business-oriented application model and constructed business model by way of business domain division-business activity division-structured description of business activity and built object-oriented logic model by iteration in line with business model construction-analysis of business unit activity-business abstraction-model integration from button to top, based on arrangement and analysis of object-oriented data models at home and abroad, combing with the research on application projects. But it should be noted that the construction of logical model requires not only a full understanding of business data and business processes but also an in-depth knowledge of object-oriented technique. It is very difficult to build logical models and we need to continue the study on the construction of logical model to complete the unified logical model in the global schema.

ACKNOWLEDGEMENT

The paper is supported by National Marine Geology Special (GZH201000501) and the Fundamental Research Funds for the Central Universities (13CX06012A).

REFERENCES

[1] WU Hong-sen. Development and Research of Object-oriented Database[J]. Computer Engineering and Applications, 1998.7:3–5.
[2] NIU Yue-hong, WANG Zhi-min. Application of OOT in modeling of GIS[J]. Science of Surveying and Mapping, 2008.10:249–251.
[3] WANG Gong-ming, GUAN Yong, ZHAO Chun-jiang, etc. Development and Research of OODB[J]. Application Research of Computers, 2006.3:1–5.
[4] ZHANG Ying. From Relational Databases to Object-oriented Database Development Overview[J]. Computer Knowledge and Technology, 2011, 7(21):5061–5062.
[5] XUN Xu-dong. Exploration of business-oriented data model application system[D]. Ocean University of China, 2008.
[6] TANG Bo. Research and Design of Sinopec E\$P Business Model and Data Element Standardization[D]. China University of Petroleum, 2009.
[7] HONG Yuan, WANG Xiu-mei, LI Can-hua, etc. Logical Design Approach for Object-Oriented Database[J]. Journal of University of Science and Technology Beijing, 2001,23(4):382–385.
[8] CHE Dun-ren, ZHOU Li-zhu, WANG Ling-chi. The Architecture of Object-Oriented Database Systems[J], 1995,6(10):599–606.
[9] ZUO Feng-chao, WANG Wen-de. Discuss of Object-Oriented Data Model[J]. Computer Engineering and Applications, 2001,16:110–112.
[10] YUN Man, JIA Wen-ju, LI Si-qiang, etc. Method of Modeling Object-Oriented Data Model[J]. Journal of daqing petroleum institute, 1998,22(4):95–97.
[11] WANG Hai-ping, GE Jun, WANG Juan. Business Process Modeling Based on POSC in Petroleum Industyr[J]. Computer System and Application, 2010,19(3):100–102.
[12] JING Rui-lin. Research on Oil and Gas Exploration and Development Data Standard System[J]. Standard Science, 2012,4:42–46.

Information Technology and Computer Application Engineering – Liu, Sung & Yao (Eds)
© *2014 Taylor & Francis Group, London, ISBN 978-1-138-00079-7*

A review of abrasive flow machining research

L.F. Yang, X.R. Zhang & W.N. Liu
Changchun University of Science and Technology, Changchun, Jilin, China

ABSTRACT: Abrasive Flow Machining (AFM), which can be applied to deburr, polish or radius surfaces and edges, is one of the most widely used abrasive-based surface finishing techniques. AFM is suitable for polishing anywhere that air, liquid, or fuel flows. Researchers have been exploring the ways to improve AFM process performance by analysing the different factors that affect the quality characteristics. The experimental and theoretical studies show that process performance can be improved considerably by proper selection of temperature, media viscosity, abrasive hardness, particles sharpness and density, workpiece hardness, pressure, piston moving speed, etc. Optimization techniques for the determination of optimum AFM condition have been contained in this article. For gaining the best quality of finishing surface in metal working industry, a multi-function intelligent control system on adjusting AFM process parameters automatically online is demanded. More research on the control of AFM process can be expected in the near future.

Keywords: Abrasive flow machining; Surface finishing; Review

1 INTRODUCTION

Modern metal working industry has developed to the situation in which several factors have to be considered, such as costs, lead time from design to production, product quality, and machining/finishing of difficult to machine materials. The most labour-intensive uncontrollable area in the manufacture of precision parts involves final machining (or finishing) operations (Walia 2009). The finishing operation sometimes may cost as high as 10–15% of the total production cost (Jain 2008). Hence, it is necessary to research finishing techniques in terms of minimizing cost and time. Further, Abrasive flow machining (AFM) as one of the most widely used abrasive-based surface finishing techniques has been studying for many years, and some progresses on AFM theory have been made. However, it is not so enough in that a multi-function intelligent control system on adjusting AFM process parameters automatically online is still not proposed. Therefore, it is of great practical and theoretical significance to review the research on AFM.

The need for high-accuracy and high-efficiency finishing of materials is making the application of abrasive finishing technologies increasingly important. In order to cater to these requirements, AFM process is gaining importance day by day (Walia 2007). AFM is used to deburr, polish or radius surfaces and edges by flowing a semisolid abrasive media over these areas (Rhoades 1991) and it is suitable for polishing anywhere that air, liquid, or fuel flows. This article attempts to comprehensively introduce AFM. With the references provided, readers may explore more by reading a particular article for detailed information. By examining the research work carried out so far, this article points out the advantages and disadvantages of AFM process, the deficiency of the present research work and the machining parameters which have a significant influence on the process performance. The last part of this article reports solutions to overcoming the disadvantages of AFM process, the follow-up research work and thereby outlines the trend for future research.

2 ABRASIVE FLOW MACHINING

Abrasive flow machining (AFM) was identified by the Extrude Hone Corporation USA in the 1960s as a method to deburr, remove recast layer, polish difficult-to-reach surfaces and edges by flowing abrasive laden viscoelastic polymer over them. AFM process was originally used in the area of aerospace for deburring and polishing critical hydraulic and fuel system components of aircraft. It uses two vertically opposed cylinders (Fig. 1), and extrudes medium with abrasive particles back and forth through a restrictive passage formed by the workpiece and the fixture. Figure 2 shows the forces acting on the workpiece surface, the axial force responsible for the removal of material in the form of microchips and the radial force responsible for the indentation of abrasives in the workpiece surface.

It is well known that the horsepower of a engine is influenced seriously by the internal surface quality of intake ports, smoother and larger intake ports make a engine gain more horsepower with a better

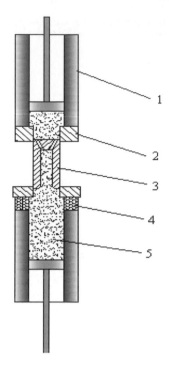

Figure 1. Schematic diagram of abrasive flow machining: 1-Medium cylinder, 2-Splitcylindrical fixture, 3-Workpiece, 4-Heating and cooling units, 5-Medium with abrasive particles.

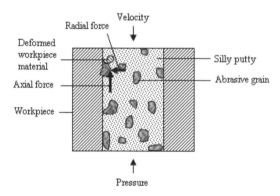

Figure 2. Types of forces acting on a grain (Gorana et al. 2004).

fuel efficiency. But it is very difficult to finish internal surface of intake ports because of its complex geometry. With the application of AFM process, it becomes easy. The AFM can process anywhere that air, liquid, or fuel flows, providing uniform, repeatable, predictable results. Almost any surfaces, for example rough cast, machined, or EDM'd surfaces, etc, typically yield a 80–90% improvement regardless of their surface complexities.

The accuracy, consistence and flexibility of abrasive flow machining make it widely used in the surface finishing of aerospace, automobile, mold and other industries. The technology was initially introduced for

critical deburring of aircraft valve bodies and spools, and now some additional AFM applications have been developed in semiconductors, medical equipment and other industries, and the processed parts and passage sizes are from gears as small as 1.5 mm (0.060 in.) in diameter and orifices as small as 0.2 mm (0.008 in.) or even smaller than 0.2 mm (0.008 in.) to splined die passages 50 mm (2 in.) in diameter. Compared with traditional manual finishing methods, it can offer high product quality with lower labor cost for its uniformity of the polished surface.

Currently, AFM technology is widely applied in the field of Automobile, Space- and aeronautics, Fine mechanics, Profile, Medical parts, Fluid technology, etc. For example, it can be used to polish the inner passage of Automotive intake and exhaust ports, common rail, fuel injectors, automotive transmission gears, precision molds, and aero-engine parts, to deburr, remove recast layer, eliminate EDM and laser machining metamorphic layer.

3 MAJOR AREAS OF AFM RESEARCH: STATE OF THE ART

AFM can be applied to finish difficult-to-reach surfaces and complex internal passages which require uniform and repeatable results. Abrasion occurs wherever the medium passes through the restrictive passage, and it will accelerate abrasive action by changing the rheological properties of the medium when it enters and passes through the restrictive passage (Rhoades 1988, Rhoades 1991). The viscosity of the medium plays a so important role in the finishing process that the quality of finishing surfaces can be controlled by selecting proper machining viscosity (Jha 2006). The AFM medium is mixed by abrasive particles and special viscoelastic polymer, and the viscosity of the viscoelastic polymer is varied when passed through the restrictive passage. Forces (axial and radial) acting during AFM have been studied and have been found to influence the reduction in surface roughness value achieved after AFM (Gorana et al. 2004). Under certain machining conditions, it has been found that the material removal can take place by ploughing (Gorana et al. 2004) other than chipping (Jain & Jain 1999). Active grain density (Gorana et al. 2004, Jain & Jain 2004) has been found to influence finishing rate and depends on the machining parameters such as abrasive mesh size, extrusion pressure, abrasive concentration and medium viscosity.

A stochastic methodology has been proposed to evaluate the interaction between spherical abrasive grains and workpiece surface (Jain & Jain 2004). Active grain density can be predicted at different concentration and mesh size, which finally control the quality of the finishing surface and machining rate. A good correlation between the predicted density and the actual results under microscope has been found. Based on the concept of shearing of peaks from the workpiece surface by flowing abrasive grains

(Jain & Jain 1999), a theoretical model has been built, which helps to simulate the expected surface after AFM. The simulated surface and experimentally generated surface after AFM are needed to be compared with each other.

Authors (Jain et al. 2001) have investigated the effects of various process parameters on the viscosity of the medium and gained some satisfactory results. It is found that the medium viscosity is influenced significantly by abrasive concentration, medium temperature and abrasive mesh size, and an increase in viscosity of the medium results in increase in material removal rate and decrease in surface roughness value or increase in finishing rate. The results of the experiment show that the generated surface roughness approaches to the "critical surface roughness" value, which will be higher for the case of higher viscosity in other given finishing conditions. However, the value will also have a upper limit in varied machining conditions. It requires further theoretical and experimental study to set these upper limits of the final generated surface after AFM.

Authors (Fang et al. 2009) have investigated the relationship between temperature and abrasive flow machining, and found that with the increase of the machining cycles, the temperature of the abrasive flow will increase and the viscosity of the medium will decrease, which will result in low machining rate. It has been studied by computational fluid dynamics (CFD) approach and the results of the study show that the abrasive particles will roll by themselves for the rising temperature which causes low machining efficiency.

A classical abrasion theory thought that the material removal is the result of accumulated plastic flow, by repeated indentation of moving abrasive particles. Based on the theory, the authors (Jain et al. 1999) have evaluated the forces acting on the workpiece surface by finite element method. They also analyzed the surface roughness produced after AFM, and the theoretical results were found to be in good agreement with the experimental results (Jain 2008).

The scratching experiments have been carried out (Gorana et al. 2006) to investigate the modes of material deformation under realistic conditions of grain-workpiece interaction, and the experimental results have shown that axial force, radial force, active grain density and grain depth of indentation, all have a significant influence on the scale of material deformation. The minimum load required for chip formation and minimum depth of indentation were found to correlate well with the mode of material deformation. The theoretical and experimental results show that the rubbing mode of material deformation dominates in the present study, however, some evidences of ploughing during AFM are also provided by Gorana et al. (2004, 2006). The authors have indicated that the weakness of the reported model could be the analysis of situation, in which some factors are failed to consider.

Parametric optimization is essential in order to achieve the best of any manufacturing process.

Authors have attempted to obtain the best of AFM process using artificial neural networks (ANN) and Taguchi method. Back-propagation neural networks have been used in the parametric optimization. Compared with all the classical optimization methods, the process optimization using neural networks can be performed in the absence of mathematical model of the process, and its accuracy depends on the accuracy of experimental observations (Jain & Jain, 2000). Mali & Manna (2010) presented the utilization of robust design-based Taguchi method for optimization of AFM parameters in case of AFM being used to finish conventionally machined cylindrical surface of Al/15 wt% SiC$_p$-MMC workpiece. The influences of AFM machining parameters on surface finish and material removal have been analyzed. Taguchi experimental design concept, L$_{18}$ ($6^1 \times 3^7$) mixed orthogonal array is used to determine the S/N ratio and optimize the AFM process parameters.

The authors (Walia et al. 2009) reported that one serious limitation of AFM process is its low productivity in terms of rate of improvement in surface roughness. They discussed improved fixturing as a technique for productivity enhancement in terms of surface roughness (Ra). A rotating centrifugal-force-generating (CFG) rod is used inside the cylindrical workpiece which provides the centrifugal force to the abrasive particles normal to the axis of workpiece. The effect of the key parameters on the performance of process has been studied. The results show that for a given improvement in Ra value, the processing time can be reduced by as much as 70–80%.

It is well known that the various machining parameters, for example abrasive concentration, extrusion pressure, abrasive mesh size, medium viscosity, etc, have a significant effect on the quality of finishing surface. However, there is no method to control the process and coordinate machining parameters properly online, and the intelligentized control system with single function is still need to be improved. The authors (Xu et al. 2011) have studied the signal measurement, transmission and control in precision process of the abrasive flow, and the measurement and closed-loop control system has been structured, which can process tasks in parallel. A softness abrasive flow monitoring method based on the acceleration vibration singal (Ji & Lan 2012) has been proposed, and a intelligent embeded real-time measurement and control platform with the capability of universal multi-channel high-speed date acquisition has been developed.

4 CONCLUSIONS

AFM process can be used to a wide range of finishing operations for complicated geometries or difficult-to-approach regions. For example, it is applied to improve the performance of high-speed automotive engines and finish difficult-to-reach surfaces in automobile, aviation, fine mechanics, tooling industry, medical components, etc.

To overcome the disadvantage of low productivity in terms of rate of improvement in surface roughness in AFM process, hybridized or improved structure is good solution, for example improved fixturing.

Further theoretical and experimental investigations to the machining conditions of temperature, media viscosity, abrasive hardness, particles sharpness and density, workpiece hardness, pressure, piston moving speed, etc, especially media viscosity, are needed in order to control the process properly.

A more realistic situation of analysis is very important for studying the modes of material deformation which helps to understand the mechanism of AFM or control the process.

For achieving the best of AFM process, parametric optimization is essential, which should be further attempted using some optimization methods. Based on the research, a excellent method that can control the process and set proper machining parameters automatically online or a multi-function intelligent control system can be proposed in the near future.

REFERENCES

Fang, L., Zhao, J., Sun, K., Zheng, D.G., Ma, D.X. 2009. Temperature as sensitive monitor for efficiency of work in abrasive flow machining. Wear 266:678–687.

Gorana, V.K., Jain, V.K., Lal, G.K. 2004. Experimental Investigation into cutting forces and active grain density during abrasive flow machining. Int. J. Machine Tools and Manufacture 44: 201–211.

Gorana, V.K., Jain, V.K., Lal, G.K. 2006. Forces prediction during material deformation in abrasive flow machining. Wear 260: 128–139.

Jain, R.K., Jain, V.K. 1999. Simulation of surface generated in abrasive flow machining process. Robotics and Computer Integrated Manufacturing 15: 403–412.

Jain, R.K., Jain, V.K. 2000. Optimum selection of machining conditions in abrasive flow machining using neural networks. J. Materials Processing Technology 108: 62–67.

Jain, R.K., Jain, V.K. 2004. Stochastic simulations of active grain density in abrasive flow machining. J. Materials Processing Technology 152: 17–22.

Jain, R.K., Jain, V.K., Dixit, P.M. 1999. Modeling of material removal and surface roughness in abrasive flow machining process. Int. J. Machine Tools and Manufacture, 39: 1903–1923.

Jain, V.K. 2008. Abrasive-Based Nano-Finishing techniques: An overview. Machining Science and Technology 12: 257–294.

Jain, V.K., Kumar, P., Behera, P.K., Jayswal, S.C. 2001. Effect of working gap and circumferential speed on the performance of magnetic abrasive finishing process. Wear 250: 384–390.

Jha, S., Jain, V.K., 2006a. Modeling and simulation of surface roughness in magnetorheological abrasive flow finishing process. Wear 26: 856–866.

Jha, S., Jain, V.K. 2006b. Nanofinishing of silicon nitride (Si_3N_4) workpieces using magnetorheological abrasive flow finishing. Intl. J. of Nanomanufacturing 1(1): 17–25.

Ji, S.M., Lan, X.H. 2012. The oriented solid-liquid two-phase soft abrasive flow embedded real-time control system. Mechanical and Electrical Engineering 29(2): 131–135. (in Chinese)

Kohut, T. 1988. Surface finishing with abrasive flow machining. Proc. of the Fourth International Aluminum Extrusion Technology Seminar. April 11–14, 1988. Washington, DC. The Aluminum Association 2: 35–43.

Mali, H.S., Manna, H. 2010. Optimum selection of abrasive flow machining conditions during fine finishing of Al/15 wt% SiC-MMC using Taguchi method. Int J Adv Manuf Technol 50:1013–1024.

Rhoades, L.J. 1988. Abrasive flow machining. Manufacturing Engineering: 75–78.

Rhoades, L.J. 1991. Abrasive flow machining: a case study. Journal of Materials Processing Technology 28:107–116.

Walia, R.S., Shan, H.S., Kumar, P. 2007. Morphology and integrity of surfaces finished by centrifugal force assisted abrasive flow machining. Int J Adv Manuf Technol 39: 1171–1179.

Walia, R.S., Shan, H.S., Kumar, P. 2009. Enhancing AFM process productivity through improved fixturing. Int J Adv Manuf Technol 44: 700–709.

Xu, J.L., Ji, S.M., Tan, D.P. 2011. A research on the measurement and control system in abrasive flow precision machining. Automation Instrumentation 32(4): 18–21. (in Chinese)

Information Technology and Computer Application Engineering – Liu, Sung & Yao (Eds)
© 2014 Taylor & Francis Group, London, ISBN 978-1-138-00079-7

Virtual design in feeding device of precoated laminating machine

XiaoHua Wang, YuanSheng Qi & YueTeng Li
Department of Mechatronic Engineering, Beijing Institute of Graphic Communication, Beijing, China

ABSTRACT: In the drive of market demand, printing equipment after developing rapidly. Pre-coating film technology is environmental protection, energy saving with wide application prospect. The application of virtual prototype technology, to key components of the pre-coated laminating machine to the paper feed mechanism (include Pre-feeder pile positioning mechanism, vice to the paper feed mechanism, universal joints, side of the block paper and wallboard etc.) improve the structural design. Through the three-dimensional modeling, assembly and disassembly, adjustment mechanism simulation, motion simulation of mechanism, the physical, mechanical performance analysis, simulation, prediction, realizes fast implementation and optimization design, improves the processing and assembly technology.

Keywords: Laminating machine; feeder; virtual prototyping

1 GENERAL INSTRUCTIONS

Laminating machine is make Printed matter surface covered with a layer of plastic film, in order to improve the gloss, firmness and water, Antifouling, wear-resistant, corrosion-resistant of the printed product after printing equipment. Widely used in Books, pictures, covers, Credentials, instructions and all kinds of paper packaging products surface decoration treatment. Laminating machine from the process aspect is divided into two kinds. Pre-coated laminating machine and Coated laminating machine. Coated laminating machine appeared earlier, mature models, low cost in use, but that glue material has Pollution and high energy consumption. Pre-coated laminating machine materials are environmental protection, simple operation. Pre-coated laminating machine in the foreign market share: 100% in The USA 90% in Europe 70% both in Japan, Korea. Our country gave the strong support for environmental protection and energy saving pre-coated film technology beginning in 2009. Pre-coated laminating machine market has a giant growth potential.

At present, has the international representative of the laminating film equipment brand include the Germany BILLHOEFER, America GBC, Korea GMP, Britain AUTOBOND, Czech KOMEI etc. Heibei Shengtian, Shanghai Oulida, Wenzhou Guangming, Ruian Huawei etc. Are the several mainly laminating machine manufacturers in China. Among them, Company GBC exhibited the 8500 "cyclone" automatic pre-coated laminating machine maximum speed of over 100 m/min. They had the most comprehensive, state-of-the-art technology. At the same time, Company GMP's Products CHALLENGER PLUS 1020 also reaches the same speed. Shanghai Oulida's Products CATL800 Outlook and CTL-DS-500 get

Circulating water heating technology and Vacuum adsorption paper separation and transport technology patent. In the field of research, Our school teacher Zhang Yang give a Innovation scheme in pre-coated laminating machine removing the powder technology[1]. Teacher Guo Junzhong against the laminating machine conveyor institutions optimized design[2], teacher Li guang use of vibration testing technology against the laminating machine pressure bonding part and wall material Suggestions for improvement[3]. Xi'an University of Technology teacher Zhang Haiyan to study the laminating machine, hot roller and heating system, Proposed a novel structure of the hot roller, and the corresponding program is designed for the temperature control system[4].

Domestic pre-coated laminating machine exist some problems. It cannot reach high degree of automation, the operator labor intensity big, wear rate is high, the speed and the accuracy is low, machine to paper adaptability is not enough and etc. At present, the domestic film laminating machine design mainly uses empirical analogy design methodology. At present, the domestic film laminating machine design mainly uses empirical analogy design methodology, and no specialized teaching material, domestic manufacturer mainly to imitation; available reference data for design is less and doesn't have a set of perfect design theory. Development mode should be changed urgently. Design is mostly in the proofreading and verification stage, there is no design innovation from the source.

This article based on the domestic actual machine as the prototype, combined with market demand. Use virtual prototyping technology to improve the structural design of the pre-coated laminating machine paper feeding part.

2 THE APPLICATION OF VIRTUAL DESIGN ON PROGRAM DESIGN

To the paper feed section of the main tasks is to periodically from paper heap-by-piece isolated the paper, and passed along to the pick roller, according to strict overlapping forms of transmission and cardboard forward transported to the former regulations. Divided into feed table lifting mechanism, pre-Feeder pile positioning mechanism, deputy to the paper feed mechanism, movable type universal joint, and side of the paper stop agencies design. Following with side of the paper stop agencies program design as an example, Descriptions virtual prototyping technology in program design application.

Side of the paper stop agencies to complete: up pendulum ready to receive the paper, on paper stop for positioning, the hem dodge paper, stop waiting for the paper passes through the four process action. Since to be coordinated with the paper feed section's others components. Decide to use the cam rod and pendulum mechanism. After initially confirmed the fulcrum location and the rod length. According to the feed paper part work circulating diagram to determine the cam profile line, Finalize the institutional diagram Figure 1.

By means of advanced engineering analysis software, we can exact test the side of the paper stopper rod displacement, velocity, acceleration values. This case adopts 3D engineering software Solidworks' Motion plug-in. Can get the visualization results easily and conveniently. By loading the model and adding the rotating motor and spring on the cam. Making completed animation to analyze. The software can be measured for any position of the member. This case detects actuator pendulum swing angular displacement, angular velocity and angular acceleration. Set the rotational speed and the spring constant of the cam, the swing arm and the axis between the torsion damping coefficient. Conduct motion Analysis. The results are shown in Figure 2.

Analyze curve, further optimization of institutions. In Motion, whether it is to modify the position, quality and parameters of the member. It can be finished quickly and intuitive. As can be seen from Figure 2(a). Cam pendulum angular acceleration in 0 to 0.2 and from 0.4 to 0.7 appears irregular movements. That can

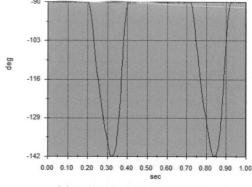

(a) Swing arm angular displacement curve

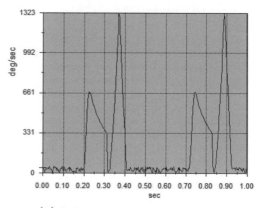

(b) Swing arm angular velocity curve

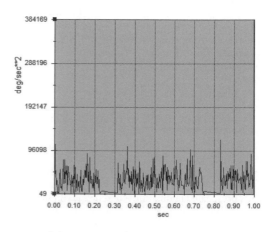

(c) Swing arm angular acceleration curve

Figure 2. Swing arm motion graph.

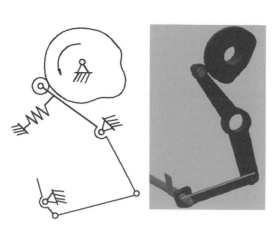

Figure 1. Side guide mechanism.

affect the paper positioning accuracy. Formed reason is due to vibration of the pendulum on the cam. It shows that the spring constant is too small. After improve the value of the parameters of the spring, and then by the analyzed. We get the movement graph in Figure 3.

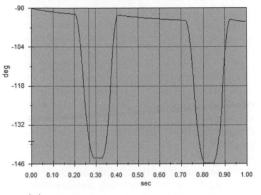

（a）revised swing arm angular displacement curve

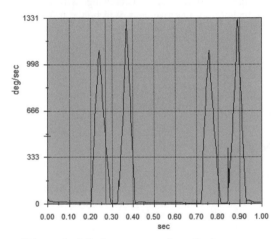

（b）revised Swing arm angular velocity curve

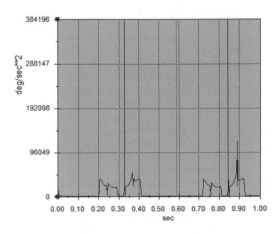

（c）revised swing arm angular acceleration curve

Figure 3. Revised swing arm motion curve.

Through the contrast can be seen by the results of the analysis of the angular acceleration. The pendulum movement is more relatively stable.

3 THE APPLICATION OF VIRTUAL DESIGN ON STRUCTURE DESIGN

In the structural design adopt modeling of three-dimensional digital parameters modeling can be applied to parts and assemblies modeling simultaneously. As SolidWorks 3D design platform for example. You can use Visual C++, VB etc. call Object Linking embedded technology OLE supplied secondary development tools API function. Create users customized dedicated SolidWorks function module. In components modeling process using parametric design can be quickly and conveniently implement modify of series Part parameters. Not only size modifications can be done, but also take advantage of the trade-off of the library feature part structure. This greatly simplifies the products' serialization design process[5]. In the assembly modeling process, can assembly relationships (coordinate) between the parts as a data variable. Applications-driven variable satisfies a predetermined condition assembly, made universal 3D design software more suitable for the characteristics of the industry. The use is more convenient, shortcut[6].

Structure design for feed paper machine is based on the existing program, combined with market demand and manufacturers improvement requests. To the feed paper table lifting mechanism, the positioning mechanism of Pre-feeding paper pile, secondary paper feed mechanism and the side of the paper stops' agencies etc. set a working cycle diagram. Proposed transmission line and structure program to operate a appearance and structure design.

3.1 Feeding device assemble

Feeder is made up of a machine frame, Feida, lifting device, pump etc. A paper feeding apparatus mainly made up by power transmission shelf, power motors, variable frequency CVT, conveyor belt, conveyor chains and pulleys, sprockets. Paper feeding apparatus is mainly made up by the sheet separating mechanism, the paper conveying mechanism, double, multi-detecting means, and the paper feed platform. Paper feeder device role is to complete the sheet-fed, which on the feeding table continuous separation and transport. For saving time to replace the paper stack. Increase pre-feeder pile positioning mechanism. When the paper which on the paper platform runs out. The paper platform will be reduced to a lowest position. Remove a paper platen and the feed table outside beams. Make the feed platform which on the pre-tidy paper trolley pushed forward until lean on the inside feeding beams. Insert the outside feeder crossbeam, which already removed. Make the feeder platform go up to the working position. Pull out the

Figure 4. Assembled feeding device.

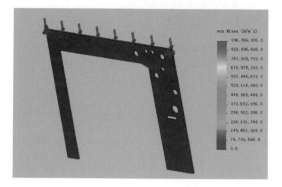

Figure 5. The wallboard stress analysis.

pre-tidy paper trolley. Fix with a bolt. Tidy the paper again.

The assembly modeling should be consistent with the actual assembly process, Finish the sub-assembly modeling first then do the final assembly.

3.2 Check analysis

After structural design complete, need to check strength of key parts. We can easily achieve with the help of the corresponding design module which in 3D design platform. In the Pre-coated lamination machine improved design. According to the manufacturers' demand make feeders' rack wall panels' material from iron castings changed to steel. After the design complete, directly use SimulationXpressStudy, a function of solidworks as stress analysis software. To do stress analyze of the wallboard transient. Give the Constraint boundary conditions of wallboard. (Fixed end), load and force. Can obtain the wallboard stress strain diagram. Graphical representation use data and color display deformation at the same time. Intuitive and easy to understand. According to these data can get further optimized. Make the design meet the working demand as well as saving the material.

4 CONCLUSION

Virtual prototyping technology is technology innovation and product design effectively auxiliary tool. Regarding the coordination of all parts of the product. For the design data management and use etc. This auxiliary tool compared to traditional two-dimensional design software, having a greater practical significance and a bright market prospects. Application of Virtual Prototype Technology is an effectively way to improve the design, manufacturing process. Through the CAE platform to determined accurately each actuator position, speed and acceleration during operation. Optimization design after determining the mechanism scheme. Use CAD platform to do three-dimensional modeling for accessory, component and complete machine. Can synchronization complete parametric design, assembly and interference detection. Check the key parts after the design is complete. I believe that with the research and development of three-dimensional model. The entire printing industry will face a great revolution. This change will not a simple technology improvement. Will be the true arrival of the era of digital printing.

ACKNOWLEDGEMENTS

Beijing Institute of graphic communication 2013 University research project 23190113079.

REFERENCES

[1] ZHANG Yang, LI Fu-yun. Study of Removing Powder Technique for Beforehand Gelatinizing [J]. Journal of Beijing Institute of Graphic Communication. 2007, 15(2):47–49

[2] GUO Jun-zhong, CAI Ji-fei. Dynamic Simulation of the Positioning and Feeding Mechanisms of Laminator Based on ADAMS [J]. Packaging Engineering. 2006, 27(4):81–83

[3] LI Guang, LI Li, ZHANG Yang. Structural Design of the Laminator Based on the Vibration Measurement [J]. Journal of Beijing Institute of Graphic Communication. 2009, 17(2):39–42

[4] ZHANG Hai-yan, HE Yan-ling. Study on the Heating Roller and Heating System of Film Laminating Machine [J]. China printing and packaging research. 2010, 2(1):37–41

[5] WANG Xiao-hua, WANG Tao. The three-dimension Parametric design of oscillating mechanism [J]. Journal of Beijing Institute of Graphic Communication. 2005, 13(3):9–11

[6] WANG Xiao-hua. Three-dimensional parametric adjustment of feed lay mechanism [J]. Packaging Engineering. 2006, 27(2):110–112

Information Technology and Computer Application Engineering – Liu, Sung & Yao (Eds)
© *2014 Taylor & Francis Group, London, ISBN 978-1-138-00079-7*

Research on the visual statement method for the passenger train plan based on GIS and system development

Hong Feng Zhu, De Wei Li & Xiao Juan Li
Beijing Jiaotong University, Beijing, China

ABSTRACT: By analyzing the characteristics of the passenger train plan, a study of the visual statement method for the passenger train plan is made. MapXtream plus C#. NET for secondary development based on the Visual Studio platform is adopted. And through the static and dynamic changes of the objects such as points, lines and areas of the GIS map in shape and color parameters, and range, pie, bar themes is the diversified expression of the passenger train plan realized, including train number, train path, departure and arriving time, train class, which provides a visual expression platform for all levels of decision-makers to support them to make a multi-dimensional assessment in the process of passenger transport products' design and adjusting.

Keywords: Visual Statement Method; Passenger Train Plan; GIS; C#.NET; MapXtream

1 INTRODUCTION

China's railway passenger transport market is gradually changing from a seller's market into a buyer's one, passengers demand more and more for the passenger transport products, for which the competition in the transport market becomes fierce. To enhance their competitiveness, the railway operators must rely on high-quality service to promote their passenger transport products. And the key to promoting the railway passenger transport products lies on making a good passenger train plan. The passenger train plan is the core of the railway passenger transport organization, which well reflects the business strategies and the quality of service of the railway passenger transport and that is the embodiment of the competition with other modes of transport.

Traditional train plan is presented to all or different levels of railway decision-makers in the form of a bivariate table so that the performance is not intuitive, in which way the data the railway departments gather cannot be made full use of for lack of multi-dimensional assessment means, which is not to the benefit of timely and effective adjusting and optimizing the passenger train plan to improve its competitiveness.

Compared to the existing statement method for the passenger train plan, the new system provides a more intuitive and interactive interface for the railway decision-makers, which is convenient for them to organize a better assessment with the features of varied thematic maps generated from the original and statistical data and other multiple visual expressions of the passenger train plan, etc.

Geographical Information System is a comprehensive discipline, which combines geography and cartography. It has been widely applied in different fields, and it is a computer system used to input, store, query, analyze and display geographic data. GIS is able to be used to visualize the passenger train plan and provide intuitive data for the staff to help them assess their decisions in different views.

2 THE VISUAL STATEMENT METHOD FOR THE PASSENGER TRAIN PLAN

To make full use of the space-time characteristics of the railway data, a common way of expressing them in the map is utilized. Map design and production should base on the characteristics of data and the relationship between the data items.

In order to realize the graphic interactive operation, a good visual design method is needed. In 1967, Bertin introduced 7 graphic variables: location, size, value, texture/grain, color, orientation and shape. Each kind of visual statement methods for the data involves at least one graphic variable mentioned above. According to the visual properties of the variables, different kinds of data need different variables to express.

In the object-oriented data structure of GIS, spatial data is usually abstracted into points, lines and areas as three types of simple objects.

2.1 *Points of geographic objects*

Points correspond to the railway stations of the railway network, which stand for the O-D and transfer stations of the passenger journeys, namely passenger flow nodes, where passenger flow emerges and vanishes. Two adjacent points identify a unique section. To visualize the points, properties such as location, size and color are utilized. While the key to visualizing the points is the visualization of the statistical data, which includes statistics of passenger flow (such as statistics of passenger flow of a station of the same period in history, statistics of floating population in the area surrounding the station in a radius of some range and so on), statistics of train departure and arriving. Themes are applied to visualize the statistical data.

2.2 *Lines of geographic objects*

Lines correspond to the railway lines of the railway network, which can be utilized to stand for the train path. A confirmed train number occupies a confirmed train path, each section is uniquely identified by two adjacent station points. The properties such as location, size and color are utilized to visualize the railway lines. Of which location makes certain of uniqueness, dynamic sizes can be utilized to show the numbers of passenger trains on the way, colors stand for the different train classes.

2.3 *Areas of geographic objects*

Areas correspond to the areas which bear the railway stations and railway lines of the railway network. Color property is utilized to distinguish the different areas.

3 SYSTEM DEVELOPMENT

3.1 *System requirements*

According to the analysis of the passenger train plan, items shown in Figure 1 are to be presented in the system.

3.2 *Development tools*

MapXtream plus C#.NET for secondary development based on the Visual Studio platform is adopted.

3.2.1 *MapXtream*
MapXtream, the main Windows software development kit of Pitney Bowes MapInfo, is utilized by the developers with the .NET development experiences to create a powerful position-enhanced desktop and client/server application. Developers can use familiar .NET programming languages to program

in this SDK, share and reuse codes between desktop and Web deployment and use standard protocols to access data of large amounts of data source and more other functions. Those mentioned above can all be achieved through MapXtream object models, which is completely managed code API developed on .NET Framework of Microsoft. Common Language Runtime (CLR) of Framework provides the basis of realizing easy development. Layered spatial data structure is the characteristics of MapXtream, which is easily used for manipulating data.

3.2.2 *C#.NET*
C#.NET, an excellent representative of the modern languages, the first component oriented language of the C/C++ language family, is the core programming language of the Microsoft.NET plan, with the characteristics of being concise, flexible, secure, object-oriented and highly compatible. C#.NET is as highly effective as Visual Basic and has no less powerful functions than Visual C++, and it has the same cross-platform characteristics as Java. C#.NET is a specially designed language for .NET, which plays an irreplaceable role in the .NET.

3.3 *System functions*

The system we developed can provide railway GIS sketch maps of various bureaus such as Shenyang Railway Bureau, Beijing Railway Bureau etc. It also provides layer exhibition query of railway lines, railway stations etc. And it can also query information about the passenger train plan based on train number, railway stations, sections etc. The system also provides the functions of zooming in and out the map. As shown in Figure 2.

According to the process of making the passenger train plan, the system has three-level users, namely three-level data structures. They are China Railway Corporation, local Railway Bureaus and Stations. As shown in Figure 3.

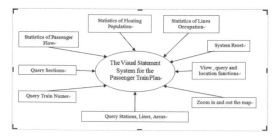

Figure 2.   System Functions.

Figure 3.   Three-level users and data structures.

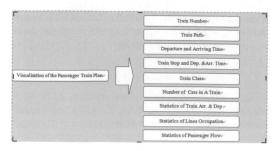

Figure 1.   System Requirements.

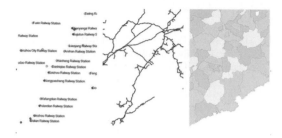

Figure 4. Map Design.

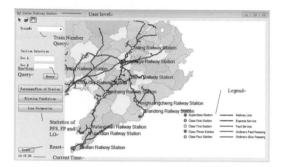

Figure 5. System Interface.

4 EXAMPLE VERIFICATION BASED ON DALIAN RAILWAY STATION

4.1 Map design

According to the spatial characteristics of MapXtream, three pieces of maps are designed to make the whole map. They are station layer, line layer and area layer. As shown in Figure 4.

4.2 System interface

According to the system functions and other aspects, the system based on Dalian Railway Station is as shown in Figure 5.

4.3 Functions presentation

Type in the *Trainno* blank with a distinct train number, then the system will present some details about this train, including train class, train path, number of cars in a train etc. As shown in Figure 6.

Type in the *Section Selection* blanks with two distinct stations, then the system will present the selected section with high light and show any train that runs through it. And the system will present the statistics of current lines' occupation in bar theme which is convenient for the decision-makers to assess the lines' capacity.

Click on the *PassengerFlow of Station* button or the *Floating Population* button, the system will present the corresponding statistical data in theme. As shown in Figure 7 and Figure 8.

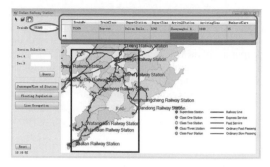

Figure 6. Train number Query.

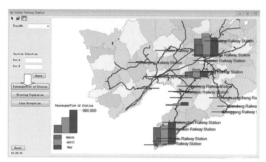

Figure 7. Bar theme.

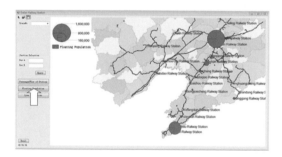

Figure 8. Pie theme.

5 SUMMARY AND FORECAST

Due to the limited time and knowledge, only party of the system is realized and can be put into use. The author provides a kind of visual statement method for the passenger train plan and makes a study of it. But a lot more work needs to be furthered, which includes:

1. The visual statement method for the passenger train plan needs to be probed into furthermore;
2. Scopes of the visual expression of the train plan need to be expanded;
3. Researches on effective index of data and passenger flow forecasting should be made;
4. Simulation and positioning technologies can be integrated into the visual statement system to make it a comprehensive one.

The visual statement for the passenger train plan involves a lot of problems, for which many scholars are engaged in this research. It's believed that with the deepening of the research and the development of other related edge disciplines, more bottleneck technologies will be broke through. The author looks forward to exchanging ideas with the counterparts of the same field and makes a progress with them together to put this research achievement into practical applications as soon as possible.

ACKNOWLEDGEMENT

This research was financially supported by the Projects in the National Science & Technology Pillar Program (Project No.:2009BAG12A10).

REFERENCES

[1] Janes Sturm. Geographical Information System of Slovenian Railways Co[OL]. www.esri.com.

[2] Hans-Jurgen Geisler. DB-Streckendaten-STREDA [J]. Eisenbahningenieur 1998, 49(12): 18–21.

[3] William C. Vantuono. Simplifying Signal Design[J]. Railway Age, 2000(8):61–62.

[4] Keith J. Watson. Use of Computer Database to Manipulate Safety Information[C]. Computers in Railways VI:93–102.

[5] Friedrich Feistle. Modern Automatic Passenger Information System[J]. Rail Engineering International Edition, 2001, (2):5–8.

[6] Software Speeds Up the Design of Italian HS Line[J]. IRJ, 2001, (9):29.

[7] GIS Based Railway Network Information System [OL]. www.esri.com.

Information Technology and Computer Application Engineering – Liu, Sung & Yao (Eds)
© *2014 Taylor & Francis Group, London, ISBN 978-1-138-00079-7*

Research on the transmission line construction risk control and management information system

Jun Luo
University of Science and Technology Beijing, Beijing, China

Yan Hui Wang
Beijing Jiaotong University, Beijing, China

Xiao Wei Du
University of Science and Technology Beijing, Beijing, China

ABSTRACT: Transmission line, with multi-point, extensive-area, long-line, is a complex construction and management process, The risk in the course of the construction of the management level is uneven, therefore result in the uneven risk management level in the construction process. Take the Electricity of Anhui province to east Huainan and Shanghai UHV AC power transmission demonstration project as the background, the great necessity and feasibility with the risk management system in the construction process has been discussed. Based on the analysis of safety influencing factors for the transmission lines engineering construction, the safety risk factor index system which contains four aspects of man – machine – environment – management has been built. Based on the on-site investigation of existing engineering risk management system and measures, the requirements of the management information system for construction risk source control has been analyzed, the service targets and the function requirements of the system has been defined. Based on above, the function structure, logical structure and network architecture of the transmission line construction risk source control management information system have been designed. At last, the system was developed by the combination of FLEX and ORACLE.

Keywords: Transmission line engineering; construction risk management; information management system

1 INTRODUCTION

Electric power industry is an important basic industry of national economy, due to the important r support for national economy development and guarantee for people's living standards. With the rapid development of national economy during the past twenty years, the development of Chinese electric power industry have maintained a rapid speed, and the average annual growth rate was up to 8.1%. As far as the beginning of 2007, 330 KV transmission line over 15000 kilometers has been built in our country, 500 KV lines have almost up to 50000 kilometers.[1]

It can be seen that China will be the largest electric power construction market of all the world in the future for a period of time. Transmission line projects is characterized as high cost, long construction period, complex technology and involving multiple units, therefore the risk of power construction project is much bigger compared with other construction projects, besides, the risk management of electric power construction project in our country is still in its infancy. With the electric power system reformation, factory departs from the nets, our country's electric power construction project investment is diverse, great changes also have taken place in enterprise ownership structure, electric power construction project risk management has gradually drawn the attention of the project participants[1–3].

Through this paper, the systematization and information method for controlling and managing of transmission lines construction was researched, in order to strengthen the risk source dynamic management and the evaluation, prediction, prevention and control of risk factors, finally achieve the purpose of reducing loss and improving earnings.

2 TARGET AND THEORETICAL BASE

2.1 Target

In terms of design, the intellectualization, platformization and integration management information system for the risk controlling was proposed in the high transmission line construction. With the systematic

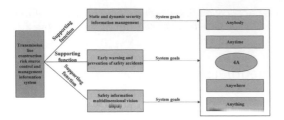

Figure 1. System designation target.

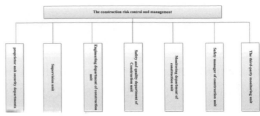

Figure 3. System service object.

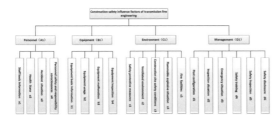

Figure 2. Transmission line engineering construction safety influence index system.

Table 1. Subordinate level 3 indicators of staff basic information.

Staff basic information (a1)	Age (a11)
	Vision (a12)
	Audition (a13)
	Constitution (a14)
	Seniority (a15)
	Professional skills level (a16)

research, sharing of the department information, conformity of scattered resources, digitization of data management, integration of security management can be realized, and eventually the intensity of safety management can be reduced and decision basis for the leadership can be provided.

With the realization of the system, managers or any other users (Anybody) with appropriate permissions, as long as to be able to access the Internet, could access to this system and cover all the management work (Anything) through the system at any time (Anytime), any place (Anywhere), eventually realize the "4A" service of the construction safety management.

2.2 Safety risk factor index system

Based on the safety influence factors analysis of the transmission line engineering, considering the needs of engineering risk control, safety factor index system has been established from four aspects of person – equipment – environment – management respects[4–6], as shown in figure 2.

In the figure, A1–D1 are Level 1 indicators, a1–d6 are Level 2 indicators, the level 2 is divided into several specific indicators as level 3 indicators, take the Staff basic information a1 as an example, the details are as in Table 1.

Figure 4. System function requirements diagrams.

3 SYSTEM REQUIREMENTS ANALYSIS

3.1 System service objects

As an application-oriented system, system services object should be firstly defined. This system is designed for transmission line construction project as a management information system, and the service object are shown in the figure 3.

3.2 System function requirement analysis

Based on the thorough investigation of engineering field, the system function requirement are definitely as given in Figure 4.

(1) Static information management

In order to conveniently deal with each program in the system, all of the information for those programs should be comprehensively managed, including design and construction scheme, construction drawing, geological exploration and environmental investigation report, risk management report, construction organization and the special solution and so on.

(2) Dynamic information management

Dynamic information is a variety of safety information with the emergence of the construction during the process of project construction. The System could provide different interfaces to input all kinds of relevant information according to different users. Therefore the system must provide a distributed application due to the various users as well as diverse office location.

(3) Early warning management

If the data is beyond the established threshold, early warning and alarm processing should be provided.

In this system, safety warning in construction risk engineering is classified as monitoring warning, patrol warning and comprehensive warning. All the warning are divided into yellow monitoring, orange warning and red monitoring according to the severity from small to large.

The "double control" indicators (change, change rate) exceeding the measurement control values of a certain proportion result in different warnings. The system should set for different warning thresholds and deal with the corresponding situation.

(4) Contingency plans management

Due to the special requirements of safety engineering, relevant special contingency plans and emergency plans should be established during the process of engineering construction, in case of an emergency to ensure the construction completion smoothly.

(5) Risk event management

No matter to launch the corresponding plan or not, safe trouble-hidden report or stop-work orders should be adopted when risk events happened.

(6) Risk source identification management

Two types of risk source should be identified and supervised during the transmission line engineering construction process. One type could be called dominant which can be found through the field reconnaissance and project design. Another type of risk source, also called recessive risk source, can be obviously discovered through a variety of monitoring data and patrol data, or through a variety of prediction model processing, with the progress of the construction work.

(7) Statistical analysis

The system need to statistically analysis all kinds of monitoring data, then present them as pie chart and histogram charts according to the time, therefore intuitively display the engineering on-site safety status.

(8) System maintenance

System should support the management for basic dictionaries, system users and its permissions.

3.3 System non-functional requirements

(1) System fluency

The operation response time for general information-input, query, electronic map operations should be less than 3 seconds, the statistical analysis of historical data should be less than 30 seconds.

(2) System openness and scalability

The system follows a modular development which should be of good extensibility, fully considering the increase of the data format in the future, and compatibility with other systems.

(3) System's convenient

User interface should be concise and beautiful, data and file input mode is flexible and fast.

(4) System safety

System should take necessary safety measures to ensure the information security, protecting information from illegal interception or illegal modification.

(5) System stability and reliability

Data of the network equipment involved in the system should be accurate and reliable, and timely add, update and backup changed data, in order to ensure the data reliability.

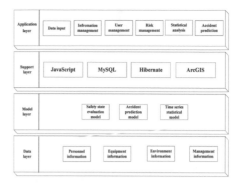

Figure 5. The system logic structure.

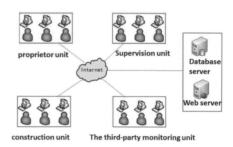

Figure 6. Systems network architecture.

Figure 7. System function structure.

4 SYSTEM OVERALL DESIGN

4.1 System logic structure

According to the functional requirements and module characteristics of the system, the system is divided into data layer, model layer, support layer and application layer, as shown in figure 5.

4.2 Systems network architecture

Due to the system involves many units and each unit does not have direct subordinate relations with others, the system network topology structure is designed as figure 6.

4.3 System function structure

Considering the requirements analysis, the system function is mainly designed as information input, risk management, emergency management, early warning alarm, statistical analysis and system maintenance and so on, as shown in figure 7.

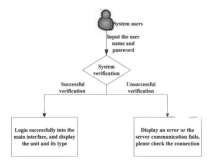

Figure 8. System login implementation logic diagram.

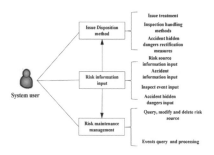

Figure 9. Systematic risk management implementation of logic diagram.

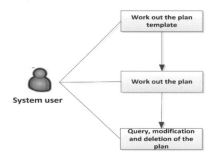

Figure 10. System emergency source management implementation of logic diagram.

5 SYSTEM IMPLEMENTATION

The system was developed by FLEX and ORACLE, and the business processes are as follows.
(1) System login (Fig. 8).
(2) Risk management (Fig. 9).
(3) Contingency plans management (Fig. 10).
(4) Early warning and alarming management (Fig. 11).

6 CONCLUSION

In this paper, the realization of the transmission lines construction risk source management and control information system, which characterizes with risk and accident management, early warning and alarm management and statistical analysis and other

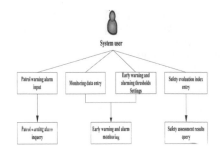

Figure 11. System warning and alarming management implementation of logic diagram.

functions, was put forward. Based on the detailed investigation of existing risk management methods on construction site and the actual management requirements, combined with the analysis of transmission line engineering construction risk, the transmission line engineering construction risk evaluation index system is constructed, besides, with the further designation of system function, physical architecture and logical architecture, supported by the eclipse platform and with the combination of FLEX + ORACLE technology, the system could realize the implementation eventually.

The implementation of transmission line construction risk source management and control information system could effectively improve the construction risk information management level, realize the dynamic management of various types of risk source in the process of engineering construction, and the evaluation of the current risk source security status, therefore decrease the accident rates and reduce the loss caused by accidents, which could provide a certain theoretical and practical guidance for similar engineering risk information managements.

REFERENCES

[1] Lv Lihui. 500 k ROM 102 line back to the power transmission and transformation project risk evaluation [D]. North China electric power university, 2009.05.
[2] Xue Lan. Helicopter transmission lines inspection system research and design based on GIS [D]. Harbin University of Science and Technology. 2008.03.
[3] Xu Donghui. Industrial and mining enterprise transmission lines management information system [D]. Nanchang University. 2006.06.
[4] Dong Jihong. Research on major hazards identification and evaluation technology [D]. Xi'an University of Architecture and Technology. 2004.06.
[5] Wang Xikui, Wu Zongzhi. Discuss on major hazards identification standard revision [J]. Journal of China Safety Science, 2007:17(1).
[6] He Tianping, ChengLing. Discussion and research on a number of major hazard identification questions [J]. Journal of China Safety Science, 2007:17(8).

Information Technology and Computer Application Engineering – Liu, Sung & Yao (Eds)
© 2014 Taylor & Francis Group, London, ISBN 978-1-138-00079-7

Recursive algorithm for circle anti-aliasing

Yin Liang Jia, Jing Gao & Bing Yang Li
College of Automation Engineering, Nanjing University of Aeronautics and Astronautics, Nanjing, China

ABSTRACT: Anti-aliasing is important for graphics and circle is no exception. Circle anti-aliasing algorithms in existence are complex and their effect does not satisfy. A recursive algorithm was explored for circle anti-aliasing. The recursive algorithm chooses two pixels closest to the ideal circle in each column and the grayscale is calculated according to the distance between the center of a pixel and the circle. The algorithm abandons two-order epsilon, reduces error, and gets the coordinates and grayscale of new pixels in the light of old pixels by recursion. The new algorithm improves circle anti-aliasing efficiency and effect.

Keywords: Circle drawing; Anti-aliasing; Grayscale; Recursive functions

1 INTRODUCTION

To show a nice picture, computer graphics is very important, and there are varied algorithms to generate graphics speedy. Raster display uses discrete pixels to show graphics while ideal graphics are sequential, so the graphics shown by raster display have serrated edges. The visual effect of these graphics is not good because of these serrated edges, and we call this phenomenon aliasing. Anti-aliasing is necessary to improve visual effect.[1]

Circle and arc are common graphics included in varied pictures, and they need anti-aliasing also. The effect and efficiency of the circle anti-aliasing algorithm are vital. Because of the two-order function of circle, the anti-aliasing algorithm is complex and existing algorithms are not satisfactory.

In this work, we explored a new algorithm for circle anti-aliasing. According to the distance between the center of a pixel and the circle, the algorithm finds the grayscale of each pixel by recursion, and simplifies calculation by abandoning two-order epsilon. This algorithm finds accurate grayscale of each pixel and the anti-aliasing effect is better.

2 RELATED WORK

Circle anti-aliasing algorithms are always based on regional sampling and look ideal circle as a pixel width graphic. The grayscale of a pixel is count according to the area overlap between circle and the pixel.[2–3]

Reference [4] explored squareness regional sampling algorithm and there are some similar algorithms.[5–7] Although these algorithms decrease aliasing of circle, these algorithms are too complex.

An algorithm was present to generate anti-aliasing circle only using integer.[8] The algorithm is very fast, but it has only 3 different grayscales. The error of grayscale is very big sometimes and the algorithm would lose some pixels in circle. Some algorithms ameliorate this algorithm.[9–10] Although the error of grayscale reduces and effect is better, these algorithm has only 7 different grayscales and may lose pixels, all the same.

An algorithm uses look-up table or retrieval to decide grayscale of pixels without floating-point.[11] If generating an anti-aliasing circle with 16 different grayscales, the algorithm is not complex. If an anti-aliasing circle has more different grayscales, the algorithm has a large amount of calculation, furthermore, the error of grayscale is high and the coordinate of some pixel may be wrong.

Reference [12] bypasses two-order epsilon to simply calculation and explores a recursion algorithm. Although abandoning floating-point, multiplication and division, the algorithm is complex because it needed too many addition, compare and shift. Even worse, the error of grayscale is big because of bypassing two-order epsilon, especially those small radius circles.

Reference [13] bypasses two-order epsilon such as reference [12], but the algorithm cut down the error by some calculation. The algorithm looks circle as line in a pixel area. If the crossover point of arc and the line between two pixels center shown in fig. 1 is J, this algorithm counts the grayscale according to the distant from pixels center to the tangent line going through J. In fig. 1, l_b decides the grayscale of the above pixel. In fact, the distant from pixels center to the ideal circle was l_a. Because of the difference between l_a and l_b, the error of grayscale was big, so the anti-aliasing circle was not good enough.

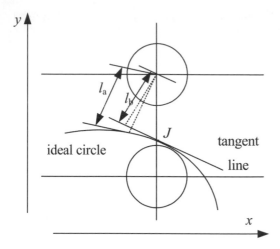

Figure 1. Method of assign grayscale assign.

The common circle anti-aliasing algorithms are deficient in either anti-aliasing effect or computational efficiency in conclusion.

3 THE CIRCLE ANTI-ALIASING ALGORITHM

Circle is a assemblage of points, which distance between these points and the centre of a circle is R. The circle has four symmetry axis: $x = 0$, $y = 0$, $x = y$ and $x = -y$, if the center of this circle is origin. Every point in a circle has 7 symmetrical points according to these symmetry axis, we only need generated eighth circle and get the other part of circle by coordinate conversion. In this paper, we only account for eighth circle from $(0, R)$ to $\left(\frac{R}{\sqrt{2}}, \frac{R}{\sqrt{2}}\right)$, which belongs to a circle whose center is origin appoint and the function is $F(x, y) = x^2 + y^2 - R^2$.

To generate a anti-aliasing circle, we chose 2 pixels in every pixel column. The 2 pixels are closest to the ideal circle than other pixels in a column. Grayscale of the pixel is proportional to the distant from pixel center to the ideal circle.

Pixel P which coordinate is (x_P, y_P) and G are closest to the ideal circle in the column which horizontal ordinate is x_P, shown in fig. 2, and we get grayscales of P and G already. The two pixels in column $x_P + 1$ are either T, M or M, B. We suppose the width of a pixel is 1 and the distant from a pixel center to the ideal circle is l. If $l = 1$, the grayscale of a pixel is 0, while the grayscale is max if $l = 0$. We suppose grayscale is g and $g \in [1, 2^n]$, so we can use function (1) to decrease grayscale as:

$$g = 2^n - 2^n \, l \qquad (1)$$

If ideal circle drills through pixel T and M such as the real arc in fig. 2, we suppose the distant from pixel

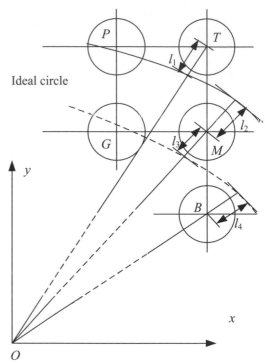

Figure 2. Some pixels in a circle.

T center to the ideal circle is l_1 and the distant of pixel M is l_2, then we can get function as:

$$(R + l_1)^2 = x_T^2 + y_T^2 = (x_P + 1)^2 + y_P^2$$

$$l_1 = \frac{x_P^2 + y_P^2 + 2x_P + 1 - R^2 - l_1^2}{2R}$$

$l_1 \leq 1$, so $\frac{l_1^2}{2R} \leq \frac{1}{2R}$. $\frac{l_1^2}{2R}$ is a two-order epsilon. We abandon $\frac{l_1^2}{2R}$ to avoid sqrt. Furthermore, we can easy give a recursion functions to get the grayscale of a pixel to simplify calculation without $\frac{l_1^2}{2R}$.

Although simplifying calculation, abandon $\frac{l_1^2}{2R}$ increases the calculation error of l_1. If $i_T = x_T^2 + y_T^2 - R^2$, we can use function (2) & (3) to decrease error as:

$$l_1 = \frac{i_T}{2R} - \frac{1}{8R} \qquad \text{if } l_1^2 \leq \frac{1}{2} \qquad (2)$$

$$l_1 = \frac{i_T}{2R} - \frac{3}{8R} \qquad \text{if } l_1^2 > \frac{1}{2} \qquad (3)$$

The error, because of abandoning $\frac{l_1^2}{2R}$, is no more than $\frac{1}{8R}$, according to the above functions. If $R \geq 16$, the grayscale error is less than 1%.

Because l_1 is positive, compare l_1^2 and $1/2$ to fine which is great, is just compare $(l_1 + R)^2$ and $(\frac{1}{\sqrt{2}} + R)^2$. Because $(l_1 + R)^2 = x_T^2 + y_T^2$, we can compare i_T and $\sqrt{2}R + \frac{1}{2}$. As the real arc in fig. 2, $i_T = i_P + 2x_P + 1$, we suppose $INCE = 2x_P + 1$ and

198

$INCNE = 2x_G - 2y_G + 2$, so $i_T = INCE + i_P$, and we calculate the increments of $INCE$ and $INCNE$ using these factions:

$$INCE_{new} = INCE_{old} + 2$$
$$INCNE_{new} = INCNE_{old} + 2$$

The same argument applies to pixel M, and we can get function as:

$$(R - l_2)^2 = x_M^2 + y_M^2 = (x_G + 1)^2 + y_G^2$$
$$l_2 = -\frac{x_G^2 + y_G^2 + 2x_G + 1 - R^2 - l_2^2}{2R}$$

Abandoning $\frac{l_2^2}{2R}$, we can use function (4) & (5) to decrease error:

$$l_2 = -\frac{i_M}{2R} + \frac{1}{8R} \qquad \text{if } l_2^2 \le \frac{1}{2} \tag{4}$$

$$l_2 = -\frac{i_M}{2R} + \frac{3}{8R} \qquad \text{if } l_2^2 > \frac{1}{2} \tag{5}$$

Comparing l_2^2 and $1/2$ to fine which is great, is just compare $(R - \frac{1}{\sqrt{2}})^2$ and $(R - l_2)^2$. Because $(R - l_2)^2 = x_M^2 + y_M^2$, we can compare $\frac{1}{2} - \sqrt{2}R$ and i_M.

As the real arc in fig. 2, we can find: $i_M = i_G + 2x_G + 1$, $i_M = INCE + i_G$, and calculate the increments of $INCE$ and $INCNE$ using these factions:

$$INCE_{new} = INCE_{old} + 2$$
$$INCNE_{new} = INCNE_{old} + 2$$

If ideal circle drills through pixel M, B such as the broken arc in fig 2, we suppose the distant from pixel M center to the ideal circle is l_3, then we can get function as:

$$(R + l_3)^2 = x_M^2 + y_M^2 = (x_G + 1)^2 + y_G^2$$
$$l_3 = \frac{x_G^2 + y_G^2 + 2x_G + 1 - R^2 - l_3^2}{2R}$$

Abandoning $\frac{l_3^2}{2R}$, we can use function (6) & (7) to decrease error:

$$l_3 = \frac{i_M}{2R} - \frac{1}{8R} \qquad \text{if } l_3^2 \le \frac{1}{2} \tag{6}$$

$$l_3 = \frac{i_M}{2R} - \frac{3}{8R} \qquad \text{if } l_3^2 > \frac{1}{2} \tag{7}$$

We can compare i_M and $\frac{1}{2} + \sqrt{2}R$. As the broken arc in fig. 2: $i_M = i_P + 2x_G + 1$, so $i_M = INCE + i_G$, and we calculate the increments of $INCE$ and $INCNE$ using these factions:

$$INCE_{new} = INCE_{old} + 2$$
$$INCNE_{new} = INCNE_{old} + 4$$

We suppose the distant from pixel B center to the ideal circle is l_4, then we can get function as:

$$(R - l_4)^2 = x_B^2 + y_B^2 = (x_G + 1)^2 + (y_G - 1)^2$$
$$l_4 = -\frac{x_G^2 + y_G^2 + 2x_G - 2y_G + 2 - R^2 - l_4^2}{2R}$$

Abandoning $\frac{l_4^2}{2R}$, we can use function (8) & (9) to decrease error:

$$l_4 = -\frac{i_B}{2R} + \frac{1}{8R} \qquad \text{if } l_4^2 \le \frac{1}{2} \tag{8}$$

$$l_4 = -\frac{i_B}{2R} + \frac{3}{8R} \qquad \text{if } l_4^2 > \frac{1}{2} \tag{9}$$

We can compare $\frac{1}{2} - \sqrt{2}R$ and i_B. As the broken arc in fig. 2, $i_B = i_G + 2x_G - 2y_G + 2$, so $i_B = INCNE + i_G$, and we calculate the increments of $INCE$ and $INCNE$ using these factions:

$$INCE_{new} = INCE_{old} + 2$$
$$INCNE_{new} = INCNE_{old} + 4$$

To find ideal circle drills through pixel T, M, such as the real arc in fig. 2, or through pixel M, B, such the broken arc, we can decide pixel M in the circle or not. If $x_M^2 + y_M^2 - R^2 < 0$, pixel M is in the circle and ideal circle drills through pixel T, M, otherwise M is out of the circle.

$$i_M = x_M^2 + y_M^2 - R^2 = (x_g + 1)^2 + y_G^2 - R^2$$
$$= x_g^2 + y_g^2 + 2x_g + 1 - R^2 = i_G + 2x_g + 1$$

The recursion functions is these: when we find the coordinate and grayscale of pixels in a new column, first, we calculate $i_M = INCE + i_G$. If $i_M < 0$, we choose pixel T, M, otherwise, we choose pixel M, B; After that, if $i_M < 0$, $i_T = INCE + i_P$, then we calculate l_1 use function (2), (3) and l_2 use function (4), (5); Otherwise, $i_B = INCNE + i_G$, we calculate l_3 use function (6), (7) and l_4 use function (8), (9); Third, we calculate each pixel grayscale using function (1); Last, we calculate the increments of $INCE$ and $INCNE$ using these factions:

$$INC_{new} = INC_{old} + 2$$
$$INCNE_{new} = INCNE_{old} + 2 \text{ if } i_M < 0$$
$$INCNE_{new} = INCNE_{old} + 4 \text{ if } i_M \ge 0$$

In first column, we choose pixel $(0, R)$ whose i is 0, and $(0, R - 1)$ whose i is $1 - 2R$, while $INCE_0 = 1$ and $INCNE_0 = 4 - 2R$.

4   ALGORITHM COMPARISON

The algorithm in this paper can generate various parameters anti-aliasing circles with different grayscales. The error is less and the effect of anti-aliasing is good. In contrast, the references [8–10] present anti-aliasing algorithm with less amounts of calculation, but there are no more than 7 different grayscales. Although reference [11–12] has more different grayscales, but the grayscale is a rough estimate, and even choose the wrong pixel. Reference [13] cuts down the grayscale error to some extent, but the error is still big and the algorithm was too complex.

The algorithm in this paper uses recursive functions to generate anti-aliasing circle. Comparing with reference [12–13], calculation results is more accurately so the anti-aliasing effect is better, and the amount of calculations is less by recursion.

5 CONCLUSION

In this paper, a recursive algorithm for circle anti-aliasing was explored. The algorithm chooses two pixels closest to the ideal circle in each column and calculates grayscale according to the distance from the center of a pixel to ideal circle. We simplify calculation by recursive functions and the algorithm calculates grayscale of each pixel accurately and the anti-aliasing effect is satisfying.

ACKNOWLEDGEMENTS

The research was supported by the Chinese Fundamental Research Funds for the Central Universities, NO. NS2012041

REFERENCES

[1] James D. Foley. (2004). *Introduction to Computer Graphics*. Beijin: China Machine Press.

[2] Xu Xiao-liang, Hong Bo. A sub-pixel regional sampling anti-aliasing algorithm based on integer coordinate. *Journal of Image and Graphics*, 2009, 14(12):2438–2442.

[3] Rokita, P. Depth-based selective antialiasing. *Journal of Graphics Tools*, 2005, 10(3):19–26.

[4] Field D. Algorithms for drawing anti aliased circles and ellipses. *Computer Vision, Graphics, and Image Processing*, 1986, 33(1):1–15.

[5] Fu, Bowen. Niu, Lianqiang. Integral algorithm for generating anti-aliasing circle based on Bresenham algorithm. China: Trans Tech Publications, 2012:490–495.

[6] Field, Dan. ALGORITHMS FOR DRAWING ANTI-ALIASED CIRCLES AND ELLIPSES. *Compute Vision Graphics Image Process*, 1986, 33(1):1–15.

[7] Niu, Lianqiang, Feng, Haiwen; Wu, Peng. Fast algorithms for generating and anti-aliased drawing circles controlled by residuals. China: Institute of Computing Technology, 2011:232–239.

[8] Wu X L, Rokne J G. Double step incremental generation of lines and circles. *Computer Vision, Graphics, and Image Processing*, 1987, 37(3): 331–344.

[9] Niu, Lian-Qiang; Shao, Zhong; Wu, Peng. Integer anti-aliased ellipse generating algorithm based on Bresenham algorithm. China: Shenyang University of Technology, 2010:316–320.

[10] Niu Yu-jing, Tang Di. Double-step anti-aliasing drawing algorithm of circle. *Computer Engineering and Applications*, 2010, 46(23):175–178.

[11] Niu Lianqiang, Shao Zhong. Unified Integral Algorithm for Anti-aliased Lines and Typical Curves. *Journal of Computer-Aided Design & Computer Graphics*, 2010, 22(8):1293–1299.

[12] Jia Yin-liang, Zhang Huan-chun, Jing Ya-zhi. Integral Algorithm for Circle Anti-aliasing. *Journal of Image and Graphics*, 2012, 17(1):130–136.

[13] Liu Jing. Implementation for Improved Algorithm of Circle Anti-aliasing. China: Mechatronics and Control Engineering, 2012:1327–1333.

Information Technology and Computer Application Engineering – Liu, Sung & Yao (Eds)
© 2014 Taylor & Francis Group, London, ISBN 978-1-138-00079-7

Study of multi-resolution modeling framework in joint operations simulation

Chen Yan Kong & Li Ju Xing
Science and Technology on Information System Engineering Key Laboratory, Nanjing, China

ABSTRACT: The Multi-Resolution Modeling (MRM) is used to study the problems in modeling and simulation of joint operations. The main methods of MRM are discussed. According to the simulation model architecture of joint operations, the MRM requirements of joint operations are analyzed according to its simulation model architecture, the design goal and the scheme are given.

Keywords: joint operations simulation; aggregate; disaggregate

1   GENERAL INSTRUCTIONS

In February 2010, the U.S. Department of Defense proposed the new "Air Sea Battle", aiming at the integration of the combat capability in areas such as air, sea, land, space and cyberspace to restrict china with integrated sea and air combat forces in Pacific. In this context, we must study the joint operations capability actively, building simulation model system for analysis and research for the joint operations.

By the way, the level of joint operations simulation is widening, the simulation objects are increasingly complex, the entities need to be described are increasing. The diversity of systems, the uncertainty of the behaviors and organizations shows the complex features of the joint operations. Modeling with single resolution is difficult to meet the changing needs. We must study the method of building multi-resolution models for the same system or process, providing different information and interfaces for different army and different level of users, establishing the interactions between those different resolution models for the multi-angle and multi-level research of joint operations.

2   THE NECESSARY OF MULTI-RESOLUTION MODELING IN JOINT OPERATIONS SIMULATION

The composition and framework of elements in joint operations are significantly different from elements in traditional operations. It is no longer simple confrontation between single platform or single system, but system confrontations behavior co-evolved by operations entities with different properties and different types. According to the level of the operations, it can be divided into strategic operation, campaign operation and tactical operation. According to army

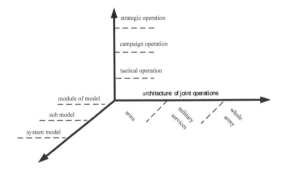

Figure 1.   Architecture of joint operations simulation.

technology, it can be divided into whole army, military services and arms, etc, as shown in Figure 1. These different level operation elements are integrated from information networks. According to objective of the operation, resources are chosen and located optimally to construct a seamless link operation system to achieve the maximization of joint operations effectiveness.

The modeling and simulation of those models is different from traditional operation simulation models in description, modeling methods, modeling emphasis and evaluation standard. The positive significances of multi-resolution modeling are mainly reflected in the following aspects.

First is complexity of the system. There are a large number of models in the joint operations simulation, single-resolution model is difficult to meet the needs of complex system which should provide proper information for different military services and man-machine interface for user of different levels. Therefore, we must analyze and handle the problem in the way of multi-level and multi-angle. We need low-resolution models to describe the microscopic properties of the system, and also need high-resolution models to describe the macroscopic properties of the system (Chen et al. 2009).

Second is flexibility of the simulation. Peoples concern different level of details about the same question. High-resolution models reveal details, low-resolution models reveal macro things, like the nature of the properties. With the continuous expansion of the scale of the simulation, different resolution models depending on the simulation needs should be combined and studied together (Liu & Huang 2004).

Third is the complexity of computation. There are lots of models running and communicating to each other in joint operations simulation. High-resolution requires lots of resources, this is unrealistic to all use high-resolution models. We should construct models in proper resolution depending on simulation purpose.

3 METHODS OF MULTI-RESOLUTION MODELING

The methods of multi-resolution modeling construct different resolution models for systems or processes, and describe them consistently, select a suitable resolution during the execution to meet the simulation needs. The purpose of multi-resolution modeling is changing resolution dynamically according to the experiment intension, giving a way to improve self-adaptability of models and simulation objects, solving the problems of the complexity in simulation and the limits of resources, simulating multi-resolution of objects for studying systems from different perspective.

Mostly, models with interactive relationship require equal resolution, i.e. aggregation models can't interact with disaggregation models, and vice versa. At present, there are some methods for multi-resolution modeling, mainly including aggregation and disaggregation, selective viewing, multi-resolution entity, IHVR, SES/MB, MOOSE, etc. These are all based on interaction with equal resolution. Three methods commonly used are introduced.

3.1 Aggregation and disaggregation

Models interact with the same resolution by disaggregating low-resolution models or aggregating high-resolution models to maintain consistency. Aggregation and disaggregation method is easier to realized with object-oriented approach. Aggregation transforms several high-resolution models to single low-resolution model, and disaggregation is exactly opposite. The method has several different forms of application: complete disaggregation, partial disaggregation, region disaggregation and pseudo disaggregation (Song 2008).

Complete disaggregation disaggregates all low-resolution models to high-resolution models which could stand for them. Complete disaggregation generates a large number of models which will consume a lot of system resources. For example, if we only want to disaggregate airplanes in a fleet. Using this method will disaggregate all fleets.

Partial disaggregation only disaggregate partly to overcome the drawback of complete disaggregation, another part will be left for interaction with aggregated models. In this application, which part to be disaggregated is the key point.

Pseudo disaggregation is used when high-resolution models do not interact directly to low-resolution models but only to some properties of low-resolution models.

Aggregation and disaggregation method consumes less resources relatively, but there are also many problems, such as temporal inconsistency, chain-disaggregation, frequent aggregation and disaggregation, delay and loss of information in model translation, etc. At the same time, the method does not support concurrent interaction, only one resolution of the model could run at one time.

3.2 Selective viewing

The idea of selective viewing method is entity in the simulation system is running in the highest resolution models. When it needs to interact with low-resolution entities, transforms high-resolution models to low-resolution models and simulates low-resolution entities.

Selective viewing method depends on the highest resolution models, only interacts with other low-resolution model at a few moments in simulation. Similar to the construction of MVC in software design patterns (Natrajan A & Tuong A N. 1995). The highest resolution model is equivalent to Model, the low resolution model transformed from highest resolution model is equivalent to View, Controller decides when and how to transform high resolution models to low resolution models. This method can reflect different perspective to the same problem.

The advantage of selective viewing method is relatively easy to achieve, the model does not change during the simulation, model consistency is easily maintained. But for complex models, this method is lack of flexibility and computationally intensive.

3.3 Multi resolution entity (MRE)

MRE method maintains running information of different resolution models, which is proposed by a research team at the University of Virginia. In the method, an entity is constituted by multi-resolution models, it interacts with different resolution models simultaneously and maintains the consistency of the internal logic. In other words, during the entire simulation, those different resolution models all exist in the entity, when the entity interacts with different resolution models, the result should be consistent. so it is necessary to maintain consistency of the interaction with different resolution models the consequent. The principle diagram is shown in Figure 2.

MRE concerns about the consistency of the multi-resolution models. As the figure shows, its consistency maintenance mainly depends on consistency

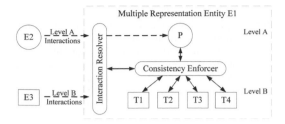

Figure 2. Principle diagram of MRE.

Figure 3. Model architecture of joint operations simulation diagram.

enforcer, consistency enforcer is constituted by attribute-dependent association graph and mapping functions. Attribute-dependent association graph describes the relationship between different resolution model properties. Mapping functions describe detail operation of the relationship. Interaction resolver maintains the consistency of properties of different resolution model to solve the concurrent interaction in MRE.

MRE is good for the development of a new multi-resolution model system. But for existing systems, it's difficult for changing greatly. And for some applications, it's not necessary for always maintaining the concurrent interaction of different resolution model.

4 JOINT OPERATIONS MODELING BASED ON MULTI-RESOLUTION

4.1 Model architecture of joint operations simulation

The confrontation experiment of joint operations simulation contains red platform, blue platform and neutral director platform (Zhang 2012). Neutral director platform is always constructed as the subsystem of red platform or blue platform because of its simple construction. Red platform and blue platform are peer-to-peer systems, so we describe the model architecture from one of it, as shown in Figure 3.

The platform mainly contains natural environment model, entity model, entity behavior and function model, assessment and analysis model and simulation foundation model, etc. And every part contains its own models. We analyze the multi-resolution modeling needs for every part in joint operations simulation.

- Multi-resolution of natural environment model
  Natural environment in joint operations simulation provides basic operating environment, including space environment, the atmospheric environment of earth's surface, terrestrial environment, marine environment, all kinds of environment can be subdivided. Numerous environment models in simulation tasks are impossible for always be running in high-resolution, so it is necessary to construct multi-resolution models for natural environment.
- Multi-resolution of entity model
  Entity models in joint operations simulation include early warning detection model, command and control model, radar countermeasures model, communications confrontation model, photoelectric confrontation model and target platform model. The most typical entity in this type like target platform, radar, etc. can be sub-divided to atomic entity and polymeric entity simulated as high-resolution and low-resolution model. Command and control (c2) may be a little special, it can be divided to joint c2, group c2, troop c2, etc. According to the level of the task and the needs of experiment system, we can construct different resolution model for different intension.
- Multi-resolution of entity behavior and function model
  Environment model and entity model are models for exist things. There also needs corresponding models for entity behavior and function in the simulation. According to multi-resolution entity model, entity behavior and function model should have appropriate resolution to support interaction of simulation data and maintain the consistency of the models, to ensure the correctness of the simulation results.
- Multi-resolution of assessment and analysis
  Assessment and analysis for simulation data is the final stage of the simulation deduction. Joint operations simulation based on multi-resolution modeling generates multi-level and many kinds of simulation output data, those data should be processed for assessment and analysis to solve different levels of simulation problems.

4.2 Multi-resolution modeling framework

As analysis above, joint operations simulation has entity-oriented model (atomic entity model and polymeric entity model), and also has parameter-oriented model (such as assessment and analysis model) (Chen 2003). So far, there does not yet exist a method which can solve all the problem of multi-resolution modeling, the tool of modeling is relative lack of too. Therefore, develop such complex system like joint operations (Mao 2012), we must build a multi-resolution modeling framework for joint operations development to deal with the relationship of hierarchical expanding models, solve the problems of aggregation and disaggregation, consistency maintenance, interactive processing etc. With the framework, model developers can focus on the logic development of models,

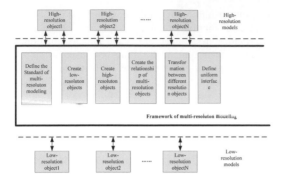

Figure 4. Multi-resolution model framework diagram.

interoperability and reusability of model could be improved. Multi-resolution modeling framework should have the following features:

- A set of standardized, realizable mode should be given. The steps of multi-resolution Object design framework should be provided.
- The base object class of multi-resolution modeling should be designed to provide public functions and properties.
- Common underlying support of multi-resolution modeling, such as aggregation and disaggregation, relationship maintenance in different resolution models, the conversion of different resolution models, consistency maintenance should be given to shield the user from details of multi-resolution modeling.
- Different run mode of multi-resolution models should be supported.

The diagram of multi-resolution model framework is shown as Figure 4. The framework mainly provides the standards of model description, the mapping relationship between different resolution models, and the transformation methods. During design and development process of multi-resolution model framework, there are some difficulties:

First is distribution of object and model in multi-resolution. Taking airplane and fleet as example, the first distribution way is putting aircraft and fleet in an object, the second way is putting aircraft and fleet in different objects, the third way is putting all airplanes in a object, putting fleet in another object. We need to consider the complexity of the computation, the reusability of modules, network performance and specific simulation application comprehensively to choose the right way.

Conversion Trigger mode of different resolution models is also very important. We can change the resolution depending on the entity action area or a specific event (Liu & Huang 2003); or using artificial trigger

such as interface trigger or manual command; or setting triggers in scenario files, triggered automatically during the simulation process.

Conversion method of different resolution models, the work of resource computation, the conflict of interaction, how to maintain the consistency, how to simulate the interaction between different resolution models are all problems we should study hardly.

5 CONCLUSION

There are many aspects which worth our further research and exploration in multi-resolution modeling. For example, the dynamic combination of different resolution modeling, we could establish a model library stored with basic multi-resolutions models according to the format, providing query, modification and execution interface. It could construct simulation systems fast and dynamically, and improve the reusability and interoperability of simulation models.

Study of multi-resolution modeling based on agent is also a hot spot. During the simulation, every object is an agent, it learns the execution environment, understand the resolution of the other models and adjust their own resolution dynamically, maintain consistency between different resolution view to solve the problems encountered in multi-resolution modeling.

REFERENCES

Bao hong LIU, and Ke di HUANG. 2004. "Multi-Resolution Modeling: Present status and trends". Journal of System Simulation 16(6):1150–1154.

Bao hong LIU and Ke di HUANG. 2003. "The Research of Multi-Resolution Modeling Based on HLA". Computer Simulation 20(3):43–71.

Chi wan, Chen. 2003. "The current situation and development of multi-resolution modeling research". Military operation research and systems engineering 3:58–61.

Cui xia Zhang, Xin Zhou and Bing Fang. 2012. "Simulaiton modeling for red and blue engagements". Command Information system and technology 3(6):10–13.

Jian-hua CHEN, Gang-qiang LI, and Tiao-ping FU. 2009. "Research of Multi-distinguish Modeling on Warship Formation Operation Simulation". Journal of System Simulation 21(22):7316–7319.

Natrajan A and Tuong A N. 1995. "To Disaggregate or Not to Disaggregate That Is Not the Question". ELECSIM95. Mar.

Ping Song. 2008. "Study on multi-resolution modeling in spatial mission simulation". Doctoral Dissertation. Xi'an institue of optics and precision mechanics chinese academy of sciences.

Shao jie Mao, Ke bo Deng and Heng Wang. 2012. "Complexity of net-centric and service-oriented C4ISR system". Command Information system and technology 3(4):1–6.

Information Technology and Computer Application Engineering – Liu, Sung & Yao (Eds)
© *2014 Taylor & Francis Group, London, ISBN 978-1-138-00079-7*

A novel approach of providing feedbacks at where a mistake occurs during solving math word problems

Yan Zhen Qu & Kyle Morton
Colorado Technical University, Colorado Springs, USA

ABSTRACT: Not having the proper help to pinpointing where the mistake occurred in solving math word problems can be frustrating for many students. In this paper we propose an interactive user input sensitive feedback model to assists students solving math word problems based on their own thinking pattern. This model uses fuzzy logic as a method to evaluate a user's input to determine if an input for each step towards a solution is correct and if that student needs help for correcting their mistakes. This model allows students to solve a math word problem by using a free form, and the immediate feedback at each step will be provided directly based on the student's input. This adaptive approach is much effective in stimulating students' confidence and learning in solving math word problems.

Keywords: E-Learning, Fuzzy Logic, On-line Tutoring, Adaptive Learning, User Input Sensitive

1 INTRODUCTION

Solving math word problems can be a challenge for many students. The inability to compute a correct solution after many attempts creates frustration and sometimes a fear of solving math word problems. The most effective approach to resolve this issue is accurately pointing out what is the mistake made by the student. Students are better served through learning from their mistakes versus building anxiety from attempting to find the solution without knowing where he or she made an error.

The use of computers in education has had an important gain at different scholastic levels in the teaching-learning process (de, de la Cruz Rivera, & Barcelo, 2009). Displaying and describing mistakes made by a student when solving a word problem gives the student more confidence and direction before making another attempt to reach a correct answer. After many failed attempts there exists a procedure in a student's mind that produces the erroneous answers (Burton & Brown, 1979; Tu, Hsu, & Wu, 2002). Especially when a student is trying to learn a new math subject, because students have difficulties in employing mathematics to solve new problems (Ferreira & Mendes, 2009). Currently, most math educational software does not have the ability to give any effective assistance at the points where student makes an error. For example, if a student enters a wrong answer to a math word problem, the current applications will only tell the student the input to the problem is incorrect without giving any help to explain why the student's answer is incorrect. We call this problem "lacking effective feedback".

Among many ways to solve this problem, the easiest also the most commonly used way is simply to provide a standard answer as the help. However, this approach does not always match with student's natural thinking process in solving a math word problem, and it also will not help students to know where the mistake occurred. In contrast we need to provide an interactive feedback model which can automatically adapt to the student's own thinking habit to provide a user input sensitive help tip to students when a mistake is discovered at any step during the process of solving a math word problem.

However, to create any solution to offer such help function, two issues need to be addressed: (a) the impossibility to predict what a student will input as the answer for each step during a math word problem solving process. The student can input anything including those inputs that may not have any relevance towards a correct solution. For example, if a math word problem has used two variables t and s, then any input does not contains t or s are very possibly not relevant to this problem; (b) the possibility to have multiple equivalent inputs for each step during a math word problem solving process. For example, "40t+12s", "(6s+20t) × 2", and "8s + 10t + 4s + 30t" as well as "(120t + 36s)/3" are all equivalent. We will use the term "input uncertainty problem" to denote the issues (a) and (b) in this paper. It is our opinion that being able to solve the input uncertainty problem is a fundamental requirement for any effective math word problem solving educational software.

In this paper we have proposed an interactive input sensitive feedback model that uses fuzzy logic to not only determine at each step if a student's input is correct or not but also provide a help tip specific to the step where the student made a mistake when computing the answer to a math word problem,. This will improve a student's procedural knowledge for solving math word problems based on their own thinking patterns, which will mitigate future errors and stimulate learning.

This paper is structured as follows. Section II presents a literature review of different methods used provide tutor students when attempting math problems. Section III is the problem statement and hypothesis. Section IV presents our model, which includes working components. Section V presents the application of our model. Lastly, section VI provides a conclusion.

2 RELATED WORK

2.1 *Intelligent tutoring*

There have been many attempts to provide an intelligent tutoring system to enhance learning. (Lu, Wu, Tu, & Hsu, 2004) proposes an intelligent tutoring system that helps teachers construct curriculum and teaching strategies by capturing a students' problem-solving process. The gathering of this data allows a teacher to understand common error types and design appropriate teaching approaches. Their model provides an ontological representation of a student's procedural knowledge that identifies the activity structure from behavioral models. Unlike our model their focus is to aid the teacher in understanding a student's weakness, whereas our model gives direct feedback and aid to the student which eliminates misinterpreting a student's problem area.

Aplusix and Treefrog are the applications that allow students to enter equations and expressions to algebra problems and offers suggestions in any given situation (Nicaud; P. M. Strickland & D. Al-Jumeily, 1999). These applications use rewrite rules instead of evaluating each expression entered by a student, which differs from our model. Also these applications are only limited to basic algebra and do not have the ability to handle math word problems.

2.2 *Fuzzy logic and learning assistants*

A multi-agent based student profiling system, which consists of fuzzy epistemic logic that represents a student's knowledge state and course content modeled by the concept of content, is proposed by (Xu, Wang, & Su, 2002). The profiling system stores a student's interactive history and learning activities into a profile which is used to generate personalized learning materials, quizzes, and advice for that student. The drawbacks of their approach are that their model does not give feedback to students when they are in need of assistances but relies on a teacher's intervention to aid students. The main purpose of this approach

is creating a personalized learning environment for an individual student. However it could cripple a student's ability to comprehend the concepts that are presented in a way different from what a student easily understands, consequently limiting that student's critical thinking.

(Cabada, Estrada, Barrientos, Velasquez, & Garc, 2008) presents an adaptive learning approach for mobile devices that allows facilitators to determine a student's learning style. Their model creates fuzzy membership functions for each type of learning style which corresponds to a set of course material and allowing their fuzzy inference process to map linguistic values to output values of learn styles. Their model only focuses on determining a student's learning style. Versus our model which addresses problem areas through using fuzzy membership function for the students' inputs during a math word problem solving process to determine where a student has had difficulty learning.

The works of (Barrón-Estrada, Zatarain-Cabada, V., R., & Hernández, 2012) takes a different approach which uses empathy via social media to determine if a student is having trouble solving a particular problem. Their affect detection systems observe and study the face, speech, conversation and other human features to detect frustration, interest, and boredom using hardware sensors like web cameras and microphones. They use fuzzy logic to evaluate cognitive and affective aspects to calculate the exercises that are presented to students. Instead of giving feedback to assist the student they provide preliminary results which does not stimulate learning.

A neural-fuzzy synergism for student modeling is proposed by (Stathacopoulou, Magoulas, & Grigoriadou, 1999). The model uses fuzzy logic to provide human-like approximate diagnosis of student's knowledge and cognitive abilities. The premise of this model is to relate student responses with appropriate knowledge and cognitive characteristics, but fails to address the issue of a student's misunderstanding of course work.

3 PROBLEM STATEMENT AND HYPOTHESIS

3.1 *Problem statement*

It was difficult for a computer based math education application to provide context-sensitive feedback to assist students to correct mistakes during solving math word problems because of the input uncertainty problem.

3.2 *Hypothesis*

Treating all the inputs made for each step in a math word problem solving process as the members of a fuzzy set will address the input uncertainty problem and enable the solution of providing students context-sensitive help tip at where the mistakes occur.

4 METHODOLOGY

4.1 A framework for solving math word problems

This input sensitive feedback model uses a math word problem framework defined by (Morton & Qu, 2013), and fuzzy logic to give users feedback when inputting their solution. The framework has the ability to interpret a math word problem, build an equation, define a process to solve the equation and provide a correct solution. The model we are proposing in this paper extends capabilities by allowing students to input their procedural solution and comparing it to the correct solution at each mathematical step using fuzzy logic to determine if the student is correct.

This model is user driven which allows the user to attempt solving the problem before giving any assistance. Our user interface is a free form design which purpose is to mimic a student writing their work on paper. This eliminates the possibilities of forcing a particular strategy upon a student and allowing a student to be creative and work individually (Pollard & Duke, 2002). Traditional mathematical teaching software only allows students to enter their answers but doesn't give students the experience of working out a problem, consequently hindering their learning ability. The first initial step is to process and create a solution to the math word problem using (Morton & Qu, 2013), which is not revealed to the student explicitly. The student's process includes interpreting a math word problem, formulating an equation, and performing each step to reach a solution. Each step is recorded on a separate line within the user's interface. The student also has the option to check their work after each input, giving them immediate assistance and keeping the user on track.

Our methodology consists of three aspects: (i) model the relevance relationship between the input made by the student and the correct solution for a step in a math word problem solving process by a fuzzy set; (ii) classify the indicators for help tips based on the level of confidence which is based on the two contiguous inputs made by the students; and (iii) special cases handling rules to interrupt any "dead loop" situation where a student is in a continued "wrong doing" such as cannot choose a correct step for a predefined "time limit" or "trial number limit".

4.2 Relevance between input and solution as a fuzzy set

Since the nature of solving a math word problem is to transform the comprehension to the problem into a math expression consists of a set of math terms, such as numbers, variables, and math operators. It is very natural for us to consider the input made by a student to a math educational software for a math word problem solving is a set of math terms. For example, a math word problem will be translated into a math equation $15x + 35 - 3x = 20$. We can consider it as a set that consists of five math terms: "$15x$", "$+35$", "$-3x$", "$=$", and "20". As we also know the equation can be further simplified to $12x + 15 = 0$ through merging "similar terms". We will call this result "*the prime expression*" of the original equation in our paper. In this example, "$12x - 15 = 0$" is the prime expression for "$15x + 35 - 3x = 20$".

Now, we can define the input for each step of a math word problem solving process as a finite set of math terms from the prime expression of the input. We denote this as C. Thus we have $C = \{c_i\}$, where c_i is the ith math term in C, $i = 1, 2, \ldots, N$; N is an integer.

Similarly, we can define the correct solution to each step of a math word problem solving process as a finite set of math terms from the prime expression of the correct solution. We denote this as S. Thus we have

$S = \{s_i\}$ where s_i is the jth math term in S, $j = 1, 2, \ldots, M$; M is an integer.

Then we can define a Fuzzy Set with the membership function shown as

$$\mu(C) = \frac{\sum_{i=1}^{N} \text{match\_count}(c_i, M)}{M} \qquad (1)$$

where match_count(c_i, M) refers to the number of math terms matched between set C and the set S; M refers to the number of math terms in S; N refers to the total number of math terms in C. Since we always have $0 \leq \sum \text{match\_count}(c_i, M) \leq M$, therefore $0 \leq \mu(C) \leq 1$.

Example 1: Assume that we have an input for one step in a math word problem solving process as "$12t + 13t + 80t = 25$", and one of the correct "solution" for that step is "$12t + 93t = 25$". The prime expression for both input and solution is "$105t - 25 = 0$". Thus we have $C = S = \{105t, -25, =, 0\}$, and $M = N = 4$, as well as $\sum \text{match\_count}(c_i, M) = 4$. Therefore we have $\mu(C) = 4/4 = 1$. This implies the student has found a right input.

Example 2: All the assumptions are the same as the Example 1 except the input has been changed to "$14t + 90 = 20$". Thus we have $C = \{14t, +70, =, 0\}$ and $S = \{105t, -25, = 0\}$, and $M = 4$, as well as $\sum \text{match\_count}(c_i, M) = 2$. Therefore we have $\mu(C) = 2/4 = 0.5$. This is normally implies the student have some idea on what will be needed but not 100% sure what he or she is doing.

Example 3: All the assumptions are the same as the Example 1 except the input has been changed to "$14t + 5y + 70$". Thus we have $C = \{14t, +5y, +70\}$ and $S = \{105t, +70, =, 0\}$, and $M = 4$, as well as $\sum \text{match\_count}(c_i, M) = 70$. Therefore we have $\mu(C) = 1/4 = 0.25$. Obviously this mostly indicates the student really doesn't know what he or she is doing.

It is worth to point out that for any step in a math word problem solving process the correct solution often can have many equivalent expressions. By introducing a Fuzzy Set based on the math terms from the prime expression of the input and the reference correct solution, we have in fact effectively resolved the input uncertainty problem.

4.3 Help tips based on the confidence level

As we have seen in three examples in previous subsection, the value of $\mu(C)$ is a good indicator to if the student have a basic grasp on what is needed to correctly resolve one step in a math word problem solving process. We can certainly take the advantage of these indicators to provide students more relevant help tips as close to the step where the difficulty occurs. To serve that purpose we will define a "confidence level classification" function as shown below.

$$\delta(\mu(C)) = \begin{cases} \text{low,} & 0 \le \mu(C) < 0.5 \\ \text{medium,} & 0.5 \le \mu(C) < 1 \\ \text{high,} & \mu(C) = 1 \end{cases} \quad (2)$$

That is we can use (2) to classify each input C into one of three confidence levels: low, medium and high based on the value of $\mu(C)$. Obviously, the help tips triggered by a low confidence on an input C will be quite different than the help tips triggered by a medium confidence on an input C because the former may imply that the student may not even understand the math word problem while the latter implies the student may just made some mistakes. We will not provide details on the specifics of the concrete help tips other than only suggesting applying this classification function as a tool to distinct the help tips. That is our model supports input sensitive feedback.

As we have already known that the high confidence level implies that the student has found a correct input to the step so that we should allow the student to move into the next step, or inform the student that he or she has successfully solved the math problem if the current step is the final step.

4.4 Special cases handling

As it is possible to get into a "dead loop" situation if the student does not correct the mistakes properly after receiving the help tips, we need to have a reliable approach to break such "dead loop". There are many ways we can achieve this goal. Below we introduce one approach that is relatively easy to do.

In the situation of a student failing to find the correct solution to a mathematical step after four attempts, a different but similar problem is presented and explained to the student to guide him or her through the math word problem. This prevents the student from entering an infinite loop of incorrect possibilities to a solution which may cause a student to become discouraged. The pseudo code to detail the algorithm of this model can be explained as following:

```
Interpret/compute word problem to build a solution
  Set solution
  SET fail_attempts TO 0
  WHILE (user_input =! solution)
    Get user input
```

Interpret and get membership value of user input
Determine the matching degree of user's current step input and prior step input

```
    If ( current_input == low || current_input ==
    medium )
      ++fail_attempts;
      Display feedback/hint
        IF ( attempts == 4 )
          Display similar example
        END-IF
      Retry/Get user's input
    END-IF
    ELSE
      Get user's next input
    END-ELSE
  END-WHILE
```

5 APPLICATION OF MODEL

To exhibit our model we apply it to a math word problem as shown below:

Two men on motorcycles started at the same time from opposite ends of a road that is 25 miles long. One motorcycle is driving 55 mph and the second motorcycle is riding at 85 mph. How long after they begin will they met?

Using the framework in (Morton & Qu, 2013) we are able to input this example and formulate a solution which is detailed in Table 1.

After the system computes a solution to the word problem the student is presented with a free form interface which allows their step-by-step procedure for solving the math word problem. In this example the feedback will be given once the students have completed entering their solution. Since the student's input is incorrect the hint was provided to help this student making a correction, which is illustrated in Figure 1. Table 2 shows the calculations that had been done by the system to help identify which step the mistake occurred. The student was then prompted to make another attempt towards reaching the correct solution. After the student has made corrections based on the suggestion provided by the system the student's solution is recalculated and more feedback is provided to the student if they still fail to find the correct answer.

Table 1. Problem Solution.

Mathematical Step	Solution of Step
1.	$25 = 55t + 85t$
2.	$25 = 140t$
3.	$140t = 25$
4.	$\dfrac{140t}{140} = \dfrac{25}{140}t$
5.	$t = \dfrac{25}{140}$
6.	$t = \dfrac{5}{28} \approx .178$

First Attempt

$$25 = 85t + 55t$$

$$140t = 25$$

$$\frac{140t}{25} = \frac{25}{25}t \quad \Leftarrow \begin{array}{l}\text{HINT!} \\ \text{Divide both sides of 140t = 25 by 140}\end{array}$$

$$t = \frac{140}{25}$$

$$t = 5.6$$

Second Attempt

$$25 = 85t + 55t$$

$$140t = 25$$

Let's try this step again!
_____
Divide both sides of 140t = 25 by 140

Figure 1. User Feedback.

Table 2. Solution Comparison.

Step	Input to Step	Solution to Step	$\mu(C)$	$\delta(\mu(C))$
1.	$25 = 85t + 55t$	$25 = 55t + 85t$	1	High
2.	$140t = 25$	$140t = 25$	1	High
3.	$\dfrac{140t}{25} = \dfrac{25}{25}t$	$\dfrac{140t}{140} = \dfrac{25}{140}t$	.5	Medium

6 CONCLUSION

Learning to comprehend, create a correct equation, and solve math word problems is a challenge for many students. We have developed a model to provide input sensitive feedback to students at each step after a student completed the data input for that step. Our method gives students the ability to make attempts solving word problems purely based on their own ideas and receive feedback when mistakes are made. Mathematics can only be learned by doing it (Al-Jumeily & Strickland, 1997). This interactive approach is more practical than the traditional write down a solution and checks your work from another source technique, which sometimes is not effective because students have a difficult time pinpointing where they made a mistake and understanding what to do to fix their mistake. Giving feedback using fuzzy logic allows the system to know exactly where an error was made and gives prompt tips to support a student's learning. This model contributes by giving students confidence when

attempting math word problems while also providing an adaptive learning environment.

REFERENCES

Barrón-Estrada, M. L., Zatarain-Cabada, R., V., J. A. B., R., F. L. C., & Hernández, Y. (2012). An Intelligent and Affective Tutoring System within a Social Network for Learning Mathematics. In J. Pavon, N. s. Duque-Mcndez & R. n. Fuentes-Fern índez (Eds.), *Advances in Artificial Intelligence – IBERAMIA 2012* (Vol. 7637, pp. 651–661): Springer Berlin Heidelberg.

Burton, R., & Brown, J. (1979). An investigation of computer coaching for informal learning activities. *International Journal of Man-Machine Studies, 11*(1), 5–24.

Cabada, R. n., Estrada, M. a., Barrientos, E., Vel?squez, M. s., & Garc, C. (2008). Multiple Intelligence Tutoring Systems for Mobile Learners. *Advanced Learning Technologies, IEEE International Conference on*, pp. 652–653.

de, L., de la Cruz Rivera, A., & Barcelo, G. (2009). An Artificial Intelligence Based Model for Algebra Education. *Networked Computing and Advanced Information Management, International Conference on*, pp. 1492–1498.

Ferreira, J. F., & Mendes, A. (2009). Students' feedback on teaching mathematics through the calculational method. *2012 Frontiers in Education Conference Proceedings*, pp. 1–6.

Lu, C.-H., Wu, S.-H., Tu, L., & Hsu, W.-L. (2004). The Design of An Intelligent Tutoring System Based on the Ontology of Procedural Knowledge. *Advanced Learning Technologies, IEEE International Conference on*, pp. 525–529.

Morton, K., & Qu, Y. (2013). A Novel Framework for Math Word Problem Solving. *International Journal of Information and Education Technology, 3*(1), pp. 88–93.

Nicaud, J.-F. Aplusix, from http://www.aplusix.com/en/

P. M. Strickland and D. Al-Jumeily, A. (1999). A Computer Algebra System for Improving Students' Manipulation Skills in Algebra. *International Journal of Computer Algebra in Mathematics Education, 6*(1), pp. 17–24.

Pollard, J., & Duke, R. (2002). A Software Design Process to Facilitate the Teaching of Mathematics. *Computers in Education, International Conference on*, p. 906.

Reyes, S. S., Pedro, S., Agapito, J., Nabos, J., Repalam, M. C., Baker, R., & Rodrigo, M. M. T. (2012). The Effects of an Interactive Software Agent on Student Affective Dynamics while Using ;an Intelligent Tutoring System. *IEEE Transactions on Affective Computing, 3*(2), pp. 224–236.

Stathacopoulou, R., Magoulas, G. D., & Grigoriadou, M. (1999). *Neural network-based fuzzy modeling of the student in intelligent tutoring systems*. Paper presented at the Neural Networks, 1999. IJCNN '99. International Joint Conference on.

Tu, L., Hsu, W.-L., & Wu, S.-H. (2002). A Cognitive Student Model – An Ontological Approach. *Computers in Education, International Conference on*, p. 111.

Xu, D., Wang, H., & Su, K. (2002). Intelligent Student Profiling with Fuzzy Models. *2013 46th Hawaii International Conference on System Sciences, 3*.

Information Technology and Computer Application Engineering – Liu, Sung & Yao (Eds)
© *2014 Taylor & Francis Group, London, ISBN 978-1-138-00079-7*

Development of a fast vibratory filtering algorithm via neural synaptic properties of facilitation and depression and its application on electrocardiogram detecting

W. Gao & F.S. Zha
State Key Laboratory of Robotics and System, Harbin Institute of Technology, China

B.Y. Song
Department of Mechanical Design, Harbin Institute of Technology, China

M.T. Li
State Key Laboratory of Robotics and System, Harbin Institute of Technology, China

ABSTRACT: In order to implement the electrocardiogram (ECG) signals detecting with high efficiency, a fast vibratory filtering algorithm via neural synapse properties of facilitation and depression is developed in this paper. The framework of filtering algorithm is based on the vibration system, and the filtering transition matrix is given depending on the neuron synapse properties. For illustration, the application on ECG signals detecting is discussed. The ECG signals are processed by developed filtering algorithm and fast wavelet transform algorithm respectively, and computation time of both methods are tested. Experiment results show that the performance of developed filtering method is better than that of fast wavelet transform algorithm. Not only in numerical performance of outputs, but also in the filtering speeds. The fast vibratory filtering algorithm via neural synapse properties of facilitation and depression can effectively detect ECG signals.

Keywords: Neural synapse; Vibration system; Fast filtering; Electrocardiogram signals; Detecting

1 INTRODUCTIONS

Theoretically, resonance, which occurs with all types of vibrations, can be used to pick out specific frequencies from a complex vibration containing many frequencies (Perlman & McCusker 1970). For vibration systems, while the frequency of external force is approximately the same as the system natural frequency, the response can produce resonance increases (Inman 2010). And towards the fixed vibration system, its natural frequency depends on the frequency of varying input signals. In other words, it means resonance could do a job of filtering out signals. However, due to the randomness of inputs, there exist many challenges on how to acquire good filtering performance. Solutions to these challenges can be traced to the research on biological neuron filtering (Forture & Rose 1997, Hutcheon & Yarom 2000, Schnupp 2006, Tucci & Raugi 2011).

Researchers pointed out that resonance, which has been found in a number of excitable cell types, is a measurable property that describes the ability of neurons to respond selectively to biological signal inputs at preferred frequencies, and the response process is achieved by a variety of synapses instantaneously (Masuda & Aihara 2003, Perrinet 2004). Synapses can increase (synaptic facilitation) or decrease (synaptic depression) markedly within milliseconds after the onset of specific temporal patterns of activity in neurons and the synaptic properties are ubiquitous among living creatures (Forture & Rose 2000, Forture & Rose 2001, Horcholle-Bossavit & Quenet 2009). The synaptic depression increases the threshold of a neuron, and the facilitation decreases it. From these research findings, we can conclude that neuron resolves the filtering problem evoked by resonance through synaptic facilitation and depression, and it can make a map of transition matrices in filtering models to improve filtering performance towards stochastic input signals. Therefore, it inspires us to probe the filtering problem of stochastic signals from the view of neural synaptic properties.

Towards one-dimensional stochastic signals, we shall consider the single degree of freedom vibration system as the basic research framework for it does not have multiple resonant frequencies. This paper is organized as follows. Section 2 presents the modeling process of developed filtering algorithm. Then Section 3 discusses the comparison experiments and the application on electrocardiogram (ECG) detecting. Finally, Section 4 gives the conclusion.

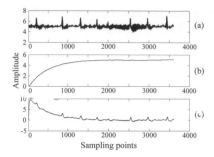

Figure 1. The response of single degree of freedom vibration system towards stochastic viable frequency input. (a) shows the stochastic viable frequency input. (b) shows the displacement response of single degree of freedom vibration system towards (a). (c) shows the velocity response of single degree of freedom vibration system towards (a).

2 METHODS

Generally, the single degree of freedom vibration system can be described as:

$$mx''(t) + cx'(t) + kx(t) = s(t) \tag{1}$$

where m, c and k are the mass, damping and stiffness coefficients of the vibration system, respectively. $x(t)$ represents the displacement at time t and $s(t)$ represents the excitation force which can be seen as the stochastic sampling inputs. For conciseness, let $x(t) = x$ and $s(t) = s$.

Firstly, we consider the response of a common vibration system with regard to stochastic inputs. Suppose m = 0.005, c = 0.5 and k = 1.0. The corresponding displacement and velocity responses towards stochastic variable frequency input are shown in Figure 1. Obviously, the displacement response in Figure 1b does not reflect the frequency property of inputs while the velocity response shown in Figure 1c indicates quite few filtering effect. Yet, we could not obtain the inputs properties. This problem is due to the unacceptable filtering transition matrices provided by the vibration system. Hence, how to improve the filtering transition matrices based on vibration system framework is the question that we shall deal with.

In order to integrate the properties of neural synapses into the single degree of freedom vibration system, we shall consider the internal state of the vibration system. The state equation can be written as:

$$\begin{cases} x' = y \\ y' = -\dfrac{kx + cy - s}{m} \end{cases} \tag{2}$$

where x is the displacement output and y is the velocity output. Solving the Equation (2) by Heun method, we have

$$\begin{cases} \Delta x = h(-\dfrac{kxh}{2m} + y - \dfrac{cyh}{2m} + \dfrac{sh}{2m}) \\ \Delta y = \dfrac{h}{2m} \left\{ -k(2 - \dfrac{hc}{m})x - [c(2 - \dfrac{hc}{m}) + kh]y + (2 - \dfrac{hc}{m})s \right\} \end{cases} \tag{3}$$

where h is the step size of Heun method. For common conditions, denote $n \in N$, and N is the set of integers. The numerical solutions would be:

$$\begin{cases} x_{n+1} = x_n + \Delta x_n \\ y_{n+1} = y_n + \Delta y_n \end{cases} \tag{4}$$

where Δx_n and Δy_n are the increments in x and y during each iteration, respectively. $0 = t_0 < t_1 < \cdots < t_n$, and $n := t_n$. Substitute Equation (3) into Equation (4). Denote

$$D = \begin{bmatrix} d_{11} & d_{12} \\ d_{21} & d_{22} \end{bmatrix} \quad R = \begin{bmatrix} r_{11} & 0 \\ r_{21} & 0 \end{bmatrix} \tag{5}$$

It yields:

$$\begin{bmatrix} x_{n+1} \\ y_{n+1} \end{bmatrix} = D \begin{bmatrix} x_n \\ y_n \end{bmatrix} + R \begin{bmatrix} s_{n+1} \\ s_{n+1} \end{bmatrix} \tag{6}$$

where D is referred as the transition matrix of the vibration system and

$$\begin{cases} d_{11} = 1 - \dfrac{kh^2}{2m} \\ d_{12} = h(\dfrac{2m - hc}{2m}) \\ d_{21} = \dfrac{hk(hc - 2m)}{2m^2} \\ d_{22} = 1 - \dfrac{h[c(2m - hc) + mhk]}{2m^2} \end{cases} \tag{7}$$

$$\begin{cases} r_{11} = \dfrac{h^2}{2m} \\ r_{21} = \dfrac{h(2m - hc)}{2m^2} \end{cases} \tag{8}$$

As shown in Figure 1, the transition matrix D does not perform well on filtering standpoint.

Considering the details mentioned above from the view of neural filtering, the stochastic inputs are depressed totally as shown in Figure 1b and Figure 1c. It just indicates one part of the neural filtering contributed by the synaptic depression, and D can be seen as the depression matrix. The instantaneous processing of neural information can be attributed to the neural synaptic properties of facilitation and depression. As referred in Forture & Rose (2001), the facilitation process mainly manifest as an information exchange process. Denote a facilitation matrix as F. If we do not consider the information loss, F can be defined as:

$$F = \begin{bmatrix} 0 & 1 \\ 1 & 0 \end{bmatrix} \tag{9}$$

The facilitation process can be modeled as:

$$\begin{bmatrix} x_{n+1} \\ y_{n+1} \end{bmatrix} = F \begin{bmatrix} x_n \\ y_n \end{bmatrix} \tag{10}$$

Thus, the transition matrix of neural filtering can be defined as:

$$H = DF = \begin{bmatrix} d_{12} & d_{11} \\ d_{22} & d_{21} \end{bmatrix} \tag{11}$$

212

Combining with the depression and the facilitation processes, we can obtain the neural filtering model based on the vibration system framework.

$$X_{n+1} = HX_n + RS_{n+1} \qquad (12)$$

where $X_{n+1} = [x_{n+1} \; y_{n+1}]^T$, and $S_{n+1} = [s_{n+1} \; s_{n+1}]^T$. Denote $k \in N$. We have:

$$\begin{cases} X_k = HX_{k-1} + RS_k \\ X_{k+1} = H^2 X_{k-1} + HRS_k + RS_{k+1} \\ \cdots \qquad \cdots \\ X_{n+k-1} = H^n X_{k-1} + H^{n-1} RS_k + H^{n-2} RS_{k+1} + \cdots + RS_{n+k-1} \end{cases} \qquad (13)$$

It yields:

$$X_{n+k-1} = H^n X_{k-1} + \sum_{i=0}^{n-1} H^{n-1-i} RS_{k+i} \qquad (14)$$

From the iteration in Equation (13), we also obtain the properties of coefficient by substituting Equation (7), (8) and (10) into Equation (13). The coefficient conditions are given as follows:

$$\begin{cases} d_{12} = 1 - r_{11} \\ d_{22} = -r_{21} \end{cases} \qquad (15)$$

Then, denote the transfer matrices as G and K. We can rewrite Equation (14) as follows:

$$\begin{bmatrix} x_n \\ y_n \end{bmatrix} = H \begin{bmatrix} x_{n-1} \\ y_{n-1} \end{bmatrix} + GK \begin{bmatrix} s_{n-k-1} \\ s_{n-k} - s_{n-k-1} \\ \vdots \\ s_n - s_{n-1} \end{bmatrix} \qquad (16)$$

where

$$G = \begin{bmatrix} r_{11} & r_{11} \\ r_{21} & r_{21} \end{bmatrix} \quad K = \begin{bmatrix} 1 & \alpha_1 & \cdots & \alpha_k \\ 0 & \beta_1 & \cdots & \beta_k \end{bmatrix} \quad H = \begin{bmatrix} 1 - r_{11} & d_{11} \\ -r_{21} & d_{21} \end{bmatrix}$$

and k is referred as the input delayed number, x_n is the filtering output and y_n is the filtering gain.

Considering the ideal condition while $k = 0$, we obtain:

$$\begin{bmatrix} x_n \\ y_n \end{bmatrix} = \begin{bmatrix} 1 - r_{11} & d_{11} \\ -r_{21} & d_{21} \end{bmatrix} \begin{bmatrix} x_{n-1} \\ y_{n-1} \end{bmatrix} + \begin{bmatrix} r_{11} & r_{11} \\ r_{21} & r_{21} \end{bmatrix} \begin{bmatrix} 1 & 0 \\ 0 & 1 \end{bmatrix} \begin{bmatrix} s_{n-1} \\ s_n - s_{n-1} \end{bmatrix} \qquad (17)$$

The values of α and β can be determined by associating Equation (12) and Equation (15). G and H are determined by the minimum error variance condition of Equation (15).

The above filtering algorithm (consists of Equation (15)) will be substantially testified in later section.

3 EXPERIMENTS AND DISCUSSIONS

To validate the proposed method and obtain applicable performance towards ECG signals, we shall investigate the properties of ECG signals first. The ECG signals are usually weak, stochastic, non-stationary and non-linear. Its frequency is quite low, about from 0.05 Hz to 100 Hz, with strong noise. In order to detect the ECG signals by the proposed method, we use the suitable chirp signals to test the feasible and the real-time properties through comparing with the fast wavelet transform (FWT) algorithm which is chosen for its high filtering speed and nonlinear processing ability. And then, we apply the developed filtering method to real ECG signals for detecting.

3.1 Comparisons with the fast wavelet transform algorithm

The comparison results are shown in Figure 2, where the adopted frequency is 1000 Hz, the sampling points are 1000, and the noise is 10 dB white Gaussian noise. Figure 2a, Figure 2b, Figure 2c and Figure 2d present the chirp signal whose frequency is from 0.05 Hz to 100 Hz, the chirp signal with 10 dB white Gaussian noise, the filtering result by the developed filtering algorithm and the filtering result by the FWT algorithm, respectively. As shown in Figure 2, both the two filtering methods can obtain acceptable outputs. In this case, we evaluate the filtering performance of the two methods by numerical testification. By concerning of Signal Noise Ratio (SNR), Root Mean Square Error (RMSE) and Signal Noise Ratio Gain (SNRG), the performance comparisons are given in Table 1. Meanwhile, the SNR of chirp signal with 10 dB white Gaussian noise is also calculated, and the value is 0.4243 dB.

The data in Table 1 indicate that the SNR and the SNRG of developed filtering algorithm are a little higher than that of FWT algorithm, while the RMSE is about 1.25 times lower than that of FWT algorithm.

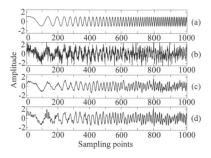

Figure 2. The comparisons between the developed filtering algorithm and the FWT algorithm. (a) shows the chirp signal whose frequency is from 0.05 Hz to 100 Hz. (b) shows chirp signal with 10 dB white Gaussian noise. (c) shows filtering result by the developed filtering algorithm. (d) shows the filtering result by the FWT algorithm.

Table 1. Performance comparison between the developed filtering algorithm and the FWT algorithm.

Filtering methods	SNR	RMSE	SNRG
The developed algorithm	14.557	0.29724	34.309
The FWT algorithm	13.531	0.37189	31.889

Table 2. The computation time (millisecond).

Filtering methods	Computation time
The developed method	0.9
The FWT algorithm	58.3

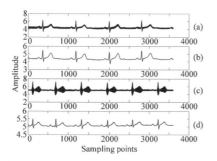

Figure 3. The testifications of actual sampled ECG signals. (a) and (c) show the actual sampled ECG signals, respectively. (b) and (d) show the detecting results by developed filtering algorithm of (a) and (c), respectively.

It means that, towards signals with strong noise whose frequency is from 0.05 Hz to 100 Hz, the developed filtering algorithm can obtain better filtering performance. Besides, we also test the computation time of the two methods in LPC2148 whose main frequency is 60 MHz. The computation time is shown in Table 2. Apparently, the filtering speed of developed filtering algorithm is about 64.7 times than that of FWT algorithm.

It can be found from the results of Figure 2, Table 1 and Table 2 that the developed filtering algorithm is better than the FWT towards the signals with strong noise whose frequency is from 0.05 Hz to 100Hz, and it is more suitable for the signals online detecting for its fast filtering speed. Therefore, we apply the developed filtering algorithm on real ECG signals detecting.

3.2 Testification on ECG signals

The ECG signals measured by a portable vital signs monitoring system is shown in Figure 3. The sampling frequency is 1000 Hz. Figure 3a and Figure 3c represent the actual sampled ECG signals. It is hard to identify and analyze the indexes of clinical diagnosis. By the developed filtering algorithm, we achieve the ECG wave detecting process successfully. Figure 3b and Figure 3d represent the filtering results of the ECG signals in Figure 3a and Figure 3c by the developed filtering algorithm.

As shown in Figure 3b and Figure 3d, the characters of ECG signals can be seen obviously, and the random noises are filtered well. Not only that but P wave, QRS complex, T wave and U wave of ECG signals are also integrity and no loss. In this case, the various ECG indexes can be obtained. Besides, for the quite high filtering speed, it can be used efficient online and obtain real-time detecting results.

4 CONCLUSIONS

In this paper, a fast filtering algorithm are proposed based on the vibration system and neural synapse properties of facilitation and depression, and apply to the real ECG signals detecting. The performance including the Signal Noise Ratio, Root Mean Square Error and the Signal Noise Ratio Gain are carefully compared between the fast wavelet transform algorithm and the developed filtering algorithm. The computation times of the two algorithms are measured in LPC2148, and the results show that the speed of developed filtering algorithm is about 64.7 times faster than that of fast wavelet transform algorithm. The effectiveness of developed filtering algorithm is further testified on the detecting of ECG signals.

ACKNOWLEDGEMENTS

This work is partially supported by National Natural Science Foundation of China (Nos. 60901074, 51075092, 61175107, 61005076), the National High Technology Research and Development Program ("863" Program) (No. 2007AA042105) of China and the Natural Science Foundation of Heilongjiang Province in China (No. E200903).

The corresponding author is M.T. Li, and the email address is skymoon.hit@gmail.com.

REFERENCES

Forture, E.J & Rose, G.J. 1997. Passive and active membrane properties contribute to the temporal filtering properties of midbrain neurons in vivo. *J. Neurosci.* 17(10): 3815–3825.

Forture, E.J & Rose, G.J. 2000. Short-time synaptic plasticity contributes to the temporal filtering of electrosensory information. *J. Neurosci.* 20(18):7122–7130.

Forture, E.J & Rose, G.J. 2001. Short-time synaptic plasticity as a temporal filter. *Trends in neurosci.* 24(7): 381–385.

Hutcheon, B. & Yarom, Y. 2000. Resonance, oscillation and the intrinsic frequency preferences of neurons. *Trends in neurosci.* 23(5):216–222.

Horcholle-Bossavit, G. & Quenet, B. 2009. Neural model of frog ventilatory rhythmogenesis. *BioSystems.* 97 (1): 35–43.

Inman, D.J. 2010. *Engineering vibrations.* New Jersey: Prentice Hall.

Masuda, N. & Aihara, K. 2003. Filtered interspike interval encoding by class II neurons. *Physics Letters A.* 311: 485–490.

Perlman, S.S. & McCusker, J.H. 1970. An adaptive resonant filter. *Proceedings of the IEEE.* 58:190–197.

Perrient, L. 2004. Emergence of filters from natural scenes in a sparse spike coding scheme. *Neurocomputing.* 58:821–826.

Schnupp, J. 2006. Auditory Filters, features, and redundant representations. *J. Neuron.* 51(3):278–280.

Tucci, M. & Raugi, M. 2011. A filter based neuron model for adaptive incremental learning of self-organizing maps. *Neurocomputing.* 74(11):1815–1822.

Information Technology and Computer Application Engineering – Liu, Sung & Yao (Eds)
© 2014 Taylor & Francis Group, London, ISBN 978-1-138-00079-7

Research and optimization on methods for reciprocal approximation

Hong Xia, Guang-Bin Wang & Bi-Cheng Xiao
School of Control and Computer Engineering, North China Electric Power University, Beijing, China

ABSTRACT: In floating-point division design, functional iteration is widely used. One of the difficulties of division with functional iteration is to get precise initial reciprocal approximation. High-accuracy initial approximation can reduce the required number of iteration and cycles of the operation. In order to get accurate reciprocal approximation with little table size, this paper presents different methods and their error analysis in detail. And an optimized symmetric table with redundant Booth encoding is discussed, which also can reduce the logic circuit area of the division operation.

Keywords: floating-point division; reciprocal approximation; functional iteration; Booth encoding

1 INSTRUCTION

The floating-point processor has an important role in high-precision calculation, image acceleration and digital signal processing. The floating-point division is the most time-consuming among four floating-point arithmetic operations, and it is an important part of the floating-point processor. In the floating-point division design, functional iteration is widely used (Flynn 1970). An important factor influencing functional iteration algorithm is the precision of initial reciprocal approximation (Fowler & Smith 1989). The precision of the initial reciprocal approximation decides the number of iteration for the final result, which influence the performance of division operation (Tang 1991, Oberman & Flynn1997). In order to generate more accurate reciprocal approximation with little table size, this paper presents different methods and their error analysis in detail. And an optimized symmetric table with redundant Booth encoding is discussed particularly, combining getting initial approximation and Booth encoding together, which can reduce the logic circuit area of the division based on functional iteration.

2 METHODS FOR FORMING RECIPROCAL APPROXIMATION

There exist many methods for forming the reciprocal approximations; the simplest way is direct look-up table that directly accesses a look-up table for the reciprocal approximation. In general, the look-up table is implemented in the form of a ROM. But obtaining a high-precision initial reciprocal approximation by a simple direct look-up table is very difficult, because the size of required table will become quite large. In order to form a more accurate initial reciprocal approximation with little table size and total area of the logic circuits, plenty of methods have been proposed.

2.1 Direct look-up table

The direct look-up table is the simplest method during the methods for reciprocal approximation. In division operation, the value can be directly used in iteration to generate infinite approximation of reciprocal, which is multiplied by the multiplicand for final quotient. The accuracy of approximation depends on the length of entry and the length of item stored in the look-up tables.

It is assumed that input operand is IEEE754 normalized $1.0 \leq b < 2.0$. A normalized operand is truncated to k bits to the right of the radix point $trunc(x) = 1.b_1b_2b_3 \ldots b_k$. These k bits are used to index a look-up table providing m output bits which are taken as the m bits after the leading bit in the $m + 1$ bits fraction reciprocal approximation $recip(x) = 0.1b_1'b_2'b_3' \ldots b_m'$ which has the range $0.5 < recip(x) \leq 1.0$ (take the case that divisor is equal to 1 as special situation). The total size of direct look-up table is:

$$Table\ Size = 2^k \times m \qquad (1)$$

In the algorithms of direct look-up table, a common algorithm is to implement a piece-constant approximation of the reciprocal function. In the algorithm, the approximation of each entry is obtained by taking the reciprocal of the mid-point between the truncated input operand and its successor, where the mid-point is $1.b_1b_2b_3 \ldots b_k1$. Then the reciprocal of the mid-point is added $2^{-(m+2)}$, finally truncated to $m + 1$ bits, producing the result in the form of $0.1b'_1b'_2b'_3 \ldots b'_m$. The item stored in table for each entry i is get from:

$$Recip_i = \lfloor 2^{m+1} \times (\frac{1}{\flat + 2^{-(k+1)}} + 2^{-(m+2)}) \rfloor / 2^{m+1} \qquad (2)$$

where $\flat = 1.b_1b_2b_3 \ldots b_k$.

DasSarma & Matula have proved that the approximation of piece-constant algorithm for forming reciprocal look-up table can minimize the maximum relative error (DasSarma & Matula 1994). The maximum relative error for reciprocal R_0 from k-bits-in, k-bits-out look-up table is bounded by:

$$|\varepsilon_r| = |R_0 - \frac{1}{\flat}| \le 1.5 \times 2^{-(k+1)} \qquad (3)$$

In addition, the reciprocal table can improve the accuracy of table output by increasing output guard bits. When $m = k + g$, where g is the output guard bits, the maximum relative error is bounded by:

$$|\varepsilon_r| \le 2^{-(k+1)}(1 + \frac{1}{2^{g+1}}) \qquad (4)$$

So, the least precision is showed for k-bits-in, $k + g$-bits-out reciprocal table:

$$k + 1 - \log_2(1 + \frac{1}{2^{g+1}}) \qquad (5)$$

The advantage of direct look-up table is that don't need complex logic circuits, the reciprocal approximation is directly stored in a look-up table. But the total table size and time delay of look-up tables grow exponentially with the accuracy of each item in tables. The method of direct look-up tables is not applicable to form high precise reciprocal, because the table size will become too large.

2.2 Linear approximation

Besides direct look-up table, linear approximation is another method for generating reciprocal approximation. Linear approximation is based on expanding a Taylor series. The value of $1/x$ is approximated by the linear function:

$$1/x \approx a_1(x_k) \times x + a_0(x_k) \qquad (6)$$

where the value of a_0 and a_1 is get from two look-up tables, then they are fed into an Multiply/Add unit, the output is the reciprocal approximation (illustrated by Figure 1).

The reason that this method exists error is that only k bits is used to index look-up tables and each item stored in tables only consist of m bits. The maximum relative error of the method is bounded by:

$$|\varepsilon_r| \le 2^{-(2k+3)} + 2^{-m} \qquad (7)$$

When m is equal to $2k + 3$, the maximum relative error of linear approximation is bounded by $2^{-(2k+2)}$,

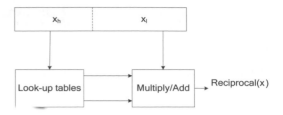

Figure 1. Linear Approximation Diagram.

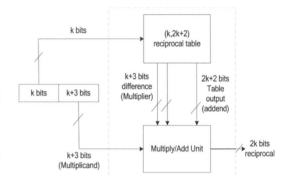

Figure 2. $2k + 3$ bits-in, $2k$ bits-out interpolation reciprocal.

which can guarantee the accuracy of $2k + 2$ bits. $m \times m$ multiplication circuit is needed to implement the method, and the total table size is $2k \times m \times 2$.

In 1997, DasSarma & Matula proposed another method of linear approximation, the method employing linear interpolation in reciprocal table (DasSarma & Matula 1997). This method just needs one table, which is implemented by employing a Multiply/Add operation. The interpolation method generates $2k$ bits precise reciprocal employing k-bit-in $2k$-bits-out look-up table and $(k + 3) \times (k + 3)$ bits multiplier.

The method is implemented by assuming that input argument is normalized $1.0 \le x < 2$. Figure 2 illustrates the method of $2k + 3$-bits-in $2k$-out reciprocal. These steps are:

1. $2k + 3$ bits of truncated input argument are partitioned into high and low part, x_h and f, whose length is k and $k + 3$ respectively.
2. high k input bits x_h index a reciprocal table which yields a $2k + 2$ bits table output $c_1(x_h)$ and implicitly the difference of successive table output $c_2(x_h) = c_1(x_h) - c_1(x_h + 1/2^k)$, which are fed into the Multiply/Add unit as addend and multiplier inputs respectively.
3. The low $k + 3$ bits part f is fed into the Multiply/Add unit as multiplicand.
4. The reciprocal is computed by operation $recip(x) = c_1(x_h) - c_2(x_h) \times f$, and the product sum is truncated to $2k$ bits.

The relative error of interpolation reciprocal results from: the rounding of the table values, the rounding of the interpolated value, and the fact that each interpolated value must represent the reciprocal for an input interval rather than an input point. We know that a table of k-bits-in $(2k+g_t)$-bits-out is needed for reciprocal of $(2k+g_i)$-bits-in $2k$-bits-out, where g_i and g_t is input guard bits and table output guard bits respectively. The interpolation error, the table discretization error, and input discretization error is defined E_∞, E_t, E_i respectively. E_∞, E_t and E_i is bounded by:

$$E_\infty \leq (\frac{1}{2} - \frac{3}{2^{k+2}} + \frac{1}{2^{2k}}) \times 2^{-2k} \qquad (8)$$

$$E_t \leq \frac{1}{2^{g_t}} \times 2^{-2k} \qquad (9)$$

$$E_i \leq \frac{1}{2^{g_i-1}} \times 2^{-2k} \qquad (10)$$

2.3 Multi-tables addition

The methods of approximate based on multi-tables addition only need look-up tables and addition operation for the initial reciprocal approximation. The method is implemented by function:

$$f(x) = a_m(x_m) + a_{m-1}(x_{m-1}) + \ldots + a_0(x_0) \qquad (11)$$

where the value of $a_i(x_i)$ $(0 \leq i \leq m)$ is stored in look-up tables $a_i (0 \leq i \leq m)$.

The methods based on multi-tables addition compared with linear approximation don't need multiplication operation, which can reduce total area of the logic circuits. DasSarma & Matula have proposed a method of bipartite reciprocal tables (DasSarma & Matula 1995). The method employs a positive table and a negative table, the output of separate tables is fed into an addition unit to generate reciprocal approximation. The method requires input argument divided into three sections, the first and second section used to index the positive table, the first and third section used to index the negative table. Figure 3 illustrates the structure of bipartite reciprocal tables.

In order to further reduce the size of look-up table, Schulte proposed Symmetric Bipartite Tables Method

(SBTM). It takes advantage of symmetry in the coefficients to reduce the size of one of the tables by a factor of two compared with the method of bipartite reciprocal tables. The method has also tight bounds on the maximum absolute error, and it also can be applicable to a wide range of function expression.

The SBTM (Schulte & Stine 1997) need to divide the input argument into three sections x_0, x_1 and x_2, whose length is n_0, n_1 and n_2 respectively. Input argument x is $x_0 + x_1 + x_2$, the length of input n is $n_0 + n_1 + n_2$. It is assumed that $0 \leq x \leq 1 - 2^{-(n0+n1+n2)}$, which results in the following condition:

$$\begin{aligned} &0 \leq x_0 \leq 1 - 2^{-n0} \\ &0 \leq x_1 \leq 2^{-n0} - 2^{-n0-n1} \\ &0 \leq x_2 \leq 2^{-n0-n1} - 2^{-n0-n1-n2} \end{aligned} \qquad (12)$$

The SBTM approximations are based on a two term Taylor series expansion of $f(x)$:

$$\begin{aligned} f(x) &= f(x_0 + x_1 + x_2) \\ &\approx f(x_0 + x_1 + \delta_2) + f'(x_0 + x_1 + \delta_2)(x_2 - \delta_2) \\ &\approx f(x_0 + x_1 + \delta_2) + f'(x_0 + \delta_1 + \delta_2)(x_2 - \delta_2) \end{aligned} \qquad (13)$$

where $\delta_1 = 2^{-n0-1} - 2^{-n0-n1-1}$, $\delta_2 = 2^{-n0-n1-1} - 2^{-n0-n1-n2-1}$. The value of $f(x_0 + x_1 + \delta_2)$ and $f'(x_0 + \delta_1 + \delta_2)(x_2 - \delta_2)$ is stored in table $a_0(x_0, x_1)$ and table $a_1(x_0, x_2)$.

By the formula (13) shows, $f'(x_0 + \delta_1 + \delta_2)(x_2 - \delta_2)$ is 1's complement of $f'(x_0 + \delta_1 + \delta_2)(\delta_2 - x_2)$. The characteristics of this symmetry can reduce 1 bit index address for the look-up table, which can reduce the table size half. In addition, because of $|x_2 - \delta_2| < 2^{-n0-n1-1}$ and $|f'(x_0 + \delta_1 + \delta_2)| < 1$, the front of value of $f'(x_0 + \delta_1 + \delta_2)(x_2 - \delta_2)$ exist at least $n_0 + n_1 + 1$ zero, which is not necessary to be stored. Figure 4 illustrates the method of symmetric bipartite tables.

The methods based on multi-tables addition can reduce the size of table space by half compared with

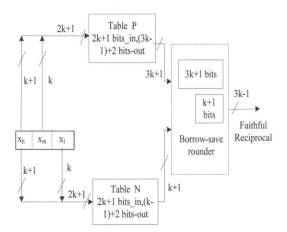

Figure 3. Bipartite Reciprocal Tables.

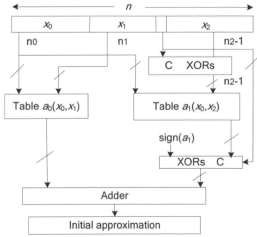

Figure 4. Symmetric Bipartite Table Method.

Table 1. Table Size for different Algorithms.

Direct look-up table	Linear approximation	Multi-table addition	
		Bipartite table	SBTM
$2^{12} \times 12$	$2^6 \times 14$	$2^9 \times 33$	$2^9 \times 25$

*SBTM: Symmetric Bipartite Tables Method.

Table 2. Comparison with SBTM for Radix 4 Booth output.

	With Redundant Booth	Symmetric tables
Gates delay	7 gates	7 gates
Gates count	17 gates	25 gates

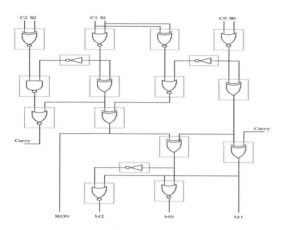

Figure 5. Redundant carry-save Booth encoder.

the method of direct look-up tables. But the size of table also increases exponentially with the required accuracy, just slower than the growth of direct look-up table method.

Table 1 shows the required table size for 13 bits reciprocal approximation (12 bits table output) for different algorithms. We know that the method of smallest table size is linear approximation, but it needs multiplication which increases circuit delay. However, the methods of bipartite tables and symmetric tables don't require of multiplication or large table size.

3 SYMMETRIC TABLES WITH REDUNDANT BOOTH ENCODING

The section 2 illustrates that SBTM doesn't need multiplication or large table size; so it is very suitable to generate initial approximation. But it also requires an addition operation for the final initial approximation, which will also increase the logic circuit area. In order to further reduce the area required for forming reciprocal approximation, this paper discusses the method of symmetric tables with redundant Booth encoding, which will encode the output of two lookup tables without addition operation.

It is assumed that the length of the output a_0 of lookup table $a_0(x_0, x_1)$ and output a_1 of $a_1(x_0, x_2)$ is p_0 and p_1 respectively. As $p_0 > p_1$, a_1 is required to be sign-extended to p_0 bits. Then a_0 and sign-extended a_1 is fed into the redundant Booth encoder circuits, producing encoded number sent to multiplier.

The Figure 5 shows the logic circuit of redundant carry-save Booth encoder (Lyu & Matula 1995). The method of symmetric table with redundant Booth encoding combines initial approximation and Booth encoding together, Table 2 illustrates the comparison of delay with SBTM, including generation of reciprocal approximation. It is concluded that the method can reduce the logic circuit area.

4 CONCLUSION

It plays an indispensible role for functional iteration algorithm to generate reciprocal approximate. The accuracy of the initial approximation is an important factor affecting the performance of division. This paper expounds different methods for forming the initial reciprocal approximation, including direct look-up table, linear approximation and multi-table addition, and their error analysis. And an optimized symmetric table with redundant Booth encoding is researched in detail, using the redundant Booth coding to encode the output of look-up tables, proved to effectively reduce the logic circuit area of the division.

REFERENCES

DasSarma, D & Matula, D.W. 1994. Measuring the Accuracy of ROM Reciprocal Tables. *IEEE Trans. Comput*: Vol. 43, No. 8, 932–940.

DasSarma, D & Matula, D.W.1995. Faithful Bipartite Rom Reciprocal Tables. *Proceedings of the 12th Symposium on Computer Arithmetic*: 17–29.

DasSarma, D & Matula, D.W. 1997. Faithful Interpolation in Reciprocal Tables. *Proceedings of 13th IEEE Symposium on Computer Arithmetic*: 82–91.

Fowler, D.L. & Smith, J.E.1989. An accurate, high speed implementation of division by reciprocal approximation. *Proceedings of the 9th IEEE Symposium on Computer Arithmetic, September*: 60–67.

Flynn, M.J 1970. On division by functional iteration. *IEEE Transactions on Computers*, vol. C-19, no. 8: 702–706.

Lyu & Matula, D.W 1995. Redundant Binary Booth Reccoding. *Proceedings of the 12th Symposium on Computer Arithmetic*: 50–57.

Oberman, S.F. & Flynn, M.J. 1997. Design issues in division and other floating-point operations. *IEEE Transactions on Computers*: 154–161.

Schulte, M & Stine. 1997. Symmetric bipartite tables for accurate function approximation. *Proceedings of the 13th IEEE Symposium on Computer Arithmetic*: 175–183.

Tang, P.T.P. 1991. Table-Lookup Algorithms for Elementary Functions and Their Error Analysis. *Proceedings of the 10th Symposium on Computer Arithmetic*: 32–236.

Information Technology and Computer Application Engineering – Liu, Sung & Yao (Eds)
© 2014 Taylor & Francis Group, London, ISBN 978-1-138-00079-7

Software trustworthiness modeling based on Interactive Markov Chains

HaiCan Peng & Feng He
Beifang University of Nationalities, Yinchuan Ningxia, China

ABSTRACT: This paper introduces a trustworthy model, and a trustworthy analysis approach for component-based software based on IMC (Interactive Markov Chains) at the architectural level. Conventional approaches are inadequate to model the behavior of a realistic software application which is likely to be made up of several interacting parts. For this reazon, we utilize characteristics of software architectural styles to assess software reliability. Moreover, a state model that synthesizes all different architectural styles embedded in the system is developed, allowing the IMC-based trusted model to be employed. Our model can be applied to software with heterogeneous architecture, can make full use of architecture design decision. At the end of this paper, part of analysis and verification results are given.

Keywords: Software trustworthiness; Software architecture; Interactive Markov Chains; State model

1 GENERAL INSTRUCTIONS

Research on trusted software is one of the branches of trusted computing. Software system is playing an increasingly important role in the information society. An error operation of software may result in economic damage and in the extreme cases cause deaths. When lives and fortunes depend on software, assurance of its quality becomes an issue of critical concern. At present, trusted degree is one of the key metrics for measuring the quality of software system. However, how to integrate the software trustworthiness into software design[7] and how to get the trusted degree as accurate as possible is still a problem in the field of trusted computing.

Conventional approaches to analyzing the performance and trustability of software system is to obtain the behavior of the whole software application and then compute the probability of a failure-free operation. Whereas, they ignore the whole software architecture, so can not generally give a credible evaluation.

Considering all the reasons above, this paper presents a research on software trustworthiness based on the software architecture and the mathematical theory of IMC. Software architecture, which describes the structure of software at an abstract level[1], consists of a set of components, connectors and configurations. Interactive Markov Chains, which combines continuous time markov chain with labelled transfer system. The reason of combining architecture with IMC has two, one is that almost every large software system has its own architecture styles, consequently we can measure the performance index as a whole. Two is IMC palces emphasis on the measurement of performance feature. In our approach, we utilize existing well-defined architectural styles and IMC theoretical research to analyze credibility of the software. The overall modeling will be introduced in section 4.1.

In the paper, we concentrate on some architectural styles which are related to state transition. These styles mainly include batch-sequential, parallel/pipe-filter, fault tolerance, and call-and-return architectural styles. we respectively build model for the four styles to measure the performance index.

The layout of the paper is as follows: Section 2 outlines the description and gives a brief overview of the IMC-based model which is modeled on the basis of the various achitectural styles. The details of our IMC-based model analysised from the architecture respect is described in section 3. Section 4 presents the overall modeling, test algorithm and test data. Conclusions and some directions for future research are given in section 5.

2 DESCRIPTION AND BRIEF OVERVIEM OF THE IMC-BASED MODEL

Our IMC-based software trusted model which is built from the architectural style respect utilizes Continuous-Time Markov Chains and Label Transition System to get performance evaluations of the system. This section requires knowledge about:

2.1 IMC (Interactive Markov Chains)

As a kind of combinable and concurrent system analysis framework, IMC combines the classical functional model-LTS and performance model-CTMC in a orthogonal way.

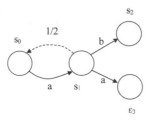

Figure 1. IMC graph.

Defination one[2]:

A IMC is a five-element-array (S, Act, $\rightarrow$, $--\rightarrow$, S_0), among it:

- S is a collection of state which is not empty.
- Act is a collection of act.
- $\rightarrow \subseteq s \times Act \times S$ is action transition relationship.
- $--\rightarrow \subseteq S \times R_{\geq 0} \times S$ is markovian transion relationship and $\forall s_1, s_2 \in S$,
- $|--\rightarrow \cap (\{S_1\} \times R \times \{S_2\})| \leq 1$. Here markovian transition is the random distribution delay transition which is indicated by exponential distribution.
- S_0 is the initial state. Here, we will choose our initial state according to the evaluation route that we selected.

In the defination one, the two transition relationship are the action transition relationship of LTS and the markovian transition relationship of CTMC respectively. As shown in Figure 1 below, there are at least the following two characteristics:

1) A state of the software system can perform multiple possible action transitions. e.g. s_1 can choose to move to state s_2 by performing the action b or to state s_3 by performing the action a. Due to the action transition and time delay is separate, the choice among these action transitions keeps the uncertain feature of the LTS[4].

2) If a state can implement action transition as well as markovian transition, such as s_1, if the happening time of action a and b are greater than the residence time of state s_1. System will execute markovian transition to state s_0 as shown in Figure 1. When performing the action transiton, the one who has the smallest happening time will be executed.

2.2 *Architecture of the software system*

Software architectural styles can be used to characterize software system by means of sharing certain common properties such as the structure of organizations, constraints, high level semantics. Over the last several decades, software architectures have attempted to deal with increasing levels of software complexity. A number of architectural styles have been identified and used to facilitate the communications among develops and to better the understanding of software systems[1]. The subject of this paper is to evaluate the trustworthiness of software based on IMC with a viewpoint of its architecture. Our ultimate aim is to analyze the

performance of the application in the given time and particular path.

Below we will give a detail demostration on the trusted model of the software with single architectural style. We can get other styles by certain modifications of these four styles.

2.3 *Performance evolution*

Performance evaluation is to show the properties of the system characteristics by certain temporal logic to depict the system properties. In our paper, we utilize a temporal logic called aCSL(action-based CSL) which is actually a variation of CSL.

Defination two:

Assume $p \in [0, 1]$, $\bowtie \in \{\leq, <, \geq, >\}$, the state formula of a CSL is produced by the following grammar: $\Phi ::= true | \Phi \wedge \Phi | \neg \Phi | P_{\bowtie p}(\varphi)$, to $t \in R_{>0}$, $A, B \in Act$, the path formula of a CSL is produced by the following grammar:

$$\varphi ::= \Phi_A U^{<t} \Phi \Big| \Phi_A U_B^{<t} \Phi$$

State formula indicates the state properties of the system, meanwhile, path formula indicates the path properties of the system[3]. The most important in the state formula is $P_{\bowtie p}(\varphi)$, which represents that the probability meeting the path formula φ is in the limited range of $\bowtie$. $U^{<t}$ in the path formula represents the time for completing a path of U is within time of t. In conclusion, we can get the performance evoluation of system like this:

- $\Phi_3 = P_{>p_3}(\Phi_{1A} U^{<t_3} \Phi_2)$; The probability eventually arriving labelled state Φ_2 is more than p_3 within the time of t_3. In addition, all the states that this path has passed must satisty the labelled state Φ_1 before arriving Φ_2, at the same time, the performed actions must satisfy the action set A.
- $\Phi_4 = P_{>p_4}(\Phi_{1A} U_B^{<t_4} \Phi_2)$; The meaning of Φ_4 is similar to Φ_3, the only difference is that the final action transition must be performed and the performed action must meet the action set B.

3 IMC-BASED TRUSTWORTHINESS MODELING FROM ARCHITECTURAL RESPECT

All above is the preparation work, below we will introduce our model detailedly. Four software architectural styles models respectively will be produced in the following four parts.

3.1 *Batch-sequential style*

Components are executed in a sequential order in this style. Choosing to any component can be probabilistic (markovian transition) or deterministic (action transition).

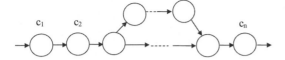

Figure 2a.   Architecture Model.

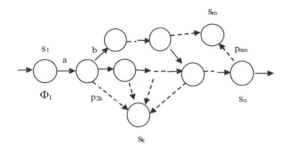

Figure 2b.   State Model.

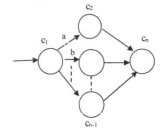

Figure 3a.   Architectural Style.

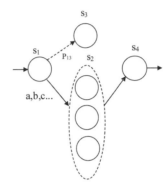

Figure 3b.   State Model.

The architecture model can be established as shown in Figure 2a, where $c_1, c_2, \ldots, c_n$ are software components. A component can just choose one path if existing more than one branchs. In this style, the system is executed from the beginning to the end in order. Therefore, the state model can be set up as follows: A state stands for the execution of a component. A transition from one state to another only through the markovian transition or the action transition. Each of action transition has a occurrence time.

The transition from the architectural style model to a state model can be viewed as a mapping of a component to a state, shown in Figure 2b, where $s_1, s_2, \ldots, s_n$ are the mapping states to components $c_1, c_2, \ldots, c_n$. In the process of mapping, we add the markovian transitions indicated by transfer rate p_{mn} and the action transitions indicated by the actions like a, b, c. ... Each of state has a residence time, if it is less than the smallest occurrence time of action from this state, system will execute the markovian transition. Besides, the probability of action transition is identified as 1.

We have introduced the performance evaluation in section 2.3. We can choose appropriate performance evaluations to analyze the performance of software. In addition, we do not give all the markovian transitions in the Figure 2b. we just choose the part of the route for instructions. As well as each state has its own labelled state which can identify some important information for it. For example, Φ_1 is the flag state of state s_1. However, we do not give all the labelled state in order to make the state model more clear. The following state models are the same situation.

3.2   *Parallel or pipe-filter style*

In this style, components often work simultaneously to finish a task. The main difference between these two styles is that parallel computation is generally in multi-processors environment, whereas pipe-filter

style occurs commonly in a single processor, multi-processes environment[1].

The state model can be set up as follows: For not existing the concurrent execution, a state represents one single component running. For existing the concurrent execution, all the concurrent execution components compose a state which represents an execution of all of the concurrent execution.

In Figure 3a, component c_1 and c_n are single component, component c_2 to c_{n-1} are running concurrently. Figure 3b is the state model with the parallel or pipe-filter architecture style.

In this style, a state maybe composed by multiple components running. Consequently, as long as it is the right action (e.g. a, b, c ...), system will transfer to the same state (e.g. s_2). When residence time is small enough, markovian transition will occur.

3.3   *Fault tolerance*

This style has a shielding ability to its own wrong actions (software errors) to a certain extent. Some specific components can compensate for the failure component by being the backup component. Apparently, the primary component and the backup components will be mapped into a state.

The state model can be modeled as follows: For a single component not having fault-tolerant function, a state represents an execution of a component. For a component with candidates, a state represents an execution of all these components. If occurence time takes long, system will execute markovian transition, such as state s_3.

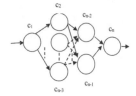

Figure 4a.   Architectural Style.

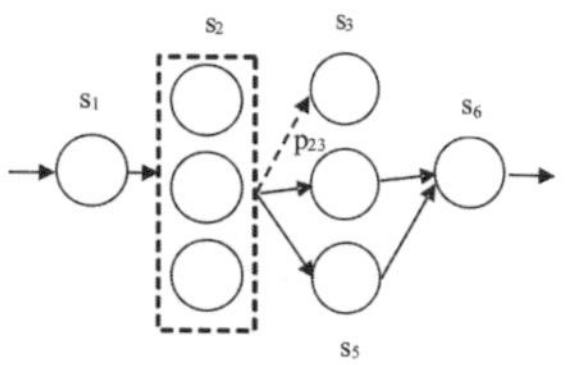

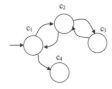

Figure 5a.   Architectural Style.

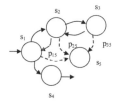

Figure 4b.   State Model.

Figure 5b.   State Model.

In Figure 4a, the components from c_3 to c_{n-3} are the backup components of c_2. There is synthetic state in both parallel/pipe-filter style and fault tolerance style. However, trigger condition is different. In the former one, the trigger action of transferring to the state (s_2) is the total trigger actions that transfers to all the concurrent components. Whereas in the latter one, the trigger action is the one transferring to the primary component (c_2). The reason is simple, system will transfer to different component when perform different action in the pipe-filter style. However, in the pipe-filter style system only performs the trigger action of transferring to original component.

3.4   Call-and-return

The dominant style in large software systems for the past 30 years has been call-and-return architecture. This kind of system shows obvious call/return relationship. Call/return style has rich connotation in the software architectural style. Here the call/return we discussed is the relationship between the components. A component can call other compoents which also can continue to call others. System will return from the last component to the first call component.

Consequently, the model can be set up as follows: As shown in Figure 5b, state s_1 is the calling state while s_2 is both a called state and a calling state and state s_3 is a called state. During the invocation, state s_2 and s_3 may be called multiple times before state s_1 gives control to the state s_4. If in a certain period of time, state s_4 still does not have the control power then the system will execute the markovian transition to state s_5.

Obviously, there is a infinite flow problem, here we will take the following approach to avoid the state explosion problem. In the Figure 5a, we just give a simple situation. In fact, this style can be a complex issues beyond imagination because of the nested or transitive call. The crux of the problem is to transfer infinite state to a finite state. Here, we choose a

simple way which can reduce the dimension of the state. When system transforms from the architectural style model to the state model, we will select n layers of nested or transitive call to set up the state model. The value of n depends on the architecture and the need for research.

4   OVERALL MODELING TESTING ALGORITHM

4.1   *Overall modeling*

We have already demonstrated the way to model based on a single architectural style and IMC in the previous sections. However, almost all of softwares have heterogeneous styles, hence the whole performance evaluation of a software system can be modeled in four steps as follows.

1) Identify the architectural styles of a system.
2) Model for each single architecture respectively.
3) Intergrate all of state models into a global state model of the system.
4) Get the performance evaluation of the system through the method of test algorithm.

4.2   *Test algorithm and test data*

The complete testing algorithm process will be given in this section. We adopt a kind of model testing algorithm based on the semantic, this approach will perform a recursive decomposition calculation for the formula needing to be detected. So we will take the recursive function Eval(Φ) as the core algorithm.

The calculation of probabilistic operator P in the state formula is the key algorithm of the IMC model. The calculation of this operator is divided into two cases: F(s,t) and G(s,t). The calculation result is the probability $\mathrm{Pr}\,ob(s_i, \varphi)$ which corresponding to every state s_i that is the starting state and satisfies φ. According to the semantics of the $P_{\rhd\lhd p}(\varphi)$, probability within the scope of $\rhd\lhd p$ is the accurate state.

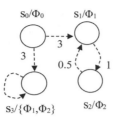

s_0/Φ_0 s_1/Φ_1

3

3 0.5 1

$s_3/\{\Phi_1,\Phi_2\}$ s_2/Φ_2

Figure 6. Component Model.

Table 1. Calculation results of properties Φ.

Prob(s_i, φ)	s_0	s_1	s_2	s_3
H = 0.1*	0	0.8086	1	0
H = 0.01*	0	0.8603	1	0
H = 0.001*	0	0.8641	1	0
H = 0.1	0	0.8540	1	0
H = 0.01	0	0.8633	1	0
H = 0.001	0	0.8645	1	0

*The datas in reference 6.

First we should calculate the state set satisfying Φ by executing Eval(Φ) function, then determine whether the initial state s_0 (or the given state) is in $Sat(\varphi)$ or not. When model test algorithm as a means of performance evaluation, we usually do not need to consider the counterexamples.

As for the two numerical iterative algorithm F(s,t) and G(s,t)[5], We here improve the algorithm G(s,t) to suit the model we have proposed. Figure 6 is adapted from reference 6 to describe a system that is based on architecture and IMC. Each state represents a component. This example is a special case which only contains markovian transitions. The nature of the each state is labelled near to it. Consequently, Sat(Φ_0) = {s_0}, Sat(Φ_1) = {s_1, s_3}, Sat(Φ_2) = {s_2, s_3}.

In reference 6, there is such a nature: $\Phi = P_{\geq 0.8}(\Phi_1 U^{\leq 2}\Phi_2)$, through our observation, there is only one qualified path, that is, a path of starting from state s_1 and ending at state s_2 within 2 unit of time. Therefore, the probability of this path can be calculated directly according to the definition as follows: $\int_0^2 e^{-1\bullet x}dx = 1 - e^{-2} \approx 0.8647$.

From the run results, as the integral step narrowing, integral results become more accurate. In our paper, although Figure 6 has another meaning, test algorithm principle is the same. For comparison, we select the same data, labelled information.

Table 1 below gives the calculation results based on three different integration step of reference 6 and our run results.

The results show that our approaches are comparatively good. Since the data is reliable, we will choose some component-based system to do further research and then give the trustworthy of a software system on the performance level.

5 CONCLUSIONS

We have presented a new approach based on architecture and IMC to model software trustworthiness from a new performance evalution perspective. In this paper, we mainly study component-based software system in order that we can focus on capturing component interactions which represents system architecture[6]. Through the analysis of heterogeneous architecturc, we have built the IMC model of each architecture and then integrate into an overall model. Finally, we will analyse the performance as required to get the trustworthy of a component-based software system. To further validate the model, we will concentrate on the experiment that applies the model to an industrial real-time component-based software system and obtain some significant promising datas in our future work.

ACKNOWLEDGEMENT

This work is supported by Innovation Project of Computer Science and Engineering Institute: An Dynamic Measurement Approach to Trustworthy Software based on Interactive Markov Chains.

REFERENCES

[1] WenLi Wang, Dai Pan & Mei Hwa Chen. 2006. Architecture-Based Software Reliability Modeling. *Journal of Systems of Software (JSS)*, Volume 79, Number 1, January 132–146.

[2] Lu Zhuang, Mian Cai & ChangXiang Shen. 2011. Trusted Dynamic Measurement Based on Interactive Markov Chains. *Journal of Computer Research and Development.* ISSN1000-1239/CN11-1777/TP 48(8):1464–1472.

[3] Chang Liu, Lian Ruan & Bin Liu et al. 2009. Heterogeneous Architecture-Based Software Reliability Model. *Computer Engineering and Applications*, 45(21):1–4.

[4] ZhenXing Xu, JinZhao Wu & JianFeng Chen. 2010. Design and Implementation of IMC-based Performance Checker. Journal of Computer Applications. 1001-9081 S1-0215-03.

[5] JinZhao Wu et al. 2007. *Interactive Markov Chain: the design, verification and evaluation of concurrent system.* Beijing: Science Press.

[6] Sherif, Y., Bojan, C. & Hany, H. A. 2004. A Scenario-Based Reliability Analysis Approach for Component-Based Software. *IEEE Transactions On Reliability.* Vol. 53, No. 4, December 2004:465–480.

[7] JunFeng Tian, Zhen Li & YuLing Liu. 2011. An Design Approach of Trustworthy Software and Its Trustworthness Evaluation. *Journal of Computer Research and Development.* ISSN 1000-1239/CN 11-1777/TP 48(8):1447–1454.

Information Technology and Computer Application Engineering – Liu, Sung & Yao (Eds)
© 2014 Taylor & Francis Group, London, ISBN 978-1-138-00079-7

Research on urban mass transit network passenger flow simulation on the basis of multi-agent

Hong Fei Yu, Yong Qin & Zi Yang Wang
Beijing Jiaotong University, Beijing, China

Bo Wang & Ming Hui Zhan
Beijing Infrastructure Investment Co., LTD., Beijing, China

ABSTRACT: In order to adapt to the situation that Urban Mass Transit (UMT) operate over network, passenger flow presents a point, line, network interaction features, the phenomenon of passenger flow accumulation have occurred time to time, passenger flow divert has become a normal work of UMT passenger flow operation, a network Passenger Flow Simulation model on the basis of multi-agent according to characteristic of UMT network operation was proposed, the deductive rules and simulation process was analyzed, and the MATLAB simulation platform was applied to realize the model and a instance analysis was carried on in this paper, which provides decision support to the UMT passenger flow divert and has great significance to improve the UMT safety management level.

Keywords: Urban Mass Transit; multi-agent; network; passenger flow; simulation

1 INTRODUCTION

With the increasing of UMT lines, the UMT operation has become from single lines to network. Instead of depending on a single station or a single line, the aggregation and dissipation of large passenger flow has been gathered into a regional and dissipate to the entire network. Under this new background, the original single-period traffic operations analysis, traffic prediction, traffic plans are no longer applicable, a new entire network collaborative persuation method for passenger flow analysis is badly needed.

Domestic and foreign scholars have realized the applicability of multi-agent in the field of transportation. Bomarius was the earliest to analyze the application of multi-agent in urban traffic in the literature [1], Cetin [2] researched on the large area traffic simulation model, Rosseti [3] put forward the dynamic network traffic model based on multi-agent system, Wei [4] introduced the microscopic traffic simulation based on multi-agent framework, Paruchuri [5] realized no organization traffic multi-agent model, verified the feasibility of multi-agent simulation. However, multi-agent modeling technology research mostly focused on the microscopic model of road traffic. To apply the multi-agent modeling technology into the UMT, a network Passenger Flow Simulation model is proposed in this paper, and the MATLAB platform is also used to realized the model.

2 ESTABLISHMENT OF THE UMT NETWORK PASSENGER FLOW MODEL ON THE BASIS OF MULTI-AGENT

2.1 *Passenger characteristics analysis and modeling*

Passengers is one of the core agent, passenger's main goal is to choose a reasonable path in complex network system, and to choose a reasonable train according to the path to arrive in the destination. In the process from origin to destination, passengers need to constantly identify the way they choose and whether or not arrive in the destination. This is a repeated process of judgment.

So the passengers can be made a formal definition like this:

Passenger : Agent =$< O, D, T, A, Ch, Tr >$

In which, O is the original station, D is the destination, T is the state, on behalf of the passenger state of get in the station, wait for the train, riding and get out of the station, and A is the action set of passengers, including aboard, get down the train and transfer, $Tr : T \times A \rightarrow T$ is the state updating function, which make the update according to the state and action. $Ch : O \times D \times T \times E \rightarrow A$ is the decision making function, on behalf of the behavior of passengers choose the reasonable path according to the OD, self-attribute, state perception, in which E is the external environment information.

In the model, passenger's behavior mainly including the following 3 kinds: get on the train and get off the train, wait for the train, transfer behavior.

2.2 Train operation characteristics analysis and modeling

In the actual operating business, the operation of the train is closely related to the passenger flow. In the simulation network, trains' actions such as running into the station, out of the station, moving, stop etc. were closely affected by other agents. Actually the train operation plan was arranged according to the day's traffic situation. So, train operation in this model should also be passive.

So the trains can be made a formal definition like this:

$$Train : Agent =< Q, S, Atrr, A, T, n >$$

In which, Q, S is the train location information, $Atrr$ is the properties set of the train including direction, train load rating, train line and so on, A is the actions set, including running and stop, T is the status set, including ready to start, driving, stop, etc.

n is the number of passengers in the train, Ch : $Q \times S \times T \times Atrr \times E \rightarrow A$ is the decision making function, shows that the train's action is based on its location information and status as well as the outside influence, $Tr : T \times E \rightarrow T$ is the status updating function, shows that the status of the train is decided by the outside environment.

Train's Q, S location information is decided by the definition of the simulation network, which is an important reference indicator of train's next move. Train obtains this information from the network after moving a distance. In this model, train's action performs step by step, train's movement is defined as uniform motion in each time step. Train's properties includes not only some natural properties such as the length, number, complement number, section number, capacity limits of the train, but also adds some properties in view of the passenger flow, which changes dynamically with the movement of the train.

2.3 UMT network characteristic analysis and modeling

The characteristics of topology network model and simulation network have been taken into account to propose the passenger flow movement simulation network, which is a refinement of the simulation network from the angle of the movement of the passenger flow. In this model, the transfer stations have been divided into two stations with a transfer path to connect with each other, and the UMT network has been divided into three layers including railway network, line and station, each layer collaborate with another.

(1) Network layer

Network agent has a simple structure, which is mainly responsible for the storage of the basic topological structure of the simulation network. The formal definition of the network agent is like this:

$$NetWork : Agent =< S, L(s), G >$$

In which, S is the node station set, including a number of stations. L is the line set, G is the correlation function, the domain of which is $L(s)$, $\forall l \in L(s)$, $\exists s \subset S$ so that $G(a) = s$.

(2) Line layer

The line agent describes the information and the basis of the structure of the line, trains are directed by the line agent, to obtain the location information and action command. Line agent can be considered as one of the key outside influence environmental factors. The formal definition of the line agent is like this:

$$Line : Agent =< s, e, Atrr, T, S, F, Tr >$$

In which, s, e represents the origin station and destination station, $Atrr$ is the properties set of the line, T is the status set, S is the station set of the line, F is the direction of the line. $Tr : T \times R \rightarrow T$ is the status updating function, which is decided by the external environment.

(3) Station layer

This simulation model is mainly aiming at the deductive process of the network passenger flow. If you consider movement of passenger flow in the station, the parameters of the passenger agent can be adjusted to change the passenger properties. The formal definition of the station agent is like this:

$$Station : Agent =< x, y, Atrr, ID, N, B, T, S, v, Tr >$$

In which, x, y is the geographical coordinates of the station, $Atrr$ is the properties set of the station, each station has its own ID, N is the ID of next station, B is the ID of the previous station, if this station is a transfer station, S is the stations correspond to the transfer station, v is the limit flow rate at the station, T is the status of the station, $Tr : T \times E \rightarrow T$ is the status updating function.

(4) Information interaction mechanism

There are a large number of passengers in the model to get on the train, get off, transfer, etc.as well as trains to run concurrently at the same time, so there is a large amount of calculation. Therefore, a high efficiency of information interaction mechanism should be considered in the model, to realize the communication between each agent. Blackboard type message mechanism is applied to a way of communication in the field of computer, with multiple intelligence that share a common area, each agent can read and write in this very area to achieve information sharing.

In the UMT passenger flow inference model, network, line and station form the main information field of the UMT system, they complete the information interaction in this field. Train and passenger also has their own information field, which has public area which the main field shown in Figure 1.

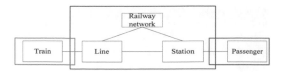

Figure 1. Interactive method of simulation model information.

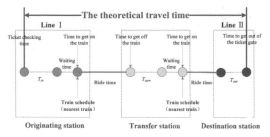

Figure 2. Assignment rules based on the passenger travel time.

Table 1. Passenger travel time classification.

Type	Content
T_{in}	Travel time from ticket gate to platform
T_{out}	Travel time to ticket gate after getting out of the train
T_{turn}	Travel time at transfer station
T_{train}	Waiting and ride time

3 SIMULATION EVOLUTION PROCESS AND PASSENGER FLOW INFERENCE RULES BASED ON THE PASSENGER TRAVEL TIME

3.1 *The establishment of passenger flow inference rules based on the passenger travel time*

Most of the passenger flow equilibrium assignment mode using time impedance as measurement has a basic assumption that each passenger can grasp the topology of the network and the time cost in each link accurately [6]. In order to obtain verifiable and reasonable routing rules, this paper put forward the passenger flow inference rules based on the passenger travel time. Statistics show that the OD of the passenger flow has a certain regularity in time depending on the different nature of the passenger flow and different travel purpose, the OD of the passenger flow has a certain regularity in time. In addition, in the process passenger flow from the station to the outbound, each passenger has to go through the station facilities (security check, ticket gate, staircase, channels, etc.), so the passenger flow in the whole process has regularity and consistency.

According to the regularity of passenger movement process, theoretically the passengers' travel time should include the following parts shown in Figure 2. The travel time of the passengers can be divided into 4 parts, T_{in}, T_{out}, T_{turn}, T_{train}, the content of which are shown in Table 1.

Theoretical travel time can be obtained by the following formula:

$$T = T_{in} + T_{out} + T_{turn} + T_{train}$$

In which, T_{in}, T_{out}, T_{train}, can be calculated through the cellular automata simulation model [7] or statistical analysis of the actual data.

When the train schedule is given, T_{train} relates to the transfer option, actually from the passenger get into the station through the ticket gate to the passenger get out of the station through the gate, the ticket card terminal records he total travel time consuming T_{travel} When the path selection of the OD pair conforms to the actual situation, the ideal travel time T calculated by the above formula can be in accordance with T_{travel} or the degree of which meets the normal distribution. So in the study, a large amount of data can be chosen as sample, and wavelet neural network, neural network and support vector machine (SVM) method etc. could be used to train the different types of passenger flow, through the fitting,

the parameters of the path selection rules can be obtained.

On the basis of the above theory, according to the a large number of field survey data, we estimate T_{in}, T_{out} and T_{turn} into a reasonable simulation step frequency time t.

$$T = T_{in} + T_{out} + T_{turn} + T_{train} \Rightarrow T = 3t + T_{train}$$

So the simulation of passenger deductive rules can be obtained as follows:

(1) The number of passenger get out of the station (t) = The number of passenger arrive in the destination;

(2) The number of passenger (t) get into the station is randomly generated in each simulation step;

(3) The number of passengers get on the train(t), the number of passenger get off from the train;

(4) Normal station:
Number of passengers waiting in the station (t) = Number of passengers waiting in the station (t − 1) + number of passengers get into the station (t) − number of passengers get on the train (t);

(5) Transfer station:
Number of passengers waiting in the station (t) = Number of passengers waiting in the station (t − 1) + number of passengers get into the station (t) − number of passengers get on the train (t) + number of passenger transfer to the next line (t − 1);

(6) Obtain the path selection parameters of each OD pair, get T_{train} through the simulation.

3.2 *Passenger flow simulation evolution process*

The passenger agent, train agent and station agent are all related to the simulation model, in which data communication only happen between station and passenger flow, the passenger flow simulation evolution process is shown in Figure 3.

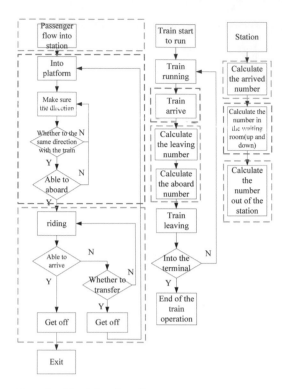

Figure 3. Simulation flow diagram.

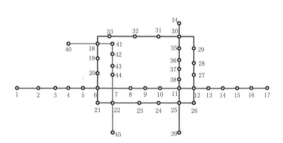

Figure 4. X City urban rail transit network.

4 THE SIMULATION RESULTS ANALYSIS

Suppose the UMT network topology of a city consists of A, B, C, D four lines and 45 stations. In which, line A is a horizontal line containing node 1 to 17, B is a ring circuit line, C containing node 34 to 39 and D containing node 40 to 45, there a 8 transfer stations in the network shown in Figure 4.

According to the simulation network structure, train operation routes for 8 directions (1~17, 17~18 inner ring 1, 18 to 19 to 33–32~18 to 33, 33 – outer ring, 39~34, 34 to 39, 40~45, 45~40). The station capacity is set at 2000, train capacity is set at 1000, interval elapsed time is 2 minutes, suppose the passenger OD obeys uniform distribution. According to the above scenario, the simulation time is set for 2 hours, all into station passenger flow value is set to unity.

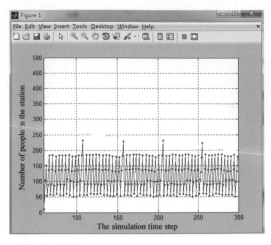

Figure 5. The results of No. 5 station.

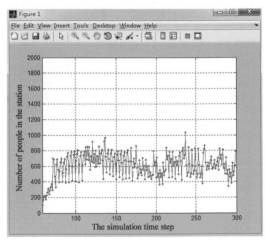

Figure 6. The results of No. 7 station (transfer station).

Through the simulation of the UMT shown in Figure 5, we can find that, when the number of passenger into the station is 47 (person/step), which means the number is 94 per minute. Under this condition, the network is operated busily, but the number of passenger in all of the stations in the network will remain below the capacity 2000.

When the number of passengers into No. 1 station increases gradually, the No. 7 station will be greatly influenced, which is shown in Figure 6. We increased the rate of No. 1 from 47 to 50, 52 and 70, what could be found is, when it reached 50, the maximum increment of No. 7 was 135; when it reached 52, the maximum increment is 464; when it reached 70, the maximum increment is 1531, and the maximum number of passengers in the station is 3460, obviously exceeds the limit of 2000. So, we can conclude that, when the number of passengers into the station reaches 94/min, a tiny increment of a station would receive great impact to the whole network.

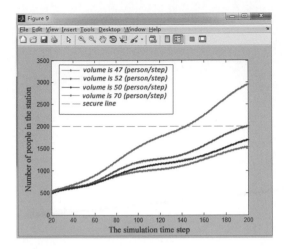

Figure 7.   Passenger flow volume of No. 7.

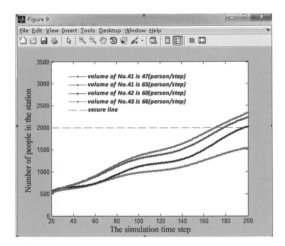

Figure 8.   Passenger flow volume of No. 7.

After filtering curve, Figure 7 shows the number of passenger into the No. 1 station is 70, 52, 50, 47, the number of passengers in No. 7 station. It clearly shows that, the unit step volume of a station can cause great impact on other stations.

Respectively increase the rate of No. 41, 42 and 43, to check the changes if No. 7, which is shown in Figure 8. We can find that network effects have big influence in No. 7 station, which will never happen in a single line. Through simulation experiments, we can infer that if the station changes at the same time, it may not have much impact on themselves, but for a transfer station as No. 7 it is likely to have a huge impact, and measures must be taken to handle it.

5   CONCLUSIONS

This paper draw lessons from the multi-agent simulation modeling methods, build the network system of UMT passenger flow simulation model, and analyzed the structure of the model and the agent's characteristics, functional properties and the communication mechanism between each agent. Through the analysis of passenger travel in urban rail transit network system, Simulation process and passenger flow inference rules based on the passenger travel time Finally the MATLAB platform was applied to realize the model, and an example was analyzed by the model which shows the the model is suitable for a network under the condition of UMT passenger flow analysis.

ACKNOWLEDGEMENTS

This work is supposed by National High Technology Research and Development Program of China (Contract No. 2012AA112401), the State Key Laboratory of Rail Traffic Control and Safety (Contract No. RCS2010ZT005), Beijing Jiaotong University and Key Project of the National Twelfth-Five Year Research Program of China (Contract No. 2011BAG01B02).

REFERENCES

[1] Bomarius F. A multi-agent approach towards modeling urban traffic scenarios. Deutches Forchungszentrum fur Kunstliche Intelligenze, PR-92-47, 1992.

[2] Cetin N, Nagel K, Raney B, et al. Large-scale multi-agent transportation simulations. Computer Physits Communications, 2002, 147(1–2):559–564A.

[3] Rossetti R J F, Liu R H. A dynamic network simulation model based on multi-agent systems. Applications of Agent Technology in Traffic and Transportation, 2005, 181–192.

[4] Wei Y Han Y Fan B. "In Agent-oriented urban traffic micro simulation system, "Industrial Technology, 2008. ICIT 2008. IEEE International Conference on, 2008, PP 1–7.

[5] Poon M.H, Wong S.C, Tong C.O, "A dynamic schedule-based model for congested transit networks. "Transportation Research Part B, 2004, 38:343–368.

[6] SI Bing-feng, MAO Bao-hua, LIU Zhi-li, "Passenger Flow Assignment Model and Algorithm for Urban Railway Traffic Network under the Condition of Seamless Transfer," Journal of the China railway society, 2007, V01, 29(12):12–18.

[7] Blue V J, Adler J L. "Modeling four-directional pedestrian flows. Traffic Flow Theory and Highway Capacity 2000-Highway Operations," Capacity, and Traffic Control, 2000, (1710):20–27.

Information Technology and Computer Application Engineering – Liu, Sung & Yao (Eds)
© 2014 Taylor & Francis Group, London, ISBN 978-1-138-00079-7

Design of fault diagnosis system for coal-bed methane gathering process and research on the fault diagnosis for compressors

Jin Su, Jian Hua Yang & Wei Lu
School of Electronic and Information Engineering, Dalian University of Technology, Dalian, China

Yi Wang & Ze Feng Lv
China Liaohe Petroleum Engineering Co., Ltd., China

ABSTRACT: The process of coal-bed methane gathering includes two important parts: coal-bed production and gas compression by booster station. So the fault often occurs in these two parts. In this article, we adopt FAM (Fuzzy Associative Memory) neural network to realize compressor fault diagnosis. This model implements intelligent fault diagnosis and self-learning of the knowledge database. Thus it significantly improves the accuracy and scalability of fault diagnosis system of the reciprocating compressor. The system has been adopted in experiment. In addition, the fault diagnosis system is based on Visual Studio 2008. NET and SQL sever 2005 to monitor operation parameters and equipment status real-timely and to realize fault management and diagnosis. The system has friendly man-machine interface and is more convenient and easy to understand the operation flow, more powerful to store data and more easy to embed fault diagnosis algorithm into it.

Keywords: Data base; Coal-bed methane gathering process; Fault diagnosis; Fuzzy; Association memory model; Self-learning

1 GENERAL INSTRUCTION

With sustained development of national economy, people's demand for energy becomes more and more strongly, but the shortage of oil and natural gas causes the contradiction between supply and demand of the energy. Now coal-bed methane extraction has alleviated the problem of energy shortage partly. The project of coal-bed methane gathering conducting in Panhe where is about 80 kms northwest of jincheng, Shanxi Province, and is under the jurisdiction of Qinshui. 200 new CBM wells are exploited in this project. The CBM gathering system consists of the well site, gas pipe network, gathering station and centralized processing booster station to complete CBM's gathering, purification, separation, pressure regulation and transmission[1]. In order to improve the utilization of coal gas and ensure fault diagnosis of the key parts timely and effectively, the fault diagnosis mainly focuses on single wellhead[2] and booster station compressor. The compressor fault diagnosis based on Fuzzy Neural Network Associative Memory Model not only has the ability of expressing fuzzy and qualitative knowledge, but also has self-learning and self-adaptive ability as well as nonlinear expression ability. In this paper, the model is applied to study the fault diagnosis of reciprocating compressors and the result shows the validity of the diagnosis and realizes the potential fault diagnosis.

In this project, we need a well-designed interface of fault diagnosis management system for monitoring the state of parameters and the fault information real-timely and the system is convenient for embedding algorithm for fault diagnosis. This article regards Visual Stdio .NET 2008 and SQL server 2005 as the development platform for the design of system to manage plenty of data. The .NET is based on the standard rules which has solved the interoperable problem with other software.

2 DESIGN SCHEME OF SOFTWARE SYSTEM

Software system can monitor the information of operation status and fault diagnosis information of the equipment real-timely and it consists of data acquisition and monitoring subsystem, database subsystem, and intelligent fault diagnosis system. The field data mainly includes two parts: one is the reciprocating compressors data transmitted by serial communication network and the other is the wellhead parameters transmitted by GPRS. The computer receives the monitoring data from the field and then stores it in the monitoring database, finally displays the operation state of the equipment on the man-machine interface. We use fault diagnosis reasoning algorithm to compare the fault diagnosis result with fault

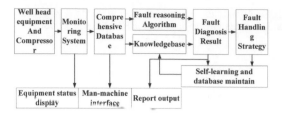

Figure 1. Function Block Diagram.

Figure 2. Function Block Diagram of Data Base.

reason between monitoring data and knowledge base sampling data and then give out the fault handling strategies.

The main function modules of the system and schematic of fault diagnosis system is shown in Figure 1.

The main functions of the system include:

(1) Monitor the parameters' state of oil wells as well as compressors and display them in the flow chart and queries the historical data through the table and curve way.
(2) Complete fault automatic detection, fault alarm, and display the fault data according to the existing knowledge base information and the field data.
(3) Manage the standard parameters of wellhead and compressor, and update the knowledge base information according to the diagnosis result to realize the management of knowledge base information.
(4) Manage the generation of the report, fault information, and count failure times.
(5) Store and update the field parameters, the knowledge base information and fault information.

2.1 The design of database system

The information in the database system is field data written by monitoring system automatically. Database system not only can query and browse the monitoring data, fault information, knowledge base information and standard parameter database and personnel information, but generate report table and then printout. Figure 2 shows the main function modules of the database system.

2.2 Main function modules of the database software system

In order to save space, query and diagnose easily, this paper designs the related table of fault diagnosis database:

(1) field data collection table for storing the related parameters of the compressor acquired through the serial communication as well as the wellhead parameters acquired through GPRS transmission. The fields in the table are: the system (well or booster station), name of the 1st grade components, parameter name, parameter values, the data acquisition time
(2) the users' information table for storing the information of registered users, the fields in the table are: username, password, name, department, accessibility, phone numbers
(3) the base information tables include standard parameters table, fault phenomenon table and fault reason table as well as rule table. Those tables will update continuously with new fault modes and improve intelligent diagnosis algorithm
(4) the historical fault information table stores the diagnosis results for querying the fault information according to the time as well as position and analyzing the report table easily, the fields in the rule base table are: the system, name of the 1st class components, name of the 2nd class components, fault reason.

3 THE RESEARCH OF FAULT DIAGNOSIS ALGORITHM WITH RECIPROCATING COMPRESSOR

The reciprocating compressor [3] has a complex structure. There are external interference factors during the normal operation and there is no correspondence between reasons and phenomena. Several faults can occur simultaneously and affect each other mutually, so fault classification is not clear. When the parameter has not reach to the threshold value, it may affect other fault in directly, so we cannot achieve the ideal diagnostic results only to rely on the threshold diagnosis. In this article, there is an accurate method relying on the fuzzy theory [4] to establish fuzzy relationship matrix about the relation between the fault and phenomena for realizing the fault diagnosis of compressor. It's hard for us to acquire the matrix, so we usually select membership function or expert experience to determine it. During the running process of equipment, we update and improve the fuzzy relation matrix depending on the fault pattern. In this paper, a method is proposed for the automatic acquisition and self-learning [5] of fuzzy relation matrix according to associative memory model based on neural network, and it has achieved good results.

Table 1.	Fault Cause.
Y1	the 1st class suction valve leakage
Y2	the 1st class exhaust valve leakage
Y3	the 2nd class suction valve leakage
Y4	the 2nd class exhaust valve leakage
Y5	the gap between cylinder body and cover in the 1st class
Y6	the gap between cylinder body and cover in the 2nd class
Y7	cooling water system fault (shortage or disruption)
Y8	the 1st class piston rod broken or blocked
Y9	the 2nd class piston rod broken or blocked
Y10	the 1st class stuffing box leakage
Y11	the 2nd class stuffing box leakage
Y12	cross head overheating
Y13	the connecting rod fracture
Y14	spindle overheat

Table 2.	Fault phenomenon.
X1	1st suction temperature rise
X2	1st exhaust temperature rise
X3	1st suction pressure rise
X4	1st suction pressure drop
X5	1st exhaust pressure rise
X6	1st exhaust pressure drop
X7	2nd suction temperature rise
X8	2nd exhaust temperature rise
X9	2nd suction pressure rise
X10	2nd suction pressure drop
X11	2nd exhaust pressure rise
X12	2nd exhaust pressure drop
X13	the intermediate cooler exhausted gas temperature rise
X14	the intermediate cooler water temperature rise
X15	the intermediate cooler pressure drop
X16	the intermediate cooler pressure rise
X17	cross head temperature rise
X18	stuffing box temperature rise
X19	piston ring temperature rise
X20	connecting rod (big) bearing temperature rise
X21	connecting rod (small) bearing temperature rise
X22	the main bearing temperature rise
X23	emissions reduction

3.1 Common fault pattern of compressor

The fault [6] of DF-5/10-40 reciprocating compressor is mainly divided into two categories: the machine thermal fault and mechanical functional fault. The main character of the former fault is the compressor exhaust problem, abnormal performance of discharge pressure and temperature pressure between different grades. The main character of latter category is overheated when the compressor operates. The causes of the two type's faults are shown in Table 1.

The fault will cause the abnormality of thermodynamic performance and dynamic performance, but the strength of performance is not the same and a more intuitive performance is the change of parameter value. It's direct and convenient to find the reason of fault diagnosis by measuring the pressure, temperature and exhaust volume.

The field parameters of the compressor are: the gas temperature, the gas pressure, temperature and pressure of the intermediate cooler water, the temperature of stuffing box, the spindle temperature, the connecting rod temperature, piston ring temperature, crosshead temperature, exhaust volume.

3.2 The establishment of fuzzy fault diagnosis model

The method of fuzzy fault diagnosis [7] is to calculate the membership degree of each fault reason depending on the membership degree of phenomena. The fuzzy relation between fault and phenomenon of compressor is given in the following:

fault phenomenon set (assuming there are m kinds of fault reasons as $x_1, x_2, x_3, \ldots, x_m$)

$$X = (\mu_{x1}, \ \mu_{x2}, \ \mu_{x3}, \ \ldots \ldots \mu_{xm}) \quad (1)$$

fault reason set (assuming there are n kinds of fault reasons as y1, y2, y3, …, yn)

$$Y = (\mu_{y1}, \ \mu_{y2}, \ \mu_{y3}, \ \ldots \ldots \mu_{yn}) \quad (2)$$

μ_{xi} ($i = 1, 2, \ldots, n$) is the membership degree when the diagnosis object has the phenomenon of μ_{yj} ($j = 1, 2, \ldots, m$), and is also the membership degree when

the fault has the fault reason of yj. Fuzzy matrix between fault phenomenon vector and fault reason vector:

$$R = \begin{Bmatrix} r_{11} & r_{12} & \cdots & r_{1n} \\ r_{21} & r_{22} & \cdots & r_{2n} \\ \cdot & \cdot & \cdot & \cdot \\ \cdot & \cdot & \cdot & \cdot \\ r_{m1} & r_{m2} & \cdots & r_{mn} \end{Bmatrix} = (r_{ij})_{m \times n} \quad (3)$$

3.3 Fuzzy reasoning

According to the sensors' detection position and field parameters, we divide the compressor fault phenomena into 23 categories as shown in Table 2.

Phenomena vector: $X = (\mu_{x1}, \ \mu_{x2}, \ \mu_{x3}, \ \ldots \ldots \mu_{x23})$,

reason vector: $Y = (\mu_{y1}, \ \mu_{y2}, \ \mu_{y3}, \ \ldots \ldots \mu_{y14})$

The data gathering in the field are accurate values, we need adopt the fuzzy distribution to transform the values of data gathered in the field into fuzzy language, namely fault phenomenon vector. Here the fault phenomenon are mostly varied values, the fuzzy distributions are partial large fuzzy distribution and partial small fuzzy distribution.

The initial value of fuzzy matrix is based on experience as well as expert statistics and is improved gradually in the practical diagnosis process through the self-learning of neural network learning mechanism.

The fuzzy relation between Y and X is $Y = X \bullet R$, ie the fuzzy relation equation between fault phenomena and fault causes. $\bullet$ is fuzzy operator represented by M. The five kinds of models are: $M(\wedge, \vee)$, $M(\bullet, \vee)$,

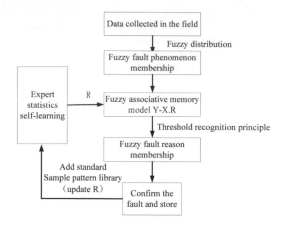

Figure 3. Fuzzy Reasoning.

$M(\wedge, \oplus)$, $M(\bullet, \oplus)$, $M(\bullet, +)$. Here we use the 5th models:

$$M(\bullet, +), y_j = \sum_{i=1}^{m} x_i r_{ij}, j = 1, 2, \ldots n .$$

Based on the fuzzy relationship model, we can get the fault reason vector. There are two fault pattern recognition methods: the maximum membership principle and the threshold limit principle. This paper uses the threshold limit principle to recognize the possible faults, and then confirms it manually. The fuzzy reasoning process is shown in Figure 3.

3.4 Self-learning principle

There are two ways to obtain and improve the fuzzy relation matrix: one is according to expert knowledge and statistical method of previous faults, the other way is that the fault diagnosis system does self-learning to modify the fuzzy relation matrix according to the initial given value to make sure that it can show the correlation [8] between fault reason and fault phenomenon more accurately. Imitating human brain neuron function makes neural network has the ability of self-learning and data processing ability. This paper attempts to combine the fuzzy logic and neural network to form fuzzy neural network, it not only can express qualitative knowledge, but also has the powerful ability of self-learning and data processing ability. Depending on the model of fuzzy neural network association memory to adjust the membership degree between the fault phenomenon and fault reason.

The self-learning samples are the fault phenomenon vector: $X = (\mu_{x1}, \mu_{x2}, \mu_{x3}, \ldots \mu_{x23})$ that includes the membership degree of each phenomena and fault reason vector: $Y = (\mu_{y1}, \mu_{y2}, \mu_{y3}, \ldots \mu_{y14})$ that includes the probabilities value of each fault reason. Supposing the number of learning sample groups is Q, self learning algorithm is shown in Figure 4.

① Initialize and give upper limit of iteration κ and precision ε

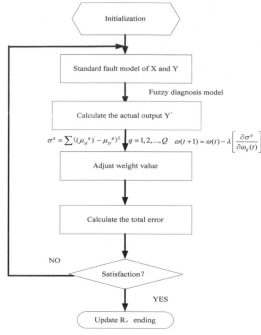

Figure 4. Self-learning.

② Give each fault phenomenon vector:

$$X^q = (\mu_{x1}{}^q, \ \mu_{x2}{}^q, \ \mu_{x3}{}^q, \ \ldots \ldots \mu_{x23}{}^q) \qquad (1)$$

and fault reason vector:

$$Y^q = (\mu_{y1}{}^q, \ \mu_{y2}{}^q, \ \mu_{y3}{}^q, \ \ldots \ldots \mu_{y14}{}^q) \qquad (2)$$

③ Calculate the associative output mode by the model operation: $M(\bullet, +)$

$$Y^{q'} = (\mu_{y1}{}^q, \ \mu_{y2}{}^q, \ \mu_{y3}{}^q, \ \ldots \ldots \mu_{y14}{}^q) \qquad (3)$$

The synthetic operation model is:
$\mu'_{yj} = \sum_{i=1}^{23} \mu_{xi} \omega_{ij}$, $j = 1, 2, \ldots, 14$, and ω_{ij} is the neural network weights, ie the value of r_{ij} in matrix of fuzzy relationship.

④ Adjust the weights value (membership), λ is proportional coefficient and it ranges in (0, 1)

$$\sigma^q = \sum ((\mu_{yj}{}^q)' - \mu_{yj}{}^q)^2 \qquad q = 1, 2, \ldots, Q \qquad (4)$$

$$\omega(t+1) = \omega(t) - \lambda \left[\frac{\partial \sigma^q}{\partial \omega_{ij}(t)} \right] \qquad (5)$$

Calculating the value of σ, $\sigma = \sum \sigma^q$ and q = q + 1, then go on the next sample, if the conditions are not satisfied, return to (3) to adjust the weight. if satisfied, it shows that we complete the adjustment of weights (membership) and achieve the correct fuzzy relation

matter, and the diagnostic performance of the system is improved.

3.5 *Diagnostic process and result analysis*

We establish the standard information base for the 23 kinds of fault phenomenon and 14 kinds of common fault reasons of compressor, the corresponding parameter to each fault are shown in table 7. Through the standard fault information and fuzzy distribution shown in Fig. 3 and Fig. 4, we obtained the fuzzy membership vector of fault phenomenon as shown in table 8. Fuzzy membership vector for the 14 fault reasons are shown in table 9, corresponding to table 8. According to the reasoning algorithm and self-learning principle, we get the relation fuzzy matrix as shown in table 3.

The learning mechanism global convergence diagram of relation fuzzy matrix is shown in Fig. 7. The test error of the fuzzy matrix is 0.001, achieving the requirement in the range of specified number (1000) of test, Epochs = 19.

We test the measured data according to the fuzzy relation matrix. The corresponding faults of the three sets of measured data are A1: 1st class suction valve leakage and a cylindergap, A2: 1st class suction valve leakage and leakage of compressor stuffingbox,

A3: the cooler water system fault. methods for identifying fuzzy model include the maximum membership principle and the threshold judgment principle. The fault of reciprocating compressor may be not a single fault, so we adopt the threshold judgment principle. According to the expert's experience, when the possibility of fault is greater than 0.6, we think that there may exist fault. For the data of the 1st and 2nd group, the diagnosis result is 1st class suction valve leakage

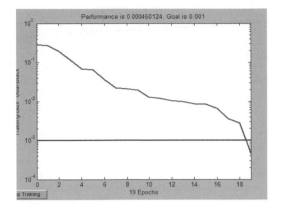

Figure 5.   Global convergence graph.

Table 3.   Testing results Fuzzy Relation Matrix by Learning.

	X1	X2	X3	X4	X5	X6	X7	X8	X9	X10	X11	X12	X13	X14
1	1.00	0.10	0.03	0.02	0.50	0.10	0.00	0.00	0.00	0.50	0.10	0.00	0.00	0.00
2	0.10	1.00	0.03	0.02	0.50	0.10	0.00	0.00	0.00	0.50	0.10	0.00	0.00	0.00
3	0.90	0.20	0.01	0.02	0.50	0.10	0.00	0.00	0.00	0.50	0.10	0.00	0.00	0.00
4	0.00	0.00	0.00	0.00	0.00	0.00	0.00	1.01	0.00	0.00	0.00	0.00	0.00	0.00
5	0.20	0.90	0.03	0.02	0.50	0.10	0.00	0.00	0.00	0.50	0.10	0.00	0.00	0.00
6	0.00	0.00	0.00	0.00	0.00	0.00	0.00	0.01	0.00	0.00	0.00	0.00	0.00	0.00
7	0.05	0.06	1.00	0.10	0.20	0.50	1.00	0.00	0.00	0.20	0.50	0.00	0.00	0.00
8	0.01	0.01	0.10	1.00	0.20	0.50	0.10	0.00	0.00	0.20	0.50	0.00	0.00	0.00
9	0.05	0.06	0.90	0.10	0.20	0.50	0.00	0.00	0.00	0.20	0.50	0.00	0.00	0.00
10	0.01	0.01	0.20	0.90	0.20	0.50	0.00	0.00	0.00	0.20	0.50	0.00	0.00	0.00
11	0.00	0.00	0.00	0.00	0.00	0.00	0.50	0.00	0.00	0.00	0.00	0.00	0.00	0.00
12	0.00	0.00	0.00	0.00	0.00	0.00	0.30	0.00	0.00	0.00	0.00	0.00	0.00	0.00
13	0.05	0.07	0.05	0.03	0.10	0.20	1.00	0.00	0.00	0.10	0.20	0.00	0.00	0.00
14	0.05	0.07	0.05	0.03	0.10	0.20	1.00	0.00	0.00	0.10	0.20	0.00	0.00	0.00
15	0.00	0.00	0.00	0.00	0.00	0.00	1.00	0.00	0.00	0.00	0.00	0.00	0.00	0.00
16	0.06	0.07	0.05	0.03	0.10	0.20	0.00	0.00	0.00	0.10	0.20	0.00	0.00	0.00
17	0.00	0.00	0.00	0.00	0.00	0.00	0.00	0.02	0.03	0.00	0.00	1.00	0.80	0.00
18	0.00	0.00	0.00	0.00	0.00	0.00	0.00	0.00	1.00	1.00	1.00	1.00	0.00	0.00
19	0.00	0.00	0.00	0.00	0.00	0.00	0.00	0.00	1.00	1.00	0.00	0.00	0.00	0.00
20	0.00	0.00	0.00	0.00	0.00	0.00	0.00	0.00	0.00	0.00	0.00	0.00	0.90	0.00
21	0.00	0.00	0.00	0.00	0.00	0.00	0.00	0.00	0.00	0.00	0.00	1.00	0.80	0.90
22	0.00	0.00	0.00	0.00	0.00	0.00	0.00	0.00	0.00	0.00	0.00	0.00	0.00	0.10
23	1.00	1.00	1.00	1.00	1.00	1.00	1.00	1.00	1.00	1.00	1.00	1.00	1.00	1.00

Table 4.   Testing result.

C1	1.0000	0.3117	0.0690	0.0408	0.6082	0.1610	0.0679	0.0561	0.0561	0.6082	0.1610	0.0000	0.0000	0.0000
C2	1.0000	0.3081	0.0822	0.0415	0.6084	0.1665	0.0805	0.0000	0.0000	0.6084	0.1665	0.0000	0.0000	0.0000
C3	0.0344	0.0456	0.2683	0.0620	0.0976	0.2220	1.0000	0.0000	0.0000	0.0976	0.2220	0.0024	0.0020	0.0022

or a cylinder gap or leakage (multiple faults) of 1st class compressor stuffing box. For the data of 3rd group, the diagnosis result is the cooler water system fault. The result of the experiments is consistent with the actual fault and indicates that the fuzzy relation matrix obtained is reasonable and the diagnosis result is effective.

4 CONCLUSION

(1) The fault diagnosis system based on .NET platform achieves good application effect by the test of simulation, field data and fault analysis and realizes fault diagnosis located in wellhead and compressor.

(2) The fuzzy neural network associative memory model (FNNAM) for the compressor fault diagnosis has achieved good effect during the self-learning of fuzzy relation matrix. The FNNAM algorithm depends on the expert experience and self-learning to update the fuzzy relation matrix, and knowledge acquisition bottleneck is solved effectively.

(3) Currently, we have only verified the existed fault. With the passage of time, the service life of the compressor is influenced by utilization times and the fault may increase, so new fault mode needs further verification.

ACKNOWLEDGEMENT

This work was supported by Major National Science and Technology Programs of china in the "Twelfth Five-Year" Plan period (2011ZX05039-3-3).

REFERENCES

[1] Sun Z, Zhang W, Hu B, et al. Geothermal field and its relation with coalbed methane distribution of the Qinshui Basin [J]. Chinese Science Bulletin, 2005, 50(1).

[2] Xiao X, Pan Y, Yu L, et al. Geological controls over coal-bed methane well production in southern Qinshui basin [J]. Journal of Coal Science and Engineering (China), 2011, 17(2).

[3] Lang W, Almbauer R A, Jajcevic D. The Piston Compressor: The Methodology of the Real-Time Condition Monitoring [J]. The International Journal of Multiphysics, 2010, 4(1):65–81.

[4] Li R, Chen J, Wu X. Fault diagnosis of rotating machinery using knowledge-based fuzzy neural network [J]. Applied Mathematics and Mechanics, 2006(1).

[5] Wang Y, Pu L, Feng J, et al. Fault Diagnosis of Compressor Valves based on Wavelet Packets and Neural Network [C]. Engineering Information Institute, 2011.

[6] Liu S, Liu S. An Efficient Expert System for Machine Fault Diagnosis [J]. International Journal of Advanced Manufacturing Technology, 2003, 21(9).

[7] Angeli C. Online expert systems for fault diagnosis in technical processes [J]. Expert Systems, 2008, 25(2).

[8] Zong-Ben Xu, Yee Leung, Xiang-Wei He. Asymmetric Bidirectional Associative Memories. IEEE Transactions on Systems, Man and Cybernetics, 1994, 24(10):1558–1564.

Information Technology and Computer Application Engineering – Liu, Sung & Yao (Eds)
© 2014 Taylor & Francis Group, London, ISBN 978-1-138-00079-7

TVBRT: A time-varying multivariate data visualization method based on radial tree

G.X. Qian
Science and Technology on Information Systems Engineering LAB, National University of Defense Technology, Changsha P.R. China

N.W. Sun
Naval Academy of Armament, Beijing, P.R. China

C. Zhang
Science and Technology on Information Systems Engineering LAB, National University of Defense Technology, Changsha, P.R. China

H.F. Liu
Naval Academy of Armament, Beijing, P.R. China

W.M. Zhang & W.D. Xiao
Science and Technology on Information Systems Engineering LAB, National University of Defense Technology, Changsha, P.R. China

ABSTRACT: The areas of social science, environmental monitoring, financial and economic, health and geographic information have generated a large amount of time-varying multivariate data. With a depth analysis of the time-varying multivariate datasets, we define "time-varying multivariate data with metric properties" and propose a layout algorithm for Radial Tree integrated metric properties called LAMPRT to demonstrate metric properties. And then, we propose another layout algorithm called LOVEBRT to exhibit the time-varying feature. These two algorithms and some human-computer interaction techniques both form the visualization method TVBRT. Case studies demonstrate the effectiveness of this method, showing that the method is more effective in exhibiting more details of the datasets. TVBRT outperforms over other methods in exhibiting data in a hierarchical and more accurate way.

Keywords: Information Visualization; Time-varying; Multivariate Data; Metric Properties; Radial Tree

1 INTRODUCTION

The areas of social science, environmental monitoring, financial and economic, health and geographic information have generated a large amount of time-varying multivariate data. With a depth analysis of the time-varying multivariate datasets, we find that some of the datasets contain only a numerical dimension (variate) and certain categorical variate, and the numerical variate is the "metric" of the categorical variates, its value indicates the specific quantity of a classification determined by the categorical variates. The numerical variate (attribute) is called "metric properties", and correspondingly, this type of datasets is called "Time-varying Multivariate Data with Metric properties (TMDM)". For Example, sales data typically includes the sales of more than one point in time in the economic field. And the data for each point in time, includes multiple variates (type of product, sales locations, sales volume, etc.). Every variate changes over time, and sales volume is the metric property.

Analysis tasks of TMDM often concern about the metric trends of multivariate data over time. If we want to visualize the trends, we need views showing the categorical variate and also the numerical variate, showing multivariate and time-varying characteristics.

Most of the existing methods for visualization of time-varying multivariate data [1][2][3][4] pieced together multivariate visualization view and time-varying visualization view. The multivariate visualization view focuses on building the ability to maintain the original multivariate metadata in low-dimensional topology visual space to assist users in the analysis of the relationship between each data item properties. The time-varying visualization view focuses on showing the evolution between data items. These two views work together to support the analysis of time-varying multivariate data. Through the comprehensive judgment of the plurality of views, we can obtain knowledge of interest. However, this type of methods not only increase the cognitive burden, but also make the analysis of variate element changes over time difficult.

Jimmy Johansson et al. [5][6] proposed a way to use the concept of the polygon rendering data between different time periods trend based on the parallel coordinate system. Such methods overcome the cognitive

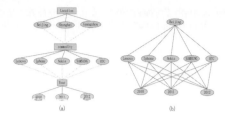

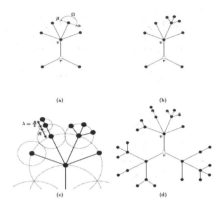

Figure 1. An Example of Sales Data.

burden to a certain extent on the mapping of the two views. But the rendering process often requires more complicated calculation, and the algorithm is inefficient.

Teng-Yok Lee et al. [7] identified trend relationship between variates based on the value changes over time. Cheng-Kai Chen et al. [8] used the sampling data relationship to approximate the relationship among the overall meteorological datasets. Conglei Shi et al. [9] used the number and relative importance of the search terms of search engine, and proposed a new visualization prototype system RankExplorer based on ThemeRiver. Such studies can reduce the number of data items need to be addressed, but can only be used to identify trends of interest or known. It is not conducive to mine implicit knowledge, and also may cause loss of information.

TMDM belong to the category of "mixed data". The mixed data visualization methods based on the existing categorical and numerical data visualization methods need to take some data conversion technology to make the data type of the element unified [10]. The data conversion process will inevitably cause the loss of information, and is not conducive to make a reasonable interpretation of the pattern found in the original dataset.

2 RELATED WORK

2.1 Time-varying multivariate data in a hierarchical way

Most variates of TMDM are categorical variates, and the dataset is actually categorical datasets without considering the metric properties. In fact, if the value of certain variates determined, the categorical datasets can be analyzed in a hierarchical way using these variates as the root. In figure 1 shows the sales data as an example. In figure 1(a) in accordance with the location – commodity – year sequence analysis from the "location" variate. After the value of "Beijing" is selected, we can conduct further analysis of the "commodity" and "year" variate, and then we get figure 1(b), which is a hierarchical view. If we select another value (not "Beijing") of the "location" variate, the hierarchical view we finally get will be different.

2.2 Radial tree

Radial Tree method using radial layout method [13], the root node is drawn at the center of the circle, and the rest of the nodes in accordance with an hierarchy of the type to be placed on the concentric circles of different radius. The depth of the node to the center of

Figure 2. Parent-centered Radial Tree view.

the circle corresponds to the distance of the node. The child nodes of the tree are arranged in concentric radial annulus wedge, each parent node has a respective wedge-shaped range, thereby eliminating the overlap of the sub-tree. The child nodes are arranged at a distance on the root node of a ring, grandchildren near the ring.

Pavlo A et al. [15] proposed the parent-centered Radial Tree layout method. Each node in the view rebuilt the coordinate system using the parent node as the center, and the child nodes are arranged with the parent node as the center of the wedge-shaped arc, as shown in Figure 2. This method can effectively avoid too much confusion in the view, and can clearly show the hierarchical relationships of the dataset.

3 TVBRT: A TIME-VARYING MULTIVARIATE DATA VISUALIZATION METHOD BASED ON RADIAL TREE

3.1 Definition

Learning from the work of SUN Yang et al. [11], we give the following definition.

Definition 1 (multivariate dataset with metric properties)

Set the given dimensional object collection $G(F) = \{F^1, F^2, L, F^m\}$ with k $(k > 0)$ dimensions, and m denotes the number of multi-dimension objects. $F^i = F^i(c_1^i, c_2^i, \ldots, c_{k-1}^i, q^i)$ $(q^i \in R, q^i \geq 0)$ is on behalf of the one-dimensional objects in the collection, and $c_1^i, c_2^i, \ldots, c_{k-1}^i$ indicate the $(k-1)$ dimensions of each object property values, which are all categorical data. q^i means the metric probability or the quantity determined by the data items $c_1^i, c_2^i, \ldots, c_{k-1}^i$. Then $G(F)$ is called a multivariate dataset with metric properties (MDMP). For of multidimensional attribute data table $G(F)$, F^i represents a record, q^i represents its metric.

Definition 2 (metric property)

Given the MDMP $G(F)$ as defined in definition 1, we define k variates of the dataset as $\{V_1, V_2, \ldots, V_{k-1}, Q\}$. V_i $(1 \leq i \leq k-1)$ are all categorical variates of the dataset, and Q is numerical variate, and any value of Q is non-negative real number. Then Q is called the metric property of $G(F)$.

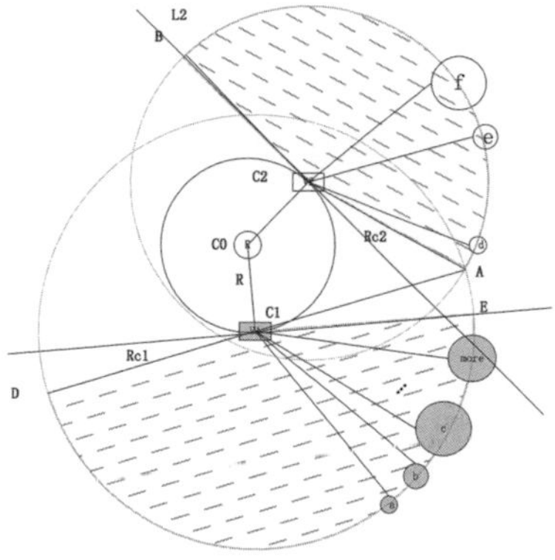

Figure 3.   LAMPRT diagram.

Definition 3 (metric of the value of categorical variate)

Set V_l $(1 \leq l \leq k - 1)$ is a variate of the MDMP $G(F)$, and $\{v_1, v_2, \ldots, v_n\}$ are n values of V_l, then the metric of any of v_i $(1 \leq i \leq n)$ can be calculated by $Q(v_i) = \sum_{j=1}^{m} q^j |_{V_l = v_i}$, meaning $Q(v_i)$ is the sum of all the metric when the value of V_l is v_i. Then $Q(v_i)$ is called the metric of v_i $(1 \leq i \leq n)$, which is one value of the categorical variate V_l.

For the convenience of the bellowing analysis we define the time-varying multivariate dataset with metric properties (TMDMP) as follows:

Definition 4 (TMDMP)

Given the MDMP $G(F)$ and k variates of the dataset as $\{V_1, V_2, \ldots, V_{k-1}, Q\}$. If the value of one of the variates V_i $(1 \leq i \leq k - 1)$ is time steps, then the dataset is time-varying. In this case, $G(F)$ is defined a time-varying multivariate dataset with metric properties (TMDMP).

3.2   LAMPRT: Layout algorithm for visualizing multivariate data with metric properties based on parent-centered radial tree

We use nodes size which is proportional to the value of metric of variates and categorical data to quantify information of the dataset, learning from some related researches [15]. We use rectangles represent the variate nodes, circles represent categorical data nodes. The variate nodes is designed as parent nodes, and the categorical data of the variate is designed as the children nodes of the parent nodes. Both the parent nodes and its children nodes are colored with the same color to maintain the unity of the cognitive, and to avoid possible confusion in the understanding.

In Figure 3, V_1 and V_2 represent two variates, a, b, c, d, e, f represent categorical data. The node size is proportional to the metric of them. C_0 is the root, and calculate the radius R of the first layer of the circle with the principle of not existing overlap among the variates in the first layer. Set C_1, C_2 two root nodes, R_1, R_2 are the radius respectively. Circle C_1 and C_2 intersect at A, F. The layout range of the categorical data, whose root nodes are C_1 and C_2, can be calculated according to the parent-centered layout method,

Algorithm 1   LAMPRT

Step 1. Calculate the metrics of the variates and categorical data, normalize them, and use the normalized value as the area of the nodes in the view. Set the area of the variates $\{V_1, V_2, \ldots, V_n\}$, the area of the categorical data $\{c_1^1, c_2^1, \ldots, c_{m1}^1, c_1^2, c_2^2, \ldots, c_{m2}^2, \ldots, c_{mn}^n\}$. n donates the numbers of the variates, mi donates the numbers of categorical value of the ith variates.

Step 2. Determine the position of the center of the circle C_0 in the view.

Step 3. Given the radius R of the first circle $R \geq 2\pi / \sum_{i=1}^{n} V_i$. Arrange the rectangular nodes on the arc in turn, and donate their position as $\{C_1, C_2, \ldots, C_n\}$.

Step 4. Set the polar coordinate of C_i (the center of a circle) (θ_i, R). Specify the angle range allowing the presence of the layer as Ω_i $(0 < \Omega_i \leq 180)$. If the angle range allowing the categorical data presence using C_i as the parent node is (α, β), then (α, β) can be calculated using the polar coordinate as follows:

$$\alpha = \theta_i + (360 - \frac{1}{2}\Omega_i)$$

$$\beta = \alpha + \Omega_i$$

Step 5. Calculate the radius R_i using C_i as the center of the circle as follows:

if the number of categorical data using C_i as parent node $== 1$ then

$$R_i = \frac{1}{2}R$$

else

$$R_i = \frac{360}{\Omega_i} \frac{1}{2\pi} \sum_{j=1}^{mi} c_j^i + r_i$$
    // r_i: correction value of the radius

end if.

Step 6. Draw an arc in the range of (α, β) using C_i as the center, R_i as the radius. Layout the nodes on the arc in accordance with the area of the nodes.

Step 7. Repeat Step 4~Step 6 until all the categorical data using $\{C_1, C_2, \ldots, C_n\}$ as the center of the circle finishing the layout process.

Step 8. Using a categorical data node as the center of the circle, repeat Step 4~Step 6, arrange the variates in turn, we can get a hierarchy view for further hierarchy information analysis.

recorded as the arc AB and DE. Then the categorical data can be arranged in the order of priority on the arc according to certain rules. The layout process can be completed by repeating the above method. Finally, we proposed the Layout Algorithm for visualizing multivariate data with Metric properties based on Parent-centered Radial Tree (LAMPRT) as is shown in Table 1.

The algorithm requires the user to specify the following parameters: the radius of the first layer circle $R \geq 2\pi / \sum_{i=1}^{n} V_i$, the angle range allowing the presence of the layer Ω_i $(0 < \Omega_i \leq 180)$, and the correction value of the second layer arc's radius r_i.

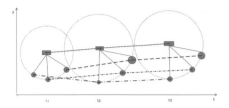

Figure 4. Radial Tree view of the trends of a single argument.

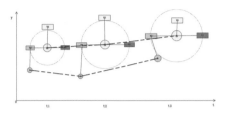

Figure 5. Radial Tree view of the trends of multiple arguments.

These parameters need to be specified according to the datasets and specific visualization process. In all, the time complexity and space complexity of the algorithm are both $O(n)$.

3.3 LOVEBRT: Layout algorithm for visualizing time-varying multivariate data based on Radial Tree

Figure 4 shows the time-varying trend of a single variate and its categorical data, the variates to be analyzed are arranged at the center of the view of each time, and the categorical data in turn arranged in the arc using the variate as the center. And the projector of each variate and categorical data on the vertical axis corresponds to the metric. The trend of the variates and categorical data are demonstrated with four different dotted lines in the figure. We can conclude apparently from the view that the increasing trend of categorical data c is the most obvious, it has the largest metric. So, we can guess c is one of the most important categorical data of the variate V_1.

Figure 5 shows time-varying analysis of multiple variates. Different variates are colored with different color (V_1 with red, V_2 with green, and V_3 with yellow). The variates to be expanded and analyzed are V_2 and V_3, which are linked by the categorical data a, and b is selected as the representative of the variate of V_2. In the view, from t_1 to t_2, the metric of a increases, b decreases, and from t_2 to t_3, the metric of a and b both increases, the increase in amplitude of b is more obvious.

Making a comparison of Figure 4 and Figure 5, we can find that Figure 5 is actually the Figure 4 expanded according to a certain categorical data. Thus, the real problem to solve of the entire layout process is firstly to complete the view of a single variate, and then expend the view in a hierarchy way, and the expended view is the view to analyse multiple variates.

Figure 6 shows the Radial Tree view of a single variate with two time steps. We first draw the Radial

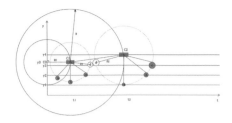

Figure 6. Radial Tree view of a single argument with two time steps.

Tree of t_1 and then the Tree of t_2, and if needed expand it according to the categorical data for the analysis of multiple variates.

On the basis of the above analysis presented and the target task, and the LAMPRT view, we proposed a Layout algOrithm for visualizing time-Varying multivariatE data Based on Radial Tree (LOVEBRT) as is shown in Table 2.

It can be seen from the description of the above algorithm that it focus on determine the center position of the follow-up period Radial Tree, the basic idea is determined by a circle with the point of intersection of the straight line, and simply need to traverse the variates and categorical data, the calculation process is not complicated, and the time and space complexity both are $O(n)$. The numbers of Radial Tree is proportional to the number of time steps. The space complexity is proportional to the number of time steps and the categorical data, therefore $O(n)$. In order to use the minimal number of computing resources in rendering of the view, our algorithm control the number of nodes that need to display.

3.4 Human-Computer Interaction of TVBRT

The Layout algorithm of LAMPRT and LOVEBRT and some human-computer interaction techniques form the visualization method TVBRT (Time-varying multivariate data Visualization Based on Radial Tree). In this section, we introduce the human-computer interaction of our method.

Step 1. Call LAMPRT, integrate the metric of variate and categorical data into Radial Tree view, and construct the Radial Tree view following the principle of showing certain numbers (10, for example) of categorical data in the view.

Step 2. Provide reasonable human-computer interaction mode of operation for each variate and the categorical data.

Step 2.1 The human-computer interaction of a certain variate includes:

(1) Click on the variate rectangle node, the variate moves to the center, the view re-layout.

(2) Double click the variate rectangle node, the variate becomes the center of the view, the view re-layout, and the variate becomes the beginning for analysis.

(3) When the number of categorical data of the variate exceeds the threshold, draw a node 'more', click on the node, then the categorical data of the variate display.

(4) Right-click on the variate rectangle node, the 'time-varying analysis' menu appears, click on the

Table 2. LOVEBRT.

Algorithm 2 LOVEBRT

Step 1. Set the time steps to draw time-varying view $\{t_1, t_2, \ldots, t_n\}$.

Step 2. Call LAMPRT, draw the initial Radial Tree view.

Step 3. Use the left corner of the view as the origin of the coordinate, draw time-metric axes yot. The metric should be normalized.

Step 4. Set r_1 the radius of the largest circular node (the categorical data) at time t_1 of the variate. Set the metric of the variate y_0, the metric of categorical data $\{V_1, V_2, \ldots V_n\}$. Denote $V_{max} = \max(V_1, V_2, \ldots, V_n)$, $V_{min} = \min(V_1, V_2, \ldots, V_n)$, then

$$r_1 = \sqrt{V_{max}/\pi}, \quad R_1 = y_0 - V_{min}.$$

Step 5. Select the node to draw a time-varying view, adjust the view according to the node type based on the Radial Tree layout as follows:

if *the node is variate* then
    reserve the selected node and its children;
if *the node is categorical data* then
    reserve the selected node and its children;
else
    reserve the selected node and its parent node;
end if

Step 6. Draw a straight line l_0 with the y axis of the shaft and cross at the point y_0, denote y_0 as C_0. Draw a circle using C_0 as the center, $R_0 = R_1 + r_1$ as the radius. The circle and the straight line l_0 intersect at C_1. Adjust the rectangular nodes to the point of C_1, and arrange the categorical data on the arc in accordance with the metric.

Step 7. Set the metric of the sub-node elements to show of t_2 $\{cy_1, cy_2, \ldots, cy_n\}$, the metric of the parent node $y_{t2,0}$. Denote $y_{max} = \max(cy_1, cy_2, \ldots, cy_n)$, $y_{min} = \min(cy_1, cy_2, \ldots, cy_n)$. Then the radius of the Radial Tree is $R_2 = y_{t2,0} - y_{min}$.

Step 8. Set r_2 the radius of the largest circular node (the categorical data) at time t_2 of the variate, then $r_2 = \sqrt{y_{max}/\pi}$. Draw a circle using C_1 as the center, $R = R_1 + R_2 + r_1 + r_2 + r_0$ (r_0 is the correction value of the radius). The point of intersection with the straight line $y_{t2,0}$ is the center of the radial diagram of t_2, recorded as C_2.

Step 9. Draw the Radial Tree using C_2 as the center, R_2 as the radius. Then, arrange the categorical data on the arc based on the metric, draw the Radial Tree of t_2.

Step 10. Repeat Step 7 ~ Step 9, until the view of entire time steps completed.

Step 11. For analysis of multiple variates, we first select a certain categorical data, and then expand the variates to be analyzed, all the variates are arranged in the same horizontal axis. Select the variate to be expanded, repeat Step 4~Step 10, until the hierarchy layer view is completed.

menu, then transferred to Step 3 and generates the variates of time-varying trends.

Step 2.2 The human-computer interaction of categorical data concludes:

(1) Click on the categorical data circular node, the node moves to the center, the view re-layout.

(2) Double click the categorical data circular node, the node becomes the center of the view, the view

re-layout, and the node becomes the beginning for analysis. At the same time, the rectangle node on behalf of all the other variates connected with the node appear, they support all interactive operations in Step 2.1.

(3) Right-click on the categorical data circular node, the 'time-varying analysis' menu appears, click on the menu, then transferred to Step 3 and generates the variates of time-varying trends.

Step 2.3 The human-computer interaction of the view concludes:

(1) Provide time steps selection window, the user can either select a specific moment in time, and generate a view of the current time, to facilitate the multivariate analysis, or a specific time period, and you can select more than one time, transferred to the Step 3, and make progressive trends analysis.

(2) Provide variates selection window, the user can select certain number of variates of interest, to generate selected variate time-varying trends, facilitate joint multivariate time-varying trend analysis.

Step 3. Call LOVEBRT, adjust the Radial Tree view to meet the analysis tasks of TMDMP. When the mouse hovers over a node, draw the line in accordance with time-varying trends. At the same time, the view is still have the human-computer interaction method as is Step 2, and the multiple Radial Tree are linkage.

4 CASE STUDY

In this section, we first introduce the dataset and the experimental environment, then verify LAMPRT algorithm and LOVEBRT algorithm for effectiveness, mine implicit knowledge of the experimental dataset with the TVBRT method.

4.1 Dataset, hardware, and software

We work with colleges admissions institution conduct our case study using the colleges admissions dataset. The dataset has accumulated in recent years, a large number of enrollment plan, including 11 variates, 40,000 records, and 8 years. The variates includes year, enrollment institutions, admissions profession, school system, education, gender, art or science, crossed category, student enrollment and so on. Each record corresponds to a specific metric property (variate), the number of admissions. The variate is numerical.

Our experimental hardware environment: Intel(R) Core(TM)2 Duo CPU P8600 2.40 GHz, 2.93 GB RAM, ATI Mobility Radeon HD 3400 Series, 320GB Disk. Software environment are: Microsoft Windows XP Professional SP3, Eclipse, MyEclipse 6.5, JDK1.6, Oracle 9i. We also use the open source software package of Prefuse [25][26].

4.2 Results of visualization

4.2.1 Result of LAMPRT visualization

Figure 7 shows the 2010 enrollment plan data, as an example. 8 variates are displayed, and we set up the number allowed to show in the view. If you want to view the other variates, click on the 'more' node. The lower-left corner of the view shows the information of the hovered node, the lower-right corner of is the search

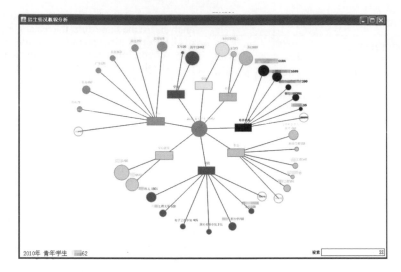

Figure 7. Results of LAMPRT Visualization.

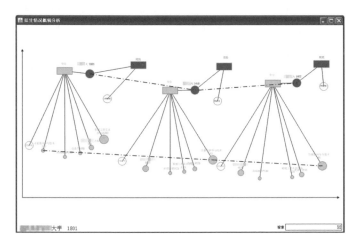

Figure 8. Results of LOVEBRT Visulization.

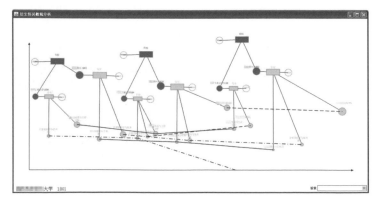

Figure 9. Result of TVBRT Visualization Method.

input box. We can intuitive see from the view that the categorical numbers of institutions, professional, training goals and student enrollment are much large, however the categorical numbers of gender, school system, education, art or science are little.

The variates in Figure 7 are colored with different colors, and the categorical data of the variates are in the same color with the variates they belong to. All the categorical data are on the arc with the variate they belong to as the center. The layout method can effectively save the view space, and can serve as the base for further analysis. Thus, we verify the effectiveness of LAMPRT.

4.2.2 Result of LOVEBRT visualization

Figure 8 shows 5 admissions profession trends using LOVEBRT of × × Technology University from 2010 to 2012, they are computer science and technology, simulation engineering, mechanical engineering and its automation, applied physics, command automation engineering. It can be seen from the view that the number of adminssions reduce significantly between 2010 and 2011, increase between 2011 and 2012. The number of admissions of the five professional research are decreasing. We can also find a very large change in the number of admissions between 2010 to 2011.

Figure 8 is a simple example, which is able to verify the validity of the algorithm. With a simple analysis, you can mine more implicit knowledge. Users can further select more institutions, professional or other variates for further analysis. The view clearly shows the time-varying characteristics, multivariate characteristics, and the relationship between the variates and the categorical data etc. It is possible to satisfactorily support the interactive hierarchical analysis.

4.2.3 Result of TVBRT visualization method

We conduct research using the variates with more categorical, the institutions and profession. For institutions, we select × × Technology University, × × Engineering University. For profession, we select computer science and technology, military command and aircraft engineering. Figure 9 shows the result of TVBRT.

From the view of Figure 9, we can at least get the following conclusions:

(1) The number of admissions two institutions in 2011, has decreased to some extent.
(2) The number of computer science and technology of × × Engineering University from 2010 to 2011 shows a downward trend, the profession is no longer enrollment in 2012.
(3) In 2011, the number of aircraft of × × Engineering University Enrollment decline, but in 2012 and increased to the original level.
(4) The enrollment number of computer science and technology show a downward trend. From 2010 to 2011, a larger decline, and in 2011 and 2012 remained stable.
(5) Starting in 2011 the two institutions to recruit military command specialty, the number of admissions in the × × Technology University was very large, remained stable in the past two years,

× × Engineering University, when the number of 2011 is not big, a substantial increase in 2012.

With a comprehensive analysis of the conclusions, we can find that there are some adjustments in the year 2011, the number of admissions of the relevant institutions, profession varying degrees of decline, indicating that in 2011 there may be some policy changes. Some profession of advantages of some institutions in the number of admissions in 2012 increase to the level before 2011, which shows the entire enrollment plan taking count of to relevant institutions the advantages of disciplines, fully illustrated the enrollment plan to be scientific and rational Universal profession (computer science and technology, as an example) began to be concentrated in certain institutions, those do not have the discipline advantage is no longer enrollment.

5 DISCUSSION

The main features of the TVBRT main features include:

(1) To the Supporting Analysis Tasks Aspect
    The method focuses on hierarchy analysis of time-varying multivariate data, can flexible mine implicit knowledge according different analysis orders of the datasets. However, the relationship of metric between variates or between categorical data is shown too little.
(2) To the Supporting Characteristics of the Dataset Aspect
    With the metric of different variates and categorical data shown in the same coordinate, the comparison of metric can be clearly demonstrated. The time-varying trends can be demonstrated. However, too many Radial Tree views may cause too much computer resourse used. It is not so useful when there are too many time steps in the dataset.
(3) To the Features of the View Aspect
    The view adjustment is very flexible and the element most concerned can be at the center of the view, elements of interest can be at more obvious location. However, the users have to specify some certain initial parameter values (as in LAMPRT), which causes some subjective factors introduced into the method.

6 CONCLUSION AND FUTURE WORK

We proposed a layout algorithm for Radial Tree integrated Metric properties called LAMPRT to demonstrate Metric properties. And then, we proposed another layout algorithm called LOVEBRT to exhibit the time-varying feature. These two algorithms and some human-computer interaction techniques both form the visualization method TVBRT. Case studies demonstrated the effectiveness of this method, showing that the method is more effective in exhibiting more details of the datasets.

Visualization methods for TVBRT will also need to conduct a study in how to be more efficient rendering view (especially when the time node is further increased, how the performance of state characteristics), to improve the efficiency of the algorithm, to further improve the human-computer interaction.

ACKNOWLEDGEMENT

This work is supported by the project of Growing Cadets Training System. Also, thanks to corresponding author PhD. Chong ZHANG.

REFERENCES

[1] Diansheng Guo, Jin Chen, Alan M, MacEachren, Ke Liao. A Visualization System for Space-Time and Multivariate Patterns (VIS-STAMP). IEEE transactions on visualization and computer graphics, vol. 12, no. 6, November/December 2006.

[2] Diansheng Guo, Ke Liao, Michael Morgan. Visualizing patterns in a global terrorism incident database. Environment and Planning B: Planning and Design 2007, volume 34, pages 767–784.

[3] Hiroshi Akiba, Kwan-Liu Ma. An Interactive Interface for Visualizing Time-Varying Multivariate Volume Data. APVIS 2007.

[4] Hiroshi Akiba, Kwan-Liu Ma. A Tri-Space Visualization Interface for Analyzing Time-Varying Multivariate Volume Data. IEEE-VGTC Symposium on Visualization 2007.

[5] Jimmy Johansson, Patric Ljung and Matthew Cooper. Depth Cues and Density in Temporal Parallel Coordinates. In Proceedings of Eurographics/ IEEE VGTC Symposium on Visualization, 35–42. Norrköping, Sweden, 2007.

[6] Jimmy Johansson. Efficient Information Visualization of Multivariate and Time-Varing Data [M]. Linköping Studies in Science and Technology Dissertations, No. 1191, 2008.

[7] Teng-Yok Lee, Han-Wei Shen. Visualization and Exploration of Temporal Trend Relationships in Multivariate Time-Varying Data. IEEE Transactions on Visualization and Computer Graphics, Vol. 15, No. 6, November/December 2009.

[8] Cheng-Kai Chen, Chali Wang, Kwan-Liu Ma, Andrew T. Wittenberg. Static Correlation Visualization for Large Time-Varying Volume Data. IEEE Pacific Visualization Symposium 2011. Hong Kong, China.

[9] Conglei Shi, Weiwei Cui, Shixia Liu, Panpan Xu, Wei Chen, and Huamin Qu. RankExplorer: Visualization of Ranking Changes in Large Time Series Data. IEEE Transactions on Visualization and Computer Graphics (InfoVis 2012). 2012.

[10] Greenacre M J. Correspondence analysis in practice[M]. Chapman& Hall, 2007.

[11] SUN Yang, Tang Jiu-Yang, Tang Da-Quan, Xiao Wei-Dong. Improved Multivariate Data Visualization Method. Journal of Software (in Chinese), Vol. 21, No. 6, June 2010, pp. 1462–1472.

[12] WANG Wei-Xin, MING Chun-Ying, WANG Hong-An, DAI Guo-Zhong. Visualization of Hierarchical Information Based on Venn Diagrams[J].

Chinese Journal of Computer (in Chinese), 2007(9):1632–1637.

[13] Xiao Wei-Dong, Sun Yang, Zhao Xiang, Zhou Cheng, Feng Xiao-Sheng. Survey on the Reaserch of Hierarchy Information Visualization [J]. Journal of Chinese Computer Systems (in Chinese), 2011(1):137–146.

[14] Michael Burch, Daniel Weishopf. Visualizing Dynamic Quantitative Data in Hierarchies-TimeEdgeTrees: Attaching Dynamic Weights to Tree Edges. IVAPP 2011.

[15] Pavlo A, Homan C, Schull J. A parent-centered radial layout algorithm for interactive graph visualization and animation. 2006.

[16] Svetlana Mansmann, Marc H. Scholl. Exploring OLAP Aggregates with Hierarchical Visualization Techniques. SAC'07, March 11–15, 2007, Seoul, Korea.

[17] Ka-Ping Yee, Danyel Fisher, Rachna Dhamija, and Marti A. Hearst. Animated exploration of dynamic graphs with radial layout. In Proceedings of the IEEE Symposium on Information Visualization, pages 43–50, 2001. URL citeseer.ist.psu.edu/article/yee01animated.html.

[18] G. Book and N. Keshary. Radial Tree graph drawing algorithm for representing large hierarchies. University of Connecticut, December 2001.

[19] Sheth N, Cai Q. Visualizaling mesh dataset using radial tree layout [R/OL]. Spring 2003 Information Visualization Class Project, Indiana University, http://iv.slis.indiana.edu/sw/papers/radialtree.pdf, 2003.

[20] Michael Douma, Grzegorz Ligierko, Ovidiu Ancuta, Pavel Gritsai, and Sean Liu. Spicy Nodes: Radial Layout Authoring for the General Public. IEEE Transactions on Visualization and Computer Graphics, Vol. 15, No. 6, November/December 2009, pp. 1089–1096.

[21] Card, S. K. Suh, B., Pendleton, B. A. Heer, J. and Bodnar, J. W. (2006). Time Tree: Exploring time changing hierarchies. In IEEE Symposium on Visual Analytics Science and Technology (VAST'06), pages 3–10.

[22] Tu, Y. and Shen, H.-W. (2007). Visualizing changes of hierarchy information using treemaps. IEEE Transactions on Visualization and Computer Graphics, 13(6):1286–1293.

[23] Burch, M., Beck, F. and Diehl, S. (2008). Timeline Trees: Visualizing sequences of transactions in information hierarchies. In International Working Conference on Advanced Visual Interfaces (AVI'08), pages 75–82.

[24] Burch, M. and Diehl, S. (2008). Time Radar Trees: Visualizing dynamic compound digraphs. Computer Graphics Forum, 27(3):823–830.

[25] http://prefuse.org

[26] Heer J, Card S K, Landay J A. Prefuse: a toolkit for interactive information visualization [C]. In: Proceedings of the SIGCHI conference on Human factors in computing systems. New York: ACM Press, 2005:421–430.

Information Technology and Computer Application Engineering – Liu, Sung & Yao (Eds)
© *2014 Taylor & Francis Group, London, ISBN 978-1-138-00079-7*

Effects of branch length and terminal impedance in multi-branch power line channel

YanHua Zheng
School of Electronic and Information Engineering, South China University of Technology, Guangzhou, China
School of Physics and Electronic Engineering, Guangzhou University, Guangzhou, China

GaoYong Luo & BingZhi Zhang
School of Physics and Electronic Engineering, Guangzhou University, Guangzhou, China

Qun Yu
School of Electronic and Information Engineering, South China University of Technology, Guangzhou, China

YongLing He
College of Medical Information Engineering, Guangdong Pharmaceutical University, Guangzhou, China

ABSTRACT: This paper presents the effects of branch length, number of branch and tap, and branch terminal impedance on the performance of indoor Low-Voltage (LV) power line channel. The two types power line structures, named the One-Tap with Multi-Branch (OTMB) and the Multi-Tap with Multi-Branch (MTMB), are studied. We simulate the transfer characteristic of two structures when the length and terminal impedance of branch vary. Our simulation results suggest that these aforementioned factors affect the amplitude and phase response of power line channel deeply. The position and number of notches and peaks in amplitude response are affected by the branch parameters and topology. This enables us to calculate the accurate channel response of the different channel branch structures in analyzing in-home LV power line grid as communication channel.

Keywords: Power Line Communication (PLC); One-Tap with Multi-Branch (OTMB); branch terminal impedance; Multi-Tap with Multi-Branch (MTMB)

1 INTRODUCTION

The LV power line as an alternative communication channel for speed over Mb/s were confirmed. However, unlike other existing and mature wired communications such as unshielded twisted pair, coaxial and optical fiber communication, the enviroment in PLC is extremly harsh. Namely, the PLC faces serious attenuation, various noises, stochastic phase shift, uncertain multipath interferences and so on. All the above factors cause austere challenge to the PLC. So, the priori of using the power line as the communication medium is familiar with the charicteristic of power line channel clearly and fully.

Fortunately, there are many researchers and academics have studied the characteristic of PLC channel. Holger Philipps analyzed the characteristics of transfer function and impulse response when the length and terminal impedance of branch is fixed and one tap with one branch configuration in PLC channel [Holger P. et al. 1999]. Manfred Zimmermann etc. provided the amplitude response, phase response and impulse response of power line with one node attaching one branch and fixed impedance [Manfred Z. et al.

2002]. H. Meng etc. presented the amplitude response and phase response in indoor power line channel [H. Meng et al. 2004]. Fabio Versolatto etc. descriped the insertion loss, phase characteristic and magnitude response in a gived three wires power line structure [Fabio V. et al. 2011]. Lars T. Berger etc. analyzed the power line transfer characteristic and offered the channel gain and impulse response in a fixed branch length and branch terminal impedance condition [Lars T. B. et al. 2009]. Anatory Justinian etc. studied the effect of the load impedance, line length, and number and length of branch in power line grid, while did not concern the number of tap and the variation of branch terminal impedance [Anatory J. et al. 2009]. From the above literature, we can know that the existed researches of transfer characteristic for power line channel mainly concern the simplest branch structure of one tap with one branch or a few taps with one branch each and the fixed length and terminal impedance of branch, while seldom study the effect of more complex branch configuration and varing length and terminal impedance of branch in power line network.

In the paper, the transmission line and chain matrix theories are exploited to analyzing transfer characteristic of the power line channel. Two kinds of configuration in power grid that is the OTMB and MTMB are studied. The amplitude response and phase response are simulated in the frequency of 1 Hz to 30 MHz. How do the length and terminal impedance of branch affect the transfer characteristic are also researched. The observations presented in this paper can be useful when designs a suitable power line communication system.

This paper is arranged as follows. The typical configuration of power line network in home and primary parameters of power cable are introduced in part 2. In part 3 and part 4, the characteristic of OTMB and MTMB power line grid are presented, respectively. The conclusion is drawn in last part 5.

2 THE POWER LINE GRID AND PRIMARY PARAMETERS

2.1 The power line grid

The actual power line grid is demonstrated in figure 1a and the corresponding abstracted structure is showing in figure 1b. From figure 1, we can see that there are many branches and sub-network structures from transmitter to receiver. There are many branches connecting to one tap. The branch can be attached by numerous other branches and loads and so on. The specification of branched line and the main trunk can be the same or not. The typical appliances in house are light, fan, computer, air-conditioner, refrigerator and TV, etc. The accessing or disconnecting of appliances will lead to the variation of branch length and branch terminal load in power line net-work. This can bring some effects to the channel characteristic.

2.2 The primary parameters

In common, the signal transmitting in the two lines of power cable and this situation can be simply considered as transmission line (TL) model.

In TL model, the two power lines can be modeled as figure 2. Here, the symbols of R, L, C and G indicate the resistance per unit length for both conductors (in Ω/m), the inductance per unit length for both conductors (in H/m), the capacitance per unit length (in F/m), the conductance per unit length (in S/m), respectively. The $v(z, t)$ and $v(z + \Delta z, t)$ are the voltages at the position of z and the position of $z + \Delta z$ at time t, while the $i(z, t)$ and $i(z + \Delta z, t)$ denote the currents at the position of z and the position of $z + \Delta z$ at time t. The Δz indicates the small distance from the place z.

The two intrinsic parameters of characteristic impedance Z_0 and transmission constant γ equalling to $\alpha + j\beta$ can be written as

$$Z_0 = \sqrt{(R + j\omega L)/(G + j\omega C)} \qquad (1)$$

$$\gamma = \sqrt{(R + j\omega L)(G + j\omega C)} = \alpha + j\beta \qquad (2)$$

where ω is the signal angular frequency transmitting in the power line; the real part α and imaginary part β of

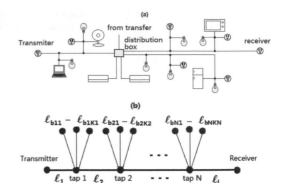

Figure 1. Common LV power line grid in house. (a) A practical structure. (b) Abstracted structure.

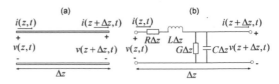

Figure 2. Power line. (a) Voltage and curent definition. (b) The lumped-element circuit.

the propagation constant γ are the attenuation constant and phase constant, respectively.

The BV4 power line with copper core of $4\,\text{mm}^2$ cross-sectional area and PVC insulator is studied.

3 OTMB CONFIGURATION

3.1 One-tap with multi-branch (OTMB) structure

As shown in figure 1b, the branches connecting to one tap are different length of l_{bi} and terminal impedance of Z_{bi} for OTMB. The K branches can be substituted for one equivalent impedance Z_{inb} attached at the tap

$$Z_{inb} = Z_{inb1} // Z_{inb2} // ... // Z_{inb(K-1)} // Z_{inbK} \qquad (3)$$

where Z_{inbk} $(k = 1, 2, \ldots, K)$ is the input impedance for kth branch at tap

$$Z_{inbk} = Z_{0k} \frac{Z_{bk} + Z_{0k}\tanh(\gamma_{bk}l_{bk})}{Z_{0k} + Z_{bk}\tanh(\gamma_{bk}l_{bk})} \qquad (4)$$

where the Z_{0k} and γ_{bk} mean the characteristic impedance and transmission constant of the kth branch, respectively.

For OTMB configuration, the TL equivalent circuit is showing in figure 3. Here, the E_S and Z_S mean the voltage and impedance of source, respectively. The l_S and l_L are the length from source to tap and from tap to load, respectively. The Z_b and Z_L define the branch terminal impedance and the load impedance, respectively.

The OTOB power line grid can be divided into three parts when used chain matrix theory. The first part T_1

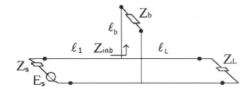

Figure 3. The TL model for one branch circuit.

is the Z_S and E_S, The second one T_2 is from source to the tap including l_1, l_b and Z_b and the third part T_3 is the right of tap including l_L and Z_L. The chain matrix of each part can be gotten

$$T_0 = \begin{bmatrix} 1, & Z_s \\ 0, & 1 \end{bmatrix} \tag{5}$$

$$T_1 = \begin{bmatrix} \cosh(\gamma_1 l_1) + \dfrac{Z_{01}\sinh(\gamma_1 l_1)}{Z_{inb}}, & Z_{01}\sinh(\gamma_1 l_1) \\ \dfrac{1}{Z_{01}\sinh(\gamma_1 l_1)} + \dfrac{\cosh(\gamma_1 l_1)}{Z_{inb}}, & \cosh(\gamma_1 l_1) \end{bmatrix} \tag{6}$$

$$T_2 = \begin{bmatrix} \cosh(\gamma_L l_L) + \dfrac{Z_{0L}\sinh(\gamma_L l_L)}{Z_L}, & Z_{0L}\sinh(\gamma_L l_L) \\ \dfrac{1}{Z_{0L}\sinh(\gamma_L l_L)} + \dfrac{\cosh(\gamma_L l_L)}{Z_L}, & \cosh(\gamma_L l_L) \end{bmatrix} \tag{7}$$

where the Z_{01} and Z_{0L} mean the characteristic impedance of the line l_1 and l_L, respectively; the γ_1 and γ_L mean the transmission constant of the line l_1 and l_L, respectively.

The chain matrix T of whole circuit can be presented

$$T = \prod_{i=0}^{2} T_i = \begin{bmatrix} T_{11}, & T_{12} \\ T_{21}, & T_{22} \end{bmatrix} \tag{8}$$

The transfer function for OTOB power grid can be gained from the relation of chain matrix and transmission line theories

$$H(f) = \frac{Z_L}{T_{11}Z_L + T_{12} + T_{21}Z_S Z_L + T_{22}Z_S} \tag{9}$$

When there are three branches at the tap, the amplitude and phase characteristic of tramsfer function for OTMB power line grid can be gained in figure 4. Both the amplitude response and phase response of transfer function for OTMB power line decrease nonlinearly with frequency increasing. The amplitude response has many notches and the minimum of notch is less than -25 dB.

In the OTMB situation, both the branch length and branch terminal impedance can affect the transfer characteristic of power line channel. The first branch l_{b1} and its terminal impedance Z_{b1} are selected as studying objects. When the l_{b1} changes from 0 to 1000 meters and the other parameters unchange, the relation of minimum and maximum amplitude response and branch length can be gained in figure 5a. Similarly, when the parameter of Z_{b1} varies from 0 Ohm to 1000 Ohms and the other values remain unchanged, the relation of minimum and maximum amplitude

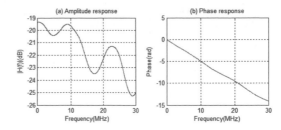

Figure 4. Transfer characteristic of one tap with three branches. (a) Amplitude characteristic. (b) Phase response.

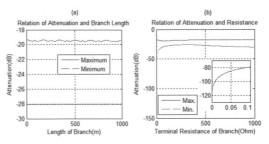

Figure 5. Minimum and maximum of transfer function in OTMB when: (a) branch length varies. (b) branch terminal impedance changes.

response and branch terminal impedance is presented in figure 5b.

In figure 5a, we can get that the maximum amplitude of transfer function vibrates and is ~-19.5 dB while the minimum one is nearly constant of ~-28 dB.

When branch length changes from 0 meter to 1000 meters. It is shown in figure 5b that the maximum amplitude response in OTMB structure is from ~-17.2 dB to ~-19.8 dB, while the minimum one increases from ~-121.5 dB to maximum ~-25.4 dB firstly but decreases soon after with the frequency increasing when terminal impedance changes from 0 Ohm to 1000 Ohms.

4 MTMB CONFIGURATION

In MTMB condition, there are N taps and K_n ($n = 1, 2, \ldots, N$) branches with n^{th} tap. The K_n branches at nth tap are different characteristic with length of l_{bnk} ($k = 1, 2, \ldots, K_n$) and terminal impedance of Z_{bnk} ($k = 1, 2, \ldots, K_n$). The input impedance of K_n branches and terminal impedance at nth tap can be synthesized to one equivalent impedance Z_{inbn}

$$Z_{inbn} = Z_{inbnK_1} // Z_{inbnK_2} // \ldots // Z_{inbn(K_{n-1})} // Z_{inbnK_n} \tag{10}$$

where the Z_{inbnk} ($k = 1, 2, \ldots, K_n$) notes the input impedance for kth branch at the nth tap

$$Z_{inbnk} = Z_{0nk} \frac{Z_{bnk} + Z_{0nk}\tanh(\gamma_{bnk}l_{bnk})}{Z_{0nk} + Z_{bnk}\tanh(\gamma_{bnk}l_{bnk})} \tag{11}$$

where the Z_{0nk} and γ_{bnk} mean the characteristic impedance and transmission constant of the kth branch at nth tap, respectively.

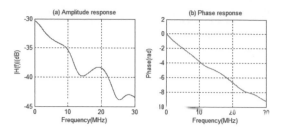

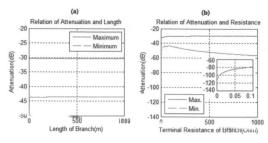

Figure 6. Transfer characteristic of MTMB. (a) Amplitude response. (b) Phase characteristic.

Figure 7. Minimum and maximum of transfer function in MTMB when: (a) branch length varies. (b) branch terminal impedance changes.

The N-tap with multi-branch circuit can be divided into $N + 2$ parts as shown in figure 1b and figure 3. The 1st part is the source of Z_S and E_S. The 2nd part is from the right of source to the left of line l_2 with line l_1 and all the lines and loads attached to 1st tap. The 3rd part is the right of 1st tap and left of line l_3 including line l_2 and all the lines and loads connected to 2nd tap and so on. The $N + 1$ part is from the right of $(N - 1)^{th}$ tap and left of line l_L including the line l_N and all the lines and loads attached to N^{th} tap. The $N + 2$ one is the right of Nth tap including the line l_L and impedance Z_L.

The chain matrices of first and last part are the same as in OTMB. The else N parts are

$$
T_{n+1} = \begin{bmatrix} \cosh(\gamma_n l_n) + \dfrac{Z_{0n}}{Z_{inbn}}, & Z_{0n} \sinh(\gamma_n l_n) \\ \dfrac{1}{Z_{0n} \sinh(\gamma_n l_n)} + \dfrac{\cosh(\gamma_n l_n)}{Z_{inbn}}, & \cosh(\gamma_n l_n) \end{bmatrix} \tag{12}
$$

where the Z_{0n} ($n = 1, 2, \ldots, N$) and γ_n mean the characteristic impedance and transmission constant of line l_n, respectively.

The expression of chain matrix and the transfer function for whole MTMB power line grid are similar to OTMB configuration showing in expressions 8 and 9.

A three taps power line grid is studied. There are two, three and four branches connected to first, second and third tap, respectively. The characteristic of amplitude response and phase response for MTMB structure is presented in figure 6. Both the amplitude and phase response in MTMB decrease nonlinearly with frequency increasing but have many notches. The minimum of amplitude response is less than -43 dB.

In the MTMB situation, the maximum and minimum of amplitude response are shown in figure 7 when branch length or branch terminal impedance changes. We select the branch l_{b11} and the branch terminal impedance Z_{b11} as researching objects.

In figure 7a, we can get that the maximum amplitude of transfer function is ~ -30.4 dB while the minimum amplitude of transfer function vibrates and is ~ -43.5 dB when branch length changes from 0 meter to 1000 meters. It is shown in figure 7b that the maximum of amplitude response for power line grid changes from ~ -31.9 dB to ~ -29.8 dB while the minimum one increases from ~ -126.3 dB to the maximum of ~ -43.9 dB and then decreases

with frequency increasing when terminal impedance changes from 0 Ohm to 1000 Ohms.

5 CONCLUSION

In this paper, we have presented an approach to compute the channel transfer function of PLC network with LV indoor scenario aided of TL and chain matrix theories. This method bases on computing the ABCD matrix via a sub-network dividing approach which is applicable to complex in-home power line network that exhibit many taps and branches. We have addressed the analytical modelling of cables with OTMB and MTMB configurations. Then, we get the amplitude response and phase response via software simulation when the branch length and branch terminal impedance change. Finally, the maximum and minimum amplitude of transfer function are presented in varing length and impedance.

ACKNOWLEDGEMENT

This work was supported by the Natural Science Foundation of China (60971039) .

REFERENCES

Anatory J. et al. 2009. The Effects of Load Impedance, Line Length, and Branches in Typical Low-Voltage Channels of the BPLC Systems of Developing Countries: Transmission-Line Analyses. *IEEE Trans. Power Del.*, 24(2):621–629.

Fabio V. et al. 2011. An MTL Theory Approach for the Simulation of MIMO Power-Line Communication Channels. *IEEE Trans. Power Del.*, 26(3):1710–1717.

H. Meng et al. 2004. Modeling of Transfer Characteristics for the Broadband Power Line Communication Channel. *IEEE Trans. Power Del.*, 19(3):1057–1064.

Holger P. et al. 1999. Modelling of Powerline Communication Channels. Proceedings of the 1999 International Symposium on Power-Line Communications, Lancaster, UK: 14–21.

Lars T. B. et al. 2009. Power Line Communication Channel Modelling Through Concatenated IIR-Filter Elements. *JOURNAL OF COMMUNICATIONS*, 4(1):41–51.

Manfred Z. et al. 2002. A multipath Model for the Powerline Channel. *IEEE Trans. on communications*, 50(4).

Development of android software update system based on web service

Ran Wang, Xiang Dong You & Qing Nan Chang
Beijing University of Posts and Telecommunications, Beijing, China

ABSTRACT: The aim of this paper is to combine WebService technology with android platform technology to develop application update system. In this update system, the Android application would interact with WebService server to get information of whether to upgrade, and display related information returned from WebService on the android device. This article consists of two parts: WebService and android application client. WebService checks the version and compares to determine whether to update this APP; Android client is responsible for downloading, installing and displaying progress bar for updating. This article also shows Handler asynchronous processing for UI updating display and Ksoap2 android open source library to parse WebService data. As the result, this solution is proved a reliable and practical way to download and update through internet after testing on different Android terminals.

Keywords: WebService; Android; App update; Handler; Ksoap2

1 INTRODUCTION

With the development of 3G/4G, mobile internet takes more important role in our life. Android has become the mainstream mobile. Apps for Android has to update regularly and continuously for bugs and optimal performance, this article introduce a new solution for updating based on WebService framework which is different from the traditional solutions like Http Service. Compared with Http protocol SOAP protocol is based on XML, it achieves an improved packing and coding way for internet transportation and enriches the date types exchanging between service and client[1].

This article describes a framework of WebService and Android client's data interaction based on the SOAP protocol. Additionally, it introduces Android asynchronous Handler mechanism, android software installation and update processes in details.

2 OVERALL ARCHITECTURE

The system mainly consists of the software update server based WebService, We can see it in Figure 1. Android client and file download server. We should start the server part to wait for client to call the Webservice method and download the new version application package, During the Android client launching time, firstly it will call the WebService's update method and upload client's current software version as the method's input value. Then the method will return

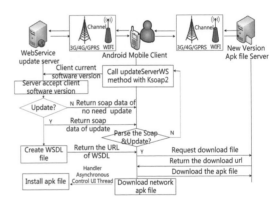

Figure 1. Overall Architecture.

SOAP data to show whether the client needs to update. If the client needs to update, WebService will return apk download address to access the latest software package. Secondly, if the client needs to update, Android main thread will show the dialog to let the users interact with and display the progress bar to show the file download condition. Thirdly, use android's Intent to install new version APK file.

3 WEBSERVICE

WebService publishes the existing applications into the open service, to eliminate the implementation

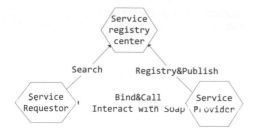

Figure 2. SOA architecture.

differences between different languages and different platforms, which will allow any platform on the Internet, the application of any location and any language to access the service. WebService uses SOA architecture[2] (Service Oriented Architecture), and mainly involves SOAP (Simple Object Access Protocol, the simple object access protocol), WSDL (Web Service description language), UDDI (Universal Description, Description and Integration). WSDL is used to describe the service, UDDI is used for service registration and lookup SOAP, located at the transport layer, is used to transmit messages between service providers and service users.

Service provider: it accepts and executes a request from a requestor, it is also an entity that can be addressable through the internet. Service provider uses WSDL and WSFL to describe its services and interface contract, then the UDDI will publish these services to a service registry to register the service interface so that the service requestor can discover and access these services in the service registry.

Service requestor: it uses the UDDI to send the queries request to service registry center to get the service interface description file, then through binding to these services it can complete the service calling function. Service requestor uses SOAP to interact with Service provider to send and get the message.

Service registry center: It contains the available services repository and uses the UDDI to register these service interfaces. It allows the service requester to find the needed Service Provider Interface.

4   SERVICE MODULE

4.1   *Service software environment*

WebService is built in JDK +Eclipse (J2ee) +Axis2 +Tomcat framework[3].

Tomcat is a Servlet container developed by Apache software fund with the support of JavaServer Page (JSP). Tomcat consists of a HTTP service, so it would run as an independent Web service.

Axis can be considered as a SOAP engine with the function of the Service client. Axis2 is also a simple standalone service which can be inserted in the Servlet engine (Tomcat).

4.2   *API for service update*

4.2.1   *Parameters for upgrading*

The version information for every Android APK is defined in AndroidManifest.xml as follow.

```
<manifest
xmlns:android="http://schemas.android.com/
apk/res/android"
  package= "com.example.updatews"
  android: versionCode ="1"
  android:versionName="1.0" >
  <uses-sdk android:minSdkVersion="8"
  android:targetSdkVersion="17" />
</manifest>
```

Android: versionCode describes the version code, android: versionName describes the version name. Version Code belongs to integral number and version-Name belongs to strings number. Normally, version-Name is shown to the APP user which is not designed for upgrade check. Obviously comparing the version-Codes is a better way for upgrading confirmation. So comparing the version code in service and the version code sent by client is preferred to be the way to confirm the upgrade.

4.2.2   *API in WebService*

Service checks the versionCode via API in WebService sent by client. It returns the Codename, Code version, downloading URL of the APK upgrade package and describes information when the upgrading is confirmed otherwise it returns the void information to the client. Specific is defined as following:

Function for input: versionCode represents the current Code

UpdateSoftwareResp updateSoftware(int version-Code);

Functions for return:

UpdateSoftwareResp = {String newVersionName, int newVersionCode, String apkName, String down-loadURL, String displayMessage}

Description: Android application device will call the function of the WebService to check if the upgrade is necessary after login.

newVersionName: name of APK

newVersionCode: version code of APK

downloadURL: URL address of APK upgrade package

displayMessage: upgrade information

4.2.3   *Comparing results*

The results of comparing newVersionCode in service and versionCode in client are as following.

5   INTRODUCTION FOR CLIENT MODULE

5.1   *Invoking Android WebService*

5.1.1   *Ksoap2*[4]

WebService is a remote invoking protocol based on SOAP, it is compatible with any OS, any languages,

Table 1. WebService newVersionCode comparison and return lists.

versioncode comparing	Service return lists
Less: upgrade is necessary	newVersionCode: version code of the new version newVersionName: version name of the new version downloadURL: URL address of APK upgrade package displayMessage: upgrade information
Equal: upgrade is necessary	newVersionCode: version code of the current version newVersionName: version name of the current version downloadURL: null displayMessage: Already the latest version
VersionCode is 0: updating error	displayMessage: Displying upating errors

any technology. On Java platform in PC client, there needs some libraries like XFirc, Axis/Axis 2, CXF to visit WebService, however, Android SDK doesn't provide the direct way to call WebService library, so it is necessary to call the WebService via the third party library like KSOAP2. Among them, KSOAP2-android.jar provides a lightweight SOAP client library for the Android platform, at the same time Android need to download the soap development kit and add to the Android project.

5.1.2 Process for calling WebService

Create a SoapObject object, specify the namespace (WSDL can be checked) and WebService method name:

SoapObject request = new SoapObject(namespace, methodName);

When the called WebService method needs paraments, the parament is set like:

request.addProperty("method name", "value")

Set SOAP to send the request information with SOAP protocol version

SoapSerializationEnvelope envelope = new SoapSerializationEnvelope(SoapEnvelope.VER11); envelope.bodyOut=request;

Send SOAP request information

envelope.setOutputSoapObject(request);

Create the Http TransportSE object, and indicates the URL of WSDL

HttpTransportSE aht = new HttpTransportSE(Uri);

Call WebService (type1:namespace+method name; type 2: Encelope Object)

aht.call(SOAP_ACTION,envelope);

Parse the returned data

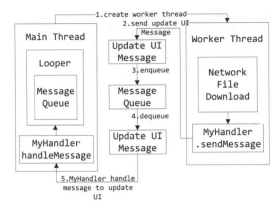

Figure 3. Handler & UI display.

SoapObject response = (SoapObject)envelope.bodyIn;

5.2 Android UI and handler module

5.2.1 Handler introduction

In Android, when an application is launched, the system creates a thread of execution for the application, called main Thread or UI Thread. Android uses the single thread model, when the app performs intensive work in response to user interaction, this model can yield poor performance. If everything is happening in UI thread, performing long operations such as network access or database queries will block the whole UI, so we must dispose them in the worker thread. Additionally, the Android UI toolkit is not thread-safe, so we must not manipulate the UI from the worker thread, we must do all manipulation in the UI thread[5].

This paper uses Handler to deal with network files downloading module. Handler is an asynchronous message processing mechanism. It allows us to perform asynchronous work on the user interface. It performs the blocking operations in a worker thread and then publishes the results on the UI thread. Handler has bridged the gap between the worker thread and the UI thread.

In Android, the system automatically creates a Looper object for the main thread and the Looper object corresponds to a MessageQueue. The MessageQueue is used to store Message performed in accordance with the FIFO (first in first out) queue principle. The Looper object opens a message loop for the thread to handle the MessageQueue.

5.2.2 Handler for upgrading UI performance

This module includes main thread, worker thread and handler. Firstly, Handler object is defined in the main thread. It performs the blocking operations in a worker thread and then publishes the results on the main thread. Secondly, the worker thread is mainly responsible for the file download from the web server. In the worker thread, we use the variable "count" to record the bytes of reading and the variable "length" to record the total bytes of the whole file. The progress bar will update the progress display through the ratio with count among length. When the handleMessage function accepts the message from the main thread's messageQuene, it will execute different disposal. When the worker thread's message is 1, Handler will call set progress function to send progressing information to main thread for displaying; when the download is finished, worker thread's returned message is 0, the main thread will terminate the processing bar widget and invoke Android's installation sub function.

5.3 Install APK of Android

After downloading the upgrade package, Android's Intent mechanism[6] will be used to call Android's installing function to reinstall the whole APK.

(1) get the downloaded APK package's name in the SDCard:
   String fileName = Environment.getExternalStorageDirectory() + apkName;
(2) get the file Uri's storage path
   Uri uri = Uri.fromFile(new File(filename));
(3) Define Intent's target and set the attribute of Action.
   Intent intent = new Intent(Intent.ACTION_VIEW);
(4) Set the attribute of Intent's Date
   intent.setDataAndType(Uri,application/vnd.android.package-archive");
(5) Initiate the application according to the attribute of Action and Date
   startActivity(intent);

6 CONCLUSION

This paper gives a framework of web Service and android client's data interaction based on the SOAP protocol. Additionally it introduces the Android asynchronous Handler mechanism and android software update process. The paper focuses on introduction process of android's calling WebService via the third KSoap2 library. It also gives an introduction about handler's sending and handling message between worker thread and main thread to display the update UI. During the test, the system is of good stability, timely response and friendly interface. Web Service based on mobile terminal data interaction has a further development on entrepreneur's distributed system industry of communication and transportation etc.

REFERENCES

[1] J.B. Bai, "Study on Integration Technologies of Building Automation Systems based on Web Services". Hainan: Shanya, 2010, pp. 68–79.
[2] C. Nizamuddin, "Comprehensive Framework for Semantic Web Service Publishing, Discovery and Automated Composition", Hangzhou: Academic, June 2006, pp. 231–245.
[3] Sun Wei-qin. Tomcat & Java Web development of technology solutions [M]. Beijing: Publishing House of Electronics Industry. 3121.
[4] Ksoap2-android, http://code.google.com/p/ksoap2-android/.
[5] Handler, http://developer.android.com/reference/android/os/Handler.html
[6] Intent and Intent Filters, http://developer.android.com/guide/components/intents-filters.html.

Information Technology and Computer Application Engineering – Liu, Sung & Yao (Eds)
© *2014 Taylor & Francis Group, London, ISBN 978-1-138-00079-7*

Spatial-temporal patterns analysis of property crime in urban district based on Moran's I and GIS

W. Ma & J.P. Ji
National Quality Inspection and Testing Center for Surveying and Mapping Products, Beijing, China

P. Chen
Institute of Policing Information and Engineering, People's Public Security University, Beijing, China

T.T. Zhao
National Geomatics Center of China, Beijing, China

ABSTRACT: Identification of outliers and clusters of crime in space is very useful to policing management. This study assessed and examined the spatial-temporal patterns of property crime which occurred in 2007 in a district of X city in China. The Moran's I was used to find spatial autocorrelation and heterogeneity of the property crime and then the results were visualized with GIS. The findings demonstrated that property crime varied by time of day and day of week, and spatial distribution of the crimes had temporal trends. The knowledge of such property can be helpful to estimate over density of property crime in city, where particular prevention and allocation of police could be assessed.

Keywords: City safety; Geographic Information System; Crime; Spatial statistic and analysis; Moran's I

1 INSTRUCTIONS

As cities have developed quickly, they have become increasingly for law enforcement departments to utilize sophisticated technology to fight crime and to increase the effectiveness of the available and limited resources. So, more extensive knowledge about crime's distribution is required.

The research about crime analysis in space had been followed for many years. Since the beginning work of Guerry and Quetelet in 19th century and proceeding with the studies of Chicago School in early of 20th century (Messner et al. 1999), people had more insight into the crime distribution features in space. Lately, the interest in crime places continued to grow, and the identification of hot spots of crime became a watershed topic in criminology. In definition, a crime hot spot is a location, or small area within an identifiable boundary, with a concentration of criminal incidents. Sherman et al had published studies to quantify what many qualitative studies had suggested, that crime in a city is highly concentrated in relatively few small areas. The study found that 3.3 percent of street addresses and intersections in Minneapolis generated 50.4 percent of all dispatched police calls for service. Similar pattern was also discovered in other cities (Pierce et al. 1988; Sherman 1992; Weisburd & Green 1994).

Lately, GIS and some algorithms were used to identify hot spots of crime in space (Hirschfield et al.

1995). The methods and technologies that have been used in hot spot identification include Kernel Density estimation, K-Means clustering and Nearest Neighbor Hierarchical spatial analysis (Levine 2002). All of them belong to point pattern analysis. Meanwhile, another equally important setting, in which the data were collected for area units of regions, had been applied in spatial analysis (Anselin 2000). Recently, the set of methods for structuring the visualization of spatial data has been referred to as exploratory spatial data analysis, or ESDA. ESDA is a collection of techniques to describe and visualize spatial distributions; identify atypical locations or spatial outliers; discover patterns of spatial association, clusters, or hot spots. Therefore, in the present paper, the spatial clustering and outliers of property crime in a district of X city in China was analyzed with ESDA. By combining descriptive graphs in an interactive computing environment, visualization was augmented through maps with hypothesis tests for spatial patterns. With application of ESDA, the utility of this approach for identifying and interpreting the spatial clustering of criminal activities was illustrated.

2 DATA AND METHOD

2.1 Study area and crime data

The district in this case is one of the largest administrative zones in X city urban area and made up of

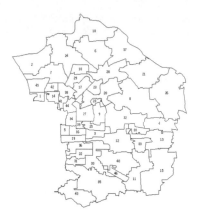

Figure 1. Map of the district and its 46 police beat zones. (Coding corresponding with Table 1).

Table 1. Name of 46 police beat zones in the districta*.

Map code	Police patrol zone	Map code	Police patrol zone
1	Anzl	24	Laigy
2	Aoyc	25	Liult
3	Balz	26	Louxz
4	Chang	27	Maizd
5	Chaows	28	Nang
6	Cuigz	29	Southl
7	Dat	30	Nanmf
8	Dongb	31	Panjy
9	Dongf	32	Pingf
10	Dongl	33	Sanjf
11	Dougz	34	Sanlt
12	Gaobd	35	Shibld
13	Guanz	36	Shuangj
14	Peaces	37	Sunh
15	Heizh	38	Taiyg
16	Hujl	39	Tuanjh
17	Huajd	40	Wangsy
18	Huangg	41	Xianghy
19	Jiagmw	42	Xiaog
20	Jiangt	43	Xiaohm
21	Jinz	44	Xinyl
22	Jins	45	Yayc
23	Jiuxq	46	Dait

* map codes match with Figure 1.

46 police beat zones (Figure 1 and Table 1). It covered the area of 455 km2. The population in the district was 2.9 million in 2007 (X City Census 2007).

The economy and population in the district is not averagely distributed. The western and southern where located many central business zones (CBD) are highly developed, and so the population concentration there is very high. But on northern and eastern, where a large area is made up of farms and rural neighborhoods, so the economy there is undeveloped and population densities are very low.

The property crime data for this analysis was collected by X City Municipal Public Security Bureau.

Detailed information on calls for police service, including crime reports, date and time of the event, address of the events, latitude and longitude coordinates of the events, and events type code were available. The data recorded information of property crime for six months (May to October on 2007) and was provided on two CDs. The entire data was loaded into a single Microsoft SQL database from which all information could be accessed.

2.2 Method

ESDA (Exploratory Spatial Data Analysis) included series principles that were relevant in analysis of spatial patterns. Among them, the methods that were taken to examine spatial features were spatial autocorrelation test and local indicators of spatial association (LISA). Spatial autocorrelation techniques test whether the distributions of events are related to each other in space, while LISA statistics assess the local association between data by comparing local averages to global averages (Anselin 1995). The more applied spatial autocorrelation test and LISA statistic are global Moran's I and local Moran's I (Tapiador & Mezo 2009; Zhang et al. 2008).

The global Moran's I index was a classic test statistic that used for spatial autocorrelation. It is a cross-product coefficient similar to a Pearson correlation coefficient and scaled to be less than one in absolute value. Global Moran's I varies between "−1.0" and "+1.0", with a high value indicating positive autocorrelation. In formal terms, let W be the weight matrix, such that $w_{ij} = 1$ if x_i and x_j are neighbors in continuity, and 0 otherwise. The Moran's I index is defined as:

$$I = \frac{n \sum_{i=1}^{n} \sum_{j=1}^{n} w_{ij}(x_i - \bar{x})(x_j - \bar{x})}{\sum_{i=1}^{n} \sum_{j=1}^{n} w_{ij} \sum_{i=1}^{n} (x_i - \bar{x})^2} \tag{1}$$

Where x_i and x_j are variables on location i and j; $\bar{x}$ is the average value of x with the sample number of n.

The local Moran's I index provid a measure of the extent to which the arrangement of values around a specific location deviated from spatial randomness. The statistic was used to identify the clustering of high and low values and outliers. The formal term of local Moran's I is

$$I_i = \frac{(x_i - \bar{x})}{S^2} \sum_j w_{ij}(x_j - \bar{x}) \tag{2}$$

Where the x_i, x_j and $\bar{x}$ have the same meaning as above. S^2 is the variance of variables x. A high positive local Moran's I value implies that the location under study has similarity high or low values as its neighbors. In this way, the locations are spatial clusters. There are two kinds of clusters, they are high-high clusters (high values in a high value neighborhood) and low-low clusters (low values in a low neighborhood). Comparatively, a high negative local Moran's I value means that the location under study is a spatial outlier. Spatial outliers are those values that are obvious different from the values of their surrounding neighbors. Similarly, spatial outliers also include two kinds. They are high-low (a high value in a low value neighborhood) and low-high (a low value in a high value neighborhood).

In this study, global and local Moran's I index were used to analyze the spatial pattern of property crime in the district. The two indexes had been used to assess spatial dependence or spatial heterogeneity in St.Louis metropolitan area by Messner and Anselin in 1999, and the results had turned out to be useful in finding the homicide pattern in local area.

2.3 Computer software

The calculation of spatial clusters and outliers was performed and visualized using the software GeoData (version 0.95i, Spatial Analysis Laboratory, 2007; Anselin, 2005).

3 TEMPORAL ANALYSIS OF CRIME BY DIFFERENT TIME UNITS

Analysis of temporal variables, particularly when linked with the micro-spatial level, will further clarify the pattern of crime (Nelson et al. 2001). For example, the robbery in places where more liquor consumption in cars and outdoors will occur late; but in the areas where near high schools, robbery tend to occur earlier (Felson & Poulsen 2003). Some work has been done to explore the crime pattern in short time period (Cettaco 2005; Brunsdon & Corcoran 2006). Thus, a property crime pattern in the district in this case was explored by day of week and time of day. This analysis revealed when the crime level reaches highest at different time units.

First, the property crime data was aggregated by days of week and plotted on the graph (Figure 2). The findings showed that most property crime concentrated on Monday (14.8%), while least property crime happened on Sunday (13.6%). Friday ranked the second day that crime took place frequently (14.5%). In addition to the frequency of crime, there was an obvious temporal trend during the week: the crime rate kept reducing from Monday to Thursday, and then suddenly increases a little on Friday, after that, it proceeded to fell off on Saturday and Sunday.

Comparatively, analysis of the property crime rate by hour in the district revealed another pattern. Figure 3 displayed crime's variation by hours of the day. According to the graph, three distinct periods of property crime were identified: between 7:00 am and 17:00 pm; between 18:00 pm and 22:00 pm; between 23:00 pm and 6:00 am. Between 7:00 am and 17:00 pm there were higher numbers of property crime, which peaked around 8:00 am. This period experienced more than 60% of the 24-hour total. While the period between 18:00 and 22:00 witnessed second higher ratio of crime rate (23.9%), and the crime rate peaked at time around

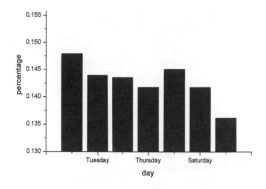

Figure 2. Property crime variation by day of week in the district.

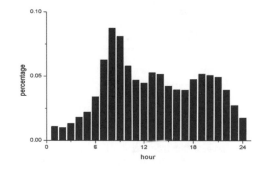

Figure 3. Property crime variation by hour of day in the district.

19:00 pm. Between 22:00 pm and 6:00 am was the period that least crime happened, only 15.3% of 24-hour property crime took place. As to the temporal trend of crime, the property crime rate in the district kept low level during the period between 22:00 pm and 6:00 am, then rose sharply to the highest of the day around 7:00 am and fell off slightly, after that, it rose to another peak level but lower comparatively around 13:00 pm and 19:00 pm, finally, crime rate reduced to the lowest level of the day after 22:00 pm.

The analysis of property crime in the district by day of week and hour of day proved the crime has a distinct weekly and daily pattern. Property crime peaked on Monday and Friday demonstrated that people changed their life pattern on the days. While hour of day, three periods of different crime level were corresponded to the stages that people were experiencing: work, evening life and sleeping.

4 SPATIAL ANALYSIS OF PROPERTY CRIME

4.1 Global analysis

The spatial pattern of property crime in the whole period of six months was investigated firstly. From this point the distribution of property crime in the long run could be assessed. The crime data was aggregated into every police beat zones and the global Moran's I index was calculated. The results in Table 2 illustrated the

Table 2. Global Moran's I index of property crime in the district for whole period.

	Moran's I	Z	p
Whole period	0.21	2.46	0.05

Table 3. Global Moran's I index of property crime in the district by week.

Days	Moran's I	Z	p
Monday	Random	–	–
Tuesday	0.22	2.77	0.01
Wednesday	Random	–	–
Thursday	0.24	3.02	0.01
Friday	0.2	2.54	0.05
Saturday	Random	–	–
Sunday	0.16	2.19	0.05

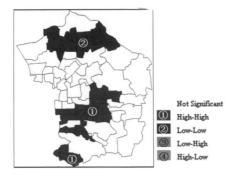

Figure 4. Spatial distribution of significant local Moran's I index for property crime in the district. ($p = 0.05$).

property crimes in the district were highly correlated. The crime in every police beat zone was influenced by their neighbors.

The spatial heterogeneity of property crime in whole period was investigated with local Moran's I index, and the results were displayed in a graph (Figure 4). Four colors represent different types of spatial pattern, and only the zones with local Moran's I index at significant level were visible. The results in Figure 4 showed that there was no outlier of property crime in whole period except some clusters. The high-high clusters concentrated on the center and southern of the district, while the low-low clusters concentrated on the northern. According to Figure 1, the three zones with low-low clusters on the northern are Cuigz, Laigy and Sunh, where located on rural places. The number of property crime happened there only took up 4.18% of the total, and crime density in these places was less than 4 incidents/km^2 on average. But to the high-high clusters, they lied on the center and south of the district which were enclosed by CBD. The percentage of the crime happened on these high-high clusters were 26.1%, and crime density reached as high as 21 incidents/km^2.

4.2 Analysis by week

For week, the property crime data was grouped into seven days and spatial patterns along the days were investigated. The findings in Table 3 showed the global spatial autocorrelation of property crime in the district was significant on Tuesday, Thursday, Friday and Sunday, but not significant on the other three days. That means on Monday, Wednesday and Saturday the property crimes were randomly distributed on police beat zones.

The pictures in Figure 5 demonstrated the results of local Moran's I index of property crime from Monday

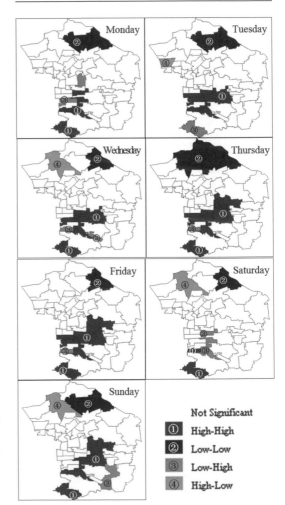

Figure 5. Spatial distribution of significant local Moran's I index for property crime in the district ($p = 0.05$).

to Sunday. Comparing to the spatial distribution in whole period in Figure 4, there was an obvious variation of the clusters and outliers during the seven days. As similar as displayed in Figure 4, the low-low clusters and high-high clusters were still located on the north and south of the district respectively. But surprisingly Pingf zone became another low-low cluster on Thursday. To the outliers, they showed different patterns. The high-low outliers, sometimes they

256

Table 4. Global Moran's I index of property crime in the district by day.

Time	Moran's I	Z	p
07:00 am–17:00 pm	0.13	1.76	0.10
18:00 pm–22:00 pm	Random	–	–
23:00 pm–06:00 am	0.15	2.13	0.05

were called hot spots, appeared on Tuesday, Wednesday, Saturday and Sunday, and all located on the northern. Among them, Laigy zone was the hot spot appeared most frequently. It turned out to be hot spot on Wednesday, Saturday and Sunday. In another way, the low-high outliers, sometimes they were called cold spots, appeared all seven days. But they were all located on the center and south of the whole district. Among them, Panjy was the most frequently appeared zone. It appeared on Wednesday, Thursday and Friday.

4.3 Analysis by day

Comparing with analysis by week, spatial analysis by day concentrated more on micro level. The global Moran's I index in Table 4 showed the property crime in zones were correlated with each other during the period between 07:00 am and 17:00 pm as well as the period between 23:00 pm and 06:00 am, but randomly distributed between 18:00 pm and 22:00 pm.

But to the local Moran's I index, the three zones on northern were still low-low clusters either in period between 18:00 pm and 22:00 pm or between 23:00 pm and 06:00 am. However, on the daytime between 07:00 am and 17:00 pm, the zones were no longer significant as low-low clusters. A small police beat zone named Yayc on the north-west corner of the district became another low-low cluster. The same case happened to the high-high clusters. The high-high clusters of property crime covered 6 zones on the center of the district between 07:00 am–17:00 pm, but decreased to only two zones on the southern during the time between 18:00 pm and 22:00 pm, then finally increased to six zones again between 23:00 and 06:00.

In addition to the clusters, the outliers also had new trends in space. According to Figure 6, the hot spot only appeared and located in the center of whole district during the period between 18:00 pm and 22:00 pm, but not significant any more in the other two periods. The cold spots appeared in three periods and distributed on the center of the whole district. The zone named Xiaohm was a high-high cluster in the whole period and by week, but turned into cold spot in period by hours.

4.4 Comparison among results from different temporal periods

For comparison of the results from the above different ways of temporal treatment, the number of significant and non-significant spatial outliers and spatial clusters were summarized in Table 5.

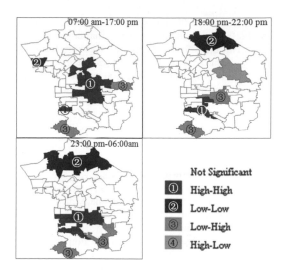

Figure 6. Spatial distribution of significant local Moran's I index for property crime in the district ($p = 0.05$).

Table 5. Comparison of numbers of non-significant and significant spatial outliers and clusters ($p = 0.05$).

	Not significant	High-High	Low-Low	Low-High	High-Low
Whole period	35	8	3	0	0
By week					
Monday	38	4	2	2	0
Tuesday	40	2	1	2	1
Wednesday	37	5	1	2	1
Thursday	35	7	4	1	0
Friday	37	7	1	1	0
Saturday	40	2	1	2	1
Sunday	37	5	2	1	1
By day					
07am–17pm	37	6	1	2	0
18pm–22pm	39	2	2	2	1
23pm–06am	34	6	3	3	0

For the whole period, only clusters with high-high and low-low patterns existed but no outliers. But for week, there was a temporal variation of spatial distribution from Monday to Sunday. From Table 5, the maximum number of high-high cluster was 7 and they appeared on Thursday and Friday. But on the same days, the number of high-low outliers was none. This phenomenon revealed that the property crimes had a trend of shifting and concentration to some certain zones on Thursday and Friday (according to Figure 5, the seven zones were Xiaohm, Balz, Gaobd, Jiangmw, Nanmf, Pingf, Shuangj), which leaded to the crime rate in other zones becoming low relatively. Another day witness this phenomenon but less obvious was Monday, 4 high-high clusters but no hotspot appeared on that day.

For the time by day, the case was similar. There were 6 high-high clusters and 0 hot spot at the time between

07:00 am and 17:00 pm as well as between 23:00 pm and 06:00 am, which showed a spatial concentration trend. But the difference was their location. Between 07:00 am and 17:00 pm, crime concentrated on suburb and rural zones, but between 23:00 pm and 06:00 am the crimes shifted to the downtown area.

5 CONCLUSION

This paper explored the temporal features of property crime in the district in Beijing, and the global and local Moran's I index were used to investigate the spatial autocorrelation and heterogeneity of the crime in different periods. The results illustrated the property crime had temporal variation by day of week and hour of day, and would shift on space along the time. The findings were helpful to estimate over density zones and make corresponding prevention strategies. In future work, the environmental factors will be considered and combined with spatial analysis to study the cause and effect of crime in space.

ACKNOWLEDGEMENT

The authors gratefully appreciate the Project of National Natural Science Foundation of China (Grant No: 41201447) and the Project of Socialization Application of National Basic Surveying and Mapping Results – Geographic Information System Testing and Quality Evaluation System Construction (National Administration of Surveying, Mapping and Geoinformation) for Supporting this work.

REFERENCES

Anselin, L. 1995. Local indication of spatial association-Lisa. *Geographical Analysis* 27(2): 93–115.

Anselin, L., Cohen, J., Cook, D., Gorr, W. & Tita, G. 2000. Spatial analysis of crime. *Criminal Justice* 4: 213–262.

Anselin L. 2005. Exploring spatial data with GeoDa™: a workbook. Spatial Analysis Laboratory, Department of Geography. *Urbana-Champaign, Urbana, IL: University of Illinois*: 61801. 226.

Brunsdon, C. & Corcoran, J. 2006. Using circular statistic to analyze time patterns in crime incidence. *Computer, Environmental and Urban Systems* 30: 300–319.

Cettaco, V. 2005. Homicide in Sao Paulo, Brazil: Assessing spatial-temporal and weather variations. *Journal of Environmental Psychology* (25): 307–321.

Felson, M. & Poulsen, E. 2003. Simple indicators of crime by time of day. *International Journal of Forecasting* 19: 595–601.

Hirschfield, A., Brown, P. & Todd, P. 1995. GIS and the analysis of spatially-referenced crime data: Experiences in Merseyside-UK. *In International Journal of Geographical Information Systems* 9: 191–210.

Levine, Ned. 2002. CrimeStat II: A Spatial Statistics Program for the Analysis of Crime Incident Locations (version 2.0). *Ned Levine & Associates: Houston, TX National Institute of Justice: Washington, DC.*

Messner, S. F., Anselin, L. & Baller, R.D. 1999. The spatial patterning of county homicide rates: an application of exploratory spatial data analysis. *Journal of Quantitative Criminology* 15(4): 423–450.

Nelson, A. L., Bromley, R. D. F. & Thomas, C.J. 2001. Identifying micro-spatial and temporal patterns of violent crime and disorder in the British city centre. *Applied Geography* 21: 249–274.

Pierce, G., Spaar, S. A. & Briggs, L. R. 1988. The character of police work: Strategic and tactical implications. *M.S. thesis, Northeastern University, Center for Applied Social Research.*

Sherman, L. W., Gartin, P. R. & Buerger. M.E. 1989. Hot spots of predatory crime: Routine activities and the criminology of place. *Criminology* 27: 27–55.

Sherman & Lawrence W. 1992. Attacking Crime: Police and Crime Control. *Crime and Justice* 15: 159–230.

Tapiador, F. J. & Mezo, J. 2009. Vote evolution in Spain, 1977–2007: a spatial analysis at the municipal scale. *Political Geography* 28: 319–328.

Weisburd, D.L., & L. Green. 1994. Defining the street level drug market. *Drugs and crime: Evaluating public policy initiatives*: 61–76.

Zhang, C. S, Luo, L., Xu, W. L. & Ledwith, V. 2008. Use of local Moran's I and GIS to identify pollution hotspots of Pb in urban soils of Galway, Ireland. *Science of the Total Environment* 398: 212–221.

Information Technology and Computer Application Engineering – Liu, Sung & Yao (Eds)
© *2014 Taylor & Francis Group, London, ISBN 978-1-138-00079-7*

Adding attributes to access delegation

Y. Li, G. Liu & H.W. Wang
Shaanxi Engineering Laboratory for Transmissions and Controls, Northwestern Polytechnical University, Xi'an, China

ABSTRACT: Access delegation regulates the process that an authorized user transfer part or total of his/her permissions to another user, which is very important in access control system especially in distributed or collaborative environment. Attribute brings salient features to access delegation such as flexibility and effectiveness. The definition of attribute is proposed that attributes selected should accord to the access control target and possess some properties. In access delegation management, the attributes consist of subject attributes, object attributes and permission attributes. As the mechanism of restricting delegation, a delegator delegate the permissions to a delegatee successfully only satisfying the delegator attribute expression, delegatee attribute expression, object attribute expression and permission expression. The definitions of the elements in the expressions are presented.

Keywords: Access Delegation, Attribute, Attribute Expression

1 INSTRUCTIONS

Access delegation regulates the process that an authorized user transfer part or total of his/her permissions to another user who is otherwise not authorized to have such access. The reverse concept of access delegation is access revocation, which deprives a group of specific access permissions that were granted previously in delegation process. The objectives of delegation are to ask someone to finish a job or invite someone to participant a collaborative work. Access control systems should support access delegation to satisfy the requirement from real world, otherwise users may perceive security as a hindrance, especially in distributed or collaborative environment. In access delegation, a *delegator* is the inviter who grants some permissions to a invitee and a *delegatee* is the invitee who receive the permissions.

There are many outputs about access delegation. The Role-Based Delegation Model (RBDM0) supporting single-step delegation is presented on the basis of flat role (E. Barka and R. Sandhu, 2000). As a subsequent research, the RBDM1 is built to take the hierarchical relationship into account (E. Barka and R. Sandhu, 2004). RDM2000 extends RBDM0 and supports delegation in presence of a role hierarchy and multi-step delegation (L.H. ZHANG, G.J. AHN, B.T. CHU, 2001). RDM2000 provides a rule based declarative language to specify and enforce delegation and revocation constraints. It introduces the *can-delegate* condition with prerequisite roles to restrict the scope of delegation. In RBDM0, RBDM1 and RDM2000, the unit of delegation is a role which means that a delegator grants all the permissions assigned to the delegation role to a delegate that

is not compliant to the least privilege rule. The Permission-Based Delegation Model (PBDM) is a family of models that extend RDM2000 with newer features (X.W. Zhang, S. Oh, and R. Sandhu, 2003). Thanks to the four types of roles with different properties, PBDM supports permission level user-to-user and role-to-role delegation and allows the security administrator to control the delegated permissions. In RDM2000 and PBDM, delegation security entirely depends on delegators and security administrators, for delegation constraints in the two models is only a prerequisite condition. The Attribute-Based Delegation Model (ABDM) with an extended delegation constraints consisting of *delegation attribute expression* and *prerequisite condition* require that delegatee must satisfy some constraints when assigned to a delegation role (C.X. Ye, Z.F. Wu, Y.Q. Fu and et al, 2006).

Nowadays, some researchers propose adding attributes to RBAC in order to encompass the benefits of attribute and role while transcend their limitations (D.R. Kuhn, E.J. Coyne and T.R. Weil, 2010; X. Jin, R. Sandhu and R. Krishnan, 2012). Delegation model can have more expressive power if introducing attributes into it. The permission depth presented in RDM2000 is an important element to realize multi-step delegation control (L.H. ZHANG, G.J. AHN, B.T. CHU, 2001). User attributes are the basis of user attribute expression in ABDM to limit the candidate scope of delegatees (C.X. Ye, Z.F. Wu, Y.Q. Fu and et al, 2006). The Temporal Constraints Based Delegation Model introduces time attribute of permission to constrain that delegatees can validate delegation permissions only on specified time or in specified period (Y.Q. Zhu, 2010).

Attribute brings salient features to access delegation and access control such as flexibility and effectiveness. However, to our best knowledge, there is no clear consensus on attribute in access delegation. We propose a definition of attribute in access control which makes contribution to powerful expression and flexible constraints of access delegation. Comparing with previous researches that focusing on subject attributes or permission attributes, attributes common used in access delegation are summarized which consist of subject attributes, object attributes and permission attributes. The feasibility of adding attributes to access delegation is proved by a logic based mechanism presented in the part 4.

2 DEFINITION OF ATTRIBUTE

An attribute is a property of a managed object that has a value. Different areas concern different sets of attributes of an object which has unlimited attributes. In access control area, an attribute can be defined as Definition 1 and the attribute function can be defined as Definition 2.

Definition 1 An attribute in access control is a variable with a codomain denoting a property of an entity in access control system, which is selected to satisfy a certain access control requirement.

Definition 2 An attribute function takes an attribute with specified value and returns a set of entities with the property expressed by the attribute with the value.

Taking E and f_e to denote an entity set and the attribute function to E respectively, and assuming that entities in E have an attribute a with a codomain $cod:\{v_1, v_2, \ldots, v_n\}$, the two properties of attributes are shown as below.

Property 1: $\cup_{v_i \in cod} f_e(a = v_i) = E$;

Property 2: $\forall v_i \in cod, \forall v_j \in cod$, if $i \neq j$,
then $f_e(a = v_i) \cap f_e(a = v_j) = \varnothing$.

The attributes selected to using in access control system should be available and agreed generally. As the reason that different areas concern different sets of attributes of an object, designers of an information system may be apt to select attributes of entities on the side of business functions, which orientation makes some access control performance not be achieved. The designers should take the access control requirements into account in the design phase of a information system in order to adding attributes to access control successfully.

3 ATTRIBUTES RELATED TO DELEGATION

The entities related to access delegation contains subjects including delegators and delegatees and permissions defined as mappings between objects and

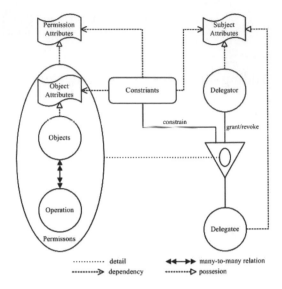

Figure 1. Process of access delegation.

operations. Fig 1 gives the process of access delegation that a delegator grants some permissions to delegatee under the restriction of constraints. Every delegator or delegatee is specified some subject attributes by the system such as secret level. Each object managed in the system is assigned to some object attributes such as secret level and owner. Permissions attributes related to delegation reflect the properties of the mappings between objects and operations. Constraints are rules on the basis of attributes to validate delegations.

The attributes of delegation management in common use are listed in the Table 1. Each attribute with a codomain is selected to achieve a management goal.

4 MECHANISM

The reason for adding attributes to access delegation is that the mechanism restricting delegation can be realized through flexible attributes assignments.

A delegator delegate the permissions to a delegatee successfully only satisfying the delegator attribute expression (*DrAE*), delegatee attribute expression (*DeAE*), object attribute expression (*OAE*) and permission expression (*PAE*) as shown below:

delegate(delegator, delegatee, permissions)
$$\leftarrow DrAE \wedge DeAE \wedge OAE \wedge PAE$$

The elements in the above expression are explained in definition 3.

Definition 3
DrAE ::= *sae* { AND *sae*}
DeAE ::= *sae* { AND *sae*}
OAE ::= *oae* { AND *oae* }
PAE ::= *pae* {AND *pae*}
sae ::= *sa oprt sav*
oae ::= *oa oprt oav*

Table 1. Attributes of delegation management in common use.

Atrrbute	Codomain	Goal
organization	all the organizations or departments managed in a system	to limit the delegations between organizations or departments
secretLevel	levels to demarcate secret grade	to combine mandatory access control rules
qualification	the titles of technical posts adopted by all the organizations	to require delegator or delegatee owning necessary abilities
project	the projects possibly need delegation to complete	to locate the objects in deferent projects
domain	all the domains	to locate the objects in deferent domain
secretLevel	levels to demarcate secret grade	to combine mandatory access control rules
owner	all users in a system	to indicate that the owner of an object have different right on the object compare with others
depth	natural number	to restrict the depth of permission in the delegation path
width	natural number	to restrict the number of users who can be granted a permission
interval	any interval	to meet the security requirements of temporality

$pae ::= pa\ oprt\ pav$

$oprt ::=$ '$<$'|'$\leq$'|'$=$'|'$\geq$'|'$>$'|'$\neq$'|'$\in$'|'$\notin$'

$sa ::=$ {subject attributes specified by system}

$oa ::=$ {object attributes specified by system}

$pa ::=$ {permission attributes specified by system}

$sav ::=$ the codomain of a subject attribute

$oav ::=$ the codomain of an object attribute

$pav ::=$ the codomain of a permission attribute

For example, Tom affiliating to organization org_A wants to delegate the permission of reading file file_A of org_A to Jack affiliating to organization org_B. Tom is a senior engineer (se) with secretLevel 3 and Jack is a junior engineer (je) with secretLevel 2. The file file_A is secretLevel 2. The delegation expression

$delegate$(Tom, Jack, (read, file_A))

$\leftarrow$(delegator $\in$ org_A)

$\wedge$(delegator.secretLevel ≥ 2)}

$\wedge$\{(delegatee $\in$ org_B) $\wedge$ (delegatee. secretLevel ≥ 2)}

$\wedge$\{object. secretLevel ≤ 2\}

$\wedge$\{(permission.depth ≤ 2) $\wedge$ permission.width ≤ 5\}

means that the delegation is valid if Tom has the original permission of reading file file_A and the current permission width less than 4, otherwise the delegation is illegal.

5 CONCLUSION

Adding attributes to access delegation can satisfy many requirement of delegation management. The definition of attribute in access control is proposed that attributes selected should accord to the access control target. The values of attributes considered in access control must be available and agreed generally. The attributes of delegation management in common use are summarized, which consist of subject attributes, object attributes and permission attributes. The mechanism restricting delegation can be realized through flexible attributes assignments.

ACKNOWLEDGEMENTS

Our work is supported by the Basic Research Foundation of Northwestern Polytechnical University (Grant No. JC201209, China) and the 111 Project (Grant No. B13044, China).

REFERENCES

E. Barka and R. Sandhu (2000): A role-based delegation model and some extensions. Proceedings of 23rd National. Proceedings of 23rd National Information Systems Security Conference (NISSC). Baltimore, USA, NIST, December, 2000.

E. Barka and R. Sandhu (2004): Role-based delegation model/hierarchical roles (RBDM1). Proceedings of the 20th Annual Computer Security Applications Conference (ACSAC'04). Tucson, Arizona, USA, IEEE Computer Society, 2004.

L.H. ZHANG, G.J. AHN, B.T. CHU (2001): A Rule-Based Framework for Role-Based Delegation. Proceedings of the 6th ACM Symposium on Access Control Models and Technologies (SACMAT), Chantilly, VA, May 3–4, 2001, pp. 153–162.

X.W. Zhang, S. Oh, and R. Sandhu (2003): PBDM: A Fexible Delegation Model in RBAC. In SACMAT '03: Proceedings of the eighth ACM symposium on Access control models and technologies, pp. 149–157. ACM Press, 2003.

C.X. Ye, Z.F. Wu, Y.Q. Fu and et al (2006): An Attribute-Based Extended Delegation Model, Journal of Computer Research and Development, 2006, 43(6), pp. 1050–1057.

D.R. Kuhn, E.J. Coyne and T.R. Weil (2010): Adding Attributes to Role-Based Access Control[C]. IEEE Computer, 2010, 43(6):79–81.

X. Jin, R. Sandhu and R. Krishnan (2012): RABAC: Role-Centric Attribute-Based Access Control[C]. Computer Network Security: 6th International Conference on Mathematical Methods, Models and Architechtures for Computer Network Security, St. Peterburg, Russia, October 17–19, 2012:84–96.

Y.Q. Zhu (2010): Temporal Constraints Based Delegation Model. Journal of ShangHai DianJi University. Feb, 2010, Vol 13, No. 1, pp. 31–38.

Information Technology and Computer Application Engineering – Liu, Sung & Yao (Eds)
© *2014 Taylor & Francis Group, London, ISBN 978-1-138-00079-7*

Unexpected passenger flow propagation in urban rail transit network based on SIR model

Miao Du, Yong Qin, Zi Yang Wang & Xiao Xia Liu
School of Traffic and Transportation, Beijing Jiaotong University, Beijing, China

Bo Wang, Ping Liang & Ming Hui Zhan
Beijing Subway Corporation, Beijing, China

ABSTRACT: The propagation and influence of unexpected passenger flow in urban rail transit network has become an urgent problem which the administrators have to face. And the research of it is very important. As a new discipline, complex network has a wide range of applications in transport field. In this paper a model on unexpected passenger flow and its propagation has been established based on SIR (Susceptive-Infected-Recovered) theory. It is found that unexpected passenger flow has impact on the urban rail transit network through simulating. The result is clear and credible.

Keywords: SIR model; Unexpected Passenger Flow; Propagation

1 INTRODUCTION

With the upgrading of China's comprehensive strength, many large-scale activities (such as large-scale cultural and sports activities, exhibitions, etc.) are frequently held in major cities, thus led to sudden accumulations and dissipation of large-scale passengers, and great impact on the urban public transport. It has become an urgent problem to make accurate analysis of the interaction between point, line and network, propagation path, scope and intensity of the unexpected passenger flow and to find the sticking point and weaknesses in the system.

The spread of outbreak peak passenger flow in urban rail transit network refers to a situation during a specific period (large-scale events, holidays, etc.), there is a high concentration of passengers while the capacity is short in some stations in a short time, leading to an overcrowd in the stations and the carriages and the following ones in the Road Network and causing series of impacts.

For the outbreak passenger flow research, foreigners in early time are mainly aimed at the traffic management experience summarized from international super games (such as the World Cup and Olympic Games, etc.). In our country the study for the theory not only starts relatively late, but also causes no enough attention.

In recent years, many scholars have started to use complex network theory to study urban rail traffic network. Latora and Marchiori [1] Introduced two new network statistical parameters the local efficiency and global efficiency, based on which he made a preliminary study on Boston subway, and found small

world characteristics in it. The small word characteristics were also found in Indian railways by lParongamaSen and SubinayDasgupta [2]. The UMT_PDSS (the Urban Mass Transit_Passenger flow Distribution Simulation System) was developed oriented to AFC mass time-sharing OD data by LuoXin [3]. Si Bing-feng and Mao Bao-hua [4] from Beijing Jiaotong University made a full consideration of the main factors impacting on passenger flow distribution, and studied the urban rail transit passenger flow distribution model and algorithm based on stochastic user equilibrium principle learning from urban road theory, Zheng Li-juan [5] established the multi-route choice probability method.

2 THE ESTABLISHMENT OF SIR MODEL ON UNEXPECTED PASSENGER FLOW PROPAGATION

The station in rail transport system may present three states: normal state, crowed state and crowded recovery state under the influence of unexpected passenger flow propagation. It just like the three states (Susceptibility state, infection state and rehabilitation state) in infectious disease model. Now, the SIR model of unexpected passenger flow propagation is as follow [6]:

$$
\begin{cases}
\dfrac{dS}{dt} = -\beta SI \\[2mm]
\dfrac{dI}{dt} = \beta SI - \delta I \\[2mm]
\dfrac{dR}{dt} = \delta I
\end{cases}
\tag{1}
$$

That is:

$$\begin{cases} \Delta S_{t+1} = -\beta S_t I_t \\ \Delta I_{t+1} = \beta S_t I_t - \delta I_t \\ \Delta R_{t+1} = \delta I_t \end{cases} \quad (2)$$

In the formula, S_t – the number of the stations which will be influenced by unexpected passenger flow at time t

I_t – the number of the stations which are crowded at time t

R_t – the number of the stations which are recovery from the crowded passenger at time t

β – crowded passenger flow propagation rate

δ – crowded Passenger flow dissipation rate

Unexpected passenger flow may happen in intermediate stations or transfer stations in the rail transport system, Unexpected passenger flow in the former ones will transmit along the stations in the rail line, while the latter will also scattered to the other lines it connects, and thus spread to the whole rail transport network.

The newly increased number of the stations which are crowed is defined as N:

$$N = N_1 \beta_L + N_2 \beta_N \quad (3)$$

In the formula, $N_1 \beta_L$ – reflect the propagation of crowed passenger in its own line

$N_2 \beta_N$ – reflect the diffusion of crowded passenger in other lines

The total number of the stations which will be influenced by crowed passenger flow can be expressed by expected value, that is:

$$E(-\Delta S) = N_1 \beta_L + N_2 \beta_N - \delta I \quad (4)$$

Base on the equation (1) and (2), we can get the following formula:

$$\Delta I_{t+1} = [N_1 \beta_{Lt} + N_2 \beta_{Nt} - \delta_t I_t] I_t - \delta_t I_t \quad (5)$$

$$I_{t+1} = I_t + \Delta I_{t+1} \quad (6)$$

In the formula, N_1 – the number of crowded stations in its own line;

N_2 – the number of crowded stations in other lines it connects;

β_{Lt} – crowded passenger flow propagation rate in its own line at time t;

β_{Nt} – crowded passenger flow propagation rate in other lines at time t;

δ_t – crowded passenger flow dissipation rate at time t.

Crowded Passenger flow propagation rate β_{Lt}, β_{Nt} and crowded Passenger flow dissipation rate δ_t are changed with time. For there are so many influential factors, we can't get the accurate expression of them. In order to make the model operable, in this paper, the changes with the passage of time in these rates are ignored. So the simplified model is obtained:

$$\Delta I_{t+1} = [N_1 \beta_L + N_2 \beta_N - \delta_t I_t] I_t - \delta_t I_t \quad (7)$$

$$I_{t+1} = I_t + \Delta I_{t+1} \quad (8)$$

3 THE PROPAGATION OF UNEXPECTED PASSENGER FLOW IN INTERCHANGE STATION

The unexpected crowd in rail transport system is mainly caused by the propagation in the transfer station, which refers to N_2 in the SIR model of unexpected passenger flow propagation. In the simulation, the value of it is 2, 4, 6, 8, 12 and 14. Despite there has no station with 10, 12 and 14 lines connected to it, these numbers are all come into consideration for the results in the simulation can make a great difference.

$I_0 = 1$ means there is only one crowded station in the network at first

$N_1 = 2$ means the crowd happens in a intermediate station

$\lambda_L = 0.4$ means there will be 0.4 stations being crowded in the line crowd happens first in unit time

$\lambda_N = 0.2$ means there will be 0.2 stations being crowded in other lines it connects in unit time

$\lambda_N = 0.2$ means there will be 0.15 stations recovered from crowd in unit time

Base on the analysis and SIR model above, we can get the simulation results:

From the figures above, when $N_2 \leq 12$, the number of passenger flow crowded stations has little increase, which suggests that the passenger flow crowd will not spread into large area; while when $N_2 \geq 14$, the number will be in wide fluctuations, which suggests that the propagation of crowd can't be controlled and it may influence the whole network.

4 UNEXPECTED PASSENGER FLOW PROPAGATION OF INITIAL SINGLE STATION AND INITIAL MULIITIPLE STATIONS IN NETEORK

In the practical network, it's probably happened that a few stations become crowded, that is the so-called initial multiples crowded stations problem. The following is the simulation on this problem.

Given conditions:

$$N_1 = 2, \ N_2 = 2, \ \lambda_L = 0.4, \ \lambda_N = 0.2, \ \delta = 0.15;$$

$$I_0 = 1, \ I_0 = 2, \ I_0 = 4, \ I_0 = 5, \ I_0 = 6, \ I_0 = 7,$$

$$I_0 = 8$$

From the figure we can come to the following conclusions:

(1) Stations recovered from crowd tend to have the same convergence with the passage of time, though the numbers of initial passenger flow crowded stations are different, the time for convergence is diverse. This is consistent with the real situation.

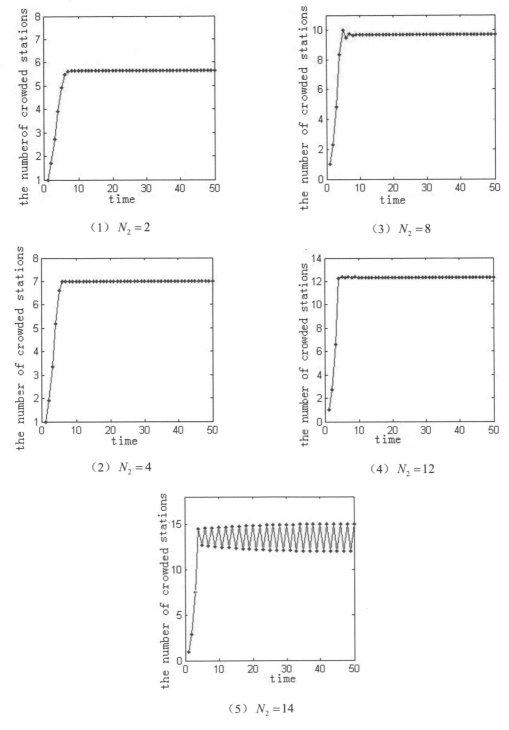

(1) $N_2 = 2$

(2) $N_2 = 4$

(3) $N_2 = 8$

(4) $N_2 = 12$

(5) $N_2 = 14$

Figure 1. The number change of passenger flow crowded stations.

(2) No matter how many crowded stations there are, the degree and sphere of influence are still the same, which means the network itself has the ability to recover from unexpected passenger flow transmission.

5 CONCLUSION

A model of unexpected passenger flow transmission in rail transport network based on SIR is set up in this paper, passenger flow transmission in station of

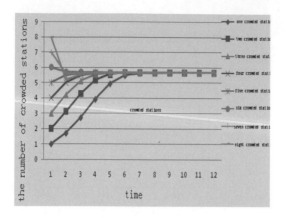

Figure 2. The number change of passenger flow in crowded station.

different connection line number is simulated. Then transmission caused by different number of stations is analysed. The conclusion shows that the network has reliability and robustness.

ACKNOWLEDGEMENT

This work is supposed by National High Technology Research and Development Program of China (Contract No. 2012AA112401), the State Key Laboratory of Rail Traffic Control and Safety (Contract No. RCS2010ZT005), Beijing Jiaotong University and Key Project of the National Twelfth-Five Year Research Program of China (Contract No. 2011BAG01B02).

REFERENCES

[1] LeBlanc. L. J, Morlok. E. IC, Pierskalla. W. P, An efficient approach to solving the road network equilibrium traffic assignment problem, Transportation Research, 1975, 9:309–318

[2] ParongamaSen, SubinayDasgupta. Small-world properties of the Indian railway network. Phys. Rev. E 67, 2003

[3] LuoXin, urban rail transit passenger flow distribution theory and simulation research based on network operation, doctoral dissertation, Tongji University, 2009

[4] Si Bing-feng, Mao Bao-hua, Passenger Flow Assignment Model and Algorithm for Urban Railway Traffic Network under the Condition of Seamless Transfer, Railway journal, 2007, V01, 29(12): 12–18

[5] Zheng Li-juan, Study of the Passenger Flow Distribution Prediction Based on URT Network Operation, master degree thesis, Tongji University, 2008

[6] LIN Guoji, JIA Xun, OURYANG QI, Predict SARS infection with the small world network model, peking university medical journal, 2003

Information Technology and Computer Application Engineering – Liu, Sung & Yao (Eds)
© *2014 Taylor & Francis Group, London, ISBN 978-1-138-00079-7*

Research on optimization of resource scheduling based on hybrid chaos particle swarm optimization

T. Wang
Beijing Institute of Technology, Beijing, China
Beijing Institute of Special Vehicles, Beijing, China

F.L. Zhang & G.F. Li
Beijing Institute of Special Vehicles, Beijing, China

ABSTRACT: Considering the optimization problem of maintenance support resource scheduling, taking the three-level service process flow as an example, basing on the model of RCPSP problem, and in view of that the particle swarm algorithm is prone to local optimization and poor searching effect in solving this problem, this paper presents the hybrid chaos particle swarm optimization, which is formed by adopting a kind of systematic random global searching method – chaos particle swarm optimization and in combination with genetic operator and refactoring particle swarm algorithm for the purpose of expanding the searching scope of the algorithm and improving the quality of the solution and enhance efficiency.

Keywords: Maintenance support; Chaos particle swarm; Hybrid particle swarm optimization

1 FOREWORD

The maintenance support "resource constrained – minimum duration" problem of special vehicle is a branch of the maintenance support resource scheduling problem, belonging to the domain of resource-constrained project scheduling problem (which is called RCPSP for short). It requires that the starting time and ending time of each activity should be arranged under the condition of meeting the task temporal constraint and resource constraint so as to achieve the minimum duration, belonging to the NP-Hard problem. This paper firstly established the solving model for the maintenance support resource scheduling problems, designed a kind of hybrid particle swarm – genetic algorithm based on double coding, and enhanced the overall convergence of the algorithm by introducing the chaos optimization technology in view of the special containment conditions for the issue of maintenance support resource scheduling.

2 PROBLEM DESCRIPTION

The maintenance support resource scheduling problem belongs to the domain of resource-constrained project scheduling problem (RCPSP). Its model can be described as (Liu & Wang, 2001): there are totally J activities in a project, j ($j = 1, 2 \ldots, J$) is to be completed by M maintainers. The duration for the task

is S_J. Each activity has two kinds of constraints, i.e. temporal constraint and resource constraint. Temporal constraint comes into being if there is precedence relationship among several activities due to technical requirement. Assuming P_j is the immediate predecessor activity set of activity j, then for any activity $i \in P_j$, if i hasn't finished, j cannot start. As to resource constraint, there are $K = 3$ categories of maintenance workers (primary workers, intermediate workers and senior workers), total allocation quantity for category k ($k = 1, 2, \ldots, K$) maintenance workers is R_k. Assuming that number of repair workers required for completing activity j is r_j, of whom, the number of category k maintenance workers is r_{kj}, performance time is d_j, starting time of activity j is ST_j, A_t is any of the moment before the activity set item which is ongoing within the time period of $(t-1, t)$ finishes, and the total amount of human resources occupied by the ongoing activity should not exceed total quantity of manpower allocation. The following conceptual mathematic model can be established (Herroelen et al. 1998):

$$Opt. \quad \min S_J \tag{1}$$

$$s.t. \quad ST_j - ST_i \geq d_i, i \in P_j \tag{2}$$

$$\sum_{j \in A_t} r_{kj} \leq R_k, t = 1, 2, \ldots, S_j, k = 1, 2, \ldots, K \tag{3}$$

Formula (1) is objective function, showing minimum duration for the task; formula (2) is constraint condition, showing the precedence relationship among the activities; formula (3) shows the resource constrained relationship and the resource usage of each stage should not exceed its maximum supply quantity.

3 THE HYBRID CHAOS PARTICLE SWARM ALGORITHM OF SOLVING RESOURCE SCHEDULING PROBLEM

3.1 Coding

Considering that the scheduling problem of special vehicle service process flow includes two subproblems, dual coding based on real number encoding (Wang & Cao, 2002) is adopted for the scheduling scheme. The first group of codes refers to the operation order of the activity, which is an array of all activities meeting the temporal constraint. The first group of codes of the chromosome come from the initial activity chain calculated based on the Regret Based Random Sampling (RBRS) algorithm, and minimum relaxation time is chosen and used by the priority rules in the sampling algorithm, which is expressed by particle swarm parameter as: D dimension search space of particle, where D is the number of activities in the target problem, and the particle population scale is N, particle $x_i = (x_{i1}, x_{i2}, \ldots, x_{iD})$, thereinto, $(i = 1, 2, \ldots, N)$, corresponding to a performable activity chain order of the problem, which can generate a feasible scheduling scheme after being decoded. The second group of codes represent the human resources type constraints of performing various activities, and for each activity in the first group of codes, we should choose randomly a number from {1, 2, 3} (respectively corresponding to primary worker, intermediate worker and senior worker) to perform Y_i times according to the human quantity required, and meanwhile, the resource constraint relationship should be satisfied. The series connection of the selection conditions of each activity can generate second group of codes of chromosome. The decoding can be made by parallel scheduling scheme, during which process, the activity operation time t is calculated according to the following formula:

$$t = t_s \frac{C_i \cdot \alpha + Z_i \cdot \eta + G_i \cdot \beta}{Y_i} \qquad (4)$$

Of which, t_s represents the operation work time of each activity in the processing procedure, C_i, Z_i and G_i represent respectively the person numbers of primary, intermediate and senior workers of performing this activity, $Y_i = C_i + Z_i + G_i$, α, η and β represents the adjustment coefficient, $\alpha/\eta = 3/2$, $\beta/\eta = 5/6$ and $\eta = 1$.

3.2 Chaos particle swarm optimization based on logistic mapping

In the basic PSO algorithm (Kennedy & Eberhart, 1995), PSO initializes a swarm of random particles.

Assuming that the particle swarm is D dimension, the number i particle is $x_i = (x_{i1}, x_{i2}, \ldots, x_{iD})$, speed is $v_i = (v_{i1}, v_{i2}, \ldots, v_{iD})$, and then the optimal solution can be obtained by tracking individual optimal value $p_i = (p_{i1}, p_{i2}, \ldots, p_{iD})$ and population optimal value $p_g = (p_{g1}, p_{g2}, \ldots, p_{gD})$ to update and iterate the particles. In the basic PSO, for each particle, its number d dimension $(1 \leq d \leq D)$ should undergo the speed and position update based on the following formula:

$$v_{id}^{k+1} = \omega \cdot v_{id}^k + c_1 r_1 (p_{id} - x_{id}^k) + c_2 r_2 (p_{gd} - x_{id}^k) \qquad (5)$$

$$x_{id}^{k+1} = x_{id}^k + v_{id}^{k+1} \qquad (6)$$

Of which, k is an iterative algebra, c_1 and c_2 are the given learning (accelerating) factors, r_1 and r_2 are random numbers evenly distributed in [0, 1] which are mutually independent, ω is inertia weight, and k is the current iterative algebra. In the basic PSO algorithm, r_2 and r_2 are random numbers, which may lead to the uncertainty of searching out all of the points in the solution space. While the chaos system is ergodic, hereby the Logistic mapping technology is introduced, and the method is as follows (Zhang et al. 2012): assume a value of generation k of r and calculate the value of generation $k + 1$ of r according to the following formula:

$$r(k+1) = \mu \cdot r(k) \cdot (1 - r(k)) \quad r(t) \in (0,1) \qquad (7)$$

Of which, μ is the control variable, whose value is 4, it is specified that $r(0) \neq \{0.25, 0.5, 0.75\}$. After the change, the system is in a chaos state, so the chaos mechanism is introduced for controlling the parameters in the particle swarm algorithm, thus increasing the overall convergence of algorithm. In each time of iteration, r_1 and r_2 are determined as $r_1(k)$ and $r_2(k)$ rather than the random numbers, and the values are determined and updated according to the Equation 7. Equation 5 is changed to:

$$v_{id}^{k+1} = \omega \cdot v_{id}^k + c_1 r_1(k)(p_{id} - x_{id}^k) + c_2 r_2(k)(p_{gd} - x_{id}^k) \quad (8)$$

3.3 Fitness function

The value of the individual fitness function should be in direct proportion to the final operation time of the chromosome to which it corresponds, and its fitness function is designed as:

$$Fitness(i) = \frac{T_{max} - T_i + \eta}{T_{max} - T_{min} + \eta} \qquad (9)$$

Of which, T_{max}, T_{min} and T_i respectively represent the maximum and minimum of the hours needed for task after the population chromosome is decoded and the activity hours needed for number i individual, η is the balance factor.

3.4 Evolutionary strategy

In the basic particle swarm algorithm, particles are updated by tracking two extreme values: individual optimal value and population optimal value, this paper introduces the genetic operators based on particle swarm algorithm and creates the hybrid chaos particle swarm optimization, which can optimize the evolutionary strategy without changing the nature of particle swarm update.

3.4.1 Selection operator

We introduce the elite strategy and sort the current particle swarm according to the high to low order of fitness and select 10% of the particles enjoying the highest fitness to enter directly into next generation of particle swarm. The characteristic of this operation is reflected in the monotone increasing of optimal solution which can realize performance enhancement generation by generation, and its drawback is that it may lead to prematurity of population, and therefore, this paper overcomes it by using the mutation operators with high probability.

3.4.2 Crossover operator

The update method for standard particles swarm is not applicable to the expression method of the first group of particles based on activity chain. Therefore, this paper introduces the crossover operator in the genetic algorithm to update the position of the particles. In the improved Equation 8, the migration distance of the current particles is determined according to the distance between the particles and the individual optimal value and population optimal value, so as to update the position of the particles. This paper introduces the single-point crossover method (Zhui et al. 2006) of genetic algorithm into the update of particles, and the particles are updated by crossing at single-point with individual optimal value and population optimal value. Setting $\omega = 1.0$, $c_1 = c_2 = D$, $pos1 = c_1 r_1(k)$ and $pos2 = c_2 r_2(k)$ in the algorithm of this paper, choose the position point $pos1$ from activity chain to cross at single point with the individual optimal value and choose the position point $pos2$ to cross at single point with population optimal value. After the above setting, the update formula of particles Equation 6 and Equation 8 are changed to:

$$x_{id}^{k+1} = x_{id}^k \oplus c_1 r_1(k) p_{id} \oplus c_2 r_2(k) p_{gd} \qquad (10)$$

Particles are updated according to the following steps:

First step: particle x_{id}^k undergoes crossover operation with individual optimal value at the position of $pos1$, and the method is as follows:

(1) Determine the cross point $pos1$ of the particles according to the random number $r_1(k)$ and the segment composed by the preceding $pos1$ activities in the x_{id}^k is J_1, and the remaining segment is J_2;

(2) Take out the part with the particle individual optimal value p_{id} same as J_2 by keeping the original order to form a new segment J_3, with the remaining party forming a new segment J_4;

(3) Combine the activities in J_1 and J_3 to form a new particle y_1, and combine the activities in J_2 and J_4 to form a new particle y_2;

(4) Calculate y_1, y_2 and fitness value, and select the particle with a high fitness as the new positions x_{id}^k and individual optimal value after update.

Second step: particle x_{id}^k undergo crossover operation with population optimal value at the position $pos2$, thereinto, the population optimal value is chosen randomly from elite population. The method is the same as the first step, i.e. form two new particles z_1 and z_2, and select the particle of high fitness as the new position x_{id}^{k+1} of the particle after the update.

For the second group of human resource distribution order, we substitute multipoint random crossover operation for the method of standard particle swarm update.

3.4.3 Mutation operator

In order to prevent the particle prematurity resulting from selection operation, the mutation operation in this paper is designed to generate some new particles to substitute for the particles in the inferior population. These new particles generate randomly by coding operation, with the same distribution characteristics as the initial population, but without integrating the genetic information of the current population. By adding newly born particles in the particle swarm, we can avoid the premature convergence of particle swarm.

3.4.4 The evolutionary strategy after the introduction of genetic operators

After the above-mentioned genetic operators are introduced, the hybrid particle swarm-genetic algorithm is formed. After the generation of the population, we sort the current particle swarm according to the high to low order of the fitness.

(1) Select 10% of the particles with the highest fitness in the population for update after the sorting based on the strategy of elite selection;

(2) Complete the update of particles of conventional population by adopting the operation of crossover operator as designed in this paper.

(3) Generate 10% of new particles through mutation operator to substitute for the inferior population in current particle swarm.

4 SOLVING THE SCHEDULING OPTIMIZATION PROBLEM OF SPECIAL VEHICLE SERVICE PROCESS FLOW

Taking the three-level service process flow of some special vehicle as example, its operation detailed information is shown in Table 1. Because of the limited space, the following table only shows 40 items therein, totally there are 79 items of complete process flow.

The table for norm of service working hours provides that the time used for three-level service is 90 working hours, i.e. if the operation is based on current 7-person service scheduling, the per person average time consumption is about 13 hours.

Table 1. Detailed information about the three-level maintenance operation.

No.	Immediate predecessor	Person number	Working hour /min	No.	Immediate predecessor	Peron number	Working hour /min
1	None	1	18	21	None	1	36
2	1	1	18	22	None	1	42
3	2, 4	1	12	23	20	2	72
4	1	1	36	24	10	1	60
5	None	1	36	25	24	1	36
6	None	1	36	26	25	1	36
7	None	1	24	27	None	1	18
8	7	1	24	28	None	1	54
9	8	1	24	29	27	1	48
10	8	2	72	30	27	1	18
11	10, 23	2	36	31	29	1	72
12	11	1	48	32	28	1	36
13	12	2	36	33	32	1	48
14	13	2	30	34	32, 33	1	48
15	26	1	24		...		
16	15	2	36	75	40	1	48
17	16	1	18	76	69 to 72	1	30
18	17	1	30	77	76	1	48
19	5, 6	1	18	78	1 to 77	2	24
20	None	1	18	79	1 to 77	2	36

Table 2. Optimal scheduling scheme of 7 intermediate workers.

No.	Starting time	Worker assignment	Ending time	No.	Starting time	Worker assignment	Ending time
27	0	A	18	41	66	F	114
1	0	B	18	29	66	D	114
45	0	C	18	30	72	E	90
47	0	D	18	8	84	A	108
40	0	E	24	43	90	E	114
39	0	F	30	4	90	B	126
28	0	G	54	75	90	G	138
7	18	C	42	74	102	C	120
50	18	A	48	69	108	A	144
59	18	D	48	70	114	F	150
58	18	B	54	71	114	D	150
73	24	E	52	72	114	E	150
67	30	F	66	46	120	C	150
20	42	C	60			...	
68	42	E	72	16	408	B, E	444
49	48	D	66	14	408	D, F	438
5	48	A	84	17	444	A	462
21	54	G	90	18	462	C	492
6	54	B	90	78	492	E, G	516
22	60	C	102	79	492	A, B	528

Parameters of algorithm are set as follows: population scale is set as 80, maximum iterations of particle swarm algorithm are set as 200, weighing coefficient $\eta = 0.8$, and proportion of elite population is 10% of total number of population, proportion of interior population is 10% of total number of population. In order to make it easier to compare with the current service process flow, we take the repair workers at the same human resource level as an example, including 7 intermediate workers of A, B, C, D, E, F and G and work out the optimal scheduling scheme, which is shown in Table 2. As the length of this paper is limited, we hereby list 40 items therein.

The scheduling results show that if performing the tasks according to the scheduling, the 7 intermediate maintenance workers will use 528 minutes to fulfill the tasks of three-level service operation, i.e. 8.8 hours, and compared with the specified 13 hours of time limit, the efficiency increases by 32.3%. If the 7 workers are primary workers and senior workers, they will respectively use 13.2 hours (792 minutes) and 7.3 hours (440 minutes).

In the above example, citing the maintenance workers at the same level mainly is for comparing with the specified figures in the regulations, however, in actual scheduling, we usually need staff the maintenance workers at three levels in a mixed way. Assuming we use two senior workers, three intermediate workers and two primary workers and solve by using the algorithm as designed in this paper, the shortest time required is 8.6 hours (515 minutes), with the efficiency increased by 33.8% as compared with the specified time limit in the regulations, and its Gantt chart of human resource

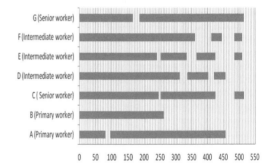

Figure 1. 7 maintenance workers mixed optimal scheduling Gantt chart.

scheduling is shown in Figure 1. We can see from the chart that the workload distributed to the primary workers is apparently less than that to the senior workers, which just conforms to the decision-making science giving precedence to superior talents.

4.1 Analysis on algorithm performance

Table 3 lists the shortest time limit obtained in solving the scheduling problems as mentioned in this paper respectively on the basis of Petri-net (Xu & Piao, 2004), genetic algorithm and the algorithm designed in this paper.

We can see from table 3 that when person number is 5 persons, the algorithm of this paper shortens the shortest hours needed for the task by 24 and 18 minutes respectively than that based on priority rule and on genetic algorithm; when person number is 7 persons, the shortest hours needed for the

Table 3. Shortest time limit based on the algorithm of this paper and the other two.

Number of workers	Petri-net based on Prior-fire rules Hours needed for the task /min	Based on genetic algorithm Hours needed for the task /min	Algorithm of this paper Hours needed for the task /min
5	672	666	648
7	558	552	528

task are shortened respectively by 30 minutes and 24 minutes; through comparison, we can see that the hybrid chaos particle swarm optimization adopted in this paper demonstrates a good performance in overall search , meanwhile, algorithm takes the actual conditions of maintenance and service operation into full consideration.

5 CONCLUSIONS

This paper adopts the dual real number encoding to design a kind of hybrid chaos particle swarm optimization. It makes use of genetic operators to rebuild the particle swarm algorithm, and meanwhile introduces chaos optimization technology to improve the overall convergence of algorithm. It takes the three-level service process flow as example, compared with the specified schedule, the results show that the efficiency increases by 32.3% and this algorithm is very effective and is superior to some existing methods.

REFERENCES

Davis, L. 1991. *Handbook of genetic algorithms*, Van Nostrand Reinhold, New York.

Herroelen, W. De Reyck B. & Demeulemeester E. 1998. Resource-constrained Project Scheduling: A survey of Recent Developments [J]. *Computers & Operations Research*, 25(4): 279–302.

Kennedy J, Eberhart R C. 1995. Particle swarm optimization [A]. In: Proceedings of the IEEE international conference on neural network [C]. Piscataway: IEEE Press, 1995: 1942–1948.

Liu Shixin, Wang Mengguang. 2001. A kind of genetic algorithm of solving the resource-constrained engineering scheduling problem [J]. *System engineering theory and practice*, 04: 24–27.

Liu Shixin. 2006. *Project optimal scheduling theory and method* [M]. Beijing: China Machine Press.

Shi Y, Eberhart R. 1998. A Modified Particle Swarm Optimizer [C]. Proc. *IEEE World Congress on Computational Intelligence*: 69–73.

Valls V, Ballestin F, Quintanilla S. 2008. A hybrid genetic algorithm for the resource-constrained project scheduling problem [J]. *European Journal of Operational Research*, 185(2): 495–508.

Wang Hong, Lin Dan, Li Mniqiang. 2005. A kind of self-adaptive genetic algorithm of solving the resource-constrained project scheduling problem [J]. *System engineering*, 23(12): 99–102.

Wang Xiaoping, Cao Liming. 2002. *Theory application and software realization of genetic algorithm* [M]. Xi'an: Xi'an Jiaotong University Press.

Xu Zongchang, Piao Yunhua. 2004. Research on a Petri-net-Based Method to Establish the Maintenance Support System [J]. *System engineering theory and practice*, 08: 141–144.

Zhang H, Li H, Tam C M. 2006. Particle swarm optimization for resource constrained project scheduling. *International Journal of Project Management*, 24(1): 83–92.

Zhang Xishan, Lian Guangyao and Yan Pengcheng. 2012. Research on maintenance support resource optimization method based on chaos genetic algorithm [J]. *Computer measurement and control*, 20(3): 741–746.

Zhui Xinggang, Wang Dingwei & Tang Jiafu. 2006. Task splitting project scheduling problem. *Journal of Northeastern University (Natural Science Press)*, 27(9): 961–964.

Information Technology and Computer Application Engineering – Liu, Sung & Yao (Eds)
© 2014 Taylor & Francis Group, London, ISBN 978-1-138-00079-7

New optimal assignment algorithm based on marginal association probability

Bin Sun

Information Center of Internet, Weifang Medical University, Weifang, Shandong, China

Qian Gao

Lijia Primary School of Fenghuang Street of Fangzi District, Weifang, Shandong, China

Xiao Ping Zhou

Information Center of Internet, Weifang Medical University, Weifang, Shandong, China

ABSTRACT: In view of the problem of long time spent of traditional optimal assignment algorithm of data association, firstly, this paper proposes a new optimal assignment algorithm of data association (OA). The new algorithm takes dynamic information to construct the model of optimal assignment algorithm and it has better association rate and better tracking performance. However, OA algorithm has a certain module error in the condition of dense targets and clutter. Therefore, on basis of OA algorithm, this paper proposes an improved OA (IOA) algorithm of data association and carries out corresponding simulation analysis. IOA algorithm makes full use of marginal association probability to replace the original statistic probability and improves the multi-target tracking performance in poor detection scenario.

Keywords: Optimal assignment; Marginal association probability; Dynamic information

1 INTRODUCTION

Pan Quan present Generalized Probability Data Association (GPDA) algorithm[1] in 2005. Compared with the typical algorithm JPDA algorithm[2–3], whether the practical associations are one-to-one or not, GPDA algorithm has the better real-time performance, less calculation and less memory. The GPDA algorithm includes the Serial Multisensor GPDA (SMS-GPDA) and the Parallel Multisensor GPDA (PMS-GPDA). Documents [3–5] have shown that the SMS-GPDA algorithm is better than the PMS-GPDA algorithm. However, because the SMS-GPDA algorithm is a cascade connection algorithm, estimate error of each sensor will be accumulated.

Multi-dimensional distribution algorithm[6–8] is a global optimal allocation algorithm and has better data association effect. But when the dimension of multi-dimensional assignment problem is equal to or bigger than three, the complexity will grow exponentially with the increase of the dimension of the assignment problem. And it is a NP-hard problem.

In order to further improve the tracking performance of the multi-target tracking system, this paper firstly put forward a new 3-dimensional (3-D) optimal assignment algorithm (OA) which is based on the dynamic information.

However, in scenario with dense targets and clutter, the new 3-D assignment algorithm model has a certain model error. This paper proposes an improved algorithm (IOA) which is based on the OA algorithm. The IOA algorithm takes use of the marginal association probability matrix between valid 3-tuple measurement combination and target tracks to replace the point-to-track association probability matrix of OA algorithm.

The simulation results have shown that when target density is relatively smaller, correct association rate and tracking performance of OA algorithm is superior to SMS-GPDA algorithm. But in dense targets and clutter environment, correct association rate and tracking performance of IOA algorithm is superior to OA algorithm.

2 MODEL DESCRIPTION OF OA ALGORITHM

Suppose that the state vector of target t follows the Normal distribution with mean value $X^t(k|k-1)$ and variance $P_t(k|k-1)$. Then the probability density function of event measurement i responding to target t is:

$$f_{it} = |2\pi S_t(k)|^{-1/2} \exp[-1/2 v_{it}'(k)S_t^{-1}(k)v_{it}(k)] \tag{1}$$

where, $v_{it}(k) = z_{it}(k) - Z_t(k|k-1)$ is the residual error vector; $S_t(k)$ is the covariance matrix of residual error of target t; Then the cluster probability statistics distance matrix can be denoted as:

$$F = [f_{it}] \quad i = 0,1,\cdots,m_k; \ t = 0,1,\cdots,T \tag{2}$$

Suppose that the point-to-track cluster probability statistics distance matrix based on three different sensors are $F_1 = [f_{it}]$, $F_2 = [f_{it}]$, $F_3 = [f_{it}]$, where i_1, i_2, i_3 are respectively measurement from sensor 1, 2 and 3. Then the basic probability value of measurement i corresponding to target t is:

$$m_{it} = f_{it} \bigg/ \sum_{t=0}^{T} f_{it} \tag{3}$$

Suppose basic probability value of measurements from different sensors corresponding to target t ($t = 0, 1, \ldots, T$) are respectively m_{1i_1}, m_{2i_2} and m_{3i_3}. And the focal elements are A_{i_1}, B_{i_2} and C_{i_3}. The inconsistent factor is $k_{i_2i_3} = \sum_{B_{i_2} \cap C_{i_3} = \phi} m_{2i_2}(B_{i_2}) \cdot m_{3i_3}(C_{i_3})$, where $A_{i_1}, B_{i_2}, C_{i_3} \subseteq U$, and U is the recognition framework of tartet t. Let

$$k_{i_1i_2i_3} = k_{i_1i_2} + k_{i_2i_3} \tag{4}$$

Then $k_{i_1i_2i_3}$ denotes the inconsistent factor of 3-tuple measurement combination $Z_{i_1i_2i_3} = \{Z_{1i_1}, Z_{2i_2}, Z_{3i_3}\}$ coming from the same target. And $K = (k_{i_1i_2i_3})_{n_1 \times n_2 \times n_3}$ is the association cost matrix. Then the data association problem based on the 3-sensor measurements can be transformed into the following 3-D assignment problem:

$$\min_{\rho_{i_1i_2i_3}} \sum_{i_1=0}^{n_1} \sum_{i_2=0}^{n_2} \sum_{i_3=0}^{n_3} k_{i_1i_2i_3} \cdot \rho_{i_1i_2i_3} \tag{5}$$

$$\begin{cases} \sum_{i_2=0}^{n_2} \sum_{i_3=0}^{n_3} \rho_{i_1i_2i_3} = 1; & i_1 = 1, 2, \cdots, n_1 \\ \sum_{i_1=0}^{n_1} \sum_{i_3=0}^{n_3} \rho_{i_1i_2i_3} = 1; & i_2 = 1, 2, \cdots, n_2 \\ \sum_{i_1=0}^{n_1} \sum_{i_2=0}^{n_2} \rho_{i_1i_2i_3} = 1; & i_3 = 1, 2, \cdots, n_3 \end{cases} \tag{6}$$

where, $\rho_{i_1i_2i_3}$ is binary association variable, if 3-tuple measurement combination is from the real target, $\rho_{i_1i_2i_3}$ is 1, otherwise $\rho_{i_1i_2i_3}$ is 0.

3 IMPROVED OPTIMAL ASSIGNMENT ALGORITHM BASED ON MARGINAL ASSOCIATION PROBABILITY (IOA)

3.1 Ideas of IOA algorithm

First, we will get multiple better solutions of the new optimal assignment problem by using Lagrange relaxation algorithm. Then all the solution components will be used to construct the set of valid 3-tuple measurement combination. Then we will use expression (7) to estimate the location (x_t, y_t) of target t:

$$\hat{x}_t = \frac{\sum_{s=1}^{S} \frac{x_s}{\sigma_{x_s}^2}}{\sum_{s=1}^{S} \frac{1}{\sigma_{x_s}^2}}, \quad \hat{y}_t = \frac{\sum_{s=1}^{S} \frac{y_s}{\sigma_{y_s}^2}}{\sum_{s=1}^{S} \frac{1}{\sigma_{y_s}^2}}, \quad \hat{z}_t = \frac{\sum_{s=1}^{S} \frac{z_s}{\sigma_{z_s}^2}}{\sum_{s=1}^{S} \frac{1}{\sigma_{z_s}^2}} \tag{7}$$

where $x_s = x_t + u_s$, $y_s = x_t + v_s$, $z_s = x_t + w_s$ are respectively the measurement components from sensor S of target t on the three coordinate axises.

Take the above estimate point as the virtual measurement corresponding to the 3-tuple measurement combination.

$$\lambda_{jt}^1(k) = \exp[-1/2 v_{jt}'(k) \cdot S_t^{-1}(k) \cdot v_{jt}(k)] \big/ \sqrt{2\pi S_t(k)} \tag{8}$$

where, $j = 0, 1, 2, \cdots, m$; $t = 0, 1, 2, \ldots, T$; m is the number of measurements following into the tracking gate at time k; T is the number of targets; $v_{jt}(k) = z_{jt}(k) - \hat{Z}_t(k|k-1)$, $z_{jt}(k)$ is the jth measurement from target t at time k, and $\hat{Z}_t(k|k-1)$ is the predicted measurement from target t at time k. $S_t(k)$ is the covariance matrix; then the corresponding point-to-track association can be denoted as

$$\Lambda^1(k) = [\lambda_{jt}^1(k)] \tag{9}$$

By solving module (5), we can get m good solutions. Components of the m good solutions, that is, components of the valid 3-tuple measurement combination, unavoidably, have repeated measurements of the same target track. So, the point-to-point assignment can't better reflect the real association situation. Therefore, under the multiple-to-multiple rules, we construct the marginal association probability between the valid 3-tuple measurement combination and the target track.

By normalizing the matrix $\Lambda^1(k) = [\lambda_{jt}^1(k)]$ according to the measurement and the target respectively, we can get new matrix elements $\xi_{jt} = \lambda_{jt} / \sum_{t=0}^{T} \lambda_{jt}$ and $\varepsilon_{jt} = \lambda_{jt} / \sum_{j=0}^{m} \lambda_{jt}$. Then the marginal association probability[1] between the valid 3-tuple measurement combination j and target t is:

$$\lambda_{jt}^2(k) = \frac{1}{C} \left(\varepsilon_{jt} \prod_{\substack{tr=0 \\ tr \neq t}}^{T} \sum_{\substack{r=0 \\ r \neq j}}^{m} \varepsilon_{rtr} + \xi_{jt} \prod_{\substack{r=0 \\ r \neq j}}^{m} \sum_{\substack{tr=0 \\ tr \neq t}}^{T} \xi_{rtr} \right) \tag{10}$$

where tr denotes the trth target; r denotes the rth 3-tuple measurement combination; c is the normalization coefficient. $\lambda_{jt}^2(k)$ is the marginal association probability between the valid 3-tuple measurement combination j and target t. So, according to the one-to-one feasibility rule, by solving the 2-D assignment problem whose association cost matrix is $\Lambda^2(k) = [\lambda_{jt}^2(k)]$, we can get the optimal point-to-track association match based on the marginal association probability.

3.2 Steps of IOA algorithm

Step 1 To get m better solutions of module (5) by using Lagrange relaxation algorithm. And then all the solution components will be used to construct the set of the valid 3-tuple measurement combination.

Step 2 According to module (8), construct cluster probability statistics distance matrix between valid 3-tuple measurement combination and target track.

Step 3 According to module (10), transform the cluster probability statistics distance matrix of step 2 to the marginal association probability matrix.

Step 4 Get the optimal point-to-track association by solving the assignment problem with cost matrix of marginal association probability matrix,

Step 5 Update the state of targets by using the optimal association result got in step 4.

4 SIMULATION ANALYSIS

Under different simulation scenarios, this section implements simulation comparisons between algorithms of SMS-GPDA and OA, algorithms of OA and IOA. Suppose that three sensors are used to track eight targets and they move at a constant speed in two-dimensional plane. The three sensors have the same distance measurement error and azimuth measurement error; The scanning internal $T = 2s$; Gate probability $P_g = 0.9997$; Detection probability $PD = 0.98$; Clutter density $\lambda = 0.01/km^2$; simulation step is 150 and the number of simulations is 50.

4.1 *Comparison between OA and SMS-GPDA algorithm*

(1) In different target distances, comparisons of Root-mean-square error (RMSE) of OA and SMS-GPDA are as shown in Figure 1 and 2.

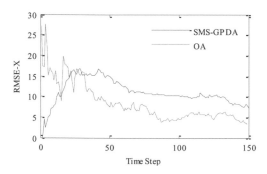

Figure 1. $d = 200\,m$, $e_r = 150\,m$, $e_\theta = 0.01\,rad$, RMSE comparison between OA and SMS-GPDA.

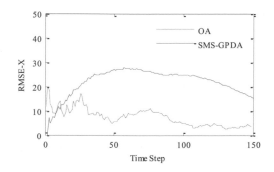

Figure 2. $d = 500\,m$, $e_r = 150\,m$, $e_\theta = 0.01\,rad$, RMSE comparison between OA and SMS-GPDA.

(2) In different target distances and measurement errors, track curve comparisons of target 1 between OA and SMS-GPDA are as shown in Figure 3 and 4.

From figure 1 and 2, we can see when target density is moderate, tracking performance of OA is superior to SMS-GPDA. From figure 3 and 4, we can see when distance of targets is smaller, tracking performance of SMS-GPDA is higher. However, when distance is bigger, tracking performance of OA is higher. This is because OA is a global optimal assignment algorithm, when target density is relatively smaller, its correct association rate is higher and tracking performance is better. SMS-GPDA is a sub-optimal data association algorithm, and it is applicable to dense target environment.

4.2 *Comparison between OA and IOA algorithm*

In different distances of targets and measurement errors, simulation comparisons of OA algorithm and IOA are as follows:

(1) Distance of targets d = 200 m.
(2) Distance between targets d = 500 m.

From figure 7 and 8, we can see when distance of targets is bigger, distance measurement error and azimuth measurement error are bigger, the multi-target tracking performance effect of IOA is superior to OA, but the increase is not large. From figure 5 and 6, we can see with the decrease of target distance, IOA is still superior to OA, and the optimization of amplitude increases. This is because in dense targets and

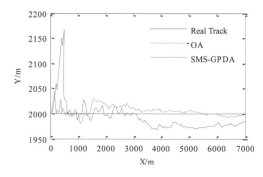

Figure 3. $d = 200\,m$, $e_r = 150\,m$, $e_\theta = 0.01\,rad$, Track comparison between OA and SMS-GPDA.

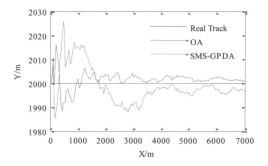

Figure 4. $d = 500\,m$, $e_r = 200\,m$, $e_\theta = 0.02\,rad$, Track comparison between OA and SMS-GPDA.

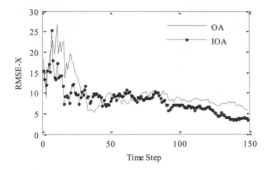

Figure 5. $e_r = 150\,\text{m}$, $e_\theta = 0.01\,\text{rad}$, RMSE comparison between OA and IOA.

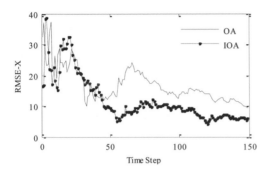

Figure 6. $e_r = 200\,\text{m}$, $e_\theta = 0.02\,\text{rad}$, RMSE comparison between OA and IOA.

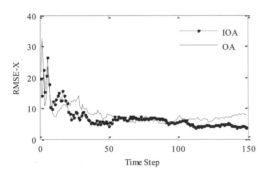

Figure 7. $e_r = 150\,\text{m}$, $e_\theta = 0.01\,\text{rad}$, RMSE comparison between OA and IOA.

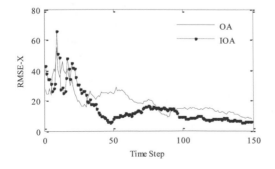

Figure 8. $e_r = 200\,\text{m}$, $e_\theta = 0.02\,\text{rad}$, RMSE comparison between OA and IOA.

clutter environment, 3-D assignment problem of data association has a certain module error, by using the traditional algorithm, the optimal solution may not correspond the exactly association. Because IOA can take full use of the useful information of marginal probability, it can reduce the influence of module error of the multi-dimension assignment problem in different extent, and improves the multi-target tracking and positioning accuracy.

5 CONCLUSIONS

In order to further improve the tracking performance of the multi-target tracking system, this paper first proposes a new OA algorithm. It is simple in principle, easy to implement. Compared with SMS-GPDA algorithm, it is also more applicable to solve the data association problem in better and secondary detection environment. In order to further improve the tracking performance in dense targets and clutter environment, this paper further proposes a new IOA algorithm. The simulation results show that, tracking performance of IOA algorithm is obviously superior to the original algorithm and has better stability. Theoretical analysis and simulation experiments have verified the feasibility and validity of the new algorithm.

REFERENCES

[1] Pan Quan, Ye Xi-ning, Zhang Hong-cai. 2005. Generalized Probability Data Association Algorithm. Acta Electronic Sinica, 33(3): 467–472.
[2] BAR-SHALOM Y, Fortmann T E. 1998. Tracking and Data Association. New York: Academic press.
[3] He You, Wang Guo-hong, Lu Da-jin, Peng Ying-ning. 2007. Multisensor Information Fusion With Applications. Beijing; Publishing House of Electronics Industry.
[4] Pan Quan, Liang Yan, Yang Feng, Cheng Yongmei. 2009. Modern Target Tracking and Information Fusion. Beijing: National Defense Industry Press.
[5] Qian Gao, Hailin Zou, Li Zhou, Lili Li. 2012. Multisensor Generalized Probability Data Association Algorithm Based on Data Compression, Journal of Computational Information Systems, 8(5): 2153–2160.
[6] Pattipati K R, Deb S, Bar-shalom Y, et al. 1992. A new relaxation algorithm and passive sensor data association. IEEE Transactions on Automatic Control, 37(1): 198–213.
[7] Deb S, Yeddanapudi M, Pattipati K, et al. 1997. A generalized S-D assignment algorithm for multisensor-multitarget state estimation. IEEE Transactions on Aerospace and Electronic Systems, 33(2): 523–537.
[8] Popp R, Pattipati K, Bar-shalom Y. 2001. m-Best S-D assignment algorithm with application to multitarget tracking. IEEE Transactions on Aerospace and Electronic Systems, (37): 22–36.

Information Technology and Computer Application Engineering – Liu, Sung & Yao (Eds)
© 2014 Taylor & Francis Group, London, ISBN 978-1-138-00079-7

Design and implementation of data transformation scheme between STEP and XML in CIMS environment

Min Zhou, Jian Hua Cao & Guo Zhang Jiang
College of Mechanical Automation, Wuhan University of Science and Technology, Wuhan, Hubei, China

ABSTRACT: CIMS (Computer Integrated Manufacturing System) is a complicated system about the manufacturing process. Product information in the internal or external system exchange process involves the compatibility among different platforms, different support environment and different data structure. XML (eXtensible Markup Language) provides an effective solution. Based on Java STEP kit JSDAI and the Eclipse development tools, a new transformation model and realization scheme between STEP and XML is put forward. The key technologies of implementation scheme are introduced, and through a concrete case, the feasibility of this scheme is verified.

Keywords: CIMS; JSDAI; STEP standard; EXPRESS; XML

1 INSTRUCTIONS

The STEP (Standard for the Exchange of Product Model Data) [1] standard is to standardizing the description of information in the whole product life cycle, and the STEP standard is the standard for the exchange of perfect product data model. It provides a neutral mechanism which does not depend on the specific system, to realize the data exchange and share. This nature of the description makes it not only suitable for the swap file, also suitable for execution and share product database and archiving. The developed countries have put the STEP standard to the industrial application. Its application can significantly reduce the exchange cost of the information in product life cycle, improve the efficiency of product development, become important foundation for standard international cooperation and participation in international competition manufacturing industry, and it is an important tool to maintain the competitiveness of enterprises.

XML (eXtensible Markup Language) is a subset of SGML (Standard Generalized Markup Language) developed by Internet Association (W3C), it inherits the main advantages of SGML, overcomes shortcomings of the structure and expansibility of the HTML [2]. Briefly speaking, XML has the following main features: Self-describing, semi-structured, Independent of platform and application, machine processable, Scalability, Wide support and so on.

XML language and STEP standard combined method under Internet environment is a new generation of information integration. Because the STEP data (Fig. 6) is a series of abstract data which is a description of product information and not readable to people, so converting product data in the STEP file to XML, can make full use of the advantages of XML to achieve the exchange and integration of product information in CIMS environment.

2 THE RESEARCH STATUS AT HOME AND ABROAD

At present, there are many domestic and foreign research institutions engaged in research work of STEP and XML data transformation. The United States of America STEP Tools Company develop a kind of XML programming technology for STEP data sharing, it is integrated in ST-Developer tools and powerful; The Fraunfofer Institute of Germany propose the model of using shared 3D viewer product collaborative design under distributed development environment [3]. Wang Min in Huazhong University of Science and Technology put forward method of indirect transformation through the corresponding mode [4]; Yin Yi in Dalian University of Technology research on the method of designing heterogeneous product information integration system, which is based on XML [5].

The existing methods mainly have the following disadvantages: 1. Platform dependencies. Many approaches are developed on Windows platform, failed to fundamentally solve the heterogeneous data sources integration problems. 2. Compatibility is poor. Many approaches can realize mutual transformation functions, but cannot be a friendly integration with other systems. 3. The charge is high. Some approaches are not open source software, Economy is poor and they are difficult to promote.

Java is general, efficient, platform independent, safe and network supported very well, at the same time, JSDAI (Java-Standard Data Access Interface)

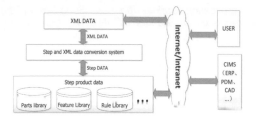

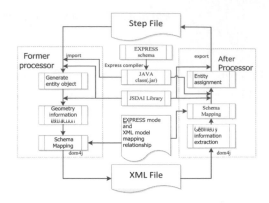

Figure 1. Overall program plans.

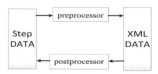

Figure 2. STEP and XML transformation diagram.

is an open source toolkit. As a result, this paper puts forward a realization scheme of mutual transformation between STEP and XML based on JSDAI. As shown in Figure 1, the transformation system is B/S architecture which is based on SSH2, the user or any other systems can access the system through the browser or the corresponding interface in WAN or LAN easily.

To realize the transformation between STEP and XML, the key lies in the realization of the two processors – preprocessor and postprocessor (Fig. 2). The processor mainly realizes the transformation from STEP data to XML data; Postprocessor mainly realizes the transformation from XML data to STEP data.

3 THE DESIGN OF DATA TRANSFORMATION SCHEME

The specific process of transformation between STEP and XML is shown in figure 3. Firstly, using EXPRESS language to prepare the corresponding file *.exp, according to the CAD software (Pro/E) to generate the STEP AP214 file, packaging the *.exp file to JAR packet through EXPRESS compiler tool, then in the jar package there is the *.exp file which corresponds to JAVA file. And then import the STEP AP214 file *.stp, JSDAI will automatically match the geometry information which is in the *.stp file and JavaBean which is generated by *.exp files, generate a set of entity object, so you can easily get the geometric information from the entity object. Finally generating the corresponding XML file according to the mapping relationship between EXPRESS mode and XML mode through the XML API – Dom4j [6], the file can be in the form of webpage, used for other users on the network, and also be used for other system. Then the function of preprocessor is completed.

After accessing the XML file, getting the geometric information of product through the Dom4j analysis, this information can be used for other business system, can also assign data to the JAVA object that in *.jar file, according to the mapping relationship between EXPRESS mode and XML mode, finally, exporting

Figure 3. Specific transformation process based on JSDAI.

the *.stp file which can be directly used for CAD software, and the function of the postprocessor is completed.

4 THE KEY TECHNOLOGY TO REALIZE THE SCHEME

To achieve the above scheme, there are two key problems need to be solved: 1) mapping between EXPRESS and JAVA; 2) mapping between EXPRESS and XML.

4.1 Mapping between EXPRESS and JAVA

Realization of the mapping between EXPRESS and JAVA, in fact is the realization of language building between EXPRESS language and Java language, the process of language binding is essentially the mapping relationship between data relationship. The EXPRESS compiler tool provided by JSDAI provides two kinds of binding mode: early binding and late binding.

Data types in EXPRESS can be divided into the following categories: Simple types, Aggregation types), Defined types, Entity types, Enumeration types, Generic types, Select types etc. Because there is only exists some basic types (such as INTEGER, REAL etc.) and the entity or entities combination type in Part21 files, so JSDAI only maps the basic type and the entity types. Entity types have many forms, below introducing the mapping relationship between EXPRESS and JAVA in the process of early binding and late.

1) Mapping rules of early binding
Through the early binding from JSDAI to EXPRESS files on the early joint editor, entities in the EXPRESS will generate some Java interface or class, these class or interface provide the corresponding get () and set () method for us to operate on its internal properties. These entities abide by the following rules during the mapping process:

• Using Java interface with the "E" prefix to express entity data types, such as "EXxx";

- Using Java class with the "C" prefix to express composite entity data types, such as "CXxx";
- Splitting the entity data types which have a plurality of leaves with the "&" symbol segmentation;
- Using Java class with the "A" prefix to express aggregate entity data types, such as "AXxx"

2) Mapping rules of late binding

Through the early binding from JSDAI to EXPRESS files on the early joint editor, entities in the EXPRESS will generate some Java interface or class, but the mapping rule of these entities in the process is different from the early binding:

- All entity data types and complex entity data types are represented by the JAVA interface EEntity;
- All aggregate entity data types are represented by the JAVA class AEntity;

The Differences between early binding and late binding are shown as follows:

- Each early binding entity is directly or indirectly inherited from EEntity, such as:
  public interface EXxx extends EEntity {...}
  public interface EYyy extends EXxx {...}
- Each early binding aggregate class is directly inherited from AEntity, such as:
  public class AXxx extends AEntity {...}

In order to facilitate the operation of the entity attributes, this paper uses early binding.

4.2 *The mapping between EXPRESS and XML Schema*

A mapping between STEP and XML mainly is the mapping between data structures, so it is the essence of the mapping relation between EXPRESS and XML Schema. The STEP Part28 standard recommend two mapping method from STEP to XML [7]: early binding and late binding. They will be discussed below, through an example.

In the early binding map, there are corresponding data types and attributes of markers and EXPRESS data model in XML. For example, for a rectangle following the EXPRESS definition, the rectangle consists of one ID and two attribute values of width and height:

```
ENTITY rectangle;
  id:STRING;
  width,height:REAL;
END_ENTITY;
```

STEPPart21 defines the instance data as follows:
5=rectangle ('r',3,4);
If using early binding, the EXPRESS model generate XML document:

```
<rectangle id='r'>
  <width>3</width>
  <height>4</height>
</rectangle>
<ENTITY id="# 5"name="rectangle">
  <attribute name="id">
```

```
    <STRING>r</STRING>
  </attribute>
  <attribute name="width">
    <REAL>3</REAL>
  </attribute>
  <attribute name="height">
    <REAl>4</REAL>
  </attribute>
</ENTITY>
```

In the late binding, XML tag does not directly correspond to the type of data in the data model of EXPRESS, but to the EXPRESS metadata object (including entities, attributes, and data type). If using late binding, the EXPRESS model generate XML document:

```
<ENTITY id="# 5"name="rectangle">
  <attribute name="id">
    <STRING>r</STRING>
  </attribute>
  <attribute name="width">
    <REAL>3</REAL>
  </attribute>
  <attribute name="height">
    <REAl>4</REAL>
  </attribute>
</ENTITY>
```

From the above comparison shows that, the mapping of late binding of EXPRESS to XML Schema is more suitable for XML switching applications that contain a plurality of EXPRESS information model, if using the strategy of early binding in such an environment, design of label for XML for each different EXPRESS model is necessary, which will make the realization complex. So the late binding mode should be used, although the early binding is more refined and simple than late-bound [8]. After the binding method of EXPRESS and XML Schema is determined, the mapping between EXPRESS and XML Schema can realize according to Part28 standard. Because the XML Schema and EXPRESS are very similar in definition of data structure, the EXPRESS tag or new defined markers can be used to describe the geometric information, and using the corresponding XML Schema to define XML document, to realize the mapping between STEP and XML. After the creation of XML Schema and determination of structure of the XML document, the entity data that is extracted from the STEP file can be converted into the corresponding XML elements. Finally the XML file can be generated according to the structure of XML document determined by Schema, and the transformation from STEP to XML is completed, and vice versa.

5 APPLICATION EXAMPLES

Below using the above scheme validate the feasibility of this project, according to the actual engineering model. As the shaft in the machinery industry is a common part, and it has a wide range of applications.

Figure 4. The Step file generated by the Pro/E.

Figure 5. Display of XML file in IE browse.

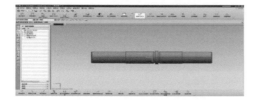

Figure 6. The model regenerated by STEP file in UG.

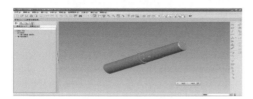

Figure 7. The model regenerated by STEP file in Pro/E.

Therefore, the cases selected axis 3D model as a data exchange.

Figure 4 is the exported STEP file which is a 3D model of the stepped shaft designed in Pro/E software displayed in the Eclipse.

As shown in Figure 5, it is the corresponding XML document generated from STEP file after the dispose through the preprocessor. Lots of Geometric information, such as the number of parts, can be obtained from it. And then the information can be used in other systems, for instance, the ERP system. So the efficiency can be improved largely.

Because XML has many advantages, so the XML file can be used by other users or system through Web. After accessing to XML documents by web transmission, the corresponding STEP file can be obtain through the post processor. Importing the STEP file into the CAD software that supports the STEP standard (UG, Pro/E) 3D model shaft can be regenerated (Fig. 6 & Fig. 7).

This instance shows the transformation scheme of STEP and XML data based on JSDAI is feasible, and the transformation of STEP and XML shows no distortion of information ensure the accuracy of the information exchange.

6 CONCLUSION

The transformation scheme of STEP and XML based on JSDAI achieves the exchange of integration product data in the CIMS environment, provides a data exchange platform for distributed and heterogeneous data sources in CIMS environment. And it also provides some corresponding access interface for other systems. it can be conveniently integrated with other system and has good scalability. However, the present research to validate the feasibility of this project is only through a relatively simple example. To really reflect its practice value, a further improved to support complex data model is needed, and a large range of related protocol in STEP standard for mapping is necessary.

ACKNOWLEDGEMENTS

This research reported in the paper is supported by the National Natural Science Foundation of China (Grant No. 71271160, No. 51205295), Innovation team project of Hubei Province (Grant No. T201102).

REFERENCES

[1] Xiankui Wang, 2008. *Manufacture technology of machinery*. Beijing: Tsinghua University Press. pp. 3–15.

[2] Hao Li, 2012. *XML and related technologies*. Beijing: Tsinghua University Press. pp. 5–10.

[3] Nikola, B. Serbedzija, 2000. Web computing framework, *Journal of System Architecture*. 20(3):1293–1306.

[4] Min Wang, Bo Wu. 2007. Application of STEP and XML in terms of product information sharing. *Journal of Huazhong University of science and technology*. 25(1):58–60.

[5] Yi Yin. 2004. Heterogeneous product information integration research in collaborative design based on XML technology. *Dalian University of Technology*. 12–14, pp. 21–23.

[6] Xinyi Zhang, 2009. *Concise XML tutorial*. Beijing: Tsinghua University Press. pp. 25–30.

[7] Qin Yang, Jianghong Han, Jiacheng Zhu, 2005. Studying on Production Planning and Scheduling for Dynamic Alliance. *Development & Innovation of Machinery & Electrical Products*. 18(21), pp. 194–199.

[8] Xiaohong Sheng, Jinzhong He, Kai Li.2009. Research on Transform file from EXPRESS Format into XML Format Based on Agile Manufacturing Product. *Modern Manufacturing Engineering*. 2009(8), pp. 27–32.

Information Technology and Computer Application Engineering – Liu, Sung & Yao (Eds)
© 2014 Taylor & Francis Group, London, ISBN 978-1-138-00079-7

An integrated framework for software product line in the collaborative environment

ZhuYun Duanmu
Science and Technology on Information Systems Engineering Laboratory, Nanjing, China

Hui Xu
Nanjing Research Institute of Electronics Engineering, Nanjing, China

ABSTRACT: Software Product Line (SPL) engineering is a popular approach which improves re-usability of software in a large number of products that share a common set of features. This article discusses in details an integrated framework for the SPL with flexible distributed architecture, and focus on the collaborative environment and parallel technology.

Keywords: software product line; collaborative environment; parallel

1 INTRODUCTION

Software product line (SPL) is a comprehensive approach to develop and evolve software products over time. An SPL is a set of software development system tools, which share some common component and manage the development process, standard and assets. These features satisfy a particular task needs, and they are developed by a group of highly re-used tools (Cai et al. 2004). Stable yet flexible architecture is the basis of a successful SPL. There are a lot SPLs at present, such as ALOAF, PLARG, Jade Bird (Xie & Yang 2000) and CTAIS, but most of current solutions focused on the component-based software engineering and paid insufficient attention to the work environment and organization pattern. In this paper (Yu et al. 2012), we will introduce an integrated framework for SPL with a distributed architecture, and focus on the collaborative environment (Xia et al. 2012).

2 OVERVIEW

In this section we describe the model of the framework, and talk about the way it works. In order to meet the need of cooperation, we make the framework a distributed system, which consists of three parts as shown in Figure 1. The server is the major facility of the system, which provides most functions of SPL such as object management, tool management, product management, message processing, etc. The tool store receives and manage the SPL tools submitted by tool producers, expose these tools to the server with a

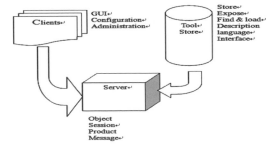

Figure 1. Architecture of SPL.

designed protocol, and it also offers some necessary services for the server to query and load the SPL tools. The client is a workspace to interact with the user. A client is mainly made of the administration interface and the working interface.

People open the administration interface in a client with proper permissions to connect the server, and request a list of SPL tools provided by the server, then select some of tools, and design the work flows for the specific software. Afterwards, people can do their jobs in the working interfaces after checking in the session in their own clients. Jobs are itemized by tools, every people takes charge of several items, get the software products from the system and work on them, then submit it to the system after finishing it in the working interface. The submittal will trigger a judgment, and one will be noticed if his requirements are satisfied, and do their works in a collaborative environment following a well arranged work flow. Moreover,

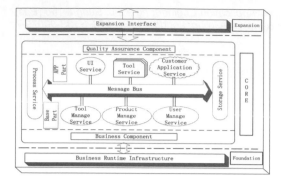

Figure 2. Detailed Structure.

administrators can monitor and control the progress in their clients. Detailed structure is shown as Figure 2.

3 SERVER

As mentioned above, the server plays the core role in the system. It communicates with the clients and the tool stores with a designed message system like a SOA infrastructure (Xu & Cao 2005) does. We can simply turn this system into an SOA/Web services (Lin & Zeng 2010) if necessary. In this case, we adopt OSGi to build our server for the benefit of its modularity, re-usability, flexibility, multitasking, PnP and the convenience to integrate into a very popular IDE (for example, Eclipse). For the following, we will discuss the services or management functions that the server should offer in details.

3.1 Object management

The kernel of the server is a container in which the instances of tools and other components run. It provides the runtime support based on OSGi (Geir 2010). As in the collaborative environment, it should be able to handle the following situations:

1) Creating objects: Because an object may come from components on the server, or be a part of the instance of an SPL tool, there should be an effective approach to locate the object to load, taking Java for example, to get the class path. There are two choices to create an object. One is on the source level, which means to obtain the source of the class (or other resources that can be used to instantiate an object) at the first stage, then to create the object. The other is to send a request to the manager of the resource for the object, leaving the responsibility of object generation to the manager, and then the manager produces the object and returns it in a proper form with some technology like Serialization.

2) Updating objects: Our system is supposed to be hot-plug-able. With a view to the frequent upgrade of some components, especially the SPL tools, and the continuity of a certain work flow, we may have to maintain same objects but in different versions.

The system should be capable to find out which version of the object instance to call.

3) Sharing objects: Since we are in a collaborative environment, there could be a great chance that people are using the same tools, which means a lot of multiplexing happen here. Here we have to deal with the traditional problems accompany with the object sharing such as resource allocation, deadlocks, synchronization, etc.

4) Destructing objects: It is not so easy to determine whether to destroy an object in consideration of the complicated concurrency, multi-version and sharing problems. Some tools will be awakened a short time since its destruction. As stated above, loading and destructing a tool may cost a lot, so here we employ a garbage collector running LRU to deal with the oscillation problem. It is hard to be the best solution, but a pragmatic one.

3.2 Session management

In a collaborative environment, an important subject is how to organize and manage the cooperation works. We introduce the idea of session to work on this topic. A session is defined as an instance of a work flow and its context. When a work flow is published, people with proper permissions can start it, which means to make an instance of the work flow. Every instance will be allocated for some necessary resources and a unique session ID which identifies this instance by the system in the first launch. A work flow can be started many times, but they are different work processes. People who want to join the work must select exactly the session ID to join the right session. The session maintains a context in its resources, which contains at least information listed below:

1) Administration: Information about the members in the session, including the roles and permissions assigned to them.

2) Catalog: Information about the component such as SPL tools and software products involved in the work, and their relationship.

3) Schedule: Information about the relationship between this session and work flows, and information about the relationship between this session and other sessions.

4) State: Information that reveals the completion of the job, and information about the running status of every component that is necessary to reproduce current state of the process.

5) Task record: Information that records the operations by job members for audit or monitoring activity.

6) Extension: Extra area to store user defined information.

In conclusion, a session can fully represent a job process. So we are able to monitor and control the job by manipulating its session data. Furthermore, with the help of session, we can also create a snapshot of the process when it is interrupted, store it and reload it whenever we want to return to the process. Besides, session is allowed to nest in order to support

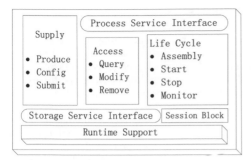

Figure 3. Session Manager.

job division. The session manager structure is shown as Figure 3.

3.3 Product management

Products refer to the artifacts produced by SPL tools or introduced by users. The product can be executable software, a module or a document, etc. Products are the data for tools to process. In our system, tools are designed to specify their inputs and outputs in the form of products, they must wait until its input products are ready, and put out the products as a return. The product management in our system is capable to meet the demands following:

1) Storage management. Solution for how to save the products, basically a proxy of the filesystem or database.
2) Access management: Permission control system used by the tool manager and session manager to judge if a request of the product can be allowed.
3) Specification management: Products are used as the inputs and outputs by SPL tools, there must be a well-formalized specification to describe them. This function ensures every product accompanying by the right specification, so that it can be tested before being passed to the tools.
4) VCS: A version control system is essential, not only for control the production of software artifacts, also for accounting and logging.

The product management guarantees that the products will be well kept and accessed, and be self-explanatory.

3.4 Message management

This system is driven by messages. There are three kinds of messages in our system. The first one is query, which is used to discover the services needed. The second one is notification, which is used to inform the listeners the happening of some events. The last one is command, which is used to request an operation. Query is the only one that can be broadcast without a specified target, and can be ignored by receivers. After broadcasting a query, the sender should set a buffer to receive multiple replies, and also have to start a timer to stop waiting in case nobody answers. Notification

is another story, senders never wait for the replies, but have to identify all the receivers in the notification. Both query and notification can be ignored by the receivers, but command must be replied by the receivers. A command always carries the necessary parameters of the operation requested and answers the return values in the reply.

We use the star topology to build the messages management for the sake of briefness and efficiency. The star topology is easy to extend both in scales and functions. When we want to introduce a new type of message, we simply upgrade the server and the clients that have to use that message. There is a message transfer and management module in the server and every source send its message to the server. All the receivers should register themselves on the server, so that they will be assigned a message box number. The server will read some meta data of the messages such as priority, classification and the receivers' box number as soon as they arrive at the server to handle in different ways. Messages will be delivered to its receiver directly with the mechanism of message box and star topology, so it reduces the difficulty of the receivers, the designer just have to get along with the message they need. Replies also are delivered by the server. Since the server controls the whole transfer of messages, we can append a filter on the server, so we have the chance to add a permission confirmation, and to make sure that vicious or mis-designed messages will be dropped and do no harm to other modules. Moreover, with the advantage of centralization of message processing, we can track the job by logging its messages, it is an effective way to diagnose in the message-driven system.

4 TOOL STORES

The SPL tools are the major participants in software productions. There are different tools in the system such as product tools, test tools, control tools, etc. Tools are designed under some certain standard established to make sure they are able to load by the framework. The tool store is just a logically-independent terminal, which can be deployed on server machines or just a single one, or even integrated with the server on the same machine. The reason for us to split it is to make the SPL tool development a independent work so that people could develop their own tools, upload it to a trusted tool store, and set the permission to control the users in a collaborative environment. An independent tool store is also more flexible when we want to upgrade and extend. It is just a collection of services provider for the server, and we do not care the details of the implementation as long as the messages could be well processed and answered. We will discuss some essential responsibility for the tool stores in the following.

4.1 Tool management

Tool management offers the service that tool providers can upload the tool and then leave the management

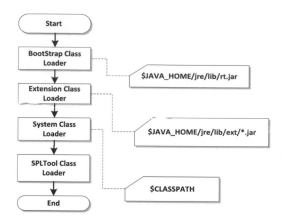

Figure 4. Tool presentation.

to the system. After reception of the tools, the store extracts and analyses its description. The description is in a designed form called TDL (Tool description Language). TDL is used by the tool designers to tell the system some important information about their product, including entry point, interfaces, inputs, outputs, parameters of functions, deployment and descriptive summaries, etc. Our framework supports TDL in two forms. One is Java annotation, the other is XML. The former is encouraged, because it is embedded in the program so that can be easily written in the stage of tool development, and so be the maintenance. TDL in XML is mainly used to describe and deploy the tools that already exist.

After finishing the analysis, the tool will be stored in a database and have its rules of competence set under the direction of deployment in the description data. Moreover, the tool store also has to support the common operations of update, deletion, modification, etc. So a friendly interface is necessary.

4.2 Tool presentation

As discussed above, the tool stores should supply the tools to the server, which means a tool store should have a mechanism to report and provide to the server the tools stored in it. With the help of TDL, the tool shows both its services and requirements. Then the tool store gathers these information and generalizes a table of its tools. There are two ways in our system to tell the server about it: to send a notification to register it actively, or to return it in the reply of a tool-discover-query passively.

After showing the tools it offers, the tool store should be capable to make these tools available. We mentioned above two ways to run the tools on the server, while another solution is to run the tools on the tool store itself, and to share the functions by ways of RMI, SOAP or XML-RPC, etc. In the case of our practice, XML-RPC is the best choice, because we always have to run the tool in the server in the complicated situation, and when in a simple situation, XML-RPC costs the least both in the design and performance. The process of tool presentation is shown as Figure 4.

5 CLIENTS

Users should be detached from the server and tool stores in most of their working time, all they can see should be the client. The client is a workspace which consists of administration interface and the working interface. In our system, we develop the client as a plugin of Eclipse.

1) Administration interface: Administration interface supplies an interface to manage and monitor the system, and more important, to design work flows by arranging activities using SPL tools. It always has to set the authority and role for the user for safety. And the workspace to design work flow is actually a special SPL tool developed with the technology of GEF.

2) Working interface: The working interface is the workplace to work on the software development according to the work flow. It is either the interface of the SPL tool or some build-in interfaces representing the data controlled by the SPL tool.

6 CONCLUSIONS

Software product line is becoming more popular due to its improvement on cost and quality of software products (DuanMu et al. 2011). A distributed SPL is significant and useful in the collaborative environment. We have built an applicable SPL with the work we have presented. The integrated framework could help software developers to design software production more effectively and conveniently. This preliminary verified the feasibility and effectiveness of this paper's work. The future work is to continue to complement the standard of tool development and the messages.

REFERENCES

CAI Yong, SANG Nan, XIONG Guang-Ze, A Case Integration Model Based on Toolbus, Computer Applications, Vol. 24, No. 3, Chengdu, Mar. 2004.
Geir Kjetil Hanssen, Opening Up Software Product Line Engineering, International Conference on Software Engineering, pp. 1–7, IEEE Computer Society, 2010.
Lin Hao, Zeng Xian-Jie, Principle and best practices on OSGi, Publishing House of Electronics Industry, 2010.
XIE Bing, YANG Fuqing, The Jade Bird Project and Its CASE Tools, Computer Engineering, Vol. 26, No. 11, Nov. 2000.
Xia Yun, Chen Hao, Liu Hui. Development Process for Large-Scale Complex Information System[J]. Command Information System and Technology, Vol. 3, No. 4, 2012.
XU Zheng-quan, CAO Li, Applying Web Services to Software Product Lines, Computer Engineering & Science, Vol. 27, No. 12, 2005.
Yu Tao, Liu Junjie, Wang Jun. Software Cost of Complex Information System[J]. Command Information System and Technology, Vol. 3, No. 4, 2012.
Zhu Yun Duanmu, Jing Jing Yan, You Fu Lee, "An Approach of Process Control in Software Product Line," greencom, pp. 139–143, 2011. IEEE/ACM International Conference on Green Computing and Communications, 2011.

Exploring Energy-Balancing Adaptive Clustering Algorithm (EBACA) in Wireless Sensor Networks (WSN)

Zhiyu Li

School of Mechatronics, Northwestern Polytechnical University, Xi'an, China

ABSTRACT: In the node-randomly-distributed clustering algorithms in WSN, it is widely accepted that the excessively centralized cluster nodes can result in unbalanced energy consumption. Section 3 of the full paper explains in some detail the results of our exploration on how to improve network load balancing with EBACA algorithm. The core of section 3 consists of (1) EBACA algorithm selects the cluster head with three parameters: node remaining energy, distance to the Sink node, node density in the cluster; (2) the communication among the cluster heads is optimized, in which selecting the path with shortest sum of Euclidean distance between the cluster heads to transmit data can lead to the least communication energy consumption. Simulation results show that EBACA algorithm, compared with Low Energy Adaptive Clustering Hierarchy (LEACH) algorithm, can effectively realize the network energy balancing and prolong the network life.

Keywords: Wireless Sensor Networks; Clustering Algorithm; Energy Balancing; Network Life

1 INTRODUCTION

In WSN clustering algorithm, the reasonable selection of cluster heads directly influences the network performance of the clustering structure. The excessive concentration of the cluster nodes will result in unbalanced energy consumption. In this paper, we put forward an Energy-Balancing Adaptive Clustering Algorithm in WSN (EBACA). EBACA algorithm selects the cluster head with three parameters: node remaining energy, distance to the Sink node, node density in the cluster. The node density in the cluster is set to be no higher than the average node density in the whole WSN, and the node with abundant remaining energy and the shortest distance to the Sink node possesses the priority of being selected as cluster head, which can make whole network energy consumption more balancing, and the network topology more reasonable. By referring to Affinity Propagation (AP) clustering algorithm, EBACA algorithm optimizes the communication mode among the cluster heads, and selects the path with shortest sum of Euclidean distance between the cluster heads to transmit data, which can minimize the communication energy consumption. Compared with Low Energy Adaptive Clustering Hierarchy (LEACH) algorithm, the simulation results show that EBACA algorithm can effectively realize the network energy balancing and prolong the network life.

2 THE RELEVANT WORK

2.1 Network model

Suppose a network consisting of N randomly distributed sensor nodes. Its application scenario is for periodic data collection (This kind of network is widely used in all kinds of monitoring fields). In this network, sensor nodes need to periodically sense the monitoring subject and transmit the collected data to the Sink node. Without loss of generality, the network model used in this paper is based on the following assumptions:

(1) The Sink node is located outside of a square observation area A. The location of sensor nodes and the Sink node will not change after the distribution and there is only one Sink node.
(2) All nodes have the same capacity and have the function of data aggregation. Each node has a unique identification (ID). The energy of sensor nodes is limited and the energy of the Sink node is not constrained.
(3) According to the distance of the receiver, the node can freely adjust its transmitting power to reduce energy consumption.
(4) The link is symmetrical. Knowing the opposite side's transmitting power, the node can calculate the approximate distance from the sender to the node itself according to the Received Signal Strength Indicator (RSSI).

2.2 Communication energy model

We adopt the same wireless communication energy consumption model as the one used in reference No. 1. Nodes transmit l bits of data to the position with the distance being d. The consumed energy $E_{Tx}(l, d)$ is made up of two parts: the transmitting circuit loss and power amplifier loss. Power amplifier loss adopts Free-space model and Multi-path Fading model respectively according to the distance between the sender and receiver, namely,

$$E_{Tx}(l,d) = \begin{cases} l*E_{elec} + l*\varepsilon_{fs}d^2, & d < d_0 \\ l*E_{elec} + l*\varepsilon_{amp}d^4, & d \geq d_0 \end{cases} \qquad (1)$$

In equation (1), E_{elec} stands for energy consumption of the transmission circuit; ε_{fs} and ε_{amp} respectively stand for the required energy when amplifying the power in the two channel models. If the transmission distance is less than the threshold d_0 (d_0 is constant). Power amplifier loss adopts Free-space model; when the transmission distance is higher than the threshold d_0, Multi-path Fading model is adopted.

3 EBACA ALGORITHM

Similar to LEACH algorithm, EBACA algorithm's implementation process is cyclical, and each cycle is divided into cluster head selection stage and inter-cluster communication stage.

3.1 Cluster head selection

Cluster head selection process is as follows:

(1) Determine whether the node v_i belongs to set B_i. If it belongs to B_i, it will participate in the subsequent cluster head selection.

(2) The determination of $T(i)$. In EBACA algorithm, in order to remain more energy, and make the node closer to the Sink node have the priority to be selected as the cluster head, the probability of the node v_i's being selected as the cluster head is determined by the equation (2):

$$T(i) = \begin{cases} \dfrac{P}{1 - P\left(r \bmod \dfrac{1}{P}\right)} \left[\rho \dfrac{E_{i\_current}}{\overline{E_i}} + (1-\rho)\dfrac{\overline{d_{i\_toSink}}}{d_{i\_toSink}} \right], & E_{i\_current} \geq E_{threshold}, v_i \in B_i \\ 0, & otherwise \end{cases} \qquad (2)$$

In equation (2), P stands for the proportion of cluster heads to all of the sensor nodes; r stands for the current round number. $E_{i\_current}$ stands for the remaining energy of the node v_i, $E_{threshold}$ stands for the threshold of the node energy; $\overline{E_i} = \sum_{i=1}^{m} E_{i\_current}/m$, m stands for the number of all nodes in set B_i, d_{i_toSink} stands for the distance between the node and the Sink node, $\overline{d_{i\_toSink}} = \sum_{i=1}^{m} d_{i\_toSink}/m$; ρ is the adaptive coefficient, and

$$\rho = \frac{1}{1 + \lambda}, \quad \lambda = \frac{E_{i\_current}}{\overline{E_i}} \qquad (3)$$

With the decrease of the node remaining energy, ρ increases within $[1/2, 1]$. This indicates that the function of remaining energy in the cluster head election process is gradually enhanced. The nodes with more remaining energy will become the cluster head[2]. Once the node is elected as the cluster head, reset $T(i)$ as 0, and this node will not be elected as cluster head in the next cluster head selection cycle.

(3) The cluster head selected in this round is not the final cluster head. Once the cluster head is selected, it will broadcast Head_Msg within the area with the radius being R. The message includes the ID number of the cluster head and the sign of the message to show this message is an announcement message. In the cluster head broadcasting stage, all of the neighbor nodes within R must continuously open the receiver to receive the Head_Msg of the cluster heads. Once the ordinary node receives the Head_Msg, it will abandon the Head_Msg coming from other cluster heads. So the nodes in the overlapped communication area of two cluster heads will be the cluster nodes of the cluster head which firstly receives the message. Then the cluster head calculates the node density D_L within its area. If D_L is higher than the average node density of the whole WSN, the cluster head selection restarts from (2). If not, the cluster head selection succeeds[3]. The purpose of this is to prevent the cluster nodes from being too concentrated, leading to the unbalanced energy consumption.

(4) After the determination of the cluster head, each cluster node makes use of CSMA (Carrier-Sense Multiple Aeeess) MAC protocol to transmit the Join_Msg to the corresponding cluster head. The Join_Msg includes the ID number of the node and the ID number of the cluster head. The cluster head receives all of the Join_Msg. The cluster head assigns each cluster node time slot in TDMA way and sends the time slot to all of the cluster nodes. This can guarantee that each cluster node transmits data only in the corresponding time slot and enters a dormant state in other time, which can help reduce the energy consumption of the node. When each node knows its time slot, it enters a stable working stage.

3.2 Inter-cluster communication stage

In the stable working stage, the cluster nodes transmit data to the cluster head, the communication between the cluster nodes and the cluster head is conducted through one-hop communication way. In this way, it is easy to dispatch each member's data transmission. After the cluster head collects the data transmitted by the cluster nodes, it makes use of data aggregation technology to process data. In the wireless communication energy consumption model, communication energy consumption is proportion to the distance. Referring to AP algorithm, if the inter-cluster communication regards the negative Euclidean distance as the measure, the maximum value shows the sum of the Euclidean distance among the cluster heads is the shortest, so the communication energy consumption is the least[4].

Referring to AP algorithm and reference No. 5, define the similarity between the cluster head H_i and the cluster head H_j as $s(i,j) = -d_{ij}$. When $i \neq j$, $s(i,j)$ stands for the possibility of the cluster head H_j being the downstream cluster head of the cluster head H_i. Define two information contents $r(i,j)$ and $a(i,j)$. $r(i,j)$ reflects the appropriateness of the cluster head H_j

being the downstream cluster head of the cluster head H_i; $a(i,j)$ reflects the appropriateness of the cluster head H_i selecting the cluster head H_j as its neighbor cluster head; as to $r(i,j) + a(i,j)$, the higher the value becomes, the more possible the cluster head H_j becomes the downstream cluster head of the cluster head H_i. The update equations of $r(i,j)$ and $a(i,j)$ are as follows:

$$r(i,j) = s(i,j) - \max_{j's.t.j' \neq j}\{a(i,j') + s(i,j')\} \tag{4}$$

$$a(i,j) = \min\left\{0, r(j,j) + \sum_{i's.t.i' \notin \{i,j\}}\max\{0, r(i',j)\}\right\} \tag{5}$$

$$a(j,j) = \sum_{i's.t.i' \neq j}\max\{0, r(i',j)\} \tag{6}$$

In the iteration process of the algorithm, through the mutual selection between the cluster head H_i and the cluster head H_j, the two information contents $r(i,j)$ and $a(i,j)$ of all cluster heads are updated continuously until the algorithm's convergence. Then the data of the cluster head H_i is assigned to the corresponding transmission path. In order to avoid algorithm shock, in the updating process of $r(i,j)$ and $a(i,j)$, the damping factor λ is introduced. The updating of iteration process is as follows:

$$R_i = (1 - \lambda)R_i + \lambda R_{i-1} \tag{7}$$

$$A_i = (1 - \lambda)A_i + \lambda A_{i-1} \tag{8}$$

4 SIMULATION EXPERIMENT AND RESULT ANALYSIS

4.1 Simulation environment setting

This paper adopts NS2 as the simulation platform. Through simulation, we compared the performance of LEACH algorithm and EBACA algorithm. The simulation experiment in this paper is based on the same experimental scene: 200 nodes are randomly deployed in the area, and the Sink node is located in the coordinates $(50, 250)$, so $d_{CH\_toSink\_max} = \sqrt{50^2 + 250^2} = 254.95$ m, and once all nodes are placed, they will not be able to be moved again. Other parameters of the simulation experiment are showed in table 1. Except the number of nodes, all parameters are the default values of LEACH algorithm source code.

4.2 Experimental results and analysis

Because WSN nodes are randomly arranged, we mainly focus on node random distribution in the simulation. Figure 1 shows the comparison of the node mortality and the communication rounds with the Sink node in the two algorithms when 200 nodes are randomly distributed and $P = 0.05$. In figure 1, under the

Table 1. The main parameters of EBACA algorithm.

Name of the parameter	Parameter value
Range size	200 m × 200 m
Primary energy	2 J
Length of the data package	500 Bytes
Band width	1 Mbit/s
$E_{threshold}$	0.01 J
d_0	86.2 m
E_{elec}	50 nJ/bit
ε_{fs}	10 nJ/bit/m²
ε_{amp}	0.0013 pJ/bit/m⁴
E_{DA}	5 nJ/bit/signal

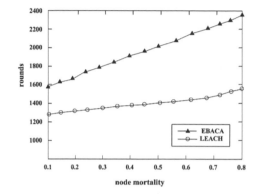

Figure 1. Comparison of network life.

condition of the same node mortality, compared with LEACH algorithm, EBACA algorithm performs better and the curve of round numbers becomes smoother with the change of node mortality. This is because in LEACH algorithm, the cluster head broadcasts the news of itself being the cluster head within the whole network, and the power attenuation of the cluster head obeys Multi-path Fading model. The communication energy consumption between the cluster nodes and the cluster head increases sharply. Under the condition of randomly distributing nodes, the node in some area is too concentrated, which can lead to enormous cluster management overhead and affect the network life. In EBACA algorithm, the node density in the cluster is set to be no higher than the average node density in the whole WSN, and the node with abundant remaining energy and the shortest distance to the Sink node possesses the priority of being selected as cluster head, which can prevent unbalanced energy consumption caused by the too concentrated nodes in the cluster. Besides, energy consumption is evenly distributed to all nodes and it can effectively avoid the early death of the node with greater energy consumption; meanwhile, EBACA algorithm selects the path set by the lowest sum of the Euclidean distance between each cluster head to transmit the data, which can minimize the energy consumption, so the network life is longer.

Figure 2 compares the total network energy consumption curves of the two algorithms along with the simulation time in $P = 0.05$. The smaller slope shows

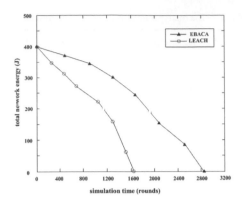

Figure 2. Comparison of network energy consumption.

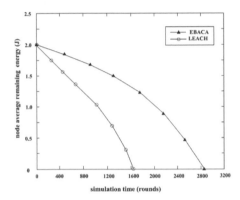

Figure 3. Comparison of node average remaining energy.

slower energy consumption rate and longer network life. The slope of EBACA algorithm is obviously less than that of LEACH algorithm.

In figure 3, node average remaining energy of EBACA algorithm has always been higher than that of LEACH algorithm, which shows that EBACA algorithm can save node energy more effectively.

Figure 4 shows the two algorithms' node remaining energy variance curve comparison along with simulation time. The node remaining energy variance of EBACA algorithm has been very low and without great change, which indicates that EBACA algorithm can effectively balance node energy. Compared with LEACH algorithm, EBACA algorithm can make the whole network energy consumption more balanceable, and the network topology formation more reasonable.

5 CONCLUSIONS

Clustering routing can highlight data aggregation status. Based on clusters, data aggregation and multiple-hop routing can better save node energy. Clustering is one of the effective ways to prolong the network life under the energy-limited condition; it not only can prolong the network life, but also is conducive to the realization of topology control and data aggregation. Aiming at the shortcomings of the existing clustering

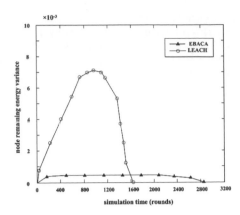

Figure 4. Comparison of node remaining energy variance.

algorithms, this paper puts forward EBACA algorithm, which is applicable to the node-randomly-distributed WSN. EBACA algorithm selects the cluster head with three parameters: node remaining energy, distance to the Sink node, node density in the cluster; meanwhile the communication among the cluster heads is optimized, in which selecting the path with shortest sum of Euclidean distance between the cluster heads to transmit data can lead to the least communication energy consumption. Simulation results and analysis show that EBACA algorithm, compared with LEACH algorithm, can effectively realize the network energy balancing, generate more stable and reasonable cluster structure, and prolong the network life.

ACKNOWLEDGEMENT

This work has been supported by the 111 Project, Grant No. B13044.

REFERENCES

[1] Heinzelman W. Application-Specific Protocol Architectures for Wireless Networks [D]. *Massachusetts Institute of Technology*, 2000.

[2] Li Zhiyu, Shi Haoshan. Exploring Load-Balancing Adaptive Clustering Algorithm (LBACA) in Wireless Sensor Networks (WSN) [J]. *Journal of Northwestern Polytechnical University,* 2009, 27(6): 822–826.

[3] BAI Enjian, GE Huayong, *et al*. A secure multi-path routing protocol for hierarchical wireless sensor networks [J]. Journal of Harbin Engineering University, 2012, 33(4): 507–511.

[4] Frey B J, Dueck D. Clustering by passing messages between data points [J]. *Science Magazine*, 2007, 315: 972–976.

[5] ZHONG Wei-min, WANG Yue-qin, *et al*. Clustering Algorithm Based on P-Changed Affinity Propagation for Heterogeneous Wireless Sensor Networks [J]. *Journal of Jiangnan University (Natural Science Edition),* 2012, 11(4): 423–427.

Information Technology and Computer Application Engineering – Liu, Sung & Yao (Eds)
© *2014 Taylor & Francis Group, London, ISBN 978-1-138-00079-7*

Face recognition using a hybrid algorithm based on improved PCA

X. Tian
Beijing University of Posts and Telecommunications, Beijing, China

M. Tian
Tongji University, Shanghai, China

ABSTRACT: In order to reduce the heavy computation and eliminate the effects of uneven illumination in face recognition based on PCA (Principal Component Analysis) method, this paper proposed an improved PCA algorithm – a new hybrid face recognition algorithm which considered the image intensification based on local means and standard deviation, which was further integrated with genetic algorithm and form a new face recognition algorithm. Experimental results show that, when tested by ORL face database, the algorithm proposed in this paper is 3.27% higher than the PCA + 2DPCA face recognition algorithm in average recognition rate.

Keywords: PCA; Genetic Algorithm; Face recognition; Pattern recognition

1 INTRODUCTION

Recent years, as a research topic valued in both theory and application, face recognition has been concerned and emphasized by more and more researchers. Various methods for face recognition emerge one after another, and PCA (Principal Component Analysis) is one of these methods (Matthew et al., 1991). Basic principle of conventional PCAs is to construct an eigenface space by using K-L transform (Ming-Hsuan et al., 2002) which can extract the main features from a face, and to obtain a set of projection coefficients by projecting the tested image onto this space, and then the recognition can be achieved by making comparisons among face images. Such a method lets the MSE (mean square error) before and after the compression minimized, and the transformed low-dimensional space has good performance in resolution. However, when processing a face image, this method extracts only global features which will be largely affected by light conditions and facial expression changes. Consequently, its recognition effect is not so good.

The improved PCA algorithm of this paper took the advantages of conventional PCA algorithms' image intensification that based on local means and variances, and can effectively reduce the impacts of uneven illumination on face recognition before making the feature extraction, thus expanding the application of PCA algorithm. Although the improved PCA algorithm performs well in lowering the dimension of the sample images, it takes insufficient account of the classificatory information of the training samples, and cannot reach the best recognition effect.

GA (Genetic Algorithm) is a random search-based method for optimization, which can seek out global optimal solution within finite algebras (Davis, 1991). It is very advantageous for which this paper has introduced GA into the new algorithm: firstly, using the improved PCA algorithm to do feature exaction; then, using GA to optimize the extracted features; finally, through experiment, such method has been proven that it not only can significantly reduce the dimension of the feature space, adapting uneven illumination, but also has fine classifying effect, hence improving the recognition rate.

2 FEATURE EXTRACTION BASED ON PCA

An $M \times N$ face image can be constituted into $D = M \times N$ column vector, where D is the dimension of the face image, namely the image space. Set n as the number of training samples, and x_i as the face vector of the jth image, then covariance matrix of the required sample is:

$$S_r = \sum_{j=1}^{n}(x_j - u)(x_j - u)^T \qquad (1)$$

where, u is the mean image vector of training samples:

$$u = \frac{1}{n}\sum_{j=1}^{n}x_j \qquad (2)$$

Make $A = [x_1-u, x_2-u, \Lambda, x_n-u]$, so:

$$S_r = AA^T \qquad (3)$$

with the dimension is $D \times D$.

According to K-L transform, eigenvector corresponding to nonzero eigenvalues of matrix AA^T

is required for new coordinate system. For the complexity of direct calculation, SVD (singular value decomposition) is employed to obtain eigenvalues and eigenvectors of AA^T by calculating those of $A^T A$.

According to the SVD theorem, given that l_i ($i = 1, 2, \Lambda, r$) is the rth nonzero eigenvalue of matrix $A^T A$, v_i the eigenvector $A^T A$ corresponding to l_i, so the orthonormal eigenvector u_i of AA^T is:

$$u_i = \frac{1}{\sqrt{l_i}} A v_i (i = 1, 2, \Lambda, r) \tag{4}$$

Then, the space of "Eigen Faces" is:

$$w = (u_1, u_2, \Lambda, u_r) \tag{5}$$

Project the training sample onto "Eigen faces" space to a set of projection vectors, which constitute the "facial recognition database". When recognizing, project each face image that needs to be detected onto the "Eigen faces" space, then, use the nearest classifier to compare it with faces of the database, so as to determine whether it belongs to the database or not, if yes, then find which one is suitable (Turk et al., 1991).

PCA is the first recognition method that is based on overall features or global visual features, and also the best compression approach that can make the MSE between original image and reconstructed image minimized (Chennamma & Lalitha, 2009). PCA is capable to get a good extraction of face features. However, it is involved in specific selection on the basis of K-L transform that cannot optimally make selection in accordance with different images, and cannot completely avoid the limitation of K-L transform. PCA ranks the eigenvalues transferred by K-L in a descending order, and chooses the relatively larger eigenvalues as the eigenvectors to constitute a subspace of the K-L Eigen space, so as to make feature extraction. In the K-L Eigen space, eigenvectors corresponding to larger eigenvalues reflect the overall trend and low frequency components of the original image; while eigenvectors corresponding to smaller eigenvalues reflect the detailed variation and high frequency components. Therefore, image's overall features extracted by PCA are reflected as the contour and grayscale of the face, which can be taken as face features, but some original information will be missed.

3 IMPROVED PCA ALGORITHM

The improved PCA algorithm of this paper took the advantages of conventional PCA algorithms' image intensification that based on local means and variances; it can effectively reduce the impacts of uneven illumination on face recognition before making feature extraction.

Given that the grayscale of an image lies in $[0, L - 1]$, r is a discrete random variable in this grayscale domain, $p(r_i)$ is probability of the grayscale r_i, so, the global mean of the image can be expressed by:

$$E_g = \sum_{i=0}^{L-1} r_i p(r_i) \tag{6}$$

Then, this image's global contrast ratio, namely the variance is:

$$\sigma_s^2 = \frac{1}{M^2} \sum_{i=1}^{M} \sum_{j=1}^{M} (x(i, j) - E_s)^2 \tag{7}$$

The brightness of an image can be measured by the mean value, and its contrast ratio can be measured by the variance. As a result, for an image that needs to be processed, its darker areas with low contrast ratio can be enhanced through comparisons between global mean value and local mean value, and between global contrast ratio and local contrast ratio, without changing the brighter areas of the image.

For an image that needs to be processed, set a M × M neighborhood $S_{(i,j)}$ with the center of $Q_{(i,j)}$, in this way, the neighborhood's mean value, which is also referred to as the local mean value, can be expressed as:

$$E_s = \frac{1}{M} \sum_{i=1}^{M} \sum_{j=1}^{M} x(i, j) \tag{8}$$

The specific procedure of image enhancement based on local mean and standard deviation goes as follows:

Determine darker areas of the image. If $E_s < k_0 E_g$, where k_0 is a positive constant number smaller than 1, meaning that the area of the image is relatively dark, and needs to be further enhanced.

Determine the areas with low contrast ratio. If the contrast ratio of an area of the image is too low, it can be considered to contain no features, and does not need to be enhanced. Therefore, areas with low contrast ratio can be assumed as: $k_1 \sigma_g < \sigma_s < k_2 \sigma_g$, and k_1, k_2 are positive constant number smaller than 1.

Make magnification to grayscale and stretch to contrast ratio of the determined areas. Processed by above approach, the algorithm for image enhancement that based on local mean value and standard deviation can be expressed as:

$$f(i, j) = \begin{cases} \lambda x(i, j) + \beta x(i, j)^\gamma, & condition\ 1 \\ x(i, j), & others \end{cases} \tag{9}$$

where, k_0, k_1, k_2 are positive constant number smaller than 1; E_s and σ_s respectively represents the global mean value and standard deviation; λ is grayscale amplification coefficient; β and γ are contrast ratio stretching coefficient. It can be known from above formula that this algorithm uses the local means and standard deviations to determine the areas which needed to be enhanced (i.e. areas with low grayscale and contrast ratio), and would not enhance the areas that do not require enhancement. The condition 1 means "$E_g < k_0$ and $k_i\sigma_g < \sigma_s < k_1\sigma_g$".

Combined with the algorithm which can enhance the image based on local mean and standard deviation, PCA algorithm is good at highlighting important parts of the face image (such as eyes, nose, mouth, etc.). As well, it can get some discriminative face features during the process of extraction, improve the recognition rate, and eliminate the effect of light factors on face recognition.

4 HYBRID ALGORITHM BASED ON IMPROVED PCA

4.1 Genetic algorithm

As a stochastic search optimization technique based on simulating the biological evolution process, Genetic algorithm (GA) is specially suitable for large-scale optimization under complex conditions. Meanwhile, the evolutionary computation takes into consideration of the natural selection of "survival of the fittest" and of some simple evolutionary operations, which lets it go beyond the limitation of searching space and other supplementary information, making the genetic algorithm not only efficient in face recognition, but also characterized with simple and available operations (Chaiyaratana et al., 1997).

In practice, face recognition is involved in a large number of features, in which exists much redundant information. Selection of face features is to choose d optimal features out of D original features (number $D > d$) to remove the redundant information, so as to achieve the best recognition accuracy. By drawing support from "survival of the fittest" of the natural evolutionary process, genetic algorithm can optimize the feature extraction and further improve the accuracy of face recognition.

4.2 Hybrid algorithm combined with the improved PCA and genetic algorithm

In this paper, genetic algorithm was applied to extract face features and optimally select from the eigenspace which has been processed by the improved PCA, to construct a new sub-eigenspace that is favorable to classification. The procedure that genetic algorithm optimizes the eigenspace is described as Figure 1.

Firstly, encode the target and map it into chromosomes. Select the length of chromosomes according to the number of eigenvectors. Here are 200 face training samples, corresponding to 200 eigenvectors, then the length of chromosomes can be selected as $L = 200$. Start an initial population of N size, with chromosomes generated randomly. Generally, N is ranged from 10 to 50 and $N = 20$ is adopted in this paper. Each chromosome corresponds to a selected eigenvector. For chromosomes, binary encoding is employed that genes of a chromosome are assigned with 1 or 0, where, 1 means to select the eigenvector, while 0 means not to.

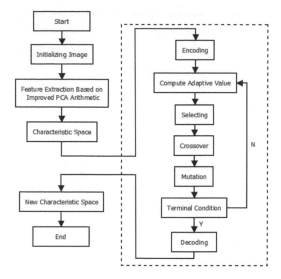

Figure 1. Genetic algorithm optimizes the eigenspace.

Then, calculate fitness of each chromosome by using the designed fitness function. Here takes the target minimized function as the fitness function, i.e. the smaller the fitness is, the better the performance of chromosome becomes. The fitness function is expressed as follows:

$$f = 1 - \frac{1}{M}\sum_{i=1}^{M}\delta_i \tag{10}$$

$$\delta_i = \begin{cases} 0, & image_i \text{ is not recognized correctly} \\ 1, & image_i \text{ is recognized correctly} \end{cases} \tag{11}$$

Thirdly, select parent chromosomes from the population of N size by using roulette wheel selection, and let them engaged in next genetic operation. Then, make new generation through crossings of the selected parent chromosomes.

Fourthly, calculate the fitness of each chromosome to determine whether it meets ending condition or not. Stop if condition is met, otherwise add the generation number by 1 and come back to step 3 to continue the genetic operation.

Fifthly, stop the genetic operation if the largest generation number is reached.

Finally, decode the optimal chromosomes to obtain the optimized subset eigenvectors. Project the face image onto such eigenspace to the projection weights, which are taken as eigenvectors of the face, and then use classifier to make recognition.

In GA, three ending conditions have been set. The first is to stop fitness function and the ending condition is: fitness $<1 \times 10^{-6}$. The second is the evolutionary algebra, i.e. the genetic operation is to be stopped when 100 generations is reached. The third one is the time which was set as 400 s for ending the genetic algorithm.

291

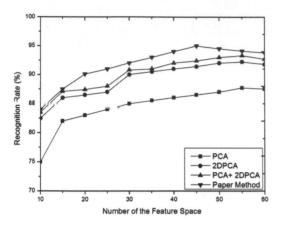

Figure 2.   Recognition Rate Relates to the Feature Space.

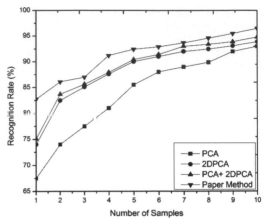

Figure 3.   Recognition Rate Relates to the Samples.

5   EXPERIMENT AND RESULTS

In the experiment, this paper made comparisons among PCA, 2DPCA, PCA+2DPCA and the proposed algorithm, to verify the effectiveness and superiority of the proposed algorithm through experimental results.

5.1   *Experimental design*

This experiment is performed based on PC (personal computer) face recognition system, where the Mathworks Matlab 2012b is used as the programming software.

In this paper, the widely used ORL face database of Cambridge University was uesd, including 40 persons and 10 images for each individual. These images, with the size of 92×112, considered changes in light, gestures and age and covers of accessories. Here, five images of each person were randomly selected as the training images, making up a training set with 200 images, and the rest 200 images were used as the testing set.

5.2   *Experimental results and analysis*

In experiment, PCA, 2DPCA, PCA+2DPCA and the proposed algorithm have been compared as shown in Figure 2 and Figure 3. Figure 2 shows the relationships between recognition rate and dimension of the eigenspace, and Figure 3 shows the relationship between recognition rate and number of the sample. It can be seen that the proposed algorithm is superior to conventional PCA. Compared to the PCA+2DPCA face recognition algorithm, the algorithm proposed in this paper is 3.27% higher in average recognition rate.

Synthesizing above experimental results comprehensively, the optimal eigenspace selection of PCA that optimize by using genetic algorithm can effectively overcome the limitation of PCA algorithm and is able to create the optimum sub-eigenspace, which improve the recognition rate.

6   CONCLUSIONS

This paper proposed a face recognition algorithm combined with the improved PCA and genetic algorithm, and elaborated how it works. This method takes the most of genetic algorithm's global search ability, and can optimize selection of eigenvectors of the obtained eigenspace to gain a sub-eigenspace which is favorable to feature classification. It can be known from the experimental results of ORL face database that, the algorithm proposed in this paper is efficient in overcoming the limitation of PCA for selecting the eigenspace and improving the recognition rate.

REFERENCES

Chaiyaratana N, Zalzala A M S. 1997. Recent developments in evolutionary and genetic algorithms: theory and applications [J]. *Genetic Algorithms in Engineering Systems: Innovations and Applications*, 1997, 20(2):270–277.

H.R. Chennamma & Lalitha Rangarajan, 2009. Robust Near-Duplicate Image Matching for Digital Image Forensics, *International Journal of Digital Crime and Forensics*, 1(3). Jul.

L. Davis, 1991, Handbook of Genetie Algorithms. New York: Van Nostrand Reinhold.

Matthew A. & Alex P. 1991. Face recognition using eigenfaces, In *Proc. IEEE Computer Society Conference on Computer Vision and Pattern Recognition.*

Ming-Hsuan YANG. Kernel Eigenfaces vs. Kernel Fisher faces: Face Recognition Using Kernel Methods. Processing IEEE International Conference on Automatic Face and Gesture Recognition. Washington D.C., 2002, 3:215–220.

Turk A, Pentland P. 1991. Face Recognition Using Eigenfaces. IEEE Vision and Modeling Group. 1991, 5(91):586–591.

Information Technology and Computer Application Engineering – Liu, Sung & Yao (Eds)
© 2014 Taylor & Francis Group, London, ISBN 978-1-138-00079-7

Exploration on practice teaching of information and computing science refering to the idea of CDIO

Guang Hui Wang, Yan Qin Zhao & Xing Hua Zhang
School of Computer and Information Engineering, Heilongjiang University of Science and Technology, Harbin, China

ABSTRACT: For the information and computing science in the characteristics of applied undergraduate colleges and universities, Practicality is strong, Operation ability request high, Then first describes the CDIO concept Analysis of the need for reform of the Practice Teaching System. Then the CDIO engineering education mode for reference on how to build up a scientific and reasonable system of practical teaching, how to cultivate applied talents with innovative spirit and practical ability are discussed, A new practical teaching system is constructed in terms of The professional characteristics. Practice shows, Practice teaching system reform based on CDIO concept effectively cultivate the students' ability of practice and innovation spirit, improve the quality of training applied talents.

Keywords: Information and computing science; CDIO; Practice teaching

1 INTRODUCTION

In 1998, the ministry of education promulgated the specialty catalogue and introduction of common colleges and universities, "information and computing science" specialty is listed as a new class of mathematics. At the same time, the ministry of education promulgated in 2012 for more than 2 years after the revision of specialty catalog and introduction of regular institutions of higher education, Clarified the information and computing science specialty in information technology, computing, operations research and mathematical foundations of control technology for the study of science majors, its main task is to study "at the core of information technology and using modern computational tools efficiently to solve the mathematical theories and methods of science and engineering". Obviously, such professional orientation and "computer science" is a difference professional, specialty information science and scientific computing as the core direction. Information science should be understood as "the information acquisition, information transmission, information processing and information control based on science", to understand this for setting information and calculation science specialized training goal and curriculum has important significance. The country set up the major of information and computation science school grew from 101 in 2001 to 500 now more than. In recent years, with the increasing number of enrollment, the profession is currently one of the largest professional national professional college of science. The specialty

of information and computing science is developing so fast, the enrollment scale is so large, the letter of development professional opportunities, more is a challenge, how to run this one major, as many believe professional workers must seriously consider the issue.

2 CDIO ENGINEERING EDUCATION MODE AND IDEA

Although the information and computing science is professional for the science of Ministry of education, but in the increasingly severe employment situation, the major of information and computation science teaching how to do that in a big project. Compared with the traditional information science, information and computing science more emphasis on the application of using the basic theory, emphasizes the cultivation of ability, professional method and technology to solve practical problems. On April 8, 2008, Heilongjiang institute of science and technology by the ministry of education determines the level of job of teaching evaluation of outstanding colleges and universities. In 2010, Heilongjiang Institute of Science and Technology was identified as the first unit construction application characteristics of Colleges of Education Department of Heilongjiang Province. At the same time, the Heilongjiang Institute of Science and Technology located in the teaching of applied engineering colleges, the opening of the professional information has the application features more distinct.

According to the guiding ideology of the "institute of technology, combined with" professional construction, corresponding the cultivation of creative spirit and practice ability and engineering quality of applied talents, become the main target of the professional personnel training, and improve the teaching quality of education at the core of the problem. In order to achieve this goal, we must carry on the reform to the practical teaching system of the specialty, build a set of theoretical teaching and supporting each other in Applied Undergraduate Talents Training System of practice teaching mode.

CDIO denotes Conceive, design, Implement and operation, is a new type of engineering education mode. It fits as a modern engineering technical personnel training education concept and methodology system, fully participate in product research and development and operation, provide students with effective links between practice and theory courses, encourage students to study in the form of active, have good development prospect and promotion value. In the exploration process of CDIO engineering education model with Chinese characteristics, Shantou University first proposed EIP-CDIO model; Tsinghua University Computer Science in two courses in the first use of the CDIO teaching method. There are including Denmark, Finland, France, South Africa, Singapore, China and other countries, including more than 20 universities join the CDIO cooperation plan, to continue to develop and perfect CDIO teaching mode. Heilongjiang Institute of Science and Technology in the summary of the school characteristics of the CDIO engineering education concept foundation school for more than sixty years experience in running a school and the domestic and foreign engineering education creates "high moral, engineering and practice of" Chinese Edition. So the professional selected reflect today's engineering education consensus CDIO education concept as the guiding ideology, and strive to cultivate scheme from continuously optimized, and constantly improve the personnel training mode; Timely integration of practice teaching resources, the practice teaching content reform, according to the features of the practice teaching link, for its science teaching process, teaching methods.

3 REFORM AND PRACTICE OF INFORMATION AND COMPUTING SCIENCE TEACHING BASED ON CDIO

3.1 *Outstanding problems that exist in the practical teaching*

In the traditional science education thought, believe in that students should have a good foundation of mathematics and mathematical thinking ability, pay more attention to mathematical theory system study. The traditional personnel training mode of theoretical and practical light weight, emphasizing the individual academic skills but ignore the team spirit, the importance of knowledge of learning and neglect the innovation culture and other problems, led to the letter of professional students are doing more teachers, graduate selection, unable to further adapt to the actual society, enterprises need.

(1) Fieldwork cognition is one of very important practical teaching link of the teaching plan of information and computing professionals, offered in the third semester. Objective is to broaden the view of the students, improve the students' understanding of the professional, let the students understand the factories and mines, enterprises and institutions of the professional knowledge and research, to determine the direction of future study and research. Due to constraints, cognition practice simply ask professional teachers to do the 2 report, to school network center, television and other campus units visit, vision does not go out, be unable to understand society really needs professional knowledge of what kind of, need to what direction.

(2) production practice, graduation practice is offered in the seventh semester, because the past system is not perfect, the lack of practice and training base, a part of the students in the room and the laboratory, and based on the experiment, the other part of the students using the "sheep" management, students find their own internship units, this has caused some aspects can not finish or complete the poor quality.

(3) Graduation design (thesis) is a professional talent training in colleges and universities in the process of one of the strongest comprehensive practice teaching link, in the eighth semester opening, is for the whole four years university learning career, a big exam is to test students' creative thinking and practical ability the important way, the quality of the pros and cons is measure whether the talents training goal and talent training quality is in line with the requirements of important standard. Previously due to find work, students after enrollment quality uneven, some teachers responsibility heart is not strong wait for a reason, as a result, the professional quality of graduation design work has been formulated with the beginning development not too conforms to the target.

3.2 *Specific measures for the reform of practice teaching based on the CDIO concept*

(1) Establishment of the Students Association for science and technology base, carry out a variety of practical teaching activities, improve students' ability.

Establishment of the professional student science base is mainly aimed at freshman to junior students, the students in their spare time to have a place to set time to form a team to determine the research direction, to analyze and solve problems. The base of good academic atmosphere, can promote the construction of style of study, correctly guide the students' interest in learning. The use of spare time to strengthen mathematical modeling, ACM training program design, actively mobilize the students to see the three northeastern provinces, the national college students' mathematical modeling competition, ACM programming competition, the national college students' mathematical knowledge competition, cultivate students' problem analysis skills, hands-on ability and innovative consciousness. Grade 2006 students Li Lisong, Li Xiaofei won the national second prize of national college students' mathematical modeling competition; Grade 2009 students Yuan Fengcheng, Zhou Niao, Li Yang won the ACM/ICPC contest bronze in heilongjiang province, northeast area the bronze medal. Also encourages students to participate in social investigation on holiday, declaration and students scientific research, organize the students to declare the Communist Youth League Committee of the Heilongjiang Institute of Science and Technology and Heilongjiang province college student scientific research project under the guidance of Teachers. From 2006 onwards, average every year 40% students to apply to the Heilongjiang Institute of Science and Technology, computer college students' scientific research project, and have certain grants. At the same time in the first 3 years continuously on students' special skills training, and encourage students to obtain Microsoft and other well-known enterprises of national qualifications certificate. These practical activities to effectively improve the analysis of problems of the professional student ability, practical ability and sense of innovation, the effect is remarkable.

(2) Strengthen the construction of practice teaching, strengthening and enterprises practice base construction.

Changing ideas, no closed schools, according to the information and computing science specialty, after careful investigation, combined with the information of enterprises and training institutions related to the professional enterprises, establish stable, practical teaching base. Beginning in 2006, we successively with the Beijing danei group, daqing careers company, Harbin Huizhi outstanding company set up a stable practice base. Cognition practice period (2 weeks) students really come into contact with the actual project, understand the specific process of project, make the students really come into contact with the actual work, understand the characteristics of the actual work and school theory knowledge to use value in practical work. At the same time, we do not regularly invited to our department under the guidance of technical personnel of enterprises in the line of work, to conduct seminar, introduce some research at the forefront of technology, make the teachers and students to keep up with the development of era of science and technology trends. The production practice, graduation practice due to a long time, all in cooperation company paid internship, through such cooperation, provides a good opportunity to practice hand for the student development and practical conditions, to develop the ability of students; on the other hand, reserve of talents for the development of enterprises, innovation technology, realize win-win school and enterprises.

(3) Strengthen the CDIO project application consciousness, reform the forms of graduation design (paper).

Graduation design (Thesis) is an important teaching link of students' scientific and systematic training or engineering problem solving ability, should be combined closely with the actual, topic selection combined with the engineering practice, combined with teachers scientific research.

To strengthen quality control, improve the project process management at the same time, the tutor team, senior title drive the young teachers, improve teachers' scientific research and professional accomplishment.

We will treat students one divides into two, first in view of the employment intention of the students, graduation design and production practice, graduation practice "trinity", continue working in the company paid internships, fully participate in the research of enterprise, combined with the actual project, with guidance competent teachers and enterprise to guide students' graduation design, do really writes really do work experience in dealing with students at the same time, training students' cooperation spirit and team spirit, improve the students' ability of independent analysis and solving practical problems, for the next step to lay a solid foundation of employment.

Such as grade 2008 student Li Pin in the internship company jingdong mall do graduation design "based on jingdong mall automation research and design of software configuration management work", it integrates the company project and graduation design topic, due to the outstanding graduation was positive, 6000 yuan a month. Then in view of the students one's deceased father grind and study, let the students participate in practical scientific research project of the teacher, to cultivate the students' scientific research consciousness in advance, teacher's scientific research and students can be organically, to let the students to help teachers scientific research work to find the actual problem, although students study may not be deep enough, but at least participate in scientific research, aware of the importance and the purpose of scientific research, in the teachers' instruction to solve the problem, so as to smoothly complete the graduation design. Such as the Yangtze university postgraduate student of grade 2008 students Wang Xiangjiu students actually participate in guiding the teacher's scientific research subject at provincial level image enhancement algorithms, and successfully completed the one sub modules, then the student told us, Postgraduate Tutor praised his participation in scientific research consciousness, practical ability. After CDIO project application consciousness training, the professional graduation design level also from qualified up to seventh in the good professional.

4 CONCLUSION

Through several years of practice, draw lessons from CDIO engineering education concept to construct the information and computing science specialty practice teaching system, benefit the students, teachers and enterprise three parties, can dramatically change the practice teaching content relatively lags behind, the relative shortage of students ability to apply situation, makes the students' innovation spirit and practice ability has improved significantly. At the same time, the construction of teachers' team also play a positive role, make the professional practice teaching with new vitality. Of course, the professional practice teaching system construction is a long-term process, there are still many problems to be studied, we will continue to improve the practice teaching system only by strengthening the reform in practice, in order to make the professional has the vitality, the Heilongjiang Institute of Science and Technology renamed the university to make due contribution.

5 FUND PROJECT

The 2012 annual task planning of Heilongjiang Education Science "Twelfth Five-Year Plan" (GBC1212076): Teaching of heilongjiang institute of science and technology research key project (2013-96).

REFERENCES

[1] The people's Republic of China Higher Education Division of the Ministry of education. College undergraduate professional directory and professional presentation (M). Higher education press, 2012.

[2] Mathematics and Statistics committee of education. On information and computing science specialty running status and report on a survey of the construction related problems [J]. Journal of university mathematics, 2003, 12(1): 1–5.

[3] Li Manli. With historical interpretation CDIO and its application [J]. Journal of tsinghua university education research, 2008, 29(5): 78–87.

[4] Cha Jianzhong. "CDIO engineering education reform strategy and the cooperation between production and internationalization [J]. Journal of China university of education, 2008(5): 16–19.

[5] Crawley E F, Malmqvist J, Ostlund S. Rethinking Engineering Education The CDIO Approach [M]. New York: Springer, 2007.

[6] Cai Xianzi, Huang Xiaohu, Wen Bin. Information and calculation science specialized practice teaching exploration of agricultural college [J]. Journal of research and exploration on room, 2011, 30(8): 235–295.

Information Technology and Computer Application Engineering – Liu, Sung & Yao (Eds)
© *2014 Taylor & Francis Group, London, ISBN 978-1-138-00079-7*

Mobile polymorphic application based on cloud computing architecture

Vanban L. Wu & Chang Heng Shao
CentLing Technologies Co. Ltd., Qingdao, Shandong, China

ABSTRACT: Mobile Polymorphic Application was a technology proposed in early 2000 for ubiquitous computing. As many applications migrate into Cloud Computing architecture, it is a natural movement to merge both the solution of Mobile Polymorphic Applications proposal with Cloud Computing architecture. As many applications migrate into Cloud Computing domain in recent years, it is worth to revisit the Mobile Polymorphic Applications proposal and merge its model into Cloud Computing architecture. In this paper, we reveal a revised model of the combined architecture, and apply the application in a Hospital Information System (HIS) to illustrate our points. The application is also applicable to other applications components of a HIS system.

Keywords: Internet of Things; Cloud Computing; SaaS; Mobile Polymorphic Application; Extensible Messaging and Presence Protocol; Hospital Information System

1 INTRODUCTION

As part of the technology evolution of Internet of Things (IoT), ubiquitous computing plays an essential role in defining and facilitating such evolution movement. Industrial and commercial sectors, education and government institutions alike are finding ways to ride on this technology curve for innovative features serving their respective communities.

In this report, we are addressing the mobility where users can transition freely applications among digital devices. A general approach to realize such needs was first proposed by Anand Ranganathan, Shiva Chetan and Roy Campbell [1] [2]. In this report, we are proposing an enhanced version of the solution where we adopt the Anand et al's proposal into a cloud computing environment, thus eliminating part of the complexity of the original framework.

To facilitate the discussion, we scale down the application of HIS to the hospital rounds system as an exemplified studying case to illustrate our points. Instead of tackle the whole HIS system, which is a formidable area to be covered in this paper, we illustrate our key points with an application component on wards, called hospital rounds system. To the readers, this writing is just an abridged version of a full report to be published in the future.

2 PAST RESEARCH AND RESULTS

To realize an application independent of its environment under which it operates, one has to segregate functional modules sensitive to an environment from parts which are common to all. In a very high logical sense, there are three major building blocks

constituting an application: input module, control module, and output module. To realize the intended mobile polymorphic objectives, Anand et al's define a breakdown of five components [1] [2]: model, presentation (generalization of view), adapter, controller and coordinator (see Figure 1). The model implements the logic of the application and exports an interface to access and manage the application's state [3]. Controllers act as input interfaces to the application and presentations as output interfaces. The adapter maps controller inputs into method calls on the model. The coordinator manages the composition of other components of the framework. More details of the framework can be found in [3].

Cloud Computing is a standard taking shape for less than ten years in IT business service offerings; it is a novel business model to deliver "computing as a service" [4] [5] [6] [7] [8] to a heterogeneous community of end-recipients. Software as a Service [9] [10] (SaaS) is a software delivery model in which software and associated data are centrally hosted on the cloud. SaaS is typically accessed by users using a thin client via a web browser. SaaS is a popular model for many

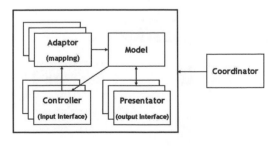

Figure 1. Mobile Polymorphic Application framework.

business applications, including collaboration, customer relationship management (CRM), management information systems (MIS), enterprise resource planning (ERP), invoicing, human resource management (HRM), content management (CM), and accounting. SaaS has been incorporated into the strategy of many leading enterprise software in U.S. Recently, the popularity of Cloud Computing technologies is also gaining grounds in the medical applications field [11].

3 PROPOSED INFRASTRUCTURE AND FRAMEWORK

3.1 *Architecture*

In this paper, we are proposing an end-to-end rounds system solution which is independent of any digital devices and OS platforms. We name it a Mobile Polymorphic System (MPS). The architecture of such system is built into a Cloud Computing environment, thus making the solution seamless and fully integrated into a HIS system.

The whole solution can be divided into five major objects: Controller, Presenter, CAdaptor, PAdaptor and Model (see Figure 2):

1. Controller

Originating device interfaces with a Controller object instance of its type to execute controls (entered by a user), including an ability to perform mobile polymorphic transfer – request to switch an application to another digital device media. The number of Controller types is proportional to the types of digital devices supported by the MPHRS system.

2. CAdaptor

A CAdaptor object acts as an interface between a Controller object and a Model object (to be described next); basically all requests from a Controller object will be translated to a set of standard form interpretable by the Model object.

3. Model

This object is a kernel component of the solution. Besides executing all intended logics of an application, it contains a key operation to bind two objects, CAdaptor and PAdaptor. It also supports a critical function of seamlessly switching applications from one digital device to the next without interrupting user's operations. The switch-over operation can be partial or full. In the partial switch-over case, major controls operating an application still reside in the originating digital device, and the terminating digital device only serves as a "pure" output media. For the later case, all operations including the display ones are switching over to the new device, i.e. a new CAdaptor object instance will also be generated to handle commands from the new device.

4. PAdaptor

The Model object interfaces with a PAdaptor object instance to generate an object instance presentable to a new digital device; all standard requests from Model object are processed by a PAdaptor object instance. PAdaptor object acts as an interface between Presentor object and Model object, it will transfer a set of customized requests interpretable by a Presenter object (to be described below) of a new terminating digital device.

5. Presenter

A Presenter object instance is responsible to execute a set of operations directed from a PAdaptor object instance to a new digital device. Like a Controller object, the number of Presenter objects is equal to the types of digital devices supported by MPHRS system.

3.2 *Layer model*

The five identified objects (Controller, Presenter, CAdaptor, PAdaptor and Model) fit perfectly well into the Cloud Computing SaaS standards [11]. Figure 3 depicts a Mobile Polymorphic Hospital Rounds System (MPHRS) in the SaaS layer. As shown in this figure, CAdaptor, Model and PAdaptor objects are parts of an integral layer above Integration and middleware. CAdaptor, Model and PAdaptor use APIs layer to interface with each Controller and Presenter object to realize operations directed by a user. Note: layers below Integration and Middleware are being simplified in Figure 3 to focus on the topic being discussed.

3.3 *Device adaptation*

As the primary goal of Mobile Polymorphic System (MPS) system is to support an environment where applications can move freely between digital devices or

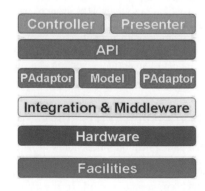

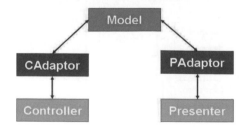

Figure 2.   Mobile Polymorphic System Structure.

Figure 3.   Mobile Polymorphic System Layer Model.

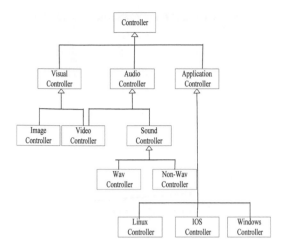

Figure 4.   Controller Hierarchy.

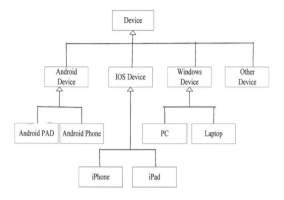

Figure 5.   Device Hierarchy.

media, the following operations will be needed when a "switch-over" attempt is issued by a user:

1. The Controller object will identify the most-suitable type of controlling operation sent by an originating device according to a device hierarchy defined in Figure 4.
2. The Model object will identify the most-suitable device to realize this request according to a device hierarchy defined in Figure 5.
3. The Presenter object will locate the most-suitable type of presenting operations according to a device hierarchy defined in Figure 6.

Using the device hierarchy as an example, the partial code implementing such structures can be written as follows:

```
<?xml version="1.0" encoding="utf-8"?>
<Device description="Top level">
  <Android provider="Google Inc">
     <AndroidPAD/>
     <AndroidPhone/>
  </ Android>
  <IOS provider="Apple Inc">
   <iPhone/>
    <iPad/>
  </ IOS >
```

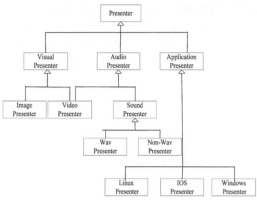

Figure 6.   Presenter Hierarchy.

```
  <Windows provider="Microsoft Inc">
     <PC/>
     <Laptop/>
  </Windows >
  <Default provider="Others">
  </ Default >
</ Device>
```

4   IMPLEMENTATION AND VERIFICATION

4.1   *Interface highlights*

In the implementation level, four primary interfaces are needed to the solution:

1. Interfaces between Presenter and PAdaptor:

   a) *PresentRequest:*
      PAdaptor calls PresentRequest to request a new terminating device to be started for an application. The interface has three input parameters including OriginatingDeviceID, TerminatingDeviceID, OriginatingDevice, BreakPointAddr and OperationType. OperationType is being used to define if a "switch" operation is intended to be partial or full. BreakPointAddr is a file address in Cloud Computing server which stores application breakpoint.

   b) *RequestComplete:*
      Presenter invokes the function to notify *PAdaptor* when the application has been migrated to terminating device successfully.

2. Interface between Controller and CAdaptor

   a) *ControlRequest:*
      Controller invokes ControlRequest to request CAdaptor to execute this operation. The interface has three input parameters including OriginatingDeviceID, TerminatingDeviceID, TerminatingDevice, BreakPointAddr and OperationType. OperationType is being used to define if a "switch" operation is being partial or full.

   b) *RequestComplete:*
      CAdaptor invokes RequestComplete to request *Controller* to end the application in the originating device.

3. Interfaces between Model and CAdaptor

    a) *ControlRequest:*
       CAdaptor invokes ControlRequest to request *Model* to identify type of a terminating device. The interface has three input parameters including OriginatingDeviceID, TerminatingDeviceID, TerminatingDevice, BreakPointAddr and OperationType. OperationType is being used to define if a "switch" operation is being partial or full.

    b) *RequestComplete:*
       Model invokes the function to notify *CAdaptor* when the application has been migrated to terminating device successfully.

4. Interface between Model and PAdaptor

    a) *PresentRequest:*
       Model invokes PresentRequest to request *PAdaptor* to present the application in the terminating device.

    b) *RequestComplete:*
       PAdaptor invokes the function to notify *Model* when the application has been migrated to terminating device successfully.

4.2 *Implementation*

When user switches application from an originating device (take Android Pad as an example) to another destination device (in this case an Andoid Phone), the procedures will be executed as follows:

1. Present populates parameters of PresentRequest and sends a request to PAdaptor:

PresentRequest(Souce_Address, Destination_Address, Android_Pad, BreakPoint_File, FULL)

2. PAdaptor triggers a presenter adaptation and sends an XMPP [12] [13] message to Model:

```
<?xml version="1.0"?>
<stream:stream xmlns:stream="http://centling.com/his"
from=" PAdaptor" to=" Model ">
 <query xmlns="PresentRequest">
   <OriginatingDeviceID> Souce_Address </OriginatingDeviceID>
   <TerminatingDeviceID> Destination_Address </TerminatingDeviceID>
   <TerminatingDeviceID> FULL</TerminatingDeviceID>
   <BreakPoint> BreakPoint_File </ BreakPoint >
   <OriginatingDevice> Android_Pad </ OriginatingDevice >
 </query>
</stream:stream>
```

3. Model performs a device adaptation and sends an XMPP message to CAdaptor:

```
<?xml version="1.0"?>
<stream:stream xmlns:stream="http://centling.com/his"
from=" Model " to=" CAdaptor ">
 <query xmlns=" ControlRequest ">
   <OriginatingDeviceID> Souce_Address </OriginatingDeviceID>
   <TerminatingDeviceID> Destination_Address </TerminatingDeviceID>
    <TerminatingDeviceID> FULL</TerminatingDeviceID>
   <TerminatingDevice> Andoid_Phone </ OriginatingDevice >
   <BreakPoint> BreakPoint_File </ BreakPoint >
 </query>
</stream:stream>
```

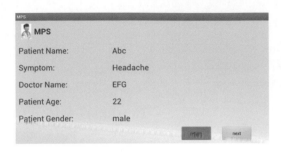

Figure 7.   Originating Device.

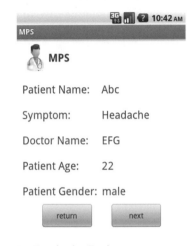

Figure 8.   Terminating Device.

4. CAdaptor initiates a controller adaptation and sends an XMPP message to Controller:

ControlRequest (Souce_Address, Destination_Address, Android_Phone, BreakPoint_File, FULL)

5. Controller receives the message loads with the application pointed by BreakPoint_File.

4.3 *Verification*

We run MPS in an Android Pad with a patient info as shown in Figure 7.

Then request to switch MRS to an Android Phone, the patient info can now be shown in the Android Phone (Figure 8).

5   CONCLUDING REMARKS

We have proposed an approach to implement a Mobile Polymorphic System with the adaptation of both Mobile Polymorphic Applications model and Cloud Computing architecture. After applied in hospital, it is proved that this solution can assist physicians and healthcare professions to focus on their primary tasks of tending patients instead of wasting their effort transferring or retrieving information once environment changes around them.

REFERENCES

[1] Anand Ranganathan, Shiva Chetan and Roy Campbell, "Mobile Polymorphic Applications in Ubiquitous Computing Environments", In Mobiquitous 2004: The First Annual International Conference on Mobile and Ubiquitous Systems: Networking and Services, August 22–25, 2004 – Boston, Massachusetts, USA.

[2] Anand Ranganathan, Shiva Chetan and Roy Campbell, "Mobile Polymorphic for Autonomic Ubiquitous Computing", To appear in Multiagent and Grid Systems – An International Journal, IOS Press.

[3] Roman, M. et al., "Application Mobility in Active Spaces", 1st International Conference on Mobile and Ubiquitous Multimedia, Oulu, Finland, Dec 11–13, 2002.

[4] Cloud Taxonomy, "Software as a Service (SaaS)": http://cloudtaxonomy.opencrowd.com/taxonomy/software-as-a-service/, April 24, 2011.

[5] CloudComputingSec: "Cloud Software as a Service (SaaS) in Cloud Computing. This is not right-Services", http://cloudcomputingsec.com/283/cloud-software-as-a-service-saas-in-cloud-computing-services.html.

[6] Barrie Sosinsky, "Cloud Computing Bible", published by Wiley Publishing, Inc. ISBN: 978-0-470-90356-8.

[7] Nick Antonopoulus, Lee Gillam editors, "Cloud Computing Principles, Systems and Applications", published by Springer, Computer Communications and Network, ISBN: 978-1-84996-240-7.

[8] Gold, N. Mohan, A., Knight, C., Munro, M., "Understanding service-oriented software[J]", IEEE Journals & Magazines, 2004, Volume: 21, Issue: 2, pp: 71–77.

[9] Dikaiakos, M.D., Katsaros, D., Mehra, P., Pallis, G., Vakali, A., Cloud Computing: Distributed Internet Computing for IT and Scientific Research[J], Internet Computing, 2009, Volume: 13, Issue: 5, pp: 10–13.

[10] Leavitt, N., Is Cloud Computing Really Ready for Prime Time[J], IEEE Journals & Magazines, 2009, Volume: 42, Issue: 1, pp: 15–20.

[11] US National Institute of Standards and Technology (NIST): http://csrc.nist.gov/groups/SNS/cloud-computing/cloud-def-v15.doc.

[12] Extensible Messaging and Presence Protocol (XMPP): Core[S], Internet Engineering Task Force RFC 6120, March 2011.

[13] Extensible Messaging and Presence Protocol (XMPP): Core[S], Internet Engineering Task Force RFC 6121, March 2011.

BIOGRAPHY

Vanban Le Wu received a B.S. degree in EECS from the National Taiwan University in 1977, an M.S. degree in EECS from the University of Illinois at Chicago in 1979, and a Ph.D. degree in EECS from the Northwestern University at Evanston, IL. in 1992. He joined AT&T Bell Laboratories, Naperville, IL. in 1982. During his 30 years professional career in the United States, he involved in various research and development activities in areas of electronic/digital/IP technologies switching, modeling network capacity, performance, reliability, and security managements for telecommunication markets. He was an adjunct professor of Beijing University in 1999, and Northwestern University in 1999–2001. His research interests include computational geometry, automata, and applied math. He was a recipient of two AT&T ISDN architecture awards in 1988 and 1992, and held a United States and Europe patent on telecommunications database accessing method in 1990. He is currently the CTO of CentLing Technologies Co. in charge of research and development departments, business strategies and investments of the company.

Changheng Shao received a B.S. degree and an M.S. degree from the Qingdao University. He had applied ten patents in mobile communication and published three IEEE technical papers. His research interests include embedded system, data mining, Internet of Things and mobile communication. He is currently a development manager of CentLing Technologies Inc., and the Innovation Team Lead of the company in charge of innovation research and prototyping tasks.

Information Technology and Computer Application Engineering – Liu, Sung & Yao (Eds)
© *2014 Taylor & Francis Group, London, ISBN 978-1-138-00079-7*

Path tracking based on H∞ suboptimal filter and fuzzy control

Y.J. Wei, X.H. Yang, W.J. Huang & M.H. Lin
Department of Mechanical and Electric Engineering, Soochow University, Suzhou, Jiangsu, China

ABSTRACT: A fuzzy predictive controller is designed for the unmanned automated vehicles. Via measuring the errors of the longitudinal displacement and vertical swing angle, the model of the vehicle is approximated by the H∞ suboptimal filter theory. The errors are also seen as the input fuzzy variables of path tracking. The output fuzzy variable represents the control gain, which regulates the swing angle of the steering wheel. Simulation results indicate a clear advantage in path prediction and tracking.

Keywords: Unmanned vehicle; H∞ suboptimal filter; Fuzzy control; Path tracking

1 INSTRUCTIONS

The unmanned automated vehicle which is also called as the wheeled mobile robot is an intelligent vehicle. It mainly depends on the computer system of intelligent autopilot to achieve unmanned control. The intelligent autopilot can make a safe and reliable driving on the road within the vehicle sensor which can control vehicles' steering and speed by perceiving the road, vehicle location and obstacle information[1].

The closed-loop system within driver-vehicle-road is recommended to investigate the smart car system, which is based on the appropriate driver model. A good driver model will implement the autopilot by processing the perceptive information of road and vehicle timely and transmitting the signal to the vehicle controller and actuator. The driver model is also used in automotive performance testing[2], eliminating fatigue and other subjective factors. On the basis of traditional control theory, McRuer did a lot of researches in the pilot model, and extended them to the car. However, these studies ignored the driver's forward-looking role. As the control of the vehicle is a typical preview control, Reddy proposed the Optimal Preview and Closed-loop Control Drive Model[3] based on the driver's forward-looking role. On this basis, Guo investigated the Optimal Preview and Curvature Model[4]. The intelligent control (such as neural networks and fuzzy control theory) have been gradually applied to the driver model, which has an advantage in the uncertain system. Hiroshi designed the Three-layer and Feed-forward Neural Network Driver Model. Based on the fuzzy control and the neural network, Michel proposed the Fuzzy-neural Network Driver Model[5], which is modelled by the test data of driver-vehicle closed loop system. The Fuzzy-neural Network Driver Model described the manipulation of the driver with fuzzy mathematics and control theory. Due to

the nonlinear and time-varying characteristics of the driver's driving, the error is existed between the model response and actual response of the input signal. In addition, in terms of the control strategy, the controllers in reference [5] only consider the prediction error of the lateral displacement without considering the yaw angle.

This fuzzy prediction algorithm of vehicle motion is motivated by the single point preview optimal controller of path tracking. With the state before the time of k, the longitudinal displacement and vertical swing angle at the time of $k+1$ are predicted by comparing the prediction error to the target locus. The errors of the state are referred to the difference between the actual and predicted values at time $k+1$, which are selected as the input fuzzy variables. The control gain of the desired control signal is inferred based on the fuzzy control theory.

2 PREDICTION MODEL OF THE VEHICLE MOTION

At time k, it is assumed that the lateral displacement of the unmanned vehicle is h_k, vertical one is ve_k and vertical swing angle is φ_k. Proposing the vehicle moves with a constant acceleration (including uniform velocity) in a very short time, the displacement of each point are $h_{k-3}, \ldots, h_{k+1}$ and instantaneous velocities are $v_{k-3}, \ldots, v_{k+1}$, as shown in figure 1. Then,

$$\begin{cases} v_{k-2} = \dfrac{v_{k-1} + v_{k-3}}{2} = \dfrac{h_{k-1} - h_{k-3}}{2T} \\[2mm] v_{k-1} = \dfrac{v_k + v_{k-2}}{2} = \dfrac{h_k - h_{k-2}}{2T} \\[2mm] v_k = \dfrac{v_{k+1} + v_{k-1}}{2} = \dfrac{h_{k+1} - h_{k-1}}{2T} \end{cases} \tag{1}$$

Figure 1. Uniformly rectilinearly accelerating.

It is concluded that

$$h_{k+1} = h_{k-3} + 2h_k - 2h_{k-2} \qquad (2)$$

from $v_k - v_{k-1} = v_{k-1} - v_{k-2}$.

Similarly,

$$ve_{k+1} = ve_{k-3} + 2ve_k - 2ve_{k-2} \qquad (3)$$

and

$$\varphi_{k+1} = \varphi_{k-3} + 2\varphi_k - 2\varphi_{k-2} \qquad (4)$$

The state variable x_k and the predictor variable $\tilde{x}_{k+1}$ of H_∞ suboptimal filter are defined as

$$x_k = \begin{bmatrix} h_k \\ h_{k-1} \\ h_{k-2} \\ h_{k-3} \end{bmatrix}, \quad \tilde{x}_{k+1} = \begin{bmatrix} \tilde{h}_{k+1|k} \\ \tilde{h}_{k|k} \\ \tilde{h}_{k-1|k} \\ \tilde{h}_{k-2|k} \end{bmatrix}$$

where $\tilde{h}_{k+1|k}$, $\tilde{h}_{k|k}$, $\tilde{h}_{k-1|k}$ and $\tilde{h}_{k-2|k}$ are respectively the predictive values of lateral displacement at time $k+1, k, k-1$ and $k-2$, according to the information before the time k. For example, $\tilde{h}_{k+1|k}$ is the lateral displacement of the car in the next moment estimated at time k. By formula (2), the system state equation can be draw as:

$$x_{k+1} = \mathbf{H}x_k + \mu_k \qquad (5)$$

in which the values of state transition matrix $\mathbf{H}$ and noise matrix μ_k are

$$\mathbf{H} = \begin{bmatrix} -2 & 0 & -2 & 1 \\ 1 & 0 & 0 & 0 \\ 0 & 1 & 0 & 0 \\ 0 & 0 & 1 & 0 \end{bmatrix}, \quad \mu_k = \begin{bmatrix} \eta_k \\ 0 \\ 0 \\ 0 \end{bmatrix}.$$

The direct measurement value z_k is defined as

$$z_k = x_k + v_k \qquad (6)$$

where

$$z_k = \begin{bmatrix} h_{k|k} & h_{k-1|k} & h_{k-2|k} & h_{k-3|k} \end{bmatrix}^T,$$

$$v_k = \begin{bmatrix} v_{k1} & v_{k2} & v_{k3} & v_{k4} \end{bmatrix}^T.$$

The actual lateral displacement of the unmanned automated vehicle at the time of $k, k-1, k-2$ and $k-3$ are respectively $h_{k|k}, h_{k-1|}, h_{k-2|k}$ and $h_{k-3|k}$.

By the formula (5) and (6), we obtain the following system:

$$\begin{cases} x_{k+1} = \mathbf{H}x_k + \mu_k \\ z_k = x_k + v_k \end{cases} \qquad (7)$$

where x_k and z_k are the state vector and the observation vector of the system, $\mathbf{H}$ is transition matrix, μ_k and v_k are measurement noise and system noise.

Based on the H_∞ suboptimal filter theory in Reference [7], the design of the H∞ suboptimal filter for system (7) is as follows:

1) $\mathbf{K}_K$ is the gain matrix of H_∞ filtering:

$$\mathbf{K_k} = \mathbf{P_k}[\mathbf{P_k} + \mathbf{R_k}]^{-1}$$

2) $\tilde{x}_{k+1}$ is the state update value:

$$\tilde{x}_{k+1} = H\tilde{x}_k + \mathbf{K_k}[z_k - H\tilde{x}_k]$$

3) $\mathbf{P}_{K+1}$ is the covariance of the estimated error:

$$\mathbf{P_{k+1}} = H\mathbf{P_k}H^T + \mathbf{Q_k} - H\mathbf{P_k}\begin{bmatrix} I & I \end{bmatrix}\mathbf{R_{hk}^{-1}}\begin{bmatrix} I & I \end{bmatrix}^T \mathbf{P_k}H^T$$

where $\mathbf{Q}_K$ and $\mathbf{R}_K$ are the variances of μ_k and v_k respectively. Otherwise

$$\mathbf{R_{hk}} = \begin{bmatrix} R_k & 0 \\ 0 & -\gamma^2 I \end{bmatrix} + \begin{bmatrix} I \\ I \end{bmatrix}\mathbf{P_k}\begin{bmatrix} I & I \end{bmatrix}, (\gamma > 0), \text{ and}$$

$$\mathbf{P_k^{-1}} + (1-\gamma^2)I > 0.$$

If $\mathbf{P}_0 = I$, then $n_k = \tilde{x}_{k+1} - H\tilde{x}_k$ and $v_k = z_k - \tilde{x}_k$.

Within the first 10 period of the running system, it is assumed that $\mathbf{Q}_{0-9} = 2I$, then $\mathbf{Q}_K$ and $\mathbf{R}_K$ are calculated online approximately. In order to save computational complexity, approximate calculation formulas are given by the following equation groups:

$$\begin{cases} M_{Q_k} = \dfrac{k-1}{k}M_{Q_{k-1}} + \dfrac{1}{k}Q_k \\ D_{Q_k} = \dfrac{k-1}{k}D_{Q_{k-1}} + \dfrac{1}{k}(Q_k - M_{Q_k})^2 \end{cases}$$

$$\begin{cases} M_{R_k} = \dfrac{k-1}{k}M_{R_{k-1}} + \dfrac{1}{k}R_k \\ D_{R_k} = \dfrac{k-1}{k}D_{R_{k-1}} + \dfrac{1}{k}(R_k - M_{R_k})^2 \end{cases}$$

Similarly, we can get the predictive values of longitudinal displacement $ve(k+1|k)$ and vertical swing angle $\varphi(k+1|k)$ at time $k+1$.

3 FUZZY CONTROLLER

Reference [6] gives single point preview optimal controller of path tracking:

$$u^0(t) = [f(t+T^*) - y_0(t+T^*)]/(T^*K)$$

where T is the preview time, K is a constant related to controlled system for the linear model parameters, f

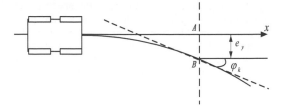

Figure 2.　Schematic diagram of predicting errors.

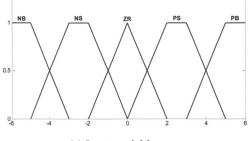

(a) Input variable e_y

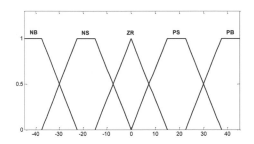

(b) Input variable e_yag

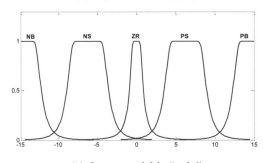

(c) Output variable "gain"

Figure 3.　Membership function definition for the input fuzzy variables.

is default track, y_0 is car's trajectory and u^0 is a control variable showing the swing angle of the steering wheel. Preview optimal control based on single-point of the linear model can be understood as the prediction error multiplied by a constant gain. Because of the car model is essentially a nonlinear model, fixed gain is clearly inappropriate. Based on the prediction error between the predicted values in chapter 2 and the target trajectory, fuzzy rules are designed to adjust the control gain dynamically.

The prediction errors of the longitudinal displacement and vertical swing angle are selected as the input fuzzy variables. As shown in Figure 2, coordinate system is established with the horizontal as x-axis and the vertical as y-axis. Input fuzzy variable $e\_y = f(\hat{x}_{k+1}) - \hat{y}_{k+1}$ means the prediction error of longitudinal displacement in the next moment (f represents the function of the target trajectory); $e\_yag = g(\hat{x}_{k+1}) - \hat{\varphi}_{k+1}$ means the prediction error of longitudinal swing angle in the next moment (g represents the function of the target angle). Output fuzzy variable 'gain' represents the control gain at time k.

The fuzzy set variables of our scheme take grade value of 'NB', 'NS', 'ZR', 'PS', 'PB' respectively. The fuzzification process is shown in figure 3(a) and (b), where e_y is positive means deviation from the direction above the desired trajectory, otherwise the opposite direction. The definition of e_yag is similar to the polar coordinate. Then output 'gain' value is through the defuzzification process, whose membership function of the fuzzy variables is also defined in Figure 3(c).

Fuzzy rules in Table 1 represent the driver should take which kind of action to keep the car going along the target path, which make the longitudinal displacement error and vertical swing angle error as small as possible. For example, when e_y and e_yag are both significantly negetive, namely, $e\_y=NB$ and $e\_yag = NB$, the car needs a larger angle to turn upward. Therefore, the gain should be a larger positive parameter, so the 'gain' is PB. Table 1 can be explained by several fuzzy *IF-THEN* rules.

Table 1 gives the fuzzy reasoning process of the controller gain, combing with the prediction error discussed above, we can get the controller design about the swing angle of the steering wheel as follows:

$u(k) = gain \times e\_y$

Table 1.　Fuzzy rule base

gain	e_yag				
e_y	NB	NS	ZR	PS	PB
NB	PB	PB	PS	PS	ZR
NS	PB	PS	PS	ZR	ZR
ZR	PS	PS	ZR	NS	NS
PS	ZR	ZR	NS	NS	NB
PB	ZR	NS	NS	PS	NB

When the controller output is positive, the steering wheel counterclockwise plays; otherwise the steering wheel clockwise plays.

4　SIMULATION EXPERIMENT

The simulation test of the unmanned vehicle path tracking is based on the CarSim combined with

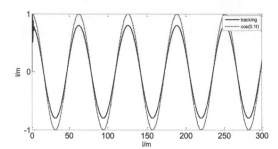

Figure 4. Path tracking simulation results.

matlab/simulink environment in this article. The CarSim software connects to the matlab/simulink interface through the S function. The target path can be expressed by the function $y = \cos(0.1x)$, where x means the lateral displacement of the vehicle's center of gravity, y means the longitudinal displacement.

Figure 4 shows in the uniform driving conditions, the tracking effect on the car with a cosine path. The dotted line represents the target path, and the solid line shows the actual path of automobile. When the target path is swing in the range of 2 m wide, path tracking average error is less than 0.2 m in less than 16.7 m/s of speed. The unmanned vehicle can track the target path with a smaller error, due to the inaccurate dynamic gain. For simple descriptions in this paper, we use less grade of the variables and fuzzy rules. In order to improve the control performance, the grade of the fuzzy set variables and the number of rules in the fuzzy rule table will be increased.

5 CONCLUSION

A novel car trajectory prediction algorithm is designed based on the single point preview control of unmanned vehicle path tracking. The preview controller, via the H∞ suboptimal filter, embodies the process of driving to the fuzzy rule base, which enables the controller to reflect the human intelligence in driving. Simulation results show that the proposed algorithm is fast and effective. The fuzzy controller can track the path well

by adjusting the control gain when the speed is not too high (16.7 m/s or less). However, at a high speed (16.7 m/s or more), error will continue to enlarge. The single direction control at high speed cannot make the vehicle track winding road. It should adjust lateral and longitudinal velocity coupling other control means.

ACKNOWLEDGEMENT

The author is indebted to the university student innovation and entrepreneurship training program for financial support.

REFERENCES

[1] Yang M. 2005. The review and prospects of unmanned automated vehicles. http://wenku.baidu.com/view/2da69b6f58fafab069dc02bc.html

[2] Hong C.W. & Shio T.W. 1996. Fuzzy control strategy design for an autopilot on automobile chassis dynamometer test stands. *Mechatronics* 6(5): 537–555.

[3] Guo K.H., Ma F.J. & Kong F.S. 2002. Drive model parameter identification of the drive-vehicle-road closed-loop system. *Automotive Engineering* 24(1): 20–24.

[4] Guan X., Gao Z.H. & Guo K.H. 2001. Driver fuzzy decision model of vehicle preview course and simulation under typical road condition. *Automotive Engineering* 23(1): 13–17.

[5] Ma J. & He Y.S. 2007. Driver model with self-adjustment fuzzy PID controller for vehicles. *Machinery & Electronics* 19(2): 35–38.

[6] MacAdan C.C. 1981. Application of an optimal preview control for simulation of closed-loop automobile driving. *IEEE Trans. Syst., Man, Cybern* 2(6): 393–399.

[7] Hassibi B., Sayed A.H. & Kailath T. 1996. Linear Estimation in Krein Space – Part II: Applications. *IEEE Transactions on Automatic Control* 41(1): 34–49.

Information Technology and Computer Application Engineering – Liu, Sung & Yao (Eds)
© *2014 Taylor & Francis Group, London, ISBN 978-1-138-00079-7*

Study hierarchical query technology based on ORACLE

Huang Ding, Cai Zhao, Yuan Zhang & Mei Wang
City College of Xi'an Jiao Tong University, Shanxi, China

ABSTRACT: Mass data has lots of hierarchical relationship with the development of information. Hierarchical relationship is analysed and discussed in deeply on the Oracle database in this paper. It is applicated in the Oracle database with a company's organizational structure.

Keywords: Oracle; Node; Hierarchical Relationship; Data; Start With

1   INTRODUCTION

The knowledge of database is more and more important with the development of network information. It is important of query to operate database. Users always apply the query not only one table or more tables but also simple query or complicated query including subquery. A lot of dates with hierarchical structure are in practical application, such as enterprise ERP (Enterprise Resource Planning), PDM (Product Data Management) and so on. However users ignore the hierarchical query with the nature of tree usually, so the paper focuses on the analysis and applications of it.

Oracle database system, a software products of distributed database, is used in client/server or browser/server system structure widely. The Oracle database has the the advantages of high efficiency, high safety performance, high reliability and rich query function and so on, so users favor it in the Information Age.

2   DATA HIERARCHY STRUCTURE

Data hierarchy structure is also called tree structure, it has one or more nodes and it must meet the following conditions:

1) A node with no parent node is called the root node, as shown in Fig. 1 node A.
2) The rest of the nodes have only one parent node and each forms a tree, as shown in Fig. 1 node B1, C1, C2, each forms a subtree.

The child node with a parent node is called brother node, as shown in Fig. 1 of the B1, B2, B3, they are brother nodes. A node that has no child is called leaf node, as shown in Fig. 1 B3, C3.

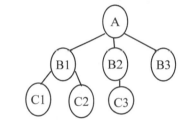

Figure 1.   Hierarchy structure.

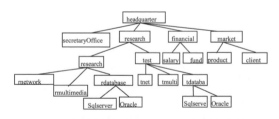

Figure 2.   A company's department hierarchy structure.

Some data have the characteristics of hierarchical structure in the Oracle database. For example, a company's department hierarchy structure is shown in Fig. 2, each department is as a node and has a parent node except the root node (headquarter), each subnode can be used as a substree.

The department information about Fig. 2 is shown in Table 1.

In order to store the department information, a table is created in the Oracle database including the department number, department name, the superior department number and so on, the SQL statement is shown in Fig. 3. The department information is stored in the table.

Table 1. Department information.

deptno	dname	dgno	deptno	dname	dgno
5000	hq	null	2110	rnetwork	2100
1000	secretaryOffice	5000	2120	rmultimedia	2100
2000	research	5000	2130	rdatabase	2100
3000	financial	5000	2131	rsqlserver	2130
4000	market	5000	2132	roracle	2130
2100	researchDlp	2000	2210	tnetwork	2200
2200	test	2000	2220	tmultimedia	2200
3100	Salary	3000	2230	tdatabase	2200
3200	funds	3000	2231	tsqlserver	2230
4100	product	4000	2232	toracle	2230
4200	client	4000			

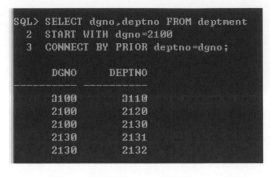

Figure 4. All the child department of the department number 2100.

```
SQL> CREATE TABLE deptment
  2  (deptno number(4) primary key,
  3  dname varchar2(20),
  4  dgno number(4)
  5  );
```

Figure 3. Created department information.

3 ORACLE HIERARCHICAL QUERY SENTENCE

Hierarchical query uses the tree traversal, the query method of tree structure is the top-down method or the bottom-up method. It also can query in the hierarchy starting point location.

3.1 Hierarchical query syntax

Oracle can be hierarchical query, syntax is as follows:

SELECT [level], column, expr..., FROM table [WHERE condition] START WITH condition CONNECT BY [PRIOR column1 = column2 |column1 = PRIOR column2];

Among them:

1. The level keyword is a "pseudo column" in order to Express the tree, such as 1 root, 2 root child, the others is the same rules.
2. It is table or view from the keyword FROM.
3. WHERE conditions restrict the rows returned by a query, but do not affect the hierarchy structure, which belong to the node truncation, but truncated nodes of the lower child is not affected by.
4. START WITH is the start node.
5. CONNECT BY PRIOR is to specify the relationship between father and son, which the position of the PRIOR doesn't have to be in CONNECT BY.

3.2 Hierarchical query restriction

Hierarchical query in Oralce database needs to pay attention to the following matters:
1. It is only one table or one view after the keyword of FROM but no keyword JOIN.

2. In order to sort, it uses keyword SIBLINGS but ORDER BY, because the hierarchical structure maybe is broken using keyword ORDER BY to sort.
3. It has subquery in START WITH expression but CONNECT BY expression.

3.3 Hierarchical query advanced application

Keyword CONNECT_BY_ISLEAF is used to query the leaf node, if the line value of 0 is not a leaf node, 1 is a leaf node.

Keyword NOCYCLE is used to avoid circular nodes in the query, circular nodes is two nodes for each child node. Keyword CONNECT_BY_ISCYCLE can get circular nodes, which value 0 is not circular and 1 is circular.

In order to sort, it uses keyword SIBLINGS, the syntax is as following:

ORDER SIBLINGS BY<expre>

This sort will protect level.

4 INSTANCE

4.1 The child nodes of any node traversal

The top-down traversal method is used to query all the child node of one node. The keyword CONNECT BY and PRIOR determine the conditions and methods of traverse, keyword PRIOR is the mean of before node, the syntax is CONNECT BY PRIOR father node=son node or CONNECT BY son node=PRIOR father node.

e.g.1: It query all the child department of the department number 2100, the SQL statement and the query results as shown in Fig. 4.

4.2 The father nodes of any node traversal

The bottom-up traversal method is used to query all the father node of one node, the syntax is CONNECT BY PRIOR son node=father node or CONNECT BY father node=PRIOR son node.

```
SQL> SELECT dgno,deptno FROM deptment
  2  START WITH deptno=2100
  3  CONNECT BY PRIOR dgno=deptno;

      DGNO     DEPTNO
---------- ----------
      2000       2100
      5000       2000
                 5000
```

Figure 5. All the father department of the department number 2000.

```
SQL> DELETE FROM deptment WHERE deptno in(
  2  SELECT deptno FROM deptment d
  3  START WITH deptno=2130
  4  CONNECT BY PRIOR deptno=dgno);
```

Figure 6. Deletes all the subordinate department of the department number 2130.

```
SQL> SELECT deptno,dname,level,PRIOR deptno as "father",
  2  CONNECT_BY_ROOT deptno as "root",
  3  DECODE(CONNECT_BY_ISLEAF,1,'It is leaf','It is not leaf') as "Is child?"
  4  FROM deptment
  5  START WITH dgno IS NULL
  6  CONNECT BY PRIOR deptno=dgno
  7  ORDER SIBLINGS BY deptno ASC;

    DEPTNO DNAME                     LEVEL     father       root Is child?
---------- ------------------- ---------- ---------- ---------- --------------
      5000 hq                           1                  5000 It is not leaf
      1000 secretaryOffice              2       5000       5000 It is leaf
      2000 research                     2       5000       5000 It is not leaf
      2100 researchDlp                  3       2000       5000 It is not leaf
      2110 rnetwork                     4       2100       5000 It is leaf
      2120 rmultimedia                  4       2100       5000 It is leaf
      2200 test                         3       2000       5000 It is not leaf
      2210 tnetwork                     4       2200       5000 It is leaf
      2220 tmultimedia                  4       2200       5000 It is leaf
      2230 tdatabase                    4       2200       5000 It is not leaf
      2231 tsqlserver                   5       2230       5000 It is leaf
      2232 toracle                      5       2230       5000 It is leaf
      3000 financial                    2       5000       5000 It is not leaf
      3100 salary                       3       3000       5000 It is leaf
      3200 funds                        3       3000       5000 It is leaf
      4000 market                       2       5000       5000 It is not leaf
      4100 product                      3       4000       5000 It is leaf
      4200 client                       3       4000       5000 It is leaf
```

Figure 7. Example 4 query statement and results.

e.g.2: It queries all the father department of the department number 2000, the SQL statement and the query results as shown in Fig. 5.

4.3 *Delete subtree*

In order to remove one department and the subordinate department, it will use delete subtree function.

e.g.3: It deletes all the subordinate department of the department number 2130, the SQL statement and the query results as shown in Fig. 6.

4.4 *Complex application*

It uses the keyword LEVEL pseudo columns to get how many the subordinate department is. It uses keyword CONNECT_BY_ISLEAF to query the leaf node, if the line value of 0 is not a leaf node, 1is a leaf node. It uses keyword CONNECT_BY_ISCYCLE to get circular nodes, which value 0 is not circular and 1 is circular. It uses keyword SIBLINGS to sort.

e.g.4: It displays information including the department number, department name, which level, the direct superior department number, the supreme department number, the SQL statement and the query results as shown in Fig. 7.

REFERENCES

- (USA) Bullock, Effective Java. Beijing: Mechanical Industry Press, 2009
- (USA) HaoSterman, Ye Naiwen, Kuang Jinyun, Du Yongping. The core technology of Java. Beijing: Mechanical Industry Press, 2008
- (USA) Bruce Eckel, Thinking in Java, hird Edition. Prentice Hall PTR, 2003
- Cui qunfa, Li Lixin, easy to learn Oracle database. Beijing: Chemical Engineering Press, 2012

Information Technology and Computer Application Engineering – Liu, Sung & Yao (Eds)
© 2014 Taylor & Francis Group, London, ISBN 978-1-138-00079-7

Design and implementation of browser/server-based intelligent decision support system for farm machinery selection

Rong Xin Zhu & Jin Yan Ju
Mechanical Engineering school, University of Science and Technology Heilongjiang, Harbin, China

ABSTRACT: Optimize the selection of Agricultural machinery system plays an important role for the run smoothly of agricultural production. To instruct selection of machine for farmers, farms, administrant department of farm machinery, and to shorten the gap of the decision – making level of agricultural machinery between China and developed countries, the Browser/Server-based intelligent decision support system for farm machinery selection has been designed and developed, refer to the domestic and foreign research results and the actual situation of China. The Browser/Server architecture was used; six major system function modules were included in the system, such as intelligent decision of machine selection, expert decision, expert information browsing, machine information browsing, geographical information input and system maintenance; the overall structure of the system was composed of the integrated components, knowledge base system, database system and model base system. SQL Server 2000 database was selected, the development of Web applications were done based on ASP.NET network theory. The COM technology integrated in Matlab 7.0 was applied to implement selection model. The more scientific selection program was proposed, it can promote normative selection for agricultural machinery and the informatization construction of agricultural machinery selection.

Keywords: Farm machinery selection; Decision support system; System design

1   GENERAL INSTRUCTIONS

In recent years, the demand of agricultural machinery increase greatly, under the impulsion of agricultural machinery subsidy policy implemented by the central and local governments. However, there is a common phenomenon of purchase blindly, because of poor information of agricultural machinery, less acquaintance with the types, performance, features, adaptability, price and maintenance of farm machinery, coupled with the lack of agricultural experts, so as to result in a huge waste of resources and funding. Agricultural machinery management departments and farmers urgent need a management information and decision support system, which can query farm machinery information, forecast and select farm machinery.

To instruct selection of machine for farmers, farms, administrant department of farm machinery, and to shorten the gap of the decision-making level of agricultural machinery between China and developed countries, the Browser/Server-based intelligent decision support system for farm machinery selection (FMSIDSS) has been designed and developed, refer to the domestic and foreign research results and the actual situation of China.

2   GOALS OF SYSTEM DESIGN

Using computer technology and communication networks technology, FMSIDSS was exploited, which is effective, efficient, reliable, friendly, easy to operate, and to suitable for actual agricultural production in China, so as to make sure effective use of agricultural machinery resources.

To provide the fundamental data for decision support system, there was established a database, which has suitable structure, lower redundancy, accurate and rich information.

The knowledge library and model base were established, which provide effective and high-quality decision information for purchasing agricultural machinery.

In order to adapt to the needs of agricultural market development, the system were maintainable and extensible.

3   ARCHITECTURE OF SYSTEM

At present the popular architecture patterns are Client/Server (C/S) and Browser/Server (B/S). Comparing the two modes between C/S and B/S, and

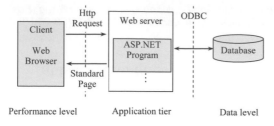

Figure 1. Architecture of system.

according to the characteristics of the system, Browser/Server/Database architecture developed from B/S was used in the system, which put main application logic on the middle layer between the client side and server end, because the upgrade and modification of software achieved in the server end, users only need to surf the Internet to meet the needs, no need to maintain the client software in this system. The architecture diagram is shown in figure 1.

Through the visual human-computer interface, the Web browser as performance level realizes the exchange of information between user and system, such as submitting data to the system, to show the user the results of decision; Database system as data level is in charge of data storage and management of database; Web server as application tier provides a variety of services components associated with application logic in decision system. The three-tier distributed architecture makes storage, handling and using of data relatively independent, the using, maintenance and renew of data easier to achieve.

4 SYSTEM FUNCTION DESIGN[1]

Considering practicality and comprehensiveness of the system, six major system function modules were included in the system, refer to the actual situation of farm machinery selection.

4.1 *Intelligent decision of machine selection module*

The intelligent decision module consists of tractor selection module and cultivated machine decision-making module.

In the tractor selection module, firstly, the inference engine matches rules in the knowledge library, based on the geographical characteristics and planting characteristics of the user area. The drive type and power range of tractor suitable for this area will be obtained in the light of these rules. Secondly, the system searches machines meeting the requirements in database, as alternative machines. Then, system brings up the performance scores of these alternative machines and evaluates performance according to appropriate selection method selected by user, the final one or several optimization machines will be proposed for decision-making reference.

In the cultivated machine decision-making module, the appropriate process was determined according to the actual cultivation conditions, after tractor selection; the type of cultivated machine was selected according to the specific job in such process, for example, planter, harvester, cultivator; then the inference engine matches rules in the knowledge library. The range of cultivated machine suitable for this area will be obtained in the light of these rules. At last, System searches alternative cultivated machines meeting the requirements in database, and brings up the performance scores of these alternative machines and evaluates performance according to appropriate selection method selected by user, the final optimization machines, such as, planters, transplanters, combine harvesters, will be proposed for decision-making reference.

4.2 *Expert decision module*

Selection methods (Fuzzy synthesize judgment, AHP, PPC) system used require the weight values of evaluation index and detailed scores for each machine that were evaluated by experts in the field agricultural machinery. In this module, agricultural machinery experts each region can visit the site, evaluate farm machine and index; it realized off-site access of scores. Expert decision module is comprised of scoring machines, scoring weight values and judgment matrix scoring.

4.3 *Expert information browsing module*

In the module, the expert information of agricultural machinery selection is reposited for users to browse, including name, gender, age, title, location, specialty, etc.

4.4 *Machine information browsing module*

The user can view the information of tractor and cultivated machine collected on the market in machine information browsing module. It includes advanced browsing and basic browsing. Through basic browsing user can view relevant information of tractor and cultivated machine domestic or imported, such as product name, manufacturers, pictures, drive type, engine power, etc. To meet the different needs of users and to achieve quick search, advanced browsing system search the suitable tractor and cultivated machine in the database, in accordance with the conditions given by the user, for example manufacturers, drive type, and so on.

4.5 *Geographical information input module*

The information of geographical features and planting characteristics in user area, required in the intelligent decision of machine selection module, was input in the geographical information input module, can be

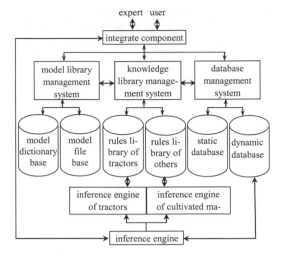

Figure 2. The structure of system.

directly displayed on the drop-down menu of intelligent decision of machine selection module, including the province, latitude, terrain features, planting characteristics, planting crops, and so on.

4.6 *System maintenance module*

System maintenance module is comprised of knowledge library management, database management, model library management, user management. Only administrator can visit the system maintenance module.

5 SYSTEM STRUCTURE DESIGN

According to functions provided by the FMSIDSS, the structure of system was comprised of the integrate component, the knowledge library system (including knowledge library, knowledge library management system and inference engine), database system (including database and database management system) and the model library system (including model library and model library management system). The structure diagram is shown in figure 2.

5.1 *Integrated component*

As the combination of the human-computer interaction system and the integrate problem system, the integrate component of the decision support system is the core part of the FMSIDSS. The major function of the integrate component is to control the FMSIDSS, which mainly obtain original data and solve problems through human-computer interactions. By integrating the knowledge library, database and model library organically, the integrate component can analyse and solve the FMSIDSS problem through controlling data flowing; starting the reference machine timely; combining modules based on requirements, and finally

supporting the machine information browsing and intelligent decision selection for the tractor and cultivated machine, and expert decision module, etc.

5.2 *Knowledge library system*

The knowledge library system of FMSIDSS system includes knowledge library, knowledge library management system and inference engine. In the knowledge library, the expert knowledge in the field of agricultural machinery selection is reposited as IF-THEN method. The knowledge library management system mainly focus on the knowledge library establishment, modification of knowledge inconsistency and incompleteness errors, and the increase and deletion of knowledge , so as to establish a high performance knowledge library. Inference engine, which adopts forward inference, can find the alternative tractor and cultivated machine suitable for users, according to the rule knowledge known already based on the initial conditions given by the user.

5.3 *Database system*

FMSIDSS database system is comprised of two parts of the database and the database management system. The database includes dynamic database and static database, which contains a total of 7 databases. The dynamic database consists of user basic information database and the intermediate information database. The static database is divided into tractor material database, cultivated machine material database, expert decision database, the user and administrator information database and case database. The database management system is in charge of the effective management and organization of database, including the data addition, deletion and modification, in order to improve the system efficiency.

5.4 *Model library system*

The FMSIDSS includes the model library and model library management system. Through comprehensive analysis of the influence factors on agricultural machine selection, based on establishment of the appraising target systems for agricultural machinery selection, the mathematical models for scientific selection were investigated for tractor and cultivated machine, including Fuzzy synthesize judgment, AHP, PPC based on GA, then build up the model library. The model library management system can not only add, delete, modify the model in the model library, but also organize and manage the model library effectively.

6 SYSTEM DEVELOPMENT ENVIRONMENT

User environment: Through configure Windows XP / Windows 7 operating system and Internet Explorer 7.0

Browser, customers can visit the Internet directly. Internet Explorer 7.0 Browser is user-friendly, extended and expanded.

Network service environment: Microsoft Windows Information Server 4.0 (IIS) was selected. The program runs on the Windows NT platform, which is stable performance, ease to operate and manage.

Database management system: SQL Server 2000 database was selected, which is relational database management system with high-performance, multi-user, large capacity, safe and reliable and seamless convergence with the windows operating system.

Professional development environment: The development of Web applications and system deployment and debugging were done based on ASP.NET network theory.

Auxiliary development environment: the COM technology integrated in Matlab 7.0 was applied to implement selection model; Photoshop CS, Dreamweaver MX and Flash MX were used for the beautification of the image processing and web design.

7 CONCLUSIONS

Decision Support System was used in the field of agricultural machinery selection, not only further to expand the scope of application of the decision support system, but also to promote the effective application of the agricultural machinery rational selection. Through utilizing a large number of detailed, credible data, and scientific selection model, the selection program accurately, efficiently and objectively was obtained.

The feature of system is that advanced mode, integral function, stable operation, high reliability, and with good interactive characteristic. The system can promote normative selection for agricultural machinery and the informatization construction of agricultural machinery selection.

In farm machinery selection process, the weight values of evaluation index and detailed scores for each machine that were evaluated by experts in the field agricultural machinery were necessary, it make selection results link to the subjective factors of experts closely, so how to avoid the impact of subjective factors on the selection results is a important issue to be resolved.

REFERENCES

Rongxin, Zhu. 2006. Study on Web-based intelligent decision support system for farm machinery selection. Northeast Agricultural University.

Jing, Du. 2003. Study on Web-based decision support system for development of agricultural mechanization. China Agricultural University.

Yidan, Bao. 2001. Study on multimedia decision support system of agricultural machinery. Journal of Zhejiang Agricultural University 27(2):187–190.

Qiang, Fu. 2003. Selection of agricultural machinery types and their optimum order based on PPC Model. Transactions of The Chinese Society of Agricultural Machinery 34(1):101–107.

Kotzabassis, Constantinos. 1991. An Integrated Farm Machinery Selection and Management System for Personal Computers. Dissertation Abstracts International 53:78–83.

Information Technology and Computer Application Engineering – Liu, Sung & Yao (Eds)
© 2014 Taylor & Francis Group, London, ISBN 978-1-138-00079-7

Analysis on earth fault and propagation characteristic of electric fire

Xin Ming Wang, Chang Zheng Zhao & Wei Gao
Shenyang Fire Research Institute of Ministry of Public Security, Shenyang, China
Key Laboratory of Fire Scene Investigation and Evidence Identification, Ministry of Public Security, Shenyang, China

ABSTRACT: In this paper, the causes of electric fires were analyzed. According to the typical low-voltage distribution system, the single-phase earth fault which occurred in the transformer's low voltage side and three-phase electric equipment was simulated by the software of matlab/simulink, the voltage and current waveforms of the main wires, three-phase and single-phase electric equipments were obtained. Based on the simulation results, the working conditions and consequences of the electric equipments under the single-phase earth fault were analyzed. Meanwhile, all of these provided the important references for the propagation characteristics of electric fires.

Keywords: Earth fault; Propagation characteristic; Electric fire; Risk analysis

1 INTRODUCTION

With the development of national economy and improvement of people's living standards, the requirement and demand of electricity becomes more than before. Meanwhile, the fires caused by the electric faults increases correspondingly, therefore leads to disastrous casualties and property losses. According to the statistics from 2002 to 2011, there were about 2 millions electric fires occurred in China and property losses went to more than 15.6 billion RMB Dollars[1].

When the wires and the electric equipments are working, short circuit, overload, over-voltage, under-voltage and leakage may occur because of the mis-using of the equipments or lines ageing[2]. In this case, if the protective devices could not cut off the fault lines immediately, the lines and equipments would be working continually under the fault current and voltage conditions, thus the abnormal energy released by the heat may ignite combustible materials, even lead to the fire. The fire which is brought by the single-phase earth fault is extrahazardous, and it is more difficult to identify the reasons for researchers[3].

In this paper, the single-phase earth fault which occurred in the typical low-voltage distribution lines were simulated and analysed, the propagation characteristics of single-phase earth fault and the mechanisms of electric fires were summed up, and the references to electric fire investigation and technical identification were provided also.

2 COMPOSITION OF TYPICAL LOW-VOLTAGE DISTRIBUTION SYSTEM

The typical low-voltage distribution system is mainly consists of the step-down transformer substation, the low-voltage distribution lines and the electric equipments.

The step-down substation mainly includes the high-voltage circuit breaker and the step-down transformer. The high-voltage circuit breaker can switch on and cut off the rated loads, and it can cut off the short-circuit currents at the same time. The low-voltage distribution lines mainly include the low-voltage circuit breakers and the transmission lines. The low-voltage circuit breakers are the key equipments, which can not only switch on and cut off the normal circuit, but also perform the fault protective functions once the wires were overloaded, short circuit or leakage. The transmission lines are set up by multi-stranded bare wires, the copper or aluminum cables with the PVC insulation and they are often adopted for the 10 kV and below lines.

It is necessary to install the neutral lines in the high or low voltage distribution systems to ensure the power supply, the electric equipments and protectors safety and reliable running. The TN-CS power supply mode is used in the low-voltage (below 1 kV) distribution systems normally in China. Its feature is that the neutral point of the power transformer is connected to the ground directly so that the single-phase loads can performe without any electric accidents.

The electric equipments in the typical buildings can be classified into two categories: the first one is three-phase equipments which must be connected to the three-phase supply, such as the three-phase motors, fan pumps, air conditioners and so on. The three-phase loads are symmetrical,and it means that the impedance values and angles of one phase must equal the other two. The other one is single-phase equipments which must be connected to anyone of three-phase supply, such as lightings and household appliances. It is difficult to make their impedance values and angles approximate equal, the loads of every phase are unsymmetrical.

3   RESULTS AND ANALYSIS ON SIMULATION

3.1   *Single-phase ground fault of the transformer' low-voltage side*

Figure 1 shows the voltage and current waveforms of the main lines while A-phase of the transformer' low-voltage side occurs the ground fault. It can be seen that the voltage and current waveforms of three-phase equipments and all single-phase equipments are similar to the main lines'. The voltage of the fault phase tends to 0, and the current is small also; while the voltage and current of non-fault phase are 1.73 times before failure. Residual current protective circuit breaker can carry out effective ground fault protection, if the three-phase circuit breaker which is equipped on the three-phase power lines can perform the functions of the breaking-phase or three-phase unbalance protection, the fault current will be cut off; otherwise, the three-phase power lines will be working

at the state of lack phase continually. In the meantime, the current of non-fault phase is 1.73 times before failure, but if the lines were runing at the state of light-load or no-load, this fault current may not reach the action value of over-current protective apparatus. Because the voltage of non-fault phase is 1.73 times before failure, the probability of insulation breakdown increases correspondingly not only for the power lines but also for the variety of electric equipments.

If the protective apparatuses are not acting and the three-phase electric equipments are working all the time, because the motors have some speed and steering already, these motors can still running normally as long as the load resistance is not too large, but the motor will generate heat serious at the state of lack phase, the winding may be burnt out due to the high temperature[4]. The single-phase electric equipments of the fault phase will stop because of the value of phase voltage is 0, while those in the non-fault phase will be working continually.

3.2   *Single-phase ground fault of three-phase equipment*

Figure 2 shows the voltage and current waveforms of the main lines while A phase of three-phase equipment occurs the ground fault. It can be seen that the voltage of fault phase tends to 0 and the current increases 3 times before failure, the voltage of non-fault phase increases 1.73 times before failure and the current increases 2 times. Residual current protective circuit breakers can carry out effective ground fault

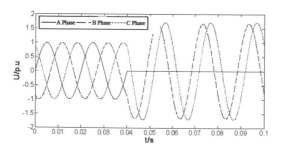

a) Voltage waveforms

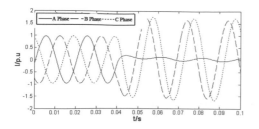

b) Current waveforms

Figure 1.   Voltage and current waveforms of the main line under single-phase ground fault of transformer' low-voltage side.

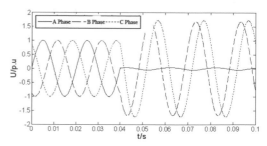

a) Voltage waveforms

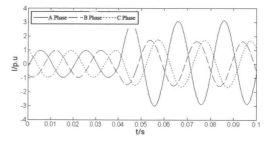

b) Current waveforms

Figure 2.   Voltage and current waveforms of main line under single-phase ground fault of three-phase equipment.

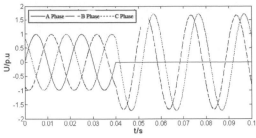

a) Voltage waveforms

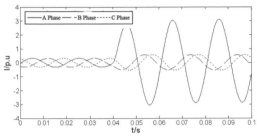

b) Current waveforms

Figure 3. Voltage and current waveforms of three-phase equipment under single-phase ground fault of itself.

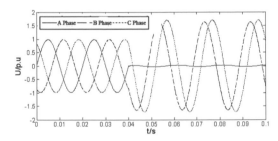

a) Voltage waveforms

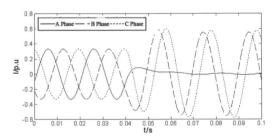

b) Current waveforms

Figure 4. Voltage and current waveforms of single-phase equipment under single-phase ground fault of three-phase equipment.

protection, if the three-phase circuit breaker which is equipped on the three-phase lines has the functions of the breaking-phase or three-phase unbalance protection, the fault current will be cut off; otherwise, the three-phase power lines will be running at the state of lack phase continually. In the meantime, if the lines are working at the state of light-load or no-load, the fault current may not reach the action value of over-current protective apparatus and so the three-phase lines will be working at the state of lack phase continually. The voltage of non-fault phase is 1.73 times before failure, and the probability of insulation breakdown increases correspondingly.

Figure 3 shows the voltage and current waveforms of the three-phase equipment while the A-phase of itself occurs ground fault. The U_A and U_B of the three-phase load are 1.73 times before failure, U_C is 0, I_A is about 6 times before failure, and I_B is about 6 times before failure as well as I_C. The A-phase' current is 6 times before failure, therefore the fault current can reach the action value of the over-current protective devices even the motors are working at the state of no-load, thus the three-phase motors will be burned out impossibly.

Figure 4 shows the voltage and current waveforms of single-phase equipment while A-phase of three-phase equipment occurs the ground fault. It can be seen in Figure 4 that the voltage and current of fault phase tends to 0, the voltage and current of non-fault phase increase 1.73 times. The result is similar to that the transformer' low side occurs single phase short circuit.

4 CONCLUSIONS

By studying the common causes of electric fires for the low-voltage distribution system and simulating the single-phase earth fault through the software of matlab/simulink, the main lines' voltage and current waveforms while one phase of the transformer' low-voltage side and three-phase equipments occurred the ground fault were obtained, the voltage and current waveforms of the three-phase and single-phase equipments while one phase of three-phase equipments occurred the earth fault were acquired also, the consequences of the electric equipments under the single-phase ground fault were analyzed, the propagation characteristics of electrical fires were summed up and the failure modes of typical building electric fires were obtained.

ACKNOWLEDGEMENT

This research was supported by the item of the strengthening police by science and technology (Grand No. 2011GABJC038).

REFERENCES

[1] Fire Department of MPS. *Fire statistical yearbook of China 2012.* Beijing: China Personnel Press.

[2] Pan Gang, Gao Wei, Zhao Chang-zheng & Di Man. Comprehensive Identification Technology about Electric Fire. *Journal of Fire Science and Technology,* 2005, 24(4):495–497.

[3] Tian Li-ying. Ground Fault Caused by the Electric Fire and Its Preventive Measures, *Journal of Taiyuan University of Technology,* 2001, 9(5): 504–506.

[4] Liu Yan. *Study on the Cause Analysis and Pattern Recognition of Electric Fire for Typical Buildings,* ShenYang: ShenYang University of Technology, 2009:31–34.

Information Technology and Computer Application Engineering – Liu, Sung & Yao (Eds)
© *2014 Taylor & Francis Group, London, ISBN 978-1-138-00079-7*

Identification of motor imagery parameters from EEG using SVM

B.L. Xu & X.X. Yin
State Key Laboratory of Robotics, Shenyang Institute of Automation (SIA), Chinese Academy of Sciences (CAS), Shenyang, P. R. China
University of Chinese Academy of Sciences, Beijing, P. R. China

Y.F. Fu
School of Automation and Information Engineering, Kunming University of Science and Technology, Kunming, P. R. China

G. Shi
State Key Laboratory of Robotics, SIA, CAS, Shenyang, P. R. China

H.Y. Li
State Key Laboratory of Robotics, SIA, CAS, Shenyang, P. R. China
School of Mechanical Engineering & Automation, Northeastern University, Shenyang, China

Z.D. Wang
Department of Advanced Robotics, Chiba Institute of Technology, Chiba, Japan

ABSTRACT: Brain Computer Interface (BCI) is a technology to control devices such as prosthetics, cars, robots and other control related applications using brain signal directly. Currently, EEG-based BCIs are mostly based on classification between different limb effectors, and only few commands can be generated directly. To increase the command number and control efficiency, both clench force parameters and clench speed parameters has been studied independently. In this paper, we try to decode speed and force hand movement imagination at the same time. Alpha band power of EEG signal is extracted as feature and then classified by Support Vector Machine (SVM). The primary result shows that speed and force motor imagery of the same hand is distinguishable, which may increase direct BCI control commands in the future.

Keywords: Brain-Computer Interface (BCI); motor imagery parameters; Support Vector Machine (SVM)

1 INTRODUCTION

Brain-computer interface (BCI) is technology to provide direct control of devices without the participation of peripheral nervous system and muscles [1]. It is first described in 1973 [2], and get great development since 1993 [3]. Currently, more and more researches have been done in this field. Generally, there are three methods to extract signals from the brain: invasive [4], half-invasive [5], and non-invasive [6].

Electroencephalograph (EEG) is the most widely used non-invasive method because its convenient and low cost. Many BCI paradigms have been developed using EEG, such as visual evoked potential (VEP) [7], steady stead visual evoked potential (SSVEP) [8], P300 [9], N200 [10] and motor imagery [11, 12], among which motor imagery is most natural and efficient to control a prosthesis device or wheelchair for stroke patients who cannot move on their self [13, 14].

However, current EEG-based BCIs are mostly based on classification between different limb effectors, and only few commands can be generated directly. Little attention has been paid to decode different ways in which a specific limb motor imagery is performed, such as force, velocity and direction. BCI command inputs could be increased by combining limb movement and task parameters, which would allow the control of prostheses or electrical stimulation of muscles more natural for functional rehabilitation of stroke patients [15–21].

Fortunately, few researches have been done to disclose the relations between EEG signals and task parameters. Do Nascimento proved that readiness potentials (RP) correlate with isometric plantar flexion torque levels, and the motor potentials (MP) and the movement-monitoring potentials (MMP) are sensitive to the different rate of torque development (RTD) [22]. Farina proposed a pattern recognition method to classify movement-related cortical potentials (MRCPs) modulated by force-related parameters using wavelet feature and support vector machine (SVM) [23]. Gu studied the discrimination between different combinations of RTD and target-torque (TT) from single-trial MRCPs during right foot imaginary tasks. Their results indicate that both TT and RTD can be classified using single-trial EEG traces [19]. On the other hand, Yuan proved that alpha, beta and gamma bands all correlated with speed decoding of imagined hand movement [15].

In this paper, we first attempt to investigate different EEG characteristics between clench speed motor imagery and clench force motor imagery, and classify EEG of the two types using support vector machine (SVM). Our results show that speed and force motor imagery is separable, which may provide more direct control commands for BCI systems.

2 EXPERIMENT DESIGN

2.1 Subjects

Six subjects aged 24~33 years (mean 26.8 years) took part in the study, including 3 male and 3 female. According to training level, they are divided into two types. One type are trained more than 3 times, while the other type are not trained and only be told how to complete the experiment. All of them are healthy subjects without sensory-motor deficits or any history of psychological disorders. The experimental procedure is approved by the Ethical Committee of the Shenyang Institute of Automation (SIA), Chinese Academy of Sciences (CAS). Written informed consent to participate in the experiment was signed by all of them.

2.2 Experiment paradigm

A single trial of the experiment is divided into four periods. A short beep is excited to remind the subjects the start of the trial. After 10 seconds baseline time, a cue is presented to inform the subjects which type of hand movement is to imagine when the cue disappears 2 seconds later. The motor imagery task sustained 10 s, followed by a random rest time between 10~12 s. The baseline time, task time and rest time is much longer than usual EEG paradigms because we acquired functional near-infrared spectroscopy (fNIRS) signals at the same time. The results of the fNIRS signal analysis can be found in our other papers [31, 32]. We will research the combined EEG-fNIRS signals in the near future.

Every subject participates in 3 sessions of the experiment, and every session includes 30 trials of right hand clench speed imagination task and 30 trials of right hand clench force imagination task.

2.3 Data acquisition

EEG signals are acquired by Neuroscan synamps2 with a sample frequency of 1000 Hz. 21 Ag/AgCl electrodes are used according to 10–20 system [33], as shown in figure 1. A1 is used for reference, and Fpz is used for ground.

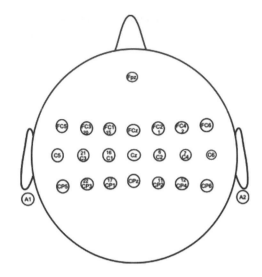

Figure 1. EEG channel locations.

3 DATA PROCESS METHODS

3.1 Preprocess

The EEG data acquired is re-referenced to common average reference first and down sampled to 250 Hz to decrease the feature dimension. Then, every channel is processed with a Laplacian filter described in Equation 1 to increase spatial resolution, where V_j is the data of channel j, and S_j is the set of 4 channels that around the channel j. A 4th order IIR Butterworth high pass filter with a cutoff frequency of 0.5 Hz is used to remove low frequency drift, and two notch filters with stop band center frequency of 50 Hz and 100 Hz are used to remove power line noise and its high-order harmonic waves.

$$V_j^{Lap} = V_j - \frac{1}{4} \sum_{k \in S_j} V_k \qquad (1)$$

3.2 Feature selection

In our research, the power of alpha band (8–12 Hz) is extracted from 21 channels are as feature space. As is shown in Figure 2 and 3, the channel spectral and maps of clench force motor imagery are different from that of clench speed motor imagery, so the alpha band power feature is reasonable.

3.3 Support vector machine

Support vector machine (SVM) is a classifier that can separate examples of two categories with a maximum margin using a small data set [24], and it is widely used for object recognition, speaker identification, face detection in images, and other pattern recognition related areas. SVM is especially appropriate for BCI application because EEG trials used for training is usually very small. In SVM, only examples

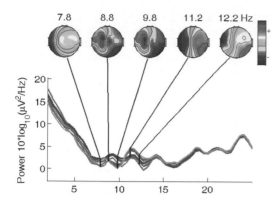

Figure 2. The channel spectral and maps of clench force motor imagery.

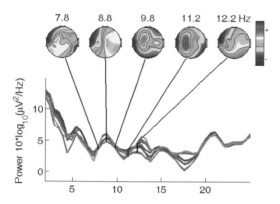

Figure 3. The channel spectral and maps of clench speed motor imagery.

that closest to the separating hyperplane are used for hyperplane orientating, and these examples are called support vectors.

SVM implementing can be described in Equation 2, where y_i is the label of example x_i with value 1 for one class and -1 for the other, w is the weight vector, and b is the offset. The margin between two classes is equal to $\frac{1}{\|w\|}$, so for maximum the margin is equivalent to minimize $\|w\|$, and also is equivalent to minimizing $\frac{1}{2}\|w\|^2$. This minimizing problem can be expressed as Equation 3 by introducing Lagrange multipliers α, which can be solved by standard quadratic programming techniques. The optimum weight vector and offset is shown in Equation 4 and 5.

$$y_i(x_i \cdot w + b) - 1 \geq 0 \ \forall_i \tag{2}$$

$$\min_{w,b} \max_{\alpha \geq 0} \left\{ \frac{1}{2}\|w\|^2 - \sum_{i=1}^{n} \alpha_i [y_i(w \cdot x_i - b) - 1] \right\} \tag{3}$$

$$w = \sum_{i=1}^{n} \alpha_i y_i x_i \tag{4}$$

$$b = \frac{1}{N_{sv}} \sum_{i=1}^{N_{su}} (w \cdot x_i - y_i) \tag{5}$$

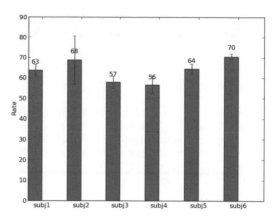

Figure 3. Classification results between clench force motor imagery and clench speed motor imagery.

For the nonlinear classification problem, the sample vector is converted into a transformed feature space by a nonlinear kernel function [25].

In our paper, the Gaussian radial basis function $K(x_i, x_j) = \exp(-\gamma \|x_i - x_j\|^2)$ is used to classify EEG feature space of clench speed motor imagery from the feature space of clench force motor imagery. The classification is implemented by libSVM [26], and the 5-fold cross-validation is used to make the results more reliable.

4 RESULTS

The classification results between speed and force motor imagery of right hand is shown in Fig 3. The mean identification accuracy is 67.65% and 59.68% for the trained subjects (subject number: 1, 2, and 6) and the no-trained subjects (subject number: 3–5), respectively, both above chance level.

Although the result is not as well as the classification accuracy between imagination of different limbs, such as left hand and right hand, it does show that decoding of different motor imagination parameters is possible.

5 DISCUSSIONS AND CONCLUSIONS

BCI is an efficient method to facilitate the rehabilitation process for stroke patients who lose voluntary limb movement. Although many EEG-based noninvasive BCI systems use motor imagery paradigm to control rehabilitation devices, their command number is very few because only different limb movement imagination is used. To increase the command number and control efficiency, both clench force parameters and clench speed parameters have been studied independently.

In this paper, we try to decode speed and force hand movement imagination at the same time, and the primary result shows it is distinguishable. Classification motor parameters of both hands will be researched in our future work, as well as more advanced feature optimization methods and classification methods.

ACKNOWLEDGEMENT

This work is supported by the National High Technology Research and Development Program of China (863 Program) under Grant No. 2012AA02A605, and National Natural Science Foundation (NNSF) of China under Grant No. 61203368, 61102014 and 61005069. The corresponding author is Li Hongyi ((hli@sia.cn)).

REFERENCES

[1] Krusienski, D.J., et al., *Critical issues in state-of-the-art brain-computer interface signal processing.* Journal of Neural Engineering, 2011. **8**(2).

[2] Vidal, J., *Toward direct brain-computer communication.* Annu Rev Biophys Bioeng., 1973. **2**: p. 157–80.

[3] G. Pfurtscheller, D.F., and J. Kalcher, *Brain-computer interface – A new communication device for handicapped persons.* J. Microcomput. Applicat., 1993. **16**: p. 293–299.

[4] Ajiboye, A.B., et al., *Prediction of Imagined Single-Joint Movements in a Person With High-Level Tetraplegia.* Ieee Transactions on Biomedical Engineering, 2012. **59**(10): p. 2755–2765.

[5] Shimoda, K., et al., *Decoding continuous three-dimensional hand trajectories from epidural electrocorticographic signals in Japanese macaques.* Journal of Neural Engineering, 2012. **9**(3).

[6] Friedrich, E.V.C., R. Scherer, and C. Neuper, *The effect of distinct mental strategies on classification performance for brain-computer interfaces.* International Journal of Psychophysiology, 2012. **84**(1): p. 86–94.

[7] Bin, G.Y., et al., *VEP-Based Brain-Computer Interfaces: Time, Frequency, and Code Modulations.* Ieee Computational Intelligence Magazine, 2009. **4**(4): p. 22–26.

[8] Ortner, R., et al., *An SSVEP BCI to Control a Hand Orthosis for Persons With Tetraplegia.* Ieee Transactions on Neural Systems and Rehabilitation Engineering, 2011. **19**(1): p. 1–5.

[9] Cecotti, H. and A. Graser, *Convolutional Neural Networks for P300 Detection with Application to Brain-Computer Interfaces.* Ieee Transactions on Pattern Analysis and Machine Intelligence, 2011. **33**(3): p. 433–445.

[10] Hong, B., et al., *N200-speller using motion-onset visual response.* Clinical Neurophysiology, 2009. **120**(9): p. 1658–1666.

[11] Vuckovic, A. and F. Sepulveda, *A two-stage four-class BCI based on imaginary movements of the left and the right wrist.* Medical Engineering & Physics, 2012. **34**(7): p. 964–971.

[12] Cano-Izquierdo, J.M., J. Ibarrola, and M. Almonacid, *Improving Motor Imagery Classification With a New BCI Design Using Neuro-Fuzzy S-dFasArt.* Ieee Transactions on Neural Systems and Rehabilitation Engineering, 2012. **20**(1): p. 2–7.

[13] Huang, D., et al., *Electroencephalography (EEG)-Based Brain-Computer Interface (BCI): A 2-D Virtual Wheelchair Control Based on Event-Related Desynchronization/Synchronization and State Control.* Ieee Transactions on Neural Systems and Rehabilitation Engineering, 2012. **20**(3): p. 379–388.

[14] Fu, Y., et al., *Direct brain-controlled robot interface technology.* Acta Automatica Sinica, 2012. **38**(8): p. 1229–1246.

[15] Yuan, H., et al., *Decoding speed of imagined hand movement from EEG,* in *2010 Annual International Conference of the Ieee Engineering in Medicine and Biology Society.* 2010, Ieee: New York. p. 142–145.

[16] Lv, J., Y. Li, and Z. Gu, *Decoding hand movement velocity from electroencephalogram signals during a drawing task.* Biomedical Engineering Online, 2010. **9**.

[17] Waldert, S., et al., *A review on directional information in neural signals for brain-machine interfaces.* Journal of Physiology – Paris, 2009. **103**(3–5): p. 244–254.

[18] Gu, Y., K. Dremstrup, and D. Farina, *Single-trial discrimination of type and speed of wrist movements from EEG recordings.* Clinical neurophysiology: official journal of the International Federation of Clinical Neurophysiology, 2009. **120**(8): p. 1596–600.

[19] Gu, Y., et al., *Identification of task parameters from movement-related cortical potentials.* Medical & Biological Engineering & Computing, 2009. **47**(12): p. 1257–1264.

[20] Waldert, S., et al., *Hand Movement Direction Decoded from MEG and EEG.* J Neurosci, 2008. **28**(4): p. 1000–1008.

[21] Boye, A.T., et al., *Identification of movement-related cortical potentials with optimized spatial filtering and principal component analysis.* Biomedical Signal Processing and Control, 2008. **3**(4): p. 300–304.

[22] do Nascimento, O.F., K.D. Nielsen, and M. Voigt, *Relationship between plantar-flexor torque generation and the magnitude of the movement-related potentials.* Exp Brain Res, 2005. **160**(2): p. 154–65.

[23] Farina, D. and O.F. Nascimento, *Optimization of wavelets for classification of movement-related cortical potentials generated by variation of force-related parameters.* Journal of neuroscience . . . , 2007. **162**: p. 357–363.

[24] Cortes, C. and V. Vapnik, *Support-Vector Networks.* Machine Learning, 1995. **20**(3): p. 273–297.

[25] Muller, K.R., et al., *An introduction to kernel-based learning algorithms.* Ieee Transactions on Neural Networks, 2001. **12**(2): p. 181–201.

[26] Chang, C. and C. Lin, *LIBSVM: a library for support vector machines.* 2001.

Information Technology and Computer Application Engineering – Liu, Sung & Yao (Eds)
© *2014 Taylor & Francis Group, London, ISBN 978-1-138-00079-7*

Pervasive real-time multi-parametric and multi-patient tele-monitoring system

Giovanna Sannino
Institute of High Performance Computing and Networking, CNR, Naples, Italy
Department of Technology. University of Naples Parthenope, Naples, Italy

Giuseppe De Pietro
Institute of High Performance Computing and Networking, CNR, Naples, Italy

ABSTRACT: Pervasive real-time tele-monitoring is a very helpful technology to remotely monitor the vital parameters of patients and ensure mobility of both patient and doctor. In the present paper we describe a pervasive multi-parametric and multi-patients tele-monitoring system designed for a cardiac department of a Hospital to process and display in a graphical user interface many physiological parameters such as Electrocardiogram, Heart Rate, Breathing Rate, SpO2, Activity, Posture and Temperature. This system is able to analyse in real time all parameters for more patients simultaneously using a rule-based DSS and if any of the vital parameters goes out of its normal range, the system generates visual and sounds alerts for a well-timed medical intervention. Finally all monitored parameters are saved in EDF files that could be used for further analysis.

Keywords: Mobile System; m-health; Wireless Monitoring; ECG; Software Architecture

1 INTRODUCTION

Because of time and cost, patients usually will not keep staying in hospital until full recovery. Instead, they typically choose to leave hospital first and take periodical subsequent visits. Therefore it becomes hard to treat chronic, especially cardiac, diseases. In recent years, related technologies have developed rapidly, making portable multi-life-parameter monitors a desirable research topic (Walker et al. 2009).

During the regular subsequent visits, patients are monitored for periods ranging from 24 to 48 hours. The use of 12-lead professional holter for this type of periodical check ups is often unnecessary, and there is the risk of keeping hospital hardware resources busy for a long period for these simple checks, instead of leaving them at the disposal of new more serious patients.

In addition, the monitoring of more patients with 12-lead Holter means that the doctors/nurses should simultaneously monitor 12 tracks for each patient, which would be difficult and tiring.

For this reason, nurses and doctors need an intelligent multi-patient and multi-parametric monitoring system that can help them to monitor patients.

This work has been partly supported by the project "Sistema avanzato per l'interpretazione e la condivisione della conoscenza in ambito sanitario A.S.K. – Health" (PON01_00850).

Pervasive real-time monitoring systems are very helpful in this type of applications.

The technological improvements pertaining to measurement and information transmission, such as wireless LAN and sensor networks, provide people with new possibilities for monitoring vital parameters through the use of wearable biomedical sensors, and give patients the freedom to move without limitation and still be under continuous monitoring and thereby allow better quality of patient care.

In this regard, our system is a multi-parametric and multi-patients tele-monitoring system which allows having a global view about the health condition of patients being monitored. The system has been made following a specific request from a team of cardiologists of Cardiac Department of Policlinico Hospital of Naples, because in this department there are a lot of chronic patients that periodically are hospitalised to be kept under monitoring.

As mentioned before, those monitoring activities have a duration from 24 h until 48 h. Nurses and doctors need an intelligent system that can help them to monitor patients because, due to the long monitoring period, the level of attention of nurses empowered to inspect in real-time the trace of ECG and the value of monitored parameters incline to diminish during the monitoring.

The main advantages of our developed system are:
1) the use of a Decision Support System in which

the formalized cardiologists' knowledge is embedded for an intelligent real-time analysis of the parameters; 2) the use of a friendly and adaptive user interface for multi-patient monitoring; 3) its ability to provide alert and alarm services; 4) no connection is required with any server because the analysis of data is embedded on a mobile device with consequent battery saving, because the mobile device of the patients shouldn't manage two connections, GPRS and Bluetooth, at the same time, but just one Bluetooth connection with the biomedical sensor.

An analysis of the state of the art in literature shows that other existing systems lack those important features presented by ours, and have also some other limitations which do not affect our system.

In fact, even some of the most interesting systems found in the literature and proposing a novel way of monitoring heart's pulse (Hata et al. 2007, Chen et al. 2008), or introducing new ways of monitoring the patient's physical activity (Jovanov et al. 2006), do not possess all the above described characteristics in conjunction.

Moreover, a number of complicate and advanced systems for in-home health monitoring were introduced (Sebestyen & Krucz 2010, Junnila 2010, Alahmad & Soh 2011). On the basis of our survey of scientific literature, we have noticed that in most cases, the current existing systems lack generality and flexibility, and are still limited by some important restrictions listed below.

Firstly, these types of pervasive health monitoring systems have been typically used only to collect patient's data. Data processing and analysis are not always in real-time, but are performed offline, which make these systems not suitable for continual monitoring and detection of abnormalities.

This problem is faced in (Sebestyen & Krucz 2010), where the CardioNET system was presented. This is a modular distributed and web-based medical application to permit remote interaction between patients and doctors. Yet, to use this system an internet connection is needed and so, sensors data cannot be analysed without sending them to a server. This is a quite limiting drawback for a lot of types of application. It should be remarked that our system, instead, does not suffer from such a limitation.

Secondly, proprietary sensors often operate as stand-alone systems and usually do not offer flexibility and integration with third-party devices. Our system, instead, is built in an open way, and new third-party sensors can be easily integrated.

Thirdly, the existing systems are rarely affordable, differently from our system which is based on wearable and easily usable devices and does not require any training for users to be employed.

Furthermore, many recent projects have been dedicated to new monitoring systems, as for example the European Union-funded research project, named 'Advanced care and alert portable tele-medical MONitor' AMON (Anliker et al. 2004), developed within

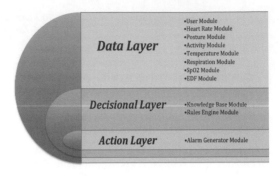

Figure 1. Pervasive Real-Time Multiparametric Monitoring System Architecture.

the 5th framework, with partners from Israel and three European countries.

The aim of the project consisted in the development of a multi-variable sensor device that can be worn like a wristwatch. The data gathered by this sensor are transmitted via GSM to a medical care centre for analysis. The patient receives real-time care at the point of need and unnecessary hospitalization can be avoided.

Some of the problems of AMON monitoring system are the used device, which appears too big to be worn everywhere and every time, the need for GSM connection to send monitoring, and the lack of a DSS to perform data analysis and to suitably activate the alert system. Also in this case, our system is free from all those drawbacks.

2 MONITORING SYSTEM ARCHITECTURE

The realized system relies on a multi-layer architecture (Sannino & De Pietro 2010), as shown in figure 1, designed to provide flexibility capacity to the system through which algorithms and sensors can be adapted to various applications and new sensors can be easily added. Each layer is independent of the upper one and of the lower one.

The Data Layer provides user interfaces and mechanisms to manage sensors data and patient information that will be processed by the Decisional Layer. This layer has the task to collect data from BioHarness device, and from Alive Pulse Oximeter device, and is responsible for computing complex parameters such as the peak of QRS or the peak of plethysmography to estimate Heart Rate and its variability. This layer has also the task to collect data from an accelerometer embedded into the BioHarness device so as to recognize the patient's posture and activity. Finally, it manages the saving of data in a specific data format, the European Data Format (EDF). The choice of EDF format has been made because it is the de-facto standard for EEG and PSG recordings in commercial equipment and multicenter research projects. It is a simple and flexible format for exchange and storage of

```
[Monitoring_rule_01:
(?m http://mH#heartRate ?hr), greaterThan(?hr,100)
-> newInstance(?abn, http://mH#abnormalCondition),
(?abn                     http://mH#conditionType
'Tachycardia'^^xsd:string),              (?abn
http://mHealthOnto#dueTo ?m) ]


[Monitoring_rule_02:
(?m http://mH#heartRate ?hr), lessThan(?hr,50) ->
newInstance(?abn, http://mH#abnormalCondition),
(?abn                     http://mH#conditionType
'Brachicardia'^^xsd:string),             (?abn
http://mHealthOnto#dueTo ?m) ]
```

Figure 2.   Example of expert's knowledge formalization.

multichannel biological and physical signals. In addition many freeware EDF viewers and EDF analysers can be found.

The Decisional Layer represents the intelligent core of the system. In this layer, data coming from the Data Layer are elaborated. Thanks to the presence of a set of rules that represent the formalization of the knowledge of experts about anomalies, this layer by means of a Decision Support System recognizes possible critical situations and determines the most suitable actions to be performed by the Action Layer. An overview of some implemented production rules is shown in Figure 2.

Finally, the Action Layer executes the actions inferred by the Decisional Layer by implementing mechanisms which produce reactions like the generation of alarms.

3   DEVICES USED

3.1   *Zephyr BioHarness BH3*

The BioHarness™ 3 is an advanced physiological monitoring device that uses Bluetooth technology to transmit data. It is small and provides a medical-grade ECG, as well as heart rate, breathing rate, and 3-axis accelerometery. The monitor could be used with the BioHarness™ strap that is a lightweight elasticised component which incorporates Zephyr Smart Fabric ECG and Breathing Rate sensors.

Data is transmitted by bluetooth to a corresponding receiver device. This will allow physiological data to be monitored using any suitably-configured mobile device with bluetooth technology, such as a laptop, phone or PDA.

3.2   *Alive Pulse Oximeter*

The Alive Pulse Oximeter is a wearable medical device which uses wireless bluetooth technology instead of cables. It reads oxygen saturation data from a sensor on the finger or earlobe, and transmits the data via bluetooth wireless technology to a mobile phone,

Figure 3.   Use Case.

PDA, laptop PC, or other Bluetooth-enabled device. The device can transmit the data in real-time or store the data for later download.

3.3   *Multi-window desktop PC*

As a server, we used a multi-window desktop pc. The hardware configuration of the server is: an Intel Core 2 Duo (3.06 GHz) CPU; 2GB RAM; Windows 7 ultimate as operating system. However, one of the key factors of the system is that a mobile device (like Tablet or PDA) can easily be used as a server provided that a suitable graphical interface is designed.

4   RESULTS

The system was designed for in-hospital monitoring applications, but it could be used also for more ubiquitous applications. The User Interface has been made following a specific request from a team of cardiologists of Cardiac Department of Policlinico Hospital of Naples to have more chronic patients under control, but thanks to the multi-layer architecture described before, it's possible to modify the user interface modifying only the User Module.

Usually, those monitoring activities have a duration from 24 h until 48 h, so the system analyses data coming from sensors on a simple pc using a DSS without any internet connection.

The knowledge used by the DSS to analyse data is formalized in collaboration with a team of expert cardiologists of the Policlinico Hospital involved in the project. A scheme of the use case is shown in figure 3.

The patient is equipped with a wearable wireless sensor instead of a classic holter and his/her information is inserted into the system. The sensor is connected to the server station and the data starts to be analysed by the system. Currently, on request, if the doctor/nurse/user wants, he/she has the possibility, using an internet connection, to send the monitored data stored in an EDF file to other doctors/specialists/medical experts to ask counsel improving diagnosis and therapy, or to store the file into an EHR (Electronic Health Record) or into a Personal Archive.

The system permits to personalize the type of monitoring for each patient, the doctor/nurse can choose

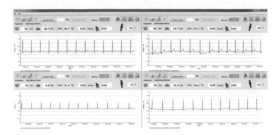

Figure 4. Example of visual interface: four patients monitored at the same time.

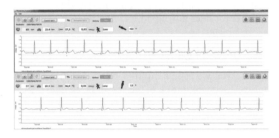

Figure 5. The system evidences when a possible dangerous situation has occurred.

what type of information he/she wants to monitor, and what type of information he/she wants to save in EDF files.

The doctors can monitor more patients in different windows using the same screen, as shown in figure 4, and they can modify the rules used by the DSS to personalize monitoring for each patient. It's possible to personalize the interface modifying the time window of the ECG signal to visualize the waveform as preferred, and finally, a mechanism is implemented to auto-reconnect sensors if these go out of bluetooth range, or if problems occur.

It's always possible to perform a zoom on a single patient, while more patients are under monitoring, and then return to visualize all monitored patients.

The goal of the system is the use of knowledge technologies that guarantee a proper application. In fact while monitoring is being performed, the DSS analyses in real time patients' data and when some possible dangerous situations are detected, the system alerts the user or the medical personnel with an acoustic sound increasing over time, underlying the involved patient in danger through a red borderline around the patient monitoring window and highlighting the parameter or the parameters at risk, as shown in figure 5.

All designed user interfaces are very simple. This choice permits all users to employ the system in an easy way, so it's possible for anyone also to install the system at home, to wear the BioHarness by himself/herself, to perform a monitoring with a real-time preliminary diagnosis and then to send the results of monitoring to their doctor for a tele-consult.

The system was tested in the laboratory of our research institute, thanks to the support of all the staff.

The system has provided excellent performance in terms of velocity in processing of data and in terms of reaction when some dangerous situations have been simulated. In the next future, the system will be installed at one of the major hospitals of Naples (Policlinico), and more accurate tests on performance will be carried out.

ACKNOWLEDGEMENTS

The authors wish to thank Dr. Aniello Minutolo and Dr. Massimo Esposito for their useful contribution in permitting us to use their developed DSS in our system (Minutolo et al. 2012).

The authors also wish to thank all the staff of Policlinico Hospital of Naples.

REFERENCES

Anliker, U. et al. 2004. AMON: a wearable multiparameter medical monitoring and alert system, Information Technology in Biomedicine, IEEE Transactions on, vol. 8, no. 4, pp. 415,427, December 2004.

Jovanov, E. et al. 2006. A WBAN System for Ambulatory Monitoring of Physical Activity and Health Status: Applications and Challenges, Engineering in Medicine and Biology Society, 27th Annual International Conference of the , vol., no., pp. 3810, 3813, 17–18, January 2006.

Hata, Y. et al. 2007. A Heart Pulse Monitoring System by Air Pressure and Ultrasonic Sensor Systems, System of Systems Engineering, 2007. SoSE '07. IEEE International Conference on, vol., no., pp. 1, 5, 16–18, April 2007.

Chen, C.M. et al. 2008. Web-based remote human pulse monitoring system with intelligent data analysis for home healthcare, Cybernetics and Intelligent Systems, 2008 IEEE Conference on, vol., no., pp. 636, 641, 21–24, September 2008.

Walker, B. A. et al. 2009. Low cost ECG monitor for developing countries, in Intelligent Sensors, Sensor Networks and Information Processing (ISSNIP), 2009 5th International Conference on, 2009, pp. 195–199.

Sebestyen, G. & Krucz, L. 2010. Remote monitoring of patients with mobile healthcare devices, Automation Quality and Testing Robotics (AQTR), 2010 IEEE International Conference on, vol. 2, no., pp. 1, 6, 28–30, May 2010.

Junnila, S. et al. 2010. Wireless, Multipurpose In-Home Health Monitoring Platform: Two Case Trials, Information Technology in Biomedicine, IEEE Transactions on, vol. 14, no. 2, pp. 447, 455, March 2010.

Sannino, G. & De Pietro, G. 2010. An Intelligent Mobile System For Cardiac Monitoring, In Proc. of IEEE Healthcom'10, Lyon, France, pp: 52–57, April 2010.

Alahmadi, A. & Soh, B. 2011. A smart approach towards a mobile e-health monitoring system architecture, Research and Innovation in Information Systems (ICRIIS), 2011 International Conference on, vol., no., pp. 1, 5, 23–24, November 2011.

Minutolo, A. et al. 2012. A Mobile Reasoning System for Supporting the Monitoring of Chronic Diseases, In Proc. of 2nd International ICST Conference on Wireless Mobile Communication and Healthcare, pp. 225–232, January 2012.

Information Technology and Computer Application Engineering – Liu, Sung & Yao (Eds)
© 2014 Taylor & Francis Group, London, ISBN 978-1-138-00079-7

Phrase table filtration based on virtual context in phrase-based statistical machine translation

Yue Yin, Yu Jie Zhang & Jin An Xu
Beijing Jiaotong University, Beijing, China

ABSTRACT: In statistical machine translation system, automatically extracted phrase table inevitably contains a large number of errors and redundant phrase pairs, which causes excessive waste of time and space in decoding and affects translation quality. In order to solve this problem, we propose a method for filtering phrase table based in which virtual context is introduced to calculate an incremental quantity in language model for score of phrase pair. By considering the maximum and minimum incremental quantity in score from the virtual context, we design a filtering strategy by re-ranking phrase pairs. We conducted experiments on NTCIR-9 data to verify the method. The experimental results show that when the size of phrase table was reduced to 47% of the original, the translation quality was improved slightly; when the size was reduced to 30% of the original, only slight decline occurred in translation quality. The experimental results indicate that this method can effectively filter out the redundant phrase pairs of the phrase table.

Keywords: phrase-based statistical machine translation, filter phrase table, virtual context

1 INTRODUCTION

Current state-of-the-art model in statistical machine translation (SMT) is hierarchical phrase-based model (Koehn et al., 2003; Zens et al., 2002; Koehn, 2004; Chiang, 2005; Chiang, 2007). Phrase table is the essential translation knowledge in SMT, which contains a large number of phrase pairs and each phrase pair consists of a source language phrase and a target language phrase. In construction of a phrase – based SMT system, a phrase table is automatically obtained from parallel corpus (Och, 2004) and target language model is trained by target monolingual corpus. This construction phase is also called training. In translation, a source sentence is segmented into phrase sequence firstly, and then each phrase is translated into target phrase by using the obtained phrase table. The target phrases are recombined to generate a target sentence. This translation phase is also called decoding and the decoding module is called decoder. The decoder will find the translation with the maximum probability from all candidate translations as final output.

For a source sentence $f_1^J = f_1 \cdots f_j \cdots f_J$, a target sentence $e_1^I = e_1 \cdots e_i \cdots e_I$, F. J. Och and H. Ney (Och and Ney, 2002) proposed maximum entropy models for SMT as follows.

$$\Pr(e_1^I \mid f_1^J)$$

$$= p_{\lambda_1}^M(e_1^I \mid f_1^J) = \frac{\exp\left[\sum_{m=1}^{M} \lambda_m h_m(e_1^I, f_1^J)\right]}{\sum_{e_1'^I}\exp\left[\sum_{m=1}^{M} \lambda_m h_m(e_1'^I, f_1^J)\right]} \quad (1)$$

$h_m(e_1^I, f_1^J)$ is feature functions, $m = 1, 2, 3 \ldots M$, is weight of feature function. Because denominator is constant, an optimum translation is selected by the following formula.

$$e_1^I = \arg\max_{e_1^I}\{\sum_{m=1}^{M} \lambda_m h_m(e_1^I, f_1^J)\} \quad (2)$$

The score of a translation is calculated by the formula (3).

$$Score = \sum_{m=1}^{M} \lambda_m h_m(e_1^I, f_1^J) \quad (3)$$

The following eight features are used in our experimental system.

(1) Phrase translation probability
(2) Inverse phrase translation probability
(3) Lexical weight: measuring the quality of word alignment inside the phrase pair
(4) Inverse lexical weight
(5) Language model
(6) Phrase penalty
(7) Word penalty
(8) Reordering

The task of decoding is to search the translation with maximum probability. (Knight, 1999) has demonstrated that decoding is a NP complete problem. A great size of phrase table will cause excessive consumption in time and space during decoding and therefore affects the practical application of machine translation system. Transplanting of such a SMT system to a moveable terminal, like PDA, is almost impractical.

To solve the problem, many researchers have proposed a lot of methods to filter phrase table and mainly focus on two types of phrase pairs, error phrase pairs and redundant phrase pairs. For filtering out error phrase pairs in phrase table, (Hua Wu and Haifeng Wang, 2007) uses log likelihood ratio and (L. Shen et al, 2008) uses dependency structure information of target phrase. For filtering out redundant phrase pairs in phrase table, (Z. He et al., 2009) proposes a method for discarding monotone composed phrase pairs.

We propose a virtual context based method aiming at filter out phrase pairs that are hardly used by decoder. In this paper such kind of phrase pairs are regarded as redundant phrases pairs. We mainly focus on phrase pairs with identical source phrase (PPISP) and compound phrase pairs (CPP). Identical source phrase pairs are those phrase pairs in which source phrases are identical and the target phrases are different. This implies that one source phrase has multiple target phrases and there probably exists redundant phrase pairs. A compound phrase pair indicates that its source phase can be decomposed into two or more sub-phrases and the phrase pairs with these sub-phrases exist in phrase table. This implies that one compound phrase pair can be replaced by sub-phrase pairs and there probably exists redundant phrase pairs.

Our method intends to find out the redundant phrase pairs. Firstly, our filter uses a log-linear model to calculate scores of phrase pairs and obtains the phrase pair with highest score and its score. Secondly, the filter uses virtual context to re-rank the phrase pairs and then discard the phrase pairs with lower ranking.

2 THE ALGORITHM OF FILTERING PHRASE TABLE

In this section, we present the method of ranking phrase pairs using log-linear model, and then describe re-ranking algorithm of exploring virtual context and strategy of discarding the redundant phrase pairs in detail.

2.1 Ranking

In order to find out phrase pairs that are hardly selected by decoder, we evaluate the quality of phrase pair in the same way as decoder. We use log-linear model to calculate score of phrase pair and consider five features, including translation probability, lexical probability, inverse translation probability, inverse lexical probability and language model, because these five feature functions are closely related to the quality of phrase pairs. The weights of features are trained by development set.

The filter ranks phrase pairs as fellows.

1) Choose phrase pairs with identical source phrase as a set S_i.

2) Calculate score of each phrase pair in S_i by formula (3) and then rank the phrase pairs. Denote the phrase pair with the highest score as S_i^H and other pairs as S_i^O, their scores as $Sc(S_i^H)$ and $Sc(S_i^O)$ respectively.

2.2 Re-ranking by virtual context

Generally speaking, the decoder prefers S_i^H to S_i^O. However, S_i^O may be selected in real decoding. The reason is explained as follows. In the above score calculation, the way of calculating language model score is not totally the same as that of decoder. The decoder takes already generated translations as context in calculating language model. However, our filter has no context information for consideration. This deficiency causes difference in ranking results between decoder and our filter when S_i^O are assigned with higher scores than S_i^H by the decoder. To compensate for this, our filter introduces virtual context information for language model calculation and then re-ranks phrase pairs. After re-ranking, phrase pairs with lower ranking are ensured to be those pairs that are hardly used in real decoding. So filtering out these phrase pairs will not affect translation quality.

We think that already generated translations in decoding will produce an incremental quantity in score from language model. If the decoder prefer S_i^O to S_i^H, the reason is that incremental quantity in score of S_i^H is lower than that of S_i^O. When the formula (4) is satisfied, the decoder will choose S_i^O.

$$Sc(S_i^O) + \Delta_{\text{Context}}(S_i^O) > Sc(S_i^H) + \Delta_{\text{Context}}(S_i^H) \qquad (4)$$

In formula (4), $\Delta_{\text{Context}}(S_i^O)$ is the incremental quantity in score of S_i^O from context in language model, $\Delta_{\text{Context}}(S_i^H)$ is the incremental quantity in score of S_i^H.

Based on the above consideration, we design a re-ranking strategy by assigning the best context to S_i^O and the worst context to S_i^H. In this way, if S_i^O is still ranked lower than S_i^H after re-ranking, we can say S_i^O is hardly selected by decoder. We call the strategy as S_i^H with the worst context vs. S_i^O with the best context.

Using virtual context to imitate the context of S_i^H makes $\Delta_{\text{Context}}(S_i^H)$ has the lowest score in language model, and the context of S_i^O makes $\Delta_{\text{Context}}(S_i^O)$ has the highest score in language model. We mark them as $\min \Delta_{\text{Context}}(S_i^H)$ and $\max \Delta_{\text{Context}}(S_i^O)$. Then re-ranking the phrase pairs based on the new score.

If $Sc(S_i^H) + \min \Delta_{\text{Context}}(S_i^H) > Sc(S_i^O) + \max \Delta_{\text{Context}}(S_i^O)$, it means S_i^O is hardly used in any case of context in decoding. The process of re-ranking and discarding is as follows.

1) For a target phrase $W_1 W_2 \ldots W_k$, W_1 and W_k are the edges of the phrase. If $W_{x1} W_{x2}$ exists in language model and $\delta(W_{x2}, W_1) = 1$, W_{x1} is the virtual context of the target phrase; if $W_{x1} W_{x2}$ exists in language model and $\delta(W_k, W_{x2}) = 1$, W_{x2} is the virtual context of the target phrase. $\delta(x, y)$ is

328

Kronecker function, if $x = y$, $\delta(x, y) = 1$; otherwise $\delta(x, y) = 0$. In addition to the above, we should consider phrase containing variables. For a target phrase $W_1 \ldots W_{m-1} \, X \, W_m \ldots W_k$, X is a variable. W_{m-1} and W_m are also the edges of the phrase. If $W_{x1} \, W_{x2}$ exists in language model and $\delta(W_{x2}, W_m) = 1$, W_{x1} is also the virtual context of the target phrase; if $W_{x1} \, W_{x2}$ exists in language model and $\delta(W_{m-1}, W_{x1}) = 1$, W_{x2} is also the virtual context of the target phrase.

2) Calculate the score of $\min \Delta_{\text{Context}}(S_i^H)$: assign a context in language model to make $\Delta_{\text{Context}}(S_i^H)$ have the lowest score, denoting the way of this calculation as worstLM.

3) Calculate the scores of other phrase pairs in S_i: assign a context in language model to make $\Delta_{\text{Context}}(S_i^O)$ have the highest score, denoting the way of this calculation as bestLM.

4) Re-rank the phrase pairs in Si by the new scores.

5) Discard the phrase pairs that have lower ranking than S_i^H.

The algorithm of re-ranking and discarding is as follows.

Input: language model; phrase pair $(e, f)_k$ in S_i; the score of phrase pair $(e, f)_k$
Output: filtered phrase table
for each phrase pairs$(e, f)_k$ in S_i^O do
  for each word f_i in f do
    if (f_i is edge)
      Sc(e, f)$_k$ += bestLM (f_i)
    end if
  end for
end for
for phrase pair$(e, f)_H$ in S_i^H do
  for each word f_i in f do
    if (f_i is edge)
      Sc(e, f)$_H$ += worstLM (f_i)
    end if
  end for
end for
rank((e, f) in S_i)
discard(the phrase pairs which has lower ranking than S_i^H)

In filtering compound phrase pairs (CPP), we denote sub-phrase pairs as minimal phrase pair. The process is as follows.

1) Calculate the score of compound phrase pair: assign a context in language model to make the incremental quantity of compound phrase pair have the highest score.

2) Calculate the score of minimal phrase pair: assign a context in language model to make the incremental quantity of minimal phrase pairs have the lowest score.

3) Discard: if the sum of the scores of minimal phrase pair is greater than the score of compound phrase pair, discard the compound phrase pair.

In this extreme case, the score of compound phrase pair is lower than the sum of the scores of minimal

Table 1. Result of filtering phrase table based on virtual context.

Filtering way	BLEU	PTS	NUM	Reminder
None(baseline)	0.2932	426 M	4733693	
PPISP	0.2938	230 M	2461777	52.42%
CPP	0.2937	309 M	3457486	73.03%
PPISP& CPP	0.2937	207 M	2225644	47.01%

phrase pairs. It means that decoder hardly uses the discarded compound phrase pairs in other case and there is almost no impact on performance of translation.

3 EXPERIMENT

3.1 Experiment setup

We use Moses (Koehn, 2004) for decoder and carry out experiment on the Chinese – English task of NTCIR-9. There are 1,000,000 sentence pairs for training set, and 2000 sentence pairs for development set and test set respectively.

We run GIZA++ (Och and Ney, 2000) on the training data in both translation direction and use heuristics method "grow-diag-final" (Koehn, Och et al, 2003) to refine word alignment results. The language model is obtained by SRI language model (Stolcke, 2002). The feature weights of translation model are turned on development set by minimum error rate training (Och, 2003). The quality of translation is evaluated by BLEU (K. Papineni et al, 2002) metric.

3.2 Experiment result

For convenience of experiment, we selected phrase pairs only if the source pairs appear in the test set and make the phrase pairs as the phrase table for filtering. Then we use the method in section 2 to filter PPISP and CPP. The changes of the phrase table and translation quality after filtering are shown in Table 1.

The first column is the way of filtering (filtering way), including PPISP, CPP and PPISP&CPP. None (baseline) is the result without filtering. The second column is the quality of translation and it is evaluated by BLEU metric; The third column is the memory of phrase table takes; The fourth column is the number of phrase pair in phrase table; the fifth column is the ratio of reminder part in phrase table. After filtering PPISP phrase pair, the number of reminder phrase pairs is 52.42% of previous number and the score of BLEU is increased 0.0006. After filtering CPP phrase pairs, the number of reminder phrase pairs is 73.03% of phrase table and the score of BLEU is increased 0.0005. After filtering CPP and PPISP phrase pairs, the number of reminder phrase pairs is only 47.01% of phrase table and the score of BLEU is also increased 0.0005. It is clear that PPISP&CPP achieves the best performance.

In order to filter more phrase pairs in phrase table, we consider filtering more PPISP phrase pair: we only

Table 2. Result of keeping phrase pairs within TOP 1–TOP 5 in filtering PPISP.

Filtering way	BLEU	PTS	NUM	Reminder
None(baseline)	0.2932	426 M	4733693	
TOP 5	0.2926	135 M	1454004	30.71%
TOP 4	0.2908	124 M	1304037	27.54%
TOP 3	0.2896	103 M	1112783	23.50%
TOP 2	0.2847	80 M	862957	18.23%
TOP 1	0.2716	48 M	521787	11.02%

keep the several top phrase pair in phrase table. In table 2, it is shown that keeping PPISP phrase pairs within Top 5 achieves the best performance. It can filter about 70% phrase pair in phrase table and the bleu score is only decreased 0.0006. It is obvious that the much less number of PPISP phrase pair keeping, the more quickly the BLEU score decreases. In our experiment, keeping top five PPISP phrase pairs is meaningful in practical application.

4 CONCLUSION

In this paper, we propose a method of filtering phrase table based on virtual context. According to the maximum and minimum incremental quantity in score from virtual context, we re-rank phrase pairs and design the filtering strategy. The experimental results show that the method can filter out 53% of the phrase table without decreasing in translation quality. When keeping phrase pairs within Top 5, the method can filter out 70% of the phrase table with only 0.0006 decreasing in BLEU score. The experiment proves the method is effective to filter out redundant phrase pair.

In the future, we will combine other information for more effective filtering method.

ACKNOWLEDGEMENT

The research work has been partially funded by the RenCai founding of Beijing Jiaotong University under Grant No. KKRC11001532.

REFERENCES

A. Stolcke. Srilm – an extensible language modeling toolkit. In Proceedings of the International Conference on Spoken language Processing, volume 2, pages 901–904, 2002.

D. Chiang. A hierarchical phrase-based model for statistical machine translation. In Proceedings of ACL 2005, pages 263–270, 2005.

D. Chiang. Hierarchical phrase-based translation. Computational Linguistics, 33(2):201–228, 2007.

F. J. Och. Minimum error rate training in statistical machine translation. In Proceedings of the 41st Annual Meeting of the Association for Computational Linguistics, pages 160–167, 2003.

F. J. Och and H. Ney. The alignment template approach to statistical machine translation [J]. Computational Linguistics, 30(4):417–449. June 2004.

F. J. Och and H. Ney. 2002. Discriminative training and maximum entropy models for statistical machine translation. In Proc. of the 40th Annual Meeting of the Association for Computational Linguistics (ACL), Philadelphia, PA, July.

F. J. Och and H. Ney. Improved statistical alignment models. In Proceedings of the 38th Annual Meeting of the Association for Computational Linguistics, pages 440–447, 2000.

K. Knight. 1999. Decoding complexity in word replacement translation models. Computational Linguistics, 25(4).

K. Papineni, S. Roukos, T. Ward, and W.-J. Zhu. Bleu: a method for automatic evaluation of machine translation. In Proceedings of the 40th Annual Meeting of the Association for Computational Linguistics, pages 311–318, 2002.

L. Shen, J. Xu, and R. Weischedel. A new string-to-dependency machine translation algorithm with a target dependency language model. In Proceedings of ACL-08: HLT, pages 577–585, Columbus, Ohio, June 2008.

Philipp Koehn, Och, F.J., and Marcu, D. (2003). Statistical Phrase-Based Translation. In Proceedings of the 2003 Human Language Technology Conference of the North American Chapter of the Association for Computational Linguistics (pp. 127–133).

Philipp Koehn. 2004. Pharaoh: a beam search decoder for phrase-based statistical machine translation models. In Proceedings of the Sixth Conference of the Association for Machine Translation in the Americas, pp. 115–124.

R. Zens, F.J. Och, H. Ney, 2002. "Phrase-Based Statistical Machine Translation". In: M. Jarke, J.Koehler, G. Lakemeyer (Eds.): KI – 2002: Advances in artificial intelligence. 25. Annual German Conference on AI, KI 2002, Vol. LNAI 2479, pp. 18–32, Springer Verlag, September 2002.

Wu, Hua and Haifeng Wang. 2007. Comparative Study of Word Alignment Heurist Based SMT. In Proc. of Machine Translation Summit XI, pages 507–514.

Z. He, Y. Meng, Y. Lj, H. Yu, and Q. Liu. Reducing SMT rule table with monolingual key phrase. In Proceedings of the ACL-IJCNLP 2009 Conference, pages 121–124, Singapore, August 2009.

Information Technology and Computer Application Engineering – Liu, Sung & Yao (Eds)
© 2014 Taylor & Francis Group, London, ISBN 978-1-138-00079-7

The method of extracting forest leaf area index based on Lidar data

Zuo Wei Huang & Shu Guang Wu
School of Architecture and Urban Planning, Hunan University of Technology, Zhuzhou, China

Jie Huang
School of Geosciences and Information-Physics, Central South University, Changsha, China

ABSTRACT: Light Detection And Ranging (LiDAR) is a new emerging active remote sensing technology in recent years, which can measure both the vertical and horizontal structure of forested areas effectively with high precision and it can accurately estimate tree height, canopy closure and above-ground biomass. The paper propose a crop fractional cover and leaf area index (LAI) retrieval method using airborne Lidar intensity of ground hits and the distance and zenith angle information contained in waveforms data. Relevant flight experiment and ground measurements in longhui area indicate that the method is reliable, the result showed that airborne LiDAR data could improve accuracy of LAI retrieval.

Keywords: point cloud, fractional cover, LAI, Forest Structural Parameters

1 INTRODUCTION

As we all known forest resource is one of the most important natural resources, which plays an more and more important role in the sustainable development of mankind. During the past years it dependent on the time-consuming and labor-intensive manual inventory methods, therefore forestry researchers are trying to find a more convenient method of investigation. Light Detection and Ranging (LiDAR) is an active remote sensing technology, which can determine the distance between the sensor and the target by emitting laser from sensors. It has been applied to forestry research in many fields. Early researches find that canopy closure is most strongly related to the penetration capability of the laser pulse, and indicates that the pulsed laser system may be used to remotely sense the vertical forest canopy profile and assess tree height. many work expose the enormous potential of laser applications in the forestry, which demonstrate that forest variables such as tree height, basal area, biomass, etc. can be estimated accurately using airborne laser scanner data (Nasset, 1997; Nelson, et al., 1997; Popescu, et al., 2002; Studies based on small footprint, high sampling density LiDAR data at individual tree level are reported later However, some studies show that the relationships between forest biophysical properties and airborne laser scanner data are different for the different geographical location, species composition, site quality, etc. In the case that remote sensing becoming more popular, the LiDAR technology has increasingly being used in the forestry as a new means of remote sensing. Explored the method of LAI retrieval based

on satellite borne LiDAR, and established the model Finally, Mapped LAI of the study area based on the model.

2 STUDY METHOD

2.1 Study site

The experimental region is the Longhui country of Hunan Province. The site is surrounded by mountains around valley, and mainly influenced by the climate in high-altitude cloud circulation. There are many crops and forest distributed on the irrigated oasis, which is on the alluvial fan and alluvial plain in the middle river.

Four forest fields on the flight were selected as the sampling fields, and were sampled on 2009-03-21. The fractional cover was measured using a Nikon Coolpix 8400 camera. Method for LAI acquisition uses the Yao's method (Yao, et al., 2010). The DSM of laser point cloud in study area is shown in Fig. 1.

According the corn growth, we chose the sample from the other fields with similar growth conditions, and then measured the leaf area by SunScan to determine LAI Each sample has precise GPS positioning data.

SunScan does not wait for the special weather conditions and it can work properly in most light conditions. Hand-held terminal PDA is simple with large storage, which can meet the needs of large data measurement. Since the sensor using a wireless transmitting and receiving technology, it makes the measurement more convenient, quick access to the large

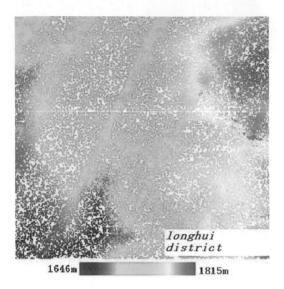

Figure 1. The DSM of laser point cloud in.

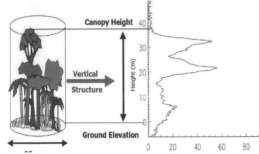

Figure 2. The principle of LiDAR Altimeter Waveform.

area of forest region LAI. Wireless transmission range 250 m, the vegetation coverage of 100 to 200 m, greatly improve the efficiency of forest LAI.

2.2 Lidar data

An airborne laser scan flight was carried out in the study area. The airborne laser scanning system (LiteMapper-5600) is developed by the German company. The laser scanner is RIEGL LMS-Q560, with laser wavelength 1550 nm, and laser pulse length 3.0 ns, and laser beam divergence of 0.5 mrad. The principle of LiDAR Altimeter Waveform is shown in Fig. 2. Lidar point cloud data uses WGS84 coordinate system, and a universal transverse Mercator (UTM) projection. The flight altitude is 600 m above the ground and the footprint diameter is about 0.35 m, which is small so that the pulse can easily get through some sparse farmland and completely reach the ground. In this paper, both point cloud and full waveform data are used. For point cloud, the system records three types of echo, first echo, last echo and single echo, with point density of 0.95 points per square meter. The full waveform data contains information about the scanning angle, time interval between launch wave and reflection wave.

Raw data contains a large number of gross errors or irrelevant information, such as data produced by local terrain mutation, reflected signal resulted from flying birds or other moving objects, or other local empty formed by no echo. In the data pre-processing, the data process software provided by equipment supplier should use artificial interactive editing to ensure the accuracy of terrain model and vegetation model generated later. The GPS carried by aircraft and DGPS on the ground positioned in real time during the flight for unifying and correcting the coordinates of data so that the data and ground data can match exactly.

3 CALCULATION OF GAP FRACTION

(1) Generating DEM: TerraScan offers the function model for classifying the ground points. The algorithm principle firstly chooses some low points, according to the area of max building to serve them as land surface points. And then, it repeatedly builds the triangle-net to gain the ground points (Guan & Li, 2009). After gaining the ground points, it uses the TerraScan software to generate and interpolate the ground points into 1 m resolution DEM.

(2) Generating the DSM: transforming the point cloud into grid, when there have multiple points, using the max value as the pixelvalue. When the pixel value equals to 0, using the average value around the pixel interpolates the grid data. And then getting 1 m resolution DSM.

(3) Generating the canopy height model (CHM) by DSM minus DEM. For a 5×5 window, if the central pixel is the minimum, let the pixel value equal to the average of the window. Furthermore, according to prior knowledge, if the CHM is less than 2.5 m, the points in the grid are the points. And then, we can get the Lidar points in the forestland.

(4) using the TerraScan to extract the ground points in the corn field. Calculating the value for every point. According to the longitude and latitude, the indensity and sensor-target distance R of every point can be gained. As usual, the points in the middle of the classification results can be considered as the pure pixels, considering the maximum likelihood classification's accuracy and features. For the huge number of points, we choose the first one hundred points.

4 RESULT VALIDATION AND ANALYSIS

In this study a few forest variables were calculated based on the investigated data, containing basal area, mean height weighted by basal area, stem wood, live branch and foliage biomass were calculated using species-specific allometric equations. The sampling data is shown in Table 1. Then the data filtering, extraction of digital terrain model from the point cloud (DTM), and then from the digital surface model

Table 1. Summary of plot data.

Forest metrics	Coniferous(n= 50)			Broad-leaved(n= 18)			Mixed(n=10)		
	Min.	Max.	Mean	Min.	Max.	Mean	Min	Max.	Mean
Lorey's mean height/m	+.600	21.500	9.800	4.200	19.300	13.700	+.200	18.100	9.800
Total above ground biomass/(Mg·ha⁻¹)	7.345	282.248	63.629	17.726	167.399	15.318	16.024	90.907	39.647
Stem biomass/(Mg·ha⁻¹)	4.233	233.513	45.098	14.206	107.682	50.077	9.804	44.045	25.125
Live branches biomass/(Mg·ha⁻¹)	1.624	45.426	14.130	2.369	45.2772	19.566	3.8998	40.322	11.222
Foliage biomass/(Mg·ha⁻¹)	0.794	21.019	4.401	0.697	14.44	5.675	1.278	8.588	3.2995

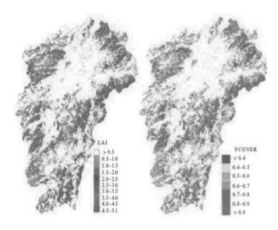

Figure 3. Results of forest canopy hight inversion.

(DSM) for eliminating the effect of DTM, get normalized digital surface model (nDSM). Data classification and extraction of vegetation and ground point under different scales of a number of ground points and vegetation points. Finally calculate the laser radar penetration index and inversion calculation on LAI.

The sampling data, LPI were calculated for each sample under various scales, LPI is the index of laser penetration, Airborne LiDAR data inversion: LAI $L = -\frac{1}{k} \ln(\frac{I}{I_0})$. L stand for the leaf area index, I was under the canopy light, I_0 is above the canopy light, k is the extinction coefficient, depending on the direction of the leaf inclination Angle and the beam of light, light beneath the canopy is available. LPI formula instead of I/I_0, (1/LPI) containing the intercept of the regression analysis with the measured LAI, inversion model based on airborne LiDAR data. This research adopts the different radius of the circular laser point cloud data resampling, sampling data are calculated respectively in different space scale of LPI, and carries on the regression analysis and the measured LAI, calculate model of the correlation coefficient (R^2) and root mean square error (RMSE), to determine the optimal sampling radius of the model. Through the forest coverage map, Results of forest canopy height inversion model.

Is shown in Fig. 3, it can be seen that the inversion result of forest coverage was consistent with the vegetation distribution in the study area. Fitted with the measured coverage data, the value of R2 was 0.9, and the error with sample data was about 0.07 to −0.08. In addition, compared to the measured results by Sheng (2001), which the forest coverage valuein sample area of Shangcun wood farm was 0.54, the difference value with the inversion results of forest coverage (0.769) was only 0.029. The inversion result and compares the result with LiDAR inversion LAI. Results show that based on airborne and spaceborne LiDAR data LAI inversion accuracy is higher than that of the corresponding optical remote sensing inversion accuracy.

REFERENCES

[1] Lim K S and Treitz P M. 2004. Estimation of above ground forest biomass from airborne discrete return laser scanner data using canopy-based quantile estimators. Scandinavian Journal of Forest Research, 19(6): 558–570

[2] Maltamo M, Eerikäinen K, Pitkänen J, Hyyppä J and Vehmas M. 2004. Estimation of timber volume and stem density based on scanning laser altimetry and expected tree size distribution functions. Remote Sensing of Environment, 90(3): 319–330

[3] Nasset E. 1997. Determination of mean tree height of forest stands using airborne laser scanner data. ISPRS Journal of Photogrammetry and Remote Sensing, 52(2): 49–56

[4] Dong L X. 2008. Estimation of the Forest Canopy Height and Above ground Biomass Based on Multiplicatel Remote Sensing in Three Gorges. Beijing: Institute of Remote Sensing Applications Chinese Academy of Sciences (CAS): 56–98

[5] Dong L X, Wu B F and Guo Z H. 2009. Estimation of forest canopy height by integrating multisensor data. Proceedings of the SPIE

Information Technology and Computer Application Engineering – Liu, Sung & Yao (Eds)
© 2014 Taylor & Francis Group, London, ISBN 978-1-138-00079-7

The research on fire spreading based on multi-agent system

Zuo Wei Huang & Xiang Chi
School of Architecture and Urban Planning, Hunan University of Technology, Zhuzhou, China

Luo Qiu
School of Geosciences and Information-Physics, Central South University, Changsha, China

ABSTRACT: The happening of forest fire has the characteristics of inevitability and chance. How can we effectively control firing disaster and make use of forest fire it is urgent to implement dynamically monitoring for forest fire, With the rapid development of spatial information technology and the increasingly artificial intelligence knowledge, multi-Agent Systems technology plays a more and more important role in conducting fire behavior forecast, this paper an interactive fire spreading simulation model based on multi-agent systems and dynamic information feedback mechanism is proposed. The model consists of a GIS (planning areas) and a multi-agent systems. The results show that this model not only better simulates fire spreading under different scenarios, but also provides solutions to fire supression action.

Keywords: MAS, Fire Spreading, building model

1   INTRODUCTION

Multi-agent is an important branch of distributed artificial intelligence. It is developed to solve the large-scale complex problems intelligently. Its basic idea is to separate large complex systems into many small autonomy systems (agents), which can communicate with each other and operate coordinately. Through cooperation with the agent's intelligent behaviors, such as interaction and collaboration, complex task can be solved [2–3]. the application of multi-agent technology in forest firing is still in the exploratory stage. Most researches are still in the concept proposed and model designed stages. Few of the them can be applied to the actual structure. scholars have studied and developed a series of fire spreading simulation models, which can be divided into several groups, including the Equation-Based Models, System Models, Statistical Models, Evolution Models, Cellular Models, Agent-Based Models and so on (Liu, et al., 2008; Chen, 2010). These simulation methods have their own advantages and disadvantages. The equation-based model regarded as a kind of static model is easy for quantitative analysis, but does not consider spatial complexity simulation of fire spreading (Parker, et al., 2003). The model has strong ability of systematic analysis, but is difficult to achieve spatial analysis. The statistical analysis method is widely used, but in the process of fire spatial decision, it is difficult to be adopted due to the difficulty of data acquisition. The evolutionary

model is a kind of effective method to solve quantitatively decision problems because of its simplicity, common use, strong robustness and the ability of parallel computing. However, it is inclined to fall into locally optimal solution when the problem is more complex. Due to the agent itself has characteristics of initiative, interactive, collaborative, reactivity, autonomy, mobility, it is more suitable for expressing complicated group behavior, in particular, have great advantages over geographical spatio-temporal related events.

2   FIRE SPREADING MODEL BASED ON MULTI-AGENT SYSTEMS

In a broad sense, the concept of Multi-agent body covers many different computational entities, these entities are able to sense the environment and effect the environment. since the method based on intelligent agent embodies the Characteristics of autonomous, adaption, robust and it easy to realize, so Multi-agent body has great advantages over some complicated structure or ill-defined tasks. Forest fire spread simulation model based on multi-agent as a discrete model, it has the following advantages: one is much room for improvement, Fire behavior model as a base module, which makes spread model independent and the change of fire behavior model does not affect the spreading model; the second is high efficiency and reliability.

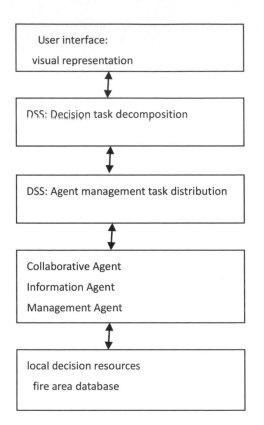

Figure 1. Hierarchical structure of the model.

2.1 The hierarchical structure of the model

The model can be divided into five levels, as shown in Fig. 1, can be expressed as the following:

$$FSDSS = < DSSM, A, UI, G, LR >$$

Where DSSM represents the decision support system for comprehensive Agent management and task distribution.

A – collection of decision-making set, $A = \{A_i, i = 1, 2, \ldots, n\}$, each decision-making agent has its own functions and duties, work together to provide decision support for fire spreading.

UI – the user interface, which is used to receive the user's decision-making and process various manual and instructions in the system.

G – attribute of agent, Including decision agent, Management Agent, Knowledge Agent, Information Agent and Collaborative Agent and so on.

LR – local decision resources, that is, including data, models, knowledge, documentation, resource collections all can aid decision-making resources and tools, it is saved in the three local libraries (model Libraries, knowledge base and database) and used by decision makers.

2.2 Model overview

Multi-Agent can be a physical entity, such as in the Cartesian task environment it operates objects work together cooperatively, which move them from one location to another location, intelligent body can also be calculation code, such as optimizing agents, they could work together to examine certain values and can effectively narrow the search space to achieve a series of small space. Regardless of the domain on agent, one thing is the same, namely, during the problem-solving processing, it conducts local physics or computing interactively in task space. In response to the different local constraints received from the task space, the agent can select and show different behavior patterns. Multi-agent is composed of multiple autonomous agents, which has problem-solving ability and interact with each other to achieve the overall goal. due to the above characteristics of multi-agent body, so it can be applied to the Forest fire spreading simulation.

The model consists of Fire spreading Decision Support System and multi-agent systems, and both systems are interacted through unified cells (Fig. 2). The agents are divided into Knowledge Agent, Information Agent and Collaborative Agent.

In order to guarantee the convergence of the model, improve the speed that agents achieve a consistent decision, and ensure that information agents can express the application requirements in time and receive feedback information, the model conforms to the following two assumed conditions.

(1) Each agent should follow a set of predefined desire rule, in which new desires can not be added and utility function cannot be dynamically changed.
(2) All agents in the model can equally and fully utilize all information of decision-making system.

3 SIMULATION EXPERIMENTS

3.1 Experiment area

The experiment area is the city of Guangzhou, which has complicated topography and plenty of lighting and heating resource. The forestland area in Guangdong has increase greatly after ten years' forestation and large-scale action of putting bare mountain to lush forest. There is a large proportion of middle age and young forest in Guangzhou; the tree population structure is haploid. In additional to dispersive natural village in suburb and frequently human being activities. so forest fire can occurs easily. To meet the requirement of forest fire prevention work which builds the city of Guangzhou into "forest surrounding city and city lies in the middle of forest" ecosystem metropolis.

3.2 Data processing

(1) The original data. Including large scale topographic map, ground control point file, fuel type distribution

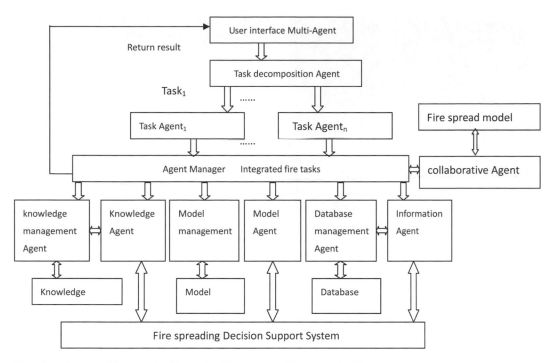

Figure 2. The general framework of interactive fire spread model based on MAS.

maps, thematic maps. While making the combustible distribution diagram, different "layers" are made by GIS map information system according to altitude, slope, soil type, annual rainfall. Which make each layer link with a certain site vegetation. Through the combination, superposition, composite of different graph, it implement the splicing of the map, and the production of various complex thematic maps for different customer.

(2) Building metadata management. Including the Gauss–Krueger projection zone, zoning, central meridian and standard parallel etc. Each subcompartment cards in forest base resources database stand for a record, in order to realize the connection of attribute database and graphics database, subcompartment database added a data item ID corresponding to the key words in graphic library, loading space database engine (Arc SDE). It also realize the management of 3D data.

(3) Data acquisition. by scanning digitizing the topographic map in GIS software. It establish digital elevation model of various forest farm, get orthographic images and region model data and attribute data, such as road, lakes, and so on. It create the DEM elevation data with reference to elevation data, which including elevation point, contour, gully line (water system network), rivers, mountain ridges, Generating DEM data mainly adopts the ArcGIS software, and using the elevation data to create TIN, TIN can be converted to DEM.

3.3 Simulation result

The model consists of a series of fire spreading related factors

$$F=(f_{1,1},\ldots f_{i,j}\ldots f_{n,n})$$

where F represents the collection of all factors, i and j represent line and row numbers of cells, each cell is a discrete part of running environment and includes environment information $S_{i,j}$ which is expressed as below:

$$S_{i,j}=(A_{i,j,1}, A_{i,j,2},\ldots A_{i,j,n}, ,R_1,R_2,B),$$

where $a_{i,j,1}, a_{i,j,2},\ldots, a_{i,j,n}$ represent land price, soil, terrain, traffic and other factors, and n is the number of factors. R_1 represents geographically bounded border in fire areas and the range of possible position of fire, which also a constraint of its movement. R_2 represents the position of road and river around the fire area, B represents the position of building and block around the fire area.

Repast is a multi-agent modeling software, which can be used to simulate the natural and social phenomenon. with its own model libraries, users can simulate and model multi-agent systems by changing various conditions set. It is particularly suitable for multiple development of complex system. Modelers can put many orders to multiple agents at the

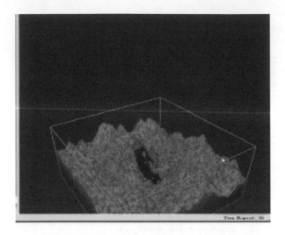

Figure 3. The simulation shape of fire spreading.

same time, it also provides some built-in adaptation functions, to support internal system dynamic model and the discrete event parallelism Operation, The final simulation shape of fire spreading is shown in Fig. 3.

ACKNOWLEDGEMENTS

The authors would like to thank the anonymous reviewers for their careful reading of this paper and for their helpful comments, This work is supported by the National Natural Science Foundation of China (No. 61273157).

REFERENCES

[1] X. F. Fan, X. W. Jiang, W. H. Huang, et al. Research of multi-agent system based satellite fault diagnosis technology. Journal of Harbin Institute of Technology, 2002, 9(3): 239–244.

[2] S. F. Yuan, X. S. Lai, X. Zhao, et al. Distributed structural health monitoring system based on smart wireless sensor and multi-agent technology. Smart Materials and Structures, 2006, 15(1):1–18.

[3] Batty M. Agent-based Technologies and GIS: Simulating Crowding, Panic and Disaster Management, in Frontiers of Geographic Information Technology [M]. Berlin: Springer-Verlay, 2006.

[4] Li X and Liu X P. 2008. Embedding sustainable development strategies in agent-based models for use as a planning tool. International Journal of Geographical Information Science, 22(1): 21–45.

[5] Ligtenberg A, Beulens A J M, Kettenis D L, Bregt A K and Wachowicz M. 2009. Simulating knowledge sharing in spatial planning: an agent-based approach. Environment and Planning B: Planning and Design, 36(4).

High-performance TV video noise reduction using adaptive filter

Shih-Chang Hsia & Cheng-Liang Tsai
Department of Electronics, National Yunlin University of Science and Technology, Taiwan

ABSTRACT: TV videos appear visible noise while the signal level becomes low. However, the current denosing methods could not remove TV noise efficiently. In this study, a new noise removal algorithm is proposed using noise detection, motion detection and adaptive filter. The five fields are referred to compute the filtering parameters and to increase the filtering ability. Experimental results show that this method can be effective to reduce the large region TV noise and keep the original TV image information. Compared with the existed approaches, the proposed method is capable of providing a better picture quality, shorter processing time than the competing algorithms.

1 INTRODUCTION

TV noise is distributed on large region display, which likes a mix of complex noises. The noise type can not be defined by a noise model. As the signal-noise ratio becomes weak, the noise level increases on display accordingly. To improve the image quality, the signal processing for noise reduction had been developed in the past literatures. Nevertheless, the noise problem for large screen TV still can not be completely resolved so far. The well-known median filter and its improved methods can provide a good feature to remove impulse noise signals [1]. However, they are not effective to reduce the large region TV noise. Recently, there are many highly efficient techniques [2–4] that discuss a new filtering approach with wide information in order to remove the noise completely. But their filtering performance is not satisfied for users since the screen noises are still existed. For large region TV noises removal, the performance of the temporal processing is generally better than that of spatial processing due to more information used. In this study, a novel adaptive algorithm is presented to improve the filtering performance based on the adaptation of temporal and spatial processing, particularly for large region TV noise.

2 PROPOSED DENOISING ALGORITHM

The adaptive algorithm is split into three parts: noise detection, motion detection and adaptive filter. The input TV videos are stored on the field memory. The motion detection used the inter field information to check the processing pixel whether is moving. The noise detection can check the noise level using intra- and inter-field information. The adaptive filter is designed to determine the filtering power based on the information of motion feature and noise level. Finally,

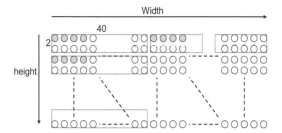

Figure 1. The block processing for intra-field noise detection.

the de-interlaced method is used to convert the field to achieve a noise-free frame.

To improve the detected accuracy, the noise detection is used to compute the amount of noise in intra- and inter- block respectively. For intra field detection, the block size used 2×40 to save the line buffer and cover large region video information, as shown in Fig. 1. As the block contains more noise, the difference of pixel and pixel becomes high. One can compute the difference of block mean variance (BV) to check the intra field noise level, by

$$BV(j,k) = \sum_{m=0}^{1} \sum_{n=0}^{39} |Mean(j,k) - f(2j+m,40k+n)| \quad (1)$$

for (j,k)th block. The mean(j,k) is the average of the first 16 pixels from the block, which can employ 2^n-bit right-shifter rather than division. If BV is high, the intra noise is higher. To reject some impossible cases, such as edge blocks, the intra noise for one field is selected by the summation of from the minimum 16 BV values over all one field, as

$$Noise_{intra-field} = \sum_{s=1}^{16} BV(s). \quad (2)$$

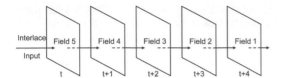

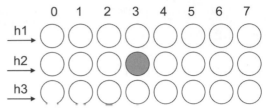

Figure 2. The five reference fields for noise detection.

● Current pixel

Figure 3. A 3×8 window for motion detection.

BV(1,2,...,16) denotes the first, second,... and sixteen minimum BV value. The selected BV must keep a suitable level to achieve a stable detection. In experiments, the BV value selected for intra noise in (2) must be satisfied two conditions: (a) The minimum BV selected must be over 100. (b) To reject too dark or brightness region, BV value limited on $20 <$ Mean(j,k) < 230.

The final intra noise adopts the summation of the results of the previous four fields, which can be expressed by

$$Noise_{intra} = (\sum_{n=1}^{4} Noise_{intra-field(n)}) >> k \qquad (3)$$

where intra-field (n) denoted the nth field, k is the number of right-shifted bit, and $k = 4$ in experiments.

To detect the temporal noise, five reference fields are employed to cover more information, as shown in Fig. 2. NTSC TV used interlaced signal that is odd/even filed alternatively. The field-5 is the current proceeded one. Two temporal differential parameters are computed with same odd or even fields by

$$Diff_1(j,i) = |f_{(t)}(j,i) - f_{(t+2)}(j,i)| \quad for\ j = 0 \sim H-1$$
$$Diff_2(j,i) = |f_{(t)}(j,i) - f_{(t+4)}(j,i)| \quad for\ i = 0 \sim W-1 \qquad (4)$$

where $f_{(t)}$, $f_{(t+2)}$ and $f_{(t+4)}$ is the current, the previous 2th, and 4th fields respectively, H and W is the height and width of field. The amount of inter noise for the current field is detected by

$$Noise_{inter}(t) = \sum_{j=0}^{H-1} \sum_{i=0}^{W-1} \begin{cases} 1, & (Diff_1(j,i) > th1 \| Diff_2(j,i) > th1) \&\& ME_C \\ 0, & otherwise \end{cases} \qquad (5)$$

where th1 is one threshold, and $th1 = 25$ in experiments. To keep the original image quality, we need to classify the noise pixel from motion pixel with motion information. The MEc is the motion feature, which can be expressed by

$$NE_C = \begin{cases} 1, & MD_{dif}(j,i) < \dfrac{Noise_{intra}}{32} \\ 0, & otherwise \end{cases} \qquad (6)$$

The various MD_{dif} is the motion differential between inter-frames, which will be described later. If the motion difference is low, but the estimate intra noise with (3) is high, the current pixel will be high probability as a noisy pixel.

The final noise level for the sampling field can be evaluated by the summation of the inter- and intra-noise, which can be given by

$$Noise_{field} = 4Noise_{inter} + Noise_{intra}. \qquad (7)$$

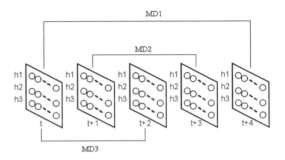

Figure 4. The computations of motion detection.

Since the level of inter noise is lower than that of intra, the amount inter noise is amplified by 4 times to balance the predicted level from the intra and inter estimation.

To detect the motion pixel, a 3×8 window is used, as shown in Fig. 3. The first check point is with the 1st and 3rd rows between the current field t and the last field $t + 4$, which can be given by

$$MD1 = \sum_{k=0}^{7} [|h1_{(t)}(k) - h1_{(t+4)}(k)| + |h3_{(t)}(k) - h3_{(t+4)}(k)|] \qquad (8)$$

where hn is the nth row pixel in the processing block. The second detection is with the 2nd row between the field $t + 1$ and $t + 3$, which can be expressed by

$$MD2 = \sum_{k=0}^{7} |h2_{(t+1)}(k) - h2_{(t+3)}(k)| . \qquad (9)$$

The last is to detect the field t and t + 2, by

$$MD3 = \sum_{k=0}^{7} |h2_{(t)}(k) - h2_{(t+2)}(k)|. \qquad (10)$$

The illustration is shown in Fig. 4. The motion deferential MD is the summation of MD1, MD2 and MD3, which can be found by

$$MD_{dif} = MD1 + MD2 + MD3. \qquad (11)$$

The MD value is used to check the current pixel whether a motion pixel compared with an adaptive threshold. The threshold is adaptive by noise level with (7) by

$$MD_{thr}(j,i) = \begin{cases} 8000/MD_{fac}, & Noise_{field} > 15000 \\ 6000/MD_{fac}, & 6000 < Noise_{field} \leq 15000 \\ NRY/MD_{fac}, & Noise_{field} \leq 6000 \end{cases} \quad (12)$$

When the noise level is high, the threshold is adaptive to high, which can avoid the error detection noisy pixel as a motion pixel. The MD_{fac} is the factor of motion detection. In experiments, $MD_{fac} = 10$ and the range of parameters with (12) can overcome the noisy pixel to effect the accuracy of motion detection.

The adaptive filter is proposed to remove the video noise based on noise detection and motion detection. If the result of noise detection is low noise level, the filtering operation can be directly skipped to avoid blurring the image. When the high noise level is found, the filter performs the temporal processing to remove noise for the still region. However, for motion region, to avoid motion object dragging and the temporal aliasing, the spatial low pass filter [1] is employed.

Generally, human eyes for luminance pixels are more sensitive than chrominance. For this, the processing domain used YUV signals. The Y- and UV-signals are luminance and chrominance components that are processed with various filtering coefficients to achieve better complexity and performance trade-off. The temporal filter for Y processing is split with three levels performing based on noise detection. If the noise level is higher, the filtering power is enhanced in order to remove noise clearly. When the noise level is the highest one, the temporal filter performs the noise removal with five fields by the weighting filter as

$$\hat{Y}_{(t)}(j,i) = Y_{(t)}(j,i) \times \frac{3}{8} + \hat{Y}_{(t+1)}(j,i) \times \frac{1}{8} + \hat{Y}_{(t+2)}(j,i) \times \frac{2}{8} +$$
$$\hat{Y}_{(t+3)}(j,i) \times \frac{1}{8} + \hat{Y}_{(t+4)}(j,i) \times \frac{1}{8} \quad (13)$$

$Y_n(j,i)$ denotes the luminance components at the (j,i) pixels in the nth field. $\hat{Y}_{(t+1)}(j,i) \ldots \hat{Y}_{(t+4)}(j,i)$ are the filtered pixels of the previous fields. In experiments, the temporal filter is performed when $Noise_{field}$ in (7) is over 3000. The processing flow can be spilt to three steps. First, the interlaced pixel is loaded to field (t) to (t+4). Second, the weighting average of various fields is found to remove the noisy pixel. Third, the filtered pixel is reloaded the field memory for the temporal recursive operation.

In a low-noise field, when $2500 < Noise_{field} < 3000$ in experiments, the filtering power is degraded. The filter employs three fields processing by

$$\hat{Y}_{(t)}(j,i) = Y_{(t)}(j,i) \times \frac{2}{4} + \hat{Y}_{(t+2)}(j,i) \times \frac{1}{4} + \hat{Y}_{(t+4)}(j,i) \times \frac{1}{4} \quad (14)$$

When the $Noise_{field} < 2500$, this field is noise free, and no filtering for this field. To improve the filtering power, the temporal filter used the recursive processing. The filtering pixel is updated to field memory for the next field processing. In (14) and (15), the pixels $\hat{Y}_{(t+1)}(j,i) \ldots \hat{Y}_{(t+4)}(j,i)$ all had been filtered.

For UV signals filtering, the noise detection can employ Y-result. When Y pixel is noise, UV signals are damaged by noise accordingly. The image color will have a little distortion. Fortunately, the eye sensitivity for UV is lower than that of Y. The filtering power can be down to save the computational complexity. The filtering operation can be done by

$$\hat{U}(\hat{V})_{(t)}(j,i) = U(V)_{(t)}(j,i) \times \frac{2}{4} + \hat{U}(\hat{V})_{(t+1)}(j,i) \times \frac{1}{4}$$
$$+ \hat{U}(\hat{V})_{(t+2)}(j,i) \times \frac{1}{4} \quad (15)$$

when the field noise is found as the field noise in (7) is over 2500 in experiments. Also, the recursive filtering is employed for UV signals processing.

3 PERFORMANCE EVALUATIONS

To simulate this adaptive filter, the interlace fields with 720×240 are sampled from NTSC TV program. We sample 5 sequences, and each sequence uses 200 fields for testing. The composite NTSC program is decoded to YUV signal. The noise detection and the motion detection used the Y signal to find the motion-level and noise-level parameters. The filtering power is dependent on the results of noise detection and motion detection. The filtered field is de-interlaced to a frame. The YUV pixels are converted to RGB pixels to show on VGA monitor.

To evaluate the filtering performance, both of the object and subject measurement are employed. In order to compute the PSNR value, Gaussian noise is artfully added to sample noise-free sequences. To show the entire frame, the de-interlace [11] is used to interpolate the missing filed. For five sequences testing, there are five algorithms used to filter the noisy sequence for comparisons. The only spatial processing [2–3] cannot remove the noisy pixel clearly. Alternatively, the quality can be improved by spatial-temporal algorithms [4–5]. Although the adaptive algorithm can achieve good filtering performance for noise removal on the sampling videos, its computational complexity is high due to motion compensation used. Table 1 lists the average PSNR after filtering, and the processing time. The PSNR value is measured with the average of 200 fields for each sequence. The processing time is evaluated with Intel Pentium 4 CPU 2.4 GHz, 2 GB DDR333 RAM. Results demonstrate that the proposed method can outperform both on PSNR and the filtering image. Our speed is close to the fast impulse filter [3], and the filtering performance is much better than [3]. The processing time of the proposed algorithm is shorter than that of adaptive ones [4–5].

Next, we directly sample the noisy sequence form TV program. The noisy images are filtered with various algorithms, and the results are shown in Fig. 5. The sampled image appears the ripple noise and like a mixing of Gaussian and ripple noise at one frame. Most of denosing algorithms are not effective to reduce

Table 1. The PSNR Performance Comparisons with Noise Filters.

		[2]	[3]	[4]	[5]	Proposed
PSNR	Video 1	29.3dB	29.5dB	31.3dB	31.2dB	31.5dB
	Video 2	26.1dB	26.0dB	27.5dB	25.8dB	28.0dB
	Video 3	28.4dB	27.8dB	30.0dB	26.5dB	30.8dB
	Video 4	27.3dB	26.6dB	28.1dB	26.5dB	28.6dB
	Video 5	29.4dB	28.1dB	30.3dB	29.5dB	30.6dB
Processing Time (per frame)		1.78 sec	0.04 sec	10.21 sec	16.32 sec	0.08 sec

(a)   (b)

(c)   (d)

(e)   (f)

Figure 5. Shows (a) Original Sampling Image; (b) Standard Median Filter [2]; (c) Non-linear Filter [3]; (d) α-trimmed Mean Filter [4]; (e) STS-DDWA Filter [5]; (f) Proposed.

TV noise. Results show that our filtering performance is better than the competing algorithms. The proposed algorithm can clearly remove most of noises, and the image quality is indeed improved.

4 CONCLUSIONS

In this paper, this low-computation filter employed the adaptation of temporal and spatial processing. The proposed noise detection and motion detection can classify the noisy pixel from motion pixel and edge pixel efficiently. The filtering power is adaptively dependent on the detected level of field noise and the motion detection. In order to remove the large range noises in TV display, five fields as reference are adopted to cover more information. Simulations show that the proposed algorithm can outperform in the object and subject measurement compared with the competing ones. With high filtering quality and low computations, the noise filter can be applied on TV video noise reduction for real-time purpose achieved better performance-complexity tradeoffs.

REFERENCES

[1] R. C. Gonzalez, R. E. Woods, Digital Image Processing, Addison Wesley, 1992.
[2] S.C. Hsia, "A fast efficient restoration algorithm for high-noise image filtering with adaptive approach", Journal of Visual Communication and Image Presentation, pp. 379–392, June, 2005.
[3] P.-Y. Chen, C.-Y. Lien, and H.-M. Chuang, "A low-cost VLSI implementation for efficient removal of impulse noise", IEEE Trans Very Large Scale Integration System, vol. 18, no. 3, pp. 473–481, Mar., 2010.
[4] V. Zlokolica, W. Philips, D. Van De Ville, "Robust Non-Linear Filtering for Video Processing", IEEE 14th International Conference on Digital Signal Processing, vol. 2, pp. 571–574, July, 2002.
[5] Kouji Miyata, Akira Taguchi, "Spatio-temporal separable data-dependent weighted average filtering for restoration of the image sequences", IEEE International Conference on Acoustics, Speech, and Signal Processing, vol. 4, pp. IV-3696–IV-3699, 2002.

Information Technology and Computer Application Engineering – Liu, Sung & Yao (Eds)
© *2014 Taylor & Francis Group, London, ISBN 978-1-138-00079-7*

Research on forecasting and pre-warning methods of social security incidents

T.T. Zhao & Y. Zhao
National Geomatics Center of China, Beijing, China

Y.L. Han
State Nuclear Security Technology Center, Beijing, China

W. Ma
National Quality Inspection and Testing Center for Surveying and Mapping Products, Beijing, China

ABSTRACT: The research on the forecast of social security incidents covers many subjects such as social science, mathematics and computer science. This paper proposes dividing forecasting and pre-warning of social security incidents into two parts: macroscopic forecast and microscopic forecast, mainly comprehensively reviews the macroscopic forecast method and systemically generalizes several forecast methods such as: time series analysis; regression analysis and data mining technique, etc. The paper also generalizes the application scope, advantages as well as disadvantages of various methods and indicates the research direction and hot spot of this field by combining the features of various types of social security incidents based on the analysis of the mathematical characteristics of each method.

Keywords: Intelligent Computing; Social Security Incident; Forecast Methods; Time Series

1 GENERAL INSTRUCTIONS

Forecasting and pre-warning methods of social security incidents can be divided into macroscopic forecast and microscopic forecast. Macroscopic forecast, generally speaking, indicates the forecast by taking the social security incidents with a certain temporal and spatial span revealing the dynamic rule of social security incidents from the perspectives of quality and quantity to provide scientific basis for social formulation of precautionary strategic measures. Microscopic forecast indicates the rule satisfied prediction of the possibility of the individual's first offense or recidivism in the future specific period by using scientific methods and making early prediction of the individual's possible first offense or recidivism under certain temporal and spatial conditions. This paper aims at systematically describing the forecasting and pre-warning methods of social security incidents and mainly systematically summarizing and introducing the macroscopic forecast method in great detail.

The research on macroscopic forecast of social security incidents is started early in the western developed countries. Macroscopic forecast can be divided into long-term forecast and short-term forecast. Long-term forecast is represented by English Derek Deadman, Richard Harries and D J Pyle, who assume that criminal activities are surely associated with social economic activities from the perspective of people's criminal mentality and predict the crime trend within certain future period through model parameter obtained by establishment of measurement economics model with social and economic indicators as the independent variables and with crime amount as the dependent variable (Harries 2003). English scholar Derek Deadman predicted property-related and theft cases of England and Wales in 2003 (Deadman 2003). They predicted the crime conditions between 1998 and 2001 through establishing the time series model and economic model by use of the crime data between 1946 and 1997 with the result that the time series model is proved to have better prediction accuracy. Furthermore, Derek Deadman and David Pyle predicted the property-related cases in UK between 1992 and 1996 by cooperatively establishing time series economic model in 1997. And also the desirable effect was achieved (Deadman & Pyle 1997).

Short-term prediction of crime is a research direction emerged recently. Wilpen Gorr and Richard Harries initially proposed "crime prediction shall be integrated into crime map technique to provide basis for judgment of crime conditions within short period in the future and decision making for the police force's patrolling and deployment" in their papers, which could be regarded as the beginning of the research into the short-term prediction of crime (Gorr et al.

2003). From the perspective of current research, short-term prediction of crime mainly adopts time series algorithm method to predict the crime conditions in future short period through the analysis of the historical crime data. Several scholars such as: Wilpen Gorr, etc. from Carnegie Mellon University performed short-term prediction of crime trend in Pittsburgh by adopting Naïve and exponential smoothing in time series algorithm in 2003. According to its result, it indicates that the ideal prediction resultcould be achieved if the data volume in time unit is kept for more than 30 (Gorr & Harries 2003).

Comparatively speaking, the research into the prediction of social security incidents is initiated late in China. The relative research was initiated in 1980s with main focus on qualitative prediction. Recently, the research is commenced by adopting qualitative scientific method. For instance, several postgraduates such as: Liu Xiaojuan, from Beijing University of Aeronautics and Astronautics applied the grey model theory to crime prediction and respectively predicted the major criminal case quantity between 1990 and 1995 of certain city and the criminal case quantity between Apr. 2006 and Sep. 2006 of certain area (Liu & Gao 2005). According to the prediction effect, it proves that the model has better applicability. In addition, in 2006, several scholars such as: Wang Zhongshen, etc. carried out long-term trend prediction of crimes between 2005 and 2010 of Beijing Chongwen district by adopting the classical long-term trend prediction model, relevant-regression model as well as Delphi method in the field of crime prediction field (Wang 2006). They suppose that economic growth, unemployment rate, gap between the rich and the poor, floating population, urban construction as well as transfer in crime, etc. are main factors influencing the crime increase trend through analysis. In 2008, several scholars such as: Yan Jun, etc. from Tsinghua University studied the mechanism of social & economic livelihood and microscopic social environmental factors in influencing the change trend of social security incidents in the course of macroscopic social development based on nearly three decades' statistical data with empirical analysis method as the main measures (Yan et al. 2009; Yan et al. 2008; Yan 2009). Several scholars such as: Chen Peng, etc. from Tsinghua University analyzed the general rule of crime activities from the angle of temporal and spatial model and predicted the incidents by adopting time series method (Chen et al. 2009).

2 OVERVIEW ON MICROSCOPIC FORECAST AND MACROSCOPIC FORECAST OF SOCIAL SECURITY INCIDENTS

The direct reason for the occurrence of social security incidents mainly focuses on human factors. But human behavior is often characterized by randomness, dynamics, complexity and relevance. Therefore, these features directly lead to uncertainty of prediction of social security incidents. However, there will always

Table 1. Comparison of microscopic forecast and macroscopic forecast of social security incidents.

Forecast of Social Security Incidents	Theoretical Basis	Information Basis	Information source
Microscopic forecast	Social studies such as: criminal psychology, criminal physiology, etc.	Information on specific incident and individual	Department of Police Information
Macroscopic forecast	Mathematics subjects such as: criminology and mathematical statistics, etc.	Historical statistical data of crimes	Statistics Department of Public Security

Forecast of Social Security Incidents	Forecasting methods	Application scope	Others
Microscopic forecast	Experience and reasoning of police personnel	Front-line police officers	Subjective but applicable
Macroscopic forecast	Reasonable mathematical model	Territory management institute	Scientific but higher requirement for data

exist some rules or some deterministic rules under the random presentation, which can make the prediction become feasible under certain probable conditions. These rules are the theoretical basis for prediction of social security incidents, required to be summarized, perfected and applied.

The forecast and pre-warning of social security incidents, generally, can be divided into microscopic forecast and macroscopic forecast. As implied by the name, microscopic forecast refers to the prediction of behavior pattern of certain individual or specific development trend from the microscopic level while macroscopic forecast refers to prediction of the development trend of total quantity of incidents occurred from the macroscopic level. For comparison of both methods, see Table 1.

3 MICROSCOPIC FORECAST OF SOCIAL SECURITY INCIDENTS

3.1 Theoretical basis for microscopic forecast

The main theoretical bases for microscopic forecast are some theories on criminology such as: criminal psychology, criminal physiology as well as criminal sociology, etc. based on observation of a lot of criminal phenomena, namely a lot of social security incidents, analyze the characteristics of different people to commit crimes with different motives and portents and summarize these rules to provide knowledge basis for future prediction of crimes (Zhang 2007).

From the angle of the individual, he will show up certain sign, incipient tendency or abnormal behavior before committing a crime. The formation of criminal mentality is a gradually development process, which develops to a certain stage, then the corresponding external behavior will occur or certain tendency of certain crime will be exposed inevitably (Mei et al. 2003). From the angle of incident, due to the demand of criminal organization for operation prior to occurrence of incidents, other series of the relevant incidents will be caused, such as: change in bank account and communication log. These omens for incidents are secluded sometimes, but these incidents are abnormal and predictable through thorough observation.

3.2 *Case study-prediction of certain terrorist attack*

On Aug. 10, 2006, the British police announced having successfully defeated a conspiracy of terrorists' attempting to blow up ten passenger aircrafts flying from Britain to America similar to 9·11 incident and arrested dozens of suspects in London.

Actually, from Nov. 2005, the MI 5 of UK started to monitor the terrorist group closely, grasped a lot of information about the group through eavesdropping and monitoring the telephone calls and tracks, personal exchanges and bank accounts and found there was a large sum of money at the terrorist suspects' accounts, which evidently were not proportional to their incomes. Further this group doesn't go its own way, the behavior of frequently transacting money in large sum came to the attention of the relevant personnel of British police and MI 5, thus great efforts were paid to monitor this group and began to monitor the phone calls of group members. Eventually, the conspiracy of terrorist attack is defeated with a complete success.

4 MACROSCOPIC FORECAST OF SOCIAL SECURITY INCIDENTS

4.1 *General process of macroscopic forecast*

The general process of macroscopic forecast: input and process initial data, select prediction algorithm, come to prediction result, select pre-warning method and give an alarm. As shown by the Figure 1.

4.2 *Comparison of classical forecast method*

4.2.1 *Time series forecasting*
To know about the surrounding world, people often perform a series of observations and forecasts according to time sequence. The future data will often be dependent on the observed value obtained at present in certain random form. The dependence of the observed value makes it possible to forecast future with past.

Time series modeling has two categories of basic hypothesis (Fan & Yao 2005):

Deterministic time series modeling hypothesis: time series is generated by a deterministic process, which often can be represented by the function f (t) of time t.

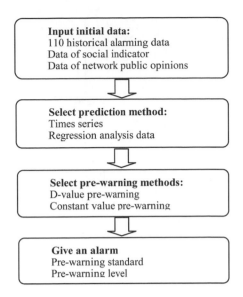

Figure 1. General process of macroscopic forecast and pre-warning of social security incidents.

each observed value of the time series is determined by the deterministic process and random factor.

Random time series modeling hypothesis: the variation process of statistic of social security incidents is random process. Time series is a sample generated by the random process. Therefore, time series is random and can be represented by random linear combination, namely using the method for analyzing random process to establish time series model.

Mainly introduce several categories of time series forecast models in accordance with the characteristics of social security incidents and application of model:

a) Winters multiplication seasonal model

As mentioned previously, for the occurrence of social security incidents is of great uncertainty, but as far as the layer of statistic is concerned, it is usually possible to find some periodicity, for instance, the statistic of Jan. and Feb. during spring festival is often a low ebb as the monthly statistic time series per year displays, but the annual statistic will often show up certain tendency (increase or decrease) during a comparatively long period, etc. In view of this, we can hypothesize the social security incident time series to be deterministic time series.

Winters multiplication seasonal model is composed of three formulas, of each formula is applicable to the three constitutional parts (steady, trend and seasonal) of the smoothing model and contains a relevant parameter, shown as below (Zhang & Cai 1993):

Level factor:

$$S_t = \alpha \frac{x_t}{I_{t-L}} + (1-\alpha)(S_{t-1} + b_{t-1}) \qquad 0 < \alpha < 1 \qquad (1)$$

Trend factor:

$$b_t = \gamma(S_t - S_{t-1}) + (1-\gamma)b_{t-1} \qquad 0 < \beta < 1 \qquad (2)$$

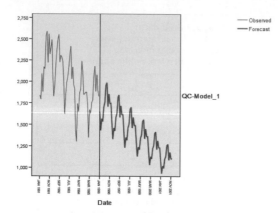

Figure 2. Seasonal forecast model example of criminal incidents in pittsburgh.

Seasonal factor:

$$I_t = \beta \frac{x_t}{S_t} + (1-\beta) I_{t-L} \qquad 0 < \gamma < 1 \qquad (3)$$

There into, L indicates the length; I indicates season-correcting coefficient.
   Predicted value:

$$F_{t+m} = (S_t + b_t m) I_{t-L+m} \qquad (4)$$

For instance: take 1 decade's data between 1991 and 2001 of Pittsburgh as the research object, use former five years' data to predict the future five years' data. Establish the prediction function Ft during the deterministic process by extracting the parameters relevant to seasonal model through the certainty similar periodicity displayed by the five-year monthly data between 1991 and 1995 and obtain the predicted data (blue line) between 1996 and 2001 through time prediction, as shown by the Figure 2.

   Due to randomness neglected by the method, this method has bad accuracy in predicting the data (such as: monthly data) with small time span but have certain accuracy in predicting the data (such as: annual data) with large time span. More, without revision of actual data, the accuracy will be worse as the prediction time is longer.

   b) ARIMA model
   ARIMA (p, d and q) model is also called autoregressive integrated moving average model. Thereinto, AR refers to autoregression, p refer to the number of autoregression terms; MA refer to moving average, q refers to the number of moving average terms; I refers to integrated, d refers to the times of difference required for transformation of non-time series into steady time series. The model is as shown as below (Bai & Yan 2007):

$$(1 - L)^d x_t = \mu + \frac{\theta(L)}{\phi(L)} \varepsilon_t \qquad (5)$$

$\phi(L)$ refers to steady autoregressive polynomial in lag operator. $\varphi(L) = 1 - \varphi_1 L - \varphi_2 L^2 - \cdots - \varphi_p L^p$.

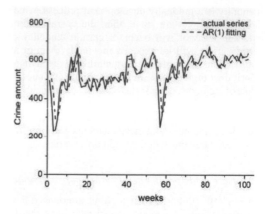

(a) By week

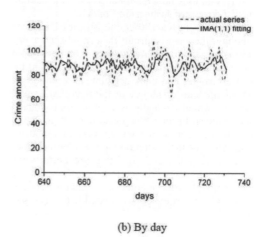

(b) By day

Figure 3. ARIMA forecast model of property related crimes in a certain city.

$\theta(L)$ refers to reversible moving average polynomial in lag operator. $\theta(L) = 1 - \theta_1 L - \theta_2 L^2 - \cdots - \theta_q L^q$.
   $(1 - L)^d$ refers to d time's difference operator.
   $(1 - L)^d = 1 - C_d^1 L + C_d^2 L^2 + \cdots + (-1)^{d-1} C_d^{d-1} L^{d-1} + (-1)^d L^d C_d^r$ refers to the combinatorial number extracted from d.
   μ refers to sequence $(x_t, t = 1, 2, \ldots, T)$ refers to the average value of after d difference, $\{\varepsilon_t\}$ refers to white noise sequences.
   The following figure is an example of applying ARIMA model to predict crimes.

4.2.2 Regression analysis

In the aforementioned time series analysis, no matter deterministic hypothesis or the random hypothesis, the basic thought is that predicting the future data by depending on the past data. But the occurrence of social security incidents is not isolated. It has profound social background, namely not only depending on the status of the past. To be definite, the occurrence of social security incidents is the result of

Table 2. Comparison of various macroscopic forecast of social security incident.

Forecast methods	Time series analysis	
	Seasonal model	ARIMA mode
Scope of application	Applicable to steady social security incidents with apparent periodicity, apparent trend such as: general public security and criminal cases, etc.	Applicable to social security incidents with strong randomness and large data volume such as: major criminal cases, etc.
Advantages and disadvantages	Advantage: simple usage. Disadvantages: narrow scope of application and not applicable to long-term prediction.	Advantage: good short-term prediction effect. Disadvantage: not applicable to long-term prediction.

Forecast methods	Regression analysis	Data mining classification and clustering techniques
Scope of application	Applicable to social security incidents with obvious social background reasons, such as: collective incident, etc.	Applicable to prediction of various social security incidents and well applicable to prediction of incidents isolated from the microscopic level
Advantages and disadvantages	Advantage: accurate prediction and applicable to long-term prediction. Disadvantage: excessive data as required.	Advantage: wide scope of application. Disadvantages: difficult and hard to popularize.

the comprehensive function of various factors. These factors are not surely the negative factors. For example, there is positive relationship between rapid economic growth and occurrence of social security incidents. Unalike as compared to the time series analysis, the regression analysis gives comprehensive consideration to the correlativity between various social factors and the occurrence of social security incidents and can further rational predict the development trend of the social security incidents. Especially, it has good prediction of the inflection point of social security situation generated by the changes in the social situation. But the time series analysis doesn't.

4.2.3 *Data mining classification and clustering techniques*

Data mining technique is an emerging technique recently, widely used to the various engineering technological fields. In a broad sense, data mining technique includes two parts: time series analysis and regression analysis. Hereby, data mining technique refers to classification and clustering techniques. The prediction of social security incidents is not just the question of time series, its occurrence will spread according to the region. For instance, the site scenes of certain criminal group successively show up relevance, represented by the continuous spread in space. The temporal and spatial prediction should be dependent on data mining technique. Additionally, the data mining technique also can summarize rules and discover knowledge in large quantity of social security incidents. For example, provided that the quantity of occurred social security incidents aggregates under the climate condition of torrential rain each time, it is able to have a rough judgment of the situation of social security when the torrential rain comes into being next time. Data mining technique is especially used to the data with complicated data such as: net work public opinion data. In brief, the scope of data mining technique application is pretty wide. There is wide prospect of the social security field. But currently, data mining technique hasn't give full play to its function, still to be studied to provide more knowledge for forecast and pre-warning of social security.

4.2.4 *Comparison of various forecast methods*

Various forecast and pre-warning methods have their own application scope, advantages and disadvantages in accordance with the mathematical characteristics and different characteristics of different social security incidents. For the specific contents, please refer to the Table 2.

5 CONCLUSION

The occurrence of social security incidents has certain sign and certain social background no matter from the microscopic level or from the macroscopic level, not only significant to the forecast and pre-warning but also feasible.

From the microscopic level, main method for forecast and pre-warning is that front-line personnel logically reasons the incident development, object refers to specific incident and individual and theoretical basis mainly refers to criminal psychology & science of criminal investigation, etc.

From the macroscopic level, the main method for forecast and pre-warning refers to establishment rational mathematical model, object refers to statistical data of incident and theoretical basis refers to mathematical subjects such as: mathematical statistics, etc. various macroscopic forecast methods have their own characteristics and scope of application.

Simple time series forecasting has good effect in short-term prediction under the guarantee of certain sample quantity, but has no effects in long-term prediction as well as prediction of reflection point.

Give consideration to the time series forecasting under the influence of comprehensive social factors can have good effects in long-term predication and prediction of reflection point, but excessive data as required and difficulty in implementation.

347

Data mining technique can perform prediction of various requirements such as: temporal and spatial prediction and can mine more value result in case of insufficient data. The method is flexible in usage but not conveniently popularized.

From the perspective of forecast and pre-warning of social security incidents, the further researches and explorations into many methods are still to be carried out and the method system is still be further perfected. The wide and in-depth application of data mining technique in forecast and pre-warning of social security incidents is very worthy of expectation.

REFERENCES

Bai, B.F. & Yan, Z.C. 2007. ARIMA model in forecasting the application of mobile communication users. *Statistical Education* 92: 41–42.

Chen, P., Shu, X.M., Yan, J. & Yuan, H.Y. 2009. Timing of criminal activities during the day. *Journal of Tsinghua University (Science and Technology)* 49: 2032–2035.

Deadman, D. 2003. Forecasting residential burglary. *International Journal of Forecasting* 19: 567–578.

Deadman, D. & Pyle, D. 1997. Forecasting recorded property crime using a time-series econometric model. *British Journal of Criminology* 37: 437–445.

Fan, J.Q. & Yao, Q.W. 2005. Nonlinear Time Series [m]. Chen, M. translation, Beijing: Higher Education Press.

Gorr, W., Olligschlaeger, A. & Thompson, Y. 2003. Short-term forecasting of crime. *International Journal of Forecasting* 19: 579–594.

Gorr, W. & Harries, R. 2003. Introduction to crime forecasting. *International Journal of Forecasting* 19: 551–555.

Harries, R. 2003. Modelling and predicting recorded property crime trends in England and Wales – a retrospective. *International Journal of Forecasting* 19: 557–566.

Liu, X.J. & Gao, L.S. 2005. Application of grey system theory to prediction of dynamic tendency of crimes. *Journal of Chinese People's Public Security University* (Social Sciences Edition) 113: 44–48.

Mei, C.Q., Wang, M. & Peng, J.H. 2003. Criminal Psychology. Law Press, China.

Wang, Z.S. 2006. Growth over the next five years in Chongwen District crime trend prediction countermeasures and actions. *Journal of Beijing People's Police College*: 88–95.

Yan, J., Yuan, H.Y., Shu, X.M. & Zhong, S.B. 2009. Optimal clustering algorithm for crime spatial aggregation states analysis. *Journal of Tsinghua University (Science and Technology)* 49: 176–178.

Yan, J., Yuan, H.Y. & Shu, X.M. 2008. Research on spatial distribution of social security incidents. *China Safety Science Journal (CSSJ)* 18: 39–42.

Yan, J. 2009. Research on the causes of social security (criminal) sase based on temporal and spatial data mining method. Tsinghua University.

Zhang, Y.H. 2007. Criminology. China Renmin University Press.

Zhang, Z.G. & Cai, Z.J. 1993. A Winter's smoothing forecasting method by forecasting seasonal factor. *Forecasting* 6: 59–62.

Information Technology and Computer Application Engineering – Liu, Sung & Yao (Eds)
© 2014 Taylor & Francis Group, London, ISBN 978-1-138-00079-7

Mixed duality of multi-objective programming

Hong Mei Miao
College of Physis and Electronic Information, Yanan University, Yanan Shaanxi, China

Xiang You Li
College of Mathematics and Computer Science, Yanan University, Yanan Shaanxi, China

ABSTRACT: In this paper, a class of multi-objective programming is considered, in which related functions are $B - (p, r, a)$-invex functions, mixed dual problem is researched, many duality theorems are proved under weaker convexity.

Keywords: $B - (p, r, a)$-invex function; multi-objective programming; mixed duality

1 INTRODUCTION

The convexity theory plays an important role in many aspects of mathematical programming. In recent years, in order to relax convexity assumption, various generalized convexity notions have been obtained. One of them is the concept of $B - (p, r)$ invexity defined by T. Antczak [1], which extended the class of B-invex functions with respect to η and b and the classes of (p, r)-invex functions with respect to η [2][3]. He proved some necessary and sufficient conditions for $B - (p, r)$-invexity and showed the relationships between the defined classes of $B - (p, r)$-invex functions and other classes of invex functions. Later Antczak defined a classes of generalized invex functions [4], that is $B - (p, r)$-pseudo-invex functions, strictly $B - (p, r)$ pseudo-invex functions, and $B - (p, r)$-quasi-invex functions, considered single objective mathematical programming problem involving $B - (p, r)$-pseudo-invex functions, $B - (p, r)$-quasi-invex functions and obtained some sufficient optimality conditions, Zengkun xu [5] considered multiobjective programming problems by a class of generalized (F, ρ) convex functions, studied mixed type dual, derived many weak, strong, strictly reverse dual conditions, Mohamed Hachimi, Brahim Aghezzaf [6] defined generalized (F, α, ρ, d)-type functions, considered multiobjective program by (F, α, ρ, d)-type functions, derived many dual conditions Xiangyou Li [7] discussed saddle-point conditions for multi-objective programming.

In this paper, we consider a class of multi-objective programming in which related functions are $B - (p, r, a)$-invex functions, $B - (p, r, a)$ quasi-invex functions and $B - (p, r, a)$-pseudo-invex functions, study mixed dual problem, obtain many duality theorems under weaker convexity.

2 DEFINITIONS

Throughout this paper, let R^n be the n-dimensional Euclidean space and R^n_+ be its non negative open subset, X *be* a nonempty open subset of R^n. For the benefit of the reader, we recall concepts of $B - (p, r, a)$-invex functions given in [5].

Definition 2.1 [5] Let $X \subset R^n$ is a nonempty open set, $u \in X$, the differentiable function $f : X \to R$ is said to be (strictly) $B - (p, r, a)$-invex function with respect to η and b at u if there exist functions $\eta : X \times X \to R^n$, $b : X \times X \to R_+, 0 \le b(.,.) \le 1, a : X \times X \to R$, for all $x \in X$, the inequalities

$$\frac{1}{r} b(x,u)(e^{r(f(x)-f(u))} - 1) \ge \frac{1}{p} \nabla f(u)(e^{p\eta(x,u)} - I) + a(x,u)$$
$(> ifx \ne u), for (p \ne 0, r \ne 0),$

$$\frac{1}{r} b(x,u)(e^{r(f(x)-f(u))} - 1) \ge \nabla f(u)\eta(x,u) + a(x,u)$$
$(> ifx \ne u), for (p \ne 0, r \ne 0),$

$$b(x,u)(f(x) - f(u)) \ge \frac{1}{p} \nabla f(u)(e^{p\eta(x,u)} - I) + a(x,u)$$
$(> ifx \ne u), for (p \ne 0, r \ne 0),$

$$b(x,u)(f(x) - f(u)) \ge \nabla f(u)\eta(x,u) + a(x,u)$$
$(> ifx \ne u), for (p = 0, r = 0),$

hold.

Function $f : X \to R$ is said to be $B - (p, r, a)$-invex function with respect to η and b on X if it is $B - (p, r, a)$-invex function with respect to the same η and b at each u on X.

Definition 2.2 [5] Let $X \subset R^n$ is a nonempty open set, $u \in X$, the differentiable function $f : X \to R$ is said to be $B-(p,r,a)$-quasi-invex function with respect to η and b at u if there exist functions

$\eta:X\times X\to R^n$, $b:X\times X\to R_+$, $0\le b(.,.)\le1$,

$a:X\times X\to R$, for all $x\in X$, the inequalities

$\dfrac{1}{r}b(x,u)(e^{r(f(x)-f(u))}-1)\le0\Rightarrow$

$\dfrac{1}{p}\nabla f(u)(e^{p\eta(x,u)}-I)+a(x,u)\le0$, $for(p\ne0,r\ne0)$,

$\dfrac{1}{r}b(x,u)(e^{r(f(x)-f(u))}-1)\le0\Rightarrow$

$\nabla f(u)\eta(x,u)+a(x,u)\le0$, $for(p=0,r\ne0)$,

$b(x,u)(f(x)-f(u))\le0\Rightarrow$

$\dfrac{1}{p}\nabla f(u)(e^{p\eta(x,u)}-I)+a(x,u)\le0$, $for(p\ne0,r=0)$,

$b(x,u)(f(x)-f(u))\le0\Rightarrow$

$\nabla f(u)\eta(x,u)+a(x,u)\le0$, $for(p=0,r=0)$,

hold.

Function $f:X\to R$ is said to be $B-(p,r,a)$-quasi-invex function with respect to η and b on X if it is $B-(p,r,a)$ quasi-invex function with respect to the same η and b at each u on X.

Definition 2.3 [5] Let $X\subset R^n$ is a nonempty open set, $u\in X$, the differentiable function $f:X\to R$ is said to be $B-(p,r,a)$-pseudo-invex function with respect to η and b at u if there exist functions

$\eta:X\times X\to R^n$, $b:X\times X\to R_+$, $0\le b(.,.)\le1$,

$a:X\times X\to R$, for all $x\in X$, the inequalities

$\dfrac{1}{p}\nabla f(u)(e^{p\eta(x,u)}-I)+a(x,u)\ge0\Rightarrow$

$\dfrac{1}{r}b(x,u)(e^{r(f(x)-f(u))}-1)\ge0$, $for(p\ne0,r\ne0)$,

$\nabla f(u)\eta(x,u)+a(x,u)\ge0\Rightarrow$

$\dfrac{1}{r}b(x,u)(e^{r(f(x)-f(u))}-1)\ge0$, $for(p=0,r\ne0)$,

$\dfrac{1}{p}\nabla f(u)(e^{p\eta(x,u)}-I)+a(x,u)\ge0\Rightarrow$

$b(x,u)(f(x)-f(u))\ge0$, $for(p\ne0,r=0)$

$\nabla f(u)\eta(x,u)+a(x,u)\ge0\Rightarrow$

$b(x,u)(f(x)-f(u))\ge0$, $for(p=0,r=0)$,

hold.

Function $f:X\to R$ is said to be $B-(p,r,a)$ pseudo- invex function with respect to η and b on X if it is $B-(p,r,a)$ pseudo-invex function with respect to the same η and b at each u on X.

Definition 2.4 [5] Let $X\subset R^n$ is a nonempty open set, $u\in X$, the differentiable function $f:X\to R$ is

said to be strictly $B-(p,r,a)$-pseudo-invex function with respect to η and b at u if there exist functions $\eta:X\times X\to R^n$, $b:X\times X\to R_+$, $0\le b(.,.)\le1$, $a:X\times X\to R$, for all $x\in X$, the inequalities

$\dfrac{1}{p}\nabla f(u)(e^{p\eta(x,u)}-I)+a(x,u)\ge0\Rightarrow$

$\dfrac{1}{r}b(x,u)(e^{r(f(x)-f(u))}-1)>0$, $for(p\ne0,r\ne0)$,

$\nabla f(u)\eta(x,u)+a(x,u)\ge0\Rightarrow$

$\dfrac{1}{r}b(x,u)(e^{r(f(x)-f(u))}-1)>0$, $for(p=0,r\ne0)$,

$\dfrac{1}{p}\nabla f(u)(e^{p\eta(x,u)}-I)+a(x,u)\ge0\Rightarrow$

$b(x,u)(f(x)-f(u))>0$, $for(p\ne0,r=0)$

$\nabla f(u)\eta(x,u)+a(x,u)\ge0\Rightarrow$

$b(x,u)(f(x)-f(u))>0$, $for(p=0,r=0)$,

hold.

Function $f:X\to R$ is said to be strictly $B-(p,r,a)$ pseudo-invex function with respect to η and b on X if it is strictly $B-(p,r,a)$-pseudo-invex function with respect to the same η and b at each u on X.

In above section, $I=(1,\ldots,1)\in R^n$, $e^{(a,\ldots,a_n)}=(e^{a_1},\ldots,e^{a_n})\in R^n$.

When $a(x,u)\ge0$, $B-(p,r,a)$-invex is $B-(p,r)$-invex, but if $a(x,u)<0$, $B-(p,r,a)$ invex may not be $B-(p,r)$-invex. Therefore, adding a parameter $a(x,u)$ means that the $B-(p,r)$ invexity may be lost.

In following section, $B-(p,r,a)$-invex functions, $B-(p,r,a)$-pseudo-invex functions and $B-(p,r,a)$-quasi-invex functions are discussed only when $p\ne0,r\ne0$, (other cases will be deal with likewise because the only changes arise from the form of inequality). We shall assume that $r>0$ (in the case when $r<0$, the direction of some of the inequalities in the proofs of theorems should be changed to the opposite one).

3 MIXED TYPE DUALITY

Let $M=\{1,\ldots,m\}$, $J_1\subseteq M$, $J_2=M/J_1$, let e be the vector of R^k whose components are all ones, We consider below vector programming

$(VP)\min\ f(x)=(f_1(x),\cdots,f_k(x))$

$s.t.\ g(x)=(g_1(x),\cdots,g_m(x))\le0,$

$\qquad x\in X\subset R^n.$

where $f_i(x):X\to R,i=1,\ldots,k,g_j(x):X\to R$, $j\in M$ are differentiable, its mixed dual grog-ramming is defined as below

$$(VD) \max \ f(y) + \sum_{j \in J_1} \mu_j g_j(y) e$$

$$\sum_{i=1}^{k} \lambda_i \nabla f_i(y) + \sum_{j=1}^{m} \mu_j \nabla g_j(y) = 0; \tag{1}$$

$$\sum_{j \in J_2} \mu_j g_j(y) \geq 0; \tag{2}$$

$$\lambda^T e = 1 \ , \lambda = (\lambda_1, \cdots, \lambda_k)^T \geq 0, \ \mu = (\mu_1, \cdots, \mu_m)^T \geq 0.$$

Note that we get a Mond-Weir duality when J_1 is empty set and a Wolfe duaity when J_2 is empty set (VD), respectively. Now we consider the mixed dual programming (VD) in which related functions are $B - (p, r, a)$-invex functions, $B - (p, r, a)$-quasi-invex functions, $B - (p, r, a)$-pseudo-invex functions.

Theorem 3.1 (Weak duality). Suppose that x is a feasible solution of (VP), (λ, μ, y) is a feasible solution of (VD), $\sum_{i=1}^{k} \lambda_i f_i + \sum_{j \in J_1} \mu_j g_j$ is $B - (p, r, a)$-invex function with respect to η and b_0 at y, $\sum_{j \in J_2} \mu_j g_j$ is $B - (p, r, c)$ quasi-invex function with respect to η and b_1 at y, $b_0(x, y) > 0$ when $x \neq y$, $a(x, y) + c(x, y) \geq 0$, then $f(x) \leqq f(y) + \sum_{j \in J_1} \mu_j g_j(y) e$ not hold.

Proof. Since $g_j(x) \leq 0, \mu_j \geq 0$, so $\sum_{j \in J_2} \mu_j g_j(x) \leq 0$, consider (2), we can get

$$\sum_{j \in J_2} \mu_j g_j(x) - \sum_{j \in J_2} \mu_j g_j(y) \leq 0, \text{ obviously}$$

$$\frac{1}{r} b_1(x, y)(e^{r(\sum_{j \in J_2} \mu_j g_j(x) - \sum_{j \in J_2} \mu_j g_j(y))} - 1) \leq 0.$$

Using $\sum_{j \in J_2} \mu_j g_j$ is $B - (p, r, c)$ quasi-invex functions. With respect to η and b_1 at y, we have

$$\frac{1}{p} \sum_{j \in J_2} \mu_j \nabla g_j(y)(e^{p\eta(x,u)} - I) + c(x,u) \leq 0, \tag{3}$$

Relating (1), (3) along with $a(x, y) + c(x, y) \geq 0$, we can get

$$\frac{1}{p} (\sum_{i=1}^{k} \lambda_i \nabla f_i(y) + \sum_{j \in J_1} \mu_j \nabla g_j(y))(e^{p\eta(x,u)} - I) +$$
$$a(x, u) \geq 0,$$

Since $\sum_{i=1}^{k} \lambda_i f_i + \sum_{j \in J_1} \mu_j g_j$ is $B - (p, r, a)$-invex function with respect to η and b_0 at y, we can get

$$\frac{1}{r} b_0(x, y)(e^{r[\sum_{i=1}^{k} \lambda_i f_i(x) + \sum_{j \in J_1} \mu_j g_j(x) - \sum_{i=1}^{k} \lambda_i f_i(y) - \sum_{j \in J_1} \mu_j g_j(y)]} - 1) \geq 0.$$

by $b_0(x, y) > 0$, we get

$$\sum_{i=1}^{k} \lambda_i f_i(x) + \sum_{j \in J_1} \mu_j g_j(x) - \sum_{i=1}^{k} \lambda_i f_i(y) - \sum_{j \in J_1} \mu_j g_j(y) \geq 0.$$

Consider $\sum_{j \in J_1} \mu_j g_j(x) \leq 0$, so we can get $f(x) \leqq f(y) + \sum_{j \in J_1} \mu_j g_j(y) e$ not hold.

Theorem 3.2 (Weak duality). Suppose that x is a feasible solution of (VP), (λ, μ, y) is a feasible solution of (VD), $\sum_{i=1}^{k} \lambda_i f_i + \sum_{j \in J_1} \mu_j g_j$ is $B - (p, r, a)$-quasi-invex function with respect to η and b_0 at y, $\sum_{j \in J_2} \mu_j g_j$ is $B - (p, r, c)$ strictly pseudo-invex functions with respect to η and b_1 at y, $b_1(x, y) > 0$ when $x \neq y, a(x, y) + c(x, y) \geq 0$, then $f(x) \leqq f(y) + \sum_{j \in J_1} \mu_j g_j(y) e$ not hold.

Proof. Suppose $f(x) \leqq f(y) + \sum_{j \in J_1} \mu_j g_j(y) e$, then there exists $\lambda \in R_+^k$, such that $\sum_{i=1}^{k} \lambda_i f_i(x) \leq \sum_{i=1}^{k} \lambda_i f_i(y) + \sum_{j \in J_1} \mu_j g_j(y)$, consider $\sum_{j \in J_1} \mu_j g_j(x) \leq 0$, so $\sum_{i=1}^{k} \lambda_i f_i(x) + \sum_{j \in J_1} \mu_j g_j(x) \leq \sum_{i=1}^{k} \lambda_i f_i(y) + \sum_{j \in J_2} \mu_j g_j(y)$. Easily get

$$\frac{1}{r} b_0(x, y)(e^{r[(\sum_{i=1}^{k} \lambda_i(f_i(x) + \sum_{j \in J_1} \mu_j g_j(x)) - (\sum_{i=1}^{k} \lambda_i(f_i(y) + \sum_{j \in J_1} \mu_j g_j(y)))]} - 1) \leq 0 \tag{4}$$

Using $\sum_{i=1}^{k} \lambda_i f_i + \sum_{j \in J_1} \mu_j g_j$ is $B - (p, r, a)$-invex functions with respect to η and b_0 at y, we have

$$\frac{1}{p} (\sum_{i=1}^{k} \lambda_i \nabla f_i(y) + \sum_{j \in J_1} \mu_j \nabla g_j(y))(e^{p\eta(x,y)} - I) + a(x, y) \leq 0.$$

Relating (1) along with $a(x, y) + c(x, y) \geq 0$, we can get

$$\frac{1}{p} \sum_{j \in J_2} \mu_j \nabla g_j(y)(e^{p\eta(x,u)} - I) + c(x, u) \geq 0,$$

also $\sum_{j \in J_2} \mu_j g_j$ is strictly $B - (p, r, c)$ pseudo-invex function with respect to η and b_1 at y, then have

$$\frac{1}{r} b_1(x, y)(e^{r(\sum_{j \in J_2} \mu_j g_j(x) - \sum_{j \in J_2} \mu_j g_j(y))} - 1) > 0. \text{So}$$

$$\sum_{j \in J_2} \mu_j g_j(x) - \sum_{j \in J_2} \mu_j g_j(y) > 0, \text{ it is a contradiction,}$$

so $f(x) \leqq f(y) + \sum_{j \in J_1} \mu_j g_j(y) e$ not hold.

Theorem 3.3 (Strong Duality). Suppose that x^0 is an efficient solution of (VP) and satisfies the Kuhn-Tucker constraint qualification for the constraints of (VP). Then there exist λ^0, μ^0 such that (λ^0, μ^0, x^0) is a feasible solution of (VD) and $\sum_{j \in J_1} \mu_j g_j(x^0) = 0$, further if weak duality holds between (VP) and (VD), then (λ^0, μ^0, x^0) is efficient for (VD).

Proof is similar to the proof of [5, Theorem 2.4].

Theorem 3.4 (strictly converse duality). Suppose that x is a feasible solution of (VP), (λ, μ, y) is a feasible solution of (VD), $\sum_{i=1}^{k} \lambda_i f_i(x) + \sum_{j \in J_1} \mu_j g_j(x) \leq \sum_{i=1}^{k} \lambda_i f_i(y) + \sum_{j \in J_1} \mu_j g_j(y)$ ($<$ if $x \neq y$), conditions of theorems 1 hold.

Then $x = y$ and y is an efficient solution of (VD).

Proof. From theorems 1, we can get (4) holded. When $x \neq y$, it's a contradiction with (ii), so $x = y$.

If x isn't an efficient solution of (VP), there exists a feasible solution of (VP) $u, u \neq x$, $\lambda \in R_+^k$ such that $\sum_{i=1}^{k} \lambda_i f_i(u) < \sum_{i=1}^{k} \lambda_i(x) = \sum_{i=1}^{k} \lambda_i f_i(y)$, by $\sum_{j=1}^{m} \mu_j g_j(u) \leq 0$ and $\sum_{j=1}^{m} \mu_j g_j(y) \geq 0$, so $[\sum_{i=1}^{k} \lambda_i f_i(u) + \sum_{j=1}^{m} \mu_j g_j(u)] - [\sum_{i=1}^{k} \lambda_i f_i(y) + \sum_{j=1}^{m} \mu_j g_j(y)] < 0$, it's a contradiction with (4), so x is an efficient solution of (VP).

4 CONCLUSION

In this paper, we introduce new classes of generalized invex functions, establish mixed dual problem of multi-objective programming, in which corresponding functions belong to the introduced classes of functions, obtain many duality conditions under weaker convexity, extend many results of [4].

Finally, duality problems of minimax fractional programming involving the introduced functions should be considered, Wolfe dual problem also should be considered in the future.

ACKNOWLEDGEMENT

This research was supported by special fund of Shaanxi Provincial high-level university building (2012sxts07), China.

REFERENCES

[1] T. Antczak, A class of B-(p,r)-invex functions and mathematical programming. J. Math. Anal. Appl. Vol. 286, pp. 187–206, 2003.

[2] C.R. Bector & C. Singh, B-vex functions. J. Optim. Theory. Appl. Vol. 71, pp. 237–253, 1991.

[3] T. Antczak, (P,r)-invex sets and functions. J. Math. Anal. Appl. Vol. 263, pp. 355–379, 2001.

[4] T. Antczak, Generalized B-(p,r)-invexity functions and nonlinear mathematical programming. Numerical functional Analysis and Optimazation. Vol. 30, pp. 1–22, 2009.

[5] Zengkun Xu, Mixed type duality in multiobjective programming problems, Journal of Math. Anal. Appl, Vol. 198(4), pp. 621–635, 1996.

[6] Mohamed Hachimi, Brahim Aghezzaf, Suffciency and duality in differentiable multi-objective programming involving generalized type I functions, Journal of Math. Anal. Appl., Vol. 296, pp. 382–392, 2004.

[7] XiangYou Li, Saddle point condition for fractional programming. Preedings of the 2012 eighth international conference on computational intelligence and security. pp. 82–85, Guang zhou. china Nov, 2012.

An efficient human face tracking method based on Camshift algorithm

T. Li* & T. Zhang
Information School, Yunnan University, Kunming, Yunnan, China

W.D. Chen & X.H. Zhang
The Transmission Cell of fourth branch of China International Telecommunication Construction
Group Design Institute Co. Ltd, Zhengzhou, Henan, China

ABSTRACT: Face tracking is a key technology in the field of face information processing, and has been widely used in such field like security and video conference, etc. But it will cause tracking error under the condition that a large class area appeared in the scene. In order to improve accuracy of face tracing in an complex environment, this paper summarizes and analyzes the current typical algorithm of face detection, the algorithms which added the typical facial feature are short in tracking region. Experiments has been tested the algorithm discussed in this paper by using improved Camshift algorithm and programming tools, and results prove it possesses higher accuracy compared to original methods.

Keywords: Face tracking; Camshift; Eyes feature; Color probability distribution image

1 INTRODUCTION

With the development of computer technology, human-computer interaction techniques have risen to unprecedented heights and new methods about it are emerging in an endless stream [1]. At present, facial tracking techniques, as a very significant value in application, is gaining attention from more and more researchers [2].

Facial tracking techniques is image processing course, ascertaining face's location and then making out it from a continuous sequence of video, but there is currently no rapid and accurate tracing method. It is difficult to trace human face in real-time video for facial dynamic change, light effect and drift caused by skin color difference.

There are many methods to trace moving objects for their high operation efficiency. Today's leading algorithm is based on Meanshift with improved Camshift which translates images into probability distribution graphs using the objective color histogram model and emerges dramatic results towards simple background setting for its less calculation so that can realize real-time tracking [3], nevertheless, it does not suitable for solving large areas of skin-based interference problems especially when the intervals between hands and face is too close which would significantly reduce the tracking accuracy. In recent years, a mass of improved algorithms are adding facial feature, such as eyes and

mouth, to original based Camshift which can solve skin-based interference at some level but bring new problems like less position area and unable to locate whole face, etc. Based on the research of improved Camshift algorithm of people of the past, the object of this paper is issue against the incomplete location problem at present and propose modified methods which use the face shape to separate the face region from other skin-color alike region to improve the problem efficiently. Under laboratory conditions, the tracking accuracy and efficiency of the improved algorithms are greatly increased.

2 GENERAL IMPROVEMENT IN CAMSHIFT ALGORITHM

2.1 *Camshift algorithm principle*

Camshift algorithm, short for Continuously Adaptive Mean Shift algorithm, is a motion-tracking algorithm based on color [4]. For the RGB color space is sensitive to the change of brightness and light, in order to reduce the effects of which on tracking performance, Camshift algorithm collect the image from RGB color space to transformed HSV color space for further processing [5]. The algorithm flow chart is shown in Figure 1.

Firstly, select the initial search window so that the window can include the whole tracking object exactly, then by sampling every pixel of H channel in the window, to get the objective hue histogram and save which as the color histogram model. Through search

*Address correspondence to: Tao Li, Information School, Yunnan University, Kunming, Yunnan Province, China

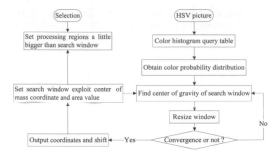

Figure 1. Flow chat of original Camshift algorithm.

Figure 2. The design sketch of typical feature algorithm added only.

in the color histogram model of every pixel in video image processing regions from tracking to get the probability of target pixel, while the probability of other regions which outside the video image processing regions is defined as 0. The video converted to target color probability distribution, also known as target color projection drawing, by the above procedure. For displayed purposes, we convert projection drawing to 8-bit gray projection drawing, if probability of pixel is 1, set its value to 255, else probability is 0 set to 0 and convert other pixels to corresponding gray value. Assuming (x,y) is pixel position of the search window and I(x,y) is the pixel value of (x,y) in the projection drawing. Define the zeroth moment M00 and first moment M01, M10 in the search window are as follows:

$$M_{00} = \sum_x \sum_y I(x, y) \tag{1}$$

$$M_{01} = \sum_x \sum_y y I(x, y) \tag{2}$$

$$M_{10} = \sum_x \sum_y x I(x, y) \tag{3}$$

Then resize the search window according to M_{00} and move the center of it to centroid, if the distance greater than presupposed fixed value, recalculate the adjusted location of the center of mass in the search window and resize the position and size of the window in next round until the distance between the center of window and centroid less than presupposed fixed value or the computational times reaching the maximum at a certain stage which can defined as under the convergent conditions then enter into the next frame of image to go on searching new targets. In new images, the final location of the center of mass in the search window and zeroth moment M_{00} of the previous frame image are used to set new position and size of the search window. Camshift tracing algorithm resizes search window adaptively in image of intra-frame and inter-frame according to previous M_{00} so that can suitably trace object dynamic deformation conditions.

2.2 *General added facial feature Camshift algorithm*

Camshift algorithm has the advantage of fast speed, little influenced by noise and can realize functions of trace but its fault is it is not suitable for solving large areas of skin-based interference problems or it is not quite right for tackling rapid movement. The existing methods for improving tracking accuracy are according to facial typical features, such as eyes and mouth, to estimate tracing regions whether is the face regions when faced greater skin-color alike region interferences [7]. Here's the general process:

(1) Use the high clustering performance of YIQ color space to define lips
(2) Use the gray level strong deviation between eyeballs and the white area around them to set threshold through which to estimate whether they are eyes.
(3) Use the characteristic triangle constituted by eyes and lips to estimate whether it is face, if not, reinitialize the search window, and the center of gravity of the characteristic triangle is the position of Camshift algorithm initialized window.

The detected design sketch according to this algorithm:

The typical feature algorithm added method can relatively accurately locates the areas between eyes and mouth but the question is that located facial areas are imperfect. Based on previous researches made by others, in order to solve this problem, this paper proposes a new tracing and locating face method based on shape features and with more positive results under laboratory conditions.

3 THE PROPOSED ALGORITHM

The improved algorithm is a human face detection method combined the feature based on the typical characteristics, and the positioning method has been improved, the results show that the method has achieved good distinction.

The method ASM (Active Shape models) and AAM (Active Appearance models) proposed by T. F. Cootes and C. J. Taylor. etc. can extract the position and shape of the main characteristics in the human face. Wherein, the method ASM is better than method AAM in the precision and accuracy of the positioning, ASM has two models, shape model and gray model, according to the representative characteristics of the model points, the method ASM search for the best position of the feature point in the target image using the gray model based on the shape model, at the same time, adjust the parameters of the method constantly according to the search results, change the shape of the model and finally, match the model to the contour of the object.

3.1 Build shape model

Before building the model, first manually detect the training set and choose n feature points from every face picture to indicate facial PDM. Each sample x_i in training set gives a $2n$ dimensional vector:

$$x_i = (x_{i1}, y_{i1}, x_{i2}, y_{i2}, \cdots, x_{in}, y_{in})^T, i = 1, 2, \cdots N \qquad (4)$$

where (x_{ij}, y_{ij}) is j-th feature point coordinate in i-th training set, n is shape model spots, N is training picture numbers.

To obtain training set model point coordinate statistic features then zoom, convert and translate them in order to reduce error in every difformity, namely, to align the training sample.

After alignment, training sample characterization facial shape are distribute in $2n$ dimensional space then analyse them by PCA and every facial shape can express below:

$$x = \bar{x} + pb \qquad (5)$$

where $\mathbf{p} = (p_1, p_2, \ldots, p_k)$ is a matrix obtained by PCA made of former kth feature vector, $\mathbf{b} = (b_1, b_2, \ldots, b_k)^T$ is arbitrary face projection in the feature space. We can get new facial shape through change project-vector b. Generally, whether we get a new shape or a facial shape through limiting variation of b. The limit formula of b is as follows:

$$D_m^2 = \sum_{h=1}^{k} (\frac{b_h^2}{\lambda_h}) \leq D_{max}^2 \qquad (6)$$

so far, we have built the shape model.

3.2 Feature point matching

Extract the gray feature of given feature point from images of training set first and give the shape initial value.

$$X_{init} = M(s_{init}, \theta_{init})[x] + X_{cinit} \qquad (7)$$

where s_{init}, θ_{init} is shape initial scale and X_{cinit} is direction and excursion parameter, these parameters can be estimated according to test results and x is mean shape.

After given facial shape initial value, we can ge the exact location of every facial feature point by using of iteration algorithm method which shown below :

Step 1: search every feature point neighbourhood in image and get the positions whose gray model are closest matching to these we find to obtain a new shape.

Step 2: According to the displacement of feature point to calculate the change of Pose Parameters and Shape Parameters in PDM model.

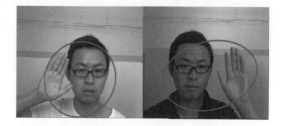

Figure 3. The design sketch of original camshift.

Step 3: Update Pose Parameters and Shape Parameters and go back to step 1 until convergence or reach certain iterations.

Feature point matching process is a process that constantly adjust the Pose Parameters and Shape Parameters actually, Pose Parameters is $s, \theta, \mathbf{X}_c$ and Shape Parameters is b.

3.3 Face area location

According to the results by rough location, it's easy to select the partial region of face by methods of gray-direction integral projection which, namely, statistical image information in a certain direction, what's more, we can get appropriate place image feature through analyze the extreme point and valley point, etc., in projection function. Generally, projection method obtains the general information distribution and has strong anti-interference ability of white noise but not enough, the projection results need further accurate calculation. Projection-direction is arbitrary and can be X-direction, Y-direction or random angle α in X-Y coordinates. These projection are all parallel, to X-direction and Y-direction, the Integral projection function are as follows:

$$H(y) = \int_{x_1}^{x_2} I(x, y) dx \qquad (8)$$

$$V(x) = \int_{y_1}^{y_2} I(x, y) dy \qquad (9)$$

Because facial image quality are varied in this paper. Multiple projection direction method is accepted to roughly locate the facial image features in order to search important facial parts position and reduce image quality dependence then finally obtain the facial areas.

4 SIMULATION RESULTS AND DISCUSSION

This algorithm is implemented in a simulation testbed based on vs2010 and opencv2.3.1, CPU is 2G, ordinary USB camera and analyzes the Camshift algorithm before and after the improvement. simulation result is shown in Figs 3–5.

Experiments show that, after a few calculations of iteration, original CamShift algorithm ascertains the

Figure 4. The design sketch of typical feature algorithm added only.

Figure 5. The design sketch of the improved Camshift algorithm.

final centroid of probability distribution graph in non-face skin-color area, thus tracking error. The search window which eyes and mouth feature algorithm added can lock key area but with incomplete tracking region. The proposed algorithm greatly improves the face tracking accuracy and avoid problems of large areas of skin-based interference or incomplete tracking region for only added typical facial feature algorithm efficiently.

5 CONCLUSION

This paper proposes a CamShift face tracing algorithm combined with facial shape and typical feature successfully matching color probability distribution to solve tracing error in large skin-color area interference and incomplete tracking region problems. In order to reduce tracing error, at the same time, ensure the real-time ability, the template matching only operate in the upper half portion of search window by simple scale of the model or turn in angle for matching. Compared with general directly matching in skin-color area algorithm, the one proposed in this paper greatly improves density gradient CamShift, and compared with the original CamShift it has higher accurancy.

In this paper, we evaluated the performance of the proposed methods using facial shape and typical feature of a tracking system and get better results; further studies should focus on face tracking in continuous video.

REFERENCES

[1] Wright P.C., Fields R.E., Harrison M.D., 2000, Analyzing human-computer interaction as distributed cognition: the resources model [J]. Human-Computer Interaction, 15(1): 1–41.

[2] Shen L., Bai L., 2006, A review on Gabor wavelets for face recognition [J]. Pattern analysis and applications, 9(2–3): 273–292.

[3] Bradski G.R., 1998, Computer vision face tracking for use in a perceptual user interface [J].

[4] Stern, H., & Efros, B. (2002, May). Adaptive color space switching for face tracking in multi-colored lighting environments. In Automatic Face and Gesture Recognition, 2002. Proceedings. Fifth IEEE International Conference on (pp. 249–254). IEEE.

[5] Boyle, M. (2001). The effects of capture conditions on the CAMSHIFT face tracker. Alberta, Canada: Department of Computer Science, University of Calgary.

[6] Li, J. H., Fu Y. S., Gao Q. H., 2008, Face tracking method based on improved CAMSHIFT algorithm [J]. Chinese journal of scientific instrument, 29(4): 55–58.

Information Technology and Computer Application Engineering – Liu, Sung & Yao (Eds)
© 2014 Taylor & Francis Group, London, ISBN 978-1-138-00079-7

The analysis of anti-noise performance of BOC modulation system

Y.F. Jia & X.D. Song
School of Aerospace Engineering, Beijing Institute of Technology, Beijing, China

ABSTRACT: BOC modulation adds a binary subcarrier on the basis of the original BPSK modulation. This article discussed the advantages of BOC modulation with respect to the traditional BPSK modulation method. Given the anti-noise performance simulation results of both of the two modulation methods, we can see that the new BOC modulation system shows a better anti-noise capacity and it has an important practical significance in compatibility and interoperability of modern GNSS navigation system.

Keywords: BOC modulation; direct spread spectrum modulation; anti-noise performance

1 RESEARCH BACKGROUND

Satellite navigation system plays a very important role in national economy, providing accurate real time position, velocity and time information for an unlimited number of users in land, sea and air on a global scale. Currently, the satellite navigation system consists of only the U.S. GPS and Russian GLONASS, while the Europe's Galileo satellite navigation will be formally put into use soon.

BOC-Binary Offset Carrier, is proposed in the process of designing the Galileo system, a new carrier modulation. In order to improve its performance due to the relatively weak GPS satellite navigation signals, the BOC modulation method is carried out, because it is able to achieve band sharing through signal spectral separation, reducing the mutual interference between the signals thus achieving a better noise immunity, and therefore BOC is very much significant to the development of satellite navigation. And this article is to verify the statements above.

2 INTRODUCTION OF DS SYSTEM AND BOC SYSTEM

2.1 The principle of generation of the spread spectrum modulation system

As is shown in Fig 2-1, direct sequence spread spectrum use a broadband spread spectrum signal on a data carrier to achieve direct modulation and bandwidth extended. The value of pseudo-code signal c(t) is ± 1, multiply the baseband signal with the pseudo-code signal in time domain, the result of which is equivalent to convolution in frequency domain, and the frequency spectrum of the signal is thus widened.

In the receiving end, the received signal along with the local oscillation signal is sent into the mixer, the output is zero-IF signal. The local pseudo-code generator generates a phase completely consistent with

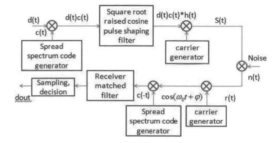

Figure 2-1. DS Spread Spectrum system block diagram.

the local pseudo-code in the transmission end. Multiply it with the output signal of the mixer in the time domain is equivalent to dispreading. For the binary signal c(t) whose value is ± 1, $c^2 = 1$, and therefore no pseudo-code component is contained after the associated dispreading, whereas for the noise added into the channel, it has only conducted the first spreading operation, so that the spectrum greatly broadened. The dispread IF signal bandwidth is relatively narrow, the interference and noise power of the filter is greatly reduced.

2.2 What BOC modulation is

Binary offset carrier (BOC) modulation technology can be seen as an improvement on the basis of traditional spread spectrum modulation. Its basic principle is adding a binary subcarrier (mainly constituted by a sine or cosine symbol function, i.e. the shape assembles sign (sin(t)) or sign (cos(t)) on the original BPSK modulation. The biggest characteristic of this form of modulation is that the main lobe of its power spectrum is split into two symmetrical parts, and according to the different selected parameter type, the distance between the two split main lobes may also be varied. Commonly expressed as BOC (m, n), wherein m represents the sub-carrier frequency, n stands for the rate of the spreading codes, the specific values are respectively m times and n times of 1.023 MHz.

The BOC modulation includes the sinusoidal modulation SinBOC, cosine-modulated CosBOC, composite modulation CBOC, the time division TMBOC and the composite carrier AltBOC. Now we are talking about SinBOC, which is similar with CosBOC.

SinBOC is a sine-BOC modulation, defined as is shown in the formula 2-1:

$$x(t) = c(t) \times sign(\sin(2\pi f_s t)) \qquad (2\text{-}1)$$

In formula 2-1, $c(t) = \sum_k c_k h(t - kT_c)$, h (t) is a non-return-to-zero waveform of which the value is 1 in $[0, T_c]$ and 0 in other cases, c_k is the k-th chip; c(t) the code sequence waveform; f_s the sub-carrier frequency; and sign () the sign function. The CosBOC modulation is the same with that of the SinBOC, only to change the $sign(\sin(2\pi f_s t))$ in the above SinBOC modulated into $sign(\cos(2\pi f_s t))$.

2.3 The generation of BOC signal

In Figure 2-2, BOC modulation signal is generated by the frequency control module, the information code generator, the spread code (PN) generator and modulation module:

(1) The frequency control module is used for controlling the other three modules, in order to achieve a variety of BOC modulated forms;
(2) The code generator is used for generating a user-defined code, which allows the user to customize the part;
(3) The spread code generator is to generate a pseudo-random code used for spreading spectrum, the pseudo-random code cycle should be long enough to adapt to the needs of the satellite navigation, and we usually select P code in GPS system;
(4) The modulating module is for completing the "spreading code modulation code" and the "spread code for square wave modulation". In figure 2-1, the modulation symbol is a multiply unit, but for BOC modulation, the modulation symbols can be replaced by a simple XOR logic instead.

Standard BOC modulation is a rectangular sub-carrier modulation. The signal is multiplied by a rectangular subcarrier, the spectrum is divided into two

parts, the expression of the offset carrier modulation signal ships is on the left and the right sides of the carrier frequency.

$$q_{n_1 T_s} s(t) = e^{-i\theta} \sum_k (-1)^k a_k q_{n_1 T_s}(t - kn_1 T_s - t_0) c_{T_s}(t - t_0)$$

$$\qquad (2\text{-}2)$$

$$= \sum_{M=0}^{n_1 - 1} (-1)^m u_{n_1 T_s}(t - mT_s)$$

In this formula, $\{a_k\}$ stands for the values of a spreading code after data modulation, $c_{T_s}(t)$ the sub-carrier, of which the cycle is $2T_s$, $u_{n_1 T_s}(t)$ the spreading symbol, time of the rectangular pulse is $n\pi$. n_1 is the number of half cycles of a subcarrier within one spread code, the values of θ and t_0 represents any offset of phase and time. When the spreading symbol is a rectangular wave, the offset carrier signal is a traditional PSK-R signal modulated by a periodic signal.

The BOC modulation is a special case of offset carrier modulation, known as BOC modulated signal or simply the BOC signal. For BOC signals, the value of $\pm 1/[2/(f_c \times 1.023 \times 10^6)(4f_s - f_c)]$ can only be $+1$ or -1, the duration of the product of the spread code and the spread symbols is nT_s. $c_T(t)$ is a rectangular pulse of which the amplitude is $+1$ or -1. For BOC modulated signals, we order

$$q_{n_1 T_s} = u_{n_1 T_s}(t - kn_1 T_s - t_0) c_{T_s}(t - t_0) = \sum_{M=0}^{n_1 - 1} (-1)^m u_{n_1 T_s}(t - mT_s) \,(2\text{-}3)$$

Formula (2–3) can be divided into two categories for whether n is even or odd.

When n is an even number:

$$s(t) = e^{-i\theta} \sum_k a_k q_{n T_s}(t - knT_s - t_0) \qquad (2\text{-}4)$$

When n is odd:

$$s(t) = e^{-i\theta} \sum_k (-1)^k a_k q_{n T_s}(t - knT_s - t_0) \qquad (2\text{-}5)$$

Because $q_{n T_s}(t)$ is a square wave of which the reference time is at time zero, and the duration is half cycle of the signal, if n is even, the mean value of $q_{n T_s}(t)$ is zero (no DC component); if n is odd, the value is not zero (DC component).

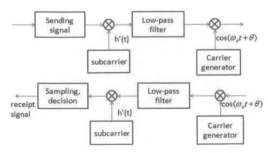

Figure 2-3. BOC modulation MATLAB simulation of flow diagram.

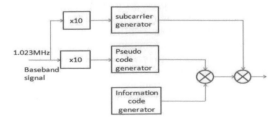

Figure 2-2. BOC signal generating block diagram.

The following is a flow diagram for BOC modulation in MATLAB simulation. There is a subcarrier modulation before spreading the spectrum signal and a subcarrier demodulation before de-spreading it:

3 ANALYSIS OF ANTI-NOISE PERFORMANCE

3.1 Noise modeling

Study on the degree of noise suppression of different signal modulating methods using MATLAB is to link the signal-to-noise ratio up with the bit error rate. First of all, we have to find out the conversion relationship of the two. To set up a discrete modeling of the above communication system, the key is the choice of the sampling rate. In order to satisfy the Nyquist sampling theorem, the sampling rate must be more than twice higher than the rate of highest frequency component of the signal:

$$f_s \geq 2[\frac{1+\alpha}{2}R_b]$$ (3-1)

So for the time being, let $f_{smp} = M \bullet R_b$ and $M > 2$. Signal sequence after discretization can be expressed as:

$$s(n) = \sum_{k=0}^{\infty} a_k g(n - kM)$$ (3-2)

The analog form signals:

$$s(t) = \sum_{k=0}^{\infty} a_k g(t - kT_s)$$ (3-3)

The contact of the two:

$$\frac{f_{smp}}{R_b} = \frac{T_s}{T_{smp}} = M$$ (3-4)

In order to study the anti-noise performance of the direct spread spectrum system and BOC system, we need to build up a relationship between the signal-to-noise ratio and the bit error rate, thus changing the values of the signal-to-noise ratio in order to get the bit error rate under different signal-to-noise ratios. So now we consider the noise link:

Construct the input signal of the receiver under the simulation environment:

$$r(n) = s(n) + x(n)$$ (3-5)

In the above formula, $x(n)$ is a Gaussian noise sequence. In MATLAB, we can control the signal and noise power, or that we can control the signal-to-noise ratio SNR (signal and noise ratio of the average power).

The form of matched filter in white Gaussian noise:

$$h(t) = g(T - t)$$ (3-6)

$$H(f) = G^*(f)e^{-j2\pi fT}$$ (3-7)

The signal-to-noise ratio of the filter's output end at moment $t = T$:

$$SNR|_{t=T} = \frac{g_h^2(T)}{E[x_h^2(T)]} = \frac{2E_g}{N_0}$$ (3-8)

All conclusions of digital communication system (any modulation mode, any signal form) under the Gaussian white noise conditions can be applied to the band-limited Gaussian noise without any change. As long as the spectral density of the band-limited Gaussian noise within the signal bandwidth is flat. If the spectral density of a band limited white Gaussian noise within the signal bandwidth is constant, then the influence of noise on a communication system is equivalent with that of the white Gaussian noise of the same spectral density. And for a band-limited Gaussian noise, if the power spectrum within the signal bandwidth B is flat, and the power spectral density of the inner band (bilateral) is $N_0/2$, then the matched filter adapted to this kind of noise is of the same form with the one matching with Gaussian white noise. In addition, for the matched filtering of limited Gaussian white noise, the signal-to-noise ratio of the optimal sampling point still meet

$$SNR|_{t=T} = \frac{g_h^2(T)}{E[x_h^2(T)]} = \frac{2E_g}{N_0}$$ (3-9)

The output of an ideal low-pass filter is the desired signal and a band-limited additive Gaussian noise. According to the matched filter principle of band-limited Gaussian noise, we know that the impact of this band-limited Gaussian noise on the system is exactly the same with that of the real white Gaussian noise. The ideal low-pass filter output is sampled to obtain a sequence of numbers. The desired signal and the band-limited Gaussian noise are both signals with limited power that can be sampled.

As mentioned before, what we can control in the simulation is the average power ratio SNR of the signal (SEQ) and noise (sequence):

$$SNR = \frac{E[s^2(n)]}{E[\hat{x}^2(n)]}$$ (3-10)

The average power of the desired signal s(t) in analog form: $S = R_b E_b$

The average power of the band-limited noise signal x(t) in analog form: $N = N_0 B$

For a band-limited zero-mean random process f(t), let f(n) be its sample sequence. If the sampling rate is greater than (or equal to) the Nyquist rate:

$$E[f^2(t)] = E[f^2(n)]$$ (3-11)

And then we will derive the relation between SNR and-EbN0:

$$SNR = \frac{E[s^2(n)]}{E[\hat{x}^2(n)]} = \frac{E_b R_b}{N_0 B} = 2\frac{E_b}{N_0}\frac{R_b}{f_{smp}} = 2\frac{kE_b}{N_0 M}$$ (3-12)

For the case of multi-band communication, it is assumed that each symbol contains k bits of information, so the similar derivation can be obtained:

$$[SNR]_{dB} = [2]_{dB} + [k]_{dB} + [\frac{E_b}{N_0}]_{dB} - [M]_{dB} \qquad (3\text{-}13)$$

Further conclusions can be applied to the spread spectrum communication system. Assumed that the spreading factor L is M times the sampling rate of the chip rate, then:

$$SNR = \frac{E[s^2(n)]}{E[\hat{x}^2(n)]} = \frac{E_b R_b}{N_0 B} = 2\frac{E_b}{N_0}\frac{R_b}{f_{smp}} = 2\frac{E_b}{N_0}\frac{1}{M}$$

$$= 2\frac{kE_b}{N_0}\frac{R_c/L}{f_{smp}} = 2k\frac{E_b}{N_0}\frac{1}{M}\frac{1}{L} \qquad (3\text{-}14)$$

$$[SNR]_{dB} = [2]_{dB} + [k]_{dB} + [\frac{E_b}{N_0}]_{dB} - [M]_{dB} - [L]_{dB} \qquad (3\text{-}15)$$

M is the ratio of sampling rate to the chip rate of the simulation. All of the above derivations are obtained in the most general conditions under which these conclusions apply to any type of modulation, any spread spectrum system and does not distinguish whether the system we study is in the baseband or the frequency band. The only condition to be met is that the (low-pass or band-pass simulation) sampling will not result in aliasing, that is, the sampling frequency is more than twice greater than that of the highest frequency component of the useful signal.

3.2 Anti-noise performance simulation in MATLAB platform

Based on the noise modeling conclusion of the previous section:

$$[SNR]_{dB} = [2]_{dB} + [k]_{dB} + [\frac{E_b}{N_0}]_{dB} - [M]_{dB} \qquad (3\text{-}16)$$

Take BOC (10, 10) as an example. The input range of signal-to-noise ratio EbN0: 3 to 8 dB. The bit error rate curves corresponding to direct spread spectrum system and BOC modulation system can be drawn as Figure 3-1.

The abscissa is the input of signal-to-noise ratio at the unit of "dB", and the vertical axis is the value of bit error rate.

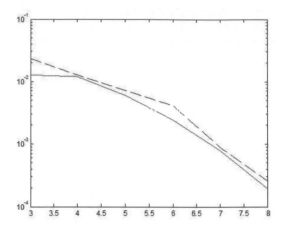

Figure 3-1. Bit error rate curves.

The dashed line "–" represents the DS system BER curve, the polygonal line represents the BOC BER curve. It can be seen from above that under the same channel conditions, the BOC modulation system show a better anti-noise performance than the original direct-sequence spread-spectrum system.

REFERENCES

Li Y. 2008. Key technology of synchronization receiving of navigation signals based on BOC modulation. 2009. Nanjing: Nanjing University of Science and Technology

Roger L., Rodger Z. & David B. 2006. Introduction to spread spectrum communication. Beijing: Beijing Electronic Industry Publisher.

Qingmin W. & Xin F. 2009. Direct modulation method and realization of BOC navigation signals. Beijing: Tsinghua University

Shibiao H. & Xiaoheng T. 2007. Realization of spread spectrum technology. Beijing: Electronic industry publisher.

Yongsong Z. & Haiyong Z. 2005. Noise immunity analysis of direct spread spectrum communication. Dalian: Dalian Naval Vessel Academy

Information Technology and Computer Application Engineering – Liu, Sung & Yao (Eds)
© 2014 Taylor & Francis Group, London, ISBN 978-1-138-00079-7

Integration platform of services and resources for water resources and environment management

Shu Yuan Li
School of Environmental Science and Engineering, Tianjin University, Tianjin, China

Jian Hua Tao
School of Mechanical Engineering, Tianjin University, Tianjin, China

ABSTRACT: To solve a series of problems of water resources and environment management in the process of system integration, an integration platform services and resources for water resources and environment management was developed. The platform based on SOA architecture. This article describes the architecture of the platform, and the architecture and function of the core components of the platform, such as the integration middleware of services and resources, service management tools and auxiliary deployment tools were introduced. The platform can be used for the rapid integration of existing application systems, data resources and services.

Keywords: SOA; Integration platform of service and resource; Water resources and environment management

1 INSTRUCTION

Information management is the basis of today's socio-economic and social development, water resources and environment management information is the foundation and an important symbol of the modernization of water management. With the economic development and the advancement of technology, the importance of water management information construction is gradually revealed. In recent years, in the process of rapid development of the water industry, information technology, construction of a large number of applications, but there have been a number of common problems, and became the bottleneck of development. These issues are as follows: dispersed sources of information, integration and sharing difficulties, duplicate construction, module reuse with difficulties and other problems, it is difficult to meet the demand of the development of water industry.

Recently, many scholars have carried out some research on information System Integration based on service-oriented architecture (SOA) (Wu et al. 2011); (Hu et al. 2011); (Contreras et al. 2008); (Umar et al. 2009); (Ma et al. 2012); (Oliveira et al. 2012). In order to solve the above problems, an integration platform services and resources for water resources and environment management was developed. The platform based on SOA, the cross-platform of J2EE enterprise architecture for building information resources service bus was adopted. The platform can be used for existing local resources and Web services resource integration, interaction, combination and management

infrastructure for distributed services, existing application systems and the development of new systems which are able to facilitate a smooth insertion to the platform, and the pooling of resources between application systems, shared, to achieve the purpose of maximum utilization of resources. This article describes the architecture and functionality of the platform, and summarizes the core techniques of the platform.

2 ARCHITECTURE OF THE PLATFORM

Integrated platform of services and resources is a set of application integration based on SOA, it is composed of middleware of integration for services and resources, access control, client components, service management tools and deployment tools. The platform allows developers to quickly integrated the old IT systems, system services, service composition reuse IT assets, improve the system's adaptability, flexibility and scalability, enabling developers to focus on real needs of the business areas, so as to improve software quality and development efficiency, and reduce the costs and risk of development. The architecture of the platform is shown in Figure 1.

Platform through different interface adapters heterogeneous resources and services of different providers for integration (WebServices, Java interfaces and classes of service standards, SQL-Mapping and OR-Mapping). The upper end of the data service bus middleware the standard WebService protocol

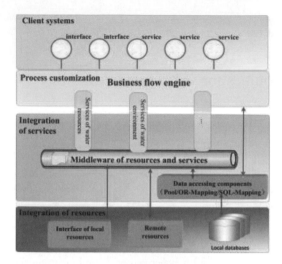

Figure 1. The architecture of the platform.

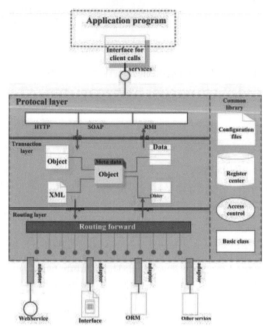

Figure 2. The architecture of the middleware.

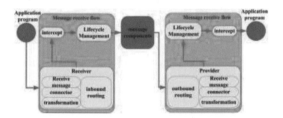

Figure 3. The message flow of the middleware.

interface to provide a unified access service. Various applications can access a service called by the client components. Different customers' heterogeneous client application through the client components connected to the adapter, the system provides integrated communications, messaging, and event infrastructure service request routing and passed to the service provider, which will build the system and the subsequent construction. The system is integrated as a whole. System through a complete set of infrastructure support for message transformation, message filtering and enhancement, Authentication Authorization and other functions, and for future system expansion, increased functionality with ease. Through the use of the platform, the system can be update without modify the client resource. The client code to call the service to independent way involved in the service location and communication protocols. Through the use of management tools, data service bus middleware management, unified, convenient and consistent service management interface.

3 MAIN COMPONENTS OF THE PLATFORM

3.1 *Middleware of integration for resources and services*

The middleware of integration for resources and services can be used to integrate existing application resources, data resources, information sharing and exchange of information; application of a uniform service access interface for the integration of data resources, data center transparent; build fast and complete. The applications system provides the infrastructure. Data Service Bus middleware divided into three layers, namely the protocol layer, the conversion layer and routing layer, including a public library classes. Protocol layer is for the application system call interface of the service interface provides a variety

of protocols, including HTTP, SOAP, RMI, and other interfaces. The conversion layer let different types of data conversion or metadata conversion. Routing layer can call forward the request to the service provider terminal. Public library classes include configuration files and their management, registry and registration services, service access control, and some other tools. The middleware architecture is shown in Figure 2. The message flow of the middleware is shown in Figure 3.

3.2 *Business workflow engine*

The business workflow engine based on the business flow configuration service scheduling, configuration management services with the business flow. Process of transfer of resources, expand the functionality of ESB, to support the process customization, interaction, combination and governance infrastructure for distributed services, and used in conjunction with the design, development and administration tools.

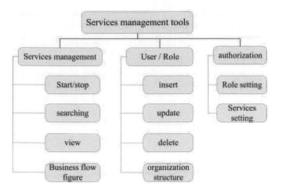

Figure 4. The main functions of the management tools.

Existing business systems and the development of new business systems are able to facilitate smooth insertion to the platform, the pooling of resources between business systems, sharing, so that different application servers work in coordination to achieve different communication between services and integration, to achieve the purpose of maximizing resource utilization.

3.3 Services management tools

Service management tools can start and stop service, publish, delete and other basic operations, as well as the definition of user roles and permissions authorized to operate, user-friendly visualization. View page form, the user can according to the service summary information, click into the details interface. The search function provided by business segment Find and Search by service name based on known information search, the user can obtain the corresponding detailed information. Service management tool service management part, by the platform service hierarchy in service functions to generate a tree list, and services related to information obtained by clicking on the node, the tabbed browsing displayed information includes: sesb business streaming service. The user clicks on the page to get a detailed description of the information, business flow service contains detailed description of the business process diagrams. The main functions of the management tools are shown in Figure 4.

3.4 Auxiliary deployment tools

The auxiliary deployment tools are special tools which can be used to simplify service deployment process, it generates Java interface, the parameter types and return value type and the corresponding configuration files for the deployment, and facilitate the development and deployment of new services. With the deployment of the middleware of integration for services and resources, it generates source code and configuration files. Auxiliary deployment tools enable WSDL file parsing Java code generation

and configuration files generated. Service resource deployment auxiliary tool consists of the follow components.

The WSDL parser: parse as described in the WSDL file for the target WebServices and production aids information object.

The JAVA code generator: Receive WSDL the parser produce the information object, the Java source code using the standard CODE MODEL generate unified style.

Simple constructor of sesb-configuration: to build a simple service in a centralized manner according to the WSDL description, and generate an XML configuration file.

The specific features are as follows:

WSDL parsing: parse WSDL file WebServices as described for the target information object and production aids.

Java code generation: Receive WSDL parser to produce the information object, the Java source code using the standard CODE MODEL generate unified style.

The configuration files generation: XML configuration file is generated based on the WSDL description.

4 CONCLUSION

Integrated platform of services and resources is composed of the middleware of integration for services and resources, service management tools and service deployment tools and other components. The platform uses standardized protocols and interfaces to integrate the existing application resources, data and resource; while providing a unified integrated resource service accessing interface, data access transparent to facilitate a variety of application integration of water resources and environment management. Using the Web Service standard protocol interface, information integration and sharing for water resources and environment management system, service support information resources to quickly build a water resources and environment management system with the management tools, client components, and deployment auxiliary tools.

REFERENCES

Contreras, M. &, Sheremetov, L. 2008.Industrial application integration using the unification approach to agent-enabled semantic SOA. *Robotics and Computer-Integrated Manufacturing* 24(5): 680–695

Hu, J. & Khalil, I. 2011.Seamless integration of dependability and security concepts in SOA: A feedback control system based framework and taxonomy. *Journal of Network and Computer Applications* 34(4): 1150–1159

Ma, J. & Yu H. 2012. Research and Implement on Application Integration Based on the Apache Synapse ESB platform. *AASRI Procedia* 1: 82–86

Oliveira, S. & Balloni, A. Information and Service-Oriented Architecture & Web Services: Enabling Integration and Organizational Agility. *Procedia Technology* 5: 141–151

Umar, A. & Zordan A. 2009. Reengineering for service oriented architectures: A strategic decision model for integration versus migration. *Journal of Systems and Software* 82(3): 448–462

Wu, Y.P. & Shu, T.T. 2011. Research on Information System Integration in Colleges Based on SOA. *Procedia Engineering* 24(1): 345–349

Information Technology and Computer Application Engineering – Liu, Sung & Yao (Eds)
© 2014 Taylor & Francis Group, London, ISBN 978-1-138-00079-7

The research of project teaching method in software development technology course

Jun Rui Liu
College of Computer Science, Northwestern Polytechnical University, Xi'an, China

Shan Jiang
National Computer Network Emergency Response Technical Team Coordination Center of China, Beijing, China

ABSTRACT: According to the deficiencies in the "software development technology" curriculum teaching, the project teaching method is suggested to use in the teaching process, and this paper has researched the practical measures and the needing attention in the using process of this teaching method. The results of the practices show that the implementation of the project teaching method can effectively improve the students' enthusiasm of learning, cultivate the students' idea of software development, and improve the students' comprehensive quality and their innovation ability.

Keywords: project teaching method; software development concept; development innovation; project selection; project task resolution; tracking and guiding

1 INSTRUCTIONS

Software development is also called software technology or software foundation, it's a computer public curriculum after the computer foundation course and the program design course for non computer majors, it has been designed to cultivate the students' idea of software development and software development technology, and to make the students to have the ability of the implementation of software engineering process and completing the scale software development. However, in practical teaching many students lack the idea of software development, can not apply the technology learned in the class to the actual software development process successfully. In view of this situation, this paper suggests that the project teaching method will be applied in software development technology course teaching. By adjusting the teaching contents and the way of practice, the advantages of the projects teaching method in the course is shown, so that the teaching effect and the quality of the students are improved.

2 THE PRESENT SITUATION OF SOFTWARE DEVELOPMENT TECHNOLOGY CURRICULUM

Based on the program design course, software development will improve the students' programming language skills, runs the concept of software engineering in the course, and hopes that the students can under the software engineering theory and complete the software development process. However, there exist the following questions in the actual teaching.

(1) The overall concept of software is lacked in theoretical course. The content of software development technology course mostly includes the software engineering, the database technology, the related technology, but these contents can't connect each other. In the book, most examples are used to explain a certain technology application method, and only includes dozens or even several lines code, while lack the description in practical software development context and unable to reflect the overall concept.

(2) The guide of the software development idea is lacking in the practice. Considering the specific executive difficulty, the experiment task of the software development technology curriculum is split into several small experiments which are designed to study each theoretical unit, and the overall concept of software development is destroyed in these small experiments, while the unified guidance of the software idea is lacking.

(3) The students' learning enthusiasm is lacking. In the study of this course, the students cannot obtain the desired software development skills, so their learning interesting cannot be mobilized.

3 THE PRACTICE OF THE PROJECT TEACHING METHOD IN SOFTWARE DEVELOPMENT TECHNOLOGY

According to the shortcoming of the software development technology course teaching described in the above, this paper discusses on the course teaching of the project teaching method to arouse students' interest in learning, to train the students' software development concept, and to cultivate the students' team spirit and cooperation ability. Therefore, the teaching activity should undertake the corresponding reform measures.

3.1 *The project teaching method in theory teaching*

First of all, the reform in the theory teaching is discussed. In order to make the curriculum to reflect the overall concept, and to cultivate the students' software development concept, when the theory teaching teachers teach, they can not only take the matter on its merits, and can not only presents the current knowledge application method. The teachers should be carefully designed a software project, and put all knowledge of the curriculum into this software development process, the teachers teach each knowledge based on the software project, so that the students is no longer learning isolated point of knowledge in the learning, but a full understanding of its actual application scene and method, thus the students is helped to cultivate the idea of software development, and to control the software development process.

3.2 *The project teaching method in practice teaching*

The project teaching method emphasizes the students' dominant position, and from the project planning, the implementation to the evaluation of final acceptance link, the students are asked to complete as the main body, so as to the students' ability to analyze and solve problems is cultivated. Therefore, the successful application of the project teaching method in practice teaching in the curriculum has a key role for the teaching effectiveness enhancing. The project teaching method in the teaching application need to do the following:

1) Assisting the students in the topic selection of the practice project. The teachers should help the students to select the project with the appropriate difficulty and the full workload based on the actual situation of the students. Only the project with the appropriate difficulty and the full workload are suitable to fulfill the students' feel in the application of the classroom knowledge, and to mobilize the students' activeness.
2) Urging the students to develop the small development team. The building of the development team is helpful for the students to cultivate the spirit of teamwork, the ability to exchange and the ability to cooperation, but also makes the teaching segment and the real enterprise project development process similar, and lets the students understand the real work of the enterprise process early, therefore the teachers should urge the students to create their own development teams.

3) Helping the students in the division of the project task.
   The project teaching method is used to guarantee that the students practice is a complete software project development process, should not be like before that each single experiment is designed only for a knowledge unit, and should help the students to divide the whole practice project into some small tasks which could be completed in each experiment, if the time is not enough, the teachers should urge the students to complete the corresponding software development tasks out of the class.
4) Tracking and guiding in the students' practice timely.
   Due to the adoption of the project teaching method in the practice, the students need to complete all development work of software, the workload is full; the difficult is greater than before. Therefore, the teachers should give timely the students the technical guidance in the process of practice, and require the students to submit the relevant documents or code data in each stage, so for the teachers, understanding the work of every student in every stage is convenient, then the teachers can prevent the students delay to the project or misunderstand the project development goals.

After the project teaching method is used, the teachers' workload will increase a lot, and they may be cannot implement the project teaching method smoothly because of their limited energy in the actual teaching. Therefore, teaching software is needed to match the application of the project teaching method. Such as, we have built a network hard disk that is convenient to the students for information management; we have open network class which can help the teachers solve the students' problems in the project development process by providing them with the video recording of class teaching, courseware information, and so on. Also, we have constructed the curriculum community, it provides the teachers and the students with the exchange platform of the course technology, the development paradigm ,the software development library and the document information to assist the students in project development; Finally, we have developed the working auxiliary scoring system to improve the teachers' work efficiency. By the assistance of this teaching software, the project teaching method could be implemented smoothly in the teaching process.

4 TEACHING EFFECTS

Through our teaching practice of the project teaching method, after the project teaching method is used in the software development technology, the course teaching has the following significant effect:

1) The students get excited. Due to the adoption of the project teaching method, the students take completing the overall project development as the study goal in the theory study and practice, they can

see how software is completed gradually, so the students' learning objectives are clear and their learning enthusiasm is very tall.

2) The students' quality and ability have been greatly improved. The project teaching method emphasizes the student as the main role in the teaching process, and the students are required to complete a whole software development which includes the project plan, the implementation, the delivery and the acceptance, then the students learn the knowledge of the whole software development process in the theoretical aspects while they can validate these knowledge in practice, so the students' abilities and qualities are greatly improved, and they have the idea of software development and a strong development capacity.

3) The students' collaboration and teamwork awareness is stronger. In practice, the teachers emphasize the students to group as a unit, and multiple person cooperate to complete the large software. In the development process, the students need to communicate and cooperate to make the project completed successfully. Therefore, the students' communication ability and team spirit of cooperation are trained in the course.

5 CONCLUSIONS

The project teaching method is an excellent teaching methods, and in the teaching process it can choose the right project as part of the course study entry point, then lead the students to explore the curriculum knowledge while in practice the theoretical knowledge imperceptibly throughout the project development process, so that it can improve the students' interest in learning the course, the students' software development philosophy is developed, and the pioneering, the innovative spirit and the team cooperation spirit are cultivated.

REFERENCES

[1] Wu Chunying The application of the case teaching method in the course of Visual Basic program design. Theory and Practice Education. 2009.9, pp. 57–58

[2] Liu Zhihong The research and practice of the project teaching method in C language teaching. Chinese adult education 2010.4, pp. 139–140

[3] Luo Xiaojuan The cultivation of college Students' occupation ability in the program design teaching. Chinese adult education 2010.9, pp. 163–164

[4] Han Junze. The design and implementation of the programming online evaluating auxiliary teaching system. Journal of Inner Mongolia Normal University (NATURAL SCIENCE EDITION) 2010.9, pp. 473–476

[5] Xu Yongxing. The design of the computer examination and automatic marking system. Open Education Research. 2005.6, pp. 80–83

Information Technology and Computer Application Engineering – Liu, Sung & Yao (Eds)
© *2014 Taylor & Francis Group, London, ISBN 978-1-138-00079-7*

Processing path study of laser cutting for sheet

Hua Qi Liang
Department of Mechanical and Electric Engineering, Anhui Jianzhu University, Hefei, China

Hui Fang Kong
Department of Foreign language Studies, Anhui Jianzhu University, Hefei, China

ABSTRACT: This essay makes further inquiries on technological treatments of laser cutting for sheet on the basis of current processing technology, which contains setting up a line of lead-in and lead-out, arranging auxiliary cutting path reasonably, determination of optimal torch path, selection of the process parameters.

Keywords: laser cutting, line of lead-in, Line of lead-out, optimal path

Compared with traditional processing technology, laser cutting for sheet has the following advantages: high quality \high efficiency \high flexibility\ and low cost. Therefore, the study on the laser cutting not only enhances the development of the processing industry but also broadens the application horizon. As we know, laser cutting for sheet is a combination of aerification and melting, which is affected by many factors such as machine tool, material and processing methods. The other factors include Laser power control, cutting speed, focal position, depth of focus, types of auxiliary gas and pressure. In fact, laser cutting technology also involves the a setting of line of lead-in and lead-out, arranging auxiliary cutting path reasonably, determination of optimal torch path, selection of the process parameters, which was neglected in some degree in the recent study. The reasons are complicated. On the one hand, no systematic study has been made on the material processing. So there is no contents of automatic processing in the automatic programming. Or there is no through study on this area. So this paper aims to study this processing technology and the application of it in the CAPP.

1   ARRANGING AUXILIARY CUTTING PATH REASONABLY

The cutting path outside of the part is called auxiliary cutting path, which is one of vital measures to guarantee the cutting quality in fine production. As we know, laser cutting makes use of heat to melt and aerify the metal. The possible uneven heat dissipation may lead to decrease of cutting quality and burning of the part. So it is important to arrange auxiliary cutting path reasonably. There are two types of auxiliary cutting path: a line of lead-in and lead-out, \circular auxiliary cutting path.

1.1   *Arranging a line of lead-in and lead-out*

To arrange a line of lead-in and lead-out, the process helps to make the part smooth in the fine production. To some metal cutting with low precision requirement, arc hole can be made on the part without such arrangement (see Figure 1a). Since arc hole has bigger diameter than kerf, it can not be made directly on

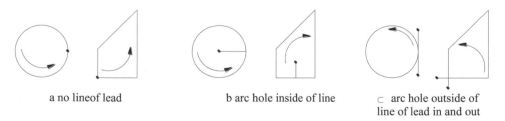

a no lineof lead   b arc hole inside of line   c arc hole outside of
                                                        line of lead in and out

Figure 1.   Setting up a line of lead-in and lead-out.

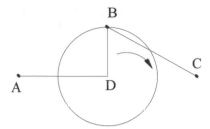

Figure 2. Straight line of lead-in and lead-out.

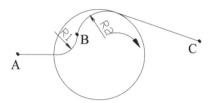

Figure 3. Arc of lead-in and lead-out.

the part itself. Instead, it should be put on the sheet to improve cutting quality. The figure 1b shows the setting up of arc hole inside of the part. The Figure 1c shows the setting up of arc hole outside of the part. The arranging a line of lead-in and lead-out can make the laser beam stable and avoid the influence of arc hole.

Both the figure b and the figure c demonstrate the straight line of lead-in and lead-out, which are used mainly in tradition processing. As is shown in Figure 2, the laser head moves from Point A to Point D, which is also arc hole. So the punch should be made at this pint. Then, the laser beam moves from Point D to Point B, processing the inside of the part. Last, the laser beam moves from point B to point C and exit from point C.

The procedure is shown clearly in the Figure 2. But this way of lead-in and lead-out has to pause twice in the production and then finish the punching and cutting. So there are speeding up and down during the processing. This may require the long duration of machine response. Meanwhile, the abrupt movement of laser head may result in discontinous cutting and influence the quality of parts. In order to solve the problem, we change the line of lead-in and lead-out into circular way and ensure the continuous cutting. The procedure is shown clearly in the Figure 3. Such adjustment not only avoid the diacontinous cutting but also make the cutting speed stable and improve the processing quality. To other shaped parts, we compare the straight line of lead-in and lead-out and circular way. The procedure is shown clearly in the Figure 4. The figure shows that such change ensure the processing quality and reduce the injury of machine tool.

1.2 *Arranging circular way of lead-in and lead-out*

The circular way of lead-in and lead-out is to finish the cutting of sharp parts without changing laser power

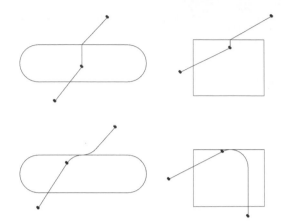

Figure 4. Copmare between line of lead-in and lead-out and arc of lead-in and lead-out.

and cutting speed. During the treatment of sharp parts, laser cutting is likely to produce Uneven heat dissipation, which can result in low precision and even the burning of parts. So the arrangement of enclosed circular cutting path can avoid such problems. During the processing, we can also realize the accuracy of parts through controlling power and speed. For instance, when the cutting approaches the sharp part, we can lower down the power and cutting speed and ensure the even distribution of heat.

2 OPTIMAL TORCH PATH AND OPTIMAL LAYOUT

Generally speaking, in the processing, many parts are put in the same sheet for the seek of efficiency. So it is necessary to find an optimal cutting path and optimal layout.

2.1 *Optimal torch path*

The productivity of laser cutting machine tool can be improved through the application of the source code program from the CAD/CAM. The optimal torch path can significantly reduce the production cycle. The optimal cutting sequence should be cut from outside to inside, from small part to big part. Nowdays, a series of commerical CAD/CAM software package has been developed to offer optimal torch path in the cutting processing.

There are many constraints on the cutting process. So common optimal algorithm can not be applied in the complicated cutting process. For instance, the cutting quality has something to do with the intricate geometric shapes. In the process of cutting acute angle groove, the over heating can lead to the decreasing quality of cutting. So the optimal torch path involves not only the shortest path but also the heating generated in the cutting process. It is difficult to cover the factors at the same time. Most of the current study focus on the change of cutting path and lowering production cost.

2.1.1 The generation of shortest cutting path

The generation of shortest cutting path can be produced with the nearest domain method. That is to say, we have discrete feature nodes, seeking to find the shortest path through the connection of feature nodes. This is a special TSP problem as well as challenging problem, which was widely studied in the optimal research. A lot of solutions have been proposed such as minimum tree weighted approximation method, multilateral transform adjustment method the nearest domain method and the GA algorithm.

So in the processing we first find the punching hole on the part and then use the algorithm to get the shortest path.

2.1.2 Influence of heating factor on the cutting path

In the process of cutting process, we have to include other factors such as overheating and then we can achieve high working effiency and satisfactory cutting quality. The heat in the processing should be minmum, reducing the risk of over heating, thermal deformation and the scrap of material. analytic function method.

2.2 Optimal layout

In the process of laser cutting, the workpiece 's layout on the sheet has great influence on the use ratio of material and the production cycle. It can help us to put as many parts as possible on the given sheet. Compared with the manual layout, automatic Layout improve the use ratio of material, reducing the production cycle. Nowdays, it is widely used in the cutting process. A lot of algorithms have been put forward such as heuristc search method, analytic function method and the human-computer interaction method. We can apply the optimal layout on the evaluation basis of these algorithms.

Meanwhile, we should also take the following factors into consideration: layout pretreatment, cutting method, cutting effiency, continuous cutting and Common cutting. We should arrange the small parts as near as possible according to some rules and then we can avoid the phenomenon of continuous lifting laser nozzle. By doing this, we can increase the service life of laser nozzle and improve the working effiency. In the process of optimal layout, the workpiece with Long side also should be put according to their long sides. So the Common side of these parts will be cut at the same time.

3 CONCLUSION

Laser cutting technology for sheet plays an important role in the industry, which influenced not only the cutting quality but also the working effiency and production cycle. The computer aided design of Laser cutting technology can help us to determine process parameters and arrange auxiliary cutting path. So this area of research will have great potential in the future.

REFERENCES

[1] Chen Bingsen, Computer aided Welding technology. Mechanical industry Publishing house 1999.
[2] Liu Yanlin, mechanics engineering of China, 1995(6)
[3] G. Hanet et al, J. Manufactur ing Processes, 1999, 1(1), 62
[4] U. Manber et al, J. Manufacturing Systems, 1998, 3(1), 81
[5] S. D. Jackso n et al, Computer in Industry, 1995, v21, 223

Information Technology and Computer Application Engineering – Liu, Sung & Yao (Eds)
© 2014 Taylor & Francis Group, London, ISBN 978-1-138-00079-7

Research on target acquisition requirements to a guidance radar of anti-missile weapon system

Qing Sun, Jian Feng Tao & Jun Liang Ji
Air and Missile Defense Institute, Air Force Engineering University, Xi'an China

ABSTRACT: Air and missile defense has now become an important task for the army. Aiming at the technology characteristics of modern radar, profound research has been processed on the acquisition course of anti-missile radar system and some specific design requirements have been proposed, which will have highly actual significance and military application value in improving the tactical and technical performance. Based on the motion characteristics and models of ballistic missile in reentry stage, the paper processes simulation work to setting measures, search time and intercept probability to interception matrix of ballistic target. Technology of broad beam tracking and multi-beam receiving has been put forward in saving search time and increasing discovery probability effectively.

Keywords: Anti-missile weapon; Guidance radar; Ballistic missile defense; Target acquisition.

1 GENERAL INSTRUCTIONS

Ballistic missile and the related technologies are now prompted developing and spreading among the world, and further research and disposition of ballistic missile defense technologies and air-defense and anti-missile system in western world, which place our country in the double threats of offense and defense ballistic missile, and it is urgent to develop the related research work. As the core of air-defense and anti-missile weapon system, the acquisition performance of guidance radar will directly influence the operational ability of the whole missile defense system. Profound research on target acquisition requirements to a guidance radar of anti-missile weapon system will supply not only technique reserve for our developing in air-defense and anti-missile weapon system, but also reference to improving defense measures and penetration ability.

2 ANALYSES ON TARGET CHARACTERISTICS OF BALLISTIC MISSILE

V-2 ballistic missile is the focus of attention which appeared in the terminal of 2nd world war, and especially, with the help of German rocket experts, American and Russian accelerate their ballistic missile development. Locating development of ballistic missile the same position with nuclear weapon, strong countries develop new type of carrier and useful load simultaneously, which becomes the important component of safety strategy.

2.1 Main features and development tendency of ballistic missile

Ballistic missile is composed by carrier rocket and bullet, and the former includes solid or liquid fuel engine and guiding and controlling parts. Having the characteristics of far range, high speed, high altitude, multi-bullet and sub-reentry, diversified penetration measures composition etc., military superpowers are devoting themselves to developing ballistic missile, which fires great progress in the weapon system and the antimissile defense system.

2.2 Target characteristics analyses

The template is used to format your paper and style the text. All margins, column widths, line spaces, and text fonts are prescribed; please do not alter them. You may note peculiarities. For example, the head margin in this template measures proportionately more than is customary. This measurement and others are deliberate, using specifications that anticipate your paper as one part of the entire proceedings, and not as an independent document. Please do not revise any of the current designations.

Being the precondition of designing a certain observation radar, target characteristics should be considered sufficiently. Ballistic missile has high reentry velocity and short flight time. As to a certain TBM with range of 120~3000 km, reentry velocity of 1.1~5 km/s, and the flight time will be 2.7~15 min. Meanwhile, ballistic missile has the characteristics of small RCS about 0.1–1 m^2 and big reentry angle. As to a certain TBM with range of 120~2500 km, the reentry angle of 44.7°~39.4° and the trajectory peak will be 30–600 km.

The beam of modern guidance radar is narrow, and the autonomous search is independent. The early-warning information should be supplied through BMEWS Ballistic Missile Early Warning System, air-base infrared system, remote ground-base early warning and detecting system. The guidance radar waits for searching in preset air space and turns into tracking after target acquisition. The coordinate dimension and precision dividend with external information source will influence arrangement strategies and numbers of beam to the searching beam, and which will determinate time and data rate of searching frame period. The reentry models to analyze the acquisition of ballistic target can be referenced in 2nd, 3th and 4th reference.

3 ANALYSES OF TARGET ACQUISITION

From the above analyses we can see that ballistic missile has become the hard-to attack target because of its characteristics. Similarly as shooting bullet with bullet, intercepting ballistic missile is harder than attacking aero dynamical target. According to the characteristics of ballistic missile and the guidance radar of anti-missile weapon system, we will analyze and discuss the target acquisition requirements from the aspects of target pointing accuracy evaluation and searching zone-setting, angular searching matrix setting and improving SNR through merging channels.

3.1 Target pointing accuracy evaluation and searching zone-setting

Target pointing accuracy has intimate relationship with the choosing acquisition zone of ballistic target. Acquisition probability equals the product of the target in acquisition zone probability P_v and target detecting probability P_d.

When the probability values of every angular resolution cell are given, the total probability of intercept can be described as (1).

$$P_a = \sum_{i=1}^{n_x} P_{vi} P_{di} \qquad (1)$$

Here, detecting probability P_{di} mainly dependents on SNR and fluctuation specification of target echo, and P_{vi} dependents on pointing accuracy.

Characteristics of range, pointing data of angular altitude and azimuth supplied by acquisition radar represent as σ_R, σ_ε and σ_β with standard deviation. Acquisition space should be chosen in the direct proportional sector areas of ΔR, $\Delta \varepsilon$ and $\Delta \beta$ with corresponding value σ in polar coordinate of acquisition radar. And so, proper filtering extrapolated models of target indicating pathway and evaluation to standard deviations of every target indicating coordinate are needed for the system to choose an appropriate acquisition matrix.

3.1.1 A definition for searching area of Gauss distribution random measurement error

When the random measurement error of exterior indicating points obey Gauss distribution, probability density functions of every dimension error can be expressed as (2).

$$\begin{cases} f(\Delta \varepsilon) = \dfrac{1}{\sqrt{2\pi}\sigma_\varepsilon} \exp\left(-\dfrac{(\Delta \varepsilon)^2}{2\sigma_\varepsilon^2}\right) \\ f(\Delta \beta) = \dfrac{1}{\sqrt{2\pi}\sigma_\beta} \exp\left(-\dfrac{(\Delta \beta)^2}{2\sigma_\beta^2}\right) \\ f(\Delta R) = \dfrac{1}{\sqrt{2\pi}\sigma_R} \exp\left(-\dfrac{(\Delta R)^2}{2\sigma_R^2}\right) \end{cases} \qquad (2)$$

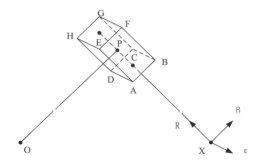

Figure 1. Acquisition space region diagram.

3σ criterion is adopted in order to set radar acquisition area appropriately and to acquire a higher intercept probability, which means acquisition area is determined by (3).

$$\begin{cases} -3\sigma_\varepsilon \le \Delta \varepsilon \le 3\sigma_\varepsilon \\ -3\sigma_\beta \le \Delta \beta \le 3\sigma_\beta \\ -3\sigma_R \le \Delta R \le 3\sigma_R \end{cases} \qquad (3)$$

Formula (4) shows the probability of target dropping-in this area.

$$P = \int_{-3\sigma_\varepsilon}^{3\sigma_\varepsilon} f(\Delta \varepsilon)d\Delta \varepsilon \int_{-3\sigma_\beta}^{3\sigma_\beta} f(\Delta \beta)d\Delta \beta \int_{-3\sigma_R}^{3\sigma_R} f(\Delta R)d\Delta R = 0.9973^3 = 99.2\% \quad (4)$$

Figure 1 gives out the acquisition area determined by σ_ε, σ_β and σ_R. Point O, X and P separately represent the home station, external information station and the coordinate of target indicating point. In the course of target searching, the space region should be covered by searching bean so as to intercept target inerrably.

From Figure 1 we can see that the shape of acquisition space region approximates cuboid and in the topocentric polar coordinate of acquisition radar, coordinates of indicting point and other peaks can be expressed as follows.

$P(\varepsilon_X, \beta_X, R_X)$,

$A(\varepsilon_X + 3\sigma_\varepsilon, \beta_X - 3\sigma_\beta, R_X - 3\sigma_R)$, $B(\varepsilon_X + 3\sigma_\varepsilon, \beta_X + 3\sigma_\beta, R_X - 3\sigma_R)$,

$C(\varepsilon_X - 3\sigma_\varepsilon, \beta_X + 3\sigma_\beta, R_X - 3\sigma_R)$, $D(\varepsilon_X - 3\sigma_\varepsilon, \beta_X - 3\sigma_\beta, R_X - 3\sigma_R)$,

$E(\varepsilon_X + 3\sigma_\varepsilon, \beta_X - 3\sigma_\beta, R_X + 3\sigma_R)$, $F(\varepsilon_X + 3\sigma_\varepsilon, \beta_X + 3\sigma_\beta, R_X + 3\sigma_R)$,

$G(\varepsilon_X - 3\sigma_\varepsilon, \beta_X + 3\sigma_\beta, R_X + 3\sigma_R)$, $H(\varepsilon_X - 3\sigma_\varepsilon, \beta_X - 3\sigma_\beta, R_X + 3\sigma_R)$.

And the length of every side can be determined by the following expression.

$$AE = BF = CG = DH = 6\sigma_R$$
$$AD = BC = EH = FG = 6R_X\sigma_\varepsilon$$
$$AB = CD = EF = GH = 6R_X\sigma_\beta \cos(\varepsilon_X)$$

3.1.2 A definition for searching area of non-Gauss distribution random measurement error

Generally, the random measurement error of exterior indicating points does not obey Gauss distribution,

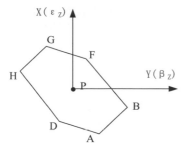

Figure 2. Acquisition matrix of fixed target.

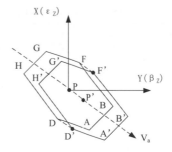

Figure 3. Acquisition matrix of moving target.

choosing of acquisition space area can be processed according to the Chebyshev inequality, which is shown in (5).

$$P\{|X - \mu| \le \varepsilon\} \ge 1 - \frac{D(X)}{\varepsilon^2} \qquad (5)$$

From (5) we can see that as to abnormal distribution, the probability will be about 91% in interval of 3σ and about 97% in interval of 6σ. So the system will need wider searching area to satisfy an authorized intercept probability when the random measurement error does not obey Gauss distribution,

3.2 Setting of angle-searching matrix

Guidance radar searches target in both angular and range dimension. In angular dimension, the radar will process serial step search according to resident period and beam width, and the step interval will just be the beam width. In range dimension, it will process parallel search, which means the radar will acquire and analyze all the echoes of detecting range in one beam resident period. We know that the analyzing time of range echo to single beam is shorter than beam resident period, so while the discussion to the questions of target acquisition matrix, acquisition time, intercept condition and so on will mainly concentrate in angular dimension.

Correspondingly, searching problem of guidance radar to a certain space area can be transformed to a two-dimension searching problem in beam section of guidance radar to the target indicating point.

3.2.1 Acquisition matrix of fixed target

The coordinates of the points are shown in Figure 2. Under the topocentric rectangular coordinates created with origin O can be calculated Using the coordinate transformation relation of topocentric coordinates to geodetic ones, and the shadow area can be obtained by (6).

$$S_S = S_{ABCD} \frac{|\overrightarrow{OP} \bullet \overrightarrow{AE}|}{|\overrightarrow{OP}| \bullet |\overrightarrow{AE}|} + S_{AEHD} \frac{|\overrightarrow{OP} \bullet \overrightarrow{AB}|}{|\overrightarrow{OP}| \bullet |\overrightarrow{AB}|} + S_{ABFE} \frac{|\overrightarrow{OP} \bullet \overrightarrow{AD}|}{|\overrightarrow{OP}| \bullet |\overrightarrow{AD}|} \qquad (6)$$

Considering there exists some superposition in searching beam, and the searching section of simple

beam or circle beam of junction center can be calculated by (7).

$$S_z = \frac{\sqrt{2}}{2} R_o \theta \cdot \frac{\sqrt{2}}{2} R_o \theta = \frac{1}{2} R_o^2 \theta^2 \qquad (7)$$

Here, R_O is the range of target to junction center and θ is the half-power beam width. The quantity of beam included in acquisition matrix can be calculated by (8).

$$N_s = \frac{S_s}{S_z} \qquad (8)$$

3.2.2 Acquisition matrix of moving target

When the target is moving, the infinite body domain will correspondingly shift to the target moving direction, and the two-dimension projection area will shift to the moving direction of target projective point. And so the two-dimension searching area will be increased. Considering the universality of calculating method to new section, it can be calculated by the following method.

Hypothesis that segment $d_k(k = 1, 2, 3 \dots 8)$ is the range between the arbitrary projective point of $A, B, \dots H$ in plane X-Y and line PP'. Then

$$S_A = 2\max(d_k)|PP'| = 2\max(d_k)V_a T_M \qquad (9)$$

Here, we will consider the approach of d_k. Firstly, we setup X-axis, Y-axis and R-axis alone the direction of ε, β and R separately. Velocity vector $V(V_X, V_Y, V_R)$ is the rays in the plane with OP as normal and its projection is the rays across origin P, shown in Figure 3.

Formula (14) gives out the projection equation.

$$V_y x - V_x y = 0 \qquad (10)$$

The coordinate of point k in plane X-Y is (X_k, Y_k), and the distance between the point and the rays.

$$d_k = \frac{|AX_k + BY_k + C|}{A^2 + B^2} = \frac{|V_y X_k - V_x Y_k|}{V_x^2 + V_y^2} \qquad (11)$$

$$N_s = \frac{S_s + S_A}{S_z} = \frac{S_s + 2\max(d_k)V_a T_M}{S_z}$$

Table 1. Simulating result of radar acquisition matrix.

N	1	2	3	4	5	6	7	8	9	10
M	9	12	4	2	0.8	0.4	0.3	0.2	0.07	0.06

N–Beam quantity for acquisition, M–Duty ratio (%).

Table 2. Simulating result of radar acquisition matrix with lower target acquisition precision.

N	1	2	3	4	5	6	7	8	9	10
M	4.5	7.6	12	20	15	4	2	2	1	1

$$= \frac{S_s}{S_Z - 2\max(d_k)V_a T_0} \tag{12}$$

In the defining field of target pointing accuracy, target velocity, radar beam width and resident period, the term of S_z-$2max(d_k)V_a T_0$ in (12) may be negative and so does the range of N_s, which means under the current circumstance, radar can not process target acquisition.

3.2.3 Simulation analyses

According to the definition of moving target acquisition matrix, radar acquisition matrix can be simulated to determine the beam quantity for acquisition. Simulating conditions are set as follows. The distance between guidance radar and acquisition radar is 300 km. Indicating precisions of range and angular are 3′, 3′ and 50 m, and the target velocity is 5 km/s. Radar beam width and resident period are 0.6° and 6 ms individually. The simulating space is defined as 1200 km × 1000 km. The result is showed in Table 1.

The simulating result shows that fewer beams can process target acquisition under the circumstance of higher precision acquisition radar (3′, 3′, 50 m), and with further depression precision, for example, (12′, 12′, 100 m), the result is showed in Table 2.

Table 2 gives out part of the result. 11 to 20 beams occupy 14% and 21 to 30 ones occupy 6%.

From the above simulation we can see that target indicating precision has obvious influence to the selection of acquisition matrix. Lower precision can not permit the system to process fast acquisition such as simple beam or four beams, and more time resource will be cost.

3.2.4 The measure of improving angular scouting speed

From the previous analyses and simulation we can see that according to the space region of search with same volume and shape, it will cost more time in close range search than in remote search if use identical beam width. And so, as to remote target, adapting the searching policy of detection and receiving with marrow beam, we can assure the scouting speed and discovery probability, and as to middle or close range target, it will improve angle sweep speed under the circumstance of fewest possible SNR loss by adapting the policy of wide beam detection and multi-beam receiving. Formula (13) and (14) individually show the relation between radar SNR and antenna gain and the relation between antenna gain and beam width.

$$s_i = \frac{P_t G_t G_r \lambda^2 \sigma}{(4\pi)^3 R^4} \tag{13}$$

$$G = \frac{\theta_0}{\theta_a \theta_e} \tag{14}$$

Here, θ_0 is a constant. Substitute antenna gain to radar receiving equation and we can get (15).

$$s_i = \frac{P_t \theta_0^2 \lambda^2 \sigma}{(4\pi)^3 (\theta_a \theta_e R^2)^2} = \frac{P_t \theta_0^2 \lambda^2 \sigma}{(4\pi)^3 (S_R)^2} \tag{15}$$

Formula (15) shows that as to arbitrary remote range R_f or close range R_n, the receive signal power can keep invariant when radar beam width changes obey the rule of $1/R$ and with invariable scanning velocity.

Further studies show that as to active array radar, multi-beam receiving technique can be adopt to increase target SNR $(R_f/R_n)^2$ times on assuring scanning velocity or increase scanning velocity $(R_f/R_n)^2$ times on assuring SNR when target is in close range. According to the acquisition period of adapting detection with wide beam and receiving with multi-beam, the same scanning velocity can be obtained when intercepting self-same target on 200 km and on 500 km and signal noise ratio of 7.96 dB can be acquired besides, or scanning velocity of 6.25 times can be increased with the same SNR acquisition.

4 CONCLUSION

Being a focus of attention nowadays in the world, the problem of ballistic missile penetration and countermeasure is still a reality to be confronted urgently in modern hi-tech warfare for our country. In this paper, profound analyses are processed on target acquisition to guidance radar system of anti-missile weapon. Simulation work is emphasized on acquisition matrix setting, search time and acquisition probability of ballistic target, and target acquisition policy of detection with wide beam and receiving with multi-beam is advanced to improve angular scouting speed and even acquisition probability to middle or close range target.

REFERENCES

Luo Qun, Zhou Wanxing, Ma Lin. The Ground Radar Manual[M]. Beijing, National Defence Industry Press, 2005.

Teng Kenan. Research on Comprehensive Evaluation Methods of Antimissile Operational Efficiency for Networked Air Defense Missile [J], Fire Control and Command Control, 2007(4):46–48

Ren Yiguang, Sha Jichang, Jing Tao. Research on Modeling and Simulation of Anti-Missile Interception for the Ship-to-Air Missile[J], Fire Control and Command Control, 2009(10):108–114.

Qi Guoqing. Theories and Technologies of close range Air-Defence and Anti-Missile [D]. Doctor's theses from Nanjing University of Science & Technology, 2006.

Information Technology and Computer Application Engineering – Liu, Sung & Yao (Eds)
© 2014 Taylor & Francis Group, London, ISBN 978-1-138-00079-7

Research on target tracking technologies to a guidance radar of anti-missile weapon system

Qing Sun & Jun Liang Ji
Air and Missile Defense Institute, Air Force Engineering University, Xi'an, China

Yong Sun
Shaanxi Radio, Film and TV Bureau, Xi'an, China

ABSTRACT: Ballistic Missile Tracking is the corn task of guidance radar in air and missile defense system. According to the technique characteristics of modern radar, deep research has been processed in this paper to target tracking technologies of anti-missile radar system and specific design requirements have been advanced. The research work will supply actual purpose and military application value in improving the technique and tactical performance of air-defense and anti-missile weapon system.

Keywords: Anti-missile weapon; Target tracking; Data transfer rate

1 GENERAL INSTRUCTIONS

Accurate tracking is the precondition and foundation for radar to recognize target correctly and intercept missile successfully. As a king of offensive weapon with advance performance, ballistic missile processes abnormal difficulty in tracking initialization for radar system because of its dramatic characteristics such as remote range, high flight speed, large acceleration and so on. Being a kind of typical non-cooperative target, unknown target type and various indeterminate factors including guided missile plate, type and magnitude of thrust, target mass and aerodynamic characteristics of target will bring great influence to radar tracking. As a special aerobat, and different from airplane and aerodynamic missile with dynamic engine, ballistic missile will be influenced more severely by gravity and aerodynamics and correct models have to be created in radar tracking. In order to increase penetration ability, new type of tactical ballistic missile often processes orbit maneuver in reentry stage or at the end of mid-section, meanwhile, the type of maneuvering ballistic is various and the mode is complex. All the above have proposed grand challenges to radar tracking.

2 GETTING STARTED: DYNAMIC SELECTION OF SIGNAL WAVEFORM IN RADAR TRACKING

In the combat of anti-missile defense system, the word track means tracking group targets, predicting bullet landing and determining defense level. A major function of tracking is data supplication, which will be used to target classification, type determination and sending target tracking report to master strategic position.

Radar signal waveform is correlative with technical indications of its sub-systems. The selection of signal waveform is determined by many factors, and the selection principle should be highest possible decrease unnecessary waveform types to simplify design and decrease unnecessary software overhead to improve the system reliability. The major factors influencing selection of radar signal waveform is multi-function and multi-mode of phase array radar. Various signal pulse width, pulse repetition frequency, instant bandwidth, pulse length, different encoded modes and its combinations are easy to change with various radar functions and working modes and which will effectively process best management to signal power of phase array radar. In searching state, it is suitable for radar to adapt signals with wide time width and narrow bandwidth to prompt echo SNR. Poor time sense of signals with narrow bandwidth after pulse compression will reduce the number of distance unit, which will reduce computation workload in signal procession. In tracking state, it is suitable to adapt signals with wide time-bandwidth product for acquiring high range accuracy and range resolution, meanwhile, the computation workload will not be increased obviously because within one radar repetition period, echoes in narrower tracking gate need to be processed, and which will insure the computation workload keep balance with searching state and signal procession be finished real-time.

3 TRACKING MODE AND DATA TRANSFER RATE ANALYSES

Anti-missile phase array radar constants the ability of tracking multiple space targets, and which processes great influence on design of phase array radar working mode, data transfer rate and tracking accuracy in both searching state and tracking state.

The ability of tracking and processing multi-batch targets depends on both the operational capability of radar controlling computer and data processing computer and the signal power supplied by radar. It needs more signal energy for tracking and irradiation to track more targets, and it is an important content for radar system design to determine an appropriate tracking target number according to the task of phase array radar.

According to the characteristics of modern operations, at least four types of tracking mode including rough tracking, intimate tracking, replica tracking and memory tracking to phase array radar of anti-missile system should be designed. In the course of target tracking, echoes may not be detected during tracking resident period because of the influence from fluctuation of signal strength, target maneuver track change, and jamming generating. Under the circumstances, it is necessary to adapt appropriate path extrapolation method to extrapolate beam direction and window position in waiting for echoes. If no echo appears during continuous several resident periods, we can judge that the target is missing and simultaneously start small range searching mode around the missing point. If the target appears after searching, then we can transfer into re-tracking and give out a mark of loss previous, and if no target appears, we can judge that thoroughly lost and give out warning message.

In this paper, Monte Carlo simulation method is adopt to establish target moving and radar tracking models, meanwhile, it will be used to simulate tracking process so as to acquire the appropriate tracking mode and data rate requirement.

Main aspects of requirements analysis to radar tracking including observational error originating from SNR, pulse width after pulse compression and lobe-half-power width, and the non-linear relation between target state vector and observation vector. Meanwhile, extended Kalman filter model is established and simulating work is processed to target tracking with various data transfer rate.

3.1 Tracking models

The observation of guidance radar to targets is processed in topocentric coordinates and the equation is as follows.

$$Z = [r, \varepsilon, \beta]^T = h(X) + V = [h_1(X), h_2(X), h_3(X)] + [v_r, v_\varepsilon, v_\beta] \quad (1)$$

Here

$$r = h_1(X) = \sqrt{x^2 + y^2 + z^2}$$
$$\varepsilon = h_2(X) = \arctan\left(\frac{z}{\sqrt{x^2 + y^2}}\right) \quad (2)$$
$$\beta = h_3(X) = \arctan\left(\frac{y}{x}\right)$$

In the above formulas, V is Gaussian noise with mean 0 and covariance R_K.

$$R_K = diag(\sigma_r^2, \sigma_\varepsilon^2, \sigma_\beta^2) \quad (3)$$

Here, σ_r, σ_ε and σ_β are the standard deviations of measuring noise to radar range, angular altitude and azimuth separately. Hypothesis that the measuring errors of each dimension submit gauss distribution with mean 0, and the corresponding standard deviations can be calculated by the following formulas.

$$\sigma_r = \frac{c\tau}{2\sqrt{2}} \cdot \frac{1}{\sqrt{S/N}} \quad (4)$$

$$\sigma_\varepsilon = \frac{\theta_\varepsilon}{1.89\sqrt{2}} \cdot \frac{1}{\sqrt{S/N}} \quad (5)$$

$$\sigma_\beta = \frac{\theta_\beta}{1.89\sqrt{2}} \cdot \frac{1}{\sqrt{S/N}} \quad (6)$$

Here, c expresses light velocity and τ does the signal width after pulse compression. θ_ε and θ_β individually are lobe-half-power widths of angular and azimuth.

The state variable transfer of target position, velocity and acceleration process in topocentric rectangular coordinates and observation process in topocentric polar coordinates, which leads to nonlinear relation between measured values and state variables, and linearization work has been done firstly to the observation for the use of extended Kalman Filter to estimate system parameters.

Adapting extended Kalman Filter, the observation equation should be expanded into a Taylor series at $\hat{X}_k$. Then

$$H(X) = \frac{\partial h}{\partial X}\Big|_{\hat{x}} = \begin{bmatrix} h_{11} & h_{12} & h_{13} & 0 & 0 & 0 \\ h_{21} & h_{22} & h_{23} & 0 & 0 & 0 \\ h_{31} & h_{32} & h_{33} & 0 & 0 & 0 \end{bmatrix} \quad (7)$$

Here,

$$h_{11} = \frac{\partial h_1}{\partial x} = \frac{x}{r}$$

$$h_{12} = \frac{\partial h_1}{\partial y} = \frac{y}{r}$$

$$h_{13} = \frac{\partial h_1}{\partial z} = \frac{z}{r}$$

$$h_{21} = \frac{\partial h_2}{\partial x} = \frac{-zx}{(x^2 + y^2 + z^2)\sqrt{x^2 + y^2}}$$

$$h_{22} = \frac{\partial h_2}{\partial y} = \frac{-zy}{(x^2 + y^2 + z^2)\sqrt{x^2 + y^2}}$$

$$h_{23} = \frac{\partial h_2}{\partial x} = \frac{\sqrt{x^2 + y^2}}{(x^2 + y^2 + z^2)}$$

$$h_{31} = \frac{-y}{x^2 + y^2}$$

$$h_{32} = \frac{x}{x^2 + y^2}$$

$$h_{33} = 0$$

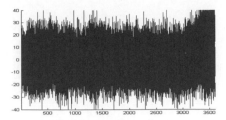

Figure 1. Range tracking error of ballistic target.

The recurrence equations of extended Kalman Filter algorithm are as follows. The step prediction equation of state variable is shown in (8).

$$\hat{X}_{K+1/K} = A\hat{X}_{K/K} + BU(\hat{X}_K) \tag{8}$$

Formula (9) is the step prediction equation of auto-covariance matrix to state variance.

$$P_{K+1/K} = AP_{K/K}A^T + \sigma^2 BB^T \tag{9}$$

Estimating equation of state variable auto-covariance matrix is shown in (10)

$$P_{K+1/K+1} = \left[P_{K+1/K}^{-1} + H_{K+1}^T R_{K+1}^{-1} H_{K+1}\right]^{-1} \tag{10}$$

Filter gain calculating equation

$$G_{K+1} = P_{K+1/K+1}H_{K+1}^T R_{K+1}^{-1} \tag{11}$$

State-variable filtering equation

$$\hat{X}_{K+1/K+1} = \hat{X}_{K+1/K} + G_{K+1}\left(Z_{K+1} - h\left(\hat{X}_{K+1/K}\right)\right) \tag{12}$$

Process noise estimating equation is shown in (13).

$$q_{K+1} = \hat{X}_{K+1/K+1} - \hat{X}_{K+1/K}$$
$$\hat{\sigma}_{K+1}^2 = \frac{K}{K+1}\hat{\sigma}_K^2 + \tag{13}$$
$$\frac{1}{\gamma(K+1)}\left[q_{K+1}^T q_{K+1} - tr\left(AP_{K/K}A^T - P_{K+1/K+1}\right)\right]$$

3.2 Simulation analysis

3.2.1 Ballistic target tracking simulation

According to the practical situation of ballistic target, simulating condition is set as follows. Hypothesis that the fire range is 3000 km, reentry velocity of ballistic target is 5000 m/s and the short-cut of airway is 100 km. Bandwidth of radar detecting signal is 2.5 MHz, time-bandwidth after pulse compression is 0.4 μs and signal noise ratio is 13.6 dB, meanwhile, radar lobe-half-power width is 0.6° and the tracking data rate is 10 Hz. The simulation result is as follows.

Figure 1 shows the tracking error curve when range RMS error equals 11 m.

When the RMS error of angular altitude equals 2.8′, the tracking error curve is shown in Figure 2.

When the RMS error of azimuth equals 2.9′, the tracking error curve is shown in Figure 3.

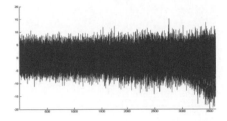

Figure 2. Angular altitude tracking error of ballistic target.

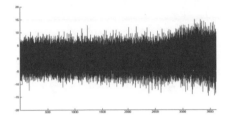

Figure 3. Azimuth tracking error of ballistic target.

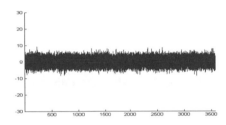

Figure 4. Range tracking error of ordinary target (with 5 Hz data rate).

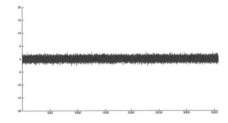

Figure 5. Angular altitude tracking error of ordinary target (with 5 Hz data rate).

3.2.2 Ordinary target tracking simulation 1 (with 5 Hz data rate)

According to ordinary target, simulating condition is set as follows. Hypothesis that the origin range is 500 km, target velocity is 700 m/s and the short-cut of airway is 30 km. Bandwidth of radar detecting signal is 2.5 MHz, time-bandwidth after pulse compression is 0.4 μs and signal noise ratio is 26 dB, meanwhile, radar lobe-half-power width is 0.6° and the tracking data rate is 5 Hz. The simulation result is as follows.

Figure 4 shows the tracking error curve when range RMS error equals 1.98 m.

When the RMS error of angular altitude equals 0.64′, the tracking error curve is shown in Figure 5.

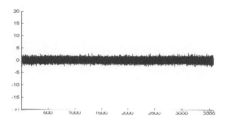

Figure 6. Azimuth tracking error of ordinary target (with 5 Hz data rate).

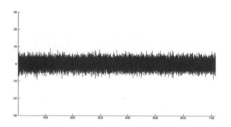

Figure 7. Range tracking error of ordinary target (with 1 Hz data rate).

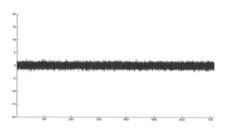

Figure 8. Angular altitude tracking error of ordinary target (with 1 Hz data rate).

When the RMS error of azimuth equals 0.64′, the tracking error curve is shown in Figure 6.

3.2.3 Ordinary target tracking simulation 2 (with 1 Hz data rate)

The simulating parameters are the same with ordinary target tracking simulation 1 expect the simulating data rate is 1 Hz.

Figure 7 shows the tracking error curve when range RMS error equals 2.2 m.

When the RMS error of angular altitude equals 0.64′, the tracking error curve is shown in Figure 8.

When RMS error of azimuth equals 1.57′, the tracking error curve is shown in Figure 9.

From the above simulating results we can see that the system can process tracking ballistic target with 10 Hz data rate and tracking ordinary target with 1 Hz for perfect tracking and 5 Hz for gross tracking separately. The results can all satisfy the tracking precision demands.

Tracking data transfer rate or its reciprocal processes great influence on multi-target tracking performance of phase array radar. Selecting tracking sample interval correctly can supply important significance to the guarantee of tracking continuity, reliability and

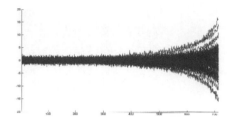

Figure 9. Azimuth tracking error of ordinary target (with 1 Hz data rate).

tracking precision. Phase array radar can not adapt high tracking data transfer rate to every tracked target for the shortage of time or signal energy. According to the materiel task and requirement, we can select different to various target based on the flexibility of antenna beam scanning. We can divided the tracked targets into several classes, and adapt different tracking sample interval to different classes. When the searching zone is small and the number of tracked target is few, tracking data transfer rate can be prompted correspondingly. The transfer of the adaptive work states mentioned above is accomplished under the control of radar controlling computer.

4 CONCLUSION

Being the core detector of air defense and anti-missile weapon system, the premise available operation of guidance radar is target acquisition, tracking and identification. Significant features under the background of ballistic missile penetration endow them with new requirements. Aiming at the technique characteristics of modern radar, tracking technologies of anti-missile radar system are further researched in this paper and specific designing requirements are prompted, which process certain virtual purpose and military application value.

REFERENCES

XIA Xinren. Present status and development of American missile defence system [J], China Astronavigation, 2007(3).

XU Feng, LI Yuanlong. Research on Technologies of Antimissile and Interception [J], Armament, Mar. 2010.

Hays D. Ground-Based Midcourse Defense(GMD). 6th Annual Small Business Day. MDA. [C/OL]. [03 Aug 2005]. http://www.mdasmall-business.com/conference/download/GMD.ppt

LIU Xing. Air and space Defense Information System and its integration technology [M], National Defence Industry Press, 2009.

TANG Yuanping. Overview of Development of World s Air Defense and Antimissile Weapon Systems [J], Aerospace Manufacturing Technology, 2010(1).

LI Weimin, XIN Yongping. Integrated Operation of Air and Missile Defense and Military System Engineering[J], Military Operations Research and Systems Engineering, 2008(04).

Information Technology and Computer Application Engineering – Liu, Sung & Yao (Eds)
© 2014 Taylor & Francis Group, London, ISBN 978-1-138-00079-7

Matlab-based small scale helicopter simulator

I. Salloum, A. Ayoubi & M. R. Khaldi
University of Balamand, Lebanon

ABSTRACT: The modeling, controlling, and emulating a small scale helicopter is presented. Helicopters are highly unstable machines that require only highly skilled pilots to fly them. They have fast response and controlling them is very complicated and need high degree of precision. Helicopters have many controllers and their behavior is monitored by an onboard panel of instruments. In this paper, a detailed mathematical model is developed, a quick overview of the Helicopter's controllers and instrumentation is stated, and a Matlab-based simulator is designed, implemented and tested. The designed simulator is composed of a 3D dynamic animation model, flight instrumentation panel, and controllers. Finally a sample run is illustrated.

Keywords: Helicopter; Matlab/Simulink; Simulator

1 INTRODUCTION

Helicopter control is a multivariable control problem with usually four outputs: the vertical velocity, the pitch angle, the roll angle, and the yaw angle. The controls used are the longitudinal and lateral cyclic command (forward, backward, left, and right). And the tail rotor and pitch collective command (up, down, and engine speed).

A large number of effective control techniques have been proposed in the literature for the helicopter flight control including robust adaptive control (Zhao et al. 2006), predictive control (Khaldi et al. 2009), state-dependent Riccati equation control (Bogdanov et al. 2003), sliding mode control (Guo et al. 2010), trajectory tracking control (Frazzoli et al. 2000; Shan et al. 2005), fuzzy control (Amaral & Crisostomo 2001) and neural network control (Tee et al. 2008). In most present works, the planned flight control technologies are focus on the linear helicopter dynamics or single-input/single-output (SISO) non-linear helicopter dynamics. To efficiently handle strongly coupled non-linearity, model uncertainties and time-varying unknown perturbations, the dynamics of helicopters must be treated as an uncertain MIMO non-linear system in the flight control design.

2 MATHEMATICAL MODEL

The fundamental equations of a helicopter motion are derived. The helicopter is treated as a rigid body having six degrees of freedom. Two frames are defined, Inertial (or fixed) and Body frames. The relationship between these two frames is derived using Euler angles. Thus, in view of Newton's laws, the helicopter's dynamics are given in (1) and (2) (Murray et al. 1994).

$$m\dot{\mathbf{v}}_B + \boldsymbol{\omega}_B \times m\mathbf{v}_B = \mathbf{f}_B$$
$$\mathbf{I}_B\dot{\boldsymbol{\omega}}_B + \boldsymbol{\omega}_B \times \mathbf{I}_B\boldsymbol{\omega}_B = \boldsymbol{\tau}_B. \tag{1}$$

The external forces $\mathbf{f}_B$ in (2) contains the fuselage aerodynamic drags in the x, y, and z- directions. Also, the main and tail rotor thrusts T and T_T, respectively, and the gravitational force mg.

$$\mathbf{f}_B = \begin{bmatrix} -D_{Fx} \\ -D_{Fy} - T_T \\ -D_{Fz} - T \end{bmatrix} + \mathbf{R}_{IB}^T \begin{bmatrix} 0 \\ 0 \\ mg \end{bmatrix} \tag{2}$$

where $\mathbf{R}_{IB}$, which can be deduced from (4), is the rotational matrix to transform terms from Inertial frame to Body frame. The external moments τ_B in (4) includes the three main directional moments M_x, M_y, and M_z as well as the torque due to the rotor hinges offset with respect to the rotor axis, and the motor torque (moment) τ_m.

$$\boldsymbol{\tau}_B = \begin{bmatrix} M_x - Tx_a \\ M_y - Ty_a \\ M_z + \tau_m \end{bmatrix} \tag{3}$$

The linear velocities with respect to the Inertial frame can be obtained as follows,

$$\begin{bmatrix} \dot{x}_I & \dot{y}_I & \dot{z}_I \end{bmatrix}^T = \mathbf{R}_{IB} \begin{bmatrix} v_x & v_y & v_z \end{bmatrix}^T \Rightarrow$$

$$\dot{x}_I = v_x \cos(\theta)\cos(\psi)$$
$$+ v_y \left(\sin(\phi)\sin(\theta)\cos(\psi) - \cos(\phi)\sin(\psi) \right)$$
$$+ v_z \left(\cos(\phi)\sin(\theta)\cos(\psi) + \sin(\phi)\sin(\psi) \right)$$

$$\dot{y}_I = v_x \cos(\theta)\sin(\psi) \tag{4}$$
$$+ v_y \left(\sin(\phi)\sin(\theta)\sin(\psi) + \cos(\phi)\cos(\psi) \right)$$
$$+ v_z \left(\cos(\phi)\sin(\theta)\sin(\psi) - \sin(\phi)\cos(\psi) \right)$$

$$\dot{z}_I = -v_x \sin(\theta) + v_y \sin(\phi)\cos(\theta) + v_z \cos(\phi)\cos(\theta)$$

where (x_I, y_I, z_I) are the helicopter's coordinate with respect to the Inertial frame. The angular velocities with respect to the Body frame after using a similar expression as in (4) can be found in terms of the angular velocities with respect to the Inertial frame as follows,

$$\omega_x = \dot{\phi} - \dot{\psi}\sin(\theta)$$
$$\omega_y = \dot{\theta}\cos(\phi) + \dot{\psi}\cos(\theta)\sin(\phi) \tag{5}$$
$$\omega_z = -\dot{\theta}\sin(\phi) + \dot{\psi}\cos(\theta)\cos(\phi)$$

where (θ, ϕ, ψ) are the Euler angles that represent the helicopter's attitude namely pitch, roll, and yaw, respectively.

From (5), the angular velocities of the helicopter with respect to the Inertial frame can be derived as follows,

$$\dot{\phi} = \omega_x + \left(\omega_y \sin(\phi) + \omega_z \cos(\phi) \right)\tan(\theta)$$
$$\dot{\theta} = \omega_y \cos(\phi) - \omega_z \sin(\phi) \tag{6}$$
$$\dot{\psi} = \left(\omega_y \sin(\phi) + \omega_z \cos(\phi) \right)/\cos(\theta)$$

Equations (1) and (2) lead to the linear acceleration with respect to the inertial frame as follows,

$$\dot{v}_x = v_y \omega_z - v_z \omega_y - \frac{D_{Fx}}{m} - g\sin(\theta)$$
$$\dot{v}_y = v_z \omega_x - v_x \omega_z - \frac{D_{Fy} + T_T}{m} + g\sin(\phi)\cos(\theta) \tag{7}$$
$$\dot{v}_z = v_x \omega_y - v_y \omega_x - \frac{D_{Fz} + T}{m} + g\cos(\theta)\cos(\phi)$$

and (1) and (3) lead to the angular acceleration with respect to the inertial frame as follows,

$$\dot{\omega}_x = \frac{\begin{bmatrix} (I_z I_y - I_{xz}^2 - I_z^2)\omega_x \omega_z + I_{xz}(I_z + I_x - I_y)\omega_x \omega_y \\ + I_z(-M_x - Ty_a) + I_{xz}(-Q - T_T d) \end{bmatrix}}{I_x I_z - I_{xz}^2}$$

$$\dot{\omega}_y = \frac{(I_z - I_x)\omega_x \omega_z + I_{xz}(\omega_z^2 - w\omega_x^2) + M_y + Tx_a}{I_y} \tag{8}$$

$$\dot{\omega}_z = \frac{\begin{bmatrix} (-I_x I_y - I_{xz}^2 - I_x^2)\omega_y \omega_x - I_{xz}(I_z + I_x - I_y)\omega_z \omega_y \\ + I_x(-Q - T_T d) + I_{xz}(-M_x + yh - Ty_a) \end{bmatrix}}{I_x I_z - I_{xz}^2}$$

where, Q is the yaw moment, d is the distance between the tail rotor axis and the helicopter's centre of gravity, h is distance between the main rotor hub and the helicopter's centre of gravity.

A component build up approach is followed to derive the helicopter's aerodynamics equations that complete the mathematical model (Etkin *et al.*, 1996).

The linear and angular positions and velocities constitute the twelve state variables that are needed to describe the behavior of the helicopter as a rigid body system. Consequently, the nonlinear coupled state space model,

$$\dot{\mathbf{x}}(t) = \mathbf{f}(\mathbf{x}, \mathbf{u})$$
$$\mathbf{y}(t) = \mathbf{g}(\mathbf{x}, \mathbf{u}) \tag{9}$$

where $\mathbf{x} = [x_I\, y_I\, z_I\, \dot{x}_I\, \dot{y}_I\, \dot{z}_I\, \phi\, \theta\, \psi\, \omega_x\, \omega_y\, \omega_z]^T$ is the state vector. Five independent inputs to the rigid body dynamics model are the main rotor thrust, the tail rotor thrust, the moments created by the main rotor around the roll and pitch axes (M_x, M_y) and the yaw moment Q. $\mathbf{u}$ and $\mathbf{y}$ are the inputs and outputs, respectively.

3 HELICOPTER FLIGHT CONTROLS

Helicopters have four basic controls used during flight. They are the collective-pitch control, throttle, the cyclic-pitch control, and the anti-torque pedals.

3.1 Collective pitch control

The collective-lever is used to change the pitch angle of the main-rotor blades collectively and is placed on the left side of the pilot's seat. When the collective pitch control is raised or lowered, the pitch-angle of the main rotor-blades increases or decreases.

3.2 Throttle control

The placement of a twist grip throttle is typically riding on the termination of the collective-lever. But some turbine helicopters have the throttles mounted on the overhead panel or on the floor in the cockpit.

3.3 Cyclic pitch control

The cyclic-pitch control may be mounted vertically between the pilot's knees or on a teetering bar from a single cyclic located in the center of the helicopter. The cyclic can pivot in all directions

3.4 Anti-torque pedals

The anti-torque pedals, located on the cabin floor by the pilot's feet, control the pitch, and therefore the thrust, of the tail rotor blades.

The main determination of the tail-rotor is to offset the effect of the torque to the main rotor. Since torque varies with changes in power, the tail-rotor thrust should also remain varied. The pitch change mechanism is connected to the pedals on the tail rotor gearbox and allows the pitch angle on the tail rotor blades to be increased or decreased.

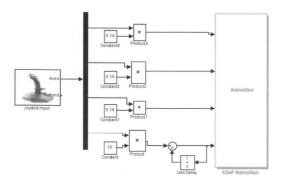

Figure 1.   Animation Model.

4   HELICOPTER SIMULATOR

4.1   *Animation model*

The animation model consists of 4 main parts which are the joystick module, demultiplexer, logic modules and animation module besides the control module, Figure 3.

The Joystick Input module provides an interaction between a physical joystick and a Simulink model that displays the virtual world associated with a Simulink 3D Animation block. Axes output ports functions in the joystick correspond to position of the axes. Buttons correspond to the number and status of buttons used in the joystick, where the digital value zero corresponds to button released and the value one corresponds to button pressed.

The constant logic modules are used to specify the sensitivity of the joystick axes and buttons controls.

The 6DOF Animation module is used to display a 3D animated view of six degrees of freedom for the motion of the helicopter displaying its trajectory and target.

4.2   *Simulation display*

The helicopter main commands are the collective, longitudinal cyclic, lateral cyclic and tail pitch. These commands are the main commands that make the helicopter able to apply different movements which are the pitch, yaw and roll.

In order to control the helicopter movement and trajectory we connected a physical Genius USB Joystick to the universal serial port of a computer.

4.2.1   *Pitch animation display in simulink*
When the joystick is moved forward the helicopter applies the forward pitch as shown in Figure 2a. When the joystick is moved backward the helicopter applies the backward pitch command as shown in Figure 2b.

4.2.2   *Yaw animation display in simulink*
Figure 3 shows when the yaw command is applied to the helicopter.

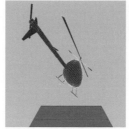 

Figure 2a. Simulink Pitch forward Display.

Figure 2b. Simulink Pitch backward Display.

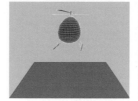 

Figure 3a. Hovering Command.

Figure 3b. 45° Yaw Command to the Right.

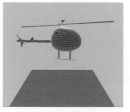 

Figure 3c. 90° Yaw Command to the Right.

Figure 3d. 145° Yaw Command to the Right.

 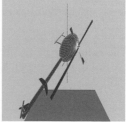

Figure 4a. Roll command to the Left side.

Figure 4b. Roll command to the Right side.

4.2.3   *Roll animation display in simulink*
When the physical joystick is moved to the left side, the helicopter will apply the roll command to the left as shown in Figure 4a. The helicopter is designed to apply 90° roll command to both sides.

4.3   *Gauges and meters*

Beside the helicopter model that is displayed in simulink animation, gauges and meters are also displayed in order to indicate the altitude, horizontal balance and direction of the plane.

Figure 5a. Altimeter low value.

Figure 5b. Helicopter at a low altitude level.

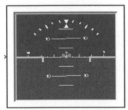

Figure 6a. Balanced Horizon Level.

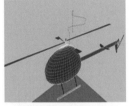

Figure 6b. Helicopter Applying forward pitch.

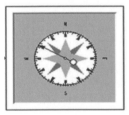

Figure 7a. Compass Indication.

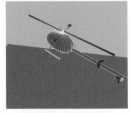

Figure 7b. Helicopter is applying yaw movement to the left.

4.3.1 *Altimeter*

The altimeter reads and displays real time altitude levels that the helicopter is at.

4.3.2 *Horizon level*

The helicopter moves forward when the joystick is pushed forward. Figure 8a shows a balanced horizon level.

4.3.3 *Compass*

Figure 7b shows the heading of the helicopter which is towards the left. At the same time the compass indicator displays a real time feedback about the correct direction of the helicopter, Figure 7a.

5 CONCLUSION

The paper has presented a small scale helicopter simulator. It presented the mathematical model and to visualize the flying course, a 3D helicopter model is created. Then a complete instrumentation display board along with controllers have developed and tested.

REFERENCES

Abiad (EL), H. & Khaldi M. R. 2006. A 3D Visualization Helicopter Model for Simulation Using Matlab and Simulink. *The 15th IASTED International Conference on Applied Simulation and Modeling*: 522–156.

Amaral T.G.B. & Crisostomo M.M. 2001. Automatic Helicopter Motion Control Using Fuzzy Logic. *Proceedings of The 10th IEEE International Conference on Fuzzy Systems*: 860–863.

Bogdanov A., Carlsson M., Harvey G., Hunt J., Kieburtz D., van der Merwe R., & Wan E. 2003. State-Dependent Riccati Equation Control of a Small Unmanned Helicopter. *AIAA Guidance, Navigation, and Control Conference and Exhibit*: 1–11.

Frazzoli E., Dahleh M., & Feron E. 2000. Trajectory Tracking Control Design for Autonomous Helicopters Using a Backstepping Algorithm, *Proceedings of American Control Conference*: 4102–4107.

Guo X., Sun W., Li Z., & Ren L. 2010. The flight control design of Mini Unmanned Helicopter on sliding mode control. *Proceedings of Control and Decision Conference*: 3245–3248.

Khaldi, M. R., Abiad (EL), H., Chamchoum, S., & Abdul Ahad, E. 2009. Plant Identification and Predictive Control of a Scaled-Model Helicopter. *IEEE International Symposium on Industrial Electronics*: 1720–1725.

Murray R., Li Z., & Sastry S. 1994, A Mathematical Introduction to Robotic Manipulation, *CRC Press*

Shan J., Liu H.-T., & Nowotny S. 2005. Synchronised Trajectory-Tracking Control of Multiple 3-DOF Experimental Helicopters. *IEE Proceeding Control Theory Applications, Vol. 152, No. 6*: 683–692.

Tee K. P., Ge S.S., & Tay, F.E.H. 2008. Adaptive Neural Network Control for Helicopters in Vertical Flight. *IEEE Transactions on Control Systems Technology, Vol. 16*: 753–762.

Zhao X., Jiang Z., Han J., & Liu G. 2006. Adaptive Robust LQR Control with the Application to the Yaw Control of Small-scale Helicopter. *Proceedings of the 2006 IEEE International Conference on Mechatronics and Automation*: 1002–1007.

Information Technology and Computer Application Engineering – Liu, Sung & Yao (Eds)
© 2014 Taylor & Francis Group, London, ISBN 978-1-138-00079-7

Design of power harmonic data acquisition system based on network

Nan Ma
China Satellite Maritime Tracking and Control Department, Jiangyin, Jiangsu, China

Jue Wang
Joint Laboratory of Ocean-Based Flight Vehicle Measurement and Control, Jiangyin, Jiangsu, China

Xiao Juan Huang & Li Li Xia
China Satellite Maritime Tracking and Control Department, Jiangyin, Jiangsu, China

ABSTRACT: Through analysis of the gathered power harmonics data, the scheme of data acquisition based on Network was put forward. According to the features of power harmonic data acquisition, the hardware design and software development were introduced in detail. After sensor and A/D converter selecting during the hardware designing, combining with using the Neport module as the network interface and the Fast Fourier Transform (FFT) data processing method by Labview software, a data acquisition system was proposed, which can realize the data acquisition and transmitting and analyzing functions.

Keywords: Power Harmonics, Data Acquisition, Network, Labview

1 INTRODUCTION

The metrical ship equips kinds of electric power devices including both linear and non-linear loads. Especially, versions of the non-linear loads are very complex and the quantity is enormous. Just considering one of the electric power devices, the harmonics generation is little. But when enormous devices work together composing an electric power harmonic generating group, the summation can be no longer neglected, which becomes the main harmonic generating source of the power system.

In order to deal with the electric power harmonics and control the harmonic contamination, precisely measuring is the precondition. In this manuscript, based on the network data acquisition by hardware circuit design and then using the harmonic analysis by Labview software, the functions such as statistics and analysis and so on were realized, which have great significance for keeping the electro web to be in motion safely.

2 POWER HARMONICS SUMMARIZATION

Harmonics are the waveform aberrations of the power system. In our power supply system, the national rules are that the alternating voltage is 220 V and the frequency of the alternating current is 50 Hz. The fundamental wave is the standard cosine wave[1], the standard voltage expression of which is shown as following:

$$u(t) = \sqrt{2}U \cos(2\pi f + \theta) \qquad (1)$$

where, represents the effective value of the supply power voltage and the value is 220 V; θ is the initial phase angle and the difference between each other is 120° in the three-phase power supply system; f is the frequency and the fundamental frequency is 50 Hz.

If the voltage does not have the waveform ahead but aberrant periodic waveforms, it can be treated as the superposition of the fundamental cosine wave with the harmonic cosine waves, the frequency of which is multiple times comparing to the fundamental cosine wave.

According to the limited voltage value of public electro web harmonic in the national standard GB/T14549-93 «Power Quality Public Electro Web Harmonic», in order to obtain the total voltage harmonics aberrant rate and each harmonic contributing rate, the nominal power voltage is needed. Detecting and decomposing the period non-cosine voltage data based on FFT, both the total voltage harmonics aberrant rate and each harmonic contributing rate can be calculated.

According to the statement before, it can be concluded that harmonic analyzing is to get the precise voltage and current signals and correctly decompose each harmonic, and then the corresponding indexes are investigated.

3 HARDWARE DESIGN

3.1 Main chips selecting

3.1.1 Ensor selecting

Using the harmonics of 220 V effective voltage as the power acquisition signals, according to the data index request of the acquisition system, the CSR Times Sensor Technique Limited Company product NCV4A voltage sensor is selected. The main technique indexes are introduced as following. The measuring range is from −375 V to +375 V (for 220 V alternating voltage, the amplitude is 311 V and there are redundancies on both hands between the measuring range), the precision is ±0.7%, nonlinearity is ≤0.1%, the response time is 13 μF, and the bandwidth is 13 KHz. The current sensor is also the company's product NT100-S(T). The electric parameters are introduced as following. The measuring range is from 0 to 200 A, the precision is ±0.7%, NP/NS is 1:1000, nonlinearity is ≤0.1%, the response time is 1 μs, and the bandwidth is between 0 and 100 KHz[2].

3.1.2 Analog to digital (A/D) converter selecting

The American MAXIM company product MAX197 with 8 channels and 12 bit resolution is selected, which has 5 MHz bandwidth and high speed data throughput as 100 kilo samples per second. The bandwidth of the needed sampling signal is 2.5 KHz. If data coding is 12 bits, according to the sample theory, the ideal data throughput is $2.5K*2*16 = 60$ Kpbs, but the data throughput of MAX197 is $100K*12 = 1200$ Kpbs under the same condition. Considering sampling not under the ideal condition, even increasing the sampling frequency three to five times high, there are still lots of redundancy. 0.15 V voltage coding width can be obtained by 12 bit resolution, which can satisfy the request of the data acquisition signals.

3.1.3 Network interface chip selecting

The NePort serial interface conversion module of the CONEXTOP Company is selected, which is a product of high integration and high performance. It affords its customers a concise network connection manner, which is cheap but of high performance for its delicate volume. The NePort product can provide 10M/100M internet interface and transmit TCP/UDP data packet to several data receiving devices at the same time. It can provide one or two serial interfaces and the bit rate can be as high as 921600bps.

3.2 Hardware circuit design

The principle sketch of the acquisition circuit is shown in Fig.1. The selected STC89C52 is the key controlling chip (MCU), the work flow of which is shown as following: the A/D data is gathered by the MCU and packed. Then the data is transmitted by the serial interface, which is received by the Neport module and then transmitted after being switched into the network data. So the data acquisition is realized. The hardware

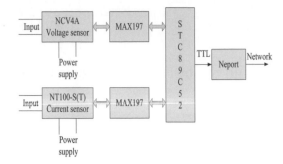

Figure 1. Principle sketch of the acquisition circuit.

design and software programming is simple and there are no unnecessary details is given.

4 HARMONIC DISTORTION ANALYSIS BASED ON LABVIEW

In this manuscript, the method of harmonic analysis is the FFT, which decomposes the acquisition data into Fourier series. And then each order harmonic are detected and judged whether they are up to the mustard or not.

4.1 Harmonic measurement

Taking the voltage as an example in the frequency domain analysis, the distortion periodic voltage can be decomposed into Fourier series as the following:

$$u(t) = \sum_{n=1}^{M} \sqrt{2} U_n \sin(n\omega t + \alpha_n) \qquad (2)$$

where ω is the angle frequency of the working frequency (fundamental frequency) with the unit rad/s, n is the harmonic order, U_n is the root mean square of the nth order harmonic voltage, a_n is the initial angle of the nth order harmonic voltage, and M is the maximum order that is determined by the degree of waveform distortion analyzing precision, the general value of which is ≤50.

The degree of sine waveform deflection caused by harmonic waveform distortion is represented as total harmonic distortion rate (THD), which can be shown as following:

$$THD_u = \frac{\sqrt{\sum_{n=2}^{M} U_n^2}}{U_1} \times 100\% \qquad (3)$$

Because the ratio between rms U_n and amplitude A_n is 2, THD can be written as following[3]:

$$\%THD = \frac{100\sqrt{A_2^2 + A_3^2 + ... + A_n^2}}{A_1} \qquad (4)$$

where A_1 is the amplitude of the fundamental harmonic, A_n is the amplitude of the nth order harmonic and n is the order.

4.2 Window smoothing technology

Time-limited signal sampling is effectively truncated the signal by rectangle window. Discrete frequency leaking effect will happen if the ratio between the rectangle length and signal period is not an integer, which is caused by the abrupt boundary conditions. Lots of high harmonics in frequency domain will be generated by the abrupt boundary conditions. The corresponding changing among the spectral in the rectangle window is that the maximum voltage of the side is higher and attenuates slower. Leaking effect results in unwanted distortion, which causes analysis error. In order to restrain the leaking error, window function is used to process the acquisition data[4]. The function of the window on the signal can be expressed as following:

$$y(t) = x(t)\omega(t) \tag{5}$$

where x(t) is the signal before being truncated, y(t) is the signal after truncated and ω(t) represents the function of the window. The substantiality of the window function is to process the signal with the weighted method. If the boundary conditions change slowly and are close to be zero gradually, the spectral leaking can be decreased even if the terminal does not synchronize when the initial signal is collected. Because the phase difference can be decreased even to be zero after multiplying the initial signal with the widow function. Hanning window is a kind of cosine window, the express of which can be written as:

$$\omega(n) = 0.5 - 0.5\cos(2\pi n / N) \tag{6}$$

The side peak values are small and attenuate quickly. But the total leaking effect of the Hanning window, which is used often for its easily gaining, is much smaller than the rectangle window. As long as dividing the selected detecting time by the signal period is an integral, the amplitudes of all the frequency that are integral times compared with the fundamental frequency are zero, which is the property of Hanning window. Because there is no leaking effect generated among the harmonics. Compared with other window functions, the leaking effect error and the calculating of the Hanning window is smaller even if the signal frequency fluctuates little.

4.3 Program design

The sketch of the program is shown in Fig. 2. Hanning window is selected for truncating process, which is used for data pretreatment analysis to reduce leaking effect. The spectral of the time domain signal is calculated based on the FFT. The converted frequency signal is analyzed and the amplitude and frequency are given in the form of array. Then the total harmonic distortion rate is calculated based on the expression (4).

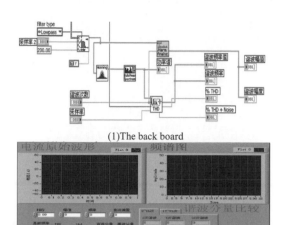

(1)The back board

(2)The front board

Figure 2. Program display.

5 CONCLUSIONS

In this manuscript, based on the power harmonic analysis, the demand of the data acquisition is proposed. Then the design of power harmonic data acquisition system based on network is investigated. Subsequently, the hardware design and software program development are fulfilled, which realizes the design function of the system. Labview provides a reliable and effective method for harmonic analysis. The properties of fundamental and harmonic waves can be obtained intuitively by analyzing distortion signal using harmonic analyzing. The measuring precision can be improved greatly by the truncated window method. So the harmonics can be analyzed and processed. Moreover the precautions can be taken on the harmonics.

ACKNOWLEDGEMENTS

I would like to express my gratitude to all those who helped me for developing this project, especially for Gaofeng PAN.

REFERENCES

[1] Wu Jingchang. 1998. Power supply system harmonic [M]. Beijing: China power press.
[2] Ma Mingjian. 2005. Data acquisition and process technology [M]. Xi'an: Xi'an Jiaotong University Press.
[3] Sun Guangjun, Zhang Jingwei. 1999. Investigation on the data acquisition and recording system [J]. Data acquisition and process.
[4] Huang Chun, Jiang Yaqun, Sun Zihua. 2001. Power harmonic detecting device based on virtual device technology [J]. Power Automation Device.

Information Technology and Computer Application Engineering – Liu, Sung & Yao (Eds)
© *2014 Taylor & Francis Group, London, ISBN 978-1-138-00079-7*

A method of on-road vehicle detection based on comprehensive feature cascade of classifier

Xiao Le Li, De Gui Xiao, Chen Xin & Huan Zhu
College of Information Science and Engineering, Hunan University

ABSTRACT: To detecting on-road vehicles rapidly and efficiently, we propose a robust method to improve the accuracy of on-road vehicle detection rate and reduce false alarm rate. Firstly, we have enhanced the feature expressive force by combining Haar-like feature with BRIEF feature. Some improvements have been done for achieving the robustness under the lighting and road conditions changes. Secondly, we have improved the performance of weak classier based on Gentle Adaboost algorithm. Experimental results show that the detection rate increased by 2.6% compared with the traditional cascade structure of classifier, and the false alarm rate reduced in some degrees.

Keywords: Vehicle Detection; BRIEF; Haar-like; Adaboost

1 INTRODUCTION

On-road vehicle detection based on computer vision is becoming more and more important in vehicle safety, especially in the vehicle assistant driving and automatic navigation system [1] [2]. However, there are many problems in vehicle detection algorithm based on computer vision, such as incapability of real-time detecting vehicle, low rate of detection, high rate of false alarm, especially with the condition of the light change. The vehicle detection rate will be greatly influenced by vehicle obscured conditions [2] [3].

Classifier performance is evaluated from three aspects, that are detection rate as high as possible, false alarm rate as low as possible and detection speed as fast as possible. These three aspects restrict each other, so we need to find a balanced point to meet the requirements of practical application. The feature extraction and classification algorithm determines the quality of the classifier. Many features have been used in current research, including the Haar-like feature [4] [5], HOG feature [6] SIFT feature, SURF feature [7], BRIEF feature [7] [8] [9], edge histogram [10], etc. For the selected these features, we need to use the appropriate classification algorithms to build robust classifiers, such as neural networks [11], support vector machine (SVM) [6], Adaboost classification algorithms [4] [5] [12] and so on.

In order to solve the problem of high false alarm rate and low detection rate of single classifier, Viola and Jones put forward to a testing framework [13] which can handle images very quickly. In this framework, the input image background region is quickly discarded by the front layers. More computing resource is used in the potential area of vehicles. The application of the simple image processing method can extract candidate regions rapidly to speed up the testing process.

However, the cascade structure demonstrated by Viola in the vehicle detection has obvious flaws [13]: On one hand, the cascade classifier which is trained by discrete Adaboost using Haar-like features can refuse simple sub-window quickly, behind a few layers, it needs more linear combination of weak classifiers to decrease the false alarm rate in complex scenarios, which easily leads to excessive learning and time-consuming. On the other hand, as rectangle feature limitations and the weakness of relatively weak classifier, the performance of each layer classifier is not greatly improved while using more weak classifiers after some layer.

In [14], the author combines Haar-like feature with HOG feature. However, introduction of HOG feature has increased the amount of the feature set and reduced the speed of detecting and training.

Ref. [7] puts forward a very simple feature named BRIEF. We improve the traditional cascade structure through three aspects. It can use much less layer to achieve a higher detection rate and lower false alarm rate: firstly, combining BRIEF feature with Haar-like feature to improve the detection rate and robustness of the classifier. Then, improve Haar-like feature's performance by reduce the Haar-like feature form.

2 THE CASCADE STRUCTURE

Constructing a cascade of classifier's main purpose is to reduce computation time while achieving increased detection performance by giving different treatments

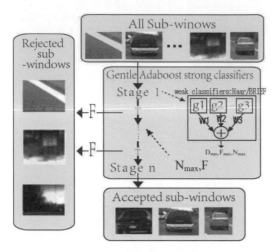

Figure 1. The proposed cascade of classifiers.

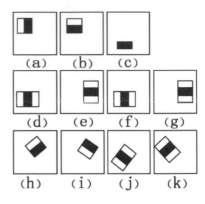

Figure 2. The Haar-like features used in this paper.

3 A CASCADE STRUCTURE DETAILS

3.1 Feature space

For vehicle detection, first of all, we should set up a large amount of training set which included a set of positive and negative samples. Then, we extract feature from the training set, which forming an n-dimensional features space. This article uses two main features: the Haar-like feature and BRIEF feature. Two types of feature descriptors generated in training, BRIEF feature have the advantage of speeding in generating and testing, all of the training set needs to turn grayscale images and gaussian filtering.

3.1.1 Haar-like features [7]

In this paper, a single-rectangular feature is used to detect vehicles which have a shadow at bottom. Because most of the gray level changes direction is vertical and horizontal, we only use edge feature and linear Feature in the Haar-like feature. All of the Haar-like feature we used in Figure 2. We will calculate single-rectangular feature, two-rectangular feature and three-rectangular feature, four-rectangular feature which not be calculated.

The Haar-like rectangle feature can be described as:

$$f = \sum_{i \in I = \{1,2,3\}} w_i \bullet \mathrm{Re}\, cSum(r_i) \tag{1}$$

In (1), w_i is the weight of the ith rectangular. RecSum(r_i) is the value of all pixels in the ith rectangle, and I is the number of rectangles feature in Haar-like feature f. Different functions can be defined as $r = (x, y, w, h, \alpha)$. (x, y) is the vertex coordinates of the left upper corner of the rectangle; (w, h) is the length and width in rectangular; α represents the rotation Angle. In order to simplify the calculation in the process of training, we use several scale features. For two rectangle features, using two feature prototype: 1×2 and 2×1, five dimensions: 1, 2, 4, 8, 16, 32 similar to prototype and scales used in the three rectangular features. The training time of strong classifiers is proportional to the number of features. Due to adding car shadow

to different complexities of input sub-window. The sub-window which passing through all of the classifiers stage that can be classified as vehicle. In this paper, the cascade structure showns in Figure 1 which is a degenerated decision tree. These sub-windows which pass through all of the stages sequentially can be regarded as a target.

Haar-like features and BRIEF features are used to train Gentle Adaboost classifier, and each weak classifier relies on a single feature. The most ideal classifier has maximum detection rate and minimum false alarm rate while using minimum computing time. Therefore, we need to find the best balance point to optimize the performance of the classifier. In practice, we identify two performance parameters to stop training of each layer of the strong classifier: F_{max}, the maximum false alarm rate, D_{min}, the minimum detection rate. Much like a decision tree, subsequent classifiers are trained by those samples which pass through all previous stages. With the purpose to reduce the false alarm rate, we used the i-1_{th} layer of negative samples to train the i$_{th}$ layer. Layers are added until the overall target F, the maximum false alarm rate of the cascade structure is met. In this way, cascade structure detection rate $D = (D_{min})^K$, K is cascade layers. This means that if we choose the strong classifier D_{min}, detection rate is 0.99, while the K=15 the finally detection rate is about 0.86.

In the first layer of Cascade structure, Adaboost classifier can use fewer features to achieve the preset target. However, classifier needs more linear combination of the weak classifiers to achieve the present target in the later layer which easily leads to excessive learning and spending more time to study. In addition, the weaker classifier is not enough to improve the performance of classifier, fixing the maximum number of each layer of weak classifier N_{max}, when increasing the number layers overall detection rate is also decreased.

features only have single-rectangular, so it can be converted into two matrix feature to calculate weights 2, and -1. The result can be calculated as:

$$f = \sum_{i \in I = \{1,2\}} w_i \bullet \mathrm{Re}\,cSum(r_i) = 2 \times \mathrm{Re}\,cSum(r) \tag{2}$$
$$+ [-1 \times \mathrm{Re}\,cSum(r)] = \mathrm{Re}\,cSum(r)$$

The integral image method [5] is employed for quickly calculate Haar-like rectangle features. Regardless of the size of the rectangle, it only needs to search four times in the integral image to find any rectangular pixel values and RecSum (r_i).

3.1.2 BRIEF [8]

As Haar-like feature is sensitive to the target shape and the change of light conditions, Haar-like feature has poor performance in vehicle detection when the image contains complex background information. BRIEF features robust in light and local geometrical changes. Both of Haar-like and BRIEF features describe local feature in image and described in the different ways.

The principle of the BRIEF descriptor is sampling the value which surrounding of all the feature points. A binary sequence of the comparison relationship is obtained among the pixel values which around the feature point. Then divide the sequence into bytes dimension and obtained the description. Each test results (binary digit) are known as a primitive, and its response calculation is as follows: the feature points around s × s pattern image preprocessed as p. Defined η as:

$$\eta(p; x, y) = \begin{cases} 1 & p(x) < p(y) \\ 0 & \text{otherwise} \end{cases} \tag{3}$$

where p (x) p (y) is the pixel intensity in a smoothed version of p, x = (u_1, v_1), y = (u_2, v_2). Choosing a set of n_d (x, y) location pairs uniquely defines as a set of binary tests. We take our BRIEF descriptor to be the n_d-dimensional bitstring.

$$D_{n_d}(p) = \sum_{i=1}^{n_d} 2^{i-1} \eta(p; x_i, y_i) \tag{4}$$

$N_d = 64$, in order to achieve the feature point recognition rate and dimension compromise. In bytes divided dimensions, namely $64/8 = 8$ dimension descriptors, a total of 8 bytes.

In order to overcome the noise affected by individual pixels, we use the gaussian smoothing pattern around the feature point as the image preprocessing to enhance describing ability of the descriptors. 9 × 9 gaussian smoothing can achieve a better effect without influencing the distribution of pixel values.

In the calculation of BRIEF descriptor, tests on the spatial distribution have the same effects on the descriptor resolution. Sampling on isotropic gaussian distribution, u = 0, $\sigma^2 = (1/25) \times S^2$ can get better features points. BRIEF descriptor computation steps are as follows:

First, Sampling under the gaussian distribution of: $G(\mu = 0, \sigma^2 = (1/25) \times s^2)$, we can get 64 descriptors of tests pair $\{\eta_1, \ldots \ldots, \eta_{64}\}$. In the process of the same target be detected, all of the test feature point patterns are carried out in accordance with the defined: $\{\eta_1, \ldots \ldots, \eta_{64}\}$. According to (3) calculate the value of every η, and then according to (4) get a binary response value, finally divided the bitstring into bytes of 8 dimension descriptors.

3.2 Weak classifiers

We train classifiers by using the method in [13]. It is very important to select a small part of important features to train the strong classifier by Adaboost algorithm. A single feature is used to train a weak classifier. The weak classifiers which only need to have a better performance than random guesses. We should choice the classification and training algorithms for Haar-like feature and BRIEF feature respectively.

We only need to determine a binary threshold for Haar-like feature. Due to high-dimensional BRIEF feature, we used a method which based on model to training the feature of BRIEF weak classifier, making the 8 dimensional feature vector turn to one dimension, and then to determine the threshold. Weak classifiers trained by this method have accuracy of the same order-of-magnitude as Haar-like weak classifiers, and the determining process of the thresholds can go with Haar-like weak classifiers' simultaneously.

3.2.1 The form of haar-like weak classifiers

We define the weak classifier for a Haar-like feature as:

$$g_{\mathrm{haar}}(x) = \begin{cases} \alpha, f_j(x) < \theta_j \\ \beta, \text{otherwise} \end{cases} \tag{5}$$

where $f_j(x)$ is the feature value and θ_j is the threshold. α and β are real numbers between $[-1, +1]$, the absolute value of which indicate the confidence of the classification results, and their signs show the classification results (negatives are non-vehicles, positives are vehicles).

3.2.2 The form of BRIEF weak classifiers

The vehicle model weighted each feature descriptor is defined as follow.

$$M_j = \sum_{i=1}^{P} \omega_i D(p)_j^i \tag{6}$$

where D_j is the feature value, defined in (2), P is the size of the sample, ω_i which determined and changed by Adaboost algorithm is the weight of each sample. One classifier calculated the distance between the feature value and the model, so a weak classifier of BRIEF feature is defined as:

$$g_{bri}(x) = \begin{cases} \alpha, d(h_j(x), m_j) < \theta_j \\ \beta, otherwise \end{cases} \tag{7}$$

391

As Haar-like weak classifier, α and β are the results of classifier. Negative values indicate non-vehicles while positives mean vehicles. $d(h_j(x), m_j)$ is the Hamming distance between BRIEF feature values to average values, θ_j is the threshold.

3.3 Adaboost classifier

The basic idea of Adaboost algorithm is repeatedly running a given weak learning algorithm on various distributions over the training data, and then combing the classifiers produced by the weak leaner into a single composite classifier. Lienhart etc. showed that Gentle Adaboost is superior to the other Adaboost algorithms on object detection tasks, while there is a lower computational complexity [4]. We adopt Gentle Adaboost to train strong classifiers through selecting a small number of good features. The weak classifiers in the iterative process are formed as: $g_j(x) = P_w(y = 1|x) - P_w(y = -1|x)$, which means the probabilities that x belongs to positives subtract that x belongs to negatives. The algorithm is as follows:

Given training examples $(x_1, y_1), \ldots, (x_n, y_n)$, where $y_i = 1$ for positive examples and $y_i = -1$ for negative examples.

Initialize weights

$$\omega_i = \begin{cases} \dfrac{1}{2\,l}, & y_i = 1 \\[2mm] \dfrac{1}{2\,m}, & y_i = -1 \end{cases} \tag{8}$$

where m and l are the number of negatives and positives respectively.

For $t = 1, \ldots, T$ (T is the number of iterations, namely the number of weak classifiers of a strong classifier):

① For each feature j, train a weak classifier $g_j \in [-1, +1]$, evaluate its error with respect to ω_i,

$$\varepsilon_j = \sum_{i=1}^{N} \omega_i^j [g_j(x_i) - y_i]^2 \tag{9}$$

② Choose the classifier g_t with the lowest error ε_t.
③ Update the strong classifier: $G(x) \leftarrow G(x) + g_t(x)$
④ Update the weights: $\omega_i \leftarrow \omega_i \exp[-y_i g_t(x_i)]$, where $i = 1, 2, \ldots, N$, and renormalize the weights.

Output the final strong classifier:

$$G(x) = sign[\sum_{t=1}^{T} g_t(x)] \tag{10}$$

After each iterate, one of the best weak classifier will be chosen due to its lowest error, and then these weak classifiers are combined to construct the final strong classifier. If the result of the combination is greater than 0, then the output of the strong classifier will be 1 as vehicle, or −1 represent non-vehicle.

Figure 3. Some training examples.

Table 1. The experiment results of the four cascade classifiers.

Feature	No. layer	Effect		
		DR (%)	FP	Time (s)
Haar-like	20	90.4	198	0.23
HOG	18	83.3	182	1.53
Haar + HOG	15	92.6	156	0.83
Haar + BRIEF	15	93.0	160	0.24

4 EXPERIMENT RESULTS

The sample of training set are collected from MIT CBCL database, computer vision database of California institute of technology and 200 samples image of vehicle which taken by myself. All of the training set contains 2516 images. All positive samples are the front or back view of vehicle which are cropped and scaled to 32 × 32 pixels and make the sample image contain the entire vehicle region and environment a small amount around information. The negative samples contain a total of 3000 images which include rocks, trees, animals, and so on. Also some images contains vehicle but is not in the path environment. Some examples from the training set are shown in Figure 3.

The testing set used in this paper from the database of Label Me from MIT, selecting 500 pictures of different background and conditions, including the 720 vehicles with the size of 360 × 240, each of them contains 1–3 cars. To test the real-time capabilities of vehicle detection, a test video was collected with size of 640 × 480 and frame rate of 30 Hz. The content of the video is collected by In-vehicle cameras which are along in the urban road. The urban road also contains the environment of tunnel, bridge under different road environment and illumination (shown in Figure 4). The effect of testing ensure that the real time detection. The smallest scanning window of testing is 32×32, dimension scaling factor is 1.1.

We determine the strong classifier parameters: $D_{min} = 0.99$, $F_{max} = 0.5$ and respectively trained four cascade classifiers. Table 1 shows the performance of four kinds of classifiers, what is tested by 500 images from the testing database of Label me from MIT.

Figure 4. The results of vehicle detection in urban traffic scene.

Haar-like + BRIEF feature has higher detection rate and lower FA compared with traditional classifiers.

Where TP stands for the number of vehicle samples detected to be vehicles; FP stands for the number of non-vehicle samples detected to be vehicle.

Figure 4 shows that the results of vehicle detection in urban traffic scenarios. Due to the using single-rectangular feature, the detection effect is robust, even in the presence of shadows.

5 CONCLUSION

In this paper, we combine BRIEF with Haar-like features in the cascade of classifiers to enhance expressive force. It makes full use of the image global information to enhance the weak classifier performance effectively. This method has a higher detection rate and it is more suitable for real-time processing.

ACKNOWLEDGEMENT

This work is supported by National Natural Science Foundation of China under grant No. 61272062.

REFERENCES

[1] Zehang Sun May 2006 On-road Vehicle Detection: a Review *IEEE Transactions on Pattern Analysis and Machine Intelligence* **28** 694–711

[2] K S Chidan and Kumar 2012 A Novel Approch for Vehicle Detection for Driver Assistance *Computer Science & Information Technology (CS & IT)* 06, pp. 39–45

[3] Sayanan S June 2010 A General Active-Learning Framework for On-Road Vehicle Recognition and Tracking *IEEE Transactions on Intelligent Transportation Systems* **11** 267–76

[4] Chensheng Sun, Jiwei Hu and Kin-Man Lam 2011 Feature subset selection for efficient AdaBoost training *2011 IEEE International Conference on Multimedia and Expo (ICME)* (Barcelona, Spain, 11–15 July 2011) pp. 1–6

[5] Xuezhi W and Fang W 2011 An Algorithm Based on Haar-like Features and Improved Adaboost Classifier for Vehicle Recognition *Acta Electronica Sinica* **39** pp. 1121–1126

[6] Xianbin C 2011 Linear SVM classification using boosting HOG features for vehicle detection in low-altitude airborne videos *2011 18th IEEE International Conference on Image Processing (ICIP)* (Brussels, 11–14 Sept. 2011) pp. 2421–2424

[7] Rublee Ethan and Rabaud Vincent Nov 2011 ORB: An Efficient Alternative to SIFT or SURF *International Conference on Computer Vision* 2011 2564–2571

[8] Michael Calonder and Vincent Lepetit 2010 BRIEF: Binary Robust Independent Elementary Features *In Computer Vision – ECCV 2010* pp. 778–792

[9] Deng Li and Wang chunhong June 2012 Fast Detection of Expanded Target Based on Binary Robust Independent Elementary Features *Chinese journal of lasers* *2012* 327–331

[10] Miyoshi T 2010 A hardware- friendly object detection algorithm based on variable-block-size directional-edge histograms *World Automation Congress (WAC)*, 2010 (Kobe, Japan, 19–23 Sept. 2010) pp. 1–6

[11] Shen Xu and Yi L. Murphey 2010 Vehicle detection using Bayesian Network Enhanced Cascades Classification (BNECC) *International Joint Conference on Neural Networks* 2010 pp. 1–6

[12] Cerri P, Gatti L, Mazzei L and Pigoni F 2010 Day and Night Pedestrian Detection Using Cascade AdaBoost System *2010 13th International IEEE Conference on Intelligent Transportation Systems (ITSC)* (Funchal, 19–22 Sept. 2010) pp. 1843–1848

[13] Viola P and Jones M May 2004 Robust Real time Object Detection *International Journal of Computer Vision* **57** 137–154

[14] Tingli Li Cascade of Classifiers Based on Multi-Feature for Road Vehicle Detection *16th International Conference on Image Processing, Computer Vision, & Pattern Recognition (IPCV'12: July 16–19, 2012, Las Vegas, USA)* pp. 724–732

Information Technology and Computer Application Engineering – Liu, Sung & Yao (Eds)
© 2014 Taylor & Francis Group, London, ISBN 978-1-138-00079-7

A loop shaped fractal antenna

W.Q. Luo & Y. Feng
University of Electronic Science and Technology, China

ABSTRACT: With the development of large-capacity high-speed communications, UWB antenna has abstracted numerous interests. Fractal antenna has been widely applied in the last 100 miles communication systems due to its particularly small size. In this paper, a new fractal antenna with loop shape was proposed. Applying the fractal theory, the antenna was designed with self similarity. It has good performance in the frequency band from 2 GHz–8 GHz. Measurements was done to verify our design.

Keywords: fractal antenna; loop; UWB

1 INTRODUCTION

The ultra-wide band (UWB) antenna has abstracted tremendous interests since the development of the high-speed and large capacity communications, especially the last 100 miles communications (Porcino & Hirt, 2003). The application of the UWB antenna were spread from the traditional shot-range communication to the Internet of Things including the Body network systems (BSN). In the UWB communication systems, the UWB antenna was considered to transmit/receive wide-band pulse in short durations. This unique feature requires the UWB antenna has uni-performance in the assigned wide frequency band (Yang & Giamalkis, 2004). Different types of UWB antennas have been proposed. As one of the branches of the UWB antenna structures, fractal antenna has been developed greatly due to its ultra wide band performance (Prasad et al, 2000).

In this paper, a fractal antenna was proposed with the design procedure was introduced. The proposed antenna has the loop shape as the outer side and the hexagon shape as the inner side. Then, the antenna was designed to have the self-similarity structure. The size of the antenna was calculated with the theory of fractal, and the parameters were optimized with the software HFSS. The suggested antenna has good performance in the frequency band from 2 GHz–8 GHz. The measured data showed a good agreement with the simulated result.

2 DESIGN PROCEDURE OF THE ANTENNA

According to the fractal theory (Azari & Rowhani, 2008), the initial value of the antenna size due to the centre frequency of the antenna should be decided at first. The antenna was designed to have the similar

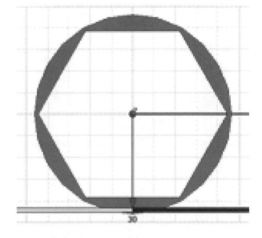

Figure 1. The first loop of the antenna.

structure of the disk antenna. So the outer round of the antenna was designed as a loop, then the inner round of the dick was inscribed by a hexagon. The structure of the first order was shown in Figure 1.

The radius of the disk was calculated as:

$$R_1 = \lambda_g = \frac{\lambda}{\sqrt{\varepsilon_{re}}} \tag{1}$$

where ε_{re} is the effective permittivity.

The second round was designed to have the outer side connected to the hexagon structure of the first round. Here we studied the bandwidth of the antenna having different orders from one order to four orders as shown in Figure 2. From this figure, we found that with the order increased, the bandwidth of the antenna got wider and wider. From this figure, it was shown that with the order increasing from 1 to 4, the

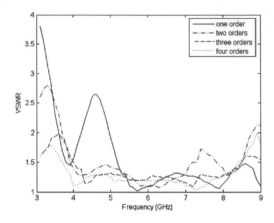

Figure 2. comparison of the VSWR of the different orders of the antenna.

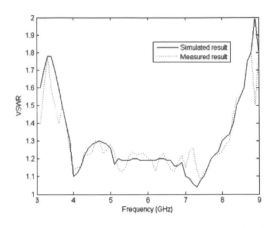

Figure 5. comparison of the VSWR of the measured reslt and the simulated result.

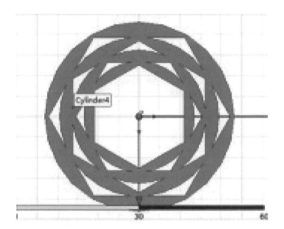

Figure 3. Prototype of the self-similarity antenna.

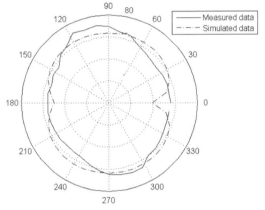

Figure 6a. Radiation of the H-plane.

Figure 4. The fabricated antenna.

3 RESULTS

The S parameter of the fabricated antenna was measured with the Agilent 8510 network analyzer shown in Figure 5. The result showed a good agreement in the concerned frequency band. And the radiation results were measurement in the microwave unreflected chamber of UESTC with the result shown in Figure 5 with the radiation pattern shown in Figure 6.

4 CONCLUSION AND FUTURE WORK

In this paper, a new fractal antenna was designed and the fabricated model was measured. The good agreement of the simulated results and measured data showed the success of the design. As this work is part of the study on the BSN antenna, we would learn more about the system response of this kind of antenna to study the feasibility of the fractal antenna in the application of BSN systems.

S11 of the antenna getting more and more flat in the frequency band from 3–9 GHz. Figure 3 shows the antenna structure with four orders.

The antenna was fabricated on the PCB board Rogers 5880 with which the related permittivity and the tangential loss are 2.2 and 0.001 respectively. The fabricated antenna was shown in Figure 4, the whole size of the antenna was 70*60 mm^2.

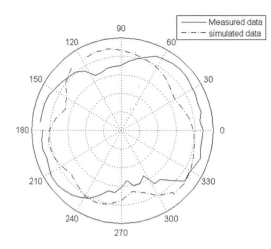

Figure 6b. The radiation patter of the E-plane.

ACKNOWLEDGEMENT

The research was supported by "The Fundamental Research Funds for the Central Universities".

REFERENCES

Azari, A. and Rowhani, J. 2008. Ultra wideband fractal microstrip antenna design. *Electromagnetics Research* 2, 7–12

Chahat, N. et al. 2011. A compact UWB antenna for on-body applications, *IEEE transactions on antenna and propagation* 59(4), 1123–1131

Porcino, D. and Hirt, W. 2003. Ultra-wideband radio technology: potential and challenges ahead, *IEEE Communication Magazine* 41(7), 66–74

Prasad, R.V.H. et al. 2000. Microstrip fractal patch antenna for multi-band communication, *Electronics Let.*, 36(14), 118

Yang, L. and Giamalkis, G. B. 2004. Ultra wide band communications, *IEEE Signal Processing Magazine*, 26–28

Information Technology and Computer Application Engineering – Liu, Sung & Yao (Eds)
© *2014 Taylor & Francis Group, London, ISBN 978-1-138-00079-7*

Estimation of parameter in a new discrete distribution analogous to Burr distribution

G. Nanjundan
Department of Statistics, Bangalore University, Bangalore, India

T. Raveendra Naika
Maharani's Science College for Women, Bangalore, India

ABSTRACT: Estimation of parameter in a new discrete distribution which is analogous to Burr distribution is discussed in this paper. The maximum likelihood and the method of moment estimators are obtained. The asymptotic normality of the moment estimator is established. The asymptotic relative efficiency of the maximum likelihood estimator over the moment estimator is computed.

Keywords: Maximum likelihood and moment estimators, Fisher information, asymptotic relative efficiency

1 INTRODUCTION

Sreehari (2010) has characterized a class of discrete distributions which turns out to be analogue of Burr (1942) family. The probability mass function (pmf) of the random variable X of the d-th distribution of the class characterized by Sreehari (2010) is

$$p(x,\theta) = \begin{cases} (x+1-\theta)\dfrac{\theta^x}{(x+1)!}, & x = 0,1,2,\ldots,0<\theta<1 \quad (1.1) \\ 0, \text{ otherwise}. \end{cases}$$

The distribution function of X is

$$F(x) = \begin{cases} 0, & x < 0 \\ 1 - \dfrac{\theta^{[x+1]}}{([x+1])!}, & x \ge 0 \end{cases} \quad (1.2)$$

We refer to this distribution as $S(d)$ distribution. The probability generating function (pgf) of X is

$$f(s) = E(s^X)$$

$$= \frac{1-(1-s)e^{\theta s}}{s}, \quad 0 < s \le 1.$$

The mean and the variance of X are respectively

$$E(X) = f'(1)$$

$$= e^\theta - 1 \qquad \text{and}$$

$$Var(X) = f''(1) + f'(1) - (f'(1))^2$$

$$= e^\theta(2\theta - e^\theta + 1).$$

It can easily be seen that E(X) > Var(X) and hence S(d) – distribution is under dispersed. The pmf of this distribution is similar to that of Poisson distribution in structure but its mean and variance are equal. Therefore the S(d) distribution can be a suitable model for the data exhibiting under dispersion and Poisson is not a good fit.

2 MAXIMUM LIKELIHOOD ESTIMATION

If $\underline{X} = (X_1, X_2, \ldots, X_n)$ is a random sample on X having the pmf specified in (1.1), then the likelihood function becomes

$$L(\theta|x) = \prod_{j=1}^{n} P(X = x_j)$$

$$L(\theta \mid x) = \prod_{j=1}^{n}(x_j + 1 - \theta)\frac{\theta^{x_j}}{(x_j+1)!}.$$

The log-likelihood function is

$$\log L(\theta \mid x) = \sum_{j=1}^{n}\log(x_j + 1 - \theta) + \sum_{j=1}^{n} x_j \log\theta - \text{constant}.$$

The likelihood equation is

$$\frac{d\log L(\theta \mid x)}{d\theta} = -\sum_{j=1}^{n}\frac{1}{x_j + 1 - \theta} + \frac{\sum_{j=1}^{n} x_j}{\theta}.$$

The maximum likelihood estimate of θ is the solution of

$$\sum_{j=1}^{n} \frac{\theta}{x_j + 1 - \theta} - n\bar{x} = 0$$

where $\bar{x} = \frac{1}{n}\sum_{j=1}^{n} x_j$. It is evident that this likelihood equation does not yield a closed form expression for the maximum likelihood estimate (MLE) of θ. Hence a numerical procedure like Newton-Raphson method can be employed to compute the MLE $\hat{\theta}_{mle}$.

3   FISHER INFORMATION

When X has the pmf specified in (1.1),

$$\log p(x, \theta) = \log(x + 1 - \theta) + x\log\theta - \log(x + 1)!$$

Also $\dfrac{d\log p}{d\theta} = \dfrac{-1}{x+1-\theta} + \dfrac{x}{\theta}$

and $\dfrac{d^2\log p}{d\theta^2} = \dfrac{-1}{(x+1-\theta)^2} - \dfrac{x}{\theta^2}.$
Hence

$$E\left[\frac{d^2\log p}{d\theta^2}\right] = -\sum_{x=0}^{\infty} \frac{1}{(x+1-\theta)^2}p(x) - \sum_{x=0}^{\infty} \frac{x}{\theta^2}p(x)$$

$$= -\sum_{x=0}^{\infty} \frac{1}{(x+1-\theta)} \frac{\theta^x}{(x+1)!} - \frac{1}{\theta}\sum_{x=0}^{\infty} \frac{\theta^{x+1}}{(x+1)!}$$

$$E\left[\frac{d^2\log p}{d\theta^2}\right] = -\sum_{x=0}^{\infty} \frac{1}{(x+1-\theta)} \frac{\theta^x}{(x+1)!} - \frac{1}{\theta^2}(e^\theta - 1).$$

The Fisher information becomes

$$I(\theta) = -E\left[\frac{d^2\log p}{d\theta^2}\right] = \sum_{x=0}^{\infty} \frac{1}{x+1-\theta} \cdot \frac{\theta^x}{(x+1)!}$$
$$+ \frac{1}{\theta^2}(e^\theta - 1)$$

The infinite series evidently converges for all values of θ in (0, 1) but its sum is not tractable. Hence it can be numerically computed correct to desired number of decimal places.

Note that

i) the support $S = \{x : p(x, \theta) > 0\}$ of X does not depend on the parameter θ
ii) The parameter space (0, 1) is an open interval.
iii) $\log p(x, \theta)$ can be differentiated thrice w.r.t. θ
iv) $\sum_{x=0}^{\infty} p(x, \theta) = 1$ is twice differentiable under the summation sign.
v) there exists a function $M(x)$ such that $\left|\frac{d^3\log p(x,\theta)}{d\theta^3}\right| \leq$
    $M(x)$ and $E[M(X)] < \infty$.

The function $M(x)$ may depend on θ. In our case,
$$\frac{d^3\log p(x,\theta)}{d\theta^3} = \frac{-2}{(x+1-\theta)^3} + \frac{2x}{\theta^3}.$$ Therefore
$$\left|\frac{d^3\log p(x,\theta)}{d\theta^3}\right| \leq \left|\frac{2}{(x+1-\theta)^3}\right| + \left|\frac{2x}{\theta^3}\right|.$$
Since $0 < \theta < 1, 0 < 1 - \theta < 1$. Hence

$$\left|\frac{d^3\log p(x,\theta)}{d\theta^3}\right| \leq \left|\frac{2}{x^3}\right| + \left|\frac{2x}{\theta^3}\right|$$

$$\leq \frac{2}{x} + \frac{2x}{\theta^3}$$

$$\leq 2 + \frac{2x}{(\theta_0 - \varepsilon)^3} = M(x),$$

$\forall\, \theta \in (\theta_0 - \varepsilon,\, \theta_0 + \varepsilon),\, \varepsilon > 0$ and $x \geq 1$.

where θ_0 is the true value of the parameter. Since the parameter space is an open interval such a neighborhood of θ_0 exists. Evidently, $E[M(X)] < \infty$.

Therefore $p(x, \theta)$ satisfies the regularity conditions of Cramer (1966) and it belongs to Cramer family. Hence if $\underline{X} = (X_1, X_2, \ldots, X_n)$ is a random sample on X having the pmf specified in (1.1) and $\hat{\theta}_{mle}$ is the MLE of θ, then

$$\sqrt{n}\left(\hat{\theta}_{mle} - \theta\right) \xrightarrow{L} N\left(0, \frac{1}{I(\theta)}\right), \text{ as } n \to \infty.$$

That is $\hat{\theta}_{mle}$ is consistently and asymptotically normal (CAN) for θ with the asymptotic variance $\frac{1}{I(\theta)}$.

4   METHOD OF MOMENT ESTIMATION

When $\underline{X} = (X_1, X_2, \ldots, X_n)$ is a random sample on X having the pmf specified in (1.1), the moment estimator of θ is the solution of the equation $e^\theta - 1 = \bar{X}_n$. Hence the moment estimator of θ is $\hat{\theta}_{mme} = \log(\bar{X}_n + 1)$.

Since the pmf of X does not belong to the exponential family, we need to establish the asymptotic normality of the moment estimator separately. The following theorem states the asymptotic normality of the moment estimator.

Theorem: If $\underline{X} = (X_1, X_2, \ldots, X_n)$ is a random sample on X having the pmf specified in (1.1), then

$$\sqrt{n}\left(\hat{\theta}_{mme} - \theta\right) \xrightarrow{L} N\left(0, e^\theta(2\theta - e^\theta + 1)\frac{1}{(\theta+1)^2}\right), \text{as } n \to \infty.$$

Proof: Since $X_1, X_2, \ldots, X_n$ are iid with $E(X) = e^\theta - 1$ and $Var(X) = e^\theta(2\theta - e^\theta + 1) < \infty$, by Levy-Lindeberg central limit theorem

$$Z_n = \sqrt{n}\left(\bar{X}_n - (e^\theta - 1)\right) \xrightarrow{L} N(0, e^\theta(2\theta - e^\theta + 1)), \text{as } n \to \infty.$$

Table 1.	ARE corresponding to various values of θ.				
θ	0.1	0.2	0.3	0.4	0.5
ARE	1.3405	2.3291	3.0257	3.8895	4.934
θ	0.6	0.7	0.8	0.9	0.95
ARE	6.1455	6.1452	7.1479	8.6441	9.0807

Take $g(x) = \log(x + 1)$. Then $g'(x) = \frac{1}{x+1}$ is non-vanishing and continuous for $0 < x < 1$. Hence by Mann-Wald theorem stated in Mukhopadhaya (2000),

$$\sqrt{n}\left(\log(\bar{X}_n + 1) - \theta\right) \xrightarrow{L} N\left(0, \frac{e^\theta(2\theta - e^\theta + 1)}{(\theta+1)^2}\right), \ as\ n \to \infty.$$

That is $\hat{\theta}_{mme}$ is CAN with the asymptotic variance

$$e^\theta(2\theta - e^\theta + 1)\frac{1}{(\theta+1)^2}.$$

The asymptotic relative efficiency of the MLE over the MME is given by

$$ARE = \frac{Asymptotic\ var\ iance\ of\ MME}{Asymptotic\ var\ iance\ of\ MLE}.$$

Since the asymptotic variance $\frac{1}{I(\theta)}$ of the MLE does not have a closed form expression, it is computed for various values of θ and the ARE is shown in Table 1.

The ARE of the MLE over the MME is uniformly greater than unity and therefore the MLE is asymptotically more efficient than the MME.

5 ILLUSTRATION

Sreehari (2010) has compared the fit of S(d) and Poisson models for a clinical trial data set. A bioequivalence study was conducted for a test drug (T) and a reference drug (R) by administering them to 144 individuals using a two period, two sequence and two treatment crossover design. The time until maximum concentration (T_{max}) was one of the characteristics observed. Let T_{it} and T_{ir} respectively denote the T_{max} values corresponding to the i-th individual for the test and the reference drugs. Take $D_i = |T_{it} - T_{ir}|$. The individuals were administered the drugs and their blood concentrations were measured just prior to medication and at time points (in hours) 1, 2, 4, 6, 8, 10, 12, 14, 16, 20, 22, 24, 30, 48, 60, 72, 96, 120 after medication. Of the 144 individuals 7 did not complete the course of treatment. The observed values of D_i for the 137 individuals are shown in Table 2.

For this observed distribution, mean $= 0.583942 >$ variance $= 0.534925$ and it exhibits under dispersion. And $\hat{\theta}_{mme} = \log(\bar{x}_n + 1) = 0.4599$. Further, it is evident that S(d) fits better than the Poisson model to this observed distribution. But S(d) model cannot be fitted to all under dispersed data. We have $0 < \theta < 1$ and

Table 2.	The observed values of $D_i = |T_{it} - T_{ir}|$.				
x	0	1	2	3	Total
Frequency	75	46	14	2	137

Table 3. The comparison of Chi-square value between S(d) and Poisson distribution.

		Expected frequency	
x	Observed frequency	S(d)	Poisson
0	75	73.9915	76.4043
1	46	48.5192	44.6157
2	14	12.268	13.0265
3	2	2.22129	2.95354
Chi-square value		**0.41111**	**0.64342**

Table 4. The data on the number of scintillations from radioactive decay of Polonium.

x	0	1	2	3	4	5	6
f	57	203	383	525	532	408	273

x	7	8	9	10	11	12	13	14
f	139	45	27	10	4	0	1	1

(x: number of scintillations, f: frequency).

hence $0 < E(X) < e - 1$. If $0 < \bar{x} < e - 1$ is violated, then it leads to $\hat{\theta}_{mme} = \log(\bar{x}_n + 1) > 1$ which is inadmissible. For example, consider the data on the number of scintillations from radioactive decay of Polonium reported by Rutherford and Geiger (1910). This data set has been reproduced in Santner and Duffy (1989).

For the observed data, mean $= 3.871549 >$ variance $= 3.694773$. This exhibits under dispersion. But $\hat{\theta}_{mme} = \log(\bar{x}_n + 1) = 1.58341 > 1$, which is inadmissible.

6 SIMULATION

Random observations on X can be simulated using the distribution function (1.2). A modest simulation study has been carried out to study the performance of the MLE and the MME of θ. Using R, 1000 samples of size 50, 100, 150 were simulated for the specified value of $\theta = 0.4$ and the estimates were computed. The MLEs were obtained by solving the likelihood equation by Newton-Raphson method and the MMEs were taken as the initial estimates. The histograms of the estimates are displayed in Figures 6.1–6.3.

These histograms give graphical evidence for the asymptotic normality of both the estimates. But the MLE approaches normality faster than the MME.

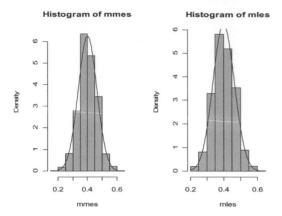

Figure 6.1. Histogram of the MMEs and the MLEs of θ based on 1000 sample of size 50 for $\theta = 0.4$.

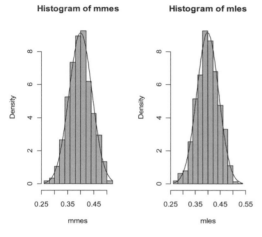

Figure 6.2. Histogram of the MMEs and the MLEs of θ based on 1000 sample of size 100 for $\theta = 0.4$.

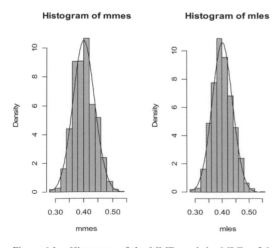

Figure 6.3. Histogram of the MMEs and the MLEs of θ based on 1000 sample of size 150 for $\theta = 0.4$.

7 DISCUSSION AND CONCLUSION

The S(d) distribution is an appropriate alternate to Poisson model when the observed data exhibit under dispersion. Though the maximum likelihood estimator of the parameter of S(d) distribution has no closed form expression, it can easily be computed by Newton-Raphson method. The moment estimator of the parameter has a closed form expression and easy to compute. Both the estimators are asymptotically normal. When computer facility is available, the MLE can be preferred to the MME since the former approaches normality faster than the later.

ACKNOWLEDGEMENT

The authors are grateful to Prof. M. Sreehari for sending a softcopy of his paper and to Professors K. Suresh Chandra and B. Chandrasekar for useful comments.

REFERENCES

Burr, I.W. 1942. Cumulative frequency functions, USA, Annals of Mathematical Statistics, pp. 215–232.
Cramer, H. 1966. Mathematical Methods of Statistics, Princeton, 11th Print, Princeton University Press.
Kale, B.K. 1999. A First Course on Parametric Infe rence, New Delhi, Narosa Publishing House.
Mukhopadhyay, N. 2000. Probability and Statistical Inference, New York, Marcel Dekker, pp. 261.
Rao, C.R. 1973. Linear Statistical Inference and Its Applications, New York, John Wiley.
Rutherford, E & Geiger, H. 1910. The probability variations in the distribution of α particles, Phil, Mag. Sixth Ser., 20, pp. 698–704.
Santner, T.J. & Duffy, D.E. 1989. The Statistical Analysis of Discrete Data, New York, Springer Verlag.
Sreehari, M. 2010. On a class of discrete distributions analogous to Burr family, Pune, Journal of Indian Statistical Association, Vol. 48, No. 1.

Information Technology and Computer Application Engineering – Liu, Sung & Yao (Eds)
© *2014 Taylor & Francis Group, London, ISBN 978-1-138-00079-7*

Application of artificial neural network on objective wearing pressure comfort evaluation model

X.L. Meng
Shanghai Institute of Visual Art, Shanghai, China

W.L. Wang
Fashion Institute of Donghua University, Shanghai, China

K. Liu
Shanghai Institute of Visual Art, Shanghai, China

W. Zhu
Shanghai University of Engineering Science, Shanghai, China

ABSTRACT: In this paper, the objective wearing pressure comfort evaluation model was established with Artificial Neural Network. Sixty fabrics with elasticity were made up into two hundred fifty eight samples to evaluate the wearing pressure comfort. Twelve females were selected as subjects. The physical parameters of fabrics were measured by Fabric Assurance by Simple Testing system or computing on through formula. The four principal factors obtained through data reduction with factor analysis method, the initial elastic modulus of fabrics and clothing allowance rate were entered the model as inputs parameters. The tightness, compression and comfort, as outputs, were exported from the model. The objective wearing pressure comfort evaluation model was established with Artificial Neural Network named Radial Basic Function Neural Network. The Pearson coefficient and the linear regression between predicted values and targets are both good. The objective wearing pressure comfort model has a practical significance.

Keywords: Artificial Neural Network; Comfort; Factor Analysis; Objective Evaluation

1 INTRODUCTION

1.1 *Type area*

As well known, many reports studied the wearing thermal and moisture comfort. Many research achievements have been applied in common life, aerospace and military etc. Recently research shows that wearing touch and pressure comfort are two main factors which have significant effect on the whole wearing comfort [1, 2]. However the study on touch and pressure comfort is a relative new field. The principles how people's touch and pressure sensation comes is unclear. The touch and pressure sensation often acquired from subjective experiment traditionally. Traditional method usually spends much time and the results could be affected by the subjects.

So in this paper an objective pressure evaluation model was established by the Artificial Neural Network.

The analysis indicates that the clothing allowance and fabric's physical properties both have correlation with wearing pressure sensation [3]. But how to find the exact relationship among them is a problem.

The Artificial Neural Network has strong capability of nonlinear mapping. Many predicted model based on ANN have in practice [4, 5]. Therefore ANN was tried to establish the objective model of pressure sensory evaluation. BP network which is based on the Error Back Propagation Training Algorithm has been widely used. But BP network often get local optimum result. To avoid the local optimization the Radial Basic Function Neural Network was adopted to create the objective evaluation model on wearing pressure sensation. The Radial Basic Function has the feature to response locally. So the accuracy of RBFNN is better. And the convergence speed is quick. In addition its structure is simple. The parameter which should be set is only one. So the human intervention on network is small. The main point is RBFNN can overcome the local optimum effectively [6].

The results show that the objective model on pressure sensory evaluation is better. The predicted values have strong positive correlation with targets. And the predicted values are close to targets. The predicted degree is up to 92% except model on comfort.

Table 1. Basic parameters of fabrics.

Parameter	Symbol	Condition	Unit
		Interpretation	
thickness	T2	0.196 kPa	mm
	T100	9.81 kPa	mm
surface thickness	ST	formula	mm
bending length	C	bending angle is 41.5°	mm
Beding rigidity	B	formula	μN · mm
elongation rate	E5	4.9 cN/cm	%
	E20	19.6 cN/cm	%
	E100	98.1 cN/cm	%
Biased elongation rate	EB5	4.9 cN/cm	%
Shearing stiffness	G	formula	N/m

Table 2. Psychological Ruler.

Sensation	Score							Sensation
loose	3	2	1	0	−1	−2	−3	tight
unconstricted	0	0	0	0	−1	−2	−3	constricted
comfortable	3	2	1	0	−1	−2	−3	discomfortable

Table 3. Specifications of Four Principal Factors.

Principal Factors	Specification		
	Symbol	Significance	Physical Property
factor1	F1	Bending and Shear	C1,C2,C3,B1,B2, B3,G
factor2	F2	Vertical and Biased Tensile	E51,E201,E1001, EB20,EB100
factor3	F3	Horizontal and Biased Tensile	E52,E202,E1002, EB5
factor4	F4	Thickness	T1,T2,ST

2 EXPERIMENTAL

2.1 Fabric and samples

Sixty fabrics with good elasticity were selected in experiment. Fabrics were sewed into two hundred and fifty eight samples with different sizes. The samples cover the whole human forearm from wrist to the upper of elbow. The difference between human body and samples was defined as clothing allowance. The ratio of clothing allowance to human body was named as allowance rate in this paper.

2.2 Fabric physical properties measurement

The stretch curves of fabrics were obtained with the YG(B)026-500 type material testing instrument. And the initial elastic modulus of each fabric was calculated according to the curve. The Fabric Assurance by Simple Testing was used to measure others fabric physical properties. The details of fabric physical properties were shown in Table 1. The vertical elongation rate was marked as E51, E201, and E1001. The horizontal elongation rate was marked as E52, E202 and E1002. The vertical, horizontal and biased bending lengths were marked as C1, C2 and C3 respectively. The biased elongation rates with 19.6 cN/cm and 98.1 cN/cm were denoted as EB20 and EB100 respectively. Then the bending rigidities with three directions were labeled as B1, B2 and B3. Thus nineteen parameters were measured or formulated with FAST system.

2.3 Subjective evaluation on pressure comfort

Twelve females who major in clothing engineering in university were selected as human subject models in the experiment. The psychological ruler was designed for sensory evaluation. The design of ruler was displayed in Table 2. The tightness, constriction and comfort were selected as the sensory factors which would evaluated by the subjects. After being trained the subjects should grade the samples on the three sensory factors when they were in dressed them.

3 THE CONSTRUCTION OF PRESS EVALUATION MODEL

3.1 The data induction

Nineteen physical parameters of fabrics were simplified with factor analysis method. Four principal factors were extracted to characterize the fabric with SPSS. The specifications of principal factors were displayed in Table 3.

3.2 The design and training of RBFNN

The Radial Basis Function Neural Network has the characteristic of local response. The model based on RBFNN can avoid the problem of local optimization in BP network.

The four principal factors, the initial elastic modulus and allowance rate were as the inputs of the sensory evaluation model by RBFNN. And the three sensations in subjective experiment were as the outputs. So the structure of RBFNN was established.

MATLAB was used to accomplish the computing of RBFNN. 'Newrbe' is the function to establish the RBFNN. Three sample sizes 110 and 160 were set to train the model of RBFNN.

Outputs are the values of objective evaluation from model. And the targets are the values of subjective evaluation from experiment.

The analysis shows that the accuracy became better when the SPREAD value is equal 0.2. The constructed model was named as RBFNN110_0.2, and RBFNN160_0.2.

Table 4. Pearson coefficient R of Models.

Sensation	RBF110	RBF160
Tightness	0.924	0.939
Constriction	0.833	0.806
Comfort	0.920	0.852

The original coding of standardize by Matlab as follows.

```
function y = StandardizeData(P)
%Standardize data
pSize = size(P);
for i = 1:pSize(1,1)
    P2(i, :) = (P(i, :) − min(P(i,: ))) / (max(P(i,:)) −
min(P(i,:)));
end
```

The original coding of RBFNN by Matlab as follows.

```
function net = ConstructRBF(P2, T2, n, P_Test)
% Construct a new RBF network

% Standardize data
P2 = StandardizeData(P2);

% Set spread factor
SPREAD = n;
net = newrbe(P2, T2,SPREAD);

% Network Simulation and Test
y = sim(net, P_Test)
```

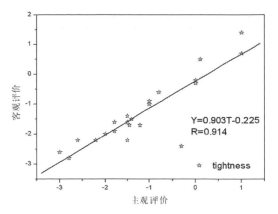

Figure 1. The regression of tightness.

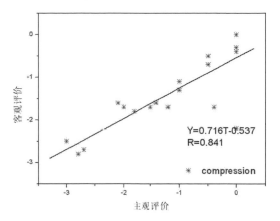

Figure 2. The regression of compression.

4 TESTING OF THE RBFNN PRESSURE COMFORT MODEL

Twenty five samples were chosen randomly from the rest as testing samples for RBFNN pressure models. Least-squares method was used to analyze the correlation between outputs and targets. R was set as the Pearson coefficient between them. If R is more than 0.8 the model was trained successfully. Otherwise the model did not pass the testing. The values of R were displayed in Table 4.

The coefficients between outputs by RBFNN and targets on tightness, constriction and comfort are all over 0.8.

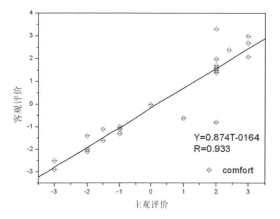

Figure 3. The regression of comfort.

5 PREDICTION OF MODEL AND CONCLUSIONS

5.1 Examining samples

Another twenty five samples which are different from the training and testing samples were picked randomly to examine the ability of objective evaluation model.

5.2 Results

When the absolute distance between outputs from model and targets is more than 0.5 the evaluation is judged as wrong. Thus the predicted accuracy rate was counted. The Pearson coefficients R between outputs and targets are obtained with SPSS.

Table 5. Relative coefficient and Accuracy Rate.

Sensation	RBFNN Model		
	index	110	160
tightness	R	0.914	0.969
	A	96%	92%
compression	R	0.841	0.974
	A	96%	92%
comfort	R	0.933	0.966
	A	68%	76%

Figure 1 to Figure 3 show the linear fitting, regression and Pearson coefficient R of tightness, compression and comfort between outputs from RBFNN110 and targets from experiment. Y means outputs, T means targets.

5.3 Subjective evaluation on pressure comfort

The predicted accuracy rate was marked as A. Pearson coefficient R and predicted accuracy rate A were displayed in Table 5.

The analysis results show that the outputs from model have strong positive correlation with the targets. The coefficient of regression on tightness, compression and comfort are all close to 1. The pots disturbed closely to the regression line. The predicted accuracy rate of tightness and compression model is over 92%.

The accuracy of comfort is lower than the other factors. The accuracy rate of comfort is 68% and 76% respectively.

Based on the above analysis the RBFNN objective evaluation model on pressure sensations is statistically significant. And the testing states that RBFNN Model has good ability of prediction in this paper.

REFERENCES

[1] Li, Y. 1998. Dimensions of sensory perceptions on next-to-skin wear in a cold environment *Journal of China Textile University*, vol. 15: 50–53.
[2] Byrne, M. S. Garden, A. P. W. & Fritz, A. M. 1993. Fibre types and end-uses: a perceptual study *Journal of Textile Institute* vol. 82:275–288.
[3] Meng, X. L. Study on Wearing touch and pressure sensation and the mathematical model of contact pressure, *unpublished*.
[4] Gao, J. 2003. The principle of the Artifical Neural Network and simulation *Peking: Mechanical Engineering Press, 2003*
[5] Zhang, L. M. 1993. The model and application of the Artifical Neural Network, *Shanghai: Fudan University Press, 1993*
[6] Ge, Z. X. & Sun, Z. Q. The thoery of the Artificial Neural Network and realization with MatlabR2007, *Peking: Electromic Enigeering Press, 200.*

Information Technology and Computer Application Engineering – Liu, Sung & Yao (Eds)
© *2014 Taylor & Francis Group, London, ISBN 978-1-138-00079-7*

Heap adjustment algorithms based on complete binary tree structure and based on array

Yuan Hu

Department of Computer Technology and Applications Qinghai University, Xining, Qinghai, China

ABSTRACT: For the given complete binary tree, the heap adjustment algorithm based on array and based on complete binary tree structure are proposed, making the data set have the nature of heap. While Useing different algorithms to achieve the same conversion, heap adjustment algorithms based on complete binary tree structure is the better algorithm by experiments, which utilizes characteristics of priority queue (delete the earliest data) and stack (delete the latest data), and conducts stack pushing on binary tree node, then accesses node successively to call glide adjustment algorithm to improve the adjustment efficiency.

Keywords: Complete binary tree; Heap; Glide adjustment; Array adjustment

1 GENERAL INSTRUCTIONS

In practical applications, data set with priority queue[1] produces convenience for various searching. Searching is to search for data elements that meet certain conditions from data set, which is one of the most common operations in data processing. For searching structure organized by different ways, the corresponding search methods are different as well. The nature of searching structure also tends to affect the searching speed. For data set exists in the form of complete binary tree[1,2] structure, adjust before searching can make data set have the nature of heap[2,3], improving searching speed. The adjusting method is concerned with the efficiency of the whole process.

At present, no records of adjusting given complete binary tree to give it the nature [3] of heap and retain the original data structure form have been found. In the process of the teaching project, there are new algorithm which could be directly written into textbooks for students having a deeper understanding of these algorithms.

For the heap adjustment of a complete binary tree, two adjustment algorithms are proposed. In the early stage, after grasping the heap adjustment[3] algorithm of array, apply this algorithm to the heap adjustment algorithm of complete binary tree to reach the purpose of adjustment.

Use linear list[3] to adjust the position of elements, and then convert into the original nonlinear list structure to replace the initial data set. This heap adjustment algorithm process based on linear list is more complex, and the complex rate of algorithm is too high. In order to reduce the algorithm's complexity in space and time, through the contrast test, the heap adjustment algorithm based on complete binary tree structure is proposed.

2 PROBLEM DESCRIPTION

2.1 *Heap adjustment algorithm based on array*

Convert a given complete binary tree accessed through a linked list[2,3] structure into a complete binary tree accessed through an array structure, and then adjust the array item positions, so as to give the nature of heap to an array set, and in the end, create a complete binary tree with the nature of heap on the basis of the above-said array.

2.2 *Heap adjustment algorithms based on complete binary tree structure*

Adjust node position directly on the given complete binary tree structure, without other intermediate conversions, only stack[1,2,3] and queue[1,2,3] are needed as carrier of intermediate record to make adjusted binary tree has the nature of heap.

3 THOUGHTS AND PROCESS OF THE ALGORITHM

Below provides the algorithm analysis with object-oriented method and C++ language, taking the adjustment of complete binary tree with keys[1,3] into complete binary tree with minimum heap property as an example, making the keys of adjusted complete

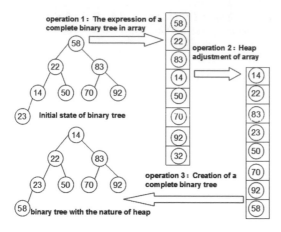

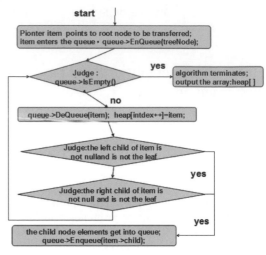

Figure 1. Steps of adjustment.

Figure 2. The expression in array.

binary tree node less than the keys of its child node around.

Use correlated variables and data structure in the algorithm:

Stack stack;//The global variable stack, used as carrier of intermediate record
template<class t>
struct TreeNode//Data structure of binary tree node
{
t date;
TreeNode<t>*Leftchild,*Rightchild;}
TreeNode head[];

3.1 Heap adjustment algorithm based on array

The whole adjustment process falls into three steps. First of all, the expression of the complete binary tree in array: convert a complete binary tree in the linked list structure into a complete binary tree in array structure. Second, heap adjustment of the array[3]: do glide adjustment to the array heap[] to give the nature of heap to the array heap[]. And in the end, the creation of a complete binary tree: create a complete binary tree on the basis of the heap[] array to replace the original complete binary tree. Figure 1 is the diagram of the three-step operation of this adjustment algorithm.

The expression of a complete binary tree in array: Basic thought: number the n nodes of a complete binary tree from top to bottom, and from the left to right in the same layer continuously in the manner of $0, 1, 2, 3, \ldots$, n; then use the queue as the intermediate carrier, and place the various nodes of the tree in the array heap[] in numerical sequence, i.e., heap[i] is the node with the number i. Figure 2 shows the algorithmic thought of the operation in a flow chart.

Heap adjustment of array: Take each item of array heap[] as the object of adjustment in turn, so that the adjusted array heap[] meets the condition of $heap[i] \leq heap[2i + 1]$ and $heap[i] \leq heap[2i + 2]$ (or $heap[i] \geq k[2i + 1]$ and $k[i] \geq k[2i + 2]$). The adjustment process can be divided into the glide adjustment and overall heapification of each node. The following

As for any given heap H_0 and H_1 and Node p, in order to get heap $H_0 \cup \{p\} \cup H_1$, we only need to take node ro and r1 as the children of p, and then conduct the glide adjustment of p.

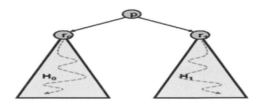

Figure 3. The overall ideology of adjustment.

code implements the glide adjustment of one node. In the meantime, during the glide adjustment of each node, its sub-tree has been converted into adjusted tree with the nature of heap. Figure 3 shows the operation thought of glide adjustment of each node from top to bottom.

The arithmetic codes implementation of glide adjustment of a node:

```
void MinHeap<t>::SiftDown(int start,int end)
{
    int i=start,j=2*i+1;
    t itme=heap[i];
    while(j<=end)
    {
    If(heap[j]>heap[j+1]&&j<end)j++;
    If(itme<heap[j])break;
    else
    {
        heap[i]=heap[j];
        i=j;
        j=2*i+1;
    }
    }
    heap[i]=itme;
}
```

408

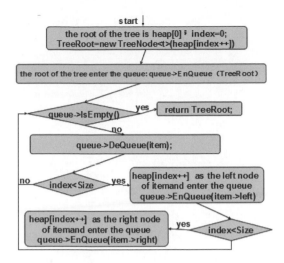

Figure 4. Creation of a complete binary tree.

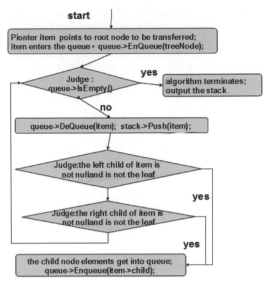

Figure 5. Stack Pushing of Binary Tree Node.

Creation of a complete binary tree: The operation is, in fact, an inverse operation of expression of a complete binary tree in array, i.e., take the items of the array heap[] as the nodes to create a complete binary tree expressed in a linked list, and after assigning consecutive numbers to the items of the binary tree from top to bottom, and from left to right in the same layer, the node with the No. i is the array item heap[i]. Figure 4 shows the algorithmic thought of the operation in a flow chart.

3.2 Heap adjustment algorithms based on complete binary tree structure

The adjustment process is divided into two steps. Firstly, conduct stack pushing on elements of binary tree node. Secondly, call glide adjustment algorithm by using pull node successively as actual parameter. When the stack is empty, heap adjustment of binary tree finishes and algorithm will come to an end. Because the pre-condition of glide adjustment algorithm is assuming the subtree is adjusted and with heap property, therefore, in the whole process of adjustment, access sequence of adjusted node must be sorted as requested before calling the glide adjustment algorithm.

Stack Pushing of Binary Tree Node Elements: Core idea of algorithm: By virtue of the queue characteristics, the node is recorded in the stake by the order of bottom-up for complete binary tree node, and right-to-left in the same layer. Here it is worth noting that leaf node[1, 3] does not need stack pushing. Figure 5 is the thought flow chart of stack pushing algorithm.

Algorithm program is as following:

```
template<class t>
void stackPush(TreeNode<t>*
&treeNode,Stack<TreeNode<t>*>*stack)
    {
    Queue<TreeNode<t>*>
```

```
*queue=new Queue<TreeNode<t>*>(100);
queue->EnQueue(treeNode);
TreeNode<t>*item;
while(queue->IsEmpty()==false)
{
queue->DeQueue(item);
stack->Push(item);//The global variable:stack
if(item->Leftchild!=NULL&&(item->Leftchild-
>Leftchild!=NULL||item->Leftchild-
>Rightchild!=NULL))
    queue->EnQueue(item->Leftchild);
if(item->Rightchild!=NULL&&(item-
>Rightchild->Leftchild!=NULL||item->Rightchild-
>Rightchild!=NULL))
    queue->EnQueue(item->Rightchild);
}}
```

Glide Adjustment Algorithm: The tree with a node as root as root is adjusted into minimum heap in the algorithm. Basic idea: Starting with the item node and adjust downwards, the precondition is to assume its subtree is adjusted and with heap property. If the key of item's left child less than the right key, then adjust along the left branch of item node, otherwise, adjust along the right branch, at this time, the local pointer variable item_child records the branch node to be adjusted. Adjustment method is to compare the item key with item_child key: if the item key is greater than the item_child key, exchanging the content of the two nodes, and then adjust the next layer of item node, that is item=item_child, repeating the glide adjustment based on item node; if item key is greater than item_child key, there is no need to conduct any operations and algorithm terminates. The final result is the node with minimum key rise to the top of the heap, and subtree with minimum heap property is formed. The core code is as following.

After stack pushing the binary tree node, conduct pull operation on, call glide adjustment algorithm with stack node successively. From part to whole, gradually enlarge the minimum heap until adjusting the whole complete binary tree into the complete binary tree with minimum heap property.

Algorithm program is as following:

```
template<class t>
void TreeNode_SiftDown(TreeNode<t>*item)
  {
  TreeNode<t>*item_child;
  while(item->Leftchild!=NULL||item->Rightchild!=NULL)
    {
    if(item->Leftchild!=NULL&&item->Rightchild==NULL)
    item_child=item->Leftchild;
    else
    if(item->Leftchild==NULL&&item->Rightchild!=NULL)
    item_child=item->Rightchild;
    else
    item_child=(item->Leftchild->date>item->Rightchild->date)?item->Rightchild:item->Leftchild;
    if(item->date<item_child->date)
    break;
    else
    {
    t it=item->date;
    item->date=item_child->date;
    item_child->date=it;
    item=item_child;
    }}}
```

Adjust the node pointed by item pointer, to make the tree with such root node possess minimum heap property. Figure 6 is the heap adjustment algorithm chart for complete binary tree whose number of node is 10.

4 ALGORITHM COMPLEXITY ANALYSIS

4.1 Algorithm complexity analysis

In the following, the complete binary tree which contains n nodes is used as analysis object, whose problem scale is n. It can be known from the properties of complete binary tree, the depth [1, 3] of binary tree with n nodes is $k = \lceil \log 2(n+1) \rceil$.

Complexity of Heap Adjustment Algorithms Based on array Structure: Space complexity: the sum of space size of queues and arrays used in this algorithm: $S(3 \lfloor n/2 \rfloor)$, the unit is size of (Tree Node).

Time complexity: push operations are conducted for n times during the expression of the items of complete binary tree in array, and after that, pop operations are conducted for n times to unload them into the array. The time complexity of the binary tree dump operation is T(2n). When calling the glide adjustment algorithm, the number of while cycle is the height of the tree

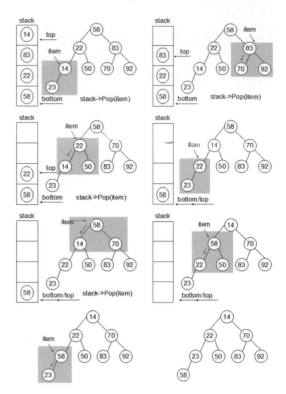

Figure 6. Glide adjustment from the bottom to top.

subtracting 1 at the maximum, and therefore the time complexity of the glide adjustment algorithm of the array is T(log 2n). Then operate each item of the array in turn, and call the glide adjustment algorithm of the array for n times. The time complexity based on the overall heap adjustment algorithm complexity of the array is T(n log 2n + 2n).

Complexity of Heap Adjustment Algorithms Based on Complete Binary Tree Structure: Space complexity: Space size of the stack and queue used in this algorithm is the same: size of $\lfloor n/2 \rfloor$ binary tree node byte.

Time complexity: when calling glide adjustment algorithm, the maximum number for while cycle will be the depth of tree minus 1, therefore, time complexity of glide adjustment algorithm will be O(log 2n). When complete binary tree elements is conducted stack pushing as well as push on operation for $\lfloor n/2 \rfloor$ times, and conducted pull operation for $\lfloor n/2 \rfloor$ times to call glide algorithm. The time complexity based on the overall heap adjustment algorithm complexity of the Complete Binary Tree Structure is T(n log 2n).

As the optimal algorithm, the heap adjustment algorithm based on complete binary tree structure has the advantages in reflecting considerable complexity of both time and space.

4.2 Comparison of algorithm tests results

Description: The heap adjustment algorithm mentioned above is referred to original algorithm there in

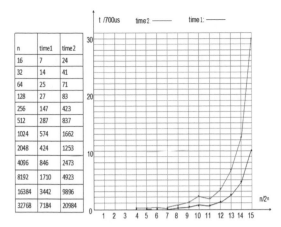

Figure 7. Charts and graphs of test results.

in linked list structure as the operation object, and proposed the heap adjustment algorithm based on complete binary tree as the best algorithm. The purpose of this paper is to show different algorithms to achieve the same conversion. These contents can be written into the textbooks, through comparative analysis, which can help students to have a deeper understanding of these algorithms. Scholars can be better appreciate the authenticity of the algorithm by these algorithms

after; the heap adjustment algorithm based on complete binary tree structure is referred to improved algorithm thereinafter.

In the experimental tests, conduct heap adjustment verification for the same complete binary tree with two methods every time, record respectively the improved algorithm execution time:time1 and original algorithm execution time:time2, unit of time is microsecond. In the following, according to the recorded data, the Figure 7 is drawn for observation and compare.

From Figure 7 we can draw a conclusion apparently: With the increasing number of binary tree node, processing rate of heap adjustment shows greater advantage. These results show that the performance of improved algorithm is better than the original algorithm.

5 CONCLUSIONS

This paper carried out the theoretic analysis, algorithm complexity analysis and comparison of test results of two algorithms, by taking the complete binary tree

ACKNOWLEDGEMENTS

This essay is partly funded by the Qinghai university education teaching and research project (No. 2012201301), based on the project-oriented of Data Structures and Algorithms courses. In addition, I want to express my gratitude to all the teachers and fellows who gave supports and advises to the essay, especially I want to express my appreciations to teacher Wang Xiao Ying and student Deng Yong Jie from the department of computer science and technology, Qinghai University.

REFERENCES

[1] Glenn W Rowe: Introduction to Data Structures and Algorithms with C++, Prentice-Hall Europe, (1997).
[2] Donald E Knuth: The Art of Computer Programming, Volume 3/Sorting and Searching. (Addison-Wesley Publishing Company, Inc. Phillipines, 1973).
[3] Ren Kun Yin: Data Structure (with object-oriented method and C++ language). Beijing: Tsinghua University Press (2007.6).

Information Technology and Computer Application Engineering – Liu, Sung & Yao (Eds)
© 2014 Taylor & Francis Group, London, ISBN 978-1-138-00079-7

A dynamic trusted measuring model based on optimum cycle

Yuan Tian
Company of Postgraduate Management, Academy of Equipment, Beijing, China

Qing Pan & Fei Wang
Department of Information Equipment, Academy of Equipment, Beijing, China

ABSTRACT: This paper presents a dynamic trusted measuring model based on optimum cycle, in order to handle the system's runtime security problems. By the method of dividing time into slices, a trust monitor will measure the critical system files dynamically, as well as the application files. It will detect the system's abnormalities and notify the user to take measures. This kind of operation is triggered by time slices and occupation rates of system resources. In this way, it can ensure the platform keeping in a safe and efficient state.

Keywords: Trusted Computing; dynamic measurement; trusted monitor

1 INTRODUCTION

With the continuous development of network technology, the computer terminals are increasingly vulnerable to hacker attacks. Trusted Computing technology provides a new idea to solve this problem. The research of trusted measurement can be divided into three categories:

The first is proposed by TCG[1], called static trusted measurement. It makes sure that the establishment of operating system environment is trusted by measuring critical system data ranging from bios to operating system. This kind of chain structure is relatively fixed. It is well compatible with existing computers.

The second is dynamic trusted measurement. The related works are as follows: Zhuang Lu[2] proposes the measurement based on extended action trace and verification based on action measurement information base. This model can distinguish the desired software behavior. Zhao Jia[3] presents the noninterference-based trusted chain model. It introduces noninterference theory into the domain of trusted computing to construct the trusted chain. Another closely related project is IBM's IMA[4]. Whenever the operating system loads a program into memory, it will validate the integrity of the program file.

The last is multiple integrity measuring of system data. After the trusted chain has established, there still exists the possibility of being attacked in the platform. Once some file was tampered, it would not be detected until rebooting. To solve the problem, hardware manufacturers have modified their microprocessors and chipsets that support multiple integrity measuring, such as Intel's LaGrande[5] and AMD's Presido project[6]. To some extend, multiple integrity

measuring is the domain of dynamic trusted measurement. However, the drawback to this approach is that we need to replace the current hardware and also, it is inflexible.

This paper present a Dynamic trusted measuring model based on optimum cycle. The trusted monitor will measure critical system data and application files periodically, with a minor effect on the platform's regular services. The model is applicable to both trusted computing platform and ordinary computing platform.

2 DESCRIPTION OF THE MODEL

After the operating system is booted, the trusted monitor firstly measure the integrity of itself, and then it takes over the execution of all executable files. The trusted monitor is responsible for two kind of jobs. One is integrity check based on trusted white list before any executable file starts to run. The other is dynamic measurement of critical system files and other executable files, during the system's runtime. It will determine whether some component is tampered.

The first job is a relatively popular technology, so this paper mainly studies the process of second job. The trusted monitor residents permanently in the memory, until the computer shuts down. The dynamic measure operation is triggered by the following conditions:

1) System's resources occupation rates fall below specified threshold values.
2) The time interval between two measurement operations reaches a specified value.

2.1 Definations

Defintion 1. Trusted monitor: mainly responsible for dynamic measurement. It can be expressed as P_{check}. Normally, P_{check} does not allow to be modified.

Defintion 2. P_{check} maintains the set of all the measurable entities, expressed as $S = \{s_i| s_i \in S, i \in N\}$. The set S includes critical system files, as well as files with the exe and dll extensions. Their expected corresponding digest values can be expressed as $H = \{h_i| h_i \in h, i \in N\}$

Defintion 3. The set of indicators, which represents system's current resources occupation rates, expressed as $E = \{e_i| e_i \in E, i \in N\}$. If $\forall e_i, \in E$, $e_i \leq \delta_i$, we say that system's resources occupation rates are low. Normally, we choose $n = 2 \cdot e_1$ represents cpu utilization; e_2 represents memory utilization.

Defintion 4. The minimal time interval between two measurement operations can be expressed as λ, and the current interval is t.

Defintion 5. At any time point when the system's running, the safe state $C = \{Trusted, Untrusted, Unknow\}$. The state function f: $s_i \times H \times e \times t \to C$.

$$f(s_i \times H \times e \times t) = \begin{cases} Trusted \\ Untrusted \\ Unknow \end{cases}$$

The conditions:

$Trusted: \forall e_i \in E, e_i \leq \delta_i \wedge t \geq \lambda \wedge Hash(s_1) = h_{s_1}$

$Untrusted: \forall e_i \in E, e_i \leq \delta_i \wedge t \geq \lambda \wedge Hash(s_1) \neq h_{s_1}$

$Unkown: \exists e_i \in E, e_i > \delta_i \vee t < \lambda$

2.2 Model description

Rule 1. At any time point, if $\exists e_i \in E$, which satisfies the in equation $e_i > \delta_i$ or if the inequality $t < \lambda$ holds, the system runs as normal.

Rule 2. Trusted monitor P_{check} dynamically validates the entities in the set S at the proper time. When it meets the condition $\forall e_i \in E, e_i \leq \delta_i \wedge t \geq \lambda$, this kind of operation will be triggered.

Rule 3. During the integrity measurement, if $\forall s_i \in S$, the inequality $Hash(s_i) \neq h_{si}$ holds, it means the integrity of s_i has been destroyed. P_{check} will display the warning information to the user.

Theorem 1. Trusted monitor can guarantee the integrity of itself.

Proof: Defintion 1 guarantees that P_{check} does not allow to be modified normally. Every time before it starts, it makes sure not to be tampered. In the trusted computing platform, P_{check} is a part of trusted chain, and the integrity measurement is done by operating system. In the ordinary platform, P_{check} measures the integrity of itself. So trusted monitor is able to guarantee the integrity of itself.

operating system environment

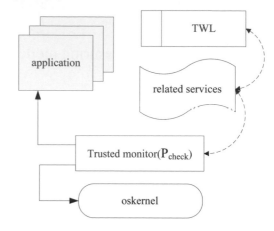

Figure 1. The implementation structure of this model.

Theorem 2. The trusted monitor will prevent the platform remaining in an untrusted state for a long time.

Proof: According to Rule 2, trusted monitor P_{check} dynamically measure the entities in the set Sat the proper time. When some element in the set S is tampered, it will be detected the next time dynamic measurement is triggered, because each element in set H does not meet the condition $Hash(s_i) = h_{si}$. The current state of the system is Untrusted. P_{check} will signal a warning to the user. Therefore the maximum length of time that the platform in an Untrusted is $\lambda + \lambda_0$. The value of λ_0 depends on the time length that the resources occupation rate is high after the minimal time interval λ.

3 IMPLEMENTATION

The implement of this model is based on ordinary platform, the workflows are as follows:

① After the booting of operating system, trusted monitor starts to run.

② The timer is started by P_{check}. When the current time interval reaches λ. P_{check} begins to monitor the system resource utilization. When $\forall e_i \in E, e_i \leq \delta_i$, P_{check} chooses some of the files in set S and calls related services to do the integrity measurement. According to Defination 5, the safe state of the platform can be calculated.

③ If the safe state is trusted, P_{check} sets the timer to zero and will switch to workflow ②. If the safe state is untrusted, it will take some corresponding measures.

The implementation structure diagram is shown in Figure 1.

The experiment is based on windows 7 operating system. The names of all the executable

Table 1. The structure of trusted white list.

file	Digest values
sms.exe	A81B48A5D6A06543ED36B7E6EA75C5E52B79DD37
user32.dll	1EEA81ADB02BFF134D2DFEBB4B4DB1AA1773674B
thunder.exe	7C46A4F8324BDD432B24DE470E3FF03CEC27A418
......	......

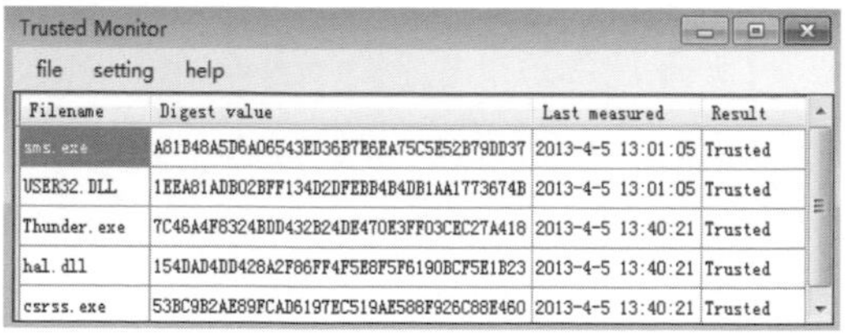

Figure 2. The interface of trusted monitor.

Figure 3. The warning interface when abnormalities are detected.

application files, names of critical system files and their expected corresponding digest values are stored in Trusted White List. We choose SHA-1 secure hashing algorithm. The critical system files include hal.dll, sms.exe, ntdetect.com, ntoskrnl.exe, csrss.exe, win32k.sys, services.exe, psxss.exe, os2ss.exe, user32.dll, ntdll.dll, gdi32.dll. A small part of TWL is listed in Table 1.

We choose the cpu and memory utilization to represents system's resources occupation rate. Trusted Monitor calls the function GetSystemTimes and GlobalMemoryStatusEx provided by windows API, to get these values above in realtime. The user is able to set time interval value to trigger integrity measurement. The trusted monitor interface is shown in Figure 2.

If any abnormality is detected, users can take measures to deal with problems, such as running setup to reinstall the application. We replace the file "thunder.exe" with the same file name and set the time interval to five minutes. After about six minutes, trusted monitor detects it and gives a warning, as shown in Figure 3.

4 ANALYSIS

4.1 Security analysis

This model will prevent the platform staying in an untrusted state.

Under the best condition, some executable file is tampered just before the dynamic measurement is triggered, so it can be detected in time.

Under the worst condition, some executable file is tampered just after a dynamic measurement finishes. The next time a dynamic measurement is triggered, this problem can be detected. So the time length that the platform stay in an untrusted state deponds on specified measurement interval and how long system resources occupation rates remain high. As for ordinary platforms, usually the system resources occupation rates are relatively low. Through setting proper threshold values and time interval, we can prevent the untrusted state timely.

4.2 Performance analysis

Any kind of integrity measurement will take system resources and then influence the system efficiency. This model validates integrity of files only when resources users need is few. In this way, it will reduce the impact on application layer in the maximum limit.

5 CONCLUSION

Recently, the risk of computer being attacked is increasing evidently. So people pay more attention to the security enhanced techniques for operating systems and application layer. This paper proposes an idea of a dynamic trusted measuring model. It gives the description of the model and discusses the concrete implementation. Furthermore, the security and performance of system is also analyzed.

REFERENCES

[1] Trusted Computing Group. 2003. *TCG PC Specific Implemetation Specification Version 1.1*. http://www.trustedcom-putinggroup.org.

[2] Zhuang Lu et al. 2010. *Software Behavior-Based Trusted Dynamic Measurement*. Journal of Wuhan University 56(2): 133–137.

[3] Zhao Jia et al. 2008. *A Noninterference-Based Trusted Chain Model*. Journal of Computer Research and Development 45(6): 974–980.

[4] Sailer R et al. 2004. *Design and implementation of a TCG-based integrity measurement architecture*. Proc of the 13th USENIX Security Symp. 223–238, 2004. Berkeley: USENIX.

[5] Intel Trusted Execution Technology. Software Development Guide Measured Launched Environment Developer's Guide. Document Number: 315168-005.2008.

[6] David Challener et al. 2008. *A Practical Guide to Trusted Computing*. Indianapolis: IBM Press.

Hierarchical ALS algorithm with constraint used in UBSS

Hong Yong Leng & Ming Jun Zhang
School of Information Science and Engineering, Xinjiang University, Urumqi, China

Wen Qiang Guo
School of Computer Science and Engineering, Xinjiang University of Finance and Economics, Urumqi, China

ABSTRACT: An improved Hierarchical Alternating Least Squares (HALS) algorithm for Non-negative Matrix Factorization (NMF) based Determinant criterion constraint has been discussed in this paper. Normally, NMF can be expressed as V = WH. In order to obtain a unique basis matrix W and weight coefficient matrix H when W is underdetermined, we always impose sparsity constraint to W and H. On the basis of HALS algorithm with sparsity constraint, we improve the method by imposing the determinant constraint to the basis matrix W. The improved algorithm is applied to solve the problem of underdetermined blind source separation when the sparsity of source signal is not sufficient. The simulation results show the effectiveness and efficiency of the improved method.

Keywords: Underdetermined BSS; NMF; HALS; Determinant criterion

1 INTRODUCTION

Blind Source Separation (BSS) as an effective signal processing technology has been widely used in the fields of remote sense, biomedical signal processing, digital communication, etc.

The Mathematical model of BSS is generally expressed as:

$$Y=AS+N \quad Y\in R^{m\times T}, A\in R^{m\times n}, S\in R^{n\times T}, N\in R^{m\times T} \quad (1)$$

where $Y = (Y_1, Y_2, \ldots, Y_m)^T$ is the mixed signal matrix observed, $S = (S_1, S_2, \ldots, S_n^T)$ is source matrix, $A = (A_1, A_2, \ldots, A_m)^n$ is mixture matrix, $N = (N_1, N_2, \ldots, N_m)^T$ is the errors of signal channel and noise.

In the past, a variety of algorithms based Independent Component Analysis (ICA) have been widely applied to the blind source separation in the case of over-determined $(m > n)$ and normal $(m = n)$. Actually, the number of sensors should be less than the number of source signals in many cases. For the mixing matrix A, there is $m < n$. The mixing matrix is ill-conditioned in this case and the matrix is defined as underdetermined. By improving the traditional algorithms and proposing new method, so as to solve the problem of underdetermined blind source separation has been becoming the focus of research.

In the field of Theoretical studies, following the ICA, Non-negative Matrix Factorization (NMF) and Sparse Component Analysis (SCA) have been applied to BSS. The development of algorithms based on the two new theories has attracted the interest of researchers.

The BSS algorithms based on SCA make use of the sparse feature to extract the signals, it is possible to achieve high separation accuracy if the signals is enough sparse.

For the first time in 2001, Bofill had used the linear clustering algorithm based SCA to solve the UBSS problem when the signals were low-dimensional and high sparsity [1]. In 2006, HE Zhao-shui proposed the K-clustering algorithm based hyper-plane, which was applied to separate the underdetermined blind signals with more dimensions and inadequate sparsity [2].

NMF is a matrix decomposition method proposed in recent years, and its mathematical model is:

$$V=WH+E \quad V\in R^{m\times T}, W\in R^{m\times n}, H\in R^{n\times T}, E\in R^{m\times T} \quad (2)$$

The meaning is that the matrix V can be expressed by the product of the base matrix W and the weighting coefficient matrix H added an error matrix E. We can also refer V as an observational data matrix, W and H can be called the factor matrix. Ignoring the error and noise, the model should be written as follows:

$$V=WH \quad (3)$$

NMF is a method to separate signals and get data through matrix factorization. NMF with non-negative constraint can be applied to the field of earth science, brain science, remote sensing. We can convert those data analysis problem into NMF optimization problem, alternating with a variety of iterative methods for solving the W and H.

Depending on the iterative method, the common NMF algorithms mainly include classic multiplicative updates algorithm [3], projected gradient algorithm [4], quasi-Newton algorithm [5], Fixed Point NMF [8], Weighted NMF [9], and so on.

In the underdetermined case, the mixing matrix is ill-conditioned, then the mixing matrix corresponding communication system is irreversible, we will get infinitely many solutions through the mixing matrix decomposition. Considering the sparsity of the signals, we can obtain the correct solution with lots of constraints if the numbers of active signals is less than the number of sensors [6]. When the NMF algorithms solve the problem of UBSS, the sparse feature of the signal is often as an important constraint [4, 5, 6, 7]. In this way, researchers have proposed many algorithms for underdetermined sparse signal separation, and some algorithms have obtained good results.

Cichochi proposed an algorithm called hierarchical alternating the least squares algorithm (HALS) [10]. Using the NMF hierarchical decomposition method had solve the problem of UBSS with 10 channels of source signals and 6 channels observation signals, in the case that the sparsity of signals is sufficient. However, the algorithm is not applicable if the sparsity of signals is inadequate. This algorithm has been improved via imposing determinant criterion constraint to estimated mixing matrix in this paper. The simulation results show that the algorithm achieves a better effect.

2 THEORY OF ALGORITHM

2.1 Determinant criterion

If P(W) is the space constructed by vector sets as $w_{*1}, \ldots w_{*i}, \ldots w_{*n}$, if W is a square matrix, then the volume of P(W) may be defined as $vol(P) = det(W)$;

If W is not square, then $vol(P) = det(WW^T)$[11].

Determinant-criterion: If P(W) is the space constructed by vectors as $w_{*1}, \ldots w_{*i}, \ldots w_{*n}$, if the value of vol(P) is minimum, and the set of vectors is unique.

As discussed above, the volume of the space constructed by matrix and value of the matrix corresponding determinant can be linked together, and via the criterion, we can find the unique vector set.

If we assume that the matrix $V \in R^{m \times T}$ is the space constructed by vector sets as $w_{*1}, \ldots w_{*i}, \ldots w_{*n}$, non-negative linear weighted vector sets as $h1*, \ldots hi*, \ldots hn*$, then

$$v_{*r} = \sum_{i=1}^{n} w_{*i} h_{ir} \qquad 0 < r < T \qquad (4)$$

If using the matrix W indicates the vectors as $w_{*1}, \ldots w_{*i}, \ldots w_{*n}$, and matrix H for the vectors as $h_{1*}, \ldots h_{i*}, \ldots h_{n*}$, then formula (4) would be another representation for formula (3)

As discussed, if we want get the only solution for W in the matrix decomposition, the process can be converted into the following optimization problem:

$$J_W(W) = min[vol(P(W))] = min[det(W)] \qquad (5)$$

If W is not square, then

$$J_W(W) = min[det(WW^T)] \qquad (6)$$

2.2 Cost function of NMF

For basic NMF algorithms, if take Euclidean distance as its cost function [7], the function can be written as follows:

$$D(V\|WH) = \frac{1}{2}\|V - WH\|_F^2 = \frac{1}{2}\sum_{i,j}[V_{ij} - (WH)_{ij}]^2 \qquad (7)$$

We always convert the NMF problem into the following optimization problem as follows:

$$F = minD(V\|WH) \quad \forall m, n, T \quad w_{mn} \geq 0 \quad h_{nT} \geq 0 \qquad (8)$$

Theoretically, via setting the F as the target and designing appropriate iterative algorithms, we can obtain the correct solution equating F close to 0. In over-determined and normal cases, the solution should be the only correct solution. But this correct result should not be the solution that we want to obtain in underdetermined case. In order to make the NMF algorithms meet all kinds of signals with different characteristics, in addition to impose non-negative constraints in the iterative process, we should also consider other constraints as sparsity, smooth, correlation.

The basis cost function can be improved as follow:

$$D(V\|WH) = \frac{1}{2}\|V - WH\|_F^2 + \alpha_W J_W(W) + \alpha_H J_H(H)$$
$$\forall m, n, T \quad w_{mn} \geq 0 \quad h_{nT} \geq 0 \quad (9)$$

where, α_W and α_H are the non-negative constraint parameter respectively correspond to W and H, $J_W(W)$ and $J_H(H)$ are constraint functions such as sparsity. $J_W(W)$ can be represented as the formula (6), then the cost function with Determinant-criterion constraint for W and sparsity constraint for H can be written as follow:

$$D(V\|WH) = \frac{1}{2}\|V - WH\|_F^2 + \alpha_W vol(P(W)) + \alpha_H \|H\|^1$$
$$\forall m, n, T \quad w_{mn} \geq 0 \quad h_{nT} \geq 0 \quad (10)$$

3 HALS ALGORITHM WITH DETERMINANT CRITERION CONSTRAINT

Instead of minimizing the global cost function (10), we can minimize the set of local functions defined as

$$D^{(j)}(V^{(j)}\| w_j h_j) = \frac{1}{2}\left\|V^{(j)} - w_j h_j\right\|_F^2 + \alpha_w^{(j)} J_W(w_j) + \alpha_H^{(j)} J_H(h_j)$$
$$\forall m, n, T \quad w_{mn} \geq 0 \quad h_{nT} \geq 0 \quad j=1,2\ldots,n \quad (11)$$

where

$$V = \sum_{r=1}^{n} w_r h_r \qquad (12)$$

$$V^{(j)} = V - \sum_{r \neq j} w_r h_r = V - WH + w_j h_j \qquad (13)$$

$w_j \in R_+^{m \times 1}$ are the columns of W, and $h_j \in R_+^{1 \times T}$ are rows of H. where $J(*) = ||*||_1$ [10], the constraint functions in (11) can be written as

$$J_W(w_j) = \Sigma_i w_{ij}, \quad J_H(h_j) = \Sigma_r h_{jr} \qquad (14)$$

At last the local cost functions with determinant criterion and sparsity constraint can be defined as

$$D^{(j)}(V^{(j)}|| w_j h_j) = \frac{1}{2}\left\|V^{(j)} - w_j h_j\right\|_F^2 + \alpha_w^{(j)} \Sigma_i w_{ij}$$
$$+ \alpha_H^{(j)} \Sigma_r h_{jr} + \alpha_{pw} vol(P(w_j w_j^T))$$
$$\forall m, n, T \quad w_{mn} \geq 0 \quad h_{nT} \geq 0 \qquad (15)$$

where

$$vol(P((w_j w_j^T)) = det(w_j w_j^T) \qquad (16)$$

$$\frac{\partial det(w_j w_j^T)}{\partial w_{ij}} = det(w_j w_j^T)\left[(w_j w_j^T)^{-1} w_j\right] \qquad (17)$$

Different from classic NMF algorithms, HALS does not use directly the gradient descent approach but rather attempt to establish iterative rules based on the KKT condition. We get the gradient of cost function as:

$$\frac{\partial D(j)(V(j)||w_j h_j)}{\partial w_j} = w_j h_j h_j^T - V^{(j)} h_j^T + \alpha_w^{(j)}$$
$$+ \alpha_p det(w_j w_j^T)\left[(w_j w_j^T)^{-1} w_j\right] \quad (18)$$

$$\frac{\partial D(j)(V(j)||w_j h_j)}{\partial h_j} = w_j^T w_j h_j - w_j^T V^{(j)} + \alpha_H^{(j)} \quad (19) \qquad (19)$$

When the gradient above approach to 0, according the two formulas above, the learning rules can be defined as

$$h_j \leftarrow \left[\frac{1}{w_j^T w_j}\left(w_j^T V^{(j)} - \alpha_H^{(j)}\right)\right]_+ \qquad (20)$$

$$w_j \leftarrow \left[\frac{1}{h_j h_j^T}(V^{(j)} h_j^T - \alpha_w^{(j)} - \alpha_{wp}\right]_+ \qquad (21)$$

where

$$\alpha_{wp} = \alpha_p det(w_j w_j^T)\left[(w_j w_j^T)^{-1} w_j\right]),$$

$[\varepsilon]_+ = \max\{\epsilon, \varepsilon\}$, ϵ is a small constant to void numerical instabilities (usually, $\epsilon = 10^{-16}$).

According to the learning rules, mixed-signal matrix V can be decomposed into mixing matrix W and the source signal matrix H through a number of iterations, and the solution would be unique.

All discussed above, we obtain a new HALS algorithm with determinant-criterion can be called DHALS.

4   SIMULATIONS AND ANALYSIS

The simulated data is got from the NMFLAB [12] which is written by A. Cichocki and R. Zdunek. The 7-channel signals with sparse insufficient are used in the experiment, and the experiment applies the improved algorithm proposed in this paper. The simulation result was analyzed in contrast to other algorithms in the toolbox in the same condition.

4.1   Evaluation of separation

The reconstruction SNR (Signal to Noise Ratio) of the separated signal and source signal is regarded as the evaluation of the separation effect, which is identical with the literature [1]. The separation evaluation is as follows.

$$SNR = 10 lg\left(\frac{||V||^2}{||V^\circ - V||^2}\right) \qquad (22)$$

V and V° are respectively the matrix of the source signal matrix and the separated source signals. Obviously the separation effect is better when the SNR value is bigger after the two matrices' normalization.

4.2   Simulation results

Some prior knowledge is needed when using the NMF algorithm to BSS. It is assumed that the source signal number has been accurately estimated.

Taking the 7 road sparsity bad signals of the NMFLAB toolbox as the simulation object, they are shown in Figure 3, the number of observed signals is 6 and the hybrid matrix is randomly generated by MATLAB. In order to reduce the amount of computation and improve the speed of computation, the formula (21) is modified as

$$w_j \leftarrow \left[\frac{1}{h_j h_j^T}\left(V^{(j)} h_j^T - \alpha_w^{(j)} - \alpha_p det(w_j w_j^T)\right)\right]_+ \qquad (23)$$

Set the values of parameters as $\alpha_H^{(j)} = 0.01$, $\alpha_w^{(j)} = 0.15$, $\alpha_p = 1$, the mixing matrix generated by NMFLAB as

$$A = \begin{pmatrix} 0.85 & 0 & 0 & 0.52 & 0 & 0 & 0 \\ 0.62 & 0.75 & 0 & 0 & 0 & 0 & 0.90 \\ 0 & 0 & 0 & 0 & 0.47 & 0.79 & 0.81 \\ 0 & 0.22 & 0.17 & 0 & 0 & 0 & 0.90 \\ 0 & 0 & 0 & 0.44 & 0.84 & 0 & 0.74 \\ 0.26 & 0 & 0.88 & 0 & 0 & 0.58 & 0 \end{pmatrix}$$

After independently running 20 times, the simulation results are as follows:

The Figure 1–3 show the result of separated source signals using the DHALS algorithm, in addition to the changes in the signal sequence and the normalized amplitude of the signals, the basic 7 separation channels are successfully separate. At the same condition, the SALS algorithm is simulated. The separation results are as follows.

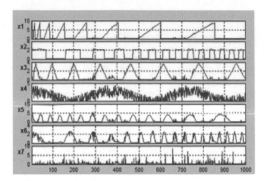

Figure 1. Estimated source matrix.

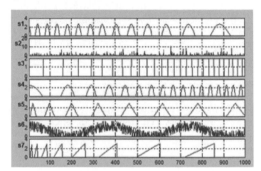

Figure 2. Source matrix.

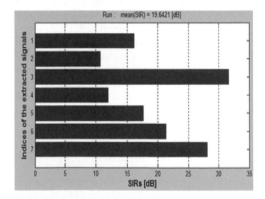

Figure 3. SNR(S).

Table 1. Comparisons between different algorithms.

Algorithm	Mean SNR(S)	Mean SNR(A)
SALS for A and S	7.7312	11.842
SALS for A, HALS for S	12.211	18.279
DHALS for A, SALS for S	14.848	29.087
DHALS for A, HALS for S	19.642	37.275

Comparing the results in table 1, it shows that the improved algorithm in this paper has a better performance, contrasting to the traditional algorithm. To impose the determinant constraint to the mixing matrix is effective.

5 CONCLUSIONS

On the basis of HALS algorithm, we impose the determinant constraint to the left matrix V separated from the NMF in this paper. The improved method is applied to UBSS in the condition of that the source signals are not sufficiently sparse. The signals were successfully separated via the new method in the simulation, and the results show that the algorithm achieves a better effect.

ACKNOWLEDGEMENT

This work was supported by the National Science Foundation of China (No. 61163066) and Scientific Research Program of the Higher Education Institution of Xinjiang (No. XJEDU2009S79).

REFERENCES

[1] Pau. Bofill, Michael. Zibulevsky, "Underdetermined blind source separation using sparse representations", Signal Processing, vol. 81, pp. 2353–2362, 2001.
[2] He zhaoshui, Xie shengli, Fu yuli. "Sparse Representation and ill-conditioned mixture blind source separation", Science in China Series E Information Sciences, vol. 36, pp. 864–879, 2006.
[3] Lee D D, Seung H S, "Learning the parts of objects by non-negative matrix factorization", Nature, vol. 401, pp. 788–791, 1999.
[4] Hoyer P O, "Non-negative matrix factorization with sparseness constraints". Journal of Machine Learning Research, vol. 5, pp. 1457–1469, 2004.
[5] R. Zdunek and A. Cichocki. "Non-negative matrix factorization with quasi-Newton optimization". Artificial Intelligence and Soft Computing – ICAISC 2006, pp. 870–879.
[6] A Cichocki, R Zdunek, Shun-ichi Amari, "Nonnegative Matrix and Tensor Factorization", IEEE signal processing magazine, vol. 25, pp. 142–145, 2008.
[7] Lee D D, Seung H S. "Algorithms for non-negative matrix factorization". Advances in Neural Information Processing Systems, vol. 13, pp. 556–562, 2001.
[8] A. Cichocki, R. Zdunek, S. Amari. "Hierarchical ALS algorithms for nonnegative matrix and 3D tensor factorization". ICA2007, LNCS 4666, pp. 169–176, 2007.
[9] Petr Tichavský, Zbynek Koldovský, "Weight Adjusted Tensor Method for Blind Separation of Underdetermined Mixtures of Nonstationary Sources", IEEE TRANSACTIONS ON SIGNAL PROCESSING, vol. 59, pp. 1037–1045, 2011.
[10] Yuanqing Li, Shun-ichi Amari, "Two Conditions for Equivalence of 0-Norm and 1-Norm Solution in Sparse Representation", IEEE Transaction on Neural Networks, vol. 21, pp. 1189–1196, 2010.
[11] R Schachtner, G Poppel, et al. "Minimum determinant constraint for non-negative matrix factorization", ICA2009, LNCS, vol. 5441, pp. 106–113.
[12] A. Cichocki, R. Zdunek, and S. Amari, "Csiszar's Divergences for Non-Negative Matrix Factorization: Family of New Algorithms", 6th International Conference on Independent Component Analysis and Blind Signal Separation, Charleston SC, USA, March 5–8, 2006 Springer LNCS 3889, pp. 32–39.

Information Technology and Computer Application Engineering – Liu, Sung & Yao (Eds)
© *2014 Taylor & Francis Group, London, ISBN 978-1-138-00079-7*

The design and implementation of University Educational Administration System with high availability

Dong Wei Guo, Yun Na Wu & Jing Ji Jin
College of Computer Science and Technology, Jilin University, Changchun, China

Zhen Yu Zou
Office of Academic Affairs, Jilin University, Changchun, China

ABSTRACT: In this paper, a high available University Educational Administration System, which refers to the Jilin University Educational Administration System, is designed and developed on the basis of the whole analysis of traditional Educational Administration System and the utilization of the modern information technology. We elaborate the system design flow, especially network topology part and web service design part comprehensively, and make an assay of the major function modules. In particular, the system security problems, such as system permission model, user authentication, etc., are presented in more detail. The new system has been in practice for two years after the accomplishment of function test and stress test, which guarantees the high-availability, stability and security of the new system. The high available system not only accelerates the pace of educational administration modernization reformation, but also provides strong support to improve the quality of educational administration.

Keywords: educational administration; informatization; high availability; educational administration system

1 INTRODUCTION

Educational Administration Management (EAM), which is the basis of various affairs in universities, plays a significant role in the university education. The EAM refer to the contents of student management, teaching management and test management, and also has the features of large amount of data, complex relationship, and various affairs. The EAM is not only the important component part that measure the stability of the teaching order, the level of teaching management and the teaching quality, but also relating to the school development and the talent cultivation. Hence, the EAM occupies a hugely important position in the college life [Cheng R.H. 2003].

In the recent years, with the development of the domestic education reformation and the introduction of the credit system manage mode, the traditional Educational Administration System (EAS) has been difficult to meet the challenges. It has been an inevitable process that developing the high available comprehensive EAS, which uses state-of-the-art information technology, automation technology and network technology, adapts to the trend of the current college reformation and accelerates the pace of educational informationization [Cheng R.H. 2003, Dai X.H. & Hu Y. 2006].

2 THE PROBLEMS IN TRADITIONAL EDUCATIONAL ADMINISTRATION SYSTEM

2.1 Function modules are insufficient

Through the practice of the traditional EAS, the author found that most modules in traditional EAS such as the test management, register management and curriculum management are basically sufficient. But there still are some problems which reduce the office efficiency, for example, the inconvenient operations, lack of data system function and so on [Meng B. 2010]. In addition, the EAS should fit the fact of each school by the way of the modules division or modules mergence, e.g. one major may across multiple campuses, and if we continue to partition modules according to different majors, will increase the teaching office personnel's workload, even the high availability feature of the EAS will also be influenced.

2.2 Accounts' security are high risk, and the permission configuration is not flexible

When we construct the EAS, the high-efficiency security and permission model is necessary, which support the EAS a more safe data basis to guarantee the access of legitimate users and the rejection of the illegal users. However, the safety management subsystems of many traditional EASs are not trustworthy so that the confidential information is easy to steal, leak, or tamper by the lawbreakers, which directly does damage to the smoothly operation work of the EAM [Sun F.X. et al. 2006].

2.3 System is not stable and concurrent number is low

EAS is a multi-users comprehensive system which provides multiple services to the educational officers, teachers and students. One of the measurement criteria of the success of our system is whether the system has long stable working time and high concurrent number. In general, the online course selection module in the traditional EAS often runs unstably or even crashes because of the concurrent operation of many students. That is why the students have dissatisfaction or complaint with the educational administration work [Wang B.H. et al. 2004].

3 THE OVERALL DESIGN OF THE JILIN UNIVERSITY EDUCATIONAL ADMINISTRATION SYSTEM

The Jilin University EAS is based on the architecture of Browser/Web Server, and also utilize RIA (Rich Internet Application) technology to implement the UI which is similar to the traditional C/S method. In the EAS, with the development process according with standard software specifications, we design the unified data structure and data model, integrate the function modules, and formulate the information transmission standard and data storage format, which ensure the data is consistent and maintainable.

3.1 Network topology

The network topology of Jilin University EAS is shown in Fig. 1.

From the Fig. 1, we can see that when the school users access the web server, they will be dynamically allocated by load balance server, which guarantees that each web server has nearly identical load to prevent one server crash due to overload. The users out of school should go through the filter layer to visit the web server; therefore, even if the outer web server is attacked by external network, EAM will still normally continue and won't be influenced directly. The common application cluster, that guarantees 7*24 services and the high availability of EAM, is made up by the application servers, one of which acts as the specialized report application server to deal with report requests only. The sharing memory server stores the cache items and the user session information to accelerate the visit. The data applied by the application clusters is from the sharing memory server or the database server. It is reasonable to adopt the magnetic

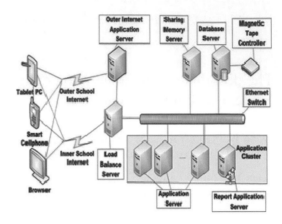

Figure 1. The network topology of Jilin University Educational Administration System.

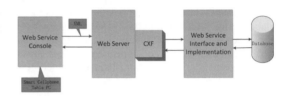

Figure 2. Workflow of web service.

tape controller to back up database data because of the high capacity and reliability.

For the cache issue, the system sets the frontend cache, which is used to cache the javascript code and css code, the 2nd-level cache for the Hibernate framework and the database cache. The three cache parts correspond to application servers, sharing memory server and database server. Utilizing this cache system, we can improve the process efficiency of data access.

3.2 Web service design

The EAS's users usually visit the website by two means: browsers and the mobile terminal. The java services and normal servlets are used to serve the browser users. But for the mobile terminal users, the system adopts the web service technique to construct the information platform, which satisfy the various mobile terminals (smart cellphone, tablet PC, etc.) and different mobile systems (like iOS, Android). As one of advanced distributed communication technologies, the web service technology is independent of platforms and utilizes XML as the information transfer medium, so it can be deployed by heterogeneous systems and have good extensibility [Erl. T. 2004, Fensel D. & Bussler C. 2002]. The workflow of web service is shown in Fig. 2. In the implementation of server-side, the apache CXF is exploited. For the part of client-side, the Android edition client of web service is accomplished with the assist of KSOAP technique so far.

3.3 System function modules

The high available EAS involves all aspects of the university teaching, mainly consists function modules of students management, teacher management, course selection management, score management, teaching plan management and so on. All these function modules are either relatively independent to constitute separate subsystem, or interacted with each other to form the complex and comprehensive EAS. In this Jilin University EAS, each module can be dynamically added or deleted when the system is online, without affecting the system's normal operations.

In the following, we present the core modules in detail.

3.3.1 Teaching plan management

Teaching work is the core axis of the university education task, and also one important standard of weighing the university whole level [Wang Y.L. & Li J.H. 2004]. The module is able to make various educational resources to contact with each other to form an organic whole, e.g. draft the semiannual course schedule for students and establish the relationship between teachers and classrooms.

The module includes the integrated planning of courses per semester, the combination pattern of various course types, the allocation of the teachers, classrooms, periods and lessons in the courses and the basic credit for each profession in per semester. These tasks are not only the fundamental protection of the complete credit system for EAM, but also endow the students with more flexible and more open management.

In addition, at the end of the teaching plan management module, the courses will be intelligently scheduled. To make the courses timetable more rational and customized, we need to set some reasonable constraints, for instance, the course schooltime should have a certain interval when the course emerge repeatedly in a week. The intelligent course schedule is able to overcome the limitations of course schedule by hands, improving the efficiency of the officers.

3.3.2 Student management

Student management refers to the overall management of student information. The module is composed by two parts: the one is students' personal basic information, and the other is important event of student, such as school entrance, school rolls, reward and punishment, school graduation, etc.

Students' personal information management is directed against the students' basic information that contains the home information, contact information, system password, and so on. The school rolls management records the multiple events of school rolls, mainly refer to the downgrade, change profession, quit school, exchange student management and profession shunt. Through the profession shunt college can uniformly distribute batch of students to different professions. Reward and punishment management supervise the students' incentives, rating certificate, penalty and so on. Graduate management reviews whether the students achieve the graduation requirement, for example, one must go through all the required courses and obtain credit over the limitation.

3.3.3 Course selection management

Course selection management is designed for students to select courses according to their semester teaching plan. This module is online and time limited, i.e. all students choose their courses independently and uniformly within a certain period, which fully reflect the freedom and equality in the study aspect for students.

In each semester, the course selection module can be divided into three stages: pre-selection, selection and back-post-selection. The three stages all have the designated date to open, so students can only make choice independently during the periods, which accords to their learning plan. Via three stages of the module is conducted orderly, we can avoid the accidental error effectively, such as wrong selection and omissive selection. In addition, there is floating timetable diagram to mark the selected courses in the course selection page. By this way, students can observe the conflict of courses selection clearly, and make arrangement of their courses more convenient.

4 THE SECURITY OF THE JILIN UNIVERSITY HIGH AVAILABLE EDUCATIONAL ADMINISTRATION SYSTEM

With the development of digital and intelligent information, security is one important standard whether the success or failure of system. Because of the shared and crucial data and the fragile school network, the security of EAS is the significant issue we have to consider.

4.1 Permission management

The permission management, as the core of system security, sets up and maintains the permission to prevent the unauthorized access of the system's resource and ensure the correct visit from legal users.

The overall development plan of permission management utilize user group to restrict the function access and set permission formula to limit the access of specific resources, thus making sure the security of the system.

The model of permission management is composed by elements of user, user group, function module, permission formulation and resource. In the EAS, permission formulation can make tight relation between the function modules and resources: when users access certain resource (search, deletion and insertion), the system will check the permission of this access to judge whether the permission formulation can pass, as that assure the safety of the resource. The user group is responsible for the relevance among users and function modules, and one user can join one or more use groups. In addition, each user group is corresponding with a collection of function modules, for instance, teacher group (one of user groups) contains score management, course management, classroom management and etc. When the user logins the EAS, the system will dynamically load the corresponding function modules according with the user groups that the user belongs to.

The above processes can be modified dynamically during the system running time to avoid the inconvenience of pre-fixed permission setting. No matter personnel changes or the varied responsibilities for the educational officers, we don't need to re-release the EAS with some modification, which results in greater availability.

4.2 *User authentication*

The purpose of user authentication is to identify the legitimacy of access from user to EAS. In general, systems detect the user name and password to decide whether the user is rightful or permissible. However, the availability and reliability of systems are relatively poor when systems only rely on the password. So it is necessary to do encryption operations on the password information by the encryption technology.

With the user authentication aspect of the high available EAS, we adopt two different strategies for different users. For normal teachers and students, in order to make sure the safety of password information, the encryption algorithm is used to transmit the password and system database without storing the password information in plaintext. For educational officer, the USBKey encryption device, the browser plugin software and the https encryption channel are utilized to encrypt the information and guarantee the transmission security in case the information is stolen or cracked by malicious persons.

4.3 *Web service safety*

With the safety of web service, the new system utilizes the WS-Security formulation to add digital signatures and encryption to the SOAP message, which guarantees the integrity and confidentiality of the message transmission and also ensures the legitimacy of the interaction behavior.

5 THE PRACTICE OF JILIN UNIVERSITY EDUCATIONAL ADMINISTRATION SYSTEM

Through the stress test and other tests, the Jilin University high available EAS, which went online at the October 2010, successfully completed the tasks of teaching evaluation, course selection and so on, having achieved the desired target. The monitor system indicates that when system is in peak and the concurrent number is higher than 30,000, the system can still provide service normally and the pressure of database is very small.

6 CONCLUSION

The high available EAS, as one method of improving the efficiency of the educational officers and enhancing the information exchange among teachers, students and educational officers, is also the specific realization of deepening the EAM reformation and achieving the EAM informationization.

The development and implement of Jilin University EAS is fit to the reality of Jilin University itself and puts the information technology into the EAM. So the high available EAS accomplishes the information standardization and resource sharing, which provides the supplementary decision-making and accelerates the pace of information processes.

REFERENCES

Cheng R.H. 2003. Business Relationships and Information System Design 0f Education and Educational Administration. *Computer Engineering*: 29(13).

Dai X.H. & Hu Y. 2006. Practice, Exploration and Reflection for Informatization of Higher Educational Administration. *Journal of Chengdu University (Natural Science)*: 25(2): 131–135.

Erl T. 2004. Service-oriented architecture: a field guide to integrating XML and web services. *Prentice Hall PTR*.

Fensel D. and Bussler C. 2002. The web service modeling framework WSMF. *Electronic Commerce Research and Applications* 1(2): 113–137.

Meng B. 2010. A Study of Educational Administration Information in Higher Institutions. *Journal of Hengshui University* 12: 88–92.

Sun F.X. et al. 2006. Design of security solution for Web-based educational administration system. *Computer Applications* 26(5): 1198–1201.

Wang B.H. et al. 2004. Research and Implementation on Networked Educational Administration System on the Basis on Scattered Campuses and the Complete Credit System. *Application Research of Computers* (5): 123–125.

Wang Y.L. & Li J.H. 2004. Applying synthetically educational administration system in educational administration management. *Computer Engineering and Design* 25(10): 1681–1684.

Fuzzy congruence and its application in fuzzy clustering

Mina Lagzian & Ali Vahidian Kamyad
Department of Applied Mathematics, Ferdowsi University of Mashhad, Mashhad, Iran

ABSTRACT: Since that the fuzzy logic be infinite value logic, it can be useful that provide a method for better data and fuzzy rules clustering. In this paper present a new method called fuzzy congruence. One of the applications of this method is clustering fuzzy numbers. In fact this method is derived from congruence in classical mathematics. In other words, this is a generalization of congruence in integers to real numbers and then to fuzzy numbers.

Keywords: fuzzy congruence, clustering, congruence, classification

1 INTRODUCTION

We know that the classical congruence is defined on integers. The module of congruence is integer too. For an integer m, there are m classes of congruence that they are partitioned integers into m parts. These classes will be determined by dividing the integers on the remaining m. these residues are $0, 1, 2, \ldots, m-1$. When any real number is assigned to the integer m, the rest is a real number in the interval $[0, m-1]$. Therefore we can extend congruence on integers to the real numbers so that any arbitrary real number x belongs to a class of congruence module m, with a membership degree. In other words we can say:

$$x \cong a \quad \text{mod}(m) \tag{1}$$

Or:

$$x - a \cong 0 \quad \text{mod}(m) \tag{2}$$

Means that a real number $x - a$ is a fuzzy congruence zero of module m, with membership degree $y(x)$ that function $y(x)$ is defined as following:

$$0 \le y(x) = \left| ((\frac{1}{m})(x - m\left[\frac{x}{m}\right]) - 1 \right| \le 1 \tag{3}$$

Therefore for every real number x, membership function of fuzzy congruence module m is defined as follows:

$$\mu_{\cong}(x) = y(x) = \left| ((\frac{1}{m})(x - m\left[\frac{x}{m}\right]) - 1 \right| \tag{4}$$

1.1 *Fuzzy congruence of two triangle fuzzy numbers*

Now we assume that $\tilde{a}$ and $\tilde{b}$ are two different triangle fuzzy numbers. We can determine congruence of this two triangle fuzzy numbers to module m, in other words we can specify:

$$\mu(\tilde{a} \cong \tilde{b}) = ? \quad \text{mod}(m) \tag{5}$$

First transform these two fuzzy numbers with an order function into two corresponding real numbers. An order function is defined as following:

$$\forall \tilde{x} = (n, \alpha, \beta)$$
$$order(\tilde{x}) = \frac{4n + (n - \alpha) + (n + \beta)}{6} \tag{6}$$

The reason of function definition is that two triangle fuzzy numbers may have overlap. By the way, two fuzzy numbers are less likely overlap and therefore the result of calculations can be achieved easier and more accurate. It is worth noting that one of the characteristics of order function is that it is linear function.

With this order function, first calculate $order(\tilde{a})$ and $order(\tilde{b})$ and we call the amount of them a_1 and b_1 respectively. By the way two gained numbers are real. Therefore we can determine fuzzy congruence of module m of any of these real numbers with zero. Means that:

$$a_1 \cong 0 \quad \text{mod}(m)$$

The membership degree of a_1 is $y(a_1)$ and

$$b_1 \cong 0 \quad \text{mod}(m)$$

The membership degree of b_1 is $y(b_1)$
Now we have:

$$a_1 - b_1 \cong 0 \quad \text{mod}(m)$$

With membership degree:

$$y(a_1 - b_1)$$

Therefore, for calculation of congruence of two triangle fuzzy numbers, we calculate congruence of difference between two fuzzy numbers.

$$order(\tilde{a} - \tilde{b}) = order(\tilde{a}) - order(\tilde{b}) = a_1 - b_1$$

The membership degree is:

$$y(a_1 - b_1) = y(c)$$

Therefore the amount of congruence of two triangle fuzzy numbers as following:

$$\mu(\tilde{a} \cong \tilde{b}) = y(c) \quad (\bmod m)$$

1.2 Clustering of fuzzy numbers

Assume that $\tilde{a}_1, \tilde{a}_2, \ldots \tilde{a}_n$ are n fuzzy numbers that we want to cluster them. First calculate the amount of $order(\tilde{a}_1), order(\tilde{a}_2), \ldots, order(\tilde{a}_b)$ with order function that defined on section 1-1 by formula number 6 and call them $a_1, a_2, \ldots, a_n$ respectively. Now we have n real numbers and we can classification this numbers into m classes that m is arbitrary number. For clustering this numbers first calculate the fuzzy congruence module m of them. (We calculate fuzzy congruence of this numbers with $0, 1, 2, \ldots, m - 1$). Therefore we have m classes of this real numbers with membership degree of fuzzy congruence of them. Now we use these classes for clustering fuzzy numbers. For any number $a_1, a_2, \ldots, a_n$ calculate minimum of membership degree in m classes of congruence then input numbers with same or nearly same amount in the same cluster. By the way we can cluster fuzzy numbers $\tilde{a}_1, \tilde{a}_2, \ldots \tilde{a}_b$.

1.3 A simple numerical example

Assume that $6.2, 4.1, 5.2, 12.4, 8.3$ are real numbers that achieved from $order(\tilde{a}_1), order(\tilde{a}_2), \ldots, order(\tilde{a}_5)$ that those are triangle fuzzy numbers.

We want classification these numbers into 3 classes. First calculate membership degree of each number in each congruence class [0], [1], [2].

These amounts are as following:

mebrship deg *rees of* $6.2 = \{0.93, 0.26, 0.6\}$

mebrship deg *rees of* $4.1 = \{0.63, 0.96, 0.3\}$

mebrship deg *rees of* $5.2 = \{0.26, 0.6, 0.93\}$

mebrship deg *rees of* $12.4 = \{0.86, 0.2, 0.53\}$

mebrship deg *rees of* $8.3 = \{0.23, 0.56, 0.9\}$

With attention of the obtained above value we can put two numbers 6.2 and 12.4 in same cluster and 8.3 and 5.2 in the same cluster too. The number 4.1 is reserved to a different cluster. Now we cluster fuzzy numbers to tree clustering.

2 CONCLUSION

Fuzzy congruence method is a simple method for clustering and classification fuzzy numbers. This method has more applications. One of these applications is in solving n degree polynomial equation that n is an arbitrary number.

REFERENCE

[1] Estivel, Jhon, translation Mirzavaziry, Majid, Fundamentals of Number Theory, Number 493, publications Ferdowsi University of Mashhad, Mashhad, Iran, 1386.

Modification proposal security analysis of RFID system based on 2nd generation security tag

Ming Zhu Lu
Department of Electrical and Mechanical Engineering, Cangzhou Normal University, Heibei, China

ABSTRACT: With the wide application of RFID, attacks upon radio frequency card have emerged and keep growing. As a result, it is becoming more and more important for the security and confidentiality of the RFID system. This paper analysed the current status of security and privacy protection of RFID system and discussed the security requirements of the RFID system. A modification proposal of security tag was proposed and its feasibility and application value were analysed.

Radio frequency identification (RFID) technology is a contactless and automatic technology. Object can be identified with relevant data obtained automatically via RF signals. As no manual operation is involved, it can be used under all kinds of severe circumstances.

Keywords: RFID; hash chain; EPC tag; 2nd generation security tag

1 RFID SYSTEM COMPOSITION

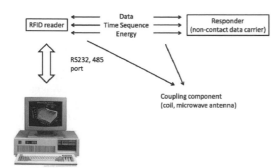

Figure 1. RFID system.

2 RFID SYSTEM SECURITY PROBLEM ANALYSES

With the wide application of RFID, all kinds of attacks against RF card have emerged and keep growing. As a result, the security and confidentiality of the RFID system is becoming more and more important. All kinds of attacks may occur to the RFID system including unintentional activities, such as faulty operation in electronic transactions, and intentional activities, such as illegal card cheating, interception and tampering of data in all kinds of transactions. Modern authentication protocol also involved the detection of the private key and appropriate algorithm can be employed for encoding. Highly-secured RFID system should keep away the following attacks including: 1) data copying/modification and unauthorised read of data carrier;

2) Insertion of external data into the inquiry range of a certain reader in order to obtain unauthorised access to buildings or free services; 3) Fraud such as the radio communication stealth and data rewrite by acting as the actual data carrier.

3 HASH CHAIN

Before discussing the new tag security model, the Hash function in the electronic product code (EPC) tag should be briefly introduced.

3.1 *Hash function*

h = H(M) is a unidirectional Hash function. Its input M is a variable length message and its output is a fixed length Hash code, H(M). Hash code is sometimes referred as message digest or Hash value. Hash code is a function of all message digits and it has fault detection capability, which means it will be changed with the alteration of either one or more message digits.

3.2 *Four properties of Hash function*

Compressibility: For a randomly input x with fixed length, the output H(x) is with fixed length as well.

Uni-directionality: With any given Hash code, h, it is impractical to find x which could fulfil the relation of H(x) = hvia calculation.

Weak collision-resistance: For any given block x, it is impractical to find a block y which could fulfil both y ≠ x and H(y) = H(x) via calculation.

Strong collision-resistance: It is impractical to to find a couple of (x, y) which could fulfil $H(y) = H(x)$ via calculation.

3.3 *Hash chain*

Hash chain is the extension and transformation of Hash function and is generated by a series of Hash functions. The generation procedure of Hash chain is as follows:

Firstly, the initial sequence S_0 is input for initialisation.

Initial sequence: S_0

First round: $H^1 = H(S_0)$

Second round: $H^2 = H(H(S_0)) = H^2(S_0)$

nth round: $H^n = H(H^{n-1}(S_0)) = H^n(S_0)$

The use of the Hash chain value is in the reverse direction, from H^n to H^1. According to the uni-directionality of Hash function, no one could deduce the previous value from the current value. For example, the value of H^{n-1} cannot be deduced from H^n and the value of each operation can only be obtained with the knowledge of the initial sequence.

Currently, the EPC tag is based on the principle of reinforced security mode tag of Hash chain. And each time when the reader visits the tag, the identifier stored in the tag will be changed automatically via two Hash functions. The tag will be in the safe mode by default and can only be visited with control authorisation after obtaining correct ID to enter the EPC mode. Read and write can then be done and identified by the reader after finishing visiting. In order to realise a security mode tag based on Hash chain, a server database B, a reader R and a tag T1 are required. The server is the only authorised entity of the traceable tag and R is the suspicious equipment without password security and Ti $(1 \leq i \leq n)$ is the read/write tag. The tag is set to security mode by default and each tag Ti stores a random initialisation ID Si which is recorded in database B. Therefore, the database stores the tag identification sequence ID1 and visits the counter k and tags the identifier.

4 2ND GENERATION SECURITY TAG PROMOTION

The default system encoding method is still the double Hash function. However, in consideration of the selectivity of encoding block, we can also employ other methods for different security levels in redundant encoding block. Take Hash as an example. EPC network and server verification is required except for the lowest level security mode SL0 and embedded system is used to realise the hardware. A unique and fixed code Ri is required for the reader/writer for this security tag model.

4.1 *Overall structure*

The system security model can be described using separate block structure. In the tag design and application, blocks are selective and can be turned on and off. In hardware design, all the blocks can be integrated and different products can be realised by disabling irrelevant blocks. This design is the prerequisite for setting tag security level system, system modification and upgrade. A design with selective and controllable function components would provide great convenience in actual design and implementation. The manufacturer could assemble the blocks based on the requirements of the customer, which is also an inexorable trend of product.

4.2 *Tag security level classification and the control authorisation of each user*

Temporarily, the lowest security level is set to SL0 and the highest is SL3. The lowest security level would turn off all security system blocks, which could achieve the maximum processing speed and the reader can work with maximum processing capability under objective security circumstances. And no EPC network and back-end server is required. The highest security level would turn on all security blocks and use the highest security strength to secure the data of the customer. A sliding button is added to the tag to control the security level (including mode switch between read and write). Although it is not profitable for the miniaturisation of the tag, yet the miniaturisation can be realised by setting the tag as uncontrollable and choosing only the required blocks. Controllable switch is obviously the best option when taking into account the versatility of design.

Key is needed for constructing the communication between EPC and reader, and following communication still involves encoding and decoding.

In the communication initialisation, the reader/writer code Ri will be added to the password character string to be verified together.

4.3 *Block function classification*

Data code verification block: In SL0 mode, the close of the block would get single-card minimum communication duration and maximise the overall communication rate of RFID system in order to deal with the condition when a large number of card, each with short length of data, exceed the system anti-collision capability.

Encoding block: Refer to EPC security model and Hash chain for details. Both card code and information will pass through this block before entering the communication block.

Password authorisation control block: card number and information segmentation. The private key and public key of this block can only read all the card information under SL0 mode and under other circumstances, part of the card number will be returned after obtaining this password at the beginning of communication. The reader will then return the data to server for verification. If verified, the 2nd level key will be given in order to get the full card number or 3rd level key will be given to read all the information and authorise to write. The highest level authorisation will not be distributed with highest security level and the tag is in read-only mode.

Key verification block: a counter and key database are contained in this block and the key is changed according to HASH function with the number of visit under control. When the number of visit, K, has reached a certain number, k, the tag will be locked and the manufacturer can add a reset button on the tag. If the password verification has reached a certain number of times, the block will add the code of the reader/writer to the blacklist and therefore the possibility of brute force attack is eliminated. Besides, more than one key verification block can be added and which block is in operation can be determined by the security level.

Bait block: According to the intelligent decoding method, if the counterpart has received large amount of information with the same tag, one might be able to get the information of the encoding mode and the operation mode of the security block. If the key verification block has judged that the wrong key is in high risk condition (the card and the server database are created using mathematical equations such as ellipse model when the initial password is created, if the model is fulfilled with wrong password, the high risk condition can be determined), the block will be started. The code of the reader/writer will be recorded and fake information will be selected and exported to keep the actual information safe.

4.4 Operation procedure

The anti-collision procedure is briefly introduced first: part of the card code will be publicised and used as the identification code for anti-collision. After finishing the anti-collision procedure, reader will start to communicate with the tag.

The system employs the RTF protocol and the electronic tag will stay in hibernate unless K is less than k and it will only respond under the reader's command. After the connection of anti-collision process has been established, the reader needs a key Ai before continuing communication with the tag. Data packet including Ri and Ai will be firstly sent to the tag. After signal demodulation, the data code verification block is called. If the block is turned off, then the communication can be continued and the reader will obtain all the authorities. If the block is turned on, then Ri will be compared with the blacklist. Correct Ri will make the block continue import Ai and wrong Ri will make the block import the number of errors. Upon reaching the predefined error number, the current Ri will be added to the blacklist and the reader/writer will be blocked. Correct Ai will enter password authorisation control block and fake information will be returned when Ai is in high risk condition and the bait block will be started. Authorisation will be allocated after Ai has entered the password authorisation control block. If it is a public key, the part of the card number of the

tag will be returned. If it is level 2, the full card number and information of limited card information will be returned. If it is the highest authorisation level, then all the information will be returned to the reader/writer. And the reader/writer could find the visiting code of the tag in the back-end server according to the card number.

The open and close of the block according to the security level can be set by the manufacturers with all blocks removable and can be turned off. The difference of security level is also indicated by the different operation modes of the block. Low security level will result in low encoding level and different encoding blocks can be used for operation. As a result, the encoding algorithm can be changed with the blocks, which provides the highest freedom of the system design.

5 PROPOSAL ANALYSIS AND FEASIBILITY

5.1 Data integrity analysis

Under normal security level, highest level password is required to write tags, which strictly remove the possibility of fake tag and data modification. Besides, the tag is set to read-only mode with highest security level, which secures the data themselves.

5.2 Forward security analysis

Due to the existence of password authorisation and the uniqueness of password, and the fact that the password will be changed each time after a visit, a series of continuous read and visit between the reader and the tag won't give the attacker the private information of the same tag or any relation between a similar category even after the private information stored in the tag is leaked. Even if the password generation mode is leaked, the attacker won't have enough times to try the password and the existence of the bait block will deal with this problem.

As a result, even if the attacker gets the encoding information of tag storage, he won't be able to go back and obtain the historic event of the tag by tracing the current information. In another word, the attacker cannot obtain the information of other tags by analysing the relation between current data and historical data.

5.3 Visit control and mutual verification

The key verification block knows how to analyse the code of the reader and the reader in the authorisation list can only be visited when it is in high security level.

5.4 Data confidentiality analysis

User information can be stored in the back-end server and there is no direct relation between tag data and user information. Therefore, the tag data don't make sense to the tag user. As a result, unauthorised readers cannot contact tag user and obtain the information even if they obtain the tag data via eavesdrop.

5.5 *Tag anonymity analysis*

As the lock up function of number of visits and block function of reader are enabled in the security block, for a series of continuous visit between reader and tag, the attacker is not able to find the same tag or any relation between tags of a similar category. In another word, no tag can be determined by multiple timescan. As separate password is required for obtaining full card number in security model, which means the encoding mode can be changed, it is impossible for attacker to identify each tag via fixed encoding information.

5.6 *Proposal feasibility*

Most of the ideas mentioned in this paper has been realised in other fields, including security level classification, password authorisation allocation technology, and therefore they can be applied in RFID security block without any technical difficulties. As a result, the versatility and applicability of the proposal make the hardware realisation valuable. However, this is only a design at system level and the following object will be realising the software and designing chip circuit.

The main disadvantage of this proposal is that the largely increasedcomplexity of the chip design, although the manufacturing cost will not be largely increased with the rapid development of the manufacturing technology. Another limitation is that the system can only be used when the network is available.

REFERENCE

Bouetm, Pujolle G.A. range-free 3-D localizationmethod for RFID tags based on virtual landmarks. Proc. of PIMRC. 2008.

Information Technology and Computer Application Engineering – Liu, Sung & Yao (Eds)
© *2014 Taylor & Francis Group, London, ISBN 978-1-138-00079-7*

Analyze the interval of street trees on campus based on the concept of low carbon—As a case of Guangxi University

Hang Li, Ying Luo, Xiu Yang, Xin Sheng Lu & Liu Di Li
College of Civil Engineering and Architecture, Guangxi University, China

ABSTRACT: This paper is based on the 45 universities in American except one McGill University (Canada). We analyze **various** carbon emission from transportation modes and simplify the equation. At last we estimate the longest interval between street trees after taking the characteristics of the local region and specify conditions of the universities where we made our survey into account.

Keywords: Street trees; Carbon emission; Interval; Guangxi University

1   PREFACE

The global warming and the melting in the Polar sides have widely aroused public attention, which are all about the emission of carbon dioxide. The carbon emission of automobiles takes up the 20% of the whole emission [1], which is well-known to every one. But there is an ignored side that the emission on campus soars increasingly.

The destination of the paper is to solve some problems about it by getting the equation which is suitable to Chinese universities, a quantitative relationship between the carbon emission of the universities and their background situation. The equation can not only cut down a big sum of fees to purchase the relevant equipment to observe some data but also be made full use of to control the longest interval from each street trees on campus, which will absorb as much carbon dioxide as possible.

Most planning projects on campus nationwide aim at setting parking places reasonably, encouraging public transportation and limiting the private cars to achieve their low carbon target. However, we can not quantitatively value what effects the following measures will bring to us or verify whether we are successful or not. Except research on Optimized Strategies of Chinese Campus Transportation under the Background of Mobilization written by Zhen Lin [2], there are not relevant papers making quantitatively surveys about the efforts some traffic measures made on campus in China.

The Campus Pedestrian and Bicycle Plans of each foreign schools (these issues are published in a regular time to guide the specific measures to finish their low carbon goals) mention the number of teachers, staffs and students who take varieties of traffic modes. On the plan, they publish the carbon dioxide let off each year like College of Charleston [3], College Dublin [4]. But they all ignore the absorption of street trees to the carbon and the relationship between carbon dioxide and the interval of each street tree.

2   SHOWING METHOD AND COLLECTING DATA

2.1   Collecting background data

Firstly, we collected information of traffic modes on campus of 45 American universities or colleges which are all in urban areas. Besides Campus Pedestrian and Bicycle Plans, we also got their background information via Internet, papers and etc. It was a pity that the data all came from schools in the United states (except McGill University, Canada). Nearly all campuses in the developing countries have not published their plans, such as Tongji in China, whose data are less than the demand. Though teachers, staffs and students may take different transportation modes to get to school or they may change the traffic modes in the way to school, we simplified the process and chose their main modes. According to the carbon dioxide, we divided the transportation into 3 species.

The private cars belong to the first type called high carbon emission. The public transportation stands for the second one called medium carbon emission. The last one is representative to the zero carbon emission such as walking and biking. We will get a result like Table 1. But it can not be guaranteed to be suitable for out country after the results are statistically meaningful.

The necessary information include: the population of campus, private car (users), bus transit, the population of campus, automotive vehicle (users), bus transit

Table 1.

Mode	kg CO2 per passenger km	Source and Assumptions
walk	0	-
cycle	0	-
bus	0.0176	Used conversion factor from Environment Canada http://www.ec.gc.ca/soerree/English/Indicators/Issues/Transpo/Tables/pttb04_e.cfm
train	0.092	Used conversion factor from Environment Canada http://www.ec.gc.ca/soerree/English/Indicators/Issues/Transpo/Tables/pttb04_e.cfm
ferry	0.0088	Used same as buses, but divided by 2 to reflect greater numbers on board (Checking with MfE)
car driver	0.22	Used conversion factor from Department for Environment, Food and Rural Affairs (DEFRA - UK) for a medium petrol car http://ww.defra.gov.uk/environment/business/envrp/gas/10.htm
car passenger	0	Assume these staff are travelling with the car drivers, so their CO2 emissions are already accounted for
motorbike	0.085	Used conversion factor from DEFRA for a small petrol car divided by 2

(CO2 Conversion Factor Table)

Table 2.

Public transit (users)	Private cars (users)	cycling+ walking (users)	% public transit	% Private car (users)	% cycling+ waking (users)
234	462	2024	7.44%	26.25%	66.31%
153	251	788			
252	1542	2882			

Picture 1.

(users), bicycle + pedestrian (users), endowment (million dollars), total area (hectare), latitude (N, degree), longitude (W, degree), student parking fees (dollars per year), school grounds maintained organically (%), CO_2 emission (kg CO_2 per passenger · km · day (the amount whose people on campus release carbon in a certain transportation mode come from the surveys done by their schools)). Some studies mentioned above came from the results that the author published.

2.2 Collecting traffic data about target school, Guangxi University

In the survey about transportation mode in Guangxi University, we tried to get the percentage of each traffic modes in Guangxi University by sampling people on campus occasionally. The sampling area is limited to a bridge connecting to the eastern and western campus, which was very representative. In order to make the conclusion more precise, the survey were chosen in various busy periods on campus that bring to us the average flow in peaking time. We observe the percentage of each traffic modes in the whole flow (Table 2).

In another survey, we should estimate how many person-times every road in Guangxi University are gone through, which is the OD Survey (Orient-Destination Survey). A road is consisted of starting point, ending point and a line between them. We select 13 sites as our starting points or ending points in Guangxi University based on two principles.

Firstly, the sites should be the significantly educational or living location.

Secondly, the sizes should be allocated evenly in the university.

The 13 sites are Western Gate, Bank Street (just outside Xiyuan Restaurant), Nanyuan Restaurant, Eastern Restaurant, Eastern Gate, Scholar Hospital outside Eastern Campus, The library, Eastern Food Market, The Hall, the Gate of Xinjian College, accessory elementary school, accessory middle school, Food Market on western campus. With all the surveys done on campus, we got 421 recycled questionaires. The total questionaires we had sent were 500. The number of recycled questionaires takes up 0.77% of the total number of people in Guangxi University (including students, teachers and staffs)

We assume that the cycle of a scholar arrangement is a week. We got the result of the OD Survey. According to the person-times that a respondents goes to a certain destination (one of the whole 13 destinations) in a week. We can estimate how many person-times one road between two destinations will have been gone through in a week. The equation is that the proportion between person-times of a certain destination and the total time-person times of the 13 destinations × the frequency of Guangxi University (it means in an average day, the times a person will go outside) × the population of Guangxi University Processing the traffic data, we drew an OD Picture (picture 1) where there are different size lines representing different frequency. It need to mention that the width of a line is not the real ones that we plan for but depends on frequency of each roads. If you ask someone how many times he get to one place from another place, he will give you an answer. But maybe there are several methods to get here to there. So we believe that people are inclined to the shortest ways which has already been there to achieve their destinations from several ways. Our simplest unit is a road from one cross to another cross. Because the length of a unit

Table 3.

		the population of campus	private car (users)	public transit (users)	cycling+walk (users)	endowment (million dollar)	whole area (hectare)	latitude (N) degree	longitude(W) degree	Parking fee (dollar per year)	% school grounds maintained organically	CO2 emission (kg CO2 per users km)
the population of campus	Pearson Correlation	1	.806(**)	.623(**)	.526(**)	0.035	0.151	-0.266	0.292	0.078	-0.109	0.191
	Sig. (2-tailed)		0	0	0	0.818	0.339	0.081	0.055	0.628	0.475	0.209
	N	45	45	45	45	45	42	44	44	41	45	45
private car (users)	Pearson Correlation	.806(**)	1	0.17	0.016	-0.117	0.087	-.534(**)	0.282	-0.145	-0.151	-0.179
	Sig. (2-tailed)	0		0.265	0.919	0.444	0.585	0	0.064	0.367	0.322	0.238
	N	45	45	45	45	45	42	44	44	41	45	45
public transit (users)	Pearson Correlation	.623(**)	0.17	1	.446(**)	0.155	0.036	0.202	0.077	.312(*)	0.028	.774(**)
	Sig. (2-tailed)	0	0.265		0.002	0.309	0.82	0.189	0.621	0.047	0.854	0
	N	45	45	45	45	45	42	44	44	41	45	45
cycling+walk (users)	Pearson Correlation	.526(**)	0.016	.446(**)	1	0.193	0.201	0.176	0.108	0.214	-0.018	0.179
	Sig. (2-tailed)	0	0.919	0.002		0.204	0.201	0.254	0.487	0.18	0.906	0.238
	N	45	45	45	45	45	42	44	44	41	45	45
endowment (million dollar)	Pearson Correlation	0.035	-0.117	0.155	0.193	1	0.033	0.083	-0.257	.668(**)	0.17	0.224
	Sig. (2-tailed)	0.818	0.444	0.309	0.204		0.836	0.591	0.093	0	0.263	0.138
	N	45	45	45	45	45	42	44	44	41	45	45
whole area (hectare)	Pearson Correlation	0.151	0.087	0.036	0.201	0.033	1	-0.167	0.032	-0.103	-0.211	0.001
	Sig. (2-tailed)	0.339	0.585	0.82	0.201	0.836		0.297	0.84	0.533	0.179	0.994
	N	42	42	42	42	42	42	41	41	39	42	42
latitude (N) degree	Pearson Correlation	-0.266	-.534(**)	0.202	0.176	0.083	-0.167	1	-0.024	0.109	0.133	0.191
	Sig. (2-tailed)	0.081	0	0.189	0.254	0.591	0.297		0.875	0.501	0.389	0.214
	N	44	44	44	44	44	41	44	44	40	44	44
longitude(W) degree	Pearson Correlation	0.292	0.282	0.077	0.108	-0.257	0.032	-0.024	1	-0.012	-0.221	-0.078
	Sig. (2-tailed)	0.055	0.064	0.621	0.487	0.093	0.84	0.875		0.943	0.149	0.614
	N	44	44	44	44	44	41	44	44	40	44	44
Parking fee (dollar per year)	Pearson Correlation	0.078	-0.145	.312(*)	0.214	.668(**)	-0.103	0.109	-0.012	1	0.222	.423(**)
	Sig. (2-tailed)	0.628	0.367	0.047	0.18	0	0.533	0.501	0.943		0.162	0.006
	N	41	41	41	41	41	39	40	40	41	41	41
% school grounds maintained organically	Pearson Correlation	-0.109	-0.151	0.028	-0.018	0.17	-0.211	0.133	-0.221	0.222	1	-0.003
	Sig. (2-tailed)	0.475	0.322	0.854	0.906	0.263	0.179	0.389	0.149	0.162		0.985
	N	45	45	45	45	45	42	44	44	41	45	45
CO2 emission (kg CO2 per users km)	Pearson Correlation	0.191	-0.179	.774(**)	0.179	0.224	0.001	0.191	-0.078	.423(**)	-0.003	1
	Sig. (2-tailed)	0.209	0.238	0	0.238	0.138	0.994	0.214	0.614	0.006	0.985	
	N	45	45	45	45	45	42	44	44	41	45	45

** Correlation is significant at the 0.01 level (2-tailed).

* Correlation is significant at the 0.05 level (2-tailed)

is positively relative to their destinations. Then we got the frequency of each units. The total frequency from one destination to the others are consisted of many simplest units. We add the frequency of each simplest unit and get the frequency of a whole road (Picture 1).

2.3 Collecting botany data about target school, Guangxi University

We assume that grass, shrubs and trees all have the ability to transform CO_2 into O_2. Now, it is in a certain interval that the street trees can all absorb the carbon dioxide that the traffic bring to us even if there are not grass or shrubs anymore. The contents of the research contain two aspects.

Firstly, take notes about street trees information about the chosen road which include tree species, crown diameters, breast-height diameters.

Secondly, measure the today's tree intervals.

3 RESULT

3.1 Background data

In the correlation analysis (Table 3), there is a positive correlation between carbon emission (kg CO_2 per passenger per km) and the population who take public traffic modes on campus ($r = 0.774$, $p = 0.000$). There is also a positive correlation between parking fees and carbon emission whose p is < 0.01.

Then we can make a stepwise regression (Table 4). We regard CO_2 emission as a dependent. The rest of

Picture 2.

the items are all independents. The entry of F is more than 0.05 and the removal is less than 0.1.

The basic equation is that $Zy = A + aX + bY + cZ$... and we get the equation.

And there are two independents entering into the equation which are the population of school and the number of people who take public transportations on campus. Then, we obtain a useful equation which reflects the relationship between the two items and the emission.

$Y = 0.04 + 5.43 \times 10^{-6} \times$ the number of people who take public transportations $-7 \times 10^{-7} \times$ the total population on campus.

3.2 The data of Guangxi University

There are 55000 people studying or working on campus. The percentage of the public transit modes is about 7% (7.44%) in the sampling survey. The users of the public transit mode might be $55000 \times 0.07 = 3850$. The figure (7%) is brought back to the linear regression which only has one unbekannte, the population on campus.

Before brought back:

The carbon emission (kg/per passenger per km per day). The users of bus transit (passengers) = 7% × the total population. The chosen site between Financial Street to Cuiyuan Road (Picture 2) is a typically suitable for we have consider two aspects.

Firstly, the site is located in the center of the school.

Secondly, frequency of this road is narrowly more than the average frequency. The equation of carbon emission is that The average emission of a person ×the frequency of this road × total population) = $70.0224 \times 55000 \times 5362.2/33110.2)/7 = 28.5$ kg per day. The 5362.2 and 33110.2 respectively stand for the person-times of respondents to the chosen site and the person-times of respondents to the whole 13 destinations. The proportion of this two number is the frequency of this road.

The street trees in the chosen place are mangoes, camphor, red bauhinias. We use a term called the fixed carbon in a given unit area on leaf surface in fall, which means that in autumn leaf surface which fixes some carbon per day per square meter depends on the species of plants. We take some indices when the leave are in

autumn as our standards because the intervals of street trees should meet the demand in nearly a year even if they work with a lowest efficiency during this period. mango ($4.16/m^2$ per day), camphor ($4.75/m^2$ per day), Red Bauhinia ($4.85/m^2$ per day), poplar ($10/m^2$ per day). According to the paper of Xinyi Wang [6], we take the capability of poplar to fix carbon and the average crown diameter, 5.8 meters as a base condition. Thus, the efficiency of a plant to fix the carbon can be respectively proportional valued as 0.55 kg per day(mango), 0.63 kg per day(camphor), 0.64 kg per day (red bauhinia). According to different number of each tree the average efficiency is 0.607 kg per day.

4 DISCUSSION

The interval of a tree = the length of a chosen road/the carbon emission of a chosen road in one day × the average efficiency to fix carbon = 0.403 km/28.5 kg per day × 0.607 kg per day = 0.00856 km = 8.56 m.

The shortest designed crown diameter is 5.8 meters which means that the interval can not be less than 5.8 meters if the street trees grow healthily. And the longest interval where each side should be planted in street trees, not just one side. The interval can be 8.56 meters even if the trees are in fall. Now, in this chosen road, today's interval is 6.5 meters. Just considering the absorption of carbon, we can extend the interval between street trees in this chosen line. It is in this way that we can estimate that the interval of street trees in every road in this school reasonably. This method, I believe, is easy to understand and strongly practical to be spread to other roads or other schools.

ACKNOWLEDGEMENT

Corresponding author: Xiu yang, lecturer, College of Civil Engineering and Architecture, Guangxi University, China.

REFERENCES

[1] World energy Outlook: submitted by International Energy Agency (2007)
[2] zhenlinz, A dissertation submitted to Tongji University in conformity with the requirements for the degree of Master of Philosophy Research on Optimized Strategies of Chinese Campus Transportation under the Background of Mobilization
[3] College of Charleston Campus Transportation Study (2011). http://sustainability.cofc.edu
[4] University College Dublin, Belfield Campus Framework Commuting Strategy 2009-2012-2015. http://www.arup.ie
[5] Xinyi Gao, yuying Wang, [J], 2011.

Information Technology and Computer Application Engineering – Liu, Sung & Yao (Eds)
© *2014 Taylor & Francis Group, London, ISBN 978-1-138-00079-7*

Digital filter window function method design based on MATLAB FDA Tool

Xin Zhao
College of Life Science and Technology, Honghe University, Mengzi, Yunnan, China

Shan Ren
Engineering College, Honghe University, Mengzi, Yunnan, China

ABSTRACT: Digital filter is one of the most important parts of digital signal processing. In practice, digital signal processing often need to limit the signal observation time interval within a certain time, choose only one period of signal that signal data will be truncated, this process is equivalent to plus window function operation to signal. In order to obtain finite unit sample response, need to truncate the infinite unit sample response sequence by window function. This paper proposes the method of using window function to design filter based on MATLAB, according to the design basic principle of digital filter.

Keywords: Digital Filter, Window function method, MATLAB, FDA Tool

1 INTRODUCTION

Digital filter is very important in signal processor, and filter design may be realized by MATLAB. There are two methods in digital filter, include FIR and IIR. Among them FIR filter can ensure strict linear phase, filter stability and less error.

At present, the methods of designing FIR Filter include window function and frequency sampling. As the window function is simple, and has closed formula to follow, so the window function method is used in this paper for designing of Digital Filter, and implement the design process though using FDA Tool-box of MATLAB.

2 WINDOW FUNCTION DESIGN METHOD OF FIR DIGITAL FILTER

Digital Filter is a signal processing devices that characteristics of certain transmission, its input and output are discrete digital signal. With digital device or a certain numerical method for processing the input signal and change its waveform or spectrum, to achieve the goal to filter out unwanted signal components[1].

The basic idea of window function design method is: firstly, given the frequency response $H_d(e^{j\omega})$ of ideal filter, and then find a group of $h(n)$, made the frequency response $H(e^{j\omega}) = \sum_{n=0}^{N-1} h(n)e^{-j\omega}$ that determined by $h(n)$ to approximate to $H_d(e^{j\omega})$. The design is in time domain. Therefore, first derived $h_d(n)$ though IFFT, namely:

$$h_d(n) = \frac{1}{2\pi} \int_{-\pi}^{\pi} H_d(e^{j\omega})e^{j\omega} d\omega \qquad (1)$$

Then use the finite length of $h(n)$ to approximate the infinite $h_d(n)$, an effective method is to cut $h_d(n)$, that is, use of finite-length window function $\omega(n)$ to intercept $h_d(n)$, namely:

$$h(n) = w(n)h_d(n) \qquad (2)$$

An ideal window function should meet the following two requirements [2]:

1. The main lobe of amplitude-frequency characteristics for window function should as narrow as possible to get a steep transition zone.
2. The maximum side-lobe of amplitude-frequency characteristics for window function should as small as possible, so that the main lobe contains as much energy as possible; this will make the spikes and ripple reduce, to obtain flat amplitude-frequency characteristics and sufficient band-stop attenuation.

Window functions that used for designing FIR Filter include: Rectangle, Triangle, Hanning, Hamming, Blackman and Kaiser Window function. Specific indexes are shown in Table 1 [3].

3 SIGNAL SPECTRUM ANALYSIS AND FILTER PARAMETERS DESIGN

Here, to a power plant measured vibration signal of a steam turbine, for example, to introduce the filter design. The speed of the steam turbine is about 3000 per minute, sampling frequency is 6400 Hz. Time and frequency domain chart of pre filtering vibration signal as shown in Figure 1.

As shown in the time domain chart, due to interfered by random noise and spikes, which affect to identify the characteristics of vibration signal.

Table 1. Window Function Index.

Window Function	Peak Amplitude of Sidelobe (dB)	Transition Zone Width	Band-stop Minimum Attenuation (dB)
Rectangle	−13	$4\pi/N$	−21
Triangle	−25	$8\pi/N$	−25
Hanning	−31	$8\pi/N$	−44
Hamming	−41	$8\pi/N$	−53
Blackman	−57	$12\pi/N$	−74
Kaiser	−57	$10\pi/N$	−80

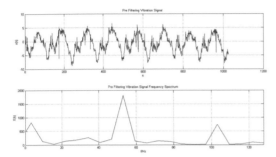

Figure 1. Time and Frequency Domain Chart of Pre Filtering Vibration Signal.

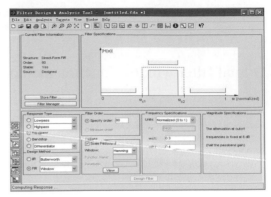

Figure 2. FDA Tool Design Interface.

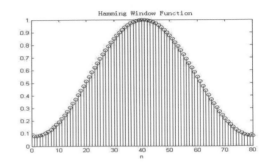

Figure 3. Hamming Window Function Oscillograph.

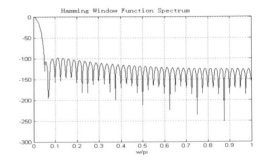

Figure 4. Hamming Window Function Spectrum.

In addition, from the frequency domain chart we can easily see that pre filtering vibration signal mixed together by the frequency domain characteristics of 4 Hz, 52 Hz, 104 Hz signal. However, the frequencies of steam turbine vibration signal roughly in the range of 50 Hz~55 Hz. So, we should design a Band-pass filter, which can filter out signals that below 50 Hz and above 55 Hz. Here, given the performance parameters of filter through normalized frequency: pass-band range is $0.3\pi \sim 0.4\pi$, minnum attenuation of stop-band is not less than −50 dB.

According to Table 1, the Hamming window and Blackman window can provide attenuation that greater than 50dB. But the Hamming window has a smaller transition zone, which has a smaller order of N. Therefore, the Hamming window was selected to complete the design. The transition zone of the filter is $\Delta\omega = \omega_{c2} - \omega_{c1} = 0.4\pi - 0.3\pi = 0.1\pi$.

As shown in Table 1, the transition zone band of filter that using Hamming window design is $\Delta\omega = 8\pi/N$, so the unit impulse response length of band-pass filter is $N = 8\pi/\Delta\omega = 80$, −3 dB pass-band cut-off frequency is $\omega = (\omega_{c2} + \omega_{c1})/2 = 0.35\pi$. Based on these performance indexes, we can design the digital filter.

4 DIGITAL FILTER FDA TOOL DESIGN

In MATLAB, the design of digital filter could implement by Filter Design & Analysis Toolbox.

According to designation index, order of the Band-pass filter is 80. In FDA Tool, we need to choice type of filter (FIR), method for designing FIR Filter (window function, frequency sampling or other), type of window function and other necessary performance index of designing filter (here, only need to input the frequency value of pass-band range) [4]. Then click the "Design Filter" that on below of the FDA Tool interface, an ideal band-pass Digital Filter could be achieved. FDA Tool design interface shown in Figure 2.

At the same time, in line with the above design index, by calling the following Hamming Window Function can achieve the oscillograph and spectrum, which shown below in figure 3, figure 4.

wls = 0.15*pi; wc1 = 0.3*pi; wc2 = 0.4*pi; wc = [wc1/pi,wc2/pi]; B = wc2-wc1; N = ceil(12/0.15);

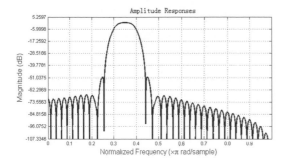

Figure 5. Magnitude Response of Band-pass Filter.

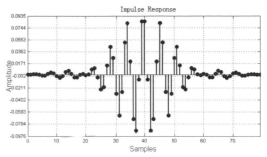

Figure 7. Band-pass Filter Impact Response.

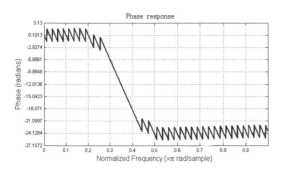

Figure 6. Phase Response of Band-pass Filter.

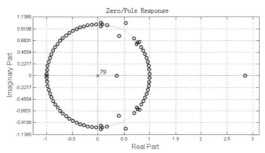

Figure 8. Band-pass Filter Zero/Pole Response.

n = 0: N-1; window = Hamming (N); [h1, w] = freqz (window, 1);

```
    figure;
    stem(window);
    xlabel ('n');
    title('HammingWindow Function');
    figure;
    plot(w/pi,20*log(abs(h1)/abs(h1(1))));
    grid;xlabel('w/pi');ylabel('Amplitude (dB)');
    title('Hamming window function spectrum');
```

Also, amplitude response and phase response of the filter could draw such as above figure 5 and figure 6 shows.

From the magnitude response curve we known, the minimum attenuation is −50 dB, to meet the design requirements; from the phase response curve can be seen, in the pass-band, phase of the filter shown a strictly linear relationship, consistent with linear phase characteristics of the FIR Filter.

Filter impact response and zero pole response as below figure 7 and figure 8 shows.

5 CONCLUSIONS

In this article, based on window function design method of Digital Filter, to a power plant measured vibration signal of a steam turbine, for example, by analyzing the composition of the vibration signal in the frequency domain, calculated and gave the performance parameters and type of Filter. Combining with the programming method and FDA Tool of MATLAB, using Hamming window function designed the band pass FIR digital filter.

This method is simple and practicable, and has certain reference value for understanding the principle and design method of Digital Filter.

REFERENCES

[1] Hongyi Zhao and Changnian Zhang: *Digital Signal Processing and Implementation Based on MATLAB.* (Chemical Industry Press, Beijing 2002), in Chinese.
[2] Peiqing Cheng: *Digital signal processing Course.* (Tsinghua University Press, Beijing 2003), in Chinese.
[3] Shengli Yan: Journal of Changchun Institute of Technology, (2003) No. 4, p. 63–65, in Chinese.
[4] Chen Liang: Machinery Design & Manufacture, (2010) No. 12, p. 87–89, in Chinese.
[5] Maoqing Li, Jie Wang and Qiang Chen: Journal of Electric Power, (2008) No. 23, p. 87–90, in Chinese.
[6] [7] Shan Ren, Xin Zhao and Wenbin Zhang: Advanced Materials Research, (2012) Vol. 490–495, p. 1867–1870, in English.
[8] Shan Ren, Xin Zhao, Jie Jiang and Dongjin Zhao: Advanced Materials Research, (2012), Vol. 203–504, p. 228–231, in English.

Information Technology and Computer Application Engineering – Liu, Sung & Yao (Eds)
© *2014 Taylor & Francis Group, London, ISBN 978-1-138-00079-7*

Window function method design and realization of high-pass digital filter based on MATLAB SPTool

An Dong Qu & Jie Min

Engineering College, Honghe University, Mengzi, Yunnan, China

ABSTRACT: Digital filter is one of the most important parts of digital signal processing. This article in accordance with the principles of window function design method, using SPTool design tools, design of a high-pass digital filter. Filtering processing for measured signal showed that filtering effect of the filter achieved the expected results.

Keywords: Window function method, High-pass Digital Filter, SPTool design method, MATLAB

1 INTRODUCTION

Digital filter is very important in signal processor, and filter design may be realized by MATLAB. There are two methods in digital filter, include FIR and IIR. FIR filter is wildly used because of it with characteristics of ensure strict linear phase, filter stability and less error.

At present, the methods of designing FIR Filter include window function, frequency sampling and equivalent ripple approximation. At the same time, these methods can also direct implementation through the FDA Tool and SPTool. The SPTool is an interactive graphic signal processing tools in MATLAB, which can be convenient to complete the signal, filter and spectrum analysis and design. In this paper, taking a steam turbine vibration signal as an example, used SPTool design and realize the vibration signal filtering.

2 WINDOW FUNCTION DESIGN METHOD OF FIR DIGITAL FILTER

Digital Filter is a signal processing devices that characteristics of certain transmission, its input and output are discrete digital signal. With digital device or a certain numerical method for processing the input signal and change its waveform or spectrum, to achieve the goal to filter out unwanted signal components [1].

The basic idea of window function design method is: firstly, given the frequency response $H_d(e^{j\omega})$ of ideal filter, and then find a group of $h(n)$, made the frequency response $H(e^{j\omega}) = \sum_{n=0}^{N-1} h(n)e^{-j\omega}$ that determined by $h(n)$ to approximate to $H_d(e^{j\omega})$. The design is in time domain. Therefore, first derived $h_d(n)$ though IFFT, namely:

$$h_d(n) = \frac{1}{2\pi} \int_{-\pi}^{\pi} H_d(e^{j\omega})e^{j\omega}d\omega \qquad (1)$$

Then use the finite length of $h(n)$ to approximate the infinite $h_d(n)$, an effective method is to cut $h_d(n)$, that is, use of finite-length window function $\omega(n)$ to intercept $h_d(n)$, namely:

$$h(n) = w(n)h_d(n) \qquad (2)$$

An ideal window function should meet the following two requirements [2]:

1. The main lobe of amplitude-frequency characteristics for window function should as narrow as possible to get a steep transition zone.
2. The maximum side-lobe of amplitude-frequency characteristics for window function should as small as possible, so that the main lobe contains as much energy as possible; this will make the spikes and ripple reduce, to obtain flat amplitude-frequency characteristics and sufficient band-stop attenuation.

Window functions that used for designing FIR Filter include: Rectangle, Triangle, Hanning, Hamming, Blackman and Kaiser Window function. Specific indexes are shown in Table 1 [3].

Table 1. Window Function Index.

Window Function	Peak Amplitude of Sidelobe (dB)	Transition Zone Width	Band-stop Minimum Attenuation (dB)
Rectangle	−13	$4\pi/N$	−21
Triangle	−25	$8\pi/N$	−25
Hanning	−31	$8\pi/N$	−44
Hamming	−41	$8\pi/N$	−53
Blackman	−57	$12\pi/N$	−74
Kaiser	−57	$10\pi/N$	−80

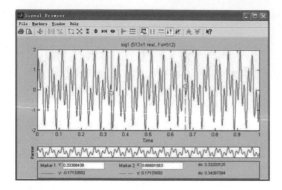

Figure 1. Time Domain Chart of Pre Filtering Vibration Signal.

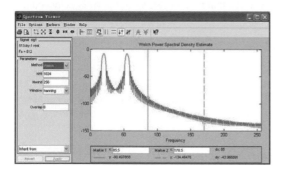

Figure 2. Frequency Domain Chart of Pre Filtering Vibration Signal.

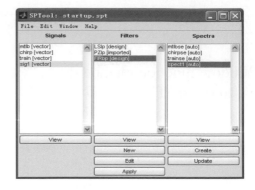

Figure 3. SPTool Work Interface.

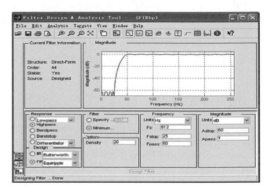

Figure 4. Filter Parameter Design Interface.

3 HIGH-PASS DIGITAL FILTER DESIGN

Here, to a power plant measured vibration signal of a steam turbine, for example, to introduce the filter design. The speed of the steam turbine is about 3000 per minute, sampling frequency is 512 Hz.

Typed "SPTool" in MATLAB command window, and opening SPTool interface. Choose the "Import" in file menu, and import the measured vibration signal and its sampling frequency, right now SPTool work interface will display a signal name for sig1, namely vibration signal. Draw time domain and frequency domain graph of the pre filtering sig1 as in figure 1 and figure 2 shows.

As shown in the time domain chart, due to interfered by random noise and spikes, which affect to identify the characteristics of vibration signal.

In addition, from the frequency domain chart we can easily see that pre filtering vibration signal mixed together by the frequency domain characteristics of 22 Hz and 55 Hz signal. However, the frequencies of steam turbine vibration signal roughly in the range of 50 Hz~55 Hz. So, we should design a High-pass filter, which can make above 50 Hz frequency part signal through.

As the chart 3 shows, design filter in SPTool work interface, choose "FIRbp[design]", and click "Edit" to modify filter parameters. Filter design interface

as the chart 4 shows. In the interface, settings filter type, method, sampling frequency, stopband cut-off frequency, pass-band cut-off frequency, stop-band attenuation and pass-band ripple, then get a high-pass filter. At this time, could observe the parameters of filter, including filter order number, stability, impact response, zero pole response, filter coefficient, etc.

After understanding the characteristic parameters of filter, we will be closed filter parameter design interface, and using filter for signal filtering.

4 MATLAB IMPLEMENT OF HIGH-PASS FILTER

Now, use high-pass filter that just design to realize vibration signal filtering.

In the SPTool work interface as shown above figure 3, choose the measured signal sig1 that has imported, and then select the "FIRbp [design]", and click "Apply" button, the system pop-up "Apply Filter" interface below figure 5 shows. This show that the system will use filter that has been designed to input signal filtering, and at the same time, the system will automatically output signal that name for sig2, which is filtered vibration signal.

As in the previous method, drawing time domain and frequency domain map of post filtering vibration signal sig2, as shown in figure 6 and figure 7.

Figure 5.   Apply Filter Dialog Box.

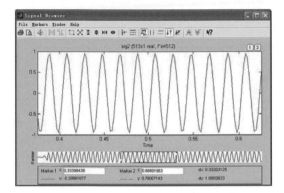

Figure 6.   Post Filtering Vibration SignalTime Domain Char.

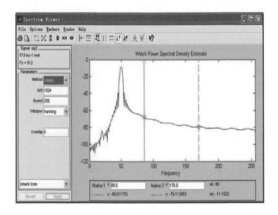

Figure 7.   Post Filtering Vibration Signal Time Domain Chart.

Magnitude and phase response of Band-pass filter are shown in Figue 8. Among them, curve 1 is magnitude response and curve 2 is phase response.

From the magnitude response curve we known, the minimum attenuation is −60 dB, to meet the design requirements; from the phase response curve can be seen, in the pass-band, phase of the filter shown a

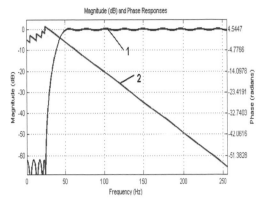

Figure 8.   High-pass Filter Amplitude and Phase Response.

strictly linear relationship, consistent with linear phase characteristics of the FIR filter.

5   CONCLUSIONS

In this article, based on window function design method of digital filter, to a power plant measured vibration signal of a steam turbine, for example, by analyzing the composition of the vibration signal in the frequency domain, calculated and gave the performance parameters and type of filter. Combining with the FDA Tool and SPTool of MATLAB, designed and implemented High-pass filter, and realized to measure signal filtering.

The results of the vibration signal show that designed filter meet the design requirements and with good performance. This method is simple and practicable, and has certain reference value for understanding the principle and design method of digital filter.

REFERENCES

[1] Hongyi Zhao and Changnian Zhang: *Digital Signal Processing and Implementation Based on MATLAB*. (Chemical Industry Press, Beijing 2002), in Chinese.
[2] Peiqing Cheng: *Digital signal processing Course*. (TSINGHUA UNIVERSITY PRESS, Beijing 2003), in Chinese.
[3] Zhonglin Huang: *Control System Calculation and Practical Training of MATLAB*. (National Defence Industry Press, Beijing 2006), in Chinese.
[4] Chen Liang: Machinery Design & Manufacture, (2010) No. 12, p. 87–89, in Chinese.
[5] Yi Shao, Yan Wen: CHINA NEW TELECOMMUNICATIONS, (2010) No. 2, p. 83–85, in Chinese.
[6] Hui Li: *Digital Signal Processing and MATLAB Implement*. (CHINA MACHINE PRESS, Beijing 2011), in Chinese.

Information Technology and Computer Application Engineering – Liu, Sung & Yao (Eds)
© *2014 Taylor & Francis Group, London, ISBN 978-1-138-00079-7*

Mathematical standard of MUSIC approach for multiple emitter location

Bao Fa Sun
Department of Computer Science and Technology, Anhui Sanlian University, Hefei, Anhui, China

ABSTRACT: To set up the mathematical standard of MUSIC approach for multiple emitter location, three new concepts are defined: linear space spanned by a matrix, distance from a vector to a subspace, inner product of a vector and a subspace. Basing on these definations, the inaccurate usage of two concepts is exacted and the wrong usage of a concept is modified. The study is useful and necessary for summarization and deeper research of MUSIC approach.

Keywords: MUSIC approach; linear space spanned by a matrix; distance from a vector to a subspace; internal product of a vector and a subspace

1 PREFACE

The literature [1] marked the appearance of the theory and technology of modern spatial spectrum estimation. In the paper, a number of new concepts were used without definition, such as matrix linear space, matrix linear subspace, internal product of a vector and a subspace, etc. People accepted these concepts and continued to use them in their essays. No one strictly defined these concepts from mathematical perspective so far.

The usage of concepts by project researchers has led to a situation: the researchers of the same field reach some kind of tacit understanding, they admit the usage of these concepts each other, don't question any longer. If a researcher has used a concept wrongly, others will circulate erroneous reports, make the same mistake continually. This case occurred on the researching process of spatial spectrum estimation. The wrong usage of concepts was recurring when the researchers interpreted the meaning of formula $d = \|a^*(\theta) \cdot E_N\|$. Therefore, it is necessary to define the concepts strictly in the project study.

Since 1980's, the theory and technology of modern spatial spectrum estimation has gotten great progress. Many researchers improved and extended MUSIC approach from different angles, so the spatial spectrum estimation has become vigorous. In the near future, there will be more and better algorithms enriching the theory and technology. However, as with any subject, the all-around, from different angles, branching, scattering research of spatial spectrum estimation will lead to synthesis of the subject. In the synthesis process, the concepts must be strictly defined.

The above analyses show that, in the complete theory system of multiple emitter direction finding, the used concepts must be strictly defined. This paper will lay the foundation for integrating work of MUSIC in future.

2 FOUNDATIONS OF MATHEMATICS

2.1 Linear space spanned by a matrix

Definition 1 Let A be a $n \times n$ real matrix, the column of A can be viewed as n-dimensional vector in linear space R^n. The linear space generated by column vectors of A is called the linear space of matrix A.

The linear space of matrix A is a subspace of R^n. If A is a full rank matrix, the column vectors of A are linearly independent, they form the base of n-dimensional linear space R^n, at this time, the linear space of matrix A is just R^n. If the matrix A is not full rank, A cannot generate the linear space R^n, the linear space of matrix A is the true subspace of R^n.

2.2 The distance from a vector to a subspace

For intuition, study an example of 3-dimensional Euclidean space. As shown in Figure 1, OXYZ is 3-dimensional Euclidean space, plane OXY is a subset of the space OXYZ, $\vec{a}$ is a vector of OXYZ. In analytic geometry, the vector is free, i.e. vector can move parallelly, so it is not suitable to define the distance of the vector $\vec{a}$ to the subspace OXY. However, in MUSIC approach, "the distance of the direction vector $a(\theta)$ to the subspace distance E_N" is mentioned, so it is necessary to define the concept "the distance of a vector to a subspace".

Definition 2 Move vector $\vec{a}$ parallelly, so that the starting point of $\vec{a}$ coincides with the coordinate origin O, and gain the vector $\vec{a}' = \{x, y, z\}$. The distance from

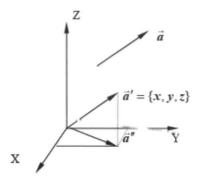

Figure 1. Distance and projection of a vector to a subspace.

point (x, y, z) to the plane OXY is called the distance of the vector $\vec{a}$ to the subspace OXY.

Two vectors are equivalent if they coincide by moving one vector parallelly. There are infinite numbers of vectors which are equivalent to $\vec{a}$. The set of the vectors equivalent to $\vec{a}$ is called an equivalence class of vector $\vec{a}$, marked as $\{\vec{a}\}$. In $\{\vec{a}\}$, take the vector $\vec{a}'$ whose starting point is the origin O as the representative of the equivalence class, calculate the distance d from finishing point of $\vec{a}'$ to OXY, and regard d as the distance of an arbitrary vector of $\{\vec{a}\}$ to OXY. The equivalence class $\{\vec{a}\}$ of a vector $\vec{a}$ determines a distance uniquely, so the distance definition is reasonable.

For Euclidean space R^n, the distance from a vector $\vec{a}$ to a subspace V can be defined similarly.

2.3 *The inner product of a vector and a subspace*

Still use Figure 1 to explain. Project $\vec{a}'$ onto the subspace OXY, and gain the projection vector $\vec{a}''$. If the starting point and finishing point coordinates of $\vec{a}$ are known, the finishing point coordinate of $\vec{a}'$ can be gotten, thus the finishing point coordinate of $\vec{a}''$ in OXY plane can be gotten.

In fact, the horizontal coordinate of the finishing point of $\vec{a}''$ is $x = \vec{a}'' \cdot \{1, 0, 0\} = \vec{a}' \cdot \{1, 0, 0\} = \vec{a} \cdot \{1, 0, 0\} = \vec{a} \cdot \vec{\alpha}^0$, where $\vec{\alpha}$ is any nonzero vector on the X axis pointing at the direction of X, $\vec{\alpha}^0$ is the unit vector with the same direction of $\vec{\alpha}$. Similarly, the vertical coordinate of the finishing point of $\vec{a}''$ is $y = \vec{a} \cdot \vec{\beta}^0$, where $\vec{\beta}$ is any nonzero vector on the Y axis pointing at the direction of Y, $\vec{\beta}^0$ is the unit vector with the same direction of $\vec{\beta}$.

Thus, the coordinates of the finishing point of the vector $\vec{a}''$ on the bases $\{1, 0, 0\}$ and $\{0, 1, 0\}$ of plane OXY are $(\vec{a} \cdot \vec{\alpha}^0, \vec{a} \cdot \vec{\beta}^0)$. In general, for any bases $\vec{\alpha}$, $\vec{\beta}$ of plane OXY, the coordinates of the finishing point of the vector $\vec{a}''$ on $\vec{\alpha}^0$, $\vec{\beta}^0$ are $(\vec{a} \cdot \vec{\alpha}^0, \vec{a} \cdot \vec{\beta}^0)$. Therefore, for the bases $\vec{\alpha}$, $\vec{\beta}$, $\vec{a}'' = \{\vec{a} \cdot \vec{\alpha}^0, \vec{a} \cdot \vec{\beta}^0\}$. In particular, if $\vec{\alpha}$, $\vec{\beta}$ are the orthogonal bases of plane OXY, the length of $\vec{a}''$ is $l = \|\vec{a}''\| = \|\{\vec{a} \cdot \vec{\alpha}^0, \vec{a} \cdot \vec{\beta}^0\}\|$.

It is easy to prove that l is uniquely identified by vector $\vec{a}$ and subspaces OXY, unrelated with the selection of orthogonal basis.

Definition 3 The inner product $\vec{a} \cdot OXY$ of vector $\vec{a}$ and subspaces OXY is a real number, satisfying $\vec{a} \cdot OXY = \|\{\vec{a} \cdot \vec{\alpha}^0, \vec{a} \cdot \vec{\beta}^0\}\|^2$, where $\vec{\alpha}$, $\vec{\beta}$ are any orthogonal bases of plane OXY.

As shown in Figure 1, let l be the length of the projection vector $\vec{a}''$ of vector $\vec{a}$ to subspaces OXY, then $l = \sqrt{\vec{a} \cdot OXY}$.

It is easy to see that the equivalence class $\{\vec{u}\}$ of each vector $\vec{a}$ uniquely identify an inner product with subspace OXY. According to definition 3, if the vector $\vec{a}$ is orthogonal with subspace OXY, the length of the projection vector of $\vec{a}$ to OXY is 0.

Similarly, the inner product of a vector and a subspace can be extended to n-dimensional Euclidean space, thus the length of the projection vector of a vector to the subspace can be gotten.

3 BRIEF INTRODUCTION TO MUSIC

Let

$S(t) = \left(s_1(t), s_2(t), \ldots, s_K(t)\right)^T$: K signals,

$N(t) = \left(n_1(t), n_2(t), \ldots, n_N(t)\right)^T$: the noise vector received by uniform linear array with N antennas,

$X(t) = \left(x_1(t), x_2(t), \ldots, x_N(t)\right)^T$: the signal vector received by uniform linear array with N antennas,

$A(\theta) = \left(a(\theta_1), a(\theta_2), \ldots, a(\theta_K)\right)$: direction matrix of signals,

$a(\theta_i) = \left(1, e^{j\frac{2\pi d}{\lambda}\sin\theta_i}, \ldots, e^{j(N-1)\frac{2\pi d}{\lambda}\sin\theta_i}\right)$:

the direction vector of the ith signal, then

$$X(t) = A(\theta)S(t) + N(t),$$
$$R_{xx} = E\left\{[X(t) - E(X(t))] \cdot [X(t) - E(X(t))]^H\right\}$$
$$= E\{X(t) \cdot X(t)^H\}$$
$$= A(\theta)E\{S(t) \cdot S^H(t)\}A^H(\theta) + E\{N(t) \cdot N^H(t)\}$$
$$= A(\theta)PA^H(\theta) + \sigma^2 I.$$

Decompose the linear space R_{xx} according to the eigenvalue, gain the signal subspace E_S and noise subspace E_N, where $a(\theta_i) \in E_S$ $(i = 1, 2, \ldots, K)$, $R_{xx} = E_S \oplus E_N$.

Spatial spectrum function is $P(\theta) = \|a^*(\theta) \cdot E_N\|^{-2}$. Scan the angle θ and observe the value of $P(\theta)$.

Since $a(\theta_i) \perp E_N (i = 1, 2, \ldots, K)$, $a^*(\theta) \cdot E_N = 0$, $P(\theta)$ have a peak when $\theta = \theta_i$. The DOA θ_i can be found according to the peak position of $P(\theta)$.

4 COMMENT ON THE CONCEPTS USED IN MUSIC APPROACH

4.1 *New concepts used in MUSIC approach*

In the MUSIC approach, several new concepts were mentioned without clear definitions, such as linear space R_{xx}, signal subspace E_S, noise subspace E_N, the distance $d = \sqrt{a^*(\theta) \cdot E_N}$ of vector $a(\theta)$ and subspace E_N, etc. According to the definitions above, the meaning of these concepts can be seen clearly. In fact, the linear space R_{xx} is the linear space of matrix R_{xx}, the signal subspace E_S is the linear space of matrix E_S, the noise subspace E_N is the linear space of matrix E_N; d^2 is the inner product of vector $a(\theta)$ and subspace E_N.

4.2 Inaccurate usage of concepts used in MUSIC approach

In the MUSIC approach, decompose the linear space R_{xx} according to the eigenvalue, get the subspace E_S and subspace E_N. E_S is named as the signal subspace and E_N is named as the noise subspace in article [1]. The naming is not accurate, the reason is as following.

Since $a(\theta_i)(i = 1, 2, \ldots, K)$ has Vandermonde matrix structure and $a(\theta_i) \neq a(\theta_j)(i \neq j)$, $a(\theta_i)(i = 1, 2, \ldots, K)$ is a set of linearly independent vectors. Since $a(\theta_i) \in E_S(i = 1, 2, \ldots, K)$, and the rank of E_S is K, $a(\theta_i)(i = 1, 2, \ldots, K)$ is the base of E_S. So $a(\theta_i)(i = 1, 2, \ldots, K)$ generates the linear space E_S, and any vector in E_S is the linear combination of the signal direction vectors $a(\theta_i)$ $(i = 1, 2, \ldots, K)$. However, the linear combination of $a(\theta_i)(i = 1, 2, \ldots, K)$ is not necessarily a signal, because the number of signal is K while the number of the linear combination of $a(\theta_i)(i = 1, 2, \ldots, K)$ is infinite. This shows that it is not accurate to name E_S as signal subspace. Accurate to say, E_S is the linear space which contains the signal direction vectors $a(\theta_i)(i = 1, 2, \ldots, K)$.

As the literature [2] explains, the signal subspace E_S contains noise, i.e. not all noise is in E_N, so it is not accurate to name E_N as noise subspace. For the director vector $a(\theta_i)(i = 1, 2, \ldots, K)$ is orthogonal with E_N, the inner product of $a(\theta_i)(i = 1, 2, \ldots, K)$ and E_N is zero, so E_N should be named as signal zero space.

4.3 Wrong usage of concepts used in MUSIC approach

Literature [1] interpreted that $d = \sqrt{a^*(\theta) \cdot E_N}$ was the Euclidean distance from the vector $a(\theta)$ to the signal subspace E_S. It isn't mistaken, but E_S is not appeared in the expression $d = \sqrt{a^*(\theta) \cdot E_N}$, so the interpretation is not obvious formally.

Literature [3] interpreted that $d = \sqrt{a^*(\theta) \cdot E_N}$ was the Euclidean distance from the vector $a(\theta)$ to the noise subspace E_N. In fact, according to the related statements of 1.3, it is visible that $a(\theta)$ is correspond to the vector $\vec{a}$, E_N is correspond to the subspace OXY. So $d = \sqrt{a^*(\theta) \cdot E_N}$ isn't the distance from $a(\theta)$ to E_N, it is the length of the projection of $a(\theta)$ to E_N. This shows that the interpretation of $d = \sqrt{a^*(\theta) \cdot E_N}$ made in literature [3] has made a mistake of concept.

5 SUMMARY

Many theory systems experience such development process. At first, a genius has a vagary, give the theory a starting point. At this time, the surging thoughts full of his head hit him, he had no leisure to clarify each concept, he merely considers keeping as intuitive as possible when using concept, not to make serious error. However, in practice, he possibly uses some concepts incorrectly, even uses some concepts wrongly. Later on, many people are attracted and follow him, these people are busy promoting the theory from different aspects, they continue to use the original concepts without careful scrutiny, and sometimes they maybe circulate erroneous reports. Finally, there is always someone to summarize and improve the theory, make it a complete theory system. He must check every detail including the usage of concepts. At this time, the existence of any fuzzy concepts is no longer allowed, the wrong usage of concepts is not allowed absolutely.

The modern spatial spectrum estimation theory has existed more than 30 years, it has experienced the former two processes [4, 5]. At present, abundant research results and further work call for a summary of the work. In order to give the mathematical decent to the upcoming summarizes, we clearly defines relevant concepts. These definitions are timely, useful and necessary.

REFERENCES

[1] Ralph O. Schmidt. 1986. Multiple Emitter Location and Signal Parameter Estimation. *IEEE. Trans on AP* 34(3): 276–280.

[2] Ralph O. Schmidt & Raymond E. Franks. 1986. Multiple Sources DF Signal Processing: An Experiment System. *IEEE. Trans on AP* 34(3): 280–287.

[3] Bencan Zhao. 1991. Spatial Spectrum Estimation DF Technology. *Detection Technology* (special issue): 2–24. (In Chinese)

[4] Sun Baofa. 2005. Visual Simulation for Determining Unidirectional DOAs. *7th ICEMI*: 692–695

[5] Sun Baofa. 2006. Simulation of Spatial Smoothing Approach for Coherent Sources DF. *1st ISTAI*: 1557–1561

Information Technology and Computer Application Engineering – Liu, Sung & Yao (Eds)
© *2014 Taylor & Francis Group, London, ISBN 978-1-138-00079-7*

Research on teaching evaluation system modeling based on UML

Xue Feng Zhao

School of Information Engineering, Binzhou Polytechnic Binzhou, China

ABSTRACT: Teaching evaluation is the core part of teaching quality assurance and monitoring system. This paper discusses the analysis and design of the teaching evaluation system for evaluation system characteristics. Beginning with the requirement analysis system models, the paper gives the teaching evaluation system modeling process based on UML.

Keywords: Teaching evaluation, Modeling UML

1 INTRODUCTION

Teaching evaluation in education administration is the core content of teaching quality assurance and supervision system. To scientifically and effectively perform teaching evaluation has significant value and realistic role in explicating educational organizations' talent cultivation objective and teaching orientation, intensifying educational reformation, and promoting the construction of teachers' group, as well as the improvement of teaching quality. However, there are still a series of technical issues existing in the teaching evaluation system, such as the method for collecting teaching evaluation data is still simplified and subjective; the actual evaluation methods are lacking of scientific nature and impartiality. All of these issues are closely related with the effectiveness of the teaching valuation system. In this paper, the advantage and ideology of UML will be employed to construct the model for teaching evaluation. On this basis, this paper will further develop the UML-based teaching evaluation system, so as to provide all teaching administrators with a handy, stable, safe and reliable teaching evaluation platform. With this platform, they will be able to conduct impartial evaluation on the quality of education, so as to promptly, effectively, comprehensively and systematically control the teaching progress.

2 UML OVERVIEW

UML (Unified Modeling Language) is a modeling language developed by Grady Booch, Jim Rumbaugh and Ivar Jacobson. The language can be used to create static structures, dynamic behaviors, and other structural models. With strong expandability and universality, the language can be applied in data modeling, business modeling, object modeling, and component modeling. As an important modeling language, UML enables developers to focus their attentions on constructing the model and structure of products, rather than thinking about which programming language or algorithm to choose. When the model has been built, it can be converted to specified programming codes by the UML tool. Thus, UML is not based on a certain language. Instead, it is a basic tool that connects and links up Java, C++, Smalltalk and other languages.

3 MODELING PROCESS OF TEACHING EVALUATION SYSTEM

Teaching evaluation refers to the comprehensive estimation conducted on the basis of theories and technologies of educational evaluation, with the aim to judge whether the teaching process and result have complied with a certain quality standard. Teaching quality evaluation is both a theoretical subject and a practical subject.

Teaching evaluation is an important and serious task, which takes teachers and their teaching activities as the evaluation objects. After the evaluation, teachers may make use of the result to know students' actual situation, to discover issues in education, to explicit the working direction of education, to reflect and improve their teaching process and effect. Through evaluation, decision-makers of schools may also get valuable references for their decision-making activities, including the referential basis for reformation and personnel decisions of the school. Teaching quality evaluation can provide stable and effective channel for the communication between teachers and students, schools and teachers, as well as schools and students. Students may also make use of the way to express their views on their teachers' instruction, the school's course design, teaching facilities, and many other aspects. On the other hand, schools and teachers are able to timely get to know students' demands and trends. Similarly, teachers may also utilize the platform to express their views and requirements on teaching system, teaching facilities and instruction budget of the school.

3.1 *User demand analysis*

Through analysis, teaching evaluation system shall include the following functions:

- Authority Administration

    Different users shall be endowed with different system operation authorities. When logging in the system, users are required to verify their ID and password, so as to ensure the security of the system. Users' password will be encrypted and stored in the database. After logging in, the logging process shall be recorded in the logging library.
- Teaching Evaluation Administration

    Teaching administrators may customize teaching evaluation models according to current teaching objectives, as well as to fulfill tasks like maintenance of teaching evaluation index library and maintenance of evaluation models. Pre-evaluated teachers' information shall be input into the system, so as to conducted corresponding evaluation, to maintain teaching evaluation data, and to process teaching evaluation. On this basis, teaching evaluation data is to be compared and analyzed.
- Query and Printing

    Teaching administrators may query teachers' evaluation results, as well as to query and print all teachers' evaluation results.
- Data Import

    Teaching evaluation system requires large amounts of basic data in teaching activities. In order to keep the data consistent, and to reduce the working load of data input, the function of importing existing basic data and evaluation data is hereby embedded. For data that schools do not have, we have also provided input options in the section of administration. The imported data can also be revised and maintained.

3.2 *Construction of case model*

Construction of case mode refers to the process of analyzing users' demands, confirming participant roles of system, extracting required functions into individual cases, describing all cases in detail, and constructing case models for the system. During this process, UML will be used to analyze all roles, to make clear that participant roles of the system include system administrator, teaching affair administrator, teacher and student. They have their respective functions and duties. System administrators are enabled to maintain the system, and to administrate users' information, evaluation indexes, evaluation items, etc. Similar contents include users' registration application, users augment, authority revision, evaluation standard modification, adding teachers, modifying evaluation contents, etc.

Non-system-administrators include teaching affair administrators, teachers and students. Teaching affair

Table 1. System Entity Class.

Name of Entity Class	Property of Entity Class
Student	Student Number, Name, Gender, Class, Department, Major
Teacher	Teacher Number, Name, Gender, Date of Birth, Post, Department
Basic Course Information	Course Number, Course Name, Course Property, Credit
Basic Performance Information	Student Number, Course Number, Semester Code, Teacher, Final Result, Ordinary Result, Comprehensive Result, Make-up Result, Repeat Result
User Logging-in	User Name, Password, Authority, Termination Date

administrators can log in the evaluation system to file up the index data. Teachers may change their ID information, query about their single-item evaluation results and comprehensive evaluation results, as well as perform evaluation on other teaching contents. Students are able to change their ID information, to find corresponding teachers according to the course they are engaged in, as well as to grade teachers in accordance with the grading standard.

3.3 *Construction of static model*

Static models describe the system from a static view. It is comprised by the mutual relation among classes, which is also called class diagram. It not only defines classes in system, representing relations among classes like association, dependence, generalization and realization, but also the inner structure of classes (properties and operations of class).

This section builds model with entity classes in the sub-system of performance administration of teachers' instruction class as the case. The rest are similar, as is shown in Table 1.

3.4 *Construction of dynamic model*

Dynamic models are mainly used to describe how the functions of system are realized. In an objective-oriented system, it is quite necessary to describe the interaction situation among objects. UML would make use of sequence diagram, collaboration diagram and state diagram to describe the interaction among objects from different dimensionalities. Sequence diagram has clear time sequence and expressions that are easy to be understood. In actual practices, it is often used to construct dynamic models. When students are going to perform teaching evaluation, they shall firstly input the user name and password for verification. If the user is legal, the teaching evaluation interface will then be displayed. At the teaching evaluation interface, students may check if the teachers and courses are correct. If the selected information is correct, they will then carry out the evaluation, and then store the results into the database.

Some non-functional demands in system can not be demonstrated with diagram. UML would utilize activity diagrams to show these dynamic properties. Activity diagram is similar to flow diagram in program design, which indicates the control flow from one activity to another. Activity diagram is a sequence for describing activities, and can be used to show conditional and concurrent behavior. Users would have to sign in the system first, i.e. to input their user ID and password for verification. Illegal users will be replied with error information. Legal user will then successful enter into the system. The first procedure is to judge whether the user information needs to be revised. If yes, the user will then enter into the page for user information revision, in which, users may modify and save their information into the database. If the user does not want to change the information, they will be directed to corresponding page according to their authority. System administrators may enter into administrators' page for teaching evaluation item administration. They may also enter into the page for user information administration, evaluation system administration, etc. Teaching affair administrator may enter into the interface of teaching evaluation, so as to perform teaching evaluation according to indexes regulated in the teaching evaluation standard. Teachers may be directed to teachers' page for query about their teaching evaluation result. They may also go to teachers' teaching evaluation interface to conduct evaluation on other teachers' teaching activity. Similarly, students will be directed to their page for teaching evaluation. All teaching evaluation results and data will be transferred to comprehensive evaluation processing center for the final evaluation, and to figure out the final result.

4 CONCLUSION

UML is a powerful and object-oriented modeling language for visual system analysis. It has employed a complete suit of mature modeling technologies, which can be widely applied in all fields. Models developed with this technology can help developer better understand business flows, so as to construct more reliable and more completed system models. Consequently, users and developers will be able to get a consistent understanding on descriptions of the same issue, so as to reduce semantic differences, and to keep the correctness of analysis. Even though, in order to make use of more advanced evaluation measures and technologies to build more scientific teaching evaluation system that complies with the actual situation of school, we will need to research and explore for a longer period in actual practice.

The evaluation system for higher vocational education is suitable for evaluation index system of different course types and evaluation entities. In the connotation of evaluation index, the system fully sticks to the objective of "customized secondary school and education". In the aspect of orientation, the system encourages all teachers to participate in higher vocation education reformation. Moreover, the system is helpful in overcoming defects brought about by the fact that "one evaluation form will be used by all members", regardless of concrete course types and evaluation subjects. By drawing support from the evaluation system for higher vocational education, teachers may timely know and correct existing issues in teaching activities, so as to improve their teaching competence and working efficiency.

REFERENCES

Carmen Cauti. Model Program: Technology Teacher Vol. 1 (2008) No. 10, p. 67–68.

Jinghua CAO, Wei JIANG, Yanzhong RAN and Fei ZHAO: Experimental Technology and Management Vol. 29 (2012) No. 11, p. 144–147.

Saifen HU: Journal of Ningbo Rdio & TV University Vol. 6 (2008) No. 3, p. 116–120.

Shane sendall, Wojtek Kozaczynski. Model Transformation: IEEE software Vol. 20 (2003) No. 5, p. 42–45.

Sobiechowska Paula, Maisch Maire: Educational Action Research Vol. 14 (2006) No. 2, p. 267–286.

Xinag-ping LIU, Xue-yan ZHAO and Jian-cheng LI: Computer Engineering Vol. 34 (2008) No. 17, p. 77–79.

Zhi-jian CHEN: Journal of Guangxi Normal University for Nationalities Vol. 27 (2010) No. 5, p. 267–286.

Information Technology and Computer Application Engineering – Liu, Sung & Yao (Eds)
© 2014 Taylor & Francis Group, London, ISBN 978-1-138-00079-7

GPU-based multi-view stereo reconstruction

Bo Ling Wang, Yan Feng Jiang, Zhen Peng & Sheng Chen Yu
Computer Science and Technology Department, North China Institute of Science and Technology, China

ABSTRACT: This paper presents a novel multi-view stereo matching algorithm, which takes account of accuracy and efficiency both. Our approach consists of two contribution point. First, we apply a gradual expansion method to the 3D points' generation process, to restrain the noise spreading in expansion; Second, we use general computing technology to accelerate Harris and DoG operator, and massive non-linear optimizations in the pipeline. The experimental results based on Middlebury benchmark data set shows that our algorithm reduces the running time of the algorithm by 80 percent, while maintain the high reconstruct accuracy and completeness.

Keywords: multi-view stereo; gradual expansion; massive non-linear optimization; GPU computing

1   INTRODUCTION

Multi-view stereo (MVS) matching reconstructs 3D geometry of objects and scenes from multiple images and camera parameters, a process known as image-based modeling and 3D photography, which is one of the fundamental problems in computer vision. MVS's potential applications include augmented reality, film and cartoon, historic preservation, target recognition, etc.

MVS has recently experienced a new era because of public available standard benchmark data set and evaluation system [1], which allows researchers to compare their algorithms with the ground truth and the state-of-the-art ones.

Seitz [2] classify the MVS algorithms into four classes: Voxel-based approaches, need a bounding box before reconstruct, its accuracy is limited by the resolution of the voxel grid; Methods based on deformable polygonal meshes, also needs a initial visual hull as a starting point; Methods based on depth map, use binocular stereo matching to generate multiple depth map and merge them to a single point cloud; Patch-based methods, use patch to optimize the 3D point, which draw more attention in recent years. In Middlebury evaluation system, the top performers are mostly depth-map methods and patch-based methods.

2   WORK RELATED

Depth-map merging method treats the problem as a global optimization problem of merging multiple depth maps. Zach [3] introduced a method to evaluate depth-map approach. Campbell [4] keeps multiple estimate value in the depth-map generation step, and use a global optimization method to decide every 3D point's best depth. The depth-map method is based on binocular stereo, each depth-map generation process is independent and parallel, which can be easily accelerated by GPU. Depth-map approach has its own constraint. It uses two images to find matching point, which is fast but also susceptible to the environment and light. It's not robust to the outdoor datasets. Also it only recovers the coordinate of the 3D point, cannot get the normal direction at the same time.

The patch-based method, such as Furukawa's PMVS [5] is almost the best performer in Middlebury. It uses multiple images to calculate one matching 3D point. But its expansion method is replicate and time-consuming. This paper is based on PMVS, uses a different expansion method to get a dense point cloud faster and more efficient; Apply GPU technology to improve the two parallel part of the pipeline.

In MVS, there are also several GPU based algorithms [3, 6]. But we notice that the most outstanding method PMVS, which achieves high accuracy, as the algorithm is complicated and cannot be completely parallelized, so far no GPU version has come out. This paper uses GPU to accelerate the most time-consuming part: feature detection, feature matching and point expansion separately, reduce the running time by about 80 percent.

3   OUR APPROACH

This section will demonstrate our approach in detail. The essential difference between our method and PMVS is in expansion part and non-linear optimization resolving method. We use a step-by-step expansion approach to restrain the noise spreading; We use Nelder-Mead, a simplex NLO algorithm which is more suitable for GPU computing to handle the NLO

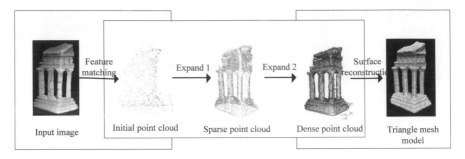

Figure 1. Overview of the proposed algorithm.

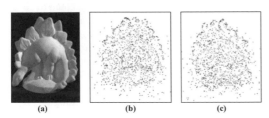

Figure 2. Initial matching point cloud of Dino48. (a) is one of the input images. (b) and (c) are two view of the initial matching point cloud.

problems. The overview of the proposed approach is illustrated in Figure 1 and summarized as follows.

3.1 Patch-based point optimization

The initial matching 3D point in feature matching is not very precise, as the epipolar matching is a weak constrain. We can't get the normal direction of the point also. To refine the 3D point with more images, we use a virtual small patch on the object surface to calculate the color similarity in images that the point is projected in. As the patch is so small (7×7 pixels in our experiment), that it can be seen as the real surface of the object. The optimize method is adjust the ordinate and normal of the patch, to make 2D projective patches' color similarity achieve maximum. We use normalized cross correlation (NCC) as the basic matching cost function. The cost function of the patch is the average of NCC of each two projective patch:

$$Ncc(p) = \frac{\sum_{i=1}^{n} \sum_{j=i+1}^{n} N(I_i, I_i, p)}{(n+1)n/2} \quad . \tag{1}$$

Compare to binocular stereo matching in depth-map merging method, Equation1 use more images to refine the point, which is more robust and steady. After the point optimization process, we get an initial 3D point cloud. The initial matching result of Dino48 are illustrated in Figure 2.

3.2 Gradual expansion

Expansion is the most time-consuming part in the whole pipeline. PMVS uses a three-times and same-grained expansion strategy, filters the noise point after each time expansion. We found that the initial matching point cloud has a lot of noise points, it's not suitable to the fine-grained expansion. The noise points will generate more noise points around them. Our improvement strategy is use different granularity in different expansion step.

First, we do a coarse-grained expansion to the initial matching point, to get the general outline of the object. Compare to the fine-grained expansion, this process can restrain the spread of noise point, and more efficient. We use the filter method in [5] to denoise the sparse point cloud. Second, we use a fine-grained expansion to the cleaner point cloud, to get the details of the object surface. The expansion result of Dino48 and Temple47 are illustrated in Figure 3.

The parameter γ in PMVS means the grid size of the image, which decides the expansion granularity. We set it 4 and 2 separately in two expansions in all our experiment.

3.3 Acceleration with GPU computing

PMVS get best result in four of six dataset on Middlebury, but execution time is also much longer. In this section, we use GPU to accelerate the three parts in PMVS separately. The experimental result shows that our GPU version PMVS reduce about 80 percent time than the original pipeline.

3.3.1 GPU based Harris and DoG

Harris and DoG are both typical SIMD (single instruction multiple data) algorithm, which is very suitable to GPU computing. There are three layers in GPU to manage threads, they are thread, block, grid. We assign each image to a grid, every 16×16 cell to a block, every pixel to a thread, and launch one thread for every pixel, as illustrated in Figure 4.

In kernel the coordinate of the pixel is calculated as:

$$x = blockIdx.x \cdot blockDim.x + threadIdx.x$$
$$y = blockIdx.y \cdot blockDim.y + threadIdx.y \tag{2}$$

where blockIdx and threadIdx are reserved words in CUDA, represent the position of block in grid and the position of thread in block. The main process of Harris and DoG is gaussian filter.

452

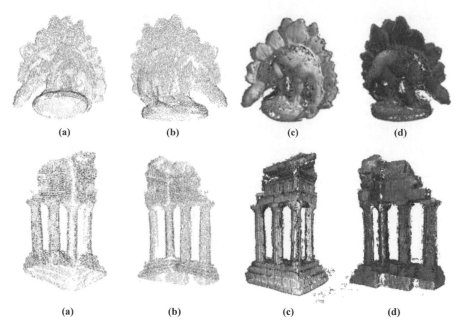

(a)   (b)   (c)   (d)

(a)   (b)   (c)   (d)

Figure 3. The point cloud of Dino48 and Temple47 in expansion step. (a) and (b) show two view of the sparse point cloud after a coarse-grained expansion. (c) and (d) show two view of the dense point.

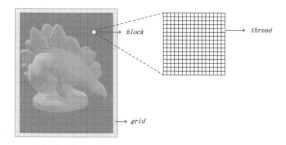

Figure 4. Thread management in feature detecting.

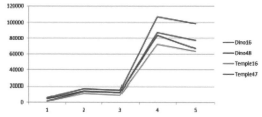

Figure 5. Number of reconstructed points after each step of the algorithm. Steps 1–5 are: Initial matching, coarse-grained expansion, filtering, fine-grained expansion, filtering.

3.3.2 GPU based massive NLOs

The 3D point optimization is a non-linear optimization (NLO) problem. In the feature matching and point expansion step, there are massive NLOs, for every point in the final point cloud is the result of a NLO problem, which is the most time-consuming process in the whole pipeline. We presents a novel strategy to handle the massive NLOs with GPU. First, we collect the dispersive NLOs together: in feature matching step, we match all the feature point through epipolar constraint, store the raw matching points; in the expansion step, we expand all the existing point, and store raw expanding points. Second, transfer the raw points into global memory (GPU), to execute the parallel NLOs. Finally, transfer the refined point to memory (CPU) to do the point selection.

Another advantage brought by GPU computing is we can get all the initial matching points and expanding point at the same time. We order these points by their NCC and select these points with a greedy strategy. Experimental result shows that parallel feature matching get less, but more accurate points.

4 EXPERIMENTAL RESULTS

We evaluate the reconstruction quality of our approach using the four benchmark Middlebury stereo data set. We run the proposed algorithm on an Intel Core i5 CPU with 3.0 GHZ and use a GeForce GTX 660Ti graphics card with 2 GB memory manufactured by NVIDIA. We apply CUDA to implement our approach on the GPU.

We track the point number after initial matching, first expansion, first filter, second expansion, second filter in our algorithm, as shown in Figure 5.

We submit our result in Middlebury evaluation system, to compare with other state-of-the-art algorithms. Table 1 shows the result compared with competitors. The accuracy and completeness of our approach, especially the Dino16 and Dino48, are competitive to the

Table 1. Reconstruction accuracy, completeness and execution time of our algorithm.

Algorithms	TempleRing			TempleSparseRing			DinoRing			DinoSparseRing		
	Acc	Com	Time	Acc	Com	Time	Acc	Com	Time	Acc	Com	Time
[5]	0.47	99.6	272:12	0.63	99.3	128:34	0.28	99.8	300:34	0.37	99.2	151:37
[6]	0.83	88.0	00:19				0.84	83.1	00:21			
[8]	0.52	98.9	21:33	0.74	96.1	11:33	0.43	98.3	22:35	0.49	97.8	12:50
[9]	0.54	99.0	02:04	0.73	94.5	01:19	0.51	94.6	08:53	0.66	89.9	03:46
[10]	0.55	99.2	60:17	0.78	95.8	25:42	0.42	98.6	43:26	0.45	99.2	20:23
Our approach	0.78	94.4	03:41	0.97	86.5	01:24	0.58	98.9	03:41	0.61	97.7	01:39

state-of-the-art algorithms, while the execution time is much shorter than non-GPU ones.

5 CONCLUSION

In this paper we present a novel multi-view stereo reconstruction algorithm. We use Harris and DoG to track the feature point, use epipolar constraint to get initial matching feature points. We introduce a gradual expansion method to restrain the noise spreading in point expansion. We apply GPU technology to our pipeline, accelerate feature detecting, feature matching, point expansion separately. Experimental results evaluated using benchmark data set on Middlebury evaluation system shows that our approach get competitive reconstruction result, and reduce the execution time of the algorithm significantly.

ACKNOWLEDGEMENT

This work is supported by Fundamental Research Funds for the Central Universities under Grant No. JSJ1202B.

REFERENCES

[1] http://vision.middlebury.edu/mview/
[2] S. M. Seitz, B. Curless, J. Diebel, D. Scharstein, and R. Szeliski. "A comparison and evaluation of multi-view stereo reconstruction algorithms". International Conference on Computer Vision. New York, 2006: 519–528.
[3] Zach, "Fast and high quality fusion of depth maps," in 3DPVT, 2008.
[4] N. Campbell, G. Vogiatzis, C. Hernandez, and R. Cipolla, "Multiple hypotheses depth-maps for multi-view stereo", European Conference on Computer Vision, Palais des Congrès Parc Chanot, 2008.
[5] Y. Furukawa, J. Ponce, "Accurate, dense and robust multiview stereopsis", Transactions on Pattern Analysis and Machine Intelligence. 2010, 32 (8):1362–1376.
[6] P. Merrell, A. Akbarzadeh, L. Wang, P. Mordohai, J. M. Frahm, R. Yang, D. Nister, and M. Pollefeys. "Real-time visibility-based fusion of depth maps". International Conference on Computer Vision. Brazil, 2007:1–8.
[7] M. Kazhdan, M. Bolitho, and H. Hoppe, "Poisson surface reconstruction", International Symposium Geometric, 2006.
[8] Y. Liu, X. Cao, Q. Dai, and W. Xu. "Continuous depth estimation for multi-view stereo". IEEE International Conference on Computer Vision. Japan, 2009.
[9] Ju Yong Chang, Haesol Park, In Kyu Park, Kyoung Mu Lee, Sang Uk Lee. "GPU-friendly multi-view stereo reconstruction using surfel representation and graph cuts". Computer Vision and Image Understanding, 2011, 115:620–634.
[10] Zaharescu, E. Boyer, and R. Horaud. Transfor Mesh: a topology-adaptive mesh-based approach to surface evolution [C]. Asian Conference on Computer Vision, Tokyo, 2007.

Information Technology and Computer Application Engineering – Liu, Sung & Yao (Eds)
© 2014 Taylor & Francis Group, London, ISBN 978-1-138-00079-7

Found the uncertainty knowledge whitch exists in the distribution of plant based on λ operator stack structure

S.Y. Song

Information Technology Engineering Institute, Yuxi Normal University, Yunnan, China

ABSTRACT: Found the uncertainty of knowledge from the data and research to quantify has been difficult, the uncertainty in the knowledge based on the concept, λ operator stack structure, and application of the operator from the actual examples found in the uncertainty of knowledge, from the results can be seen that the algorithm is more reliable knowledge of uncertainty.

Keywords: Data mining; Uncertainty; λ Operator; Stack structure

1   INTRODUCTION

Today is a data flood era, although we acquire knowledge from a large amount of data for decision-making and forecasting work, and use the knowledge gained from the huge data obtained knowledge is scarce, due to objective things in the real world or uncertainty phenomenon led to most of the people in the understanding of the field of information and knowledge inaccurate, knowledge truly is, and will always be uncertain

2   KNOWLEDGE DISCOVERY

2.1   *Knowledge discovery task*

Summary: summary of the data and summarized. Traditional easiest Summary method is to calculate the statistical value of the summed value on the individual fields of the database, mean, variance value, or represented by the histograms, pie charts and other graphical way.

Clustering: According to the different characteristics of the data, divided into different classes, are unsupervised learning [1].

Category: model based on the classification of the data collection classification, about a given object is classified in a class. Classification (Classification) is a very important task in knowledge discovery, is a supervised learning (machine learning appellation).

Deviation analysis: the basic idea is to find the difference between the observed results with reference to the amount. Abnormal, can cause people to pay more attention to special circumstances.

Modelling: A mathematical model constructed to describe an activity or state (such as Bayesian model).

2.2   *The problem of uncertainty*

Randomness: statistical regularity of probability theory to reveal a random phenomenon.

Fuzziness: using fuzzy sets and rough sets to reveal blur regularity [2]. Randomness and fuzziness is uncertainty, uncertainty can be seen as a special case of uncertainty.

2.3   *Knowledge discovery*

Knowledge discovery methods: traditional methods (regression analysis, cluster analysis, etc.); fuzzy set method [4]; rough set methods and machine learning (rule induction, decision trees, case-based reasoning, support vector machines, neural networks, Bayesian belief networks etc.). The following describes the application of rough set method.

3   APPLICATION OF ROUGH SET

3.1   *Knowledge discovery*

An effective tool for rough set incomplete information as a handle imprecise, inconsistent, incomplete, on the one hand, thanks to his mathematical basis of mature, does not require a priori knowledge [5]; On the one hand lies in its ease of use. As the rough set theory created purpose, and the starting point of the study is directly on the data analysis and reasoning hidden knowledge, reveals the potential law, therefore, is a natural method of data mining, or knowledge discovery, it is based on probability theory data mining method based on fuzzy theory of data mining methods and evidence-based theory of data mining methods for dealing with uncertainty problem theory comparison, the most significant difference is that it does not need to provide the required processing data collection any prior knowledge of the outside and highly complementary with the theory of uncertainty (especially fuzzy theory). Rough set-based methods can be used: approximate reasoning, information retrieval, machine learning and data mining. These concepts are defined in "A Fuzzy Knowledge Matching Method Based on Variable Weight IDM", 2011, Applied Mechanics and Materials, 71-78, 2424 and "Based on Bayesian Network of Artificial Fish of Auditory System Research", publish in Advanced Materials Research.

Table 1. The case history.

Patient	Headache A1	Muscle pain A2	Temperature A3	Cold A4
e1	y	y	Normal	n
e2	y	y	High	y
e3	y	y	Very high	y
e4	n	y	Normal	n
e5	n	n	High	n
e6	n	y	Very high	y
e7	n	n	High	y
e8	n	y	Very high	n

3.2 Basic concept

Assume U: Non-empty set of objects, called the domain. R: Equivalence relation on U, and it has the following properties [3].

Reflexivity: (a, a) ∈ R; Symmetry: if (a, b) ∈ R, then (b, a) ∈ R; Transitive: if (a, b) ∈ R, (b, c) ∈ R, then (a, c) ∈ R

Divided by the equivalence relation R is defined on U, each divided into blocks called equivalence classes. U/R are defined as the R exported equivalence class. [x]R Is defined as the equivalence class containing the object x.

Example 1: Assume U = {x1, x2…x8} is the building block set.

R1: Colors (red, yellow, blue), R2: Shape (square, circle, triangle), R3: Volume (large and small).

Then:

U/R1 = {Red(x1, x2, x7), blue(x4, x5), yellow(x3, x6, x8)}, U/R2 = {Circle (x1, x5), square (x2, x6), Triangle (x3, x4, x7, x8)}, U/R3 = {big(x2, x7, x8), small (x1, x3, x4, x5, x6)}

Let U is the domain = {R1, R2, R3} Equivalence relation on U cluster, then {R1, R2, R3}, the cross is also the equivalence relation, Denoted Ind(R).

Definition 1: Assume R = {R1…Rn}. If U/ind(R) = U/ind(R − {Ri}), then Ri Called redundancy relations on R.

Definition 2: Assume P, Q is two equivalence relations, and Q ⊆ P. If no redundant relationships in Q and U/ind (P) = U/ind (Q), Q is a reduction of P.

Definition 3: Assume P and Q is two equivalence relations. Q depends on P, dnoted P ⇒ Q, only U/ind (P) ⊆ U/ind (Q).

Definition 4: Q dependence of P denoted d(P ⇒ Q), d(P⇒Q) = | POSP(Q)|/|U|, among POSP(Q) call P about Q positive domain,I.e. P is the object package in Q,|U | is object umber in U,| POSP(Q)|, P is the number of objects wrapped in Q.

3.3 Application of attribute reduction

Example 2: provided the medical records table as shown in table 1.(refer with table 1)

U = (e1, e2, …, e8) is object sets (patients), A = {A1, …, A4} is properties on U, C = (A1, A2, A3)

called condition properties, D = (A4) is deice properties.

then:

U/A1= {(e1, e2, e3), (e4, e5, e6, e7, e8)}
U/A2= {(e1, e2, e3, e4, e6, e8), (e5, e7)}
U/A3= {(e1, e4), (e2, e5, e7), (e3, e6, e8)}
U/ind(A1,A2,A3)={(e1),(e2),(e3),(e4),(e5,e7),(e6 ,e8)}
U/ind(A1,A3)={(e1),(e2),(e3),(e4),(e5,e7),(e6,e8)}

(A1, A3) is reduction of C; we can delete properties of table A2. (Refer with table 2)

4 THE APPLICATION OF ATTRIBUTE DEPENDENCE

The dependence of the decision attribute on condition attribute is a measure of the degree of importance of the condition attributes [6]. When the degree of importance of the evaluation of an attribute, based on the degree of importance of the attributes (single attribute), but also based on the attributes of this property and other properties set the degree of importance. Currently, the vast majority of the research literature on the dependence, limited to solving a single attribute dependency, single attribute dependence 0 properties do not contribute to the decision-making table, the reduction will discard them. However, through research that attributes to delete single attribute dependence 0, tends to create knowledge discard. Therefore, the study of the dependence of the two or more attributes of a set of attributes has more important meaning [7]. Clustering, association, identify and associate degrees superimposed practical examples of research on plant distribution. Clustering, fuzzy clustering and other methods can be used, and here we are mainly concerned about the associate degree determined and related degree superimposed.

4.1 Correlation degree

Example 3: Plant Distribution Table (refer with table 2).

According to Table 2, Sample sets T = {t1, t2, …, t8}, Attribute sets C = {A1, A2, A3} is condition

Table 2. Plant distribution table.

	Plant	Root A1	Leaf A2	Subjects A3	Mean annual temperature B1	Annual precipitation B2	Elevation B3
t1	Orchid	Aerial root a1	Strip b1	Herbal c1	[11,20] e1	[900,1500] f1	[1000,2000] g1
t2	Fichus	Aerial root a1	Big leaf b2	Woody c2	[14,22] e1	[800,1400] f1	[900,1500] g1
t3	Tequila	Deep root a2	Fleshy b3	Herbal c1	[20,26] e1	[350,600] f2	[700,1000] g1
t4	Garlic	Deep root a2	Fleshy b3	Shrub c3	[15,24] e1	[800,1500] f1	[200,500] g2
t5	Calligonum	Deep root a2	Lobular b4	Shrub c3	[9,14] e2	[350,550] f2	[120,400] g2
t6	Calligonium	Deep root a2	Lobular b4	Shrub c3	[−3.10] e3	[300,450] f2	[2000,3000] g3
t7	Wild taro	Shallow root a3	Big leaf b2	Herbal c1	[8,14] e2	[2000,3000] f3	[600,1000] g1
t8	Water lily	Shallow root a3	Big leaf b2	Herbal c1	[14,20] e1	[800,1500] f1	[1000,2000] g1

attributes = {B1, B2, B3} is the determine attributes of set, range of A1 is dom(A1) = {a1, a2, a3}, range of A2 is dom (A2) = {b1, b2, b3, b4}, range of A3 is dom(A3) = {c1, c2, c3 }.

4.2 The divide of T based Ai

Visually, Attribute-based A1 can be divided T into {(t1,t2), (t3,t4,t5,t6), (t7,t8)}, Based Ai T can divided into IND(Ai). The IND(Ai) in the K-th equivalence class referred to as Ajk, IND(A1) = {A11 (t1,t2), A12 (t3,t4,t5,t6), A13 (t7,t8) }. Ajk and Bnm intersection denoted Pos(Ajk,Bnm), The number of objects denoted|(Pos(Ajk,Bnm))| In Pos(Ajk,Bnm), then:

IND(A1)={A11(t1,t2),A12(t3,t4,t5,t6),A13(t7,t8)}
IND(A2)={A21(t1),A22(t2,t7,t8),A23(t3,t4),A24(t5,t6)}

IND(B1)={B11(t1,t2,t3,t4,t7),B12(t5,t7),B13(t6)}
IND(B2)={B21(t1,t2,t4,t8),B22(t3,t5,t6),B23(t7)}
IND(B3)={B31(t1,t2,t3, t7,t8),B32(t4,t5),B33(t6)}
Pos (A11, B11) = {t1, t2} | (Pos (A11, B11))|=2
Pos (A11, B12) = { Φ } | (Pos (A11, B12))|=0

4.3 Associate degree

Assume a1 as aerial root, f1 as semi-humid root, f2 as drought, f3 as wet. Visually, a1 Only life in f3 area, a1 relationship on humid environment. a1 life in f3 area, f2 and f1 while living area, The a1 hasn't relationship on wet environment. (Refer with table 3)

When we study f1:

The closeness of f2 relative to f1 noted SP(f1, f2), SP(f1, f2) are weights of f2 relative to f1, noted w12, Of course W11 = SP(f1, f1) = 1.

Definition 5 (associate degree): The associate degree of Ajk for Bnm are noted λ(Ajk,Bnm), λ(Ajk,Bnm) = |(pos(Ajk,Bn1) |*w1 + |(Pos(Ajk,Bn2) |*w2 + ... + | (pos(Ajk,Bnm))|*wm]/|(Ajk)|.

For λ(A12,B22), among A12 = a2, B22 = f2, (The relationship of deep-rooted plants (a2) with less precipitation (f2), shown in table 3.

Table 3. Deep-rooted plants with less precipitation (f2).

	Plant	Root A1	Annual precipitation B2
t1	Orchid	Aerial root a1	[900,1500] f1
t2	Fichus	Aerial root a1	[800,1400] f1
t3	Tequila	Deep root a2	[350,600] f2
t4	Garlic	Deep root a2	[800,1500] f1
t5	Calligonum	Deep root a2	[350,550] f2
t6	Calligonum	Deep root a2	[300,450] f2
t7	Wild taro	Shallow root a3	[2000,3000] f3
t8	Saliva lotus	Shallow root a3	[800,1500] f1

Assume w1 = SP(f1, f2) = 0.4, w2 = SP(f2, f2) = 1, w3 = SP(f3,f2) = 0, Can be seen from Table 3 A2 located in the f2 region there are three objects, there is one object located in f1 region, there is zero in f3 region.

Then λ(A12,B22) = | (Pos(a2,f1)|*w1 + | (Pos(a2, f2)| *w2 + |(Pos(a2,f3)|* w3]/4 = (1*0.4+3*1 + 0*0)/4 =3.4/4 = 0.85

4.4 Folding operator of λ

Associate degree λ ∈ [0,1], would like to extend it to [−1,1], assume μ(Aij,Bnm) = (λ(Ajk,Bnm)−0.5)/0.5. That λ to pan and zoom. $0 < μ \leq 1$ said the Aij suitable environment Bnm; $−1 \leq μ < 0$ said Aij not suitable environment Bnm; $μ = 0$ said Aij does not relationship for environment Bnm.

$$μ_1 \oplus μ_2 = μ_2 \oplus μ_1 \quad (μ_1 \oplus μ_2) \oplus μ_3 = μ_1 \oplus (μ_2 \oplus μ_3)$$

μ also called as the associated degrees, λ folding operator shall meet the following properties:

$$if \ μ_1 < 0, μ_2 < 0 \quad then \quad μ_1 \oplus μ_2 < \min(μ_1, μ_2)$$
$$if \ μ_1 > 0, μ_2 > 0 \quad then \quad μ_1 \oplus μ_2 > \max(μ_1, μ_2)$$

$$if \ |μ_1| = 1 \quad then \quad μ_1 \oplus μ_2 = 1$$

Table 4. Cannas basic attribute.

Cannas		
root	leaf	shape
Deep of root a2	Big leaf b2	Herbal c1

Table 5. Cannas roots with precipitation.

The relationship between the root and precipitation			
Deep-rooted	f1	f2	f3
	1	3	0

Table 6. Cannas adapt to the environment (refer with table 6).

Environment		
The annual average temperature B1	Annual precipitation B2	Elevation B3
[16,21] e1	[300,500] f2	[2000,3000] g3

Table 7. Cannas superposition operator results.

μ	e1	f2	g3	results
Deep root a2	−0.09	0.7	0.8	0.93
Big leaf b2	0.22	−1	−1	−1
Herbal c1	0.34	−0.55	−1	−1

4.5 Constructed superimposed operator of λ

Definition 6: We called $\mu_1 \oplus \mu_2 = \mu_1 + \mu_2 + sig(\mu_1 * \mu_2)$ as superimposed opera to on μ_1 and μ_2.
Among

$$sig = \begin{cases} 1 & (\mu_1 < 0 \wedge \mu_2 < 0) or (\mu_1 > 0 \wedge \mu_2 < 0 \wedge |\mu_1| < |\mu_2|) \\ -1 & (\mu_1 > 0 \wedge \mu_2 > 0) or (\mu_1 > 0 \wedge \mu_2 < 0 \wedge |\mu_1| > |\mu_2|) \\ 0 & (\mu_1 > 0 \wedge \mu_2 < 0) \wedge |\mu_1| = |\mu_2| \end{cases}$$

The instance: set Cannas basic attributes such as table 4, according to previous knowledge of the relationship between the deep roots and precipitation are shown in table 5, predicted Cannas adapt to the environment. (refer with table 4 and table 5)

λ (a2, f2) = (1*0.4+3*1+0*0)/4 = 3.4/4 = 0.85
μ (a2, f2) = 0.7, from table 7 we can get Cannas adapt to the environment in Table 6.

5 SUMMARY

As can be seen from Table 7, cannas λ stack result was 0.93, higher confidence. Uncertainty of artificial intelligence is a major breakthrough in the future research directions. Although there are many uncertainties in knowledge representation and processing methods, but some of these methods are still under development, some practical problems need further study. The superposition algorithm proposed in this paper, I hope to be able to provide a reference to the uncertain real researchers.

REFERENCES

Wallerstein, Wang Bing translated the uncertainty of knowledge [M], Shandong University Press, 2006-1-1.
the Li Fan experts system uncertainty [M] Meteorological Press, 1992.
Wang Daoping etc. Study on the Classification and Disposal of Uncertain Knowledge in Intelligent Fault Diagnosis Systems [C]. Pro. of the 3th WCIIC & A. 2000, Heifei, PRChina.
Wu Quanyuan, Liu Jiangning, artificial intelligence and expert systems [M] National Defense Science and Technology University Press, 1995.
Wu Bo, Ma Yuxiang the expert system [M] Beijing Institute of Technology Press, 2001.
Yu Ji, and fault diagnosis expert system [M] Metallurgical Industry Press, 1991 The wood just Bo the management on behalf of the logic
[Japan] end of Du grapefruit Stone, Sun Zhongyuan translation [M] China Renmin University Press, 1983.
John F. Sowa. Conceptual Structure. UK: ADDISON_welslely, 1984.
Had yellow phosphorus rough set theory and its application [M] Chongqing University Press, 1998.

Information Technology and Computer Application Engineering – Liu, Sung & Yao (Eds)
© *2014 Taylor & Francis Group, London, ISBN 978-1-138-00079-7*

Band-pass digital filter window function method design and realization based on MATLAB SPTool

Xin Zhao
College of Life Science and Technology, Honghe University, Mengzi, Yunnan, China

Shan Ren
Engineering College, Honghe University, Mengzi, Yunnan, China

ABSTRACT: In the digital signal processing process, the length of signal that we can get is limited, so should be infinite long time sequence limit in a certain interval, only choose one of the segment signal to carry on the analysis, namely cut off the signal, and this process is equivalent to add a window function to the signal is, using the window function truncation infinite unit sampling response sequence. This article in accordance with the principles of window function design method, using SPTool design tools, design of a high-pass digital filter. Filtering processing for measured signal showed that filtering effect of the filter achieved the expected results.

Keywords: Window function method; Band-pass Digital Filter; SPTool Design Method; MATLAB

1 INTRODUCTION

Digital filter is very important in signal processor, and filter design may be realized by MATLAB. There are two methods in digital filter, include FIR and IIR. FIR filter is wildly used because of it with characteristics of ensure strict linear phase, filter stability and less error.

At present, the methods of designing FIR Filter include window function, frequency sampling and equivalent ripple approximation. At the same time, these methods can also direct implementation through the FDA Tool and SPTool. The SPTool is an interactive graphic signal processing tools in MATLAB, which can be convenient to complete the signal, filter and spectrum analysis and design. In this paper, taking a steam turbine vibration signal as an example, used SPTool design and realize the vibration signal filtering.

2 WINDOW FUNCTION DESIGN METHOD OF FIR DIGITAL FILTER

Digital Filter is a signal processing devices that characterics of certain transmission, its input and output are discrete digital signal. With digital device or a certain numerical method for processing the input signal and change its waveform or spectrum, to achieve the goal to filter out unwanted signal components [1].

The basic idea of window function design method is: firstly, given the frequency response $H_d(e^{j\omega})$ of

ideal filter, and then find a group of $h(n)$, made the frequency response $H(e^{j\omega}) = \sum_{n=0}^{N-1} h(n)e^{-j\omega}$ that determined by $h(n)$ to approximate to $H_d(e^{j\omega})$. The design is in time domain. Therefore, first derived $h_d(n)$ though IFFT, namely:

$$h_d(n) = \frac{1}{2\pi} \int_{-\pi}^{\pi} H_d(e^{j\omega})e^{j\omega}d\omega \qquad (1)$$

Then use the finite length of $h(n)$ to approximate the infinite $h_d(n)$, an effective method is to cut $h_d(n)$, that is, use of finite-length window function $\omega(n)$ to intercept $h_d(n)$, namely:

$$h(n) = w(n)h_d(n) \qquad (2)$$

An ideal window function should meet the following two requirements [2]:

1. The main lobe of amplitude-frequency characteristics for window function should as narrow as possible to get a steep transition zone.
2. The maximum side-lobe of amplitude-frequency characteristics for window function should as small as possible, so that the main lobe contains as much energy as possible; this will make the spikes and ripple reduce, to obtain flat amplitude-frequency characteristics and sufficient band-stop attenuation.

Window functions that used for designing FIR Filter include: Rectangle, Triangle, Hanning, Hamming, Blackman and Kaiser Window function.

3 BAND-PASS DIGITAL FILTER DESIGN

Here, to a power plant measured vibration signal of a steam turbine, for example, to introduce the filter design. The speed of the steam turbine is about 3000 per minute, sampling frequency is 1024 Hz.

Typed "SPTool" in MATLAB command window, and opening SPTool interface. Choose the "Import" in file menu, and import the measured vibration signal and its sampling frequency, right now SPTool work interface will display a signal name for sig1, namely vibration signal. Draw time domain and frequency domain graph of the pre filtering sig1 as in figure 1 and figure 2 shows.

As shown in the time domain chart, due to interfered by random noise and spikes, which affect to identify the characteristics of vibration signal.

In addition, from the frequency domain chart we can easily see that pre filtering vibration signal mixed together by the frequency domain characteristics of 5 Hz, 52 Hz and 105 Hz signal. However, the frequencies of steam turbine vibration signal roughly in the range of 50 Hz~55 Hz. So, we should design a Band-pass filter, which can filter out signals that below 50 Hz and above 55 Hz.

As the chart 3 shows, design filter in SPTool work interface, choose "FIRbp[design]", and click "Edit" to modify filter parameters. Filter design interface as the chart 4 shows. In the interface, settings filter type, method, sampling frequency, stopband cut-off frequency, pass-band cut-off frequency, stop-band attenuation and pass-band ripple, then get a high-pass filter. At this time, could observe the parameters of filter, including filter order number, stability, impact response, zero pole response, filter coefficient, etc.

After understanding the characteristic parameters of filter, we will be closed filter parameter design interface, and using filter for signal filtering.

4 MATLAB IMPLEMENT OF BAND-PASS FILTER

Now, use high-pass filter that just design to realize vibration signal filtering.

In the SPTool work interface as shown above figure 3, choose the measured signal sig1 that has imported, and then select the "FIRbp [design]", and click "Apply" button, the system pop-up "Apply Filter" interface below figure 5 shows. This show that the system will use filter that has been designed to input signal filtering, and at the same time, the system will automatically output signal that name for sig2, which is filtered vibration signal.

As in the previous method, drawing time domain and frequency domain map of post filtering vibration signal sig2, as shown in figure 6 and figure 7.

Magnitude and phase response of Band-pass filter are shown in Figue 8. Among them, curve 1 is magnitude response and curve 2 is phase response.

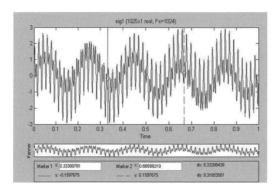

Figure 1. Time domain chart of pre filtering vibration signal.

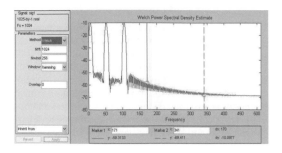

Figure 2. Frequency domain chart of pre filtering vibration signal.

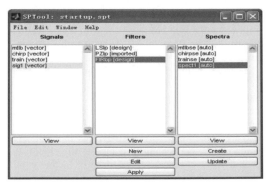

Figure 3. SPTool work interface.

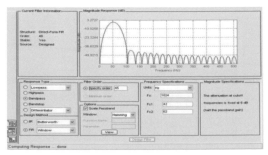

Figure 4. Filter parameter design interface.

From the magnitude response curve we known, in the pass band main lobe valve is more narrow, the level of energy concentration in the main lobe is relatively good, therefore can obtain higher frequency resolution, to meet the design requirements that mentioned above; from the phase response curve can be seen, in the pass-band, phase of the filter shown a strictly linear relationship, consistent with linear phase characteristics of the FIR filter.

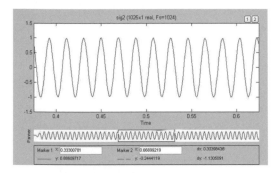

Figure 5.   Apply filter dialog box.

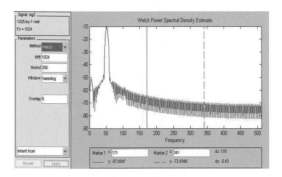

Figure 6.   Post filtering vibration signaltime domain char.

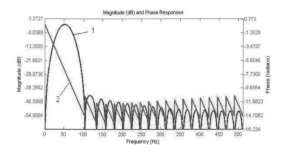

Figure 7.   Post filtering vibration signal time domain chart.

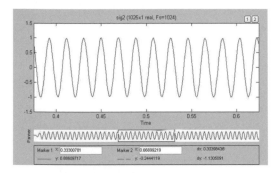

Figure 8.   High-pass filter amplitude and phase response.

5   CONCLUSIONS

In this article, based on window function design method of digital filter, to a power plant measured vibration signal of a steam turbine, for example, by analyzing the composition of the vibration signal in the frequency domain, calculated and gave the performance parameters and type of filter. Combining with the FDA Tool and SPTool of MATLAB, designed and implemented Band-pass filter, and realized to measure signal filtering.

The results of the vibration signal show that designed filter meet the design requirements and with good performance. This method is simple and practicable, and has certain reference value for understanding the principle and design method of digital filter.Photographs should be with good contrast and on glossy paper. Photographic reproductions cut from books or journals, photocopies of photographs and screened photographs are unacceptable. The proceedings will be printed in black only. For this reason avoid the use of colour in figures and photographs. Colour is also nearly always unnecessary for scientific work.

PREFERENCES

[1] Hongyi Zhao and Changnian Zhang: *Digital Signal Processing and Implementation Based on MATLAB*. (Chemical Industry Press, Beijing 2002), in Chinese.

[2] Peiqing Cheng: *Digital signal processing Course*. (TSINGHUA UNIVERSITY PRESS, Beijing 2003), in Chinese.

[3] Zhonglin Huang: *Control System Calculation and Practical Training of MATLAB*. (National Defence Industry Press, Beijing 2006), in Chinese.

[4] Chen Liang: Machinery Design & Manufacture, (2010) No. 12, p. 87–89, in Chinese.

[5] Yi Shao, Yan Wen: CHINA NEW TELECOMMUNICATIONS, (2010) No. 2, p. 83–85, in Chinese.

[6] Jihong Zhu, Longsheng Huang and Wenyi Zhou: Science Mosaic, (2007) No. 11, p. 225–227, in Chinese.

[7] Shan Ren, Xin Zhao and Wenbin Zhang: Advanced Materials Research, (2012) Vol: 490–495, p. 1867–1870, in English.

[8] Shan Ren, Xin Zhao, Jie Jiang and Dongjin Zhao: Advanced Materials Research, (2012) Vol: 203–504, p. 228–231, in English.

Information Technology and Computer Application Engineering – Liu, Sung & Yao (Eds)
© 2014 Taylor & Francis Group, London, ISBN 978-1-138-00079-7

The new lightweight encryption mechanism for large media signal processing system in global content delivery network

Juseung Heo, Chulwoo Park & Keecheon Kim
Department of Computer Science and Engineering, Konkuk University, Seoul, Republic of Korea

Kisang Ok
KT Corporation, Seoul, Republic of Korea

ABSTRACT: GCDN (Global Content Delivery Networks) is supposed to provide contents quickly through the nearest cache server which can reduce the load of a cache server as the user's contents request. Recently, high-capacity media stream services have been increasing because of explosive demand, but these are only speedy consideration ones mostly. These services have no security mechanism which is extremely critical on their works in contents delivery network. For this reason, If the whole media stream data is encrypted for the security consideration, it is not only damages features of GCDN services characteristic (fast providing of contents), but also wastes network resources inefficiently. Therefore, this research uses Selective Encryption, one of the stream encryption techniques. Additionally, we implement the new lightweight encryption algorithm for the Large Media Signal Processing System (LMSPS). We simulate and compare between the new lightweight encryption algorithm and the previous encryption algorithm.

Keywords: Global Content Delivery Network; Encryption Algorithm; Security System; Large Media Signal Processing System

1 INTRODUCTION

As the development of media transmission technology, various forms of streaming services have been increased, especially high quality video streaming service is being provided in various ways. Additionally, multimedia data transmission has become widespread with the high performance improvement devices which due to performance improvement of contents delivery network and prevalence of smart phones and other electronic devices. In this respect, those are necessary for the media service to protect copyright and classified intelligence which transmit in contents delivery network. (Fig. 1) shows the structure of the Large Media Signal Processing System (LMSPS) and security services in GCDN.

Encryption algorithm is the most important algorithm to embody the LMSPS. Even if we used excellent encryption technique, the resource wasting of the low-power electronic devices like mobile phone will be increasing and the quality of media images will be decreasing when the performance of encryption algorithm was low. But, If we lightweight the encryption algorithm in the optimum level, it brings many advantages such as an effective use of resources, a reduction of network traffic and a reduction of cache server overload. According to this, this research presents the designs and implementations of previous encryption algorithm and the lightweight encryption algorithm with the round reduction to adapt a multimedia streaming security system and simulation. (Boscovicm, D et al. 2011) (Seyyedi, S.M.Y et al. 2011) (Spagna, S et al. 2013)

2 MOTIVATION

Previous streaming security encryption techniques has some disadvantages such as, low security because of less encrypted data rating, low speed because of too many encrypt calculates to many encryption data, and there is risk that a quasi-random number which used for encryption process can be exposed because

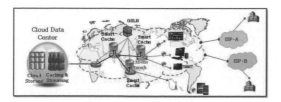

Figure 1. LMSPS in GCDN.

they were not using the key. Therefore, the large media stream service generally uses selective encryption algorithm which is one of the many streaming media encryption techniques (KISA 1999). Selective encryption technique analyzes images or videos, and then encrypts important parts based on the analyzed data. It means, compared to the way of encryption all data, it reduces resources consumption and can protect the data efficiently. To encrypt videos, the selective encryption uses three ways defined as before compression, during compression, and after compression to classify the relational consideration of compression technique. And i-fream, DCT coefficient, and header information can be used in this technique. Although selective encryption Tech is used, mobile device performance will be decreased without changing of real encryption algorithm because mobile devices use low electronic process that it is not adaptive to use previous algorithms. Therefore, this research presents the design and implementation for large media signal processing system construction (Boscovic, D et al. 2011) (Jie, Ding et al. 2011) (Sangmin, Lee et al. 2007).

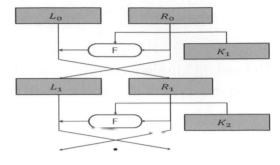

Figure 2. NL-SEED.

Table 1. Key creation algorithm.

Key Creation Algorithm

if (i=1; i<=8; i++){
$K_{i,0} <= G(A+C=K_{ci+1})$;
$K_{i,1} <= G(B-D+K_{ci-1})$;
if(i%2==1) $A\|\|I <= (A\|\|B)^{>>8}$;
}else $C\|\|I <= (C\|\|D)^{<<8}$;

3 IMPLEMENTATION OF NEW LIGHTWEIGHT ENCRYPTION ALGORITHM FOR LMSPS

Our algorithm implementation method is to make a lightweight seed encryption algorithm which reduces the round time. When we make the encryption key based on controlling the rounding times, we can reduce the performance of each round time by reducing the original SEED algorithm round time (KISA 1999). In this research, we define this new algorithm called New Lightweight SEED (NL-SEED) algorithm and implement as follows.

NL-SEED Algorithm which is composed of Feistel structure repeatedly creates Lr and Rr of ciphertext from the tbit L0 and tbit R0 of Plaintext which is 128 bit size in each 8 rounds. Repeatedly structure means plaintext block performs encryption process time in a few rounds, round $i(1 \leq i \leq r)$ means that Li = Ri1 and Ri = Li1exor f(Ri1,Ki) which are needed each subkey 'Ki' which is main input is directed by encryption key 'K' as a main input to perform (Li1,Ri1) > (Li,Ri) transform function. The round condition of previous SEED algorithm is not less than 2^{128} bit which are calculation complex, plaintext and ciphertext pair of BruteForce Attack, in addition, that should satisfy the condition of efficiency. Creating Key algorithm has a fixed rule every 2 round to make subkey. The input data to make subkey is same at the round data of (8 + i) and i. That means that round 1 and round 9 create equal data using same input data. Furthermore, it is rare impossible attacking encryption paper using the relationship between subkeys after five rounds.

Therefore, we reduced 16 round times to 8 round times that guarantee relatively safe performance with improving encrypt speed compare the previous seed

```
17
18    /*G function*/
19    #define SEED_ROUND(L0, L1, R0, R1) ₩
20    { ₩
21        T1 ^= T0; ₩
22        T1 = SS0[B0(T1)] ^ SS1[B1(T1)] ^ SS2[B2(T1)] ^ SS3[B3(T1)]; ₩
23        T0 += T1; ₩
24        T0 = SS0[B0(T0)] ^ SS1[B1(T0)] ^ SS2[B2(T0)] ^ SS3[B3(T0)]; ₩
25        T1 += T0; ₩
26        T1 = SS0[B0(T1)] ^ SS1[B1(T1)] ^ SS2[B2(T1)] ^ SS3[B3(T1)]; ₩
27        T0 += T1; ₩
28        L0 ^= T0; ₩
29        L1 ^= T1; ₩
30    }
```

Figure 3a. G function code.

algorithm. NLSEED algorithm follows such as procedures that 128 bit plaintext block processes eight round times inputting data which is composed of 64 bit size of each 8 round keys, we can get a 128 bit ciphertext block after these procedure. (Fig. 2) shows the structure of NL-SEED.

Key creation algorithm of NL-SEED algorithm (Table 1), reducing previous 16 round to 8 round, is created that 64 bit encryption key divides right 32 bit and left 32 bit then work by turns 16 bit right and 16 bit left rotary translation makes 32bit result data which is applied arithmetical operation and G function.

As (Figs. 3a,b), G function which uses two 16 bit Sbox are nonlinear transformation of each input bit that likes previous SEED algorithm makes a result data.

The result data processes a rotational displacement and then makes output data. In other word, G function input data (32 bit) is separated four 16 bit block of X3||X2||X1||X0 which Sbox block, S2||S1||S2||S1, is adapted in the order to create Y3||Y2||Y1||Y0 which are processed 8 bit left rotational displacement and get Z3||Z2||Z1||Z0 which are composed four 16 bit block.

$$Y_3 = S_2(X_3), Y_2 = S_1(X_2), Y_1 = S_2(X_1), Y_0 = S_1(X_0)$$

$$Z_3 = (Y_0 \& m_3) \oplus (Y_1 \& m_0) \oplus (Y_2 \& m_1) \oplus (Y_3 \& m_2)$$
$$Z_2 = (Y_0 \& m_2) \oplus (Y_1 \& m_3) \oplus (Y_2 \& m_0) \oplus (Y_3 \& m_1)$$
$$Z_1 = (Y_0 \& m_1) \oplus (Y_1 \& m_2) \oplus (Y_2 \& m_3) \oplus (Y_3 \& m_0)$$
$$Z_0 = (Y_0 \& m_0) \oplus (Y_1 \& m_1) \oplus (Y_2 \& m_2) \oplus (Y_3 \& m_3)$$

$$(m_0 = 0\text{xfc}, m_1 = 0\text{xf3}, m_2 = 0\text{xcf},\ m_3 = 0\text{x3f})$$

Figure 3b.   G function Algorithm.

```
32   /*F function*/
33   #define SEED_ENC(L0, L1, R0, R1)
34   {
35       T0 = R0 ^ *(k++);
36       T1 = R1 ^ *(k++);
37       SEED_ROUND(L0, L1, R0, R1);
38   }
39
40   #define SEED_DEC(L0, L1, R0, R1)
41   {
42       T1 = R1 ^ *(k--);
43       T0 = R0 ^ *(k--);
44       SEED_ROUND(L0, L1, R0, R1);
45   }
46
47   #define SEED_ROTR
48   {
49       t = A;
50       A = (A >> 8) ^ (B << 24);
51       B = (B >> 8) ^ (t << 24);
52   }
53
54   #define SEED_ROTL
55   {
56       t = C;
57       C = (C << 8) ^ (D >> 24);
58       D = (D << 8) ^ (t >> 24);
59   }
60
61   #define SEED_KEYROUND
62   {
63       T0 = A + C - *kc;
64       T1 = B - D + *(kc++);
65       *(k++) = SS0[B0(T0)] ^ SS1[B1(T0)] ^ SS2[B2(T0)] ^ SS3[B3(T0)];
66       *(k++) = SS0[B0(T1)] ^ SS1[B1(T1)] ^ SS2[B2(T1)] ^ SS3[B3(T1)];
67   }
68
```

Figure 4a.   F function code.

$$C' = G[G[G\{((C \oplus K_{i,0}) \oplus (D \oplus K_{j,1})\} \boxplus (C \oplus K_{i,0})] \boxplus$$
$$G\{\{(C \oplus K_{i,0}) \oplus (D \oplus K_{j,1})\} \boxplus$$
$$G[G\{(C \oplus K_{i,0}) \oplus (D \oplus K_{j,1})\} \boxplus (C \oplus K_{i,0})]$$
$$D' = G[G[G\{(C \oplus K_{i,0}) \oplus (D \oplus K_{j,1})\} \boxplus (C \oplus K_{i,0})] \boxplus$$
$$G\{\{(C \oplus K_{i,0}) \oplus (D \oplus K_{j,1})\}\ ]$$

Figure 4b.   F function Algorithm.

As (Figs. 4a,b), Feistel algorithm which is given 128 bit size composed the F function.

F function has two input data to make two output data of 64 bit. In other word, we define F function input data which are block C, D of 64 bit and key data Ki (i.e. Ki = Ki,0,Ki,1) during the encrypt processing and get output data C',D' which are the result output data.

Table 2.   Performance evaluation.

Number	SEED E/D (Mbps)	NL-SEED E/D (Mbps)
1	20.0/20.5	32.1/32.8
2	22.4/22.9	33.5/33.9
3	20.7/21.1	32.9/33.1
4	21.1/21.8	33.1/33.9
5	20.1/23.8	31.2/31.7
6	22.6/22.9	33.8/34.2
7	21.1/23.7	32.7/33.1
8	22.8/23.1	33.9/34.1
9	22.2/23.8	34.2/34.8
10	22.1/22.2	33.4/34.1
Average	21.51/22.58	33.08/33.57

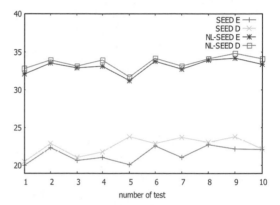

Figure 5a.   Performance evaluation.

4   PERFORMANCE EVALUATION

In this section, we apply to NL-SEED algorithm implementation code to simulate. The Table 2 shows the result which we tested using NL-SEED algorithm.

(Fig. 5a) shows test data in the graph. As graph shows, performance of lightweight algorithm is much better than previous algorithm. In addition, our incremental algorithm performance shows that our encrypt performance is improved as follows $33.08 - 21.51/21.51 * 100 = 53.78\%$ and decrypt performance is also improved as follows $33.57 - 22.58/22.58 * 100 = 48.67\%$ (Fig. 5b).

PSNR (Peak Signal-to-Noise Ratio) is the ratio between the maximum possible power of a signal and the power of corrupting noise that affects the fidelity of its representation. We can easily calculate by

PSNR=10*log{10}[(255*255)/MSE], MSE=1/n^2*SUM{y=1}^{N}[f(x,y)]^2]]

without considering about the possible power of a signal*.

(Fig. 6a) is an original media** and (Fig. 6b) is an encrypt media. According to the PSNR analysis (Fig. 6a) and (Fig. 6b), we make (Fig. 6c) which is the result of the difference between (Fig. 6a) and (Fig. 6b).

Performance rate

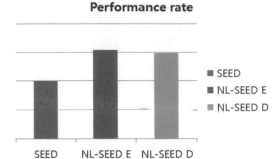

- SEED
- NL-SEED E
- NL-SEED D

SEED   NL-SEED E   NL-SEED D

Figure 5b.   Performance rate.

Figure 6a.   Original media.   Figure 6b.   Encyrption media.

Figure 6c.   PSNR.

5   CONCLUSION

These days, the encryption system of contents delivery network has increased electrical power because of the increase growth of large media service. Thus, we designed and implemented a new lightweight algorithm which is extremely efficient for large media streaming transfer system in global contents delivery network environment. Our implementation algorithm increases about 53.78% (Encrypt) and 48.67% (Decrypt) performance rates more than the previous algorithms. Furthermore, by adapting selective algorithm in PSNR with analysis result, it is proved that our algorithm is very powerful and secured in large media signal processing system. Our future research will be extended to the secure mechanism using our NL-SEED algorithm for large data handling and load distribution system in global contents delivery network.

ACKNOWLEDGEMENT

This research was supported by the IT R&D program of MKE/KEIT [10041910, Development of global cloud delivery platform that can reduce video traffic up to 50%].

REFERENCES

Boscovic D. & Vakil F. & Dautovic S. & Tosic M. 2011. Pervasive wireless CDN for greening video streaming to mobile devices. MIPRO, 2011 Proceedings of the 34th International Convention: 629–636.
Jie Ding. & Ning Li. 2011. A Distributed Adaptation Management Framework in Content Delivery Netowkrs. WiCOM: 1–4.
KISA. 1999. 128bit block encryption algorithm standard SEED, TTAS. KO-12.0004
Sangmin Lee. & Keecheon Kim. & Yonggeun Won. & Chaetae Im. 2007. An Effective Encryption Mechanism using Selective algorithm for Video Telephony. ICUT:210–214.
Seyyedi SMY. & Akbari B. 2011. Hybrid CDN-P2P architectures for live video streaming: Comparative study of connected and unconnected meshes. Computer Networks and Distributed Systems (CNDS):175–180.
Shangming Zhu. & Min Jiang. & Qimin Dang. 2011. Research and implementation on mixed Content Delivery Network. CSSS:1533–1536.
Spagna S. & Liebsch M. & Baldessari R. & Niccolini S. & Schmid S. & Garroppo R. & Ozawa K. & Awano J. 2013. Design principles of an operator-owned highly distributed content delivery network. Communications Magazine IEEE Vol 51, Issue 4: 132–140.

NOTE

*http://en.wikipedia.org/wiki/PSNR
**http://www.dreamworksanimation.com

Information Technology and Computer Application Engineering – Liu, Sung & Yao (Eds)
© 2014 Taylor & Francis Group, London, ISBN 978-1-138-00079-7

Objectionable information detection based on video content

S. Tang & W.Q. Hu

School of Electro-Information Engineering, Xuchang University, Xuchang, China

ABSTRACT: Preventing the spread of objectionable videos on the network is an arduous task, this paper attempts to do a preliminary work in this field, a method of detection the objectionable sensitive based on video content is proposed. First step, the video data is segmented, and the key-frames are extracted from video shot. Next, the characteristics of the video information are extracted from the key-frames. Finally, objectionable information judging mechanism based on video content is established, and video data is detected by Support Vector Machine. The experimental results demonstrate that the proposed method is feasible and effective.

Keywords: Video Processing; Feature Extraction; Information Detection; Support Vector Machine

1 INTRODUCTION

With the rapid development of the Internet and video processing technology, network videos increase dramatically, pornographic and violent camera lens also are mixed in, resulting in a significant negative impact on people, especially minors. Preventing objectionable video content to spread on the network will become an important task of purifying the Internet. Need to increase to more effective monitoring and control of network video content. Obscene information detection based on video content gradually becomes a new research focus.

Research efforts for filtering objectionable videos can be found in the recent literatures. Wang Qian et al. [1] Selected the key frame of videos based on motion analysis using the three-dimensional structure tensor, then detected skin color in each key frame based color model, finally applied the video estimation algorithm to estimate objectionable degree in videos. LV Li et al. [2] presented an optical flow based key-frame selection algorithm for detecting objectionable video. Chang-Yul Kim et al. [3] proposed an automatic system for filtering obscene video sequences based on segmentation. Seungmin Lee et al. [4] presented a multilevel hierarchical system for detecting objectionable videos. An Hongxin and Liu Yanmin [5] proposed an objectionable video detecting algorithm based on motion features. [1] and [2] detected the skin color information through the key-frame images of videos to judge the objectionable of the videos, likely to cause miscarriage of justice because of using less information on the frame images. [3] and [5] analyzed the obscene of the video based on motion characteristics, the detection would lose accuracy when a small video screen motion or still. [4] was constructed multilayer grading system, and tested each frame image, the complexity was relatively high degree.

In this paper, the videos were segmented and key-frame images were extracted, and analyzed the contents of the key-frame images, including the colors, textures and shapes. Then the judging mechanism was constructed for objectionable information based video contents, finally, the objectionable information was detected by Support Vector Machine. Experimental results have demonstrated that the proposed method was effective to determine the objectionable information of the videos.

2 BASIC FRAMEWORK OF THE DETECTION METHOD

In this paper, the framework of the detection method can be divided into 3 phases, namely, video processing, feature Extraction and objectionable information judgment. As shown in Figure 1.

3 DETAIL OF NOVEL FRAMEWORK

3.1 *Video processing*

The video data has a certain implied hierarchy and temporal redundancy, to detect the objectionable information based video content, the video data must be structured. Camera shot is the basic physical component units of a video, it can reduce the complexity of the video processing that the video is divided into camera shots. Key-frame is an image extracted from a camera shot which can describe the main content of the shot, according to the complexity of the content, one or several key-frames is used to describe the shot.

In this paper, according to the characteristics of objectionable information detection based video content and speeding up the video processing, the method

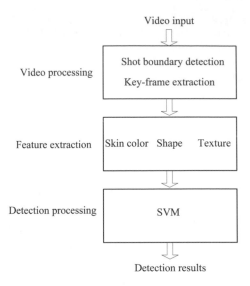

Figure 1.　Framework of detection method.

3.2.2　Feature extraction of color mask image

Pornographic sensitive key-frame images and non-Pornographic sensitive key-frame images will exhibit different characteristics after skin color model segmentation and morphological operations, such as the proportion of skin color, the area of the skin color region, the number of color connected regions and so on. By analyzing a large number of Pornographic sensitive key-frame images, the following five features extraction as skin color characteristics to judge the objectionable sensitive videos

① the number of the skin color connected regions (N).
② the ratio of the skin color pixel to whole image pixel (P_i).
③ the ratio of the skin color region to skin color rectangular area (P_r).
④ the ratio of the maximum skin color area to skin color circumscribing rectangle area (P_{rmax}).
⑤ the ratio of the maximum skin color connected region to whole image (P_{imax}).

3.2.3　Skin texture feature extraction

The real skin regions are usually smooth, there is no obvious texture features, at least there is no obvious visual perception texture features. Therefore, the texture detection and analysis for key-frame images can effectively distinguish skin color regions and non-skin color regions, to improve the accuracy of the skin color detection.

Gray Level Co-occurrence Matrix (GLCM) is an effective method for the analysis of texture characteristics that developed in recent years, which reflects the gray correlation of any two points in the image. The frequently used metrics of GLCM include four indicators: local consistency, contrast, energy and correlation. Experimental results [8] show that the indicators of contrast and energy are the most appropriate to reflect the overall contour texture measure indicators of the skin color region and non-skin color region.

of [6] is used to detect the video shot boundary, and the first frame image of the shot is selected as a key-frame.

3.2　Feature extraction of the key-frame image

The features of color, texture and shape are the lowest level characteristics of the video, also are the most significant characteristics of video frame image. The semantic features of the video can be comprehended through analyzing these characteristics of the key-frame images. Pornography sensitive video contains a lot of bare skin, to make full use of the features of the key-frame images, establishes a proper judgment mechanism, and detects the objectionable information of the video.

3.2.1　Skin color segmentation of the key-frame image

The key-frame images can be succinctly expressed the main content of the video shot, pornography sensitive video contains a lot of bare skin. Therefore, the skin color regions are segmented from the key-frame images can be a good characteristic to descript the pornographic sensitive video.

The study [7] showed that the skin color pixel has good separability in the color space HSV and YCgCr. Therefore, the pixel is a skin color pixel if a pixel is satisfying Eq. 1 and Eq. 2.

$$0 < H < 0.1 \quad \text{or} \quad 0.9 < H < 1 \qquad (1)$$

$$\begin{cases} Cr = -Cg + 258 \\ Cr = -Cg + 275 \\ Cg = 95 \\ Cg = 135 \end{cases} \qquad (2)$$

3.3　Objectionable information detection based on Support Vector Machine

3.3.1　Objectionable information judgment mechanism based the video content

Most of the key-frame images must be pornographic pictures if a video content contains objectionable information, in this paper, this feature is using to detect the objectionable sensitive of the video content. When detecting the objectionable sensitive of a video content, 8–10 key-frame images are extracted from the Candidate video slips as input, if output of the discriminant function $f(x)$ is +1, the sample is considered objectionable key-frame image, if the output is −1, the sample is non-objectionable key-frame image.

3.3.2　Experimental data preparation

A small database is established in this paper, which contains 40 objectionable sensitive video clips and 60 non-objectionable sensitive video clips, length of the video clips ranges from 1 minute to 5 minutes.

Table 1. Parts sample data of video key-frames.

No.	N	P_i	P_r	P_{rmax}	P_{imax}	f_{con}	f_{ene}	label
1	2	0.6565	0.9016	0.9210	0.6515	0.4784	0.2411	+1
2	2	0.6104	0.7136	0.7112	0.6069	0.7645	0.1995	+1
3	1	0.5964	0.6645	0.6645	0.5964	0.7088	0.2109	+1
4	8	0.1574	0.1574	0.7168	0.0968	0.3907	0.6904	−1
5	1	0.3793	0.5619	0.5619	0.3790	0.2733	0.4030	−1
6	3	0.1076	0.3513	0.7618	0.0634	0.1858	0.7841	−1

15 objectionable sensitive video clips and 25 non-objectionable sensitive video clips are selected from the database as sample data, the others are as test data.

A feature vector is constituted by 5 characteristics of the skin color region and 2 characteristics of skin texture (contrast (f_{con}) and energy (f_{ene})) to determine the objectionable sensitive based video content. The key-frame extracted from the objectionable sensitive video clips is labeled +1, and the key-frame extracted from the non-objectionable sensitive video clips is labeled −1, thus the sample set $\{(x_i, y_i), x_i \in R^7, y_i = \pm 1, i = 1, \ldots, n\}$ is obtained. Parts sample data extracted from the key-frame images as shown in Table 1.

3.3.3 Objectionable information detection based on Support Vector Machine

Support Vector Machine (SVM) was developed by Vapnik et al. as a method for learning linear, through the use of kernels, non-linear rules. In this paper, simulation is conducted by libSVM which was developed by Pro. Zhiren Lin of the National Taiwan University, and the radial basis function is selected as kernel function.

3.4 Experiment results

Process of objectionable information detection based on SVM is divided into SVM learning and SVM testing. In the learning process, the sample data have to normalize, and each vector of the characteristics is normalized to between [−1, 1], and the best parameters C = 32 and gamma = 2 are obtained by Cross-validation. In the testing process, the test data are detected by the SVM learning machine. 21 objectionable sensitive video clips and 30 non-objectionable sensitive video clips are correctly detected from the test data, total correct detection rate is 85%.

Experimental results have shown that the proposed method was effective to determine the objectionable information of the videos. But there was still a few of false detection, false detection of the non-objectionable sensitive video clip was caused by the close range, the reason was that People's facial skin in the entire image frame was in a higher proportion.

4 CONCLUSIONS

A method of objectionable information detection based on video content was proposed in this paper, according to the characteristics of the pornographic video, the detection judgment mechanism was established, and the SVM was used to detect the objectionable sensitive of the video content. Experimental results show that the proposed method could achieve a high detection rate with a relatively low false positive rate.

REFERENCES

[1] Q. Wang, W.M. Hu & T.N. Tan. 2005. Detecting objectionable videos. *Acta Automatica Sinica* 31(1): 280–286.

[2] L. Lv, S.T. Yang & S.N. Lu. 2007. Objectionable videos detection algorithm based on optical flow. *ComputerEngineering*. 33(12): 220–222.

[3] Chang-Yul Kim, O. Kwon, W. Kim & S. Choi. 2008. Automatic system for filtering obscene video. *10th International Conference on Advanced Communication Technology*, VOLS I-III–Innovations Toward Future Networks and Services:1435–1438.

[4] L. Seungmin, S. Woochul & K. Sehun. 2009. Hiearchical system for objectionable video detection. *Transactions on Consumer Electronics* Vol. 2: 677–684.

[5] AN Hongxin & Liu Yanmin. 2010. Objectionable video detection algorithm based on motion features. *Microcomputer Applications* 31(7):32–37.

[6] S. Tang, Y. Chai & H.P. Yin. 2011. Rapid method for adaptive shot boundary detection. *Application Research of Computer* 28(11):4380–4382.

[7] S. Tang. 2013. Human Face Detection Method Based on Skin Color Model. *Advanced Materials Research* Vols.706–708: 1877–1881.

[8] H. Guo & H.T. Huo. 2010.Research on the application of Gray Level Co-occurrence Matrix for skin texture detection. Journal of Image and Graphics 15(7):1074–1078.

Information Technology and Computer Application Engineering – Liu, Sung & Yao (Eds)
© 2014 Taylor & Francis Group, London, ISBN 978-1-138-00079-7

Association of education with daily life practice research in computer education

Zhi Fang Tu
Wuhan University of Technology Huaxia College, Hubei Wuhan, China

ABSTRACT: With the constant progress of science and technology, human life has developed an inseparable relationship with computer education. In this study, we have transferred the common sense into computer education, simplifying the complicated computer education. For instance, we have used racetracks to explain the break and continue statements of the C language, associated writing love letters with writing computer programs, used cups to analyze the computer storage mode "stack", used highway to explain computer-bit computing, and moved the isosceles right-angled triangle to write traversal of binary tree. All of this makes computer education easy to be understood, improves the efficiency and quality of education, and also enhances the learners' interest and enthusiasm, which is helpful for memorizing computer knowledge. Research conclusion suggests that we should make a further study of the cross connection all subject education, explore the similar knowledge. By broadening learners' thinking, using familiar knowledge from other subjects to analyze the computer knowledge learning to enhance study efficiency.

Keywords: Education research; Computer education; Life

With the constant progress of technology, computer has become an indispensable part of human life. In computer education, how to simplify the abstruse knowledge, to popularize the strange knowledge, and to improve the efficiency and quality of learners, are worth exploration and practice. In this study, we have transferred the common sense of life into computer education, by analyzing the profound truth contained in the common sense of life, compared it with computer knowledge, which makes the computer education vivid, and also enable learners to understand and easy to remember.

1   THEORETICAL BASIS

In 1897, the famous American philosopher and educator John Dewey proposed "Education is life" theory, which requested connection between the school education and students' social life, students' life contact, focused on life's impact on education. He also believed that getting involved in real social life, was the correct way to develop one's body and mind. Chinese educator and thinker Mr. Tao Xingzhi said: "life education is original, self-support, and necessary. The fundamental significance of education is the change of life. Life is ever-changing, so life ever bears the educational significance."[1] He put forward "life education theory", supporting that "life is education", "society is school", "teaching, learning and doing should be

integrate", which are derived from John Dewey education theory, demanding education to come out of the classroom and books, to emphasize social life more than books. The famous Chinese writer and educator, Mr. Ye Shengtao also said: "The textbooks can only serve as the basis of teaching, teachers should make the best of the textbooks to teach students and to benefit them." Knowledge is generated in real life, combining with the characteristics of the educational content, according to the rules and the current level of students' cognition, breaking the restrictions of the textbook, starting with the life stories that students are familiar with and the living examples that students are interested in, so teachers can absorb and introduce information of social life for comparative analysis, and propose unique and, easy way of education. In this way, it can not only make a very boring educational content alive, and interest students, but also make students truly understand the knowledge around them, that is closely linked with life, which in turn pushes to actively engage in learning.[2]

2   RESEARCH AND PRACTICE

"Knowledge comes from life, and is again applied to life." Computer enters human life, then how can the computer education approach the life?[3] This study connected social life and book knowledge, the common life experience and the computer education, to

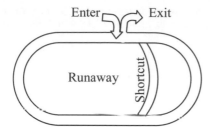

Figure 1.   Continue statement with runway diagrams.

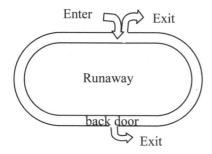

Figure 2.   Break statement with runway diagrams.

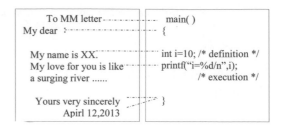

Figure 3.   Diagram of the C language function with a love letter.

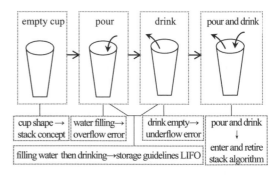

Figure 4.   Process of cup profiling stack knowledge.

achieve the objective of using familiar knowledge to explain the strange knowledge.

2.1   Case study one: The connection of break and continue statements in the C language and the stadium runway

Comparing two looping statements in C language, the break statement means the early termination of loop, the continue statement means the early termination of the current round of circulation. In this study, we associate them with the Marathon race, which requires sometimes going through some stadium; which requires athletes to run laps into the stadium. Assuming that there are "shortcut" in the stadium runway, taking the "shortcut" resembles ending prematurely counting laps, which is similar to continue statement; Assuming that there are gymnasiums "back door" at the gymnasium, if you go out of the "back door" and then advance ending counting laps at the gymnasium, it is similar with break statement. Shown in Figure 1 and Figure 2.

2.2   Case study two: The connection of C language functions and love letters

When teaching the C language function, we asked learners whether they can write function or not. For beginners, perhaps they can copy or imitate, but it is difficult for them to understand and master. Just imagine, even if you can not write a function, you may be able to write a love letter, The introduction of a "love letter" makes teaching atmosphere active. Associating it with C language function makes the learners feel even more vivid. Shown in Figure 3.

2.3   Case study three: The connection between the stack and the cup

Stack as a data structure, is a special linear form that can be only inserted and deleted at one end. It stores data according to the principle of LIFO. The first incoming data is pushed onto the stack end, and the last data stores in the top of the stack. When data is needed reading, it pops up data from the top of the stack (the last data is first read out). In this study, we used the cup to explain the stack. By analyzing the stack from pouring and drinking water action, the stack concept, operations and overflow error can be fully explained. Shown in Figure 4.

2.4   Case study four: The connection of bitwise "<<" (left-shift) or ">>" (right-shift) and the highway

In many computer programming languages (for example: C, C++, Java, JavaScript, Pascal), "<<" stands for the left-shift operator (equivalent to 'shl'), which is a binary operator, combined with direction from left to right. ">>" stands for the right-shift operator (equivalent to "shr"), which is opposed to the "<<" operator. In order to prevent students confuse the two operations, we introduce the lanes of the highway to explain. Figure 5 ① means that the car moves to the left lane, and according to common sense, the general ① lane is fast speed lane, at the same time, through "<<" operator calculating, it makes the original data larger, the two are very similar; Figure 5 ② means that the car moves to the right lane, according to common sense, the general ② lane is slow speed lane, at the same time, through

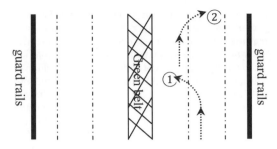

Figure 5.   The diagram of between bitwise "<<" (left-shift) or ">>" (right-shift) and the highway.

"<<" operator calculating, it makes the original data smaller, the two are very similar. On the highway, if shifted left too much, will hit the middle of the green belt; if shifted right too much, vehicles will hit the roadside guard rails. That has a natural similarity with the overflow error of "<<" or ">>".

2.5   Case study five: The connection of binary tree and isosceles right-angled triangle

In computer science, the binary tree is a tree displayed in a certain way that has a maximum of two subtrees at each node. Usually the root of the subtree is called the left subtree and right subtree. The binary tree is often used as a binary search tree and binary heap or binary sort tree. Traversal is one of the most important operations in the binary tree, is the basis for other operations on binary trees. The so-called traversal is the search along a route, followed by each node in the tree once and only do a visit. In this study, we take the isosceles right-angled triangle as a model, the vertex of two fork tree roots, at the end of two points for the left and right subtree, from top to bottom, from left to right isosceles triangle, to solve the preorder traversal, inorder traversal and postorder traversal. Following the preorder written for example, as shown in Figure 6 ("□" represents a space).

3   RESULTS AND ANALYSIS

3.1   Education connected with daily life can enhance students' interest in learning computer

The introduction of common sense in computer learning and teaching, can effectively motivate the students learning, guide the students to actively explore, enliven the classroom education atmosphere and stir the students' enthusiasm. In the survey with the students, the students mostly felt that introduction of the life into the computer learning, makes learning not so difficult to imagine, and improve students' learning interest. In the association of writing a love letter with writing program relations, especially when it comes to the fashionable terms in our daily life "my love for you is like the surging river …", students all burst into laughing; In the connection of C language "&&" (logical AND) and "||" (logical OR) relational operators, the

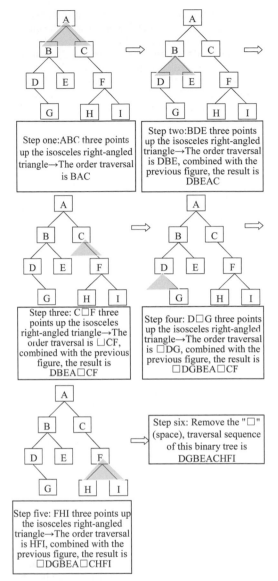

Figure 6.   Mobile isosceles right triangle to write a binary tree traversal sequence.

introduction of "Duguqiubai" and "Quanzhenqizhi" martial arts figures in television shows to explain, makes the students listen with relish, and the knowledge very difficult to grasp appear to be easy to understand; while connecting stack and cups teachers, suddenly asked: "when water is absorbed in the cup with a straw inserted, which part of the water is firstly sucked?" which stimulates the interest of the students to explore knowledge and develops students' creative thinking.

3.2   Education connected with daily life can improve the effect of computer teaching

Using simple example in life to explain complex computer knowledge, can help students easily grasp the knowledge and simplify the complicated knowledge for teachers under the premise of ensuring the quality of teaching, and also can help improve the efficiency of learning and teaching. The study found that teaching binary tree traversal, according to the argument of professional books, 80% of the students can understand, however, only 30% of the students can solve problems next class; The introduction of the isosceles right-angled triangle for binary tree traversal written, 100% of the students can understand and can solve problems. Even in NCRE training, some non-computer-science major students can understand and master it.

3.3 Education connected with daily life is conducive to students' memory of the computer knowledge

There are, in computer science, many rules, similar and contradictory knowledge, it is difficult for students to distinguish them. However, by analyzing the knowledge with the common sense students can, not only easily grasp it, but also keep it in mind for a long time. The study found that, if the concept of the highway has been transferred to explain the "<<" (left-shift) and ">>" (right-shift) operation, 80% of the students still remember it after one year. Similarly, associating software life cycle, with a person's, childhood, youth and aging characteristics in teaching, the students can easily understand it and does not need to learn it by heart as well. When teaching the degree of the tree, teachers introduce the human body, "fat or thin" to explain the degree of the tree, and "height" to explain the depth of the tree. In this way, students do not need to learn by heart and will not feel confused.

4 CONCLUSIONS AND SUGGESTIONS

4.1 The impact of connecting education with daily life on teaching computer science

Transferring common sense in life to computer education makes the abstract computer knowledge become concrete, visible, tangible. Organic combination of book knowledge and life knowledge, and the analysis of similarity computer education activities and real-life activities, can stimulate students' creative thinking, and enable students to learn about computers in a happy and relaxed atmosphere. Association of education with daily life can help develop students' self-learning ability and practical ability[4]. The scientific analysis of living resources, and the transfer of them to computer learning, enable students to fully recognize that knowledge is "from life, but above life"; so that students understand that each knowledge point is reflected in the life.

4.2 Teaching implications on computer science teaching

Using the familiar life example to explain the new computer knowledge enlightens us that we can try using familiar, and mastered knowledge of other disciplines to explain computer knowledge. The study found that the understanding of computer loop structure, the introduction of the circuit diagram of the physics, can make complex computer problems be easily solved and that introducing the C language "&&" (logical AND) and "||" (logical OR) relational operator concept into mathematics model, using the concept of interval axis to explain, can make the two kinds of relationships operator concept become clear. It can be seen that the cross-cutting nature of the disciplinary education is worthy of further exploration, and that the relevance of knowledge is worthy of further study.

REFERENCES

[1] Tao Xingzhi, 1985. Tao Xingzhi Complete Works (Volume II) (ed.) *China: Hunan Education Publishing House.*

[2] Classroom Teaching life, 2011. *http://www.zxxk.com/ArticleInfo.aspx?InfoID=161373.*

[3] Yan Yongjian 2007. Computer education "life" in the end. *Journal of Fujian Institute of Education,* (12).

[4] Sun Shanshan, 2009. The teaching of computer basis "education of life". *Profession,* (2).

Information Technology and Computer Application Engineering – Liu, Sung & Yao (Eds)
© 2014 Taylor & Francis Group, London, ISBN 978-1-138-00079-7

Multiple attribute decision making based congestion control algorithm for wireless sensor networks

Y. Sun, M. Li & Q. Wang
School of Electrical and Electronic Engineering, North China Electric Power University, Beijing, China

ABSTRACT: It is not sensible to try to reduce the incoming traffic into the network during congestion occurrence when important data packets are transmitted in wireless sensor networks. This will reduce the fidelity of observed by applications and influence the performance of the network. Aiming at this issue, a congestion control algorithm based on multiple attribute decision making theory (CCAMA) was proposed. CCAMA considers the nodes residual energy, congestion level, hop number from the original path and channel occupancy ratio as the bases for selecting next-hop, and builds a next-hop selection model based on multiple attribute decision making theory. Selecting the optimum next-hop according to the next-hop selection model and then establishing the distributary routing path to relieve congestion effectively. Simulation results show that CCAMA can increase the global throughput and reduce transmission latency, which achieves efficient congestion control and improves the performance of network.

Keywords: wireless sensor networks; congestion control; channel occupancy ratio; multiple attribute decision making; distributary routing path

1 INSTRUCTIONS

Wireless sensor networks (WSN) is a distributed information system, which consists of one or more sinks and large number of sensor nodes. The sensor nodes collect perception data and send it to the sink through the way of multi-hop[1]. There are several reasons which can lead to local or even global congestion in wireless sensor networks: Densely deployment of sensor nodes, resource-constrained node, multi-hop communication method and many-to-one convergence property[2]. Congestion always results in packet losses, transmission collision and an increase in transmission latency[3], which severely affecting the network performance. Therefore, how to control congestion effectively is a problem need to be solved urgently, and congestion control research is getting increasingly concerned.

In recent years, lots of work is going on in congestion control for wireless sensor networks. C. Wan proposed an energy efficient congestion control algorithm CODA (congestion detection and avoidance)[4] which provides basis for putting forward new congestion scheme. It adopts three mechanisms as follows: (1) monitoring mechanism based on the receiver, (2) open loop hop-by-hop backpressure, (3) closed-loop multi-source regulation. According to the characteristics of sensor network that sink nodes only care about source nodes information, ESRT (Event-to-sink reliable transport in wireless sensor networks)[5] divided the network into five states based on the situation of node buffer changes, and configured it to approach the optimal network operation. However, ESRT can't deal with the temporary congestion well, and does not apply to applications of large scale network. Jaewon Kang proposed an adaptive resource control scheme (ARC)[6] which create multiple paths to share the traffic to alleviate congestion, but it does not accurately reveal the relationship between node attributes and the next-hop node selection. The typical algorithms above provide the basis for proposing new algorithms. Paper [7] adopts different rate adjustment schemes according to the current queue length and the queue length change rate, so as to satisfy the demands of different QoS in wireless sensor network application. Literature [8] introduces RED active queue management algorithm, and control the length of queue by discarding the packets or reducing node rate. However, when network needs to transport important data packets, it is irrational to reduce source traffic by restraining upstream node's rate and it will reduce the monitoring fidelity of sensor network. Because the data generated during the crisis state are of great importance, often critical to the applications.

Based on the above analysis, in order to relieve congestion in wireless sensor network and ensure the monitoring fidelity of network at the same time, a congestion control algorithm based on multiple attribute decision making theory was proposed. CCAMA increases the capacity of network by building the

Table 1. An example neighbor table.

Neighbor ID	Congestion level	Residual energy (J)	Hop number from original path	Channel occupancy ratio
12	0.6	2	1	40%
2	0.4	5	3	30%
23	0.5	4	2	70%
34	0.3	6	4	30%

distributary routing path to relieve congestion and tears down the path as soon as the congestion has relieved.

2 CONGESTION CONTROL ALGORITHM BASED ON MULTIPLE ATTRIBUTE DECISION MAKING THEORY

2.1 Problem description

Reducing source traffic during a crisis state is undesirable since it will reduce the monitoring accuracy level of wireless sensor networks. Congestion is also transient and sensor nodes are deployed densely in sensor networks, therefore it can adjust the resource provisioning by establishing distributary routing path to increase the capacity of network when congestion occurrence. The existing network layer resource control algorithm proposed the node who own high residual energy, low congestion level and modest hop number from the original path will be selected as the next-hop to build a distributary routing path, but the algorithm didn't reveal the relationship between the three factors and the selection of next-hop precisely. Meanwhile, the algorithm ignored the influence of channel occupancy ratio for selecting next-hop. In view of the problems above, a new algorithm to select next-hop based on multiple attribute decision making was proposed, which can enhance the performance of the wireless sensor networks further.

CCAMA consists of four stages, i.e. congestion detection, distributary routing path construction, stable transmission and distributary routing path deconstruction. Algorithm flowchart is shown in Fig. 1. In the network, each node maintains a neighbor nodes information table, and the information can be got by monitoring RTS frame from neighbor nodes. The table includes the unique identification number ID, residual energy, congestion level, hop number from the original path and channel occupancy ratio of neighbor nodes within the communication range. An example neighbor table is shown in Table 1.

2.2 Congestion detection

At this stage, nodes calculate its own buffer occupancy ratio c periodically. When the congestion level exceeds

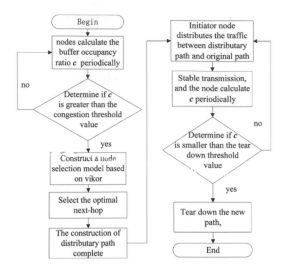

Figure 1. The flowchart of CCAMA.

a predefined threshold c_t, it indicates that congestion is about to happen. At this time, the first upstream node in the original path whose congestion level is below the congestion threshold will become the initiator and start to create distributary routing path.

2.3 Distributary routing path construction

Assume that there are n neighbor nodes within the communication range of the initiator node. The initiator node can get the information about its neighbor nodes by the information table, like identification number ID, residual energy, congestion level, hop number from the original path and channel occupancy ratio, etc. Choosing node with high residual energy as the next-hop can prevent the distributary routing path from failing due to exhaustion of node energy. Congestion level, i.e. buffer occupancy ratio, indicates the node current degree of congestion. Try to choose the node with low congestion level as next-hop to avoid new congestion occurs after the construction of new path. The distributary routing path can neither be too far away from the original path nor be too close to the original path. Because if the path is too close to the congestion area, it doesn't help, but it will further interfere with the already bad congestion. On the other hand, if the path is too far from the original one, then the packet delivery latency and the energy consumption will increase. Therefore, hop number from the original path should be considered for selecting next-hop. Channel occupancy ratio reflects the busy degree of wireless channel. Choosing the node with low channel occupancy ratio as the next-hop can make the node easily access to the channel to forward packets. In order to choose the optimum next-hop, CCAMA constructs mathematic model based on VIKOR multiple attribute decision making theory, VIKOR[9] method makes the benefits of all the attributes maximum, while keep the individual attributes regret value minimum. According

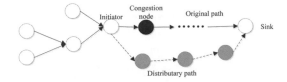

Figure 2. Distributary path creation.

to the decision value, CCAMA selects the optimum next-hop so as to build a distributary routing path. The creation of distributary path is shown as Fig. 2.

2.3.1 Pretreatment of node attributes

The state of node i is defined as $D_i = \{x_{i1}, x_{i2}, x_{i3}, x_{i4}\}$, and $x_{i1}, x_{i2}, x_{i3}, x_{i4}$ denotes residual energy, congestion level, hop number from the original path and channel occupancy ratio respectively. The comprehensive measurement of node is denoted by M_i, which is used to measure the state of node. The initiator node first eliminates the neighbor nodes which has already congested, and regards the no congestion nodes as the candidate nodes, assumes that there are m candidate nodes. Calculate the absolute deviation of hop number from the original path of each candidate node, i.e. the difference value between hop number from original path of each candidate node and the average hop number, so as to bring them into formula for calculation. Because the measuring units of the four attributes values are inconsistent, so the attributes values should be standardized and then generate the standardization matrix, as (1) shows.

$$S = \left(s_{ij}\right)_{m \times 4} = \begin{pmatrix} s_{11} & s_{12} & s_{13} & s_{14} \\ s_{21} & s_{22} & s_{23} & s_{24} \\ \cdots & \cdots & \cdots & \cdots \\ s_{m1} & s_{m2} & s_{m3} & s_{m4} \end{pmatrix} \quad (1)$$

In the formula, s_{ij} denotes normalized values of node attribute, and the normalization process is shown in (2).

$$s_{ij} = \frac{x_{ij}}{\sum_{i=1}^{m} x_{ij}} \quad (j = 1,2,3,4) \quad (2)$$

The residual energy of node is the higher the better, so it is the benefit criteria J_1, while the other attributes belong to the cost criteria J_2. According to (3) and (4), the best value s_j^+ and the worst value s_j^- of each attribute can be got. If one associates all s_j^+, one will have the optimal combination, which will be the ideal node.

$$s_j^+ = \begin{cases} \max\{s_{1j}, s_{2j}, \ldots, s_{mj}\}, j \in J_1 \\ \min\{s_{1j}, s_{2j}, \ldots, s_{mj}\}, j \in J_2 \end{cases} \quad (3)$$

$$s_j^- = \begin{cases} \min\{s_{1j}, s_{2j}, \ldots, s_{mj}\}, j \in J_1 \\ \max\{s_{1j}, s_{2j}, \ldots, s_{mj}\}, j \in J_2 \end{cases} \quad (4)$$

2.3.2 The attributes beneficial of node

When selecting the next-hop, we hope that there is an ideal node. But in practice, the possibility of the existence of such node is small. Under normal circumstances, the attributes values ??of the candidate nodes are not all the best. S_i is the attributes benefit of node which denotes the distance from each candidate node to the ideal node (best combination). The smaller the S_i, the more close to the best attributes value. It is presented by weighted sums formula, as (5) shows:

$$S_i = \sum_{j=1}^{4} w_j \left(s_j^+ - s_{ij}\right)/\left(s_j^+ - s_j^-\right) \quad (5)$$

In the formula (5), w_j denotes the weight coefficient of attributes. In this paper, congestion level and residual energy are considered more important, because node no congestion and having enough energy are important in distributary routing path construction. So the weight is set as follows:

$$w_1 = 0.3, w_2 = 0.3, w_3 = 0.2, w_4 = 0.2;$$

2.3.3 The comprehensive measurement of node

The smaller the S_i, the more close to the ideal node. However, there may be such a situation that the node is still not the optimum next-hop when S_i is the smallest. For example, the node's residual energy is very low, while the other three attributes values are wonderful which makes S_i is the smallest. Choosing the node as the next-hop may have a certain risk. Therefore, it is necessary to measure whether the risk value has affected its performance as the next-hop node. By (6)–(7) to compute comprehensive measurement of node, and its physical meaning is: not merely considering the S_i of the node when selecting next-hop but also taking the risk value of the node into account. Choosing the node whose attributes benefit S_i is good and the attributes values relatively balanced as the next-hop by calculating comprehensive measurement M_i. The smaller the M_i means that the node attributes are more close to the best attributes values, and the risk value is smaller at the same time, so the comprehensive performance is better. The formula is as follows:

$$R_i = \max\{w_j \left(s_j^+ - s_{ij}\right)/\left(s_j^+ - s_j^-\right)\} \quad j = 1,2,3,4 \quad (6)$$

$$M_i = \lambda\left(S_i - S^+\right)/\left(S^- - S^+\right) + (1 - \lambda)\left(R_i - R^+\right)/\left(R^- - R^+\right) \quad (7)$$

where, $S^+ = \min S_i$, $S^- = \max S_i$, $R^+ = \min R_i$, $R^- = \max R_i$. $\lambda = 0.5$, namely calculating the comprehensive measurement in a balanced compromised way.

Comparing the candidate nodes' M_i, which is the smaller the better, then the node with the minimum M_i is selected as the optimum next-hop. Using the same methods to look for the next hop until a new path has been created.

Table 2.   Simulation parameters.

Parameters	Value
Simulation area	500 m × 500 m
Communication range	50 m
MAC protocol	802.11DCF
Buffer size	60 packets
Ct	0.8
Simulation time	3 min

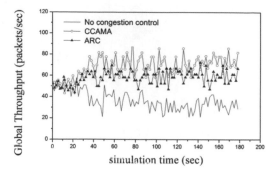

Figure 3.   The global throughput.

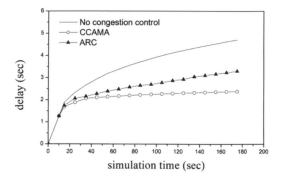

Figure 4.   The end-to-end delay.

2.4 *Stable transmission and path de-construction*

After distributary routing path is established, the initiator node become a traffic dispatcher who will distribute the traffic between distributary routing path and original path. The network comes into the stage of stable transmission. If the congestion level of nodes in the original path is lower than the path tear down threshold, then the initiator node tears down the distributary routing path, and all the traffic will be transmitted through the original path.

3 EXPERIMENT AND ANALYSIS

In order to verify the validity of CCAMA, this paper designed a corresponding simulation experiment on the OPNET 14.5A network platform. The simulation parameters are shown in Table 2. The experiment simulates the state of no congestion control, ARC and CCAMA protocol. We compared the global throughput and end-to-end delay two properties, and the results as in Table 2.

Fig. 3 shows the throughput of the three situations. The global throughput is the number of packets received by sink node per second. It can be seen from the graph, as the time increases, the throughput of no congestion control decreases, this follows because it occurs congestion in network. As a result, both the retransmission times and package lose number increase, and the global throughput reduces. However, the throughput of ARC is greater than the case without congestion control, because it begins to create a distributary path as soon as nodes have detected congestion, and the distributary path can undertake a part of traffic. CCAMA takes channel occupancy ratio into consideration to choose the next-hop which can make the node access channel easily and improve the forwarding data rate indirectly. At the same time, CCAMA chooses the next-hop based on multiple attribute decision making theory. Therefore, compared to ARC, the throughput of CCAMA improves about 16.7%.

The end to end delay is compared with the state of no congestion control, ARC and CCAMA. The result is as shown in Fig. 4. With the increasing of time, the latency of the three situations become higher. This is due to the accumulation of data packets in the buffer and even leads to congestion. Under the no congestion

control situation, the number of both drop packets and the retransmission packets increase, so the end to end delay is higher. Compared with no congestion control, the end-to-end delay of ARC and CCAMA reduce by 33.3% and 46.7% respectively. This is because both of the schemes take corresponding measures to alleviate congestion. CCAMA construct node selection model based on multiple attribute decision making theory reasonably, so the performance of the CCAMA path is superior to the path of ARC. As a result, the delay of CCAMA is smaller.

4 CONCLUSION

In order to relieve congestion in WSN effectively and not reduce source traffic during a crisis state of network. We proposed a new algorithm based on multiple attribute decision making to adjust the resource provisioning in sensor networks, i.e. CCAMA. CCAMA considers the nodes residual energy, congestion level, hop number from the original path and channel occupancy ratio as the bases for selecting next-hop, and selects the next-hop based on multiple attribute decision making theory so as to build a distributary routing path. The distributary routing path can share the load, so the capacity of the network is increase. The simulation results show that the CCAMA can relieve congestion effectively and improve the performance of network.

478

ACKNOWLEDGEMENT

The work was supported by a grant from the National Science and Technology Major Project of China (No. 2010ZX03006-005-01).

REFERENCES

[1] Sun Guodong, Liao Minghong, Qiu Shuo. A Congestion Control Scheme in Wireless Sensor Networks [J]. Journal of Electronics & Information Technology, 2008, 30(10):2494–2498.

[2] Sun Limin, Li Bo, Zhou Xinyun. Congestion control techniques for wireless sensor networks [J]. Research and development on computer, 2008, (12): 63–72.

[3] Zhang Yupeng, Liu Kai, Wang Guangxue. Cross-Layer Congestion Control for Wireless Sensor Networks [J]. Acta Electronica Sinica, 2011, 39(10):2258–2262.

[4] Wan C, Eisenman S, Campbell A. CODA: congestion detection and avoidance in sensor networks [C]. Proc. of the 1st International Conference on Embedded Networked Sensor Systems. Los Angeles: ACM Press, 2003:266–279.

[5] Y Sankarasubramaniam, O B Akan, I F Akyidiz. ESRT: Event-to-sink reliable transport in wireless sensor networks [C]. Proc of the 4th ACM Int'1 Symp on Mobile Ad Hoc Networking and Computing. ACMPress, 2003. 177–188.

[6] Jaewon Kang, Badri Nath, Yanyong Zhang. Adaptive resource control scheme to alleviate congestion in sensor networks [C]. Proc of the 3st IEEE Workshop on Broadband Advanced Sensor Networks (BASENETS). IEEE Communications Society, 2004.

[7] Liang Lulu, Gao Deyun, Qin Yajuan. A Reliable Transport Protocol for Urgent Information in Wireless Sensor Networks [J]. Journal of Electronics & Information Technology. 2012, 34(1):95–100.

[8] Li Luwei, Yang Hongyong. Congestion Control Strategy Based on RED Algorithm in Wireless Sensor Network [J]. Computer Simulation, 2012, 29(3):13–16.

[9] Li Qingsheng, Liu Sifeng, Fang Zhigeng. Stochastic VIKOR method based on prospect theory [J]. Computer Engineering and Applications, 2012, 48(30):1–4.

Information Technology and Computer Application Engineering – Liu, Sung & Yao (Eds)
© *2014 Taylor & Francis Group, London, ISBN 978-1-138-00079-7*

Realization of the voice acquisition and secrecy transmission system based on DSP5416

Shi Qiang Gao & Ze Zhang

College of Electronic Information Engineering, Inner Mongolia University, Hohhot, China

ABSTRACT: In this article, the design of speech signal processing system based on digital signal processor of TMS320VC5416 and the hardware platform is constructed utilizing TMS320VC5416 processors and TLV320AIC23 voice chip. Then some functions have realized in this speech signal processing system in debugging, which are audio collection, secrecy transmission and sending out. And the chip of signal processing encrypts or unbinds the voice signal via the arithmetic of recycling XOR. The testing results shows that this algorithm has high precision and strong anti-jamming feature and obtains the satisfying result.

Keywords: Data collection and analysis; DSP; TLV320AIC23; TMS320VC5416; secrecy transmission

1 INTRODUCTION

Data communication is an important way of information transmission and mutual communication in modern informational society. However, with the booming of data communication, there appears the problem of data compromised, such as the frequency occurrence of information gotten by illegal methods and the database information being stolen. Thus the secrecy communication becomes an important issue. Secrecy communication, as a way of communication in which the details of communication are hidden, plays a significant role in preventing information from stolen. The technology of data encryption is an important means of secrecy communication[4].

Digital signal processor is a dedicated processor which is suitable to high-speed real-time processing of digital signal. Its chief application is the realization of various digital signal processing algorithms in real-time. In nowadays' digital era, DSP has become the basic device of Communication, Computer, Consumer Electronics and other fields[1]. DSP can realize data encryption in secrecy communication faster and more efficiently. TMS320VC5416 processor, which is suitable for the speech signal processing and has been practiced, together with the special phonetic acquisition template of TLV320AIC23, consists of the voice system of audio collection, processing and playing. This thesis successfully realizes the acquisition and playback of the voice signal, and carries out the secrecy transmission of the voice signal.

2 DESIGN OF THE SYSTEM HARDWARE

The structure block diagram of system is consisted of two templates shows in figure 2.1. One is signal

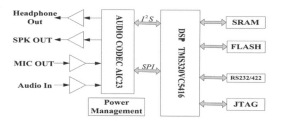

Figure 2.1. Structure block diagram of the system.

processing part which is based on DSP, and the other part is speech acquisition processing. Two parts together implement the system functions. According to the characteristics of audio signal processing application, the core of DSP processing system adopts the TMS320VC5416 chip of C54x series of TI company, and the speech acquisition processing part adopts AIC23 integrated speech chip, and DSP plays an integral part in the whole system.

2.1 *DSP control signal processing part*

Signal control processing part mainly implements on the function platform of VC5416. VC5416 is a 16 bit fixed-point high performance DSP of TI company, and is the third generation chip of TMS320VC54x series. Because of its low power consumption and high performance, the separate data and instruction space provide the chip with high capacity of parallel operation and the access of instructions and data at the same time in a single cycle. With highly optimized set of instructions, the chip has a high operation speed and the chip itself has rich on chip memory resources and a variety of on-chip peripherals. Therefore, it is widely applied in engineering, especially in speech signal processing

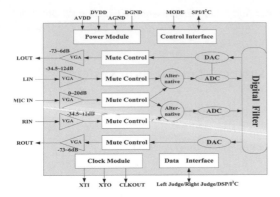

Figure 2.2. Basic structure and main interface of TLV320AIC23.

Table 1. Internal configuration register of AIC23.

Address	Register
0000000	Left sound channel input control
0000001	Right sound channel input control
0000010	Left earphone channel control
0000011	Right earphone channel control
0000100	Analog audio channel control
0000101	Digital audio channel control
0000110	Start the control
0000111	Digital audio formats
0001000	Sampling rate speed control
0001001	Digital interface activation
0001111	Initialization register

and secrecy communication applications, satisfying the requirement of system design[8].

2.2 Speech acquisition processing part[7]

The core of speech acquisition processing unit is Codec chip TLV320AIC23 (hereinafter referred to as AIC23), which is a high-quality stereo audio codec of TI company. Its internal integration, analog-digital converter (ADCs) and digital-to-analog converters (DACs), they all adopt the multi-digit Sigma-Delta technology of oversampling digital interpolation filter. Its data word lengths are 16, 20, 24, 32 bit and supports sampling frequency ranging from 8 kHz to 96 kHz. The signal-to-noise ratio of ADC and DAC can be up to 90 dB and 100 dB respectively. It has a built-in headphone amplifier output and supports two kinds of input mode, MIC and LINE IN. It also has programmable gain control of input and output. Moreover, AIC23 has the characteristic of low power consumption, with power being only 3 mW in the playback mode power, even less than 15 uW in power saving mode. Therefore, AIC23 becomes the ideal choice in the field of digital audio applications. The basic structure and main interface of TLV320AIC23 are shown in figure 2.2.

AIC23 has these main features: it is a kind of high performance audio codec chip; compatible with the McBSP of DSP chip of TI company through software control; can handle audio data input and output, stereo linear input and various audio input (stereo line input and microphone input) through the programmable audio interface that is compatible with McBSP; with mute function analog volume control function; with a high performance linear headphone amplifier; elastic management of power supply under software control; has the industry-specific minimum packaging; suitable for portable audio player and recorder; with many programmable features; can edit the device's registers through the control of interface; can control registers to set and configure the registers functions bits (0 or 1) through data serial transmission and control its work.

The peripheral interfaces of AIC23 mainly include three parts as follows:

a) Digital Audio Interface

BCLK: the clock signal of digital audio interface, when AIC23 is the secondary pattern, the clock is generated by DSP; when the AIC23 is primarily mode, the clock is produced by AIC23

LRCIN: the frame synchronization clock signal in the direction of DAC digital audio interface

LRCOUT: the frame synchronization clock signal in the direction of ADC digital audio interface

DIN: the signals of DAC direction given by serial data input of I^2S format

DOUT: the signals of DSP direction given from DAC output by serial data input I^2S format

This part of the interface signal can achieve seamless connection with McBSP (multi-channel buffered serial interface) of DSP. The McBSP's receive clock and BCLK signal of AIC23 are both provided by AIC23, in which AIC is the family equipment while DSP is the secondary device.

b) The Voice Signal Input Interface

MICBIAS: microphone's bias voltage input, usually 3/4 VDD

MICIN: microphone's input, the default is five times the gain

LLINEIN: linear input of left sound channel

RLINEIN: linear input of right sound channel

c) The Configuration Interface of AIC23

SDIN: serial data input signal of console port

SCLK: serial data clock signal of console port

CS#: input latch/address selection signal of console port

Before normal work, AIC23 configures its internal register through the configuration interfaces. The first 7 bits of the words have been sent before mean the register address, while the last 9 bits mean the contents of configuration register. Internal configuration register of AIC23 is shown in Table 1.

2.3 The interface design of AIC23 and VC5416

The connection of VC5416 and AIC23 is shown in figure 2.3. The TLV320AIC23 clock is supplied by

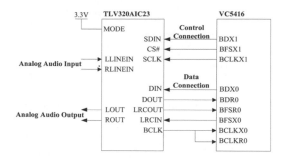

Figure 2.3.   The connection of VC5416 and AIC23.

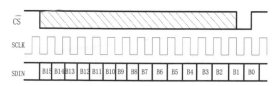

Figure 2.4.   SPI timing.

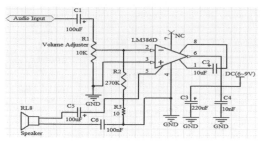

Figure 2.5.   Audio amplifier circuit.

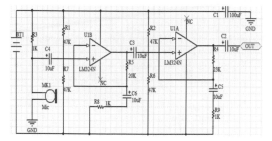

Figure 2.6.   MIC circuit.

the external crystals; its frequency is 12 MHz; its control connection interfaces with McBSP1 of DSP. VC5416 can make necessary configuration towards AIC23 through this interface. The part of data connection achieves the sending and receiving of voice data along with the McBSP1 of DSP[3][10].

In figure 2.3, MODE pin decides the mode of control connection. The connection of high voltage level in this system means the use of SPI mode. SPI timing is shown in figure 2.4.

The characteristic of this mode is that only when the chip select signal is effective, can it latch the data. In the 16 control words of serial input, B[15~9] is the register address and B[8~0] is the register data to be written. It is necessary to set register before system initialization so that the device can work normally. AIC23 supports four digital audio interface patterns: Right-Justified, Left-Justified, I^2S format and DSP format. In this system, the connection between AIC23 and the serial port 0 of VC5416 is the DSP format.

Two McBSPs, Multi-channel Buffered Serial Ports, of VC5416 configure I^2S mode and SPI mode respectively. McBSP0 is the sending and receiving data port; McBSP1 as the control port writes control word to codec. AIC23 as the secondary device sets as Slave mode while VC5416 sets Master mode. The codec sampling rate sets as 8 KHz.

The data encryption communication of voice acquisition module and DSP module mainly are accomplished through the McBSP interface of VC5416. The structure of McBSP is consists of a data channel and a control channel. Data channel sends and receives data while the function of control channel include the generation of internal clock, the generation of frame synchronization signal, the control of these signals and multi-channel choice, etc. Audio signal will be a string of digital signals after the high-precision and high-speed ADC conversion of TLV320AIC23 then input to the input buffer RAM. And then audio signal is

added into the internal of TMS320VC5416 by processing algorithms to dispose in a high speed processing. The processed audio signal can be stored, then input to the high-precision and high-speed AIC23 DAC converter. Then output it after being reverted into analog voice signal.

2.4   *The peripheral circuit design of DSP*

The purpose of the system hardware design is to identify and monitor whether the voice signal is under secrecy transmission or not. So to make DSP work, there need to make peripheral interface circuit achieve the input of the adopted voice signal and the output of the processing results. The following are the designs of hardware circuits of input module and output module.

2.4.1   *Audio amplifier circuit*
Audio amplifier circuit is the back-end processing module and can validate the restudy of speech signals and the effect of encryption. The design of hardware circuit is shown in figure 2.5.

2.4.2   *MIC circuit*
MIC circuit is the front-end speech input module of the voice acquisition module AIC23. It mainly complete the voice input of human-computer interaction module. The typical circuit is shown in figure 2.6.

3   DESIGN OF THE SYSTEM SOFTWARE

The program of the system software design mainly includes the main program and subprogram. The main program firstly completes the initialization of

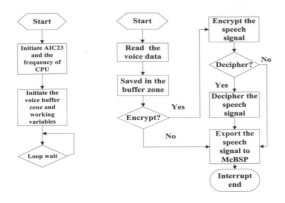

Figure 3.3. Audio signal secure communication processes.

Figure 3.4. Audio signal encryption process.

Figure 3.1. The flow chart of main program and secrecy communication module subprogram.

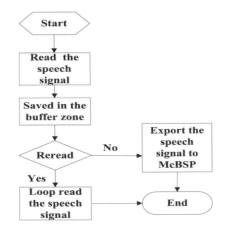

Figure 3.2. Flow chart of voice signal acquisition module.

VC5416 McBSP0, McBSP1, DMA and AIC23, then after the interrupt respond of CPU, enters the interrupt service routine and loops to wait. Subprogram includes voice acquisition module subprogram, voice encryption module subroutine and human-computer interaction module subprogram[9]. The flow chart of main program and secrecy communication module subprogram are shown in figure 3.1.

3.1 The software design of the voice signal acquisition module

This module is mainly to complete speech signal acquisition, compression, signal denoising and codec, that is to say read the stored speech signal in cycle on the basis of the original order being not disturbed. It can change the store content through PS/2 keyboard and can also make it stop reread. The flow chart is shown in figure 3.2.

3.2 The software design of the voice encryption and decryption algorithm[2]

Audio signal secure communication system based on DSP can conduct the audio signal at the sending end with encrypt processing and restore the original

audio after decryption at the receiving end, making eavesdropper not understand when transmitting in channel[1][4]. Corresponding control and information tips are required at this time, so are the data buffering and storage. The program of application system consists of two modules: the sending end and the receiving end. The former is responsible for the original audio signal input, real-time encryption and sending. The latter is responsible for receiving and encrypting audio signal and restoring by decrypting. There are DSPs at the sending and receiving end, which are responsible for dealing with these two modules respectively. The overall process is shown in figure 3.3.

There are many encryption algorithms used for secure communications. However, because of the particularity of audio signal, this system uses cyclic XOR to achieve data encryption. Its principle is to xor the original data and the previous encrypted data so as to form the current encrypted data, and loop until all data are processed. The algorithm formula of cyclic xor encryption is shown as follows[4]:

$$y(n) = x(n) \quad xor \quad y(n-1) \tag{1}$$

When $n = 0$, $y(-1)$ can take a value randomly, 0 for example. Although the security of this algorithm is lower than that of the chaos algorithm, it has the characteristics of simplicity, fast encryption processing.

In order to ensure that the audio signals can restore the original after decryption, decryption algorithm adopts the inverse operation of the encryption algorithm. The decryption algorithm formula of cyclic xor is shown as follows:

$$x(n) = y(n) \quad xor \quad y(n-1) \tag{2}$$

The audio signal encryption encrypts the original speech stored in the FLASH so as to make the newly-formed encrypted signals not understandable, then stores the encrypted signals in the FLASH area and exports to the audio codec for playback audition. Specific encryption flow chart is presented in figure 3.4.

The decryption function of audio signal is to read encrypted audio from FLASH and decrypt it by

Figure 3.5.   Audio signal decryption process.

Figure 3.6.   Flow chart of PS/2 keyboard processing program.

decryption algorithm and then output to audio codec equipment. Specific decryption flow chart is shown in figure 3.5.

3.3   *The software design of the human-computer interaction module*

The main human-computer interaction platform of this system is PS/2 keyboard of 17 key. It can achieve the real-time control through the keyboard. PS/2 keyboard perform a bidirectional synchronous serial protocol, i.e. every time send a bit data on the data line then each of the sent pulse on the clock line is read. Keyboard can send data to the host and the host can send data to the device, but the host always has priority in the bus. The host can inhibit from keyboard communication at any time, as long as we pull the data line low. PS/2 keyboard processing program flow chart is shown in figure 3.6.

4   THE FUNCTIONAL VERIFICATION AND TEST RESULTS OF THE SYSTEM

The system selects the CD quality music output by computer music software in the testing process as test speech signal[11]~[12]. The sampling frequency of setting up data to capture buffer in the CCS environment is 1 Hz, the test results are shown in figure 4.1.

Figure 4.1(a), (b), (c), (d) reflects the time domain, frequency domain of the original speech signal and the corresponding encrypted speech signal. From figure 4.1(c), (d) can be seen that after encryption not only the frequency domain signal and amplitude becomes a "dwarf", but also frequency domain component become abundant. We can see the obvious differences from the vision and the actual hearing audition also illustrates that the encryption processing is valid. It is unable to correctly understand the meaning of encrypted voice signal. And what we can see from the test results of experiment 4.1(e) & (f) that the decryption module of this system can completely decrypt the encrypted audio signal into the original audio signal. If we want to further improve the security of encryption and decryption algorithms, we can improve the length of cyclic xor, i.e. choose a number of consecutive signal points then xor them to realize.

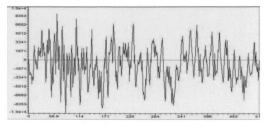

Figure 4.1a.   Original speech signal in time domain.

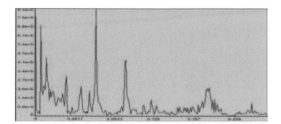

Figure 4.1b.   Original speech signal frequency domain.

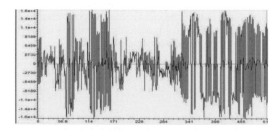

Figure 4.1c.   Speech signal in time domain after encryption.

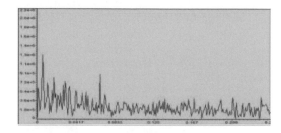

Figure 4.1d.   Speech signal frequency domain after encryption.

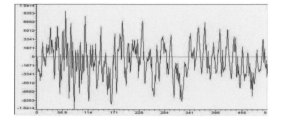

Figure 4.1e.   Speech signal in time domain after decryption.

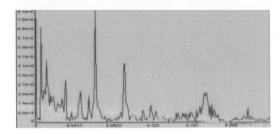

Figure 4.1f. Speech signal frequency domain after decryption.

5 CONCLUSION

Based on the characteristics of digital signal processing chip TMS320VC5416 of TI company and the audio codec chip TLV320AIC23, by writing applications and debugging software and hardware in the CCS integrated development environment, this thesis designs and implements a speech secure communication system which can encrypt or decrypt audio signal through cyclic xor algorithm. The system test results show that it can meet the requirements of real-time signal processing and can be used as a universal platform for voice signal processing algorithms research and real-time implementation[5]~[6].

REFERENCES

[1] Sanjit K.Mitra. 2011. Digital Signal Processing—A Computer-Based Approach, Fourth Edition [M]. Publishing house of electronics industry.

[2] Chen Chaoyang. 2004. Process Speech Communication System Based on DSP Algorithm Implementation [J]. Computer & Digital Engineering. Vol. 32 (2004)6. p. 73.

[3] Arindam Sanyall. 2007. An Efficient Time Domain Speech Compression Technique and Hardware Implementation on TMS320C5416 Digital Signal Processor [J]. IEEE – ICSCN 2007, MIT Campus, Anna University, Chennai, India. Feb. 22–24, 2007. pp. 26–29.

[4] Cheng zhaoming. 2004. The Audio System Design and Processing Algorithm Based on DSP Implementation [J]. Ship Electronic Engineering. Vol. 143. (2004)5.

[5] Zhou changlin. Based on the DSP Voice Recorders and Digital Implementation of the Echo [J]. Electronic Measurement Technology, Vol. 30 (2007)8.

[6] Zhu Z Q, Howe D. Halbach. 2001. Permanent Magnet Machines and Applications: a Review [J]. IEEE Proceedings Electronics Power, 148(4):299–308.

[7] Texas Instruments Inc. 2001. TLV320AIC23EVM: Evaluation Module for the TLV320AI23 Codec and the TLV320AIC23.

[8] Audio DAC Users Guide [EB/OL]. SLEU003. http://www. ti.com.

[9] Texas Instruments Inc. 2005. TMS320VC5416 Fixed-Point Digital Signal Processor Data Manual [EB/OL]. SPRS095O. http://www.ti.com.

[10] Chen Hucheng, An Qi, Wang Yanfang, Chen Jiaqin. 2000. Digital Voice Processing System Based on High-speed Multi DSP [J]. Proceedings of the 3rd World Congress on Intelligent Control and Automation June 28–July 2, Hefei, P.R. China.

[11] Fariha Muzaffar, Bushra Mohsin, Farah NazLecturer Farooq Jawed. DSP Implementation of Voice Recognition Using Dynamic Time Warping Algorithm [J]. Department Of Electronics Engineering, NED University of Engineering & Technology, Karachi.

[12] Yi Kefei, Hu Qingfeng. 2011. Real-time Multitask Scheduling Kernel Design Based on DSP [J]. Telecommunication Engineering, 2011 (4):135–137.

[13] Zhou Changlin, Xiao Ganfeng, Wang Yudong. 2007. Based on DSP of g.729 a Decoding Real-time Implementation [J]. Chinese Journal of Scientific Instrument, 28 (10), 1911–1915.

Information Technology and Computer Application Engineering – Liu, Sung & Yao (Eds)
© 2014 Taylor & Francis Group, London, ISBN 978-1-138-00079-7

The design of restricted domain automatic question answering system based on question base

Zheng Gong & Dan Zhang
College of Computer Science of Inner Mongolia University, Huhhot, China

ABSTRACT: In this paper, taking computer network course as an example, we design a Chinese question answering system against this course based on Frequently-Asked Question (FAQ). The system mainly focuses on the construction of domain knowledge, question comprehension, sentence similarity calculation method and so on. In the construction of domain knowledge base part, we design the curriculum knowledge table structure, the storage mode of FAQ and FAQ pretreatment; in question comprehension part, we mainly discuss Chinese word segmentation, keyword extraction and expansion, problem classification methods etc; in sentence similarity calculation part, we adopt semantic-based similarity calculation method.

Keywords: question answering system; FAQ; question comprehension; similarity calculation

1 INTRODUCTION

In view of the particularity of the knowledge in restricted domain, this paper revolves the theories and technologies of QA system to study, including the construction of domain knowledge base of computer network curriculum, frequently-asked question (FAQ) collection and organization, problem classification, sentence similarity calculation and answer extraction. We discuss the following aspects: (1) According to the characteristics of domain knowledge, we design domain knowledge base presentation and construction method by means of ontological thinking; (2) We improve the efficiency of automatic question answering system by means of the FAQ collection, presentation and classification; (3) Using domain knowledge characteristics and problem classification rules, we design the classification methods of question sentence in domain knowledge in order to reduce the search scope of candidate problems and improve the accuracy of locating answers; (4) We improve the matching precision of question sentences by studying the similarity calculation method of Chinese question sentences in restricted domain.

2 SYSTEM DESIGN

The system mainly includes four parts: question comprehension, information retrieval, answer extraction and domain knowledge base construction. The overall design flow of question answering system is shown in Figure 1.

2.1 *Question comprehension*

The question comprehension part is to extract the core meaning expressed by the user's question sentences, so that it can better and more accurately match the user's

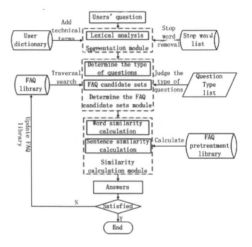

Figure 1. System Flowchart.

question in the following processing part in order to improve the performance of the system.

2.1.1 *Chinese word segmentation*
This paper uses the forward feature maximum match algorithm to segment word to the question sentence. We use ICTCLAS (Institute of Computing Technology Chinese Lexical Analysis System) word segmentation system developed by the Institute of Computing Technology, Academia Sinica in specific word segmentation process. Because this paper is focusing on question answering system for knowledge of computer network curriculum, while ICTCLAS system can't segment correctly to some computer network technical terms (in this paper, computer network technical terms is also called curriculum keywords), such as simulation data, cyclic redundancy check, TCP/IP protocol

etc. This article makes certain improvements to ICT-CLAS system in order to solve this problem: create a user-defined dictionary and add some technical terms and English abbreviation of computer network curriculum to this dictionary to ensure segment course keywords correctly.

2.1.2 Problem classification

Problem classification is a very important part of question comprehension, having significant effect on whether the question answering system handle the problem correctly in the following processes, and is also a key factor in making answer extraction strategy, playing a direct role to the accuracy of answers. For example: "the definition of simulation data?", if it can determine this question sentence belonging to conceptual type, then we can formulate an answer extraction strategy: extract the corresponding answer from all conceptual type questions, which effectively reduce the search space to the answers to improve the accuracy of answers and the search speed.

Because the method of judging question type is not ideal only by interrogative. For example, "What is the code element?" We can determine this question sentence belonging to conceptual type by "What" and curricular keywords "code element". While for the problem, "What is the function of the network card?", "What are the advantages of optical fiber?" We determine the question type not only by "what" and courses Keywords but also by question focus such as "function" and "advantages". For this problem, this paper proposes a question classification method of combining interrogative words and question focus.

In this paper, according to question characteristics of users asking frequently in the computer network curricular in conjunction with Chinese own grammar, we divide the questions into eight types. Classification rule is shown in Table 1.

2.1.3 Keywords extraction

The composition of keywords can be nouns, verbs, restrictive adverbs and so on. However, a large number of corpus statistics find that not all words in question sentence can become keywords. Many interrogative words, modal particles, auxiliary word etc appear more frequently in the document, but the amount of information it contains is very low, which almost has no effect on the subsequent retrieval, so we called these words stop words.

The function of stop words removal is implemented by a stop words list in this paper, which is created manually. Stop words list include: (1) auxiliaries, function words such as: "de", "ne", "le"; (2) words that more frequently appearing, but contributing little to the meaning of the sentence. For example: "please", "may"; (3) punctuations that cannot be used to form words.

2.1.4 Keywords expansion

Some keywords in the answer sentence are not keywords appearing in the user's question sentences, but synonym or abbreviation of these keywords. For example, user's question is "analog data definition?"

Table 1. Table type styles.

Number	Question type	Interrogative combination rules	Examples
1	conceptual type	What's/ definition / concept	What's the computer network?
2	principle type	principles/ ideas/ algorithms	What is the working principle of continuous ARQ protocol?
3	method type	how/how to /how about	How to prevent network congestion?
4	enumeration type	which/which type/which kind	Which hard devices are for transmission data?
5	reason type	why/reason	Why would produce errors?
6	functional type	What + function/effect	What is the primary function of the network layer?
7	relationship type	connection/ relationship/ difference	What's the relationship between signal rate and modulation rate?
8	other types	other complex problems	What are the detailed steps of dividing the IP subnet?

"analog data" is used in the user's question sentence, while "analog signal" is used in the answers that are stored in the database. In order to match such synonyms successfully, we need to expand these keywords. Expanding keywords properly can enhance the recall rate of the system, but if improperly, it will lead to the accuracy rate of the system dropping. Therefore, the special attention should be paid to keyword expansion in the general question answering system. The usual question answering system uses some certain semantic resources to expand the keywords. We can use common semantic dictionary, such as HowNet, Synonyms Dictionary and so on. However, common dictionary is not suitable for restricted domain question answering system studied in this paper. Therefore, in this paper, we design a curriculum key word synonym table for "Computer network" curriculum to carry on the synonym expansion for the curriculum key word. For example, "the definition of frequency division multiplexing? "After segmentation process, we obtain the key word set "the frequency division multiplexing, definition", while after the keywords synonymy expansion, we obtain the set "the frequency division multiplexing, frequency division multiplexing, definition".

2.2 Information retrieval

Information retrieval section plays a key role in question answering system, which completes the refinement function from the vast amounts of information to accurate information. Information retrieval is divided into two parts: document retrieval and segment retrieval.

2.2.1 Determine FAQ Candidate Set

In this paper, the process of determining the FAQ candidate set is the document retrieval process. Search those questions that are consistent with the type of user's questions to constitute FAQ candidate sets. The significance of determining FAQ candidate set is that it can reduce the question search scope greatly and improve the system performance. For example, a user's question "analog data definition?", This problem has been identified conceptual type in question classification process, then all conceptual type questions in the FAQ can be extracted to form FAQ candidate set

2.2.2 Word similarity calculation

In this paper, we adopt word semantic similarity calculation algorithm based on HowNet for word similarity calculation problem. This system combined with WordSimilarity software, which is developed by the Institute of Computing Technology, Academia Sinica, carries on semantic similarity computation between keyword sets of user questions and keyword sets of FAQ pretreatment library.

2.2.3 Sentence similarity calculation

We can calculate the similarity between two sentences after obtaining the similarity between words. We assume that there are two sentences A and Q, sentence Q contains words $Q_1, Q_2, \ldots, Q_m$, and sentence A contains words $A_1, A_2, \ldots, A_n$, then the word similarity between $Q_i(1 \leq i \leq m)$ and $A_j(1 \leq j \leq n)$ is $Sim(Q_i, A_j)$. Semantic similarity between Q and A is:

$$Similarity(Q, A) = \left(\frac{\sum_{i=1}^{m} q_i}{m} + \frac{\sum_{j=1}^{n} a_j}{n} \right) / 2 \qquad (1)$$

Among which,

$$q_i = \max(Sim(Q_i, A_1), Sim(Q_i, A_2), L, Sim(Q_i, A_n))$$
$$a_j = \max(Sim(A_j, Q_1), Sim(A_j, Q_2), L, Sim(A_j, Q_m))$$

The processing steps of information retrieval part are:

Step1: Identify the FAQ candidate set;
Step2: Carry on semantic similarity computation between the keyword set of user's questions and keyword set of FAQ pretreatment library;
Step3: Calculate sentence similarity set by formula (1) after obtaining word similarity.

2.3 Answer extraction

Different question answering system returns different result depends on answer extraction strategies.

There are two main kinds of answer extraction strategies: one is that related answers are sorted in reverse order according to the similarity values, then the first n answers are regarded as the search results and returned to the users, and the final answers are chosen by the users. Another is setting threshold theta, and all answers whose similarity value greater than theta are returned to the users as reference.

We use the first method—answer extraction— in the system design of this paper and set the value five in the actual running process in the system, that is to say the first five answers to question with the highest similarity value are returned to the user.

2.4 Domain knowledge base construction

Domain knowledge base in this paper, that is, computer network curricular knowledge base, is established under the guidance of ontology idea. We design domain knowledge base which is suitable for the question answering system in this paper by referring to design ideas of reference [6] in conjunction with the actual situation of computer network course. In this article domain knowledge library mainly includes: curriculum knowledge table, FAQ table, FAQ pretreatment table, similar words and expressions table, question type table.

Curriculum knowledge table is obtained by numbering chapter, section and knowledge point of computer network course.

FAQ table is an important component of question answering system. When user asks a question, we can directly search answers to your question in the FAQ table. At present, many question answering systems are implemented based on the FAQ table, especially the answering system used in specific areas. In this paper, questions in the FAQ table mainly come from "Computer network" teaching material, collection of test questions and network collection. The FAQ pretreatment table is a data table resulting from the preprocessing of the FAQ table. Preprocessing refers to carrying on word segmentation, stop words removal and determining the types of problems to all question set of FAQ table, finally resulting the question keyword set. The FAQ table pretreatment has an important influence on execution efficiency of word similarity calculation module. Similar word list mainly records the similarity between synonyms or near synonyms. Similarity value is represented by a numeric value between 0 and 1. Higher values indicate that the meaning of two words are more closer, thus the higher substitution level. Question type list includes eight types of problems that users often ask in the "Computer Network" courses.

3 SYSTEM TEST

The system uses a combination of interrogative and question focus classification method. As a result of complexity and changeableness of Chinese question, it is impossible to exhaust all the questioning methods. The system lists rules as much as possible. We give the

Table 2. Test results of question classification.

The type of questions	The number of questions	The number of correctly classified	classification accuracy
conceptual type	25	25	100%
principle type	20	20	100%
method type	10	9	90%
enumeration type	30	28	93.3%
reason type	10	10	100%
functional type	10	10	100%
Relationship type	15	14	93.3%
total	120	116	96.7%

Table 3. Test results of QA system.

The type of questions	The number of questions	The number of correct answers	Accuracy
conceptual type	26	26	100%
principle type	18	16	88.9%
method type	15	8	53.3%
enumeration type	35	24	68.6%
reason type	10	8	80%
functional type	18	15	83.3%
relationship type	24	17	70.8%
Total	146	114	78.1%

test results of question classification for 120 questions. The test results of problem classification are shown in Table 2.

The classification method used by this system has reached the accuracy rate of 96.7%, which is an ideal result in general. For example, "How many kinds are the hub divided into according to the function?", which is an enumeration type, but the sentence contains the question focus "function", and therefore this sentence is assigned wrongly to a function type. For this kind of question with characteristics of two types, we write program in the system. FAQ candidate set will be identified as collection of two types of problems so as to avoid the problem being divided into the wrong classification and leading to an incorrect answer.

The test sets used by this system are random problems from the computer network courses, and we use different questioning methods for the identical question to carry on the test. There are 146 question sentences in total. The test results of question answering system are shown in Table 3.

By carrying on the experimental test to the test set of computer network course, we finally obtain that the accuracy rate of this question answering system is 78.1%. The experimental results are more ideal, indicating that this system has a certain practicality. The accuracy is higher in this system for the questions of conceptual type, principle type, reason type and functional type, which reaches above 80%. The reason is that questioning methods of these four kinds of questions are unitary comparing with the target

question, and there always appearing the same or similar keywords in the user's question sentence, and the question focus is also easy to determine, therefore we have obtained the good experimental results; The accuracy is relatively low for the questions of method type and enumeration type, which is below 80%. The reason is that questioning methods of these two kinds of questions are variable and keywords are not very clear. So the system for answering such questions may be a little more difficult and the experimental results are not very ideal.

4 CONCLUSION

This paper implements an automatic question answering system for the knowledge of computer network course. System can automatically search candidate question set in FAQ (Frequently-Asked Question) library, according to the natural language input of students about questions of computer network course, calculating sentence similarity and returning the matched answers to the user. This system enables students to access information related to the current study, which is not limited by time and space, through question answering systems over the network, in conjunction with their own studies, thus deepening the understanding and grasping of the relevant knowledge to improve learning efficiency; Teachers obtain the widespread problems from the students and punctually reflect it to the teaching activities to help students understand and master the knowledge involved by those issues, thus improving the quality of teaching.

ACKNOWLEDGEMENT

This Work is Supported by Natural Science Foundation Project of Inner Mongolia, Grant No. 2013MS0902.

REFERENCES

Kang Chen, Xiaozhong Fan, Jie Liu, etc. 2008. Chinese Question Analysis and Research of Restricted Domain Question Answering System. Computer Engineering, 34(10), 25–27.

Qun Liu, Huaping Zhang, Hongkui Yu. 2004. Chinese Lexical Analysis Using Cascaded Hidden Markov Model. Computer Research and Development 41(8), 1421–1429.

Qun Liu, Sujian Li, 2002. Word Semantic Similarity Calculation Based on Hownet. // Taibei: Third Chinese Lexical Semantics Workshop, 8–15.

Shifu Zheng, Ting Li, Bing Qin. 2002. Overview of Question-Answering. Journal of Chinese Information Processing, 6(6), 46–52.

Shuxi Wang. 2005. Question Answering System: Core Technology, Application]. Computer Engineering and Applications 41(18), 1–3.

Xiang Li, 2010. Research and Application of Curriculum Knowledge-Based Question Answering System. Dalian Maritime University, 16–23.

Youzheng Wu, Jun Zhao. 2005. Research on Question Answering & Evaluation: A Survey. Journal of Chinese Information Processing 19(3), 1–13.

Information Technology and Computer Application Engineering – Liu, Sung & Yao (Eds)
© 2014 Taylor & Francis Group, London, ISBN 978-1-138-00079-7

Rationality analysis of Chinese innovative education

Yi Hui Pan
Jiangxi Science and Technology Normal University, Nanchang, Jiangxi, P.R. China

Ke Qin Tu
NanChang Taohua School, Nanchang, Jiangxi, P.R. China

Xi Mei Wu
NanChang Foreign Language School, Nanchang, Jiangxi, P.R. China

ABSTRACT: In the context of our time, innovations become the driving force. With the developing of economic globalization and Chinese-Western cultural exchange and penetration, we are increasingly focused on keeping the pace with the times on quality-oriented education, and adapt to this rapidly changing society. International advocate of national rejuvenation through science and education, society also needs creative talents, so we need innovation in Chinese education, not only the requirements of times, but also the requirements of Chinese education itself.

Keywords: Chinese education; innovation; rational appeal

Chinese is a foundational subject. It is the basic knowledge before we met other disciplines. It also taught us to recognize the society and to know everything around us. At first the knowledge is emotional, but with the continuous development of society information, and the requirement of times, Chinese education was being asked to combine with age, integrate into the society, so as not to be pedantic or on the situation of being eliminated. Then for Chinese innovative education, how to treat and innovated by a rational way?

The purpose of education is to make people become smart and sensible, to learn people skills and the ability to survive in society. Chinese innovative education is a basic part.

1 ESTABLISH A SENSE OF INNOVATION AND MAKE CHINESE PLAY THE MAIN ROLE

Chinese is what we grew up mostly has been implementing the learning about a subject, because the wide range of the content and the deepening expansion of the basic knowledge, bring many difficulties to Chinese education. Such as some teachers thought teaching Chinese waste a lot of time and energy, some students feel Chinese is very difficult to learn, thinking never fit the questions, or during Chinese text, they don't know what the best answer is and so on. All above are all because of the lack of innovation in Chinese education. It is weakening the students' efficiency and students' interest in the classroom.

In the past, the mode of "a question, an answer" confined students' divergent thinking, and also obstructed the cultivation of their creative consciousness. Thus, it's a must to bring up their creative consciousness in Chinese teaching. Although the class requires both teachers and students to participate, students are the main part of the teaching process. Teachers should pay attention to cultivating students' thinking ability, actively mobilize students' activeness, and take students as developing subject. Then, the Chinese teaching can emerge its significance.

2 PEOPLE-ORIENTED, EXTENDING INDIVIDUALIZED CHINESE INSTRUCTION

Education in the 21st century pays more attention to people-centered, to fully mobilize the enthusiasm and initiative. Students now receive an education is not only to learn some skills to seek future development, but also a manifestation of personal value and progress with society. The Chinese teaching is more to teach students how to learn, how to care for others, how to survive and how to better create. This is what we call people-oriented, putting people on the important position, promising students into young hopefuls to be innovative talents in the new era.

The past Chinese teaching emphasized too much on the overall solemnity and required common more than personality, which ignored the human personality charm. That not only affected the students' ability to acquire knowledge, but also hindered the development of the student's personality thinking, making their creative desire can't be released. Thus it reminded that we should think about the liberation of the student nature, to give students enough free space, so that students will communicate with teachers in teaching, and do freelance writing. After that, students' charismatic personality can be fully demonstrated, and the individualized Chinese teaching can get a better implementation.

3 ASSEMBLING SPREAD TO BE WHOLE, STRESSING THE WHOLE, EXPLORING THE LAW OF CHINESE

The aim of Chinese teaching is to teach students how to use language. Life is the best teacher. The width of society is synchronous with Chinese teaching. This requires us to comprehend and use language in everyday life, expand the serviceable range and function, to deepen and expand language knowledge. On the other hand, the arranging of the teaching material has level and a system. We should pay attention to carding the events, characters in teaching material, sum up its internal rules so that students can raise the irrigation of rational knowledge on the basis of perceptual knowledge, in order to do better learning. We should pay attention to the integrity of the Chinese teaching and optimize the structure of teaching, guiding students to look at the problem from a holistic point of view, perceiving the Chinese teaching.

4 INSTITUTE OF OPEN COMMUNICATION, EXPANDING CHINESE TEACHING SYSTEM

We do not require students to think completely consistent with the books. If paying too much emphasis on the integrity and certainty of the book knowledge, but ignoring its development, Chinese teaching will be completely confined within the inherent mode. At the time of giving great importance to the integrity of the Chinese teaching, we cannot ignore its development.

Today, the western open teaching has become a trend of the times, and we should do open teaching with the pace of the world. By this, the knowledge system of Chinese teaching is possible to get expanding and deepening, and students' sense of innovation can be possible to be inspired, then the learning space will expand and be more free and easy. This is a great benefit for improving student literacy and lifelong learning ability. What we want is a free and an enjoyable learning space where students can freely learn, and you can easily crash out of the spark of thinking, and then the fruits of innovation will be sweeter.

5 INSTITUTE OF DYNAMIC TEACHING, FULLY EMERGING CHINESE CONTEXT

We are often quietly facing the books to learn when learning an article or appreciating some verses. This static approach of learning in the new requirements of

the times is required to be changed. The development of modern information technology provides a good platform for dynamic teaching, such as the application of multimedia technology in Chinese teaching. From simple static teaching to a lively and dynamic teaching, all the senses are mobilized to deepen the perception of the language knowledge context.

6 ABLE TO LEARN TO THINK, INITIATING AWARENESS BY ASKING, AND ENHANCING THE DESIRE FOR KNOWLEDGE

Some book knowledge is easy to understand; especially Chinese teaching in the classroom, for example, a simple background introduce of the characters, as well as the narrative of life experiences. The question is to answer how this person's individual behavior formed. Some students read over, but never get the hidden factors in the context of its age; you do not know where to start to answer questions above. The historical background is a large environment, in particular times the special behavior must be found a potential cause in the background, regardless of the reason is not the main reason. So during the process of learning, students need well digested completely understood.

In the classroom teachers need to guide students to ask questions, eliminate some students' inferior and fear mentality which because of afraid to ask questions or are not good at expression of their own problems. Of course, in solving the problem needs both of teachers and students to participate in interactive communication to find out the best answer. Give students enough confidence and certainly encourage students to ask questions, so as to create a good learning atmosphere.

7 FOCUS ON THE QUALITY OF CHINESE TEACHING, ENHANCING STUDENTS' CHINESE PROFICIENCY

In the past, we were learning in accordance with the book, chapter by chapter, for the purpose of enabling students to have sufficient understanding of the knowledge to answer questions. But the development of the society is far faster than we imagine, the rapid changes and exchanges of information makes the simple knowledge learned from the books cannot solve life's doubts. So the teachers cannot judge the depth of students' knowledge levels by the level of student achievement, coupled with new knowledge emergences in the new era; with the rapid development of network media, channels to access knowledge is increasing; the way of learning has undergone tremendous changes.

In the past, teachers try all the ways to make the students learn the taught courses, but now we should use "know to learn" to replace the "learn to know" to equip students with the methods and strategies of Chinese learning, so that in the future when encounter different knowledge doubts, they can solve the mystery by the learned. This quality assurance can continuously stimulate the students' sense of innovation, deepen the quality of students training, and try to nurture students to be the successor of a new era.

8 INSTITUTE OF IMAGINATION AND ASSOCIATION, TO EXTEND THE TIME AND SPACE OF CHINESE

Why to build associate awareness? Students have endless imagination, which is an important factor for the realization of Chinese education innovation. A celebrity once said, knowledge is limited, but the imagination is infinite, it is like the wings of a bird, which can take you fly everywhere. Chinese is always been an exchange and transfer of emotion. If imagination strangled in the classroom, the exchange between emotions cannot be extended, and also cannot crash a wonderful sparkle. The building of related awareness adds vigor to innovative education of Chinese teaching.

9 LEARN TO THINK INDEPENDENTLY, LEARN TO CRITIQUE, AND THINK BACK ON CHINESE TEACHING

The criticism is one of the important features of innovative thinking. The understanding of things should not just stay on the surface. Sometimes, things seem correct may have error point hidden inside. Only judging things with a critical view, they can more clearly and more correctly know and understand these things, so that to better understand themselves. So in Chinese innovative education, it needs to learn to think back of the deficiencies and shortcomings in the Chinese education, find solutions in reflection to promote the development and progress of Chinese innovative education.

10 HAVING A SENSE OF LIFE, ENRICHING THE LIFE OF THE CHINESE

Education has always been a serious problem. It's not to say that education is in regular sequence, but that must be treated seriously and piously. The Chinese in daily life and the Chinese in the books cannot be separated, so as to fully enjoy the colorful world Chinese brings us.

11 EMPHASIS ON CHINESE TEACHING AND LEARNING ENVIRONMENT, AND CREATING A RELAXED ATMOSPHERE

Interaction of students and teachers are usually on the class in the classroom, and the collision of two magnetic fields requires a right learning environment just as a carrier. Now teaching has new tricks, and learning atmosphere has naturally undergone great changes, while the constant should be that no matter in what occasion, such as classrooms, lawns, parks, the learning atmosphere must be pleasant and relaxed. This will be a good help to Chinese innovative education.

12 CONCLUSION

The cultivation of students' innovative consciousness and ability is a major area of education in the context of the new era. The mentioned above provides rational support for Chinese innovative education. Language education will be more flexible, and the quality education progress will also across a big step.

REFERENCES

[1] Fan Chunjin. Reconstruction: the rational pursuit of implementation of creative education in Chinese [J]. Culture and Education Resources, 2009(36).
[2] Zhu Aijun, Wei Xiaoyan. The discussion on language teaching innovation [J]. Selections of Small writers (Teaching Exchanges), 2012(12).
[3] Wu Xiaochun, Cui Xiaoling. Discussions on the secondary language education's basic positioning and innovative fusion [J]. Wireless Internet Technology (Basic Education), 2011(1).
[4] Tao Dongfeng. Secondary language education, the task is not to cultivate genius [J]. Language Teaching in Middle School, 2011(6).
[5] Liu Fen. Explore the value orientation of the secondary language education, based on the new curriculum reform [J]. New Curriculum Research, 2010(10).
[6] Huang Yanhui. Language teachers in the new curriculum positioning [J]. Reference for Middle School Education, 2011 (15).

Information Technology and Computer Application Engineering – Liu, Sung & Yao (Eds)
© *2014 Taylor & Francis Group, London, ISBN 978-1-138-00079-7*

How to enhance the employability of the graduates majored in economic

Yi Hui Pan & Jie Ming Cao
Jiangxi Science and Technology Normal University, Nanchang, Jiangxi, P.R.China

Hong Gui Jiang
East China Jiaotong University, Nanchang, Jiangxi, P.R. China

Ke Qin Tu
NanChang Taohua School, Nanchang, Jiangxi, P.R. China

ABSTRACT: With the quality education became white-hot, collages and universities keep increasing enrollment students, bring the increasing pressure of graduates' employment, the problem of graduates' employment has become increasingly prominent. In this economic globalization era, the rapid consumption information and continuous developing economic provide a large number of employment information to the graduates, but still there are many economic class graduates is difficult to find an ideal job. This paper simple analyzed the employment problem of economics graduates, and put forward some countermeasures and suggestions.

Keywords: Economic categories; graduates employment; ability training

The employment problem of graduates has increasingly become the focus of attention of the community. No matter nation or school, including parents, regard this phenomenon as the most important problem. Solving the problem that graduates are hard to get a job is not only related to the future of the students themselves, which reflects their personal values in the social value, but also an important step for the country to adhere to the scientific concept of development and promote sustainable development. The employment problem of graduates has been thought to be hard to solve by general publics. Solving the problem is related to the healthy development of people's livelihood and the public mind. In the follow, it proposes several possible advices to enhance the employability of graduates majored in economics.

1 IMPROVING THE QUALITY TRAINING OF GRADUATES THEMSELVES, ALL-ROUND DEVELOPMENT

In recent years, many college students are increasingly concerned about those popular majors and consider that only the popular majors can help them to find a good job, which is not really. No matter what major is, only when students' own efforts to learn, constantly strive and practice to strengthen their ability to master the professional knowledge, it will be possible to find a satisfactory job. Well, how should the economics graduates enhance their self-cultivation?

1.1 *Mastering professional knowledge, raising professional skills, developing professionalism*

The economic majors are mostly liberal arts students. Liberal arts students are mostly with more active thinking and pleasant personality, and relatively master a wide range of professional knowledge, thus to be approved by society. However, in the university, there is a notable difference between knowing a profession and mastering a profession. In university, most of the courses are focused on theoretical research. They want the book knowledge can be actually useful in their own hands. It would have to go in-depth studying those formulas to master the trick. And then, they need to join some economic activities or competitions to exercise their abilities. By this, it can integrate theory with practice. Our society requires the ones are able to have the ability to do things, but not the students only delved into books.

1.2 *To guide students participated in the social practice, doing more exercises, lay a good foundation for the future employment.*

We cannot shut the students up in the ivory tower, for they must learn to understand the ever-changing information and social developing status, to recognize their position in the world, in order to find the opportunities. They need to learn how to live in the society, broaden their social circle, learn to correctly deal with some unexpected events and to enhance the ability to deal with conflicts, so that they can best fit the society.

1.3 Strengthen the psychological quality of graduates, and training them improves the ability of working under pressure, to make the graduate has a strong heart

There are many factors that influence a person doing thing, but the psychological quality level is a crucial factor to affect the efficiency. Some graduates often say that the reason why they cannot find the ideal job is that the external environment is too harsh or the people are too demanding; some even become negative and depressed because they cannot afford to repeatedly rebuff, Reason is still the relatively fragile heart. Economics graduates need to learn to participate in social practice, unremitting efforts to enhance their own psychological qualities. So that in facing the problems like looking for work or a dispute in the workplace they will not feel panic, deal with it smoothly.

1.4 Straighten employment mentality, keep a positive attitude.

When economic class graduates look for jobs, it shows out that they cannot endure hardship, inferiority complex, feel disappointed on wages and work carelessly. Graduation season, in addition to the students whom decided to continue studying or going abroad, most students are busy looking for a work. University recruitment and social recruitment provide many employment channels. Some students went to several job fairs, feeling hard getting a job, then lost confidence. When graduated, they find they get nothing. Some graduates thought that some work is to do hard labor, he is a college student why tortures himself? This is a huge mistake on the understanding. Any industry to have its existing value, everything is difficult at the start, when economics graduate looking for work they must Straighten employment mentality, keep a positive attitude and then they can catch the opportunity to find an ideal job .

On the other hand, graduates do not blindness feeling superiority on urban and regional issues. Some graduates in the economically developed cities or coastal open areas, they have more social contact opportunities than the other places. They can learn more about the development of the society, and they have more learning opportunities. All of the above made they thought they had a greater advantage in looking for work, then a little smug, so as to miss the opportunity to seek a good job, and at last they would be extreme regret.

2 LEARN TO START A BUSINESS, PRACTICE BRAVELY, AND KEEP THE SPIRIT OF INNOVATION AND HARD WORK.

In current economic times, people's need not only a manifestation of personal value, but also the interpretation of the social value. Our social needs of high-end talent, professionals, encourage employment, but encourage more positive entrepreneurial. This requires that you must have excellent skills and ability to innovate, requiring students to bravely explore, and actively develop new roads, which have great influence in stimulate students' potential aspect. Let innovative undertaking promoting employment, so as to deal with the difficult problem of employment of economics major graduates.

3 GOVERNMENT AND SOCIETY PROVIDE GOOD SUPPORT FOR THE EMPLOYMENT OF GRADUATES

3.1 Create a healthy employment Atmosphere

When graduation season, most of students busy attending job fairs, and some even ask their relatives and friends to help looking for a job. After these measures, the graduates who don't find the job will feel sad and great pressure. At this time, we should promote a healthy and positive atmosphere, and guide the graduates to keep a positive state when seek work.

3.2 Create green employment channel

The academic diploma of the economic graduates is different, but this is not related to the students' skills directly. Government should clear the channels, guide enterprises employing graduates by the correct way, do not look through colored glasses, to be a reasonable employer, the specification employer.

3.3 Guide enterprises to reasonable allocation of human resources

In fact, the economy has many aspects, the economic graduates also can serve variety kinds of positions, such as marketing, accounting, financial analysis and so on, small and medium enterprises should be a reasonable allocation of these graduates, to provide them with internships and exercise opportunity to establish effective partnerships with colleges and universities, to achieve a mutual win-win situation, so as to solve the difficult problem of employment of economics graduates.

4 SCHOOL CREATES MOMENTUM FOR GRADUATES TO FIND JOBS

4.1 Adhering to the characteristic teaching, being not scripted

Economy courses are mostly linked with the economy, some boring and some hard to understand. Teachers should pay attention to using effective methods to guide students in the characteristic teaching. It's necessary to pay attention to both the relevance of the economic profession and the feasibility of the occupation; it needs to focus on the grasp of students' professional knowledge, but also pay attention to practicing the graduates' comprehensive ability to respond to the community, so as to effectively create the conditions for graduates to get jobs.

4.2 Strengthen the exercise of the graduates' reserve capacity

The ability in addition to the professional knowledge and skills is the reserve capacity. For graduates in the ivory tower, it is very important to master extra skills besides their major. Companies want the person can create effective for them and help their considerable development. It's impossible for them to employ one who doesn't know any other skill but the knowledge of the books. So, for economics graduates, it is essential to master some other skills besides their major.

4.3 Improve the work ethics

Whether a graduate is able to find an ideal job or have good workplace and life after work has a great relationship with his work ethics. If an economic graduate has good work ethics, being honesty and trustworthiness, favoring others, with the spirits of sacrifice and dedication and knowing to serve others, then he will get the favor of the interviewer, and after working will get an important position from the boss. Thus, it is significant to have good work ethics.

4.4 Give good pre-employment guidance in all aspects to graduates

Even for the graduates who have exercised sometime in the community, it can't guarantee there is an adequate understanding of employment, quite apart from the inexperienced graduate-to-be. This time the teachers' employment guidance and expansion will be a beacon on the future road for the graduate-to-be. Such as the problems that should be noted on the signing of the employment agreement in the Labor Contract Law, reasonable choice of job objective, some factors that need to consider in an interview and interview skills, the popularity of this knowledge will provide a favorable of conditions to help graduates successfully join the employment team.

5 CONCLUSION

In summary, the problem that graduates are hard to get satisfied job is not incapable of being improved. In this era that the economy is globalization and graduates are common, in order to find an ideal job, starting their career, graduates must arm themselves on all fronts to enhance their image, do preparatory work before seeking a job, adapt to the wave of seeking jobs and improve the ability of surviving. The series of preparatory graduates does before seeking a job is not achieved overnight. This is a continuous process of accumulation, and would be gradually done. Therefore, in the course of a university, students should consciously develop their own abilities and literacy in different aspects. It will be a great benefit to the road development in the future of the graduates.

REFERENCES

[1] Ren Aizhen. The cultivation of core competitiveness of vocational college graduates—research based on Five vocational colleges in Changzhou Science and Education Town[J]. Education Research Monthly, 2012(9).

[2] Liu Wenwen. Exploration of effective initiatives to enhance the vocational college graduates core competitiveness[J]. Journal of Liaoning Administration College, 2011,13(3).

[3] Zuo Diansheng, Li Zhaozhi. Talk about how to break the employment plight of Economics and Management graduates[J]. Youth & Juvenile Research—Journal of Shandong Province Youth League, 2009(4).

[4] Zhao Bo. Strategy to enhance the core competitiveness in Independent Colleges Graduate Employment[J]. Modern Education Science (Higher Education Research), 2010(4).

[5] Xu Haoxiang. Investigation and Analysis of the Economic Class University Graduates Employment Intentions—Take 2011 Nanjing University of Finance and Economics School Graduates as the Example[J]. Journal of Puyang Vocational and Technical College, 2012, 25(6).

[6] Peng Bin. Graduates Psychological Problems and Solutions—take coastal areas popular professional as the example[J]. Management Observer, 2012(9).

Information Technology and Computer Application Engineering – Liu, Sung & Yao (Eds)
© *2014 Taylor & Francis Group, London, ISBN 978-1-138-00079-7*

Employment situation of graduates' discussion about economics-related course in colleges and universities

Yi Hui Pan, Li Tian & Wei Bao
Jiangxi Science and Technology Normal University, Nanchang, Jiangxi P.R.China

Ke Qin Tu
NanChang Taohua School, Nanchang, Jiangxi, P.R. China

ABSTRACT: Employment of graduates is the big issue with national and social attention. Under the perfect employment, each graduate can find the suitable job. This is significant to the nation, enterprise and the graduate. However, the imperfect world economy holds the economic crisis since 2009. It affects the society employment, especially in economics and finance. Therefore, the employment of economics major graduates is austerity. This article will analyze the employment of economics major graduates in colleges and universities. The provided reasonable suggestion can develop the full-employment of economics major graduates.

Keywords: Colleges and universities; economy; graduates; employment

The current world economy is in a resurgent moment. The American economy continues this depressed condition. The European debt crisis has great influence. Japan is under the financial crisis. Moreover, the proud electronics manufacturing is at a lower bet after the tsunami damage. Chinese economy is the best through the world. China is the leader of the world economy, which brings global economy recovery. However, the domestic employment condition is un-optimistic. Especially the economics major graduates. This is a challenged major. Good emulation is everything, bad emulation is nothing. We need to analysis the economics major graduates in dispassionate heart, provide guidance to the students and promote the full-employment of economics major graduates.

1 EMPLOYMENT SITUATION ANALYSIS OF CURRENT ECONOMICS MAJOR GRADUATES

An economics major includes economics, international economics and trade, finance, risk management and insurance, cameralistics, environmental resources and development economics. Some parts particular emphasis on the macroeconomic study and some parts particular emphasis on microcosmic study. The further employment direction and filed will be different based on this condition. After graduation, the economics major graduates will work in a government agency, bank, scientific research institution, industry and commerce, and the management of financial enterprise. Totally speaking, it is a high-end filed with good wages

and welfare. Our marketing economic development and international trade increase bring the great requirement of economy talents. However, it is an opportunity as well as the challenge.

Based on the past years' employment of graduates, southeastern coastal areas have developed economy, and the economic enterprises can provide sufficient positions. The employment condition in China's South-East coastal areas is better than other areas. From academy conditions, the national brand colleges such as '211', '985' economics major graduates will have better employment. Some of them will go to the government institutions, some of them work in transnational enterprises. However, the common colleges and universities take a dim view of employment condition. The graduates take part in the works, which has no relationship with their majors or at home unemployed. Four years' college life means waste of time with two certifications to be neet generation.

Research the graduate supply of economic management, most colleges and universities increase or expand the enrollment, and most provinces and cities build more colleges with economic management major. Even some non-economic management university and comprehensive colleges set up this kind of junior college classes based on the hot talent requirement of finance, international trade, finance and accounting, marketing management, and economic law. Some adult colleges, spare-time universities, and universities for employees take the part to establish the economic majors, build training course everywhere. This condition will make a jump of graduate quantity without quality assurance. All these schools break their own talent market. From the graduate requirement, their employment is under the condition of supply exceeds demand. That is all we have known about the 'law of value'. The years' of expanding enrollment create this embarrassing condition.

On the whole, the current economic major graduates have a serious employment situation. They cannot obtain full-employment. This kind of long-term condition will create idle labor forces, national resource wastage. There might arouse social contradictions even influence national stability and development.

2 INSUFFICIENCY EMPLOYMENT EVALUATION OF ECONOMICS MAJOR GRADUATES

The reasons have various aspects. They can divide into objective factor and subjective factor. We will detailed introduce the reasons in the following.

The first one is subjective factor. The subjective factor includes personal character, ability, quality, and professional level. These are the basic factors to influence the employment.

Most students have no learning during the college period. They have no professional knowledge, insufficient knowledge storage and even has no basic theory of economics. This condition cannot satisfy the enterprise requirement when they are facing the employment. Although some students have perfect knowledge storage with hard working during the school time, they have bad ability. The current required talents need professional knowledge and necessary ability. Especially the finance industry requires high competencies of student ability, analytical ability, associates' ability, anti-pressure ability, and creation ability. The comprehensive ability combines professional knowledge will simply obtain a good job. In addition, personal character is necessary. From the qualities we can evaluate the industry, because it is displayed from the inner quality. Politics have sagacious quality and businessman has smart quality. The economics major graduates need financier quality and the brain to evaluate economics. Each graduate has to correctly analyze the personal quality, and model the suit quality for you.

Characters influence a person's life. Although this is exaggerated, it can express some definite principles. The optimistic character is popular in the job market. We need to have perfect character, continuously make up the weakness, build good natures, support positive attitude, speak less and work more with concise language. Moreover, the easy-going, principle, action determination and strong orientation are necessary all the same.

The second one is the objective factor. The unemployment of economics major graduates has objective factors and subjective factors. The objective reasons have: national policy, school education, economic condition at home and board, and enterprise employment system.

It is no doubt that the economic crisis in 2009 brought great influence. All the graduates received an employment impact. Especially the economics major graduates. The international trade develops more and more quickly and increases close relationship with other countries. The economy crisis started from credit financing. Large amount of banks goes out of business, Greece and Iceland faced economy bankrupt. Our financial industry expressed great impact at the same condition. Moreover, the transnational enterprises bankrupt increase domestic graduates difficulty of employment. Our nation has less power to express policy function and strategy adjustment leads the employment graduates flock together without reasonable guidance. Nobody wants to go the west area and everybody wants to go the east area. The colleges and universities didn't provide employment guidance, and only pay attention to the fake employment rate. The modern enterprises do not provide opportunities to the graduates, which only employ the experienced people. This condition aggravates the employment difficulty to the graduating students. Here we propose the social companies can supply more chances to the graduating students.

3 SUGGESTIONS TO IMPROVE EMPLOYMENT OF ECONOMICS MAJOR GRADUATES

Employment is a significant livelihood project as well as the important measurement to maintain social stability. This project needs concerted effort among the nation, enterprise, school, and graduates. It will promote full-employment of economics major graduates, and realize personal value and social value.

In the first place, the economics major graduates need to develop the personal profession abilities, promote comprehensive quality and change the old employment opinion. College students have to work hard on cultural knowledge learning, grasp professional technology, emulate well on various courses, and cultivate perfect self-learning and management ability. They can joint into the social practice, part-time job. This method can increase social practice experiences and earn some money. They will understand the laborious of parents, encourage the good study. At the same time, expand social circle is necessary. It is important to learn about different types of people. Moreover, adjust mentality, do not forget one and underestimate one's own capability, and convert the employment idea. The students need to establish correct and positive value, view of life, view of employment, and keep the positive psychology. Work hard down to earth, look forward to the bright future, continuously increase personal professional quality and comprehensive quality is the best way to be a perfect compound talent.

In the second, we need positive guidance from the nation, and the colleges should implement the function as a bridge. Since we cancel the work distribution policy, the talent free mobility is beneficial to the economic development. However, the nation has insufficient guidance to the economics major graduates. The nation needs to keep employment channel communication with policy guidance. At the same, ensure the employment fairness. Each graduate will has the employment chance, and each student will get success chance. The college and universities should implement the function as a bridge. During the school time, we need to enhance ideological education work, infuse correct view of employment. When the students are facing the graduation, the colleges should build job fair with the employers, and integrate relative graduates with the employer requires. Meanwhile, the daily education needs to pay attention to the practical ability training. The college students have grandiose aims but puny abilities. It is necessary to build social practical actions, send them to the enterprise training, and improve their abilities.

In the third, the enterprises need to provide more chances for the graduates and increase their sense of social responsibility. The economics major graduates have less working experiences and most enterprises play it low upon this during the recruitment. This condition develops the difficulty of the graduate employment. Most students have no chance to take the interview. In my opinion, this is quite unfair. Everyone has the right of equal employment and the personal ability cannot decide by experiences. This is wastage to both enterprise and the graduates. Therefore, the enterprises should give up the old perception, grow full of confidence in the young people and create an equitable employment environment.

4 SUMMARY

Employment problem is the livelihood issues. The perfect employment solution can stimulate domestic demand, which has significant meaning to maintain social stability. The current employment situation of economics graduates is not optimistic for the national economy affect and independent employment misunderstandings. Moreover, it is also caused by school guidance absence and knowledge level disqualification of the students. If we want to solve the employment problem and ensure the full employment of graduates, the concerted effort is necessary. National policy support, school guidance of graduate employment, and enterprise opportunities providing is important. Moreover, every student continually promotes ability and quantity, research jobs with placid mentality and realizes the unified personal value and social value.

REFERENCES

[1] Zeng Xiang. Graduated employment problems and strategy of economics-related course in colleges and universities [J]. Vocational & Technical Education Forum. 2010 (06).

[2] Bao Dihong. Employment evaluation of finance and economics major [J]. Occupational Planning. 2011 (09).

[3] Yang Yiyong. Employment situation, employment environment and employment measurement after joining into WTO [J]. Economic Tribune. 2009 (06).

[4] Zhang Yun. Problem of employment evaluation about economical talents in colleges and universities [J]. Journal of Changchun University of Science and Technology. 2012 (09).

[5] Li Haifeng. An employment situation evaluation under the new situation [J]. Journal of Lanzhou University. 2011 (12).

[6] Cao Zhen. Occupational guidance of economic management students [J]. New Culture Daily. 2009(06).

Information Technology and Computer Application Engineering – Liu, Sung & Yao (Eds)
© 2014 Taylor & Francis Group, London, ISBN 978-1-138-00079-7

Network resource development and application in college Chinese education

Bin Ying Wu & Li Fang
Jiangxi Science and Technology Normal University, Nanchang, Jiangxi P.R. China

Shi Qiang Chen
East China Jiaotong University, Nanchang, Jiangxi P.R. China

ABSTRACT: Information network has been an unstoppable tendency. Under the new rapid network development, Chinese teaching has new development space. How to get rid of the traditional teaching mode constraint, the positive development and network resource usage even expand Chinese teaching space has been one assignable issue of Chinese teaching. When we are learning to network, it is necessary to select useful information for improving Chinese teaching quality.

Keywords: Chinese learning; network technology; conformity utilization

Chinese teaching obtains new method in the information network period. It changes the passivity learning mode into active exploration. This is beneficial for the students to research and collect information by themselves and improve their innovation. However, the various and abundant network resources have difficulty to select the useful information for Chinese learning. In order to improve Chinese learning ability, it is worth to have deepened analysis and research.

At the very beginning, we will provide simple explanations for the development and utilization inevitability of Chinese teaching network resources.

1 CONFORM TO THE TIMES

More and more people use the internet to obtain new information. The network has been the irreversible mode in our daily life. In the modern period of information networks, ever filed influenced by the network no matter in economy and polity. The Chinese teaching has no exception. We need to clear one point that people cannot be separated from the daily matured network society. We have to learn to use network resource when we are teaching Chinese for promoting the teaching quantity.

2 INEVITABLE REQUIREMENT OF NEW COURSE REFORMATORY

During the modern Chinese reformatory, lots of new theories provided. One bright spot is 'curriculum resource'. This article clearly points out each course resource development and utilization during the Chinese teaching process.

For the various reasons, we have no emphasis attitude to the curriculum resource of some knowledge contents. This makes the waste of course resources. In order to develop new course reformatory, we need to increase the understanding and take the full advantage of these resources.

The traditional teaching mode support single and simple information acquisition. The modern network popularity fortifies our knowledge. The development and utilization of course resources provide a convenient approach under this background.

2.1 Network resource is the important origin of Chinese teaching course in colleges and universities

The internet has abundant resources. Lots of people will browse and quote the information from the network. If we input the poetry name, it will pop up countless information about this poetry. Information can share all over the world, which provides convenient condition to learn Chinese. Knowledge and abilities on the network include various custom cultures of different nationalities. This information can be treated in Chinese course reference. Otherwise, the course resource on the internet suits Chinese teaching approach of the new period. The traditional Chinese teaching cannot satisfy the professor requirement and course revolution demand.

2.2 The network provides chances to develop and utilize Chinese course resources

The perfect learning environment places significant position of teaching. The network technique development provides a possibility to study in the perfect

Chinese learning environment. Students can learn the knowledge through the virtual environment, develop knowledge cognition and conclusion, and understand the knowledge construction. Information Network has no limitation in time and space. Many materials can be found from reading mode or listening mode. For example, the traditional poetry appreciation and analysis has limited materials. The modern students can select useful resources through the network anywhere. This is convenient to the student learning.

2.3 Chinese course development and utilization on the network expands the course implementation

The Chinese generation and development cannot be separated from specific society. Lots of knowledge and emotion come from our daily life. Organic combination of life and Chinese can fully understand the Chinese and make out our daily life, even the independent soul. Network places important character for knowledge expanding and deepen researching. The development and utilization of Chinese course has significant meaning to inspect network culture development.

2.4 Chinese course reformatory needs development and utilization of network course resources

The course and resource have a close relationship. Not course resource will have few courses. It doesn't mean the course is the course resource. They have differences. The course resource has a larger range than the course.

Not all the course resources can be put into the courses. It is only an alternative material. If we want to use these materials in the course, we need to manage it for suiting to teaching requirement.

3 THE CYBER SOURCE AND UNIVERSITY CHINESE TEACHING ORGANIC COMBINATION

Since network resource development and utilization is imperative in college Chinese teaching, how can we organic combine network resource and Chinese teaching?

3.1 Bring network resource to expand Chinese learning space

The abundant Chinese resource can increase their learning interests. Students can positive joint into the practical course and fully mobilize their subjective initiative. Although, the network resource provides convenient to Chinese teaching resources, it will bring a huge database in front of the students. The resource integration and application work out their learning ability, and makes them deepen their modern Chinese teaching.

3.2 Use network resource and develop Chinese course study

The network brings great changes to students and teachers in colleges and universities. The perfect Chinese teacher needs wild knowledge and understands the teaching method and take full advantage of this course. With the timely development, Chinese teachers should stay close to the time, advance with the times, even understand the way to promote classroom learning.

College Chinese is not the simple cognitive activity. It is the affective activity. The modern Chinese teaching can use the network and multimedia to create the best learning environment. It can adjust students' thirst for knowledge and deepen their learning experiences.

The multimedia technique application can effectively express writer thinking of each paper. Students can understand more about the papers. Multimedia technology has sound and picture. The design creates learning environment, which is hard to realize in the traditional class with one teacher and one mouth.

3.3 Focus on network resource selection, ensure teaching quantity

The network provides a large amount of information and knowledge to us. The perfect utilization is helpful for expanding Chinese teaching contents. Here is the question that not all the network information is suited for Chinese teaching. Unhealthy resources are everywhere. Therefore, we need to seriously select resources before using them.

Teachers should guide students to access some scientific websites when facing some Chinese course questions. It will take much time and spirit to find the necessary information. Moreover, it will increase the learning effect. The balance between autonomously online learning and course missions is emphasized to be thought.

From the network technique application, we can find out some errors. For example, the text association and imagination with multiple course wares will influence student thinking. If we depend on the network and explain the simple meaning through the network, it is unnecessary. Moreover, multimedia technology has limitation in Chinese teaching.

During the Chinese teaching process, we need to find out convenience and benefit of multimedia network technology and pay attention to the limitation. We need to take full consideration of how to correct use these resources. Network education should teach students in accordance of their aptitude. We can select key information through different files. Effectively combines modern information technology and students learning can full express network advantages, promote teaching quantity, and even train their autonomous learning ability.

4 CONCLUSION

The information globalization makes reasonable development and utilization of network resources in college Chinese teaching. It changes the old passive learning in the modern initiative study. Students can fully express their activeness. They increase their Chinese learning interests and develop the learning quantity. Under the information globalization, Chinese teachers need to follow the time changes in a positive condition. Moreover, it is necessary to express distinctive features while keeping the characteristics. The Chinese teaching will not degenerate and fall behind. Therefore, we need to take full advantage of these network resources and support powerful back up to college Chinese teaching.

REFERENCES

[1] Jia Hui. Admission against interest-discussion of network resource application during Chinese teaching [J]. Education Teaching Forum, 2011(11).

[2] Dai Jing. Rational use of network resource for Chinese classroom teaching—discussion of network resources development and utilization in Chinese teaching [J]. New Curriculum (Research Version) 2010(3).

[3] Zhang Hexin. Discussion of resource development and utilization inevitability about Chinese network course [J]. The Languge Theacher's Friend, 2010(3).

[4] Zhao Bingli. Network resource utilization in Chinese teaching. [J]. Science Weekly C, 2012(4).

[5] Zheng Yunhong. Discussion of network resource utilization in Chinese teaching [J]. Dajiang Weekly (Forum), 2011(4).

[6] Huang Hongjiao. Brief discussion of college Chinese teaching based on network [J]. The Science Education Article Collects, 2012(28).

Information Technology and Computer Application Engineering – Liu, Sung & Yao (Eds)
© 2014 Taylor & Francis Group, London, ISBN 978-1-138-00079-7

Brief discussion of students' vocational counsel idea and practice pattern

Xuan Ping Luo

Jiangxi Science and Technology Normal University, Nanchang, Jiangxi P.R. China

ABSTRACT: It is one of the most difficult problems of the modern undergraduates when they are facing the graduation. Most students have no idea about the graduation. They don't know what to do, where should go, and which kind of job is perfect to themselves. It is emphasized to develop series education work about vocational counsel principle and practice pattern.

Keywords: Vocational counsel principle; practice pattern; students;

Appropriate and reasonable guidance idea can help students to find their good qualities, disadvantages, what can be done well, and support the decision and confidence. At the same time, the perfect practice pattern can provide guidance information to the students, which includes practice problems and other questions during the further work. It is very important to establish reasonable guidance idea and practice pattern for Chinese graduates' employment. However, our vocational counsel still stands in the traditional idea. It means schools only guide students when they are selecting jobs. This kind of graduation guidance has a limitation that cannot provide effective direction and ignore the students' professional ability development. Vocational counseling places important function to the graduate employment. Therefore, some colleges need to perfect vocational counsel idea and practice patterns.

1 EMPHASIS MEANING OF REASONABLE EMPLOYMENT IDEA AND PRACTICE PATTERN OF UNDERGRADUATES

The graduation research shows, most school vocational counsel have no clear attitude. They only care about the good jobs. However, the institutions have many students, less quality. Especially the teachers are lack of interaction and practical experiences. Students cannot absorb the meaning. Therefore, it is very important to build perfect, reasonable training of employment idea and practice pattern. It can help students to select better jobs and train their physiology quantity and personal comprehensive development.

2 PROBLEMS OF VOCATIONAL COUNSEL AND PRACTICE PATTERN OF CHINA'S UNDERGRADUATES

2.1 *Misguide in job selection and employment*

America is one of the earliest countries to develop employment guidance. In America, some colleges have one theory. It is a Parsons Character — factor matching theory, which is the most basic theory among various theories. This theory points out one important idea: job selection is the matching among personal mentality, physiology character and job requirement to people. However, this theory has not reached further practice and research for the misunderstanding that people believe vocational counsel is job selection. In China, most schools provide some kind of job selection classes when they are going or facing undergraduate. The being believed help ignores the students' personal skill development and has no obvious assistance to the further employment.

2.2 *Colleges provide insufficient practice to the students*

Some Chinese colleges only pay attention to the professional knowledge of students. They ignore the practical manipulative ability. It will lead students have low quality with high requirement. The current enterprises only emphasize in proficiency and calculate the new employer proficiency. They avoid spending money and time to train and guide the new employers. What they need is to create proficiency once they enter the company with practical experience and manual ability. It is hard to calculate the practical ability of undergraduates. Therefore, this is one of the biggest malpractices during the employment.

2.3 Employment mentality problems of undergraduates

When facing the graduation, most students will feel scary. They fear cannot find good jobs, stand on the society. They feel confused to find what kind of job. A major job, or change the other jobs. Which job is suited for them? All the above problems are normal to the undergraduates. For most of them, undergraduate is out of work. The pressure of facing to the work, society, further life of independent exist is heavy to the undergraduates. Some students with bad mentality might think the positive side and even commit suicide to solve all the problems. However, death cannot solve any question. They should face what they should be. Train a good graduation mentality is very important during college life.

3 HOW TO OPERATE VOCATIONAL COUNSEL IDEA AND PRACTICE PATTERN TEACHING

Develop vocational counsel idea and practice pattern teaching to college students has no time to delay. The undergraduate of higher education related to the students' personal development and higher education image and position. Moreover, it has emphasis influence to the 'popular' realization of our higher education. The current condition shows vocational counsel is the bridge between teaching input realization and efficiency production. It is the premise condition for teaching recovery. The employment rate is one of the most important conditions to balance teaching quality. The college employment rate influences enrollment tendency, and the school development. Therefore, college guidance of vocational counsel and practice pattern becomes more and more important.

3.1 Employment idea determination

Employment guidance places all around the education period. It should be a long term planning without limitation in time and objects. Students should start the reasonable vocational counsel since they step into the colleges. Aiming at different stages, different students need to follow the mission teaching guidance. Prepare the necessary mentality, knowledge and skill and students can improve their further adoption when taking a job.

Vocational counseling helps students to find their advantages and disadvantages. Students can ensure the development direction, create correct employment ideas and avoid following blindly during the employment.

Ideology education is important during the period of vocational counsel. It can correct student mentality no matter in life and study. The combination of employment idea and ideology education can teach student ideology and employment knowledge.

3.2 Establishment of practical pattern

The practice during school life is important to the graduate employment. The perfect practice pattern can help students to find the suitable job and train their ability to develop. The college vocational counsel needs to under the theory guidance and develop in the direction of standardization and efficiency. It is necessary to deepen understanding and study of theory knowledge during the continue research and practice. Moreover, working method and experience summarization is emphasized to improve practical work level. Based on practical experience, the writer believes the practice pattern is very important.

3.3 The practical work needs to separate into different periods

Vocational counsel works through the whole college period. Different ages grasp different knowledge. Therefore, it is necessary to develop relative practice in the different stages. During the continue practice and learning, students can understand better about the textbook.

The freshman has less profession knowledge and they need to focus on the college life adaption. Realize college character and learning meaning, how to better understand profession knowledge, and try to plan and define their future is very important.

The sophomore year of college life is the most important period. No matter the major courses learning and further planning, the students need to under details as much as possible. In the sophomore year of college life, students need to store and expand the professional basic and viewpoints, evaluate personal advantages and disadvantages, make self-perfection, and try to plan the further occupation.

In the junior year of college life, students have much understanding about the profession knowledge. They need to combine personal character (mentality and physiology) and personal ability to develop in the professional field in order to train the suitable quality for further development.

In the senior year of college life, students nearly learned all the major knowledge. It is necessary to train them suitable profession value, information service, employment form, and discussion skill and mentality quality. Moreover, systematic guidance on job selection during the employment period is very important.

3.4 Enroll vocational counsel into educational planning, develop classroom teaching

We enroll vocational counsel into educational planning can make students understand more about employment. Vocational counseling class should develop from basic vocational knowledge, self-cognition, and choice of occupation. The vocation basic knowledge includes vocational introduction, vocational development tendency, and various requirements for the employers.

Self-cognition includes personal character of mind and body, superiority and defect. It helps students to operate self-evaluation and train the employment confidence. Choice of occupation includes the series problems such as graduate procedure, employment regulation, skill of job selection, and how to adapt the further operating post and so on.

4 SUMMARY

The current employment situation shows some colleges have been realized their disadvantages of students' employment. They absorb different vocational counsel and practice pattern for increasing atten-

tion and education to the student employment. These universities will create a perfect employment environment, develop employment rate, and provide contribution to the national employment.

REFERENCES

[1] Xiu Hua. Undergraduates' vocational counsel's theory and practice patter [A]. Fujian Tribune, 2005(9)
[2] Tang Yu. Innovation theory and practice strategy research of modern undergraduates'vocational counsel [J]. Policy & Scientific Consult, 2008(18)

Information Technology and Computer Application Engineering – Liu, Sung & Yao (Eds)
© *2014 Taylor & Francis Group, London, ISBN 978-1-138-00079-7*

The application of multi-agent technology in multi-motor control system

J.D. Yang, H. Wang & X.W. Han
Institute of Information Engineering, Shenyang University, Shenyang, China

ABSTRACT: This paper introduces the concept and characteristics of multi-agent and analyses the problems existing in the multi-motor control system. According to the characteristics of multi-motor coordinated control and welding production technology, we discussed the internal relations between the multi-agent system and the multi-motor coordinated control system. Then, we established the structure and mathematical models for the multi-agent system of transverse seam submerged arc welding machines' multi-motor coordinated control system.

Keywords: Multi-agent technology; Transverse seam submerged arc welding machine; Multi-motor control system

1   INTRODUCTION

With the increasing demand for welding automation, the welding automation technology is developing more and more integrating, intelligent and flexible. Consequently, there is still much room for improvement in the automation of current transverse seam submerged arc welding machine.

Currently, multi-motor synchronous control technology mainly includes parallel control, master-slave control, cross-coupling control, virtual-shaft control and deviation coupling control[1]–[5]. Including the traditional PID control method, people tend to choose more advanced control algorithm, such as neural network control, digital PID control, fuzzy control, sliding mode variable structure and H^∞ control. As can be seen from the research conclusions, multi-motor synchronous coordinated control technologies are still at an early stage of development. While, there are several key issues in the control of high performance coordination driven of multi-motor[6].

(1) The nonlinearity of AC motor
(2) The boundedness of motor output torque
(3) The mismatch between multi-motors
(4) The uncertainty of the load
(5) The nonlinearity and uncertainty of the tracking trajectory curve
(6) The feeding speed of trajectory curve

In recent years, in the field of distributed artificial intelligence, multi-agent system (MAS) has become the main content and research hot spot[7]. It is a powerful thoughtway and tool to complex system analysis and design.Multi-agent system (MAS) is made up of several Agents. As a branch of the distributed artificial intelligence, it can make the logically and physically dispersed systems become parallel and coordinate to solve problems. MAS controls through multiple agents. It isn't a single large and complex system but it is divided functionally into several agent systems according to the requirement of the control system. They can coordinate and communicate with each other to complete the control task of whole complex system. This system is not only a general distributed system which is characterized by resource sharing, high reliability, strong real-time performance and simplicity of expanding, but also has strong flexibility and robustness which are manifested by the mutual coordination and collaboration between each agent to solve large-scale and complex problems.

In the multi-motor control system of transverse seam submerged arc welding machine, each running motor is considered as an agent. Through the mutual coordination among multiple agents, the multi-motor system of transverse seam submerged arc welding machine will run coordinately, which is of great significance for improving the automation level of welding equipments. A typical agent includes the local database, reasoning and decision module, communication module and motion or behavior control module (Fig. 1).

2   INTERNATIONAL RELATIONS BETWEEN MULTI-MOTOR COORDINATED CONTROL SYSTEM AND MAS

In the multi-motor control system of transverse seam submerged arc welding machine, every motor needs to run coordinately with one another at every procedure, including the information exchange of flux recovery, wire welding, swing of welding torch and the

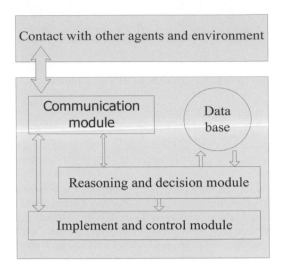

Figure 1. The typical structure of an agent.

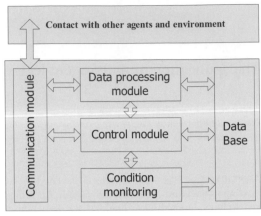

Figure 2. The structure model of single motor agent.

running of welding tractor. Consequently, the coordination among intelligent equipment, control software and the system architecture of the whole multi-motor control system is particularly important.

The fundamental principle of transforming the multi-motor coordinated control system into MAS is to make the system simple to control conveniently. Simple interactions between agents can greatly reduce the complexity of the manufacturing system. There are 4 aspects of theory and technology of MAS which could be drawn lessons from when we build the multi-agent system model of multi-motor coordinated control system.

(1) The agent formal modeling method of MAS can be used to construct the agent model of multi-motor control system or other equipment.
(2) The topological structure and organization methods of MAS can be applied to the agent system framework of multi-motor coordinated control.
(3) The communication protocols and methods of MAS provide available reference for the design of agents' communication standards and protocols as well as implementing scheme in multi-motor coordinated control system.
(4) The consultation and negotiation strategy of MAS can be employed in the multi-motor coordinated control agent system.

3 THE STRUCTURAL MODEL AND MATHEMATICAL MODEL OF MAS FOR MOTOR

3.1 The structure model of MAS for motor

A running motor can be considered as a single agent whose structural model can be divided into 5 modules (Fig. 2).

(1) Database: It is mainly to store the data associated with the motor running and the received information.
(2) Communication module: It is mainly responsible for the communication, data sending and receiving between task agent and other agents.
(3) Data processing module: It mainly acts to process the data received by communication module, and the processing results are sent to the database and control module respectively.
(4) Control module: On the basis of data from data processing module, database data and the motor running status, control module serves mainly to control the operation of motors reasonably following a certain decision strategy and algorithm and send the results to other agents by the communication interface.
(5) Monitoring module for the motor operation status: The main function of monitoring module is to monitor the running status of motors and send the data to database and control module.

Based on the MAS structure, multiple single motor agents constitute a transverse seam submerged arc welding machine multi-motor system (Fig.3) whose structure model not only shares the properties of single motor agent but also has the following characteristics.

(1) The complexity of the system reduces greatly due to the transformation of working mechanism from multi-motor control to a single motor agent running independently or multi-motor agent working collaboratively.
(2) This system model has strong adaptability to the conditions which are difficult to predict exactly in the production process. At the same time, the system model has high flexibility, that is, it is easy to increase or reduce agents when the resources or projects change.
(3) Owing to the autonomy of agent, each one can be modified separately without affecting the other agents, which results in the efficiency improvement of system operation. However, each agent

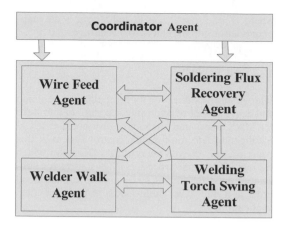

Figure 3. The structure model of multi-motor MAS.

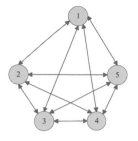

Figure 4. Directed graph of multi-motor MAS.

$$l_{ij} = \begin{cases} \sum_{j=1,2,\cdots,n} a_{ij}, i = j \\ -a_{ij}, i = j \end{cases} \quad (II)$$

In a directed graph, if there is a node whose information can be transmitted to any node in the system, then the directed graph contains a directed spanning tree. The MAS structure of multi-motor in Figure 3 is a directed graph (Fig. 4).

The motion equation of brushless DC motor is as follow

$$T_e - T_L - B\omega = J\frac{d\omega}{dt} \quad (1)$$

where T_e = electromagnetic torque; T_L = load torque; B = damping coefficient; ω = mechanical rotational speed of motor; and J = rotational inertia of the motor.

If make $\omega_i = R^q$ represent the rotational speed of the ith vertex v_i and meet $\dot\omega = f(\omega_i, u_i)$, then the network system of multi-motor's running can be represented by the two-tuples (G, ω) in which $\omega = \left(\omega_1^T, \omega_2^T, \ldots, \omega_n^T\right)^T$. Therefore, the state equation of system is as follow.

$$\dot\omega = F(\omega, u) = \frac{1}{J}\left(T_e - T_L - B\omega\right) \quad (2)$$

If all the i and j meet the following formula (3), the multi-agent system achieve consistency, in other words, the multiple motors operate synchronously.

$$\lim_{t\to\infty}\left\|\omega_i(t) - \omega_j(t)\right\| = 0 \quad (3)$$

From the formula (1), we can get the following solution.

$$\omega = \frac{t}{J}\left(T_e - T_L\right) + Ce^{-\frac{B}{J}t} \quad (4)$$

When t tends to infinity,

$$\omega = \frac{t}{J}\left(T_e - T_L\right) \quad (5)$$

If we drag it in the formula (3), then

$$\lim_{t\to\infty}\left\|\omega_i(t) - \omega_j(t)\right\| = \lim_{t\to\infty}\left\|\frac{t}{J}\left(T_e - T_L\right) - \frac{t}{J}\left(T_e - T_L\right)\right\|$$

has incomplete ability of information and problem solving. What's more, there is no overall control.
(4) This system model can be extended to other advanced manufacturing systems without modification of the model structure.

3.2 The mathematical model of MAS for motor

In order to achieve a coordinated operation of multiple motors, the control process in multi-motor system can be abstracted as the solving of the consistency of multiple motors' running status. Graph theory is an important tool for analysis of the consistency problems. It is used to represent the process of information exchange among agents in MAS. A multi-agent system can be expressed by graph $G = (V, E)$ in which the node set $V = [1, 2, \ldots, n]$ represents the independent agent and the edge set $E \subseteq \{(i, j) : i, j \in V, i \neq j\}$ represents the information exchange between them. If and only if $(v_i - v_j) \in E \Leftrightarrow (v_j - v_i) \in E$, the graph is undirected. The neighbor set of the ith agent can be denoted by $N_i = \{v_j \in V : (v_i, v_j) \in E\}$.

In graph theory, the adjacency matrix A(as formula (I)) is usually used to describe the relationship among element nodes, in which the elements are defined as follows. If the jth node has an edge point to the ith node, when $a_{ij} > 0, \forall i \neq j$, a_{ij} means the weight of edge (i, j). Otherwise, $a_{ij} = 0$ and the diagonal elements $a_{ii} = 0$. In particular, when the value a_{ij} is not only 0 and 1, the graph of this sort is called weighted graph. The Laplace matrix L is another matrix to describe the topological structure of graph G. It has a similar structure with the adjacency matrix. The elements are defined as L (as formula (II)). For undirected graph, the Laplace matrix L is a symmetric matrix.

$$A = \begin{bmatrix} a_{11} & \cdot & \cdot & \cdot & a_{ni} \\ \cdot & \cdot & & & \cdot \\ \cdot & & \cdot & & \cdot \\ \cdot & & & \cdot & \cdot \\ a_{1n} & \cdot & \cdot & \cdot & a_{nn} \end{bmatrix} \quad (I)$$

Because the formula meets the condition of formula (3), multiple motors of MAS can achieve the synchronous operation.

In this paper, only to consider the speed of brushless dc motor. In other words the system is a first-order multi-agent system. In order to achieve the coordination of five motor running. If we make $\dot{\omega} = u_i (1, 2, \ldots, 5)$ and utilize the control law[8] as formula (6), the whole system can be described as matrix (6).

$$u_i = -\sum_{j=1}^{n} a_{ij}\left(\omega_i - \omega_j\right), \quad (i = 1, \cdots, 5) \qquad (6)$$

$$\begin{cases} \dot{W} = U \\ U = -LW \end{cases} \qquad (7)$$

where $W = [\omega_1, \omega_2, \ldots, \omega_5]^T$ and $L = [l_{ij}]$. At the same time, we can get the adjacency matrix A and Laplace matrix L from Fig. 4 by definition.

$$A = \begin{bmatrix} 0 & 1 & 1 & 1 & 1 \\ 1 & 0 & 1 & 1 & 1 \\ 1 & 1 & 0 & 1 & 1 \\ 1 & 1 & 1 & 0 & 1 \\ 1 & 1 & 1 & 1 & 0 \end{bmatrix} \quad L = \begin{bmatrix} 5 & -1 & -1 & -1 & -1 \\ -1 & 5 & -1 & -1 & -1 \\ -1 & -1 & 5 & -1 & -1 \\ -1 & -1 & -1 & 5 & -1 \\ -1 & -1 & -1 & -1 & 5 \end{bmatrix}$$

3.3 Model description

In a distributed multi-agent cooperation control system with 5 brushless DC motor, we can consider mechanical speed as the only variable and the others constant in the seeking of multiple motors' coordinated operation, which can be derived from the dynamic equation (1). Namely, we should consider the system as a first-order MAS. To achieve the coordinated operation of 5 motors, we must make each motor's status change on the basis of the status of neighbor motors and itself. Therefore, the system is autonomous to ensure that each or all status variables of every agent tend to equal.

this model it adopts the distributed control law, and each motor state changes depends on its and its neighbors current state, so the system is an autonomous system, make some or other quantity of state or all quantity of state tend to be equal of all the agents. The dynamic model can be expressed by following formula.

$$\dot{W} = \frac{1}{J}\begin{bmatrix} T_e \\ T_L \\ B \end{bmatrix}^T \begin{bmatrix} 1 & 0 & 0 \\ 0 & -1 & 0 \\ 0 & 0 & -\omega \end{bmatrix} = f(\omega, u)$$

where $W = [\omega_1, \omega_2 \ldots, \omega_5]^T$ and $f(\omega, u)$ is the mapping relationship of motors' status change.

4 CONCLUSIONS

It provides a new direction for the application, research and development of multi-motor control to introduce the MAS theory and methods of distributed artificial intelligence into multi-motor control system. In this paper we only discuss the brushless DC motor which we regard as a first-order agent. Thus, there is a substantial need for further study aiming at the second order and high level multi-motor MAS in practical application.

REFERENCES

[1] Masayoshi, Tomizuka & Hu, J-S & Chiu, T-C & Kamano, Takuya. 1992. Synchronization of two motion control axes under adaptive feed forward control. Journal of Dynamic Systems Measurement and Control 114(2): 196–203.

[2] Yoram, Koren. 1980. Cross-coupled biaxial computer control for manufacturing systems. Journal of Dynamic Systems Measurement and Control 102(4): 265–272.

[3] Robb, G. Anderson & Andrew, J. Meyer & M. Anibal Valenzuela & Robert, D. Lorenz. 2001. Web machine coordinated motion control via electronic line-shafting. IEEE Transactions on Industry Applications 37(1): 247–254.

[4] Francisco, J. Perez-Pinal & Gerardo, Calderon & Araujo-Vargas, I. 2003. Relative coupling strategy. In Electric Machines and Drives Conference, 2003. IEMDC 2003. IEEE International (2): 1162–1166.

[5] Francisco, J. Perez-Pinal & Ciro, Nunze & Ricardo, Alvarez & Ilse, Cervantes. 2004. Comparison of multi-motor synchronization techniques. In Industrial Electronics Society, 2004. IECON 2004. 30th Annual Conference of IEEE (2): 1670–1675.

[6] Wenhuan, Sun & Shanmei, Cheng & Xiaoxiang, Wang & Yi Qin. 1999. The development of multi-motor coordinated control. Electric Drive (6):51–55.

[7] Dumodn, Y. & Roche, C. 2000. Formal specification of a multi-agent system architecture for manufacture: the contribution of the π-calculus. Journal of Materials Processing Technology 107(1):209–215.

[8] Olfati-Saber, Reza & Richard, M. Murray. 2004. Consensus problems in networks of agents with switching topology and time-delays. IEEE Transactions on Automatic Control 49(9): 1520–1533.

Information Technology and Computer Application Engineering – Liu, Sung & Yao (Eds)
© 2014 Taylor & Francis Group, London, ISBN 978-1-138-00079-7

A study on micropayment system based on self-updatable two-dimensional hash chain

Yan Ling Huo
Department of Information Engineering, Xingtai Polytechnic College, Xingtai, Hebei, China

Hai Ying Li
Department of Information Science and Technology, Xingtai University, Xingtai, Hebei, China

Hai Bin Wang
Department of Information Engineering, Xingtai Polytechnic College, Xingtai, Hebei, China

ABSTRACT: In view of the existing problem of micropayment systems, that is, users need to use the existing nodes to combine the electronic cash, and the efficiency of initializing the Hash chain is low, the thought of the binary tree is used to present a new micropayment system based on self-updatable two-dimensional Hash chain. It values specific evaluation to the node, thus the user can complete the transaction by Log2N times Hash calculation (N is the total amount of electronic cash users spend). It also uses the self-updatable method of the multi-Hash chain to avoid the reinitializtion, so as to promote the efficiency of micropayment system greatly.

Keywords: micropayment; Hash chain; self-updatable; one-time signature

1 INTRODUCTION

With the rapid development of Internet and mobile communication, more and more people begin to buy electronic products through the Internet and mobile phone, usually, only small electronic cash payment is used in shopping, for example, when browsing certain webpage, the users only need to pay a few cents to businesses, if we still use the original macro payment mode in the transaction, it will not embody the characteristics of micropayment. If we use the RSA signature technology to guarantee the security of the transaction, according to reference [2], we can see that the Hash function is 100 times faster than the RSA signature, and about 1000 times faster than the RSA signature generation, so the transaction efficiency will be low in the use of RSA to ensure the security of transactions.

Relative to the macro payment, Micro-payment has the characteristics of small denominations payment and high efficiency, and the Hash function exactly has the characteristics of high efficiency and security, so the Hash function is used to ensure the safety of the transaction. Hash chain is proposed by Lamport [1], although it was originally designed to protect the one-time password from wiretapping and play backing, it was used to micropayment schemes soon for the high efficiency and the similar technology of the public key. Rivest and Shamir [5] proposed the Payword scheme, in this scheme, the users carried out Hash calculation many times and then formed a Hash sequence.

Each payment of the users needs to submit values in the Hash sequence to businesses in reverse order of the Hash chain calculation. But the scheme above is based on the one-dimension, when the users consume with the micropayment system, if each node represents ten cents, then one dollar will need ten times hash computation to complete the transaction. Considering this shortcoming of one-dimensional Hash, Quan Son Nguyen [2] proposed a scheme of multi-dimensional Hash chain based on RSA algorithm, the micropayment system contains a variety of denominations and the transaction efficiency has improved, but there is a problem that users need to use existing nodes to combine the electronic cash, and the efficiency of reinitializing the Hash chain is low. Liu Yining and Li Hongwei [3] pointed out that the scheme of Quan Son Nguyen is not feasible, and they put forward a new Hash structure, but the two-dimensional Hash chain proposed by Liu Yining and Li Hongwei does not have the property of self-update, so when the Hash chain is depleted, only the public key signature technology is available to initialize Hash chain, which resulted the low efficiency of the micropayment. In all of the micropayment system above, the electronic cash needs to be combined in accordance with the existing denominations when transactions happen. In the micropayment system based on multi-dimensional, the Hash chain needs to be initialized when it is depleted, and the public key signature technology which has low efficiency will be used in the initialization.

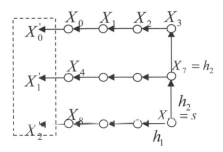

Figure 1. The structure of 2D Hash chain.

This article valued specific evaluation to two-dimensional Hash chain proposed by Liu Yining and Li Hongwei [3] with the binary tree thought, and borrowed the structure of renewable Hash chain proposed by Zhao Yuanchao and Li Daoben [4], which made the micropayment system based on 2D Hash structure not only can automatically update the multi-dimensional Hash chains, but also can complete the transactions between the users and businesses with the help of binary tree thought. During the use of a multi-dimensional Hash chain, the public key signature technology was needed only once, so it can reduce the additional cost caused by updating the Hash chain.

2 PRELIMINARY KNOWLEDGE

2.1 Symbols

h, h_1 and h_2 represent the hash function, the specific use of SHA-256 is viewed as secure Hash function, $\|$ represents the connection string, and r_1, r_2 and r_i represents random number.

2.2 The introduction of the 2D Hash chain

Hash chain is obtained by a recursive algorithm of a public function h which is called "one-way Hash function." Among them, an arbitrary length (or a pre-determined maximum length range) bit string will be mapped to a fixed length bit string by h. Moreover, h has 3 properties: (1) given x, easy to calculate $h(x)$; (2) given $h(x)$, the calculation of x is not feasible; (3) we find two value x and y, and $x \neq y$, the calculation of $h(x) = h(y)$ is not feasible.

When two-dimensional Hash chain was constructed, a random seed s was selected firstly, then repeat the Hash calculation of s by Hash function h_2, and generate seeds X_7, X_3, finally using Hash function h_1, h_2 to generate root which is corresponding to seed. Figure 1 is the structure:

Three root generated in Figure 1, $X_0' = h_1^4(h_2^2(s))$, $X_1' = h_1^4(h_2(s))$ and $X_2' = h_1^4(s)$. Thus users expose X_0, X_1 etc. to businesses, and allow businesses to test the correctness of transactions.

2.3 One-time signature technology

One-time signature was firstly proposed by Lamport [1], the main idea is: sign on 1bit message: select two random numbers x_0 and x_1 as a private key representing '0' and '1' accordingly: $y_0 = h(x_0)$ and $y_1 = h(x_1)$ are calculated by one-way Hash function, and y_0 and y_1 are released as a public key; when signing on 1 bit message, if the message is '0', then open x_0, the receiver of the message calculate $h(x_0)$, and make a comparison between the numerical results and y_0, if the two are the same it is verified; if the message is '1', then open x_1, the receiver of the message calculate $h(x_1)$, and make a comparison between the numerical results and y_1, if the two are the same it is verified. Signature on n message were required to prepare $2nx$ and y

Merkle improved the method above: only one random number $x_i(i = 1, 2, \ldots, n)$ is generated for each bit of n bit, calculate $y_i = h(x_i)$ $(i = 1, 2, \ldots, n)$, and release all y_i; when the i bit is '1', open x_i, if it is '0', no value is opened. In order to prevent the recipient who received '1' from pretending to receive '0', the signer also need to sign on the number of '0' in the messages, and the signing method is just the same as it is on the message body. Therefore, according to Merkle, the signature on n bit were required to prepare $n + \lceil \log_2(n) \rceil$ x and y.

2.4 The self-updatable property of multi-dimensional Hash chain

The self-updatable property of multi-dimensional Hash chain refers to: the users randomly select 1 seed s_1, then generate $M - 1$ seed, they are, $s_2 = h_2(s_1), \ldots, s_{M-1} = h_2^{M-2}(s_1)$ and $s_M = h_2^{M-1}(s_1)$. And then generate M root accordingly, they are, $h_1^4(s_1), h_1^4(h_2(s_1)), \ldots, h_1^4(h_2^{M-2}(s_1))$ and $h_1^4(h_2^{M-1}(s_1))$. At the same time, prepare a pair of one-time signature key example to the message with length of L bit, concatenation including $L + \lceil \log_2(L) \rceil$ cascade connection S_U of random numbers and $L + \lceil \log_2(L) \rceil$ cascade connection P_U of the Hash function value corresponding to the random number, S_U could be looked at as private key elements of one-time signature and P_U could be looked at as public key elements of one-time signature. The flag, $h_1^4(h_2^{M-1}(s_1))$, $\ldots, h_1^4(h_2(s_1)), h_1^4(s_1), CR = h(flag, h_1^4(h^{M-1}(s_1)), \ldots h_1^4(h_2(s_1)), h_1^4(s_1), h(P_U))$ the flag is a M bit long binary string, the i is used to mark whether the start send the i Hash chain to the business, the correctness of data is tested by CR. Now the users can use the Hash chain like normal transactions. When the Hash chain is depleted, for example, the node in the i and $i - 1$ chain is depleted, then the number of Hash chain needed to be started is two. The users need to choose 2 random number r_1 and r_2, and make the 2 random number hash with s_i and s_{i-1} respectively, then get $s_i' = h_2(r_1\|s_i)$ and $s_{i-1}' = h_2(r_1\|s_{i-1})$, moreover, generate a pair of new one-time signature key S_U' and P_U'. Then calculation of new composite root: $CR' = h(flag', s_i', s_{i-1}', h(P_U'))$, now, users send P_U (not P_U') to the business, the opening part required by CR' signature is S_U (not S_U') and CR'. The business verified P_U by $h(P_U)$ received in last startup, and verified CR' by the opening part of S_U. Of course, $flag'$, $h_2(r_1\|s_i)$, $h_2(r_1\|s_{i-1})$ and $h(P_U')$ should

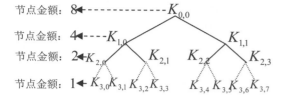

节点金额: 8◄- - - - - - - - - - - - $K_{0,0}$

节点金额: 4◄- - - -$K_{1,0}$ $K_{1,1}$

节点金额: 2◄$K_{2,0}$ $K_{2,1}$ $K_{2,2}$ $K_{2,3}$

节点金额: 1◄ $K_{3,0}$ $K_{3,1}$ $K_{3,2}$ $K_{3,3}$ $K_{3,4}$ $K_{3,5}$ $K_{3,6}$ $K_{3,7}$

Figure 2. The binary tree description of e-cash.

be sent to businesses, and businesses will test the correctness of data by calculating CR'. Then the users can use the new generated chain in consumptions without initializing of all chain.

2.5 *The thought of binary tree*

Divisibility of e-cash is shown in Figure 2, we can know from the Figure 2 that the electronic cash 80 cents could be shown in the way of electronic cash 40 cents, 20 cents and the 10 cents, and this is also the binary display of electronic cash 80 cents. So when the users spend e-cash in certain denomination, the binary show will be used first to represent the electronic cash, and then spend e-cash in the layer where the binary of that bit is 1. For example, the user wants to spend electronic cash 30 cents, the binary of 3 is displayed as 101, and then the user will know that the nodes in second layer and fourth layer could be used to complete the transaction.

This paper values specific evaluation to the node in two-dimensional Hash chain proposed by Liu Yining and Li Hongwei, borrowing the thought of binary tree. Thus, the users can divide e-cash according to the thought of binary tree, because any number in the computer is used in binary representation, then it is very fast to use the computer to determine how many times the hash computation is needed. That is, the users only need to carry out Log_2N Hash calculation to complete transaction.

3 THE STRUCTURE OF TWO-DIMENSIONAL HASH CHAIN AFTER THE ASSIGNMENT.

The assignment to the two-dimensional Hash chain makes each node in Hash chain represent electronic cash, $X_0 = 1$ represents the electronic cash 10 cents. The two-dimensional Hash chain after the assignment is shown in Figure 3, $X_0' = h_1^4(h_2^2(X_N))$ (X_N is the random seed of X_{11} selection) is the root node of the first line, $X_1' = h_1^4(h_2(X_N))$ is the root node of the second line, and $X_2' = h_1^4(X_N)$ is the root node of the third line. The dashed parts in the diagram are the root node which will be sent to the business to verify rather than the Payword value used in micropayment system.

The process of two-dimensional Hash chain applied in micropayment is as follows:

(1) The users choose X randomly and make it a secret, and calculate three roots correspondingly $h_1^4(h_2^2(X_N))$, $h_1^4(h_2(X_N))$, $h_1^4(X_N)$, at the same time,

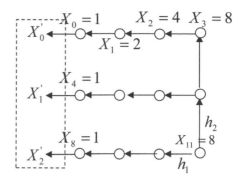

Figure 3. The two-dimensional Hash chain after the assignment.

prepare $L + \lceil \log_2(L) \rceil$ cascade stage S_U of random number to the message with length of L bit, and prepare $L + \lceil \log_2(L) \rceil$ cascade stage P_U of Hash function value corresponding to the random number. Actually, the instance of the one-time signature key is not truly used to the Hash chain initiation, but for the next regeneration of Hash chain. The users need to send flag, $h_1^4(h_2^2(X_N))$, $h_1^4(h_2(X_N))$, $h_1^4(X_N)$, $CR = h(flag, h_1^4(h_2^2(X_N))$, $h_1^4(h_2(X_N))$, $h_1^4(X_N)$, $h(P_U))$ ① to the businesses, where *flag* is a 3bit long binary string, the i bit of *flag* is used to mark whether the i Hash chain is sent to businesses in this startup, because it is the first start of the Hash chain, so the value of flag must be 1 series, namely 111, it means all Hash chains are started, after receiving, the businesses test the correctness of data by CR.

(2) When he user sends first request of spending 20 cents, the amount is described by binary representation as 10, when the bit is 1, it means the node in this bit is spent by consumer. because the second bit is 1, the user sends $(X_1, 1)$ and $h(10||X_1||r_1)$ to businesses, at the same time, the user sends $h(10||X_1||r_1)$ to the bank which is used to test the message sent by businesses to see whether businesses fake users' spending. When businesses received it, just need to verify $h_1^2(X_1) = X_0'$ to judge whether the request is issued by the users. Of course, businesses need to record each consumption of users in the database, namely: the binary description of the amount of money 10, $(X_1, 1)$ and $h(10||X_1||r_1)$, finally, in the last settlement between business and bank, these are sent to the bank, the bank checks the correctness of data, then the final macro payment will be used in the transaction between businesses and users.

(3) When the nodes in the chain of users are depleted, for example, the nodes in second chain are depleted, and then the number of Hash chain which needs to be started is one. The user needs to choose 1 random numbers r_i, then hash the random number and the original seed and get new seeds $X_7 = h_2(r_i||X_7)$ and generates a pair of new one-time signature key examples S_U' and P_U'. Then

517

the calculation of new composite root: $CR' = h(flag', h_2(r_i||X_7), h(P'_U))$② Now the users send to P_U (not P'_U) to the business, the opening part required for CR' signature is S_U (not S'_U) and CR'. The business verified P_U by $h(P_U)$ received in last startup, and verified CR' by the opening part of S_U. Of course, $flag', X_7, h(P'_U)$ should be sent to businesses, and businesses will test the correctness of data by calculating CR'. Then the users can use the new generated chain in consumptions without initializing of all chain.

In this case, when users spend small denomination e-cash N, Hash calculation can be used to complete the transaction, instead of using the existing denominations to combine the electronic cash of users. At the same time, because of the self-updatable property of the multi-dimensional Hash chain, users will be able to update the Hash chain without using the public key cryptography technology when the Hash chain is depleted. So, as long as the user uses a public key signature technology to generate all the chain, the chain can be self renewal and available for users in consumption, which improve the efficiency of the system.

4 PERFORMANCE ANALYSIS

This paper aimed to use the Hash function to ensure the characteristics of micropayment, that is, small denomination and high efficiency.

4.1 *Analysis of correctness*

The correctness of the system is validated by the data sent to business, it also verified the correctness of equality ① and the equality ②. equality ① and the equality ② are realized by the Hash function, and this is correct, so businesses will be able to carry out Hash with users' data, and contrast it with the results of equality ① and the equality ②, if they are the equal one, it means the data from users is correct, then transactions are dealt with, if they are not the equal one, the businesses will refuse the transactions.

4.2 *The unforgeability*

Because the Hash function is used in sending date from users to businesses as well as the business to the bank, then anyone who wants to forge users or businesses must find the counter image of Hash function, however it is impossible, we can know this by the properties of Hash function, so this system has unforgeability.

4.3 *Analysis of efficiency*

Due to the specific assignment of the existing two-dimensional Hash chain, the small denomination electronic cash needs little calculation done by Hash. Moreover, the initialization of the Hash chain will be done efficiently with the self-update technology, and the public-key cryptography technology is no longer needed.

5 SUMMARY

In this paper, the two-dimensional Hash chain proposed by Liu Yining and Li Hongwei was improved, so that the user can take the Log_2N hash computation in consumption, and the two-dimensional Hash chain can do the self-update, thus, when the Hash chain is exhausted, the users' Hash chain can update by itself without using the public key signature so as to improve the efficiency of the micropayment system.

REFERENCES

[1] Lamport L. Password authentication with insecure communication. Communications of the ACM, 1981, 24(11): 770—772.
[2] Quan S.N. Multi-dimensional hash chains and application to micropayment schemes [C]. Proc. of International Workshop on Coding and Cryptography, Bergen, Norway, 2005.
[3] Yining Liu, Hongwei Li, Jinbing Tian. Payword Scheme Based on Two-dimensional Hash Chain. Computer Engineering 2006, 32(23): 34-36.
[4] Yuanchao Zhao, Daoben Li.A novel Construction of renewable Hash chain Journal ofElectronics & Information Technology. 2006, 28(2): 299–302.

Information Technology and Computer Application Engineering – Liu, Sung & Yao (Eds)
© 2014 Taylor & Francis Group, London, ISBN 978-1-138-00079-7

Formulating of recipe for fireproof coatings via DOE based on statistical software package

Jian Hu, Zhi Bin Wang, Wen Zhou & Yan Rong Dou
Sichuan Vocational College of Chemical Technology, Luzhou, China

ABSTRACT: Many experiments in research and development in the paint preparation involve mixture components. These are experiments with mixtures in which the experimental factors are the components of a mixture and the response variable depends on the relative proportion of each components, but not on the absolute amount of the mixture. This paper presents a D-optimal design methodology for computer assisted experimental design for fireproof coating formulations involve mixture components, exemplifies the benefits of using Design of Experiments (DOE) together with statistical software package to facilitate the formulating of recipe for structural steelworks. Goal of this paper is to encourage greater utilization of these techniques in paint preparation research and development.

Keywords: Computer Assisted Experimental Design; Design of Experiments; Formulating of recipe; Fireproof Coating

1 GENERAL INTRODUCTIONS

1.1 *Mixture experiments*

Many products are mixtures of several components. A mixture experiment involves mixing the components in various proportions within an experimental region and observing the response for each mixture (Piepel et al. 2009). Experiments to study mixtures of several components occur commonly in a variety of coatings applications. The relative proportion of each components included in the composite is thought to influence the overall product characteristics and hence is of importance.

1.2 *Fireproof coatings*

The use of fireproof coatings is one of the easiest, one of the oldest and one of the most efficient ways to protect materials against fire (Vandersall 1971). Some responses considered in fireproof coatings applications include fire-resisting time, corrosion resisting property, water-resistance and rheological behavior etc. The proportions can be measured as proportions by weight or volume, depending on which is most natural to the process. Fireproof coating formulations are commonly comprised of many components, including active fire-retardant ingredients as well as inactive coalescent which influence general product characteristics and improve construction performance. Finding the optimal combination of components to produce a desirable product is made more efficient by the effective use of designed experiments and, multivariate analysis of the resulting data based on statistical software package (Jian et al. 2012). In this paper, we consider different models for the mixture components and select an appropriate design for a particular model, the analysis suitable for the experiment resulting from the design will be considered.

2 PREPARATION OF FIREPROOF COATINGS

In our prior experiment, fireproof coatings are usually composed of three active ingredients: an acid source (generally ammonium polyphosphate-APP), a carbon source (such as pentaerythritol-PER) and a blowing agent (most often melamine-MEL) linked together by a binder. In additional, nano-LDH is used as a new generation of flame-retardant fillers and the nano-TiO_2 as a nanometer filler in the formulation. APP, MEL, PER, LDH, TiO_2 and silicone-acrylate emulsion and water were mixed by high-speed disperse mixer. The prepared coatings were applied onto the surface of a steel plate (Q235 carbon steel, 15 cm × 7 cm × 6 mm) and dried at room temperature. This process was repeated 10–15 times until the dry film thickness of 3 ± 0.1 mm was obtained.

3 PERFORMANCE TESTING OF FIREPROOF COATINGS

3.1 *Fire protection test*

Fire protection test was an examination of heat insulation of flame-retardant coatings. The coated plates

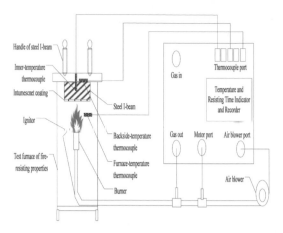

Figure 1. Experimental apparatus for fire protection test.

were exposed to gas blowlamp flame whose temperature increased in accordance with the standard temperature–time curve (ISO 834 curve) (Duquesne et al. 2005). Three thermocouples were placed on the backside of the test plate and the back temperatures of the plate were recorded. When the average of three back temperatures of the test plate reached 300°C, the time was defined as fire-resisting time.

3.2 Experimental apparatus

The experimental apparatus of measuring fire-resisting time is shown in Figure 1. To detect steel substrate temperature, the inner-temperature thermocouple tip was set in the centre of the steel I-beam and at half the thickness, thus ensuring good thermal contact of the thermocouple with the steel plate and providing a reliable average temperature of steel along its thickness.

4 FUNDAMENTALS OF MIXTURE EXPERIMENT

4.1 Multi-component mixture equations

Mixture experiments, of which the property studied depends on the proportions of the components present, but not on the amount of the mixture. In a q-component mixture ($q \geq 3$) let x_i be the proportion (by volume, or by weight, or by moles, etc.) of the jth component in the mixture, when expressed as fractions of a mixture, the proportions are nonnegative and sum to one or unity, so that

$$x_i \geq 0 (i = 1,2,...,q), x_1 + x_2 + ... + x_q = 1 \quad (1)$$

The factor space is thus a regular $(q-1)$-dimensional simplex (triangle for q = 3, tetrahedron for q = 4) (Cornell 2011).

4.2 Response surface methods

As with standard response surface methods, common model choices are based on lower-order polynomials

to model the relationship between the factors and the response μ. However, the constraint in Equation 1 leads to some modification of the basic models to allow for unique estimation of the model parameters. The standard first-order linear model used in response surface methods, see for Equation 2 below:

$$\mu = \beta_0 + \beta_1 x_1 + \quad + \beta_k x_k \quad (2)$$

does not have unique estimates for the parameters, $\beta_0, \beta_1, ..., \beta_k$. The preferred solution to this overparameterized model is to remove the intercept term, β_0, since its usual interpretation as the value of the response when all factors are set to zero has no meaning under the constraint in Equation 1. Hence the first-order model for mixtures has the form below:

$$\mu = \beta_1 x_1 + ... + \beta_k x_k \quad (3)$$

and the second-order medel where β_{ii} is the curvature term of independent variable

$$\mu = \beta_0 + \sum_{i=1}^{k} \beta_i x_i + \sum_{i=1}^{k} \beta_{ii} x_i^2 + \sum_{\substack{i=1 \\ i<j}}^{k} \sum_{j=1}^{k} \beta_{ij} x_i x_j + \varepsilon \quad (4)$$

needs to be modified to compensate for the constraint in Equation 1. The pure quadratic terms, $\beta_{ii} x_i^2$, are redundant, and hence can be removed from the model, leading to the general form,

$$\mu = \sum_{i=1}^{k} \beta_i x_i + \sum_{\substack{i=1 \\ i<j}}^{k} \sum_{j=1}^{k} \beta_{ij} x_i x_j + \varepsilon \quad (5)$$

and β_{ij} is the interaction coefficient between variables x_i and x_j. k is the number of factors and x_i are the coded variables. (ε) represents the statistical random error in (μ) that are mutually independent in the statistical sense, often assuming it to have a normal distribution with mean zero and common variance σ^2 (Myers 2009).

5 MULTIVARIATE DESIGN VIA DOE

Design of Experiment (DOE) is a structured, organized method that is used to determine the relationship between the different components affecting a process and the output of that process (response) through observance of forced changes made methodically as directed by systematic tables. There are four interrelated steps in building a DOE (Trifkovic1 et al. 2010): 1) Defining an objective of the study, e.g., better understanding of the system, sorting out important variables, or finding an optimum response. 2) Defining variables that will be manipulated during the experiments (components) and their levels or ranges of variation. 3) Defining variables that will be measured to describe the outcome of the experimental runs (responses).

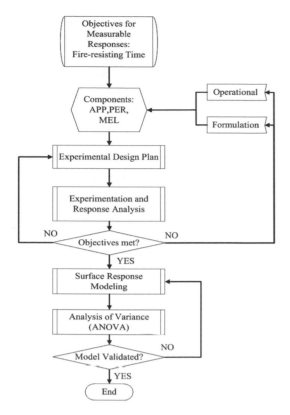

Figure 2.   Flow diagram of DOE.

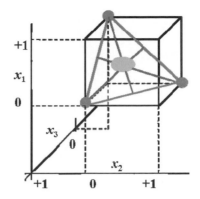

Figure 3.   Three-dimensiona restricted design space for 3-component mixture.

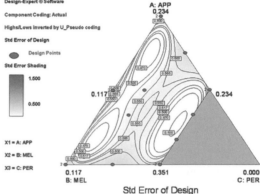

Figure 4.   Design and standard error of D-optimal design for the example.

4) Choosing one standard design that is compatible with the objective, number of components, and precision of measurements and has a reasonable cost. Figure 2 summarizes the DOE approach used in this study. The DOE objective is to obtain optimal fire protective properties of coatings, operational components are related to the protective properties of the coatings.

6   A CASE OF MIXTURE EXPERIMENT

In paint industry, the fireproof coatings are to be formulated mainly from six ingredients, the silicone-acrylate emulsion resin as a binder which is fixed at 31.5% of the mixture, the nano-TiO_2 as a nanometer filler and nano-LDHs as nano-flame retardant which are both fixed at 5% respectively, a APP x_1 (used to be an acid source), a PER x_2 (used to be a carbon source), and a MEL x_3 (used to be a blowing agent). Since the resin a binder and nano-TiO_2 and nano-LDHs have fixed percentage in the mixture, no experimentation with these components is required. Figure 3 shows the general design space for the three variable components.

Typically we wish to obtain some measure of pure error and lack of fit for the process. Pure error measures how much variability is expected if the same experimental set-up is run multiple times. Lack of fit

checks the adequacy of the model and provides feedback as to whether additional terms are required in the model. The statistical software package, Design-Expert® 8.0 (Stat-Ease, Inc.), can construct optimal designs for a variety of constrained and unconstrained mixture regions and models.

If we did not have constraints on the three variable components beyond their total proportions, the ideal design for the second-order model would be the simplex design. To obtain this design, under the 'Mixture' option, we would request a simplex centroid design with three components. Since we have the components fixed at 41.5% we would set the remaining variable components to sum to 0.585. Using the recommended sample size in Design-Expert®, we allocate four degrees of freedom to lack-of-fit testing and four degrees of freedom for estimating pure error.

However in our case, there are additional constraints on the ingredients, which leads to a much more restrictive region of experimentation. Figure 4 shows the design that was obtained by requesting a *D-Optimal* design from Design-Expert® with 13 runs for the restricted region. Again, we use the recommended sample size, which allows for assessment of both lack of fit and pure error.

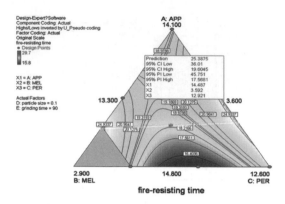

Figure 5. Contour plot of the reduced fitted model with particle size 0.1 mm and grinding time 90 min.

Two approaches are considered for building the design under the *Combined option*. The first, an *User defined approach*, that user can specify which mixture blends to run and which process point to run. A second approach is to generate an *Optimal design* in the combined mixture–process space. For a typical approach selected we generate a D-optimal design in the combined mixture–process space in Design-Expert® with the recommended number of lack-of-fit and replicate runs (5 and 5), for a total of 34 runs.

In Figure 5, process variables are set at their low levels. A numerical optimization can be performed to find the mixture–process combination with the highest fire-resisting time values. In this case, we could find the maximum graphically. A coating formulation with 14.5% APP, 3.6% MEL, 12.9% PER, and particle size 0.1 mm will yield a fireproof coating with a predicted fire-resisting time of 25.38.

7 CONCLUSIONS

Experiments involving mixture components of fire retardants arise frequently in the fireproof coatings development. A statistical experimental design is an effective approach to these problems. However, because of the specialized nature of experimental problems involving mixtures, standard designs such as orthogonal design are often ineffective. We presented a example of multivariate design based on D-optimal design and illustrated how experimental designs for these problems can be developed. Modern computer software such as statistical software package can greatly facilitate the design, execution and subsequent analysis of these experiments. This design is especially practical when the investigator is faced with large number of factors and is unsure which settings are likely to produce optimal or near optimum responses.

REFERENCES

Cornell, J. A. 2011. A retrospective view of mixture experiments. *Quality Engineering* 23(4): 315–331.
Duquesne, S. et al. 2004. Intumescent paints: fire protective coatings for metallic substrates. *Surface & Coatings Technology* 180–181: 302–307
Jian, H. 2012. Mixture experiments design including interactions with process variables for applications in paint preparation. *Advanced Materials Research* 418–420: 965–971.
Myers, R.H. & Montgomery, D.C. 2009. *Response surface methodology: process and product optimization using designed experiments*. New York: John Wiley & Sons.
Piepel, G.F. & Cooley, S.K. 2009. Automated method for reducing scheffé linear mixture experiment models. *Quality Technology & Quantitative Management* 6(3): 255–270.
Trifkovic, M. et al. 2010. Experimental and statistical study of the effects of material properties, curing agents, and process variables on the production of thermoplastic vulcanizates. *Journal of Applied Polymer Science* 118: 764–777.
Vandersall, H.L. 1971. Intumescent coating systems, their development and chemistry. *J. Fire & Flammability* 2: 97–140.

Information Technology and Computer Application Engineering – Liu, Sung & Yao (Eds)
© 2014 Taylor & Francis Group, London, ISBN 978-1-138-00079-7

A new synthesis method of N-ethyl-m-methyoxy aniline

Lin Mei, Jiu Wang, Shi Yuan Song & Yong Gang Shi
Department of Petro-chemistry, LEU of PLA, Chongqing, China

Yu Hong Sun
Department of Basic Science, LEU of PLA, Chongqing, China

Ke Zhang
Foreign Language College, Chongqing University, Chongqing, China

ABSTRACT: N-ethyl-m-methyoxy Aniline was synthesized by two reaction steps, and the yield has been risen to 48.8%, while m-aminophenol was used as material, ethyl bromide as ethyl reactant, dimethylsulfate as methyl reactant, and synthesis process was optimized by means of changing solvent and feeding appropriate phase-transfer catalyst etc. The synthesis process has less reaction steps, and its material is cheap and can be purchased easily. The synthesis reaction condition is extremely moderate and the method is very simple. Thus it provides scientific basis for industrial manufacturing.

Keywords: synthesize; m-aminophenol; dimethylsulfate; N-ethyl-m-methyoxy Aniline

1   GENERAL INSTRUCTIONS

In recent years, nitrogenous drug intermediates are growing rapidly that can be synthesized many sorts of drug. N-ethyl-anisidine is an important drug intermediates which can be widely used in intermediates of pharmaceutical industy, surfactant and textile auxiliary. Now N-alkylation method always adopt, but the resource of its initial raw materials are difficult to obtain, Production costs is high, meanwhile prcess route is high temperatue and high pressure, and response condition is harsh. This resarch is based on the refference [1–4] and cheap raw material of m-aminophenol, after its experience of two reaction steps as N-enthylize and O-methylation, under extremly temperate reaction condition, object production was gained. Its chemical reaction mechanism is as below:

2   EXPERIMENT

2.1   *Instruments and reagents*

Fu Liye infrared spectrometer 550 II (USA, LiGaoni); counter balance TM-100B; JJ-1 precision electric mixer; electric heating constant temperature water bath (SenXin experimental apparatus Co., Ltd. Shanghai); ammonia phenol (CP, Beijing Chemical Plant); dibromoethane (CP, Shanghai Chemical Reagent Factory); dimethyl sulfate (CP, Xinhua Active material Research Institute).

2.2   *The experimental procedure*

2.2.1   *N-ethyl-m-aminophenol synthesis*

Add 1.50 g of m-ammoniophenol and 5 mL of solvent water in a 50 mL round-bottomed flask, then add bromoethane according to 1:1.1 mole ratio, add a small amount of polyethylene glycol, control temperature $30°C \pm 0.5°C$ with continuously stirred, after reaction for 4 h, stewing, pass the night for use.

2.2.2   *N-ethyl-m-methyoxy Aniline synthesis*

Add 10% sodium hydrate into the stewed aqueous solution till solution shows neutral, then heated it to 80°C, remove unreactive bromoethane, after 15 min later, add appropriate sodium dodecyl benzene sulfonate, and dropwise dimethyl sulfate, when reaction end, solution shows red. Add concentrated hydrochloric acid till pH is 2–3, add brand new Sodium nitrite solution, get yellow solid matters, filtered it, add 3% dilute hydrochloric acid 4 mL into solid matters, heated it to 65°C, then can obtained water-immiscible brownish red oily liquids.

2.2.3   *Production purification*

Stewing and layering Previous obtained liquid, water scrubbing, drying organic layer with anhydrous sodium sulfate, vacuum distillation, collecting 198°–200°/700 mmHg fraction, then as object product

Table 1. Synthesis yield with different mole ratio of amine and an alkylating agent (12 h, 30°C).

Mole ratio	Synthesis yield
1:1.0	43.8%
1:1.1	48.8%
1:1.2	48.9%
1:1.3	47.5%

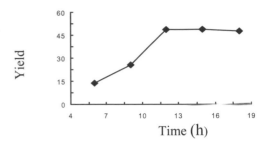

Figure 1. Reaction time of synthesis yield diagram (molar ratio 1:1.1, 30°C).

Table 2. Solvent and reaction yield (12 h, molar ratio 1:1.1, 30°C).

Solvent	Yield
Water	37.7%
Water + Polyethylene glycol	48.8%
Ethyl alcohol	40.8%
Pyridine	41.1%

was obtained. Total yield can reach 48.8% (according to m-aminophenol calculation).

3 RESULT AND DISCUSSION

3.1 Optimization of experiment process

3.1.1 N-ethyl-m-aminophenol synthes

N-alkylation process, by primary amine regeneration into secondary amines react further finally can stay in a tertiary amine stage, to get a secondary amine, we must use the excess amine to suppress generation of tertiary amine. However, in N-alkylation process, the synthesis of secondary amine reaction temperature at 100°C, amine and an alkylating agent in the molar ratio of 1:1 or greater. While the generation of tertiary amine synthesis need have concentrated sulfuric acid as catalyst condition, the same alkylating agent, high temperature and high pressure condition of excessive and alkylating agent. Table 1 shows different molar ratio of reaction synthesis yield with polyethylene glycol as a phase transfer catalyst, amine and an alkylating agent.

We can see from table 1 that, along with the primary amine and an alkylating agent mole ratio increases, the synthetic yield increased slightly, but not significantly. Considering the organic volatile, this experiment set amine and alkylating agent raw material mixture ratio is 1:1.1.

3.1.2 The influence of reaction time

Organic chemistry reaction time is long, by experiment, when N-ethylation reaction time between ethyl bromide and m-ammonia phenol is more than 12 h, we can achieve high yield, see as figure 1 below. From the graph 1 shows, as the reaction time increases, the increased rate of synthesis, when reaction time was 12 h, the yield is almost constant. So we selected 12 h.

3.1.3 The influence of different solvents

Experiment of water, ethanol and pyridine as solvents are compared. Table 2 shows, choose a different solvent to target will impact molecules yield. When the water is selected as solvent, in the N-ethylation of addition of polyethylene glycol as a phase transfer catalyst in methylation reaction, adding the surface active agent SDS, synthesis yield is the highest, up to 48.8%. So we selected water as solvent.

3.1.4 Phase transfer catalysts and the effect of the surface active agent

The reaction in two phases, with bromoethane and sulfuric acid two ester does not dissolve in water, can be in the N-ethylation reaction of adding phase transfer catalyst, in methylation reaction, adding surfactant SDS, to improve the reaction yield. This experiment selects the easy, convenient use, easy processing of waste polyethylene glycol as a phase transfer catalyst. Table 3 shows yield has risen to highest 48.8%.

3.2 Methylation reaction

The methylation reaction was completed in two steps reaction, with sulfuric acid two ester as methylation reagent, The first step: N-ethyl phenol with 10% sodium hydroxide solution after the reaction solution form negative ion, phenol, reaction temperature 80°C. The second step: reaction temperature is about 100°C, dropping to two methyl sulfate solution, sulfuric acid two ester on a methyl phenol negative ion oxygen atoms, generating a target molecule, the sulfuric acid two ester is to generate a CH_3OSO_2Na, sodium salt of methyl phenol and another negative ion binding, a N-ethyl room methoxy aniline. The results show that: when adding SDS, extension reaction time 1–2 times.

3.3 Optimization reaction conditions of N-ethyl methoxy synthetic

N-ethylation reaction: water as solvent, adding polyethylene glycol as a phase transfer catalyst, at 30°C, 12 h, reaction molar ratio 1:1.1; methylation reaction: 80°C–100°C, SDS as surfactant, $(CH_3O)_2SO_2$ was add mode.

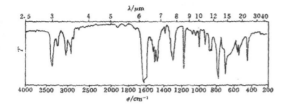

Figure 2. Infrared spectrum of purification sample.

3.4 *Infrared spectrum information expression*

Apply the purification of the brown red sample in liquid paraffin mixing, using infrared scanning, the infrared spectrum was shown in the figure 2.

4 CONCLUSIONS

The synthesis experiment by changing the solvent and adding proper surfactant, phase transfer catalyst, optimizing the synthesis process, the two step reaction for the synthesis of N-ethyl-m-methoxy aniline, can yield up to 48.8%. The process have less reaction steps, easily available raw materials, temper reaction condition, simplify experimental method, it has a certain reference value, also a notable synthesis route.

REFERENCES

[1] Shinohara Akira, Wada Nobuhide, Tokurage Yukio, et al. Process for producing aminophenol ether: DE, 2944030[P], 1980-05-08

[2] Akira Shinohara, Shimizu, Nobuhide Wada, Yukio Tokurage, et al. Process for producing aminophenol ether: US, 4231963[P], 1980-11-04

[3] Kimura Kazuo, Shimizu Hiroshi, Usui Masahiro. Production of aminophenol ether: JP, 2000 244[P], 1990-01-05

[4] Oyoshi Hajime. Production of aminophenol alkyl ether: JP, 1242562[P], 1989-09-27

Information Technology and Computer Application Engineering – Liu, Sung & Yao (Eds)
© *2014 Taylor & Francis Group, London, ISBN 978-1-138-00079-7*

Research and application of BP neural network algorithm in license plate recognition

J.X. Wang
Department of Computer, Hebei Institute of Architecture and Civil Engineering, Zhangjiakou, China

Y.L. Wang
Department of Mathematics and Physics, Hebei Institute of Architecture and Civil Engineering, Zhangjiakou, China

J.G. Zhao
Department of Computer, Hebei Institute of Architecture and Civil Engineering, Zhangjiakou, China

ABSTRACT: In the vehicle license plate automatic recognition system, it is difficult to identify the character due to natural or sampling factors. It makes the original rules of printed character distortion. BP neural network has been widely used to identify the license plate. How to improve the accuracy and speed of the license plate recognition is the most fundamental problem of license plate recognition system. In this paper we use momentum factor and adaptive learning rate to improve the traditional BP network. The algorithm is simple, high recognition rate and applied to character recognition in a variety of high-noise environments.

Keywords: License Plate Recognition; Character Recognition; BP neural network; Feature Extraction

1 INTRODUCTION

Currently, the license plate recognition technology has been used in the parking systems and traffic monitoring systems in some developed countries. In all-weather conditions the recognition accuracy is more than 95%. The license plate recognition technology development is slower in China. Although we achieved some results, but still remain in the experimental stage. Although the current laboratory recognition accuracy of 90%, in all-weather conditions recognition accuracy is less than 85%, far less than the actual application requirements. At present, the license plate recognition mainly has the following types of identification methods: template matching method, the statistical feature matching method and the neural network identification method. Template matching method of regular character recognition rate is relatively high, but in the case of the character deformation, the identification capability is limited. The feature statistical matching method in practical applications, when the characters appeared missing or fuzzy, the recognition effect is not ideal. The neural network recognition can effectively identify high resolution license plate. It has a strong ability of classification and fault tolerance. A lot of character recognitions in vehicle license are based on neural networks. The BP neural network is the most widely used network algorithm. BP neural network algorithm has some shortcomings, such as the slow convergence, local optimal, it is difficult

to determine the number of hidden layer nodes and the training process oscillation. For the above, this paper has made some improvements in the selection of the momentum factor, the learning rate improvements and modifications of the network weights. It can effectively prevent network may fall into the local minimum, and can adaptively adjust the learning rate, accelerate the convergence speed, avoid oscillation.

2 LICENSE PLATE RECOGNITION PROCESS

License plate recognition system includes a video input, image preprocessing, license plate coarse positioning, precise positioning license plate character segmentation, character recognition, and several other parts[1]. Collected in the natural environment, the image quality is not very good. It must be the image a series of image preprocessing denoising, boundary enhancement. In the plate positioning process, it may be a certain rotation angle on the license plate, so that its position parallel to the X axis. At last its location and size are normalized. Locate the position of the license plate, it is necessary to the license plate character segmentation. We put all the characters of the license plate area is divided, and then identify the character segmentation. Finally the recognized character is transferred to the decision-making and management algorithms. This paper focuses on how to improve BP algorithm, in order to better identify the characters.

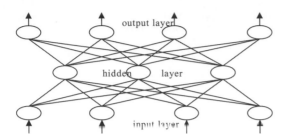

Figure 1. BP neural network structure.

3 IMPROVED BP NETWORK LICENSE PLATE RECOGNITION

3.1 *BP neural network algorithm model*

The basic principle of BP neural network model to process information is[2]: The input signal X_i through the intermediate node (hidden layer point) to the output node, after the non-linear transformation to generate an output signal Y_k. The every sample of network training consisted of input vector X and the expected output t. The deviation between the network output value Y and the desired output value t, by adjusting the input nodes and hidden layer node weights W_{ij}, and the output node between the weights and thresholds of T_{jk}, the error decreased along the gradient direction. After repeated training, it is identified the determination and minimum error corresponding to the network parameters (the weights and thresholds), training is stopped. BP network structure as shown in Figure 1.

BP neural network model includes its input-output model, role function model, error calculation models and self-learning model. Hidden layer node output model such as Equation 1, the output node outputs model such as Equation 2. Where f is a non-linear function, q is a nerve cell threshold.

$$O_j = f(\sum W_{ij} * X_i - q_j) \qquad (1)$$

$$Y_k = f(\sum T_{jk} * O_j - q_k) \qquad (2)$$

Role functions reflect lower input pulse intensity stimulation of the upper node function. It is also known as the stimulus function. The general value is (0, 1). The role function model is shown as Equation 3.

$$f(x) = 1/(1 + e^x) \qquad (3)$$

Error calculation model is a function of reflect neural network output and calculate the error between the desired output function. Shown as Equation 4.

$$E_p = \frac{1}{2}\sum(t_{pi} - O_{pi})^2 \qquad (4)$$

t_{pi} is a node to the desired output value, O_{pi} is a node to calculate output value.

Neural network learning process is connected to the lower level node and the node of the upper layer between the weight matrix. W_{ij} is set and the error correction process. Learning models such as Equation 5:

$$\Delta W_{ij}(n+1) = h \times \Phi_i \times O_{j+a} \times \Delta W_{ij}(n) \qquad (5)$$

3.2 *BP network model defect analysis and optimization strategy*

(1) Learning factor h optimization such as Equation 6. Using method of changing step length based on output error automatically adjusts the size of learning factor, which can reduce the number of iterations and speed up the convergence rate. A is the adjustment step and it is between 0~1 value.

$$h = h + a \times (E_p(n) - E_p(n-1))/E_p(n) \qquad (6)$$

(2) The number of hidden nodes of much larger impact on network performance. When the number of hidden layer nodes is too much, it can cause network learning time is too long and even can't convergence. When the number of hidden nodes is too small, network fault tolerable ability is poor. Use the stepwise regression analysis method and the parameters of the test of significance it can dynamically delete some of linear correlation of hidden nodes. When ownership of the node is the starting point for the next layer of nodes and threshold values are falling in an area (Usually take ±0.1, ±0.05 range), the node can be deleted. Best of hidden nodes L is reference to Equation 7. In the equation, m is the number of input node; n is the number of output nodes; c is in the range of 1~10 constants.

$$L = (m+n)1/2 + c \qquad (7)$$

(3) We use multiple regression analysis to deal with the neural network input and output parameters. Delete the strong correlation between input and output parameters, to reduce the input and output nodes. In this paper, character normalization into 32 × 16 size, output only 10 digital, 26 characters and 34 Chinese characters. So the input layer node number is 512, the output layer node number is 70.

(4) Algorithm optimization. Because the BP algorithm using the gradient descent method, which is easy to fall into local minimum and longer training time. We use based on the biological immune mechanism can both global search and can avoid premature convergence of immune genetic algorithm IGA to replace the traditional BP algorithm to overcome this shortcoming[3].

Before the system is running, shooting from the road to a large number of the license plate data (including normal operation and normal operation) as the content of the training. It uses a certain input and a desired output to modify the network weights through the BP algorithm. After the running, also can according to the specific situation of scene learning, with the extended ANN memory knowledge quantity.

3.3 *Further adjustment of BP neural network*

Because the BP neural network algorithm itself has some shortcomings, such as slow convergence, local optimum, it is difficult to determine hidden nodes and training process often oscillate. This paper uses the momentum factor and adaptive learning rate to improve it[4]. Based on the gradient descent algorithm we use the momentum factor a ($0 < a < 1$). Weight calculation is shown as Equation 8. Wherein, D(k) represents the negative gradient.

$$w(k + 1) = w(k) + n((1 - a)D(k) + aD(k - 1)) \qquad (8)$$

Thus, in the network weight modification process, take into account not only the role of the error on the gradient, but also consider the impact of the trend on the error surface, which can effectively avoid network falling into local minimum. In BP neural network training process, the adaptive learning rate, accelerate the convergence speed and avoid oscillation.

4 THE ANALYSIS AND COMPARISON OF THE EXPERIMENTAL RESULTS

This paper studies the license plate recognition system is based on the road monitoring screen, captured images of the toll stations collect. First, image preprocessing, license plate positioning, character segmentation, and then get character owned by one into a 32×16 size[5]. Each pixel is the feature points in the input. There are 26 characters, 10 digits and 34 characters in the output, so the input layer node number is 512. The output layer adopts coding mode and needs 7 nodes. These data are divided into two groups: one as a training sample set, as a standard template; the other is used as a test sample set. Apply VC++6.0 and Matlab7.0 software development. The specific function of the system is shown in Fig. 2. In Fig. 2, the left side is a typical license plate. It is the result of the image after the pretreatment and license plate positioning. On the right, there is a license plate identification and management system. After recognizing the license plate, it can search the car and all of the owner's information through networking. It can provide sufficient resources to decision makers to quickly and accurately make decisions for the vehicle.

In order to test real-time license plate recognition of this article, we compare traditional BP network and the algorithm convergence speed. Conclusion of this paper is the license plate recognition algorithm convergence speed is accelerated apparently, and the oscillation is very small. It shows that the traditional BP network by using additional momentum factor, adaptive learning rate is improved, improved the recognition speed, to meet the needs of modern intelligent traffic management and real time characteristics.

Figure 2. Recognition and management system functions of vehicle license.

Table 1. Comparison of the four algorithms recognition success rate.

Character type	Template matching	Traditional BP network	Support vector machine	The algorithm
Chinese character	77%	86%	85%	93%
English characters	81%	89%	86%	95%
Arabic numbers	82%	90%	89%	96%

In order to test the accuracy of license plate recognition of this article, the template matching algorithm, the traditional BP neural network algorithm, support vector machine as a comparison model and the algorithm. We use the license plate recognition correct rate as evaluation criteria. Several algorithms to identify the success rate as shown in Table 1. From Table 1, the proposed algorithm recognition accuracy rate is higher than the traditional BP neural network algorithm, the template matching algorithm and support vector machine. The comparison results show that the proposed license plate recognition algorithm can effectively identify the license plate characters, and faster learning and recognition.

5 SUMMARY

Automatic license plate recognition technology is a very important technology in modern intelligent transportation systems, and it becomes a research hotspot in recent years. This paper analyzes some shortcomings of the BP neural network in license plate recognition applications, and the shortcomings have been improved. It focuses on the construction and improvement of the BP neural network. Finally it tests the accuracy, reliability and timeliness of verification through the simulation. The simulation results show that the improved BP neural network license plate recognition algorithm to further improve the correct rate of license plate recognition, accelerate the recognition speed and real-time monitoring of the passing vehicles. The good performance of the system has a wide range of applications in modern intelligent transportation systems.

REFERENCES

[1] Danian Zheng, Yannan Zhao, Jiaxin Wang. An efficient method of license plate location[J]. Pattern Recognition Letters, 2005, 26(26): 2431–2438

[2] S.Z. Wang, H.M. Lee. Detection and recognition of license plate characters with different appearances[C]. Proc. Confi. Intell. Transp. Syst. 2003, 979–984

[3] Danlan Zheng, Yannan Zhao, Jiaxin Wang. An efficient method of license plate location[J]. Pattern Recognition Letters. 2005(26): 2431–2438

[4] Vahid Abolghasemi, Alireze Ahmadyfard. An Edge-based Color-aided method for license plate detection[J]. Image Vision computing, 2008, 23(10)

[5] Sergios Theodoridis. Pattern Recognition[M]. Beijing: China Machine Press, 2003.

Information Technology and Computer Application Engineering – Liu, Sung & Yao (Eds)
© *2014 Taylor & Francis Group, London, ISBN 978-1-138-00079-7*

The research of improved Sobel-image edge detection algorithm

Y.L. Wang
Department of Computer, Hebei Institute of Architecture and Civil Engineering, Zhangjiakou, China

J.X. Wang
Department of Mathematics and Physics, Hebei Institute of Architecture and Civil Engineering, Zhangjiakou, China

H.D. Wang
Library, Hebei Institute of Architecture and Civil Engineering, Zhangjiakou, China

ABSTRACT: Sobel operator is easy to achieve in the space, and it can produce a better effect of edge detection. But it is more sensitive to noise, and the extracted edges are thicker. In order to overcome these disadvantages, this paper proposes an improved algorithm. The algorithm improves and reorganizes the median filtering algorithm and Sobel algorithm based on them. Simulation results show that: the improved algorithm has a strong ability to inhibit the image noise. It can also refine the extracted edges, and extracts the edge of the positioning accuracy, good continuity.

Keywords: edge detection; edge detailed; improved Sobel operator; improved median filter

1 INTRODUCTION

Edge refers to the local image gray scale changes the most significant part, it exists in the target and background, goals and objectives, between area and area. The edge has two characteristics: direction and magnitude. Along the edge to the pixel values change relatively flat; and along the vertical edges towards, the values have more violent changes. This drastic step shape may be present, may also present the shape of slope.

The substance of the edge detection is the use of an algorithm to extract a boundary line between the objects in the image and background. We define the edge of the boundary of the image gray area changed dramatically. How to quickly and accurately extract the image edge information has been a research hotspot, the classical Sobel algorithm is one of this kind. Although the Sobel algorithm in many fields has been widely used because of a small amount of calculation and fast speed, it is capable of detecting the direction of the noise immunity is low, edge extraction is not precise enough and so on, which also bring limitations to its use. Therefore, this paper puts forward a kind of simple and feasible improvement method, it makes the detection of the direction and accuracy have been better to improve noise immunity be enhanced.

2 VIDEO DATA ANALYSIS

The commonly used edge detection algorithm is using the edge of one or two order derivative rules to detect the edge. In many gradient operators, the most classic gradient operator is Sobel operator[1]. Sobel operator is also a gradient magnitude. Each pixel of the digital image of {f(i, j)}, examine its upper, lower, left and right neighbor point gray weighted difference, the right of the neighboring points close thereto is big. Accordingly, the definition of Sobel gradient operator is $M = \sqrt{s_x^2 + s_y^2}$. Like other gradient operator, S_x and S_y can be used to achieve the convolution operator. As shown in Equation 1:

$$S_x = \begin{bmatrix} -1 & 0 & 1 \\ -2 & 0 & 2 \\ -1 & 0 & 1 \end{bmatrix} \qquad S_y = \begin{bmatrix} -1 & -2 & -1 \\ 0 & 0 & 0 \\ 1 & 2 & 1 \end{bmatrix} \qquad (1)$$

We appropriate threshold for Th1, make the following judgments: If $M \geq$ th1, then the point is an edge point; and if $M <$ th1, the point is not the edge point. The operator is focus on the center pixels close to the template. Because of the relatively large changes in the brightness of the image near the edges of, you can put those in the neighborhood gray level exceeds a certain value pixels as edge points.

3 IMPROVEMENT OF THE SOBEL OPERATOR

3.1 Background modeling

Assuming the digital image to be processed contains L gradation, X denotes the noise image, and Y represents

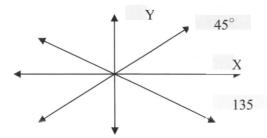

Figure 1. 4 directions of filtering window.

the image denoising. The $x(n)$ represents a pixel gray value at the position of the noise images $n = [n_1, n_2]$. $W(n)$ represents a filter denoising window that n is the center of the $(2N + 1) \times (2N + 1)$ size. This filtering window contains $(2N + 1) \times (2N + 1)$ pixels.

R(n) is the set of the pixel value in the filter window are arranged in accordance with the descending order of the obtained. Defined as follows Equation 2

$$R(n) = \{r_1(n), r_2(n), \ldots, r_{(2N+1)\times(2N+1)}(n)\} \quad (2)$$

The $r_1(n)$ is a maximum value of the pixel in the filter window, and $r_{(2N+1)*(2N+1)}(n)$ is a minimum value of pixels in the filter window. According to the above definition of a variety of local statistical properties, we can test each of the pixel noise detection. For the detected noise pixel, we use the improved median denoising algorithm for correction. Otherwise, the pixel is not any treatment. The denoising algorithm output using the Equation 3 calculated[2].

$$y(n) = (1 - a(n)) \cdot x(n) + a(n) \cdot m(n) \quad (3)$$

where m(n) is the n position noise instead of value, that n location within the filter window median point sorting, $m(n) = mid(R(n))$, $a(n)$ is switch variable. When n location at the pixel is determined to be noise, let $a(n) = 1$; else, let $a(n) = 0$. Defined the following Equation 4 below:

$$a(n) = \begin{cases} 1, & if\ x(n) = r_1(n)\ \ and\ \ (d_1 > th2, d_2 > th2, d_3 > th2, d_4 > th2) \\ 1, & if\ x(n) = r_{(2N+1)\times(2N+1)}(n)\ \ and\ \ (d_1 > th2, d_2 > th2, d_3 > th2, d_4 > th2) \\ 0, & other \end{cases} \quad (4)$$

Here, $x(n)$ is the pixel value of the filter window centre point, $r_1(n)$ is the maximum value of the filter window centre point, and $r_{(2N+1)*(2N+1)}(n)$ is the minimum value. D_1, d_2, d_3, and d_4 are respectively the variance of the x-direction, y-direction, the direction of 45 degrees, 135 degrees in filtering window at $x(n)$ as the centre. As shown in Figure 1. The th2 is an adjustable threshold value of the parameter. If x(n) is in the filtering window pixel values of the maximum or minimum values, and its four direction pixel variances are greater than the threshold value th2, the point is considered not in on the edge, it is the noise. So $a(n) = 1$, else $a(n) = 0$. Using this algorithm to remove noise, it can be able to improve retention of the edge portions of the image.

3.2 The change of Sobel direction convolution operators

To be able to more accurately describe the image edge points, reduce the impact of noise on the test results and improve noise immunity operator, we reconstruct four of the size of 5×5 convolution operator[3]. As shown in Figure 1 and Equation 5. S_x is the horizontal direction, S_y is the vertical direction, S_{45} is a 45° direction and S_{135} is a 135° direction.

$$S_x = \begin{bmatrix} 2 & 3 & 0 & -3 & -2 \\ 3 & 4 & 0 & -4 & -3 \\ 6 & 6 & 0 & -6 & -6 \\ 3 & 4 & 0 & -4 & -3 \\ 2 & 3 & 0 & -3 & -2 \end{bmatrix} \quad S_y = \begin{bmatrix} 2 & 3 & 6 & 3 & 2 \\ 3 & 4 & 6 & 4 & 3 \\ 0 & 0 & 0 & 0 & 0 \\ -3 & -4 & -6 & -4 & -3 \\ -2 & -3 & -6 & -3 & -2 \end{bmatrix}$$

$$S_{45} = \begin{bmatrix} 0 & -2 & -3 & -2 & -6 \\ 2 & 0 & -4 & -6 & -2 \\ 3 & 4 & 0 & -4 & -3 \\ 2 & 6 & 4 & 0 & -2 \\ 6 & 2 & 3 & 2 & 0 \end{bmatrix} \quad S_{135} = \begin{bmatrix} -6 & -2 & -3 & -2 & 0 \\ -2 & -6 & -4 & 0 & 2 \\ -3 & -4 & 0 & 4 & 2 \\ -2 & 0 & 4 & 6 & 2 \\ 0 & 2 & 3 & 2 & 6 \end{bmatrix} \quad (5)$$

3.3 The thinning processing of gradient image edge

The value calculated by the above template is the gradient image of the image corresponding, the edges appear in the gradient where the value of the larger. If we want to extract the edge of the image must be set the threshold value of the binary.

The curve is the shape of the roof at the edges and extracted from the edge width is related to the threshold values. Each threshold corresponds to an edge pixel width value. When choosing appropriate threshold it can reduce the edge width, which can achieve the purpose of thinned edge. It is difficult to find suitable threshold if directly on the gradient image threshold values of two. This makes the edge detection is difficult to meet the requirements. Based on the above discussion, we can find the pixels in the gradient image in the neighborhood of (m, n) a maximum value Max(m, n). We set the local threshold th3 according to the Max(m, n).

We will gradient images of the two values according to the gradient value and the threshold, to achieve purpose of edge thinning. According to the gradient computation of image edge detection image such as Equation 6 and Equation 7 shown:

$$th3 = a \times Max(m, n) \quad (6)$$

$$edge(m, n) = \begin{cases} 1 & grade(m, n) > th3 \ \ and \ \ grade(m, n) > th1 \\ 0 & other \end{cases} \quad (7)$$

wherein: edge(m, n) is obtained edge detection images; edge(m, n) is the value of the gradient corresponding to the gradient image pixel (m, n); th1 is the minimum threshold of the whole image edge detection. As long as the gradient value is greater than the threshold value th1, and the direction of the point of the direction of the edge points, it can be judged as an edge point. The smaller Th1, the more abundant edge details is. The greater Th1, the more obvious the main outlines of the image, but it reduced the edge detail. Max(m, n) is the maximum gradient value of the gradient image (m, n) in the eight directions. A is the control factor, $0 < a < 1$. By selecting different value of a, we can obtain the highest threshold th3 of the different local edge detection, so as to control the width of edge. This no only highlights the edges of the image, thins the edges, but also enhances the edge effect and convenient two values of threshold selection. In the edge detection algorithm, the global minimum threshold selection of Th1 is always a difficult problem in image processing. In the gray images, background and foreground share statistical weight tend to be large. Therefore, in the histogram of the gradient image, most of the gradient values are concentrated in the low gradient zone. But the real edge in the gradient values of the larger places, in a histogram is a decay curve. So it can be characterized by analysis of the decay curve, set the appropriated threshold value to filter out the non-edge regions retain the edges of the image[4].

4 IMPROVED ALGORITHMS TO ACHIEVE THE SPECIFIC PROCESS

Step (1) Determine the denoising threshold Th2, use the 3.1 improved median denoising algorithms for denoising.

Step (2) Use the Equation 6 and Equation 7 the 4 directions template image of X was calculated point by point, and taking the maximum gradient images in the new gray value. Maximum value corresponding to the template for the direction of pixel edge direction, finally obtain the gradient image grade(m, n).

Step (3) Determine the global minimum threshold Th1. If the gradient image pixel gray value greater than or equal to Th1, and the direction for the edge points in the direction, we can determine the point of edge points. Otherwise, it is a non-edge point[5]. The smaller Th1, the more details edges have. The greater th1, the main contours of the image, but the reduced edge detail. According to the Equation 6 we calculate the local maximum threshold th3. The th3's size represents the thickness of the edge line. The edges of the control factor "a" general values of 0.5 to 0.9.

Step (4) According to the Equation 7 to calculate the edge extracted image, and ultimately the edges of the image of the original is detected.

5 EXPERIMENTAL RESULTS AND COMPARATIVE ANALYSIS

We use VC6.0 to do simulation program. In this experiment, the th1 value is 80. If you want to extract more details, th1 value is smaller. The th2 value is 30. Denoising algorithm is used 5X5 convolution mask. The a values of 0.7 and 0.9. If you want the edge width is greater, then a is smaller. When a values of 0.9, the value of the two image edge width can reach one to two pixels. In order to conduct comparative analysis,

(a) Original image   (b) Sobel operator

Figure 2. Sobel gradient operator in original image.

(a)0.02 Salt and pepper noise   (b) Sobel operator

Figure 3. Sobel gradient operator to add noise.

(a)Weighted smooth denoise   (b)Sobel operator

Figure 4. Sobel gradient operator after denoising.

(a)a=0.7 Results   (b)a=0.9 Results

Figure 5. The results of this paper algorithm.

we operate the traditional Sobel gradient operatoron on the normal image, image noise and weighted smoothing denoising images. As shown in Figure 2, Figure 3 and Figure 4. The Figure 2(b) is the traditional Sobel gradient operator on the noise-free original Figure 2(a). The results show more clearly. Figure 3(b) is the Sobel gradient operator on the Figure 3(a) which adds 0.02 salt and pepper noise. Figure 4(a) is the image of Figure 3(a) after the weighted smoothing denoising. The Figure 4(b) is the traditional Sobel gradient operator on Figure 4(a). The result shows that the traditional Sobel operator is very sensitive to the noise and the processed image is very fuzzy. Figure 5(a) and

Figure 5(b) are the algorithm processing on the add noise Figure 2. In Figure 5(a) the threshold value a is 0.7; in Figure 5(b) the a is 0.9. The results show that the algorithm processing the image noise is very clear, and it can effectively control the thickness of the edge extraction by use the size of the threshold value a. It is be found, the anti-noise ability of traditional Sobel gradient operator is very weak. Using this algorithm for noise image edge extraction, the effect is very good and much better than the conventional Sobel gradient operator. Shown in Figure 5.

6 SUMMARY

From the above analysis and the experimental results we can see: the improved gradient edge detection algorithm in this paper, it overcomes the Sobel operator edge detection that is rough edges and noise-sensitive shortcoming. The algorithm has the advantage of extracted fine edge and strong resistance to noise. It is a simple and effective edge detection algorithm.

REFERENCES

[1] Chen Dali. Research on De-noise Algorithms in Digital Image Processing, Shenyang: Northeastern University Doctoral Dissertation 2008.5.

[2] Gonzalez R C, Richard E W. Digital image processing [M], Beijing: Electronic Industry Press, 2002, 241–242.

[3] Han Jiandong, Xiong Jianyong, Yin Chao. Algorithm of adaptive image fuzzy-entropy segmentation based on wavelet analysis[J]. Infrared Technology, 2004, (26) 3; 29–31.

[4] Li Jie, Tang Xing-ke, Jiang Yan-jun. Comparing study of some edge detection algorithms[J]. Formation Technology, 2007, 38(9):106–108.

[5] Kwan H K, Cai Y. Fuzzy filters for noisy image filtering [A], Proceedings of the IEEE International Symposium on Circuits and Systems, 2002:672–675.

Information Technology and Computer Application Engineering – Liu, Sung & Yao (Eds)
© *2014 Taylor & Francis Group, London, ISBN 978-1-138-00079-7*

Integral inequalities of Hermite-Hadamard type for functions whose 3rd derivatives are (α, m)-convex

Ling Chun
College of Mathematics, Inner Mongolia University for Nationalities, Tongliao City, Inner Mongolia Autonomous Region, China

ABSTRACT: In the paper, the author establish some new Hermite-Hadamard type inequalities for functions whose 3rd derivatives are (α, m)-convex.

Keywords: Hermite-Hadamard's Integral Inequality; (α, m)-Convex Function; Hölder Inequality

1   INTRODUCTION

Let $f : I \subseteq \mathbb{R} \to \mathbb{R}$ be a convex function on I and $a, b \in I$ with $a < b$. The inequality

$$f\left(\frac{a+b}{2}\right) \le \frac{1}{b-a} \int_a^b f(x)\mathrm{d}x \le \frac{f(a)+f(b)}{2}. \qquad (1.1)$$

This inequality is well known in the literature as Hermite-Hadamard's inequality for convex mappings.

Definition 1.1 A function $f : I \subseteq \mathbb{R} = (-\infty, +\infty) \to \mathbb{R}$ is said to be convex if

$$f(tx+(1-t)y) \le tf(x)+(1-t)f(y) \qquad (1.2)$$

holds for all $x, y \in I$ and $t \in [0, 1]$.

Definition 1.2 A function $f : [0, b] \to \mathbb{R}$ is said to be m-convex if

$$f(tx+m(1-t)y) \le tf(x)+m(1-t)f(y) \qquad (1.3)$$

holds for all $x, y \in [0, b], t \in [0, 1]$ and $m \in (0, 1]$.

Definition 1.3 A function $f : [0, b] \to \mathbb{R}$ is said to be (α, m)-convex if

$$f(tx+m(1-t)y) \le t^\alpha f(x)+m(1-t^\alpha)f(y) \qquad (1.4)$$

is valid for all $x, y \in [0, b]$ and $t \in [0, 1]$, and $(\alpha, m) \in (0, 1]^2$.

2   A LEMMA

In order to prove our main results, we need the following lemma.

Lemma 2.1 Let $f : I \subset \mathbb{R} \to \mathbb{R}$ be a three times differentiable mapping on I°, where $a, b \in I^\circ$ with $a < b$. If $f''' \in L[a, b]$, then

$$I(f) := \frac{f(a)+f(b)}{2} - \frac{1}{b-a} \int_a^b f(x)\mathrm{d}x - \frac{(b-a)^2}{12} f''(a)$$

$$= \frac{(b-a)^3}{12} \int_0^1 t^2(3-2t)f'''(ta+(1-t)b)\mathrm{d}t. \qquad (2.1)$$

Proof. By integrating by part, we have

$$\int_0^1 t^2(3-2t)f'''(ta+(1-t)b)\mathrm{d}t$$

$$= -\frac{1}{b-a} \int_0^1 t^2(3-2t)\mathrm{d}f''(ta+(1-t)b)$$

$$= -\frac{1}{b-a}f''(a) - \frac{6}{(b-a)^2} \int_0^1 (t-t^2)\mathrm{d}f'(ta+(1-t)b)$$

$$= -\frac{1}{b-a}f''(a) + \frac{6}{(b-a)^3}[f(a)+f(b)]$$

$$\quad - \frac{12}{(b-a)^3} \int_0^1 f(ta+(1-t)b)\mathrm{d}t$$

$$= -\frac{1}{b-a}f''(a) + \frac{6}{(b-a)^3}[f(a)+f(b)]$$

$$\quad - \frac{12}{(b-a)^4} \int_a^b f(x)\mathrm{d}x$$

The proof lemma 2.1 is thus proved.

3   HERMITE-HADAMARD'S TYPE INEQUALITIES FOR (α, m)-CONVEX FUNCTIONS

Theorem 3.1 Let $f : I \subseteq \mathbb{R}_0 \to \mathbb{R}$ be a three times differentiable function on I°, such that $f''' \in L[a, b]$ for

$a, b \in I$ with $a < b$, if $|f'''|^q$ is (α, m)-convex on $[a, b]$ for $(\alpha, m) \in (0, 1]^2$ and $q \geq 1$, then

$$|I(f)| \leq \frac{(b-a)^3}{24} \left(\frac{1}{(\alpha+3)(\alpha+4)} \right)^{1/q}$$

$$\times \left(2(\alpha+6)|f''(a)|^q + m\alpha(\alpha+5)\left|f''(\frac{b}{m})\right|^q \right)^{1/q} \quad (3.1)$$

Proof. From Lemma 2.1, using the well known Hölder's inequality, we have

$$|I(f)| \leq \frac{(b-a)^3}{12} \int_0^1 t^2(3-2t)|f''(ta+(1-t)b)|dt$$

$$\leq \frac{(b-a)^3}{12} \left(\int_0^1 t^2(3-2t)dt \right)^{1-1/q}$$

$$\times \left(\int_0^1 t^2(3-2t)|f'''(ta+(1-t)b)|^q dt \right)^{1/q}$$

$$\leq \frac{(b-a)^3}{12} \left(\int_0^1 t^2(3-2t)dt \right)^{1-1/q} \left\{ \int_0^1 t^2(3-2t) \right.$$

$$\times \left[t^\alpha |f''(a)|^q + m(1-t^\alpha)\left|f''(\frac{b}{m})\right|^q \right] dt \right\}^{1/q}$$

$$= \frac{(b-a)^3}{24} \left(\frac{1}{(\alpha+3)(\alpha+4)} \right)^{1/q} \left(2(\alpha+6)|f''(a)|^q \right.$$

$$\left. + m\alpha(\alpha+5)\left|f''(\frac{b}{m})\right|^q \right)^{1/q}$$

The proof of Theorem 3.1 is complete.

Theorem 3.2 Let $f : I \subseteq \mathbb{R}_0 \to \mathbb{R}$ be a three times differentiable function on I°, such that $f''' \in L[a, b]$ for $a, b \in I$ with $a < b$, if $|f'''|^q$ is (α, m)-convex on $[a, b]$ for $(\alpha, m) \in (0, 1]^2$ and $q > 1$, then

$$|I(f)| \leq \frac{(b-a)^3}{12} \left(\frac{3^{p+1}-1}{2(p+1)} \right)^{1/p} \left(\frac{1}{(2q+\alpha+1)(2q+1)} \right)^{1/q}$$

$$\times \left((2q+1)|f'''(a)|^q + \alpha m\left|f'''(\frac{b}{m})\right|^q \right)^{1/q} \quad (3.2)$$

where $1/p + 1/q = 1$.

Proof. By Lemma 2.1 and the (α, m)-convexity of $|f'''(x)|$ on $[a, b]$ and by Hölder's inequality, we get

$$|I(f)| \leq \frac{(b-a)^3}{12} \int_0^1 t^2(3-2t)|f''(ta+(1-t)b)|dt$$

$$\leq \frac{(b-a)^3}{12} \left(\int_0^1 (3-2t)^p dt \right)^{1/p}$$

$$\times \left(\int_0^1 t^{2q} |f'''(ta+(1-t)b)|^q dt \right)^{1/q}$$

$$\leq \frac{(b-a)^3}{12} \left(\int_0^1 (3-2t)^p dt \right)^{1/p}$$

$$\left\{ \int_0^1 t^{2q} \left[t^\alpha |f'''(a)|^q + m(1-t^\alpha)\left|f'''(\frac{b}{m})\right|^q \right] dt \right\}^{1/q}$$

$$= \frac{(b-a)^3}{12} \left(\frac{3^{p+1}-1}{2(p+1)} \right)^{1/p} \left(\frac{1}{(2q+\alpha+1)(2q+1)} \right)^{1/q}$$

$$\times \left((2q+1)|f'''(a)|^q + \alpha m\left|f'''(\frac{b}{m})\right|^q \right)^{1/q}.$$

The proof of Theorem 3.2 is complete.

Theorem 3.3 Under conditions of theorem 3.2, then

$$|I(f)| \leq \frac{(b-a)^3}{12} \left(\frac{3^{p+3}-2p^2-12p-19}{4(p+1)(p+2)(p+3)} \right)^{1/p}$$

$$\times \left(\frac{1}{3(\alpha+3)} \right)^{1/q} \left(3|f'''(a)|^q + m\alpha\left|f'''(\frac{b}{m})\right|^q \right)^{1/q}. \quad (3.3)$$

Proof. Using Lemma 2.1, by Hölder's inequality, we have

$$|I(f)| \leq \frac{(b-a)^3}{12} \int_0^1 t^2(3-2t)|f''(ta+(1-t)b)|dt$$

$$\leq \frac{(b-a)^3}{12} \left(\int_0^1 t^2(3-2t)^p dt \right)^{1/p}$$

$$\times \left(\int_0^1 t^2 |f'''(ta+(1-t)b)|^q dt \right)^{1/q}$$

$$\leq \frac{(b-a)^3}{12} \left(\int_0^1 t^2(3-2t)^p dt \right)^{1/p}$$

$$\times \left\{ \int_0^1 t^2 \left[t^\alpha |f'''(a)|^q + m(1-t^\alpha)\left|f''(\frac{b}{m})\right|^q \right] dt \right\}^{1/q}$$

$$= \frac{(b-a)^3}{12} \left(\frac{3^{p+3}-2p^2-12p-19}{4(p+1)(p+2)(p+3)} \right)^{1/p} \left(\frac{1}{3(\alpha+3)} \right)^{1/q}$$

$$\times \left(3|f'''(a)|^q + m\alpha\left|f'''(\frac{b}{m})\right|^q \right)^{1/q}.$$

The proof of Theorem 3.3 is complete.

Theorem 3.4 Under conditions of theorem 3.2, then

$$|I(f)| \leq \frac{(b-a)^3}{12} \left(\frac{p+2}{(2p+1)(p+1)} \right)^{1/p} \left(\frac{1}{(\alpha+1)(\alpha+2)} \right)^{1/q}$$

$$\times \left((\alpha+4)|f'''(a)|^q + m\alpha(2\alpha+5)\left|f'''(\frac{b}{m})\right|^q \right)^{1/q}. \quad (3.4)$$

Proof. By Lemma 2.1, using Hölder's inequality, we have

$$|I(f)| \leq \frac{(b-a)^3}{12} \int_0^1 t^2(3-2t)|f''(ta+(1-t)b)|dt$$

$$\leq \frac{(b-a)^3}{12}\left(\int_0^1 t^{2p}(3-2t)\mathrm{d}t\right)^{1/p}$$

$$\times\left(\int_0^1 (3-2t)\left|f'''(ta+(1-t)b)\right|^q \mathrm{d}t\right)^{1/q}$$

$$\leq \frac{(b-a)^3}{12}\left(\int_0^1 t^{2p}(3-2t)\mathrm{d}t\right)^{1/p}$$

$$\times\left\{\int_0^1 (3-2t)\left[t^\alpha\left|f'''(a)\right|^q + m(1-t^\alpha)\left|f'''(\frac{b}{m})\right|^q\right]\mathrm{d}t\right\}^{1/q}$$

$$= \frac{(b-a)^3}{12}\left(\frac{p+2}{(2p+1)(p+1)}\right)^{1/p}\left(\frac{1}{(\alpha+1)(\alpha+2)}\right)^{1/q}$$

$$\times\left((\alpha+4)\left|f'''(a)\right|^q + m\alpha(2\alpha+5)\left|f'''(\frac{b}{m})\right|^q\right)^{1/q}.$$

The proof of Theorem 3.4 is complete.

Theorem 3.5. Under conditions of theorem 3.2, then

$$|I(f)| \leq \frac{(b-a)^3}{12}\left(\frac{3^{P+2}-2p-5}{4(p+1)(P+2)}\right)^{1/p}$$

$$\times\left(\frac{1}{(q+2)(\alpha+q+2)}\right)^{1/q} \tag{3.5}$$

$$\times\left((q+2)\left|f'''(a)\right|^q + \alpha m\left|f'''(\frac{b}{m})\right|^q\right)^{1/q}.$$

Proof. From Lemma 2.1, using the well known Hölder's inequality, we have

$$|I(f)| \leq \frac{(b-a)^3}{12}\int_0^1 t^2(3-2t)\left|f'''(ta+(1-t)b)\right|\mathrm{d}t$$

$$\leq \frac{(b-a)^3}{12}\left(\int_0^1 t(3-2t)^p\mathrm{d}t\right)^{1/p}$$

$$\times\left(\int_0^1 t^{q+1}\left|f'''(ta+(1-t)b)\right|^q \mathrm{d}t\right)^{1/q}$$

$$\leq \frac{(b-a)^3}{12}\left(\int_0^1 t(3-2t)^p\mathrm{d}t\right)^{1/p}$$

$$\times\left\{\int_0^1 t^{q+1}\left[t^\alpha\left|f'''(a)\right|^q + m(1-t^\alpha)\left|f'''(\frac{b}{m})\right|^q\right]\mathrm{d}t\right\}^{1/q}$$

$$\leq \frac{(b-a)^3}{12}\left(\frac{3^{P+2}-2p-5}{4(p+1)(P+2)}\right)^{1/p}\left(\frac{1}{(q+2)(\alpha+q+2)}\right)^{1/q}$$

$$\times\left((q+2)\left|f'''(a)\right|^q + \alpha m\left|f'''(\frac{b}{m})\right|^q\right)^{1/q}.$$

The proof of Theorem 3.5 is complete.

ACKNOWLEDGEMENTS

The author was supported by Science Research Funding of Inner Mongolia University for Nationalities under Grant No. NMD1225.

REFERENCES

M.K. Bakula, M.E. Ozdemir and J. Pecaric, Hadamard type inequalities for m-convex and (α, m)-convex functions, *J. Inequal. Pure and Appl. Math.*, 9 (4) (2008), Article 96. Available online at http://www.emis.de/journals/JIPAM/article1032.html

S.S. Dragomir and G.H. Toader, Some inequalities for m-convex functions, *Studia Univ. Babeş-Bolyai, Math.*, 38 (1) (1993), 21–28.

S.S. Dragomir and R.P. Agarwal, "Two Inequalities for Differentiable Mappings and Applications to Special Means of Real Numbers and to Trapezoidal Formula," Applied Mathematics Letters, Vol. 11, No. 5, 1998, pp. 91–95. doi:10.1016/S0893-9659(98)00086-X

M.E. Özdemir, M. Avci, E. Set, On some inequalities of Hermite–Hadamard type via m-convexity, *Appl. Math. Lett.* 23 (9) (2010) 1065–1070.

Shu-Hong Wang, Bo-Yan Xi, and Feng Qi, Some new inequalities of Hermite-Hadamard type for n-time differentiable functions which are m-convex, *Analysis (Munich)* 32 (2012), no. 3, 247–262; Available online at http://dx.doi.org/10.1524/anly.2012.1167

Information Technology and Computer Application Engineering – Liu, Sung & Yao (Eds)
© 2014 Taylor & Francis Group, London, ISBN 978-1-138-00079-7

The research of immunization strategy on complex networks

Xin Yi Chen, Jing Jiang, Tao Jiang, Jian Hua Xia & Jin Xi Zhang
China Institute of Minorities Information Technology, Northwest University for Nationalities, Lanzhou, Gansu, China

ABSTRACT: Two viruses spreading mechanism were used to the nodes in complex networks based on SIS propagation model: the same probability of infection and the different probability infection, different immunization strategies implemented in the network with the same topology to simulate the spread of the virus with different immunization process, and perform the same immunization strategy in the different network topologies. The experimental results showed that: The node probability of infection on maximum degree immunization strategy which proposed by the authors of this article is between the target immunization strategy and the acquaintance immunization strategy, and comparing with the target immunization strategy, it is relatively low-cost.

Keywords: Complex networks; SIS propagation model; the probability of infection; the maximum degree of immunization strategy

1 INTRODUCTION

With the rapid development of computer technology, the computer plays an important role in our daily life and production, a combination of computer technology and communication technology make networks play an increasingly important role in daily life and production, the network has become human communication and the channels of getting information. Network makes people's day-to-day activities more and more convenient; it also brings a lot of new problems: the spread of computer viruses in the network more and more rampant, if not control, computer virus will make human life and production cause huge losses. More than a decade into the 21st century, infectious diseases on make A series of provocative on human society itself: the epidemic of SARS in 2003, the spread of the avian influenza virus (H5N1) in 2005, H1N1 in 2009 and so on. With the global integration process, convenient transportation promotes the communication of people around the world, and convenient the human activities, but also it speeds up the transmission of the virus. Whether in the social network of human life, or in a computer network, how to stop the spread of the virus or slow down the rate of the spread of the virus, which are one of the hot issues in today's complex network research.

The research of the network model in the time of the 1990s in foreign countries. The research of network model process through the regular networks, random networks and complex networks three important stages of development. In the stage of regular network, according to some completely rule; connect each node of the network, such as the global coupling network, the nearest-neighbor coupling network and the star network. In the stage of random network, the network model is thought as a completely random evolution, such as the ER random graph model proposed by Erdös and Rényi [1]. In the complex network research stage, by analysis and simulation of the real network, the formation of the network model is not completely rule and is not completely random, and they are between regular and randomly. Through the empirical research the scholar found that the complex network have a common characteristic, namely small-world and scale-free properties. Watts and Strogtz introduced small-world network model by "random reconnection" the regular network which is called WS small-world model [2]. Barabási and Albert in the United States promoted a scale-free network model, which also known as the BA model [3].

In the studying of the propagation model, scholars abstract the status of individuals in the population as some typical states, including: Susceptible, Infected and Removed. The naming of the general model of infection is named by the transition between these several states. Individuals from the susceptible state (Susceptible) infected (Infected) returning to a healthy state and with the antibody is no longer infectious (Removed), which known as the SIR model; If the individuals are cured after being infected through back to the susceptible status (Susceptible) which still have the possibility of infected, this process is known as SIS model; from susceptible state to the infection status (infected) is known as the SI model [4,5]. In this paper, the SIS virus propagation model was used to simulate the spread of the virus in a complex network, by comparing the effect of different networks of which the largest immunization strategy is used and find the best network which fit using the maximum degree immunization strategy very much; by comparing different immunization strategies implemented in the same topology network, comparing the immunization effect of random immunization, target immunization and acquaintances immunization and the maximum immunization immune effect.

2 THE PROPOSED OF THE MAXIMUM DEGREE IMMUNIZATION STRATEGY

2.1 The development of immunization strategy

Selecting some nodes randomly in the network, and treat each node equally no matter the nodes' degree is large or small, which is the idea of random immunization strategy.

The critical value of random immunization strategy is [6]:

$$g_c = 1 - \frac{\lambda_c}{\lambda} \quad (1)$$

Implemented random immunization strategy get the immunization critical value in scale-free networks is:

$$g_c = 1 - \frac{1}{\lambda} \frac{<k>}{<k^2>} \quad (2)$$

$<k>$ is assumed to as the average degree of the network. λ is the effective rate of transmission, λ_c is the epidemic threshold. Obviously, random immunization strategy is not fit the scale-free network model which is similar to many real systems. If take the random immunization strategy in scale-free networks, in order to eventually eliminate the virus, almost all of the nodes in the network should immune. According to the scale-free network have robustness against random attacks and have vulnerability on deliberate attacks, choose a small amount of degree node to immune (target immunization strategy [7]) is very feasible. Although the target immunization strategy is very effective for scale-free networks, this approach need to know the global information of the network, at least to know the degree of each node, so that the current maximum degree node can be find before the implementation of immunization. For many large and complex and ever-changing Complex networks, obtaining information about each node in the network is difficult. Based on this, Cohen proposed acquaintance immunization strategy [8]: Choose the ratio of p from the network of n nodes randomly, selected from the selected nodes randomly and immune its neighbor, which takes full use of the characteristics of scale-free networks, the probability of large degree node being selected are much

higher than the small degree nodes. The author of this article proposed the maximum degree immunization strategy is based on the acquaintance immunization strategy and the target immunization strategy, and is combining both of their advantages, the acquaintance immunization strategy is based on the local network structure characteristics, the neighbors of the selected node is immune, though its immunization effects is not good as the target immunization which is based on the global, the acquaintance immunization avoid knowing the node's degree of the network, but the accuracy of acquaintance immunization still need to be improved.

2.2 The maximum degree immunization strategy

The maximum immunization strategy is based on the ideological control from the local node of acquaintance immunization strategy, the basic idea is: selected $m = np$ nodes randomly from n nodes in the and calculate the degree of neighbors of each selected node, and select the neighbor node which has the maximum degree. This article uses the idea of the adjacency matrix to define a scale-free network connection array. The diagonal position number indicates the degree of the node, if n_i, n_j are connected, $n_i n_j$ corresponding array position -1, then the corresponding array position of $n_j n_i$ is -1; If there is no connection, the position of the corresponding array $n_i n_j$ and $n_j n_i$ are 0.

The maximum degree immunization strategy algorithm steps are as follows:

Step 1: Initialize the scale-free network topology, and construct a network containing n = 1000 nodes.
Step 2: Generate a random probability p, select m=np nodes from the network and added to the array Infecting.
Step 3: Simulate the virus spread in the networks use the m selected nodes, and the implementation of the maximum immunization strategy:

① Choose the node n_i from array infecting $(i = 1, 2, \ldots, m)$, And calculate the degree of its neighbor node, immune the neighbor node with the maximum degree, namely the biggest neighbor node where the row and column elements are set to 0, add n_i to array ImuneNod, update the network topology.
② Repeated to immune the nodes neighbors in the array infecting in accordance with the step ① until all nodes in the array of Infecting performed the immunization of this article.

Step 4: Generate a network after this immunization strategy is generated after immunization.

According to the rate of the spread of virus $\lambda = p/u$ [9], in the formula p is the rate of infected nodes, and u is the probability of recovery. In the experiment set $\lambda = 0.4$, in the experimental simulation, make the virus spread spontaneously in the network.

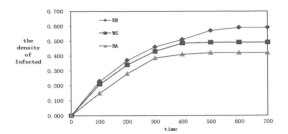

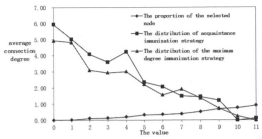

Figure 1. The immunization effect of maximum degree immunization strategy in ER random graph, the WS small-world network model and the network of BA scale-free network model.

Figure 2. The average connection degree comparison of the maximum degree immunization strategy and the acquaintance immunization strategy.

3 THE SIMULATION OF IMMUNIZATION STRATEGY

3.1 Immunization effect of the maximum degree immunization strategy in different networks

Define the infection density as the proportion of infected nodes in the network nodes as follows:

$$\rho = \frac{n}{N} \tag{3}$$

Among them ρ means the infection density and n for the infected nodes, N is the total number of nodes in the network.

Respectively, In accordance with the ER random graph, the WS small-world network model and BA scale-free network model to use the maximum degree immunization strategy within the stipulated time, the changes of infected nodes proportion shown in Figure 1: the maximum degree immunization strategy is more suitable for the BA scale-free networks.

3.2 The immunization simulation in scale-free networks

3.2.1 The average connection degree
The nodes' average connection degree in the network is an important parameter of a network, which reflects the average connectivity of the network, in general, the greater the average degree of connection, the better of the networks' average connection degree. It is calculated as follows:

$$D_{avg} = \frac{1}{n} \sum_{n=1}^{n} d_n \tag{4}$$

Here d_n represents the degree of the node n.

In the same topology of scale-free network with the same initial nodes, implement the acquaintance immunization strategy and maximum degree immunization strategy separately and figure out the network nodes distribution which was shown as follows in Figure 2. In the same network topology, by comparing the two strategist's average connection degree in varying proportions nodes in network. In the same proportion and with the same selected nodes in scale-free

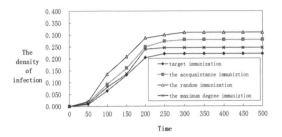

Figure 3. In the same probability of infection transmission of the virus by using four immunization strategies in Scale-free network.

networks, it's obviously that the maximum immunization strategy in the selected node ratio near $p < 0.6$, the average connection degree is less than implement acquaintances immunization strategy; When the selected proportion is $p > 0.6$, the effects of the maximum degree immunization strategy and the acquaintance immunization strategy are similar. Therefore, when select fewer nodes to immune, the effect of maximum immunization degree are better than the acquaintance immunization strategy.

3.2.2 The density of infection
In the case of ignoring the difference between the nodes in scale-free network that all nodes have the same probability of infection, 5% nodes are selected at random as infected nodes, simulate the spread of the virus in the network by the SIS propagation model, implement random immunization strategy, the acquaintance immunization strategy, the maximum degree immunization strategy and the target immunization strategy, do 25 times average value of the experimental results, the figure as below can be characterized the network node infected virus. From the figure we can be seen that the density of infection when implement the maximum degree immunization strategy in the network is higher than implement the target immunization, but is lower than the effects of acquaintance immunization strategy and the random immunization strategy. Compared with the effect of the random immunization strategy, use the strategy of acquaintance immunization, the target immunization and the maximum degree immunization strategy are easier to achieve the balanced state.

4 CONCLUSION

The immunization strategy effects of scale-free networks are validated: Ignoring the difference between the nodes, the average connection degree is lower when using the maximum degree immunization strategy network than using acquaintance immunization strategy in scale-free networks. By ignoring the difference of the nodes and considering the difference among nodes, the trends in the spread of the virus in scale-free networks are researched in this paper. The simulation results show that, in the two mechanisms, The density of maximum degree immunization is lower than the acquaintance immunization, but is higher than the target immunization, the effect of maximum degree immunization is between the effect of the acquaintance immunization and the target immunization, the maximum immunization strategy provides new ideas and solutions for using the limited funds to control the virus spread in the networks. In the study of the next job, introducing the parameters of initial infectious node as well as the location of the node to study.

For the virus spreading in the network, the initial infection node is the big degree node or a small degree node will also affect the rate of diffusion of the virus and the number of nodes eventually be infected in the complex network.

REFERENCES

[1] Erdös P, Rényi A. On the evolution of random graphs. Publ. Math. Inst. Hung. Acad. Sci., 1960, (05): 17~60.

[2] Watts D J, Strogatz S H. Collective Dynamics of Small-world Networks. Nature [J], 1998, 393(6684): 440–442.

[3] Barabási A L, Albert R. Emergence of scaling in random networks [J]. Science, 1999, 286 (543): 509–512.

[4] Pastor-Satorras R, Vespingnani A. Epidemic spreading in scale-free networks. Phy. Rev. Lett., 2001, 86(4): 3200~3203.

[5] Bailey N T J. The Mathematical Theory of Infectious Diseases and Its Applications. New York: Hafner Press, 1975.

[6] Gomez-Gardenes J, Echenique P, Moreno Y. Immunization of real complex communication networks [J]. European Physical Journal B, 2006, 49(2): 259–264.

[7] Cohen R, Havlin S, Ben-avraham D. Efficient immunization strategies for computer networks and populations [J]. Physical Review Letters, 2003, 91(24): 247901–247901.

[8] Pastor Satorras R, Vespignani A. Epidemics and immunization in scale-free networks [A]. Handbook of Graphs and Networks: Fran the Gename to the Internet [C]. Berlin: Wiley-VCH, 2002: 113–132.

[9] LI X, Chen G R. A local-world evolving network model [J]. Physical A: Statistical Mechanics and its Applications, 2003, (328): 274–286.

Information Technology and Computer Application Engineering – Liu, Sung & Yao (Eds)
© 2014 Taylor & Francis Group, London, ISBN 978-1-138-00079-7

Expert system based on testing knowledge management

Xiang Lei Zhao & Hong Ying Lu
School of Computer and Information Technology, Beijing Jiaotong University, Beijing China

ABSTRACT: Software testing is highly dependent on testers experience, skills and corresponding quality. To establish a knowledge management system will provide significant help to software testing. Also, the testing automation becomes more and more popular. First, a knowledge management system is built by using testers' experience and testing knowledge in which testers provide the most fundamental knowledge content, summarize knowledge content, provide summary information and preliminary screen knowledge then analysts assess knowledge content, classify knowledge, and establish a knowledge index to establish knowledge management system. Then, expert system knowledge base gains knowledge from the knowledge management system. The knowledge base accompanying with the inference engine, interpreter and conversational consists the expert system based on testing knowledge management with the key technology of knowledge classification, knowledge evaluation and knowledge representation. Using the expert system brings a great progress for software testing automation.

Keywords: Software testing; Knowledge management; Expert system.

1 INSTRUCTIONS

As a part of information management, knowledge management has been a key of information management and becomes more and more important. In order to improve the storage capacity of the company's internal knowledge and liquidity, more and more companies are trying to establish firm-specific knowledge management system.

Software testing, as a human-driven activity, has a strong dependence of knowledge. Therefore, to establish knowledge management system to summarize knowledge from test projects is an effective measure for software testing to improve the knowledge of the testers and ensure the quality of software testing.

At the same time, knowledge management system with a wealth of knowledge is the cornerstone of the testing automation. By classifying and abstracting knowledge, we can build the knowledge base and then complete the expert system based on testing knowledge management.

The knowledge management in software testing organizations and expert system based on testing knowledge management will be introduced in the next in detail.

2 KNOWLEDGE MANAGEMENT IN SOFTWARE TESTING ORGANISATIONS

Software testing is any activity aimed at evaluating an attribute or capability of a program or system and determining that it meets its required results (Hetzel

William C. 1988). As a knowledge-driven activity, software testing is highly dependent on tester's experience, skills and corresponding quality. At the same time, the experience of software testing is varied. For example, the experience of the test cases design in function testing, the knowledge of testing and the monitoring results analysis in performance testing, the knowledge of testing requirements analysis phase and so on. Testers use the experience and knowledge to guide the testing activities in the testing project. Therefore using effective knowledge management mechanisms in the testing organization and establishing perfect knowledge management system will provide significant help for the test. In the testing organizations, effective knowledge management can improve the testing knowledge's reuse rate, avoid the loss of test knowledge, improve the circulation of knowledge.

A perfect test knowledge management system consists mainly of four parts:

1) Knowledge acquisition part. Knowledge acquisition part is the key of knowledge management system because the knowledge for knowledge management system comes from knowledge acquisition part. Knowledge acquisition part gains the basic knowledge from the testing organization, screening and abstraction, and provides a summary of information about the corresponding knowledge, so as to convey knowledge to the knowledge base of the management system.

2) Knowledge assessment part. The main task of knowledge assessment part is to assess the gained

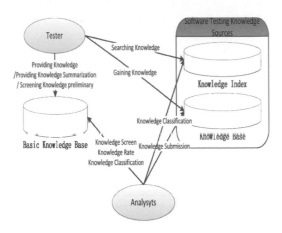

Figure 1. Knowledge management system.

knowledge and delineate the weight of knowledge. Assessing the knowledge and delineating the weight of knowledge will provide guidance for the retrieval and application of knowledge. Knowledge can be assessed from different aspects, such as the level of knowledge citation and knowledge Raised level.

3) Knowledge classification retrieval part. Knowledge reuse is the fundamental purpose of knowledge management system. Testing-related knowledge is varied, so effective classification of knowledge is good for testers to retrieve knowledge and apply knowledge.

4) Knowledge storage part. Knowledge storage part is the basis of knowledge management system. The knowledge storage structure of the knowledge management system determines the stability of the knowledge management system. The preservation and accumulation of knowledge is also one of the purpose of knowledge management system.

Figure 1 reveals the work process of the testing knowledge management system.

Testers are the operating body of knowledge acquisition system and the source of knowledge. By summing up the experience in test projects, they provide the most fundamental knowledge content for knowledge management system. At the same time, testers should summarize knowledge content and provide summary information then preliminary screen knowledge by searching for knowledge in knowledge management system to avoid duplicate knowledge. Analysts are the operating body in knowledge assessment part and classification knowledge retrieval part, they will assess knowledge content, classify knowledge, and establish a knowledge index and other related work. Analysts are managers of knowledge database. They should build and maintain knowledge databases and knowledge indexing library by corresponding knowledge screening assessment tools provided by the system.

3 EXPERT SYSTEM BASED ON TESTING KNOWLEDGE MANAGEMENT

Expert system is a program system with expert level problem solving in a specific area. It is a computer system that emulates the decision-making ability of a human expert (Jackson Peter 1998). It can use experience and expertise accumulated by experts effectively to solve the problem that expert does, by simulation expert's thought. Expert systems are designed to solve complex problems by reasoning about knowledge, like an expert, and not by following the procedure of a developer as is the case in conventional programming (Regina Barzilay et al. 1998). The first expert systems were created in the 1970s and then proliferated in the 1980s (Cornelius T. Leondes. 2002). Expert systems were among the first truly successful forms of AI software.

An expert system has a unique structure, different from traditional computer programming. It is divided into two parts, one fixed, independent of the expert system: the inference engine, and one variable: the knowledge base. To run an expert system, the engine reasons about the knowledge base like a human. In the 80s a third part appeared: a dialog interface to communicate with users. This ability to conduct a conversation with users was later called "conversational".

Depending on the application purposes, the expert system can be divided into: 1. Diagnostic expert system: system based on the observation and analysis of the symptoms, derive out the cause of symptoms and troubleshooting methods. 2. Interpreted expert system: system to explain the deep structure or internal environment based on surface information. 3. Predictive expert system: system to forecast the future according to the status. 4. Design expert system: system design products according to the given product. 5. Decision-making expert system: system to choose a feasible option from comprehensive evaluation. 6. Planning expert system: system to develop an action plan. 7. Teaching expert system: system with the ability to support teaching.

Software testing is a complicated process, according to the type of test purpose and the test method, software testing can be divided into several different types. Also, a test process can be divided into different test stages. According to different testing types and the testing stages, the expert system gives corresponding guidance which is of great help for the software test automation development. For example, in the design phase of function testing test case or the performance testing test scenarios, we can build an expert system to design case using case designing experience and knowledge. For the analysis phase of the testing and monitoring results of the performance test, we can build a performance test results diagnostic expert system by using correlation analysis knowledge. It can be seen that the expert system will greatly improve the automation of the testing process.

The basic structure of the general expert system is shown in Figure 2. The inference engine based on the

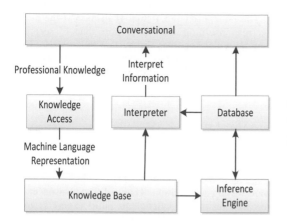

Figure 2. Expert system structure.

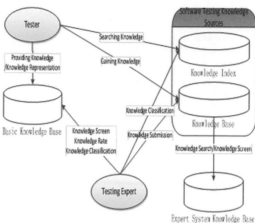

Figure 3. Expert System Knowledge Base.

corresponding knowledge in the knowledge base, reasons the data in the consolidated database and records the reasoning process so as to achieve the purpose of using machines instead of experts.

For expert system, the most important part is expert knowledge base. The quality of knowledge in expert knowledge base largely determines the quality of the expert system itself. However, the knowledge base is different in nature with the knowledge management system knowledge base. First of all, the knowledge base of knowledge management system is storing as a form of natural language knowledge, system users access to text form of knowledge and apply knowledge in practice while knowledge in the expert system knowledge base should be stored as two forms that one is formal language and the other one is natural language, because the knowledge of the expert system not only requires to interact with human, but also required to interact with the machine. In addition, the knowledge base of knowledge management system is quite complex and huge while the knowledge in the expert system is an aspect about the testing process. Also it is relatively simple to build the expert system.

The expert system knowledge base is essentially different from the knowledge management system knowledge base. But the knowledge management system can provide a lot of knowledge for the expert system knowledge base. By extracting the appropriate classification of natural language knowledge and formalizing the knowledge, it will provide basic knowledge for the expert system knowledge base.

Figure 3 describes the basic flow envisage to build expert system using knowledge in the knowledge management system. Testers give further formal representation when the knowledge is provided, so as to provide two forms of knowledge representation in knowledge base. The expert system builders can retrieve the appropriate types of knowledge, to gain a large number of knowledge about how to build the expert system. Builders make the knowledge in the knowledge base dump to expert knowledge base by screening the knowledge, serving for the corresponding type of expert system.

4 KEY TECHNOLOGY IN BUILDING EXPERT SYSTEM KNOWLEDGE BASE

4.1 Knowledge classification and evaluation

Knowledge classification and evaluation is good for the building of the knowledge base of expert system. As expert system often faces part of the testing process, so classification of software testing knowledge can improve the efficiency of the expert system knowledge base to be built. For example, according to whether the properties of the internal structure of the software are concerned, software testing knowledge can be divided into: white-box testing knowledge, black-box testing knowledge, gray-box testing knowledge. Different knowledge based on knowledge characteristics can belong to various knowledge types. Meanwhile, the assessment of knowledge will give reference for the knowledge choosing. If there is conflicting knowledge in the expert system, analysts choose for knowledge based on assessment coefficient of the knowledge.

4.2 Knowledge representation

How to represent and store knowledge in the knowledge base is the key to build an expert system (Russell Stuart J. et al. 2003). The main knowledge representation methods in Artificial Intelligence are: first-order predicate logic representation, generation representation, frame representation, semantic grid representation, object-oriented representation.

Generation can be used to represent the reasoning knowledge and the basic rule is IF P THEN Q (Allen Newell et al. 1972). Among it, P is a precondition for generation, used to indicate whether the conditions of generation are available while Q is a set of conclusions or actions used to indicate what the conclusion should be or operation should do when P is satisfied. P or Q may be a mathematical expression or a natural language. Uncertainty rules knowledge in the knowledge base can be used $P \rightarrow Q$ (credibility), IF P THEN Q (credibility) to represent. Formulation using tuple as follows: (object, attribute, value), or (relation, object 1, object 2). For example: "IF the identity of the germ is not known with certainty AND the germ is gram-positive AND the morphology of the organism is "rod" AND the germ is aerobic THEN there is a strong probability (0.8) that the germ is of type enterobacte-riacae" (Buchanan B.G. et al. 1984). This formulation has the advantage of speaking in everyday language which is very rare in computer science (a classic program is coded), so knowledge with the form of natural language in management system knowledge can be expressed by generation representation, then stored in expert knowledge base.

In the performance testing analysis process, the expert system is often used. For example, there exists knowledge in the knowledge base of testing knowledge management system with the following documentation form:

CPU may become a bottleneck omen in the system:

1) slow response time;
2) zero percent idle CPU;
3) high percent CPU usage.

Memory may become a system bottleneck omen

1) small free memory;
2) high memory paging rate.

As described above, in most text representation, the knowledge is described in the form of uncertainty, such as "high", "slow" and some fuzzy meaning words. Therefore, in the description of generation, knowledge provider should give the specific representation of knowledge and provide credibility of uncertainty rules knowledge.

For example:
Rule 1:
E1: (CPU, average response time, >15 s)
H: (become, CPU, system bottleneck)

The generation If E1 then H means "if the average response time of CPU is more than 15 seconds then the CPU becomes the system bottleneck". The expression of uncertainty knowledge is the basis to make knowledge be used by machine to build expert system.

5 CONCLUSION

With the development of software testing, the software testing automation demand has become increasingly. Building expert system will greatly enhance the software testing automation. Using different aspects of software testing knowledge to build expert system based on the testing knowledge management will bring great progress to testing automation.

REFERENCES

Allen Newell & Herbert Alexander Simon. 1972. *Human problem solving*. Prentice Hall.
Buchanan B.G. & Shortliffe E.H. 1984. *Rule Based Expert Systems: The MYCIN Experiments of the Stanford Heuristic Programming Project*. Addison-Wesley.
Cornelius T. Leondes. 2002. *Expert systems: the technology of knowledge management and decision making for the 21st century*: 1–22.
Hetzel William C. 1988. *The Complete Guide to Software Testing, 2nd ed*. Wellesley.
Jackson Peter. 1998. *Introduction To Expert Systems (3 ed.)*. Addison Wesley.
Regina Barzilay, Daryl McCullough, Owen Rambow, Jonathan DeCristofaro, Tanya Korelsky & Benoit Lavoie. 1998. *A new approach to expert system explanations*.
Russell Stuart J. & Norvig Peter. 2003. *Artificial Intelligence: A Modern Approach (2nd ed.)*. Prentice Hall.

Information Technology and Computer Application Engineering – Liu, Sung & Yao (Eds)
© *2014 Taylor & Francis Group, London, ISBN 978-1-138-00079-7*

Research of a new model about the age reduction of repairable systems

Jin Qiang Liang, Jiong Sun & Kai Liu
Office of Research and Development, Naval University of Engineering, China

Yi Zhang
Department of Weapon Engineering, Naval University of Engineering, China

ABSTRACT: Study about the phenomenon of excessive repair during the maintenance of repairable systems. The weakness of the traditional method and age reduction mechanism about excessive repair is analyzed. Meanwhile, its shortcoming of is modified. Then, a newly nonlinear decreasing model about the age reduction of repairable system is established, and is simplified on the situation of periodically preventive maintenance. Finally, An example is given, and the results of simulation proved the superiority and feasibility of the models. Besides, some advices about maintenance is given according to the results of simulation.

Keywords: Repairable System; Excessive Repair; Periodically Preventive Maintenance; Non-linear Decreasing; Age Reduction

1 INSTRUCTIONS

Studying about the variation of repairable systems after maintaining has always been an important work of Reliability and Maintainability and Supportability (RMS). In 1979, Mailk firstly proposed the conception of improvement factor [1], in order to describe the effect of maintenance on the failure rate of systems. In 1998, I.T. Dedopoulos and Y. Smeers [2] developed the Mailk's theories, and put forward the age reduction factor-η, which can be use to express this effect through the age reduction of systems and has been wildly used. Nowadays, most of researchers use a stationary η so as to simplify the calculation and research [3] [4] [5]. However, the age reduction of systems is actually not stationary and will decrease with the increase of maintenance, which shows us a phenomenon of excessive repair.

Thus, according to the weakness of traditional methods, this paper modifies the traditional age reduction mechanism, and establishs a newly nonlinear decreasing model about the age reduction of repairable system.

2 WEAKNESS OF TRADITIONAL AGE-REDUCTION METHODS

Nowadays, we use both age reduction factor (η) and equivalent age (e) to explicate the failure rule of repairable system.

Table 1. Age reduction factor (η) of different kinds of maintenance.

Different kinds of maintenance	Perfect maintenance	Imperfect maintenance	Minimal repair	Worst repair
Age reduction factor (η)	$\eta = 1$	$0 < \eta < 1$	$\eta = 0$	$\eta < 0$

The definition of age reduction factor (η) is the degree of the age reduction after every maintenance, which also means the degree of "rejuvenation".

According to different kinds of maintenance, age reduction factor (η) is different. As listed in table 1, it can be divided into four kinds:

The equivalent age (e) indicates an equivalent age of systems after every maintenance. It expresses the connection between the actual age and reduced age, shows us the effect of maintenance on the age of systems.

Then, their connections can be written as follows:

$$e_1 = t_1 - \tau_1 \quad , \quad \tau_1 = \eta_1 \cdot T_1;$$
$$e_2 = t_2 - \tau_2 \quad , \quad \tau_2 = \eta_2 \cdot T_2;$$
$$\vdots$$
$$e_n = t_n - \tau_n \quad , \quad \tau_n = \eta_n \cdot T_n; \qquad (1)$$

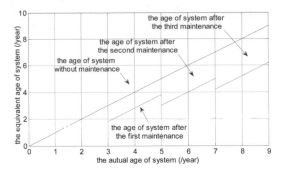

Figure 1. The age of systems after every maintenance.

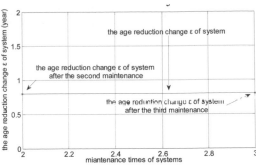

Figure 2. The age reduction change ε between any twice maintenance.

Then, the age of systems after every maintenance can be written as follows:

$$
\begin{cases}
e = t - \tau_i \\
\tau_i = \eta_i \cdot T_i \\
i = 1, 2, \cdots, n, \cdots \\
t \in [0, \infty) \\
e \in [0, \infty)
\end{cases}
\tag{2}
$$

Where i is the frequencies of maintenance, T_i is the specific time of every maintenance. τ_i is the reduction age of every maintenance, η_i is the age reduction factor of every maintenance.

As can be seen from Fig. 1 and equation (2), the reduced age τ_i after every maintenance only depends on the age reduction factor (η_i) and the specific time (T_i).

While if we assume the age reduction factor is stationary (it also means every η_i is equal) as most researchers do, a new conclusion can be draw as follows.

Firstly, we make a definition that the age reduction change between any twice maintenance is ε, which describes the further effect of next maintenance.

The equation can be written as follows:

$$
\varepsilon_{ij} = \tau_i - \tau_j = \eta \cdot (t_i - t_j) = \eta \cdot T_{ij}
\tag{3}
$$

And the equivalent age (e) can be written as follows:

$$
\begin{aligned}
e_i &= e_j + (1-\eta)T_{ij} = e_j + T_{ij} - \varepsilon_{ij} = (1-\eta) \cdot t_i \\
&= (1-\eta)(t_j + T_{ij}) \\
& \quad i > j, i = 1, 2, \cdots, j = 1, 2, \cdots
\end{aligned}
\tag{4}
$$

where i and j is frequencies of maintenance, T_{ij} is the maintenance cycle which means the period between the ith maintenance and the jth maintenance.

Specially, if the maintenance cycle T_{ij} is equal ($T_{ij} = T_0$), the change ε can be a constant ($\varepsilon_0 = \eta \cdot T_0$). As shown in Fig. 2, $\varepsilon_{21} = \varepsilon_{32} = \eta T_0 = \varepsilon_0$.

Apparently, traditional methods can represent the age reduction of systems after maintenance, but it neglects the influence of maintenance times because of the assumption of equal η. During the actual maintenance, we always face the situation that the effect of

maintenance becomes less and less with maintenance times increasing, it shows us a phenomenon of excessive repair. At this moment, it must waste our time and expenses to keep on maintenance.

Thus, according to the weakness of traditional methods, a new age reduction model is established.

3 A NON-LINEAR DECREASING AGE REDUCTION MODEL

Firstly, to modify the traditional definition of η as follows:

Age reduction factor η is the degree of the age reduction after the first maintenance, which also means the degree of "first rejuvenation". It indicates that η only depends on the techniques of maintenance, and has nothing to do with the maintenance times or the actual age.

Then, to make an definition that increments of the age reduction change ε is U and decrements of ε is D, which both express the change of ε.

Therefore, the new age reduction mechanism can be presented as follows:

$$
e_n = t - \tau_n \ , \quad \tau_n = \varepsilon_{n(n-1)} + \tau_{n-1} \ , \quad \varepsilon_{n(n-1)} = U_n + D_n .
$$

the new mechanism expresss that the effect of maintenance on systems is not single but comprehensive, it contains all of the earlier and current maintenance.

And the change of effect reflect on the ε, which includes increments and decrements. Apparently, the ε is changeable and relate to η, maintenance cycle T and maintenance times n.

Actually, as time went by and increase of maintenance times, their connection can be listed in table 2.

According to traditional methods, $\varepsilon = g(\eta, T) = \eta \cdot T$, and $\frac{\partial g}{\partial \eta} > 0$, $\frac{\partial g}{\partial T} > 0$. Then, as the new age reduction mechanism indicated, $\varepsilon = f(\eta, T, n)$, and $\frac{\partial f}{\partial \eta} > 0, \frac{\partial f}{\partial T} > 0, \frac{\partial f}{\partial n} < 0$. The boundary conditions is written as follows:

i) Initial value is 0:

$$
\varepsilon\big|_{\eta=0} = \varepsilon\big|_{T=0} = \varepsilon\big|_{n=0} = 0 ;
$$

Table 2. The connection among ε, η, T and n.

	η	T	n
ε	+	+	−

*"+" means that dependent variables will increase (decrease) with independent variables increasing (decreasing). "−" means that dependent variables will increase (decrease) with independent variables decreasing (increasing).

ii) there is a final value:

$$\varepsilon = f(\eta, T, \mathrm{n})\big|_{n \to A} = 0, \quad A \text{ exists or } A \to +\infty ;$$

iii) Nonnegative: $\varepsilon \geq 0$.

To combine the traditional methods and the new one, the connection among ε, η, T and n can be expressed as follows:

$$\varepsilon = f(\eta, T, \mathrm{n}) \propto \eta ;$$

$$\varepsilon = f(\eta, T, \mathrm{n}) \propto T ;$$

$$\varepsilon = f(\eta, T, \mathrm{n}) \propto \frac{1}{n} .$$

Thus, the new function of age reduction change ε is written as follows:

$$\varepsilon = f(\eta, T, \mathrm{n}) = \eta \cdot T \cdot f_1(n)$$

Where $f_1(n)$ is the decreasing function of age reduction change ε, and $\frac{\partial f}{\partial n} = \eta \cdot T \cdot \frac{df_1}{dn} < 0$. The non-linear decreasing function of ε means that ε will decrease as the maintenance times increase, and the decrements is a not constant, it's a non-linear gradual change, scilicet, $\frac{df_1(n)}{dn} = -\frac{1}{n^2}$.

Then, the age reduction change ε is written as follows:

$$\varepsilon_{21} = \frac{\eta_2 \cdot T_{21}}{2} ;$$
$$\vdots \qquad\qquad (5)$$
$$\varepsilon_{n(n-1)} = \frac{\eta_n \cdot T_{n(n-1)}}{n} .$$

Where, n is the maintenance times, and $n = 2, 3, 4, \ldots$. Of course, equation (5) is satisfied with the boundary conditions.

Then, the reduced age τ_i can be written as follows:

$$\tau_1 = \eta_1 \cdot T_1 ;$$
$$\tau_2 = \tau_1 + \frac{\eta_2 \cdot T_{21}}{2} ;$$
$$\vdots \qquad\qquad (6)$$
$$\tau_n = \eta_1 \cdot T_1 + \frac{\eta_2 \cdot T_{21}}{2} + \cdots + \frac{\eta_n \cdot T_{n(n-1)}}{n}$$

The age of systems after the nth maintenance can be written as follows:

$$e_n = t - \left(\eta_1 \cdot T_1 + \frac{\eta_2 \cdot T_{21}}{2} + \cdots + \frac{\eta_n \cdot T_{n(n-1)}}{n} \right) \qquad (7)$$

Then, according to equation (5), (6), (7), a new age-reduction model can be written as follows:

$$\begin{cases} e = t - \tau_i \\ \tau_i = \eta_1 \cdot T_1 + \frac{\eta_2 \cdot T_{21}}{2} + \cdots + \frac{\eta_i \cdot T_{i(i-1)}}{i} \\ \varepsilon_{n(n-1)} = \frac{\eta_i \cdot T_{i(i-1)}}{i} \\ i = 1, 2, \cdots, n, \cdots \\ t \in [0, \infty) \\ e \in [0, \infty) \end{cases} \qquad (8)$$

4 THE SIMPLIFIED AGE-REDUCTION MODEL BASE ON PERIODICALLY PREVENTIVE MAINTENANCE

The meaning of periodically preventive maintenance [6][7] is that we don't maintain the systems until it has went through the maintenance cycle T, and T is a constant.

Then, we make some assumptions as follows:

i) If the system breakdown during the maintenance cycle T, we will repair it by minimal repair [8], which can't change the age of systems.

ii) The time spent on preventive maintenance or minimal repair is lift out.

iii) To assume that the techniques of every maintenance is identical, then, every η is equal ($\eta = \eta_1 = \eta_2 = \ldots$).

Then, the model based on periodically preventive maintenance is presented as follows:

$$\begin{cases} e = t - \tau_i \\ \tau_i = \eta \cdot T_1 + \eta \cdot T \cdot \sum_{m=1}^{i} \frac{1}{m} - \eta \cdot T \\ \varepsilon_{i(i-1)} = \frac{\eta \cdot T}{i} \\ i = 1, 2, \cdots, n, \cdots \\ t \in [0, \infty) \\ e \in [0, \infty) \end{cases} \qquad (9)$$

5 COMPUTATIONAL EXAMPLE AND ANALYSIS

In order to illustrate the feasibility and effectiveness of the model, we take certain repairable system for example and simulate [9] it by the traditional and new

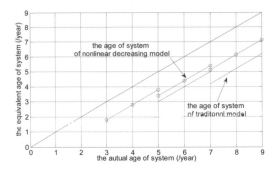

Figure 3. The age of systems after every maintenance.

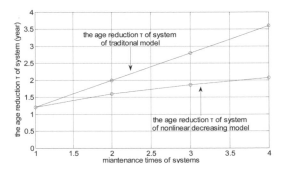

Figure 4. The age reduction τ of systems after every maintenance.

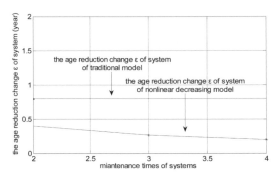

Figure 5. The age reduction change ε of systems after every maintenance.

approaches respectively. The parameters are set as follows: the maintenance cycle is 2 years, age reduction factor η is 0.4, the maintenance times is 3, and the system has been firstly maintained in the third year. The results of simulation are presented as follows:

As shown in fig. 3, both of the traditional and new models can describe the age reduction of systems, the difference is that the new one express the age reduction of systems decrease with the increase of maintenance, which shows us a phenomenon of excessive repair.

As shown in fig. 4 and fig. 5, we compare the description of age reduction τ and age reduction change ε between the two models. Both of them can illustrate the variation of τ and ε, the difference is that new model indicate the increase of age reduction τ is varied from large to little, and trended to steady. Besides, to the age reduction change ε, the new model proposes that ε will decrease with maintenance times increasing and be trended to zero at last.

All in all, according to the result of simulation, the age reduction of the system has become slight after the third maintenance. Then, the maintenance depot should consider how to improve η with adjusting their techniques.

6 CONCLUSIONS

As the result of simulation given above, the new model can preferably describe the phenomenon of excessive repair during maintenance, and modify the weakness of traditional methods, which has made a good influence on the improvement of maintenance.

REFERENCES

[1] Chan, J. K. & Shaw, L., Modeling repairable system with failure rate on age and maintenance, IEEE Trans. Reliab., 1993, 42: 566–571.
[2] L.T. Dedopoulos, Y. Smeers. An age reduction approach for finite horizon optimization of preventive maintenance for single units subject to random failures. Computers ind. Engng, 1998 Vol. 34, No. 3: 643–654.
[3] Mahmood Shafiee, Ming J. Zuo. Adapting An Age-Reduction Model to Extend The Useful-Life Duration. The 3rd International Conference on Maintenance Engineering. 2012, P. 1057–1061.
[4] Wei Peng. Diesel Repair Analysis Modeling and Application. University of Electronic Science and Technology. 2008.
[5] Peng Sun, Shaohui Chen, Caiqin Zhang. LCC Assessment on the Best Life Cycle of Substation Equipments. Southern Power System Technology. 2011, 5(6): 96–100.
[6] Zhaoyong Mao, Baowei Song, Peng Tu. Optimization of Preventive Maintenance Period based on GA . Fire Control and Command Control. 2006, 31(11): 73–75.
[7] Yuncong Yang, Ping He. Flexible cycle preventive maintenance model based on reliability constraint. Journal of Huanggang Normal University. 2012, 32(3): 10–14.
[8] Gu Wei, Chu Jianxin. "Fault information model and maintenance cycle forecasting for ship's power system", Reliability and Maintainability Symposium, 2002. Proceedings. Annual, 2002. p. 445–449.
[9] Pin Zhou, Xinfang Zhao. Analysis of MATLAB statistics. Beijing: National Defense Industry Press, 2009: 270–283.

Information Technology and Computer Application Engineering – Liu, Sung & Yao (Eds)
© 2014 Taylor & Francis Group, London, ISBN 978-1-138-00079-7

Face detection method based on skin color and template matching

Q.H. Liu, B. Li & M. Lin
Shanghai University, Shanghai, China

ABSTRACT: In order to solve the face detection problem of color image under complex background, this paper proposes a novel method combined with skin color detection and template matching. Firstly, after comparing skin segmentation effects of some color spaces, YCgCb color space is confirmed to skin detection. Then using region merging rules makes the separated face affected by accessories to merge. Further by template match searches the faces in face candidate regions and confirms the detected face. Non-face regions are gotten rid of at the same time. Experiments show that this method is versatile and effective.

Keywords: YCgCb color space; face detection; region merging; template matching

1 INSTRUCTION

Face detection is to determine whether there are any faces in the given image and video, and if exist, return the size, pose and range of each face [1]. Face detection is the first step of any face processing system, which is not only applied to face recognition, also can be used in many fields of vehicle safety driving, vision inspection, digital video processing and so on. Face detection is a challenging task and the detection process is affected by many factors, such as the difference of human's appearance, occlusions on the face, imaging angle and light conditions, something similar to face color in nature. Therefore, face detection has been one of the hot researches by scholars at home and abroad.

For a single image, the techniques to detect face can be classified into four categories [1]: knowledge-based method, feature invariant approaches, template matching methods, appearance-based methods, including some classical algorithms of face detection. For example, Sung and Poggio [2] proposed based on sample learning algorithm to solve frontal face detection without rotating under the complex background. Osuna [3] et al. put SVM (support vector machine) into use for face detection by classification to distinct 'face' and 'non-face'. Viola and Jones [4] used Adaboost algorithm combined with Cascade algorithm to realize the real-time and efficient face detection. Hsu [5] proposed face detection algorithm based on skin color model under complex background.

Comparing with the commonly used algorithm for face detection, such as SVM, Adaboost, neural network and so on, they all require a large number of training samples. The detection complexity and time-consuming are increased greatly. The face detection method based on skin color with the advantages as a small amount of calculation, rapid processing speed, can be used as a preliminary screening method to detect face. But because of the influence of illumination, background complexity and occlusions on face, missing or error detection could appear only by skin color detection method. Therefore it is necessary to utilize the other method to further verify. This paper presents a face detection method combined with skin color and template matching.

2 SKIN COLOR DETECTION

2.1 Selection of color space

Skin color as one of the most obvious features of face, can be expressed in a variety of color spaces, such as RGB, rgb, YUV, YCbCr, HIS, YIQ, Lab [6]. The difference of human's skin color is mainly caused by brightness, but less affected by chrominance. So the color space which is not affected by luminance is preferred. YCbCr color space has the characteristic of chrominance and luminance separated, and each component of YCbCr is a non-linear relationship. YCbCr is a color space that is suitable for skin color segmentation. YCgCr color space proposed by De Dios [7] and YCgCb color space in [8] have the same characteristic of YCbCr color space; also can be applied to skin color detection.

In order to determine the most suitable color space for skin color segmentation in this paper, 150 color images contains faces are selected. From these images, 200 pieces of skin sample including face skin and body skin are cut manually. The total number of skin color pixels is 15992880 which consist of skin color training samples. All of the skin color samples are converted to YCbCr, YCgCr and YCgCb color space from RGB

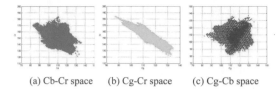

(a) Cb-Cr space   (b) Cg-Cr space   (c) Cg-Cb space

Figure 1. Space distribution of skin color.

(a) Original image

(b) Detection result of YCbCr

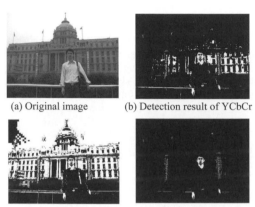

(a) Original image   (b) Detection result of YCbCr

(c) Detection result of YCgCr (d) Detection result of YCgCb

Figure 3. Multiplayer detection results in three color spaces.

(c) Detection result of YCgCr (d) Detection result of YCgCb

Figure 2. Single detection results in three color spaces.

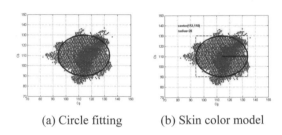

(a) Circle fitting   (b) Skin color model

Figure 4. Skin color model of circle fitting in Cg-Cb.

respectively. The sample pixels gather respectively in three color spaces to form two-dimensional space distribution figures. As is shown in Figure 1, three color spaces have the different clustering effects and all of the clustering effects are very good.

According to the distribution model of skin color clustering, the color range of each chrominance component in three color spaces is preliminarily identified. In YCbCr color space, select $[85 < Cb < 135]$, $[125 < Cr < 180]$ to meet skin color regions. In YCgCr color space, select $[125 < Cr < 180]$, $[90 < Cg < 135]$ to meet skin color regions. In YCgCb color space, select $[90 < Cg < 135]$, $[80 < Cb < 135]$ to meet skin color regions. Using above skin color models detects 50 color images contain human faces. Comparing the detection results in every color spaces, it is concluded that the best color segmentation result is in YCgCb color space. Most of non-skin regions in the background are separated. As is shown in Figures 2–3. So this paper determines to detect skin in YCgCb color space firstly.

2.2 Establishment of skin color model in YCgCb

When YCgCb color space is confirmed to detect skin color, the clustering shape is observed. As is shown in Figure 1c, the shape that is composed of all of the sample pixels clustering is similar to a circle in Cg-Cb color space. Skin color model is established by using a circle to fit the boundary of clustering shape. See

Figure 4. Mathematical expression is the best method to show skin color model. See Equation 1 below:

$$\frac{(x - 112)^2 + (y - 110)^2}{20^2} \leq 1$$
$$\begin{cases} x = 20 \cos \theta + 112 \\ y = 20 \sin \theta + 110 \end{cases} \qquad (1)$$

After many times fitting and calculation, when the center of skin color fitting boundary circle locates at (112, 110), 20 for radius size, the boundary fitting effect is best. It ensures almost clustering points are inside the internal of the circle. If the values of pixels in the inspected image meet to Equation 1, namely the pixels fall into the internal circle in Cg-Cb plane, the pixels can be judged as skin, otherwise as non-skin.

2.3 Morphological operations

Based on circle skin color model in YCgCb color space, the color image can convert into binary image which contains two kinds of colors as white and black. The white regions temporarily are regarded as skin perhaps including non-skin, and all of the black regions belong to non-skin. In order to effectively remove the noise and interference, while preserve the useful information of image, a series of morphological operations are processed. Firstly some small noises in the background are removed by median filtering. Next, with

(a) Original image   (b) Detection result of YCgCb

(c) Morphological operation result (d) Label candidate regions

Figure 5. Processing of face segmentation.

open and close operation, erosion and dilation operation, many complete connected regions are formed. All the connected regions can be regarded as the candidate region of face.

2.4 Region merging

Although skin color is consistent in the same face, some accessories such as glasses, beard and hair would affect face to be detection. There is a great difference between color characteristic of these accessories and skin color. Morphological processing sometimes divides face into multiple regions because of the above reason. As is shown in Figure 5, the third person from left is divided into two parts because of glasses. In order to solve this problem, this paper uses region merging [9].

This paper mainly discusses the face is divided into upper and lower two parts by glasses or beard. Some candidate regions would be merged based on two rules below:

(1) Color consistency: The chrominance of same face is similar usually. The mean of Cg component and the mean of Cb component are calculated in YCgCb color space with two adjacent regions. If the in equations $|\overline{C_{g1}} - \overline{C_{g2}}| \le T_g$, $|\overline{C_{b1}} - \overline{C_{b2}}| \le T_b$ are satisfied, where T_g, T_b are certain threshold, it can conclude that the two regions are similar.

(2) Location adjacency: Just consider two neighboring regions up and down. Suppose there are two regions where W_1, W_2, h_1, h_2 are width and height of region 1 and region 2.

$d_x d_y$ are respectively the center distance of x direction and y direction.

If the in equations $d_y < \frac{h_1+h_2}{2} + \Delta_y$ and $d_x < \frac{W_1+W_2}{4}$ are satisfied at the same time, it is considered region 1 and region 2 are the same region.

As is shown in Figure 5d, the rectangles marked 3 and 4 meet the above rules, so they are adjacent regions. The two regions can be merged to one region by the minimum enclosing rectangle of them. The next template matching is applied to the merged region

(a) Template matching result   (b) Final result

Figure 6. Detection result.

to verify face. And the candidate regions of face are decreased.

3 TEMPLATE MATCHING

After skin color segmentation and region merging, the candidate regions of face are obtained. But it is possible that some non-face regions exist such as skin of body or something in the background similar to skin color. In order to locate the face regions accurately and exclude the non-face regions, template matching method is applied further. Template matching is one of the methods to search target in the image. The correlation degree is judged by the template and target. This paper calculates the correlation between face template and every candidate regions of face to select the scale and location of true face.

Face template is constructed by the mean value of 30 faces, without any accessories such as glasses, beard and hair. The average face size is 50×50. Face template must adapt to different size and direction of face. Firstly, the length and width, centroid and rotation angle of candidate region is calculated. Next, face template is adjusted by this information to reach the best matching effect. Refer to Equation 2 below, the correlation coefficient is calculated. If the correlation coefficient is more than 0.4, the region is a face, otherwise excluded. Template matching result see Figure 6.

$$c = \frac{\sum_m \sum_n (A_{mn} - \overline{A})(B_{mn} - \overline{B})}{\sqrt{\left[\sum_m \sum_n (A_{mn} - \overline{A})^2\right]\left[\sum_m \sum_n (B_{mn} - \overline{B})^2\right]}} \quad (2)$$

where c = correlation coefficient; A_{mn} = face template matrix; $\overline{A}$ = average of face template matrix; B_{mn} = candidate region matrix and $\overline{B}$ = average of candidate region matrix.

4 EXPRIMENTAL RESULT

In order to prove the face detection method of this paper whether has the versatility and effectiveness, 200 color images under non-uniform background that mainly contain multiple person are selected to construct test set. Some images come from camera taking and some images come from Internet downloading.

(a) Five people   (b) Thirteen people

Figure 7.   Partial face detection results.

Table 1.   Statistical data.

Samples numbers	Face	hits	misses	Precision
1	5	5	0	100%
2	8	8	0	100%
3	15	15	0	100%
4	24	22	2	91.7%
5	37	34	3	91.9%
Total	89	84	5	94.4%

Table 2.   Contrastive analysis result.

Method	Based on skin color only	This paper
Detection ratio	87.5%	93.5%
False ratio	8.2%	6.0%

Faces of test set have various kinds of position, scale, orientation and facial expression. The partial experimental results are shown in Figure 7. The statistical data is shown in Table 1. Comparing the experimental result of this paper's method with the detection method based on skin color only, the contrastive analysis result is shown in Table 2. The experimental result shows that the method combined with skin color and template matching has high detection rate and low false acceptance rate. And this paper's detection effect is better than that method only based on skin color.

5   CONCLUSION

In this paper, a face detection method for color images using skin color detection and template matching is proposed. The detection process includes selection of skin color space, skin color segmentation, morphological operation, region merging and template matching. In YCgCb color space, by circle skin color model, skin and non-skin are separated. Morphological operation

help to simplify calculation and form the best face candidate regions. Especially the problem of face that is segmented because of accessories is solved by region merging successfully. And then template matching method is used for further face confirmation. Experiments show this method has better detection effects for face detection under non-uniform background. It can be applied to face recognition field and other face process fields.

There still exit some problems need to solve. Especially something in the background similar to skin which is connecting with face would enlarge the face range and interfere with the detection of face. In further works, the above problem is worth to research. Moreover some other face detection methods can be combined with the method of this paper to enhance the detection rate and reduce the false detection rate.

REFERENCES

[1]   Yang M.H. & Kriegman D.J. & Ahuja N. 2002. Detecting faces in images: A survey. *IEEE Trans Pattern Analysis and Machine Intelligence.* 24(1): 34–58.

[2]   Sung K. & Poggio T. 1998. Example-based learning for view based human face detection. *IEEE Trans Pattern Analysis and Machine Intelligence.* 20(1):39–51.

[3]   Osuna E. & Freund R. & Girosi F. 1997. Training support vector machines: An application to face detection. *Proc. IEEE Conf. on Computer Vision and Pattern Recognition*: 130–136.

[4]   Viola P., Jones M., Robust real time object detection, Technical Report, CRL 2001/01, Compaq Cambridge Research Laboratory, February 2001.

[5]   Hsu Rein-Lien, Mohamed, Jain Anil K. 2002. Face detection in color images, *IEEE Trans. on Pattern Analysis and Machine Intelligence.* 24(5):696–706.

[6]   Liang, L.H & Ai, H.Z. & Zhang, B. 2002. A survey of human face detection. *Chinese J. Computers* 25(5): 449–458.

[7]   de Dios J.J. & Garcia N. 2003. Face detection based on a new color space YCgCr. *Proceedings of International Conference on Image Processing*: 909–912.

[8]   Zhang, Z.Z. & Shi, Y.Q. 2009. Skin color detecting based on clustering in YCgCb color space under complicated background. *International Conference on Information Technology and Computer Science*: 410–413.

[9]   Ai, H.Z. & Liang, L.H. & Xu, G.Y. 2001. Face detection based on skin color and template. *Journal of software* 12(12):1784–1792.

Information Technology and Computer Application Engineering – Liu, Sung & Yao (Eds)
© 2014 Taylor & Francis Group, London, ISBN 978-1-138-00079-7

Image-based crack detection and properties retrieval for high-speed railway bridge

D.F. Zhang & N. Zhang
School of Computer and Information Technology, Beijing Jiaotong University, Beijing, China

ABSTRACT: Recently, image processing technology has been developed into the inspection of bridges to detect cracks. However, there are still some difficult problems unresolved, for example, noises in the concrete bridge pictures are very similar to cracks, may be even more dominant than cracks. In this paper, an improved method to handle this situation is proposed. A modified percolation model is used to detect cracks automatically, generating a crack map for automatic measurement. Then, a distance transform is employed to get the information of distance and store it into black pixels. Finally, a thinning algorithm is applied to the crack map to get the skeletons of cracks and extract data stored in pixels. The performance of the proposed method is evaluated by receiver operating characteristics analysis. Experimental results show that the proposed method can simulate the real data correctly.

Keywords: crack detection; properties retrieval; crack width; automatic measurement

1 INSTRUCTION

With the development of High-speed rail, the demand of maintenance of High-speed Railway Bridge is further increased. In the bridge inspection, it's critical to retrieve the crack properties precisely. Usually, the crack is measured manually by specialists. However, this lacks efficiency and objectivity which are needed for quantitative analysis. Besides, it can be very dangerous for people because the bridges of high-speed railway are usually very high. Consequently, some new detection methods must be developed to improve the inspections.

Automatic crack detection is a promising research point and has been rapidly developed in recent years. Abdel-Qader et al. (Abdel-Qader et al. 2003) gives a comparison of the effectiveness of crack detection methods for the images of a bridge surface. Among the four common used methods, fast Haar transform, fast Fourier transform, Sobel and Canny, it's concluded that the fast Haar transform is most reliable. Y. Fujita et al. (Fujita et al. 2006) proposed a method to detect cracks using a substraction preprocessing to remove slight variation, as well as a Hessian matrix filter to emphasize the line structures. Moreover, threshold processing is applied to the image to separate cracks from the background. Je-Keun Oh et al. (Oh et al. 2009) proposed a robotic system for inspecting the safety status of bridges using computer vision. T. Yamaguchi et al. (Yamaguchi & Hashimoto 2006) (Yamaguchi, Nakamura, & Hashimoto 2008)

(Yamaguchi & Hashimoto 2007) (Yamaguchi, Nakamura, Saegusa, et al. 2008) (Yamaguchi & Hashimoto 2010) proposed an automated crack detection method using percolation model. Z. Zhu et al. (Zhu et al. 2011) proposed a method to retrieve concrete crack properties for automated post-earthquake structural safety evaluation. However, the methods discussed above mainly focus on the locating crack points, neglecting the properties retrieval from detection. Besides, the situations of High-speed Railway Bridge should be taken into account. Because of the shade of the beams under the bridge, bottom of the bridge pictures are usually cannot be not well observed. Textures and noises can be even more dominant than the slight cracks. For this situation, a modified method based on the percolation model (Yamaguchi & Hashimoto 2006) (Yamaguchi, Nakamura, & Hashimoto 2008) (Yamaguchi & Hashimoto 2007) (Yamaguchi, Nakamura, Saegusa, et al. 2008) (Yamaguchi & Hashimoto 2010) is proposed in this paper. A mask is used to scan black points to run percolation processing, while a binary image is generated at the same time for automatic measurement. This procedure will be discussed in the following part. With the binary image, some morphology methods (Zhu et al. 2011) are applied to the image to get crack properties. The camera calibration is not concerned in this paper, we focus on the automatic calculation of crack properties and assume that the calibration properties are already known. 30 real bridge pictures are used to test the effectiveness of the proposed method.

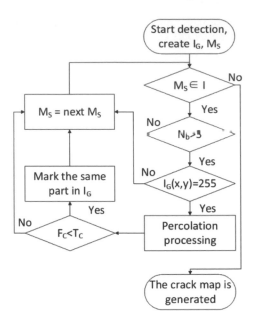

Figure 1. Method Overview.

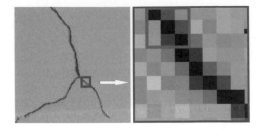

Figure 2. The Process of Scanning.

2 CRACK DETECTION

The accuracy of the crack detection is a crucial consideration which affects the measurement results largely. So far, there doesn't exist perfect solutions for crack detection, because some slight cracks are inevitable lost in image processing while removing the noise with filters. For the safety of the bridge, obvious cracks must be detected and maintained immediately. In the inspection of the bridge, the maximum crack width is the most important element which determines whether the bridge should be repaired or not. That's not say the slight cracks are not harmful but just emphasize that the obvious ones are more important to be measured. A reasonable hypothesis is proposed that obvious cracks in pictures always have a cluster of some lowest intensity points. Based on this hypothesis we use a modified percolation model to detect cracks. In this paper, 256-level gray scale pictures are used in the crack detection. And color pictures will be converted to gray scale. Contrasts of all pictures are improved in order to stretch out the intensity range using histogram equalization.

2.1 Method overview

C Figure 1 shows the flowchart of the crack detection process. The detailed steps of the proposed algorithm are given as follows.

1. When image I is opened for check, a separate image I_G with the same size of the input image is created, then the intensity of all positions of the new image is set as: $I_G(x, y) = 255$, which means that I_G is a pure white image. And a 3*3 mask is used to scan the original image I, we call it M_S, which denotes that mask for scanning.

2. Check the mask M_S to find that whether it's in the region of image I or not. If the answer is yes continue the process, and if the answer is no, it means that the detection is over and the image I_G has been marked. The generated image I_G is the crack map in the proposed method.

3. Check the intensities of pixels in mask M_S, pixel $p_i \in M_S$, i ranges from (1, 9), and N_b is the number of pixels where $I_p = 0$. When $N_b \geq 3$ the process continues, and when $N_b < 3$ the mask M_S moves to scan next part. The threshold value 3 is chosen because we consider the crack that needs to be measured has a 3*3 mask that contains more than 3 pixels with the lowest intensity 0. A diagrammatic sketch is given in Figure 2. The percolation process will cover other pixels of the crack and this will be discussed in the following section. Some very slight cracks those have only grey color will be ignored as we discussed before. This scanning is performed to save time because the number of the percolation process will be decreased drastically in automatic crack detection.

4. Check the intensity of pixels in I_G, for pixel $p_i \in M_S$ in image I, assume its coordinate value is $I(x, y)$, its corresponding pixel in I_G will be $I_G(x, y)$. When $I_G(x, y) = 255$ it means that the corresponding part in image I is not percolated and the percolation processing will start. On the contrary, the percolation processing will be skipped to save computation time.

5. Percolation processing is performed. The detail will be discussed in next section.

6. Check the percolated part with F_c (Yamaguchi & Hashimoto 2006). Compare F_c with the threshold value T_c, where F_c denotes the circularity of the percolated part, it is expressed by the equation

$$F = \frac{4 \cdot C_{count}}{\pi \cdot C_{max}^2} \qquad (1)$$

where C_{count} is the number of pixels in the percolated cluster and C_{max} is the maximum length of the cluster. The value of F ranges from 0 to 1. When F is close to 1, it means that the shape of the cluster is similar to a circle, while the value F of a crack is close to 0. The quality is used to classify the dark pixels in pictures into two parts, cracks and non-cracks. If $F_c \geq T_c$, mask M_S moves to the next, if $F_c < T_c$, the corresponding part in image I_G will be marked, by setting the intensity value $I_G(x, y) = 0$.

Once the mark process is completed, the mask M_S moves to the next. The loops will continue unless the scanning is completed.

2.2 *Percolation process*

Percolation is the process of a liquid slowly passing through a filter. It is a nature phenomenon that developed into the use of computer science. This model is very effective to describe the procedure of expanding. Water poured at crack boundaries always makes its way to fill the cracks, while water poured on a concrete surface will be spread evenly as a circle. T. Yamaguchi et al. (Yamaguchi & Hashimoto 2006) proposed a method to detect cracks using image-processing in 2006 and developed it in next few years with some improvements. The accuracy of crack location (Yamaguchi, Nakamura, Saegusa, et al. 2008) and the computing time (Yamaguchi, Nakamura, & Hashimoto 2008) (Yamaguchi & Hashimoto 2010) are both improved. In this paper the crack detection method is modified from the percolation model. Basically the intensities of crack pixels are easily distinguished from non-crack pixels, which is the foundation of the percolation method. As discussed before, the percolation process is started when some conditions are satisfied. And the center pixel in the mask M_S is chosen to be the seed point. The details of the percolation procedure are described as below. (Yamaguchi & Hashimoto 2010)

1. The seed point p_s selected to initiate the percolation is included in the percolated part D_p, the 8 neighboring regions of D_p are defined as the candidate region D_c. Threshold T is set to the value of the initial pixel brightness $I(p_s)$.
2. In the region of D_c, check the intensities of pixels, those satisfied the equation $(I_p)_{p \in D_c} \leq T$ are percolated and included in D_p. If such pixels do not exist and the value of threshold $T < 30$, the least bright pixel in D_c is included in D_p, and the threshold T will be updated using the equation:

$$T = \max_{p \in D_p}(I_p) \qquad (2)$$

This operation is performed to involve the sub-pixels of cracks which have higher intensities. At the same time the maximum value of T should be limited, otherwise the percolation process will grow into non-crack regions. This maximum value T_{max} is determined by experiments.
3. As the region D_p expands, new pixels neighboring to D_p are included in D_c. The process of percolation continues until there are no pixels having lower intensities in D_c than those in D_p and $T = T_{max}$.
4. The circularity F_c of D_p will be calculated after the percolation completed using equation (1), and following operations are performed to generate the final binary image I_G as discussed in the step 6 of method overview.

This modified percolation process is quite different from the original method. Firstly, the proposed

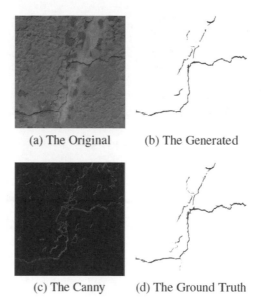

(a) The Original   (b) The Generated

(c) The Canny   (d) The Ground Truth

Figure 3.   The Comparison of Different Methods.

method performs the percolation process only in the area that satisfied some conditions as given in the method overview, in order to suppress interference of noises and reduce computation time. Secondly, the parameter w used to accelerate the percolation (Yamaguchi, Nakamura, Saegusa, et al. 2008) is not used in the proposed method, the region D_p slowly expands by including the darker pixels neighboring to it, like water flows downhill, and to void a local minima, a back-flow way is designed, the least bright pixel in D_c is included in D_p. And T_{max} is designed to prevent the percolation part expands to the non-crack area. The proposed method works effectively on finding cracks. Figure 3 shows the original picture, the generated crack map, detection using canny edge detector and the ground truth image created manually. Figure 3(b) generated by the proposed method can simulate the ground truth image quite precisely.

3   PROPERTIES RETRIEVAL

When the crack map is generated, properties of cracks should be retrieved from it to evaluate the safety of the bridge. The properties in the investigation include crack length, maximum width, and average width and so on. The maximum width is a very crucial data (Lee et al. 2008). A thinning processing is performed to get the length of the crack and a distance transform is applied to the picture to get the maximum width of the crack in the picture.

3.1 *Distance transform*

A distance transform is also known as distance map or distance field. It is an operator normally applied to binary images. The result of the transform is a

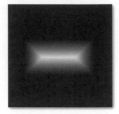

Figure 6. An Example of Thinning.

Figure 4. The Input and Output of Distance Transform.

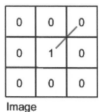

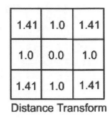

0	0	0
0	1	0
0	0	0

Image

1.41	1.0	1.41
1.0	0.0	1.0
1.41	1.0	1.41

Distance Transform

Figure 5. Euclidean Distance Transform.

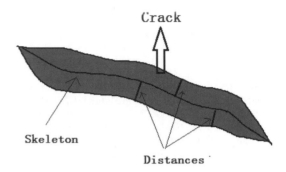

Figure 7. The Reconstruction of a Crack.

gray scale image that looks similar to the input image, except that the intensities of points inside foreground regions are changed to show the distance to the closest boundary from each point (Fisher et al.), as shown in Figure 4.

In this paper, the Euclidean metric is employed in the distance transform, which means that the distance used in procedure is the straight-line distance between two pixels. The Euclidean distance between point p and q is the length of the line segment connecting them. Figure 5 shows an example of Euclidean distance transform.

This method is very potential to store the width information in the skeleton pixels. The transform function calculates the distance between each pixel that is set to off (0, the black pixels in I_G) and the nearest nonzero pixel for crack maps. Different from Figure 5, the distance information would be stored in the "black pixels" (only the shortest distance would be saved). When the distance transform is completed, the black pixels in the crack map would contain a data of its distance information to the closest boundary. The retrieval of information would be completed by thinning process.

3.2 Thinning process

Generally the width of a crack is the distance between its two edges, the measuring direction should be perpendicular to the orientation of the crack (Nishikawa et al. 2012). In fact it is difficult to determine the actual orientation automatically by computer vision. The skeleton of a crack is used instead of the orientation. Thinning process is performed to get the skeleton of a crack. In digital image processing, morphological skeleton is a skeleton (or medial axis) representation of a shape or binary image, computed by means of morphological operators. Thinning is the transformation of a digital image into a simplified, but topologically

equivalent image, it is obvious that thinning produces a sort of skeleton (Fisher et al.) (Fisher et al.). Figure 6 gives an example of the thinning process.

After the procedure of the distance transform, the information of a crack shape is stored in the skeleton pixels of a crack. Figure 7 shows the reconstruction of a crack using the skeleton points and the distance values from the skeleton to the crack boundaries (Zhu et al. 2011). The doubled results of the data retrieved from skeleton is the width of a crack. Obviously, the average of the doubled results will be the average width of the cracks, and the maximum number will be the maximum width, the length of the skeleton will be the length of the crack. The data obtained from the properties retrieval is the number of pixels, rather than the actual width. Further unit conversion is performed using the results of camera calibration, which is not the concern of this paper.

4 RESULTS ANALYSIS

The results analysis includes two parts, the precision of the crack location and the measurement errors of the results. For precision analysis, a ground truth image of the original image is created manually. The generated crack map is compared to the ground truth image, and the Receiver Operating Characteristics (ROC) analysis is performed (Hutchinson & Chen 2006). ROC curve is graphical plot which illustrates the performance of a binary classifier system as its discrimination threshold is varied. The same method is used in references (Fujita et al. 2006) (Fujita & Hamamoto 2010) (Yamaguchi, Nakamura, Saegusa, et al. 2008).

The evaluation of the detection results can be consider a two-class prediction problem, in which the outcomes are labeled either as positive (P, black pixels are real cracks) or negative (N, white pixels are

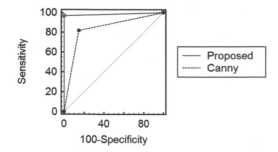

Figure 8. The ROC Curve of Proposed Method and Canny.

Table 1. Experiment Results of Properties Retrieval.

Image Num.	Results	Real Data	Error
Image 1	0.0619	0.1	−0.0381
Image 2	0.1728	0.15	0.0228
Image 3	0.3376	0.35	−0.0124
Image 4	0.1033	0.12	−0.0167
Image 5	0.2976	0.27	0.0276
Image 6	0.1920	0.17	0.0220
Image 7	0.0641	0.12	−0.0559
Image 8	0.0130	0.05	−0.0370
Image 9	0.2130	0.2	0.0130
Image 10	0.0166	0.05	−0.0334
Image 11	0.2602	0.27	−0.0098
Image 12	0.1591	0.17	−0.0109
Image 13	0.1148	0.13	−0.0152
Image 14	0.2212	0.22	0.0012
Image 15	0.1026	0.1	0.0026
Image 16	0.1913	0.15	0.0413
Image 17	0.1563	0.12	0.0363
Image 18	0.1933	0.17	0.0233
Image 19	0.2495	0.26	−0.0105
Image 20	0.3713	0.35	0.0213
Image 21	0.1692	0.16	0.0092
Image 22	0.2085	0.24	−0.0315
Image 23	0.2275	0.18	0.0475
Image 24	0.1854	0.17	0.0154
Image 25	0.2068	0.2	0.0068
Image 26	0.1558	0.13	0.0258
Image 27	0.1531	0.14	0.0131
Image 28	0.0329	0.08	−0.0471
Image 29	0.3119	0.32	−0.0081
Image 30	0.0627	0.07	−0.0073

not cracks). There are four possible outcomes from a crack map. If a black pixel from the crack map is on a real crack, then denote as true positive (TP); however if the actual position of the pixel is not on a crack then it is identified to be a false positive (FP). Conversely, a true negative (TN) is the situation that white pixel in the crack map is not on a crack, and false negative (FN) is when the white pixel is on a crack.

Name the generated crack map I_G and the ground truth image I_T, then TP can be regarded as the intersection of the two sets of pixels satisfied $I_p = 0$ (I_p means that the intensity of p is 0) in I_G and I_T, which can be represented as $(p_{(Ip=0)} \in I_G) \cap (p_{(Ip=0)} \in I_T)$. Similarly FP can be represented as $(p_{(Ip=0)} \in I_G) \cap (p_{(Ip=255)} \in I_T)$. P and N will be $(p_{(Ip=0)} \in I_T)$ and $(p_{(Ip=255)} \in I_T)$.

To draw a ROC curve, only the True Positive Rate (TPR) and False Positive Rate (FPR) are needed (as functions of some classifier parameter). TPR and FPR can be represented as:

$$TPR = \frac{TP}{P} = \frac{(p_{(Ip=0)} \in I_G) \cap (p_{(Ip=0)} \in I_T)}{(p_{(Ip=0)} \in I_T)}$$

$$FPR = \frac{FP}{N} = \frac{(p_{(Ip=0)} \in I_G) \cap (p_{(Ip=255)} \in I_T)}{(p_{(Ip=255)} \in I_T)}$$

The ROC curve generated using Figure 3(b) and Figure 3(d) shows in Figure 8. The closer a result is to the upper left corner, the better it predicts. The curve indicates that the proposed method works effectively on finding cracks, and avoid taking noises as cracks.

The experiment results of the maximum widths and the real data measured manually are given in Table 1. The unit of all numbers is millimeter (mm). The proposed method finds out all cracks those are larger than 0.15 mm, while the safety standard is 0.2 mm. The results shows that the measurement error is very small, the proposed method can simulate the real data correctly.

5 CONCLUSION

The inspection of High-speed Railway Bridge is very important for the safety of the operation of the High-speed rail system. And manually inspection is burdensome and impossible. Therefore, a solution with

modified percolation model is proposed in this paper. By this way, the crack detection can be performed automatically. Besides, a binary image is generated to simplify the measurement procedure. Though some information of slight cracks is lost, the dangerous cracks are kept in the generated images with no noisy parts. It becomes much easier to get properties of cracks from clean picture without disturbs than present methods. And experimental results show that the proposed method can simulate the real data effectively.

ACKNOWLEDGEMENT

The study is supported by National High Technology Research and Development Program of China (2012AA040912).

REFERENCES

Abdel-Qader I, Abudayyeh O, Kelly ME. 2003. Analysis of edge-detection techniques for crack identification in bridges. *Journal of Computing in Civil Engineering*. 17:255–263.

Fisher R, Perkins S, Walker A, Wolfart E. Morphology – Distance Transform [Internet]. [cited 2013a Apr 27].

Fisher R, Perkins S, Walker A, Wolfart E. Morphology – Skeletonization/Medial Axis Transform [Internet]. [cited 2013b Apr 27].

Fisher R, Perkins S, Walker A, Wolfart E. Morphology – Thinning [Internet]. [cited 2013c Apr 27].

Fujita Y, Hamamoto Y. 2010. A robust automatic crack detection method from noisy concrete surfaces. *Machine Vision and Applications* [Internet]. [cited 2013 Apr 21]; 22:245–254.

Fujita Y, Mitani Y, Hamamoto Y. 2006. A method for crack detection on a concrete structure. In: *Pattern Recognition, 2006 ICPR 2006 18th International Conference on. Vol. 3.* [place unknown]: IEEE; p. 901–904.

Hutchinson TC, Chen Z. 2006. Improved image analysis for evaluating concrete damage. *Journal of Computing in Civil Engineering* [Internet]. 20:210–216.

Lee JH, Lee JM, Park JW, Moon YS. 2008. Efficient algorithms for automatic detection of cracks on a concrete bridge. In: *The 23rd International Technical Conference on Circuits/Systems, Computers and Communications.* [place unknown]; p. 1213–1216.

Nishikawa T, Yoshida J, Sugiyama T, Fujino Y. 2012. Concrete crack detection by multiple sequential image filtering. *Computer-Aided Civil and Infrastructure Engineering* [Internet]. 27:29–47.

Oh J-K, Jang G, Oh S, Lee JH, Yi B-J, Moon YS, Lee JS, Choi Y. 2009. Bridge inspection robot system with machine vision. *Automation in Construction* [Internet]. 18:929–941.

Yamaguchi T, Hashimoto S. 2006. Automated crack detection for concrete surface image using percolation model and edge information. In: *IEEE Industrial Electronics, IECON 2006-32nd Annual Conference on.* [place unknown]: IEEE; p. 3355–3360.

Yamaguchi T, Hashimoto S. 2007. Practical image measurement of crack width for real concrete structure. *IEEJ Transactions on Electronics, Information and Systems.* 127:605–614.

Yamaguchi T, Hashimoto S. 2010. Fast crack detection method for large-size concrete surface images using percolation-based image processing. *Machine Vision and Applications* [Internet]. 21:797–809.

Yamaguchi T, Nakamura S, Hashimoto S. 2008. An efficient crack detection method using percolation-based image processing. In: *Industrial Electronics and Applications, 2008 ICIEA 2008 3rd IEEE Conference on.* [place unknown]: IEEE; p. 1875–1880.

Yamaguchi T, Nakamura S, Saegusa R, Hashimoto S. 2008. Image-Based Crack Detection for Real Concrete Surfaces. *IEEJ Transactions on Electrical and Electronic Engineering.* 3:128–135.

Zhu Z, German S, Brilakis I. 2011. Visual retrieval of concrete crack properties for automated post-earthquake structural safety evaluation. *Automation in Construction* [Internet]. 20:874–883.

Information Technology and Computer Application Engineering – Liu, Sung & Yao (Eds)
© 2014 Taylor & Francis Group, London, ISBN 978-1-138-00079-7

The research of video retrieval technology

J.X. Wang
Department of Computer, Hebei Institute of Architecture and Civil Engineering, Zhangjiakou, China

Y.L. Wang & Z.H. Ma
Department of Mathematics and Physics, Hebei Institute of Architecture and Civil Engineering, Zhangjiakou, China

ABSTRACT: With the continuous development of information technology and the rapid growth of the video information, how to effectively organize, manage, express and retrieve video data has become a hot issue in video retrieval research field. In this paper, it discusses the background of the video image, extraction of dynamic target, index and sort the target image features, and video retrieval and browsing. The simulation experiment has achieved good results to test video clips with different characteristics. It is proved that the validity of the research methods through comparison with other methods test results.

Keywords: Establish the background image; Dynamic target detection; Image feature extraction; Video retrieval

1 INTRODUCTION

With the increasing number of surveillance cameras in the city, the video data are rapid growth. How to quickly obtain the information required by the user has become a key issue of intelligent video surveillance system. In surveillance applications, people are concerned on only a smaller proportion of the target picture, instead of the features of the whole image. In the retrieval, people more expect to find occurrences of the same target at different times and different locations. Usually, the target is moving rather than stationary. Therefore, the existing Internet video retrieval methods are usually difficult to be applied to surveillance video retrieval. The research of moving target discovery provides feasibility and support for retrieval, which is for the moving target object.

With the construction and development of communication technology and network technique, network monitoring terminal becomes a popular trend. The monitoring terminal around the city will form a three-dimensional interactive network in the futher. Object-oriented video retrieval technology can search for video content between terminals and give full play to the function and role of network. It is also can quickly search and location of a target in the city.

2 VIDEO DATA ANALYSIS

In order to quiry the video database with the content-based, we first to construct to facilitate retrieval of video structure. A video sequence is composed of frames, is the camcra to a group of consecutive image frames in the process of closing down from the open. In practical application, because between the frames of images in the same shot content redundancy degree, image can be selected to reflect main information content in the lens as key frame. Users to browse all image frames in a scene is very time consuming, so commonly used key frame technology for fast browsing. Key frame is the most important representative in the lens of one or more images of representative. Based on the complexity of the content of the lens, it can be one or more key frames extracted from a lens.

3 DATA COLLECTION

The system automatically analyzes the new collection of video data, extracts dynamic target information, and then adds to the database. Feature vector is calculated to target small picture at query time. We regard each video as a document, a small picture as a word. Then the document is a combination of a series of words, with a group of corresponding feature vector to represent. Finally, the set of documents to establish the inverted index provide fast retrieval[1]. First we use background modeling algorithm to video moving object extraction.

3.1 *Background modeling*

Multi-frame average method background extraction algorithm error is relatively large, the statistical histogram background extraction algorithm's time and space, and the size of the error depends on the size of the interval with the split range and background to match the degree of adaptationso in practical applications often can not obtain satisfactory results.

Multi-frame average method background extraction's algorithm error is relatively large, and the statistical histogram background extraction has large time complexity and space algorithm. The size of the error depends on the size of the interval with the split range and background to match the degree of adaptation, so in practical application, it usually can not get satisfactory results. Based on clustering thought time adaptive background extraction and updating algorithm is a background modeling algorithm after in-depth study[2]. Through the experiment, it has prominent effect. The process of the algorithm is to remove the same position in the N frame image for each pixel value, and then find the mean and variance. If the variance is greater than a certain threshold, then remove the pixel value that is farthest with the mean. Then it is for the mean and variance again. Until the remaining variance of pixel value is less than a specific threshold, the mean is the pixel of background value. Then take the next pixel value until all the pixel values of the image background extraction is completed. The algorithm steps are as follows.

Step 1: Sort point with a position on the N-frame image pixel gray value. Read the gray N-frame in video streams with a position of image pixel values. These data are in increasing order, into a pre-cleared (Temp) array. At the beginning, k = 1, x = 1, y = 1. With the Equation 1 expressed as:

$$Temp(k) = \begin{cases} Temp(k-1) & if \quad Image_K(x,y) < Temp(k) \\ Image_K(x,y) & if \quad Image_K(x,y) \geq Temp(k) \end{cases}$$
$$k = k+1 \qquad\qquad if \quad k < N \qquad (1)$$

In the formula (1), TEMP (k) means that the the array Temp deposited ascending ordering K gray value data. $1 \leq k \leq N$. Image$_k$ (x, y) denotes the gray value of (x, y) at the k frame image. $1 \leq x \leq Height$ (Image row height). $1 \leq y \leq Width$ (Image column width). It is a total of N frames.

Step 2: Calculating mean and variance of the pixel at image sequences (x, y) point. Initialize v low = 1, high = N. With the Equation 2 expressed as:

$$Average(x,y) = \frac{1}{high-low+1} \sum_{k=low}^{high} Temp(k)$$
$$D(x,y) = \sum_{k=low}^{high} (Temp(k) - Average(x,y))^2 \qquad (2)$$

Average(x, y) is the average gradation of the video image sequence at the point (x, y). And D(x, y) is the variance. The low point to a minimum value has not removed in the Temp array; the high point to a maximum value has not removed in the Temp array.

Step 3: Comparison of the variance and threshold. If the variance is greater than the threshold, comparing of the first and last data value in the Temp with the

Average(x, y) distance value, and then remove that a large distance. Repeat Step 2 and Step 3. If the variance is less than the threshold, the average value of the pixel is the background value.

When $D(x,y) > TH$, using Equation 3 is expressed as:

$$low = low+1 \quad if \quad Average(x,y) > \frac{1}{2}(Temp(low)+Temp(high))$$
$$high = high-1 \quad else \qquad\qquad (3)$$
$$goto \qquad step2$$

When $D(x,y) \leq TH$, using Equation 4 is expressed as:

$$B\,Image(x,y) = Average(x,y) \qquad (4)$$

B Image (x, y) is a video image sequence (x, y) at the background gray value. In this paper the TH threshold value is 50.

Step 4: Move a row or column unit, repeat Step 1, Step 2, Step 3 until the entire images are calculated in the background pixel value B Image (x, y). With the Equation 5 expressed as:

$$y = y+1 \quad goto \quad step1 \qquad if \quad y < width \qquad (5)$$
$$x = x+1, y=1 \quad goto \quad step1 \quad if \quad x < height \quad and \quad y = width$$

X represents the number of rows in the image pixel; y represents the number of columns in the image pixel. Width is the width of the image and height is the height of the image.

3.2 Moving object extraction

The video has significant changes in light (sun, light) and other dynamic background. But these video clips have almost the same background scene. If we use content-based network image (video) retrieval methods directly, because of the highly similar between these videos, we can't return the correct results. By moving target detection techniques, however, we eliminate the adverse effects of the same background on the retrieval results by moving objects in the scene to retrieve the object.

After we have the background image, the current frame and background image subtraction, regional T$_i$ can get moving target. Specific algorithms such as Equation 6:

$$T_i(x,y) = \begin{cases} 1 & if \; | I_i(x,y) - B(x,y)| \geq TH \\ 0 & otherwise \end{cases} \qquad (6)$$

Due to the influence of noise and disturbance, there are some holes and isolated noise target image by using the algorithm directly. We can use some mathematical morphology method to split the two value image processing. Then we use the median filter to remove noise present in the image of the isolated, by morphological dilation fill voids present in the target area. Finally, using the treatment after the template image to intercept the target area scene images, and achieved good results.

4 DATA FEATURE EXTRACTION

Researchers are usually selected to measure certain characteristics of video video content discontinuity, such as color, motion vector features, edge feature, etc.

Usually the surveillance video is not very clear, and has a larger amount of data. While the color feature of the process is more convenient, has good resistance. So we choose the image color characteristics to be related to retrieval[3]. Commonly used color space with the RGB color space, the HSV color space, the HSI color space. The HIS color space is based on the human visual system, to describe a color from three angles of hue, saturation and brightness, accords with people's perception and discrimination to the color. So we select HIS color space to extract the color histogram.

For a given image, we first use the Equation 7–9 to turn the image from the RGB color space into HIS color space.

$$H = \begin{cases} \arccos\left(\dfrac{(R-G)+(R-B)}{2\sqrt{(R-G)(R-G)+(R-B)(G-B)}}\right), & R \neq G或R \neq B \\ 2\pi - \arccos\left(\dfrac{(R-G)+(R-B)}{2\sqrt{(R-G)(R-G)+(R-B)(G-B)}}\right), & B > G \end{cases} \quad (7)$$

$$S = MAX(R,G,B) - MIN(R,G,B) \quad (8)$$

$$I = \frac{R+G+B}{3} \quad (9)$$

Because there are many kinds of color conversion, direct calculation takes time and plenty of storage space. So we calculate the advanced sex of the proper color quantization, which can save a large amount of space and improve operational efficiency. Color histogram feature completely discard the space structure color distribution, the visual difference images may have the same feature vector. For example, one wearing a red coat and black trousers man and a red pants in a black coat man should have no similar or similarity is very low, it is necessary to distinguish between the two. But the difference between the color histogram may be very similar. In order to deal with this situation, we divide the image into several sub image block, and then extract the color histogram respectively. Because the main target of the database in the video is pedestrians, we follow the structural characteristics of people to 0.3:0.4:0.3 ratio of the image is divided into upper, middle and lower three. Then we extract color feature vector. Finally, we image the three blocks of color histogram is normalized, and the characteristics of the various parts together, to form a multi-dimensional color histogram as the characterization of the original image and to retrieve.

5 INDEXING AND RANKING DATA CHARACTERISTICS

Inverted table is a more mature and efficient indexing technology. It has been widely used in text retrieval

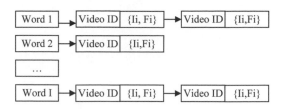

Figure 1. System uses the inverted index structure.

based on the webpage, especially suitable for large-scale data index. Object-oriented video retrieval system also uses the inverted index table. Wherein, each index entry includes a video number and the number of pairs of the target image and the frame number. It records the word appearing in the video image of the target entry of the number and the destination of which is in the video frame.

Analogy with the text document retrieval, video segment is equivalent to the document, the characteristics of the target image is equivalent lexical items. But the image has an almost infinite variety of possibility, we must find a limited number of dictionary to describe all of the video documentation. The process of moving target detection and feature quantization is equivalent to the text of the steps in the index entry. After a comparative analysis of various algorithms, we use the Hash algorithm for dictionary quantified[4]. As shown in Figure 1.

6 SYSTEM FUTURE TREND

At present, the system only using the color feature of video retrieval, it also needs to make the function and performance of the upgrade and expansion in many ways in the future. It includes the following[5]: the fusion of multiple features, face recognition, retrieval based on local feature. One of the future research directions is the description of the event study method and index for abnormal events.

REFERENCES

[1] C.D. Manning, P. Raghavan, and H. Schutze, Introduction to Information Retrieval. ISBN: 978-7-115-23424-7, 2010.9.

[2] E. Erdem, and S. Dubuisson, and I. Bloch. Fragments Based Tracking with Adaptive Cue Integration [J]. Computer Vision and Image Understanding, 116(7): 827–841, 2012.

[3] J. Huang, and S.R. Kumar, and M. Mitra, and W.J. Zhu, and R, Zabih, Image indexing using color correlograms [C]. IEEE Computer Conference on Computer Vision and Pattern Recognition, pp. 762–768, 1997.

[4] Y. Weiss, A. Torralba, and R. Fergus. Spectral Hashing [C]. NIPS 2008.

[5] N. Dalai and B. Triggs. Histograms of oriented gradients for human detection [C]. In Proc. CVPR, 2005, pp. 886–893.

Information Technology and Computer Application Engineering – Liu, Sung & Yao (Eds)
© 2014 Taylor & Francis Group, London, ISBN 978-1-138-00079-7

Water quality remote retrieve model based on Neural Network for dispersed water source in typical hilly area

X.J. Long & Y. Ye
College of Resources and Environment, Southwest University, Chongqing, China

C.M. Zhang
College of Water Resource and Hydropower & State Key Library of Hydraulics and Mountain River Engineering, Sichuan University, Chengdu, China

ABSTRACT: Based on TM image data and synchronization of the measured data, this manuscript constructs the quantitative inversion model of water quantity parameters, in view of distributed water resource in Yongchuan distinct. In order to compare the inversion effect, BP ANN model and multiple linear regression model are applied. Experimental results show that (1) The water quality parameters have different correlation with each wave band. CS and NH_3-N reflect evidence relativity to TM band 1–4. And Chl, SD and COD_{mn} just prominent relate to TM band 6. (2) This paper presents a valuable trail of comparative analysis. The application of remote sensing technology to estimate water quality can get more information, efficiently. And coulping with BP ANN can supply better prediction data than traditional regression model. (3) For long term monitoring, BP ANN couple with remote sensing inversion can be more effort and economical.

Keywords: In order to Dispersed water resources; Water quality retrieving; Remote sensing image; BP Neural Networks Model

1 GENERAL INSTRUCTIONS

Water quality monitoring is the main basis of water quality assessment and water pollution prevention and control. Accurate and efficient water quality monitoring for water conservation is very important. Water quality of traditional monitoring needs point-wise collection and quantitative chemical analysis. This method not only requires plentiful of labour, material and financial resources, but also inefficient. Therefore, it is more and more difficult to satisfy social development with backward techniques. Since the 1970s, the development of satelite remote sensing bring new approach to this industry [Wang et al. 2012]. Remote sensing technology, by contrast, has the following advantages which are much more excellent than traditional methods, such as wide coverage, temporal synchronization, low cost, real-time acquisition and convenient monitor for long term and dynamically [Yu et al. 2008]. With the progress of remote sensing technology theory, method research and applications, direct or indirect monitoring content have been involved Chlorophyll, suspended solids, transparency, water tempreture, dissolved organic matter and so on [Dekker 2002, Gu et al. 2007, Giardino & Pepe 2001]. These studies also demonstrated the feasibility of remote sensing inversion.

Remote sensing inversion is a nonlinear process. Nonlinear approaching ability of artificial neural network for the implementation provides a powerful technical support. And the coupling model have been applied by water quality parameter retrieval [Chen 2003, Shi et al. 2006, Zhao et al. 2009]. This manuscript, based on the TM image of Landsat-5 with 1–4 waveband and 6 wave band and measurment data of decentralized water resources, constructs the BP neural network inversion model for water quality parameter prediction. Moreover, in order to quicken learning outcome, the core parameter of BP neural network that is learning rate is determined as gradient method. And this study is aimed at technical supports to local sustainable utilization of water resource.

2 METHODS

2.1 Research area

This study was conducted at yongchuan district, Chongqing, $105°37'37''\sim106°05'06''$W, $28°56'16''\sim29°34'23''$N. This area was typical hilly area in southeast of Sichuan basin (Fig. 1). And more than 70% of this distinct was hills. Because of that, there were 133 reservoirs which ensured supply for all walks

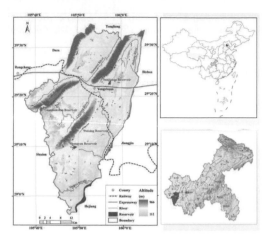

Figure 1. Location of study area.

of life in this county. Climatologically, the area is humid subtropical climate zone, and is characterized by large seasonal fluctuations in temperature and rainfall. Mean annual temperature is about 17.7°C. Mean annual precipitation is about 1049 mm. And nearly 80% of the annual rainfall occurs from April to September.

2.2 Approach

At present, a lot of remote sensing data of inland water body mainly rely on the landsat spectral sensor. Such as Landsat TM/ETM, SPOT HRV, CBERS CCD, EO-1 ASTER and so on. Relatively speaking, the TM image for remote sensing retrieval obtained perfect results, especially on chlorophyll a (Chl), suspended solids (ss), clarity (cla) and yellow substance [Ekstrand 1992, Lei et al. 2004]. This study collected the TM image of Landsat-5 with 7 wave band on 25th, March, 2005. The satellite orbit was 128, and width of cloth was 40. The TM image data quality was better because of without cloud cover influence. Among these remote sensing data, spatial resolution of all selected wave band pixels were 30 m × 30 m, except wave band 6 whose pixel was 120 m × 120 m. Pretreatment progress were contained of radiation correction, geometric correction, spectral enhancement, wave band selection and image cropping. In order to contrastive analyze the inversion results, ground monitoring data with almost the same time were also collected. In view of the pollutant distribution characteristics of the study area, the data acquisition were 12 reservoir which of the main water quantity parameters were chlorophyll a(Chl), concentration of suspended substances (CS), NH_3-N, water transparency (SD) and chemical oxygen demand (COD).

3 CORRELATION ANALYSIS

In order to confirm the correlation between selected wave band and the measured data of water quality

Table 1. Correlation coefficients between the reflectivity of bands and the water quality parameters.

	$\ln b_1$	$\ln b_2$	$\ln b_3$	$\ln b_4$	$\ln b_6$
$\ln$ CS	0.93*	0.93*	0.92*	0.98**	−0.01
$\ln$ NH_3-N	0.93*	0.93*	0.92*	0.98**	0.31
$\ln$ SD	−0.43	−0.06	−0.48	−0.26	0.96*
$\ln$ Chl	0.55	0.18	0.58	0.36	−0.95*
$\ln$ COD_{mn}	−0.3	−0.29	−0.51	−0.19	0.94*

*$p < 0.05$; **$p < 0.01$; b1, b2, b3, b4 & b6 indicate TM band of blue, green, red, near-infrared and thermal infrared, separately.

parameters, correlation analysis of all data with logarithmic form in this manuscript were carried out by software DPS. The results are show in Table 1. And the analysis of outcome indicates that CS and NH_3-N reflected evidence relativity to TM band 1–4. However, Chl, SD and CODmn were just prominent related to TM band 6.

4 INVERSION MODEL OF WATER QUALITY PARAMETERS

4.1 Multiple linear regression model

According to the relation between water quality parameter and waters spectral response, multispectral multiple linear regression models, based on TM wave bands, were obtained as follows [Ye et al. 2012]:

$$\text{Chl-a} = 37.2 - 1.37b_1 - 1.08b_2 + 1.7b_3 + 3.66b_4 \quad (1)$$

$$\text{Cs} = 28.99 - 0.39b_1 - 2.04b_2 - 2.04b_3^3 + 0.34b_4 \quad (2)$$

$$\text{Ln}(\text{SD}) = -0.28\text{Ln}(b_3) + 0.58 \quad (3)$$

$$NH_3\text{-N} = -1.18 - 0.14b_1 - 0.22b_2 - 0.24b_3 + 0.18b_4 \quad (4)$$

$$\text{CODmn} = 12.74 - 0.11b_1 - 0.14b_2 + 0.35b_3 + 0.14b_4 - 0.04b_6 \quad (5)$$

where b_1, b_2, b_3, b_4 and b_6 were remote sensing reflectivity of TM wave band 1, 2, 3, 4 and 6, separately.

4.2 BP neural networks model

Artificial Neural Networks (ANN), by means of mathematical methods, which simulate human brain character and thinking ability, was nonlinear information handling system [FECIT 2005, Deng et al. 2006]. After more than half a century of development, because of various network structure and algorithm, ANN theory was also gradually developed comparatively perfect system. However, the BP ANN was the most wide application model in all ANN models. This is due to several benefits such as self-adaption, self-organization, self-learning, non-locality and non-convexity [Zhang et al. 2005, Ling et al. 2007]. The BP ANN, in this paper, were consist of input layer,

Table 2. The comparison results of retrieved CS [mg/L].

Sequence	Measured value	Prediction value	
		Multiple linear regression models	BP ANN
1	10	8.014706	10.0384
2	11.94	8.014706	11.7773
3	12.97	11.07353	12.2809
4	9.97	12.02941	9.9391
MAD		2.466618	0.2303
R^2		0.3202	0.9892

Table 3. The comparison results of retrieved SD [mg/L].

Sequence	Measured value	Prediction value	
		Multiple linear regression models	BP ANN
1	0.549	0.59085	0.5509
2	0.601	0.620915	0.6018
3	0.634	0.585621	0.6341
4	0.638	0.684967	0.6697
MAD		0.039278	0.008625
R^2		0.6296	0.9287

hidden layer and output layer. In the training process, the TM wave bands with good correlation (namely, wave band 1–4 and 6) were selected as input layer data, and the corresponding measured data were confirmed as output value. The neuron of hidden layer was intended to be 3 after training effect contrast. That is to say, the BP network structure was 5-3-1. Furthermore, excitation function of hidden layer was S function"Sigmoid", export function was "purelin", and training function was self-adaption momentum gradient method "traingdm". Owning to fundamental theory of the BP neural network, training process demands that the more sample the better outcome. In the case of limited data, K-fold cross-training method was used to improve training generalization ability of this study [Zhao et al. 2009]. Grouping of sample data were as follow: the training data were divided into four group by random, three groups of random selected data were used to training by turns and the rest group data was served as confirmation. In the training process, the training parameter and network connections that were in the group of minimum test error were provided to prediction. Moreover, before training, it was necessary to normalization all the data. According to the training, the initial learning rate, dynamic parameter, Sigmoid parameter, permissible error and maximum iterations were confirmed as 0.1, 0.7, 0.9, 0.00001 and 2000, separately.

5 RESULT ANALYSIS

The TM remote sensing images, in the last ten day of March, 2008, were used to inverse water quality parameters which were composed of CS, NH_3-N, Chl, SD and COD. Based on inversion results and measured data, the BP ANN model was build up. As a contrast, the simulation outcome of multiple linear regression model was given at the same time. In addition, mean absolute deviation (MAD) and R^2 were taking out to verify the inversion accuracy. The matching and testing consequence were show in the following tables.

Comparing the results between simulation values and measured values, the basic law is as follows: prediction values of BP ANN are more closely to measured data than multiple linear regression model.

Table 4. The comparison results of retrieved NH_3-N [mg/L].

Sequence	Measured value	Prediction value	
		Multiple linear regression models	BP ANN
1	0.361	0.821229	0.4222
2	0.477	0.673184	0.5824
3	0.678	0.717877	0.7616
4	0.682	0.691341	0.7780
MAD		0.176408	0.086548
R		0.9281	0.9529

Table 5. The comparison results of retrieved Chl [mg/L].

Sequence	Measured value	Prediction value	
		Multiple linear regression models	BP ANN
1	14.99	14.505	15.1404
2	15.94	14.52	15.7131
3	15.83	16.2	16.0997
4	16.06	16.815	16.4047
MAD		0.7575	0.247925
R^2		0.9256	0.9706

Table 6. The comparison results of retrieved COD_{mn} [mg/L].

Sequence	Measured value	Prediction value	
		Multiple linear regression models	BP ANN
1	10	8.014706	10.0384
2	11.94	8.014706	11.7773
3	12.97	11.07353	12.2809
4	9.97	12.02941	9.9391
MAD		2.466618	0.230265
R^2	10	0.3202	0.9892

And the model fitting results of BP ANN were much better than multiple linear regression model. No matter the correlation index of MAD or R^2 were point out that BP ANN given better simulation result. Through comparative analysis, we can consider that inversion result, in the study area, combining remote sensing image with BPANN can better reflect the status of water quality parameters.

6 CONCLUSIONS

Based on TM image data, the quantitative inversion models of water quantity parameters with distributed water resource in Yongchuan distinct were constructed. That is including two models, namely, multiple linear regression model and BP ANN model. This paper presents a valuable trail of comparative analysis. The application of remote sensing technology to estimate water quality can survive more information, efficiently. And coulping with BP ANN can supply better prediction data than traditional regression model. Especially, for long term monitoring, BP ANN based remote sensing inversion can be more effort and economical.

ACKNOWLEDGEMENT

This work was financially supported by the PHD fund of Southwest University under Contract NO. SWU112045.

REFERENCES

Chen, L. 2003. A study of applying genetic programming to reservoir trophic state evaluation using remote sensor data. *International Journal of Remote Sensing* 24(11): 2265–2275.

Dekker, A.G. 2002. Analytical algorithms for lake water TSM estimation for retrospective analyses of TM and SPOT sensor data. *Int J Remote Sens* 23(1):15–35.

Deng, J.X. & Li, S.S.2006. A New Method of Ascertaining Radial Basis Function Network Parameter. *Microprocessors* 4:48–52.

Ekstrand, S. 1992. Landsat TM based quantification of Chlorolphyll-a during algae blooms in coastal waters. *Int J Remote Sensing* 13(10):1913–1926.

FECIT Info-Tech. 2005. *Neural Networks & Matlab Application*. Beijing: Publishing Housing of Electronics Industry

Gu, L. & Zhang, Y.C. 2007. Study on Retrieval of Chlorophyll-a Concentration by Remote Sensing in Taihu Lake. *Science and Management* 32(6):25–29.

Giardino, C. & Pepe, M. 2001. Detecting Chlonophyll, Secchi Disk Depth and Surface Temperature in a Sub-alpine Lake Using Landsat Imagery. *The Science of the Total Environment* 268:19–29.

Lei, K. & Zheng, B.H. 2004. Monitoring the surface water quality of Taihu Lake based on the data of CBERS-1. *ACTA SCIENTIAE CIRCUMSTANTIAE* 24(3):376–380.

Ling, L.H. & Gui, F.L. 2007. Forecasting and analysis of water demand in Jiangxi Province in the near future. *Mathematics in Practice and Theory* 37(22):42–47.

Shi, A.Y. & Xu, L.Z. 2006. A Neural Network Model for Water Quality Retrievals Using Knowledge and Remotesensed Image. *Journal of Image and Graphics* 11(4): 521–528.

Wang, H. & Zhao, D.Z. 2012. Advance in remote sensing of water quality. *MARINE ENVIRONMENTAL SCIENCE* 31:285–288.

Yu, D.H. & Wang, Y.H. 2008. Research progress in remote sensing technology for inland water quality monitoring. *CHINA WATER & WASTEWATER* 24(22):12–16.

Ye, Y. & Ren, Y.G. 2012. Preliminary Study on water quality monitoring of the dispersed water source Based on remote sensing technology. *Journal of Anhui Ahri.Sci* 40(2):1206–1208.

Zhang, X.F. & Guo, X.R. 2005. Prediction of urban water demand in Tangshan City with BP neutral network method. *Journal of safety and environment* 5(5): 95–98.

Zhao, Y.Q. & Wang, X.L. 2009. Jiang Sai. Study on neural network model for weihe river water quality retrieving using remote sensing image. *REMOTE SENSING TECHNOLOGY AND APPLICATION* 24(1):63–67.

Information Technology and Computer Application Engineering – Liu, Sung & Yao (Eds)
© 2014 Taylor & Francis Group, London, ISBN 978-1-138-00079-7

Sensor network coverage problem research based on rail transport of dangerous goods

X.Z. Han & N. Zhang

College of Computer and Information Technology, Beijing JiaoTong University, Beijing, China

ABSTRACT: This paper researches on node coverage problem of the wire-less sensor network in the course of transportation of dangerous goods in railway. Under the condition of completely covering the monitoring area and neighbor nodes connected, an optimal node covering method which makes use of regular polyhedron space subdivision is proposed to achieve complete coverage with the least number of nodes.

Keywords: Wireless sensor network; Node coverage; Railway

1 INTRODUCTION

With the development of economy, the demand for transport of dangerous goods has been increasing. Railway accounts for more than 36% of total amount national transport of dangerous goods. Because of the special nature of the dangerous goods and environmental uncertainty in railway transportation, accidents happen easily, cause huge casualties, economic losses and environmental pollution to the country (Xi, Xian-ning, Ru-yan, & Guo-qiang, 2008). Transport of dangerous goods vehicles are generally divided into the tanker, container, caravan, there are many different kinds of transport of dangerous goods. This paper selected container loaded with calcium carbide as the research object.

The length, width and height of train container are 15.5 m, 2.8 m, 2.8 m, the weight and load are 24 t, 60 t. The container is considered as a cuboid space rules, the lower space of the container place the dangerous goods calcium carbide. Because the calcium carbide easily release acetylene gas with water or damp air, near the window or door or gaps of the container, the dangerous goods easily dampened to release acetylene gas. Acetylene gas is lighter than air, it is likely to concentrate in the container upper space. When acetylene gas reaches a certain concentration, it is easy to explosion. Sensor nodes should be placed in the cuboid space which is formed by the upper layer of the container, and focus on monitoring the concentration of acetylene gas in the upper container space.

2 APPLICATION OF WIRELESS SENSOR NETWORK IN THE RAILWAY TRANSPORT OF DANGEROUS GOODS

Wireless sensor networks (WSN), which have its own superiority on regional multi-point detection,

self-organizing network and achieving real-time monitoring and transmission of information. It has great significance to be used in transit monitoring of transport dangerous goods. The problem of node coverage is very important, which needs to be solved in wireless sensor networks. Reasonable and effective node deployment can improve the accuracy and comprehensiveness of the monitoring results, it can also prolong the network lifetime through the coordination controlling (Li-ping, Zhi, & You-xian, 2006). Paper (Huang, Tseng, & Lo, 2004) proposes an efficient polynomial-time algorithm, whose goal is to determine whether every point in the service area of the sensor network is covered by at least ∂ sensors. A coverage algorithm using specified 3D network model is proposed and it solves the network sensor density and redundancy problems (Lei, Wenyu, & Peng, 2007). In the paper (Akkaya & Newell, 2009), it proposes a distributed node deployment scheme which can increase the initial network coverage in an iterative basis. Paper (Ammari & Das, 2010) proposes the Reuleaux tetrahedron model to characterize k-coverage of a 3D field and investigate the corresponding minimum sensor spatial density.

Currently, node deployment methods are often based on experience in the process of railway transport of dangerous goods (Figure 1), there are no specific methods and it can easily lead to monitoring loopholes. On the one hand, current research methods of node deployment based on wireless sensor network, are aiming to the two-dimensional plane. Node deployment methods on three-dimensional space are still relatively lacking. On the other hand, the environment of the railway transport of dangerous goods is very special, compared with the monitoring of the wild area, sensor nodes have smaller range of sensing, the communication distance between the nodes is also relatively closer and monitoring space is limited in the railway application. Therefore, the common node deployment

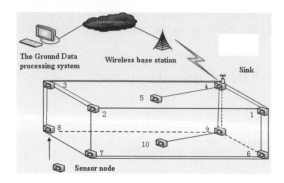

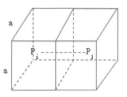

The Ground Data processing system

Wireless base station

Sink

3 4 5 2 1 8 9 10 7 6

Sensor node

Figure 1. Application of wireless sensor network in the railway.

methods are not suitable for the process of monitoring of the railway transport of dangerous goods.

In this research, it is aiming to monitor the safety condition of dangerous goods in rail transport process. A three-dimensional sensor network node deployment method with using the regular polyhedral space split is proposed. This method analyzes the spatial characteristics of cube, regular triangular prism and regular hexagonal prism, respectively. Under the condition of completely covering the monitoring area and neighbor nodes connected, the maximum effective volume of nodes can be got according to the different communication radius. Finally an optimal node covering method is proposed.

3 NETWORK MODEL AND DEFINITIONS

3.1 *Network model*

Assuming that sensor nodes deployed in the three-dimensional space and the position basic won't change, the static three-dimensional sensing network space is Ω, $P = \{P_1, P_2, \ldots, P_n\}$ is the limited sensor node collection. It is assumed that the wireless sensor network has the following properties:

1) The sensing radius and communication radius of sensor nodes are R_S, R_C, communication model and sensing model are spheroidal distribution.
2) The sensor node is located in the center of the sphere.
3) Each nodes in the sensor network has the same sensing radius and communication radius.
4) The communication ball and sensing ball of the node P_i is denoted by $C(P_i)$ and $S(P_i)$.

3.2 *Definitions*

Definition 3.2.1: Any point in the three-dimensional space is sensed at least by a sensor node, then there exists the node $P_i \in P$, such that $d(a, P_i) \leq R_S$, $d(a, P_i)$ is represented as the Euclidean distance between the points a and node P_i.

Definition 3.2.2: In the network space Ω, there are two sensor nodes P_i and P_j, P_i, $P_j \in P$. If

$S(P_i) \cap S(P_j) \neq \Phi$ and $d(P_i, P_j) \leq R_C$, So called nodes P_i and P_j Neighbor Node. $d(P_i, P_j)$ is represented as the Euclidean distance between the node P_i and node P_j.

Definition 3.2.3: Any node of P_i corresponding Voronoi cell is denoted by $VC(P_i)$, by the distance to node P_i smaller than the distance to the other node P_j ($i \neq j$). Assume that any point q is located at the corresponding Voronoi cell of node P_i, if and only if for any, have $d(q, P_i) < d(q, P_j)$ ($i \neq j$).

Definition 3.2.4: The node of P_i corresponding Voronoi cell volume is called the volume of unit node, which is denoted by $V(P_i)$.

According to the related definitions, each sensor node in the network model is a sphere. However, because the sphere is not continuous seamless non-overlapping coverage in three-dimensional space, so three-dimensional space can not be independently division by sphere. The cube, regular triangular prism and regular hexagonal prism can be composed of independent filling three-dimensional polyhedron (Alam & Haas, 2006). Sensor nodes deployed in the center of the polyhedron, Voronoi cell formed by the node sensing ball is the polyhedron. Because the collection of Voronoi cell has space subdivision, it only need each node to fully cover its corresponding polyhedron, and makes the volume of the polyhedron as big as possible, then the nodes in the sensor network can use the minimum number of nodes to fully cover the entire area of the three-dimensional space.

4 NODE OPTIMAL COVERAGE

Because the collection of Voronoi cell has space subdivision, through analysis and study on the spatial characteristics of regular polyhedron (cube, regular triangular prism, regular hexagonal prism), the constraints can be got under the condition of complete coverage and neighbor nodes connected. Further, the maximum effective volume of the node can be calculated and the optimal coverage mode of sensor nodes are obtained.

4.1 *Cube*

In sensor network space, Voronoi cell of Node sensing ball formed is cube. Nodes deployed in the center of the cube. The distance between two neighbor nodes is equal to the side length of each cube (Figure 2), assume that each side length of a cube is 'a', then the volume

Figure 2. Adjacent cube.

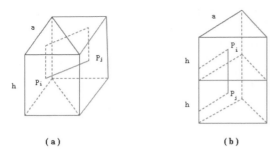

(a)   (b)

Figure 3.   Adjacent regular triangular prism.

of unit node is $V = a^3$. In order to reach each sensing ball formed by the node of sensor completely covering the "Voronoi" cell which it formed, it must make node to each vertex of the cube of the distance $d \leq R_S$. After calculation $d = \sqrt{3}a/2 \leq R_S$. At the same time, in order to ensure the connectivity between neighbor nodes, the distance between neighbor nodes is the cube of side length a, it must satisfy $a \leq R_C$. After the above analysis, getting the constraints for

$$\begin{cases} 0 < d \leq 2R_s/\sqrt{3}a \\ 0 < a \leq R_C \end{cases}$$

Finally obtaining maximum effective volume

$$V_{max} = \begin{cases} R_C{}^3 & 0 < \dfrac{R_C}{R_S} < \dfrac{2}{\sqrt{3}} \\ \dfrac{8}{3\sqrt{3}} R_s{}^3 & \dfrac{R_C}{R_S} \geq \dfrac{2}{\sqrt{3}} \end{cases}$$

4.2   Regular triangular prism

In sensor network space, Voronoi cell of Node sensing ball formed is regular triangular prism. Nodes deployed in the center of the regular triangular prism. The distance between two neighbor nodes is which the distance between the neighbor two regular triangular prisms (Figure 3). Assuming the side side of the bottom surface of the prism is 'a', the height is 'h', then the volume of unit node is $V = \sqrt{3}a^2h/4$. In order to reach each sensing ball formed by the node of sensor completely covering the Voronoi cell which it formed, we must enable the distance (d) between each node to the regular triangular prism vertex to $d \leq R_S$, calculated as $d = \sqrt{(h^2/4 + a^2/3)}$. The distance between two adjacent left and right prism center (Figure 3a) can be calculated as $d_1(P_i, P_j) = \sqrt{3}a/3$, the distance between two adjacent up and down prism center (Figure 3b) can be calculated as $d_2(P_i, P_j) = h$, in order to ensure the connectivity between neighbor nodes, it must be $\sqrt{3}a/3 \leq R_C$ and $h \leq R_C$. After the above analysis, getting the constraints for

$$\begin{cases} \sqrt{h^2/4 + a^2/3} \leq R_s \\ \sqrt{3}a/3 \leq R_c \\ h \leq R_c \end{cases}$$

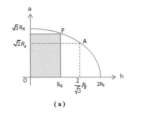
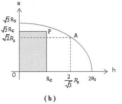

(a)   (b)

Figure 4.   Constraint relations.

In order to obtain the maximum effective volume of the unit node, the problem is converted to find the maximum value of 'V' under the above constraints.

On the basis of $V = \sqrt{3}a^2h/4$, it is known unit volume (V) of a node is the function of a and h, and has a positive correlation with them. Figure 4 shows the constraints need to meet, the coordinates of the point (h, a) in the plane coordinate system must be within the area of the ellipse in the first quadrant. When "V" to obtain the maximum, point (h, a) must be on the elliptical boundary, or on the boundary of the rectangular area composed by the abscissa and ordinate of the point (h, a), that can be proved by reduction to absurdity. It would be dividing to two cases and discussed in next.

i: Assuming that the maximum point is on the elliptical boundary (Figure 4a). If

$$V(h) = \frac{\sqrt{3}}{4}a^2h \tag{1}$$

By the equation

$$\sqrt{h^2/4 + a^2/3} = R_S \tag{2}$$

Solve 'a' into equation (1) then

$$V(h) = \frac{3\sqrt{3}}{4}\left(R_S{}^2h - \frac{h^3}{4}\right) \tag{3}$$

Derivative of V(h), get

$$V'(h) = \frac{3\sqrt{3}}{4}R_S{}^2 - \frac{9\sqrt{3}}{16}h^2 \tag{4}$$

If $V'(h) = 0$ then $h = 2R_S/\sqrt{3}$, $a = \sqrt{2}R_S$. On the condition of $h \in (0, 2R_S/\sqrt{3})$ the function V(h) is increasing, on the condition of $h \in [2R_S/\sqrt{3}, 2R_S)$ the function V(h) is decreasing. According to the constraints $0 < h \leq R_C$, on the condition of $2R_S/\sqrt{3} \leq R_C$, 'V' to obtain the maximum value at $h = 2R_S/\sqrt{3}$, after the calculation $V_{max} = R_S^3$.

On the condition of $2R_S/\sqrt{3} > R_C$, 'V' to obtain the maximum value at $h = R_C$, substituting into equation (2), solve $a = \sqrt{(3R_S^2 - 3R_C^2/4)}$. According to the constraints $\sqrt{3}a/3 \leq R_C$ in the case of $\sqrt{(3R_S^2 - 3R_C^2/4)} \leq R_C$,

$$V_{max} = \frac{3\sqrt{3}}{4}\left(R_S{}^2R_C - \frac{R_C{}^3}{4}\right) \tag{5}$$

ii: Assuming maximum points obtained in the elliptic (Figure 4b), namely in the vertex of the area which is

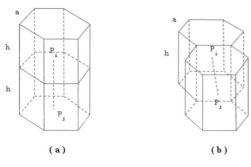

(a)   **(b)**

Figure 5.   Adjacent regular hexagonal prism.

formed by the point (h, a) of the abscissa and ordinate. Right now $R_C < 2R_S/\sqrt{5}$ the maximum of 'a' is $\sqrt{3}R_C$, and the maximum of 'h' is R_C. Easy to get $V_{max} = 3\sqrt{3}R_C^3/4$.

In conclusion, 'V_{max}' max can be calculate as

$$V_{max} = \begin{cases} R_s^3 & R_c \ge \dfrac{2}{\sqrt{3}}R_s \\ \dfrac{3\sqrt{3}}{4}\left(R_s^2 R_c - \dfrac{R_c^3}{4}\right) & \dfrac{2}{\sqrt{5}}R_s \le R_c < \dfrac{2}{\sqrt{3}}R_s \\ \dfrac{3\sqrt{3}}{4}R_c^3 & 0 < R_c < \dfrac{2}{\sqrt{5}}R_s \end{cases}$$

4.3   Regular hexagonal prism

In sensor network space, Voronoi cell of node sensing ball formed is regular hexagonal prism. Nodes deployed in the center of the regular hexagonal prism. The distance between two neighbor nodes is which the distance between the neighbor two regular hexagonal prisms (Figure 5). Assuming the side side of the bottom surface of the prism is 'a', the height is 'h', then the volume of unit node is $V = 3\sqrt{3}a^2h/2$. In order to reach each sensing ball formed by the node of sensor completely covering the Voronoi cell which it formed, we must enable the distance (d) between each node to the regular triangular prism vertex to $d \le R_S$, calculated as $d = \sqrt{(a^2 + h^2/4)}$. The distance between two adjacent left and right prism center (Figure 5a) can be calculated as $d_1(P_i, P_j) = \sqrt{3}a$, the distance between two adjacent up and down prism center (Figure 5a) can be calculated as $d_2(P_i, P_j) = h$. According to the method which described in the regular triangular prism, the constraints can be got as

$$\begin{cases} \sqrt{a^2 + h^2/4} \le R_S \\ \sqrt{3}\,a \le R_C \\ h \le R_C \end{cases}$$

Finally obtaining maximum effective volume

$$V_{max} = \begin{cases} 2R_s^3 & R_c \ge \sqrt{2}R_s \\ R_c^2\sqrt{3R_s^2 - R_c^2} & \sqrt{\dfrac{12}{7}}R_s \le R_c < \sqrt{2}R_s \\ \dfrac{\sqrt{3}}{2}R_c^3 & 0 < R_c < \sqrt{\dfrac{12}{7}}R_s \end{cases}$$

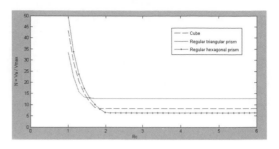

Figure 6.   Change curve of nodes and communication radius.

Table 1.   Ten sets of data selected from the curve.

R_C	N_C	N_T	N_H
1.0	43.4	33.4093	50.1140
1.1821	26.2685	20.2215	30.3323
1.3643	17.0882	13.7211	19.7318
1.5185	12.3945	13.1466	14.3119
1.7287	8.4005	12.8593	9.7001
1.7427	8.3523	12.8592	9.4680
2.1211	8.3523	12.8592	6.4296
2.9059	8.3523	12.8592	6.4296
4.0130	8.3523	12.8592	6.4296
4.9939	8.3523	12.8592	6.4296

5   ANALYSIS OF SIMULATION EXPERIMENT

Through the analysis of cube, regular triangular prism and regular hexagonal prism in the previous chapter, the maximum effective volume of the node can be got under the condition of complete coverage and neighbor nodes connected. Assuming that volume of the monitoring area is 'V_a', the required number of sensor nodes is N under the premise of without considering the boundary effect, $N = V_a/V_{max}$. In the course of transportation of dangerous goods in railway, in order to get the optimal deployment mode of the sensor nodes in the container freight trains, it can analyse the performance of the sensor nodes in a deployment mode of cube, regular triangular prism, regular hexagonal prism through Matlab simulation software. In the process of simulation experiment, assuming the sensor network monitoring area remaining is the upper region of space after the accumulation of goods for container trains, if $V_a = 15.5\,\text{m} * 2.8\,\text{m} * 1\,\text{m}$ and sensing radius of the sensor node $R_S = 1.5\,\text{m}$, range of the communication radius is $1\,\text{m} \le R_C \le 15\,\text{m}$.

Figure 6 shows the number of sensor nodes changing curve according to different polyhedral deployment with different communication radius. Table 1 shows ten sets of data are randomly selected from small to large in the curve of Figure 6 in accordance with the communication radius. N_C, N_T, N_H, separately represent the required number of nodes in the deployment mode of cube, regular triangular prism and regular hexagonal prism. It can be seen the number of nodes required does not change when the communication radius of a certain value, under three different deployment modes from the Figure 6 and Table 1. The following conclusions can be got after experiment:

1) On the condition of $1 \leq R_C < 1.485$, $N_T < N_C < N_H$. The number of nodes are required at least under the deployment mode of regular triangular prism. The number of nodes are required at least under the deployment mode of regular triangular prism.

2) On the condition of $1.485 \leq R_C < 1.566$, $N_C \leq N_T < N_H$. The number of nodes are required at least under the deployment mode of cube.

3) On the condition of $1.566 \leq R_C < 1.817$, $N_C < N_H \leq N_T$. The number of nodes are required at least under the deployment mode of cube.

4) On the condition of $R_C \geq 1.817$, $N_H \leq N_C < N_T$. The number of nodes are required at least under the deployment mode of regular hexagonal prism.

5) On the condition of $R_C \geq 1.7427$, $N_C = 8.3523$; on the condition of $R_C \geq 1.7287$, $N_T = 12.8592$; on the condition of $R_C \geq 2.1211$, $N_C = 6.4296$; on the condition of $R_C \geq 1.927$, $6.4296 \leq N_H \leq 7$.

It can be drawn from the above analysis, in the course of transportation of dangerous goods in railway, the sensor nodes can achieve the optimal coverage under the deployment mode of regular hexagonal prism. Communication radius can be selected as $R_c = 2$ m, the number of sensor nodes is needed for $N = 7$. In this case, On the one hand can reduce the node transmission power and save energy, on the other hand can take advantage of the least number of nodes to achieve the purpose of completely covering the monitoring area.

6 CONCLUSION

This paper analyzes regular polyhedron spatial characteristics, an node covering method is proposed, which is suitable for the transportation of railway dangerous goods. Through the simulation experiments it is to obtain the least number of nodes and suitable communication radius, under the condition of completely covering the monitoring area and neighbor nodes connected. On the one hand this method can save energy, on the other hand it can improve the coverage rate.

ACKNOWLEDGEMENT

The study is supported by National High Technology Research and Development Program of China (2012AA040912).

REFERENCES

Akkaya, K., & Newell, A. (2009). Self-deployment of sensors for maximized coverage in underwater acoustic sensor networks. *Computer Communications, 32*(7), 1233–1244.

Alam, S. M., & Haas, Z. J. (2006). Coverage and connectivity in three-dimensional networks. *Proceedings of the 12th annual international conference on Mobile computing and networking* (pp. 346–357).

Ammari, H. M., & Das, S. K. (2010). A Study of k-coverage and measures of connectivity in 3D wireless sensor networks. *Computers, IEEE Transactions on, 59*(2), 243 257.

Huang, C.-F., Tseng, Y.-C., & Lo, L.-C. (2004). The coverage problem in three-dimensional wireless sensor networks. *Global Telecommunications Conference, 2004. GLOBECOM'04. IEEE* (Vol. 5, pp. 3182–3186).

Lei, R., Wenyu, L., & Peng, G. (2007). A coverage algorithm for three-dimensional large-scale sensor network. *Intelligent Signal Processing and Communication Systems, 2007. ISPACS 2007. International Symposium on* (pp. 420–423).

Li-ping, L., Zhi, W., & You-xian, S. (2006). Survey on Coverage in Wireless Sensor Networks Deployment. *JOURNAL OF ELECTRONICS AND INFORMATION TECHNOLOGY, 28*(9), 1752–1757.

Xi, L. I., Xian-ning, Z. H. U., Ru-yan, Z., & Guo-qiang, C. A. I. (2008). How to Identify and Control the Unsafe Factors in Dangerous Goods Transportation. *LOGISTICS TECHNOLOGY, 27*(9), 17–19.

Information Technology and Computer Application Engineering – Liu, Sung & Yao (Eds)
© *2014 Taylor & Francis Group, London, ISBN 978-1-138-00079-7*

An improved Camshift algorithm based on occlusion and scale variation

Ying Ying Yue

School of Information Science and Engineering, Yuxi Normal University, Yuxi, China

ABSTRACT: At present, target tracking under complex scenes is still a difficult problem in the field of computer vision For occlusion and the scale change of targets which are two kinds of complex scenes, an improved Camshift algorithm is proposed in this paper. Firstly, the improved algorithm adopts block tracking to handle occlusion problem. Secondly, considering that targets template should update when scale changes but not when occlusion, adaptive template-updating mechanism is proposed. Finally, the geometric histogram, which is robust to scale variation, is introduced to describe the moving targets. The experimental results show that, the improved algorithm can well adapt to scale change of targets, and also has good anticonclusion ability, can achieve a good tracking effect.

Keywords: Targets tracking; Camshift; Occlusion; Scale variation; Geometry histogram; Adaptive template-updating mechanism

1 INTRODUCTION

Target tracking is key problem in the field of computer vision. At present, there are many widely used targets tracking algorithm: Kalman filter [1], Camshift [2,3] algorithm and particle filtering method [1,4]. Among them, Camshift works effectively in simple scenes, but easily lose tracking targets when target scale changes. In addition, many improved tracking algorithm have been proposed, such as [5], solved the half-occlusion problem, but don't work in full-occlusion and has low robustness when targets scale changes. Such as [6], improved the tracking robustness of scale-changing, but can't use when occlusion occurs.

2 RELATED WORK

To sum up, achieving robust tracking effect when occlusion and target scale changes, is still a difficult problem. In order to solve these two problems, this paper proposes an improved Camshift algorithm, which uses the adaptive template-updating mechanism.

In improved algorithm, firstly, geometry histogram is combined with color histogram of traditional Camshift for target scale variations. Secondly, for total-occlusion, based on timing mechanism, out-scene targets and temporarily disappear targets are respectively tracked. Finally, for half-occlusion, block tracking and partially matching principle is used to ensure correct tracking results. The experimental results show that, the algorithm can adapt to the scale changes of objects, has certain anti-occlusion ability, can achieve robust tracking effect.

3 BASIC PRINCIPLE OF TRADITIONAL CAMSHITM ALGORITHM [TCA]

Camshift, a tracking algorithm based on color histogram, is currently one of the most widely used tracking algorithms. It has 3 main parts: calculation of back projection, Mean Shift algorithm [7] and Camshift operation. The basal steps of Camshift are as follows.

(1) To initialize the search area with the whole image area;
(2) To initialize the size and position of Track Window;
(3) To calculate the probability distribution of color in Track Window;
(4) To obtain a new size and position of the Track Window with Mean Shift algorithm;
(5) To set the size and position of Track Window in the next frame with the calculated results of the fourth step, and jump to the third step.

By the iteration operations in such way, object tracking can be realized. TCA, which is based on color histogram, can track targets effectively in simple scenes, but robustness is greatly affected when occlusion and scale variation occurs. Therefore, the tracking performance of TCA is still need to be improved.

4 IMPROVED ALGORITHM

TCA, which uses color histogram as target template, has following problems: (1) when there is occlusion, color-based TCA will lose the target. (2) TCA, which is based on whole template update mechanism, will

Figure 1.   Divide target into 16 blocks.

make template updating mistakes because of half-occlusion; (3) TCA, which contains color information but lost the geometry information of targets, continuous tracking of target is interfaced by scale variation. Aiming at above problems, improved algorithm maintained the excellent characteristics of TCA, and put forward three measures. For (1), the block tracking is introduced to deal with half-occlusion problem. For (2), adaptive template updating mechanism is proposed to avoid template updating mistakes. For (3), geometry histogram, which is robust to scale variation, is combined to describe target template for accurate and robust tracking performance. Adaptive template update mechanism, block tracking and geometry histogram are described as follows.

4.1   Block tracking [5]

In this paper, target is divided into n blocks. According to several experimental results, the optimum value of n is 16, as shown in the figure 1. Color-based template of TCA can be expressed as formula (1). Pt and W is serial number and weight value of each block, and their initial value is 1. In addition, $b = 1, 2, \ldots, B$.

$$C_{p_t} = (h_{p_t}(b), w_{p_t})_{p_t = p_1, p_2, \cdots, p_n} \tag{1}$$

When tracking, firstly, similarity measurement is calculated between target template and candidate region in blocks. The central coordinates of Pt (block in target template) and Qt (block in candidate region) are defined as y and qt:y separately. Then matching distance of histogram is as formula (2) and (3). Finally, similarity degree of candidate target and template is described by linear weighted method as formula (4).

$$S(y) = \sum_{p_t = p_1}^{p_n} d(p_t, q_{t:y}) w_{p_t} \tag{2}$$

$$d(p_t, q_{t:y}) = \sum_{b=1}^{B} \frac{(h_{p_t}(b) - h_{q_{t:y}}(b))^2}{h_{p_t}(b) + h_{q_{t:y}}(b)} \tag{3}$$

After calculating the similarity for each corresponding blocks, similarity vector D is obtained as formula (4).

$$D = \{d(p_1, q_{1:y}), d(p_2, q_{2:y}), \cdots, d(p_n, q_{n:y})\} \tag{4}$$

4.2   Geometric histogram [6]

Geometry histogram, which uses the statistical characteristics of geometric relationship of vertices and

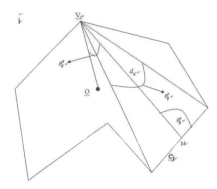

Figure 2.   Relationship between the vertices and edges.

edges as structure information of the graph, is simple in structure and has high matching accuracy for scale variation.

Assume $G = (V, E)$, $V = (v_1, v_2, v_3, \ldots, v_n)$, $E = (e_1, e_2, \ldots, e_m)$ respectively as a collection of vertices and edges of the graph, $F(v_i)$ as the characteristic matrix of vertex v_i. Considering location feature of every vertex is determined by other edges, $F(v_i)$ can be seen as a function of E and described as $F(v_i) = f(e_1, e_2, \ldots, e_m)$.

Relationship between v_i and e_j can be expressed as formula (5). d_{ij} is the distance of v_i and M (midpoint of e_j), θ_{ij}^e is clockwise angle of e_j and L (line from M to v_i), θ_{ij}^d is clockwise angle between L and line from v_i to O (the center point of graph), θ_{ij}^c is angle between v_i and two ends of e_j.

$$S(v_i) = \begin{pmatrix} \overline{d_i} \\ \theta_i^e \\ \theta_i^d \\ \theta_i^c \end{pmatrix} = \begin{pmatrix} \overline{d_{i1}} & \overline{d_{i2}} & \overline{d_{i3}} & \cdots & \overline{d_{im}} \\ \theta_{i1}^e & \theta_{i2}^e & \theta_{i3}^e & \cdots & \theta_{im}^e \\ \theta_{i1}^d & \theta_{i2}^d & \theta_{i3}^d & \cdots & \theta_{im}^d \\ \theta_{i1}^c & \theta_{i2}^c & \theta_{i3}^c & \cdots & \theta_{im}^c \end{pmatrix}, \overline{d_{ij}} = \frac{d_{ij}}{\sum_{i=1}^{n} \sum_{j=1}^{m} d_{ij}} \tag{5}$$

We define $S = \{a_i | i = 1, 2, \ldots |S|\}$ as a data set, and then define H (S, X) as histogram description of S as formula (6). In formula (6), $X = \{x_i | x_{i-1} < x_i, i = 1, 2, \ldots, m\}$ and $a_j \in [x_0, x_m](j = 1, 2, \ldots, |S|)$, h_i is the number of data elements of S falls into the interval $([x_{i-1}, x_i])$.

$$H(S, X) = \left(\frac{h_1}{|S|}, \frac{h_2}{|S|}, \cdots, \frac{h_m}{|S|} \right) \tag{6}$$

According to the description of the histogram above, matrix H(S(vi),X) can be defined as histogram characteristic matrix of vertices, which is used as vertex features, as shown in formula (7). $X_i(i = 1, 2, 3, 4)$ is parameter vector of histogram.

$$F(v_i) = f_i(e_1, e_2, \cdots, e_m) = H(S(v_i), X) = \begin{pmatrix} H(\overline{d_i}, X_1) \\ H(\theta_i^e, X_2) \\ H(\theta_i^d, X_3) \\ H(\theta_i^c, X_4) \end{pmatrix} \tag{7}$$

There are two sets of data, S1 and S2, as showed in formula (8), and their histogram description are as formula (9), then weighted distance can be defined as formula (10).

$$S_1 = \{a_i\}_{i=1}^N, S_2 = \{a_i\}_{i=1}^M \tag{8}$$

$$H(S_1, X) = \{h_1, h_2, \cdots h_m\}, H(S_2, X) = \{h_1', h_2', \cdots h_m'\} \tag{9}$$

$$d(H(S_1, X), H(S_2, X)) = \sum_{i=1}^m w_i \left| h_i - h_i' \right| \tag{10}$$

$$w_i = \cos\left(\frac{h_i + h_i'}{\alpha \max_i (h_i + h_i')} \frac{\pi}{2}\right) \tag{11}$$

Distance matrix of $G = (V, E)$ and $G' = (V', E')$ can be expressed as formula (12). Finally, similarity between candidate target and template is defined as formula (13). In formula, k_i and w_{ij} are the weight coefficient.

$$C(G, G') = \begin{pmatrix} c(v_1, v_1') & \cdots & c(v_1, v_m') \\ \vdots & & \vdots \\ c(v_n, v_1') & \cdots & c(v_n, v_m') \end{pmatrix}, c(v_i, v_j') = \sum_{i=1}^4 k_i d(H_i, H_i') \tag{12}$$

$$Q(y) = \sum_{i=1}^n \sum_{j=1}^m c(v_i, v_j') w_{ij} \tag{13}$$

4.3 Adaptive template updating mechanism

Template updating is one of the key and difficult problems in target tracking. TCA updates template fully when there is gray variation, not suit for occlusion and scale variation, eventually lead to tracking failure. In order to solve problems above, we need to distinguish these two special scenes and using different update mechanism respectively.

The same point: occlusion and scale variation are all result in gray changes within the target region.

Different point:

1) Different causes. Occlusion is mostly because that target area is covered by background obstacle. But the scale variation is caused by changes of angle or distance between target and camera.
2) Different treatment. Gray changes caused by occlusion can't update into template, but gray change caused by scale variation shall be updated into template timely.

Aiming at half-occlusion, full-occlusion and scale variation, an adaptive local template updating mechanism is introduced. For occlusion, templates are not updated. But for scale variation, template should be updated timely. These three scenes above are distinguished by A (block matching proportion) adaptively. When A is relatively small, there are three possible situations: occlusion, scale variation, and out-scene target, If occlusion is defined as scale variation falsely, template is updated and target losing will be caused, Conversely, template is not update and tracking error will happen, but target will not completely lost. Based on the above analysis, improved algorithm gives priority to occlusion.

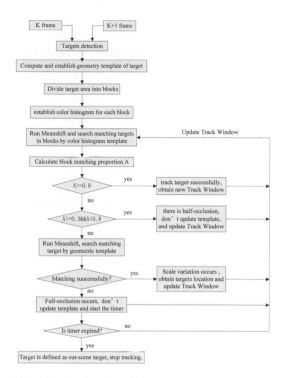

Figure 3. The flowchart of improved algorithm.

A is considered as criterion, and after repeated experiments, the criterion is set as follows. When A is greater than 0.8, we define that target is tracked successfully; If A is between 0.3 and 0.8, we define that target is half-occluded, and then search the target without template updating; If A is less than 0.3, we define that there are full-occlusion and scale variation, start the timer, and search targets based on geometry histogram. If there are matching target, scale variation occurs and template should be updated. Otherwise, there is full-occlusion without updating template. When timer is beyond the threshold, we define that target is out-scene, and stop tracking.

4.4 Algorithm steps and flowchart

Steps of the improved algorithm are as follows and showed in figure 3.

(1) Targets detection. Obtain the initial target position, and initialize the size and position of Track Window.
(2) Compute geometry histogram within the target region, establish target geometry template.
(3) Divide target area into blocks, and establish color histogram for each block.
(4) Run Meanshift, and search matching targets in blocks within Track Window by color histogram template.
(5) Calculate block matching proportion A, and update template adaptively according to value of A.

 A >= 0.8: track target successfully, obtain new Track Window, turn to step (4);

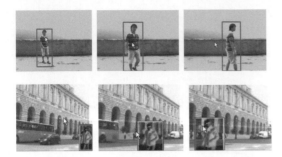

Figure 4.   Tracking result under scale variation.

Figure 5.   Tracking result under occlusion.

A >= 0.3&&A < 0.8: there is half-occlusion, don't update template, and update Track Window, turn to step (4);

A < 0.3: there may be full-occlusion or scale variation.

(6) Run Meanshift, use geometric template to search matching target in Track Window. If success, we think that scale variation occurs , then obtain targets location and update Track Window, turn to step (4); If failure, we think that full-occlusion occurs, and then don't update template, start the timer, turn to step (4);

(7) If the timer expires, the corresponding target is defined as out-scene target, stop tracking.

5   ANALYSIS OF EXPERIMENTAL RESULTS

In order to verify the validity of the improved algorithm proposed in this paper, we made three experiments under two scenes, occlusion and scale variation. Experimental video are obtained outdoor, and size is 320*240. The results are shown as follows.

In Figure 4, target turned around and walked from far to near, which leads to scale variation, but the target is still tracked accurately by improved algorithm. In Figure 5, because of fixed objects in the scene, the target is occluded to some extent, but the target still can be continuously, effectively tracked. In Figure 6, the experimental comparison of TCA and improved algorithm is showed, and we can find that improved algorithm has better, more accurate tracking effect than TCA. The experimental results show that, the improved algorithm can well adapt to scale variation, and also has good anti-occlusion ability, can maintain a good tracking effect.

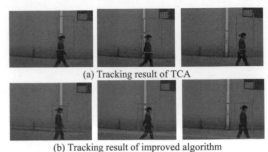

(a) Tracking result of TCA

(b) Tracking result of improved algorithm

Figure 6.   Comparative tracking result.

(1) Tracking result of the improved algorithm under scale variation.

(2) Tracking result of the improved algorithm under occlusion.

(3) Comparative tracking result of improved algorithm and TCA.

6   CONCLUSION AND FUTURE WORK

An improved Camshift algorithm based on block tracking and geometry histogram, which can distinguish occlusion and scale variation and update template adaptively, is presented in this paper. The experimental results show that, the algorithm can maintain a good tracking effect when there are occlusion and scale variation. However, the algorithm still has some problems, such as: 1) occlusion between targets is not considered; 2) target itself changes are not solved, which requires a combination of face recognition. These problems need to be further studied.

REFERENCES

[1]   Ristic B, Arulampalam S, Gordon N. 2004. Beyond the Kalman Filter: Particle Filters for Tracking Applications [M]. Boston, London: Artech House.

[2]   ZHANG Hong-zhi, ZHANG Jin-huan, YUE Hui, HUANG Shi-lin. 2006. Object tracking algorithm based on CamShift [J]. Computer Engineering and Design 27(11):108–110.

[3]   Comaniciu Dorin, Ramesh Visvanathan, Meer Peter. 2003. Kernel-based object tracking [J]. IEEE Transactions on Pattern Analysis and Machine Intelligence 25(5):564-577.

[4]   Nummiaro Katja, Koller-meier Esther, Van Gool Luc. 2003. Object Tracking with an Adaptive Color-Based Particle Filter [C]. Proceedings of the 24th DAGM Symposium on Pattern Recognition. London, UK: Springer-Verlag: 591–599.

[5]   QI Meibin, ZHANG Li, JIANG Jiangguo, WU Hui. 2011. Target template update method in fragment tracking[J]. Journal of Image and Graphics 16(6):976–982.

[6]   TANG Jin, JIANG Bo, LUO Bin, GUO Yutang. 2011. Graph geometry relation histogram and application in graph matching [J]. Journal of Image and Graphics 16(7):1234–1240.

[7]   Comaniciu Dorin, Ramesh Visvanathan, Meer Peter. 2003. Kernel-based object tracking [J]. IEEE Transactions on Pattern Analysis and Machine Intelligence 25(5):564–577.

Information Technology and Computer Application Engineering – Liu, Sung & Yao (Eds)
© *2014 Taylor & Francis Group, London, ISBN 978-1-138-00079-7*

A model of software reliability with actual fault removal efficiency

T. Hong

School of Electronic and Information Engineering, Tongji University, Shanghai, China

ABSTRACT: Addressing at the problem about the uncertainty of the fault removal efficiency in the existing software reliability growth models, a new improved model based on the rate of removing faults, which incorporates linear decrease-rate of removing faults with exponent decrease-rate of removing faults, is proposed in this paper. Meanwhile, the fault introduction is also considered in the new model. By comparing to the existing models and analyzing simulation results, the new proposed model is proved to be more suitable for the actual software reliability process.

Keywords: Software reliability model; fault removal efficiency; nonhomogeneous Poisson process; decrease-rate.

1 INTRODUCTION

With the rapid development of computing technology and complexity of software systems, the assurance of software quality becomes a critical concern. Software reliability is cited by many developers as one of the most important features of software products, which means the probability of failure-free software operation for a specified period of time in a specified environment. Many factors including the scale of software, operating environment, software architecture, software testing may have significant impact on the software reliability.

Researchers have conducted lots of activities in software reliability engineering. A proliferation of software reliability models have been proposed as they attempt to understand the characteristics of software fail and try to quantify software reliability. In another word, they focus their attention on the problem of quantifying the software reliability by using software reliability growth models (SRGM). Software reliability models based on non-homogeneous Poisson process (NHPP) have been the most important and applied most widely in SRGM, they are frequently used in stochastic simulations to model nonstationary point processes.

Up to now, lots of improved NHPP models have been proposed. Goel and Okumoto proposed the G-O model which was the earliest NHPP model in 1979. It ignored both the fault introduction and the fault removal efficiency, the fault removal time was also overlooked. This model was too ideal to meet the actual software development process. Reference [5] proposed a model, which thought over the fault introduction but assumed the detected faults were immediately removed. Reference [6] formulated the hypothesis that the fault removal efficiency was a constant and the fault introduction was ignored. Reference [7] considered both the fault introduction and fault removal efficiency, but it assumed the fault removal efficiency was a constant. Reference [8] proposed a model considered the linear decrease-rate of removing faults, Reference [9] assumed the exponent decrease-rate of removing faults in the model. These researchers focused their attention on the imperfect debugging phenomenon in the software engineering, but they could not consider fault removal efficiency reasonably, so their models could not satisfy actual situations.

In fact, the imperfect debugging phenomenon includes two sections. The first section is that it exists the fault removal time, and the fault removal efficiency is not always 100%, experiment results show that the fault removal efficiency is 15%~50% in the phase of unit testing, 25%~40% in the phase of integration testing and 25%~55% in the phase of system testing, so it should be a function of the time. The second section is that we should also consider the fault introduction when we are removing faults, and the number of total errors should also be a function of the time when fault occurs.

In this paper, we propose a new software reliability model that discusses the assumption, which incorporates linear decrease-rate of removing faults with exponent decrease-rate of removing faults. At the same time, we also take the fault introduction situation into account. The paper is organized as follows. Section 2 presents a conventional NHPP SRGM. Section 3 deduces the new model. The analysis and simulation of the proposed model are investigated in Section 4. Finally, the conclusions are given in Section 5.

2 NON-HOMOGENEOUS POISSON PROCESS MODEL

As the basis of the paper, we give the notations of the traditional NHPP.

$N(t)$: Counting process for the total number of faults in $[0, t)$.

$m(t)$: Expected number of software failures by time t, $m(t) = E[N(t)]$.

a: Total number of faults in the software.

b: Fault detection rate per fault.

$p(t)$: Fault removal efficiency function.

$x(t)$: Total number of faults detected and successfully removed by time t.

MTTF: Mean time to failure.

The conventional G-O model can be expressed as (1),

$$m(t + \Delta t) = b(a(t) - x(t))\Delta t + o(\Delta t). \tag{1}$$

then we can deduce:

$$\frac{dm(t)}{dt} = b(a(t) - x(t)). \tag{2}$$

$$\frac{dx(t)}{dt} = p(t)\frac{dm(t)}{dt}. \tag{3}$$

Through differential (2) and (3), we have

$$\frac{dx(t)}{dt} = b \cdot p(t) \cdot (a(t) - x(t)). \tag{4}$$

The marginal conditions for the differential (2) and (3) are as follows:

$$m(0) = 0. \tag{5}$$

$$x(0) = 0. \tag{6}$$

Hence, the total number of faults detected and successfully removed by time t is

$$x(t) = a(1 - \exp(-b\int_0^t p(\tau)d\tau)). \tag{7}$$

Therefore, the explicit expression of the expected number of faults by time t can be obtained as follows:

$$m(t) = ab\int_0^t \exp(-b\int_0^v p(\tau)d\tau)dv. \tag{8}$$

The total number of detected faults $N(t)$ follows a Possion distribution with parameter $m(t)$. It can be denoted as:

$$P\{N(t) = n\} = \exp(-m(t)) \cdot (m(t))^n / n!, n = 0,1,2\dots. \tag{9}$$

The reliability function can be expressed as:

$$R(x|t) = P\{N(t+x) - N(t) = 0\} = \exp[-(m(t+x) - m(t))]. \tag{10}$$

The traditional NHPP model treats the fault introduction rate a and the fault detection rate b as a constant. The fault removal rate $p(t)$ is uncertain in the model. These hypotheses do not meet the demand of the actual software development process. So we should concern on these issues in the new model.

3 A NOVEL SOFTWARE RELIABILITY MODEL

The key of setting up software reliability model is to meet the actual situations. In this new model, we consider the related functions to fit the process of software as much as possible.

3.1 The fault detection rate

In fact, the rate of fault detection is changed with properties of the fault and the process of development. The rate is similar to the growth function of S, as follows:

$$b(t) = \frac{b^2 t}{1 + bt}, 0 < b \le 1. \tag{11}$$

In this paper, we consider b as a constant.

3.2 The fault introduction rate

Cite the model based on constant-rate of fault introduction, we have

$$a(t) = a_0 \cdot (1 + \delta t). \tag{12}$$

$\delta > 0$, the values of α decide the probability of fault introduction.

3.3 The fault removal rate

In generally, the failure rate of software may be higher if there are more errors in the software. Based on this theory, scholars made three simple but important modes, such as modes based on constant-rate of removing faults, linear decrease-rate of removing faults and exponent decrease-rate of removing faults. However, none of them are valid at all times and there is no unique model which can perform well for all situations.

In fact, the fault removal efficiency can not be expressed by a constant function. It can be affected by the complexity of software, the complexity of fault, operating environment, standard of tester etc. In the beginning of removing fault, the fault removal efficiency remains a high level. With the process of test, the easily found faults are removed and the removing process becomes harder. When the test process reaches a phase, the fault removal efficiency will stay a low level and remain stable.

So, considering more factors, we give mathematical expressions:

$$\begin{cases} \dfrac{dE_r(t)}{dt} = \dfrac{u}{1 + \alpha \cdot e^{\beta t}} \\[2ex] E_r(t) = a_0 - \dfrac{u \cdot \ln(\alpha + 1)}{\beta} + \dfrac{u \cdot \ln(\alpha + e^{-\beta t})}{\beta} \\[2ex] p(t) = \dfrac{u}{1 + \alpha \cdot e^{\beta t}} \\[2ex] MTTF = \dfrac{1}{k \cdot \left[a_0 + \dfrac{u \cdot \ln(\alpha + e^{-\beta t})}{\beta} - \dfrac{u \cdot \ln(\alpha + 1)}{\beta} \right]} \end{cases} \tag{13}$$

Substituting (12) and (13) into (7) and (8), we obtain

$$x(t) = a_0(1+\delta t)\left[1 - \left(\frac{\alpha + e^{-\beta t}}{\alpha + 1}\right)^{bu/\beta}\right]. \quad (14)$$

$$m(t) = a_0 \cdot (1+\delta t) \cdot b \int_0^t \left(\frac{\alpha + e^{-\beta t}}{\alpha + 1}\right)^{bu/\beta} d\tau. \quad (15)$$

We can estimate the parameters a_0, δ, b, α, β, u by maximum likelihood estimation and least square method.

4 PERFORMANCES AND SIMULATION

In this paper, the performance of the newly improved NHPP model is evaluated by using the sum of squared error (SSE) and R-square.

The sum of squared error is the criterion that sum up the squares of the residuals of the actual data and the mean value function ($m(t)$) of each model in terms of the number of actual faults at any time points. The SSE function can be expressed as follows:

$$SSE = \sum_{i=1}^{n} (y_i - \hat{m}(t_i))^2$$

where y_i is the total number of faults observed at time t_i according to the testing data, $\hat{m}(t_i)$ is the estimated cumulative number of faults at time t_i.

Another criterion we use is R-square, which can be expressed as follows:

$$R\text{-square} = \frac{\sum_{i=1}^{n}(\hat{m}(t_i) - \bar{y})^2}{\sum_{i=1}^{n}(y_i - \bar{y})^2}$$

where $\bar{y}$ is the mean number of faults observed.

4.1 Compared models

To make the experiment convinced, we choice three models as the compared models.

Model I: G-O model. We can obtain the $m(t)$ of G-O model

$$m(t) = a(1 - \exp(-bt))$$

Model II: The model proposed by Zhang in 2003, the $m(t)$ of this model can be expressed as follows

$$m(t) = \frac{a}{p-\beta}\left[1 - \left(\frac{(1+\alpha)e^{-bt}}{1+\alpha e^{-bt}}\right)^{\frac{c(p-\beta)}{b}}\right]$$

where the fault removal rate p is constant, β is the rate of fault introduction.

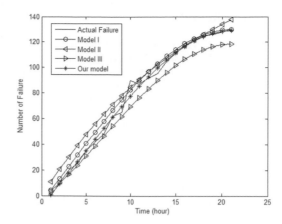

Figure 1. Comparison of goodness-of-fit for Four Different Models.

Table 1. Comparison of goodness-of-fit and predictive power of NHPP models.

Model	SSE (fitted)	SSE (predicted)	R-square
Model I	2479	204.8	0.9827
Model II	5212.7	326.7	0.9153
Model III	4315.4	254.6	0.949
Our model	1432.8	107.5	0.9931

Model III: The model proposed by Xie Jing-Yan, the $m(t)$ of this model can be expressed as follows

$$m(t) = b \cdot \left[\frac{a_0(1+kt)}{k-b} + \frac{\alpha}{b+k}\left(\frac{(1+kt)^2}{2k} - \frac{(1+kt)^{1-\frac{b}{k}} - 1}{k-b}\right)\right]$$

where the fault removal rate $p(t) = 1/(1+kt)$, α is the rate of fault introduction, b is the rate of fault detection.

4.2 Simulation

In this section, we use the data set that is documented in Reference [7] to examine the goodness-of-fit and predictive power of the improved model.

There are totally 136 faults reported, and the TBF is extremely long from the 122nd fault to the 123rd fault, and the TBFs after the 123rd fault increases tremendously. It implies that the reliability grows, and the system becomes stabilized. Here, we use the first 122 data points for the goodness-of-fit evaluation and the remaining data points for the predictive power test.

Figure 1 shows the new proposed model has a best goodness-of-fit and predictive power.

The SSE and R-square values for goodness-of-fit and prediction are listed in Table 1.

Table 1 shows that for the improved model, the SSE value for prediction is the lowest and the R-square value is the highest among all models.

5 CONCLUSIONS

In this paper, we discuss a new software reliability model, which is based on the rate of removing faults incorporates linear decrease-rate of removing faults with exponent decrease-rate of removing faults, considering the fault introduction. By using two sets of practical data and the comparison of performance between the improved NHPP model and classical NHPP models. The results show that this model can improve both the descriptive and the predictive ability of a model.

ACKNOWLEDGEMENT

This work is supported by the Open foundation of the State Key Laboratory of Rail Traffic Control and Safety (Contract No. RCS2010K007), Beijing Jiaotong University.

REFERENCES

[1] Musa JD. 1999. Software Reliability Engineering. New York: McGraw Hill.
[2] Lyu M. 1996. Handbook on Software Reliability Engineering. New York: McGraw Hill.
[3] Goel AL. & Okumoto K. 1979. Time-Dependent error-detection rate model for software and other performance measures. IEEE Trans. On Reliability, 28:206–211.
[4] Swapna S G. 2007. Architecture-based Software Reliability Analysis: Overview and Limitations [J]. IEEE Transactions on Dependable and Secure Computing, 4(1):32–40.
[5] Pham H. & Nordmann L. & Zhang XM. 1999. A general imperfect-software-debugging model with S-shaped fault-detection rate. IEEE Trans. on Reliability, 48(2):169–175.
[6] Li CZ. & Yue XG. 2005. A NHPP software reliability growth model incorporating test coverage and fault removal efficiency. Electronics Quality, (3):34–36.
[7] Zhang XM. & Teng XL. 2003. Considering fault removal efficiency in software reliability assessment. IEEE Trans. on Systems, Man, and Cybernetics: Systems and Humans, 33(1):114–119.
[8] Xie JY. & An JX. 2010. NHPP Software Reliability Growth Model Considering Imperfect Debugging. 21(5):942–949.
[9] Zheng L. & Shen YL. 2010. A Model of Software-Reliability with Imperfect Fault Correction Process.
[10] Huang C Y. & Lyu M R. 2003. A unified scheme of some nonhomogenous Poisson process models for software reliability estimation [J]. IEEE Transaction on Software Engineering, 29(3): 261–269.
[11] Ledoux J. 2003. Software reliability modeling [M]. New York: Springer.
[12] Xue Y. & Nan S. 2008. An Improved NHPP Model with Time-Varying Fault Removal Delay. Journal of the electronic science and technology of china.
[13] Liu HW.& Yang XZ. 2005. Qu F. A framework for NHPP software reliability growth models. Computer Engineering & Science, 27(4):1,2,18.
[14] Ossmane Krini. 2012. A New Method to Detect and Correct the Critical Errors and Determine the Software-Reliability in Critical Software-System. Journal of Physics.
[15] Hou YZ. & Yu BY. 2011. A novel software reliability model with the decrease-rate of removing errors. Proceedings of the 2011 International Conference on Machine Learning and Cybernetics.
[16] Zhang X. & Teng X. 2003. "Considering fault removal efficiency in software reliability assessment," IEEE Trans. on Systems Man and Cybernetics Part A: Systems and Humans, vol. 33, no. 1, pp. 114–119.

Information Technology and Computer Application Engineering – Liu, Sung & Yao (Eds)
© 2014 Taylor & Francis Group, London, ISBN 978-1-138-00079-7

A brief analysis into e-commerce website mode of the domestic luxury

Lian Lu

Nanjing Institute of Industry Technology, Nanjing, China

ABSTRACT: With the current prosperity of the luxury market, this paper focuses on e-commerce market situation of China's luxury, disintegrates all the parts during the process of the luxury's e-commerce online shopping, analyzes the three targets and three links of the current e-commerce websites at the material level and finds out its appearing relevant problems. This paper also delves deeper and puts forward the online shopping mode at the spiritual level in order to bring some benefit to Chinese e-commerce businesses.

Keywords: Luxury; E-commerce; Mode; Analysis

1 INTRODUCTION

China's consumers have an indissoluble bond with the luxury from the Pierre Cardin in the eighties and nineties to the current Louis Vuitton in their minds.

According to the statistics of WLA, the annual total consumption amounts of China's luxury market has reached 12.6 trillion (without private planes, yachts and limousines), accounting for 28 percent of the global shares in 2011, and is still increasing year by year [1]

Consistent with the prosperous luxury market in China, the domestic luxury e-commerce websites have been online successively in the two past years and have entered public view in a high-profile way. In particular, nearly ten websites including VIPStore, Fclubcn, XIU, Vipshop and so on have won VC's favor and gained more than 500 million investment capital.

However, not for long, the luxury website Vipshop encountered a break during its listing in the NYSE on March, 2012. The former chairman as well as CEO of VIPKU was being dismissed; the pay back gate of Wooha was reported; the redundancy news of Shangpin and XIU was coming out successively. The luxury businesses that flourished in 2011 are suffering asperity at present [2]

In addition, the global famous luxury websites that enjoy high market shares in developed countries such as Europe and America rolled out Chinese web pages in quick succession. For example, the yoox in Italy rolled out www.thecorner.com.cn; the net-a-porter in Britain detruded www.theoutnet.cn to grab Chinese luxury businesses' markets.

With the domestic strife and foreign aggression, how to promote the domestic e-commerce market development healthily is a great problem for luxury business enterprises now.

2 THE ANALYSIS INTO LUXURY E-COMMERCE WEBSITE'S PROBLEMS

The VIPStore's banner ad in the baidu search engine is "the VIPStore is a discount website for top-level luxury; the international brands has started from 90 percent off; fashionable watches, brands clothing, brand-name bags and top-level cosmetics and so on have captured the fashion and led the luxurious fashion in the year of 2012" [3].

The Shangpin's banner ad in the baidu search engine is "the Shangpin luxury website is licensed exclusively by international bands; 100 percent of quality goods; try first, pay later; free home delivery; creative commercial mode and excellent service experience has won the trust of medium and high customers" [4].

After so many years' penetration into the industries and the relevant analysis on the market research of the current domestic luxury e-commerce websites including Vipshop, Shangpin, ihaveu and VIPStore etc, the front-line writers who have been working in the luxury e-commerce enable to disintegrate the space structural forms of various parts and links in the current luxury e-commerce websites and decomposed it into material level and spiritual level. The first level, material level, is the basic structural form of the current most luxury e-commerce websites (see Figure 1). There are three targets in the space including websites, things (luxury)£and people (buyers) in this level. Three targets link and connect each other through payment, logistics and after-sales service. The buyers pay to the websites through the online payment platform; the front-line of website inform the warehouse to deliver the luxury to buyer by logistics facility; the buyer can solve the problems through the after-sales link on the website to complete the procedure of online luxury purchasing.

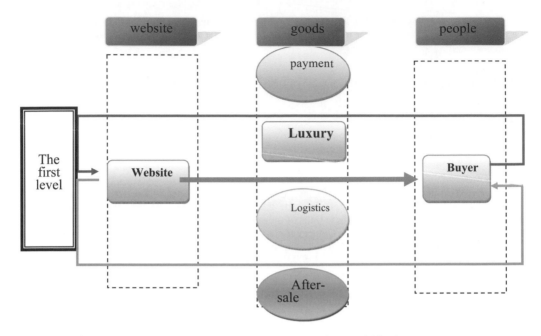

Figure 1.   The first level: the architectural pattern of luxury online shopping material level.

During this level, we found the following relevant problems through the analysis on the three targets and three links.

2.1   *Website problems*

The differentiation among websites is not obvious enough. When you open lots of luxury websites, you may have the similar feelings. For example, the websites' home pages of the Fifth Avenue, Vipshop and Meici have a larger similarity. They are almost the same things. What's more, Vipshop and Shangpin both detruded the discount promotion.

2.2   *Commodity problems*

(1) Commodity supply

Commodity supply is always the damage of domestic luxury e-commerce, which has hindered the e-commerce website development. Up to now, domestic e-commerce websites mainly have three types of commodity supplies. At first, buy out and purchase the commodity from the brands' domestic factor and franchiser. Secondary, purchase the commodity during the overseas discount period or purchase the out-of-season ones from the overseas discount stores and OUTLET. At last, purchase parts of them in the overseas exclusive stores. The above three methods do not purchase commodity from the brands, so all the websites encounter a common authorization problem.

Due to the authorization problems, the China's e-commerce websites including Jing Dong Mall received the "roll call" of the international renowned luxury brands on March, 2012.

According to the claim of "crystal" luxury brands Swarovski, they have not authorized any websites in China regions to sell Swarovski products. After that, luxury brands such as LV, GUCCI, Coach, PRADA come forward to claim that they have not authorized any websites in China to sell their products too. [5]

(2) Price difference

The commodity appreciation accredits to the brands while the brand appreciation accredits to consumer's feeling for brands. We find that the luxury never participates in the discount activity and its price is increasing year by year; however, the consumer's enthusiasm will not decrease with the higher price of luxury. Therefore, the pricing of luxury brand businesses becomes more confident and firm about the price. Generally speaking, luxury brand businesses have strict pricing system. But the luxury businesses' "attracting" price will definitely shock the brand owners to cause the increasing conflict between the businesses and brand owners. At the same time, the recent euro's weakness leads a large price difference between the domestic exclusive shops' price and businesses' price, which delights the fashionable buyers and the businesses' sellers but depresses the luxury brand businesses and domestic agents.

(3) Quality assurance

The luxury may inevitably carry some flaws, because parts of the commodity sources of luxury e-commerce websites are out-of-season ones or discount ones. Though the buyers bought it at a preferential price, they can not experience the luxurious feelings or it may ruin the nice brand image in buyer's minds. Meanwhile, buyers can not obtain the maintenance from the domestic brand luxury exclusive shops for those commodities.

2.3 Buyer's problems

The buyers who purchase luxury from e-commerce websites are usually white collars or gold collar workers. They are familiar with the online resources and have certain cognition for internet. Meanwhile, they have a fabulous income. However, according to the actual survey, it is found that those customers do not have high viscosity and seldom have repeated purchases. By the statistics, the annual times of a stable luxury consumer's online shopping will not surpass ten times, and ten times are being divided up by a series of professional luxury e-commerce websites such as Shangpin, Vipstore, XIU and Fifth Avenue etc. [6]

2.4 Online payment problems

At present, the sales of luxury businesses websites almost connect with the third party payment platform through its own system. The third party payment platforms include Unionpay, Cyberbanking, alipay, Tenpay, lakala, fast money etc.

Fclubcn brought about a slow refund and unsound user experience due to the website system problems recently, which caused buyers' extreme dissatisfaction. So the buyer safeguarded rights of online shopping, which also had a negative impact on the websites.

2.5 Logistics problem

Network sales have to face buyers from all over the word. Currently, luxury e-commerce has three main logistics modes: first, it depends on external Logistics Distribution Company or so-called the Third Party Logistics [7]; second, logistics channel cooperates with express company; third, extremely few of e-commerce with relatively sufficient capital creates logistics on their own. Presently, most of websites are by means of external Logistics Distribution Company.

Support value. Security is the first problem during process of luxury logistics. And what the express company launches is the security of support value. But as to support value, problems usually happen during logistics process: first, express company is not willing to receive orders of commodity with excessive amount of money; second, breakage of package as well as quality problems happens in uninsured commodity. Therefore, luxury support value is beneficial to both luxury e-commerce and express company. If the luxury with high value is to be insured, commodity's invaluable luxury property will naturally show, which is also likely to suffer taking the wrong one; on the other hand, it is natural that the support value of commodity with high value is many times higher than express fee. Take the fifty-thousand RMB bag for example, if it is delivered by sf-express, the express fee is within 30 RMB

including that of bag. However, if express fee is 0.5 percent of commodity value, then, the support value is 250 RMB. It can be seen that the logistics cost is not low. So luxury e-commerce should take the high cost of natural logistics chain into consideration.

2.6 After-sales problems

Luxury e-commerce websites' supply of goods is relatively special and its channels are diverse. Most of commodity hasn't been authorized by brand owners; besides, some merchants may purchase goods from foreign OUTLET or the goods on sale. Some goods have defects and have not brand authorization certification. Thus, whether the luxury shopped online is authentic or false could not be inspected. Consequently, purchasers can not experience after-sale service, like maintenance, in domestic brand exclusive shops.

3 NEW MODEL STUDY ON LUXURY E-COMMERCE WEBSITE

Based on years' investigation and researches, the author has analyzed each part of three links of above three targets. There are diverse problems and situations in the sales mode in the first level of luxury e-commerce. If they are solved one-on-one, the problems can not be settled down fundamentally. It is just like that treat the head when the head aches and treat the feet when the feet ache.

3.1 Brand license, honest sales

The soul of luxury is brand; authorized brand is the premise of luxury e-commerce. Only that can eliminate customers' worry to buy the fake and enable them to buy things safely. Consequently, the credit problems are solved and e-commerce sales will become bigger and stronger. Because of its special charm, each luxury item plays a role in spreading brand cultural deposits and delivering brand value, which makes luxury have so many followers. The authorized e-commerce website builds a bridge between brand and buyers by means of luxury.

As it is shown in the mode of second level in Figure 2, there is not only the luxury between websites and buyers. In turn, websites convey the brand culture and brand values to buyers through luxury, bringing buyers remarkable brand experience; once buyers have enjoyed the "special treatment", they will be closely around brand with their special feelings. No matter whether it is a new product or a discount activity, buyers trust brand and follow the brand. Thus, websites can interact with buyers to form a big circle. In this big circle, they continuously contact with each other and communicate with each other. At the very start, there are interactions between them, then, they benefit from each other, creating the maximum value in this circulation.

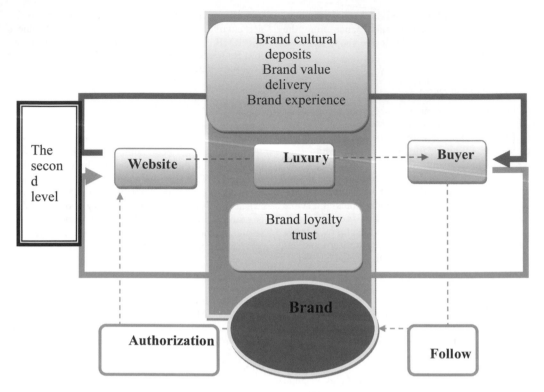

Figure 2. The second level: architectural pattern of luxury online shopping's spiritual level.

3.2 Spreading brand culture, delivering brand value and increasing brand experience through websites

Luxury e-commerce merchants have to survive healthily. Particularly, they need to be authorized via brand owners which have already registered in China. But their "attractive" price must destroy the price system of luxury brand owners. As merchants, it is important for them to know how to cooperate with brand owners delightfully and manage their own "base" respectively. Rather than damage the markets that brand owners possess. E-commerce merchants need acting as disseminators of luxury brand and being the protectors of brand owners. In a word, merchants will obtain win-win when they are favored by brand owners [8].

With regards to luxury which has not yet entered into Chinese inland market, new brand can be known and recognized by domestic masses by means of "brand story" and "brand introduction". Promoting by aiming at the fixed user group is time saving and has a significant effect. It is a new channel of promoting new brand and the beginning of cooperation between e-commerce merchants and brand owners.

With the spread of brand by e-commerce on the website and implantation of new culture, brand owners build a good strategic relationship with e-commerce, which can ensure the true or false of commodity. With the help of brand owners, users' attention and customer experience can be finished through brand platform. Meanwhile, customers enjoy after-sale service, like return or replacement of goods and maintenance, equal to the service in exclusive shops [9]. Photographs should be with good contrast and on glossy paper. Photographic reproductions cut from books or journals, photocopies of photographs and screened photographs are unacceptable. The proceedings will be printed in black only. For this reason avoid the use of colour in figures and photographs. Colour is also nearly always unnecessary for scientific work.

3.3 Buyers' trust and loyalty to brand

A shopping experience is the basic motivation of repeat purchase. How to attract a new buyer to the website, look through it and make purchasing behavior is a better reason for e-commerce merchants to survive after they obtain brand owners' authorization. When experience luxury shopping online, they have enjoyed the following processes: brand introduction, cultural influence, preferential price, three-dimensional or even video display products, one-on-one trend, specific introduction of use and maintenance, sincere after-sale problems solving and door-to-door service by Smart. Buyers have already realized influence from brand. Such heart to heart communication enables buyers to be around brand. Naturally, it is also recognition of e-commerce websites. Buyers' trust produces the website bonding. It is obvious that the enthusiasm has been inspired. Both e-commerce merchants and brand owners look forward to seeing the rise of websites' popularity and increase of volume.

Starting from 2010, luxury e-commerce websites reluctantly maintain the likelihood through several venture capitals. After several shuffle from 2011 to 2012, if luxury e-commerce websites want to live in a truly healthy and ecological way, they need to dig out the connotation of luxury, a special good. Besides, they need to treat spiritual level of luxury shopping online as their way of life. Due to cultural heritage and influential spread, luxury markets make commodity as links to closely connect brand, websites and buyers with each other. E-commerce merchants obtain large sales; meanwhile, buyers enjoy first-rate service and have the sense of belonging of equivalent values. So more buyers are willing to accept them, trust them and be loyal to them; with the increase of new buyers, the circulation consisting of buyers, websites and brand can be magnified all the way. As a result, buyers, websites and brand can benefit from each other. Current e-commerce merchants need to find their own breakthrough to stand out in the present fierce fighting.

REFERENCES

[1] Chinese Luxury Ten-year Report Launched by World Luxury Association [N]. World luxury association. 2012.

[2] Revealing the behind secret on luxury websites' going public [N]. New wealth, 2012.

[3] http://www.baidu.com/s?wd=%BC%D1%C6%B7%CD%F8&tn=sogouie_dg

[4] http://www.baidu.com/s?wd=%C9%D0%C6%B7%CD%F8&tn=sogouie_dg

[5] Burst of Listing in Vipshop, The compatibility of Luxury Websites' marketing Mode [N]. Chinese E-commerce Research Center, 2012.

[6] X. Zha, Y. Guo. Game. Analysis of Cooperative Behaviors Based on Third-party Logistics. The Fifth Wuhan International Conference on E-business. 1145–1151.

[7] Huang CC, Liang WY, et al. "The agent-based negotiation process for B2C e-commerce", Expert Systems with Applications, 2010, 37(1): 348–359.

[8] Zhao Jiayi. "New Breakthrough" in Luxury Online Shopping. China Internet Weekly [J]. 2012. 4.

[9] Jun Tan, Zhongchun MI. Game Analysis of Consumer Decision-making and Return Policy in Electronic Commerce. Journal of Computational Information Systems, 2009, 5(2):701–708.

Information Technology and Computer Application Engineering – Liu, Sung & Yao (Eds)
© 2014 Taylor & Francis Group, London, ISBN 978-1-138-00079-7

Advertising matching model based on user interests and social relationship

Jia Hui Wang, Bing Fei Ren & Fang Zhao
Software Engineering School, Beijing University of Posts and Telecommunications, Beijing, China

Yu Nong Lin
International School, Beijing University of Posts and Telecommunications, Beijing, China

ABSTRACT: Recently, Location Based Advertising (LBA), which integrates mobile advertising with location based service (LBS), has become the new form of advertisement, bringing new experience to users. However, it results in bad advertising effect and user experience, which is attributed to the existing LBA applications' advertisement pushing only according to user location. In this report, we aim to study the advertising matching model based on the user interests and social relationship which are mined through social network platform. And these two dimensions help to increase the matching degree between advertising and potential clients, and to some extent, improve the advertising effect and user experience.

Keywords: Advertising; User Interests; Social Relationship; LBS

1   INSTRUCTIONS

LBS locates users to provide relevant services through the cooperation of mobile terminal and mobile network. With all these years' efforts, its technique becomes mature and is applied widely. It involves life services, social network, electronic commerce and other aspects, which bring convenience to people's life. LBA, based on the geographic location, is a kind of service vertically combined with advertising.

Nowadays, social media, like Facebook, Twitter or Weibo, rise rapidly and have a great amount of users. These social platforms provide users with services of publishing, sharing and receiving information, and also present user preference and social relationship. Traditional advertising spent a lot of money on "pushing" ads through traditional media, while social platforms create an excellent chance to promote the products by "pulling" strategy with less costs.

The effectiveness of advertisement distribution highly relies on well understanding the preferences information of the target users [1]. Besides, if the users do not consume right after seeing the ads, it means a loss of potential consumers. An advertisement which is based on location and meets interests of users can attract users to consume in no time, which creates a win-win mode for both consumer and provider.

Take advantage of LBS and social platform, a matching model based on user interests and social relationship is proposed to raise matching degree between ads content and user. Meanwhile, it combines advertisement targeting and pulling marketing strategy, improves the efficiency of advertising and is convenient for user consumption.

2   MODEL PRESENTATION

Merchants may prefer a platform that can promote a product to consumers in a timely, effective, and low cost way. Similarly, consumers would prefer to receive relevant and useful promotions [2]. For example, a scenario in which a well-known pizza restaurant in a shopping mall wishes to promote its popular but high profit margin triple sausage pizzas to draw the attention from nearby customers before the peak lunchtime hour. The promotion may be limited to ten pizzas selling between 11:30 and 12:00. The advertisement should be easily prepared by the owner and distributed via a platform to the mobile phones of target customers who are nearby or whose preference is pizza [2].

In this study, the model based on user interests and social relationship is a platform connecting advertiser and user. It manages advertising information provided by advertisers, mine user interests and relationship by data from social platforms, and send target user corresponding advertisements through matching degree of preferences, location and social relationship. The main framework of model, presented in Figure 1, contains three modules: Ads Management, User Character and Ads Intelligent Pushing. These modules interact with user and databases. And Ads Intelligent Pushing module interacts with Ads Management module and User Character Management module.

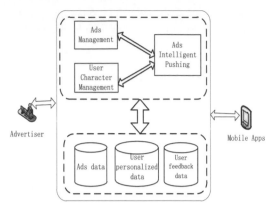

Figure 1.   Main Framework of Model.

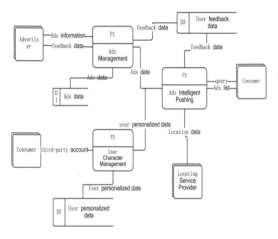

Figure 2.   Detailed representation of Model in DFD.

Figure 2 presents interactive data stream chart between inner and outer environment of the model. All mobile advertisements are provided by advertisers through the Advertisement Manager (P1) and are stored in the database (D1). As soon as user logs in the model via the third party social platform account, User Character Management (P2) is activated. Data from platform is used to analyze users' preferences and relevant user with high similarity. Then the results will be restored in the database (D2). When user sends a query instruction, Ads Intelligent Pushing (P3) is activated, matching degree between user and ads are calculated by specific algorithm through Ads data and user personalized data. And the feedback of user is stored in the database (D3).

2.1   Ads Management

Ads Management module provides advertiser with an interface to publish ads. Advertiser can fill in the relevant information of ads, e.g. category, banner, start time, end time, location, tab and so on. At the same time, Ads Management module is responsible for deleting overdue Ads in Ads Data Database as well

as send correlative user response in Feedback Data Database to advertiser on a regular basis.

2.2   User Character Management

User Character classification is an important issue in online marketing. By pre-classifying customers and offering personalized recommendation services or products it helps to improve customer satisfaction and target marketing [3]. Main function of User Character Management: (1) User Preferences Analysis; (2) User Relationship Analysis.

2.2.1   User Preferences Analysis

User Preferences Analysis is mainly used to abstract user keywords which demonstrates user preferences. Through API provided by social network platform, a document is summarized from information published by the user. Keywords are discovered from that document. Keyword abstraction is based on TF/IDF algorithm, the formula is:

$$w_j = TF_j \times IDF \qquad (1)$$

In formula, w_j is the weight of Term$_j$ in document; TF$_j$ stands for the frequency Term$_j$ appears in current document; DF represents the frequency Term$_j$ in all documents and IDF is the reciprocal, IDF $= \log\left(\frac{|D|}{DF_j}\right)$ in general case, which D is the number of document.

2.2.2   User relationship analysis

In practical applications, it is difficult to discover user preferences when his publishing and setting information is not enough. So it's hard to make precise ads-user match only with preferences dimension. At the moment, the system will analyze social relationship of user i on platform and discover a user group j of similar activities. The main procedure of User Relationship Analysis is: (1) Calculate concern degree of user group j to user i; (2) Figure out similarity degree between user group j and user i with Cosine similarity algorithm.

i) Concern Degree
According to research, if user i interacts frequently with user j, they probably have something in common, or user j is interested in user i. Therefore, the products or ads that user i likes are likely to draw user j's attention. In the experiment, the system uses formula (2) to calculate concern degree.

$$\text{CD(i,j)} = E_{ij} \times \sum_{t=1}^{T} \frac{|\Phi_{response(i,j,t)}| + |\Phi_{repost(i,j,t)}|}{|\Phi_{post(i,t)}|} \qquad (2)$$

where CD(i,j) stands for concern degree from user j to user i in period T. E_{ij} equals to α if edge j to i exists while edge i to j doesn't exist; E_{ij} equals to 2α if an edge from node i to j exists and an edge from node j to node i exists, too. $\Phi_{post(i,t)}$ stands for the set of the messages posted by user j at the time period t. $\Phi_{response(i,j,t)}$ stands for response from user j to user i

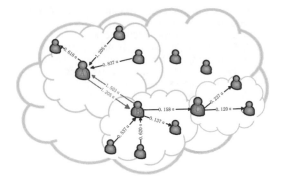

Figure 3. User Direct Weight Graph.

during t period, $\Phi_{repost(i,j,t)}$ stands for forwarding from user j to user i during t period.

Figure 3 is part of sampling result from data of over 2000 users in Sina Weibo, Inc. In the figure, user A has bi-directional edge with user B the knot between user A and user B is bi-directional, which means A and B have mutual concern. The weight of edge shows the degree. The concern degree is 1.205α from A to B and 1.503α from B to A.

ii) Similarity Degree

$$SD(i,j) = \cos\theta = \frac{I \cdot J}{\|I\| \cdot \|J\|} \qquad (3)$$

where user i keyword vector $I = \{I_1, I_2, \ldots, I_n\}$, and user j keyword vector $J = \{J_1, J_2, \ldots, J_n\}$.

2.3 Ads intelligent Pushing

Ads Intelligent Pushing (AIP) is based on the ideas of ItemCF and UserCF. As soon as user location is in the corresponding domain, prioritize the matching between user and ads by preferences, which means pushing products that conforms to their past preferences according to the ItemCF; if user information on home page is not enough to abstract keywords precisely, then pushing ads depending on his friends' ads pushing history, which means pushing products which users of similar interests preferred according to UserCF.

Figure 4 displays interaction data stream chart between inner environments of the Ads Intelligent Pushing module. The module is combined with 3 submodules: Ads searching (P3.1), Ads Matching (P3.2) and Rank Manager (P3.3). When query from user is received, Ads Searching is activated. It searches ads in the scope of 200 meters centered with user location and sends information to Ads Matching. Then Ads Matching will work out the matching degree of these ads and clients and send ads of high matching degree to Rank Manager, who will sequence the information and send user the sequential list.

2.3.1 Advertisement matching

The matching between consumers and advertisements involves a number of factors. There is a number of

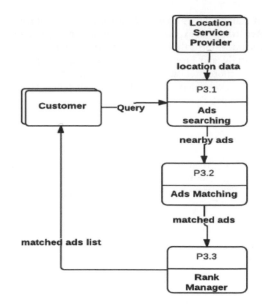

Figure 4. Detailed representation of AIP in DFD.

attributes for ranking advertisements, such as distance, discount, tabs, price and many others. The system uses algorithm below to match ads and user interests.

i) User preference can be presented by vector:

$$P = (P_1 : w_1, P_2 : w_2, \ldots, P_n : w_n)$$

where P_k stands for keyword of user preference, stands for weight. Generally, the frontier keyword has higher weight. Semantics of keyword user interested are often connected with information scopes and information resource models.

ii) Ads resources can be presented by vector:

$$A = (A_1 : w_1, A_2 : w_2, \ldots, A_m : w_m)$$

where A_k stands for the attribute of advertisement, w_m stands for weight of the attributes.

iii) Matching of User interests and ads attributes:
Matching degree of user interests and ads attributes is measured by correlation. The algorithm is showed below:

$$R = \sum_{j=0}^{m} (w_j \times \sum_{k=0}^{n} F_{kj}) \qquad (4)$$

where F_{kj} stands for the frequency P_k appearing in A_j, w_j stands for weight of A_j, $j = 1, \ldots, m$, According to the formula, relevancies of user interests and contents of ads nearby can be figured out. Then the result is ranked. Ads with higher relevancies are placed ahead [4].

3 CONCLUSION

In this report, a model based on user interests and social relationship is has been discussed. The key point of the model is the precision of discovering user interests and social relationship, as well as the precision of matching user characteristics with advertisements. According to experiments, the model has favorable accuracy. In further study, more efforts will be made to study and promote how to discover user preferences more precisely, so that higher accuracy of user preferences discovering and user-ads matching can be achieved.

REFERENCES

[1] S. Ha, Helping online customers decide through Web personalization, IEEE Intelligent.

[2] Kai Li a, Timon C. Du b, Building a targeted mobile advertising system for location-based services, Decision Support Systems 54 (2012) 1–8.

[3] A. Bagherjeiran, R. Parekh, Combining behavioral and social network data for online advertising, Proceedings of the 2008 IEEE International Conference on Data Mining Workshops, Pisa, Italy, 2008, pp. 837–846.

[4] Zhang Shiming et al, Design and Realization of Resource Retrieval Algorithm Based on User's Interest Model Matching, Computer Applications and Software, 2009.

Information Technology and Computer Application Engineering – Liu, Sung & Yao (Eds)
© 2014 Taylor & Francis Group, London, ISBN 978-1-138-00079-7

Research and design of Chinese-English corpus alignment

Cheng Ying Chi, Bo Zhang & Yin Qi

School of Software Engineering University of Science and Technology, Liaoning Anshan, Liaoning, China

ABSTRACT: Sentence alignment is most important for Chinese-English bilingual corpus alignment. This paper analyzes the features of the length-based method and the lexical-information method. First, the N-Path method is proposed which combines length and lexical-information. This method avoids the insufficiency of length-based method and limitations of lexical-information methods. Secondly, we can get expanded result aggregate and improve the alignment effect using the N-Path alignment method. Finally, we can confirm the best one from the result aggregate by using synonym-expansion alignment. The experiment result shows that our methods can be applied to general text processing and improves alignment accuracy.

Keywords: Sentence alignment; N-Path; alignment method; synonym-expansion

1 INTRODUCTION

Bilingual corpora are built to acquire language information. Language corpora are important in machine translation[1], bilingual dictionary compiling[2], and cross-language information retrieval[3], etc. How corpora are processed is determined by their applications. Units of bilingual alignment include paragraph, sentence, phrase, and word. Various applications may choose different alignment unit. However, sentence alignment is the most commonly used.

As an important link in text alignment, sentence alignment has been studied by many researchers [6]. Compared with studies on English-French, English-German alignment, Chinese-English alignment has a lot of work to do. Brown [7] and Gale [8] implemented bilingual alignment according to sentence length relation. The former used the number of words as length; while the latter used bytes as the length of sentence. They have succeeded in English-French alignment. But Gale pointed out that length relation does not suit sentence alignment between languages of different families. Howerver, LIU Xin, etc. [9] proved with their experiment that length information can be applied to Chinese-English sentence alignment. Key[11] proposed an alignment method for English-French sentence alignment using vocabulary distribution information. Wu [13], used length and vocabulary distribution in their experiments on Chinese-Japanese sentence alignment, and discovered that the mixed method is superior to either the length method or the vocabulary method..

In this paper, we propose a new method based on sentence length and N-path, that is, using N paths to implement sentence alignment. The basic idea is finding all possible meaningful sentence pairs according to the alignment model, and constructing a non-circled digraph using these pair information. All pairs thus generated are represented in the graph as directed edges, and each edge a label indicating its score of alignment. Through analysis of all the paths from the starting point to the end point, each path is assigned a score, thus generating the desired path information. If there are path overlap, we merge them into the first path. The set of N paths form the alignment result set.

2 N-PATH ALIGNMENT

2.1 N-Path alignment model

For quick matching in text alignment, we use context-free grammar, and suppose that all alignment pairs are indenpendant of each other. This leads to the evaluation of alignment $W(E|C)$ formula as follows,

$$W(E \mid C) = \prod_{i=1}^{m} W(P(c_{k-l}, c_k) P(e_{h-r}, e_h)) \tag{1}$$

Using this formula, we get the score of edge $<c_k, e_h>$ to be $W(P(c_{k-l}, c_k) P(e_{h-r}, e_h))$, and the score of the path to be $W(E|C)$. Let NRS be the set of N alignment sequences generated by the N-path method. When $N = 1$, the result is the exact alignment. This converts the problem from N-path text alignment to finding NRS of a non-circled digraph. We can use the greedy method to solve this problem. In fact, it is a simple extention of Dijkstra's algorithm. This method records the predecessor position of current Chinese sentence, this is, the position of previous English sentence. At the same time, it records relevant Chinese information within the same set. For an English sentence, its predecessor position should be recorded, that is, the position

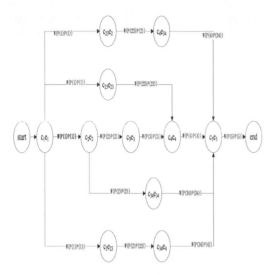

Figure 1.   NRS text alignment result set.

of previous Chinese sentence. And, at the same time, records relevant English information within the same set. In this way, a newly generated Chinese sentence and English sentence generate sentence pair information. If there are more than one path on the same length, we need to store predecessor information. This can be implemented by backtracking to obtain the resulting NRS. Possible NRS alignments are shown in Figure 1.

After the execution of the alignment algorithm, we get N text alignment paths. If the paths are the same, we get the alignment information. Otherwise, we have to apply the vocabulary alignment method to obtain the alignment path. The major reason for the differences of alignment information is that there are mutual disturbances between sentence lengths and vocabulary information. In that case, we only consider the vocabulary information for the scoring. Compared with the length information, vocabulary takes higher weight.

2.2   Evaluation of sentence alignment

In the above model, $W(P(c_{k-l}, c_k)P(e_{h-r}, e_h))$ is the score value of Chinese-English sentence alignment based on both length and vocabulary information. The computation formula is as follows,

$$W(P(c_{k-l}, c_k)P(e_{h-r}, e_h)) = W_a \times SimLen_i + W_b \times SimWordLen_i + W_c \times SimWord_i \quad (2)$$

where $SimLen_i$, $SimWordLen_i$, $SimWord_i$ are sentence length value, vocabulary length value, and vocabulary value, respectively. W_a, W_b, W_c are their weight, respectively, and according to our experiment on large corpus, are assigned 0.8, 1.0 & 1.2, respectively.

2.3   Sentence length evaluation

Character length ratio ByteValue, word lengthratio, and vocabulary length ratio ItemValue are obtained from their length ByteLen, WordLen, and ItemLen,

respectively. Here, we need to compute the lenght similarity using the following vector space method.

$$SimLen = \frac{\sum_{i=1}^{n} Value(i) \times StdValue(i)}{\sqrt{\sum_{i=1}^{n} Value(i)^2 \times \sum_{i=1}^{n} StdValue(i)^2}} \quad (3)$$

where, Value is the sentence length sequence, and StdValue is the sentence length ratio sequence. The resulting Sim is the measure of similarity of sentence length. The standard sequence is obtained from corpus training.

2.4   Sentence length information evaluation

In the process of computing sentence similarity, by statistical computation, we obtain Chinese and English alignment vocabulary sequences. And we use the following formula to compute the vocabulary length value.

$$SimWordLen = \frac{\sum_{i=0} CWordLen(i) + \sum_{j=0} EWordLen(j)}{\sum_{i=0} L(Ci) + \sum_{j=0} L(Ej)} \quad (4)$$

where, CwordLen, EwordLen are Chinese and English alignment vocabulary length sequence, respectivel; and L(Ci) and L(Ei) are Chinese and English sentence lengths, respectively.

2.5   Vocabulary information evaluation

Vocabulary alignment evaluation is based on Chinese-English correspongding vocabulary pairs. The core idea is to compute the alignment vocabulary occurrences. The evaluation function is as follows.,

$$SimWord = \frac{\sum_{i=0} (W(Ci) + W(Ej)) \times P(CE_{ij})}{\sum_{i=0} W(Ci) + \sum_{j=0} W(Ej)} \quad (5)$$

where, W(Ci) and W(Ei) are Chinese and English weights, respectively. P(CEi) is the probability of Chinese vocabulary Ci corresponding English vocabulary Ej. Chinese and English weights computation takes the form of TF*IDF.

3   SYNONYM EXPANSION

In this paper, we modified the vocabulary alignment evaluation function. Using sysnonym expansion, a list of Chinese keyword sysnonyms can be obtained. These helps us find the corresponding English information. By looking up a Chinese-English dictionary,

Table 1. 5 experiments.

	Precision	Recall
Run1	18.46%	82.83%
Run2	59.20%	50.10%
Run3	69.09%	74.94%
Run4	73.48%	84.57%
Run5	76.26%	85.28%

a new vocabulary alignment evaluation function is formulated.

$$W(C_i, ExtC_i) = TF(C_i, ExtC_i) \times IDF(C_i, ExtC_i) \quad (6)$$

where

$$TF(C_i, ExtC_i) = \frac{ExtTF(C_i, ExtC_i)}{TF(C_i)} \quad (7)$$

$$IDF(C_i, ExtC_i) = \frac{DF(C_i)}{ExtDF(C_i, ExtC_i)} \quad (8)$$

$TF(C_i)$ indicates the number of occurrences of C_i in the corpus sentences. And $ExtTF(C_i, ExtC_i)$ indcates the number of expansions of C_i in the corpus sentences. $ExtDF(C_i, ExtC_i)$ indicates the number of passages C_i is expanded to $ExtC_i$.

In this way, we can obtain vocabulary alignment information the does no exist in Chinese-English dictionaries. When the frequency of a keyword reaches the alignment frequency, it can be considered to be a correct alignment pair, and will be put into the keyword dictionary and assigned corresponding weight. At the same time, the evaluation value is updated as,

$$W_{new}(P(c_{k-l}, c_k) P(e_{h-l}, e_h)) = W(P(c_{k-l}, c_k) P(e_{h-l}, e_h)) + W_d \times SimExtWord \quad (9)$$

Please do not revise any of the current designations.

4 EXPERIMENT AND CONCLUSION

For the test set, two major values are observed.

$$precision = \frac{number-of-correctly-aligned-sentence-pairs}{number-of-output-sentence-pairs} \quad (10)$$

$$recall = \frac{number-of-correctly-aligned-sentence-pairs}{number-of-corpus-sentence-pairs} \quad (11)$$

We have designed 5 experiments. Run1 uses length alignment, Run2 uses vocabulary alignment, Run3 uses length and vocabulary mixed method, Run4 uses N-Path model with dynamic programming, and Run5 uses N-Path model combined with synonym expansion.

Sentence length alignment heavily relies on Chinese-English translation rules. This makes it hard to recognize the sentence length relation. Only those sentences that comply with the rules are alighned. Vocabulary alignment can effectively obtain the relation of corresponding sentences, but cannot recognize those sentences that have no prominent vocabulary alignment value. The mixed method make use of both methods, and shows better result. The proposed methods (Run4) raised the precision from 69.09% to 73.4%, and raise the recall from 74.94 to 84.57%; and Run5 shows even better result.

Errors occur where there is no 1-1 correspondance, which needs to be further studied.

REFERENCES

[1] Dolan W. B., J. Pinkhan and S. D. Richardson. The Microsoft Research Machine Translation [C]// AMTA 2002: 237–239.

[2] CHEN BoXin, DU LiMin. Alignment of Single Source Words and Target Multi-word Units from Parallel Corpus [J]. Journal of Chinese Information Processing, 2002.1.

[3] Chen A., Gry F. C. Translation term weighting and combining translation resources in cross-language retrieval[C]//TREC 2001.

[4] Gey F. C., A. Chen, M. K. Buckland and R. R. Larson. Translingual vocabulary mappings for multilingual information access[C]//SIGIR 2002: 455–456.

[5] Morre R. C.. Fast and accurate sentence alignment of bilingual corpora[C]//ATMTA2002: 135–144.

[6] Li Weigang LiuTing Wangzhen Li Sheng. Research of Paragraph Realignment of Bilingual Corpus. The Eighth National Conference on Computational Linguistics (JSCL-2003) proceedings, 2003.

[7] Brown, P. F., Lai, J. C., and Mercer, R. L. Aligning Sentences in Parallel Corpora. In Proceedings of the 29th Annual Meeting of the Association for Computational Linguistics (ACL'91), 1991: 169–176.

[8] Gale, W. A., and Church, K. W.1991. A Program for Aligning Sentences in Bilingual Corpora. In Proceedings of the Association for Computational Linguistics (ACL'91): 177–184.

[9] LIU Xin, ZHOU Ming, ZHU Sheng-Huo, HUANG Chang-Ning. Aligning Sentences In Parallel Corpora Using Self-Extracted Lexical Information. Chinese Journal of Computer, 1998.8.

[10] Simard, M., Foster, G. F., and Isabelle, P. 1992. Using Cognates to Align Sentences in Vilingual Corpora. In Proceedings of the Fourth Inernational Conference on Theoretical and Methodological Issues in Machine Translation (TMI'92):67–81.

[11] Kay, M., Roscheisen, M. 1993. Text-Translation Alignment. Computational Lingustic, 23(2): 313–343.

[12] Chen, S. F. 1993. Aligning Sentences in Bilingual Corpora Using Lexical Information. In Proceedings of the 31st Annual Meeting of the Association for Computational Linguistics (ACL'93): 9–16.

[13] Wu, D. Aligning a Parallel English-Chinese Corpus Statistically With Lexical Criteria. In Proceedings of the 32nd Annual Meeting of The ACM.

Information Technology and Computer Application Engineering – Liu, Sung & Yao (Eds)
© 2014 Taylor & Francis Group, London, ISBN 978-1-138-00079-7

Microwave-induced thermoacoustic imaging – A new method for breast tumor detection

Q.Y. Long & Q. Lv
Thyroid Breast Surgery, West China Hospital of Sichuan University, Chengdu, Sichuan, China

Y. Gao, L. Huang, W.Z. Qi, D. Wu, J.Y. Xu & J. Rong
School of Physical Electronics, University of Electronic Science and Technology of China, Chengdu

H.B. Jiang
School of Physical Electronics, University of Electronic Science and Technology of China, Chengdu
Department of Biomedical Engineering, University of Florida, Gainesville, Florida

ABSTRACT: In this paper, a new biomedical imaging technology called thermoacoustic imaging (TAI) for breast tumor detection is presented. We describe in detail the TAI imaging system developed, and validate the system using breast tissue-mimicking phantoms and human breast tumor specimens. The results obtained indicate that TAI can be a candidate for in vivo noninvasive imaging of breast tumor with high sensitivity and high contrast.

Keywords: Breast tumor; Early detection; Microwave; Thermoacoustic imaging

1 INTRODUCTIONS

Breast cancer is currently the second leading cause of cancer deaths in women and is the most common cancer among women. According to the Word Health Organization reported, >1.2 million people will be diagnosed with breast cancer each year worldwide, and approximately 500,000 of them died. inChinese women, around 27% of female breast cancers are linked to largely modifiable lifestyle and environmental factors. There is an rapidly increasing trend for breast cancer incidence and mortality in our country. Research shows average annual growth rate is about 2% higher than that of the developed countries and is increasing at an annual rate of 3%. Breast cancer had already been the primacy malignant tumor of the female in some big cities such as Peking, Tianjin and Shanghai.

Now, for the early detection and diagnosis of breast cancer, there are many ways such as Mammary gland of color doppler ultrasound, mammography, electrical impedance tomography (EIT), magnetic resonance imaging (MRI) and microwave imaging technology etc. [1, 2].

For the early detection and diagnosis of breast cancer, mammography is long considered the most reliable and cost-effective method. However, because of its false-negative rate of misdiagnosis through the X-ray of the breast tissue limits its development and application [3–6].

The reported sensitivity of mammography for fatty breasts ranges from 80% to 92% compared to 30% to 69% for dense breasts. Premenopausal women mostly with fiber and is based on dense glandular breast cystic breast disease main ingredients Premenopausal women mostly with fbrocystic mammary gland disease, whose breast is primarily composed of the dense glands. So there will be a high misdiagnosis rate in a breast X-ray detection of diseased tissue. Compared with the postmenopausal women, Which will cause more damage to young females with breast cancer, and the progression of the disease in young women also will be faster. Another issue that deserves concern, breast X-ray examination of the positive predictive value is very low [7–9]. In order to verify canceration, leading to further do a variety of biopsy experiments, And thus increasing the number of medical tests and economic costs. Therefore developing a new breast cancer detection method to make up for the insufficient of the traditional mammogram is very urgent.

Besides the mammogram, conventional imaging techniques such as ultrasound imaging (USI), computed tomography (CT) and magnetic resonance imaging (MRI), are also being used for the detection of breast cancer [10]. Which imaging mechanism for the changes brought about by the X-ray examination of the breast or replace. In these imaging mechanism examination of the breast, MRI can bring about changes or can be used as replacement of the X-ray examination of the breast. MRI is the most extensively studied and is considered the most potential to become a means of detection of breast cancer. According to a study of 821 women, MRI gives a sensitivity of about 88% and a specificity of 68%, whether the breast tissue is dense or not [11]. A similar study for the detection of high-risk breast cancer groups also reveals the same results that MRI has higher sensitivity in the detection of breast cancer [12]. Although breast MRI has a high sensitivity of around 80 to 100 percent, but whose's specificity is not high and expensive. So still without catholicity.

Compared with the traditional imaging mechanism, microwave tomography is an emerging imaging modality, and shows the unique advantages. In addition to the use of non-ionizing radiation and wthout the tested breastst squeezing and stereotypes. Microwave imaging for breast cancer in contrast is much higher than the existing imaging modalities. The microwave imaging results showed that malignant lesions of the breast tissue and normal breast tissue conductivity ratio of 6:1[13–15], and the conductivity contrast ratio between malignant lesions of the breast tissue and the fat-based normal breast tissue is up to 10:1. The enhancement of contrast is considered to be the result of an increase of the water content and the particle concentration in the tumor tissue [16]. Changes in water content and particle concentration can lead to change in the local electrical conductivity and the dielectric constant. The study of malignant tumor of breast cancer indicates that permittivity and conductivity is associated with the vascular distribution, cellular structures and the substrate specificity. Therefore, imaging in water content and particle concentration contribute to the analysis of many parameters of malignant tumors. Previous study had revealed that water content in breast cancer tissue is higher than the water content in the benign lesions or normal tissue. Accordingly, how much of the water content is an important parameter in diagnosis of breast cancer. But other recent studies indicted that tumor at the early growth stage will increase in local blood content, with the increase of the conductivity of organization [17]. This means that the use of microwave imaging for hemoglobin can be achieved, because the parameters of hemoglobin play a potentially decisive role in the diagnosis of breast cancer. Numerous clinical studies indicate the hemoglobin content in breast cancer tissue is always higher than the normal tissue near the tumor [18–21]. According the above analysis, microwave imaging can improve the sensitivity and specificity in breast cancer screening, and also has the potential for structure and functional information imaging.

The clinical validation by many research groups reveals the strong application value of the microwave imaging in detection of breast cancer [22–25]. Because of the inherent limitations, the resolution of microwave imaging is only in the order of 10 mm, which limited its application [26–27]. Thermoacoustic tomography (TAT) has been developed to overcome the limitations of the poor resolution [27–29]. Compared with the pure microwave imaging, the resolution of which up to 1 mm or even better [30]. Prior clinical trials have demonstrated that TAT has great potential in early breast cancer diagnosis [30–32]. So microwave-induced thermo-acoustic tomography (MITAT) has received more and more concerns in biologic tissue imaging field.

Microwave-Induced Thermo-Acoustic Tomography system is discussed in this paper, combined with the experimental analysis of phantoms and human breast carcinoma specimens, the system has the potential in the daily census of breast cancer.

2 EXPERIMENTAL SYSTEM

Microwave-induced thermoacoustic tomography experimental system uses a single probe to collect the data by the circular scanning method. The frequency of the the microwave source is 3 GHz (peak power: $\geq$70 kW; the pulse width: 0.75 μs; designed by Chengdu Jinjiang Electronic System Engineering Co., Ltd). Microwave through a horn antenna at the end of the waveguide coupled to the biological tissues, so as to induce thermal acoustic, thereby generating thermal acoustic. In the experiment, microwave power density irradiated onto the sample surface is less than 10 mW/cm^2, which is less than the security irradiation power(provisions: 20 mW/cm^2 at 3.0 GHz), which ensures that the security of the experiment and clinical application. Ultrasound transducer used in the system, the center frequency is 2.25 MHz, bandwidth is 60%, which can obtain a spatial resolution of approximately 0.5 mm. In order to realize the coupling of microwave transmission and ultrasonic signal, oil is used as the coupling liquid in the experiment. The transducer controlled by the step-motor (ZOLIX INSTRUMENTS CO. LTD, RSA100, resolution: 0.00125°) rotates around the centerline of the antenna, the step angle of rotation is 2°, each data acquisition takes about 3 minutes. First the collected signals are amplified by a pre-amplifier. then via the data acquisition card (DAQ) the signal converted into digital signal, and stored in the computer. The entire data collection process is controlled by the LabVIEW program.

3 EXPERIMENTAL STUDY OF THE PHANTOMS

In order to verify the feasibility of the experimental system for breast imaging, We have produced a phantom with muscle to simulate breast lesions approximately [27]. According to electrodynamics, the microwave absorption coefficient can be expressed as:

$$\alpha = \omega\sqrt{\mu\varepsilon}\left[\frac{1}{2}\left(\sqrt{1+\frac{\sigma^2}{\varepsilon^2\omega^2}}-1\right)\right]^{\frac{1}{2}} \quad (1)$$

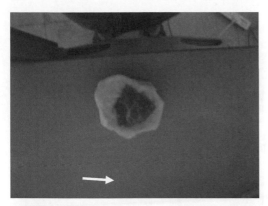

10mm

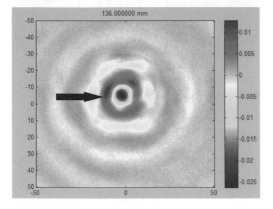

Figure 1. Upper: Picture of the phantom. lower: recovered by thermoacoustic imaging (colarbar unit: a.u, indicated relaitive Mw absorption).

detection, and supply experiment evidence. Description of each specimen's nature and dielectric properties can be found in Table 1. All the three human breast tumor specimens were acquired from West China Hospital, Sichuan University. After these imaging experiments, the specimen was rendered to West China Hospital, Sichuan University. The results of the experiments are as in Table 1.

From the above three groups of experiments for slices of human breast cancer, we can arrive at the conclusion that microwave-induced thermal acoustic imaging for breast cancer can easily distinguish cancerous tissue from normal tissue, and has higher imaging contrast. Because, water content and the particle concentration in the cancerous tissue is higher than that in the normal tissues, which can result in the fairly large differences between the dielectric properties of cancerous tissue and that of normal tissues, and eventually lead to the difference of absorption of microwave, and then image contrast in the reconstructed image is relatively large. The high sensitivity of thermoacoustic tomography can be used in the early detection of breast cancer.

5 DISCUSSION AND CONCLUSION

Many of the techniques developed for medical imaging is expected to be used for clinical diagnosis. This paper points out: when breast cancer appears, water and the type of ion content in the tissue will increase, compared with the normal breast tissues, which would result in significant changes in the electrical properties of cancerous breast tissues. This remarkable change would result in the difference of microwave absorption. This difference in the thermoacoustic recovered images indicted that the signal strength, and thus thermoacoustic tomography can easily detect the tumor lesions in the breast. High contrast and high resolution images can be obtained in breast cancer detection by thermoacoustic tomography in terms of theory and experimental validation.

In this paper, thermoacoustic imaging reconstructed the phantoms and specimens of human breast cancer, preliminary results showed that thermoacoustic imaging has the potential for high-contrast and resolution breast cancer imaging. Meanwhile, these results fully demonstrate that thermoacoustic imaging has potential

where ω is the angular frequency of microwave, μ is the permeability, ε is the relative dielectric constant, and σ is the conductivity. Usually, the permeability is a constant in the biological tissues, so the relative dielectric constant and conductivity are taken into account. Porcine fat tissue simulate the normal breast and muscle as the tumor mimicking tissues.

Based on the above experimental result, we have the following conclusions: the system deliver great image quality in the imaging for the simulated phantom of breast cancer, the object size obtained from thermoacoustic recovered was in good agreement with the actual size of the object. Meanwhile, the contrast between fat and muscle is very high.

4 THE STUDY OF HUMAN BREAST SPECIMENS

In order to further study the ability of the microwave thermal acoustic imaging system for breast cancer

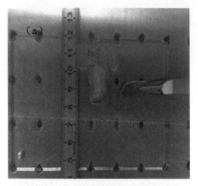

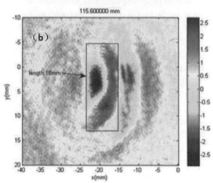

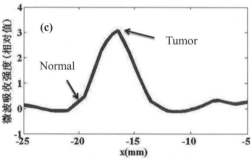

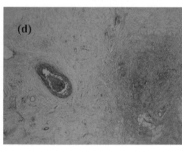

Figure 2. (a) Picture of the specimen; (b) reconstructed image with thermoacoustic imaging; (c) when y = 3 mm, relative intensity of microwave absorption on x-axis; (d) pathological analysis of the specimen showed in (a).

as a means of the day-to-day census of breast cancer detection, and the results of these experiments is important to promote this imaging modality used in clinical research of breast cancer.

Figure 3. Left: the second slice of the human breast cancer; Right: reconstructed image with thermoacoustic imaging.

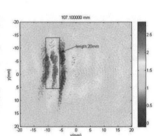

Figure 4. Left: the third slice of the human breast cancer; Right: reconstructed image with thermoacoustic imaging.

REFERENCES

[1] KD Paulsen, P.M. Meaney, 2006. Alternative Breast Imaging. *Springer Press*.
[2] SK Moore. 2001. Better breast cancer detection. *Spectrum, IEEE*, 38(5):50–54.
[3] Carney PA, Miglioretti DL, Yankaskas BC, et al. 2003. Individual and combined effects of age, breast density, and hormone replacement therapy use on the accuracy of screening mammography, *ANNALS OF INTERNAL MEDICINE*, 138 (3): 168–175.
[4] Sickles EA, Miglioretti DL, Ballard-Barbash R, et al, 2005. Performance benchmarks for diagnostic mammography, *Radiology*, 235 (3): 775–790.
[5] Boyd NF, Lockwood GA, Byng JW, et al. 1998. Mammographic densities and breast cancer risk, *CANCER EPIDEMIOLOGY BIOMARKERS & PREVENTION*, 7(12): 1133–1144
[6] LAM Stephen, MACAULAY Calum, LERICHE Jean C. PALCIC Branko, 2000. Detection and localization of early lung cancer by fluorescence bronchoscopy. *American Cancer Society International Conference on Prevention and Early Diagnosis of Lung Cancer*, 89 (11): 2468–2473.
[7] R. White, TJ. Halperin, JA. Olson Jr, MS. Soo, et al. 2000. Stereotactic Core-Needle Breast Biopsy by Surgeons. *Annals of Surgery*, 232(4): 542–548.
[8] Doyle AJ, Murray KA, Nelson EW, 1995. Selective Use of Image-Guided Large-Core Needle Biopsy of the Breast: Accuracy and Cost-Effectiveness. *Am J Roentgenol*, 165:281–284.
[9] Liberman L, Feng TL, Dershaw DD, 1998. US-guided Core Breast Biopsy: Use and Cost-effectiveness, *Radiology*, 208:717–723.
[10] M. Sabel, H. Aichinger, 1996. Recent developments in breast imaging. *Phys Med Biol* 41(3): 315–368.
[11] Bluemke DA, Gatsonis CA, Chen MH, et al. 2004. Magnetic resonance imaging of the breast prior to

biopsy. *JAMA-JOURNAL OF THE AMERICAN MEDICAL ASSOCIATION* 292 (22): 2735–2742.

[12] Warner E, Plewes DB, Hill KA, et al. 2004. Surveillance of BRCA1 and BRCA2 mutation carriers with magnetic resonance imaging, ultrasound, mammography, and clinical breast examination. *JAMA-JOURNAL OF THE AMERICAN MEDICAL ASSOCIATION,* 292 (11): 1317–1325.

[13] W.T. Joines, Y. Zhang, C. Li, R. Jirtle, 1994. The measured electrical properties of normal and malignant human tissues from 50 to 900 MHz, *Med Phys*, 21(4): 547–550.

[14] S.S. Chaudhary, R.K. Mishra, A. Swarup J. Thomas. 1984. Dielectric properties of normal and malignant human breast tissues at radiowave and microwave frequencies. *Indian J Biochem. Biophys*, 21(1):76–79.

[15] A.J. Surowiec, S. Stuchly, J. Barr, A. Swarup, 1998. Dielectric properties of breast carcinoma and the surrounding tissues. *IEEE Trans. Biomed. Eng.* 35(4): 257–263.

[16] S.P. Poplack, K. Paulsen, A. Hartov, et al. 2004. Electromagnetic breast imaging: Average tissue property values in women with negative clinical findings, *Radiology,* 231(2): 571–580.

[17] Poplack SP, Tosteson TD, Wells WA, et al. 2007. Electromagnetic breast imaging: Results of a pilot study in women with abnormal mammograms, *Radiology*, 243(2):350–359.

[18] Zhu Q, Cronin EB, Currier AA, et al. 2005. Benign versus malignant breast masses: Optical differentiation with US-guided optical imaging reconstruction, *Radiology*, 237 (1): 57–66.

[19] Chance B, Nioka S, Zhang J, et al. 2005. Breast cancer detection based on incremental biochemical and physiological properties of breast cancers: A six-year, two-site study. *Academic Radiology*, 12(8): 925–933.

[20] Pogue, B. W., McBride, T. O., Osterman, S., Poplack, S., et al. 2001. Quantitative hemoglobin tomography with diffuse near-infrared spectroscopy: pilot results in the breast. *Radiology*, 218(1) 261–266.

[21] Cerussi A, Hsiang D, Shah N, et al. 2007. Predicting response to breast cancer neoadjuvant chemotherapy using diffuse optical spectroscopy. *Proceedings of the National Academy of Sciences of the United Sates of America*, 104 (10): 4014–4019.

[22] Meaney, P.M., Fanning, M.W., Dun Li, et al. 2000. A clinical prototype for active microwave imaging of the breast, *IEEE Transactions on Microwave Theory and Techniques*, 48(10):1841–1853.

[23] Meaney PM, Demidenko E, Yagnamurthy NK, et al. 2001. A two-stage microwave image reconstruction procedure for improved internal feature extraction, *Med Phys*, 28(11):2358–2369.

[24] H. Jiang, C. Li, D. Pearlstone, L. Fajardo. 2005. Ultrasound-guided microwave imaging of breast cancer: tissue Phantom and pilot clinical experiments, *Med Phys*, 32(8): 2528–2535.

[25] Poplack SP, Tosteson TD, Wells WA, et al. 2007. Electromagnetic breast imaging: Results of a pilot study in women with abnormal mammograms, *Radiology*, 243(2): 350–359.

[26] Fear EC, Hagness SC, Meaney PM, 2002. Enhancing breast tumor detection with near-field imaging, *Microwave Magazine, IEEE*, 3(1):48–56.

[27] M. Pramanik, G. Ku, C.H. Li, & Lihong V. Wang. 2008. Design and evaluation of a novel breast cancer detection system combining both thermoacoustic (TA) and photoacoustic (PA) tomography. *Med Phys*, 35(6): 2218–2223.

[28] Kruger RA, Kopecky KK, Aisen AM, et al. 1999. Thermoacoustic CT with radio waves: A medical imaging paradigm, *Radiology*, 211(1): 275–278.

[29] RA Kruger et al. 1999. Thermoacoustic CT-Technical considerations, *Med. Phys*, 26(9): 1832–1837.

[30] Ku G, Fornage BD, Jin X, Xu M, Hunt KK, Wang LV, 2005. Thermoacoustic and photoacoustic tomography of thick biological tissues toward breast imaging, *Technol Cancer Res Treat,* 4(5):559–566.

[31] Robert A. Kruger, William L. Kiser, Jr., A. P. Romilly, Phyllis Scmidt 2001. Thermoacoustic CT of the breast: Pilot study observations, *Proc. SPIE*, 4256, 1–5.

[32] R.A. Kruger et al. 2002. Thermoacoustic CT. Proc. SPIE 4682, 521–525.

Information Technology and Computer Application Engineering – Liu, Sung & Yao (Eds)
© *2014 Taylor & Francis Group, London, ISBN 978-1-138-00079-7*

The evaluation model about the trust between enterprise and university in industrial technology innovation strategy alliance

Zhan Zhang, Jing Zhang & Wei Bai
School of Economics Management, Shenyang University of Chemical Technology Shenyang, China

ABSTRACT: The article aims at setting up the evaluation model about the trust between enterprise and university in Industrial technology innovation strategy alliance. In my opinion, three factors influence the trust, namely the selection of alliance partners, the governance of alliance, the incentive and restraining mechanism of alliance. The fuzzy evaluation method can deal with above-mentioned datum well, and provide analytical tools for improving the levels of trust between enterprise and university in alliance, so as to promote the stable development of industrial technology innovation strategy alliance.

Keywords: Industrial technology innovation strategy alliance, trust, fuzzy evaluation method

1 THE BACKGROUND

The rise of Industrial technology innovation strategy alliance has changed the boundary of traditional enterprise, but not all the alliances can obtain expected performance. The report showed that the failure rate of alliances is 40%–70%[1]. And the failure was mostly attributed to the lack of mutual trust[2]. The cooperation relationship between alliance partners is essentially based on the commitment for the future behavior. The commitment can be regulated in public, also can be reached by tacit ways. In alliances, only mutual trust and all the parties keep their promises, can the alliances obtain due performances and enhance the competitive advantage of the parties, so as to lay a solid foundation for common development. In the industrial technology innovation strategy alliance, the trust relationship between enterprises and university is no exception. Therefore, studying the factors that influence the trust is very necessary.

2 FACTORS INFLUENCE THE TRUST

In my opinion, three factors influence the trust, namely the selection of alliance partners, the governance of alliance, the incentive and restraining mechanism of alliance.

2.1 *The selection of alliance partners*

$$Y_1 = F(X_{11}, X_{12}, X_{13}, X_{14}, X_{15}, X_{16}) \qquad (1)$$

X_{11}: Interdependence and complementary degree about resources and ability between enterprise and university. The mutual dependence degree will strengthen the mutual trust between enterprise and university.

X_{12}: Whether the strength of enterprise and university is nearing or not. If the strength is nearing, the partners will be confident in cooperation.

X_{13}: Whether the enterprise and university have clear willingness to co-operate or not. Clear goals will contribute to the long-term existence of trust.

X_{14}: Whether the reputation of enterprise and university is good or not. Good reputations can strength the degree of trust.

X_{15}: The degree of cultural differences between enterprise and university. A lesser degree of cultural differences can strengthen the continuity and consistency of partners' behavior; prevent mutual trust from some unnecessary interference.[3]

X_{16}: Whether the strategic goals in the areas of cooperation between enterprise and university are close to or not. Similar strategic goals will make their cooperation more closely.

2.2 *The governance of alliance*

$$Y_2 = F(X_{21}, X_{22}, X_{23}, X_{24}, X_{25}, X_{26}) \qquad (2)$$

X_{21}: The degree of communication between enterprises' high-level and university's top-level. The communication between the top can directly influence both relations, and reduce the asymmetry information.

X_{22}: The tolerance to the risk of enterprise and university. Having a certain degree of tolerance to the risk can enhance the trust between enterprise and university.

X_{23}: Whether the oral agreement between enterprise and university is frequent or not. Oral agreement can accumulate the trust relationship between enterprise and university.

X_{24}: Whether the dispute and conflict can be dealt with in time or not. If the contradiction can be coordinated timely, the gap and strangeness between enterprise and university will be eliminated easily.

X_{25}: Whether the secrets between enterprise and university are under protection or not. Once the secrets can't be protected well, there will be some problems. For example, the trust relationship between alliance partners will be weakened a lot and the two sides will also reduce trust to partners for their damaged interests.

X_{26}: Whether the alliance has formal and regular operating report or not. Good mutual supervision helps to avoid speculative behavior and unnecessary conflict.

2.3 The incentive and restraining mechanism of alliance

$$Y_3 = F(X_{31}, X_{32}, X_{33}, X_{34}, X_{35}, X_{36}) \quad (3)$$

X_{31}: Whether the alliance run fairly or not. Sense of unfairness will lead to the tension of alliance and the bursting of trust foundation.

X_{32}: Whether the costs accord with the benefits. If the costs match the benefits well, the enterprise and university will maintain long-term trust relationship.

X_{33}: If enterprise and university are satisfied with ever cooperation. The effect of cooperation has a decisive influence on partners' next action.

X_{34}: Whether the incentive mechanism for partners who participate in cooperation actively is perfect or not. Perfect incentive mechanism affects the enthusiasm of alliance partners to a certain extent. [4]

X_{35}: The degree of constrain to enterprise and university. Effective constraint system can raise cost of cheating; improve enterprise and university's faith degree objectively.

X_{36}: The complete sanction. The threats from the complete sanction will restrain enterprise and university's behavior effectively; force them to keep their promises and cooperate with each other sincerely. That helps to build up good relationships of trust.

3 THE EVALUATION MODEL OF TRUST

3.1 The basic principle and characteristics of the Fuzzy Evaluation Method

The Fuzzy Comprehensive Evaluation Method is a comprehensive evaluation method which based on the fuzzy mathematics. It converts the qualitative evaluation into quantitative evaluation according to the membership degree theory of fuzzy mathematics. Using fuzzy mathematics, it makes a overall evaluation about things or objects which are restricted by various factors. It has characteristics with clear results and strong systematicness, can solve the fuzzy and hard-to-quantify problems well and is suitable for solving all kinds of uncertain problems. [5]

The basic procedures of the method: First, setting up evaluation indexes set A $\{a_1, a_2, ..., a_n\}$ based on selected a series of evaluation indexes, and giving a weight R $\{r_1, r_2, ..., r_n\}$ to every evaluation index. Second setting up the evaluation set U $\{u_1, u_2, ..., u_n\}$, then evaluating every index by the relevant experts. Finally calculating A × R.

The advantages of this method are to avoid subjective capriciousness of the qualitative evaluation method; and to overcome the disadvantages of scoring method. The disadvantages of this method are too complicated; and hard to operate. The Fuzzy Comprehensive Evaluation Method can solve the fuzziness of comprehensive evaluation well, thus the method has been widely used in many fields.[6]

3.2 The evaluation process of trust between the enterprise and the university

First, set up the evaluation indexes set. According to the factors that influence the trust between enterprise and university, the trust evaluation indexes are divided into three first level indexes, which are the selection of alliance partners, the governance of alliance, and the incentive and restraining mechanism of alliance. Every first level index involves six second level indexes. The evaluation indexes set T is composed by these indexes. T = T$\{y_i\}$, $y_i = \{x_{ij}\}$ (i = 1, 2, 3; j = 1,2,3,4,5,6).

Second, give a weight to every index. Given the weight of y_i to the T is a_1, a_2, a_3 respectively, and then the weight matrix is A (a_1, a_2, a_3). Similarly, Given the weight of y_{ij} to y_i is B_{ij}, then $B_i = \{B_{ij}\}$, i = 1, 2, 3; j = 1, 2, 3, 4, 5, 6.

Third, set up the comment set and the fuzzy evaluation matrix. For setting up evaluation indexes set, we can choose ten or so to constitute evaluation group, then the evaluation group can evaluate every factor in the way of registration according to the specific conditions of different alliances, Lastly, Setting up the comment set c. Namely: c = $\{c_i\}$ = $\{c_1, c_2, c_3, c_4, c_5, c_6\}$ = {biggest, bigger, big, small, smaller, smallest}. According to the established evaluation set, setting up the Ti' fuzzy evaluation matrix W_i (i = 1, 2, 3, 4, 5, 6), considering alone every secondary index's weight to every comment, drawing the corresponding numerical.

Fourth, set up the fuzzy evaluation set of the first level index. That is $R_i = B_i \times W_i$, then we can obtain the Fuzzy comprehensive evaluation set: $R_i = (r_{i1}, r_{i2}, r_{i3})$ (i = 1, 2, 3), then we can obtain the transposed matrix of R(R_1, R_2, R_3).

Fifth, set up the fuzzy evaluation matrix T of the evaluation object. T = A × R = (a_1, a_2, a_3) × (R_1, R_2, R_3), T = (d_1, d_2, d_3)

Sixth, draw an evaluation conclusion by normalization. Because the sum of d_1, d_2, d_3 is not necessarily equal to 1 so we need to deal with that by normalization. Namely, $d'_i = d_i / \sum d_i$, thus obtains d' = (d'_1, d'_2, d'_3), corresponding y_i respectively. According to the maximum subordination principle, we think that the possibility of y_i is d'_i, corresponding to the evaluation of the trust between enterprise and university in the industrial technology innovation strategy alliance.

4 THE PROSPECT OF THE MODEL APPLICATION

We can get the weight of above indexes by many methods, such as experts' method and Delphi method.

The prospect of data sources: Delphi method. Namely expert consultation method, for solving the problem of every level indicator's weight, we can ask experts of industrial technology innovation strategy alliance to put forward opinions or views, then we conclude experts' opinions and new ideas, and feedback the concluded results to the experts again anonymously. We will not do that repeatedly until the opinions tend to be relatively concentrated, consistent, and reliable[7]. The questionnaire survey method. This method is that extending questionnaire to the related person of enterprise and university, the related person can provide data, grade every index, and put forward suggestions to perfect the index system [8].

The prospect of the model application: After collecting data, we can analyze the rationality of every evaluation index, and how the indexes influence the trust between the enterprise and university of every alliance by using the model this paper provides. If the alliance has high degree of trust, we can study its experiences further. If the alliance has low degree of trust, we can discuss the factors that lead to the low trust by using the model this paper provides, and then put forward corresponding measures to promote steady development of industrial technology innovation strategy alliance.

ACKNOWLEDGEMENT

National Social Science Fund, China (12BJY071) supports this paper. The Ministry of education of Humanities and social science research fund plan, China (12YJA790203). Hall of Liaoning province science and technology, China (2011401017). Science and Technology Bureau of Shenyang City, Liaoning province, China (F11-263-5-26). The Reform Project of Education Department of Liaoning Province, China (200914116). Education Department of Liaoning Province, China (W2101341), Liaoning regular schools of higher learning combination of production of innovative pilot test of the construction project staggered results.

REFERENCES

[1] Barber. The logic and limits of trust [M]. New Brunswick, NJ: Rutgers University Press (2007).

[2] DASTK, TENGBS. Trust, control, and risk in strategic alliances: An integrated framework [J]. Organization Studies (2010), P. 22.

[3] Thorelli. H.B. Networks between market sand hierarchies [J]. Strategic Management Journal (2010), P. 37–511.

[4] Shapiro, D. H. A preliminary study of long-term mediators: Goals, effects, religious orientation, cognitions. Journal of Transpersonal Psychology (2009), P. 23–39.

[5] Dale E.Zand. The Leadership Triad: Knowledge Trust and Power [M]. Oxford Univ. Press (2007).

[6] Baughn, C.C. Protecting Intellectual Capital in International Alliances [J]. Journal of World Business (2007), P. 103–177

[7] Lewis, J. D. & Weigert, A. Trust as a social reality [J]. Social Forces, 2006, Gulati, Ranjay. Alliance and networks [J]. Strategic Management Journal (2008), P. 293.

[8] Fishbein, M.I. Ajzen. Belief, Attitude, Intention and Behavior: An Introduction to Theory and Research. Addison-Wesley [J]. Reading. MA.

Information Technology and Computer Application Engineering – Liu, Sung & Yao (Eds)
© *2014 Taylor & Francis Group, London, ISBN 978-1-138-00079-7*

The financial performance evaluation of the electric power enterprise based on entropy weight and grey correlation degree

S.S. Guo & F.W. Yang

North China Electric Power University, Baoding, China

ABSTRACT: In view of the problem that using subjective weighting method to determine index's weight leads to the result of performance evaluation has strong subjectivity, this paper combine the entropy weight method and the grey correlation method together to evaluate the financial performance of the electric power enterprise. We use the entropy weight method to determine index's comprehensive weight and according to the principle of grey correlation theory calculate the correlation degree. We choose 8 electric power enterprises as examples and use this method analyze their financial performance level. The result is objective and can provide directions for power enterprise's development.

Keywords: Financial performance; Entropy weight method; Grey correlation degree; Indicator; Power

1 GENERAL INSTRUCTIONS

Power industry as the basic industry play an important role in promoting the development of national economy and society, the electric power industry's well running have a closely relationship with national economic security, social stability and People's Daily life (Liu 2003, Li & Jiang 2011). To achieve the goal of operation, enterprises usually adopt suitable financial performance evaluation index system, use scientific and reasonable method, according to a unified performance evaluation standard, to make an objective scientific and fair financial performance evaluation of the enterprise's performance during a certain period. Making financial performance evaluation on electric power enterprise A's during a certain period can provide reference and guidance for power enterprise and help them make right decisions, promoting a better development of power enterprise and making our power supply more stable and cheaper.

2 THE FINANCIAL PERFORMANCE EVALUATION METHOD OF POWER ENTERPRISE

2.1 *Build the performance evaluation index system*

Some scholars evaluate operating performance of the electric power enterprise from four aspects, that is, financial benefits, operation ability, solvency and development (Wang 2006, Yu 2011, Zhang & Dong 2012. Wang & Li 2012). Some scholars use factor analysis method to select indicators (Shu & Pei 2008, Yang & Li 2009, Xing & Ma2011). There are still some scholars directly select a number of financial indicators to evaluate performance (Xu & Dong 2009). This paper's performance evaluation index system is as follows. Our index system has four one class index, they are profitability (X_1), solvency (X_2), development ability (X_3) and operation capability (X_4). The profitability has three secondary indexes, Rate of Return on Sale (X_{11}), Return on total asset (X_{12}) and Return on Equity (X_{13}). The solvency has three secondary indexes, current ratio (X_{21}), Quick ratio (X_{22}) and ratio of liabilities to assets (X_{23}). The development capacity has three secondary indexes, sales increase rate (X_{31}) and Rate of capital accumulation (X_{32}). The operation ability has three secondary indexes, total assets turnover (X_{41}), Current asset turnover (X_{42}) and accounts receivable turnover (X_{43}).

2.2 *Financial performance evaluation system used to determine the index's weight*

There are subjective and objective methods to determine the weight of index. The subjective method including the Delphi method and analytic hierarchy process (AHP), etc., and objective method including multiple objective programming method and entropy weight method, and so on. Entropy weight method determines the entropy values of the indicators according to the amount of information each index provides, based on this determine the index's weight. We determine the weight of one class index based on Enterprise's integrated performance evaluation indexes and weights table and using the entropy weight method determine the weight of secondary index, making the weight we got more objective and more reasonable. Suppose there are n evaluation index, evaluation matrix $A = a_{ij}$ $(i = 1,2,3,...m; \ j = 1,2,3,...n)$, n represents there are n index, m represents the amount of enterprise. The specific steps of entropy weight method are as follows:

Use the following formula converse the suitability indexes:

$$c_{ij} = \frac{1}{|a_{ij} - c|} \quad (1)$$

where c represents the index's suitable value.

Standardize the indicators according to the following forluma:

$$r_{ij} = \frac{a_{ij} - \min(a_{ij})}{\max(a_{ij}) - \min(a_{ij})} \quad (2)$$

and if c_{ij} is suitability index, $a_{ij} = c_{ij}$.

Calculate index j's entropy e_j according to the formula:

$$e_j = -\frac{1}{\ln m} \sum_{i=1}^{m} \frac{r_{ij}}{\sum_{i=1}^{m} r_{ij}} \ln \frac{r_{ij}}{\sum_{i=1}^{m} r_{ij}} \quad (3)$$

The weight w_j of index j:

$$w_j = \frac{1 - e_j}{\sum_{i=1}^{m} (1 - e_j)} \quad (4)$$

2.3 Calculated the grey correlation degree

The steps of grey correlation degree are as follows:

Decide the reference sequence C, $C = (c_1, c_2, ...c_n)$, and $c_i = \max(r_{i1}, r_{i2}, ...r_{im})$, where $i = 1, 2,...n$.

According to the formula of grey correlation calculate the Correlation coefficient ε_i between comparative sequence R_j and C:

$$\varepsilon(R_j, C) = \frac{\min_i \min_j |r_{ij} - c_i| + \rho \max_i \max_j |r_{ij} - c_i|}{|r_{ij} - c_i| + \rho \max_i \max_j |r_{ij} - c_i|} \quad (5)$$

Where ρ is the distinguish coefficient.

Calculate correlation R between objects Rj and C according to the formula:

$$R = \sum_{j=1}^{n} \omega_j \varepsilon(R_j, C) \quad (6)$$

The valuation criterion. The performance level of the electric power enterprise can be divided into five grades. If the correlation degree is lower than 0.3, the financial performance level is lower, if the correlation degree is between 0.3 and 0.5, the performance level is low, if the correlation degree is between 0.5 and 0.7, the level is average, if the degree is between 0.7 and 0.9, the level is high and if the correlation degree is between 0.9 and 1.0, the level is high.

3 ANALYSIS ON EXAMPLES

3.1 Companies and datas

Power is essential to the national economy development, we choose eight power production enterprises from power listed enterprise. Multiple is the unit of measurement for X21 and X22, times is the unit of measurement for X42 and X43, percent is the unit of measurement for other indexes. A represents HUADIAN POWER INTERNATIONAL CORPORATION LIMITED, B represents DATANG International Power Generation Company, C represents SDIC HUAJING Power Holdings co., ltd, D represents SP POWER DEVELOPMENT co., ltd, E represents HUANENG Power International, F represents GUANGDONG BAOLIHUA NEW ENERGY STOCK CO., LTD, G represents XINJIANG TIANFU THERMOELECTRIC CO., LTD, H represents GUANGXI GUIDONG ELECTRIC POWER CO., LTD. Each company's financial data in 2012 is listed in order, that is $X_{11}, X_{12}, X_{13}, X_{21}, X_{22}, X_{23}, X_{31}, X_{32}, X_{41}, X_{42}, X_{43}$. A=(3.22, 1.22, 6.93, 0.30, 0.24, 83.20, 9.17, 16.91, 0.38, 3.94, 10.31), B = (7.96, 2.40, 11.54, 0.42, 0.34, 79.17, 7.21, 11.72, 0.30, 2.94, 7.82), C = (8.45, 1.49, 8.54, 0.41, 0.38, 82.54, 9.97, 17.13, 0.18, 1.75, 8.43), D = (12.00, 2.59, 14.59, 0.29, 0.24, 75.24, 10.14, 27.72, 0.29, 3.02, 8.82), E = (5.11, 2.68, 11.14, 0.39, 0.31, 74.73, 0.41, 11.62, 0.52, 3.85, 9.00), F = (11.42, 5.11, 12.64, 1.33, 0.79, 59.81, 12.39, 15.86, 0.45, 1.67, 7.37), G = (8.89, 3.13, 10.70, 1.77, 1.62, 72.73, 21.68, 4.98, 0.35, 0.78, 8.45), H = (2.76, 1.84, 4.11, 0.99, 0.92, 53.69, 103.64, 58.87, 0.67, 2.90, 10.52).

3.2 Determine the weight of index

According to the enterprise's comprehensive performance evaluation index and weights table, we determine the weight of profitability is 0.34, the weight of solvency is 0.22, the weight of development ability is 0.22 and the weight of operation ability is 0.22. The suitable values for current ratio, quick ratio and ratio of liabilities to assets are 2.00, 1.00 and 50%. By conversing the suitable indexes and using the entropy weight method, we get the comprehensive weight for the indexes, $\omega = $ (0.1220, 0.1417, 0.0763, 0.0748, 0.0774, 0.0678, 0.1381, 0.0819, 0.0747, 0.0630, 0.0823).

3.3 Grey correlation degree

Make $\rho = 0.5$, according to the formula we can get the gray correlation degree for each company. $\varepsilon_A = $ (0.3448, 0.3333, 0.4062, 0.3335, 0.3333, 0.3333, 0.3533, 0.3910, 0.4579, 1.0000, 0.8823), $\varepsilon_B = $ (0.5334, 0.4178, 0.6321, 0.3362, 0.3373, 0.3372, 0.3486, 0.3637, 0.3984, 0.6124, 0.3684), $\varepsilon_C = $ (0.5658, 0.3495, 0.4641, 0.3360, 0.3393, 0.3339, 0.3553, 0.3923, 0.3333, 0.4191, 0.4297), $\varepsilon_D = $ (1.0000, 0.4356, 1.0000, 0.3333, 0.3333, 0.3423, 0.3557, 0.4638, 0.3920, 0.6320, 0.4809), $\varepsilon_E = $ (0.4014, 0.4446, 0.6030, 0.3355,

608

0.3360, 0.3431, 0.3333, 0.3632, 0.6203, 0.9461, 0.5089), $\varepsilon_F = (0.8885, 1.0000, 0.7288, 0.3972, 0.4195, 0.4160, 0.3613, 0.3852, 0.5269, 0.4104, 0.3333)$. $\varepsilon_G = (0.5977, 0.4955, 0.5739, 1.0000, 0.3393, 0.3466, 0.3864, 0.3333, 0.4336, 0.3333, 0.4321)$, $\varepsilon_H = (0.3333, 0.3730, 0.3333, 0.3591, 1.0000, 1.0000, 1.0000, 1.0000, 1.0000, 0.6030, 1.0000)$. R0.4232, 0.3958, 0.5298, 0.4559, 0.5707, 0.4811, 0.7060). Based on the grey correlation degree of the eight power listed enterprise, we can get the company's performance order. Their performance sequence from higher to lower is H, F, D, G, E, A, B and C. By compare their grey correlation degree with the evaluation criterion, we can see that only H's financial performance is high, D and F's financial performance is on the average and the other power companies in the eight examples is low. From this we can get the information that their financial performance is low on the whole and the power listed companies need to analyze their own financial condition from four aspects, profitability, solvency, development capacity and operation ability so as to get a better development in the future and promote our country's power energy's stable development.

4 SUMMARY

Combine the entropy weight method and grey relation analysis method together to evaluate the power listed company's financial performance from four aspects, profitability, solvency and development capability and operation ability, can effectively avoid the disadvantages of analysis based on single aspect index and accurately evaluate the enterprise's position in the peer and reflect the electric power enterprise's comprehensive strength. It provide theoretical guidance for power enterprises finding problems exists in the operation and management and analyzing the gap between advanced enterprises, which plays an important role in the stable development and great progress of the power listed company and our power energy.

REFERENCES

Liu, Z.Y. 2003. Speed up development, strengthen cooperation, continuous innovation. *China Power Enterprise Management* 5: 15–16.

Li, S.Q. & Jiang, D.L. 2011. The research on the application of performance evaluation in electric power enterprise senior management. *Review of Economic Research* 29: 69–70.

Wang, G.N. 2006. The explore on Electric power listed company's performance evaluation system. *Science & Technology Ecnony Market* 12: 22.

Yu, X.Q. 2011. The research on electric power enterprise's financial performance comprehensive evaluation.*Securities & Futures of China* 2: 80–81.

Zhang, G. & Dong, Y. 2012. The financial performance evaluation of electric power listed company based on the grey correlation analysis. *Securities & Futures of China* 8: 28–29.

Wang, Y. & Li, J. 2012. The management performance evaluation of power listed company based on analytic hierarchy process. *Business* 4: 110–111.

Shu, Y.Y. & Pei, X.W. 2008. The financial performance evaluation of power listed companies based on factor analysis. *Industrial & Science Tribune* 9: 164–166.

Yang, X.H. & Li, J.H. 2009. The research on our country's electric power listed company's operating performance evaluation based on the factor cluster analysis. *Chinese business* 9: 5–6.

Xing, H. & Ma, H.Q. 2011. The operating performance evaluation of electric power listed companies based on the factor analysis method. *Communication of Finance and Accounting* 8: 28–29.

Xu. X.H. & Dong, H. 2009. The operating performance evaluation on coal listed companies based on the improved fuzzy clustering. *Science and Technology Innovation Herald* 4: 190.

Information Technology and Computer Application Engineering – Liu, Sung & Yao (Eds)
© *2014 Taylor & Francis Group, London, ISBN 978-1-138-00079-7*

A long-range power-law correlations analysis in Taiwan's stock market

Peng Hsiang Kao
Department of Food & Beverage Management, China University of Science and Technology, Taipei City, Taiwan

Huei Huang Chen
Department of Information Management, Tatung University, Taiwan

Yu Bin Chiu
Department of Information Management, China University of Technology, Taiwan

Yi Lin Huang
Department of Computer Science & Engineering, Tatung University, Taiwan

Shih Chih Chen
Department of Accounting Information, Southern Taiwan University of Science and Technology, Taiwan

ABSTRACT: In time series analysis, there are many statistic models; some models could estimate long memory. A new function for analyzing time series is Detrended Fluctuation Analysis (DFA), which was originally developed for finding long-rage power-law correlation in DNA sequences. This study applies DFA to Taiwan stock market for three categories of data: TAIEX (Taiwan Stock Exchange Capitalization Weighted Stock Index), the group indices aggregated from individual stock indices, and individual stock indices. The results show that long memory exits in most listed companies of Taiwan stock market for the cases when $\alpha \neq 0.5$. However, the correlations detected from aggregated data series do not imply the correlation of original data series. The findings show that the correlations detected from main index do not imply the same correlation of group indices and individual stock indices, but there are greater than half of group indices and individual stock indices following the same correlation with the main index.

Keywords: Time Series Analysis, Detrended Fluctuation Analysis (DFA), TAIEX (Taiwan Stock Exchange Capitalization Weighted Stock Index)

1 INTRODUCTION

Time series analysis has been used to various fields for many years. Methods to analyze time series could be applied in different domains, such as Hurst method that was originally developed for water resource researches and also used in finance researches, it also apply in long memory analysis (Hurst, 1951). Due to the data from the existence of long memory in financial time series, it would be a good reference for stock market investment.

Most previous researches on stock market using DFA focused on the main indices of each nation to detect long-range power-law correlation. They all proved that DFA used to estimate long memory in stock markets. However, detecting long-range correlations in main index only adapt for long-term investment in future market. On the other hand, long-term investors want to invest individual companies, also can use the detecting long-range correlations in individual stock indices.

The main index and group indices are aggregated from individual stock indices. If the main index is analyzed as persistent correlation, what correlations of group indices and individual stock indices will be? Will the three categories of indices follow the same correlation or there has no relation among them? Therefore, this study will use DFA to analyzed Taiwan stock market with the three categories of data: TAIEX (Taiwan Stock Exchange Capitalization Weighted Stock Index), the group indices, and individual stock indices).

Many researchers using DFA focused on the main indices of different nation to detect long-range power-law correlation on stock market. This method has been applied to financial time series analysis in recent years (Liu et al, 1997; Pilar, 2001; Ivanova, 1999). It also was important applying DFA to analyze Taiwan stock market. There are three categories of data: TAIEX (Taiwan Stock Exchange Capitalization Weighted Stock Index), the group indices aggregated from individual stock indices, and individual stock indices. Using DFA to analyze those data can understand the long-range power-law correlation exist or not in Taiwan stock market.

2 RESEARCH METHOD

Weron (2002) had tested these three methods on samples drawn from Gaussian white noise. The DFA method is more accurate on estimating the scaling exponent. Another similar work (Hu et al, 2001) compared the performance of the DFA method with Hurst method and show that the DFA is a superior method to quantify the correlation of noisy signals. There are many non-economic factors (noisy signal) will influence the stock market in short-term period. So, the DFA can work better then Hurst method, because DFA can avoid spurious detection.

This study chose DFA method. For this research purpose, it substituted the l in Equations (2.6) to (2.7) for t. The variable t indicates the time. Then it redefine the Equation (2.6) as $F_d^2(t) = 2, k = 0,1,2,k, (-1)$ or $F_d(t) = 0,1,2,k,(-1)$ (3.1) and redefine the Equation (2.7) as $F_d(t) \sim t^a$ (3.2). This study will use the new definitions in our work.

It will analyze the Taiwan stock market for daily data of 3 categories of indices: 1 main index, 19 group indices, and 667 individual stock indices. It will use the time series of the main index starting on January 1991 and extending to May 2004 and use the time series of group indices starting on January 1995 and extending to May 2004. The ranges of third category of data are not described here because they listed on the stock market started on different dates.

2.1 DFA Method

DFA method was well built as C program by Peng, and this program is free and could be downloaded from reference (http://physionet.incor.usp.br/physiotools/dfa/dfa-1.htm). After downloading the program, this study modified it for our purpose. This study rewrote the function input ( ) and added a lsm ( ) function to fit the results from function dfa ( ) and estimate the scaling exponent. Then it used Microsoft Visual C++ to compile this program and generated a DOS program.

The original program has some options and it will give brief description as following:

- −d k: Detrend the data using a polynomial of degree k(1: linear, 2: quadratic, etc.). The default value option is linear detrending.
- −h : Print a usage summary and exit.
- −l $minbox$: Set the smallest box width. The default, and the minimum allowed value for minbox, is $2k + 2$ (where k is determined by the −d option, see above).
- s : Perform a sliding window DFA (measure the fluctuations using all possible boxes at each box size). By default, fluctuations are measured using non-overlapping boxes only. Using the −s option will make the calculation much slower.
- u $maxbox$: Set the largest box width. The default, and the maximum allowed value for maxbox, is one-fourth the length of the input series.

Options −u and −l are both follow a number, if the given numbers are logically conflict t(e.g. minbox > maxbox), the program will automatically swap the value of them. And if the given numbers are out of range of the time series, the program will automatically set the default value for them. Because all of our time series data don't have the same number of data points, it uses the default values for the two options. As to option −d and −s, it also use default values for them for the original idea of DFA method.

The study modified program will automatically read the external data described from the database of ez Chart (http://www.ezchart.com.tw) and generate the scaling exponent and the time holding the exponent of each time series

3 ANALYSIS RESULT

3.1 Comparison – main index vs. group index

The result shows the scaling exponent for main index and group index. It can be observed that the main index TAIEX shows a persistent power-law, and 12 group indices also have the persistent correlation, 4 group indices are antipersistent, and others are uncorrelated series. By the results, it found that not all of the group indices follow the same correlation with the main index (12 group indices follow persistent correlation, 4 group indices follow antipersistent correlation and 3 group indices are random correlation), but the number of persistent correlation of group indices is grater than half of all group indices. So the results show that more than half of all group indices follow the same correlation with the main index.

3.2 Comparison – group index vs. individual stock index

The result shows the scaling exponent for individual indices of Appliance & Cable. There are 5 individual indices follow the persistent correlation, 10 individual stock indices follow the antipersistent correlation. Comparing the results with group indices of Appliance & Cable, it can observe that all individual stock indices of Appliance & Cable don't follow the same correlation with group indices Appliance & Cable (5 individual stock indices follow persistent correlation, 10 individual stock indices follow antipersistent correlation) and there are greater than half of individual stock indices of Appliance & Cable following the different correlations from group index of Appliance & Cable. This phenomenon can also be found in group indices of Cement, Chemical, Steel Iron, Construction, Tourism, Department Stores and Other with their individual indices. So the results show that there is no obvious relation found between group indices and individual stock indices.

3.3 Comparison – main index vs. individual stock index

Finally, it compares the correlations between main index and individual indices. There are 355 individual stock indices following the persistent correlation, 275 individual stock indices following the antipersistent correlation and 37 individual stock indices following the random correlation. There are greater than half of individual stock indices following the same correlation with main index. So the results say that more than half of individual stock indices follow the same correlation with the main index.

4   CONCLUSION

In this study, DFA method to detect long-range power-law correlations in the main index, group indices and individual stock indices and the result show that the long-range power-law correlations exist in the main index as similar as the previous researches and the long memory also exist in most listed companies of Taiwan stock market for the $\alpha \neq 0.5$.

Besides, the correlation detected from main index was different with the correlation detected from group indices and individual stock indices. However, there are half of group indices and individual stock indices as similar as detected from main index. It also found that there is no obvious relation between group indices and individual stock indices with their detected correlations. It also construct experimental result tables of DFA method of all individual stock indices and theses result is referable for long-term investment in listed companies of Taiwan stock market.

The DFA method can also substitute quadratic fit or higher order polynomial fit for linear fit $y(n)t$. This will be able to give different views for further researcher to detect long-range correlation in financial time series. Moreover, the persistent type series means that the past signals have a positive increment, the future signals will be expected as positive and the antipersistent type series means that the decreasing signals are in the past, the increasing signals will be in the future. According to the tow phenomenon, a forecasting system can be developed while how to implement it will be in the future works.

REFERENCES

[1] Anderson T. W. *The Statistical Analysis of Time Series*, John Wiley and Sons, Inc. (1971), in press.

[2] Detrended Fluctuation Analysis (DFA), information on *http://www.physionet.org/tutorials/fmnc/ node5.html*

[3] Hurst, H., E.: *Long-term Storage Capacity of Reservoirs*. Trans. Am. Soc. Civ. Eng., 116, (1951), p. 770–779.

[4] Hu, K., Ivanov P.C., Chen Z., Carpena P. and Stanley H.E.: *Effect of trends on detrended fluctuation analysis*. Phys. Rev. E, 64, (2001), p.1–19.

[5] Ivanova, K.: *Toward a phase diagram for stocks*. Physica A, 270, (1999), p.567–577

[6] Liu, Y., Cizeau, P., Meyer, M., Peng, C.K. and Stanley, H.E.: *Correlations in Economic Time Series*. Physica A, 245, (1997), p. 437–440.

[7] Pandit, S.M. and Wu, S.M.: *Time Series and System Analysis with Applications*, edtied by John Wiley and Sons, New York. (1983), in press.

[8] Grau-Carles, P.: *Long-range Power-law Correlations in Stock Returns*. Physica A, 299(3–4), (2001), p. 521–527.

[9] Peng, C.K., Buldyrev, S.V., S. Simons H., M., Stanley, H.E., and Goldberger, A.L.: *Mosaic Organization of DNA Nucleotides*. Phys. Rev. E 49, (1994). P. 1685–1689.

[10] Weron, R.: *Estimating long-range dependence: Finite sample properties and confidence intervals*. Physica A, 312(1–2), (2002). p. 285–299.

[11] Shumway, R.H. and Stoffer, D.S.: Time Series Analysis and Its Applications. New York: Springer, (2000), in press.

[12] Information on http://reylab.bidmc.harvard.edu/DynaDx/

Information Technology and Computer Application Engineering – Liu, Sung & Yao (Eds)
© 2014 Taylor & Francis Group, London, ISBN 978-1-138-00079-7

Analysis on the usage characteristics of electrical equipment of Taiwan's convenience stores

Po Yen Kuo, Jen Chi Fu & Chen Hao Jhuo
Jifeng E. Rd., Wufeng District, Taichung, Taiwan, R.O.C.

ABSTRACT: This study aims to explore the composition and characteristics of power consumption of convenience storesregarding electric equipment. The electricity monitoring system was applied to conduct long time measurement and recording of the equipments in convenience stores. The application characteristics of the equipments were discussed. Convenience stores can be divided into four types, including "affected by season, climate and business hours", "affected by season and climate but not by business hours", "not affected by season, climate but affected by business hours" and "not affected by season, climate or business hours" according to the application characteristics of the various equipment in convenience stores. The findings can provide an understanding of electricity consumption rush hour and serve as references to the development of the future energy saving strategies.

Keywords: Convenience stores, Total electricity consumption, Usage characteristics

1 BACKGROUND AND PURPOSES

Most studies on the energy saving of convenience stores in Taiwan focus on the energy-saving technology or power consumption of single equipment. Studies on the overall power consumption of convenience stores are limited in the discussion of the annual EUI (electricity consumption per unit area), The composition of the energy consumption of convenience stores should be analyzed and studied in multiple aspects. For the comprehensive in-depth discussion of the energy consumption and equipment application characteristics of convenience stores in Taiwan, this study attempts to achieve the following purposes through fundamental data statistics and real time electricity monitoring: (1) to understand the application status and characteristics of various equipments in convenience stores by using the year-long 24-hour uninterrupted electricity monitoring system to understand the application status and characteristics, (2) to categorize the electric equipments of the convenience stores by application characteristics as a reference for the subsequent development of energy saving strategies.

2 LITERATURE REVIEW AND RESEARCH METHOD

The review on recent domestic and foreign studies showed that, the "Convenience Store Energy Saving Technical Manual" issued by Bureau of Energy, Ministry of Economic Affairs (2001) simply focused on the energy saving technical operation of convenience stores. The study "An Analysis on the Effect of Refrigeration System on the Energy Consumption of Buildings of General Merchandise – Taking Convenience Stores for Example" conducted by Li (2003) used power-monitoring system to understand that refrigeration system is a critical factor affecting the energy use of convenience stores. The study conducted by Kuo (2010) investigated the overall energy consumption of convenience stores according to energy saving management and technology to take in to account the influence of refrigerating system, air conditioner, lighting, architectural space, power management and use management. "Green Convenience Store Grading Certification and Incentive Improvement Plan" increased the diversity and comprehensiveness of studies concerning convenience stores. However, there is still a lack of studies on the usage characteristics of electrical equipment which is most significantly correlated with the energy use of convenience stores. Therefore, this study intended to use the following methods to conduct a preliminary investigation and have a better understand.

Electric equipment electricity monitoring: as for various electric equipments of convenience stores, by using digital electrical meter, this study conducted the long term monitoring and observation on the installed electricity system to monitoring the data in real time by wireless network to analyze the electricity consumption composition and electric equipment application characteristics of the convenience stores, in an attempt to find the ways to reduce energy consumption and enhance the efficiency of equipment.

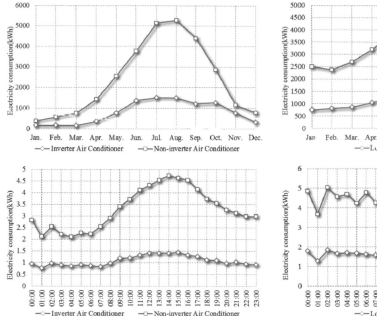

Figure 1. Statistics of monthly electricity consumption of air conditioner (top: monthly, bottom: hourly).

Figure 2. Statistics of monthly electricity consumption of combined compressor (top: monthly, bottom: hourly).

3 Research Procedures and Results

Analysis on the application characteristics of various equipments in convenience stores. There are various types of equipment in convenience stores. In addition to the statistical analysis of five main categories by function including "freezing and refrigeration", "air conditioner", " lamps", "heat source" and "others". Those equipments in convenience stores have different peak and non-peak periods due to different seasons and time periods. Knowing the characteristics could contribute to energy saving. Therefore, this study further developed for equipment use models by "season and climate", and "business hours" to explore the differences in electricity consumption of the electric equipment. The knowledge of the timing and application characteristics of the equipment could serve as a reference to energy saving improvement of the industry.

Affected by season, climate and business hour. As shown in Figure 1, for the 24-hour, all-weather air-conditioned environment of convenience stores, regardless of using the inverter or non-inverter air conditioner, as shown in the monthly statistic diagram of the year, the electricity consumption gradually increases from April until November when it goes back under the average value of the annual electricity consumption. As shown in the hourly statistical diagram of the day, the electricity consumption gradually increases from 8 am to 8 pm when it goes back to the average value of the day. This suggests that the electricity consumption of air conditioners differ greatly

in summer and the noon period of time of the day as compared to other time periods. Moreover, the electricity curve changes of the non-inverter air conditioner changes more significantly as compared with inverter air conditioner as reflected by the electricity consumption. The average electricity consumption gap between the inverter and the non-inverter air conditioners is 32%.

Affected by season, climate but not affected by business hour. To maintainthe temperature for the fresh food and drinks, the combined compressor powering the freezing and refrigeration equipment should run or stand by for 24 hours in all weather condiions. Hence, as shown in the hourly statistical diagram of the day in Figure 2, 5-ton plus and 5-ton minus combined compressor have no significant electricity consumption differences. However, under the effect of summer drinks and hot sales of snacks, for both the 5-ton plus or 5-ton minus combined compressor, the electricity consumption increases apparently from April until November when it goes back under the average annual electricity power consumption, suggesting that the combined compressor is affected by season and climate but not affected by business hour.

Not affected by season, climate but affected by business hours. In recent years, the wave of cheap coffee has made the coffee machine one of the basic equipment of the convenience stores. By using the monthly electricity consumption of the year statistical diagram as shown in Figure 3 to explore the equipment application characteristics, it can be found that

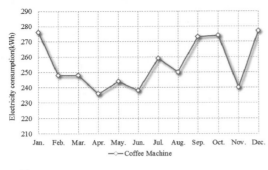

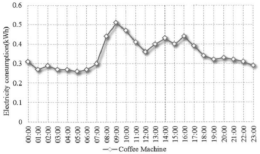

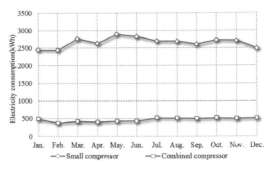

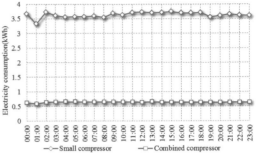

Figure 3. Statistics of monthly electricity consumption of coffee machine (top: monthly, bottom: hourly).

Figure 4. Statistics of monthly electricity consumption of open refrigerator (top: monthly, bottom: hourly).

period from September to January of the next year when winter approaches is the peak hour of the coffee machine. However, the electricity consumption in the peak period and the off-peak period of April ~ July have a gap of only 40 (kWh/month), suggesting that the application characteristics of the coffee machine are not affected by season and climate. It can be found from the hourly electricity consumption statistics diagram that the period from 8 o'clock in the morning to 6 o'clock in the afternoon is the best hour of coffee sales, in particular, the sales is best during the period from nine o'clock in the morning to four o'clock in the afternoon. It is inferred that it is closely related to the drinking of coffee for refreshing and relieving of the stress in class and afternoon tea.

Not affected by season, climate or business hour. The open refrigerator to store drinks and snacks provides a rapid and convenient shopping environment. The cooling measure of open cooling cycle relies on the long time operation and standby of the compressor to maintain the temperature and prevent food from going rotten. Hence, according to the electricity consumption statistic diagrams in Figure 4, the electricity statistical curves have no apparent changes. In particular, the electricity consumption curve of the open refrigerator using the combined compressor is more stable and seems to be a straight line. The total electricity consumption of it is different from the open refrigerator using the single small compressor by about 50% to 60%.

4 CONCLUSIONS

According to the application characteristics of the various equipment in convenience stores, they can be categorized into equipment "affected by season, climate, and business hour" (e.g., air conditioner, signboard and arcade lamps, water dispenser), "affected by season, climate but not affected by business hour" (e.g., combined compressor, freezer, combination refrigerator, cabinet refrigerator, horizontal refrigerator, smoothies machine), "not affected by season, climate but affected by business hours" (e.g., coffee machine, microwave oven, Oden machine, electric cookers) and "not affected by season, climate or business hour" (e.g., open refrigerator, indoor lamps, hot dog machine).

REFERENCES

[1] B.Y. Kuo, 2010, "Green Convenience Store Grading Certification and Incentive Improvement Plan", Architecture and Building Research Institute, Ministry of The Interior
[2] K.P. Li, 2003, "An Analysis on the Effect of Refrigeration System on the Energy Consumption of Buildings of General Merchandise–Taking Convenience Stores for Example", Architectural Institute of the ROC

Information Technology and Computer Application Engineering – Liu, Sung & Yao (Eds)
© 2014 Taylor & Francis Group, London, ISBN 978-1-138-00079-7

Case study of an intelligent green store based on NFC technology

D.Y. Sha
Department of Technology Management, Chung Hua University, Taiwan

Der Baau Perng
Department of Applied Informatics and Multimedia, Asia University, Taiwan

Guo Liang Lai
Department of Industrial Engineering and Management, National Chiao Tung University, Taiwan

ABSTRACT: This study adopts an empirical case studythat focuses on improving the process of products exhibit and sale in the store using NFC technology. Through intelligent interactive guide-shopping system proposed by this research, this case shows how to reduce the cost of products exhibition and simplify the sales process based on the NFC technology. Furthermore, this shopping flow not only enhances the shopping efficiency for consumers but also saves energy in the stores with less display space and marketing costs. The stores could also collect shopping information in order to realize the consumers' behavior and then offer better service.

Keywords: Near Field Communication (NFC), Radio Frequency Identification (RFID), wireless network

1 INTRODUCTION

The high costs of products display in the stores usually hinder the expansion plan of retail stores. How to save cost and simplify the sales process and let customers browse the commodity they want to buy speedily are very important. The store should provide good services that are closer to a customer's demands and therefore satisfy the customer's demands. However, some literatures pointed out that the methods to obtaining information of consumer behavior are still very rare [1]. Due to the concerns mentioned above, Taiwan's Kang Siang Technology Company (KS Technology)had developed the digital shopping system to solve the problems. The system is named the intelligent interactive guide-shopping system. KS Technology had combined near field communication (NFC) technology and radio frequency identification (RFID) technology for the application of this system. The paper will try to use case study methods to discuss how this shopping system can enhances the shopping efficiency for consumers but also saves energy in the stores with less display space and marketing costs. The stores could also collect shopping information in order to realize the consumers' behavior and then offer better service.

2 LITERATURE REVIEW

Recently, not only the RFID technology had been applied in retail industry but also NFC, such as browsing commodity and offering payment while shopping. The rapid improvement of RFID technology, including mobile RFID and NFC applications, had influenced one consumer behavior in our daily life [2][3][4]. Furthermore, retailers have shown a great interest in RFID tags as a means to improve supply chain efficiency and customer relationship management [5].

Besides, good service quality means that customers are satisfied with the service they use and that they are willing to enhance their loyalty [6]. In fact, offering good service quality is the indispensable key for their competition. To enhance store service quality, researchers confirmed that store service quality can be affected with the help of information technology [7][8]. Information technology not only offers retail stores shopping assistant function, but also obtains information about a customer's demand as well. Through NFC and RFID technology in the intelligent interactive guide-shopping system proposed by KS Technology, this study will demonstrate how this shopping system can effectively reduce the cost and space of products display and simplify the sales process. Furthermore, various studies are being actively researched about the mobile payment system through these NFC devices. It could enhance the shopping process and save the time for payment.

3 RESEARCH METHOD

This study adopts an empirical case study to illustrate the behavior information integration system. The case study method is a qualitative study method. Through

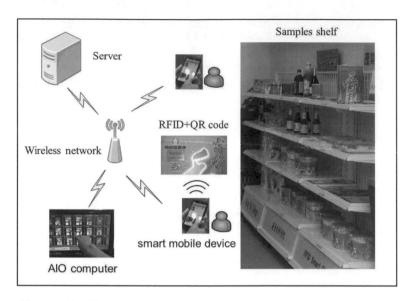

Figure 1. The architecture of intelligent green stores.

observation, in-depth analysis and summarized results of a case, one can have more detailed understanding of an issue or a phenomenon. The knowledge obtained from understanding can help to provide suggestions for other cases or similar issues existing in the knowledge field. With regard to applicable research, Yin [9] believed that any issue that was still at its exploratory stage could be investigated with the case study method. In other words, a case study is suitable for exploring the "why" or "how" of an issue.

Hence, in terms of the steps of a case study, the investigation has three stages. The first stage is to explain the basis and theory for the development of the intelligent interactive guide-shopping system. The second stage is to explain the technology used for the construction of this system, and the third stage is to explain the equipment and operation required for this system. These steps included the design, planning and completion of the information technology and its introduction into an experimental store for testing. In addition to conducting interviews, during this time period, the researchers also performed observations to understand the structure of the overall technology and the result after it was introduced into the store. The research results obtained from this kind of case study are believed to contribute to the information system being used in the store, and are also valuable for further development of more advanced technology in the future.

4 CASE STUDY

KS Technology began a government program in March 2010 to try to facilitate the development of a set of information systems that can effectively obtain and analyze customer behavior information. This system was denominated the intelligent interactive guide-shopping system. The design of this system is based on the combination of powerful databases operating in a server, smart mobile device with NFC and WiFi. Using RFID technology, every sample product has a unique tag that incorporates simple product information. Customers can use smart mobile device or AIO computer to obtain product information by scanning the product tag. Based on the relevant information, the customer can decide to buy or not. Using this technology, retail stores can also plan their stocking needs, marketing, and customer service more effectively, even improve the process of products display and sales. Figure 1 illustrates the architecture of intelligent green stores.

NFC technology is a standard communication protocol for very short-range radio frequency recognition. This protocol enables the user to safely exchange all types of information between two electronic devices when puts the two electronic devices close to one another. Beside, this protocol can also conduct information processing and integration and then information transmission. NFC is developed on the basis of RFID frequency recognition technology. RFID uses Tag to store identification data, uses radio frequency waves to transmit information and uses non-contact manners to repeatedly read and write records, which causes the rapid development of RFID in the product supply chain. NFC is a much stronger element frame compared to pure RFID. NFC contains a security chip, an NFC communications chip and a sensor antenna. These characteristics of NFC make the devices capable of performing non-contact point-to-point transmission, reading or writing into non-contact cards or RFID tags and even mimicking the non-contact cards and being used as Contactless Smart Cards. Currently, the most extensive and promising application of NFC is its complete Contactless Smart Card function

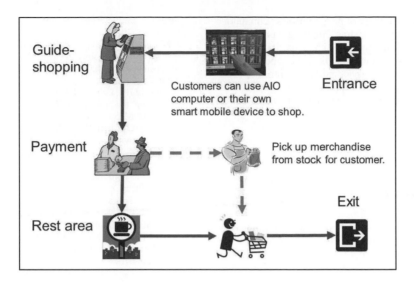

Figure 2. The shopping flow of intelligent green stores.

and its ability to integrate with consumer electronics. The application frequency of NFC technology is HF 13.56 MHz, so this frequency is suitable for short-range communications. This technology simplifies the entire recognition process and makes the communication among electronic devices more direct, safer and clearer.

In the intelligent interactive guide-shopping system, KS Technology uses the function of mobile NFC devices to read RFID tags with the purpose of providing relevant services. At the present time, the setup for most RFID systems is to use a fixed RFID reader to read the information within an RFID tag and then to transfer data inside the tag to the server information system through computer terminals. All equipment and applications above during shopping processes in experiment store is showed in figure 2. When customers come in the store, they could use AIO computer or use their own mobile device to browse and purchase the goods they want to buy including payment. Hence, this mobile device must have a bi-directional interaction function, which can assist the customer to learn about products and make purchases including payment. Hence, this shopping flow not only enhances the shopping efficiency for consumers but also saves energy in retail stores with less display space and marketing costs.

While concerning the bi-directional interacting ability, mobile NFC device must have RFID reader capacity and must be able to provide information services and record customer behavior information. An RFID label tag should be applied to every product, with the product recognition information written on the tags. When the customer holds up mobile NFC device to approach the product tag, the product recognition information can be detected immediately.

However, the bi-directional interaction is not completed with only the above contact, which means using the reader function of the mobile NFC device and posting label tags on products alone is inadequate. Therefore, KS Technology has added a multi-function purchasing software set into the intelligent interactive guide-shopping system to act as an interface for bi-directional interaction. This software set is connected to the product database and the customer database in the store server. Before or after the customer enters the store, the customer must use the internet function within the mobile NFC device to send signals and to connect to the network. Then the customer can enter the store website to download and install the multi-function purchasing software provided by the store.

While finish shopping, the customer can immediately send out a purchasing request through the shopping function of the multi-function purchasing software. At that time, the customer decides the payment method (credit card internet payment or cash payment) while the purchasing request is transferred to the server system. In fact, the most recent NFC mobile phone can add an e-wallet function. At the time of product pickup, the customer only demands to pay through e-wallet, which makes the payment method even more convenient. Therefore, the NFC mobile phone plays an important role in the introduction of this customer information integration technology.

5 SUMMARY AND CONCLUSIONS

Because the introduction of the intelligent interactive guide-shopping system breaks the bottleneck of previous store information system, KS Technology received acclaims during the physical experiment in the store. There is no doubt that this system will reduce the space of products display. So that the stores can reduce the cost of products display and simplify the sales process based on the technology of NFC to browse commodities and make payment while shopping. Furthermore, this shopping flow not only enhances the shopping efficiency for consumers but also saves energy in the stores with less display space and marketing costs. The stores could also collect shopping information in order to realize the consumers' behavior and then offer better service.

REFERENCES

[1] F. Wang, M. Head, "How can the web help build customer relationship? An empirical study in E-tailing. Information & Management", Vol. 44, No. 1, pp.115–129, 2007.

[2] M. C. Lee, C. C. Lee, "Stable Tag Identification in Mobile RFID Systems", Journal of Convergence Information Technology, Vol. 7, No. 7, pp. 312–322, 2012.

[3] X. Yu, X. Guo, Y. Feng, Q. Yao, "A RFID-based Collaborative Process Management Approach for Real-time Food Quality Management", Advances in Information Sciences and Service Sciences, Vol. 4, No. 10, pp. 34–43, 2012.

[4] J. N. Luo, M. H. Yang, "Mobile RFID Mutual Authentication and Ownership Transfer", International Journal of Advancements in Computing Technology, Vol. 4, No. 7, pp. 28–40, 2012.

[5] Hardgrave BC, Langford S, Waller MA, Miller R (2008) Measuring the impact of RFID on out of stocks at Wal-Mart. MIS Quart Executive, 7:113–124.

[6] J. J. Jiang, G. Klein, D. Tesch, H. G. Chen, "Closing the user and provider service quality gap", Communications of the ACM, Vol. 46, No. 1, pp. 72–76, 2003.

[7] G. S. Irene, B. C. Gloria, R. M. María-Eugenia, "Information and communication technology in retailing: A cross-industry comparison", Journal of Retailing and Consumer Services, Vol. 16, No. 1, pp. 232–238, 2009.

[8] C. Loebbecke, J. H. Kronen, T. Jelassi, "The role of information technology in retailing: the case of supporting fashion purchasing at a European department store chain", Journal of Strategic Information Systems, Vol. 5, No. 1, pp. 67–78, 1996.

[9] R. K. Yin, Case Study Research: Design and Methods. Fourth Edition. London: Sage Publications, 2008.

Information Technology and Computer Application Engineering – Liu, Sung & Yao (Eds)
© *2014 Taylor & Francis Group, London, ISBN 978-1-138-00079-7*

An RNS based perfectly optimal data allocation scheme

Chih Ying Chen
Department of Communications, Feng Chia University, Taichung, Taiwan

Chih Yen Yang
Program in Electrical and Communications Engineering, Feng Chia University, Taichung, Taiwan

Hsiu Feng Lin
Institude of Information Engineering, Feng Chia University, Taichung, Taiwan

ABSTRACT: In this paper, we present a data allocation scheme using a Residue Number System (RNS) representation of integers. This allocation can be used in a partial match query system to make the data retrieval perfectly optimal in a data base system under certain commonly occurred condition. The RNS method is also suitable for parallel nearest neighbor searching.

Keywords: multidisk data allocation, partial match query system, perfectly optimal performance.

1 INTRODUCTION

Data base systems are used frequently in many knowledge fields. For example, the Civil Engineering Database (CEDB) of American Society of Civil Engineers (ASCE), or data bases of other organizations, provides access to bibliographic records of publications. Data records are usually distributed over several storage units (like disks) in order to improve the response performance for queries.

Given a set of multi attribute data, a good way of allocating data among multiple disks for facilitating partial match retrieval can also be served as an effective declustering technique. Suppose further that each data assumes integer attribute values. Based on the RNS (Residue Number System) representation of integers that produces a linear ordering of the records, data records may be effectively allocated among disks so that the data retrieval achieves perfect optimality. In Section 2, we describe the problem of data allocation among disks. In section 3, we propose the RNS allocation method, and show that it is perfectly optimal for partial match retrieval under certain conditions that are commonly occurred in practice. In section 4, we give a modification of the allocation to handle the data records that don't satisfy the above condition. Section 5 is the conclusion.

2 THE MULTIDISK DATA ALLOCATION PROBLEM

2.1 Data allocation

The multidisk data allocation problem for partial match retrieval was first proposed by Du and Sobolewski [1] in 1982 and studied subsequently by many researchers [1–17]. It can be formally stated as follows [1,6–7,14]:

We are given a set of N-attribute records, $N \geq 1$, and an m-disk system, $m \geq 2$, our job is to allocate the records among the m disks in such a way that the average response time, over all possible partial match queries, is minimized (i.e., the disk access concurrency is maximized).

By an N-attribute record (data) we mean a record which is characterized by N attribute values. By a partial match query (PMQ), we mean an access specification of the form $q = (a_1, a_2, \ldots, a_N)$, where each a_i, $1 \leq i \leq N$, is either a specified value belonging to the domain D_i of the i-th attribute or is "*" which denotes a don't care condition (i.e., it can be any value in D_i) [1]. For instance, $q = (a, *, b)$ denotes a PMQ to retrieve the records with the first attribute a, the third attribute b, and the second attribute arbitrarily from some set of 3-attribute records.

In the multidisk data allocation problem, it is assumed that the m disks can be accessed independently. Also, it is assumed that the retrieval of one record takes one unit of time. Therefore, the time taken to respond to a query can be simply measured in terms of the maximum number of records needed to be accessed on a particular disk. Accordingly, let $R(q)$ denote the set of all qualifying records for a query q. Then a lower bound to the response time of q is $\lceil |R(q)|/m \rceil$, where m is the total number of available disks.

2.2 *The perfectly optimal data allocation*

In the multidisk data allocation problem, an allocation method which minimizes the average response time over all possible queries is called an *optimal* allocation method [1]. An allocation method is strictly optimal if, for any query q, the $|R(q)|$ qualifying records are distributed uniformly to the m disks such that the work load to each disk is balanced in response to q and whence the lower bound $\lceil |R(q)|/m \rceil$ to the response time of q is achieved.

If, in addition, a strictly optimal allocation method also assigns all records uniformly among the disks, meaning that the number of records that are assigned on each disk differs only one at most, it is called a *perfectly optimal* allocation method [11]. Therefore, a perfectly optimal allocation method is superior to all other allocation methods and is the most desired approach to the multidisk data allocation problem in real applications.

Nevertheless, it has been shown in [15,17] that, for certain types of PMQs, strictly optimal allocation doesn't exist in the general case, perfect optimality is difficult to achieve in general. Accordingly, all previous solutions, including ours, are near optimal or heuristic in the general case. Optimal solutions can be obtained only under certain conditions [1–15,18]. For examples, the method suggested by [5] does not guarantee any perfect optimality, the perfect optimalities given in [6–7,13] are limited to two-disk systems only, the perfect optimalities given in [3–4,6,12–13,16] are limited to a very special case where $|D_1| = |D_2| = \cdots = |D_N| = 2$ only (D_i denotes the domain of the i-th attribute), the perfect optimalities given in [2,9,11] are limited to the special cases where $|D_1| = |D_2| = \ldots = |D_N| = k$, $k > 1$, and the perfect optimalities given in [2] are limited to the special case where m (the total number of available disks) is a power of an integer. Accordingly, the problem of perfectly optimally allocating a set of records among multiple disks is still very challenging at present.

Based upon the RNS (Residue Number System) representation of integers, in the next section, we shall propose a new data allocation method, called the RNS allocation method, and show that it is perfectly optimal under certain conditions that are more flexible than each of the previous solutions.

3 THE RNS DATA ALLOCATION SCHEME AND ITS OPTIMALITY PROPERTY

3.1 *The RNS data allocation scheme*

Let there be a set of N-attribute records $F = \{(b_1, b_2, \ldots, b_N) | 0 \le b_i \le m_i - 1$ for $1 \le i \le N\}$, where $m_1, m_2, \ldots, m_N$ are pairwise relatively prime integers. Let $NB = m_1 m_2 \ldots m_N$ denote the total number of records, and $m \ge 2$ denote the total number of available disks. Consider a record $(b_1, b_2, \ldots, b_N)$ in F. Since $m_1, m_2, \ldots, m_N$ are pairwise relatively prime and $0 \le b_i \le m_i - 1$ for $1 \le i \le N$, $(b_1, b_2, \ldots, b_N)$ can be served as the unique RNS representation [18, Section 4.7] of some integer, denoted as $x_{b_1 b_2 \ldots b_N}$, in the range [0,NB] by using the set of moduli $\{m_1, m_2, \ldots, m_N\}$. That is $b_i \equiv x_{b_1 b_2 \ldots b_N}$ (mod m_i) for $1 \le i \le N$. And in this way all records in F are aligned into a linearly ordered list according to the order of their integer correspondences in [0,NB]. Consider, for illustration, an example where $N = 3$, $m_1 = 2$, $m_2 = 3$ and $m_3 = 5$. Table 1 depicts the one-to-one correspondence between each record (b_1, b_2, b_3) in F and the corresponding integer $x_{b_1 b_2 b_3}$ in [0,30].

It is interesting to observe, from Table 1, that the integer correspondences of all qualifying records of any PMQ for F show up periodically in the range [0,30].

Theorem 3.1 For any PMQ, the RNS integer correspondences of all qualifying records show up periodically.

Proof: Let $q_{i_1 i_2 \ldots i_n}$ be a PMQ for which the i_j-th key, $1 \le j \le n$, are specified and other keys are unspecified. Suppose $(b_1, b_2, \ldots, b_N)$ and $(b'_1, b'_2, \ldots, b'_N)$ are two distinct records qualified by $q_{i_1 i_2 \ldots i_n}$. Then $b_{i_j} = b'_{i_j}$ for $1 \le j \le n$, or equivalently, $x_{b_1 b_2 \ldots b_N} \equiv x_{b'_1 b'_2 \ldots b'_N}$ (mod m_{i_j}) for $1 \le j \le n$. Since $m_{i_1}, m_{i_2}, \ldots, m_{i_n}$ are pairwise relatively prime, we have $x_{b_1 b_2 \ldots b_N} \equiv x_{b'_1 b'_2 \ldots b'_N}$ (mod $p_{i_1 i_2 \ldots i_n}$), where $p_{i_1 i_2 \ldots i_n} = m_{i_1} m_{i_2} \ldots m_{i_n}$. This says that the integer correspondences of the records qualified by $q_{i_1 i_2 \ldots i_n}$ show up periodically in the range [0, NB] with period $p_{i_1 i_2 \ldots i_n}$ Q.E.D.

Another intriguing observation from Table 1 is that the components of two consecutive records (b_1, b_2, b_3) and (b'_1, b'_2, b'_3) for which $x_{b'_1 b'_2 b'_3} = x_{b_1 b_2 b_3} + 1$ satisfy $b'_i \equiv b_i + 1$ (mod m_i) for $1 \le i \le 3$.

Theorem 3.2 Let $(b_1, b_2, \ldots b_N)$ and $(b'_1, b'_2, \ldots b'_N)$ be two records such that $x_{b'_1 b'_2 \ldots b'_N} = x_{b_1 b_2 \ldots b_N} + 1$. Then $b'_i \equiv b_i + 1$ (mod m_i) for $1 \le i \le N$.

Proof: Since $b'_i \equiv x_{b'_1 b'_2 \ldots b'_N}$ (mod m_i) and $b_i \equiv x_{b_1 b_2 \ldots b_N}$ (mod m_i) for $1 \le i \le N$, $x_{b'_1 b'_2 \ldots b'_N} = x_{b_1 b_2 \ldots b_N} + 1$ implies that $b'_i \equiv x_{b'_1 b'_2 \ldots b'_N} \equiv x_{b_1 b_2 \ldots b_N} + 1 \equiv b_i + 1$ (mod m_i) for $1 \le i \le N$. Q.E.D.

Based upon the correspondence properties stated in Theorem 3.1 and Theorem 3.2, we propose a new data allocation scheme, called the RNS data allocation scheme as follows.

Table 1. The 1-1 Correspondence between Integers in [0, 30] and Records in F.

Integer $x_{b_1 b_2 b_3}$	Record (b_1, b_2, b_3)	Integer $x_{b_1 b_2 b_3}$	Record (b_1, b_2, b_3)	Integer $x_{b_1 b_2 b_3}$	Record (b_1, b_2, b_3)
0	(0,0,0)	10	(0,1,0)	20	(0,2,0)
1	(1,1,1)	11	(1,2,1)	21	(1,0,1)
2	(0,2,2)	12	(0,0,2)	22	(0,1,2)
3	(1,0,3)	13	(1,1,3)	23	(1,2,3)
4	(0,1,4)	14	(0,2,4)	24	(0,0,4)
5	(1,2,0)	15	(1,0,0)	25	(1,1,0)
6	(0,0,1)	16	(0,1,1)	26	(0,2,1)
7	(1,1,2)	17	(1,2,2)	27	(1,0,2)
8	(0,2,3)	18	(0,0,3)	28	(0,1,3)
9	(1,0,4)	19	(1,1,4)	29	(1,2,4)

Algorithm 3.1 The RNS Data Allocation Scheme

Input: A set of records $F = \{(b_1, b_2, \ldots, b_N)$ | $0 \leq b_i \leq m_i - 1$ for $1 \leq i \leq N\}$, where m_1, m_2, ..., m_N are pairwise relatively prime; and m disks, $m \geq 2$.

Output: The allocation of the records in F onto m disks.

Steps:
1. Compute ℓ and r such that $NB = m_1 m_2 \ldots m_N = m\ell + r$, where $0 \leq r < m$.
2. Assign each record $(b_1, b_2, \ldots, b_N)$ in F to disk i if $i\ell + i \leq x_{b_1 b_2 \ldots b_N} \leq (i+1)\ell + i$, for $0 \leq i \leq r - 1$; disk j if $j\ell + r \leq x_{b_1 b_2 \ldots b_N} \leq (j+1)\ell + r - 1$, for $r \leq j \leq m - 1$. (3.1)

If all records $(b_1, b_2, \ldots, b_N)$ in F are aligned linearly according to the ascending order of $x_{b_1 b_2 \ldots b_N}$, then the RNS allocation scheme assigns the first $\ell + 1$ records onto disk 0, the second $\ell + 1$ records onto disk 1,...,the r-th $\ell + 1$ records onto disk $(r - 1)$, respectively, and then takes turns to assigns the next ℓ records onto disk j for $r \leq j \leq m - 1$. It is easy to determine the record $(b'_1, b'_2, \ldots, b'_N)$ next to a given record $(b_1, b_2, \ldots, b_N)$ in the linearly ordered list of records by Theorem 3.2.

Example 3.1 If $N = 3$, $m_1 = 2$, $m_2 = 3$, $m_3 = 5$ and $m = 4$. Then $\ell = 7$ and $r = 2$. Table 2 depicts the assignment of all records among 4 disks by using the RNS allocation scheme given by Algorithm 3.1.

3.2 The perfect optimality of the RNS allocation scheme

The RNS allocation scheme is perfectly optimal in Example 3.1. In fact, the perfect optimality of the RNS allocation scheme is true in general and can be formally established as follows.

Theorem 3.3 Let $F = \{(b_1, b_2, \ldots, b_N)$ | $0 \leq b_i \leq m_i - 1$ for $1 \leq i \leq N\}$ be a set of records, where m_1, $m_2, \ldots, m_N$ are pairwise relatively prime. Let $NB = m_1 m_2 \ldots m_N$ denote the total number of records, and $m \geq 2$ be the number of available disks. Then the RNS allocation scheme is perfectly optimal for allocating F among m disks.

Table 2. he Assignment of 30 Records among 4 Disks by Using the RNS Allocation Scheme

Disk 0	Disk 1	Disk 2	Disk 3
(0,0,0)	(0,2,3)	(0,1,1)	(1,2,3)
(1,1,1)	(1,0,4)	(1,2,2)	(0,0,4)
(0,2,2)	(0,1,0)	(0,0,3)	(1,0,0)
(1,0,3)	(1,2,1)	(1,1,4)	(0,2,1)
(0,1,4)	(0,0,2)	(0,2,0)	(1,0,2)
(1,2,0)	(1,1,3)	(1,0,1)	(0,1,3)
(0,0,1)	(0,2,4)	(0,1,2)	(1,2,4)
(1,1,2)	(1,1,0)		

Proof: Suppose $NB = m\ell + r$, where $0 \leq r < m$, and let n_i, $0 \leq i \leq m - 1$, denote the number of records in F that are assigned on disk i by the RNS allocation scheme. Then, according to (3.1), we have $n_0 = n_1 = \cdots = n_{r-1} = \ell + 1$ and $n_r = n_{r+1} = \ldots = n_{m-1} = \ell$. Therefore, all records in F are uniformly distributed among m disks by the RNS scheme. Next, let q be any PMQ, R(q) be the set of all records qualified by q, and p be the period of occurrence of the RNS integer correspondences of all records in R(q), where the set of used moduli is $\{m_1, m_2, \ldots, m_N\}$. Further, let α_i denote the number of buckets in R(q) that are assigned on disk i by the RNS scheme. Then it is obvious that $\alpha_i = 0$ or 1, $0 \leq i \leq m - 1$, if $p > \ell$. In this case, the RNS scheme is strictly optimal for q. For the case where $0 < p \leq \ell$, we must have $\alpha_i \geq \lfloor \ell/p \rfloor$ and $(\alpha_i - 1) \times p + 1 \leq \ell + 1$. This implies that $\lfloor \ell/p \rfloor \leq \alpha_i \leq \lfloor \ell/p \rfloor + 1 = \lceil \ell/p \rceil$. Accordingly, the RNS scheme is also strictly optimal for q in this case. Consequently, we have shown that the RNS scheme is perfectly optimal for allocating F among m disks. Q.E.D.

It is worth pointing out that the perfect optimality of the RNS allocation method is far more flexible than that of each of the previous solutions (as we have stated in Section 2.2) even though it is still somewhat strict when the number of attributes is large in real applications.

Table 3. The assignment of 36 records among 4 disks.

Disk 0	Disk 1	Disk 2	Disk 3
$(0,0,0)^1$	$(0,2,3)^1$	$(0,1,1)^1$	$(1,2,3)^1$
$(1,1,1)^1$	$(1,0,4)^1$	$(1,2,2)^1$	$(0,0,4)^1$
$(0,2,2)^1$	$(0,1,0)^1$	$(0,0,3)^1$	$(1,1,0)^1$
$(1,0,3)^1$	$(1,2,1)^1$	$(1,1,4)^1$	$(0,2,1)^1$
$(0,1,4)^1$	$(0,0,2)^1$	$(0,2,0)^1$	$(1,0,2)^1$
$(1,2,0)^1$	$(1,1,3)^1$	$(1,0,1)^1$	$(0,1,3)^1$
$(0,0,1)^1$	$(0,2,4)^1$	$(0,1,2)^1$	$(1,2,4)^1$
$(1,1,2)^1$	$(1,0,0)^1$	$(0,0,5)^2$	$(0,2,5)^2$
$(0,1,5)^2$	$(1,2,5)^2$	$(1,1,5)^2$	$(1,0,5)^2$

In addition, let us call two multiattribute records a pair of k-th nearest neighbors if the Hamming distance between the corresponding attributes $\leq k$. Since the RNS method assigns every subset of k-th nearest neighbors uniformly among independently accessible disks respectively, we would also like to emphasize here that our method is also suitable for parallel nearest neighbor searching.

4 A MODIFIED RNS DATA ALLOCATION SCHEME

Let $F = \{(b_1, b_2, \ldots, b_N) \mid 0 \leq b_i \leq m_i - 1$ for $1 \leq i \leq N\}$ be a set of records, where $m_1, m_2, \ldots, m_N$ are not pairwise relatively prime. Then there is no one to one correspondence between the records in F and the integers in $[0, NB]$, where $NB = m_1 m_2 \ldots m_N$ is the number of records. If we may decompose an integer m_k in the set $S = \{m_1, m_2, \ldots, m_N\}$ as $m_k = m_k^1 + m_k^2$ such that both the set of integers $S_1 = \{m_1, \ldots, m_k^1, \ldots, m_N\}$ and the set of integers $S_2 = \{m_1, \ldots, m_k^2, \ldots, m_N\}$ are pairwise relatively prime, then we still can apply RNS approach to the two sets F_1 and F_2 of records corresponding to the two sets S_1 and S_2 of integers respectively. The performance of this application is not perfectly optimal in general, but close to perfect optimality usually.

Example 4.1 Consider a data allocation with $N = 3$, $m_1 = 2$, $m_2 = 3$, $m_3 = 6$ and $m = 4$. Table 3 shows an assignment of the records among 4 disks.

The sets of integers $S_1 = \{2, 3, 5\}$ and $S_2 = \{2, 3, 1\}$ are both pairwise relatively prime. We apply RNS data allocation to both the set F_1 and the set F_2 of data records., where F_1 is the set of records mentioned in Example 3.1, and $F_2 = \{(0,0,5), (1,1,5), (0,2,5), (1,0,5), (0,1,5), (1,2,5)\}$. The uppercase 1's that appear in the notations of records denote the records in F_1, and the uppercase 2's denote the records in F_2. We may check to see that the response time of most of the queries achieve their lower bounds. For example, the query $(0, 0, *)$ has response time $2 = \lceil 6/4 \rceil$, and the query $(1, *, 4)$ has response time $1 = \lceil 3/4 \rceil$, Hence the approach has a performance close to perfect optimality.

5 CONCLUSIONS

In this paper, we have concerned with the multidisk data allocation problem for partial match retrieval. Based upon using the well-known RNS representation of integers to produce a linear ordering of the data, we have proposed a new data allocation scheme, called the RNS allocation scheme. It has been shown that the RNS allocation method guarantees perfectly optimal response time and load-balance performance if the domain sizes of all attributes are pairwise relatively prime. In addition, the RNS method is also suitable for parallel nearest neighbor searching because it assigns every subset of k-th nearest neighbors uniformly among independently accessible disks respectively. Finally, how to solve the general multidisk data allocation problem through an effective declustering technique is interesting and challenging. This remains to be one of our future research topics.

REFERENCES

[1] H. C. Du and J. S. Sobolewski, "Disk Allocation for Cartesian Product Files on Multiple Disk Systems," *ACM Trans. Database Systems*, vol. 7, no. 1, pp. 82–101, 1982.

[2] K. A. S. Abdel-Ghaffar and A. El. Abbadi, "Optimal Disk Allocation for Partial Match Queries," *ACM Trans. Database Systems*, vol. 18, no. 1, pp. 132–156, 1993.

[3] M. Y. Chan, "Multidisk File Design: An Analysis of Folding Buckets to Disks," *BIT*, vol. 24, pp. 262–268, 1984.

[4] M. Y. Chan, "A Note on Redundant Disk Allocation," *IPL*, vol. 20, pp. 121–123, 1985

[5] C. C. Chang, "Application of Principal Component Analysis to Multidisk Concurrent Accessing," *BIT*, vol. 28, pp. 205–214, 1988.

[6] C. C. Chang and C. Y. Chen, "Gray Code as a Declustering Scheme for Concurrent Disk Retrieval," *Information Science and Engineering*, vol. 13, no. 2, pp. 177–188, 1987.

[7] C. C. Chang and C. Y. Chen, "Symbolic Gray Code as a Data Allocation Scheme for Two-disk Systems," *The Computer Journal*, U. K., vol. 35, no. 3, pp. 299–305, 1992.

[8] C. C. Chang and J. C. Shieh, "On the Complexity of File Allocation Problem," *Proc. Int'l Conf. Foundation of Data Organization*, Kyoto, Japan, pp. 113–115, May 1985.

[9] C. Y. Chen and H. F. Lin, "Optimality Criteria of the Disk Modulo Allocation Method for Cartesian Product Files," *BIT*, vol. 31, pp. 566–575, 1991.

[10] C. Y. Chen, H. F. Lin, R. C. T. Lee and C. C. Chang, "Redundant MKH Files Design among Multiple Disks for Concurrent Partial Match Retrieval," *Journal of Systems and software*, vol. 35, pp. 199–207, 1996.

[11] C. Y. Chen, H. F. Lin, C. C. Chang and R. C. T. Lee, "Optimal Bucket Allocation Design of

K-ary MKH Files for Partial Match Retrieval," *IEEE Trans. Knowledge and Data Engineering*, vol. 9, no. 1, pp. 148–159, 1997.

[12] H. C. Du, "Disk Allocation Methods for Binary Cartesian Product Files," *BIT*, vol. 26, pp. 138–147, 1986.

[13] C. Faloutsos and D. Metaxas, "Disk Allocation Methods Using Error Correcting Codes," *IEEE Trans. Computers*, vol. 40, no. 8, pp. 907–914, 1991.

[14] M. F. Fang, R. C. T. Lee, and C. C. Chang, "The Idea of Declustering and Its Applications," *Proc. 12th Int'l Conf. VLDB*, Kyoto, Japan, pp. 181–188, Aug. 1986.

[15] M. H. Kim and S. Pramanik, "Optimal File Distribution for Partial Match Retrieval," *Proc. ACM-SIGMOD Conf.*, pp. 173–182, 1988.

[16] D. E. Knuth, *The Art of Computer Programming*, volume 2: Seminumerical Algorithms. Addison-Wesley, 1969 Second edition, 1981.

[17] Y. Y. Sung, "Performance Analysis of Disk Allocation Method for Cartesian Product Files," *IEEE Trans. Software Engineering*, vol. 13, no. 9, pp. 1,018–1,026, 1987.

[18] R. L. Graham, D. E. Knuth and O. Patashnik, *Concrete Mathematics*, Addison-Wesley, 1989.

Information Technology and Computer Application Engineering – Liu, Sung & Yao (Eds)
© 2014 Taylor & Francis Group, London, ISBN 978-1-138-00079-7

Design and development of interactive-whiteboard system based on teaching system

Long Fei Liu & Qi Shan Zhang
Department of Electronics and Information Engineering, Beijing University of Aeronautics and Astronautics, Beijing, P.R. China

ABSTRACT: In this paper, for the characteristics of the teaching system application, we select a hardware implementation of the whiteboard based on infrared sensor, and for devices to complete software system. Based on frequent user interaction and information necessary to conduct more real-time features, we choose the C/S structure mode. The GDI + SDK, TCP/IP Socket and a new fuzzy recognition algorithm are used in the system to achieve the main system functions. Test result shows the system has the higher recognition rate, and the effect is much more obviously.

Keywords: E-Teaching; infrared sensor; Interactive-Whiteboard; C/S Frame

1 INTRODUCTION

Interactive white board, as the replacement of the traditional white board, is a electronic sensor board with the interaction between human and computer as well as human and human to realize the paper-less teaching and working. The development of the electronic white board was based on the micro-electronic technique, computer technique and electro-communication technique. In order to expand the teaching space, make fully use of the e-Education, an interactive white board system is designed in this article. It helps to combine the theoretical teaching with the visual teaching, fully stimulate the student's different senses to improve the teaching quality and efficiency.

Interactive white board has different classifications due to the different standards. Based on the precise positioning technique, the interactive white board in the market can be classified into electromagnetic induction technology whiteboard, resistor film technology whiteboard, infrared induction technology whiteboard, ultrasonic technology whiteboard, and CCD scan technology whiteboard etc. In general, the whiteboard based on infrared sensor has advantages of accurate positioning, quick reaction, not fearing cut and long life-span. It doesn't need a special writing pen. Instead, teachers can use their fingers or a pointer directly writing on it, making the large-area white board production become possible. Thus the infrared induction technology whiteboard becomes one of the most popular and ideal aids in the teaching industry, due to its well-rounded performance. Also, since all the products have its own flaws, some manufacturers may combine two or even more techniques together,

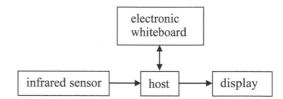

Figure 1. Structure of the system.

for example to produce the infrared and ultrasonic technology whiteboard etc, in order to improve the disadvantage of the single technique.

2 SYSTEM CONFIGURATION AND WORKING PRINCIPLE

2.1 *The system structure and main functions*

To meet the requirement of teaching on electronic whiteboard system, this paper designed an interactive electronic whiteboard system based on infrared induction. Fig. 1 shows the structure of the system.

The system combined the field teaching with network teaching, using plasma TVs as the main body of the display, associating with infrared scan positioning device, IPC (industrial computer), etc to consist of the interactive and multi-functional teaching system.

The main functions are: traditional MCAI functions like electronic courseware, web browsing, teaching video playback, etc; others like screen notation, painting graphics, textures, graphics editing or a certain part key shows; The system can capture the contents of the screen, make full screen snapshots, regional snapshots and save; set up material warehouse which has centralized saving and management, to convenient the teacher by realizing network sharing, so that teachers can call out their materials no matter where they teach; this direct network interconnection can realize the campus LAN or remote network teaching; according to teachers' non-standard graphing like plotting a circle, it can recognize the common geometric graphs and help to correct, etc.

2.2 The hardware structure of the system

The hardware platform of the system is mainly composed of infrared induction device, host and plasma TV.

Among them, the infrared induction device in the electronic whiteboard is used to realize accurate positioning part. The infrared light-emitting/receiving diodes are distributed densely around the display area, generating horizontal and vertical infrared to form a plane-scan mesh. When the infrared is blocked by the object inside the area, it can locate the abscissa and ordinate of the object and send the position to PC to realize positioning.

So far, the display device mainly use a projector coordinating with a white board and due to its fuzziness under sun light or shade problem when writing. The system uses a sixty inch plasma TV as a display screen.

Powerful graphics card and windows operation systems are chosen for the host.

2.3 The design of software

The software mainly contains two subsystems: local sub-system and network sharing sub-system respectively. It can be divided into the following modules:

(1) Stand-alone system: Complete the basic requirements of teachers. The concrete is divided into the following five modules:

- Graphics module: Brush function realizes writing on the interactive infrared touchpad system. The system supports fluorescent pen, pencil and other common writing mode. The thickness and color of the pen can be set. It also supports line, circle, oval, rectangular, square and other geometric graphics drawing. Graphic color, line width, linear and infilling can be set and the handwriting can be erased on the interactive infrared touch system.
- Texture module: Can copy and paste the pictures in JPG or BMP on the screen, change the size and edit them.
- Re-edit module: All the objects can be re-edited to change the size or transparency; cut, copy and paste; move, rotate, overturn and mirror image.
- Image interception module: Can capture the content on the screen and do region snapshot, full-screen snapshot or object snapshot to a certain area and save the snapshot.

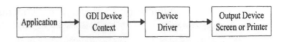

Figure 2. GDI working process.

- Other function module: mask(used to cover a certain area) and system mode(mouse mode, whiteboard mode, writing screen mode, etc.) change, undo and redo etc.

(2) Network interactive module: The network interconnection can realize the sharing of graphs, operations and teaching resources to make the campus LAN and remote network teaching come true.

(3) Graphics recognition module: It can recognize the teacher's non-standard graphing and help to correct.

3 THE REALIZATION OF THE FUNCTION

3.1 Single computer system

In the realization of the system, we directly call the functions from GDI (Graphics Device Interface) to easily create, output various needed Graphics, text and operation in the screen, printers and other output devices. It can automatically change the output from the program into the output from the hardware without considering the hardware and device driver, so that the designer of program doesn't need to be with the hardware which makes the job quick and convenient. Its way of working is shown in Fig. 2:

3.1.1 Graphics module
Graphics module is the most frequently-used module in this system, including the function of painting line, circle, ellipse, rectangle, square and changing their line shape, line width and fill color, etc. Before we drawing, a brush object must be first created or selected. MFC provides a encapsulated Cpen class, which simplified the use of brush objects. We can use pen as an object for processing by calling some functions to create the user-defined pen including functions like: CreatPen(); CreatPenIndirect(), etc which are capsulated in the Cpen. Then we can set the pen attribute, draw different graphs or delete the pen.

3.1.2 Texture module
When using the electronic whiteboard, users often need to give a description about their picture. Therefore, we provide attribute table and attribute page here by using the CPropertySheet and Page capsulated in the MFC to call the functions of AddPage (&m_properpage1) and AddPage(&m_properpage2) to display the two newly created attribute pages. Use CTreeCtrl in MFC to call the corresponding interface function to display the pictures in the disk directory or files. It is similar to the browsing module in explorer, showing the thumbnail of JPG or BMP, which is convenient for users to choose and process.

3.1.3 *Re-edit module*

When using the electronic whiteboard, teachers often need to re-edit and revise the previously drawn pictures, for example: delete, move or change the size or position etc. Therefore, we bring in CPtrAffay to save the graphic properties and set up CGraph to save the properties of graphs and pens.

3.1.4 *Image interception module*

In order to know the students' ideas in the future, the teacher often need to save their courseware. Therefore, we call the function Copybitmap() and change it into BMP format to save the DIB file in the disc according to image file head, bitmap information head, color table and bitmaps image data.

3.2 *Network interactive module*

We select a client/server (C/S) network model and use TCP/IP protocol in the electronic whiteboard system. The main function of the system's server is to transmit data, including graphic data transmission module, image and file transferring module. We establish connection with the server in the client based on TCP/IP, use the CFileSocket class to realize graphic data transmission, and use RLE, LZW, Rice and Hoffman coding algorithm to compress and transmit the large-capacity image files.

3.3 *Graphics recognition module*

To solve the problem of non-standard drawing during the teaching, we introduce graphic fuzzy recognition method based on geometric property which combines statistics with geometry to recognize the common used graphs and correct them.

Therefore, we studied the frequently-used graphs and signs, for example: 6 common graphs like straight line, rectangle, circle, ellipse, diamond, triangle and 5 signs like delete, undo, copy, move, wave type. In view of the graphics' geometry characteristics, we program the recognition and repainting steps according to the hand-painted graph input, sketches before processing, primitive recognition, primitive beautification, etc.

After the studies above, we can identify straight line, circle, triangle, ellipse, rectangular and their respective rotary graphics. The testing shows the recognition time is less than 0.6 s, the recognition rate is above 95%, and the recognition effect is much more obviously.

4 THE APPLICATION OF THE SYSTEM

Interactive electronic whiteboard inherits the normal blackboard's advantages, and integrates with modern multimedia teaching technology, to realize the combination between traditional and modern education. At present, the electronic white board technology is mature in foreign markets. Our country has also begun the investigation on the experiment of electronic whiteboard application now. In Beijing, white board experiments have successively been held in five middle school and primary school. Through a few semesters' experiment, teachers and students fully affirmed this teaching method. Therefore, the electronic whiteboard, as a modern teaching method, will be an important aid in the class of quality-oriented education. With the development of computer, network and processing technique, the application of electronic white board in teaching will increase widely and step into a new stage.

REFERENCES

[1] Chen J, Chen W. VC++ Network advanced programming [M]. Beijing: People's Post and Telecommunication Publishing House, 2004.

[2] Liu F Q. Multimedia technology and application [M]. Beijing: People's Post and Telecommunication Press, 2002.

[3] Zhang J H, Tan Q P. Multimedia whiteboard design and application [J]. Computer engineering and application, 2003(3):130–132.

[4] Yuan Z, Chen X C. Design and implementation of whiteboard module [J]. Computer engineering, 2001(4): 15–18.

[5] He J, Xie S W. Socket based on TIP/IP network communication model [J]. Computer application research, 2001(3):54–58.

[6] Manuel J. Fonseca, Joaquim A. Jorge. Using Fuzzy Logic to Recognize Geometric Shapes Interactively [J]. IEEE Press (2000).

Information Technology and Computer Application Engineering – Liu, Sung & Yao (Eds)
© 2014 Taylor & Francis Group, London, ISBN 978-1-138-00079-7

Similarity assessment method for CBR in a multi-dimensional scenario space system

Y.L. Feng, S.B. Zhong, Y. Liu & H. Zhang
Institute of Public Safety Research, Department of Engineering Physics, Tsinghua University, Beijing, China

ABSTRACT: Similarity assessment is fundamental to Case-Based Reasoning (CBR). This paper first proposes a new method of building a well quantified case library which is called the Multi-Dimensional Scenario Space (MDSS) system. A new similarity assessment method is developed in order to get more precise results.

Keywords: Case-Based Reasoning; Multi-Dimensional Scenario Space system; Similarity assessment method

1 INTRODUCTION

Case-based reasoning (CBR) is an artificial intelligence method used in many fields. The fundamental idea of CBR is that similar problems may have similar solutions. A CBR cycle always contains 4 steps, also known as '4-REs'[1]: Retrieval, Reuse, Revise, and Retain. Case retrieval is the first, also the most important step which aims at retrieving the most similar cases to the unsolved case from the case library.

Similarity assessment has been a hot topic in CBR related studies in recent year, and different methods have been developed. However, as cases becoming more and more complex, the similarity assessment is getting more and more difficult. This paper developed a method of defining and calculating similarity with the multi-dimensional scenario system for CBR, which shows more efficiency and precision. In this method, a certain case is defined as a collection of scenarios, and a scenario is described as a point in a multi-dimensional scenario space and the cases can be split into small scenarios in this space. The similarity assessment method between scenarios is developed which can helps in finding more precise results.

2 MULTI-DIMENSIOAL SCENARIO SPACE SYSTEM

2.1 Case library building

A case library is always essential when applying CBR in practice. Take cases of oil tank fire emergency system as an example. When the fire department wants to efficiently put out the fire, many factors should be taken into account, such as the number of the oil tanks, the type of the oil, whether there is an oil depot around, and the wind direction etc. However, the question lies that how to properly and efficiently record the factors.

In a multi-dimensional scenario space, a certain scenario is defined as a point where the factors are served as axis of coordinates (dimensions), and the cases are collection of scenarios[2]. The dimensions are in great detail so that the case can be well defined without omitting any important information. The values in the dimensions can be of different types, discrete, continuous or a mixture of the two. For example, when defining the dimension describing the number of oil tanks, the values can simply be. However, considering that the differences between 101 tank and 102 tanks is surely much smaller than the differences between 1 and 2 tanks because when there are large number of tanks, one more tank won't affect the measures taken to put up the fire that much. Denote as the number of oil tanks, as a threshold above which the influence of increasing number of oil tanks starts to fade. The values in this dimension may be defined as follows:

$$X_i = \begin{cases} i & \text{, if } i \leq k \\ k + exp(i - k) & , otherwise \end{cases} \quad (1)$$

Here we use the exponential function to describe the fading process, some other functions can also be used if necessary. The definition of other dimensions can be defined in the same way, i.e. first find the characters of the dimension, then get some function to describe the dimension.

[1] A. Aamodt, E. Plaza. 1994. Case-Based Reasoning: Foundational Issues, Methodological Variations, and System Approaches. AI Communications. IOS Press, Vol. 7: 1, pp. 39–59.

[2] Yi Liu, Yulin Feng, Hui Zhang, Rui Yang, Lili Zheng, Study on Multi-dimensional Scenario-space Method for Case-Based Reasoning, The 10th International Conference on Cybernetics and Information Technologies, Systems and Applications, July 9–12, 2013, Orlando, Florida, USA.

Using the method above, we can build a new case library in which the cases are split into well quantified dimensions.

2.2 Scenario extracting from the case library

By defining the multi-dimensional space, the case libraries can be transformed as a well quantified database for future use. However, new problem of the database is that as the cases becoming more and more complex, the databases can be very huge.

To split the case into small scenarios is necessary. For a fire emergency case, the measures taken may be decided on the natural information, the severity of the fire, the basic information of the oil tanks, etc. Each can be defined as a small scenario extracted from the specific cases. By defining the scenarios, the solutions for a case can also be described as a composition of solutions to each scenarios, which leads to more convenient since the searching for similar scenario process is much more efficient than the process of searching for similar cases.

One of the advantages of the multi-dimensional space system is that we can easily define the scenarios by selecting necessary dimensions together, i.e. the scenario is formed as a combination of several dimensions. Therefore, the scenario is also defined as a point in a multi-dimensional space, which is defined as the Multi-Dimensional Scenario Space system (MDSS) [3]. The similarity assessment method for MDSS system is developed below.

3 SIMILARITY ASSESSMENT METHOD

3.1 Existing weighing sum method

Similarity assessment method, as a key point in the CBR methods, has been studied by many researchers. The most frequently used method is the 'Weighing Sum Method'. In this method, the similarity of each dimension of the scenarios is derived first (here denoted as SIMs). The overall similarity between two scenarios can be a weighting sum of the similarities of the different dimensions. The analytic hierarchy process (AHP) method and the Delphi method are commonly used for deriving the weights of each dimension.

When calculating the weighting sum, the sum can be in the form of a t-norm which is expressed as below:

$$Similarity = (\sum_k w_k \times SIM_k^t)^{1/t} \qquad (2)$$

[3] Yi Liu, Yulin Feng, Hui Zhang, Rui Yang, Lili Zheng, Study on Multi-dimensional Scenario-space Method for Case-Based Reasoning, The 10th International Conference on Cybernetics and Information Technologies, Systems and Applications, July 9–12, 2013, Orlando, Florida, USA.

Table 1. Scenario formation and weight definition.

	d_1	d_2	d_3	d_4	d_5
weight	0.05	0.10	0.15	0.30	0.40
S_0	0.5	0.5	0.5	0.5	0.5
S_1	0.5	/	/	/	/
S_2	/	/	0.5	/	1
S_3	0.4	0.4	0.6	/	/

* The symbol '/' means that this scenario doesn't have information for this dimension.
** The values of the dimensions are normalized, i.e. the value are in the range of 0–1.

Another commonly used form is expressed as below:

$$Similarity = 1 - [\sum_k w_k \times (1 - SIM_k)^t]^{1/t} \qquad (3)$$

Where similarly = the overall similarity between two scenarios; w_k = weight of each dimension in the unsolved scenario; SIM_k = similarity of each dimension between the unsolved scenario and the existing scenario; t = the order in norm.

For both Eq. (2) and (3), only the similarities of the common dimensions are calculated, other dimensions are omitted from calculation. The latter form is to define the differences of each dimension first, then calculate the overall differences. Though the two forms are calculated in different ways, both of the two forms can be seen as a weighing sum of the similarities of each dimension.

The weighing sum method has been studied by many, either in how to define the weights of each dimension, or in defining the form of the overall similarity process, and some very complex models have been develop inorder to make the calculation result more precise [4].

One fundamental requirement for this method is that the scenarios should have enough common dimensions for the calculation to be effective. But when it comes to the similarity estimation of two scenarios, which don't have enough common dimensions, the results can be tricky using Equation (2) or (3).

Below is an example indicating the shortcomings of Equation (2) and (3). In this example, the unsolved scenario (Denoted as S_0) has 5 dimensions (Denoted as d_1, d_2, d_3, d_4 and d_5, S_1, S_2, S_3, are three scenarios from the MDSS system. The values and the weights of the dimensions are listed below:

Comparing these scenarios. We can see that scenario S_3 is more similar to scenario S_0 than S_1, S_2 because S_3 and S_0 have more common dimensions and the values in the dimensions are of fewer differences. Though S_2 has got a dimension d_5 in common with S_0, which has a weight of 0.4. But the values are far different which may have no reference value at all.

[4] Li, F. 2007. Research on Intelligent Decision Techniques Based on Optimized Case-Based Reasoning. Doctoral Dissertation, HeFei University of Technology.

Table 2. Results of Similarity Calculation.

	S_1	S_1	S_3
Formula (2)	0.22	0.50	0.49
Formula (3)	1.00	0.68	0.94

Here we define the similarity measure of each dimension as:

$$SIM_k = 1 - (X_{0k} - X_{ik}) \qquad (4)$$

Apply Eq. (2) and (3) separately to calculate the similarity of S_0 and the others. The results are as below:

When applying Eq. (2), one can find that the result is greatly dependent on the weight, which makes S_2 has a bigger similarity than S_3. When Eq. (3) is used, S_1 is calculated as most similar, which is far beyond our expectations.

Therefore, new similarity assessment methods need to be developed to get more precise results.

3.2 New similarity assessment method with MDSS

The number of common dimensions becomes an important factor in defining the similarity when the number of common dimensions is very small. Scenario with only one dimension can be of little reference value though the unsolved scenario and the existing scenario shares the same value of dimension. Some thresholds are considered in this method for better accuracy.

3.2.1 Definition of parameters
The parameters used in this method is listed below:

a) Distance of dimensions (D_{si})

In the MDSS system, scenarios are clarified by dimensions. Since each scenario is described as a point in the MDSS system, calculating the distance between two scenarios is more convenient. The values of each dimensions in MDSS system is well quantified, which simplifies the calculation. In addition, the normalization of each dimension is also essential since the ranges of the dimensions may vary a lot.

As stated above, the scenarios are defined as points in the MDSS system. The distance between two points can be easily calculated as below:

$$Ds_i = (X_{0i} - X_{ki})/(Max_i - Min_i) \qquad (5)$$

where X_{0i} = value of the ith dimension of the unsolved scenario; X_{ki} = value of the ith dimension of the kth scenario in the MDSS system; Max_i, Min = the maximum and the minimum value of the ith dimension.

Here the denominator is used for normalization.

b) Number of common dimensions (Count)

The common dimensions refer to the ones which appears both in the unsolved scenario and the existing scenarios. To be more precise, the word 'common' needs to be redefined with thresholds.

For a dimension whose values vary from 0 to 1, the threshold (d_i) can be set as the difference between two values shouldn't be more than 0.5 or this dimension won't be counted. The threshold of each dimension can be defined separately. When the values in the dimensions are normalized, the thresholds can be presented as percentages. If the threshold is regarded as a global variable, different d_i should be the same.

Thus, the parameter Count should be regarded as the number of dimensions which is not only shared by the unsolved scenario and the existing scenario, but also the difference between the two values shouldn't exceed the threshold set before.

c) Weights of dimensions (w_i):

Weights of each dimension is still used, but not just for calculating the weighing sum. The sum of weights of common dimensions is also playing an important role.

If the summation of weights of a scenario is too small, the scenario should also be omitted. When defining the threshold, some factors should be taken into consideration. First is the requirement of accuracy. The threshold should be bigger to be more accurate. Second is the number of existing scenarios in the MDSS system. If there are not enough scenarios, the threshold cannot be to larger, or only few scenarios can be found.

The effect of the weights of common dimensions can be set as defining a new function $I_{SW>T}$, which is set as below:

$$I_{SW>T} = \begin{cases} 1 & , \text{ if } SW > T \\ 0 & , \text{ otherwise} \end{cases} \qquad (6)$$

Where SW = summation of weights of common dimensions; T = the relative threshold below which the scenario will be omitted.

3.2.2 Definition of the method
In the MDSS system, the overall similarity is expressed as a function inversely related to the distance between two points. The distance of two scenarios can be defined as a weighing sum of the common dimensions, which is similar to existing methods. But in the MDSS system, the dimensions are quantified more efficiently, so the calculation process can be efficient. The distance between two scenarios can be expressed as below:

$$Distance = \left(\sum_k w_k \times Dis_k^2 \right)^{1/2} \qquad (7)$$

The '2 − norm' form (i.e. the Euclidean distance) is used for calculating the distance. Other orders of norm can also be used here, such as the '1 − norm' form '∞ − norm' form, etc.

Now, we can define the formula for similarity assessment method as below:

$$Similarity = f(Count) * f(Distance) \\ * f(\text{Sum of Weight}) \qquad (8)$$

Here the symbol '*' doesn't simply mean 'Multiple (×)', it's a symbol indicating the composite function of the three parts. The detailed function of the three parts in defined as below:

a) The function f(Sum of weight) has been defined before, which is Eq. (8). Thus:

$$f(\text{Sum of weight}) = I_{SW>T}$$

b) The functions f(Distance) and f(count)

Though *Count* and *Distance* are not exactly independent, the relation isn't strong enough to need a very complex function to combine them together. We can describe the function as below:

$$f(Count) * f(Distance)$$
$$= count^m \times (1 - Distance)^n \quad (9)$$

Since the similarity is positively associated with *Count* and negatively associated with *Distance* (i.e. positively associated with $(1 - Distance)$), the basic bound of the parameters is below:

$$m > 0, n > 0 \quad (10)$$

The parameters can be get using the existing MDSS system. The process can be described as below:

Step 1: Manually set a series of pairs of (m, n). Monte Carlo methods may be used.

Step 2: Randomly select a scenario in the MDSS system.

Step 3: Find some scenarios related to this scenario, these scenarios may not be so similar, but it's okay.

Step 4: Sort the scenarios derived in 'Step 2' by their similarities to the scenario got in 'Step 1'. Here the results are get by manual analysis.

Step 5: Run tests to find possible results for m, n. Find the most appropriate pairs of (m, n). The control variable method can also be used here to simplify the process.

Step 6: Repeat 'Step 2' to 'Step 5', till a most appropriate pair of (m, n) are derived. If not, repeat 'Step 1' to 'Step 5'.

After defining the parameters, the overall equation can be expressed as:

$$Similarity = count^m \times Distance^n \times I_{SW>T} \quad (11)$$

The methods of defining each parameter have been proposed before.

Let $m = 1$, $n = 1$, $T = 0.03$, formula (14) can be expressed as below:

$$Similarity = Count \times (1 - Distance) \times I_{SW>0.03} \quad (12)$$

Using the data in Table to test the accuracy, the results are listed in the table below:

Comparing to Eq. (2) and Eq. (3), this method is much more accurate. The results shows clearly which the most similar scenario is and the least one.

Table 3. New Results of Similarity Calculation

	S_1	S_2	S_3
Eq. (2)	0.22	0.50	0.49
Eq. (3)	1.00	0.68	0.94
Eq. (15)	1.00	1.36	2.82

4 CONCLUSIONS

Case retrieval is a key step in case-based reasoning (CBR). To efficiently retrieval similar cases, a well-defined case base is essential. This paper first proposes a new method of building the case library, naming multi-dimensional scenario space (MDSS) system which is well quantified. The cases are described in this system for future study. To simplify the process of similarity assessment, scenarios are extracted from the system by combining relevant dimensions together.

The similarity assessment method for the MDSS system is developed in this paper. In this method the number of 'common' dimensions and some thresholds are applied to filter scenarios of little reference value, thus makes the results more accurate. Methods of defining each parameter is also presented in this paper in detail. The example shows that the method developed in this paper provides more accurate results in similarity assessment in CBR.

Both the methods of developing a MDSS-based case library and the similarity assessment method can be applied in many areas for future studies.

ACKNOWLEDGEMENT

This work is funded by the National Natural Science Foundation of China (NO.91024032, NO.91224008, NO.70601015, NO.70833003), the Project of the National Science & Technology Pillar Program (NO. 2011BAK07B02), and the Project of the Key Laboratory of Firefighting and Rescuing Technology of Ministry of Public Security (NO. KF2011002).

REFERENCES

Aamodt, A., & Plaza, E. (1994). Case-based reasoning: Foundational issues, methodological variations, and system approaches. *AI communications*, 7(1), 39–59.

Behbahani, M., Saghaee, A., & Noorossana, R. (2012). A case-based reasoning system development for statistical process control: Case representation and retrieval. *Computers & Industrial Engineering*.

Bergmann, R., & Gil, Y. (2012). Similarity assessment and efficient retrieval of semantic workflows. *Information Systems*.

Li, F. 2007. Research on Intelligent Decision Techniques Based on Optimized Case-Based Reasoning. *Doctoral Dissertation*, HeFei University of Technology.

Yi Liu, Yulin Feng, Hui Zhang, Rui Yang, Lili Zheng, Study on Multi-dimensional Scenario-space Method for Case-Based Reasoning, *The 10th International Conference on Cybernetics and Information Technologies, Systems and Applications*, July 9–12, 2013, Orlando, Florida, USA.

Information Technology and Computer Application Engineering – Liu, Sung & Yao (Eds)
© *2014 Taylor & Francis Group, London, ISBN 978-1-138-00079-7*

Simulation analysis and optimization on unidirectional wheel vibration of McPherson front suspension with ADAMS software

Guang Shuo Xin, Wei Zhou & Yun Zhou

East China University of Science and Technology, Mechanical Engineering Department, Shanghai

ABSTRACT: This research models multi-body dynamic assembly of Macpherson front suspension with Adams software based on the parameters of Santana and in accordance with regular rules sets up parameters of front suspension and motion. Simulation analysis is conducted on unidirectional wheel vibration and in line with simulation results the changing law of orienting key suspension in the proceeding of wheel vibration is analyzed. Evaluate existing rationality and corresponding deficiencies of set suspension, if any, judging from its parameter variation range and accordingly set forth certain suggestions and improvements on optimization. Re-utilize module Adams/Insight to ameliorate the hard point value which has impacts on suspension performance. And then observe the effects to wheels caused by substituting elastic constraining elements for rigid ones.

1 GENERAL INSTRUCTIONS

This research sets the front suspension of Santana as prototype and simulates the McPherson front suspension with Adams software. Systematical parameter setting and simulated analysis of unidirectional vibration will be done before optimizing the suspension parameter by module Adams/Insight.

2 BRIEFING ON SIMULATION SOFTWARE

Adams/Car can be applied to establish and test both subsystem and assembly system, also practicable for module Insight to design complicated simulation tests for mechanical performance evaluation. Combined Insight and Car modules are capable of analyzing feature optimizing and sensitivity of front suspensions.

3 ESTABLISHED MODEL OF MCPHERSON FRONT SUSPENSION AND SIMULATION ANALYSIS

3.1 *Modeling on simulated prototype of McPherson independent front suspension [1]*

Select the template of simulated prototype of McPherson independent suspension in the Adams/Car data base, build up its subsystem by using the template. Add the suspension subsystem and test board with front steering system to the McPherson assembly system. The model is shown below:

3.2 *Setting suspension parameters*

This research exemplifies Santana 2000-GSi-1.8-MT' free boiling point (state standard II) as full vehicle assembly for parameter setting.

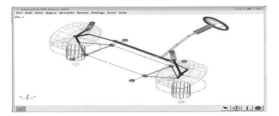

Figure 3.1. model of front suspension assembly system.

The Unloaded Radius: type of tires is 195/60R14, and the unloaded radius is calculated as: $14 \times 25.4/2 + 195 \times 0.6 = 294.8$.

The stiffness: it is calculated that when the tire pressure is between 210 kpa and 250 kpa (common referred as 2.1 catties of air pressure and 2.5 catties of air pressure), the radial stiffness ranges from 190–210 (unit: N/mm). The set stiffness in this research is 200.

Wheel Mass: Most tires weight from 5–15 kg each, choose the intermediate value 10kg in this research

Sprung Mass: generally the sprung mass is taken 90% of the entire car weight which is 1089 kg after calculation

GC height: the GC height is calculated approximately 700 mm according to the body height.

Wheel base: 2656 mm

Drive ratio: This car is a front drive vehicle, and its drive ration is 100%.

Brake ratio: Calculated with correlated brake data, the front to back brake coefficient ratio is 0.67.

Make both right and left wheels vibrate up and down to simulate the common work condition with a scope of −50~50 mm within the weight limitation for sprung mass when the automobile is fully loaded.

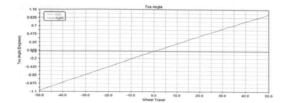

Figure 3.2. Changing curve of toe-in angle with wheel vibration.

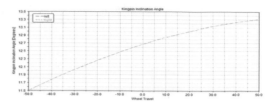

Figure 3.4. Changing curve of king pin inclination angle with wheel vibration.

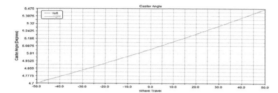

Figure 3.3. Changing curve of king pin caster angle with wheel vibration.

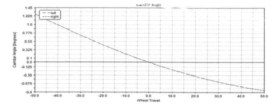

Figure 3.5. Changing curve of front-wheel camber angle with wheel vibration.

3.3 Two-wheeled unidirectional vibration simulation analysis

Start the bidirectional vibration simulation, and analyze the variation tendency in wheel alignment parameters and its effects on suspension performance.

(1) Toe-in angle: excessive variation ranges will add rolling resistance between tires and the ground surface and in turn simultaneously influence on straight line driving stability. Thus the designing principle applied is to minimize the variation range during wheel vibration. The picture below illustrates the changing curve when front toe-in angles vibrate in pace with wheels in the same direction. Judging from the figures listed in the picture we can conclude that, when wheels vibrate in the same direction to 50 mm, the toe-in angle varies from 1.0°–1.0375°, larger than the ideal variation range. Furthermore, in the aforesaid wheel vibration proceeding, the rake ratio is supposed to be negative.

(2) King pin caster angle: King pin caster angle can form stable aligning torque to restrain the braking nod. But excessively large king pin caster angle may cause hard steering. Seen from the picture below, the king pin caster angle ranges from 4.71°–5.475° compared with the variation value at static balance(the value of king pin caster angle at static balance is 5.049°) being −0.34°–0.426°, the variable quantity is 0.765° relatively larger than the value at static balance.

(3) King pin inclination angle: king pin inclination angle can lessen king pin offset and in turn alleviates the impact force transmitted from deflecting roller to steering wheel. In the process of wheel vibration, possible excessively large variation range of the king pin inclination angle may lead to hard steering which results in tire wear.

Telling from the picture, it is evident that the king pin inclination angle ranges from 11.5°–13.31°, a variable quantity of −1.168°–0.630° compared with the value at static balance(the value at static balance is 12.68°). The overall ranging scope is slightly larger being 1.798°.

(4) Front-wheel camber angle: Front-wheel camber angle can locate the direction and guarantee the automobile is driving along the straight line. When fully loaded, the axle will distort in a certain degree due to the weight, which cambers the wheels inwards. In addition, the vertical counterforce passing through road surface to wheels along axial force of the wheel hub makes the hub fall to the small bearings of outer end, which increase load bearing on the outer small bearings as well as clamp nuts. Therefore, to even tire wear and reduce weight the bearings uphold outside the wheel hub, a certain camber angle must be added to the wheels in advance. In the same time, wheels fits more to arched road surface with camber angles. However, variable range of the camber angle is not appropriate if large beyond limitation, otherwise uneven wear may be appear in wheels. The picture below shows the changing curve of king pin inclination angle with wheel vibration. We can tell from it that the variable quantity of the king pin camber angle is −0.744°–1.281° with the overall variation being 2.028°. It's a slightly big figure and optimization should be applied.

(5) Suspension stiffness: In the process of wheel upper vibrating to 50 mm, the suspension stiffness varies a little and is steady in driving. But in the lower vibrating to 50 mm, the stiffness strengthens in a short period due to the buffering stopper.

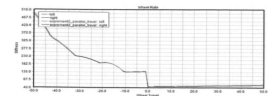

Figure 3.6. Changing curve of suspension stiffness with wheel vibration.

4 OPTIMIZATION AND ANALYSIS OF SUSPENSION PERFORMANCE PARAMETER

4.1 *Confirmation on optimized variation and objectives*

After data-analyzing the parameters, select the five aligning parameters in module Insight as optimized objectives, namely front-wheel camber angle, front-wheel toe-in angle, king pin inclination angle and suspension stiffness. Afterwards choose the suspension hard point parameter: front hinge point, spherical hinge point and back hinge point of the lower swinging arm, outer and inner supporting points of steering tie rod as design variables to conduct analysis and optimization.

4.2 *Adams/Insight applied in optimization*

After establishing the optimized objectives, select 15 coordinate values of 5 points as independent variables for analysis with each value ranging within 10%. Set up systematic work matrix for module Insight iterative calculation in 215. After the calculation, the interactive webpage will save the optimized results as dynamic data by native functions of Insight. Later the interactive webpage is utilized for parameter sensitivity analysis of independent variables. The analysis results are as follows:

Select 9 hard point values that have major impacts on target variables to re-optimize through 512 times of partial iterative calculations until the optimized dependent variables are gained. Modify the initial hard point coordinates, then carry out two-wheeled unidirectional vibration experiment, and compare the consequences with pre-optimized results for analysis.

4.3 *Comparisons between pre-optimized results and post-optimized results*

Toe-in angle before and after optimization: as is shown in the picture, the variable range of toe-in angle is conspicuously lessened in unidirectional vibration after Insight optimization with the value of variation being 0.9°. And in the upper vibration the toe-in angle becomes toe-out angle which transfers the steering characteristic of the automobile.

King pin caster angle before and after optimization: telling from the picture that the varied quantity of king pin caster angle is about 0.65°. This value is respectively smaller than that before optimization and the

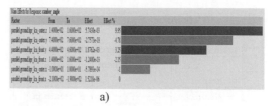

a)

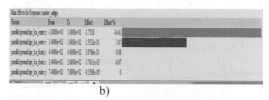

b)

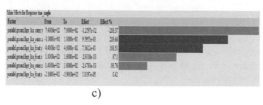

c)

Figure 4.1. Pensitivity analysis on interactive web pages.

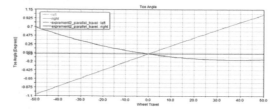

Figure 4.2. Hard point modification values.

Figure 4.3. Toe-in angle before and after optimization comparison.

value at balance is reduced hence reinforce steering stability of the automobile.

King pin inclination angle before and after optimization comparison: the value at balance of king pin inclination angle varies after optimization with the varied rang being 0.8°. Such alternation lightened the steering operation.

Front-wheel camber angle before and after optimization: Judging from the figure that the varied quantity of wheel camber angle is −0.518°−1.125°. As is shown in the picture, the variation decreases than that before optimization but not close enough to the ideal expectations. But such alternation is so far the

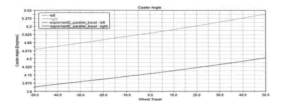

Figure 4.4. King pin caster angle before and after optimization comparison.

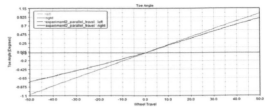

Figure 5.1. Toe-in angle comparison in kinematics state and complying state.

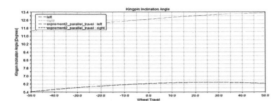

Figure 4.5. King pin inclination angle before and after optimization.

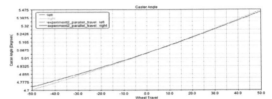

Figure 5.2. King pin caster angle comparison in kinematics state and complying state

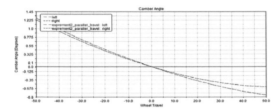

Figure 4.6. Front-wheel camber angle before and after optimization comparison.

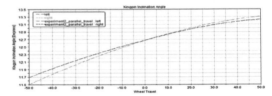

Figure 5.3. King pin inclination angle comparison in kinematics state and complying state

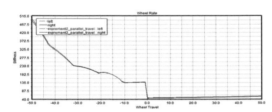

Figure 4.7. Suspension stiffness before and after optimization comparison.

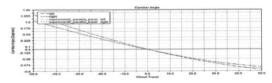

Figure 5.4. Front-wheel camber angle comparison in kinematics state and complying state

optimal result taking the optimized variable limitation into consideration.

Suspension stiffness before and after optimization comparison: decreasing tendency of the suspension stiffness is improved in a narrow range.

5 SIMULATION ANALYSIS ON ELASTIC KINEMATICS OF MCPHERSON SUSPENSION

5.1 *Simulation analysis on McPherson suspension in motional state and elastic kinematics*

Stiffness constraint is needed in kinematics analysis in module Adams, while in the complying analysis, one should transfer the analyzing mode from kinematics to complying state and then conduct simulation experiment by setting original parameters. After the simulation, draw the two results in the reprocessing window simultaneously and make comparison by analysis.

(1) Toe-in angle: varied range decreases to a certain amount in elastic simulation, the suspension performance is optimized.
(2) Kin pin cas gter angle: the varied range of king pin caster angle is not considerable which explains that elastic constraint is not influential to king pin caster angle
(3) King pin inclination angle: the varied range of king pin inclination angle is not considerable which explains that elastic constraint is not influential to king pin inclination angle.
(4) Front-wheel camber angle: Partially decline in the varied range of front-wheel camber angle can effectively prevent uneven wheel wear.

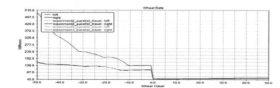

Figure 5.5. suspension stiffness comparison in kinematic-sstate and complying state.

(5) Suspension stiffness: The initial suspension stiffness decreases due to elastic constraint. However, it drops dramatically in lower 50 mm vibration.

6 CONCLUSION

Wheel unidirectional vibration simulation in McPherson suspension is carried out with Adams/Car software. Insight is applied in optimizing hard point coordinates of suspension which decline impacts wheel vibration lays on wheel aligning parameters. In the meantime, the kinematics characteristics of suspension are optimized by replacing stiffness constraint with elastic components.

REFERENCES

[1] Fan Chengjian, Xiong Guangming, Zhou Mingfei Application and Improvement in Stimulated Prototype Software MSC. ADAMS[M], Beijing: Mechanical Working Press, 2006, 287–292.

[2] Ge Feng, Qu Xuechun, Yang Shenghui, Chen Dingkun, Zhu Yuan Relation between Tire Pressure and Automobile Performance and Optimal Choices[J], Academic Journal of Military Communications Institute, 2009, 11(1): 54–55.

[3] Fang Fei Investigations into Operational Stability on Automobiles with McPherson Independent Front Suspensions[D], Wu Han: Wu Han University of Technology, 2006.

[4] Chen Jun Technological and Engineering Case Analysis[M], Beijing: China Water Conservancy and Hydropower Press, 2008, 155–156.

[5] Mao Nanhai, Ning Xiaobin, Xie Weidong Kinematics Stimulation and Optimization of Automobile Front Suspensions Based on Adams[J], Mechanical Manufacturing, 2010, 48(533):12–15.

Information Technology and Computer Application Engineering – Liu, Sung & Yao (Eds)
© 2014 Taylor & Francis Group, London, ISBN 978-1-138-00079-7

A study on the impact of e-HR on Chinese firms

X.L. Wen & C.H. Chen
School of Business Administration, South China University of Technology, Guangzhou, China

ABSTRACT: Information Technology (IT) provides Human Resource (HR) functions with the opportunity to create new avenues for contributing to organizational success. Electronic Human Resource (e-HR) is viewed as an umbrella term covering the integration of IT and HR, aimed at creating value for targeted firms. This paper firstly introduced the principle of e-HR, discussed the IT research and e-HR. Then, this study identified the challenges associated with the implementation and maintenance of e-HR in Chinese firms. By the case study method, the paper also attempted to understand the extent of e-HR practiced in China by investigating the impact of e-HR on Chinese firms from several important aspects. Lastly, some suggestions to improve the functions of e-HR in Chinese firms in the future were given.

Keywords: Information Technology (IT); Electronic Human Resource (e-HR); Chinese firms

1 INTRODUCTION

With the latest advanced information technology (IT) offering the potential to streamline many human resource (HR) functions, a lot of firms have been utilizing IT to design and deliver their HR practices. HR is traditionally considered as the planned and consistent approach to the management of firms for the most valued possessions of firms: human capital. It employs people, develops their capacities, utilizes, maintains and compensates their services with the key job requirement (Bondarouk & Ruel, 2009).

It's not surprising that traditional HR processes have been propelled into some entirely new directions in digital age's global networking timeframe. IT has helped modify many HR processes including human resource planning, recruitment, selection, performance management, work flow, training, and compensation. Moreover, most of the large firms have used Web-based recruiting systems, and implemented Web-based training programs in the digital era. These new systems have enabled HR professionals to provide better service to all of their stakeholders and reduced the administrative burden in different kind of firms (Stone & Dulebohn, 2013).

In China, there is also obvious evidence showing that HR technology budgets in Chinese firms and sales of the HR technologies were increase fast over the last decade. Although administrative e-HR was still the most popular application, Chinese firms have broadened the scope of HR applications. Firms reported an increasing use of strategic applications like talent acquisition services, performance management, or compensation management. Financial investments in e-HR in 2011 already showed a slight preference towards strategic applications. Chinese firms

estimated their investments in simple management reporting tools as no less than 5% for an up-coming year, while they planned to increase their budgets, for example, in career planning tools and competence management more than ever before.

Based on the review of some key aspects of the e-HR, this study firstly hopes to identify the challenges associated with the implementation and maintenance of e-HR in Chinese firms. Secondly, we will explore the extent of e-HR practiced in Chinese firms, including large state-own enterprises, small and medium sized private firms, etc. The essay will focus on several main areas of e-HR, which are believed to have a significant impact on the competitiveness of firms, such as: e-recruitment, e-training and development, e-performance appraisal, and communication, etc. It is believed that the feasibility of implementing e-HR is very much dependent on the national characters, so the researchers also discuss the main issues that could captured the changing trends of HR environment and practices in China in a timely manner.

2 PRINCIPLE OF E-HR

2.1 *Conception of e-HR*

Researchers have not standardized a definition of e-HR yet. Different perspectives (IT and HR) fall under a common label, despite there being no common terminology set in which to create and test ideas, constructs, or concepts. Strohmeier(2007) defined e-HR as the (planning, implementation and) application of IT for both networking and supporting at least two individual or collective actors in their shared performing of HR activities. Bondarouk & Ruel(2009) defined e-HR as an umbrella term covering all possible integration mechanisms and contents between HR and IT aiming at creating value within and across organizations for targeted employees and management. It's believed that this definition suggests an integration of four aspects of e-HR: content, implementation, targeted employees and managers and consequences.

2.2 Information technology research and e-HR

E-HR practitioner and suppliers often assert that internet-based technological innovations are important in realizing the outcomes of successful e-HR implementation. Marler & Fisher (2013) have argued three perspectives on technology in e-HR research: technological determinism, organizational imperative, and technology as a process. In the first and earliest stream of literature, IT is seen as a distinct entity that interacts with various aspects of the organization. IT is an independent variable having a range of effects at different levels of analysis on multiple organizational outcomes as the dependent variable. This research stream takes a deterministic perspective, in that IT is a causal factor that is expected to create predictable, theoretically-determined consequences. The second broad stream of research focuses on dynamic interactions between people and IT over time. It is less deterministic, viewing IT as part of a complex process through which organizing is accomplished. IT is no longer a discrete entity and thus no longer a quantifiable independent or dependent variable. Instead, IT is emergent and not fully determinate. In the third theoretic paradigm, IT is not an objective external force that has deterministic impacts on organizational structure, but is an entity, resource or an organizational process that emerges from operating in a particular organizational context.

2.3 The impact of e-HR on the human resource management function

E-HR enable managers to access relevant information and data, conduct analyses, make decisions, and communicate with others even without consulting an HR professional. Generally, as might be expected, e-HR systems are thought to provide a number of key benefits to firms, such as (a) enhance HR efficiency, (b) reduce costs, (c) decrease administrative burdens, (d) facilitate HR planning, and (e) allow HR professionals to become a strategic or business partner in firms. On the other hand, in spite of these benefits, there are concerns that e-HR focus primarily on efficiency and cost containment rather than enhance the effectiveness of HR processes (e.g., selection systems). So, e-HR may potential to affect effectiveness of human resources function: (a) have an adverse impact on members of some protected groups (e.g., older job applicants), (b) have the potential to invade personal privacy. Given these potential benefits and unintended consequences, it is needed to enable e-HR professionals to design systems that meet the goals of firms and all stakeholders (Mark & Steve, 2003).

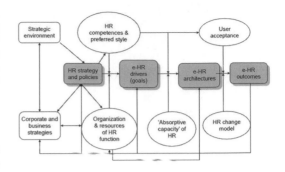

Figure 1. A model of e-HR.

2.4 A model of e-HR

Martin & Reddington (2010) produced a model explaining the links between HR strategies, e-HR strategic drivers, e-HR technologies and e-HR outcomes (see Figure 1). In the model, the HR strategies and policies of an organization interact with its strategic environment, corporate and business strategies, often in a complex, recursive manner. Outside-in approaches to strategy stress the linear and hierarchical relationships in which HR strategies are deemed to flow from key corporate and business strategies as second or even third order strategies.

3 CHALLENGES OF E-HR IN CHINESE FIRMS IN DIGITAL AGE

Led by rapid e-HR developments in the professional HR world, Chinese firms have been accelerating their attempts to understand the phenomenon of e-HR and its multilevel consequences within and across firms. According to Varma & Gopal (2011), some of the major challenges for implementing e-HR have been identified (see Table 1). So, before embarking on an e-HR journey, Chinese firms should answer strategic questions to adopt these new challenges. Especially for those traditional Chinese firms, the challenges may be more serious. These firms as compared to those in the IT sector are having a tougher time getting people to speed up on a new e-HR system.

4 THE PRESENT AND FUTURE STATUS OF E-HR IN CHINESE FIRMS

In the last decade, the application of IT to HR in Chinese firms is believed to change the role of the HR function; while at the same time, HR allows technology to develop its full potential. On the other side, due to the increasing market demand, HR software have become the second largest management software in Chinese market(just after financial software). There are more and more e-HR software suppliers in China, such as foreign software suppliers-SAP, ORACLE and domestic companies – YONYOU, KINGDEE and EASTERSOFT, etc. Thus, to what extent has e-HR already been implemented in Chinese firms?

Table 1. Challenges of e-HR Chinese firms faced.	
Type of challenge	Description
1. Cost implications	Chinese firms must e-enable only those operations that are vital, essential or desirable.
2. Aligning the e-HR system with the business requirements	Management must consider the e-HR stance on control to ensure it will meet the business requirements of the company.
3. Security of the information generated	A Chinese firm needs to ensure that outsiders or competitors should not access the information.
4. Managing the data	Managing the huge amount of data generated through e-HR is a relatively new challenge for Chinese firms.
5. 'Overkill' and loss of the 'human touch'	It should not be the case that in a bid to be techno-savvy Chinese firms neglect the human side.
6. Customization to be taken up in the right perspective	Customizations can also be costly and maintaining and upgrading customizations can be cumbersome.
7. Training the users a crucial issue	Training the users is many a time a long drawn out process, as many people do not find them to be user friendly.
8. The Return on Investment (ROI) on an e-HR project to be justified	The e-HR initiative should align itself with the overall HR and IT strategy and ultimately, with the business strategy to ensure ROI in Chinese firms.
9. E-HR to function along with other systems to be successful	Firms wanting to integrate their value chains with the business activities of their suppliers, business partners and customers typically have to implement systems other than e-HR.
10. Continuous monitoring and feedback	Continuous monitoring and feedback are critical for the success of any e-HR in Chinese firms.

Table 2. A typical sample of Chinese firms which implemented e-HR.	
Facet of e-HR	Content
E-recruitment	An online portal is used in Company A to attract potential candidates to join the organization all over China. People first must upload their CV online. After this stage, depending on the position, applicants must take electronic tests to ensure they are appropriate for the job opening. Company A's electronic recruitment process also facilitates the recording and monitoring of personnel data.
E-training	Company A is an organization with multiple different locations nationwide, so it uses electronic forms of training. It helps to standardize training options in all branches. One of the programs is the "Training and Development Program" and it is done online. It's reported that the most suitable trainings in the company for online training are theoretical or instructional programs.
E-learning	E-learning is also used to teach some personal skills without the need of an instructor or a physical session. For instance, Company A uses audio books and self-learning software commonly in order to teach employees presentations skills or how to approach audiences if they have to give speeches.
E-performance Appraisal	Digitalized forms of performance evaluations are implemented in Company A to monitor employees. These forms help the company to keep a measure of performance of its staff and the different departments within the organization. The company's electronic performance appraisal system is reliable and enables management to give evaluations in a faster way.

4.1 A typical sample of Chinese firms which implemented e-HR

Company A, founded in 1999, is a Chinese Internet-based businesses Group which makes it easy for anyone to buy or sell online anywhere in the world. Since its inception, it has developed leading businesses in consumer e-commerce, online payment, business-to-business marketplaces and cloud computing, reaching Internet users in more than 240 countries and regions. Company A consists of 25 business units and is focused on fostering the development of an open, collaborative and prosperous e-commerce ecosystem.

In 2005, company A began implementing components of an e-HR system (named People Soft Enterprise Human Capital Management) supplied by Oracle to build a unified electronically human resources management platform within the Group people. Since then, surveys have shown that the e-HR systems introduced in company A improve efficiency and service delivery, increase the strategic orientation of the HR function, improve standardization and organizational image and empower managers (see Table 2).

4.2 The present status of e-HR in Chinese firms

The other typical samples of Chinese firms which implemented e-HR, such as China Vanke Co.,

Ltd, which introduced a SAP e- HR system since 2005,

Fujian Nanping Nanfu Battery Co., Ltd, which used a YONYOU e-HRM system since 2007, and Suzhou Huasu Plastics Co., Ltd.etc, which adopted a Kayang e- HR system in 2010, include large state-own enterprises, small and medium sized private enterprises. It is also suggested clearly that there are six main goals (no matter realization or not) for these firms introducing e-HRM: a) operational and efficiency, b) service delivery) manager empowerment; d) strategic orientation; e) standardization and f) organizational image. It is seemed that the characters of firms, such as size or ownership don't have influence on the adoption of HR practices in Chinese firms.

4.3 *The future directions of e-HR of Chinese firms*

Are Chinese firms keeping pace with advances of advanced techniques? Is the present status of e-HR implemented in Chinese firms reflecting the changing trends in the HR environment and practices in China? What are some of the areas Chinese firms should focus on in the future? Based on the research of Cooke(2009), there may some important issues that practitioners and scholar should pay much more attention: (1) the changing role of the HR function; (2) strategic role of HR and firm performance; (3) work-life balance, diversity management, employee well-beings, and employee engagement; (4) variable pay systems and stock options; (5) single HR issues; (6) employees' voice and representation, union officials' work; (7) the role of HRM in corporate social responsibility; (8) HR of Chinese MNCs (multi-national corporations).

5 DISCCUSION AND CONCLUSION

In the new digital age, Chinese firms are exposed to the effects of IT changes and there is no choice to stay immune to these changes. IT is responsible for dramatic changes through such revolutionary concepts like e-HR. This paper gives some samples of Chinese firms which facing challenge of e-HR to achieve their targets.

Any e-HR installation keeping all the challenges in mind can take a Chinese firm a long way towards success. Increased transparency in functions and a total systems approach of e-HR has facilitated better control by top management. In the future, for Chinese firms, firstly, HR practitioners must play a down to business

role, to create an effective e-statement, standardizing and centralizing HR administration in a fast response service center, assessing and ensuring the flexibility of the e-HR technology tools. Secondly, the realization of improved efficiency and effectiveness in Chinese firms is dependent on the design and implementation of the system, and increased effectiveness and involvement in delivering the business strategy may depend on appropriate redeployment and up-skilling of HR staff. Thirdly, to enhance the transformation of traditional HR practices to e-HR in Chinese firms, the government may also have an increasing instrumental role to play.

All in all, this study showed that a large number of Chinese firms are no longer surprised by e-HR and are ready to invest in it further. We see new steps in the practice of e-HR caused by (or due to) recent IT and organizational developments. In short, the transformation of Chinese firms from traditional HR to e-HR in the near future is rather encouraging.

REFERENCES

Bondarouk, T.V. & Ruel, H.J.M. 2009. Electronic human resource management: challenges in the digital era. The International Journal of Human Resource Management 20(3): 505–514.

Cooke. F L. 2009. A decade of transformation of HRM in China: a review of literature and suggestions for future studies Asia Pacific Journal of Human Resources 47(1): 640.

Leda, P. Eleanna, G. & Nancy, P. 2010. Adoption of electronic systems in HRM: is national background of the firm relevant? New Technology, Work and Employment 25(3): 253–269.

Marler, J. H. &Fisher, S. L. 2013. An evidence-based review of e-HRM and strategic human resource management. Human Resource Management Review 23: 18–26.

Mark, L. L. & Steve, M. 2003. The impact of e-HR on the human resource management functiong. Journal of Labor Research 24(3): 365–378.

Martin, G. & Reddington, M. 2010. Theorizing the links between e-HR and strategic HRM: a model, case illustration and reflections. The International Journal of Human Resource Management 21(10): 1553–1574.

Stone, D.L. & Dulebohn, J.H. 2013. Emerging issues in theory and research on electronic human resource management (eHRM). Human Resource Management Review 23: 1–5.

Strohmeier, S. 2007. Research in e-HRM: review and implications. Human Resource Management Review 17: 19–37.

Varma, S. & Gopal, R. 2011. The implications of implementing electronic-human resource management (E-HRM) systems in companies. Journal of Information Systems and Communication 2(1): 10–29.

Information Technology and Computer Application Engineering – Liu, Sung & Yao (Eds)
© 2014 Taylor & Francis Group, London, ISBN 978-1-138-00079-7

A longitudinal voxel-based morphometry study of GM atrophy progression

Li Ho Tseng & Chia Yi Chou
Behavioral and Cognitive Electrophysiology Laboratory, Department of Environmental Science and Occupational Safety and Hygiene, Tajen University, Pingtung County, Taiwan

The Alzheimer's Disease Neuroimaging Initiative
Alzheimer's Disease Neuroimaging Initiative, Indianapolis, Indiana, United States

ABSTRACT: Longitudinal neuroimaging measurements have received substantial attention for their ability to track disease progression. Patients with Mild Cognitive Impairment (MCI), who have been proposed to represent a high risk of developing Alzheimer's Disease (AD). This study examined brain atrophy in patients by using longitudinal MCI data. We hypothesized that the atrophy of the Gray Matter (GM) was affected by the neurodegenerative process; thus, the Voxel-Based Morphometry (VBM) method was applied to analyze GM volumes in 16 Healthy Control (HC) patients, 16 MCI stable (MCI-S) patients and 16 MCI convert patients (MCI-C) who degenerated from MCI to probable Alzheimer's Disease (AD). The results showed that the earliest differences in atrophy progression occur in the right Middle Temporal Gyrus (MTG) in the MCI-C compared to the HC and MCI-S during baseline observations. This suggests that VBM analysis assisted in causing a significant GM atrophy difference in right MTG atrophy in MCI-C patients.

Keywords: Mild Cognitive Impairment (MCI); Voxel-Based Morphometry (VBM); Brain atrophy; Middle Temporal Gyrus (MTG)

1 INTRODUCTION

In the past 10 years, medical professionals have been challenged when using magnetic resonance imaging (MRI) to characterize brain changes and elucidate the organization of brains with lesions. In such cases, VBM is often used [1, 2]. Systematic meta-analyses of VBM studies have identified decreasing GM volumes in AD patients [3]. Voxel-based morphometry is a powerful method for detecting longitudinal brain morphometric changes [4]. In 2006, more than 26.6 million people were affected by AD [5]. Mild cognitive impairment has been proposed as a transitional stage between normal aging and a diagnosis of clinically probable AD [6]. Previous studies have examined brain atrophy associated with progressive diseases [7–10]. Age-associated GM loss occurs predominantly in the frontal, cingulate, insular, and parietal brain regions [11]. Using the VBM method, numerous studies have indicated that MCI and probable AD are associated with the atrophy of GM, primarily in the parahippocampal gyrus, medial temporal lobe, insula, and thalamus [12–14]. Research groups have recently studied AD and MCI from various perspectives, attempting to understand the pathogenesis and discover effective therapies, such as the effect of the *APOE* genotype. A longitudinal study showed that *APOE* ε4 positive HC patients demonstrated an accelerated rate of atrophy relative to ε4 negative participants, but only in hippocampal volume [3].

Gray matter changes identified using VBM remain inadequately understood, especially when used to predict if patients will progress to AD in the future. Voxel-based morphometry analysis is frequently used to examine GM changes for neuropsychiatric conditions such as schizophrenia [15] and dementia [16]. Earlier studies have revealed that hippocampal atrophy is a strong predictor of progression from MCI to AD [17–19]. In this study, we examined the GM atrophy patterns of MCI patients compared to patterns in a cognitively healthy group with a baseline evaluation and follow-ups at 12, 24, and 36 months. We hypothesized that the atrophy of brain regions was affected by the neurodegenerative process; thus, neurodegenerative diseases are associated with atrophy in particular brain regions. These analyses allowed researchers to understand more clearly the atrophy patterns associated with different pathologies, especially in progressive MCI patients However, VBM analysis has seldom been used for longitudinal data in studies on MCI [3, 20]. Therefore, this study provides a comprehensive account of the GM loss patterns in a group of well-characterized patients with MCI compared with healthy aged control patients.

2 MATERIALS AND METHODS

The Alzheimer's disease neuroimaging initiative (ANDI) database (http://www.loni.ucla.edu/ADNI/) provided the patients for this study. Sixteen HC

Table 1.   Demographic characteristics of the participants.

	HC (n = 16)	MCI-S (n = 16)	MCI-C (n = 16)	P-value	Significant pair comparisons
Baseline age	75.3 ± 5.0	74.8 ± 6.0	73.6 ± 9.3	NS	None
Age range	65–85	65–84	55–87		
Gender (male, female)	8, 8	11, 5	11, 5		
Baseline MMSE	30.0 ± 0.0	26.7 ± 1.7	25.4 ± 1.1		HC > MCI-S,
12-month change in MMSE	29.6 ± 0.5	25.9 ± 2.0	23.7 ± 2.0	P < 0.01	HC > MCI-C
24-month change in MMSE	29.6 ± 0.7	25.1 ± 2.4	20.2 ± 3.7		MCI-S > MCI-C
36-month change in MMSE	29.9 ± 0.3	23.2 ± 1.1	14.8 ± 3.9		
48-month change in MMSE	29.3 ± 1.2	22.9 ± 1.5			

Data are given as mean ± standard deviation.
Key: HC, healthy control; MCI-C, converters from mild cognitive impairment to probable AD; MCI-S, mild cognitive impairment-stable; MMSE, Mini Mental State Examination; NS, nonsignificant.

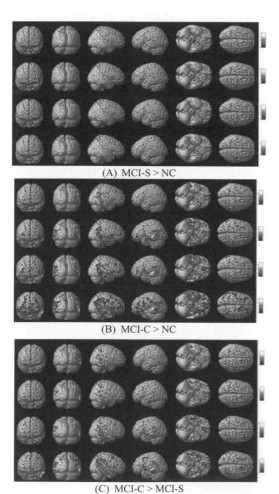

(A) MCI-S > NC

(B) MCI-C > NC

(C) MCI-C > MCI-S

Figure 1.   Group differences in pattern of reduction in GM. Group differences in atrophy progression reflected by reduction in GM density from baseline to 1 year, 2 years, and 3 years (from top to bottom) in the Alzheimer's Disease Neuroimaging Initiative (ADNI) cohort (n = 48; 16 HC, 16 MCI-S, 16 MCI-C); (A) NC > MCI-S (B) NC > MCI-C (C) MCI-S > MCI-C. Interaction contrasts are displayed at a threshold of P < 0.001 (uncorrected) with a minimum cluster size (k) = 25 voxels.

patients, 16 MCI-S patients, and 16 MCI-C patients who converted from MCI to probable AD were included (as defined in ADNI 1). The baseline, 12-month, 24-month, and 36-month follow-up observations were performed for each patient. Table 1 lists the demographic information.

This study included standard T1-weighted MRI scans that were acquired using volumetric 3D magnetization prepared rapid acquisition gradient echo (MPRAGE) data sets. All structural MRI brain scans were obtained from multiple ADNI sites using 1.5 Tesla MRI scanners from General Electric Healthcare and Siemens Medical Solutions. All scans were downloaded in an NIFTI format. For detailed information regarding the MRI experiments, please refer to the ADNI website: (http://www.adni-info.org/Home.aspx).

The functional data were processed and analyzed using statistical parametric mapping software (SPM8; Wellcome Trust Centre for Neuroimaging, London; http://www.fil.ion.ucl.ac.uk/spm; [21]), and all analyses were executed in Matlab. VBM analysis was performed using a VBM8 toolbox (http://dbm.neuro.uni-jena.de/vbm/). Voxel-based morphometry is a powerful method for the noninvasive study of human brain functions. It is an unbiased whole-brain analysis method that assists in examining voxel comparisons of local concentrations of GM [9].

The GM volume of each patient was obtained using the following steps: First, to preserve the brain volume for each patient, longitudinal data were collected at 4 time points, and all MRI were bias-corrected and segmented to extract the GM, white matter (WM), and cerebrospinal fluid (CSF). Second, a high-dimensional DARTEL normalization was used as the default spatial normalization. Gray matter maps were normalized to Montreal Neurological Institute (MNI) atlas space as 1 mm × 1 mm × 1 mm voxels. Segmented images were smoothed using an isotropic 8 mm full width at half maximum (FWHM) Gaussian kernel [22]. Furthermore, Jacobian modulation was applied to the tissue class segments for nonlinear normalization. Third, specific differences in GM among the HC, MCI-S and MCI-C patients were statistically assessed using

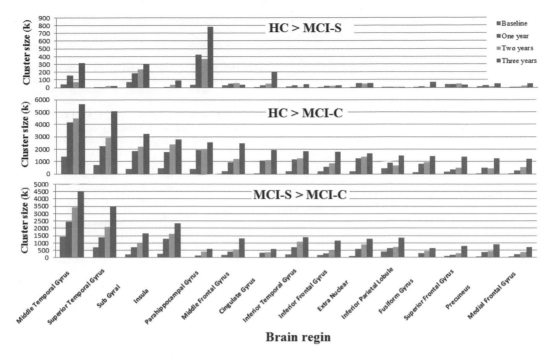

Brain regin

Figure 2.   Group differences in GM reductions of comparisons ($P < 0.01$) between HC, MCI-S and MCI-C. Based on the Talairach Deamon atlas at the level 3 labels, were used to define the regions for analyzing VOIs.

two-sample *t*-tests. Additionally, an absolute threshold mask of 0.1 was used. The family-wise error (FWE) and *P*-value statistics ($P < .001$) were used for the voxel-level analysis. The anatomical brain regions were selected for analysis based on Volumes of Interest (VOIs). Based on the Talairach Deamon atlas, level 3 labels were used to define the regions [23, 24]. A nonlinear transformations approach was used to convert the MNI coordinates to Talairach coordinates. This longitudinal study analyzed and compared the smoothed GM images from these 2 groups.

3   RESULTS

Table 1 shows the demographic characteristics of the HC, MCI-S and MCI-C patient groups. There were no significant differences in age or handedness between the groups. The MCI-C patients scored significantly lower than the HC and MCI-S patients on the MMSE at all stages of life ($P < 0.01$).

The topography of GM atrophy among the baseline, 12-month, 24-month and 36-month observations was revealed in our study, and is illustrated by the surface rendering in Figure 1. The earliest differences in GM atrophy progression occurred in the right MTG in the MCI-C patients compared to the HC and MCI-S patients in baseline observations. The MCI-C patients showed significant atrophy in GM volume in the superior temporal gyrus, sub gyral, insula, parahippocampal gyrus, middle frontal gyrus and cingulate

gyrus regions, which are shown in Figure 2. The MCI-S patients showed that the earliest differences in GM atrophy progression occurred in the parahippocampal gyrus.

4   DISCUSSION

This study investigated whether GM atrophy predicted the outcome in patients with MCI-C. Right MTG atrophy was found to be a significant pathological indicator, even in patients at the earliest stages of MCI-C. However, our finding of significantly greater and earlier right MTG atrophy in MCI-C relative to MCI-S has not been discussed exhaustively [3, 20]. Cabeza et al (2000) showed that the MTG is involved in several cognitive processes including language and semantic memory processing [25]. Other studies have indicated that the MTG is involved in episodic memory performance, including verbal memory encoding [26], recognition retrieval [27], and multimodal sensory integration [28]. Moreover, previous results have shown the obvious GM atrophy in the parahippocampal gyrus of MCI and AD patients [29]. However, relatively few studies have reported parahippocampal gyrus atrophies in MCI-S patients that appeared so early and so extensively. Investigating the GM atrophy effects on HC and MCI-S patients is essential.

Significant progressive right MTG atrophy may indicate the presence of underlying pathologic processes in patients with cognitive impairment.

A plausible explanation is that right MTG atrophy indicates a decline in cognitive and memory functions, thus causing the brain function to deteriorate from a stable state to a relatively unstable state. In addition, MCI-C patients with a decreased GM volume have an increased risk of progression toward dementia. Therefore, the results might indicate that asymmetric atrophies in the progression of MCI-C patients are related to the decline of cognitive functioning

This study was subject to several limitations. First, because of the small sample size these results should be interpreted with caution. Second, the numbers and genders of the MCI patients were inadequately matched. Third, future studies should use MRI acquired using a higher magnetic intensity. These limitations affected the generalizability of our findings.

5 CONCLUSION

A significant difference was noted among clinically well-characterized patients in this study, which enabled examining GM atrophy as patients progressed from MCI-S to MCI-C. We identified atrophy in the right MTG of patients with MCI-C and compared them with HC and MCI-S patients. Substantially more atrophy in the right MTG was present during the baseline observation of MCI-C patients. This study suggests that VBM can be used to map the progression of GM loss in patients with MCI-C. These results provide significant insight into the pathogenesis of MCI.

ACKNOWLEDGEMENTS

The authors are grateful for the Alzheimer's Disease Neuroimaging Initiative (ADNI) database. Data used in preparation of this article were obtained from the Alzheimer's Disease Neuroimaging Initiative (ADNI) database (http://adni.loni.ucla.edu/). As such, the investigators within the ADNI contributed to the design and implementation of ADNI and/or provided data but did not participate in analysis or writing of this report. A complete listing of ADNI investigators can be found at: http://adni.loni.ucla.edu/wp-content/uploads/how_to_apply/ADNI_Acknowledgement_List.pdf.

The authors also would like to thank the National Science Council, Taiwan for financially supporting this study under contract number NSC 101-2410-H-127-001.

REFERENCES

[1] Ashburner, J., Friston, K.J. 2000. Voxel-based morphometry—the methods, *Neuroimage* 11(6): 805–821.

[2] Gaser, C. 2012. Voxel-based morphometry toolbox (VBM8). Available at: http://dbm.neuro.uni-jena.de.

[3] Risacher, S.L., Shen, L., West, J.D., Kim, S., McDonald, B.C., Beckett, L.A., Harvey, D.J., Jack,

C.R., Jr., Weiner, M.W., Saykin, A.J. 2010. Longitudinal MRI atrophy biomarkers: relationship to conversion in the ADNI cohort, *Neurobiology of Aging* 31(8): 1401–1418.

[4] Whitwell, J.L. 2009. Voxel-based morphometry: an automated technique for assessing structural changes in the brain, *The Journal of Neuroscience* 29(31): 9661–9664.

[5] Brookmeyer, R., Johnson, E., Ziegler-Graham, K., Arrighi, H.M. 2007. Forecasting the global burden of Alzheimer's disease, *Alzheimer's and Dementia* 3(3): 186–191.

[6] Gauthier, S., Reisberg, B., Zaudig, M., Petersen, R.C., Ritchie, K., Broich, K., Belleville, S., Brodaty, H., Bennett, D., Chertkow, H. 2006. Mild cognitive impairment, *The Lancet* 367(9518): 1262–1270.

[7] Matsuda, H., Kitayama, N., Ohnishi, T., Asada, T., Nakano, S., Sakamoto, S., Imabayashi, E., Katoh, A. 2002. Longitudinal evaluation of both morphologic and functional changes in the same individuals with Alzheimer's disease, *Journal of Nuclear Medicine* 43(3): 304–311.

[8] Dickerson, B.C., Goncharova, I., Sullivan, M., Forchetti, C., Wilson, R., Bennett, D., Beckett, L., deToledo-Morrell, L. 2001. MRI-derived entorhinal and hippocampal atrophy in incipient and very mild Alzheimer's disease, *Neurobiology of Aging* 22(5): 747–754.

[9] Good, C.D., Johnsrude, I.S., Ashburner, J., Henson, R.N.A., Fristen, K., Frackowiak, R.S.J., A voxel-based morphometric study of ageing in 465 normal adult human brains. Secondary 2002. *A voxel-based morphometric study of ageing in 465 normal adult human brains*, IEEE, p. 16 pp.

[10] Frisoni, G., Testa, C., Zorzan, A., Sabattoli, F., Beltramello, A., Soininen, H., Laakso, M. 2002. Detection of grey matter loss in mild Alzheimer's disease with voxel based morphometry, *Journal of Neurology, Neurosurgery and Psychiatry* 73(6): 657–664.

[11] Resnick, S.M., Pham, D.L., Kraut, M.A., Zonderman, A.B., Davatzikos, C. 2003. Longitudinal magnetic resonance imaging studies of older adults: a shrinking brain, *The Journal of Neuroscience* 23(8): 3295–3301.

[12] Karas, G.B., Scheltens, P., Rombouts, S.A., Visser, P.J., van Schijndel, R.A., Fox, N.C., Barkhof, F. 2004. Global and local gray matter loss in mild cognitive impairment and Alzheimer's disease, *Neuroimage* 23(2): 708–716.

[13] Visser, P.J., Scheltens, P., Verhey, F.R.J., Schmand, B., Launer, L.J., Jolles, J., Jonker, C. 1999. Medial temporal lobe atrophy and memory dysfunction as predictors for dementia in subjects with mild cognitive impairment, *Journal of Neurology* 246(6): 477–485.

[14] Scheltens, P., Leys, D., Barkhof, F., Huglo, D., Weinstein, H., Vermersch, P., Kuiper, M., Steinling, M., Wolters, E.C., Valk, J. 1992. Atrophy of medial temporal lobes on MRI in" probable" Alzheimer's disease and normal ageing: diagnostic value and neuropsychological correlates, *Journal of Neurology, Neurosurgery and Psychiatry* 55(10): 967–972.

[15] Wright, I., Ellison, Z., Sharma, T., Friston, K., Murray, R., McGuire, P. 1999. Mapping of grey matter changes in schizophrenia, *Schizophrenia Research* 35(1): 1–14.

[16] Mummery, C.J., Patterson, K., Price, C., Ashburner, J., Frackowiak, R., Hodges, J.R. 2000. A voxel-based morphometry study of semantic dementia: relationship

between temporal lobe atrophy and semantic memory, *Annals of Neurology* 47(1): 36–45.

[17] Fox, N., Warrington, E., Freeborough, P., Hartikainen, P., Kennedy, A., Stevens, J., Rossor, M.N. 1996. Presymptomatic hippocampal atrophy in Alzheimer's disease, *Brain* 119(6): 2001–2007.

[18] Whitwell, J.L., Przybelski, S.A., Weigand, S.D., Knopman, D.S., Boeve, B.F., Petersen, R.C., Jack Jr, C.R. 2007. 3D maps from multiple MRI illustrate changing atrophy patterns as subjects progress from mild cognitive impairment to Alzheimer's disease, *Brain* 130(7): 1777–1786.

[19] Teipel, S., Bayer, W., Alexander, G., Bokde, A., Zebuhr, Y., Teichberg, D., Müller-Spahn, F., Schapiro, M., Möller, H., Rapoport, S. 2003. Regional pattern of hippocampus and corpus callosum atrophy in Alzheimer's disease in relation to dementia severity: evidence for early neocortical degeneration, *Neurobiology of Aging* 24(1): 85–94.

[20] Chetelat, G., Landeau, B., Eustache, F., Mezenge, F., Viader, F., de la Sayette, V., Desgranges, B., Baron, J.C. 2005. Using voxel-based morphometry to map the structural changes associated with rapid conversion in MCI: a longitudinal MRI study, *Neuroimage* 27(4): 934–946.

[21] Friston, K.J., Holmes, A.P., Worsley, K.J., Poline, J.P., Frith, C.D., Frackowiak, R.S.J. 1994. Statistical parametric maps in functional imaging: a general linear approach, *Human Brain Mapping* 2(4): 189–210.

[22] Karas, G., Burton, E., Rombouts, S., Van Schijndel, R., O'Brien, J., Scheltens, P., McKeith, I., Williams, D., Ballard, C., Barkhof, F. 2003. A comprehensive study of gray matter loss in patients with Alzheimer's

disease using optimized voxel-based morphometry, *Neuroimage* 18(4): 895–907.

[23] Lancaster, J., Rainey, L., Summerlin, J., Freitas, C., Fox, P., Evans, A., Toga, A., Mazziotta, J. 1997. Automated labeling of the human brain: a preliminary report on the development and evaluation of a forward-transform method, *Human Brain Mapping* 5(4): 238.

[24] Pakhomov, S. 2006. MSU - MNI Space Utility. Available at: http://www.ihb.spb.ru/~pet_lab/MSU/MSUMain.html.

[25] Cabeza, R., Nyberg, L. 2000. Imaging cognition II: An empirical review of 275 PET and fMRI studies, *Journal of Cognitive Neuroscience* 12(1): 1–47.

[26] Leube, D.T., Weis, S., Freymann, K., Erb, M., Jessen, F., Heun, R., Grodd, W., Kircher, T.T. 2008. Neural correlates of verbal episodic memory in patients with MCI and Alzheimer's disease—a VBM study, *International Journal of Geriatric Psychiatry* 23(11): 1114–1118.

[27] Ojemann, G., Schoenfield-McNeill, J., Corina, D. 2001. Anatomic subdivisions in human temporal cortical neuronal activity related to recent verbal memory, *Nature Neuroscience* 5(1): 64–71.

[28] Onitsuka, T., Shenton, M.E., Salisbury, D.F., Dickey, C.C., Kasai, K., Toner, S.K., Frumin, M., Kikinis, R., Jolesz, F.A., McCarley, R.W. 2004. Middle and inferior temporal gyrus gray matter volume abnormalities in chronic schizophrenia: an MRI study, *The American journal of psychiatry* 161(9): 1603.

[29] Yao, Z., Hu, B., Zhao, L., Liang, C., Analysis of gray matter in AD patients and MCI subjects based voxel-based morphometry. Brain Informatics, Springer, 2011, pp. 209–217.

Information Technology and Computer Application Engineering – Liu, Sung & Yao (Eds)
© 2014 Taylor & Francis Group, London, ISBN 978-1-138-00079-7

Application of thermoacoustic computed tomography to joint imaging

X.Y. Jing, J. Rong, Y. Gao, L. Huang & T.T. Li
School of Physical Electronics, University of Electonic Science and Technology of China,
Chengdu, Sichuang, China

X.C. Zhong
School of Physical Science and Technology, Southwest Jiaotong University, Chengdu, Sichuan, China

ABSTRACT: This paper presents a preliminary experimental study on joint imaging in vitro using thermoacoustic computed tomography (TACT). In this pilot study, the experiments were conducted on chicken claw joints in vitro, and a delay-and-sum algorithm was used to reconstruct the two-dimensional (2D) thermoacoustic images. Experimental results suggest that TACT has the potential to be a useful tool for joints imaging except for breast cancer imaging and further to be a new modality for diagnosis of osteoarthritis and other bone joint diseases.

Keywords: thermoacoustic computed tomography; joints; experimental study.

1   INTRODUCTION

When electromagnetic energy is delivered into biological tissue, a portion of the energy is absorbed by the tissue and converted to heat. A temperature gradient is then produced by the heating based on the energy absorption pattern, and subsequently ultrasonic waves are generated through thermal expansion. This is called thermoacoustic effects (Jin 2007, Foster & Finch 1974, Gutfeld 1980). Thermoacoustic computed tomography (TACT) is a new imaging modality based on thermoacoustic effects. In recent years, there is an increasing interest in studying it and it has attracted the attentions of researchers from various fields.

Other imaging methods relating different biological parameters have both advantages and limitations, such as X-ray, ultrasound imaging, positron emission tomography (PET), magnetic resonance imaging (MRI), etc. The ultrasound imaging has a relatively high spatial resolution, but limited to image contrast. Traditional X-ray and positron emission tomography (PET) are either ionizing or radioactive, and the expensive cost of MRI prevents its use as a routine screening tool (Jin, 2007). TACT can provide some unique physical and chemical information of biological tissue due to his special principle.

TACT is a noninvasive and non-ionizing imaging modality as photoacoustic tomography (PAT), which is also based on thermoacoustic effects. PAT has been successfully applied in imaging vascular structure in the tissue and doing functional brain imaging in small

animals (Kolkman et al. 2003, Wang et al. 2003). Nevertheless, its applications are limited by the penetration depth of the laser lights in the visual light region. The Penetration depth of microwave-induced TACT is much larger than near-infrared PAT. Besides, different from PAT related to the optical absorption coefficient of biological tissues, TACT is related to conductivity and relative permittivity of tissues, which can obtain different thermal properties of the tissue.

In this experimental study, we explore potential application for TACT in imaging joint based on the different thermal properties of tissues

2   MATERIALS AND METHODS

2.1   *Thermoacoustic computed tomography systerm*

The schematic of the experimental setup is shown in Figure 1. The central frequency of the microwave generator (S-band microwave source) is 3 GHz. The pulse width of the microwave source is $0.75\,\mu s$, which means that ultrasonic waves up to ~ 1 MHz were generated, thus providing spatial resolution on the millimeter scale.

The experiment was conducted in a plastic container filled with transformer oil, which acted as coupling medium due to its weak microwave absorption and high ultrasonic penetrating ability. To absorb the leaking microwave, outside the transformer oil container,

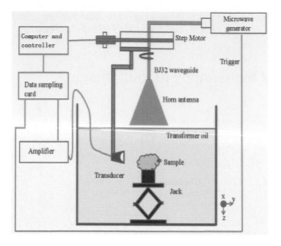

Figure 1. *Schematic of the thermoacoustic computed tomography systerm.*

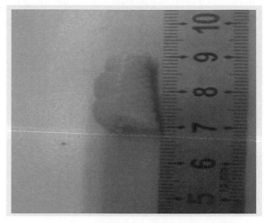

Figure 2. Photograph of the chicken claw joint.

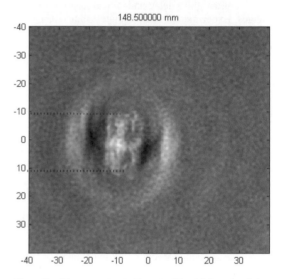

Figure 3. The reconstructed image of the chicken claw joint. The units in both x axis and y axis are mm.

there is another plastic container filled with water, which is not showed in Figure 1. The microwave energy was delivered into the biological tissue through an air-filled pyramidal horn antenna. The tissue sample was immersed in the oil and placed on a base in the X-Y plane under the opening of the antenna as shown in Figure 1.

The ultrasonic receiver was an unfocused transducer (V323, Panametrics NDT Inc.), with a central frequency of 2.25 MHz and a diameter of 6 mm. This receiver was immersed in the oil and scanned around the Z axis, and its axis was aligned with the center of samples. The transducer, driven by the computer and the controllers (MC600-2B, Zolix Instruments Co., Ltd.) to scan around the sample, detected the thermoacoustic signals in the imaging plane at each scanning position. A pulse amplifier received the signals from the transducer and transmitted the amplified signals to a data sampling card. One set of data at 180 different positions was taken when the receiver moves over 360 degrees. After 180 series of data were recorded, the TACT image was reconstructed with the delay-and-sum algorithm (Xing, 2007).

2.2 The experimental sample

In the experiment, a joint with ∼2 cm long, which was cut from a chicken claw bought from supermarket, was used as the experimental sample. The photograph of the chicken claw joint is shown in Figure 2.

3 RESULTS AND DISCUSSION

Existing reconstruction algorithms for TACT are based on the assumption that the acoustic properties in the tissue are homogeneous, biological tissue, however, has heterogeneous acoustic properties, which lead to distortion and blurring of the reconstructed images.

Besides, the imperfection existing in the delay-and-sum algorithm results in artifact. That all can be seen from Figure 3 and Figure 4. Compared to the physical object in Figure 2, the reconstructed image can well correspond with it, from whatever the physical size (∼2 cm), the structure and so on. The difference between Figure 4 and Figure 3 depend on the different image display function in the algorithm. From the Figure 4, the structure of the joint like skin, articular cavity, cartilage can be imaged very well. The reflected signals from the test sample are related to the complex relative permittivity, thus, the more water, the higher signal. However, the water in the articular cavity of the sample in this experiment is partial losing, which lead to its microwave absorbing ability is lower than cartilage. That can be clearly reflected from Figure 4 and Figure 5.

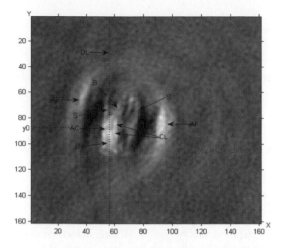

Figure 4. The reconstructed image of the chicken claw joint. B: bone; AF: artifact; S: skin; AC: articular cavity; CL: cartilage; DL: dotted line. X-coordinate and Y-coordinate are the element subscript corresponding to image matrix.

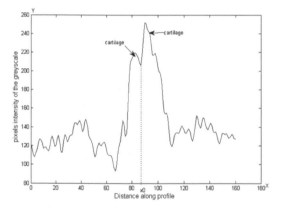

Figure 5. Pixels intensity of the greyscale along DL dotted line in Figure 4. The x0 in this figure is equal to y0 in Figure 4.

4 CONCLUSION

In summary, this study represents the first attempt to apply the thermoacoustic computed tomography to joint imaging. The physical size, the structure like skin, articular cavity, cartilage can be well correspond to the physical object. The signal intensity of various parts from the test sample, which is reflected by the pixels intensity along a line, well corresponds to the difference of dielectric properties in the tissue. Thus, TACT has the potential to become a useful tool for joint imaging and it may be a new diagnosing method for osteoarthritis and other bone joint diseases in the future.

ACKNOWLEDGEMENT

Thanks Prof. Jian Rong for giving me some significant advice, and he is the corresponding author of this paper. Thanks Yi Gao, Dr. Lin Huang, Tingting Li and the other members of the teaching and research section for giving me a lot of help in experiments.

REFERENCES

Jin, X. 2007. *Microwave induced thermoacoustic tomography: applications and corrections for the effects of acoustic heterogeneities.* Texas A&M University.

Foster, K. R. & Finch, E. D. 1974. Microwave hearing: evidence for thermoacoustic auditory stimulation by pulsed microwaves. *Science* 185(147), 256–258.

Gutfeld, R. T. V. 1980. Thermoelastic generation of elastic waves for non-destructive testing and medical diagnostics. *Ultrasonics.* 18(4), 175–181.

Bowen, T. 1981. Radiation-induced thermoacoustic soft tissue imaging. *Proc. IEEE Ultrason. Symp.* 2, 817–822.

Olsen, R. G.1982. Generation of acoustic images from the absorption of pulsed microwave energy. *in Acoustic Imaging, edited by J. P. Powers* (Plenum Publishing, New York), pp. 53–59.

Lin, J. C. & Chan, K. H.1984. Microwave thermoelastic tissue imaging–system design. *IEEE Trans. Microwave Theory Tech* 32, 854–860.

Oraevsky, A. Esenaliev, R. Jacques, S. & F. Tittel.1995. Laser optoacoustic tomography for medical diagnostics – principles. *Proc. of SPIE* 2676, 22–31.

Kolkman, R. G. M., Hondebrink, E., Steenbergen, W. & de Mul, F. F. M. 2003. In vivo photoacoustic imaging of blood vessels using an extreme-narrow aperture sensor. *IEEE J. Sel. Top. Quant.* 9, 343–346.

Wang, X. Pang, Y. Ku, G. Xie, X. Stoica, G. & Wang, L. V. 2003. Non-invasive laserinduced photoacoustic tomography for structural and functional imaging of the brain in vivo. *Nat. Biotechnol.* 21, 803–806.

Xing, D. & Xiang, L. Z. 2007. Photoacoustic and Thermoacoustic Imaging Application in Cancer Early Detection and Treatment Monitoring. *Proc. of SPIE Vol.* 6826, 68260B-1-8.

Information Technology and Computer Application Engineering – Liu, Sung & Yao (Eds)
© 2014 Taylor & Francis Group, London, ISBN 978-1-138-00079-7

Photoacoustic imaging of mouse brain using ultrasonic transducers with different central frequencies

T.T. Li, J. Rong, L. Huang, B.Z. Chen & X.Y. Jin
School of Physical Electronics, University of Electonic Science and Technology of China, Chengdu, Sichuan, China

X.C. Zhong
School of Physical Science and Technology, Southwest Jiaotong University, Chengdu, Sichuan, China

H.B. Jiang
J. Crayton Pruitt Family Department of Biomedical Engineering, University of Florida, Gainesville, Florida, USA

ABSTRACT: A circular scanning photoacoustic imaging system was utilized to image the blood vessels of a mouse brain. The ultrasonic transducers with central frequencies of 2.25 MHz and 5 MHz, respectively, were used. The photoacoustic signals as well as the recovered images of mouse brain were compared among the two different transducers. The results show that the vascular structure of the brain is imaged in higher resolution using the 5 MHz transducer, which implies that it is important to choose a transducer with proper central frequency for a particular biological tissue. This study demonstrates that photoacoustic imaging has the potential to become a new modality for functional brain imaging.

Keywords: Photoacoustic imaging; Transducers; Blood vessels of mouse brain

1 INSTRUCTIONS

Brain injury, brain tumors, stroke, dementia and epilepsy severely affected people's lives. Existing clinical brain imaging techniques contained X-ray computed tomography, magneticresonance imaging (MRI), ultrasonography and positron emission tomography (PET) (Elwell & Beard, 2005) can provide structure or functional information of the brain, and however, they all have limitations. X-ray computed tomography is expensive and moreover it uses ionizing radiation. MRI requires high cost of maintenance and demands a strong magnetic field (Wang, 2008). Ultrasonography has strong attenuation due to echo reflection of skull. PET injects radioactive substance into the human body, which will lead to irreversible damage to the body. Thus, it's urgent to develop a novel, noninvasive and high-resolution imaging technology for brain imaging.

Photoacoustic image tomography (PAT) employs a pulsed laser as the excitation source. Because different biological tissues have different coefficient absorption, which generate ultrasonic waves, also known as photoacoustic signals, with a bandwidth ranging from 20 KHz to 20 MHz (Tan, Xing, Wang, Zeng & Yin, 2005) after absorbing the laser energy. The ultrasonic transducer feeds these signals into computer. Using a reconstruction algorithm, the absorbed laser energy distribution can be reconstructed. Therefore, PAT has both high optical contrast and high ultrasonic resolution (Ku, Wang, Xie, Stoica & Wang, 2005). Many groups have made an in-depth study of brain imaging using PAT system, e.g. Geng Ku et al. imaged tumor angiogenesis in the rat brains in vivo; Xueding Wang et al. used a single transducer to image the structural and functional information of the brain in vivo successfully (Wang, Pang, Ku, Xie, Stoica & Wang, 2003); Yang Sihua et al. combined the photoacoustic and thermoacoustic imaging modalities, achieved functional imaging of rapid foreign body localization and tissue damage detection (Yang, Xing & Xiang, 2007). These results promote the development of PAT and facilitate the combination of PAT with other imaging modalities meanwhile.

2 EXPERIMENTAL SETUP, ANIMAL MODEL AND IMAGING RECONSTRUCTION

2.1 Experimental setup

The schematic of experimental setup is shown in Figure 1. A Q-switched Nd:YAG laser emits laser light at a wavelength of 532 nm, of which the pulse with is 6 ns and the pulse repetition rate is 1 Hz. The direction of

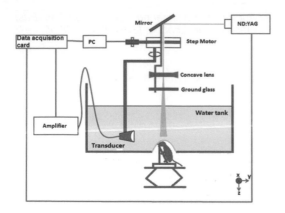

Figure 1. Schematic of experimental setup.

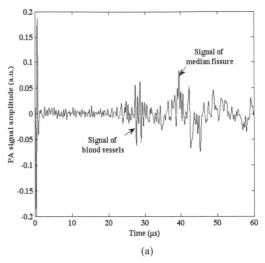

(a)

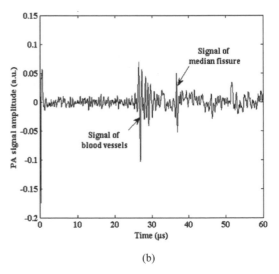

(b)

Figure 2. Comparison of PA signal amplitudes acquired with (a) 2.25 MHz ultrasonic transducer and (b) 5 MHz ultrasonic transducer.

light path is changed by 90 degrees by the mirror. The laser beam is expanded by a concave lens. To obtain uniform laser beam, a ground glass is used. Then, the homogeneous beam illuminates the mouse head with scalp removed but skull intact from its top. Two transducers with 2.25, 5 MHz central frequencies will be used. The diameter of receiving area is 0.8 mm. The transducer is fixed on a step motor which is controlled by the computer. During the experiment, the transducer rotate around the mouse brain 360 degrees with a step size of 2 degrees. The entire received signals are sent to the amplifier and then transmitted into PCI-4732 data acquisition card. After the scanning, the delay and sum algorithm is adopted for image reconstruction.

2.2 Animal model

A Kunming mouse weighted 28g was used as the animal model. The mouse was sacrificed by intraperitoneally injecting overdose chloral hydrate solution (the concentration is 10%). Before imaging, the hair on this mouse was removed using depilatory paste. The skull was exposed by removing the scalp. After coated with a layer of ultrasound coupling agent on the head, the mouse was placed in a homemade mounting bracket. The water tank was covered with a piece of polyethylene film that would not affect the light propagation. A circular opening is at the bottom of the water tank with a diameter of 5 cm, from which the mouse head can rise up to the water tank. Adjust the relative position between the mouse head and the transducer scanning plane so that both of them are in the same level. During the experiment two ultrasound transducers with central frequencies of 2.25 MHz and 5 MHz were employed.

3 RESULTS

Figures 2(a) and 2(b) show the photoacoustic signals (PA) amplitude as a function of time measured by the 2.25 MHz and 5 MHz transducers respectively. The data acquisition processes of two transducers were under the same conditions. Figure 2(a) shows that the signal of blood vessels is less than 2(b). Moreover, the number of peak signal of blood vessels in 2(a) is fewer than in 2(b), which implies that only three blood vessels are visible using 2.25 MHz while five are in sight using 5 MHz, as it will be proved in Figure 3. Theoretically, huge biological tissues generate low frequencies more while tiny tissues generate more high frequencies when they are excited by the laser light (Wang, 2008). It means that the 5 MHz transducers are more sensitive to PA signals generated by tiny tissues.

Figure 3(a) and 3(b) demonstrate two reconstruction images of the superficial layer of the mouse brain acquired with the 2.25 MHz and 5 MHz ultrasonic transducers respectively. Figure 3(c) is the scalp-removed photograph of the mouse brain acquired after the PAT experiment. The outline of left and right hemi encephalon are both visible in 3(a) and 3(b). Because

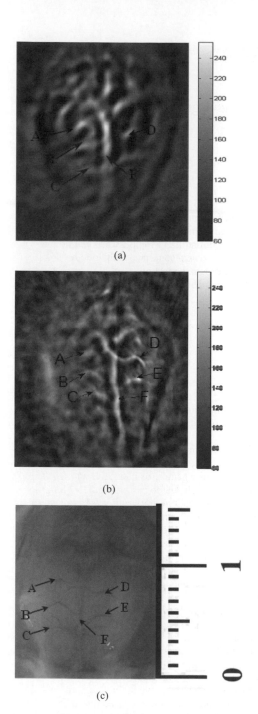

(a)

(b)

(c)

Figure 3. Photoacoustic images of the mouse brain acquired with (a) 2.25 MHz ultrasonic transducer and (b) 5 MHz ultrasonic transducer. (c) Scalp-removed photograph of the mouse brain. A,B,C,D,E, blood vessels; F, median fissure.

high frequency signal sharps the boundary of image and low frequency signal lead to image smoothing, the median fissure and blood vessels are distinct in 3(b) while they are blurred in 3(a). It's obvious that blood vessels D and E are clearer in 3(a) than in 3(b). The blood vessels C and E in 3(a) can hardly

be distinguished. In general, both reconstructed PAT images match well with the photograph picture.

4 CONCLUSION AND DISCUSSION

This research employed a circular scanning photoacoustic imaging system and successfully imaged the mouse brain, of which the outline of the brain, median fissure and the vascular structure of the blood vessels were clearly showed using the 5 MHz transducer. In addition, the PA amplitude received by the 2.25 MHz and 5 MHz transducers as well as the reconstruction images were compared. Different sizes of biological tissues result in different frequencies. Large biological tissues generate low frequencies more while tiny tissues generate higher frequencies when they are excited by laser light. The results show that the 5 MHz transducer are more sensitive to tiny tissues, which implies that it is important to choose a corresponding transducer with proper central frequency for a particular biological tissue. PAT is expected to become a new modality for functional brain imaging.

Because the current PAT system uses a single-element ultrasonic transducer, the total scanning time of 4.5 min are so long that it's impossible to achieve real-time photoacoustic imaging. In our future work, by employing an transducer array, the real-time imaging could be achieved. Furthermore, it's urgent to carry out the experiment noninvasively with scalp intact.

ACKNOWLEDGEMENTS

This research was carried out under the guidance of professor Huabei Jiang, who is the corresponding author of this paper.

REFERENCES

Clare Elwell & Paul Beard, 2005. Shedding light on the brain, *NIR news,* 16(7): 28–30.

Geng Ku, Xueding Wang, Xueyi Xie, George Stoica & Lihong V. Wang, 2005. Imaging of tumor angiogenesis in rat brains in vivo by photoacoustic tomography, *Applied Optics,* 44(5): 770–775.

Lihong V.Wang, 2008. Prospects of photoacoustic tomography, *Medical Physics,* 35(12): 5758–5767.

Lihong V. Wang, 2008. Tutorial on Photoacoustic Microscopy and Computed Tomography, *IEEE Journal of Selected Topics in Quantumelectronics,* 14(1): 171–179.

Tan Yi, Xing Da, Wang Yi, Zeng Yaguang & Yin Bangzheng, 2005. Influence of Bandwidth of Ultraonic Transducer on Photoacoustic Imaging, *ACTA OPTICA SINICA,* 25(1): 40–44.

Xueding Wang, Yongjiang Pang, Geng Ku, Xueyi Xie, George Stoica & Lihong V.Wang, 2003. Noninvasive laser-induced photoacoustic tomography for structural and functional in vivo imaging of the brain, *Nature Biotechnology,* 21(7): 803–806.

Yang Sihua, Xingda & Xiang liangzhong, 2007. Brain structure and functional imaging based on photoacoustic and thermoacoustic technology, *Science in China,* 37: 101–109.

Information Technology and Computer Application Engineering – Liu, Sung & Yao (Eds)
© *2014 Taylor & Francis Group, London, ISBN 978-1-138-00079-7*

The photoacoustic computed tomography of bones and joints using the system based on PCI4732

X.C. Zhong
School of Physical Science and Technology, Southwest Jiaotong University, Chengdu, Sichuan, China

X.Y. Jing, L. Huang, J. Rong, T.T. Li & Y. Gao
School of Physical Electronics, University of Electonic Science and Technology of China, Chengdu, Sichuang, China

ABSTRACT: We present a preliminary experimental study in two-dimensional (2D) photoacoustic tomography reconstruction of bones and joints using a photoacoustic imaging system based on PCI4732 with high positioning accuracy. In this pilot study, the experiments were conducted on in vitro chicken bone joints, which were embedded in larger cylindrical background phantoms. The photoacoustic images were reconstructed with a delay-and-sum algorithm. The position, the physical size and the structure shape of the joints can be imaged clearly with this system, and can well correspond with those of actual joints compared to the real objects. This study suggests that photoacoustic imaging system based on PCI4732 with high positioning accuracy has the potential to become a novel noninvasive bone and joint imaging system and further a useful tool for diagnosis of osteoarthritis and other bone joint diseases.

Keywords: photoacoustic tomography; bones and joints; two-dimensional; imaging reconstruction, PCI4732.

1 INTRODUCTION

Photoacoustic tomography (PAT) is a novel noninvasive imaging method of biological tissues, which is different from other imaging methods. It irradiates biological samples with short-pulse laser, biological tissues thermally expand in result of absorbing laser energy and produces ultrasonic pressure which is also referred to as photoacoustic signal, the ultrasonic transducer detects ultrasonic signals in all directions, then the light absorption distribution of biological tissues can be reconstructed with the detected signals by corresponding image reconstruction algorithms. This method combines the advantages of pure optical imaging and pure ultrasonic imaging, which has high resolution and high contrast.

In the PAT study of bones and joints, particularly in the PAT research of osteoarthritis of the fingers, has been relatively mature in foreign countries such as American. In China, it is still at its preliminary stage. At abroad, they have combined photoacoustic tomography system and optical tomography system to detect osteoarthritis in the finger joints with array detector, and have made multispectral quantitative analysis.

The photoacoustic imaging system based on PCI4732 has been used in brain imaging and vascular imaging, etc, which has been achieved relative success. However, the imaging of bones and joints with this PAT system has not been attempted. So, we want to do a pilot study in this aspect with this system.

In this paper, we use the photoacoustic imaging system based on PCI4732 and delay-and-sum algorithm, which are different from those of published articles in bone and joint imaging, to acquire the photoacoustic signals of in vitro chicken bone joints and to reconstruct images.

2 MATERIALS AND METHODS

2.1 *Photoacoustic imaging system based on PCI473*

Photoacoustic imaging system based on PCI4732 mainly consists of pulsed tunable laser (Nd: YAG), a step rotary table and its driver, an ultrasonic detector, a super low-noise preamplifier, a PCI4732 acquisition card and a computer, etc. In this system, the wavelength of laser pulse output by the laser is 532 nm, the pulse repetition frequency is 1–10 Hz adjustable, the pulse width is 6 ns. The laser beam is spread out by fiber and through concave mirror and ground glass before irradiating the experimental sample. The effect of concave mirror and ground glass is expanding and making light uniform respectively. The experimental sample thermally expands in result of absorbing laser energy and produces ultrasonic signal, the ultrasonic signals are detected by ultrasonic transducer and amplified by preamplifier, then the signals are collected by PCI4732 card before sending to the computer. The

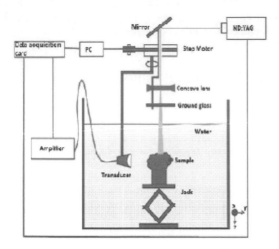

Figure 1. Schematic of the photoacoustic imaging system based on PCI4732.

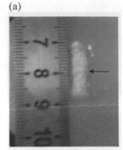

Figure 2. (a) Photograph of the bone and joint experimental sample 1. (b) Reconstructed photoacoustic image. The black arrow points to the joint and the red arrow points to the articular cavity.

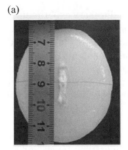

Figure 3. (a) Photograph of the bone and joint experimental sample 2. The black hair was placed on the joint. (b) Reconstructed photoacoustic image. The red arrow point to the articular cavity and the yellow arrow point to the hair.

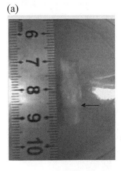

Figure 4. (a) Photograph of the bone and joint from in vitro chicken claws, which was only removed the skin. (b) Reconstructed photoacoustic image. The black arrow points to the joint and the red arrow points to the articular cavity.

whole process is controlled by LabVIEW programs. The photoacoustic image of the experimental sample is reconstructed by delay-and-sum algorithm with acquired photoacoustic signals.

2.2 The experimental samples

A. The bone and joint experimental sample 1. The bone and joint tissue phantom was simply made of water and agar powder, among which the agar serves as the coagulator. We got one fresh joint from the in vitro raw chicken claw, removing its skin and flesh, blood vessels and the other tissues attached to the bones so as to keep only the joint and the bones at both sides. The bone joint was embedded in a larger cylindrical background phantom.

B. The bone and joint experimental sample 2 was the bone and joint experimental sample 1 with a single hair which was placed at the joint as a marker.

C. The bone and joint experimental sample 3. The bone and joint tissue phantom was simply made of water and agar powder, among which the agar serves as the coagulator. We got one fresh joint from the raw chicken claw in vitro, which was only removed its skin. The bone joint was embedded in a larger cylindrical solid phantom.

D. The bone and joint experimental sample 4 was the bone and joint experimental sample 3 with a single hair which was placed at the joint as a marker.

The LabVIEW program controlled the PAT system as show in Figure 1 to rotate 360° scanning for the bone and joint experimental samples and to acquire photoacoustic signals. The image was reconstructed using a delay-and-sum algorithm with the acquired signals.

3 RESULTS AND DISCUSSION

The reconstruction images as can be seen from above, can well correspond with the real objects both in terms of physical size and structure shape. We know from previous articles that the optical absorption coefficient of bones is higher than that of articular cavities which mainly consist of cartilage and synovial fluid. So, we infer the places where the arrows point at Figure 2(b) and Figure 4(b) are the articular cavities and the high optical absorption objects at both sides are bones. The chicken bone joint in experimental sample 2 is the

(a)  (b)

Figure 5. (a) Photograph of the bone and joint experimental sample 4. The black hair was placed on the joint place. (b) the reconstructed photoacoustic mage.

same bone joint used in experimental sample 1. Similarly, the chicken bone joint in experimental sample 4 is the same bone joint used in experimental sample 3. The hairs in the Figure 3(a) and Figure 5(a) act as the markers of joints. The Figure 3(b) and Figure 5(b), are presented as the evidence that the places where the arrows point at Figure 2(b) and Figure 4(b) are the articular cavities. So, the position of the articular cavity can be imaged correctly by this system. What's more, the physical size and the structure shape of the reconstructed joints can be well in agreement with the anatomical structure. By making progress in reconstruction algorithm or improving the ability of doing experiments, the tissues in articular cavities may be reconstructed clearly. The changes in joint tissues and joint thickness, which are compared with normal joints, usually serve as the symbols of early osteoarthritis.

4 CONCLUSION

We present our pilot study on two-dimensional (2D) photoacoustic tomography of bones and joints using a photoacoustic imaging system based on PCI4732 with high positioning accuracy. From the reconstruction results of photoacoustic images, we can see that the position, the physical size and the structure shape of in vitro bone joints can be imaged clearly by the PAT system based on PCI4732. In future, we will try to conduct experiments on in vivo joints. By making progress in reconstruction algorithm or improving the ability of doing experiments, the tissues in articular cavities may be reconstructed clearly. The changes in

joint tissues and joint thickness, which are compared with normal joints, usually serve as the symbols of the early osteoarthritis, so the PAT system based on PCI4732 has the potentiality to be a useful tool for the early diagnosis of osteoarthritis.

ACKNOWLEDGEMENT

Thanks Prof. Jian Rong for giving me some significant advice, and he is the corresponding author of this paper. Thanks Xiangyu Jing, Dr. Lin Huang, Tingting Li, Yi Gao and the other members of the teaching and research section for giving me a lot of help in experiments.

REFERENCES

Chen, B. Rong, J. Zhong, X. Huang, L. Jiang, H. 2012. Design of photoacoustic imaging system with high positioning accuracy based on PCI4732. *J. Sou. J. Univ.* 47: 0258–2724.

Sun, Y. Sobel, E. Jiang, H. 2009. Quantitative three-dimensional photoacoustic tomography of the finger joints: an in vivo study. *J. Biomed. Opt.* 14: 064002-1-5.

Sun, Y. Sobel, E. Jiang, H. 2010. In vivo detection of osteoarthritis in the hand with three-dimensional photoacoustic tomography. *Proc. SPIE,* 7548: 75484H-1-6.

Sun, Y. Sobel, E. Jiang, S. H. 2011. First assessment of three-dimensional quantitative photoacoustic tomography for in vivo detection of osteoarthritis in the finger joints. *Med. Phys.* 38: 4009–4017.

Tan, Y. 2010. The Production methods of photoacoustic coupled fluid. *Sci. Tec. Infor.* (6): 15.

Xu, Y. Lftimia, N. Jiang,H. Key,L. Ly. Bolster, Mar. B. 2002. Three-dimensional diffuse optical tomography of bones and joints. *J. Biomed.Opt.* 7: 88–92.

Xiao, J. Yao, L. Sun, Y. Sobel, E. S. He, J. Jiang, H. 2010. Quantitative two-dimensional photoacoustic tomography of osteoarthritis in the finger joints. *Opt. Express* 18: 14359–14365.

Xiao, J. He, J. 2010. Multispectral quantitative photoacoustic imaging of osteoarthritis in finger joints. *Appl. Opt.* 49: 5721–5727.

Xiao, J. Tang, J. Lu, G. Chen, Z. 2011. Functional Photoacoustic Imaging Of Osteoarthritis In The Finger Joints. *BMEI 2011,* 1: 299–303.

Yuan, Z. Zhang, Q. Sobel, E. Jiang, H. 2007. Three-dimensional diffuse optical tomography of osteoarthritis: initial results in the finger joints. *J. Biomed. Opt.* 12: 034001-1-11.

Zhang, Q. Yuan, Z. Sobel, E. S. Jiang, H. 2009. Three-dimensional diffuse optical tomography of osteoarthritis: A study of 38 finger joints. *Proc. SPIE,* 7166: 71660K-1-5.

Information Technology and Computer Application Engineering – Liu, Sung & Yao (Eds)
© 2014 Taylor & Francis Group, London, ISBN 978-1-138-00079-7

A heuristic rule mining algorithm based on inner cognitive mechanism

Bing Ru Yang, Wen Bin Qian, Hui Li & Yong Hong Xie
School of Computer and Communication Engineering, University of Science and Technology Beijing, China
Beijing Key Laboratory of Knowledge Engineering for Materials Science, Beijing, China

ABSTRACT: At present, most researchers are concentrated on mining algorithms with high performance and high scalability in different kinds of databases. However, there are still some problems can't be solved by current structure models and techniques. From the cross view of multi-subject, the authors discuss in detail the Relation between Reasoning Category and Accessible Category, on this basis, the authors present a heuristic rule mining algorithm based on inner cognitive mechanism, the inner cognitive mechanism is double bases cooperating mechanism, where double bases are database and knowledge base.

Keywords: data mining; rule extraction; structure model; inner cognitive mechanism

1 INTRODUCTION

We have found some propositions on this subject when having comprehensible investigation corpus such as: G. Piatesky-Shapiro developed a knowledge discovery platform–KDW and proposed the idea of "assisting the focus in initial discovery by adopting the field knowledge and finite searching" [1]; Two operations, consolidation and link formation, which complement the usual machine learning techniques that use similarity-based clustering to discover classifications, are proposed as essential components of KDD systems for certain applications [2]. Sarabjot S. Anand, based on evidence theory, developed the general frame of data mining—ESD and proposed that "users" transcendental knowledge and the known knowledge can be coupled into the process of knowledge discovery" [3]; However, none of them involved the theoretical foundation of Knowledge Discovery, not a bit of reaching the height of cognition mechanism and running system, namely there are still on the stage of scholar thoughts. In the view of philosophy, a database show the quantity of knowledge base, and knowledge base shows the quality of database. Only the research of quantity and quality, rather than one aspect only reflecting part side of nature of matters, will reflect the nature of Knowledge Discovery better. Database and knowledge base distinguish as well as connect each other, besides under certain conditions they have some counterpart relation [4,5]. The specific relation between database and knowledge is exactly the start of the construction of double bases cooperating mechanism. Furthermore, the KDD algorithms mentioned only focus according to man-machine interaction, can embody the cognitive independence of

system, so they are unable to reflect novelty and validity of KDD.

In past years, rule mining techniques are applied to decision support system and expert system field [6–10], a gate reminder scheme. The identity decision is reached by two approaches is developed [7]. A decision support system for RME design evaluation is presented, and Intelligent IDS using fuzzy rough set based C4.5 classification algorithm to improve the detection accuracy is proposed [9]. Multiple methods of discretization and reduction of the data sets obtained from measurements are tested to find effective combinations for diagnostic purposes.

2 BASIC THEORIES

2.1 *Reasoning category*

In the universe of discourse X, there is a set of knowledge nodes N among each of which there exist intrinsic reasoning relation. in the universe of discourse X, knowledge node n_1 = high temperature, n_2 = high pressure. If there is an intrinsic rule in universe of discourse X—"If the temperature is high, then the pressure is high", then there exists the reasoning relation from knowledge node n_1 to n_2: $n_1 \rightarrow n_2$ or $r(n_1, n_2) = r$. It means that if the temperature is chosen from the numerical sub-domain of "high temperature", the pressure must be in the numerical sub-domain of "high pressure"; another intrinsic rule is that proposition of "if temperature is high, pressure is high" is untenable, i.e. $n_1^\times \rightarrow n_2$, or $r(n_1, n_2) = \phi$. It means that there exists at least one phenomenon in which temperature is high while pressure is not high.

Definition 1: We call the reasoning relation from knowledge node n_1 to n_2: $n_1 \rightarrow n_2$ a positive rule and $n_1^\times \rightarrow n_2$ a negative rule. n_1 marks the beginning of knowledge node, while n_2 the ultimate of it. Both the positive rules and the negative rules are gathered into the rule set of universe of discourse X.

Definition 2: The primitive knowledge base is composed of the knowledge node set N in universe of discourse X and the set of intrinsic rules between knowledge nodes. It is marked down as $K^p(X)$.

Primitive knowledge base veritably shows whether there exists the reasoning relation among knowledge node in primitive knowledge base $K^p(X)$, $r(n_1, n_2) = r$ (the inference relation exists) if in universe of discourse X $n_1 \rightarrow n_2$. Based on this, in given universe of discourse X, only the primitive knowledge base is certain. What's more, the negative rule set can be decided by the positive rule set, so if inference category is determined, primitive knowledge base is determined.

2.2 Accessibility category

Definition 3: Suppose y is an element in numerical domain Boolean algebra $<\gamma, \Re(\gamma)>$, $\psi(y)$ is the proper set of y, $n = f^{-1}(y)$ is the corresponding knowledge node of y. Then we say that the tuple u in database $\Re(X)$ satisfies y (or satisfies n), if

$$a(u) \in \psi(y) \tag{1}$$

Which is marked down as u/y or u/n.

The data sub-class structure originates from the primitive data base is called the primitive data sub-class structure; the data sub-class structure originates from the complete data base is called the complete data sub-class structure.

Definition 4: On the data sub-class structure set $<Y, \Re(X)>$ of universe of discourse X, we can establish, between elements, the accessibility relation "$\propto$": $<y_1, \Re(y_1)> \propto <d_2, \Re(y_2)>$ iff $\Re(y_1) \subseteq \Re(y_2)$. If in $<Y, \Re(X)>$, there is no accessibility relation from the element $<y_1, \Re(y_1)>$ to the element $<y_2, \Re(y_2)>$, we say that there is inaccessibility relation from $<y_1, \Re(y_1)>$ to $<y_2, \Re(y_2)>$. All accessibility relations constitute the accessibility relation set; all inaccessibility relations constitute the inaccessibility relation set.

Theorem 1: the data sub-class structure set $<D, \Re(X)>$ of universe of discourse X and the accessible relations "$\propto$" among its elements can create a category. (Proof omitted)

The category constructed by the data sub-class structure set $<D, \Re(X)>$ and the accessible relations among its elements are called the data sub-class structure accessibility category of universe of discourse X and is marked down as $C_\propto <D, \Re(X)>$. Accordingly, primitive data sub-class structure accessibility category is marked down as $C_\propto <D, \Re^p(X)>$; complete sub-class structure accessibility category is marked down as $C_\propto <D, \Re^c(X)>$. Obviously, these 3 categories can be solely determined by their corresponding databases.

Lemma 1: There are functors between the inference category $C_r(N)$ of universe of discourse X and the accessibility category $C_\propto <\gamma, \Re(\gamma)>$ of data sub-class structure.

Proof: First, let's establish the one to one mapping from knowledge node set N to the data sub-class structure set $<D, \Re(X)>$ of universe of discourse X

$$F_0 = gf^1: N \rightarrow <\gamma, \Re(\gamma)>, \tag{2}$$

The meaning of F_0 will remain unchanged if we change data sub-class structure set into the primitive data sub-class structure set $<D, \Re^p(X)>$ or complete data sub-class structure set $C_\propto <\gamma, \Re^c(\gamma)>$.

Theorem 2: (Structural Corresponding Theorem): The reasoning category $C_r(N)$ of universe of discourse X and the accessibility category $C_\propto <D, \Re^c(X)>$ of complete data sub-class structure are equal.

Proof: Suppose the meaning of functor (F_O, F_H) is as what is described in Lemma 1. According to the proof of Lemma 1, we can know that F_0 is a one to one mapping; therefore there exists F_O^{-1}. Now we will prove that F_H is also a one to one mapping.

Choose an arbitrary morphism $(F_O(m) \propto F_O(n))$ from $C_\propto <\gamma, \Re^c(X)>$, and we will prove that $m \rightarrow n$. Conversely, if the above proof is untenable, we will prove that $m^\times \rightarrow n$. According to the definition of complete database $\Re^c(X)$, there exists at least one tuple u, which makes u/m and u/¬n, i.e. $u \in \Re(f^{-1}(m))$ but $u \notin \Re(f^{-1}(n))$, i.e. $\Re(f^-(m)) \subseteq (f^{-1}(n))$ is untenable and therefore $F_O(m) \propto F_O(n)$ is untenable. This is contradictory to the supposition that $(F_O(m) \propto F_O(n))$ is a morphism. Therefore, $m \rightarrow n$ and F_H^{-1} exists.

It is easy to prove that (F_O^{-1}, F_H^{-1}) is a functor from $C_\propto <D, \Re^c(X)>$ to $C_r(N)$. Therefore, $C_r(N)$ and $C_\propto <D, \Re^c(X)>$ are equivalent. Proof accomplished.

Theorem 2 is significant. Although primitive knowledge base exists objectively, we can't know its content. What we can do is to infer its content through its specific expression, such as complete data sub-class structure base.

We have discussed the five algebra systems and their relations deduced by universe of discourse X. Three of them are formally determined by the attribute of universe of discourse and the partition of the numerical domain. The other two categories are concretely determined by the intrinsic relations between the attributes in the universe of discourse, with contents concerned. Furthermore, Theorem 2 presents the 1-1 mapping between knowledge nodes of knowledge sub-base and layers of corresponding data sub-class structure.

2.3 The method for measuring dual rule intensity

1) Rule Intensity
Definition 5: In X, the data sub-class structure $<n, \Re(n)>$ of knowledge node n has $S(n)$ number of elements. The number of tuples in the database is $S(\Re)$. Then for a piece of rule $r = (n \rightarrow k)$, $S(n)/S(\Re)$ is called the pre-support of rule r being marked down

as $S_f(r)$, and $S(k)/S(\Re)$ is called post-support of rule r being marked down as $S_b(R)$. $S(n \wedge k)/S(n)$ is called the confidence of rule r and is marked down as $C(r)$. The ratio of confidence to post-support is called the entropy and is marked down as $\gamma(r)$.

Any function that satisfies definition 5 can be regular intensity function, and we should choose one that can be easily described and computed and is smooth, for example

$$\Gamma(S_f, \gamma) = S_f \cdot R(\gamma, 0) = S_f \cdot \gamma \cdot e^{\frac{r^2}{2}} \tag{3}$$

Where $R(\gamma, 0)$ is Rayleigh distribution function.

2) Discussion of Special Cases

It is very significant to define the rule intensity. This will not only provide clear priority level to rules before being presented to the users. It is also significant in the partition of data sub- domain. In definition 5, there is a hidden hypothesis: post-support is not 0. But in fact, there exists some rules whose post-support is 0. When there is such a case, we cannot apply definition 5 directly. We adopt three schemes here.

The first scheme is based on information diffusion hypothesis. For a given rule $r = (n \rightarrow k)$, if $S(k) = 0$, then $S(n \wedge k) = 0$, therefore, both $S_f(r)$ and $C(r)$ are 0. The type of entropy ratio $\gamma(r) = C(r)/S_f(r)$ is "0/0". So we should diffuse the information contend in the sample of other space into these two areas: $d = \psi(f^{-1}(k))$ and $e = \psi(f^{-1}(n \wedge k))$. Then, we establish, on the following s-dimension Euclidean space $(-\infty, +\infty)^s$, parabolic partial differential equation:

$$\frac{\partial u}{\partial t} - a^2 \Delta u = f$$

$$\Delta u = \frac{\partial^2}{\partial x_1^2} + \frac{\partial^2 u}{\partial x_2^2} + \cdots + \frac{\partial^2 u}{\partial x_n^2} \tag{4}$$

where, $u(x_1, x_2, \ldots, x_s, t)$ is temperature function (information intensity function), Δu is s-dimension Laplacian, a_2 $a^2 = k/c\rho, f = f_0/c$, k is heat exchange constant, c is specific heat exchange constant, ρ is density, f is heat resource (information resource) function $f(x_1, x_2, \ldots, x_s, \varepsilon)$.

After confirming original value and threshold, we can then solve the partial differential equation, and get $u(x_1, x_2, \ldots, x_s, t, \varepsilon)$. Then, if supposing $t \rightarrow +\infty$, we can get a function $u_\infty(x_1, x_2, \ldots, x_s, \varepsilon)$ in one space.

$$u_\infty(x_1, x_2, \ldots, x_s, \varepsilon) = \lim_{t \to \infty} u(x_1, x_2, \ldots, x_s, t, \varepsilon) \tag{5}$$

Then, we will do the integral of function $u_\infty(x_1, x_2, \ldots, x_s, \varepsilon)$ in s-dimension space $d_1 = \psi(f^{-1}(n \wedge k))$, $d_2 = \psi(f^{-1}(n))$, $d_3 = \psi(f^{-1}(\Re))$ and $d_4 = \psi(f^{-1}(k))$.

$$I_i(\varepsilon) = \int_{d_i} u_\infty(x_1, x_2, \ldots x_s, \varepsilon), i = 1, 2, 3, 4 \tag{6}$$

Finally, we can get the intensity of rule $r = (n \rightarrow k)$:

$$\Gamma(r) = \lim_{\varepsilon \to \infty} \frac{I_1(\varepsilon) * I_3(\varepsilon)}{I_2(\varepsilon) * I_4(\varepsilon)} \tag{7}$$

The second scheme is a reduced one of the first scheme. Using n source to represent the enormous heat source. Then we can easily get the results according to methods similar to the first schema.

The third scheme is a reduced one. It will compute the arithmetic mean of all the known rule intensity. When "0/0", the intensity is the multiplication of this arithmetic mean and a fixed parameter. The rule intensity function demonstrates the importance of rules from both sides, positively and negatively. If the entropy ratio of rule is more than 1, the rule is a positive one. Otherwise, it is a negative one. When the entropy ratio is 1, the rule is a common one.

3 RULE MINING ALGORITHM BASED ON INNER COGNITIVE MECHANISM

3.1 Purpose and function of heuristic coordinator

The main purpose of heuristic coordinator is to provide another approach for the focus on the process of data mining. In classic KD, the focus is directed by the interest of the user and KD will mine accordingly. But if it mines only along this direction, a large amount of potential and may-be-useful information will be neglected by the user. To help KD find more useful information, make up for the limitation of user and domain experts and enhance the self-recognition ability of the machine, we constructed the heuristic coordinator.

The function of heuristic coordinator is shown as follows. Given basic knowledge base is constructed based on attributes, by searching the non-association of knowledge nodes in knowledge base, the coordinator can find knowledge shortage, produce mining suggestion, and activate corresponding data sub-class in real database. Thus directional mining will be equipped to the system. To find knowledge shortage and the superiority of knowledge, we use the association intensity of causal rules among knowledge nodes. Association intensity of causal rules is a triple, denoted as $\pi(H, E) = <\alpha, \beta, \gamma>, \alpha = CF(E)*P(E), \beta = CF(H, E), \gamma = CF(H)*P(H)$. The three formulas include all the random uncertain and fuzzy uncertain information, which can be of help for the research of relation and evaluation of rules. Here, $CF(E)$ is the confidence of premise, $P(E)$ is the prior probability of premise, CF(H,E) is the confidence of rule, CF(H) is the confidence of result, and $P(H)$ is the prior probability of result.

After storing the elementary knowledge into the base, the reasoning arcs from reduced starting knowledge node to terminal knowledge node will be partially constructed. But there are still some arcs that haven't been constructed; we cannot only rely on the experts' experience and the degree of interest in the process of mining. We should establish a kind of searching strategy and search the deficiency of the knowledge by the machine itself. The order of searching should be based on the intensity of the to-be-mined rules.

667

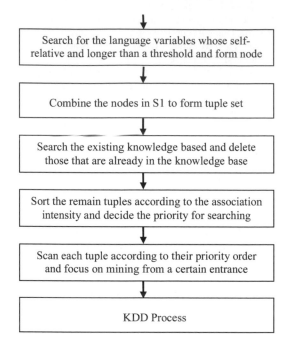

Figure 1. Heuristic rule mining framework flow chart.

3.2 Heuristic rule mining framework

Heuristic coordinator is realized by heuristic rule mining framework, the basis of which is pan-isomorphic theory and structural corresponding theorem; the flow chart of the framework is shown in Fig. 2.

By using Heuristic Coordinator and hyper-graph, redundancy and subordination, recycling will be precisely defined and eliminated as soon as possible in real time. Thus only the most likely be knowledge hypotheses will be evaluated, reducing the quantity of evaluation at the largest. In real expert system, the probability of accepting hypothesis is very low, and lots of hypothesis is redundant. So by utilizing Heuristic coordinator, the efficiency of KDD is improved.

4 CONCLUSION

In view of many rule mining algorithms have not regarded KDD as the cognitive complicated system to study its inner rules, and have not considered designing procedures to concretely implement the direct mining process of the knowledge base; some produced hypothesis rules are repeated and redundant with intrinsic knowledge in the knowledge bases. Double bases cooperating mechanism, one of inner mechanisms, can solve these problems fundamentally. We present a heuristic rule mining algorithm based on inner cognitive mechanism, the proposed algorithm can find knowledge shortage and the superiority of knowledge, and can reflect novelty and validity in KDD.

ACKNOWLEDGEMENT

This work was supported by the National Natural Science Foundation of China (No.61175048); the Key Project of Ministry of Science and Technology of China (No.2010IM020900); the 2012 Ladder Plan Project of Beijing Key Laboratory of Knowledge Engineering for Materials Science (No.Z121101002 812005).

REFERENCES

Piatesky-Shapiro, G. and Mathe C. 1992. Knowledge Discovery Workbench for Exploring Business Databases, *International Journal of Intelligence Systems*, 7(7): 675–686.

Henry G. Goldberg, Ted E. Senator. 1995. Restructuring Databases for Knowledge Discovery by Consolidation and Link Formation, *the First International Conference on Knowledge Discovery & Data Mining*, Montreal, AAAI Press, pp. 1–6.

Sarabjot S. Anand, David A. Bell, John G. Hughes. 1996. EDM: A General Framework for Data Mining Based on Evidence Theory, *Data Knowledge Engineering*, 18(3): 189–223.

Liu D., Li T.L., Ruan D. and Zou W. 2009. An incremental approach for inducing knowledge from dynamic information systems, *Fundamenta Informaticae*, 94(3):245–260.

Bingru Yang. 2002. The inner mechanism of knowledge discovery, *Journal of University of Science and Technology Beijing*. 24(3):345–351.

Bingru Yang, et al. 2009. KAAPRO: An approach of protein secondary structure prediction based on KDD* in the compound pyramid prediction model, *Expert Systems with Applications*, 36(5):9000–9006.

Kyoung-Yun Kim, Jihoon Kim. 2011. Design Decision Support System toward Environment Sustainability in Reusable Medical Equipment, *the AAAI 2011 Spring Symposium*, pp. 74–77.

Jaisankar N, Ganapathy S, A. Kannan. 2012 Intelligent Intrusion Detection System Using Fuzzy Rough Set Based C4.5 Algorithm, *the International Conference on Advances in Computing, Communications and Informatics*, Chennai, T Nadu, India, pp. 596–601.

Neil Mac Parthalain, Qiang Shen. 2010. A distance measure approach to exploring rough set boundary region for attribute reduction, *IEEE Transactions on Knowledge and Data Engineering*, 22(3): 305–317.

Information Technology and Computer Application Engineering – Liu, Sung & Yao (Eds)
© 2014 Taylor & Francis Group, London, ISBN 978-1-138-00079-7

Research on construction of digital garment automatic pattern system

Yong Qing Yang

School of Art Design, Shandong Polytechnic University, Jinan, China

ABSTRACT: Using mathematical functions to simulate clothing structure curve, complete digitalization of garment prototype structure modify; Various constraints in the method broken down into basic constraint type, simplify the constraint relationships between entities in the pattern, set up the association between shape characteristics and structural characteristics; provide basic research for the development of digital automatic pattern system.

Keywords: digital; pattern; data base; garment CAD; pattern design; block pattern

If garment pattern design system wants to realize automatic and intelligent pattern making, not only need to use parameterized pattern design method, but also need with the aid of professional technology and experience, and be able to complete express corresponding structure pattern style of certain types clothing. But, there is no perfect database in almost all the existing systems. So this paper will design the perfect garment pattern database.

1 RATIONALITY OF DEVELOPING GARMENT DATABASE

The garment PCAD (Pattern Computer Aided Design)'s intelligent development present situation stay only on the operating functions and tools used, the development and establishment of garment pattern database are still in its early stage. The design and development of garment pattern database can provides abundant experience and knowledge to pattern designer effectively. Accumulate various clothing styles and typical patterns of various parts through garment CAD digital technology, and form database files, so that system can invoke and modify those database files at any time. According to this assumption, pattern database can be embedded in PCAD system be used as Expert Knowledge Database.

2 GARMENT PATTERN DATABASE DESIGN

Garment pattern database is Taxonomic Database. Because of the need to build a comparative systemic digital pattern library, the basis of classification design of database requires not only to meet the needs of the

pattern design of the basic style, but also have the flexible ability to change of the design.

2.1 *The classification of the database*

The general assumption of database structure can be showed by Figure 1.

Garment pattern in general is divided into three series: block pattern, basic style and change style components; Because there is a big difference in the structure of men and women, so clothes can be divided to men's and women's based on gender; men's clothes can be divided into the tops and bottoms; man tops divided into some basic styles such as: shirt, jacket, vest, suits, coats, etc; then according to the change of structure of the tops components, there have shoulder line, partition line, sleeve, collar, pocket, etc; finally, also can be divided for each widget. We can build a garment pattern database which is full, orderly and

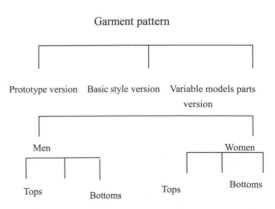

Figure 1. Template database tree diagram.

infinitely expandable if design the database on the basis of such idea.

2.2 Establish a database pattern size table

Size is the precondition of clothing pattern making. Due to there are many kinds of size for human body, so provide for the pattern size in the database come from intermediate size (women is 160/84A, men is 170/88A) in the largest proportion of human. Establish basic size table usually include across chest, waistline, hip, neck, shoulder, back length, across back, clothing length, sleeve length, skirt length, trouser length, hip height, body rise, design parameter (the parameter that control of the change of lengthwise shape) and other major anthropometric dimensions.

2.3 Garment block pattern data structure

Block pattern is the foundation of the clothing structure design, its structure is the most simple and contain the information of the most important site of the human body, strive to the pattern has the most wide coverage. Broadly speaking, prototype also includes the simplest structure pattern in all varieties of clothes. So the date in block pattern database come from two aspects: one is basic block pattern for men and women, another is the simplest block pattern in some basic style categories, often referred to as the "name of style category + prototype". Such as women suit prototype, men's jacket prototype, dress prototype, trousers prototype, etc.

2.4 Structure of garment parts date

Garment parts data is used for change the garment style. Pattern structure will change when garment parts style change. Garment parts structure is the foundation of clothing pattern design. Take the change of collar structure for example, when a style file is established and needs to change the collar style, only need to call-out the parts database and select the desired type. When the selection is done, enter the variables such as neck width, collar length, collar point angle, etc, then the new collar pattern is generate and save in the collar parts database at the same time.

2.5 Pattern making method based on parametric design

The data in garment pattern database must recycle, namely achieve the function that pattern will generate automatically when replace the parts data, the pattern making method of parametric design is more applicable. Parametric design refers to a mode that a set of parameters to define geometry dimensions and appoint the relationship of the size. It can be provided to designer for geometric modeling design, and the figure can be updated automatically when the size of geometry had change. The graph description of garment pattern can be divided into graphic topology relation, graphic geometric parameters and

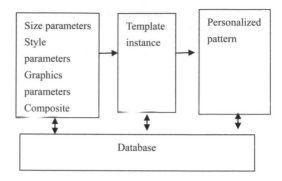

Figure 2.   Garment pattern parametric design principle.

the relationship between those geometric parameters and graphic structure parameters. Pattern's topology structure can be regard as the rule of pattern making; pattern's geometry parameters form the geometry information, such as the coordinates of point; the structure parameters of garment pattern can be summarized as the following four types:

1) The parameters to define the dimensions which refer to the size that listed from pattern design software's standard measurement chart.
2) The parameters to define the style which refers to a variable that meet the requirements of garment style and process design prescribed by the operators themselves in the design process, such as degree of looseness, dart measure and so on.
3) Define the parameters of drawing object properties. Length, width and other properties of graphics object can be defined as variable parameters by operators. For example, define the armhole curve as variable, in order to make the sleeve sample as a reference.
4) Complex parameter means when the above three kinds of simple parameters are difficult to express, and the mathematical or logical operators can be used to connect these simple parameters together to form parameter expressions. The essence of the pattern parametric design is record pattern making (graphic topology relation) process in the program, and uses a set of parameter constraint a set of pattern structure size sequence. There has a correspondence between parameters and the control size of patterns, when give different parameters of value, original pattern will become the new pattern. (See Figure 2)

3   THE USE OF THE PATTERN DATABASE

Establishing scientific and systematic database in the process of garment CAD pattern making, only need use the combined pattern element method, by changing the main parameters data, using parametric design drawing method to established a big frame shape, change the control parameters of structure line or add new structure line, the garment pattern we need can be generate in a short time. Though this method, pattern designer can enrich their pattern library in the process of pattern design, so that the process of garment pattern design will more effectively, diversified, freely and circularly.

There has part of data is no formula can depend on in the process of pattern making, such as the positions of pockets, the positions of the design dividing line and the size of pleat measurement, etc. These data in the use of computer tools to multi-code make pattern generally take a fixed number, so the fixed number must transform a formula which rely on use the size in specification as a parameter then involved in the actual pattern making applications. The basic principle of conversion is: count horizontal dividing line according to the clothing length proportion, count vertical dividing line according to the proportion of chest.

Data of database can not in strict according with the classification of garment structure to store at sometimes, and must be adjusted appropriately according to the characteristics of working sketch, special processing for special circumstances. Take female suit lapel style for example, lapel can be divided into two categories, have collar stand and no collar stand. Their curves are very similar in graphics expression, and the main distinction is the difference in neckline and placket. Then, the placket can be treated as the internal lines of garment pieces. So the lapel which either has or no collar stand can be classified as one type of structure, pattern making only need to define some parameters such as the curves curvature and tilt angle of top collar of lapel.

4 INTEGRATED DEVELOPMENT OF BUSINESS COMPONENTS UNFER THE APPAREL INDUSTRY CHAIN PLATFORM INTEGRATION FRAMEWORK

Components of basic service type, common interface type and platform management type make up kernel components of the apparel industry chain collaborative work platform, providing the basis for development of other components, especially business ones. When developing business components, kernel components of the platform can be called as needed to accomplish acquisition of alliance, district and cooperative relationship information on the platform. Business components integrated development mode based on kernel interface of the apparel industry chain collaborative work platform is as shown in Figure 3.

The kernel interface components and business-style ones of the platform can be called as needed during design and development of business components, so as to simplify development of business components and realize integration with the platform. When finished, business components can be tested under the integrated environmental test framework which is a kernel running program deployed in testing environment. The procedure provides basic functions such as business component registration and loading, user authorization

Solution 1		Solution 2
Business component	Business component management	Alliance management
Business component		User management
Business component		Authority management
Component library		Daily management
	Common interface	Cooperative relationship management
		Operation monitoring

Figure 3. Industry Chain Collaborative Work Platform Based on SaaS.

and login, etc. Business components can be registered and tested in the integrated testing environment.

Among them, components of basic service type mainly provide some general basic services like message and short message management; components of common interface type are interface service provided by kernel of the platform, mainly including user identity interface, alliance message interface, etc., and they can be called by business components to obtain relevant information for development and realize integrated development under the platform architecture; components of platform management type mainly includes those of user, enterprise, alliance, cooperative relationship and authority management type; components of sales type are mainly used for accomplishing collaborative sales business of the platform, and those of purchase type are mainly used for accomplishing collaborative purchase business of the platform; As a collaborative work platform supporting apparel industry chain enterprise cooperation, the apparel industry chain platform demands dynamic expansibility in system structure. Component technology is effective means to solve system reconfiguration and dynamic expansion. It simplifies development and maintenance of the platform through component reusing and assembly based on components. The paper carries out study on platform component model, component classification management based on XML as well as development and debugging of business components under the platform integration framework.

5 CONCLUSION

This article conducts the research to garment pattern database, presents the structure classification basis of establishing garment pattern database, and analyzes the content of data structure so as to form data by applying parameter design graphics, state the circular usage rule of database and handling method of database for special data.

REFERENCES

[1] F.F. Zheng: Three-dimensional Parametric Standard Parts Research and Implementation. Journal of Computer-Aided Design & Computer Graphics. 1999, 11(3) pp. 218–220.

[2] L.Y. Li, H.Z. Zhang and Q.R. Han: Garment CAD Theory and Application, (China Textile Press, China, 1997), p. 6269.

[3] B.Z. Tan, B.Y. Bang and C.Z. Chen: Based on object-oriented group technology CAD systems Research. Journal of Computer-Aided Design & Computer Graphics. 1999, 11 (5), pp. 437–440.

[4] X.X. Meng: Parametric design study Journal of Computer-Aided Design & Computer Graphics. 2002 (11), pp. 1087–1090.

[5] D.P. Zuo, Z.W. Huang: Parameters of the Garment structure of CAD research. Journal of Beijing Institute of Clothing Technology, 2003, 60(1).

[6] D.W. Ralston: Principles of Artificial Intelligence and Expert Systems Development (Shanghai Jiaotong University Press, China 1991). J.Q. Chen, T.X. Yuan and L.Z. Ge: translation. pp. 103–109.

[7] W.B, Zhang: Clothing Technology, Structure and Design volumes (China Textile Press, China, 1997) pp. 3–4.

[8] G.D. Lu, Z.Q. Wu and J.H. Jin: Serialization Graph Generated Automatically Formming Method. Journal of Engineering Graphics, 1992(2), pp. 42–48.

[9] L.Y. Li, H.Z. Zhang and Q.R. Han: Garment CAD Principles and Applications (China Textile Press, China, 1997).

[10] Q. Zhang: 21st–Garment CAD to Enter the Era of Intelligent, Journal of Chinese and Foreign Sewing Equipment, 1999, 21(5).

Information Technology and Computer Application Engineering – Liu, Sung & Yao (Eds)
© 2014 Taylor & Francis Group, London, ISBN 978-1-138-00079-7

The search of build ERP system platform of brand clothing enterprise

Yong Qing Yang

School of Art Design, Shandong Polytechnic University, Jinan, China

ABSTRACT: This article proceeds from the necessity of the brand clothing enterprise to establish and implement ERP architecture platform, then analyses some problems in the ERP application, discuss on platform build and method research systems of ERP architecture, in order to provide reference for brand clothing enterprises to implement ERP successfully.

Keywords: brand clothing enterprise; supply chain; design principle; ERP system

ERP (Enterprise Resources Planning) is software of integrated management which aimed at manpower, financial, logistics and information resources within the enterprise, is the most advanced management mode in manufacturing enterprises currently.

1 THE NECESSITY OF THE BRAND CLOTHING ENTERPRISE TO ESTABLISH AND IMPLEMENT ERP SYSTEM

ERP system is one of the most high-profile management modes in the present China. In today's global economic integration, Chinese brand clothing enterprise introduction and implementation of ERP has its special importance. The core of the ERP management idea is to achieve the effective management of supply chain, namely the customer demand, enterprise internal manufacture as well as supplier's manufacture resources conformity in the same place, has form a network of supply and demand of the system building, and to effectively manage all links in the supply chain. These links include: orders, purchasing, inventory, planning, manufacturing, quality control, transportation, distribution, service and maintenance, financial management, personnel management, project management, etc.

1) Help enterprises to optimize the internal management and realize management innovation.

 Many brand clothing enterprises in China have some problems on the management concept and management mode, such as system implementation is not strict, post setting repeat, human resources is not reasonable, financial information false and Co-ordination not in place and etc. These problems seriously affected the scale expansion and efficiency improvement of the brand clothing enterprises, than make the low market competition of enterprises. Through the introduction and implementation of the ERP system, can make the brand clothing enterprises to increase resource utilization rate and market sensitivity, enhance the strain capacity of brand clothing enterprise to market, and achieve the innovation of enterprise management.

2) Improve the operation efficiency of the enterprise

 There is a repeated phenomenon that a lot of brand clothing enterprise in the department and the post setting, more enterprises have many unreasonable business process because some old management concept. It is easy to appear the phenomenon of buck-passing between each department when the division of responsibilities is indeterminate, seriously affect the enterprise's operation efficiency. Implementing ERP system, brand clothing enterprises will be in-depth analysis and optimize the process for all departments, and break down the responsibilities, rights and interests into every link. With the establishment of system, channel dredging and implementation of management, the efficiency will be promoted. Figure 1 is one of the apparel supply chain architecture under the large-scale production and sales network model provide by brand clothing enterprise. It consists of purchasing, product design, production, inventory, sales and etc, able to exert its flexible and efficient management performance.

2 THE SYSTEM CONSTRUCTION UNDER THE BRAND CLOTHING ENTERPRISE ERP SYSTEMATIC EVALUATION OF INTERNAL CONTROL

ERP system's goal is through the close cooperation between trading partners, provide maximum value and

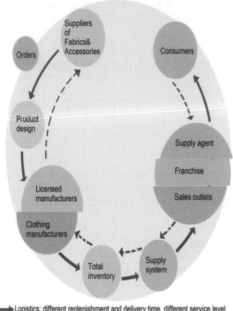

➤ Logistics: different replenishment and delivery time, different service level

⇢ Fund flow, order flow: demand diversification, different replenishment frequency

Figure 1.

best service with minimal costs and charges. ERP system's internal control design is the pilot work of internal control build, and its rationality will directly determine the construction effect and running effect of internal control, it is the foundation of the construction of internal control.

1) Systemic evaluation of internal control of clothing ERP system

Brand clothing enterprise internal control system is an organic system consisting of many factors. There have mutually affective and restrictive relation between each element. Elements combination for the system, system as an organic link of the whole, gets the new characteristic that each part hasn't had before, and this new feature is a result of interaction of elements, whole system and external. So, if the brand clothing enterprise internal control is incomplete, lack of systematic, like that organic connection between each element will cease to exist, lead to the overall functionality greatly reduced.

2) The adequacy evaluation of the internal control

When design the internal control of brand clothing enterprise, the first step is set the goal, it is the starting point and endpoint of control activities, internal control must revolve around the goal of brand clothing enterprise, each control element and control measures should served the target, and aimed at each goal of internal control evaluate its reliability. Whether the control measures used by internal control system adequacy for achieve the goal depends on if the internal control's goal

can be realized. If the internal control goal can be achieved, then the brand clothing enterprise the internal control system is effective.

3 THE EFFECT OF IMPLEMENT ERP SYSTEM OF BRAND CLOTHING ENTERPRISE

In the enterprises' large-scale operations, optimizing the entire supply chain from manufacture to consumer, and use the ERP system to achieve real-time information technology statistics and management, can effectively improve the "loose" link in enterprise large-scale operation. More importantly, the optimized supply chain is unblocked and efficient, and it is the important means to increase the size of brand scale competition strength. The implement ERP system of brand clothing enterprises can create huge benefits to the enterprise. Build the internal control is to reduce the possibility loss, and the design and implementation of internal control need pay a certain cost, so brand clothing enterprise must focus on the relation between loss that may reduce by implement internal control and implementation cost, and the control cost compared with the control gains must be economic. Or the control link, procedure and method of the enterprise internal control must be necessary and reasonable, parts of duplicating and overlapping can not be existed.

The implement ERP of brand clothing enterprises is destined to become the mainstream of promote the enterprise management level. How to more effectively implement ERP system, and take the corresponding measure aim at the problems existing in the ERP implementation? First, formulate the implementation plans, prepare for the economic and human before the implementation. The implementation of ERP is a high investment project which accompanied by certain risk, need to made all the preparations before its implementation. Second is the professional staff training system. "human" is a key of success of system, as a result, if the enterprise want to do a good work of implement the ERP system, must make good ERP knowledge training for employees at first. The training is not only aimed to the operation and management of each module involved in the ERP system, but also to set up the correct understanding of ERP for staffs, let staffs convinced the system from the heart. Third, strengthen the supervision of system.

4 CONCLUSION

"There is no end in ERP construction". ERP project's implementing in brand clothing enterprise means update of management concept and reengineering of management process, the corresponding internal control system is bound to be constantly modify and perfect with the use of the ERP system. In the entire life cycle of ERP system, brand clothing enterprise should constantly self-test and self-evaluation for the effect of ERP internal control, and change and expand the internal control system continuously.

FUND PROGRAM

This is college humanities and social science research project of Shandong province, project number (J09WB01).

REFERENCES

[1] Stevenson, W. J. *Production and Operation Management*[M]. Q zhang, & J zhang, (Trs). Beijing: China Machine Press, 2000.

[2] Z. Q. Wan, H. J. Song. *Garment Production Management*[M]. Beijing: China Textile Press, 2004.

[3] X. H. Liu. *Fashion Marketing*[M]. Beijing: China Textile Press, 2003.

[4] Kale, V. *SAP Software Implementation*[M]. Y. Zhu, (Trs). Beijing: China Renmin University Press, 2003.

Information Technology and Computer Application Engineering – Liu, Sung & Yao (Eds)
© 2014 Taylor & Francis Group, London, ISBN 978-1-138-00079-7

BIM technology applications explore in the Beijing No. 4 high school at Changyang campus of mechanical and electrical design

Y.D. Cai
Beijing University of Civil Engineering and Architecture, Beijing

D.G. Dong
Beijing Institute of Architectural Design, Beijing

D.Y. Li
Beijing University of Civil Engineering and Architecture, Beijing

J. Zhang
Beijing Institute of Architectural Design, Beijing

ABSTRACT: Through actual case of Beijing No. 4 high school at Changyang campus, this paper presents electromechanical professional system. The system was analyzed by building the electromechanical professional building information model. Based on analysis of project management files, product, professional collaboration and other difficulties of the BIM technology application in the project practice, and practical application experience finally was summarized.

Keywords: BIM, Building Information Modeling, Project document, database, Professional collaboration

1 INTRODUCTION

With the development of science and technology, information technology is increasingly dominant the society. In the field of construction engineering, architectural design has run to the Building Information Model (BIM) from the original hand-painted drawings to CAD drawings. The information technologies play a particularly important role.

BIM is building information model created by digital technology, which covers a large number of related information in the entire life cycle. BIM provides a way of thinking or ideas. As the core of information to answer the problems encountered in the all life cycle by this workflow.

According to related statistics, from 2002 to the present, the application of BIM technology in Europe and the United States and other developed counties probably account for about 30 percent of the project. Although our BIM application is still in the primary stage, the academic and practical application in the field has been recognized to the importance.

This paper taking Beijing No. 4 high school at Changyang campus as an example, describes electromechanical professional system, and to try and explore its application of BIM technology.

2 THE ENGINEERING GENERAL SITUATION

Changyao middle school project locate in Changyao town, Fangshan District, Beijing. The campus planning a total construction area is $55484\,m^2$. The construction category is multilayer civil construction. Its main building height is 24 m. It composes of two parts of the teaching building and dormitory. Teaching building has 6 floors above ground and one level underground. Student dormitory has 4 floors and 1 basement. The architectural effect is shown in Figure 1.

Figure 1. Architectural effect diagram.

Table 1. Heating and air conditioning design.

Type of building	Program
Office, conference, classrooms	Multi-split air conditioning system + radiator heating system
Dormitory	Radiator heating system, the reserved split air conditioner installation condition
Auditorium	Constant air volume and all-air heat recovery conditioning system (the transitional quarter or public health emergency program startup system can run 70% fresh air)
Gymnasium, lecture hall, indoor activity space	Radiator heating system
Auditorium foyer	Constant air volume and all-air heat recovery conditioning system (the transitional quarter or public health emergency program startup system can run full fresh air) Radiant floor heating on duty
Restaurant	Fan coil unit + circulation wind air conditioning + fresh air system

3 ELECTRICAL AND PROFESSIONAL DESIGN

3.1 HVAC system

3.1.1 Cold and heat source

A lecture hall, hall, restaurants use ground-source heat pump system as cold and heat source.

The B1 layer cold and hot source equipment room set two ground-source heat pump units, single rated refrigeration capacity is 289KW, and the heating capacity is 288 KW. Refrigeration conditions chilled water inlet and outlet temperature is 12/7°, the water-side temperature is 30°/35°. In the heating conditions, the hot water temperature is 45/40°, the temperature of the water side is 8/4°. Using vertical double-U-shaped buried tube heat transfer holes amount to 144, spacing of 5 m, depth of 120 m, below ground in the playground.

In winter heating, the regional boiler room provide a hot water to the classrooms, offices and dormitories area, the water temperature is 95/70°. And secondary water after heat exchanger, temperature is 85/60°. The borne of summer cooling loads use a multi-split air conditioning system.

3.1.2 Heating and air conditioning design

Some system classification is shown in Table 1.

3.2 Water supply and drainage system

3.2.1 Domestic water system

Indoor domestic water supply system has two areas in building vertical. The underground layer and interlayer is low area, set up a set of flow for 25 m³/h, head for 50 m frequency conversion water supply equipment and a 45 m³ water tank, using municipal water direct supply. On the ground floor to the fifth floor is high zone, using frequency conversion water supply equipment pressurizes water.

3.2.2 Domestic hot water

A primary heat source is the solar-thermal systems. While standby heat source adopt regional boiler room heat-supply pipe network in the winter, and use gas water heater heating in the summer. The domestic hot water preparation, water supplement, recycling facilities and other equipment are all located in the student dormitory basement heat transfer room. The domestic hot water heat consumption per hour is 142.6 KW, providing 60° hot water for a shower.

3.2.3 Drainage

Indoor drainage system uses the way of dirt waste water confluence. In the outside, confluence treated by septic tank back to outdoor municipal sewage pipe network. Sewage of the kitchen are processed through grease treatment, discharging into the nearest neighboring municipal sewage pipeline.

3.2.4 Rainwater collection and utilization

Roof rain is the within drainage system. The rain of the playground and basketball court after drainage tubes to collect the rainwater by drain, into the rainwater collection pool, processing for water filling water, street flushing water, green land irrigation, etc. The teaching area outdoor sidewalk use permeable practice to reduce discharge rainwater and infiltration into the underground reservoir, the excess rainwater through the inlet for storm water to the municipal rainwater pipe network.

3.2.5 Fire water supply

The indoor set the fire hydrant system, pipelines into a ring-shaped layout. Underground garage, auditorium set up automatic sprinkler system.

3.3 Electric system

3.3.1 Power supply

This engineering load rating is the secondary load of users. The important load, firefighting equipment power supply are based on double-circuit power supply, at the end of each other.

3.3.2 Fire detection

Automatic fire alarm system is designed according to a class fire rating, using overall protection mode. Fire control room set on basement level one. Less than 15 KW motor started directly, greater than 15 KW motor adopt reduce voltage startup.

3.3.3 Light

Activity room, office, classroom, laboratory, garage, substation, weak current distribution room and other function rooms adopt high photosynthetic efficiency of the three-band fluorescent lamp and low loss of high-frequency electronic ballast or energy-saving electromagnetic ballasts with high efficiency lighting.

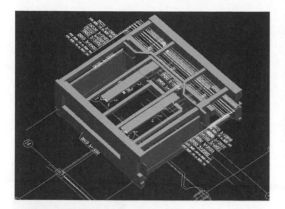

Figure 2.   Classroom model.

Figure 4.   Basement model.

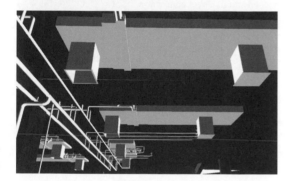

Figure 3.   Restaurant model.

4   BIM PRACTICES

4.1   *Software selection*

BIM software is widely used both at home and abroad at present, such as Autodesk's the Revit, Graphsoft's the ArichCAD and so on. MagiCAD of this selection is based on CAD platform software. Its main reason is the use of CAD platform, for two-dimensional operating habits of the people easy to learn and use.

4.2   *Working method*

Apply BIM technology mainly for Layer 2 standard classrooms, a restaurant, a basement garage and power station area. In view of the project area is not large, the only distinction made into a separate file system, including air systems, water systems, water supply systems, drainage systems, sprinkler systems, fire hydrant systems, electrical systems.

4.2.1   *Function analysis*

In accordance with the requirements of the architecture, use an integrated three-dimensional data model, to analyze local scene.

4.2.2   *Modeling and optimization*

The design results in all the major coordination on the basis of checking and inspection.

4.2.3   *Detailed design*

Detect professional collision problems and optimize the layout of integrated pipeline. Try to run the debugger, and generate collision report.

4.3   *Use of software*

4.3.1   *Project document*

In professional collaboration, electromechanical common to use a project documents, collaborate together to complete the design task.

For a new project, you need to create a new project management documents. First and very important is the need to establish the concept of the floors, in order to accurately determine the floor and the origin of installation. Secondly, you can improve project documents in the process of the project. To set the project file of parameters, can also reconfigure reasonably according to the different each project, each professional staff work style, work habits. For a similar project, there is no need to create a new project management file, and can choose to modify the existing project document.

It is worth noting that the project management documents recommended by a special person. This is because, when increase the setting of a project management file, if the non-management staff increase or modify the settings by self, then professional collaboration will use different project management documents.

4.3.2   *Application of product*

Products in the database should have the right size and accurate information, reflecting the core of the building information model. Although the product is continuing to improve, it does not meet the situation. At this moment, there are three ways to solve it.

The first method is editing the product. Second, edit block. Third, you can seek the help of software manufacturers.

Here to underground refrigeration as an example, the ground source heat pump water chiller CH 1, rated at 289 KW cooling capacity, rated heat is 288 KW, control size is 3600 mm long, width is 1000 mm, and

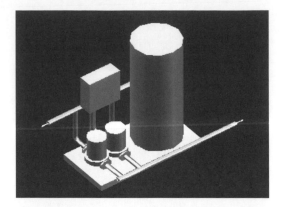

Figure 5.   Constant pressure replenishment equipment.

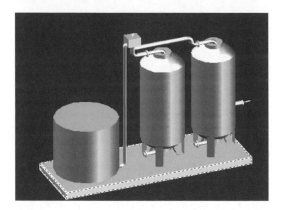

Figure 6.   Softened water device.

Table 2.   BIM technology problems and the solution.

Problem	Solution	Graphic expression
Slope	When it comes to the slope of pipes, the adjustment is too much trouble. If there is a break, need to be repainted.	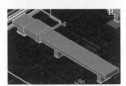
Reducer tee piece	Now, software support variable diameter tee in the contour of the premise.	
Intelligent connection	No-interface. Direct connection.	
Valve	Cannot use the copy command of this valve. Because at the time of installation, make the pipe insertion point. Copy the branch command can be used.	

height is 1700 mm. Did not meet the size of the chiller, we edited the product to complete production. The constant pressure replenishment equipment and softened water devices were not in the product. Therefore, we adopted the method of block instead temporarily. Constant pressure replenishment equipment is shown in Figure 5. Softened water device is shown in Figure 6.

4.3.3   *Problem and solution*
The following are the problems in the process of BIM practices, solution and graphic expression. BIM technology problems and the solution are listed in Table 2.

4.3.4   *Problem and solution*
We should improve attention in the pipeline comprehensive: every professional should be communication timely to complete a preliminary comprehensive as early as possible. Professional collaboration, not only refers to the comprehensive of mechanical and electrical major, it is more important with the construction of walls, doors and windows, floor, etc., and the structure of the foundation, beams, plate, etc.

In this case, architecture used Revit to design, while engineering chose MagiCAD for CAD. Both belong to the BIM technology, just the format conversion between different software. When cooperate

with civil engineering model, people should export the single-layer 3D and 2D DWG format, and coordination with electrical and mechanical model, in which the two-dimensional construction model mainly is to determine horizontal location of the electrical and mechanical model, 3D building model is aimed to determine vertical position.

4.3.5   *Post processing*
The Navisworks software with the model of Revit, MagiCAD is linked to the post-production work, which introduced at this stage after the completion of the building information model. Navisworks software is the real-time 3D visualization tool software. It relates to the field of architectural design, engineering, construction, mechanical design, etc.

In this case, we will import model to Navisworks. According to the different systems, such as wall, column, air conditioning wind system, electrical system, etc., create selection should be given real material. Navisworks software can make the professions better coordinate through the visual, detecting the collision of various professional problems, optimizing comprehensive pipeline layout, trying running debugging, Collision effects are shown in Figure 7 and 8.

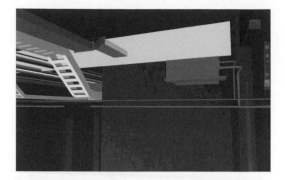

Figure 7. Original design basement.

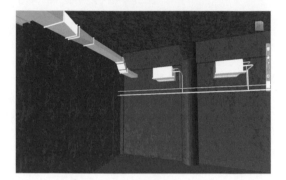

Figure 8. Modified design basement.

Collision report identifies collision. The picture is the hard collision of architectural structure and air system, which distance is -0.12 m. The professional pipeline is various and intensive in the basement, so that the net space cannot reach requirements. Coordinate by BIM technology to make it possible to meet the space requirements in a number of professional.

5 SUMMARIES

In the whole process of implementation of BIM, we aimed at every step to combine BIM information with construction better, from the initial analysis to post-processing of the project.

First, the visual advantages of BIM technology help to enhance communication.

Visual advantages of BIM technology can enhance professional communication. Through constructing building information model, let more people understand as soon as possible and grasp the essence of different professional design content (Li and Yang 2012).

Second, information advantages of BIM technology will help improve the information precision.

BIM technology can replenish information at any time, formed a powerful information resource library gradually. We can be convenient, fast and efficient check information and coordinate work.

Third, BIM technology is more emphasis on the importance of professional collaboration.

Professional coordination is the core work of the entire design process, to avoid exchanging the content constantly. While overcoming a lot of troubles caused by the information exchange not in time, so that we are able to view real-time changes in the relevant profession and timely identify problems and correct mistakes.

Fourth, using of BIM technology to improve efficiency.

We should focus on the entire project cycle and play high efficiency on BIM technology. Due to the incomplete product library, lack some non-standard connection, etc., may cause a slight drop in efficiency in the beginning stage. With the understanding of BIM technology more deeply, operate more skilled, the software open more perfect, BIM technology advantage will be more stand out. In terms of design, what you see is what you get is more advantageous to design optimization, reducing the unreasonable design. The construction can improve the efficiency, reduce the construction period and cost savings benefits. In the aspect of operations, can observe work in detail, and improve the efficiency of the engineering control greatly.

Fifth, BIM technology applications affect the design of traditional workflow.

The two-dimensional drawing workflow may not apply to BIM technology. BIM technology is more emphasis on improving the accuracy of primary drawing, reducing due to lack of communication and brought a lot of modification work.

Sixth, BIM technology requires increasing professional content continuously.

BIM technology more emphasis on the multifaceted development of talent, you need to understand and comprehend the different professional design content and some BIM software.

Seventh, BIM technology can be combined with green building for research.

With information as the core, Using BIM technology establish building information model, extending application of other auxiliary analysis software, such as energy analysis software, airflow analysis software, operation management software and so on, to optimize design. BIM technology and related auxiliary software should have a good combination and operability of the practice of engineering projects, to realize the construction of information technology.

REFERENCES

Li H.Y. & Yang Y.C. 2012. The applications of BIM in architecture & structure designing of "Haitangwan mangrove seven-star hotel project". The *Second international conference on "Application of BIM technology in design, construction and real estate cooperation" China, 8 November 2012. Beijing:22–31.*

Information Technology and Computer Application Engineering – Liu, Sung & Yao (Eds)
© 2014 Taylor & Francis Group, London, ISBN 978-1-138-00079-7

Analysis of the features of religious language from the development of socialism

J. Li, H. Yin & W.J. Gu

Faculty Of Mechanical And Electrical Engineering, Kunming University of Science and Technology, Kunming, Yunnan, P. R. China

ABSTRACT: Religious language expresses religious thoughts which can be understand in a broad as well as a narrow sense. It plays a critical role in the national formation and social development. Religious language is different from daily language, philosophical language, the language of science and art. The features of religious language are independent, special and regional. Our religion is under the conditions of socialism existence and activity. So it must accord with the socialist society, the creation of a harmonious society.

Keywords: Religion; Religious Language; Features; Social harmony

1 GENERAL INSTRUCTIONS

Religion is a human characteristic. China is a multi-religious country, as Chinese religious including Buddhism, Taoism, Islam, Catholicism and Protestantism. According to incomplete statistics, there are various religious faiths in China, more than 100 million religious people, 85000 sites for religious activities, about 30 million religious teaching people, more than 3,000 religious organizations. Religious groups also organized 74 religious schools to train religious personnel.

President Jintao Hu once pointed out that, under the new historical conditions, we must adhere to the Marxist stand, viewpoint and method. A comprehensive understanding of the objective reality of religion in a socialist society will exist for a long time. A comprehensive understanding of the religious issue with political, economic, cultural, ethnic factors intertwined complex situation, a comprehensive understanding of the special status of the religious factor in the contradictions among the people, efforts to explore and master the law of the religion itself, and constantly improve the level of religious work [1]. This shows understanding of religion is very important and of full practical significance, the dialectical analysis of the language of religion is to know and understand the basis of religion.

2 RELIGIOUS LANGUAGE AND ITS CLASSIFICATION

Human language is the most important communication tool. It is the direct reality of ideas as a tool for thinking. Language is a special kind of social phenomenon. It emerged with the emergence of society, and developed with the social development, treating for various social classes equally [2]. It can be called religious language when the everyday language used in the description or thinking of the religious phenomenon. Religious language (Religious Language) refers to a variety of religious language in the form of expression of their beliefs and religious sentiments.

Religious language can be divided into broad and narrow. Generalized language of religion includes the language used to communicate with each other, as well as the physical symbols and actions, behavior symbols in religious classics and religious activities. It means that generalized language includes religious entity etiquette, entity objects, such as murals, in addition to the language itself. Implement, robes the monks wearing, the cross Christians wearing all belong to the generalized symbol of religious language.

Narrow religious language refer only to the language used in the field of religion. It is the use of religious texts, the relatively secular language in terms of the ultra-secular language.It has a specific concept, scope and Language chain used to express and interpretation of religious doctrine, religious rules as well as metaphysical religious ideology.

A lot of words are created in religion, which constitute the religious discourse to depict the presence and perfect nature. This language does not have the characteristics of reasoning, experience and empirical, with great symbolic meaning and inspiration. These symbolic and apocalyptic religious terminologies constitute a religious language features. So can religious language called "symbolic language" [3].

3 DIFFERENCES BETWEEN RELIGIOUS LANGUAGE AND OTHER LANGUAGES

3.1 *Difference between religious language and everyday language*

Firstly, everyday language and religious language are largely similarity. Secondly, there is a great difference between the religious language and everyday language because of religious language to be able to give new meaning to the everyday language [4]. The original semantics and experience of everyday language are the basis of religious language. The vocabulary in religious language often contains more profound symbolic significance, and broader coverage.

3.2 *Difference between religious language and philosophical language*

Philosophy is an understanding of nature, society and life, comprehend and understand. It is not difficult to understand if look back at the history of philosophy. The opening of ancient Greeks is the thinking of natural phenomena. The first emerged philosophy is also called natural philosophy. They questioned the basis of the world's existence and the structure of the world. Then they gave answers to these questions. Though the numerous answers are not uniform, it really is progress of the human mind to thinking on these issues. Their philosophical rationale does not seem to meet limited to natural areas. It finally shift the focus on human, as Socratic "know themself" is the best example. Objects of philosophical reflection gradually enriched the lives of people such as how to happiness, what kind of society is the ideal society, et al. Political, ethical, moral, legal, art, language, and et al have become the objects of philosophical reflection. In this way, philosophy greatly enriched the spiritual life of mankind, and plays an important role in the spiritual life of mankind.

3.3 *Difference between religious language and scientific language*

Scientific language is the language of logic, reasoning. Scientific language is the empirical language. It is based on everyday language. And scientific language is clear, precise directions variant, until completely exclude all incidental symbolic meaning. Religion is not logical reasoning, scientific proof to reveal the nature and law of nature and society. Religious language is different from scientific language as it never use the concept, judgment, reasoning and logical form to express. But it has a unique different logic of scientific language the language logic [5].

Einstein once said that, science without religion is blind; religion without science is lame. Scientific has the uncertainty characteristic. And thus can become a shared understanding of man. Any healthy religion and philosophy advocates science, although people of different philosophical have different views on the role of science and its ultimate meaning. In short, science is certain, but it will always be partial. Religion has two connotations. First, it provides a stone for confused religious people to rely on, which essentially safeguards the basic human family relations order. Second, it contains the knowledge of the truth (this philosophy) and witnessed the way of the truth (this is not the same with the philosophy). In fact, the only consciously or unconsciously witness of truth through religious methods, are likely to become a philosopher. People cannot enter the kingdom of Philosophy through the accumulation of knowledge and thinking.

3.4 *Difference between religious language and artistic language*

The famous American philosopher Nelson Goodman Professor said in the book Artistic Language that, the language of art is the system of symbols used in the various disciplines of the arts creation. It is the substance media for the artist to generate their conception and work. Artistic language is the symbol system for art of different categories or different artists. Art is the image to reflect the typical life ideology, which including literature, painting, sculpture, architecture, music, and so on. Religion and art developed together from the origin [6]. Their relationship is close, as religions rely on a certain artistic language to performance, through the application of certain artistic language to develop, disseminate. The arts also develop because of heavy use of religious language and produce a large number of works, becoming a large part of the most important human treasure.

4 THE INDEPENDENCE OF THE RELIGIOUS LANGUAGE

Religious language is based on everyday language, attached to everyday language. But there is a certain degree of independence. For example, Buddha, God, Allah, Lohan, Road long, monks, novices terms, etc. For the title of death, religion has a special language, and different from everyday language, including Taoism Feather, soaring, passed away and so on. Feather, adult wings to take this means changes after soaring immortality. Taoists call old and sick to death as Feather. Nirvana is a transliteration of the Sanskrit, meaning death. Perishable refers to the highest state of Buddhist. That is, to become a Buddha, passed away.

5 THE PARTICULARITY OF RELIGIOUS LANGUAGE

The particularity of religious language is that it tries to describe the transcendent reality mystical experience by human language. So the experience is beyond the day-to-day experience of the human. And thus the expression of the exact and validity often become a problem. Religious philosophers and theologians who tried to resolve this problem take religious language as analogies and symbolic language. Analogical language refers to religious language to describe the content, not the day-to-day language can grasp. People express the similarity between objects through religious language description of the object and the language of daily life to understanding its meaning through association and analogy. Such language is often used to describe the nature and attributes of God metaphysical proposition. Religious language of symbolic will be deemed to convey the transcendental real meaning of symbols, of the ineffable object speaking, is a manifestation of symbolic significance, rather than empirical description.

Tillich thought that the religious symbol has four characteristics. Image, the visible symbol vividly expressed not visible reality. Cognitive is the recipient's insight into its meaning. Inner strength, that is the ultimate concern of the strength of the carrier, and its inner strength cannot be separated. The characteristics widely accepted as a society that is a symbol of the symbolic expression of religious belief is accepted as a religious community as a whole, a symbol of the groups become available public exchange of symbols, which has contributed to the mutual understanding of the members of the groups in the religious beliefs.

Tillich believes that the in general religious symbols unconditionally beyond the field of the human mind. It is the manifestation of that point implicit in the religious activities of the ultimate reality. "Analogy and symbolic language" legitimacy has been questioned by the logical positivist philosopher. It may be pointed out that the particularity where such particularity is precisely its legitimacy.

6 RELIGION AND ITS ADAPTION TO SOCIALIST SOCIETY

China is a socialist country. Our religion is the existence and activities under the conditions of socialism. It must adapt to socialist society. This is both a socialist society the objective requirements of our religion, our religion itself the objective requirements. We are building socialism with Chinese characteristics, in line with the fundamental interests of the broad masses of the people, including the masses of believers, including political basis for doing religious work.

Actively guide religion and socialist society to adapt, not asking for religious and religious believers to give up religious belief, but for their love for the motherland, support the socialist system, and to uphold the leadership of the Communist Party, and to comply with national laws, regulations and policies; requirements they engaged in religious activities should be subordinated to and serve the highest interests of the country and the interests of the nation as a whole; support their efforts to comply with the requirements of social progress interpretation of religious doctrine; religious laws and regulations within the scope of activities, religion and socialism compatible central element. Carry forward the patriotism in the fine tradition and religion of the positive factors to guide the religious, inhibition of the negative factors that unite the religious believers to actively participate in the modernization, which is adapted to the specific requirements of the religion and socialism.

7 SUMMARY

President Jintao Hu pointed out that understand and deal with religious issues correctly, earnestly relates to the party and the country overall relationship, social harmony and stability, the relationship of building a moderately prosperous society in process, and the relationship between the development of the cause of socialism with Chinese characteristics. From such a strategic height, we fully understand the new situation of the importance of religious work. Religious work, the key is to know and understand the religious thought. And religious language is an expression of religious thought. It must be properly understood and to understand the language of religion, under socialist conditions, in order to better play the positive role of religion, as social harmony and building blocks.

ACKNOWLEDGEMENTS

This work was financially supported by the Project of Science and Technology Department in Yunnan Province under Grant No. KKSY201201051.

REFERENCES

Hu Jintao. 2007. Implement the Party's religious approach to religious work under the new situation on December 19, Source: Xinhua *http://www.xinhuanet.com/?utm_source=weibolife*
Cihai. 1979. Microprinting this. Shanghai *Lexicographical Publishing House*, 890–892.
Wei Bohui. 2010. *The key to open the soul of the believer – symbolic of religious language of religion in China*, 01: 45–47.
Hu Zixin. 2004. *Religious language Preliminary Beijing Second Foreign Language Institute.* 6: 25–26.
High Yangtze River. 1992. *Relationship between language and religion preliminary study. Yunnan Normal University: Philosophy and Social Sciences*, 05: 88–89.
Peng Ziqiang. 2008. *Introduction to religious studies. Religious Culture Press.* 330 is –333.
Ji Xianlin. 2007. Buddhist Q15. *On Zhonghua.* 25–47.
Yang Xuezheng. 1993. Yunnan Province of the three major world religions – geographical religious Comparative Study. *Yunnan People's Publishing* House, 2–5.

Information Technology and Computer Application Engineering – Liu, Sung & Yao (Eds)
© 2014 Taylor & Francis Group, London, ISBN 978-1-138-00079-7

Research and realization of a hardware eliminating echo technology on building talkback system

S.H. Tong

College of Computer, Chongqing College of Electronic Engineering, Chongqing, China

ABSTRACT: This paper researches a hardware eliminating echo technology, which applied to audio and video transmission system in building talkback. Echo Cancellation chip 34801 is added to the audio and video transmission circuit. A reasonable power amplifier circuit is designed in its external. Both the filtered echo, but also to ensure the quality of the audio signal transmission. The design method is simple and practical, low cost, has good market prospects.

Keywords: Echo; Hardware eliminating echo; Power amplification

1 INTRODUCTION

The visual-talk and simple-talk are widely used in building talkback. But regardless of the use of visual-talk or simple-talk during transmission, audio after several feedbacks, echo is caused by a variety of reasons. Almost not any echo eliminating function in market besides some special switch. This paper researches a simple hardware eliminating echo technology, and it is realized in visual-talk system.

2 HARDWARE ELIMINATING ECHO TECHNOLOGY

Communication echo mainly includes two kinds in modern voice communication. A technique called Acoustic Echo, which its reason is various. Such as a hands-free telephone function will produce an echo, or the voice of the speaker after repeated feedback spread to the microphone cause echo, etc. The results of it due to lead their own spoken words are heard, and great impact to the quality of voice communication (Yilizhong etc. 2007). Another called Line Echo, and the reasons for its cause is the mismatch of physical electronic circuit 2/4 line. Reflection of the signal is caused, its results almost like the Acoustic Echo's (Chenliangyin etc, 2005). ADSL Modem and switch have 2/4 line conversion of electronic circuits (Fengzhe, 2007). The circuit mismatch will lead to a small part of the signal be fed back. Through the speakers, echo will be produced. If switch without any echo eliminating function, echo will be produced through voice communication. Almost not any echo eliminating function in market besides some special switch.

There are a lot of echo cancellation technologies and it is divided into Hardware Echo Cancellation technology and Software Echo Cancellation technology (Chentao etc, 2010). Hardware Echo Cancellation mainly eliminate echo through the application of related circuit of the chip design to inhibit and attenuate the echo signal, its effect is more obvious. Another way to eliminate echo signal is designing the software algorithm (Kyung-Ae Cha, 2004). Hardware Echo Cancellation is used in this paper to eliminating echo in visual-talk.

Hardware Echo Cancellation designed circuit is mainly in order to improve the quality of an audio transmission, and to eliminate echo and line echo of Talkback process from the speaker to the handset. It is mainly designed the echo generating principle, the echo cancellation chip and the echo cancellation circuit.

3 VIDEO INPUT, AUDIO INPUT AND OUTPUT AND HARDWARE ECHO CANCELLATION CIRCUIT DESIGN

3.1 *Echo Cancellation chip 34801*

Echo canceller chip is the 34018, which has a microphone amplifier, audio power amplifier, transmit and receive attenuator, the background noise level monitoring system. 34018 also integrates adjusting power for the requirement of chip internal and external circuit, and it has the following characteristics: In a monolithic chip includes a desired level detector and attenuation controller; Can be a long time stability monitoring for background noise level; The power and reference voltage adjusting circuit is designed on the circuit; A circuit control work and waiting state;

With the linear volume control function; The working temperature is $-20°-60°$.

3.2 Echo elimination principle

The background noise level monitoring system of 34018 is the main application to eliminate the echo. It can make a judgment on transmission and reception of signal and background noise, and eliminates noise according to some characteristics of the audio. In this paper, the working principle is as follows: In visual-talk process, the acoustic echo is caused due to the horn's vocal of the door phone. The original sound and echo is compared by the background noise level monitoring system of 34018. Compared with the original sound, the echo level is relatively weak, so background noise level monitoring system will start the attenuator to the attenuation of the echo signal. Audio signal eliminates acoustic echo before the A/D conversion, reducing the difficulty of the echo cancellation. Line echo cancellation is divided into two types. When the other party's voice came over, the echo signal and the other original signals are compared by the background noise level monitoring system. Because the line echo is partial feedback, so line echo level is lower than the other party's original sound, the echo signal is judged, then the attenuation of echo. When others not talking, the background noise level monitoring system is judged not echo, here is the 34018 half-duplex characteristics to eliminate. When the line echo feedback from the switch back, due to 34018 is a half-duplex chip, feedback sound is not over when original sound in the transmission, so line echo is eliminated flexibly. Although some properties were sacrificed, the effect of visual-talk is not affected overall.

3.3 Echo Cancellation circuit design

Audio and video signals input and output are the key of visual-talk. In design, CCD camera is used in the video input, which is Tripod Technology 1/4 SHARP camera. Its output is S-VIDEO video signal, and the working voltage is 8–12 V. Audio input is from the microphone input, then after 34018 hardware echo cancellation. Audio output is also from the output of the 34018. The audio signal from the UDA1345 transmission input to the 34018 hardware echo cancellation, then output to the power amplification circuit for amplification, and through the horn to sound reduction. Video input circuit as shown in Figure 1.

This circuit mainly realizes the control of camera, Camera module output have three lines, which were the power cord,groundwire and data line,and connect the J8 interface, From in Fig. 1, we can see that the control camera is through the control of the camera power supply to carry on the control, V_CONTROL connect with SCM P6 interface, when PC6 interface for high level, there is a positive pressure differential between triode 8050 base and emitter, because 8050 is NPN transistor, so 8050 conduction, the triode collector and emitter of differential pressure will be very small, because the

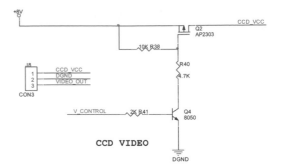

Figure 1. Video input circuit principle diagram.

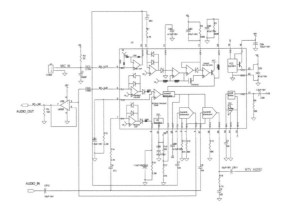

Figure 2. Audio input and Hardware Echo Cancellation circuit.

partial pressure between R40 and R38 make MOS tube AP2303 exist pressure, and is negative voltage, AP2303 is P channel enhanced MOS tubes, Before its conduction between the drain-source plus a negative voltage, so AP2303MOS tube conduction, CCD_VCC terminal voltage is 8V, the camera work; When PC6 interfacs for low level, the triode 8050 base and emitter electrode does not exist pressure, 8050 is in the OFF state, Triode collector and emitter is equivalent to break, there is no pressure differential between the P2303 drain-source, So AP2303 cutoff, CCD_VCC terminal voltage to 0V, the camera dose not work. Thus, when visual-talk to give CCD_VCC high level, make the camera work, other times do not work, So, greatly reduces the power consumption, and in accordance with the design principles and requirements. audio input and Hardware Echo Cancellation circuit as shown in Figure 2.

As can be seen from Fig. 2, Audio input signal from MIC_IN ,Hardware Echo Cancellation by 34018, After processing output through AUDO_OUT port, in this circuit is designed in an output buffer device for output, Output to UDA1345 for A/D conversion; The audio signal output from GM8120 input to UDA1345 on D/A conversion, input to AUDO_IN elimination echo processing, Through the MTV_AUDO output to

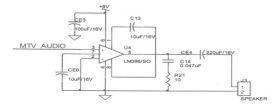

Figure 3. Audio power amplifier circuit.

the power amplifying circuit for processing, power amplification circuit as shown in Figure 3.

As can be seen from Fig. 3, This power amplifier circuit used in the design of the power amplifier LM386, through the feet 1 and feet 8 collocation can increase the gain to 200, the input voltage range 4–12V, work when the current consumption of 4 mA, and the distortion is very low. This circuit uses a gain for the 200, between feet 1 and feet 8 access a 10 uF capacitance, the power supply voltage is 8 v, audio signal from the third foot input, after LM386 for power amplifier, through the 5 feet for output, and then drive speakers, the sound reduction.

4 CONCLUSIONS

This paper studies the hardware eliminating echo technology, Specific study the principle of the echo, echo cancellation chip, echo cancellation principle and echo cancellation of the circuit design. This paper adopts 34018 echo cancellation chip, the power amplifier cir-

cuit design is reasonable in its peripherals, not only filter out the echoes, but also to ensure the quality of the audio signal transmission. After testing, audio performance obvious improved. This kind of hardware echo cancellation technology, no software programming, design principle of simple, in building talkback system or other applications has a good advantage and development prospect.

ACKNOWLEDGEMENT

I would like to express my gratitude to all those who have helped me during the writing of this paper. I gratefully acknowledge the help of my wife. I do appreciate her patience, encouragement, and professional instructions during my paper writing.

REFERENCES

Chenliangyin etc, 2005. The Design and Implementation of the smart home system[J]. Xinjiang electricity journal, 22: 108–111.
Chentao etc, 2010. The linux driver design of signal generator. Process Automation Instrumentation[J], 37: 88–91.
Fengzhe, 2007. The ARM926 based MPEG-4 video decoding system design[J]. Video technology application and engineering, 31: 85–87.
Kyung-Ae Cha, 2004.Content Complexity Adaptation for MPEG-4 Audio-visual scene[J]. IEEE Transactions onConsumer Electronics, 50: 760–765.
Yilizhong etc. 2007. On the research about the visual talkback system in the application of well-off residential[J]. Low Temperature Architecture Technology, 3:27–28.

Information Technology and Computer Application Engineering – Liu, Sung & Yao (Eds)
© 2014 Taylor & Francis Group, London, ISBN 978-1-138-00079-7

Deformation analysis using Danish approach

Q. Dai & Y.Y. Chang
Chongqing Water Resources and Electric Engineering College, China

ABSTRACT: Deformation surveying is a kind of technique to monitor engineering structure. Network adjustment and deformation monitoring are main activities of deformation surveying. Geodetic networks are estimated by the method of Least-Squares Estimation (LSE) and the 'goodness' of the network is determined by precision analysis based upon the covariance matrix of the estimated parameters. The main purpose of this study is to analyze the deformation by using Danish methods. Furthermore, a data set is acquired from field work observation and analyzed by the program developed in this study to verify the practice of Danish methods in deformation analysis.

Keywords: Deformation analysis, Danish approach, LSE, MATLAB, AutoCAD VBA

1 INTRODUCTION

Owing to the development of the various engineering technique, a considerable number of modern constructions will become more and more common in our life. However, all the constructions are supposed to be deformed during its operating time, and the safety of constructions is related to human's life critically. So the study of monitoring object deformation has already become very important in Geomatics science nowadays. The work of deformation detection and network analysis can represent object movement accurately. And it can evaluate the safety of the construction object. The stability of deformation monitoring network is the most important issue. Therefore, it is important for us to find suitable methods to perform adjustment and data analysis. A part from that, processing surveying data is tedious work, and an efficient programming is required for it.

The aim of this research is to analyze the stability of monitoring network using robust method. A correct interpretation of displacement is directly related to safety of engineering structures and human life. Today many equipments and data processing software are provided by commercial company. And admittedly suitable analysis methods and accurate surveying equipment are important in deformation surveying work.

2 DEFORMATION METHODOLOGY

Deformation Survey play a significant role in modern engineering surveying. Deformation results are directly relevant to the safety of human life. Displacement measurement and deformation monitoring have an important role for the design, assessment and control of large engineering structures and in the predicting and preventing of ground disaster. In general deformation detection normally consists of design, measurement and analysis stage.

A complete geodetic deformation analysis normally has several procedures. They are per-analysis of geodetic network, data acquisition, network adjustment computation, pillar stability analysis, trend analysis, modeling of the deformation and physical interpretation. And the least squares adjustment is a common adjustment method for geodetic network adjustment, and it could solve all the surveying problems.

2.1 Danish method

Different robust methods have different weight schemes. For Danish method, the weight matrix is defined as W=1(k=1) for all the common points in the first iteration transformation. After the first iteration, the weight matrix is define for gross error detection. This conception also can be used for deformation analysis. And the iterative procedure continues until the absolute differences between the successive transformed displacements of all the common points.

2.2 Deformation graphical presentation

In order to interpret the displacement of the points, a graphical presentation in the form of displacement vector and error ellipse is used. The displacement vector and covariance are numerical form, it is hard to interpret. And error ellipse which can be visualized is used for graphical representation of results. For 2D network, the coordinate differences between two epochs do not always caused by point movement for

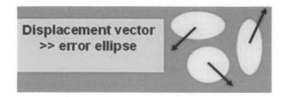

Figure 1. Deformation is significant.

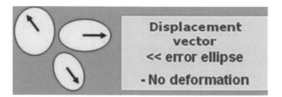

Figure 2. Deformation is not significant.

sometimes they are caused by error of measurement. In order to investigate the reason of the differences, the significance analysis of displacement should be carried out. A method of using the error ellipse to analyze the significance of displacement of points is presented based on the error ellipse method which is used to analyze plane points deformation, Fig. 1 and Fig. 2 show using error ellipse to represent deformation significance in 2D network case.

3 IMPLEMENTATION

This program is developed using Visual Basic 6.0 and AutoCAD VBA respectively. There are six modules in this program. Observation data module is design to insert the observation data and perform degree unit exchange. And adjustment module is developed by MATLAB. The deformation analysis module can perform deformation detection. After programming of all the calculation, graphic representation module is design to plot network, error ellipse and displacement vector. This module is developed by AutoCAD VBA. In the end, result and print module can link to EXCEL to show all the result information. The design flowchart is shown in Figure 3.

After the adjustment computation, deformation analysis is performed by Danish Method. Point 1, 2, 3 and 6 are assumed as datum points here. Danish Method is used for determining the movement points in the network. And the test is processed according to $a = 0.05$. X and Y are point coordinates, while Danish Method determine the displacement in X axis (Dx), Y axis (Dy) and displacement vector (Disp.). FTAB in table shows value from Fisher distribution and Fcom are calculated from covariance matrix in single point test. The summary of the deformation analysis by Danish Method is shown in Table 1. And the displacement vector and error ellipse are plotted in AutoCAD program shown in Figure 4. And user can use AutoCAD command to measure values of displacement vector.

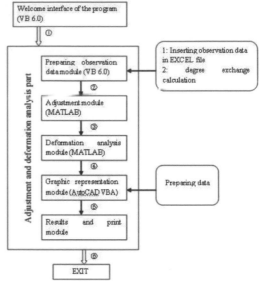

Figure 3. Flowchart of program design.

Table 1. Deformation analysis by Danish Method for data set three.

DETECTION		DEFORMATION		BY			
DANISH-----------------------							
Test on Variance Ratio ***							
DF1=		27					
DF2=		22					
SIGNIFICANCE LEVEL=		0.05					
POOLED VARIANCE							
FACTOR=		1.15060022					
Test on Variance Ratio=		P	A	S	S		
SINGLE POINT TEST ***							
M=	2						
DF=	49						
SIGNIFICANCE	LEVEL =	0.05					
STN	DX (m)	DY (m)	DISP.VECT (m)	FCOM	FTAB		INFO
1	-0.0002	-0.0005	0.0005	0.25827956	3.18658235	0	
2	0.0004	-0.0003	0.0005	0.20817349	3.18658235	0	1=MOVE
3	-0.0002	0.0004	0.0004	0.12303831	3.18658235	0	
4	0.0790	0.0004	0.0790	1194.28997	3.18658235	1	
5	0.0013	0.0017	0.0022	0.97333857	3.18658235	0	0=STABLE
6	0.0102	-0.0254	0.0273	135.86864	3.18658235	1	

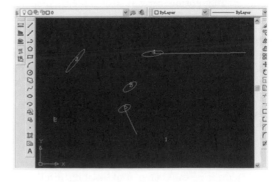

Figure 4. Plotting displacement vector and error ellipse by AutoCAD VBA program.

4 CONCLUSION AND RECOMMENDATION

Deformation of datum point is very important in deformation analysis. After performing LSE adjustment for each epoch, Danish Method is used to obtain the displacement vector, which shows that this method is practical for deformation analysis. And plotting displacement vector and error ellipse by VBA program developed in this study make the deformation analysis results more directly. In field work, only 2D observations are required, because a precise leveling is required to determine the movement in vertical dimension.

Input file and output file are Office Excel format. For future work, Microsoft Office Excel VBA can be used to develop new function for input and output file, which can make the program more powerful. Most of the computation program is developed by MATLAB. But the program for adjustment computation can not compiler to executable file. In future work, Visual Basic 6.0 is recommended for programming.

REFERENCES

Caspary, W.F. (1987): Concepts of Network and Deformation Analysis, 1st. ed., School of Surveying, The University of New South Wales, Monograph 11, Kensington, N.S.W.

Caspary, W.F., and H. Borutta. 1987. Robust estimation in deformation models, Survey Review, Vol. 29, No. 223, pp. 29–45.

Choudhury, M., Harvey. B. and Rizos, Ch. (2010). A survey of techniques and algorithms in deformation monitoring applications and the use of the Locata technology for such applications. University of New South Wales, Australia.

Chen, Y.Q. (1983). Analysis of Deformation Surveys-A Generalized Method. Department of Surveying Engineering Technical Report No. 94. University of New Brunswick, Fredericton.

Chen, Y.Q., A. Chrzanowski, and J.M. Secord. 1990. A strategy for the analysis of the stability of reference points in deformation surveys, CISM JOURNAL ACSGC, Vol. 44, No. 2, pp. 141–149.

CRC Press LLC (2003). Graphics and GUIs with MATLAB. 3rd. ed.

Idris, K.m. & Setan, H. (2008). A Geodetic Deformation Survey to Monitor the Behavior of a Concrete Slab During Its Axial Compression Testing, Malaysia. Proceedings, international city event on Civil Engineering 2008 (ICCE08), Kuantan, 12–14 May 2008.

Grontmij, Move3, How To Combined TPS and GPS adjustment Version 4.0.2, Retrieved on May 7, 2010, from fttp://www.MOVE3.com

Information Technology and Computer Application Engineering – Liu, Sung & Yao (Eds)
© 2014 Taylor & Francis Group, London, ISBN 978-1-138-00079-7

Research on the anti-grade trip system in coal mine high-voltage grid based on WAMS

Z.J. Wang & Y.W. Wang
Department of Mechanical Electronic & Information Engineering, China University of Mining & Technology (Beijing), Beijing, China

ABSTRACT: Grade-skip trip incident happens occasionally in coal mine high-voltage grid. In order to solve this problem, WAMS based on fiber communication network is introeuced. In this paper, the composition and basic structure of WAMS, as well as how such technology is applied to the anti-grade trip system in coal mine high-voltage grid is introduced. Taking one of the renovation projects in Shanxi province as example, the methods on the connection of GPS and phasor measurement units, the communication style between the control center, phasor measurement unit and intelligent controller are detailed described.

Keywords: WAMS; Anti-grade Trip System; Coal Mine High-voltage Grid

1 INTRODUCTION

Grade-skip trip incident is very common in coal mine high-voltage grid, but never be solved completely. It will not only expand the power off area, and also may cause gas accumulation because of the shutdown of fans. The main reason causing grade-skip trip is that both the fault feeder and the superior substation incoming line will receive fault signal when fault happens such as short circuit fault and single-phase earth fault, but fail to choose the right feeder switch to trip. Taking short circuit fault as example, optical fiber differential protection criterion can be used to solve grade-skip trip problem, but there are two requirements needed to be met. One is all the data should sampled synchronously, and the other is the trip decision should be made according to all the data in the whole power system. And WAMS (wide area measurement system) will just meet these two requirements.

2 WAMS INTRODUCTION

2.1 *WAMS*

It originates from the wide area requirements of time and space to the power system. Such system collects the real-time information with PMU (Phasor Measurement Unit) as the primary unit, and uploads the information to the control center by communication system, to achieve monitoring the whole system (Cheng K. 2008).

In recent years, WAMS and applications based on such system have developed rapidly in the United States, Japan, France and other countries and regions; at the same time the major grids in our country have implemented wide area measurement system in different levels on the power supply system (Qiao X.Y. 2008, Meng Z.Q. 2009). WAMS is widely used in real-time monitoring of power system state, power system disturbance records, step-out protection, energy management system (EMS), grid stability control and other fields.

2.2 *PMU structure principle*

The roles played by the PMU include the original data collection, synchronization phase calculation and synchronization phase sending in the WAMS system. PMU is composed of the microprocessor, GPS receiver, the signal transmission module and communication modules (Hu Y.S. & Zhang M. 2007), the typical structural principle is shown in Figure 1.

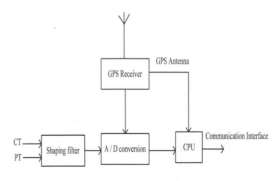

Figure 1. The typical architecture of PMU.

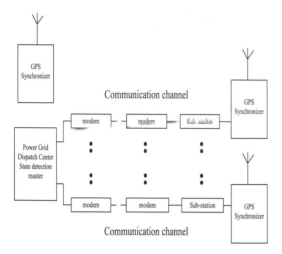

Figure 2. WAMS Structure 1.

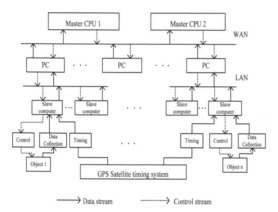

→ Data stream   ·····> Control stream

Figure 3. WAMS Structure 2.

2.3 Composition of WAMS

Wide area measurement system consists of four parts (Zhang J.Y. 2006, Adamiak M.G. et al. 2006), as shown in Figures 2–3.

GPS. GPS satellites provide global timing signal.

PMU. PMU device based on GPS signal uniform sampling, using Discrete Fourier Algorithm calculate the phasor data and add the time scale.

Communication Network. It uses high-capacity fiber-optic network as the backbone of wide-area communication network.

Control Centre. It collects all the information in the PMU of the wide-area power system, and stores, filters, displays the data, as well as runs on the system safety assessment, forms protect and control program.

In Figure 2, the master station of WAMS is located in the total control center, and sub-stations are monitoring points of the power angle. While in Figure 3, the host computer and the slave computer constitute a basic set of PMU. Compared to the first mode, the

communication system of this kind adopts LAN and WAN communication technology, so that the whole system can be more powerful and therefore is more reasonable.

Throughout the phasor measurement unit and high-speed communications all over the whole network, WAMS system can achieve wide-area synchronized data sampling and send the real-time status to the control center.

Compared to the traditional methods, GPS has many advantages such as: high precision (ns), range, do not need access to contact, free from geographic and climatic conditions, etc., it is the ideal method to uniform grid time, and will be more and more used in the field of electric power system.

3 ANTI-GRADE TRIP SYSTEM BASED ON WAMS

3.1 Renovation introduction

So far, one of the anti-grade trip systems based on WAMS is built up in Shanxi province, to avoid the appearance of grade-skip trip, so as to reduce power cut off range and time.

According to the spot power supply structure and renovation requirement, there are three substations that need to be protected to avoid grade-skip trip, which are the ground substation, the central substation underground, the third substation. In the paper, we define the three substations as substation A, substation B and substation C.

3.2 Renovation scheme

GPS. The GPS antenna is set up on the roof of the control center, which makes sure the GPS server can get precise time signal from the GPS satellites. The GPS is connected to the PMU in substation A, B, C and control center by single mode fiber (SMF) to achieve the time synchronization of the whole system. All the PMU in the system will get pulse per second (PPS) from the GPS by SMF to adjust current time, which makes good preparation for the synchronous data sampling of all the intelligent controllers.

PMU. The system consists of four phasor measurement units, being fixed respectively in substation A, B, C and control center. Here we define the phasor measurement unit in control center as Head Unit, and others as Part Units.

Obviously all the Units will get PPS from GPS by SMF as mentioned above. The Part Units will guide the intelligent controllers to sample data synchronously, and get corresponding calculation results from them, such as the phase current, phase voltage, zero sequence current, and zero sequence voltage. Then all the data is send to the Head Unit in the control center, where the final decision is made according to the data. The final decision will be then send to all the Part Units, and therefore all the intelligent controllers will determine

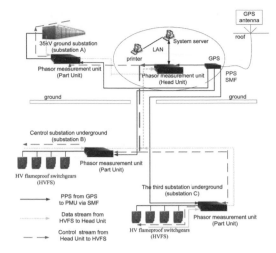

Figure 4.   Renovation schematic diagram of the project.

whether to trip or not according to the order they receive.

Communication Network. In this renovation project, we adopt point to point (PTP) fiber communication network between the phasor measurement units.

Control Center. The control centers consist of system server, GPS server, Head Unit and printer. The communication protocol in this system is made according to IEC 61850 standard, so we can easily realize telemeter, tele-signaling, tele-control and tele-regulation in the protection system.

Until now, most of the intelligent controllers in the HV flameproof switchgears (HVFS) have been replaced, and we can receive all the information we had expected, with all the tele-operations such as tele-control operate successfully.

The whole renovation schematic diagram of the project is shown in Figure 4.

4   FAULT TEST

Until now, the project has completed the renovation of substation C, most of substation B, and the control center on ground. The renovation of substation A will be left to the last part because of its importance on power supply reliability. We can now control the HVFS reformed to trip, to get power, and read the setting values, change the setting values. We can also get the waveform of each HVFS reformed.

During the renovation, we did several times short circuit fault test in substation C, and every time the protection system chose the right feeder in substation C according to optical fiber differential protection criterion, which achieve the goal we expected. Certainly, more kinds of fault tests will be made after the whole renovation project is completed.

5   CONCLUSIONS

Grad-skip trip incidents have been very common in coal mine power supply system. It will not only expand the power off area, and also may cause gas accumulation because of the shutdown of fans. WAMS technology has been widely used in power systems for its outstanding merits, especially in ground power system. In this paper, this technology is applied to construct anti-grade trip protection system. Combine with the renovation project in Shanxi province, the construction of communication network and equipment configuration is detailed described. The main innovation point of the project is the communication protocol based on IEC 61850. According to the renovation results so far, the anti-grade trip protection system is running very well.

ACKNOWLEDGEMENT

This paper is based upon a research project (grant number 2012YJ11) supported by the special funds for basic research business expenses of central university.

REFERENCES

Adamiak M.G. et al. 2006. Wide area Protection-technology and infrastructures[J]. *IEEE Transaction on Power Delivery*, 21(2):601–609.
Cheng K. 2008. Research on line voltage stability index and its assessment method based on WAMS. *North China Electric Power University*. Beijing.
Hu Y.S. & Zhang M. 2007. Applications of WAMS in Power System. *Electrotechnical*. 5:19–21.
Meng Z.Q. 2009. Assessment on security of Power System based on WAMS. *North China Electric Power University*. Hebei.
Qiao X.Y. 2008. Research on the parameter identification of Synchronous Machine based on WAMS. *North China Electric Power University*. Beijing.
Zhang J.Y. 2006. Research on Power System transient stability security and defense in the Observation and Control of Wide Area. *Tsinghua University*. Beijing.

Study on the radiosensitivity of AT cells with HPRT gene mutation

J. Liu, K.J. Dai, C.H. Zheng & L. Chen
Changzhou Tumor Hospital, Soochow University, Changzhou

Z.H. Lu & X.F. Zhou
Dongping People Hospital, Taian

ABSTRACT: It is to investigate the high radiosensitivity of fibroblast cell lines AT5BIVA (AT cells) from the skin of the patients with ataxia-telangiectasia (AT). With the fibroblast cell lines GM0639 (GM cells) as control, hypoxanthine phospho-ribosyltransferase (hprt) locus mutation analysis and Cytokinesis-Block micronucleus method (CBMN) are employed. After AT cells and GM cells were irradiated by 0, 1, 2, 3 and 4 Gy of $^{60}Co\gamma$ ray, the differences of hprt gene mutation frequency (hprt MF), micronuclear frequency (MNF) and micronucleated cell frequency (MNCF) between AT and GM cells were observed, and curve fittings were separately conducted. At different dose points, AT cells have significant higher hprtMF, MNF and MNCF than GM cells do, and the difference is of statistical significance ($P < 0.01$); AT and GM cells' hprtMF, MNF and MNCF are positively correlated with irradiation dose, which can be fitting as dose effect linear equation $Y = a + bX$. The results indicate that in the patients with A-T, the radiosensitivity of AT cells is remarkably higher than that of GM cells, being high radiosensitivity.

Keywords: Ataxia-telangiectasia, high radiosensitivity, Hypoxanthine phospho-ribosyltransferase, Cytokinesis-Block micronucleus method

Ataxia-telangiectasia (A-T) is an autosome recessive genetic disease, being characteristic of the special clinical performances like progressive cerebellar degeneration, high radiosensitivity, immunodeficiency, genomic instability, early-maturing, high incidence rate of tumor. AT mutation (ATM) gene protein encoding involves in cell cycle regulation, DNA damage signaling conduction and repair and the maintenance of chromosome stability. When cells are ionizing radiated, the signal conduction paths mediated by ATM kinase are activated so as to identify and repair the damaged DNA. In the fibroblast of AT patients, the loss of ATM kinase activity caused by ATM gene mutation results in AT cells cannot be repaired the DNA damages caused by damage factors like irradiation, expressing high readiosensitivity and chromosomal instability [Lavin 1997, Meyn 1999]. We adopted hypoxanthine phospho-ribosyltransferase (hprt) locus mutation analysis and Cytokinesis-Block micronucleus method (CBMN) to study the high radiosensitivity of AT cells. Hprt gene locates in the end of X chromosome q, being a locus which is sensitive to ionizing radiation, so it is easy and rapid to detect the gene locus damages of genetic material caused by ionizing radiation [Mazin 2009, Czaja 2012]. Cytokinesis-Block micronucleus method was put forward by Fenech et al [Fenech 1985] in 1985, which can simply and easily measure chromosomal damage.

1 EXPERIMENTAL MATERIALS AND METHODS

1.1 Reagents and instruments

DMEM medium (American Invitrogen company), calf serum (Shanghai Huamei Biotechnology Company), cytochalasin B (Cyt-B) (American Sigma), 6-TG (Sigma), dimethyl sulfoxide (Shanghai Shangliu Chemical Co., Ltd) and microscope (Olympus).

1.2 Cell line and culture

Human fibroblast cell line AT5BIVA and GM0639 (presented by Germany GSF National environment and health research Centre for Radiation Biology Laboratory) were incubated in DMEM medium respectively (containing 15% calf serum, 100 U/ml of penicillin and 100 U/ml of streptomycin) in incubator with 5% CO_2 at 37°C.

1.3 Ionizing radiation

AT and GM cells at exponential growth phase were irradiated by $^{60}Co\gamma$ ray at room temperature, and the irradiation doses were separately 0, 1, 2, 3 and 4 Gy (dose rate 1.0 Gy/min), and the culture was continued after irradiation.

1.4 hprt gene mutation detection

The fourth generation cells after irradiation were taken, 2 bottles for each dose point. One of them was added into 6-TG, and the final concentration was 1×10^{-5} mol/ml. At the 60 h of culture, each bottle was added into Cyt-B, making the final concentration 4 μg/ml. Continue to culture until 84 hours. 0.25% trypsin was used to gather cells, which was centrifuged at 1500 rpm for 8 min, and the upper medium was removed, and then 3ml of pre-warmed 0.15 M KCl at 37°C was added into for hypotonic treatment. Fresh fixed liquid (methanol: glacial acetic acid = 3:1) was used for fixation. Conduct conventional drip piece, 5% Giemsa staining and microscopic examination. In 6-TG culture solution, the number of dikaryocyte or coenocyte among 1000 cells (including those pressing each other, tangent and separation dikaryocyte or coenocyte) is divided by the number of dikaryocyte or coenocyte among 1000 cells in the culture solution without 6-TG, and the result is the hprt gene mutation frequency(‰)[Torres 2007].

1.5 Micronucleus detection

As for AT and GM cells at exponential growth phase, after irradiated by $^{60}Co\gamma$ ray, culture solution was changed and added into cytochalasin B (the final concentration is 4 μg/ml), and then cultured for another 24 hours. 0.25% trypsin was used to gather cells, which were centrifuged at 1500 rpm for 8 min. The upper culture medium was removed, and then 3 ml pre-warmed 0.15 mol/l KCl of 37°C was added into for hypotonic treatment. Fresh fixed liquid (methanol: glacial acetic acid = 3:1) was used for fixation. Conduct conventional drip piece and prepare complete cell micronuclei tablets after Giemsa staining. Observe and count: ① dikaryocyte: large cell body with 2 cell nucleus, which is caused by the first division. The chromatin is meticulous and loose, and most nucleolus is clear and visible with rich endochylema. ② the identification standard of micronucleus: micronucleus must be in the endochylema of dikaryocyte, being separated from the main nucleus. If any overlapping or being tangent, their complete nuclear membrane should be visible; micronucleus presents to be round or oval, with a diameter less than one third to one fifth of main nucleus; without any refractivity, staining is similar or a little slighter than main nucleus. ③ Cell analysis: with high power lens (400×), count the micronucleus and the cells with micronucleus among 3000 dikaryocytes, and express it with micronucleus rate and micronucleated cell rate (‰) [Gutierrez 2003, Wang 1997].

1.6 Statistical method

SPSS 11.0 is used to analyze data, while the relations between hprt MF, MNF, MNCF and dose are analyzed by linear regression and correlation, and the pair comparison was conducted by t or u-test.

Table 1. hprtMF of AT and GM cells post-irradiation of $^{60}CO\gamma$-ray ($\bar{x}\pm s$).

Dose (Gy)	Hprt MF(‰)	
	AT	GM
0	0.348 ± 0.015	$0.195 \pm 0.019^{\#\#}$
1	$0.559 \pm 0.031^{**}$	$0.288 \pm 0.017^{**\#\#}$
2	$0.715 \pm 0.045^{**}$	$0.417 \pm 0.023^{**\#\#}$
3	$0.876 \pm 0.054^{**}$	$0.518 \pm 0.039^{**\#\#}$
4	$0.901 \pm 0.077^{**}$	$0.624 \pm 0.053^{**\#\#}$

n = 6; 1, 2, 3 and 4 Gy compare with 0 Gy respectively in AT and GM, **P < 0.01;
AT compares with GM at the same dose points, ##P < 0.01

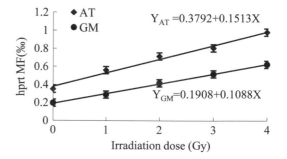

Figure 1. Relations of Hprt MF in AT and GM cells with irradiation dose.

2 RESULTS

2.1 After being irradiated by $^{60}Co\gamma$ ray, the hprtMF comparison of AT and GM cells

After being irradiated by 60COγ ray, the hprtMF of AT and GM cells change along with doses, which is shown in table 1. It shows that at the same dose point, after being irradiated by 0, 1, 2, 3 and 4 Gy of $^{60}Co\gamma$ ray, AT cells have a significantly higher hprt MF than GM cells do, and the difference is of statistical significance (P < 0.01); thehprtMF of AT and GM cells increase when dose increases, and the differences are also remarkably and statistical significance (P < 0.01), and the tendency is indicated by figure 1. The correlation analysis manifests that both cells' hprtMFs are positively correlated with irradiation dose, and their linear regression equation are separately fitting as follows:

AT cells'hprt MF(‰): $Y_{AT} = 0.3792 + 0.1513X$, r = 0.992, $R^2 = 0.984$ (P < 0.01);
GM cells'hprt MF(‰): $Y_{GM} = 0.1908 + 0.1088X$, r = 0.998, $R^2 = 0.998$ (P < 0.01),
X stands for irradiation dose (Gy), and Y stands for hprt gene mutation frequency (‰).

2.2 After being irradiated by $^{60}Co\gamma$ ray, the distribution of AT and GM cellmicronucleus

After the irradiation of 60Coγray, the distribution of AT and GM cell micronucleus is shown by table 2.

Table 2. Micronucleus distribution of AT cells post-irradiation of $^{60}CO\gamma$-ray.

Dose (Gy)	Observed cells	AT				
		Uni-MN	Bi-MN	Tris-MN	Tetra-MN	Five-MN
0	3000	51	2	0	0	0
1	3000	146	16	7	0	0
2	3000	220	38	12	0	0
3	3000	465	225	70	15	0
4	3000	520	231	111	50	4

MN: micronuclei

Table 3. Micronucleus distribution of GM cells post-irradiation of $^{60}CO\gamma$-ray.

Dose (Gy)	Observed cells	GM				
		Uni-MN	Bi-MN	Tris-MN	Tetra-MN	Five-MN
0	3000	39	0	0	0	0
1	3000	80	4	3	0	0
2	3000	133	6	4	0	0
3	3000	144	13	7	0	0
4	3000	181	25	18	0	0

MN: micronuclei

Table 2 & 3 show that after being irradiated by 0, 1, 2, 3 and 4 Gy of $^{60}CO\gamma$ ray, there is an increase in the micronucleus number among the micronucleated cell type of AT and GM cells and different micronucleated cell types, as the irradiation dose increases; there is possible micronucleated cells with monokaryon, two-, three-, four- or five-karyon in AT cells, while there are only micronucleated cells with monokaryon, two- or three-karyon; the micronucleated number of AT cells and GM cells and micronucleated cells numbers rise when irradiation dose increases; at the same dose point, the numbers of AT cells' micronucleus and micronucleated cells greater than those of GM cells.

2.3 After being irradiated by $^{60}CO\gamma$ ray, the comparison of AT and GM cells' MNF and MNCF

After being irradiated by $60CO\gamma$ ray, the comparison of MNF and MNCF of AT and GM cells is shown in table 4 & 5. It shows that after being irradiated by 0, 1, 2, 3 and 4 Gy of $^{60}CO\gamma$ ray, AT cells have a significantly higher MNF and MNCF than GM cells do ($P < 0.01$), and the MNF and MNCF of AT and GM cells grow when dose increases. The relations are indicated by figure 2 and figure 3. The correlation analysis manifests that both cells' MNF and MNCFs are positively correlated with irradiation dose and their linear regression equation are separately fitting as follows:

AT MNF: $Y_{AT} = -42.8 + 131.7X$ $r = 0.952$ $R^2 = 0.906$
    ($P < 0.05$).

Table 4. MNF of AT and GM cells post-irradiation of $^{60}CO\gamma$-ray ($\bar{x} \pm s$).

Dose (Gy)	Observed cells	MNF (‰)	
		AT	GM
0	3000	18.33 ± 2.45	13.00 ± 2.07
1	3000	66.33 ± 4.54**	32.33 ± 3.23**##
2	3000	110.67 ± 5.73**	52.33 ± 4.07**##
3	3000	395.00 ± 8.93**	65.67 ± 4.52**##
4	3000	512.33 ± 9.13**	95.00 ± 5.35**##

1, 2, 3 and 4 Gy compare with 0 Gy respectively in AT and GM, **$P < 0.01$;
AT compares with GM at the same dose points, ##$P < 0.01$

Table 5. MNCF of AT and GM cells post-irradiation of $^{60}CO\gamma$-ray ($\bar{x} \pm s$).

Dose Gy	Observed cells	MNCF (‰)	
		AT	GM
0	3000	17.67 ± 2.41	13.00 ± 2.07
1	3000	56.33 ± 4.21**	29.00 ± 3.06**##
2	3000	90.00 ± 5.22**	47.67 ± 3.89**##
3	3000	258.33 ± 7.99**	55.33 ± 4.17**##
4	3000	302.33 ± 8.39**	74.67 ± 4.80**##

1, 2, 3 and 4 Gy compare with 0 Gy respectively in AT and GM, **$P < 0.01$;
AT compares with GM at the same dose points, ##$P < 0.01$

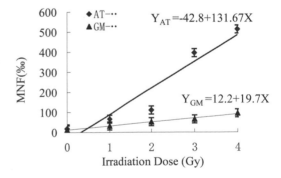

Figure 2. Relations of MNF in AT and GM cells with irradiation dose.

GM MNF: $Y_{GM} = 12.2 + 19.7X$ $r = 0.994$ $R^2 = 0.988$
    ($P < 0.01$).
AT MNCF: $Y_{AT} = -9.3 + 77.1X$ $r = 0.959$ $R^2 = 0.920$
    ($P < 0.05$).
GM MNCF: $Y_{GM} = 1.4 + 15.0X$ $r = 0.994$ $R^2 = 0.988$
    ($P < 0.01$).

Y stands for MFN or MNCF (‰), while X is irradiation dose(Gy). As shown from the previous equations, the linear regression equation slope of AT cells' MNF and MNCF is obviously greater than GM cells, and the difference is of statistical significance ($P < 0.05=$, indicating that the increasing tendency of AT cells'

701

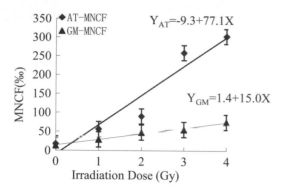

Figure 3. Relations of MNCF in AT and GM cells with irradiation dose.

MNF and MNCF along with the increase of dose is faster than GM cells.

3 DISCUSSION

AT is caused by the mutation of ATM genes. ATM genes encode a protein kinase which is high homologous with the catalytic domain of phosphatidylinositol 3-kinase (PI-3K). This kinase regulates cell cycle prosecution point through phosphorylating different target proteins in different cell cycles, and plays a vital role in DNA damage signal conduction and repair, apoptosis and chromosome stability. Genomic instability is the major pathological phenotype of A-T cells, which is expressed by high radiositivity and the chromosome spontaneous aberration rate in vivo and vitro, chromosome aberration rate induced by irradiation and the abnormal increase of loci injury of genetic material caused by irradiation [Khanna 2001]. By detecting the differences of hprt MF of AT and GM cells after being irradiated by 0, 1, 2, 3 and 4 Gy of $60CO\gamma$ ray, we found that at the same dose point, the hprt MF of AT cells is significantly higher than that of GM cells, and the difference is of statistical significance (P < 0.01); thehprt MF of AT and GM cell rates grow with the increase of dose. The correlation analysis indicates that both cells' hprt MF is positively correlated with irradiation dose. Based on the linear relation of hprt MF and irradiation dose given by Jilan Shi [10], we make it fitting to be linear regression equation in the form of $y = a + bx$. The equation indicates that AT cells' hprt MF's increasing tendency is faster than that of GM cells when the dose increases, and the hprt MF can be employed to measure individual radiosensity [Wang 1997].

From this standpoint, AT cells have a higher radiosensitivity than GM cells. Meanwhile, with Cytokinesis-Block micronucleus method, the injury of chromosome was detected after AT and GM cells' being irradiated by rays. We found that after being irradiated by 0, 1, 2, 3 and 4 Gy of $60CO\gamma$ ray, there may be various micronucleated cells in AT cells, including micronucleated cells with monokaryon, two-, three-, four- or five-karyon, while there are only micronucleated cells with monokaryon, two- or three-karyon

in GM cells; at the same dose point, themicronucleus number of AT cells' various micronucleated cells is higher than GM cells. The MNF and MNCF of AT cells are significantly higher than that of GM cells (P<0.01, and the MNF and MNCF of AT cells and GM cells increase along with dose. Accordingly, we based on the four mathematical models ($y = a + bx$, $y = a + bx + cx^2$, $y = kxn$, $y = u + cx^2$)provided by WHO and referred to YushuBai [Torres 2007]'s related statistical method to conduct curve fitting, and also tested regression coefficient and fitting degree, finally chose $y = a + bx$ as the best regression equation. The linear regression equation indicates that the linear regression equation slopes of MNF and MNCF of AT cells are greater than those of GM cells, and the differences are of statistical significance (p < 0.05), which demonstrates that AT cells have an obvious faster increasing tendency than GM cells. Micronucleus is caused by chromosomal aberration, reflecting individual radiosensitivity [Wang 1997]. Therefore, these research results manifest that at the same dose point, the degree of AT cells' chromosomal damage is greater than that of GM cells, and AT cells' radiosensitivity is significantly greater than that of GM cells.

In general, when eucaryotic cells are ionizing radiated, there are some effects like cell cycle arrest, enhanced repair of DNA damage and apoptosis, and the response of AT cells to ionizing radiation is different. When AT cells are ionizing radiated at G1 phase, p53 protein accumulation significantly delays; meanwhile, some downstream genes of p53 in signal conduction pathways (such as WAF1, gadd45, mdm2 and p21) also have a decreased expression level or a delayed presentation [Khanna 1998]. When AT cells are ionizing radiated, there is no delay of S phase progress, so as to reduce DNA synthesis and the replication of damaged DNA; after irradiation, p34cdc2 tyrosine residues' phosphorylation reduces, while the activation cyclin B/cdc2 protein kinase also make cells lose G2/M block, so cells cannot stagnate before karyokinesis for long, causing that cell damage not to be repaired. Abnormal protein expression of ATM in AT cells stops p53 phosphorylation induced by ionizing radiation, and thus weakening or dalaying p53's regulation, inhibiting the synthesis of damaged DNA, blocking DNA's repair, finally leading to the abnormal increase of loci injury of genetic material [Khanna 1998, Enoch 1995]. That is why AT cells have much higher MNF, MNCF and hprtMF after ionizing radiated than normal GM cells.

ACKNOWLEDGEMENTS

This work is supported by Innovative Project for Graduate Students of Jiangsu Province (CXLX11_0082), Scientific Program of Changzhou (CE20125026 and ZD201005), Special funds of Jiangsu province scientific innovation and achievements transformation (BL2012046) and The natural science foundation of Jiangsu Province (BK2009102).

REFERENCES

Czaja AJ. Nonstandard drugs and feasible new interventions for autoimmune hepatitis: part I. Inflamm Allergy Drug Targets. 2012; 11(5): 337–50.

Enoch T, Norbury C. Cellular responses to DNA damage: cell-cycle checkpoints, apoptosis and the roles of p53 and ATM. Trends Biochem Sci. 1995 Oct; 20(10): 426–30.

Fenech M, Morley AA. Measurement of micronuclei in lymphocytes. Mutation Research. 1985, 147: 29.

Gutierrez-Enriquez S, Hall J. Use of the cytokinesis-block micronucleus assay to measure radiation-induced chromosome damage in lymphoblastoid cell lines [J]. Mutat Res. 2003 Feb 5;535(1): 1–13.

Khanna KK, Keating KE, Kozlov S et al, ATM associates with and phosphorylates p53: mapping the region of interaction [J]. Nat Genet, 1998; 20(4): 398–400.

Khanna KK, Lavin MF, Jackson SP et al. ATM, a central controller of cellular responses to DNA damage. Cell Death Differ. 2001 Nov; 8(11): 1052–65.

Lavin MF, Shiloh Y. The genetic defect in ataxia telangiectasia. Annu. Rev. Genet. 1997, 15: 177–202.

Meyn MS. Ataxia telangiectasia and cellular responses to DNA damage. Cancer Res. 1995, 55: 5991–6001.

Mazin AL. Suicidal function of DNA methylation in age-related genome disintegration. Ageing Res Rev. 2009; 8(4): 314–27.

Progress in hprt Mutation Assay and Its Application in Radiation Biology. NUCLEAR PHYSICS REVIEW. 2008, 25(3): 294–298.

Torres RJ, Puig JG. Hypoxanthine-guanine phosophoribosyl-transferase (HPRT) deficiency: Lesch-Nyhan syndrome. Orphanet J Rare Dis. 2007 (8); 2: 48.

Wang Jixian. Radiobiological dosimetry [M]. Beijing: Atomic Energy Press. 1997, 6: 65–75, 216–235.

Information Technology and Computer Application Engineering – Liu, Sung & Yao (Eds)
© 2014 Taylor & Francis Group, London, ISBN 978-1-138-00079-7

Implementation of single-phase leakage fault line selection in ungrounded coal mine grid based on WAMS

Z.J. Wang & Y.W. Wang
Department of Mechanical Electronic & Information Engineering, China University of Mining & Technology (Beijing), Beijing, China

ABSTRACT: Single-phase leakage fault cannot be solved by one single substation, especially in coal mine high-voltage gird. In order to solve this fault, the implementation of single-phase leakage fault line selection based on WAMS is introduced combine with one renovation project in Shanxi province. In the system, the final fault line will be decided by the PMU in control center on ground, to avoid causing grade-skip trip. A new method, without sampling zero sequence, to select fault line by comparing the zero sequence current of feeders is introduced. The real part of normal feeders' ratio is positive, otherwise negative. Simulations show the effectiveness of this method.

Keywords: Single-phase Leakage; Fault Line Selection; Coal Mine Power Supply System; WAMS

1 INTRODUCTION

Single-phase leakage fault is one of the most common line faults in coal mine power supply system. It will easily cause grade-skip trip because the switch in upper substation don't have enough wait time. So when there is single-phase leakage fault happened in the coal mine grid, the switch of fault line and that in upper level substation will both trip, which leads to grade-skip trip. While with the application of WAMS (wide area measurement system) in coal mine grid, the control center will judge the fault line according to all the information in the grid, and send trip instruction only to fault switch. The biggest advantages of WAMS are data sampling synchronously in whole grid and decisions made by all the information received in control center.

WAMS consists of four parts (Cheng K. 2008), and collects the real-time information with PMU (Phasor Measurement Unit) as the primary unit, and uploads the information to the control center by communication system, to achieve monitoring the whole system (Zhang J.Y. 2006, Adamiak M.G. et al. 2006).

In recent years, WAMS and applications based on such system have developed rapidly in the United States, Japan, France and other countries and regions; at the same time the major grids in our country have implemented wide area measurement system in different levels on the power supply system (Qiao X.Y. 2008, Meng Z.Q. 2009), especially in the ground substation construction of integrated automation.

2 FAULT LINE SELECTION SYSTEM BASED ON WAMS

So far, one renovation project based on WAMS is built up in Shanxi province, to solve single-phase leakage problem. The whole renovation schematic diagram of the project is shown in Figure 1.

GPS. The GPS antenna is set up on the roof of the control center, which makes sure the GPS server can get precise time signal from the GPS satellites. The GPS is connected to the PMU in substation A, B, C and control center by single mode fiber (SMF) to achieve the time synchronization of the whole system. All the PMU in the system will get pulse per second (PPS) from the GPS by SMF to adjust current time, which makes good preparation for the synchronous data sampling of all the intelligent controllers.

PMU. The system consists of four phasor measurement units, being fixed respectively in substation A, B, C and control center. Here we define the phasor measurement unit in control center as Head Unit, and others as Part Units.

Obviously all the Units will get PPS from GPS by SMF as mentioned above. The Part Units will guide the intelligent controllers to sample data synchronously, and get corresponding calculation results from them, such as zero sequence current, and zero sequence voltage. Then all the data is send to the Head Unit in the control center, where the final decision is made according to the data. The final decision will be then send to all the Part Units, and therefore all the intelligent

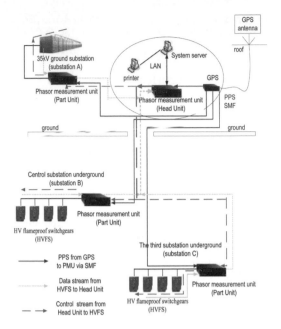

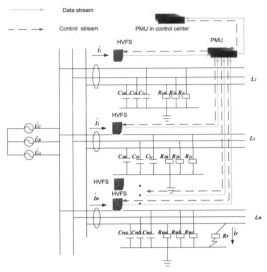

Figure 2.   Single-phase leakage fault line selection model.

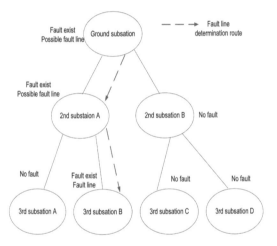

Figure 1.   Renovation schematic diagram of the project.

controllers will determine whether to trip or not according to the order they receive.

Communication Network. In this renovation project, we adopt point to point (PTP) fiber communication network between the PMU.

Control Center. The control center consists of system server, GPS server, Head Unit and printer. The communication protocol in this system is made according to IEC 61850 standard, so we can easily realize telemeter, tele-signaling, tele-control and tele-regulation in the protection system.

Figure 3.   Fault line determination schematic diagram.

3 FAULT LINE SELECTION PRINCIPLE

Single-phase leakage fault model is shown in Figure 2. C_{ia}, C_{ib} and C_{ic} are distributed capacitances of feeder i, R_{ia}, R_{ib}, R_{ic} are insulation resistances of feeder i; I_i are zero sequence current of feeder i. Here define feeder g as the fault feeder and

$$\dot{I}_{ij} = \frac{\dot{I}_i}{\dot{I}_j} = I_{ijR} + jI_{ijM}, (i=1,2,3\cdots n) \qquad (1)$$

where I_{ijR}, I_{ijM} are the real and imaginary parts of I_{ij}. Then we have:

$$I_{giR} = \frac{-(\frac{1}{R_i}\sum_{j=1}^{n}\frac{1}{R_j}) - \omega^2 C_i \sum_{j=1}^{n} C_j}{\frac{1}{R_i^2} + \omega^2 C_i^2} < 0 \quad i,j=1,2,3\cdots n \qquad (2)$$

and $i,j \neq g$

$$I_{ijR} = \frac{\frac{1}{R_i R_j} + \omega^2 C_i C_j}{\frac{1}{R_j^2} + \omega^2 C_j^2} > 0 \qquad i,j=1,2,3\cdots n \quad \text{and}$$
$$i \neq j \neq g \qquad (3)$$

4 FAULT LINE DETERMINATION

All the integrated controllers in the HVFS will sample the zero sequence current $3I_0$ synchronously according to the PPS from GPS, and send $3I_0$ to the PMU in the substation. PMU will choose the possible fault line according to Equation 2 and Equation 3, and send the fault status of the feeders in the substation to the PMU in the control center. The fault line will finally be decided according to all the possible fault lines, as Figure 3 shows.

Table 1. fault line selection simulations in different cases

I_i [A]	$Rr = 1\,k\Omega$	$C_1/C_2/C_3 =$ 0.3/0.3/0.3	I_{ij} results	Fault line
(a) $Rr = 1000\,\Omega$ and $C_1/C_2/C_3 = 0.3/0.3/0.3$				
I_1	8.47 − 23.6i		$I_{12R} = 1.00 > 0$	3
I_2	8.47 − 23.63i		$I_{31R} = -2.33 < 0$	
I_3	230.8 + 144.9i			
(b) $Rr = 11\,k\Omega$ and $C_1/C_2/C_3 = 0.3/0.9/0.6$				
I_1	0.04 − 3.76i		$I_{12R} = 0.90 > 0$	3
I_2	0.37 − 11.26i		$I_{31R} = -4.35 < 0$	
I_3	39.40 + 16.76i			

5 FAULT TEST

In order to verify the effectiveness of the fault line selection approach, simulations in different cases are done. Line 3 is chosen as the fault line during the simulation, and the simulation result is shown in Table 1.

From the simulation results in table 1, we can get the real part of the comparison between normal lines is positive, with $I_{12R} = 1$ and $I_{12R} = 0.9$ respectively, the real part of the comparison between fault line and normal line is negative, with $I_{31R} = -2.33$ and $I_{31R} = -4.35$ respectively. All the simulation results in different grounding resistance and distributed capacitors show that the fault line is line 3, which squares with the fact.

6 CONCLUSIONS

Single-phase leakage fault line selection system based on WAMS could guarantee all the data sampled synchronously, and the final fault line will be decided according to the fault line determination route and all the status of switches. In ungrounded coal mine grid, the fault line can be selected by the comparing the zero sequence current of feeders and the real part of the ratio between normal feeders is positive, while the real part of the ratio between normal feeder and fault feeder is negative. Renovation project being implemented in Shanxi province and simulation test prove the effectiveness of this fault line selection implementation.

ACKNOWLEDGEMENT

This paper is based upon a research project (grant number 2012YJ11) supported by the special funds for basic research business expenses of central university.

REFERENCES

Adamiak M.G. et al. 2006. Wide area Protection- technology and infrastructures[J]. *IEEE Transaction on Power Delivery*, 21(2):601–609

Cheng K. 2008. Research on line voltage stability index and its assessment method based on WAMS. *North China Electric Power University*. Beijing.

Hu Y.S. & Zhang M. 2007. Applications of WAMS in Power System. *Electrotechnical*. 5:19–21.

Meng Z.Q. 2009. Assessment on security of Power System based on WAMS. *North China Electric Power University*. Hebei.

Qiao X.Y. 2008. Research on the parameter identification of Synchronous Machine based on WAMS. *North China Electric Power University*. Beijing.

Zhang J.Y. 2006. Research on Power System transient stability security and defense in the Observation and Control of Wide Area. *Tsinghua University*. Beijing.

Information Technology and Computer Application Engineering – Liu, Sung & Yao (Eds)
© *2014 Taylor & Francis Group, London, ISBN 978-1-138-00079-7*

An ant colony optimization approach to chemical equilibrium calculations of complex system problem

Shang Yong Li

Sichuan Electromechanical Institute of Vocation and Technology Panzhihua Sichuan, China

ABSTRACT: This paper deals with the problem of continuous function optimization in chemical equilibrium calculations of complex system. It addresses this problem by means of the development of an ant colony optimization-based. This new algorithm, here named as continuous function ant colony optimization is composed of the global and local search algorithm. The results show this new approach can be used to solve the problem efficiently and stably.

Keywords: ACO; CFACO; Chemical equilibrium; styling; Algorithm

1 INTRODUCTION

In 1991, the ant colony optimization (ACO) algorithm presented by Dorigo M et al, and later extended by several studies. The algorithm is essentially a new type of global random search algorithm, which easy to combined with other methods and applied to distributed computer system. Currently, it has become a hot spot in artificial intelligence field. The research has penetrated into many applications from the initial one-dimensional algorithm to solve the static optimization problem to solve multi-dimensional development of dynamic combinatorial optimization problems. The research has attracted many scholars to the various studies because of excellent performance and great development potential.

On the basis of these research results, different categories of problems have been studied using ACO, for example, travelling salesman problem (TSP), the sequential ordering problem (SOP) and the capacitated vehicle routing problem (CVRP). It is still difficult for researchers to solve in this context of combinational problems.

The ACO algorithm is mainly applied to solve discrete combinatorial optimization problems, not suitable for solving continuous space problem. Because ACO in discrete space, the amount of information retained and changes in the selection of the optimal solution is through the distribution of discrete point-like approach to the solution. In continuous space optimization problem, the solution space is a regional, rather than a discrete set of points. In order to solve continuous function optimization problem (CFO), the algorithm must be changed in the way of information retained, optimization methods and ant colony road strategy. The proposed algorithm developed in this paper is used to solve CFO problem on the basis of improved optimization strategies. It well be presented in further sections, is here named continuous function ant colony optimization (CFACO).

The paper deals with the chemical equilibrium calculations of complex system problems use ACO. The chemical equilibrium calculations of complex system (CECCS) are mainly to solve the composition of each component, when the system reached chemical equilibrium. According to equilibrium theory, chemical equilibrium can be expressed as a system Gibbs free energy minimization of the system. Thus, the problem is transformed to solve the minimum of a continuous function in the certain limit conditions. The CFACO is used to solve it.

To present the CFACO algorithm used to solve the CECCS problem, this paper is divided as follows.

Section II presents the formulation of the CECCS problem considered in this paper; section III presents ACO basic concepts, which is used as a basis for the development of the proposed algorithm; section IV is devoted to describing the proposal of the CFACO algorithm; section V presents and analyses the results of an calculation example of CECCS; and section VI presents the conclusion and final remarks.

2 PROBLEM FORMULATION

In this section, we describe the problem formulation and notions used in the paper.

2.1 *CECCS problem formulation and notions*

From the mathematical point of view, the Gibbs free energy minimization method of CECCS can be considered to the objective function minimization problem in the linear constraints. Therefore, we need to first establish the objective function and constraints of the forms of expression, as (1) shows.

$$\begin{cases} \min G(n_i) = \sum_{i=1}^{N} n_i G_i \\ Subject\ to : \sum_{i=1}^{N} a_{ie} n_i = B_e \left(n_i \geq 0; e = 1, 2, ..., M \right) \end{cases} \quad (1)$$

The problem approached here is defined as: let M presented elements, N are components, G is the total Gibbs free energy of the system. Presented partial molar Gibbs free energy of the component i, presented the total number of moles of e element, presented subatomic number of the element in the composition i of the chemical formula, presented the mole number of composition i and its value is non-negative.

2.2 Constrained optimization problem conversion

As (1) shows, The CECCS problem is a constrained optimization problem. It must convert unconstrained optimization in order to facilitate the calculation of using CFACO. Using the penalty function method, the constrained problem can be simply converted to an unconstrained one, as (2) shows.

$$F(n_i) = f(n_i) + Zp(n_i) \quad (2)$$

where $f(n_i)$ is objective optimization function, Z is the common penalty parameter, $p(n_i)$ is constraint functions and its value is non-negative.

If the value of does not meet the constraints, $p(n_i)$ will be get a large value. When the constraints are more severe damage, the greater the value, so that the greater the value of, Conversely, When the constraint is satisfied, regardless of how much to take the penalty factor Z, the objective function is no longer affected by the penalty function, which is the penalty factor and the penalty term of the function. Now, we consider convert (1) to be an unconstrained function in (3).

$$\min F(n_i) = \sum_{i=1}^{N} n_i G_i + Z \sum_{e=1}^{M} \left(\sum_{i=1}^{N} a_{ie} n_i - B_e \right)^2 \quad (3)$$

In the following section, the CFACO algorithm used to solve this problem will be shown.

3 ANT COLONY OPTIMIZATION

Biology studies have shown that, although the ability of individual ants is very limited, but the groups composed of multiple ants have the ability that can find the shortest path between food and ant home. This ability relies on the release of volatile secretions (pheromone) in its path through [1][2]. When the ants on the path forward, it will choose path what will be

through according to pheromone on the path. The probability of choosing a path will be proportional to the intensity of pheromone on the path [3][4][5]. Therefore, the collective behavior of ant colony actually constitutes a positive feedback phenomenon of learning information [6]. A path through the more ants, ants choose the path behind the more likely. Ant between individuals seeks through this exchange of information leading to the shortest path to food.

ACO algorithm is that using artificial ants to simulate the behavior of real ants seeking for foods in order to solve optimization problems. The rule of releasing and updating pheromones must be designed for analog ants. Gambardella and Dorigo define as transition rule for ant colony system (ACS) show in (4).

$$p_{ij}^k(t) = \begin{cases} \dfrac{\left[\tau_{ij}(t)\right]^\alpha \bullet \left[\eta_{ij}(t)\right]^\beta}{\sum_{s \subset allowed_k} \left[\tau_{is}(t)\right]^\alpha \bullet \left[\eta_{is}(t)\right]^\beta} & , if\ j \in allowed_k \\ 0 & , otherwise\ j \notin allowed_k \end{cases} \quad (4)$$

i and j are not visited node, p is the path choice probability of i to j, τ is the ant pheromone of i to j, η is the heuristic function. s is a set of nodes to allow ant access [7].

4 CFACO ALGORITHM

Because the basic ant colony algorithm to solve the discrete optimization problem domain, CECCS is continuous global optimization problem, its solution space is not a discrete set of points, but a "regional". The objective function of CECCS problem is multivariable nonlinear function and the optimal solution value of each variable vary widely. So, in a complex system of chemical equilibrium calculations using ant colony optimization algorithm to solve the continuous problem domain, not simply to do the discretization the continuous domain, the need to improve on the basic ant colony algorithm to adapt to the continuous domain non-linear function optimization problems solving.

4.1 Continuous solution domain decomposition methods

In the whole solution domain, according to the number of ants, randomly generated a number of points. Each ant is placed on these points. And this point is the center, forming a number of radiuses R of the circular area. Every ant searches the optimal solution in their respective circular area.

4.2 The whole thought of the CFACO algorithm

First, every ant searches the optimal solution in their respective circular area, then the overall evaluation.

It is to be sorted According to the objective function value, optimal ant in its area to continue the search. Other ants moved to neighborhood where the ant searches objective function value than it better. When the Direction of movement has multiple goals, decision-making rely on probability function. The CFACO algorithm is shown in Algorithm 1.

1. initialize
2. each ant is positioned on the initial node randomly
3. **repeat** each execution is called iteration
4. **repeat** each execution is called step
5. each ant searches the optimal solution in their respective circular area
6. **until** all the ants find the optimal solution in their area
7. sorting and evaluation
8. updating pheromone intensity
9. a new research area is assigned for all the ants
10. **until** the stop criteria is satisfied

In the following sections, the key aspects of the implementation of both stages of the CFACO algorithm will be discussed.

1) The Global search algorithm of CFACO: Global search algorithm aimed to avoid the algorithm falling into local optimal solution, when it falling, we can choose a strategy to enable it to jump out of the local optimal solution. Each ant global search strategy is shown in (5)(6).

$$
act(k) = \begin{cases} Global\ Search & ,if\ F\left(n_i^k\right) > F_{avg}\ and\ q < q_0 \\ p_{ks} = \dfrac{\tau(s)e^{\frac{\eta}{d_{ks}}}}{\sum \tau(s)e^{\frac{\eta}{d_{ks}}}} & ,otherwise \end{cases} \tag{5}
$$

$$
F_{avg} = \frac{\sum\limits_{k=1}^{m} F\left(n_i^k\right)}{m} \tag{6}
$$

where $F(n_i^k)$ is minimum of function in local search by ant k; F_{avg} is average of m ants; q_0 is a global search probability $(0 \leq q_0 \leq 1)$; q is Ant K path choice probability $(0 < q < 1)$; p_{ks} is path choice probability from ant k to ant s $(F(n_i^k) > F(n_i^s))$; $\tau(s)$ is pheromone of ant s, it reflects that the pheromone of ant s have effect on ant k when it selects path; $d_{ks} = |F(n_i^k) - F(n_i^s)|$ and $d_{ks} > 0$; d_{ks} is distance value from ant k to ant s; $\eta = |F(n_i^k) - F_{avg}|$; $e_{ks}^{-\eta/d}$ is expectation factor of path selection, d_{ks} is larger, the probability of ant k moving to ant s is greater.

2) The local search algorithm: Local search is designed to make the ant k to the identified domain value as the center, in the radius r of the neighborhood to get the best solution precision value. The rules are as follows:

Ant k initial point is n_i^k, initial value is $n_{i\,init}^k$, neighborhood is $[n_{i\,init}^k - r,\ n_{i\,init}^k + r]$, $n_{i\,new}^k$ is new value

that the ant reached in neighborhood. The local search rules is shown (7).

$$
n_i^k = \begin{cases} n_{i\,new}^k & ,if\ F\left(n_{i\,new}^k\right) < F\left(n_i^k\right) \\ n_i^k & ,otherwise \end{cases} \tag{7}
$$

3) The ant pheromone updating rule: The ant pheromone will be updated at the end of each global search in the algorithm, each ant represented as a function of the solution, so the ant pheromone should not only represent the quality of solutions, and can reflect other ant attraction degree. The ant s pheromone updating rule is shown in (8).

$$
\tau(s) = \rho\tau(s) + \sum_{k=1}^{m_{k\to s}} e^{-\frac{F\left(n_i^k\right) - F\left(n_i^{best}\right)}{F_{avg} - F\left(n_i^{best}\right)}} + \tau_0 e^{\frac{F\left(n_i^k\right) - F\left(n_i^{best}\right)}{F_{avg} - F\left(n_i^{best}\right)}} \tag{8}
$$

$F(n_i^{best})$ is the current global optimum function value; $F(n_i^k)$ is function value of all of ant that move to ant k; τ_0 is initial value of ant pheromone ($\tau_0 = 6$); ρ is Pheromone evaporation factor ($\rho = 0.9$); $m_{k\to s}$ is ant number of moving to ant s.

Clearly, from (8) It can be seen, the more excellent function value that ant s get, the greater the amount of pheromone growth. Ant amount of moving to ant k is more, the greater the amount of pheromone growth. Therefore, it is a good reflection of the ant path selection results and the pros and cons of the current search path.

4) The convergence condition of CFACO: Because of the randomness of ant colony algorithm, the global optimal solution can be obtained after several iterations. The convergence condition of CFACO is as follow as shown.

(1) The Ant with the global optimal solution should not change and the value of the global optimal solution should no longer change or a very small change.
(2) Most of the ant colony should be moved to a nearby neighborhood of the ant that has obtained a global optima solution.

5 COMPUTATIONAL RESULTS

In order to verify the results of iterative calculation is stable, we have the same system under the same conditions repeated calculations to verify the optimal objective function value of stability. The best case and worst case of Iterative process have conducted a comparative analysis.

Calculation Example: Water-gas reaction

Initial condition: 3 mol C + 1 mol H_2O + 1 mol O_2

Equilibrium conditions: Temperature is 1000 K; pressure is 1 atm.

The data from Table 1 is shown, either in best or worst case, the algorithm of CFACO can be very good at or near to the global optimal value. The algorithm has good stability.

711

Table 1. Iteration of the best and worst results comparison table.

	Reference Data	The Best	The Worst
Gibbs free energy	−1142684.31675	−1142679.80351	−1142604.21961
Number of iterations		1705	2587

6 FINAL REMARKS

In this paper, the Gibbs free energy minimization principles and penalty function method, the chemical equilibrium calculations of complex system converted to unrestricted continuous function optimization problems, to facilitate the use of ant colony algorithm.

Use zoning approach to solving domain is divided into several circular areas, and improved ant colony optimization algorithm, so that space can be used to solve continuous optimization problems.

In the improved ant colony algorithm, the algorithm is divided into two phases, the first stage, using a local search algorithm to find the local optimal value, in order to avoid premature algorithm into a local optimal solution; the second stage, the design of global search strategy, the faster the algorithm to jump out of local optimum, and ultimately get the global optimal solution.

An example of chemical equilibrium calculation of complex system is tested using improved ant colony algorithm, the test results show that the algorithm is feasible and effective.

REFERENCES

[1] Bonabeau E, Dorigo M, Theraulaz G. "Inspiration for optimization from social insect behavior," Nature, vol. 406, 2000, pp. 39–42

[2] Jackson D E, Holcombe M, Ratnieks F L W. "Trail geometry gives polarity to ant foraging networks," Nature, vol. 432, 2004, pp. 907–909

[3] Gambardella L M, Dorigo M. "Solving symmetric and asymmetric TSPs by ant colonies," Proccedings of the IEEE International Conference on Evolutionary Computation, 1996, pp. 622–627

[4] Katja V, Ann N. "Colonies of learning automata," IEEE Transactions on Systems, Man, and Cybernetics-Part B, vol. 32, 2002, pp. 772–780

[5] Dorigo M, Di Caro G. "Ant colony optimization: a new meta-beuristic," Proceedings of the 1999 Congress on Evolutionary Computation, 1999, pp. 1470–1477

[6] Dorigo M, Gambardella L M. "Ant colonies for the traveling salesman problem," Bio Systems, vol. 43, 1997, pp. 73–81

[7] Dorigo M, Bonabeau E, theraulaz G. "Ant algorithms and stigmergy," Future Generation Computer Systems, vol. 16, 2000, pp. 851–871

Information Technology and Computer Application Engineering – Liu, Sung & Yao (Eds)
© 2014 Taylor & Francis Group, London, ISBN 978-1-138-00079-7

Finite element analysis of aircraft skin clamping deformation based on the technology of vacuum adsorption

Da Wei Wu, Hong Min Cui, Ying Chun Liu & Wei Ji
Key Laboratory of Fundamental Science for National Defense of Aeronautical Digital Manufacturing Process, Shenyang Aerospace University, Shenyang, China

ABSTRACT: As the aircraft skin is less rigid and it requires a higher degree on the surface roughness of the aircraft skin, therefore it can easily have the problem of clamping deformation in the process of assembly. In this paper, the aluminum alloy of aircraft skin is regarded as a typical example, and the vacuum adsorption technology is adopted in the aircraft skin clamping. Vacuum suction cups for different thickness of the aircraft skin producing the local deformation were simulated by the finite element method under the condition of different vacuum pressures. The results showed that different thickness of the aircraft skin should have different vacuum pressures so as to effectively control the deformation of aircraft skin during assembly process, thereby increasing the surface roughness of the aircraft skin.

Keywords: Aircraft skin; Clamping deformation; Vacuum adsorption technology; Finite element simulation

1 INSTRUCTIONS

Modern aircraft skin materials are usually of high-strength aluminum alloy, carbon fiber reinforced aluminum alloy composites and other composite materials, and has the characteristics of big structure size, poor rigidity, and high positioning accuracy requirements. Because the aircraft skin is in the form of thin-walled open surfaces and the entire part is in a flexible form of state, the clamping assembly process is difficult and it often requires the fixture design department to design more fixture system to assemble. The traditional six-point positioning principle is not suitable for this type of thin elastic stiffness and poor surface profile of freeform parts[1], as it seriously affects the quality and efficiency of assembly. Therefore, under the conditions of the elastomeric surface positioning, the aircraft skin should be in contact with the surface in tooling. The finite deformation is a guiding principle which cannot be ignored in clamping system designing and analyzing[2], to rationally determine the size of the clamping force and the position of points, in order to make the aircraft skin small deformation, thereby to enhance the accuracy of the entire aircraft skin assembly. Therefore, the clamping system should ensure the accurate positioning of aircraft skin, maintain the stability of aircraft skin, in order to control the aircraft skin deformation within a certain range.

In response to the above problems, based on the study of flexible fixture clamping technology abroad,

combined with our process of aircraft assembly features, analyzed the clamping deformation condition of aircraft skin clamping system based on the vacuum adsorption technology, it can provide a theoretical reference for the deformation control of aircraft in the assembly process, to ensure the accuracy of the whole assembly. In this paper, according to the 7075 aluminum alloy aircraft skin Material, the author analyze the finite element of aircraft skin affected in different status of the clamping process.

2 VACUUM ADSORPTION CLAMPING TECHNOLOGY

We have got a new philosophy through the study of assembly—by the type of vacuum sucker flexible fixture can generate adsorption lattice to be consistent with product appearance accurately and adsorb aircraft skin accurately and solidly[3]. The type of contact of the structure of the suction cup with the aircraft skin is the use of surface and surface contact. The positioning, support, clamping during the whole process are accomplished at one time. The location and stay of aircraft skin in the space are not only relying on 6-DOFs, but also rely on the entire surface constituted by each column on the flexible fixture. Figure 1 shows the work form of flexible fixture clamping the aircraft skin. The pillars form the same curvature of curved surface compared with the aircraft skin in the figure. If we want to complete this idea, we can use two

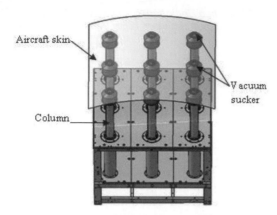

Figure 1. The type of vacuum sucker flexible fixture.

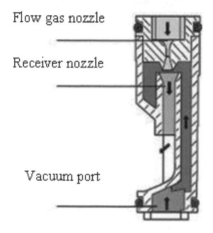

Figure 2. Vacuum generator structure principle diagrams.

methods—rigid and flexible. If we adopt the method of rigidity, we will have significant limitations. Different surface will need to produce a different column, so this design adopts the method of flexible. The vacuum sucker can form different curvature of curved surface shape by changing the column height, so that the flexible fixture can adapt to different shape of the aircraft skin. the flexible fixture is universal—reducing the production cost and improving the assembly precision.

The application of vacuum system in the flexible fixture mainly take advantage of vacuum to aircraft skin adsorb on the vacuum suction cups. Its adsorption force is adjustable, no creasing and easy mounting-dismounting work piece. Vacuum sucker used for positioning and clamping of the aircraft skin is according to the venturi principle[4]. The structure and principles of the vacuum generator is shown in Figure 2.

Vacuum generator includes a flow gas nozzle (Venturi nozzle) and one receiver nozzle at least, which is connected to the vacuum chuck. Through the high vacuum produced by the vacuum generator, making the suction disc connected with the vacuum generator produce a negative pressure. The negative pressure

makes the aircraft skin adsorb on the surface of the vacuum chuck type. Compressed air pass into the vacuum generator and pass through the narrow flow venturi nozzle, the airflow is accelerated to 5 times of the speed of sound. At the entrances to the receiver nozzle has a short gap. When the compressed air from the venturi nozzle passes through this gap, its volume rapids expansion and generates a suction effect. So in the output port of the device (vacuum port) forms a vacuum. Accordingly, make the suction disc connected with the vacuum generator produce a adsorption force and effect on the aircraft skin.

3 FINITE ELEMENT MODEL AND ANALYSIS

In this paper, a single sucker was mainly considered by the deformation of aircraft skin.

Application Examples: Now, the model of before assembling the aircraft fuselage skin is analyzed. The 7075 aluminum alloy aircraft skin Material, Length × width = 200 mm × 100 mm, the cylindrical radius of curvature R = 400 mm, E is the elastic modulus of 71,000 MPa, Poisson's ratio of 0.33 μ. Assembling the skin, the vacuum chuck 9 is utilized simultaneously. The Assembly style was shown in Figure 1, moreover, the structure of a single vacuum suction cups skinned was shown in Figure 4. According to standard suction cup, suction cups are selected to Φ60.

Combined with the shape and material aircraft skin and the shape of sucker are chosen to high-load-type sucker, sucker is made of silicone rubber.

3.1 Finite element mode

(1) Geometry
The model of Part solid was created through the three-dimensional CAD software of CATIA, and then it was imported into the Hyperworks. 7075 aircraft skin material, E is the elastic modulus of 71,000 MPa, Poisson's ratio of 0.33 μ, the density is 2.7 g/cm³. Model is defined element type of Pshell, it was divided into 832 unit overall by control the cell size.

(2) Constraints imposed
The contact type was applied by Elastic contact, and the degrees of positioning holes were secured.

(3) Suction force applied
The size of the adsorption force was changed by vacuum pressure; a closed chamber can be constituted of suction nozzle and the aircraft skin. Therefore, when vacuum extraction in a closed room; adsorption is mainly distributed in the skin of vacuum suction cups. Finite element model is shown in Figure 3.

3.2 Results and analysis solution

At different vacuum pressure adsorption of different thicknesses aircraft skin deformation was calculated

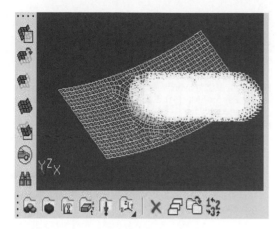

Figure 3.   Finite element model.

by using Hyperworks, the displacement and deformation cloud of aircraft skin can being viewed through Hyperview. Excel was used to process the data, which was obtained by finite element analysis. The main contents in this paper are shown as: the deformation of aircraft skin were affected by adsorption force in different thickness of aircraft skin in the process of assembly, According to the actual situation, the vacuum pressure is generally satisfactory between −40 kPa and −60 kPa, Therefore, this paper were adopted by −40 kPa, −45 kPa, −50 kPa, −55 kPa, −60 kPa, respectively to the thickness of d = 1 mm, 1.5 mm, 2 mm, 2.5 mm, 3 mm of the aircraft skin adsorption. Its relationship between Maximum deformation of the aircraft skin and the vacuum pressure are shown as below.

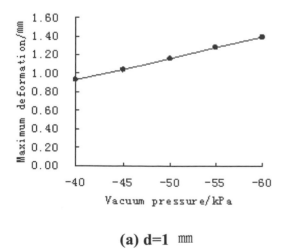

(a) d=1 mm

When the aircraft skin thickness must be, the maximum deformation is nearly linear rule change with the increases of vacuum pressure. Therefore, in the process of vacuum adsorption, the vacuum suction force was regulated by switching regulator the minimum vacuum pressure should be adopted so that the

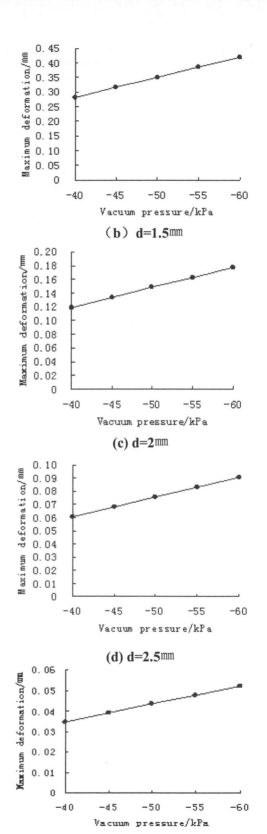

（b）d=1.5mm

(c) d=2mm

(d) d=2.5mm

(e) d= 3 mm

smallest deformation of t aircraft skin in the condition of the aircraft skin being guaranteed to adsorptions.

4 CONCLUSIONS

As a new kind of aircraft assembly technology, namely vacuum sucker type flexible fixture system, vacuum system in regulating switch is used to adjust the vacuum pressure and adjust the adsorption force, to avoid the adsorption force is too large and cause skin deformation or the adsorption force is smaller and it can not stick to skin, make the whole assembly efficiency and quality to rise, to provide theoretical reference about controlling the aircraft skin deformation in the assembly process.

REFERENCES

[1] Men Yanwu, Zhou Kai. The study of flexible positioning method of free curved thin walled workpiece [J]. Manufacturing Technology & Machine Tool, 2008(10): 113–117.

[2] Nee A.Y.C, Whybrew K., Senthil K.A. Advanced fixture design for FMS [M]. London: Springer-Verlag London Limited, 1995.

[3] Zhu Yaoxiang, Rong Yiming. The development of flexible fixture and computer aided fixture design technology [J]. Manufacturing Technology & Machine Tool, 2006(8):5–8.

[4] Zheng, Xiaoyan, Que Lijuan, etc. The technology and its application of the location and flexible assembly of Rocket skin and wall plate [J]. Missile & Space Launch Technology, 2012(2):47–53.

Information Technology and Computer Application Engineering – Liu, Sung & Yao (Eds)
© 2014 Taylor & Francis Group, London, ISBN 978-1-138-00079-7

EDZL schedulability analysis on performance asymmetric multiprocessors

Wu Qi Li, Peng Wu & Hyunmin Yoon
Department of Electrical and Computer Engineering, Hanyang University, Seoul, Korea

Minsoo Ryu
Department of Computer Science and Engineering, Hanyang University, Seoul, Korea

ABSTRACT: EDZL (Earliest Deadline Zero Laxity) scheduling policy is a hybrid policy that combines EDF (Earliest Deadline First) and LLF (Least Laxity First) policies. Previous work showed that EDZL performs better than EDF in a sense that EDZL can always schedule task sets that EDF can schedule on identical multiprocessors. In this paper, we apply EDZL to performance asymmetric multiprocessors that may have different speeds and present an effective schedulability test. Our EDZL for performance asymmetric multiprocessors slightly differs from the original EDZL algorithm. Our EDZL chooses the fastest available processor at the moment of scheduling in a way similar to that of global EDF for performance asymmetric multiprocessors. With this EDZL scheduling, we derive an effective schedulability condition. To our knowledge, this is the first schedulability analysis on EDZL for performance asymmetric multiprocessors.

Keywords: Performance asymmetric multiprocessor; Sporadic task system; EDZL algorithm; Schedulability analysis

1 INTRODUCTION

EDZL (earliest deadline zero laxity) scheduling policy is a hybrid policy that combines EDF (earliest deadline first) and LLF (least laxity first) policies. EDZL normally gives higher priorities to tasks with earlier deadlines as EDF does, but elevates task priorities to the highest one whenever they have zero laxity. Previous work (Cirinei et al. 2007, Lee et al. 2012, Park et al. 2005, Baker et al. 2008) showed that EDZL performs better than EDF in a sense that EDZL can always schedule task sets that EDF can schedule on identical multiprocessors.

In this paper, we apply EDZL to performance asymmetric multiprocessors that may have different speeds and present an effective schedulability test. Our EDZL for performance asymmetric multiprocessors slightly differs from the original EDZL algorithm. Our EDZL chooses the fastest available processor at the moment of scheduling in a way similar to that of global EDF for performance asymmetric multiprocessors (Baruah et al. 2008). With this EDZL scheduling, we derive an effective schedulability condition. To our knowledge, this is the first schedulability analysis on EDZL for performance asymmetric multiprocessors.

2 SYSTEM MODEL AND SCHEDULING ALGORITHM

A sporadic task τ_i is defined by (C_i, D_i, P_i), where C_i is a worst-case execution time, D_i is a relative deadline, P_i is a minimum inter-arrival time, and $C_i \leq D_i \leq P_i$. A sporadic task set is defined by $T = \{\tau_1, \ldots, \tau_n\}$ where all tasks are assumed to be independent. As each sporadic task generates an infinite sequence of requests with a minimum inter-arrival time, we use $\tau_{i,j}$ to denote the j_{th} request and refer to such an individual request as a job. We use $A_{i,j}$ to denote the arrival time of $\tau_{i,j}$. We consider m performance asymmetric multiprocessors and use $\pi = (s_1, \ldots, s_m)$ to denote the set of their processing speeds in a non-decreasing order, i.e., $s_k \geq s_{k+1}$ for all k, where s_k is the k_{th} processor's speed. For notational convenience, we assume that the highest speed $s_1 = 1$ and there can be r processors that have the highest speed. Let Y_p be the cumulative speed of the first p fastest processors.

$$Y_p = \sum_{k=1}^{p} s_k$$

Let δ_i be the density of task τ_i, i.e., the ratio of its worst-case computation time to its relative deadline. We use $\delta_{max}(T)$ to denote the largest density of task set T.

Based on the above system model, our EDZL scheduling for performance asymmetric multiprocessors works as follows. First, it computes the laxity for each job with respect to the fastest processor. Second, when it finds tasks that have zero laxity, it gives them the highest priority. Otherwise, it gives priorities following the original EDF policy. Third, it assigns tasks to the fastest possible processors available at the moment of scheduling.

3 SCHEDULABILITY ANALYSIS

In this section, we describe our schedulability condition for EDZL on performance asymmetric multiprocessors. To this end, we begin with an important observation that if the number of jobs that have zero or negative laxity is always less than or equal to the number of fastest processors, then the given sporadic task system is schedulable. The following theorem proves this observation.

Theorem 1: For performance asymmetric multiprocessors $\pi = (s_1, \ldots, s_m)$ with $s_1 = \ldots = s_r = 1$ and $s_{r+1} < 1$, sporadic task set $T = \{\tau_1, \ldots, \tau_n\}$ is not schedulable by EDZL if and only if there exists a time point t_c when the number of zero laxity jobs is greater than r.

Proof: We first prove the "if" part. Suppose that the number of zero laxity jobs is greater than r at time t_c. Since there are only r fastest processors, some zero laxity tasks must either run on processors whose speeds are smaller than 1 or wait until processors are available. Since the laxity values are computed with respect to the fastest processor, those tasks will miss their deadlines.

We now prove the "only if" part using proof by contradiction. Suppose that the task set is not schedulable and the number of zero laxity tasks always does not exceed r. Let t_x be a task that misses its deadline. For t_x to miss its deadline, there are two cases to consider when the laxity of t_x becomes zero. The first case is where t_x is assigned to a processor whose speed is less than 1. This case is only possible when the r fastest processors are busy running other jobs. Since our EDZL always assign zero laxity tasks to fastest processors unless the number of zero laxity tasks does not exceed r, this leads to a contradiction. The second case is where t_x must wait until any processor becomes idle. This is also impossible since our EDZL always gives the highest priority to zero laxity tasks. □

Corollary 1: For performance asymmetric multiprocessors $\pi = (s_1, \ldots, s_m)$ with $s_1 = \ldots = s_r = 1$ and $s_{r+1} < 1$, sporadic task set $T = \{\tau_1, \ldots, \tau_n\}$ is schedulable by EDZL if and only if the number of zero laxity jobs always does not exceed r.

Based on Corollary 1, we now describe our schedulability condition. We first derive a necessary condition for a task to become a zero-laxity task, called a critical task. We then use this necessary condition in finding non-critical tasks that always have positive laxity values. If the number of such non-critical tasks is greater than $m - r$, i.e., the number of critical tasks does not exceed r, we can conclude using Corollary 1 that the given sporadic task system is schedulable.

We derive the necessary condition for task τ_i to become a critical task in three steps; (1) deriving a lower bound on the total amount of CPU time demand during $[t_o, A_{i,j} + D_i)$ for certain time point $t_o \leq A_{i,j}$ in subsection 3.1,(2) deriving an upper bound on the total amount of CPU time demand during $[t_o, A_{i,j} + D_i)$ for

in subsection 3.2,(3) combining the lower bound and upper bound into a necessary condition for task τ_i to become a critical task in subsection 3.3.

Note that we only consider tasks have higher priorities than task τ_i when we derive the lower and upper bounds since lower priority tasks cannot affect the scheduling of task τ_i.

3.1 Lower bound

For a given task set $T = \{\tau_1, \ldots, \tau_n\}$, let $W(t_p, t_q)$ be the total amount of work done in $[t_p, t_q)$ and $R(t_p, t_q)$ be the total amount of processor time demand $[t_p, t_q)$ required by T.

Consider job $\tau_{i,j}$ arriving at time $A_{i,j}$ that becomes a critical task in $[A_{i,j}, A_{i,j} + D_i)$. Since $\tau_{i,j}$ has zero laxity at some time point in $[A_{i,j}, A_{i,j} + D_i)$, it cannot complete before $A_{i,j} + D_i$, but either completes its execution at $A_{i,j} + D_i$ or misses its deadline $A_{i,j} + D_i$. Let B_v be the total duration over $[A_{i,j}, A_{i,j} + D_i)$ for which exactly v processors are busy. We can then write

$$W(A_{i,j}, A_{i,j} + D_i) = \sum_{v=1}^{m} Y_v B_v. \tag{1}$$

Since $\sum_{v=1}^{m} Y_v B_v = Y_m B_m + \sum_{v=1}^{m-1} Y_v B_v$ and $Y_m B_m = Y_m(D_i - \sum_{v=1}^{m-1} B_v)$, we can write

$$W(A_{i,j}, A_{i,j} + D_i) = Y_m D_i - \sum_{v=1}^{m-1}(Y_m - Y_v)B_v$$
$$= Y_m D_i - \sum_{v=1}^{m-1} \frac{(Y_m - Y_v)}{s_v} s_v B_v.$$

Let $\lambda(\pi) = \max_{0 \leq i \leq m} \sum_{j=i+1}^{m} s_j / s_i$ (Funk et al. 2001). It then immediately follows that $\lambda(\pi) \geq Y_m - Y_v / s_v$. Thus, we have

$$W(A_{i,j}, A_{i,j} + D_i) \geq Y_m D_i - \sum_{v=1}^{m-1} \lambda(\pi) s_v B_v.$$

Due to the work conserving property of EDZL, $\tau_{i,j}$ must always be running in $[A_{i,j}, A_{i,j} + D_i)$ except when all the processors are busy. Thus, the processor time that $\tau_{i,j}$ receives during $[A_{i,j}, A_{i,j} + D_i)$ cannot be less than $\sum_{v=1}^{m-1} s_v B_v$, which must not exceed C_i, i.e., $C_i \geq \sum_{v=1}^{m-1} s_v B_v$. Therefore, we can write

$$W(A_{i,j}, A_{i,j} + D_i) \geq Y_m D_i - \lambda(\pi) C_i$$

and

$$\frac{W(A_{i,j}, A_{i,j} + D_i)}{D_i} \geq Y_m - \lambda(\pi) \delta_i.$$

Let $\mu_i = Y_m - \lambda(\pi) \delta_i$. Then we have

$$W(A_{i,j}, A_{i,j} + D_i) \geq \mu_i D_i. \tag{2}$$

Note that the total amount of work processed can never exceed the amount of processing time demand. Thus, we have

$$R(A_{i,j}, A_{i,j} + D_i) \geq W(A_{i,j}, A_{i,j} + D_i).$$

It immediately follows that

$$R(A_{i,j}, A_{i,j} + D_i)/D_i \geq \mu_i.$$

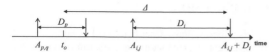

Figure 1. A job of task τ_k arrives at $A_{i,j}$ and miss its deadline at time-instant $A_{i,j} + D_i$.

Based on this, we now obtain anlower bound of $R(t_o, A_{i,j} + D_i)$, where t_o is the earliest time point such that $t \leq A_{i,j}$ and $R(t_o, A_{i,j} + D_i)/(A_{i,j} + D_i - t_o) \geq \mu_i$. Note that we can always find such a time point since at least $A_{i,j}$ satisfies $t \leq A_{i,j}$ and $R(t_o, A_{i,j} + D_i)/(A_{i,j} + D_i - t_o) \geq \mu_i$. Let $\Delta = A_{i,j} + D_i - t_o$, then we finally have

$$R(t_o, A_{i,j} + D_i) \geq \mu_i \Delta. \tag{3}$$

3.2 Upper bound

In this section, we obtain an upper bound of $R(t_o, A_{i,j} + D_i)$.

To compute the upper bound of $R(t_o, A_{i,j} + D_i)$, we need to consider two types of jobs, carry-in jobs that arrive before t_o and regular jobs that arrive during $[t_o, A_{i,j} + D_i)$.

We first compute the contribution made by carry-in jobs by: (1)determiningan upper bound on the number of carry-in jobs, (2) determining an upper bound on the remaining processor time demand for each carry in job, and (3) combining the per-job bounds to obtain an upper bound on the total remaining processor time demand.

Let ξ be an arbitrarily small positive number. From the definition of t_o, it follows that $W(t_o - \xi, A_{i,j} + D_i) < \mu_i(\xi + \Delta)$ and $W(t_o, A_{i,j} + D_i) \geq \mu_i \Delta$. Therefore, the work processed during $[t_o - \zeta, t_o)$ is less than $\mu_i \xi$. Since $\mu_i \leq Y_m$, it follows that some processor is idle during $[t_o - \xi, t_o)$. This implies that all the carry-in jobs are runningin this interval and that the number of carry-in jobs is the same as the number of busy processors. Let ω be thenumber of busy processors over $[t_o - \zeta, t_o)$ and Ω bean upper bound on ω Thus, we have

$$\Omega = max\{\omega : S_\omega < \mu_i\}. \tag{4}$$

Now consider the remaining processor time demand of each carry-in job. Let us consider a carry-in job $\tau_{p,q}$ arriving at time $A_{p,q}$. Let $\Phi_{p,q} = t_o - A_{p,q}$. From the definition of t_o, it follows that $R(A_{p,q}, A_{i,j} + D_i) < \mu_i(\Delta + \Phi_{p,q})$ and $R(t_o, A_{i,j} + D_i) \geq \mu_i \Delta$. Let $C'_{p,q}$ be the processing time that $\tau_{p,q}$ receives in $[A_{p,q}, t_o)$. By use the sameapproach as computing inequality (1). Let B'_v be the total duration over $[A_{p,q}, t_o)$ for which exactly v processors are busy. Thus, the total amount of work processed in $[A_{p,q}, t_o)$ is $Y_m \Phi_{p,q} - \sum_{v=1}^{m-1}(Y_m - Y_v)B'_v$. Since We can also express the total amount of work during the same interval by $R(A_{p,q}, A_{i,j} + D_i) - R(t_o, A_{i,j} + D_i)$, we can write

$$R(A_{p,q}, A_{i,j} + D_i) - R(t_o, A_{i,j} + D_i)$$
$$= Y_m \Phi_{p,q} - \sum_{v=1}^{m-1}(Y_m - Y_v)B'_v.$$

Since $R(A_{p,q}, A_{i,j} + D_i) < \mu_i(\Delta + \Phi_{p,q})$ and $(t_o, A_{i,j} + D_i) \geq \mu_i \Delta$, it immediately follows that

$$\mu_i(\Delta + \Phi_{p,q}) - \mu_i \Delta$$

$$> Y_m \Phi_{p,q} - \sum_{v=1}^{m-1}(Y_m - Y_v)B'_v.$$

Thus, we have

$$\mu_i \Phi_{p,q} > Y_m \Phi_{p,q} - \sum_{v=1}^{m-1}(Y_m - Y_v)B'_v. \tag{5}$$

Since

$$\sum_{v=1}^{m-1}(Y_m - Y_v)B'_v = \sum_{v=1}^{m-1}\frac{(Y_m - Y_v)}{S_v}S_v B'_v$$

and $\lambda(\pi) = max_{0 \leq i \leq m}\frac{\sum_{j=i+1}^{m}S_j}{S_i}$ and $\lambda(\pi) \geq \frac{Y_m - Y_v}{S_v}$, we can write

$$\sum_{v=1}^{m-1}(Y_m - Y_v)B'_v \leq \sum_{v=1}^{m-1}\lambda(\pi)S_v B'_v.$$

Since $C'_{p,q} \geq \sum_{v=1}^{m-1}S_v B'_v$

$$\sum_{v=1}^{m-1}(Y_m - Y_v)B'_v \leq \lambda(\pi)C'_{p,q}.$$

By applying this to Eq. (5), we have

$$\mu_i \Phi_{p,q} > Y_m \Phi_{p,q} - \lambda(\pi)C'_{p,q}.$$

Using $\mu_i = Y_m - \lambda(\pi)\delta_i$, it follows that

$$(Y_m - \lambda(\pi)\delta_i)\Phi_{p,q} > Y_m \Phi_{p,q} - \lambda(\pi)C'_{p,q},$$

resulting in

$$C'_{p,q} > \delta_i \Phi_{p,q}. \tag{6}$$

Note that $C_p = D_p \delta_p$ and that the remaining processing time demand of the carry-in job of task $\tau_{p,q}$ is $C_p - C'_{p,q}$. Through arithmetic manipulation of Eq. (6), we can write

$$C_p - C'_{p,q} < D_p \delta_p - \delta_i \Phi_{p,q}.$$

Since the absolute deadline of this carry-in job is no later than time $A_{i,j} + D_i$, $D_p - \Phi_{p,q} \leq \Delta$. Also, $C_p - C'_{p,q}$ cannot be greaterthan Δ. Therefore, when$\delta_p \leq \delta_i$,

$$C_p - C'_{p,q} < D_p \delta_p - \delta_i \Phi_{p,q}$$
$$< \delta_i(D_p - \Phi_{p,q})$$
$$< \delta_i \Delta,$$

otherwise ($\delta_p > \delta_i$)

$$C_p - C'_{p,q} \leq \Delta.$$

Let $\sigma_p \Delta$ be the upper bound of remaining processing time demand of carry-in job $\tau_{p,q}$. If $\delta_p \leq \delta_i$,

$\sigma_p = \delta_i$, else $\sigma_p = 1$. Let T_C be the set of carry-in jobs $\tau_{p,q}$. Since the number of carry-in jobs are upper bounded by Ω as shown above, the size of T_C is also bounded by Ω The upper bound of remaining processing time demand of all carry-in jobs is $\sum_{T_C} \sigma_p \Delta$. Note that it is hard to identify carry-in jobs $\tau_{p,q}$ for real schedulability analysis. Due to this limitation, we simply choose Ω tasks that have the highest density values from the task set $T - \{\tau_i\}$.

Now let us consider the upper bound of the execution requirement of regular jobs. Baruah *et al.* showed that the demand bound function $DBF(\tau_i, I)$ bounds the maximum cumulative processing time demand caused by τ_i within any time interval of length I.

$$DBF(\tau_i, I) = max\left(0, \left(\left\lfloor \frac{I - D_i}{P_i} \right\rfloor + 1\right) C_i\right).$$

We define $LOAD$(T) as follows.

$$LOAD(T) = max_{I>0}\left(\frac{\sum_{\tau_i \in T} DBF(\tau_i, I)}{I}\right).$$

The total processing time demand of all regular jobs can then be upper bounded by $\Delta \times LOAD$(T). As a result, an upper bound of total processing time demand caused by all regular and carry-in jobs can be expressed by

$$R(t_o, A_{i,j} + D_i) \le LOAD(T)\Delta + \sum_{T_C} \sigma_p \Delta. \qquad (7)$$

3.3 Schedulability test

Recall that to has been chosen such that $R(t_o, A_{i,j} + D_i)/(A_{i,j} + D_i - t_o) \ge \mu_i$. Since $\Delta = A_{i,j} + D_i - t_o$, We can rewrite Eq. (7) as follows

$$LOAD(T)\Delta + \sum_{T_C} \sigma_p \Delta \ge \mu_i \Delta,$$

and thus we have

$$LOAD(T) + \sum_{T_C} \sigma_p \ge \mu_i. \qquad (8)$$

Now we describe a sufficient schedulability test for EDZL.

Theorem 2: A sporadic task system T $= \{\tau_1, \ldots, \tau_n\}$, is schedulable by EDZL on performance asymmetric multiprocessor $\pi = (s_1, \ldots, s_m)$, and $s_1 = \ldots = s_r = 1$, $s_{r+1} < 1$, if there are no more than r different tasks τ_i hold the following condition.

$$LOAD(T) + \sum_{T_C} \sigma_p \ge \mu_i. \qquad (9)$$

Proof: From the foregoing discussion, $LOAD$(T) is the upper bound of the load of all regular jobs $\sum_{T_C} \sigma_p$ is the upper bound of the load of all carry-in jobs. μ_i is the lower bound of the load of all jobs. And we know if a job of a task τ_i gets zero laxity, the inequality $LOAD$(T) $+ \sum_{T_C} \sigma_p \ge \mu_i$ must hold. By combining it with Theorem 1, a task set is schedulable by EDZL if there are at most r tasks with zero laxity jobs. Then we get theorem 2 $\qquad \square$

4 CONCLUSIONS

In this paper we have described EDZL scheduling and analysis for performance asymmetric multiprocessors. To our knowledge, this is the first schedulability analysis for EDZL scheduling upon performance asymmetric multiprocessors. However, there still remain many issues about EDZL. First, there are many ways of applying EDZL to performance asymmetric multiprocessors. For example, our EDZL approach always chooses the fastest processors while we may choose other slow processors whose speed is sufficiently enough to meet the required deadlines. Second, our EDZL schedulability analysis can also be improved in many ways. In this paper, we made some pessimistic assumptions while deriving the schedulability condition. In the future work, we may relax those pessimistic assumptions so that we can obtain more accurate schedulability tests.

ACKNOWLEDGEMENTS

This work was supported partly by Seoul Creative Human Development Program (HM120006), partly by Mid-career Researcher Program through NRF (National Research Foundation) grant funded by the MEST (Ministry of Education, Science and Technology) (NRF-2011-0015997), partly by the IT R&D Program of MKE/KEIT [10035708, "The Development of CPS (Cyber-Physical Systems) Core Technologies for High Confidential Autonomic Control Software"], and partly by the MSIP(Ministry of Science, ICT&Future Planning), Korea, under the CITRC(Convergence Information Technology Research Center) support program (NIPA-2013-H0401-13-1008) supervised by the NIPA (National IT Industry Promotion Agency), and partly by Business for Cooperative R&D between Industry, Academy, and Research Institute funded Korea Small and Medium Business Administration in 2013 (Grants No. 00045488).

REFERENCES

Baker, T.P. & Cirinei, M. & Bertogna, M. 2008. EDZL scheduling analysis. Real-Time Syst: 264–289.

Baruah, S. & Goossens, J. 2008. The EDF Scheduling of Sporadic Task Systems on Uniform Multiprocessors. Real-Time Systems Symposium: 367–374.

Cirinei, M. & Baker, T.P. 2007. EDZL scheduling analysis. In Proceedings of the Euro Micro Conference on Real-Time Systems: 9–18.

Funk, S. & Goossens, J. & Baruah, S. 2001. On-Line Scheduling on Uniform Multiprocessors. 22nd IEEE Real-Time Systems Symposium: 183.

Lee, J. & Shin, I. 2012. EDZL Schedulability Analysis in Real-Time Multi-Core Scheduling. IEEE Transactions on software engineering.

Park, M. & Cho, S. & Cho, Y. 2005. Comparison of Deadline-Based Scheduling Algorithms for Periodic Real-Time Tasks on Multiprocessor. IEICE-Transactions on Information and Systems Volume E88-D Issue 3: 658–661.

Information Technology and Computer Application Engineering – Liu, Sung & Yao (Eds)
© *2014 Taylor & Francis Group, London, ISBN 978-1-138-00079-7*

Interactive game design of axles upon augmented reality

J.Y. Chao
Graduate School of Curriculum and Instructional Communications Technology,
National Taipei University of Education Taipei, Taiwan

J.Y. Chen
Department of Electronic Engineering, Ming Chuan University Taoyuan, Taiwan

C.H. Liu
Department of Mechatronic Technology, National Taiwan Normal University, Taipei, Taiwan

C.K. Yang
Department of Computer Science and Information Engineering, Ming Chuan University Taoyuan, Taiwan

K.F. Lu
Department of Electronic Engineering, Ming Chuan University Taoyuan, Taiwan

ABSTRACT: This study developed an interactive game of axles upon Augmented Reality. Using different force positions, the game design includes two scientific concepts, which are "time-consuming and easy-to-use" and "time-saving and difficult-to-use", of axles. In order to match the characteristic of game, the game situation was designed by drawing water from the well. By using video camera of reasonable prices and writing valid video program, the interaction was enhanced by using a large number of marks of AR Toolkit and Kinect sensory system with interface of expensive images. Interactive game developed by this study can be applied to instruction of characteristics of axle and popular science activities. It will certainly enhance learning effectiveness.

Keywords: Augmented Reality; Interactive Game Design; Image Processing

1 INTRODUCTION

Augmented Reality is the technique different from Virtual Reality. By Virtual Reality, users are indulged in virtual space created by Virtual Reality. However, Augmented Reality is the overlapping between reality and virtual information. Thus, in reality, users see virtual image information to expand realistic space. Azuma(1997) suggested that Augmented Reality combines reality and virtual world and is based on immediate interaction. Thus, Augmented Reality can be applied in different fields. For instance, Kuchera (2007) designed card game by Augmented Reality and users can see the invisible objects in daily lives; McKenzie designed the book different from the past. By Augmented Reality, invisible characters and objects are shown in the book. In addition, users can see the characters' actions and sound in the book to increase reading interest and efficiency (McKenzie & Darnell, 2003).

Some scholar suggested that by Augmented Reality, learning can be advantageous. For instance, in 2006, many universities introduced games of Augmented Reality in high schools in Boston. When interacting with virtual characters, students must deal with the

questions of mathematics and natural science in order to pass and move to the next position. Hannes combined reality and virtual world of Augmented Reality to help students learn 3D objects in order to strengthen students' concept of space (Kaufmann, 2002).

In addition, according to Hogle(1996), there are four advantages by including game in learning: 1. It can trigger internal motive and enhance interest; 2. It keeps the memory; 3. Practice and feedback; 4. High-level thinking Games are included in instruction and learners can continuously ponder on the problems in games to find the solution and solve the problems. It is the best learning method.

Therefore, by Augmented Reality and image processing technique, this study designs and develops interactive game of axles which can be applied to instruction to enhance students' scientific concept and learning effectiveness.

2 SYSTEM FRAMEWORK

In order to accomplish the goal of "education through entertainment", this study is divided into two parts. The first is to fulfill interactive game of Augmented

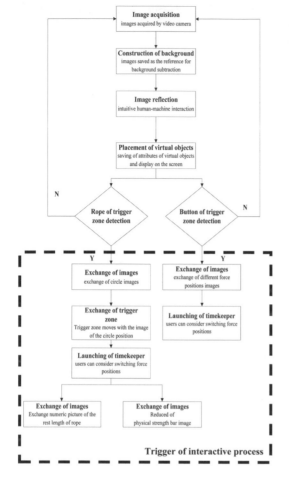

Figure 1. Image processing process of Augmented Reality.

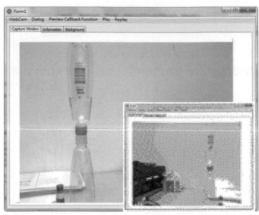

Figure 2. Contrast between image without reflection and image in reality.

Figure 3. Contrast between image with reflection and image in reality.

Reality by image processing technique. The second is the design of game situations. Scientific knowledge is introduced in situation of game to reinforce learning effectiveness.

2.1 Augmented reality

This study treats video camera as the input of image, acquires the operators' images and shows virtual objects on computer screens. Video camera at reasonable price will enhance users' purchase and intention to use (Lee, Wang & Wu 2012). Image processing process of Augmented Reality is shown in the following figure.

As to image processing of Augmented Reality, in order to detect the change of images, we adopt background subtraction to save the detected images as background and then compare each frame shot by video camera with the background.

For human-machine interaction, image reflection should be practiced on images acquired. At the beginning, the original images acquired will resemble the

reality. In other words, when right hand is raised, the character in the image will also raise the right hand. Thus, the image is not natural and intuitive (Figure 2, Figure 3) and it indirectly influences flexibility of interactive game.

2.2 Situation of game

Wheel and axle in the game rotate together. Force can be pass to axle from wheel. According to the positions of force, force can be passed from axle to wheel. When force is on the wheel, it saves the energy, but consumes time. However, with force on the axle, it takes the energy.

Based on the situation to draw water from the well, the researcher combines "time-saving and difficult-to-use" and "time-consuming and easy-to-use" of axle. Users can change force position in the game to highlight the effects of "time-saving and difficult-to-use" and "time-consuming and easy-to-use". By design of

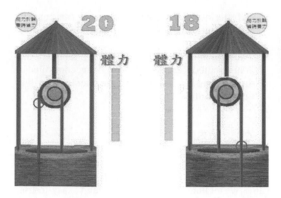

Figure 4.   Interface of game.

Figure 6.   Rope of trigger zone.

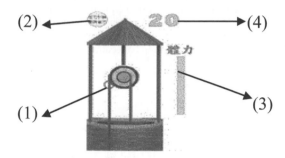

Figure 5.   Interface of game.

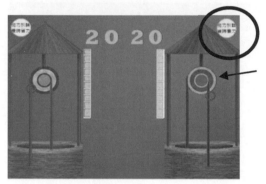

Figure 7.   Button of trigger zone.

physical strength parameter and parameter of the rest rope length, through physical strength consumption and change of the rest rope length of rope pulling (rope on wheel or axle is pulled), users find that force on wheel or axle is different. In addition, in order to increase the fun, the game allows two people to compete with the video at the same time to see who will be the one to acquire the bucket upon the conditions of physical strength parameter and the rest rope length. In the game, force positions should be changed sometimes; otherwise, water cannot be acquired (total force on wheel, in comparison to axle, will be more time consuming. However, with force on axle, physical strength will drop rapidly and players cannot immediately pass the level), as shown in Figure 4.

Interface of game provides two trigger zones, physical strength bar and the rest rope length, as shown in Figure 5.

(1) Rope of trigger zone
When users touch the circle, they move the hand downward. Thus, the hand moves with the circle to the bottom. The circle will move to the top. The action is repeated as pulling the rope downward.

(2) Button of trigger zone
In the game, users can switch force on wheel or axle all the time. Force on wheel is time-consuming and

easy-to-use. Force on axle is time-saving and difficult-to-use. In order to avoid the increase of delay at touch caused by rapid switch time, users can consider switching force positions.

(3) Physical strength bar
When the rope is triggered to pull the well, physical strength bar will be reduced. Users will significantly perceive "time-saving and difficult-to-use" and "time-consuming and easy-to-use". When the rest physical strength is insufficient, the rope cannot be pulled and the screen will be still. Physical strength will recovered when users do not have any actions.

(4) The rest length of rope
In the screen, the figures will be deducted with the rest rope length. The initial value is 20. When the value is deducted as 0, it means the bucket is pulled. The user will be the winner of interactive game. It will thus increase the playfulness of the game.

The figure below shows two people to compete with the video at the same time

According to game characteristics analyzed by Garris (2002), this study designs game situation matching the following:

(1) Fantastic: by drawing water from the well, users experience the situation in life.

Figure 8. Physical strength bar.

Figure 9. The rest length of rope.

Figure 10. Two people to compete with the video.

(2) Rules/goals: in the game, there are the rules such as the rest rope length, physical strength bar and game control. Goals are based on game and thus learners learn scientific concept of axles.

(3) Sensory stimulus: when users play the game, they will change the images by force on wheel or axle to have sensory stimulus.

(4) Challenging: the game relies on the brain, limbs and thinking to win. It controls interactive game by limbs

(5) Mysterious: in the game, learners can consider switching force positions to control the information to win.

(6) Controllable: the game is based on interaction of Augmented Reality instead of being controlled by mouse and keyboard. Thus, the interaction will be more intuitive.

3 CONCLUSION

By image processing techniques, this study fulfills game of axles by Augmented Reality. In the situation of game, we include "difficult-to-use" and "easy-to-use" of axles. According to analysis of Garris, game of this study is fantastic, has rules/goals, sensory stimulus, challenging, mysterious and controllable. The game can be applied to promotion of popular science and teaching tool in instruction of Nature and Life Technology for elementary school students. By playing the game, learners' learning concentration and interest will certainly be enhanced.

ACKNOWLEDGEMENTS

This study was financially sponsored by the National Science Council under Grant No. NSC101-2511-S-130-001 and NSC101-2221-E-130 -025

REFERENCES

Azuma, R. 1997. A Survey of Augmented Reality, *Teleoperators and Virtual Environments* 6(4): 355–185.

Kuchera, B. 2007. *Games with strong online components outsell the competition*, Retrieved September 6, 2007, from http://arstechnica.com/news.ars/post/20070906-games-with-strong-online-components-outsell-the-competition.html.

McKenzie, J. & Darnell, D. 2003.*The Eye Magic Book: A Report into Augmented Reality Storytelling in the Context of a Children's Workshop 2003*, Retrieved March 17, 2008, from http://www.hitlabnz.org/fileman_store/2004-eyeMagic_workship.pdf.

Kaufmann, H. 2002. Comstruct3D: an augmented reality application for mathematics and geometry education, *ACM international conference on Multimedia*: 656–657.

Hogle, J. G. 1996. Considering Games as Cognitive Tools: In Search of Effective "Edutainment", *University of Georgia Department of Instructional Technology*. Retrieved November 15, 2011, from http://twinpinefarm.com/pdfs/games.pdf.

Lee, J. S., Wang S. R. & Wu P. P. 2012. Input Interface Design via Corner Features, *International Journal of Science and Engineering* 2(4): 49–55.

Garris, R., Ahlers, R., & Driskell, J. E. 2002. Games, motivation, and learning: A research and practice model, *Simulation and Gaming* 33(4): 441–467.

Information Technology and Computer Application Engineering – Liu, Sung & Yao (Eds)
© *2014 Taylor & Francis Group, London, ISBN 978-1-138-00079-7*

Measures and construction methods of service-oriented network platform architecture

Jing Tian
School of Education Science and Management, Yunnan Normal University, Kunming, China

Yun Zeng
Department of Engineering Mechanics, Kunming University of Science and Technology, Kunming, China

ABSTRACT: Service-oriented network platform is the hub position in the social and economic operation, which has become an indispensable part of people's daily life and socio-economic. Based on the UNIX system philosophy, the paper design a set of core tools including service network system kernel of the dynamic optimization, data storage, network optimization and relationship mining, e system creates maintenance tools, system services growth tools, visualization tools, etc. Architectural approaches are given about application system of Service-oriented network platform. And Main measures are put forward, which is the re-design of the architecture of the platform; redesigned platform ecosystem value chain; adjust the organizational structure and business processes.

Keywords: Service-oriented; Network platform; Construction method; Measurement

1 INSTRUCTIONS

Enterprises rely on services provided to them by other parties for the realization of their own service offers. The interconnections in terms of services offered and required by enterprises shape complex webs called service network. Furthermore, service network is not phenomena occurring only among distinct businesses. In fact, Service Network (SN) exists as well inside enterprises because of the interplay among business units, divisions, departments, etc [1]. SN are at the crossroad of many different and converging disciplines, each approaching the topics from a different point of view and focusing on different aspects [2]. SN system's goal is to build a comprehensive service management platform, which support higher recall rate and higher precision service discovery mechanism on the basis of the function of the traditional service registry, and support automatic/manual combination of Web services. In addition, as a service management platform the service network system provides a set of complete solutions to support re-development of the service network, and makes possible to develop a large number of Web services-based systems. With the rapid development of the thought of service-oriented architecture (SOA) and software as a service (SaaS), a large number of public Web services have emerged. UDDI (Universal Description, Discovery and Integration) provides a mechanism of service registry and discovery, but UDDI service discovery model is too crude,

and can not provide effective support for automatic service composition. The lack of platform and large-scale service management specification made incomplete to SOA. Moreover, the current SOA infrastructure is lack of supporting automated processing of Web services. Build new SOA infrastructure is becoming more and more important.

2 SYSTEM FRAMEWORK OF SERVICE NETWORK PLATFORM

In the state of the service networks come under many names service value networks [3], service ecosystems, service system, Approaches to service networks with an economy focus are mainly concerned with the creation of value [4]. On the contrary, the business communities investigate the structure of organizations in service networks and the related business models. The SOA and Business Process Management communities focus on the technology to realize and operate service networks and automate the business processes that take place inside them. The SOA communities focus on the technology to realize and operate service networks and automate the business processes that take place inside them.

As can be seen from the SN system design goals, there are four role of service network platform, which are managers, contributors, Web users and application program interfaces users. The contributors are who provide services, and provide relationship attributes amendments on the network. Due to the extensive of the user, for the guarantee the spirituality and reliability of the system, the design of the service network system follows the design philosophy of UNIX systems, which are micro-kernel and peripheral tool. The service network system is combined by a small kernel and a large number of peripheral tools. That small kernel can guarantee the maintenance of the service network system is relatively simple, and help to improve the stability, but also to ensure the scalability of the system of the service network. SN system has a relatively large number of peripheral tools to together forming a service network platform. The design of tools adhere the principle of single and excellent, that each tool can well done a single function. Under the guidance of its design principles, the system is divided into several modules.

2.1 Kernel module

Due to the semantic characteristics of the service, the system kernel structure is designed around service network body. Each service and its various attributes are reflected as an instance of the service network ontology concept. Meanwhile, in order to improve the efficiency of the kernel data, there is a buffer layer in kernel data. SN body operation process, the buffer layer data manipulation process, and guarantee the synchronization of information between the two threads are formed main portion of the service network kernel. SN kernel provides some operating kernel interface, which these information operations interfaces become a bridge of communication between the services network core application sequence and service network kernel. At the same time, there are four important processes in service network kernel, which are consistency validation process, service relationship mining process, the service network ontology buffer layer synchronization process, and process scheduling system.

2.2 Create module

SN created program is used to create, delete, backup and recovery service network system, Start background process, Initialize the entire core tools independent program operating environment. Service network create a program to provide the client, so administrator with the management permission of the service network system can remotely control service network system.

2.3 Active growth module

SN can rely on the module to the increase in service initiative. Service network targeted at large-scale Internet Web services management platform that a large number of Web services is coming from is a key problem. In fact, the current Internet littered with a large

number of Web services, so system development service active growth module and service registration module, which two core modules can solve the problem of the source of service. Automatic growth modules fact are Web services reptiles, they can search WSDL files scattered on the Internet.

2.4 Service registration module

Corresponding to the active growth Module of service network, service network registration module is called passive growth module. Its design is also similar to the active growth modules, and shares core layer architecture with active growth modules. The module provides the user interface, which can submit Web service by the users of the service network through the following three ways: First, it provides the URL of the Web service description file. The user needs to fill in the appropriate URL in the service network registration system, the system will automatically go to download the description file and analyze it to extract the corresponding added to the service network; Secondly, users upload Web service description file. The user needs to upload service description file in the service network registration system, the system analyzes description file of the user, and extract the corresponding added to the service network; Finally, users provide service details form. Some of the services provided by the user may not be the standard Web services, Web service description file does not exist at this time. At this point, the user can fill in the corresponding values in the system to provide service details form, the system will analyze the user's input, extract the appropriate content added to the service network.

2.5 Visualization module

The module is achieve display and editing of service network based on service network core API, which can help User queries editorial services and service relationships in a graphical manner.

2.6 Service search engine

SN search engine provide Web interface for the user. Web users can be sent service queries to the service network search engine; SN system queries the attributes of each service, the user request and Semantic distance of service attributes. The results are returned to visual presentation tool of the service network, then intuitive results presented to the user. Finally, the SN system provides a set of open APIs, which can easily development of new tools package to meet the changing needs. One of the primary goals of SONP is to provide a common, unified, and flexible control environment that can support multiple types of services and management applications over multiple types of transport. There are three critical characteristics as follow: Architectural Layering; Open Services Interface/API; Distributed Network Intelligence. The concept of architectural layering is central to SONP environments. First and foremost, SONP cleanly separate service/session controls from the underlying transport elements. An open development environment based on an Application Programming Interface will enable service providers, third party application developers, and potentially end users to create and introduce applications quickly and seamlessly. Network intelligence can be distributed to the most suitable locations in the network or, if appropriate, to the CPE.

3 ARCHITECTURAL APPROACH OF SERVICE-ORIENTED NETWORK PLATFORM APPLICATION SYSTEM

Service oriented network platform development cycle is constructed, Deployment, Test, release, which its core is to build. Web service building includes the definition of WSDL, development corresponding component, implementing one or more proxy class. In the development, the order of creation these documents and source code is determined according to the project development requirements and development tools, and the development order is decided by abstraction hierarchy of development components. WSDL document describes a Web service, which more abstract than the components. Therefore, we will look for the WSDL Web service as the top floor, and the corresponding realization component as the bottom layer, and use the following several methods to implement Web service.

3.1 The top-down method

Firstly, application system services interface need be defined, then services metadata are extracted, and Step by step realization of each WSDL defined interface component. In the development process, WSDL document mapping is became corresponding Web service realization templates by tools. In using top-down development method, it is the key that are carefully designed to the entire application system metadata, and analyzed business process in detail, refined system required services interface, designed WSDL document. This model development steps are as follows: analysis business process, extracting metadata; XML describe the metadata; the metadata for XSD modeling, XSD document; the XSD document mapping database Schema; the design of the service WSDL document; and through the WSDL document class framework, using the template; The traditional program development tool for the development of specific implementation class; Deployment; test.

3.2 The bottom-up method

This method start concrete implementation component of service, and gradually progress to the direction of the service interface definition developed. This model is the most important development Method, because this architecture can be simple and quick derived from existing components or application to

Web services. On the Windows platform, WSDL document can be generated from the existing COM components. Using the bottom-up model to develop Web services provider, follow these steps: analysis of business processes, the extracted metadata; using XML to describe the metadata; metadata XSD modeling, generate XSD document; analysis of business processes, extract metadata; metadata mapping for Database Schema; custom component interface; generate the WSDL document components; deploy the Web services; test.

4 DESIGN MEASURES OF SERVICE-ORIENTED NETWORK PLATFORM

The major measures of Service-oriented network platform designed cover three aspects, which are redesign the architecture of the platform, and redesign the value chain of the platform ecosystem, adjusting the organizational structure and business processes. The main goal of service-oriented network platform is to improve the function and operating efficiency of the network platform, and value chain redesign goal is to rationalize the relationship between the interests and values delivery mechanism. To make a real implementation and the role of these re-design measures, platform internal organizational structure and business processes must be adjusted accordingly.

4.1 Re-design platform architecture

The platform are constitute of components (such as hardware, software, services) and rules (standards, protocols, policies and contracts). In accordance with the role of the various elements of the platform running the platform architecture can be divided into three levels: (1) the infrastructure layer, which is platform running physical conditions composed mainly by computers, network equipment and other hardware infrastructure; (2) Component communication protocol layer, which provides a variety of automated functions and services, and mainly constituted by the software, services, standards and protocols, the platform; (3) Interaction rules layer, which contains policies and contract, and provisions the parties involved rights, obligations and actions that can be taken. The redesign of the architecture is to redefine constitute the platform components and rules. Improved new platform technology affects some components of the platform architecture. Platform enterprise can gradual upgrade and update those components. Breakthrough technology impacts platform architecture that platform enterprise should be a fundamental rethinking of the technical characteristics of the platform architecture and competitiveness. If you want to switch to the new technology, it is necessary to formulate a complete plan of switching, which platform components are classified and clearly switched priorities to ensure that the platform switch while the platform is still normal.

4.2 Re-design platform ecosystem value chain

Based on the re-engineering strategy, the platform enterprise redesigns the value chain of the ecosystem of the platform in accordance with the following steps to: First, designer need clear information, capital and material between the platform user flowing' direction, path and size. Second, based on user type, designer need determine the role of platform user, and determine the value of the relationship between the platform and the user. Third, designer need clear charges and provide subsidies to different types users based on a certain of principle.

4.3 To adjust the organizational structure and business processes

After the network platform redesign, the internal organizational structure and business processes to be done corresponding adjustment. Business units need to be re-divided, the department's staffing and work responsibilities need to be adjusted, need to improve communication and orchestration between departments. In addition, according to the re-design platform features and services design business processes, overall design principle is to minimize user aspects of the platform and improve operating friendliness.

5 SUMMARY

With the rapid development of the network, Internet and Intranet development have become the focus of software development. Internet and Intranet applications development experience three stages: the technology of the initial use is HTTP / HTML, which is static Web Development; second stage is the application of the Java language. Due to the cross-platform features of Java itself, so that software developers can use a single development language and programming environment to develop application systems on various platforms; the third stage is the rise of XML, which provides a standard data encapsulation technology, making the exchange of data across a variety of platforms, operating systems and development tools. Java and XML is a unique advantage in cross-platform and data exchange, but it still can not solve the integration problems of different systems, different application data and application logic. The reason for this problem is that each system is highly independent and closed. Therefore, we need research the measures and construction methods of service-oriented network platform architecture. Based on the existing SN model, service network ontology is built for large-scale management and automated process of Web services, which introduced Semantic Web services specifications and related technologies to support the automatic processing, and design and Implemented service network system, completed management functions of a large-scale service. As Web services are loosely coupled, it needs to be improved in transaction processing, security for general applications. We recommend using a bottom-up design, so that developers can use traditional means to building systems, and can give full play to the advantages of Web services data and application integration. In particular, the future research will need to investigate the optimization of aspects of service networks stemming from both the technical and economical perspectives.

ACKNOWLEDGEMENT

The research reported here is financially supported by the Humanity and Social Science Funds of Education Ministry of China Grant NO. 10YJC880112. Part works is supported by the National Natural Science Foundation of China under Grant No. 51179079.

REFERENCES

[1] Olha Danylevych, Dimka Karastoyanova, Frank Leymann. Service Networks Modelling: An SOA & BPM Standpoint. Journal of Universal Computer Science, vol. 16, no. 13 (2010), pp.1668–1693.
[2] Paul P. Maglio, Stephen L. Vargo, Nathan Caswell, and Jim Spohrer. "The service system is the basic abstraction of service science." Inf. Syst. E-Business Management, 7(4), 2009, pp. 395–406.
[3] Benjamin Blau, Jan Kramer, Tobias Conte, and Clemens van Dinther. "Service Value Networks." In IEEE International Conference on Ecommerce Technology, 2009, pp. 194–201.
[4] Alain Biem and Nathan Caswell. "A value network model for strategic analysis." In HICSS, IEEE Computer Society, 2008, pp. 361.

Information Technology and Computer Application Engineering – Liu, Sung & Yao (Eds)
© *2014 Taylor & Francis Group, London, ISBN 978-1-138-00079-7*

A research of assembly technique-oriented three-dimensional assembly model file structure tree information restructuring method

Ying Chun Liu, Jun Chao Yuan & DaWei Wu
Shenyang Aerospace University, Shenyang China

ABSTRACT: This paper presents a for assembly process of 3D assembly model file tree structure information recombination method, which makes the information process reengineering is done entirely in 3Denvironment,to abandon the traditional two-dimensional file tree structure information based on recombinant forms, improves the efficiency of process.

Keywords: 3D model; assembly process; structure tree; Information reorganization

1 IN 3D ENVIRONMENT ASSEMBLY INFORMATION EXPRESSION

Currently, three-dimensional CAD software has been widely used in China. The thought of product design has significant changes. From the parts to the assembly design is entirely based on the completion of the three-dimensional environment. In the three-dimensional CAD design environment, the assembly product information expressed in the form of tree structure. Centering on the component object, all the product design data, process data and production data organization in three-dimensional structure of a tree. In different periods of building product structure tree, the main source of information is the contents of the 3 d assembly information. Assembly is composed of a series of component model assembly components structure tree plus with words of annotation and attribute data. Assembly model and part model in the form of numbered layers of father-child nodes reflected in the structure tree.

2 IN 3D ENVIRONMENT 3D MODEL OF TREE STRUCTURE INFORMATION REORGANIZATION

BOM is one of the most important concepts in MRP system, bill of material (Bill of Material, BOM), is a word with a broader meaning, it is all products, semi-finished products, raw materials, WIP, blank, parts and auxiliary production related materials referred to as many species such as EBOM, BOM, PBOM, MBOM, there is a close relationship between them.

There are many BOM in the whole life cycle of products, and BOM is used in this article mainly has

EBOM and PBOM as well as EBOM and the PBOM transformation and so on:

(1) EBOM is the product design department the product data which designs according to the product design request produces, such as product materials, names, product structure and other information;

(2) PBOM is based on the EBOM data process design department, which design the product assembly data, such as product assembly unit partition, set assembly, assembly station, specify the assembly process, etc.

(3) BOM tree was started in the product design department, design department design produce EBOM, Then process design department produces PBOM structure tree in the base of EBOM structure tree, Finally passes to the PBOM structure tree the assembly production department to instruct the production. Transformation process between EBOM and PBOM.

2.1 *Analysis of EBOM product data structure design*

The structure of EBOM tree is by design department in accordance with the specific requirements of customers design production, which includes the product list, name, structure information, the final EBOM data form the three-dimensional model obtained from the design department. The EBOM data is the base of product assembly process design and manufacturing, the data source is the only.

EBOM, EBOM date is produced by the design department, according to the requirement of design, in terms of product design, product designer generate

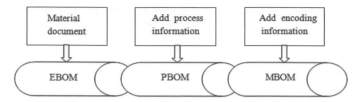

Figure 1.   The relationship between the different BOM.

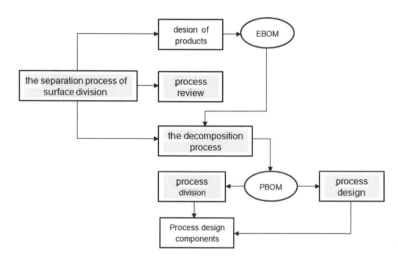

Figure 2.   The transformation between EBOM and PBOM.

the information such as product name, structure and so on. This paper mainly studies type of parent and child EBOM, which is mainly composed of the paren and the child and child to the parent parts assembly number this three parts to express. Mainly uses the paren and the child to identify, its mainly advantage is: under the condition of the parts borrowed from each other, all the parts need statement only once, and once it has been defined, you can use in many place, so it can reduce the data redundancy of BOM, indeed it improve the BOM data consistency. Its data relations mainly includes: ① assembly property relationship (used to represent the product and parts processing and assembly relationship and the relationship between the assembly number); ② the natural property relationships (used to indicate the nature of the products and parts, including size, material, weight, type of production, etc.); ③attached design information (including part of sketches, reference surface, heat treatment, etc.)

2.2   *Analysis of PBOM data structure of the assembly process*

PBOM, assembly process BOM, according to the data as the basis, process design department to generate BOM is PBOM, it is lay down process planning, divide process, position information, and specify the assembly process, process design for all parts. PBOM is on the basis of the EBOM, according to the characteristics of the enterprise technological process, and then to formation the product components, assembly parts and manufacturing technique of the final product, but also determine the components of cutting tool in the BOM module, frock clamp, auxiliary means and processing equipment technique information. In the general case, the PBOM and EBOM is different, because of the technology department according to the technological equipment of enterprise's condition, and thus modify of product parts of assembly relationship and material composition of the ingredients.

3   THREE-DIMENSIONAL MODEL OF ASSEMBLY FILE TREE INFORMATION REORGANIZATION FOR ASSEMBLY PROCESS IMPLEMENTATION

Based on COM component object model technology development tree information restructuring tool, that the inherited COM is language independent, transparent process, simplifying the COM low-level details, more extensive application.

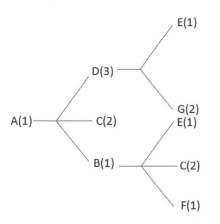

Father and son BOM data		
Parent code	Child code	Assembly number
NULL	A	1
A	B	1
A	C	2
A	D	3
B	E	1
B	C	2
B	F	1
D	E	1
D	G	2

Figure 3. EBOM product structure tree.

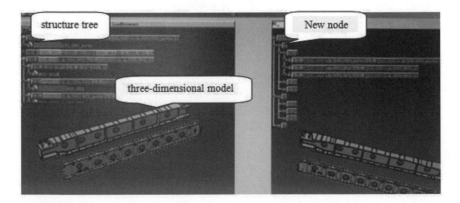

Figure 4. Structure tree information restructuring tool.

In the restructuring tool, the main display structure tree to restructuring, and on the right is the tree structure of new generation, restructured by the way, information attached to the parts model, and transfer with the number change parts.

4 CONCLUSION

Through the research of 3D model file tree structure information recombination method to assembly process, realize the complete 3D assembly file tree structure information recombination in three-dimensional environment, ensure the consistency of data sources, the synchronization process design, cancelled the process personnel information reorganization based on the traditional 2D assembly file, make the process easy to understand the assembly process, shorten the product development cycle, reduce the cost of development.

REFERENCES

[1] Wei Zhiqiang, Wang Xiankui, Wu Dan, et al. BOM multi-view mapping of product based on a single data source [J]. Journal of Tsinghua University (Science and Technology), 2002, 42(6): 802–805 (in Chinese).
[2] Yuan Jianguo, Zhang Guojun, Li Peigen. The key technology research of BOM-oriented manufacturing information system integration [J] . Application Research of Computer, 2004, (4): 39–41 (in Chinese).
[3] Xie Guiliang, J Iang Lin, Sun Shudong. Component based model of works hop management information system [J]. Computer Engineering and Applications, 2002, 38 (18): 16–17 (in Chinese).

A master-slave immune clone selection algorithm and its application

Jia Ma
School of Economic and Management Shenyang Aerospace University, Shenyang china,

ABSTRACT: Based on the immune regulating mechanism of biological immune system, a kind of Master-Slave Immune Clone Selection Algorithm (MSICSA). The algorithm uses multi population strategy, through the immigration and emigration operation, update information between population and maintain the diversity of population. Experimental results showed that the proposed MSICSA can effectively improve the basic immune clonal selection algorithm to solve the shortcomings of large-scale optimization problems, with excellent global convergence and stability.

Keywords: Immune clone selection algorithm; Clone selection; Master-slave; Immigration and Emigration

1   INTRODUCTION

In recent years, artificial immune system and its engineering application has gradually become the intelligent information system of popular research direction [1], biological immune system is a highly evolved, complex function system, it can adaptive to identify and eliminate the effects of foreign body to enter the body and has the learning, memory and self-adaptive ability, maintain the stability of the environment in the body. Immunity clone selection algorithm (ICSA), from the tradition of clonal selection principle [2], by De Castro is put forward based on the principle of clonal selection of artificial immune algorithm [3], the algorithm has been in solving such as combinatorial optimization, intelligent optimization and production scheduling problems and Shows the strong data search capability.

This article puts forward the improved immune clonal selection algorithm - Master-slave immune clone selection algorithm (MSICSA), the algorithm is introducing a masterslave structure, this structure actually uses a multigroup strategy [4] Multigroup strategy can independently update each species, and indirectly exchange information between each species. It both can maintain the same number of the species, and can strengthen to exchange information between the species and the son species, increase the diversity of species, so as to prevent effectively algorithm in local optimum. The algorithm will be applied to task allocation problem, the simulation results show that this algorithm is practicable and effective[5].

2   MASTER-SLAVE IMMUNE CLONE SELECTION ALGORITHM

Immune clonal selection algorithm is for a single population to be modified [6], the target is that improving overall quality in every iteration of the species But along with the increase of number of iterations for a single population increase more and more slowly the entire population is likely to converge to the local advantages. In order to overcome the shortcomings, this paper presents a masterslave structure this structure actually uses multi population strategy Algorithm for the population and the sub populationrespectively the selection, cloning and variation operation through the three operation can improve the quality of the antibodies in each species In addition to for the population and the sub population above the operation, also puts forward the ingoing and emigration operation populations to adapt to handle high antibody values moved into the main population, the main populations of poor move out to the child populations antibodies, such already can maintain the same number of population and can strengthen population and the sub populationson between information exchange increase the diversity of population so as to effectively prevent algorithm in local optimum [7] The main characteristic of this algorithm is between the son of population and the population update operation is independent only after each to finish the upgrade, just exchange information, also generally is to move out and operation.

2.1   The improvement of Son population

For a subset of group As (s = 1,2, ..., S), The following will be presented on the sub population improved seven steps.

(1) Initial population Ps: randomly generated n_s antibody initial to from population Ps.
(2) Choice: select n_c adapted to the highest antibody from Ps.
(3) Clone: the selected n adaptive values the highest antibody were cloned. The number of each antibody were cloned will contact with its affinity

In short, the k solution vector affinity calculation are described below:

$$A_k = \frac{\sum_{j \neq k} \exp(\|a_k - a_j\|)}{\sum_{k=1}^{n} \sum_{j \neq k} \exp(\|a_k - a_j\|)} \qquad (1)$$

Here, A_k Represents k solution vectors affinity, $\|a_k - a_j\|$ represents between antibody K and antibody J the euclidean distance. $\exp(\|a_k - a_j\|)$ represents on the $\|a_k - a_j\|$ index type conversion, the conversion can be amplified between K and J distance.

If cloned more affinity low antibody, most clones are relatively close to each other after the antibody, the diversity of the population can not be guaranteed. Therefore, low affinity antibody clone number should be reduced, while the number of the high affinity antibody clone should be increased. Cloning of n_c antibody set of definitions for C. The number of each antibody clone can be calculated according to $N_c = \sum_{i=1}^{n} round(\beta \cdot N / i)$

(4) Get n clones after, then they are going to small probability of variation. The k antibody mutation rate is set to $p_k = 1 - A_k$. Variation after the n_c clones set of definitions for C'. This paper used a method based on the chaotic mutation operation, in this chaotic sequence mutation operation can acquire a mature antibody population (C*). The mutation operation can be described as follows:

$$m_j = \underline{m}_j + c \times (m_j - \underline{m}_j) \qquad (2)$$

Here, m_j represents a clone of j attributes, is also the solution vector j dimensional component. $\underline{m}_j$ represents the j dimensional component lower bound, $\overline{m}_j$ represents the j dimensional component upper bound. c represents a chaotic sequence [7], Detailed description is as follows.

$$c_{n+1} = a \times c_n \times (1 - c_n) \qquad (3)$$

Here, a represents a coefficient, and is set to 4. c_n represents the n iteration sequence value, while the c_{n+1} represent the $n+1$ iteration sequence value. Attention is required, $c_n \in (0, 1)$, $c_0 \in (0, 1)$ and $c_0 \neq \{0.0, 0.25, 0.75, 1.0\}$. The advantages of the chaotic sequence has is, it is in constant change in the next state, and cannot be predicted, This increase in search of randomness. This paper adopts $c_0 = 0.23$

(5) From the set C and C' select a number of fitness the highest antibody to M memory population.

(6) Choose the highest value $30\% \times n_s$ antibodies from M to update the population Ps fitness minimum quantity of antibody.

(7) $k = k + 1$, and judging whether to reach a maximum number of iterations $k = k + 1$. If the condition is satisfied, then Stop calculation, Otherwise, go to step (2).

2.2 The improvement of the population

The population improvement and the son of population of exactly the same steps. On the population A, Select n_m adaptive values the highest antibody. Then, to be chosen for this N antibody were cloned, after the cloning of all the antibody variation. The main population of mutation operation is still chaotic sequences form, it is helpful to improve the algorithm to search the solution space. Then, from after the cloning antibody and mutated antibodies to choose a number of fitness the highest antibody to memory consists of population, from memory to choose the highest population of $30\% \times n_m$ antibodies to update the population Pm fitness minimum quantity of antibody [8]. Finally, to determine whether the maximum number of iterations K_m. If the condition is satisfied, then stop the improved main population, otherwise, repeat the above steps in the operation.

2.3 Immigration and emigration

The operation of Immigration and emigration is the son of population and the population between update operation, only after independently finish the upgrade to exchange information. For Immigration, from each sub population to choose a fitness value of the highest antibody, and it will be moved to the main population [9]. Therefore, the amount of antibody moved to the main populations as $S(S$ for the number of sub population). In order to maintain a constant number between the son of population and the population, it must move out the main population. Select S antibody that adaptability in the value of the worst from the main population, they will be moved out to the son of population. Each sub population was randomly assigned to a poor antibody. Such, the main population has retained many fitness of high antibody. Through continuous immigration and emigration operation, primary antibodies in population quality has been continuously improved [10].

2.4 Master-slave Immune Clone Selection Algorithm

According to the above algorithm description, the process of Master-slave Immune Clone Selection Algorithm can be described as shown in figure 1.

3 CONCLUSION

The analysis of this paper is based on the principle of the immune clone selection algorithm, put forward the improvement strategy based on master-slave structure, constitute a type of immune clone selection algorithm. Through immigration and emigration operation can strengthen the exchange of information between the sub population and the host population, increase the diversity of population, so as to effectively prevent the algorithm being trapped in local optimal solution.

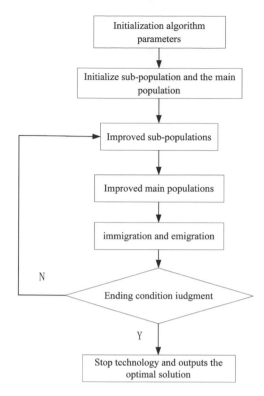

Figure 1. Procedure of MSICSA.

REFERENCES

[1] V. Chaudhary, J.K. Aggarwal. A generalized scheme for mapping parallel algorithms[J]. IEEE Transactions on Parallel and Distributed Systems, 1993, 4(3): 328–346.

[2] Kacem I., Hammadi S., Borne P. Pareto-optimality approach for flexible job-shop scheduling problems: Hybridization of evolutionary algorithms and fuzzy logic[J]. Mathematics and Computers in Simulation, 2002, 60: 245–276.

[3] Timmis J, et al. Data analysis with artificial immune systems[J]. Proc. of IEEE Int. Conf. SMC. Tokyo, Japan. 1999, 21:245–252.

[4] Jiao Li-cheng, Wang Lei. A novel genetic algorithm based on immunity[J]. IEEE Transactions on Systems, Man, and Cybernetics, 2000, 30(5): 552–561.

[5] D. Castro, V. Zuben. Learning and Optimization Using the Clonal Selection Principle [J], IEEE Transaction on Evolutionary Computation. 2002, 6(3): 239–251.

[6] De Castro L.N, Von Zuben F.J. Learning and optimization using the clonal selecting principle[J]. IEEE Transactions on Evolutionary Computation, 2002, 6(3): 239–251.

[7] Yin, P.Y., Yu, S.S., Wang, P.P., Wang, Y.T. A hybrid particle swarm optimization algorithm for optimal task assignment in distributed systems[J]. Computer Standards & Interfaces, 2006, 28(4): 441–450.

[8] Maoguo Gong, Licheng Jiao, Lining Zhang. Baldwinian learning in clonal selection algorithm for optimization[J]. Information Sciences, 2010, 180(8): 1218–1236.

[9] Leandro N De Castro, Fernando J Von Zuben. Learning and Optimization Using the Clonal Selection Principle[J]. IEEE Transactions on Evolutionary Computation, Special Issue on Aritifical Immune System, 2002, 6(3):239–251.

[10] Bilal Alatas. Chaotic harmony search algorithms [J]. Applied Mathematics and Computation, 2010, 216(9): 2687–2699.

Information Technology and Computer Application Engineering – Liu, Sung & Yao (Eds)
© 2014 Taylor & Francis Group, London, ISBN 978-1-138-00079-7

A KVM-based elastic Cloud Computing framework

Xing Zhang, Wei Nong Wu & Qing Zeng
Chongqing Electric Power Information and Communication Branch Company, Chongqing, China

Peng Liu
Information & Network Management Center, North China Electric Power University, Baoding, China

ABSTRACT: Cloud Computing would combine with all IT resources and provide infrastructure as a service (IaaS) to users by virtualization technology, and users could obtain resources on demand. Therefore, monitoring and collecting the information of resources are the basis of all Cloud Computing environments to achieve dynamic resources management based on the running status of resources and users' requirements. An elastic Cloud Computing structure based on KVM virtualization technology was proposed to improve the utilization of system resources and Cloud Computing environments automation management level.

Keywords: Cloud Computing; KVM; workload balance; Live virtual machine migration

1 INTRODUCTION

As an emerging network computing model, Cloud Computing could distribute computing tasks on a large number of IT resources, and provide all application systems with appropriate computing resources, storage resources, network resources, and a variety of software services. Cloud Computing through the delivery model of IaaS(IT as a Service), would greatly improve application deployment speed and cut down expenses. The resources of Cloud Computing were infinitely expandable, and could be dynamically used based on user's demand. There are three types of service model in Cloud Computing (Foster I et al. 2008, Zhu,Y et al. 2012): SaaS (Software as a Service), PaaS (Platform as a Service) and IaaS (Infrastructure as a Service). IaaS should provide a variety of IT resources with dynamically allocation to meet all users' requests, and the resources are shared by multiple users. But actual demands of users were often dynamic, which would cause the irrational distribution of resources. In addition, Cloud Computing should schedule resources based on the actual workload status of applications and system resources. To this end, many IT companies are building Cloud Computing services with advanced technology for all levels and areas and have launched the corresponding Cloud Computing framework solution.

Cloud Computing uses virtualization technology as IaaS infrastructure of platform, and virtualization technology would provide an effective solution for resource management in Cloud Computing. The services providers of Cloud Computing have provided all kinds of virtual machine instances for user with different processing capability. Such as Amazon EC2 (Amazon EC2), users can choose Small, Large, Extra Large number of different levels virtual machine instances according to their needs, and can be independently leased to the state of the virtual machine to start, stop, turn off and other operations. But user's actual needs are dynamic changing according to continuous development of applications and services, and need to acquire greater flexibility and control of the virtual machine resources rented from Cloud Computing to reduce the waste of resources and user's costs. When needs of applications increased or the demands for resources decreased, users could select appropriate virtual machine to migrate or release unnecessary resources.

Therefore, this paper proposed a flexible, dynamic allocation of elastic Cloud Computing platform architecture used KVM (KVM, Deshane, T et al. 2008)virtualization technology as the core part, according to the load conditions of user's applications and services, to achieve reasonable optimization of resources in a cloud computing environment and keep load balancing of Cloud Computing. The main advantage of this architecture including: (1) The scalability dynamic scheduling of virtual machine according to the conditions of applications and services; (2) Provide users with more flexible management solutions of Cloud Computing resources.

2 KVM VIRTUALIZATION TECHNOLOGY

KVM is an open source virtualization solution based on Linux ring and the core idea is adding virtual

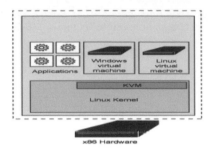

Figure 1. KVM architecture.

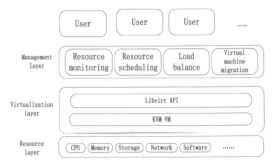

Figure 2. Cloud Computing Framework.

machine management module on the basis of the Linux kernel to reuse the impeccable functions of Linux kernel, including process scheduling, memory management, IO management and other functions, to making it being a hypervisor with virtualization technology supporting. KVM is not a complete simulator, just is a kernel plug-in virtualization capabilities, and its specific simulator need to use QEMU. A virtual machine is a thread in KVM, and can be managed by the Linux process management. User can learn the virtual machine startup command with QEMU-KVM and can also enter corresponding control commands. KVM is a full virtualization solution based on hardware-assisted virtualization (Intel's VT-x or AMD-V) technology, and allows that client operation system would not been modified to run directly on KVM virtual machine with independent virtual hardware resources: network card, disk, graphics adapter. In addition, KVM supports virtualization nested, which means that user could create a virtual machine in a KVM virtual machine. KVM architecture is shown in Figure 1.

3 ELASTIC CLOUD COMPUTING FRAMEWORK DESIGN

The Figure 2 describes the framework of elastic Cloud Computing in detail, and this framework was mainly divided into three layers from bottom to top: resource layer, virtualization layer and management layer The framework would integrate all IT resources into the resources pool with basic storage and computing capabilities as standardized services, and combine with servers, storage systems, switches, routers, and other systems through networks for users to process the workloads of application components or high performance computing applications.

3.1 The resource layer

In the bottom of framework, resource layer was consisted of IT hardware and software resources Hardware resources constituted entire resources pool of Cloud Computing infrastructure. Users could use virtualized IT resources, including compute, storage and network resources as service to deploy and run their applications on operation system. The cloud infrastructure was transparent to users, but they could control

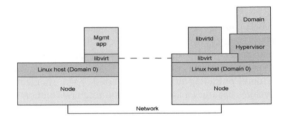

Figure 3. Libvirt working diagram.

the operating system, storage, and applications with selecting appropriate network components.

3.2 The virtualization layer

The virtualization layer is mainly based on the KVM virtualization technology to combine hardware resources pool, and it is the fundamental component of Cloud Computing data center to achieve logical resources pool with servers, storages and networks Resources pool architecture are mainly two models: one is the high availability architecture more commonly used in the cloud data center, and the other one is the independent host server architecture with low-cost but lower availability. Highly available architecture is a hypervisor cluster formed by more than one physical server node. Those cluster nodes connect to a centralized shared storage with host bus adapters iSCSI adapters and storage network, and use a number of high-speed Ethernet to connect to the core switch of data center. Independent host-server architecture is a simple pattern of Cloud Computing data center. The architecture runs a number of virtual machines over the running hypervisor host servers.

Virtual machine management layer manages routine affairs of virtual machines through the Libvirt (Bolte, M et al. 2010) API. The Libvirt API is a set of virtualization management API with supporting Xen, KVM, etc. The monitoring module achieve the following functions through Libvirt: start, stop, pause, save, restoration, migration and access to the running status and resources usage information of physical hosts and virtual machines. Its control model was shown in Fag.3.

3.3 The management layer

3.3.1 Resource monitoring

Resource monitoring used two modes to monitor the running status of resources: (1) active mode. The resource monitoring component and virtual machine monitor acquisition in work node would gather the running information of virtual machine, and send them to the master node initiatively (2) passive mode. The master node needs to send a request to the work node to acquire those monitoring information.

In Cloud Computing, resource monitor is real-time and needs the pollingstrategy: periodic query pattern or event-driven pattern. The periodic query pattern is that the work node would send those monitoring information to the master node at intervals or the master node sends a request to the work nodes for collecting their monitoring information every once in a while. Event-driven pattern is that work nodes would produce a series of events, and each event would trigger the corresponding of collector for checking the status of resources monitoring with compared with the last checked. When the change between two events was beyond threshold value, work nodes would send monitoring information actively or passively

3.3.2 Resource scheduling

Cloud Computing providers have their own resource allocation and scheduling mode, but there are no uniform standards and norms. In Cloud Computing, The efficiency of resource allocation would affect the overall performance of cloud. Because of the difference between Cloud Computing and grid computing, the resource allocation and scheduling algorithm of gird computing could not work effectively So considering Cloud Computing platform environment and requirements of users, this paper introduced the heuristic algorithms to resource schedule in Cloud Computing (Liu S et al. 2013, Fang J.M. et al. 2012) and task scheduling (Lin J.M. et al. 2011) based on genetic algorithm for assigning appropriate resources and remaining fairs for user tasks.

3.3.3 Workload balancing

In running process of physical hosts, workload would change all the time. Some hosts would have a light workload, while others were heavy, which would affect the implementation of virtual machine seriously The workload between physical hosts should achieve dynamic balancing with appropriate load balancing strategy to improve system performance. There was a virtual machine layer in the Cloud Computing architecture and virtual machine could be viewed as a stand-alone host with software achievement to keep workload balancing by virtual machine migrating. The temporarily interrupt latency would affect the performances of virtual machines when virtual machines migrating, which must be considered.

3.3.4 Live virtual machine migration

Resource Scheduling module would determine whether virtual machine should be migrated and which virtual machines would be selected to migrate according to the status of resources using and pre-set policies (Meng X.G. et al. 2012) Taking into account the differences of applications running on virtual machines in Cloud Computing, the workloads of physical hosts and virtual machines would jump high or low. It would add unnecessary overhead and not achieve workload balancing purpose according to the monitoring peak value of some point. To avoid this situation, this paper made the following improvements: when the workload value monitored was beyond the threshold value, it should monitor the next N value, and then would reallocate resources or virtual machine migrate when there were more than M values exceeded the threshold value. M/N represents the different scheduling strategy. If M/N = 0, it means that user adapt a more aggressive scheduling policy with migrating virtual machine immediately when there was a peak value on virtual machine scheduling; Conversely, if the proportion of M/N value was closer to 100%, then user would adopt a more conservative scheduling policy and virtual machine migrate after several peak values.

The selection of migration destination host should consider two factors: response speed and processing power. According to ping response time of each physical host, as well as the physical host CPU and memory, there was a weight value, which means that the physical host is the target host with biggest value. This algorithm can reflect the status of each physical host. After virtual machine migration done, the network load and processing capacity of physical host with biggest weight value would decline, and its value would also decline. In this process, the physical host following the host with biggest weight value would become the target host, and improve its resource occupancy rate to keep workload balance of the cluster as a whole.

4 SUMMARY

This paper proposed a Cloud Computing architecture based on KVM virtualization technology with analysis of current Cloud Computing technology. This architecture includes three layers, and each layer had a detailed introduces in paper. The Cloud Computing platform framework proposed a solution to meet different user's flexibility needs, adapt to the different applications and services of users, and improve automated management of Cloud Computing platform.

REFERENCES

Amazon EC2. http://aws.amazon.com/ec2/.
Bolte, M. Sievers, M. & Birkenheuer G. 2010. Non-intrusive Virtualization Management using libvirt. *Design, Automation & Test in Europe Conference & Exhibition:* 574–579.
Deshane, T. Shepherd Z. & Matthews, J.N. 2008. Quantitative Comparison of Xen and KVM. Xen Summit *Xen Summit:* 1–2.
Fang Jin-ming. 2012. Cloud Computing based on NSGAI-Ivirtual resource scheduling algorithm. *Computer Engineering and Design*: 1454–1455.

Foster I. Zhao, Y. Raicu, I. & Lu, S.Y. 2008. Cloud Computing and Grid Computing 360-Degree Compared. *Grid Computing Environments Workshop:* 3–4.

KVM. http://www.linux-kvm.org/page/HOWTO

LI Jian-feng, PENG Jian. 2011. Task scheduling algorithm based on improved genetic algorithm in cloud computing environment. *Journal of Computer Applications:* 184–186.

LIU, S. Sai, L. Xu, R.W. Lin, R. CHEN Tao. 2013. Research of cloud computing resource scheduling model. *Computer Engineering & Science:* 49–49.

Meng XQ, Pappas V, Zhang L.2010.Improving the Scalability of Data Center Networks with Traffic-aware Virtual Machine Placement. *INFOCOM:* 1–2.

Zhu, Y.C. Liu, P. & Wang, J.T. 2012. Enterprisedata security r esearch in public cloud computing. *Applied Mechanics and Materials:* 435–438.

Zhu, Y.C. Liu, P. & Wang, J.T. 2012. Cloud security research in cloud computing. *Applied Mechanics and Materials:* 415–419.

Information Technology and Computer Application Engineering – Liu, Sung & Yao (Eds)
© 2014 Taylor & Francis Group, London, ISBN 978-1-138-00079-7

Based on computer technology of the automobile logistics informationization

Yun Fei Lv, Xiao Quan Li & Ming Wei Zuo
Automotive Engineering Department, ChengDe Petroleum College, ChengDe, China

ABSTRACT: With the continuous development of automobile logistics, automotive logistics informatization construction based on computer technology is more and more attention for the enterprise. Automobile logistics enterprise should make full use of information technology to development of modern automobile logistics, realize accurate link of automobile logistics activities a series of links, to provide users with convenient and efficient service. This paper analyzes the current situation of the construction of automobile logistics informationization in our country, and put forward the strategies and measures to strengthen the construction of automobile logistics enterprise informatization. To further reduce the cost of automobile logistics process, to achieve maximize the value of society as a whole.

Keywords: computer technology, automobile logistics, information systems, current situation, strategy

1 INTRODUCTION

In 2012 China's automobile production and sales more than 19 million vehicles hit a record high, to refresh the record again, in the world for four consecutive years. According to experts predict that market demand about 20.8 million cars in this year, annual sales of 20.65 million cars, the growth rate of around 7%. Total auto industry has entered a higher stage of steady development, for the development of automobile logistics industry provides a huge opportunity. "Automobile logistics management, information construction has become a global consensus on automobile logistics enterprises. On the one hand, the automotive logistics quality depends on the information, logistics services relying on information, and actively use computer technology based on data collection, sorting, mining and analysis, for the logistics enterprises at all levels within the organization staff to provide information, and improve enterprise decision-making ability, speed up the decision, to ensure that the decision-making accuracy, at the same time realize the remote management inside the enterprise. On the other side, and provide effective information for automobile logistics enterprises external users, share sales, inventory and other business data, common category analysis and management, promote foreign service levels.

2 THE PRESENT SITUATION OF THE AUTOMOBILE LOGISTICS ENTERPRISE INFORMATION CONSTRUCTION

At present, China's automobile logistics enterprise informatization level is low, generally, without perfect information platform, information service industry overall level is not high, the service function is not perfect, the information statistical dispersion, messy, not system, a lot of statistical information is still in a blank state, which creates a high error rate, slow information transfer and the low efficiency of management. Mainly manifested in the following aspects:

2.1 The deviation of understanding of automobile logistics informationization

Enterprises of automobile logistics informationization awareness deviation is reflected in: (1) understanding of the importance of automobile logistics informationization to promote enterprise development. This respect, many automobile logistics enterprises, especially small and medium-sized enterprises (smes) on the understanding is still in vacant state. Within the enterprise, the understanding of the leadership needs to improve more, business leaders are not generally known to the informationization is the head of engineering", the leadership of the value or not is a matter of implementation effect and the success or failure. (2) the difficulty of auto logistics enterprise informatization project construction estimate shortage, to the understanding of the process of informatization construction, mode and means there is a deviation. Some enterprises think that as long as you use the computer instead of manual processing information is the implementation of information management, the information construction as just simply buy a computer and a two sets of computer software, is only by the technical department or departments to undertake informatization. Without considering the transformation and development of business, management and technology. Automobile logistics informationization must accompany to the enterprise huge changes to the way the operation process and management.

2.2 Low degree of standardization of automobile logistics informationization

Automobile logistics activities including transportation, warehousing, distribution, and many other links, for logistics information system on the supply chain of each link connected into a whole, which requires the coding, the file format, data interface, EDI, GPS and other related code to achieve standardization, to eliminate the different between auto logistics enterprise information communication obstacles. And our country because of the lack of auto logistics information standard, the basis of the interface of different information systems become a bottleneck restricting the development of information technology, automobile logistics enterprises in the processing order, sometimes data exchange to seven or eight different oriented mode, hinder the integration of information.

2.3 Automobile logistics information system is inefficient

In the automobile logistics supply chain, each node in the supply chain between enterprises demand forecasting, inventory status, production planning, etc. Are all the important data of supply chain management, the data points between different supply chain organization, to do effectively and quickly respond to user needs, must to transfer information in real time. To do this, you need to model of supply chain information system to make corresponding change, through system integration to make the supply chain inventory data in real time, fast delivery. But now many of the enterprise information system is not well integrated, when supplier need to understand the needs of the user information, often get the delay information and inaccurate information. Which caused the inaccurate delivery status data.

2.4 Automobile logistics enterprises lack of cooperation and coordination

Automobile logistics supply chain as a whole, the need to co-ordinate activities, if you don't consider the upstream enterprise and downstream enterprise inventory status. Will not be able to get the best operation result, make the whole supply chain can be coordinated according to user requirements, to form a more reasonable relationship between supply and demand, adapt to the complex and changeable market environment.

3 AUTO LOGISTICS ENTERPRISE INFORMATIZATION STRATEGY

Country should be worked out as soon as possible to promote our social and automobile logistics enterprise informatization development plan, to speed up the pace of China's economic and social informatization development, from the overall improve the level of application of modern information technology in our country, for the automotive logistics development to create a good technical environment and market demand. At the same time should be in line with the overall planning, step-by-step implementation, our strengths, our pragmatic policy, accelerate the construction of automobile logistics information infrastructure, both to meet the needs of the development of social informatization, and has good scalability and compatibility.

3.1 Promotion of automotive logistics standardization construction

To speed up the popularization and application of advanced applicable technology, automobile logistics should be widely used in standardization, seriation and standardization of transportation, storage, loading and unloading, handling, packaging machinery facilities, bar code technology, etc. Automotive logistics standardization construction mainly includes material container in my homework with the standardization and standardization of coding. If established the material container with the standard requirements, suppliers can to different vehicle manufacturer use the same packaging and transport, material requirements of the society as a whole to reduce a lot of positive role. Should consider the factors are: auto parts of the packing and shipping requirements; The volume size of transport vehicles; To match the size of the material container has its own; Is advantageous for the mechanized transporting, stacking and so on. Coding and standardization is the premise of the Chinese automobile logistics industry informatization, the unity of the material code and standardization for the serial number of the same material transfer between different companies and identify, shorten the supply chain process time.

3.2 Auto logistics area should support and encourage the development and application of information technology

Countries insist on resource sharing, the application leading, market-oriented, step by step, gradually optimized principle, encourage and support related to automobile logistics information technology and related equipment development and research activities, one is in the national key scientific research project of automobile logistics technology, especially the automobile logistics information technology application in the field of research and development efforts.

2 it is to encourage and support enterprises use information technology to improve automobile logistics management and operation means, such as using state funds, technical transformation to encourage circulation enterprises and manufacturing enterprises of automobile logistics information system construction, the introduction of various kinds of advanced automobile logistics management software and systems, etc. Automobile logistics enterprise logistics informatization construction should begin from the problems of enterprise survival and development, seize the key link in the process of automobile logistics management, tightly around improve enterprise core competitiveness key breakthrough, out of a low cost, quick effect. 3 it is to support information communication and information sharing between upstream and downstream automobile logistics enterprise network construction and management innovation, reduce and even eliminate the "information island" between enterprises and enterprises, automobile logistics information resources sharing, constantly improve the level of auto logistics enterprise logistics information construction and application, in order to optimize process to improve the whole logistics chain, with emphasis on the efficiency and reduce the cost of each link, gradually formed in the field of professional logistics public information platform.

3.3 To speed up the related to the automotive logistics industry informationization management departments

Automobile logistics informationization construction and development, needs the government management department to support. Actively promote government management informatization, will be conducive to the smooth development of automobile logistics informationization. To speed up the logistics related industry management information construction should start from two aspects. On the one hand, to promote the administrative departments of electronic government affairs, the full realization of the government departments at all levels of government to deal with electronic. On the other hand, we need to promote the department of the government online project. Through the implementation of the government online project, is advantageous to the reform of government management mode and workflow, improve office efficiency, enhance the government and enterprises, social interaction and communication, realize the government information resources sharing, provide guidance and services for the development of automobile logistics enterprise, ensure the smooth progress of automobile logistics activities and the automobile logistics enterprise smooth operation.

3.4 Strengthen automobile logistics information technology application in the field of demonstration and guidance

At present our country has some internal auto logistics management level of higher production and circulation enterprises, also have a batch of rapid development of advanced automobile logistics enterprises, these enterprises have many experiences and lessons in application of information technology, should seriously review, better use of information technology provides more demonstrations. In addition, the relevant departments also can consider to set up auto logistics information technology application and promotion of demonstration project or base, and explore automobile logistics established between upstream and downstream enterprises ways and means of information sharing mechanism, accelerate the information technology application in the field of automobile logistics and promotion.

3.5 Strengthen the training and cultivation of talents

As many of the information technology is widely used in the field of automobile logistics and the improvement of auto logistics enterprise information intensive, automobile logistics staff level of knowledge and skill levels also change accordingly. It is the cultivation of logistics talents and automobile logistics personnel training is put forward higher requirements. Therefore, strengthen the cultivation of information technology talents and auto logistics practitioners knowledge and skills training, information technology is completely change backward key automotive logistics field of information technology. Therefore, the government should broaden the channels of education and training, encourage the industry association, the enterprises and colleges and universities to carry out various and multi-level training work, and to speed up cultivating automobile logistics information technology in the field of research and development personnel.

3.6 Strengthen the automobile logistics supply chain management, integration of resources

Automobile logistics supply chain management is a kind of integrated management thought and method, is the automobile logistics in the supply chain cash flow, information flow, business flow, etc. To plan, organize, coordinate and control the integration of management process. Automobile logistics supply chain management is the basic concept of competition based on cooperation of beliefs, it is able to plan jointly by sharing information and improve the efficiency of the entire vehicle logistics system, the automobile logistics channels from a loosely connected group of independent enterprises, become a kind of alliance is committed to improve efficiency and increase competitiveness. Automobile logistics supply chain management mainly through the control and coordination of supply chain node enterprise behavior, to reduce the system cost, improve product quality, improve service levels and other purposes, so as to comprehensively improve the comprehensive competitiveness of the automobile logistics supply chain system.

REFERENCES

1. The national development and reform commission, bureau of economic operation, modern logistics center of nankai university. The China's modern logistics development report (2008) [M]. Electronic industry press. 2008.8
2. The China federation of logistics and purchasing, logistics association in China. China's logistics development report (2007–2008) [M]. China press. 2008.5
3. The national development and reform commission, bureau of economic operation, modern logistics center of nankai university. The China's modern logistics development report (2007) [M]. Machinery industry press. 2007.8
4. Lu Hui. Information technology help the fifth party logistics [J]. Science and technology association BBS. 2009, 6
5. Lili. Logistics information platform construction and application [D]. Wuhan. Ph.D. Dissertation, university of science and technology, 2006

Information Technology and Computer Application Engineering – Liu, Sung & Yao (Eds)
© 2014 Taylor & Francis Group, London, ISBN 978-1-138-00079-7

Research on evaluation of e-commerce website based on method of hybrid TOPSIS group decision-making

J.M. Li, X.D. Hu, J.X. Cheng & R. Zhang
School of Information Engineering, AnHui XinHua Univeristy, Hefei, China

ABSTRACT: Due to the fuzzy of the evaluation information, a kind of E-business websites evaluation under uncertainty group decision-making is researched in this paper. The real number, interval number, triangle fuzzy value and linguistic are used to express and aggregate the evaluation information. A new group evaluation method based on TOPSIS (Technique for Order Preference by Similarity to an Ideal Solution) and hybrid indicators is presented to determine the order of the E-business websites. Finally, a numerical example is used to illustrate the validity and simplicity method.

Keywords: website; evaluation; TOPSIS; multi-attribute decision-making; group decision-making

1 INSTRUCTIONS

With the rapid development of information technology, computer network technology and Internet, e-commerce market also have developed rapidly. E-commerce sites have emerged in large numbers and have the booming trend. Market competition become more and more severity. However, the evaluation of e-commerce website becomes an important foundation work of the development of e-commerce, which is also a hot problem in the field of e-commerce research. At the same time, it provides a scientific basis for the development and utilization of e-commerce information resources.

Domestic and foreign experts and scholars put forward a variety of evaluation index system and evaluation methods on the evaluation of e-commerce sites [1-4]. Wang wei-jun[1] have analysis and studied the evaluation method, content and index system of e-commerce; Chang Jinling[2] provides a heuristic algorithm to evaluate the website based the usability guidelines as a basis for the evaluation; Zhang ling[3] provides analytic hierarchy method based on the correlations considering the correlation between the evaluation indicators of e-commerce website. Fu Li-fang [4] who considers the uncertainty of the evaluation of e-commerce sites, the comprehensive evaluation method of e-commerce sites is proposed based on the possible degree. On the basis of the traditional TOPSIS method, these methods have their own characteristics and advantages, but there are some shortcomings, especially in the evaluation process of the e-commerce sites, ignoring the subjective evaluation of experts and customers, which does not reflect the competitiveness and management ability of true electronic Business Website.

2 PRELIMINARIES

2.1 *Interval number*

Definition 1[5] In this section, we briefly review basic concepts of interval numbers, [5] represented a generalized interval numbers number $\tilde{a}$ as $\tilde{a} = [a^L, a^U] = \{x | a^L \le x \le a^U, a^L, a^U \in R\}$, where a^L and a^U are real values when $\tilde{a} = [a^L, a^U] = a$, $\tilde{a}$ is constant value. The arithmetic operations of interval number $\tilde{A} = [a^L, a^U]$ and $\tilde{B} = [b^L, b^U]$, as follows:

(1) $\tilde{A} + \tilde{B} = [a^L, a^U] + [b^L, b^U] = [a^L + b^U, a^L + b^U]$;
(2) $\beta A = [\beta a^L, \beta a^U]$, where $\beta \ge 0$ if $\beta = 0$, then $\beta \tilde{a} = 0$;
(3) $\tilde{A} = \tilde{B}$ when $a^L = b^L$ and $a^U = b^U$.

Definition 2[6]. In this section, [6] presented a distance measure between interval-valued numbers. Let $\tilde{A}$ and $\tilde{B}$ be two interval-valued numbers, where $\tilde{A} = [a^L, a^U]$ and $\tilde{B} = [b^L, b^U]$. Then, the degree of distance between interval-valued numbers $\tilde{A}$ and $\tilde{B}$ is calculated as follows:

(1) $d_p(\tilde{A}, \tilde{B}) = [(a^L - b^L)^p + (a^U - b^U)^p]^{1/p}$, where $p \ge 1$;
(2) $d_1(\tilde{A}, \tilde{B}) = |a^L - b^L| + |a^U - b^U|$ where $p = 1$, it calls $d_1(\tilde{A}, \tilde{B})$ as Hamming distance;
(3) $d_2(\tilde{A}, \tilde{B}) = \sqrt{(a^L - b^L)^2 + (a^U - b^U)^2}$, where $p = 2$, it calls $d_2(\tilde{A}, \tilde{B})$ as Euclidean distance;

2.2 *Linguistic*

Definition 3[7] In this section, [7] introduced a finite and totally ordered discrete linguistic label Set $S = \{s_0, s_1, s_2, \cdots, s_g\}$, whose cardinality value is odd and must be small enough so as not to impose useless

Table 1. Transform between linguistic and triangular fuzzy number.

Type of language	Rank	Range of value
Triangular fuzzy	Perfect	[0.8,0.9,1]
Triangular fuzzy	Very good	[0.7,0.8,0.9]
Triangular fuzzy	Good	[0.5,0.6,0.7]
Triangular fuzzy	Medium	[0.4,0.5,0.6]
Triangular fuzzy	Bad	[0.3,0.4,0.5]
Triangular fuzzy	Very bad	[0.2,0.3,0.4]
Triangular fuzzy	None	[0.1,0.2,0.3]

precision on the experts, and it must also be rich enough in order to allow a discrimination of the performances of each object in a limited number of grades, such as 7 and 9, the limit of cardinality is 11 or not more than 13, where g is a positive integer, s_i represents a possible value for a linguistic label, and it requires that:

(1) The set is ordered: $S_i \geq S_j$ if $i \geq j$;
(2) The negation operator is defined: $Neg(s_i) = s_{g-i}$;

For example, a set of seven linguistic labels S could be

$$S = \{s_1 = none, s_2 = very\ bad, s_3 = bad, s_4 = medium, s_5 = good, s_6 = very\ good,\}$$

This paper has transformed linguistic indicators into triangular fuzzy number, as shown in Table 1.

2.3 Triangular fuzzy number

Definition 4[8]. In this section, we briefly review basic concepts of triangular fuzzy number (TFNs), [8] represented a triangular fuzzy number $\tilde{P}$ as $\tilde{P} = (a^L, a^M, a^U)$, whose membership function is given by

$$\mu_{\tilde{p}}(x) = \begin{cases} x - a^L / (a^M - a^L), & a^L \leq x \leq a^M \\ x - a^U / (a^M - a^U), & a^M \leq x \leq a^U \\ 0, & other \end{cases}$$

where a^M is the mean of $\tilde{P}$, a^L and a^U are the lower and upper limits of $\tilde{P}$, respectively.

Obviously, if $a^L = a^M = a^U$ then the TFN $\tilde{P} = (a^L, a^M, a^U)$ is reduced to a real number. Conversely, real numbers are easily rewritten as TFNs. Thus, the TFN can be flexible to represent various semantics of uncertainty such as ill-quantity.

$\tilde{P} = (a^L, a^M, a^U)$ is called a non-negative TFN if $a^L > 0$ and $a^U > 0$. Let $\tilde{A} = [a^L, a^M, a^U]$ and $\tilde{B} = [b^L, b^M, b^U]$ be two TFNs. Then, arithmetical operations can be expressed as follows:

(1) $\tilde{A} + \tilde{B} = [a^L, a^M, a^U] + [b^L, b^M, b^U] = [a^L + b^L, a^M + b^M, a^U + b^U]$;
(2) $\lambda \tilde{A} = [\lambda a^L, \lambda a^M, \lambda a^U]$, where $\lambda \geq 0$;

Definition 5[9]. In this section, [9] presented a distance measure between triangular fuzzy number. Let $\tilde{A}$ and $\tilde{B}$ be two triangular fuzzy numbers,

where $\tilde{A} = [a^L, a^M, a^U]$ and $\tilde{B} = [b^L, b^M, b^U]$. Then, the degree of distance between triangular fuzzy numbers $\tilde{A}$ and $\tilde{B}$ is calculated as follows:

$$d(\tilde{A}, \tilde{B}) = \sqrt{\frac{(a^L - b^L)^2 + (a^M - b^M)^2 + (a^U - b^U)^2}{3}}$$

3 HYBRID INDICATOR DECISION-MAKING EVALUATION METHOD BASED ON TOPSIS

For a specified decision problem or decision alternative, different decision – maker usually give different estimations or judgments over a set of evaluation criteria. Assume that the set of alternatives $X = \{X_1, X_2, \ldots, X_n\}$, and the set of criteria $C = \{c_j | j = 1, 2, \ldots, m\}$, in the decision expert group E joint by k expert $e_1, e_2, \ldots, e_t$, the set of expert weight $w = \{w_1, w_2, \ldots, w_n\}$ and $w_k \in [0, 1]$, each expert evaluates n alternatives $X_1, X_2 \ldots, X_n$, where a_{ij}^k $(i = 1, 2, \ldots, n; j = 1, 2, \ldots, m; k = 1, 2, \ldots, t)$ refers to the evaluation value to alternative X_i based on criteria c_j by expert e_k, and m, n and t are the maximum numbers of criteria, alternatives, and experts, respectively.

The procedures of calculation for the hybrid indicator decision-making evaluation method can be described as follows:

Step1: Normalization of evaluation matrix. Accordingly, the normalized evaluation matrix $A' = (a'_{ij})_{n \times m}$ can be obtained from Eq. (1–4).

When a_{ij} is precise value, J_1 and J_2 are positive and negative indicator respectively.

$$a'_{ij} = \begin{cases} a_{ij} / \max(a_{ij}), & j \in J1 \\ \min(a_{ij}) / a_{ij}, & j \in J2 \end{cases} \quad (1)$$

When a_{ij} is interval value, J_1 and J_2 are positive and negative indicator respectively.

$$\begin{cases} a_{ij}'^L = a_{ij}^L / \sum_{i=1}^n a_{ij}^U \\ a_{ij}'^U = a_{ij}^U / \sum_{i=1}^n a_{ij}^L \end{cases} j \in J1 \quad \begin{cases} a_{ij}'^L = (1/a_{ij}^L) / \sum_{i=1}^n (1/a_{ij}^U) \\ a_{ij}'^U = (1/a_{ij}^U) / \sum_{i=1}^n (1/a_{ij}^L) \end{cases} j \in J2 \quad (2)$$

When a_{ij} is linguistic value, linguistic value need to transform into triangular fuzzy number, J_1 and J_2 are positive and negative indicators respectively.

$$\begin{cases} a_{ij}'^L = a_{ij}^L / \sum_{i=1}^n a_{ij}^U \\ a_{ij}'^M = a_{ij}^M / \sum_{i=1}^n a_{ij}^M \quad j \in J1 \\ a_{ij}'^U = a_{ij}^U / \sum_{i=1}^n a_{ij}^L \end{cases}$$

$$\begin{cases} a_{ij}'^L = (1/a_{ij}^L) / \sum_{i=1}^n 1/a_{ij}^U \\ a_{ij}'^M = (1/a_{ij}^M) / \sum_{i=1}^n 1/a_{ij}^M \quad j \in J2 \\ a_{ij}'^U = (1/a_{ij}^U) / \sum_{i=1}^n (1/a_{ij}^L) \end{cases} \quad (3)$$

Step2: Aggregation every expert evaluation matrix. Different expert evaluation matrix $A'_k = (a''^{(k)}_{ij})_{n \times m}$ are aggregated into a group evaluation matrix $R = (r_{ij})_{n \times m}$ can be obtained from Eq. (4) based on WA operator:

$$r_{ij} = WA_w (a^{(1)}_{ij}, a^{(2)}_{ij}, \cdots, a^{(t)}_{ij}) = w_1 a^{(1)}_{ij} \oplus w_2 a^{(2)}_{ij} \oplus \cdots \oplus w_t a^{(t)}_{ij} \quad (4)$$

Step3: Compute optimistic and pessimistic evaluation values for the j th evaluation criterion. For each criterion c_j, optimistic and pessimistic values are $r^+ = (r_1^+, r_2^+, \cdots, r_m^+)$, $r^- = (r_1^-, r_2^-, \cdots, r_m^-)$ respectively, defined as follows:

When a_{ij} is precise value, J_1 and J_2 are positive and cost indicator respectively.

$$r_j^+ = \max_i(r_{ij}), j \in J1, \quad r_j^- = \min_i(r_{ij}), j \in J2 \quad (5)$$

When a_{ij} is interval value, J_1 and J_2 are positive and cost indicator respectively.

$$r_j^+ = [r_j^{+^L}, r_j^{+^U}] = [\max_i(r_{ij}^L), \max_i(r_{ij}^U)], j \in J1 \quad (6)$$

$$r_j^- = [r_j^{-^L}, r_j^{-^U}] = [\min_i(r_{ij}^L), \min_i(r_{ij}^U)], j \in J2 \quad (7)$$

When a_{ij} is triangular fuzzy number, J_1 and J_2 is positive and cost indicators respectively.

$$r_j^+ = [r_j^{+^L}, r_j^{+^M}, r_j^{+^U}] = [\max_i(r_{ij}^L), \max_i(r_{ij}^M), \max_i(r_{ij}^U)], j \in J1 \quad (8)$$

$$r_j^- = [r_j^{-^L}, r_j^{-^M}, r_j^{-^U}] = [\min_i(r_{ij}^L), \min_i(r_{ij}^M), \min_i(r_{ij}^{U'})], j \in J2 \quad (9)$$

Step4: Distance computation between criteria values and optimistic/pessimistic values. Using optimistic/pessimistic values, and definition (2) and (5), the distance between attribute values of the jth criteria and optimistic/pessimistic values of the alternative can be calculated by

$$D_i^+ = \sqrt{\sum_{j=1}^m [d(r_{ij}, Y^+)]^2} \quad (10)$$

$$D_i^- = \sqrt{\sum_{j=1}^m [d(r_{ij}, Y^-)]^2} \quad (11)$$

Step5: The relative closeness coefficient (CC) measurement for each alternative X_i with respect to the ideal solutions. In the distance-based method, the measure of dispersion for the ith alternative is expressed as:

$$CC_i = D_i^+ / D_i^+ + D_i^-$$
where $0 \le CC_i \le 1, i = 1, 2, \ldots, n$. $\quad (12)$

where $0 \le CC_i \le 1, i = 1, 2, \ldots, n$.

The larger value of CC indicates that an alternative is closer to optimistic solution and farther from pessimistic solution simultaneously. Therefore, the ranking order of all the alternatives can be determined according to the descending order of CC values. The most preferred alternative is the one with the highest value.

4 ILLUSTRATIVE EXAMPLE

In this section, in order to demonstrate the calculation process of the proposed approach, an evaluation example of e-commerce sites is provided. Selecting the seven evaluation criteria of e-commerce sites, such as the business model innovation c_1, website technical ability c_2, the popularity of web applications c_3, business level c_4, visits c_5, the quality of information and services c_6, website development prospects c_7, the four appliance products of e-commerce website (X_1, X_2, X_3, X_4), four decision experts (e_1, e_2, e_3, e_4), give four evaluation matrix (A_1, A_2, A_3, A_4) of the e-commerce website sites by the evaluation of expert, and let the expert weights $w = (0.15, 0.25, 0.25, 0.35)$, this example's linguistic indicator is seven particle size of the linguistic evaluation set, as shown in Table 1 above.

Four experts give themselves evaluation matrix A_1, A_2, A_3, A_4 to alternative X_1, X_2, X_3, X_4 with respect to the seven evaluation criteria, the normalized evaluation matrix $A' = (a'_{ij})_{n \times m}$ can be obtained from Eq. (1–3), and group evaluation matrix $R = (r_{ij})_{n \times m}$ can be obtained from Eq. (4).

$$A_2 = \begin{bmatrix} 90 & [50,56] & \text{very good} & [4.8,5.8] & 85 & [0.90,0.92,0.95] & \text{very good} \\ 85 & [35,45] & \text{good} & [4.5,5.5] & 75 & [0.89,0.90,0.93] & \text{medium} \\ 85 & [48,58] & \text{very good} & [5.0,5.7] & 90 & [0.84,0.86,0.90] & \text{good} \\ 80 & [40,45] & \text{medium} & [4.3,5.3] & 80 & [0.91,0.93,0.95] & \text{medium} \end{bmatrix}$$

$$A_2 = \begin{bmatrix} 90 & [50,56] & \text{very good} & [4.8,5.8] & 85 & [0.90,0.92,0.95] & \text{very good} \\ 85 & [35,45] & \text{good} & [4.5,5.5] & 75 & [0.89,0.90,0.93] & \text{medium} \\ 85 & [48,58] & \text{very good} & [5.0,5.7] & 90 & [0.84,0.86,0.90] & \text{good} \\ 80 & [40,45] & \text{medium} & [4.3,5.3] & 80 & [0.91,0.93,0.95] & \text{medium} \end{bmatrix}$$

$$A_3 = \begin{bmatrix} 85 & [50,55] & \text{good} & [4.5,5.5] & 85 & [0.91,0.94,0.95] & \text{very good} \\ 85 & [35,45] & \text{good} & [4.3,5.3] & 75 & [0.90,0.92,0.95] & \text{good} \\ 90 & [47,57] & \text{very good} & [4.7,5.4] & 90 & [0.91,0.94,0.97] & \text{good} \\ 80 & [40,45] & \text{medium} & [4.2,5.0] & 80 & [0.85,0.88,0.90] & \text{medium} \end{bmatrix}$$

$$A_4 = \begin{bmatrix} 90 & [48,56] & \text{very good} & [4.4,5.2] & 85 & [0.92,0.96,0.99] & \text{good} \\ 80 & [35,45] & \text{medium} & [4.6,5.6] & 75 & [0.90,0.92,0.95] & \text{medium} \\ 85 & [50,55] & \text{very good} & [4.8,5.5] & 90 & [0.92,0.94,0.97] & \text{very good} \\ 85 & [37,45] & \text{good} & [4.3,5.0] & 80 & [0.86,0.89,0.93] & \text{medium} \end{bmatrix}$$

$$R = \begin{bmatrix} 0.976 & [0.250,0.325] & [0.206,0.276,0.366] & [0.213,0.303] & 0.94 & [0.236,0.255,0.275] & [0.219,0.293,0.397] \\ 0.915 & [0.170,0.250] & [0.149,0.212,0.292] & [0.203,0.297] & 0.83 & [0.237,0.253,0.267] & [0.150,0.217,0.306] \\ 0.964 & [0.245,0.332] & [0.234,0.308,0.406] & [0.226,0.307] & 1.00 & [0.236,0.250,0.267] & [0.216,0.289,0.392] \\ 0.915 & [0.189,0.260] & [0.149,0.211,0.292] & [0.200,0.288] & 0.89 & [0.228,0.245,0.260] & [0.138,0.204,0.289] \end{bmatrix}$$

$r^+ =[0.976 \ [0.25,0.332] \ [0.234,0.308,0.406] \ [0.226,0.307] \ 1.0 \ [0.237,0.255,0.275] \ [0.219,0.293,0.397]]$

$r^- =[0.915 \ [0.17,0.250] \ [0.149,0.211,0.292] \ [0.200,0.288] \ 0.8 \ [0.228,0.245,0.260] \ [0.138,0.204,0.289]]$

2) Compute optimistic and pessimistic evaluation values for the evaluation criterion based on Eq. (5–9).

3) Compute distance between criteria values and optimistic/pessimistic values based on definition (2) and (5) and Eq. (10–11).

$D_1^+ = 0.0/1$, $D_2^+ = 0.250$, $D_3^+ = 0.015$, $D_4^+ = 0.211$

$D_1^- = 0.204$, $D_2^- = 0.019$, $D_3^- = 0.251$, $D_4^- = 0.064$

4) Compute the relative closeness coefficient (CC) measurement for each alternative X_i with respect to the ideal solutions. In the distance-based method, the measure of dispersion for the ith alternative is expressed as

$CC_1 = 0.257$, $CC_2 = 0.931$, $CC_3 = 0.055$, $CC_4 = 0.768$

$$CC_2 > CC_4 > CC_1 > CC_3$$

Therefore, we can see that the order of rating among four e-commerce website is $X_2 \succ X_4 \succ X_1 \succ X_3$, where "$\succ$" indicates the relation "preferred to"

5 CONCLUSION

The evaluation of e-commerce sites is not only a very important issue, but also is a complex problem for the customer, site builders and enterprise. In the fact, decision-makers usually use single evaluation form to reflect the evaluation information which can not accurately describe the objective and subjective information of decision-makers, so this paper propose a new multiple attributes group decision evaluation method based hybrid TOPSIS, in the complete known weights information of expert, and the form of attribute values including interval numbers, real numbers, triangular fuzzy number for the evaluation of e-commerce site, which can scientific and intuitive reflect the overall strength and competitiveness of the e-commerce site.

REFERENCES

[1] Wang Weijun. Analysis and Comment on Study and Application of EC Websites Evaluation [J]. Information Science, 2004, 22(2): 1484–1486.

[2] Chang Jinling,Xia Guoping. Usability Evaluation of B2C E-Commerce Web Site [J]. Journal of the China Society for Scientific and Technical Information, 2005, 24(2): 238–242.

[3] Zhang Ling, Zhou Dequn. Study on Hierarchical Judgment Method with Interdependence and Its Application to Web Evaluation [J]. Journal of the China Society for Scientific and Technical Information, 2007, 26(5): 699–703.

[4] Fu Li-fang, Feng Yuqiang, Liu Ke-xing. Research on the Synthetic Evaluating Model for E-Business Website under Indeterminate Information Conditions [J]. Information Science, 2007, 25(9): 1423–1426.

[5] Wu Jiang, Huang Deng-shi. The Uniform Methods for Interval Number Preference Information in Multi-Attribute Decision Making [J]. Systems Engineering Theory Methodology Applications, 2003, 12(4): 359–362.

[6] Xu Ze-shui. Algorithm for priority of fuzzy complementary judgment matrix [J]. Journal of Systems Engineering, 2001,16(4): 311–314.

[7] Herrera F, Herrera-Viedma. Aggregation operators for linguistic weighted information [J]. IEEE Transactions on Systems, Man, and Cybernetics, 1997, 27:646–656.

[8] Van Laarhoven P J M, Pedrycz W. A fuzzy extension of Saaty's priority theory [J]. Fuzzy Sets and Systems, 1983, 11(2): 199–227.

[9] Chen C T. Extensions of the TOPSIS for group decision-making under fuzzy environment [J].Fuzzy Sets and Systems, 2000, 114(1): 1–9.

Information Technology and Computer Application Engineering – Liu, Sung & Yao (Eds)
© 2014 Taylor & Francis Group, London, ISBN 978-1-138-00079-7

Research of portable heart sound recognition instrument based on embedded computer

Y.T. Wang, H.Y. Yu, X.H. Yang & Q.F. Meng
School of Information Science and Engineering, University of Jinan, Jinan, China
Shandong Provincial Key Laboratory of Network based Intelligent Computing, Jinan, China

ABSTRACT: The present status of heart sound diagnosis technology was introduced in the paper. A new system based on embedded computer was given to solve the drawbacks of traditional heard sound analyzer. This system adopted an autocorrelation segmentation algorithm to implement the accurate positioning of heart sounds signal period and overcame the traditional problem that only aimed at particular period in heart sound localization. It made the heart sounds of different period had better adaptability and less complexity. Besides, on the basis of extracting MFCC characteristic parameters, we proposed the recognition method based on DTW to improve the stability of the heart sound signal classification. The experimental results show that the average recognition rate reachs 88.5%. Its performance is superior to other heart sound recognition systems. Also, the instrument has the advantages such as small volume, low cost, intelligence and convenient to use.

Keywords: Heart sound; Autocorrelation coefficient; Embedded computer; Dynamic Time Warping; Mel Frequency Cepstrum Coefficient

1   INTRODUCTION

Heart sound is reflection of the mechanical movement in heart and cardiovascular system, which contains the physiological and pathological information in each part of heart and their interactions. Before cardiovascular disease can generate clinical and pathological changes, there are diagnostic information in heart sound such as noise and distortion. Thus, we can prevent and find disease before it occurs. In recent years, the cardiovascular disease rate is higher and higher. The efficient collection and processing of heart sound data is the basis of heart disease detection. Currently, heart sound signal recognition is a hot research both in China and abroad.

However, there are no mature products on the market and the heart sound analysis system still exists many shortcomings. The present directions of researches and developments mianly include the following aspects. The first tendency is miniaturization. The traditional system is based on PC, it is uneasy to carry and inconvenient to test for it is big and heavy. The second tendency is intelligentization. Because the heart sound signal is very complex and no accurate algorithm, heavy work is necessary in the field. The third tendency is the remote communication of system. With the development of communication technology, how to transmit ECG signals to the remote area is a hotspot.

In conclusion, there are some immature aspects in the analysis and detection of heart sounds signal, which need to be enhanced. Therefore, a portable heart sound analysis instrument based on embedded computer is designed in the experiment, which has the advantages such as small volume, low cost, real-time, intelligence, etc. It is of great significance to the detection and prevention of cardiovascular disease.

2   THE EXPERIMENT CIRCUIT

The embedded heart sound diagnosis system is mainly composed of heart sound signal acquisition module, microprocessor module, USB interface module, storage module, JTAG debug module, human-computer interaction module. The system compositions are shown in Figure 1.

This system chooses the embedded S3C2440 microprocessor, which is produced in SAMSUNG corporation. It is mainly dedicated to handheld devices, and its advantages are low-power and high-speed processing. In order to reduce system cost, S3C2440 uses the following components: ARM920T core, 0.13 Um cmos unit and complex storage unit. It has the features like small power consumption, simple and stable design, so it is suitable for products that need high-quality power. It adopts new bus architecture.

The specific descriptions of each module functions are shown as follows.

a) Signal collecting module includes sensors, filter circuits, amplified circuits and A/D converting

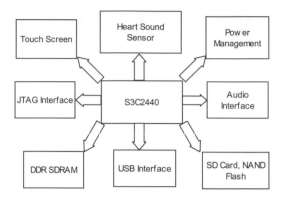

Figure 1. Composition diagram of embedded heart sound recognition system.

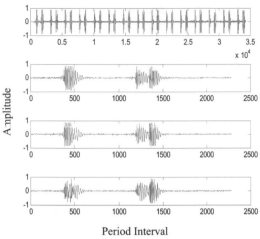

Figure 2. The second heart sound segmentation.

circuits. Through filtering, amplifying and A/D converting, signals collected by sensor enter the main system.

b) Main control chip module adopts S3C2440 microprocessor. It is the core of the entire system. On the one hand, it can control heart sound sensors to collect heart sound signals, establish sample bank. On the other hand, it can control the circuit to implement data storage, USB communication, LCD display. Further, it can deal with feature extraction, recognition algorithm and matching.

c) USB interfaces module is the inner integrated part of S3C2440 microprocessor. It makes the analyzer more flexible and convenient. Users can communicate between USB and PC. Besides, data can be got via USB, which facilitate storage and processing.

d) Storage module mainly stores heart sound data, operating procedures and algorithm code.

e) Human-computer interaction module includes seven-inch touch screen, buttons and related control circuits. Through pressing buttons or touching screen, users can control the system operation, and check the signal waveform or the results.

3 THE SOFTWARE IMPLEMENTATION

3.1 Heart sound segmentation

Traditional segmentation algorithm mainly refer to ECG or carotid artery signal. This method increases the burden of the hardware and software. Heart sound is a kind of non-stationary signal, but traditional algorithm process it as stationary one, which may lead to significant errors. The traditional algorithm need the high-quality signal and is sensitive to noise. It can only locate the approximate positions of S1 and S2, nor can distinguish systolic and diastolic time ranges, or precisely locate heart sound signal periods.

An autocorrelation algorithm is proposed in this paper. It can overcome the present problem of only aiming at particular period in the heart sound positioning field, which make different period heart sounds have better adaptability and less complexity.

The autocorrelation coefficient of random signal x(t) can be denoted as

$$\rho_{x_{t1}x_{t2}} = \frac{\sigma_{x_{t1}x_{t2}}}{\sigma_{x_{t1}}\sigma_{x_{t2}}} = \frac{|R_x(t_1,t_2)|^2}{R_x(t_1,t_1)R_x(t_2,t_2)} \quad (1)$$

Where, $R_x(t_1,t_2)$ is the autocorrelation function between time t_1 and t_2. According to Cauchy-Schwarz inequality, the absolute value of autocorrelation coefficient is not more than one. It is just a ratio without unit name, its positive or negative sign denotes the autocorrelation direction, and the value represents autocorrelation degree.

Taking the segmentation of Second Heart Sound Split as an example, we accurately locate its interval position by adopting the autocorrelation coefficient, which is shown as Figure 2.

In which, we can see that heart sound period interval can be accurately positioned by adapting the autocorrelation coefficient method for heart sound segmentation. Therefore, the period can be extracted completely. In the experiment, we select 45 segmentation signals from five types. Among them, 38 heart sound signals can be precisely positioned. The accurate rate of positioning is 94%.

3.2 Title, author and affiliation frame

The human auditory is a special nonlinear system. Its sensitivity to different frequency signal is a logarithm. MFCC can simulate the characteristics of how ears process the sound signal. So as characteristic parameter of heart sound, MFCC can represent the information of heart sound signals.

The calculation of MFCC is based on 'bark', it can convert to linear frequency according to following formula.

$$f_{mel} = 2595\log_{10}(1+\frac{f}{700}) \quad (2)$$

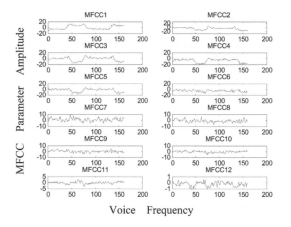

Figure 3.　MFCC parameters of tricuspid regurgitation.

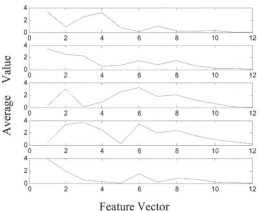

Figure 4.　Average value of 12 order Mel-frequency cepstral coefficient.

The process of MFCC characteristic parameters extraction can be divided into several steps such as framing, windowing, DCT transform, Mel spectrum calculation. In the experiment, we select each frame from heart sounds signals.Where, the sampling sequence points and the number of filter is 256 and 24 respectively. Thus, a frame with 12-order MFCC characteristic parameters is shown as Figure 3.

Where, abscissa represents voice frequency and ordinate represents amplitude. The characteristic value plays a significant role in the pattern recognition. The selections of them are directly related to the accuracy of the results, thus, selected characteristic values should have distinguishing ability, stability, and independence. In this system, we divide each 1s signal into approximately 500 frames, then extract MFCC coefficient from each frame and finally get 500*24 matrix. Inputing too much data is a heavy workload. We use the follwing formula to process MFCC coefficient.

$$x_j = \frac{\sum_{i=1}^{n} |d_{ji}|}{n} \qquad (3)$$

In which, x_j represents average value of j-order MFCC coefficient, d_{ji} represents j-order one, and n is the total order number. After the processing above, we get 12 feature vectors , which greatly reduces the calculational work and is shown as Figure 4.

3.3 Feature matching

The classification algorithm DTW belongs to the field of audio signal processing. It is introduced into the recognition of heart sound signals. On the basis of feature extraction, we processing it to improve the stability of the classification.

The heartbeat period is various from person to person, so the length of testing and reference template frame is not commonly equal. While DTW algorithm requires the same frame length and shift of both the testing and reference template, so testing template

need to be mapped to reference template by a certain method and then the distance between them can be calculated.

According to DTW algorithm, testing and reference templates are commonly labeled by rectangular coordinate system. Usually, the testing template frames $n = 1 \sim N$ are labeled by the horizontal axis and the reference template frames $m=1 \sim M$ are labeled by the vertical axis in the rectangular coordinate system. Each grid intersection in the two-dimensional coordinate system represents the intersection of testing template frame and reference template frame. If the distance between two points in this system is defined as d[n,m], then the distance between testing template and reference template can be shown below.

$$D[N, M] = \sum_{n=m=1}^{N} d[n, m] \qquad (4)$$

DTW adopts the first and the last frames of testing template as its corresponding points in reference templates.This method can find a best path function, which minimize the distance from N to M. The formula is shown as follows.

$$D[N, M] = \min_{\substack{n_i=1 \\ m_i=\phi(n_i)\in\eta}} \sum^{N} D[n_i, m_i] \qquad (5)$$

Among them, the minimum distance means the maximum similarity. So the heart sound type that corresponds to the reference template is the final recognition result.

Some heart sounds are not easy to find in actual experiment due to their particularity. Therefore, all the signals were from Robert Laboratory. We selected ten typical heart sounds such as Normal Heart Sound, First Heart Sound Split, Second Heart Sound Split, Tricuspid Incompetence, Fourth Heart Sound Enhancement and Continuous Murmur, etc. The reference samples and test samples of every type are twenty.

Table 1. Heart Sound Recognition Result.

Type	Recognition		
	Result	Rate	Time
Normal Heart Sound	19	95%	1.50s
First Heart Sound Split	17	85%	1.73s
Mitral Stenosis	18	90%	1.37s
Tricuspid Incompetence	17	85%	1.60s
Fourth Heart Sound Enhancement	16	80%	1.42s
Continuous Murmur	19	95%	1.31s
Summation Gallop	18	90%	1.58s
Active Regurgitation	19	95%	1.40s
Third Heart Sound Enhancement	18	90%	0.97s
Active Valve Closed-insufficiency	18	80%	1.56s

In the recognition process, we firstly preprocess it and then extract MFCC and calculate its features, getting a set of characteristic sequence points. Finally, comparing characteristic sequence points with heart sound to get recognition result, which is shown as Table 1.

From the results above, it can be concluded that:

a) the correct recognition rate based on S3C2440 can reach 88.5%.
b) the heart sound system test 20 samples respectively and the average time is 1.44s, which basically achieves real-time effect.

4 CONCLUSION

Hardware design adopts S3C2440 microprocessor as the core of the embedded heart sound diagnostic system. Because S3C2440 has powerful arithmetic function, the real-time ability and accuracy of this system can be guaranteed.

Software design adopts an autocorrelation heart segmentation algorithm. It implements the accurate positioning of heart sounds signal period. Also, it overcomes the problem of only aiming at particular period in heart sound positioning field. In the experiment, we combine the dynamic time programming algorithm in acoustic field and MFCC coefficient together to recognize the period of heart sound signal, which achieves satisfactory purpose.

Portable heart sounds analyzer based on embedded computer can break the limitations from uncertain factors in traditional auscultation. It also provides better analytical methods of aiming at the clinical research. This system lays solid foundation for the clinical auxiliary diagnosis and is of great significance in both and economy society.

ACKNOWLEDGEMENT

Project supported by the National Natural Science Foundation of China (Grant No. 61201428), the Natural Science Foundation of Shandong Province, China (Grant No. ZR2010FQ020).

REFERENCES

Ahlstrom, C. & Lanne, T. & Ask, P. & Johansson, A. 2008. A method for accurate localization of the first heart sound and possible applications. *Physiological Measurement* 29(3): 117–122.

Chung, Y.J. 2007. Classification of continuous heart sound signals using the ergodic hidden markov model. *Pattern Recognition and Image Analysis Lecture Notes in Computer Science* 24(2): 563–570.

Guo, X.M. & Lin, H.J. & Xiao, S.Z. 2010. Complexity in the application of heart sound analysis. *Journal of Instrumentation* 31(2): 259–263.

Ji, A. & Guo, X.G. & Xiao, S.Z. 2006. The research progress of heart sound detector.*Modern Scientific Instruments* 8(6): 34–36.

Kao, C.W. & Wei, C. & J. Liu. 2009. Automatic heart sound analysis with short-time fourier transform and support vector machines. *IEEE International Midwest Symposium on Circuits and Systems* 23(3): 188–191.

Wang, X.P. & Liu, C. & Li, Y. 2010. Heart sound segmentation algorithm based on higher order shannon entropy. *Journal of Jilin University (Engineering Edition)* 40(5): 1433–1437.

Wang, Y.T. & Li, B. & Jiang, X.Q. 2009. Speaker Recognition Based on Dynamic MFCC Parameters. *IEEE International Conference on Image Analysis and Signal Processing* (1): 101–106.

Zhang, X.G. & He, W. & Zhou, J. & Li, J. 2007. Study on portable heart sound analyzer based on embedded system. *Journal of Instrumentation* 28(2): 303–307.

Zhan, M. & Han, Y. & Wei,Z. 2012. Heart sound feature extraction based on S transformation. *Journal of Vibration AND SHOCK*. 31(20): 179–184.

Zhao, Z.D. & Zhao, Z.J. & Zhang, C. 2004. Heartsound auto-segmentation research. *Space Medicine and Medical Engineering* 17(6): 452–456.

Zhou, J. &Yang,Y.M. & He, W. 2005. Research on heart sound analysis and feature extraction method. *Chinese Journal of Biomedical Engineering* 24(6): 685–689.

Information Technology and Computer Application Engineering – Liu, Sung & Yao (Eds)
© *2014 Taylor & Francis Group, London, ISBN 978-1-138-00079-7*

Analysis on the causes of Chinese vegetable exports to Japan under the background of positive list system – Based on the CMS model

H. Pang
College of Economics, Shenyang University, Shenyang China

M. Zhou
College of Economics and Management, Shenyang Agricultural University, Shenyang China

ABSTRACT: Based on the CMS model, this paper analyzed the changes, causes and inherent mechanism of Chinese vegetable exports to Japan during 1997–2011 under the background of Japanese positive list system. Analysis showed both export quantity and value in 2006 continued to decline after the implementation the system; The demand effect on Chinese vegetable exports to Japan was improved, and the contribution rate increased gradually; the structure effect had a negative action on Chinese vegetable exports to Japan, and the inhibitory action increased gradually; the competitiveness effect played a major role in affecting Chinese vegetable exports to Japan, but it brought negative effect on of ability of earning foreign exchange of Chinese exports at the same time.

Keywords: Vegetable product; Positive list system; CMS model; Influencing factor

1 INTRODUCTION

Chinese vegetables export has a relatively high concentration, and according to the data from UNCOMTRADE, Japan has always been the largest importer of Chinese vegetables. Due to the lack of natural resources, less arable land as well as the aging problem of agricultural labor force, a considerable part of Japanese vegetables need to import from abroad. According to UN statistics, Japan is the largest in Asia and the world's third largest vegetable import country[1]. While China has geographical and cost advantages and low labor costs, as well as low transportation cost, which made Japan become an important export market for Chinese vegetables.

However, since 2001, Japanese inspection and quarantine standards on Chinese vegetable products is strengthened ceaselessly, especially the "positive list system", which was published on May 29, 2006, caused Chinese vegetables become one of products with higher export risks. Moreover, the abuse of chemical fertilizer and pesticide in vegetables planting in our country and the very big disparity between our food sanitation standard and international standards, made the food security situation in our country was not good enough. The proportion of Chinese vegetable exports to Japan occupying Chinese total exports volume decreased year by year after 2001, dropped from 55.05% in 2000 to 18.54% in 2009, with an average decrease of 4.1%. Then, under the background of the positive list system, What changes had

taken place in Chinese vegetables exports to Japan? what caused this significant growth? The answers to these questions may provide some policy references to improve the competitiveness of Chinese vegetable products. In view of this, this paper analyzed effect degree of various factors affecting Chinese vegetable exports to Japan based on the CMS model.

2 CHANGES IN EXPORT QUANTITY AND VALUE OF CHINESE VEGETABLES EXPORTING TO JAPAN

2.1 Changes of export quantity

The overall growth of Chinese vegetable exports to Japan before 2006, the exports in 2005 reached 1.73 million tons, which was 4 times that of 1992. A yearly progressive decrease of exports began in 2006, and exports amounted to 1.6924 million tons in 2006, a decrease of 2.5% over the previous year. What's more, the export volume in 2009 dropped to 1.147 million tons, 588700 tons less than 2005.

Fresh vegetables[1] and processing vegetables' exports accounted for an overwhelming proportion

---

[1] We use Chen Yongfu's (2001)[2] classification method for reference, and divided vegetables into four types: fresh vegetables, frozen vegetables, dehydrated vegetables and processing vegetables (the original classification also includes vegetable juice, but we leave it out because the export volume of Chinese vegetable juice to Japan is too small).

in the total export volume, and they accounted for 73.21% of total exports in 2009; the proportion of dehydrated vegetables was minimum, only around 3%. In addition, the export volume of fresh vegetables reached the maximum in 2005, and the number of 2006 was 714200 tons, a decrease of 7.7% over the previous year; the export volume of frozen vegetables and processing vegetables both reached the maximum in 2006. By 2009, the export volume of fresh, frozen and processing vegetables was 411000 tons, 268200 tons and 428800 tons. Dehydrated vegetables' export volume was relatively stable at 40000–50000 tons.

2.2 Changes of export value

Chinese total exports of vegetables to Japan before 2006 showed a rising trend, the export value amounted to 1.734 billion dollars in 2006. However, the trend turned downward after 2006 and the export value in 2007 was 1.5781 billion dollars, a decrease of 9% over the previous year. Exports in 2008 continued to decline to 1.434 billion dollars, a decrease of 9.1% over the previous year, then it recovered somewhat in 2009 and the number was 1.4862 billion dollars and but still below 2007 levels.

The ratio of processing vegetables' export value occupying the total export was maximal, basically above 30%, and that of dehydrated vegetables was minimum, basically below 20%. The other two products' proportion were between 20%–30%.

3 CONSTANT MARKET SHARE MODEL

Constant Market Share model (CMS) is used to reflect the influencing direction and degree of elements affecting a country's export change. The basic assumption is that, if the competitiveness of a country's product remain unchanged, its market share of the world market should hold the line. The model was first proposed by Tysznski in 1951, and after several amendments and improvement, it has become one of the most important models that study the reason of foreign trade growth and the change of competitiveness of export products.

The assumption of Constant Market Share model is that the market share of the product of a country in a certain period of time is constant, and based on this assumption, the actual changes of a country's exports are due to the change of product demand, export structure or the competitiveness of products. Therefore, according to this model, the growth of a country's certain product exports to another country can be chalked up to three main aspects: (1) the demand of importing country for this product grow faster (demand effect); (2) the exports mainly concentrated in the fast-growing species (structure effect); (3) the country can compete effectively with other exporters (competitiveness effect).

According to the modified CMS model proposed by Leamer and Stern (1970), we suppose $X_{(t)}$ as the export value of Chinese vegetables to Japan in period t; $X_{i(t)}$ indicates the export value of vegetable product i (fresh vegetables, frozen vegetables, dehydrated vegetables and processing vegetables) exported from China to Japan in period t; $X_{(0)}$ indicates the export value of Chinese vegetables to Japan in the base period; $X_{i(0)}$ indicates the export value of vegetable product i exported from China to Japan in the base period; ΔQ indicates the change of vegetable exports from base period to period t; ΔQ_i indicates the change of vegetable product i's exports from base period to period t; m indicates the growth rate of Japanese vegetable import from base period to period t; m_i indicates the growth rate of Japanese vegetable product i's import from base period to period t. Therefore, export changes in vegetable product i can be expressed as:

$$\Delta Q_i = X_{i(t)} - X_{i(0)} = m_i X_{i(0)} + X_{i(t)} - X_{i(0)} - m_i X_{i(0)} \quad (1)$$

Then, the change of Chinese vegetable exports to Japan can be expressed as follows:

$$\begin{aligned}
\Delta Q &= X_{(t)} - X_{(0)} = \sum X_{i(t)} - \sum X_{i(0)} = \sum m_i X_{i(0)} + \sum X_{i(t)} - \sum X_{i(0)} - \sum m_i X_{i(0)} \\
&= m \sum X_{i(0)} + \sum m_i X_{i(0)} + \sum X_{i(t)} - \sum X_{i(0)} - m \sum X_{i(0)} - \sum m_i X_{i(0)} \\
&= m \sum X_{i(0)} + \sum \left[(m_i - m) X_{i(0)} \right] + \sum \left[X_{i(t)} - X_{i(0)} - m_i X_{i(0)} \right] \\
&= m X_{(0)} + \sum \left[(m_i - m) X_{i(0)} \right] + \sum \left[X_{i(t)} - X_{i(0)} - m_i X_{i(0)} \right]
\end{aligned} \quad (2)$$

demand effect $= m X_{(0)}$, it shows the export growth of the Chinese vegetables brought by Japanese vegetables import volume growth; structure effect $= \sum \left[(m_i - m) X_{i(0)} \right]$, if this effect was positive, it indicates Chinese vegetable exports focused on the varieties that has a faster growth in Japanese imports, and negative effect shows the other way round;

competitiveness effect $= \sum \left[X_{i(t)} - X_{i(0)} - m_i X_{i(0)} \right]$

4 ANALYSIS ON THE EFFECT DECOMPOSITION OF CHINESE VEGETABLE EXPORTS TO JAPAN

In this paper, the research period is from 1997–2011, and we divide it into three phases. The first phase is 1997–2000, the reason why we choose 1997 as the starting point is that the Asian financial crisis happened in 1997, the study results of 1997–2000 may objectively reflect the situation before China joined in the WTO; the second phase is 2001–2005, the export of Chinese vegetable to Japan in this phase increased fairly fast; the third phase started in 2006 after the implementation of positive list system, Chinese vegetables exports to Japan presented the drop. In this section, according to the above formula, we calculated the demand effect, structure effect and competitiveness effect on the change of total export volume respectively during the course of the three phases. According to the principle of CMS model, we put the

Table 1. The effect decomposition of Chinese vegetable exports to Japan in 1997–2011.

Effect decomposition	Phase 1 to phase 2		Phase 2 to phase 3	
	value	ratio	value	ratio
Total effect	28.057	100	21.883	100
Demand effect	8.924	31.81	10.123	46.26
Structure effect	−2.935	−10.46	−3.931	−17.96
Competitiveness effect	22.096	78.75	15.691	71.7

years with similar data characteristics into one group, and use the average value of data of these years for calculation. The source data originates from COM-TRADE database. The calculation results are shown in the following table:

4.1 Demand effect

From the calculating results, the demand effect of both the first stage and the second stage on Chinese vegetable exports to Japan was improved, and the contribution rate increased gradually. In the first stage, the vegetable trade scales expanded in Japan, which promoted Chinese vegetable exports to Japan under the condition of keeping the original share. The effect was 89.2 millions dollars and the contribution ratio reached 31.81%; In the second stage, the expansion of trade volume of Japanese vegetable products continued to produce positive effect on the growth of Chinese vegetables exports, the effect was 101.2 millions dollars and contribution ratio was 46.26%, a 14.45% growth compared with the previous stage. The reasons are as follows. although Japanese government has taken a series of measures to protect domestic vegetable industry, in recent years, Japanese domestic vegetables sown areas and agricultural labors were running down steadily. As well as the aging problem of agricultural labor force, Japanese vegetables industry shrank unceasingly and vegetable export has continued unabated. Since late 1990s, Japanese vegetable self-sufficiency rate decreased year by year, the rate was 84% in 1997–2000 and 82% in 2001–2005, and further dropped to 80% in 2006–2011, while Japanese vegetable imports increased from 2.52 billion dollars in 1997–2000 to 2.733 billion dollars in 2001–2005, and further amounted to 2.94 billion dollars in 2006–2011. In the context of continued expansion of Japanese vegetable imports, even the market share of our country's vegetables in Japanese market declined, the increasing demand of Japanese imports would also lead to rising exports of Chinese vegetables, this explains well why the demand effect of Chinese vegetable products increased on gross theoretically. The contribution of fresh and frozen vegetables was larger, while the imports of Japanese dehydrated vegetables decreased, which inhibited Chinese vegetables export to increase.

4.2 Structure effect

Based on the analysis results of CMS model, the structure effect of both the first stage and the second stage on Chinese vegetable exports to Japan was inhibitory, and the inhibitory action increased gradually. That is, under the condition of keeping the original share, the structure changes of Japanese vegetable imports led to the reduction of Chinese vegetable exports to Japan. In the first stage, the structure effect was −29.4 million dollars and the contribution ratio was −10.46%; In the second stage, the structure effect was −39.3 million dollars and the contribution ratio was −17.96%; this showed that export structure of Chinese vegetables to Japan was not so reasonable. The adjustment of importing structure of Japanese vegetables changed the proportion of all kinds of vegetables' imports occupying the total imports. The proportion of fresh and frozen vegetables rose and the proportion of dehydrated vegetables and processing vegetables dropped. The proportion of Japanese fresh and frozen vegetables imports occupying the total imports increased from 30% and 35% in 1997–2000 to 34.5% and 37% in 2006–2011. While fresh and frozen vegetables have always been lack of advantage among Chinese vegetable products exporting to Japan, and the processing vegetables accounted for a large proportion in Japanese total imports, that is to say, Chinese vegetable exports focused on the varieties that has a slower growth in Japanese imports, which inhibited the increase of total exports of Chinese vegetables. Since 2006, owning to the implementation of positive list system, Chinese exports of fresh and frozen vegetables has faced more barriers, making these two kinds of vegetables' disadvantage much more prominent, while Japanese import of these two kinds of vegetables increased, which caused the negative structure effect contribution rate increase by 7.5%.

4.3 Competitiveness effect

As shown in table 1, the competitiveness effect in both the first stage and the second stage was the main factor on to promote the increase of Chinese vegetable exports to Japan, the effect of two stages were 221 and 156.9 million dollars and the contribution rate were 78.75% and 71.7%. However, we found the contribution rate reduced by 7.05%. There is a strong relationship between the competitiveness advantage and the price advantage of Chinese vegetables. Taking fresh vegetables for example, the average export price of Chinese fresh vegetables fell from 278 yen per kilogram in1988 to 69 yen per kilogram in 2005[3], and the price advantage made our country vegetables' market share in the Japanese market continue to increase. In recent years, the problem of pesticide residues exceeding the standard happened in Japanese imported agricultural products making Japanese consumers doubt on the safety and quality of food. Price has not been the first thing that Japanese consumers need think about when they purchase agricultural products, but the food safety. Therefore, Japanese MHLW continued to strengthen the inspection of pesticide residues in imported vegetables and other agricultural products, especially after the implementation of positive list system. Thus, due to the change of core part of the competitiveness, promoting Chinese vegetable exports by lowering the prices in the future may gradually won't work well.

5 CONCLUSION

This paper firstly analyzed the changes in export quantity and value of Chinese vegetables exporting to Japan and found that both export quantity and value in 2006 continued to decline after the implementation Japanese positive list system; Secondly, this paper discussed the influencing factors of Chinese vegetable exports to Japan through the Constant Market Share model. From the empirical analysis results, we found that because of the expansion of Japanese vegetable imports, the demand effect on Chinese vegetable exports to Japan was improved, and the contribution rate increased gradually; as a result of the implementation of positive list system, Chinese exports of fresh and frozen vegetables was blocked, while the import of these two kinds of vegetables grew faster in Japan, which made the structure effect have a negative action on Chinese vegetable exports to Japan, and the inhibitory action increased gradually; the competitiveness effect played a major role in affecting Chinese vegetable exports to Japan, however, the competitiveness effect counteracted the negative effect structure effect, it brought negative effect on of ability of earning foreign exchange of Chinese exports at the same time. The author analyzed the terms of trade of Chinese vegetables before and found that the net terms of trade in the research period of the Chinese vegetable products were deteriorating,

but the income terms of trade during the period were improved. This phenomenon should arouse our attention, especially after the implementation of the positive list system, Chinese export of vegetables already had a powerful competitiveness, and it is necessary to change the core content of the export competitiveness. we should improve the quality and added value of the products in order to promote the ability of earning foreign exchange of Chinese vegetable exports. Perhaps, on the other hand, the implementation of the positive list system may provide the opportunity for Chinese vegetables to improve their net terms of trade.

ACKNOWLEDGEMENTS

Mi Zhou is a corresponding author. This research is jointly funded by the National Natural Science Foundation of China (71203146) and supported by Program for Shenyang Agricultural University youth fund (20111020).

REFERENCES

Wang Bailing, Wang Yexiao. 2010. Analysis of the Effects of the Positive List System on Chinese Vegetable Exports. *Contemporary Economy of Japan*(1): 27–29.

Chen Yongfu, He Xiurong. 2001. Analysis on the cause and countermeasure of vegetable trade war between China and Japan. *Problem of Agricultural Economy* (8): 44–47.

Liu Yazhao, Wang Xiuqing. 2007. Analysis of import market of Japanese fresh vegetables and its demand elasticity. *Journal of Agrotechnical Economics* (2): 31–36.

Huang Guansheng. 2007. Study on agricultural products' TBT and its countermeasure. *Problem of Agricultural Economy* (5):18–22.

Zhang Xiaodi, Li Xiaozhong. 2004. Study on the double effects of TBT on the export of our country's agricultural products. *Management World* (6): 26–33.

Information Technology and Computer Application Engineering – Liu, Sung & Yao (Eds)
© 2014 Taylor & Francis Group, London, ISBN 978-1-138-00079-7

Studies on informational practice of higher education teachers' teaching skills

Hong Yun Chen & Chuan Jie Cheng

School of Chemistry and Chemical Engineering, Jiangxi Science & Technology Normal University,
Nanchang, China

ABSTRACT: Based on the problems existing in the practice of teaching skills in normal universities, effective teaching approach and evaluation method including teaching preparation, teaching process and teaching evaluation are constructed by combining information technology, information source, micro-teaching, education practice and curriculum. And all these will finally promote student's efficient learning as well as optimized teaching.

Keywords: Informationization; Teaching Skills; Micro-teaching; Educational Practice

1 ESTABLISHMENT OF TEACHING AIMS OF TEACHING SKILLS TRAINING FOR NORMAL SCHOOL STUDENTS IN THE NEW ERA

The teaching aims of teaching skills training of normal school students in the new era should be established based on the current situation of teaching skills so as to cultivate fundamental teaching skills of normal school students under the background of new curriculum standard (Zhao, C.L. 2010). The teaching skills aims can be classified as writing skills on the blackboard, demonstration skills in classes, teaching design skills, introducing skills in classes, teaching language skills, questioning and explanation skills, media application skills, and learning design skills. The above eight aspects of teaching aims can be finished by three stages: teaching analysis skills, teaching design skills, teaching performance and evaluation skills (Zhang, W.H. et al. 2011). Formation of teaching skills can be also classified into four stages: demonstration-cognitive stage, imitation-decomposition stage, setting-fixing process and automation-reaction system. Constant development of informationization and deepening of new curriculum, extensive application of information technology in teaching, modification of new learning, teaching and evaluation thought, and all this will require improvement of teaching skills of normal school students. Informational teaching design skills, modern class teaching skills, organizing and instructing out-of-class activities skills, informational teaching evaluation skills, media application skills, and so on are all new teaching skills which are adapted to the new era.

2 OPTIMIZATION OF MICRO-TEACHING

2.1 *Increasing new teaching skills training that are adapted to new curriculum of fundamental education*

In addition to teaching theory, teaching organizing skills as well as media application skills, normal school students should have application skills, lessons talking about skills of modern education technology, evaluation skills adapted to new thoughts, etc. This will make normal school students able to informational teaching design. The major contents of informational teaching design include informational lessons plan, electronic work examples of students, evaluation tool (gauge), supporting materials and unit plan (Zhong, Z.X. 2006).

2.2 *Optimization of teaching process based on net curriculum*

Many platforms have been established in schools such as net teaching platform, high-quality curriculum construction platform, research teaching platform, major and curriculum construction platform, teaching source managing platform, and so on, in order to further improve construction of informational cultivation and teaching source, and the construction of informational teaching platform for promoting personal learning of students. Net teaching curriculum solved the problem that time of micro-teaching training is insufficient. The construction of microteaching network course website includes theoretical knowledge of teaching skills, excellent lesson plans, expert video case, representative data left by previous normal school students, and teaching reflection etc.

2.3 To create a microteaching environment based on web

Now many practical teaching modes need the network environment, and part of training courses were taken in computer network classrooms, such as Web Quest teaching, network for research teaching. Teaching skills training was realized by network tools (Li, B.H. 2009).

3 CONSTRUCTION OF NEW MODEL FOR MICRO-TEACHING AND EDUCATION TRAINING

In our country, conventional micro-teaching has five basic steps, i.e., theory learning, watching and demonstration, preparation of lesson plans, teaching practice and evaluation feedback. In recent years, lots of schools generally require talking about lessons and rehearsal lessons when hiring teachers, and the two aspects are consistent with micro-teaching both in form and content, so we can construct micro-teaching adaptable to talking about lessons and rehearsal lessons (Cheng, J.B. 2009). New model should be established by combining micro-teaching and teaching practice to solve the problem that real classroom teaching is not enough. The new mode of training is divided into two stages, the first stage is mainly individual learning teaching skills, the second stage is to complete a "whole lesson" teaching after the first stage. By the second phase of the study and practice, students can master teaching basic skills and will become qualified teaching personnel.

3.1 Learning-observing-practice-evaluation

First, normal college students study teaching skills under the guidance of teachers, and study the usual aspects of classroom teaching and teaching philosophy, including the guidance of teachers and the students' self learning. Students first need to master some basic theoretical knowledge, then, standard lesson plans of micro-teaching training are prepared for students as a reference to write plans. Study of teaching skills is the base of teaching skills training(Sun, X.M. 2009). Then, these student watch the expert teaching videos and attend a real class in teaching base. Here, observing and studying play a role of bridging the past and the future. They can feel the real class, become real understanding of teaching skills, and also improve the adaptability to real classes.

Single teaching skill training is performed on the basis of study and observation, which will make theory be transformed into practice and let students experience real teaching. Finally, teaching skills are assessed according to practice, including assessment by teachers and mutual assessment by their own classmates. This mutual assessment will help them to improve themselves by discussion, learning and even criticizing. Micro-equipment and campus net must be fully applied to improve the quality of teaching. Exercises and evaluation can be done in microteaching classroom, while study, observing, the following evaluation can be conducted through teaching skill training site, which will help to solve the problem of less micro-classrooms and limitation of time and space.

3.2 Preparing lessons-talking about lessons-evaluation-teaching- observation – rethinking

Normal school students firstly need to study textbooks to form whole teaching thoughts for class preparation training after they are proficient with single teaching skill, then they should elaborate the whole teaching design for class talking training according to class preparation plan, which is assessed by other students. After that, rehearsal teaching is conducted, followed by study and observation in middle schools. These students should also compare lessons of experts and those of themselves. In this model, rehearsal teaching is the core, while talking about lessons is the key stage. Class preparation, evaluation and consideration can be finished on net, while talking about lessons and teaching in micro-teaching classrooms, which will combine micro-teaching classrooms and webs to solve problems such as less micro-teaching classrooms, less instruction teachers and less training time.

By the stage of teaching practice and research, especially for the new curriculum reform, normal school students have learned more knowledge and accumulated practical knowledge, which will help them to be more confident to practice and improve understanding of education, and be able to go beyond the limitations of classroom practice to think about the issues and actions.

4 INFORMATIONAL MANAGEMENT OF TEACHING SKILLS TRAINING

4.1 Construction of training web blog

Education blog is a good network form of teaching and research. Dynamic information of training can be published and achievements of the members can be exhibited. Normal school students, instruction teachers and middle school teachers of training base can organize to form a community to communicate with each other. On this platform, you can expand more to exchange knowledge of teaching skills for specific teaching problems.

4.2 Construction of electronic study files of training

Electronic study files have two kinds of forms: one is the folder form, the other is network form. The files include research, lecture listening records in secondary schools, lesson plans, classroom instruction, extracurricular activities and other information based on individual student unit. The footprints of normal school students in the course of training can be truly recorded by the management of electronic portfolio, and the records will urge students to carry out self-evaluation, mutual evaluation, and reflection on learning methods. They also develop students' learning autonomy and self-confidence to promote the teaching skills. Teachers can grasp the teaching skills of normal school students, which can be used as the basis of the evaluation of the training process.

5 MODIFICATION OF EVALUATION METHOD OF TEACHING SKILLS TRAINING

5.1 Process evaluation based on electronic portfolio

Individual learning room for each collaborative group can be provided by using network collaborative learning platform, and evaluation space of the group's electronic file folders is established. Thus, every teacher and student may enter their individual space by their account password. Students can input learning thoughts, observation diary, record of investigation process, personal learning plan, etc. at any time. They can also read their own or other students' contents of the folders, and give the corresponding evaluation. The system can also timely announce the index scores of the group, evaluation review, expected targets and performance ranking, and so on (Guo, L. & Xu, X.D. 2003). We should not only emphasize single works of students in electronic files, but also stress the importance of entire learning process evaluation and reflection when we evaluate by electronic learning files.

5.2 Evaluation of teaching skills training on the basis of electronic rubric

The rubric is a structured quantitative evaluation criteria, which requires four evaluation criteria from 0 (lowest) to 4 (highest) based on multiple aspects of evaluation criteria. There two kinds of rubrics, one is on about results, and the other is process. Electronic rubric should be made by full use of information technology to achieve the automatic, timely, and scientific evaluation of teaching skills.

6 PRACTICAL SIGNIFICANCE OF INFORMATIONAL TEACHING SKILLS TRAINING OF NORMAL UNIVERSITIES

6.1 Teaching skills training and informationization of evaluation overcome the disadvantages of cultivation methods of teaching skills

Because of the problems such as single method for teaching skills training, less training time and lack of process evaluation of teaching skills, information

technology, information resources and curriculum are combined to construct effective teaching methods and evaluation methods, to promote teaching optimization and make use of advantages of teachers and teaching methods. It will also help to make up time and space limitations of microteaching, to overcome less training time and achieved good results in practice.

6.2 Compensate the drawbacks of inadequate cooperation with middle schools and integrate with the new curriculum reform

Due to the problems that teaching skills training is out of the actual situation of the middle schools, normal school students are given chances to enter middle school classrooms, to experience the real teaching, and to understand the new curriculum reform. This will help the students to exercise their teaching skills, to enhance classroom adaptability in the actual teaching process. The students should also modify their teaching design after the end of each teaching activities to further improve their teaching skills.

6.3 Promote the construction of school-based specialty courses to improve the overall quality of normal school students

Based on the situation that instruction teachers are not enough and modern teaching skills are not emphasized, information technology as tool, environment and resource is incorporated into detailed disciplinary teaching and each aspect of learning including teaching preparation, class teaching process and teaching evaluation, which ultimately promote effective learning of students. The characteristics of normal colleges are demonstrated, normal education is healthy developed, and all this will help to cultivate qualified teachers of the people.

ACKNOWLEDGEMENTS

Higher school teaching reform project of the Provincial Education Department of Jiangxi "Research and practice of model of higher school chemistry teaching skills training on the basis of information technology"(No: JXJG-11-13-22); the Twelfth-Five-Year-Plan project of Jiangxi Provincial Education Science (No: 11YB115).

REFERENCES

Cheng, J.B. 2009. The Improved Microteaching Pattern of Getting Used to the Training of the Tryout Teaching and Classroom Teaching Skills. *Journal of Modern Educational Technology*, Vol. 8: 112–115.
Guo, L. & Xu, X.D. 2003. Design of E-Portfolio learning evaluation based on inter-school cooperation of web. *Journal of China Educational Technology*, Vol. 12: 72–75.

Li, B.H. 2009. Construction of multiple cultivation modes of education and teaching skills of normal school students on the basis of information technology.*Journal of China Adult Education*, Vol. 23: 126–127.

Sun, X.M. 2009. Discussion on the Training of Teaching Technical Ability in the Micro Situation. *Journal of Education and Teaching Research,* Vol. 11: 12–14.

Zhang, W.H. et al. 2011. Research on Curriculum Integration of Information Technology and Teaching Skills of Chemistry and its Practice. *Journal of Higher Education of Sciences*, Vol. 06: 130–134.

Zhao, C.L. 2010. Practical Research on Training Mode of Teachers' Education Teaching Skills. *Yan Bian University.*

Zhong, Z.X. 2006. *Informational teaching mode.* Beijing: Beijing Normal University Press.

Information Technology and Computer Application Engineering – Liu, Sung & Yao (Eds)
© *2014 Taylor & Francis Group, London, ISBN 978-1-138-00079-7*

Effect of body stiffness on the swimming performance of a robotic fish

Z. Cui & H.Z. Jiang
School of Mechatronics Engineering, Harbin Institute of Technology, Harbin, China

ABSTRACT: Previous biological experiments show that the fish has a super swimming performance when the driven frequency is closed to the natural frequency. Inspired by that, a biomimetic fish based on continuous compliant visco-elastic body is presented, and the body dynamics deduced from the Bernoulli-Euller beam equation is also proposed. More important, we provide a proposition to determine the body stiffness and the driven frequency, namely, a robotic fish with different body stiffness $E_1 I_1$, $E_2 I_2$ obtain a better swimming performance at the corresponding driven frequency ω_1, ω_2, then there exist this relationship: $E_2 I_2 : E_1 I_1 = \omega_2^2 : \omega_1^2$. Further, it is proved by the experiments of robotic fishes with different body stiffness. The experiment results show that the swimming performance of robotic fish is largely dependent on the body stiffness and the driven frequency.

Keywords: robotic fish, stiffness, driven frequency

1 INTRODUCTION

Fish have attracted the interest of researchers due to their superior swimming ability compared to man-made devices. Traditionally, the body motions on fish robots have been implemented using complex mechanisms which employ several discrete, stiff components. As a result, several actuators are required, along with sophisticated controls[1, 2]. However, this design focuses on the spinal motion, ignoring the impacts of the muscles. Alternatively, Valdivia y Alvarado have proposed a compliant robotic fish made by the visco-elastic material [2, 3]. Although the resulting device is relative simpler and more robust, the research about the effects of body stiffness on the swimming performance has remained incomplete.

Besides, J.E. Colgate et.al [4, 5] revealed that the fish use their muscles to stiffen their bodies for improving the swimming performance. When the driven frequency was close to the natural frequency, the swimming fish had the minimum of negative work. Further, the natural frequency of a fish is determined by its elastic components, such as muscle, skin, muscle tendon, collagenous fiber and the skeletal system. So to reach a super swimming performance, the body stiffness of robotic fish should be adjusted to match with the driven frequency, or the driven frequency should be changed to match with the natural frequency of the robotic fish.

The understanding of these physical principles behind fish swimming has in turn motivated efforts to replicate such performance. In this paper we propose a biomimetic fish based on continuous compliant visco-elastic body, and expect to obtain the relationship between the body stiffness and the driven frequency by analysis of the body dynamics.

2 EXPLOITING NATURAL VIBRATIONS OF THE SLENDER BODY

Based on the mechanics of vibration and Lighthill's elongated body theory, this section briefly describes the approach used to build the mathematical model of the vibrations of slender body and proposes the relationship between the body stiffness and the driven frequency.

The Fig.1 displays the top view of a fish-like slender body that is exited by a concentrated harmonic moment of magnitde M and frequency ω located at a distance $x = a$ from the anterior end. The geometry and material properties are defined by cross-section area $A(x)$, second moment of inertia $I(x)$, modulus of elasticity $E(x)$, viscosity $\mu(x)$, and density $\rho(x)$, which are all function of x. The forces per unit length acting

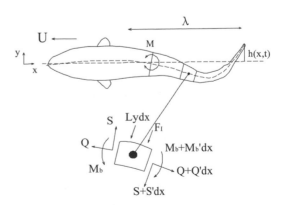

Figure 1. The analysis of the forces per unit length acting on a body element.

on a body element dx include: the longitudinal tension $Q(x, t)$, the shear force $S(x, t)$, a bending moment $M_b(x, t)$, the lateral force $L_y(x, t)$ exerted by the liquid medium and the inertia force F_I exerted by the body mass.

It is assumed the body is composed of an anisotropic material. The resulting lateral motions $h(x, t)$ of the body centerline are assumed to lie in the xy plane. For small centerline deflections, the plane cross-sections initially perpendicular to the body centerline remain plane and perpendicular to the neutral axis during bending. So the balance of forces in the y-direction yields:

$$-F_I = -\rho A dx \frac{\partial^2 h}{\partial t^2} = S - \left(S + \frac{\partial S}{\partial x} dx\right) + L_y(x)dx \quad (1)$$

And the moment equilibrium yields:

$$-\left(M_b + \frac{\partial M_b}{\partial x} dx\right) + M_b - L_y(x)dx\frac{dx}{2} + F_s dx = 0 \quad (2)$$

The total bending moment M_b is a combination of the actuation moment $M(x, t)$ and the moment due to the body resistance to bending. Energy storage and dissipation are presented through the visco-elasticity of the materials used. Therefore, assuming a uniaxial stress configuration, the simplest constitutive relationship for elastic and viscous stresses are respectively, as Eq.(3), Eq.(4) shown.

$$M_e = \int_A \sigma_e y dA = EI \frac{\partial^2 h}{\partial x^2} \quad (3)$$

$$M_v = \int_A \sigma_v y dA = \mu I \frac{\partial}{\partial t}\left(\frac{\partial^2 h}{\partial x^2}\right) \quad (4)$$

The total bending moment can be found by adding all the components,

$$M_b = M - M_e - M_v \quad (5)$$

According to Lighthill's elongated body theory and the biological observation, the lateral force $L_y(x, t)$ can be further simplified approximately [3], as Eq. (6) shown, m_a is the added mass.

$$L_y \sim 1.11 m_a \frac{\partial^2 h}{\partial t^2} \quad (6)$$

Combined all the Equation above, the body dynamics deduced from the Bernoulli-Euler beam equation can be expressed as Eq.(7) shown.

$$\frac{\partial^2}{\partial x^2}\left(M - EI \frac{\partial^2 h}{\partial x^2} + \mu I \frac{\partial}{\partial t}\left(\frac{\partial^2 h}{\partial x^2}\right)\right) = (m_a + \rho A)\frac{\partial^2 h}{\partial t^2} \quad (7)$$

where the first term in the left hand side is the input moment, the second term represents capacitive energy storage, and the third term respesents energy dissipation through viscosity. The terms in the right hand

side respresent the fluid resistance and the body inertia respectively.

Previous biological experiments show the stiffness of fish muscles plays a primary role in the swimming performance. So it is reasonable to ignore the effect of the viscosity when analysing the natural vibrations of the fish body, as Eq.(8) shown.

$$\frac{\partial^2}{\partial x^2}\left(-EI \frac{\partial^2 h}{\partial x^2}\right) = (m_a + \rho A)\frac{\partial^2 h}{\partial t^2} \quad (8)$$

The solution of the difference equation (8) has the form like Eq.(9) [6].

$$h(x, t) = \varphi(x)q(t) \quad (9)$$

Assumed that:

$$\frac{\ddot{q}(t)}{q(t)} = -\frac{EI\varphi(x)^{(4)}}{(\rho A + m_a)\varphi(x)} = -\omega^2 \quad (10)$$

So the Eq.(10) can be further writen as Eq. (11) shown.

$$\begin{cases} \ddot{q}(t) + \omega^2 q(t) = 0 \cdots\cdots(A) \\ \varphi(x)^{(4)} - \beta^4 \varphi(x) = 0 \cdots\cdots(B) \end{cases} \quad (11)$$

With

$$\beta^4 = \frac{(\rho A + m_a)}{EI}\omega^2 \quad (12)$$

In Eq. (11), the first one (A) is a form of free vibration, and the solution is determined by the initial condition. The second one (B) is sloved by the eulerian method. Based on the boundary conditions (free-free end), the resonant frequency equation is obtained by Eq. (13).

$$\cos(\beta l)ch(\beta l) - 1 = 0 \quad (13)$$

With

$$\beta_i l \approx (i + 0.5)\pi (i \geq 2) \quad (14)$$

The natural vibrations of the slender body satisfies the following condition, as Eq. (15) shown.

$$\frac{(i + 0.5)\pi}{l^4(\rho A + m_a)} = \frac{\omega_i^2}{EI} (i = 1, 2, 3...) \quad (15)$$

For the robotic fish, the length l, the density ρ, the added mass m_a are constant values, so the terms in the left hand side is just a constant, mainly determined by the lateral force and the inertia force. Therefore, Eq.(15) can be deduced the expression $EI \sim \omega^2$, so it can be described as: the stiffness EI has a direct proportion to the square of the natural frequency of the slender fish.

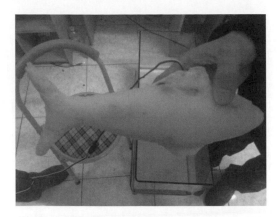

Figure 2.   The experimental model of the compliant robotic fish.

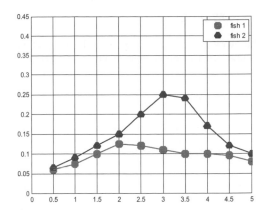

Figure 3.   Swimming velocity measurements versus swimming frequency.

Combined the biological evidences mentioned above, the proposition can be expressed as: when the robotic fish has a super swimming performance, the stiffness EI should be changed as the ratio of the square of the driven frequency ω^2. In other words, to match with the different driven frequency, the robotic fish should be adjusted its body stiffness EI, and the adjustment discipline is expressed in Eq.(16).

$$\frac{(EI)_1}{(EI)_2} = \frac{\omega_1^2}{\omega_2^2} \qquad (16)$$

where $\omega_1 =$ the driven frequency; $\omega_2 =$ another driven frequency; $(EI)_1 =$ the body stiffness of the robotic fish when the natural frequency is ω_1; $(EI)_2 =$ the body stiffness of the robotic fish when the natural frequency is ω_2;

This proposition is very useful, and with huge potential payoffs, to design the compliant robotic fish by changing the body stiffness to match with the different driven frequency so that to achieve a better swimming performance.

3   THE EXPERIMENTAL RESULTS

In the experiment of robotic fish, we focus on the swimming velocity with respect to the body stiffness. As Fig.2 shown, the design of biomimetic fish is result from the features for thunniform type swimmers, and the fish model consists compliant body, actuation mechanism and control hardware. The motions of the robotic fish are mainly relied on the material property. In this paper, we present two prototypes. The first prototype which we called fish 1 is made of a single very soft silicone material with hardness shore 10A ($E_1 = 37000\,N/m^2$). The second prototype, or fish 2, is made of silicone material with hardness shore 28A ($E_2 = 95000\,N/m^2$). Both prototypes have the same features except for the stiffness.

During steady rectilinear locomotion, a robotic fish swimming forward by pitching its tail. Changed the driven frequency in the interval of 0.5 Hz from 0.5 Hz

to 5 Hz, the robotic fish swims in different speeds. Fig.3 displays the swimming speed versus different driven frequency with the certain stiffness. The robotic fish 1 and fish 2 respectively obtain the maximum forward speeds at the driven frequency 2 Hz and 3 Hz. As expected, when the driven frequency in tune with the natural frequency, the forward speed is obvious higher than other driven frequencies. The maximum forward speed of fish 2 is about 0.25 BL/s. Due to the same shape of both fish1 and fish 2, the ratio of body stiffness (EI) is determined by the elasticity modulus, so it equals 0.421(40000/95000). For another, the ratio of the driven frequency is approximately equals to 0.667 (2/3), or equals to the square root of the body stiffness ratio 0.649. Based on the experiment results, the proposition to determine the body stiffness and the driven frequency is proved. Meanwhile, the pattern of speed curve in Fig.3 is concordant with the experiment done in MIT ([2] [3] [7]).

4   SUMMARY

The work in this paper was mainly focused on the swimming performance of robotic fish in terms of the body stiffness and the dirven frequency. Based on the analysis of a slender body dynamics, the proposition can be described as: when the robotic fish has a super swimming performance, the body stiffness EI should be changed as the ratio of the square of the driven frequency ω^2. In particular, for a better swimming performance, the robotic fish made by the visco-elastic material is required to match the driven frequency with the natural frequency by choosing different metrials or adjusting the driven frequency. Further, the experiments of robotic fish with two different stiffness proved it, showing the swimming performance is largely dependent on the body stiffness and the driven frequency. It has a great guiding significance to design a robotic fish to achieve a better swimming performance by matching the body stiffness with the driven frequency.

ACKNOWLEDGEMENTS

This work is financially supported by the National Natural Science Foundation of China (No.51275127).

REFERENCES

[1] J.E. Colgate, K.M. Lynch, Mechanics and control of swimming: a review, IEEE Journal of Oceanic Engineering, 29 (3) (2004) 660–673

[2] B.P. Alvarado, P.V.Y. Youcef-Toumi, K. Techet, Swimming performance of a biomimetic compliant fish-like robot. Experiments in Fluids 47 (2009) 927–939.

[3] P.V. Alvarado, Design of biomimetic compliant devices for locomotion in liquid environments. PhD thesis, Massachusetts Institute of Technology, 2007.

[4] J.H. Long, JR., Muscles, Elastic Energy, and the Dynamics of Body Stiffness in Swimming Eels, American Zoologist.38 (1998) 771–792.

[5] J.H. Long, JR. and K.S. Nipper, The importance of body stiffness in undulatory propulsion, American Zoologist. 36(1996) 678–694.

[6] Timoshenko, S., Young, D. H., and Weaver, W. Jr., 1974, Vibration Problems in Engineering, Wiley, New York.

[7] Valdivia y Alvarado and Youcef-Toumi, Performance of Machines with Flexible Bodies Designed for Biomimetic Locomotion in Liquid Environments, Proc. IEEE. International Conference on Robotics and Automation, Barcelona, Spain, (2005) April 18–22.

Information Technology and Computer Application Engineering – Liu, Sung & Yao (Eds)
© 2014 Taylor & Francis Group, London, ISBN 978-1-138-00079-7

The supply and demand equilibrium of highly specific human capital

Xue Lin Wang

Department of Studies, Civil Aviation Flight University of China, Guanghan, Sichuan, China

ABSTRACT: This paper presents a concept of highly specific human capital which has the nature of bilateral monopoly. We must spend large sums of money and long time to get highly specific human capital and we must accumulate its value in work experience. The supply and demand of highly specific human capital is almost inelastic and its price is determined by the equilibrium of given enterprises. So the highly specific human capital has special properties on fluidity, risk control and macro-regulation of supply and demand. At last this paper analyzes some problems which are a kind of special highly specific human capital.

Keywords: Highly specific human capital, Market equilibrium, Enterprise equilibrium

1 INTRODUCTION

Human capital be seen as a special factor of production since Adam Smith has aroused the attention of researchers. But it was not until the middle of last century since the concept was formally proposed by Theodore Schultz, he believes that Human capital is the output of knowledge, techniques embodied upon workers as well as the capacities presented. Therefore, the human capital of workers can be acquired by learning and there may be a big difference. These different elements of human capital are bound to have different characteristics, and showing the unique laws in the market transactions. This article will preliminary explore a special kind of human capital——the basic characteristics and the law of the supply and demand equilibrium of highly specific human capital.

2 HIGHLY SPECIFIC HUMAN CAPITAL

Williamson recognize firstly the specificity of the assets, which refers to the asset extent which can be redeployed for other purposes, or the extent which used by others without loss of the value of production[1].Yang Rui Long further explained it as a persistent investment. [2] Luo Pin Liang studies assets specificity combined with specific human capital [3]. However, all the previous studies emphasized the unilateral monopoly from client to agent only. We will further define highly specificity as bilateral monopoly both client and agent. It is to say, not only there will be of no value if clients are away from agents, but also the process of production cannot be finished if agents are away from clients. So, highly specific human capital refers to bilateral monopoly of the human capital and physical capital.

2.1 *The value of highly specific human capital*

Schultz summarized the contents of the human capital investment into five areas[4]. The above five aspects commonly condensed the value of human capital. The value of the formation of highly specific human capital also depends on the above aspects, and more, it has three special characteristics. The first is that highly specific human capital requires relatively large investments. The second is that the time to form highly specific human capital is longer than the other human capital. The third is that highly specific human capital is depended on heavily the process of production.

2.2 *The property rights of highly specific human capital*

The property rights of human capital are a bundle of rights as well as other assets. These rights can be decomposed and human capital is traded in the market through the decomposition of property rights. Normally, human capital forms complex employment relationship by separating from each other of four capabilities. However, the capability of the ownership of the highly specific human capital may be further divided because of large investment. When the people can not afford the large cost of education and training of highly specific human capital, other units or organizations will share it. The result is that the investors who bear the value of formation of human capital and the workers share the ownership of human capital. Such ownership further impacts the

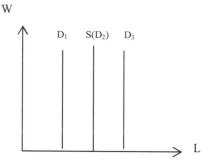

Figure 1.

other bundles of ownership, which makes highly specific human capital transactions more complex than general human capital.

3 EQUILIBRIUM OF HIGHLY SPECIFIC HUMAN CAPITAL: BASIC MODEL

The bilateral monopoly structure of highly specific human capital brings the special equilibrium structure in the short-term. This structure is perfectly inelastic in supply and demand (fig.1). In the factor market, the demand for highly specific human capital depends on the buyer's product scale which is scheduled in short-term.. Thus, the demand curve D is a vertical line which parallel to the longitudinal axis. The specific performance as D_1 or D_2 or D_3 depends on different situations. In the other aspect, the supply cannot be replaced. Therefore, the supply only expressed as the stock of given time. If labor intensity is constant, the supply curve is a vertical line parallel to the longitudinal axis.

The curves of supply and demand of the highly specific human capital will almost not intersect in short-term. Then, the balance amount depends on the short side of supply and demand. There are three cases $D > S$ or $S > D$ or $D = S$, and supply and demand situation are precisely equal in the third case only. And the probability of this case is very low and extremely unstable. Father more, the equilibrium price cannot be determined in this market. When the curves of supply and demand do not intersect, the idea of pricing based on the balance of supply and demand of market will fail, and then the price must be determined separately.

4 EQUILIBRIUM OF HIGHLY SPECIFIC HUMAN CAPITAL: COMPETITION MODEL

In different amount of supply and demand have different equilibrium outcome. For the clients, the sole criterion for signing is MRP = MFC, that means the marginal product value equal to its marginal factor cost [5]. The supply curve for individual clients is determined by MFC and upward-sloping. MFC is independent of each client in the same market, so

different clients making employment decisions will be face the same MFC and change MRP. MRP is downward-sloping, that means a reduce benefit from each labor capital. It further decided that each employment owe the substances status, the production technology level and management skill levels. General, the more advanced of material, the higher the level of production technology and management skills, MRP is greater.

4.1 When the supply is equal to the demand in the market (D = S)

A firm will hire L_i which is the equilibrium quantity, according the profit-maximizing law (fig.2). (1or i or x stands for one of the agents, n is the total agents, the same below).Obviously, for the two different clients i_1 and i_x, the value of employment is not equal ($VMP_i \neq VMP_x$), so the equilibrium price is also not the same ($W_i \neq W_x$). Because $\sum_{i=1}^{n} L_i = S = D$ in this market, different wages will inevitably bring labor and capital to flow, until the entire client's needs meet the supply of agents.

4.2 When the demand is greater than the supply (D > S)

Every client is still decides hire quantity Li according to the profit-maximizing law (fig.3). At this time, $\sum_{i=1}^{n} L_i = D > S$. Under the guidance of different price W, the labor will flow which lead to result that the high-W client's needs satisfied and the low-W client's needs can not be achieved eventually. How much the low-W client's needs can not be achieved depends on the gap between D and S. Of course, the premise of providing high W is high MRP, but a low W client may not be low MRP. Eventually the companies with the lowest MRP can not produce due to lack of labors, the market survival laws spontaneously role. In fig.3, the minimum MRP company obtains, the gap $\triangle L_i$ between Li and the theoretically optimal labor Lin' ($\triangle L_i = L_i' - L_i$).

4.3 When the demand is less than the supply (D < S)

All clients' need will be met entirely, even the condition the minimum MRP principal's who provide the lowest W (fig.4). Because of $\sum_{i=1}^{n} L_i = D < S$, some people can not be hired. This time each client face a perfectly competitive labor markets, they can hire satisfactory labor by only giving the minimum wage. Perhaps, the lowest wage may Less than the minimum production efficiency clients' W in the other market. It is because that highly specific human capital can not find other similar alternative jobs, so they can only be engaged in the simplest work which just depends on physical strength. The outputs of the simple work are very low which causes a great waste in the investment of highly specific human capital. From the perspective of highly investment, it is huge risk to invest highly specific human capital when D < S.

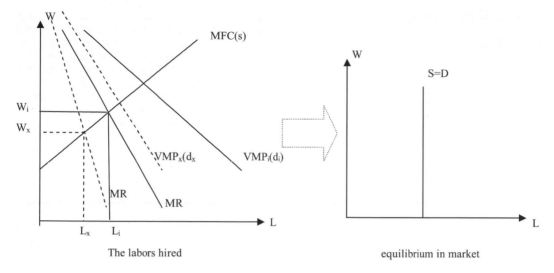

The labors hired

equilibrium in market

Figure 2. D = S.

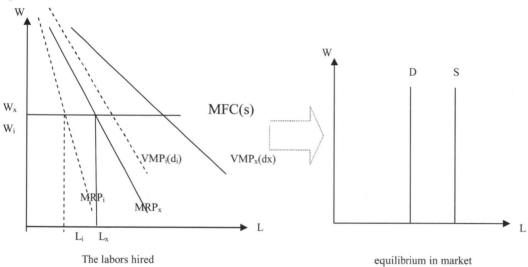

The labors hired

equilibrium in market

Figure 3. D < S.

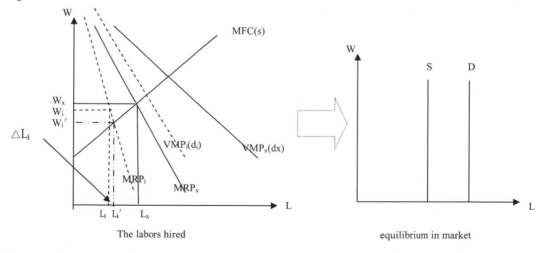

The labors hired

equilibrium in market

Figure 4. D > S.

5 SOME CONCLUSIONS

5.1 It is important for highly specific human capital to flow

If it flows freely, when $D = S$, supply meets demand; When $D > S$, the most inefficient client will exit. However, if it can not flow freely, the results will change. Whatever $D = S$ or $D > S$, the nature of the highly specific human capital may offer a chance for clients to control the supply. When the client is inefficient, the phenomenon of "bad money drives out good money" appears.

5.2 The behaviors who invest the highly specialized human capital face high risk

When $S > D$, highly specific human capital which can flow freely will be surplus. In the absence of corresponding alternative works, excess human capital can only get the opportunity of simple labor. Statically, in this period, the surplus human capital lost corresponding income which match with investment. Dynamically, due to the dependent on productive process, the loss temporarily may mean the loss forever, which will lead to the annihilation of all investments. Therefore, the investor must be ability to invest in, to bear risk and to predict market. Obviously, most individual is difficult to meet the three conditions at the same time. The clients will be better to be considered because of its finance and information advantage. No matter who is the investor, the risk is here objectively, which lead to the underinvestment in highly specialized human capital. When the investors and the human capital carrier (worker) separates, some more complex problems will be inevitably appeared, such as the human capital property right, the residual distribution and the mobility of highly specialized human capital.

5.3 It is importance for the highly specific human capital to macro-control supply and demand

The blade balanced which highly specific human capital supply and demand is precisely equal is an accidental situation. The imbalance between supply and demand is likely to have adverse consequences. If the demand is determined by the enterprises according to market demand, inadequate investment will lead to lack, which bring out the bottleneck of industrial development. And the over-investment damages the benefits of investors. Therefore, to macro grasp the supply and demand is very important. When the demand is a planned expansion, we can guarantee to supply complete from the macro level. Thus, we need a third-party which separate with highly specific human capital supply and demand. The agency needs to grasp the information of demand, provide the platform for the transfer and act as a reservoir. From the task which needs to be completed by the third-party, a pure market organization can not be qualified. The job to collect information and pre-invest highly specific human capital need the supporting of authority and financial, which may not be done directly by the government.

REFERENCES

[1] Williamson, O. E., The Economic Institu-tions of Capitalism. Free Press, New York, 1985.
[2] Yang Ruilong ,Yang Qijing, Specificity, Exclusiveness and Enterprise Institution, Economic Research Journal, 2001(3), pp.3–11.
[3] Luo Pin liang,Si Chun Lin, Studies on incentives for investment in specific human capital, Journal of Manegement Sciences In China, 2001(2), pp. 19–24.
[4] Theodore W. Schultz, Investment in Human Capital: The Role of Education and of Research, New York: Free Press, 1971, p.31.
[5] Huang Yajun, Microeconomics, Bei Jing :Higher Education Press, pp.177–195.

Information Technology and Computer Application Engineering – Liu, Sung & Yao (Eds)
© *2014 Taylor & Francis Group, London, ISBN 978-1-138-00079-7*

Analysis to the influence factors of consumption behaviors of cruise tourism in China

Guo Dong Yan, Hong Wang & Ling Qiu
School of Management, Shanghai University of Engineering Science, China

Jian Cheng Kang
Urban Ecology and Environment Research Center, Shanghai Normal University, China

ABSTRACT: Cruise tourism market presents fast development in China which due to the influence of particularly factors, aim to provide reference for the government, based on relevant study outcomes to discuss the characteristics as follows: The consumption willing of Chinese cruise tourists increase obviously but mainly on low price. The outbound travel of Chinese citizens is mainly short distant, while middle and long distant destinations increase fast. The route demand of domestic cruise increases obviously and the domestic tourists' first choice is the Mediterranean Sea and the Far East areas. But domestic tourists knew about cruise tourism not long ago and the understanding to cruise tourism needs to be further improved.

Keywords: Cruise Tourism; Consumption Behaviors; Influence Factors

1 GENERAL INSTRUCTIONS

1.1 *The cruise market development in Asian region has great potential*

The cruise market development in Asian region has great potential. In past 8 years, the cruise tourist in Asian region has been tripled. In 2010, Singapore, Thailand, Malaysia and Indonesia have benefited from ASEAN (the Association of Southeast Asian Nations) cruise economy remarkably. Asia and Pacific region has become an important engine to the increase of global cruise market in the future. During January to April 2010, Asian tourism market has increased by 12% simultaneously, which is 5% higher than global average. It's estimated that until 2020, the population of cruise tourist globally will reach 40 million. Hence, Asian cruise market will be further enlarged (Cruise Lines International Association, 2009~2010 China Cruise Industry Development Report).

1.2 *The cruise industry in China was regarded as the major drive*

The cruise industry in China began from 1976. Although started late, it presented good development tendency and was regarded as the major drive for the recovery of Asian tourism market. In 2001, there were 8,000 outbound Chinese cruise tourists and there were 790,000 outbound cruise tourists from Mainland China in 2010 (excluding Hong Kong, Macao and Taiwan), an increase of 20.1% than that in 2009.

There were 462,000 foreign cruise tourists to China, an increase of 15.5% (China Communications and Transportation Association – Cruise & Yacht Industry Association, 2010). In 2010, Mainland China has received 223 international cruises, an increase of 42.9%. Cruise tourists to China increase distinctly. It's estimated that by 2020, China may become the largest tourism destination in the world and attract more international cruises mooring (China Communications and Transportation Association – Cruise & Yacht Industry Association, 2010-2011 China Cruise Industry Development Report; Cheng Juehao, 2004).

1.3 *Shanghai will become the terminal cruise port in East Asian region*

According to the Development Planning of Cruise Industry in the 12th Five-Year Plan of Shanghai, by the end of the "12th Five-Year Plan", Shanghai will become the terminal cruise port in East Asian region and also one of the three largest international cruise centers in the Asia and Pacific region after Singapore and Hong Kong. Now world top three cruise groups have established branches in Shanghai and opened regional cruise tourism routes regarding Shanghai as the home port, which has accelerated the development of cruise economy in Shanghai. In the respect of the number of cruises received, there were 36 international cruises visited Shanghai in 2002 and 80 visiting cruises and cruises of home port totally in 2009, an increase of 33.3% than that in 2008 (China Ports and

Harbors Association – China Cruise Terminal Association, 2010). In 2010, since Shanghai Expo was held successfully, there were 177 international cruises received only by Shanghai Port International Passenger Transport Center and around 200 cruises received in Shanghai in 2011. It's estimated that there would be 100 cruises received in the home port in 2012, 500 outbound cruises in Shanghai by the end of the "12th Five-Year Plan" and 5~8 cruises regarding Shanghai as the home port base, as for the population of tourists, the population of the outbound cruise tourist from Shanghai increased from 100,000 in 2007 to 340,000 in 2010. Based on the information about the routes of cruise corporations confirmed, the population of cruise tourist would have reached 300,000 in 2012. By the end of the "12th Five-Year Plan", cruise tourists will reach 1~1.2 million by estimation. The establishment of Disneyland will further accelerate the increase of cruise and tourist of Shanghai.

Cruise tourism market in China presents fast development, which has promoted the gradual perfection of related studies. However, what are the factors influencing cruise tourism consumption, how to effectively promote the development of cruise market and how to gradually meet increasing demand of cruise tourism have become significant topics which need to be solved urgently. Hence, based on relevant study outcomes, the Paper tries to discuss the major factors influencing the consumption behaviors of cruise tourism and learns the characteristics of the consumption behaviors of cruise tourism, so as to promote the sustainable development of cruise tourism and provide basis for the government to make administrative policies.

2 MAJOR FACTORS INFLUENCING CONSUMPTION BEHAVIORS OF CRUISE TOURISM

2.1 Consumption level

As high-end tourism product the average consumption level of cruise tourists is about 50% higher than that of non-cruise tourists. A region possesses the basic condition of developing cruise tourism economy when the GDP per capita is above USD 6,000. The top four countries of global consumption are China, Russia, Japan and the USA. The consumption level of China is high, which has driven the increase of the consumption of cruise tourism in some certain degree. In 2010, China has already become the third largest tourism destination and the fourth largest consumption country of outbound tourism worldwide. In 2011, the revenue from domestic tourism of China is around RMB 130 million Yuan, an increase of 12%; the foreign exchange revenue from foreign tourists is USD 50 billion, an increase of 8%; the expenditure on outbound tourism is USD 5.5 billion, an increase of 14%. In such situation, the consumption demand of cruise tourism becomes more and more distinct. It's

worthy noticing that, above 50% non-cruise tourists in China are willing to accept cruise product under RMB 5,000 Yuan (Zheng Hui, 2009; Tang Zhaoyu, 2010; Wu Xianfu, Chen Pan, 2011). In 2009, the consumption per capita of the transit cruise tourists in Hong Kong was HKD 1,670 (Cruise Lines International Association, 2009~2010 China Cruise Industry Development Report). The cruise tourism consumption in China still features low price. How to enhance the consumption on high-end cruise tourism products in China is the key of improving the overall level.

2.2 Consumption willing

Consumption willing directly influences the decision-making of cruise tourism consumption. Chinese citizens of outbound tourism often go to short-distant destinations, mainly North America, Africa, Hong Kong, Macao, Taiwan, Japan, Korea and ASEAN countries. In last few years, middle and long-distant tourism increased fast, which has driven the increase of the consumption willing of cruise tourism. In 2007, the percentage of Chinese tourists who have taken cruise tourism is 10.40% (Zheng Hui, 2009). In 2010, the study aiming at the tourists in 3 regions of Guangxi shows that, the percentage is 21.25% (Wu Xianfu, Chen Pan, 2011). The consumption willing of cruise tourism increased distinctly. It's notable that, there is certain difference on the demand preference between cruise tourists and non-cruise tourists. In 2010, the study aiming at the tourists in 3 regions of Guangxi shows that, the top three attracting factors of cruise tourism are leisure, romance and convenience (Wu Xianfu, Chen Pan, 2011). In the same year, the study on the tourists of Costa Romantica shows that, the top three are plenty activities, onshore sightseeing and generous catering. Hence, non-cruise tourists prefer cozy and romantic circumstances, while cruise tourists focus on cruise experience. The demand difference between the two categories of tourists is obvious. How to bridge the gap of consumption willing between the two categories of tourists is an important precondition of popularizing cruise tourism consumption.

2.3 Leisure time

Leisure time is a necessity to realize the consumption behavior of cruise tourism. The cruise tourism products at present are mainly global travel of around 100 days, 60-day ocean travel, 6~7-dayregional travel and 2~3-day nearshore travel. During 2006-2012, most of Chinese cruise routes last for 5~7 days and there are less middle and long-term routes above 7 days. Among them, Chinese tourists accepted 6~8-day cruise tourism products were the most in 2007, accounted for 38.52% (Zheng Hui, 2009). In 2011, there were 68 home port routes in China (including Hong Kong), which were all under 7 days and mostly 3~5 days. It's extremely distinct that the home port routes of Chinese cruise show short-distance tendency. There is also certain difference on time demand between the tourists in different regions. In 2010, tourists from Guangzhou accepted shortest cruise tourism products, 3~4 days, and are willing to travel in July and August. Among tourists from Shanghai, 71.84% accept cruise tourism of above 5 days and are willing to travel in May. Hence, it's the highlight for cruise companies and travel agencies to undertake route design and product promotion in their next step aiming at the above difference on leisure time.

2.4 Route demand

The first choice of tourism destinations of Chinese tourists are the Mediterranean Sea and the Far East areas. The Caribbean Sea and North Europe rank the second. The percentage of accepting the route to the Mediterranean Sea is 40.93% (Zheng Hui, 2009), since France, Spain, Italy and Greek islands along the route can attract a large number of tourists who are willing for honeymoon tourism. During 2009–2011, the international cruise received by Mainland China increased from 156 to 262. Among them, the cruises started from our coastal areas increase from 80 to 142 and cruises visited our coastal cities increased from 76 to 120 (CCYIA, 2011). Cruise route demand increase obviously. Meanwhile, the demand of Chinese tourists to international route exceeds the demand of foreign tourists to China. There is difference on international routes between tourists from different regions, which is presented that tourists from Guangzhou prefer the Middle East route, those from Beijing prefer the Mediterranean Sea route and those from Shanghai prefer China-Japan-Korea and Southeast Asia routes. Among them, 66.02% of Shanghai tourists accept China, Japan, Korea and Southeast Asia routes (Tang Zhaoyu, 2010). This is because Shanghai is relatively nearer to Japan, Korea and Southeast Asia. Moreover, 5-day Japan and Korea routes have promoted relevant routes. It's notable that, among the tourists on Costa Romantica started from Shanghai in 2010, 44.66% prefer routes to Japan and Korea and almost 50% of them prefer 6–8 day tourism. Therefore, routes to Japan and Korea shall be prolonged to meet the demand of tourists.

2.5 Information demand characteristics

Cruise tourism developed late in China and tourists knew about cruise tourism not long ago. In 2007, only 39.04% of the tourists heard about cruise tourism 3 years ago, 14.27% heard about it for the first time, so the understanding of tourists to cruise tourism needs to be further improved. Furthermore, in the respect of information acquisition channel, in 2007, the top three channels that domestic tourists obtain information about cruise tourism are relatives and friends' introduction, TV & broadcast, newspaper, magazine & book. Internet and travel agencies shall strengthen their publication of cruise tourism (Zheng Hui, 2009). It's worthy noticing that, there is distinct difference on the percentage of information obtaining channels

between cruise tourists and non-cruise tourists. Take Shanghai as an example, in 2010, the tourists on Costa obtained information about cruises mainly through internet, travel agencies and friends and relatives. While non-cruise tourists are mainly through internet, TV & broadcast, friends and relatives. Comparing the two, the gap of percentage on travel agencies' introduction is the most distinct: that of cruise tourists is 40.07% more than that of non-cruise tourists. The gap on friends and relatives' introduction ranks the second: that of cruise tourists is 19.42% more than that of non-cruise tourists. Gap on TV & broadcast ranks the third: that of cruise tourists is 14.62% less than that of non-cruise tourists. This shows that, cruise tourists prefer to obtain information through internet and travel agencies and non-cruise tourists prefer to obtain information through internet and TV & broadcast.

3 CONCLUSION

3.1 The consumption willing of Chinese cruise tourists increase obviously

In terms of consumption willing, the consumption willing of Chinese cruise tourists increase obviously, which provides the cruise tourism market great potential. But the consumption of domestic cruise tourism is mainly on low price. Therefore, market promotion shall be accelerated to attract more consumers to accept high-end cruise tourism products, improve the consumption level of cruise tourism, bridge the gap on consumption willing between cruise tourists and non-cruise tourists and promote the popularization of cruise tourism consumption.

3.2 The outbound travel of Chinese citizens is mainly short distant

In terms of leisure time, the outbound travel of Chinese citizens is mainly short distant, while middle and long distant destinations increase fast. The short-distance tendency of home port cruise route is extremely distinct. Tourists from Guangzhou need cruise tourism of shorter period than those from Shanghai, so route design shall be done aiming at the demand features of tourists.

3.3 The route demand of domestic cruise increases obviously

In terms of route demand, the route demand of domestic cruise increases obviously. The demand of tourists to international routes exceeds that of foreign tourists visiting our country. Moreover, domestic tourists' first choice is the Mediterranean Sea and the Far East areas, followed by the Caribbean Sea and North Europe. Tourists from Guangzhou prefer to the Middle East, those from Beijing prefer to the Mediterranean Sea and those from Shanghai prefer the routes of China-Japan-Korea and Southeast Asia. Route design shall be done as per regional route demand and preference, e.g., to prolong the routes to Japan and Korea.

3.4 *Domestic tourists knew about cruise tourism not long ago*

In terms of information demand features, domestic tourists knew about cruise tourism not long ago and the understanding to cruise tourism needs to be further improved. Domestic cruise tourists focus more on cruise experience and non-cruise tourists prefer cozy and romantic circumstances. Publication through internet and travel agencies shall be strengthened aiming at cruise tourists and that through internet and TV & broadcast shall be strengthened aiming at non-cruise tourists.

ACKNOWLEDGEMENTS

This work was financially supported by the Connotation Construction Project of the Twelfth Five-Year Guideline for the Shanghai local undergraduate universities: Construction of Research Bases for Intelligent Management Engineering of Modern traffic and Key Discipline Construction of Decision Support System for modern public transportation and the Public Decision Consulting research(0852011XKZY15), Funding scheme for training young teachers in Colleges and universities in Shanghai: Characteristics of cruise tourism demand and marketing strategy in Shanghai(ZZGJD12032).We are grateful for their valuable comments.

REFERENCES

[1] Cheng Juehao & Gao Xin in: on the Development of Global Cruise Tourism Market. World Shipping (2004)
[2] China Communications and Transportation Association. Cruise & Yacht Industry Association, 2010–2011 China Cruise Industry Development Report, Shanghai (2012)
[3] CLIA, 2009 Cruise Market Profile Study[R]. Cruise Lines International Association, New York
[4] Cruise Lines International Association, 2009~2010 China Cruise Industry Development Report
[5] Tang Zhaoyu Study of Development of Shanghai Cruise Tourism professional master's dissertation of East china normal university (2010)
[6] Wu Xianfu & Chen Pan in: Research of Consumption Prospects and Development Strategy of Domestic Cruise Tourism. Journal of Anhui Agricultural Sciences (2011)
[7] Zheng Hui. Research on the Development Strategy of Cruise Tourism Products Based on the Demand of Domestic tourist Master Degree Thesis of China Ocean University (2009)

Information Technology and Computer Application Engineering – Liu, Sung & Yao (Eds)
© 2014 Taylor & Francis Group, London, ISBN 978-1-138-00079-7

Research and implementation of terrestrial test for underwater acoustic detection device under multi-target environment

Xi Chen
Department of Weaponry Engineering, Naval University of Engineering, Wuhan, China

Li Rui
PLA, Qingdao, China

ABSTRACT: Along with the continuous improvement of multi-target detection ability of acoustic detection device with high speed motion platform, it is particularly important to the research of terrestrial test for underwater acoustic detection device under multi-target environment. In this paper the multi-target echo model is presented through the exploitation of the various target models of echoes received by the underwater acoustic detection device with high speed motion platform and the numerical simulations are also presented. Furthermore, a terrestrial test system for underwater acoustic detection device under multi-target environment is designed. The experiment of the terrestrial test system for underwater acoustic detection device shows that it can simulate the multi-target echoes on sea effectively.

Keywords: underwater acoustic detection device; multi-target; echo; terrestrial test

1 INTRODUCTION

The terrestrial test of underwater acoustic detection device with high speed motion platform means to pull the whole under acoustic detection device or part of it in the simulation circuit under laboratory conditions and process the analog simulation test on complex underwater environment, various artificial jamming targets, various point targets, various volume targets, different target detection function and target identification function. The object of this kind of terrestrial test is to build a terrestrially semi-object simulation testing system for underwater acoustic detection device, which can simulate the testing platform environmentat sea of the acoustic detection device with high-speed platform and finally achieves the qualitative and quantitative simulating test of the underwater acoustic detection device[1].

2 MULTI-TARGET ECHO MODEL

Shown as in Figure 1, suppose that an underwater acoustic detection device work at active mode in a sea area with certain sea-condition. The coverage of this underwater acoustic detection device is R = 1000 m and the flare angle of self-navigation operating sector is $\theta = 70°$. Under the assumption that SNR is constant, an echo includes 4 targets information, i.e. x = 4. Target A and target B are at far field, which both are point

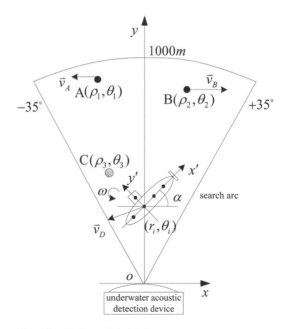

Figure 1. Multi-target Model.

target. Target D is at near field, which is volume target. Target C is wideband noise jamming released by target D, which is jamming target.

The multi-target model is shown in Figure 1. xoy is the rectangular coordinate system, whose origin is the acoustics centre of the underwater acoustic detection device and y-coordinate is the axle wire of the underwater acoustic detection device. $x'o'y'$ is volume target coordinate system, whose origin is the target centre. The angle between $x'o'y'$ and xoy is α. The coordinate of each point target and the centre coordinate of volume target are given. $\vec{v}_A$, $\vec{v}_B$ are the linear velocity of point targetswith respect to the underwater acoustic detection device, whose vector directionsare shown as in Figure 1. $\vec{v}_D$ is the linear velocity of the volume target centre with respect to the underwater acoustic detection device, whose vector direction is also shown as in Figure 1. The volume target includes 5 scatters which rotate around the centre of the volume target and the angular velocity is ω. The coordinates of 5 scatters are as follows:

(-20, 0);   (0, 20);   (0, 0);
(20, 0);   (40, 0);

Then the coordinate of centre o' at xoy can be written as:

$$(o_x, o_y) = (\rho_D \cos\theta_D, \rho_D \sin\theta_D)$$

Also the coordinates of 5 scatters on volume target at xoy are:

$$\begin{pmatrix} x_i \\ y_i \end{pmatrix} = \begin{pmatrix} \cos\alpha & -\sin\alpha \\ \sin\alpha & \cos\alpha \end{pmatrix} \begin{pmatrix} x_i' \\ y_i' \end{pmatrix} + \begin{pmatrix} o_x \\ o_y \end{pmatrix}$$

The distance between scatters and origin o is $\rho_i = \sqrt{x_i^2 + y_i^2}$ and the velocity of each scatter with respect to xoy is $\vec{v}_i = \vec{v}_D + \vec{r}_i \times \vec{\omega}$, where $r_i = \sqrt{x_i'^2 + y_i'^2}$, which is the vector radius between each scatter and the geometric centre of the volume target.

Project $\vec{v}_i$ to the direction at which each scatter connects to origin, then the radial velocity v_i of each scatter with respect to origin can be obtained. If the range and radial velocity of each target with respect to underwater acoustic detection device and the range and radial velocity of each scatter on volume target with respect to the origin of xoy coordinate system are all known, the echo signal model can be established according to their relative position.

Let transmitting signal $f(t)$ is pure-tone pulse CW signal with rectangle envelope and the processing method of other signal form is the same. The center frequency of $f(t)$ is f_0, bandwidth is BW and the pulse width is T, i.e.

$$f(t) = \cos(2\pi f_0 t) \qquad t \in [0, T]$$

$\omega_0 = 2\pi f_0$ is the center angular frequency, i.e. carrier frequency. The Doppler frequency shifts of target A and target B are $\omega_{dA} = -\frac{2v_A}{c}\omega_0$ and $\omega_{dB} = -\frac{2v_B}{c}\omega_0$ respectively. The Doppler frequency shifts of the scatters on volume target are:

$$\omega_{di} = -\frac{2v_i}{c}\omega_0$$

Then, the narrow echo signals of scatters on target A, target B and target D are:

$$s_A(t) = \tilde{a}_A f(t - \tau_A) e^{jw_{dA}(t-\tau_A)} \qquad t \in [0, T]$$
$$s_B(t) = \tilde{a}_B f(t - \tau_B) e^{jw_{dB}(t-\tau_B)} \qquad t \in [0, T]$$
$$s_i(t) = \tilde{a}_i f(t - \tau_i) e^{jw_{di}(t-\tau_i)} \qquad t \in [0, T]$$

Where $\tilde{a}_A$ and $\tilde{a}_B$ are the attenuation factors of target A and target B respectively. $\tilde{a}_i$ is the attenuation factors of each scatter on target D, which is not only related to the propagation loss, but also is related to random amplitude fading.

The time delay of target A and target B are $\tau_A = \frac{2\rho_A}{c}$ and $\tau_B = \frac{2\rho_B}{c}$ respectively, the time delay of each scatter on target is $\tau_i = \frac{2\rho_i}{c}$.

After obtaining the echo signal $s_i(t)$ of each scatter on target D, align the data of $s_i(t)$ by computing the different time delay and then the echo signal of target D after superposition can be obtained, i.e.

$$s_D(t) = \sum_{i=1}^{5} s_i(t) = \sum_{i=1}^{5} \tilde{a}_i f(t - \tau_i) e^{jw_{di}(t-\tau_i)}$$

Finally, the multi-target echo signal can be written as:

$$\begin{aligned} s(t) &= s_A(t) + s_B(t) + s_D(t) \\ &= \tilde{a}_A f(t - \tau_A) e^{jw_{dA}(t-\tau_A)} \\ &\quad + \tilde{a}_B f(t - \tau_B) e^{jw_{dB}(t-\tau_B)} \\ &\quad + \sum_{i=1}^{5} \tilde{a}_i f(t - \tau_i) e^{jw_{di}(t-\tau_i)} \end{aligned}$$

Add the echo signal, noise signal, reverberation signal and the jamming signal of wideband noise jammer together, the self-navigation echo signal of multi-target in the context of this paper can be obtained. The simulation result is shown as in Figure 2, where the frequency of transmitting signal is 30 KHz, bandwidth is 50 Hz and pulse width is 25 ms.

In Figure 2a, 2 point targets and a volume target can be distinguished obviously and the amplitude

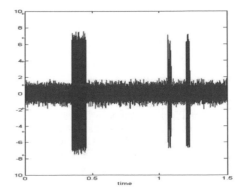

Figure 2a.   Multi-target echo without jamming.

and width of the volume target is obviously higher than those of point targets at far field. Because of the selection of signal pulse width, the sub-echoes of the scatters on volume target are interfered with each other, so the sub-echo of each scatter can't be distinguished and target response becomes steady state response. By calculating the time delay of each target in Figure 2, the distance from each target to origin is conformed to the premise on this paper. In Figure 2b, since the wideband noise jamming covers the target echoes, the target echoes can't be distinguished obviously.

3 SIMULATION PLATFORM DESIGN

The functional block diagram of the terrestrial test for underwater acoustic detection device is shown as in Figure 3. The system includes waveform acquisition and playback module based on PXI bus, acoustic interconnection device, PXI controller, PXI case and monitor, etc. The system operates at acoustic input mode. Simulating impedance matching material in addition certain pre-stress, the interconnection transducer can mechanically couple the transducer array of underwater acoustic detection device exactly. The operating principle is that, after system has started

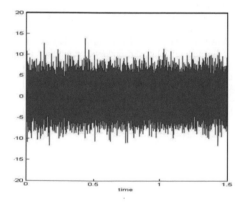

Figure 2b. Multi-target echo with wideband noise jamming.

up, the transmitting signal from underwater acoustic detection device is converted into input signal fed into trigger circuit via matching material and interconnection transducer 1. The input signal is used as system synchronous signal. The multi-target echo data and wideband noise jamming data are written into waveform playback module via PXI bus. According to the received synchronous signal, set the parameters of echo signal, such as frequency f_0, pulse width τ, repetition period and the decibels of signal attenuation. After the superposition of above signals, the superposition signal pass D/A and attenuation and is sent to interconnection transducer. Now the electric signal is converted into acoustic signal. The acoustic signal is sent to the transducer array via matching material and form the testing and simulating environment[61,62]. The underwater acoustic detection device can analyze and process the received simulating echo signals to obtain the multi-target information.

According to certain tactical premise, the parameters of target echo, jamming signal and background environment can be generated and controlled by control computer.

4 RESULT OF TERRESTRIAL TEST AND CONCLUSION

Set the parameters of echo signal, which are frequency f_0, pulse width τ, repetition period and the decibels of signal attenuation, and start up the system. The output waveform of analogue amplifier and filter circuit of underwater acoustic detection device on oscillograph is shown as in Figure 4. It is the multi-target echo signal waveform real-time displayed of a experiment. The center frequency of transmitting signal is 30.025 kHz, the bandwidth of transmitting signal is 50Hz and its pulse width is 25 ms. In Figure 4, 2 point targets and one volume target can be distinguished. The distance between the center of volume target and the underwater acoustic detection device is 300 m and the distances between point targets and the underwater acoustic detection device are 800 m and 900 m respectively. The output waveform of amplifying and

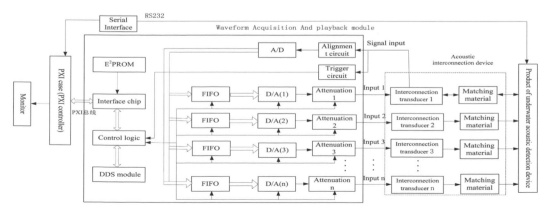

Figure 3. Functional Diagram of Terrestrial Test For Underwater Acoustic Detection Device.

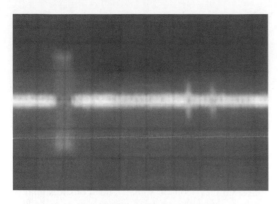

Figure 4a. Multi-target echo with no jamming.

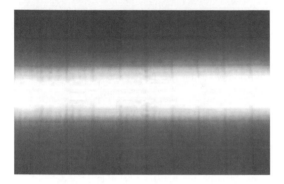

Figure 4b. Multi-target echo with wideband noise.

filtering circuit is almost the same of the simulating waveform and the distance between each target and the origin is conformed to the assumption presented on this paper.

By analyzing various target models in echoes received by underwater acoustic detection device with high speed, the multi-target model and its simulating result is presented. The test platform of acoustic detection device operating at sea is formed through acoustic interconnection device and the modeling data is written into waveform playback module by PXI bus. The matching test of acoustic detection device verifies the validity of the multi-target module and the effectiveness of the simulating method.

REFERENCES

[1] The 705 Research Institute compilation. Modern Torpedo[R]. KunMing: Torpedo Magazine, 1985, 57–58.
[2] ZHAO Guo-tong, GU Zhao-yang, LI Jian. 1996. Predictive study of foreign naval weapon development between 2000 and 2020[R]. Beijing: Naval equipment demonstration research center of science and Technology Information Research, 158–167.
[3] ZHOU Tao, YUAN Bing-cheng. 1996. Study of Torpedo Sound Input Target Simulator[J]. Journal of Naval University of Engineering, (4): 59–66.
[4] SHI Yang, YANG Wei-sheng, REN Zhang. 1997. A real time control semi physical simulation system for Torpedo [J]. Ship Engineering, (6): 47–49.
[5] WANG Xin-xiao, HUANG Jian-guo, ZHANG Yong-feng, YAN Wei. 2006. Simulation Research on Target Echo in Underwater Acoustic Detection System[J]. PIEZOELECTECTRICS & ACOUSTOOPTICS, Vol.28(3): 370–372.
[6] ZHANG Lijie, HUANG Jianguo, ZHANG Qunfei, LEI Kaizhuo. 2006. An underwater multi-source signal generator[J]. COMPUTER MEASUREMENT & CONTROL, 14(3): 365–373.

The game analysis of the lack of credit of the construction enterprises

Y.P. Wu
Jinhua Polytechnic, Jinhua, China

Z.J. Kong
Hongyu Building, Limited Liability Company, Jinhua, China

ABSTRACT: This article, by making use of the theory of game analysis and imitating the credit game theory in the field of construction, tries to establish the game analysis model, so as to determine the root reasons why the credit system is absent from the construction enterprises, and proposes the countermeasure and arrangement.

Keywords: construction enterprises credit the reasons of lack of credit theory of games

1 INTRODUCTION

The rapid development of the construction industry has brought about a series of issues, such as project quality, arrears of pay, commercial bribery and so on. All these problems can be considered as lack of credibility of an enterprise. With China's accession to WTO, credibility becomes a vital iss- ue. If it cannot be solved, our construction companies are likely to stay in an inferior status.

This paper, from the analysis of game models based on the condition of good and lack of credit system and theoretical respect, tries to propose the essential reasons of lack of credit in construction field and major measures to solve the credit situation in present construction market.

2 GAME ANALYSIS OF LACK OF CREDIT IN CONSTRUCTION COMPANY

Analysis of game model for the participants: The construction unit A, The construction company B. We suppose that the construction unit A (hereinafter referred to as A) is the purchaser of building products and construction companies B (hereinafter referred to as B) is a producer of building pro- ducts. A entrusts B to produce the construction products. In the building process of the following models, we make simulation setup of the respective participants' benefit in the game by simulating the situation of credit game.

2.1 *The game model's establishment and analysis of imperfect credit system in construction enterprise*

2.1.1 *Rules of the game*
1) Because of lack of credit system, A and B do not know each other's credit. Suppose the search cost A needs to pay is D_1 and A requires B to prepay C_1. The probability of prepay and not-prepay is q and 1-q respectively. Because of the market competition, the probability of not-prepay is hardly e- ists; 2) Project contract payments are C2. The probabilities of A's default are m and 1-m; 3) performance bond provided to A by B is C_3; 4) A's yield rate from default is i_1, which can be considered as the bank interest rate; while B's loss rate resulted from A's default is i_2. $i_1 > i_2$ is always right because B will default subcontractors, material suppliers and migrant workers and so on; 5) Suppose A and B are both rational and both ofthem are trying to maximize their profits.

2.1.2 *Model calculation*

(1) A utility analysis of the construction unit A
  a. the effectiveness of construction unit A's choosing default

$$U_{11}=[C_2+(C_2+C_1+C_3)\times i_1+C_1]\times q$$
$$=(C_1+C_2)(i_1+1)+C_3i_1$$

  b. the effectiveness of construction unit A's choosing no-default

$$U_{12}=(-C_1-C_2)\times q=-C_1q -C_2q$$
Result: U11> U12.

(2) Utility Analysis of Construction Enterprise B
  a. the effectiveness of construction enterprise B when A chooses default

$$U_{21}=[-(C_2+C_1+C_3)\times i_2-C_2-C_1]\times q+[-(C_2+C_3)\times i_2-C_2](1-q)= -(C_1q+C_2)(i_2+1)-C_3i_2$$

  b. The effectiveness of construction enterprise B when A chooses no-default

$$U_{22}=(C_1+C_2)\times q+ C_2\times(1-q)=C_1+C_2$$
Result: U22> U21.

Table 1. Game matrix.

		Construction unit A's default payments m	Construction unit A's No default payments 1-m
Construction enterprises B	Prepay q	$-(C_2+C_1+C_3) \times i_2$-$C_2$-$C_1$, $C_2+(C_1+C_2+C_3) \times i_1+C_1$	C_1+C_2, C_1-C_2

Note: The formula for each column of the first represents the effectiveness of the contractor, the second represents the effectiveness theowner

It proves: on the condition of lack of credit system in the construction field, A's effectiveness when choosing default is maximum, meanwhile, B benefits the worst. Therefore, B is bound to choose no credit.

2.2 The game model's establishment and analysis of perfect credit system in construction enterprise

2.2.1 Rules of the Game

Because of the perfect credit system, A and B's credit information are relatively transparent. In addition, the transaction will have an effect on the future ones. Thus, the game rules will be the following:

1) A's search cost is D2. Because of the relatively good credit rating system, So D2 can be considered zero. 2) Because the credit policy is relatively perfect, the requirement for A's providing the amount of guarantee of payment is C1. The probability of providing guarantee is p, while the opposite p robability is1-p. 3) A requires B to prepay C2. The probabilities of B's prepay and no-prepay are q and 1-q; 4) construction contract price is C3, the probabilities of A's default and no-default are m a- nd1-m; If A defaults payments, A will lose the guarantee C1; A does not default, then A gets C1; 5) the default may cause B to resort the legal proceedings. The litigation cost is C4. The probability of litigation is n, while the opposite probability 1-n; 6) A's yield rate from default is i1, while B's loss rate resulted from A's default is i2, i1> i2; 6) Suppose A and B are both rational and both of them are trying to maximize their profits [2].

2.2.2 Assumptions

Suppose the law is just, B can win the lawsuit if B lodges an appeal when A default. If B wins the lawsuit, B can obtain C1 paid by A and A should be fined Ca and compensate for B's loss $(C_2+C_3) \times i_2$. If B doesn't lodge an appeal, A will refuse to pay and benefit from the default.

2.2.3 Model calculation

(1) If construction unit A guarantees default, B will choose to litigation when A defaults.
   a. The effectiveness of construction unit A choosing default

U_{11}=[-C_1-C_a-(C_3+C_2)×i_2-C_3+(C_3+C_2)×i_1]×q+[-C_1-C_a-C_3×i_2-C_3+C_3×i_1](1-q)

=-C_1-C_a-C_3-C_2×(i_2-i_1)×q-C_3×(i_2-i_1)

b. The effectiveness of construction unit A choosing no default

U_{12}=(C_1-C_3-C_2)×q+(C_1-C_3)×(1-q)=C_1-C_2×q-C_3。

Proven: U12 > U11.
   (2) If construction unit A provides guarantee, B will not appeal to the lawsuit even A defaults
   a. the effectiveness of construction unit A chooses default

U_{21}=[-C_1+C_3+(C_3+C_2)×i_1+C_2]×q+[- C_1+C_3+ C_3×i_1](1-q)=-C_1 +C_2×(i_1+1)×q+C_3×(1+i_1)。

b. The effectiveness of construction unit A chooses no default

U_{22}=(C_1-C_3-C_2)×q+(C_1-C_3)×(1-q)=C_1-C_2×q-C_3。

Proven: When C1 > (C3 + C2q) (2 + i1)/2, U21 > U22.
   (3) If construction unit A provides no guarantee, B will appeal to the lawsuit when A defaults.
   a. The effectiveness of construction unit A chooses default

U_{31}=[-C_a-(C_3+C_2)×i_2-C_3+(C_3+C_2)×i_1]×q+[-C_a-C_3×i_2-C_3+C_3×i_1]×(1-q)

= -C_a-C_3-C_2×(i_2-i_1)×q-C_3×(i_2-i_1)。

b. The effectiveness of construction unit A chooses no default

U_{32}=(-C_3-C_2)×q+(-C_3)×(1-q)=-C_2×q-C_3。

Proven: U32 > U31. It proves that: When a landlord fails to provide guarantee, he will still choose no default.
   Of course, if construction unit A provides no guarantee, B will appeal to the lawsuit when A defaults.

2.3 Analysis

We can see from the result of calculation based on game analysis

1) In the condition of imperfect credit system, A benefits a lot from default. Without discreditable cost, namely, no credit in the process of transaction, A will maximize its profit. Meanwhile, B will benefit the worst. Thus, B is also likely to disobey the

Table 2. Game matrix.

		Construction unit A outstanding payments m		Construction unit A No outstanding payments $1-m$	
		Owners guarantee p	The owners do not guarantee $1-p$	Owners guarantee p	The owners do not guarantee $1-p$
Construction enterprises B Litigation n	Advancer q	$C_1 + C_a + (C_3 + C_2) \times i_2 + C_3 - C_4,$ $-C_1 - C_a - (C_3 + C_2) \times i_2 - C_3 + (C_3 + C_2) \times i_1$	$C_a + (C_3 + C_2) \times i_2 + C_3 - C_4,$ $-C_a - (C_3 + C_2) \times i_2 - C_3 + (C_3 + C_2) \times i_1$	–	–
	Not Advancer $1-q$	$C_1 + C_a + C_3 \times i_2 + C_3 - C_4,$ $-C_1 - C_a - C_3 \times i_2 - C_3 + C_3 \times i_1$	$C_a + C_3 \times i_2 + C_3 - C_4,$ $-C_a - C_3 \times i_2 - C_3 + C_3 \times i_1$	–	–
Construction enterprises B	Advancer q	$C_1 - (C_3 + C_2) \times i_2 C_3 - C_2,$ $-C_1 + C_3 + (C_3 + C_2) \times i_1 + C_2$	$-(C_3 + C_2) \times i_2 C_3 - C_2,$ $C_3 + (C_3 + C_2) \times i_1 + C_2$	$C_3 + C_2 - C_1,$ $C_1 - C_3 - C_2$	$C_3 + C_2,$ $-C_3 - C_2$
No proceedings $1-n$	Not Advancer $1-q$	$C_1 - C_3 \times i_2 - C_3,$ $-C_1 + C_3 + C_3 \times i_1$	$-C_3 \times i_2 - C_3$ $C_3 + C_3 \times i_1$	$C_3 - C_1,$ $C_1 - C_3$	$C_3,$ $-C_3$

Note: The formula for each column of the first represents the effectiveness of the contractor, the second represents the utility of owner.

rule or doesn't finish the project. Thus, in the process of transaction, both parties are likely to protect their own profits by means of discredit, which may worsen the whole industry's credit and lower the efficiency as well as enhance the cost in this field.

2) If the credit system is complete and perfect, A must provide payment guarantee. Under this circumstance, the cost will be enhanced if A defaults. In addition, if the credit system is improved, the law will be perfect. Thus, A will pay for much if default exists. Besides, the amount should be ruled by the law exactly.

3) We also can understand from the game analy sis mentioned above that one of the leading standards of whether a credit system is soundornot is the information's transparency. However,one of the important factors of low transparen cy at present is the imperfect credit rating system.

3 REASONS FPR THE LACK OF CREDIT SYSTEM

Based on the above game analysis, the author summarizes the reasons for the lack of credit in the current construction market as follows:

1. People's weak sense of credit
   In the time when Chinese economy and society is developing very quickly, and when the mate rial base of the society is becoming more abun dant, we are faced with the growing creditcrisis. Actions in bad faith and unlawful practices can be usually seen : swindling, deception, tax evasion, and so on. It is more alarming that this abnormal social phenomenon is like a plague spread among the population.
2. Imperfect credit evaluation system
   The author believes that the basic reason for weak sense of credit among people is the low cost of

dishonesty, while with the respect of institution, it is caused by the imperfect credit evaluation system and low information transparency.

3. Imperfect legal system also contributes to the loss of Honesty
   Currently, honesty still remains in morality, and there is no effective link with the legal form. However, observing the present situation, although rather complete laws and regulations have been laid out in China, such as "Construction Law", "Security Law", etc., these regulations are quite loose in safeguard honesty. Therefore an effective law system to ensure honesty has not been formed. Moreover, the relevant departments have been disapproved for many years due to their weak enforcement of judgments, low settlement rate and difficult implementation. Seen form the above model, if the legal system is sound, most people would choose litigation, and then the construction unit A would increase the cost of dishonesty.

4. Immature project risk management system in the building industry
   Compared with developed countries, risk management of the project in China is still in its infancy. Viewed from the current overall situation, the scope of project insurance limited, project warranty basically blank. The immature project risk management system needs to be further strengthened so as to promote the establishment of credit system in the building industry

5. Analysis of the asymmetrical construction market information
   The asymmetrical information among the transaction parties in the construction market transaction is observed from the following aspects:

   (1) Asymmetrical information between government supervision departments and proprietors;
   (2) Asymmetrical information between proprietors and project supervisors;

(3) Asymmetrical information between proprietors and construction contractors.

4 MEASURES AND SUGGESTIONS ON THE ESTABLISHMENT OF CREDIT SYSTEM IN CONSTRUTION INDUSTRY

Based on the analysis of the above models and reasons, I think we should focus on the following issues:

1. Enhance the publicity and education

   To create a credit economy, we must first establish a credit concept. Whether it is a legal entity or individual citizen, he should establish trustworthy public images and social consciousness valuing honesty and despising dishonesty,which will be achieved by various publicity, education and typical models and by facilitating the credit education, scientific research and training.The publicity and education of credit concept, credit awareness, and credit ethics should deserve great attention all the way from primary school to university. Meanwhile, great importance should be attached to the inservice education of credit management, especially to that of enterprises' core leaders. This is the fundamental condition for the effective operation of the credit management system.

2. Establish a perfect credit evaluation system

   Firstly, a unified credit information database and information platform should be established; then, it is essential to build up a set of comprehensive credit evaluation systems as soon as possible. For example, we can use the AHP method to determine credit evaluation coefficient or the Delphi method to determine the weight of each rating coefficient. Finally, we should set up a group of credit evaluation agencies such as Standard &Poor's, Moody's Inves tors Service. If the credit evaluation system of construction enterprises is relatively perfect, we will curtail search costs and increase information transparency and law enforcement, thus saving social costs.

3. Improve the law of credit in construction industry and enhance law enforcement.

   China should improve the law related to credit, which includes two aspects: First, improving the law related to credit evaluation such as problems concerning credit data collection and data opening up, so as to enhance our information transparency. Second, strengthening the law enforcement and punishment for those untrustworthy groups and facilitate governmental departments' supervision while improving individua ls' law awareness.

4. Enhance the risk prevention awareness and contract management of enterprises.

With the construction market integrating with the international standards, market competition has become increasingly stiff. According to developed countries' experience, credit enterprises enjoy obvious advantages in competition; therefore, construction enterprises must increase their credit awareness. However, construction industry is a highlyrisky industry,which means construction enterprises must increase their risk prevention awareness as well, i.e. acts like defaulting project money and breakingcontracts should be avoided, requiring contractors regulating their behavior according to the law while doing business, construction enterprises fully evaluating project risks and don't contract projects discriminately when the funds cannot be put into effect and don't promise to construction contract without advance payment. This will not only eliminates the dangers of defaul ting project money at the source, but also avoids contractors involving into illegitimate competition.

5 CONCLUSION

The construction of the credit system in construction industry is an important measure to rectify and regulate the order of construction market and ensure the construction industry's reform and development. It is an integral part of the whole nation's construction of credit system instead of a simple industry management system. As a result, establishing a credit system in construction industry is an imminent social problem. Due to the limited space, this article only analyses major reasons and measures of the lacking of credit theoretically. As a matter of fact, credit system construction requires a long and arduous process of research.

REFERENCES

[1] Kuang Kaicui,Ma Ziqiang. Study the Credit management institution in the construction industry[J]. Journal of Tongji University (Social sciences edition), 2004,15(5):114–119.

[2] Xing Xiuqing, Niu Bingkun. The game analysis of credit problems in construction field[J]. Shanxi architecture, 2007, (23):205–206.

[3] Huai Xianfeng, Wang Xiaolei. Construction project supervision and analysis of information asymmetry[J]. Construction management modernization 2004, (5): 45–47.

[4] Wang Mengjun. Construction market credit mechanism and credit system construction research[D]. Zhongnan University, 2004.

Information Technology and Computer Application Engineering – Liu, Sung & Yao (Eds)
© *2014 Taylor & Francis Group, London, ISBN 978-1-138-00079-7*

The presence of materials in the permeable transparency space based on time-dimension

Xie Qing Sunny & Mei Feng Lu
Shanghai Publishing and Printing College, Shanghai, China

ABSTRACT: Based on time-dimension, modern space appears either transparent or permeable in and with surroundings by means of construction and presence of materials. This article tries to make an analysis on the presence of materials in permeable transparency space based on time-dimension and push forward the idea of using materials as medium to express spatio-temporal spirit.

Keywords: the presence of materials, time-dimension, sensory perception, place, phenomenology

With the domination of modernism fading away, various styles and genres became popular, such as post-modernism, surface and tectonic architecture. People found there's no accepted space model which could solve the problems arise in space design. Under this situation, one universally accepted concept emerged: personal perception and specific experience of architecture is the origin of design inspiration, and it is ultimately what architecture is all about. Modern space design is a process which is directly based on space sensory perception: it highlights the involvement of human being as main body, the experience accumulated in space based on time-dimension, the space value percepted from movement in it and the experience got from different view angles. Its property of layered construction, transparency and spatio-temporal synaesthesie makes contribution to the name "permeable transparency space". The following are two cases from which we can have further understanding of the presence of materials in space.

Take AVEFENIX Fire Station as an example, which is in Mexico city by BGP Architecture Agency in 2006. Seen from outside, the entire building appears like a metal house suspending in the air. With its mass form got decompounded, what is left in the city is its image. Seen from inside, except small numbers of auxiliary functional spaces are enclosed, the floor is fully released. Various functions are arranged in a large open space. According to blueprint, holes are arranged as center of the floor. These holes vary greatly in size and scatter freely in public space. The space which is enclosed by partitions makes use of large amount of permeable material as wall surface. The use of light strengthens flow ability and transparency of space. (refer with: picture 1,2) Especially its public stairs, it is a public transport space. With natural light led in from the top, it makes the inner steel structure visible. Thus, the whole building looks perfect both visually and functionally.

Take Wang Shu's Tengtou Pavilion in Expo as another example. The building is a plain vacant rectangle with no windows in length and with holes in width. The main material used is bamboo plate, molded concrete and recycled bricks and tiles. Inside, there are rooms at the bottom and gardens on the top. People walking up along the winding path will look up slightly. Looking through from south to north, doorways of various sizes appear laminated, and this forms a changeable and far-reaching landscape. With movement of viewer, there appear corridors, terraces or open spaces, with vegetable in the garden and plants on the roof. Thus there are varied and rich space forms presented before viewer. In this plain looking rectangle space, viewer will experience wonderful landscape from different view angles and locations. (refer with: picture 3,4)

At first sight, these two buildings are totally different. One represents advanced technology, while the other stands for local design; one makes use of glass titanium LED, the other employs recycled bricks and tiles as main material; one functions as a fire station, the other is a rural exhibition. But after further study, it is not difficult to find common properties of these two buildings. They are as followings: first, as public space, they both make use of materials which reflects features of surroundings to construct spaces, relatively independent function corresponds to relatively independent single space. Though one is western, the other is eastern, these two buildings both possess the visual feature of transparency, lamination and multi-viewpoint. It is material that constructs the space order of multi-layers. Transparency and permeating phenomenon come into being between single spaces, between space and surroundings, and this interprets the concept of non-linearity spatio-temporality, presenting the feature of 4D. It is not hard for us to find that it makes breakthrough in single viewpoint and it laminates space perception which is got from different locations. The perception comes not only from the enclosed space but also from space leap. There is time perception left from history in the space, and this experience perception and emotional wakening are the results of the application of various material and its various ways of presence.

In space construction, material plays 3 roles: surface, structure and space. The design of modern public space commits itself to the application and presence of material which can function all the 3 roles at the same time. On one hand, the use of permeable material gets effect of overlying and with time passing by material brings perception of memory leap, telling stories of space and arousing synaesthesie; on the other hand, the application of stress rule and logic construction creates multi-view interior and exterior environment by decomposing and developing 3D space, and the combination of visual sight and light design promotes the fusion of interiors and exteriors. Temporality which is expressed by material itself and design construction brings us unusual space image.

Different materials present featured space look, which is either rustic or fancy. Take Tengtou Pavilion as an example, the only country venue in Expo. Its walls are built by recycled bricks and tiles, bamboo plate and molded concrete, which convey pure local flavor. Fire Station in Mexico city is totally different. Its surface is made by metallic titanium, which conveys the architecture look of high efficiency, tense and seriousness. The metal exterior finish of Fire Station creates a space surface which reflects city environment, and this makes viewer blurred with the boundary of concrete building and its surroundings. The application of metallic titanium creates a reflective, transparent or translucent scenery. The scenery is fused with modern city image, and this highlights its modernity and leading position. Tengtou Pavilion makes use of recycled materials, with centuries-old tiles as its walls, bamboo plate as weaving texture and drawing tact as a way to present ancient perception. That makes this new building in 08 Expo a modern one with history memories of south China. On one hand, super abundance of materials reduces the visual recognition. Compared to traditional materials such as bricks and stones, bamboo plate, overlapping permeable glass and metallic titanium bring about a trance-like state to viewer. On the other hand, the rich texture, weaving texture and permeable effect enhance space sensitivity. It is temporality represented in materials that promotes space perception in building, and that makes it symbolic. Tengtou Pavilion is a typical case. The temporality represented by materials brings about memory leap and embodies ancient artistic conception that traditional art seeks and the age value that the 19th century western art seeks.

If above is only visual image of building, the following is the presence of material in space construction. In modern rectangle building, the partition and transparency of wall board and interaction between wall boards are the basic rules which are applied to an enclosed box. "Once the box is decomposed, its panel is not a constituent part of limited space any more, it composes a space which is flowing, syncretic and continuous kinetic . with the participation of time (4D), classism static space is replaced by a kinetic one." (refer with: table 1,) Motion on a single level is the effect by this way. This doesn't break the basic concept of "layer" in traditional building space, it only makes freelance and open facade come true. But this is only a start. After that, the construction of open space and sequence space get further developed by the way of 4D. Using time as links, a narrative space with development, climax and twist and turns is created on a non-horizontal surface. With the breakthrough of the concept of "layer", spaces can be overlapped without visually spoiling one another, thus a wider space hierarchy is created. Based on the development of technology, with the spatiality and sensuality of materials being developed, and the form of vertical angle and plane being decomposed, the space of double curved surface can also be created easily (for example, the interior of Guangzhou Opera House by GRG).

In Wang Shu's square space, there are decomposed wall boards, irregular terraces and winding paths. Based on the image of landscape painting, doors are opened in interior walls as camera aperture. So in the cubic space, there is lamination of masking transparency and doorways with landscape image. (refer with: picture 5) Based on 3-meter construction module, space layout is presented with multi-view centers and scenery spots, with decorating greenings on the location of varied terraces to attract view sight. With movement and change of view angle, viewer experiences fantastic garden scenery. Then, let's have a closer study on Fire Station. Public stairs which goes through entire building has similar function of insertion as Steven Holl's , (refer with: table 2) and it promotes space interestingness. The stairs opens at the top, and its space view center is created by transmitting form which is enclosed by colored organic glass. It echoes with the round opening transmitting ceiling cut by enclosed curve on floor panel. Stairs is set in building interior, with part of it connecting with the space on traditional first and second floor and part of it opening to architecturalization of interior atrium. That forms the climax of interior atrium space. With application of armored concrete, the interior space is architecturalized, forming windows and terraces, among which are stainless steel pipes functioning as firefighters' slider, and transmitting cylinder enclosed by glass and steel. Thus interior and exterior spaces blend in each other, with a result that they are open to and shared with each other. With viewer's movement and change of view sight, sceneries which are successive or trimming are presented before viewer, forming space stories independent of novels and films. Along time dimension, transparent space model covers the experience accumulated both in space and material, containing more non-linear spatio-temporality and non-objective perception.

Space has its aspect not only of perception but also of materiality. And material is used as medium to present human being's accumulated experience of the world. Though modernism German Venus is visually transparent, the marbled texture and delicate glass metal texture in space exist only as wallboards to define space. It is merely a surface phenomenon. In Tengdou Pavilion, plain material exists in the form of building mass in which human being lives and survives. Wall –facing material with brick and clay texture blends itself in the rural daily activities. Material is part of the place. Space and environment perception meet and blend in each other. According to Professor David Leather barrow from University of Pennsylvania, material and space place are interdependent, like clothes to body. Specific behavior and habit concise and solidify to form place. There is convergence in the understanding of space all over the world. It exists in the perception of large-scan time in buildings - just like the sigh we make when we stand in front of An then Temple- and existence experience based on the routines in "living space".

For example, the fire station in Mexico city shares some common features with eastern Buddhist space in aspect of trade-off when dealing with routines. The fire station has two functions: fire station and firefighters' library. In the building, Public space and auxiliary space are put together on a plane. Public properties and fire fighting function, data query and office routine are relatively separate but visually open. The separation is created by means of simple sub region. A large amount of natural light is led into interior by means of cutting and plugging holes. Its frame which separates it from exterior and ternary form division which separates the interior spaces, especially the arrangement of atrium are exactly the same as Tadao Audo's Asuma House. (refer with: picture 6) Seen from usual view single, they are both closed spaces. But when seen beyond view angle of 1.7 meters, approximately 1/3 open space (atrium) can be found to take in spring, summer, autumn and winter. And it connects fixed functional spaces at both ends. When decomposing the frame of walls between neighboring spaces, atrium provides spaces for transportation, lighting, airing and scenery. In Asuma House, garden function and place spirit of traditional Japanese family life get extended by means of opening the space. (refer with: table 3) The same is the design of Fire Station which trades off the surroundings in a Zen type. Functional spaces are negative-space. It forms empty space which is positive and contains human being's activities. It broadens space dimension to the maximum and contains surroundings in space. Fire Station trades off simplified surroundings, functions as the public platform in the city, and solidifies place spirit of city story. It also conveys non-objective spatio-temporal perception of surroundings and seasons.

SUMMARY

Following the space concept in western architecture development, space in building has been developed from a closed one to an open and interpenetrating one; from a mass enclosure one to a spiritual one. It stimulates emergence of the concept of place. Modern Architecture Phenomenology admits the existence of space value in human being's consciousness and experience and it is independent of concrete building. By analyzing cases of Fire Station in Mexico city and Wang Shu's Tengtou Pavilion, this article explores the existence of experience body which is accumulated based on time-dimension and its perception of the world. It put emphasis on the aspects presented by material. What's more, it sheds new light on the idea that design of modern public space has convergence in multi-view, multi-order and multi-perception. It values the co-existence of artificial space and natural environment. The author hopes that this article will do its contribution to the exploration of Chinese culture and spirit in local design.

References to a book

[1] (Italy) Bruno Zevi: The Language of Modern Architecture (Architecture Industry press, China 1986), p. 34.
[2] J.P. Chen: A Narrative Building — Steven Holl's Studies Series, Architect (2004) No. 10, p. 94.
[3] (Britain) D. Scott: Ji shao zhu yi yu chan zong (Architecture Industry press, China 2002), p. 20.

References to photos

[P1], [P2] (Germany) M. Galindo: 1000 Architecture of Americas (Huazhong University of Science and Technology, China 2008) P676.
[P3], [P4] S. Wang: Shanghai World Expo Ninbo Tengtou Pavilion, World Architecture (2012) No.05, p. 110.
[P5] S. Wang: The view of Section—Ninbo Tengtou Pavilion, Journal of Architecture (2010) No.05, p. 129.
[P6] (Japan) Tadao Audo: Tadao Audo lun Jianzhu, (Architecture Industry press, China 2003), p. 140.

APPENDIX

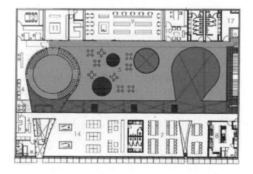

Picture 1. Fire Station.

Picture 2.　Ninbo Tengtou Pavilion.

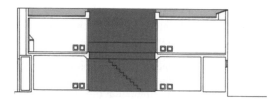

Picture 3.　Tadao Audo's, Asuma House.

Picture 4.　Shanghai World Expo.

Picture 5.　Ninbo Tengtou Pavilio.

Information Technology and Computer Application Engineering – Liu, Sung & Yao (Eds)
© 2014 Taylor & Francis Group, London, ISBN 978-1-138-00079-7

Monomial form approach to the construction and the conversion for curves in CAGD

Natasha Dejdumrong
King Mongkut's University of Technology Thonburi, Bangkok, Thailand

ABSTRACT: Their basis functions (polynomials) or via recursive algorithms. An efficient means for representing curves can be introduced by using their monomial forms. Monomial functions of B'ezier curves were investigated by Faux and Pratt [1], Mortenson [2], and Chang [3]. Nevertheless, monomial functions for other types of CAGD curves are nonexistent. This work proposes monomial functions for Said-Ball, Wang-Ball, DP, Dejdumrong and NB1 curves. These monomial functions will be extremely useful and far less computational intensive in the curve-conversion procedures.

Keywords: Monomial Matrix, Bézier Curves, Said-Ball Curves, Wang-Ball Curves, DP Curves.

1 INTRODUCTION

In geometric modeling and design for civil engineering and architectural drafting, it is compulsory tomake use of curve segments in surface and solid modeling. Nowadays, the curve modeling schemehas been based on mathematical representations of the Bézier-Bernstein, B-Spline and NURB curves.In the recent years, such curves have gained enormous popularity because they are appropriate formodeling, and have quickly become the curves for a wide range of graphics applications. However,these curves are usually represented in terms of mathematical equations that consume much computationaltime. This is problematic for facilitating interactive CAD/CAM applications. In this research,we are particularly concerned with the Bézier curves. Compared to others, Bézier curves have beenused as the basic foundation of the curve representation for the longest time.

During the late 80s, the Said-Ball(1989) [7] and Wang-Ball(1987) [8] polynomials were independentlydeveloped as the two alternative forms for curve modeling. They are generalizations of thecubic Ball polynomial scheme [4][5][6] proposed in 1974. The effectiveness of Said-Ball and Wang-Ball curves was explained in the work by Hu et al. [10] in 1996. In 2000, they were introduced astwo of the fastest and most efficient methods in evaluating Bézier curves [13]. This is particularly truewhen transformed intoWang-Ball control points. The computational time for a Bézier curve from theWang-Ball control points by theWang algorithm is linear. Due to the difficulty of the conversion from Bézier into Wang-Ball control points,

however, the transformation is formulated in improper formsfor coding in most programming languages. Moreover, they can cause slow and erroneous computationsbecause of the underlying complexities. Thus, the formulae have not been widely used eventhough they are more efficient in theory.

Lately, there have been several attempts to introduce of Bézier-like curve representations byproposing new polynomial bases and new algorithms. Such examples are Delgado-Peña (2003) [9], B1 of Wu (2005) [11][12] and Dejdumrong (2008)[14][15]. In this group, the Delgado-Peña and Dejdumrong curves possess linear computational algorithms while the NB1 curve has quadratic computationalcomplexity. However, the new algorithms are still not suitable for CAD applications due totheir complex formulae.

A simple approach to represent polynomial curves is provided in the form of monomial matricesbecause it is more convenient to code and implement matrix operations than to solve for symboliccomputations. The monomial form of Bézier curves was first investigated by Faux and Pratt [1], Mortenson [3], and Chang [2] although it was obvious that the formula for Bézier curve was verysimple. To date, however, no monomial functions for other curve types with complex formulae hasbeen suggested. This paper presents monomial functions for the Said-Ball, Wang-Ball, DP, Dejdumrong and NB1 curves. The proposed functions are less complicated and more efficient for constructingcurves and surfaces in CAD Applications. Further, the conversions among polynomial curves can bereadily obtained from their monomial matrices.

2 THE CONSTRUCTION OF THE CURVES IN CAGD USING MONOMIAL FORMS

If conventional polynomial equations are used in the calculation of points on curves, it will be a big challenge to write a program to support the symbolic computations. This paper introduces the curve monomials to make coding more efficient and less computational intensive. Hence, it is necessary to define monomial forms for each curve in the following propositions.

Proposition 1 (Bézier Monomial Form) [16] A Bézier curve of degree n, denoted by $B^n(t)$, with a row matrix of $n + 1$ control points, denoted by $G = \{P_i\}(i = 0)^n$, can be written in terms of the powerbasis form of a column matrix, denoted by, $T = \{t^j\}(j = 0)^n$, as follows:

$$B^n(t) = G \cdot B^n \cdot T =$$
$$[P_0 P_1 \ldots P_n] \begin{bmatrix} m_{0,0} & m_{0,1} & \cdots & m_{0,n} \\ m_{1,0} & m_{1,1} & \cdots & m_{1,n} \\ \vdots & \vdots & \ddots & \vdots \\ m_{n,0} & m_{n,1} & \cdots & m_{n,n} \end{bmatrix} \begin{bmatrix} t^0 \\ t^1 \\ \vdots \\ t^n \end{bmatrix}, \quad (1)$$

where B^n is a Bernstein monomial matrix of the element $m_{i,j}$ and $m_{i,j} = (-1)^{j-i} \binom{n}{j}\binom{j}{i}$.

Proposition 2 (Said-Ball Monomial Form) An nth-degree Said-Ball curve, denoted by $S^n(t)$, given by $n + 1$ control points, denoted by $G = \{P_i\}_{i=0}^n$, can be expressed in power basis form as follows:

$$S^n(t) = G \cdot S^n \cdot T =$$
$$[P_0 P_1 \ldots P_n] \begin{bmatrix} s_{0,0} & s_{0,1} & \cdots & s_{0,n} \\ s_{1,0} & s_{1,1} & \cdots & s_{1,n} \\ \vdots & \vdots & \ddots & \vdots \\ s_{n,0} & s_{n,1} & \cdots & s_{n,n} \end{bmatrix} \begin{bmatrix} t^0 \\ t^1 \\ \vdots \\ t^n \end{bmatrix}, \quad (2)$$

where S^n is a Said-Ball monomial matrix of the element $s_{i,j}$ and

$$s_{i,j} = \begin{cases} (-1)^{j-i}\binom{i+\lfloor\frac{n}{2}\rfloor}{i}\binom{\lfloor\frac{n}{2}\rfloor+1}{j-i}, & \text{for } 0 \le i \le \lfloor\frac{n}{2}\rfloor - 1, \\ (-1)^{j-i}\binom{n}{i}\binom{i}{j-i}, & \text{for } i = \frac{n}{2}, \\ (-1)^{j-\lfloor\frac{n}{2}\rfloor-1}\binom{\lfloor\frac{n}{2}\rfloor+n-i}{n-1}\binom{n-i}{j-\lfloor\frac{n}{2}\rfloor-1}, & \text{for } \lfloor\frac{n}{2}\rfloor + 1 \le i \le n \end{cases} \quad (3)$$

Proposition 3 (Wang-Ball Monomial Form) A Wang-Ball curve, denoted by $A^n(t)$, provided with $n + 1$ control points, denoted by $G = \{P_i\}_{i=0}^n$, can be shown as

$$A^n(t) = G \cdot A^n \cdot T =$$
$$[P_0 P_1 \ldots P_n] \begin{bmatrix} a_{0,0} & a_{0,1} & \cdots & a_{0,n} \\ a_{1,0} & a_{1,1} & \cdots & a_{1,n} \\ \vdots & \vdots & \ddots & \vdots \\ a_{n,0} & a_{n,1} & \cdots & a_{n,n} \end{bmatrix} \begin{bmatrix} t^0 \\ t^1 \\ \vdots \\ t^n \end{bmatrix}, \quad (4)$$

where A^n is a Wang-Ball monomial matrix of the element $a_{i,j}$ and

$$a_{i,j} = \begin{cases} (-1)^{j-i}2^i\binom{i+2}{j-i}, & \text{for } 0 \le i \le \lfloor\frac{n}{2}\rfloor - 1, \\ (-1)^{j-i}2^i\binom{n-i}{j-i}, & \text{for } i = \lfloor\frac{n}{2}\rfloor, \\ (-1)^{j-i}2^{n-i}\binom{n-i}{j-i}, & \text{for } i = \lceil\frac{n}{2}\rceil, \\ (-1)^{j-n+i}2^{n-i}\binom{n-i}{j-n+i-2}, & \text{for } \lceil\frac{n}{2}\rceil + 1 \le i \le n \end{cases} \quad (5)$$

Proposition 4 (DP Monomial Form) An nth-degree DP curve, denoted by $C^n(t)$, given by a set of $n + 1$ control points, denoted by $G = \{P_i\}_{i=0}^n$, can be formulated in power basis form by

$$C^n(t) = G \cdot C^n \cdot T =$$
$$[P_0 P_1 \ldots P_n] \begin{bmatrix} c_{0,0} & c_{0,1} & \cdots & c_{0,n} \\ c_{1,0} & c_{1,1} & \cdots & c_{1,n} \\ \vdots & \vdots & \ddots & \vdots \\ c_{n,0} & c_{n,1} & \cdots & c_{n,n} \end{bmatrix} \begin{bmatrix} t^0 \\ t^1 \\ \vdots \\ t^n \end{bmatrix}, \quad (6)$$

where C^n is a DP monomial matrix of the element $c_{i,j}$ and

$$c_{i,j} = \begin{cases} (-1)^j\binom{n}{t}, & \text{for } i = 0, \\ (-1)^{j-1}\binom{n-i}{j-i}, & \text{for } 0 < i \le \lfloor\frac{n}{2}\rfloor - 1, \\ (-1)^{j-1}(n-2i)\binom{i+1}{j-i}, & \\ +\left(\frac{1}{2}\right)^{n-2i}\left(\binom{0}{j}-\binom{0}{j-i-1}-(-1)^j\binom{i+1}{j}\right), & \text{for } i = \lfloor\frac{n}{2}\rfloor, \\ (-1)^{j-n+i}(n-2i)\binom{1}{j-n+i-1}, & \\ +\left(\frac{1}{2}\right)^{2i-n}\left(\binom{0}{j}-\binom{0}{j-n+i-1}-(-1)^j\binom{n-i+1}{j}\right), & \text{for } i = \lceil\frac{n}{2}\rceil, \\ (-1)^{j-i}\binom{1}{j-i}, & \text{for } \lceil\frac{n}{2}\rceil + 1 \le i \le n-1 \\ \binom{0}{j-n}, & \text{for } i = n \end{cases} \quad (7)$$

Proposition 5 (Dejdumrong Monomial Form) A Dejdumrong curve of degree n, denoted by $D^n(t)$, with $n + 1$ control points, denoted by $G = \{P_i\}_{i=0}^n$, can be computed by

$$D^n(t) = G \cdot D^n \cdot T =$$
$$[P_0 P_1 \ldots P_n] \begin{bmatrix} d_{0,0} & d_{0,1} & \cdots & d_{0,n} \\ d_{1,0} & d_{1,1} & \cdots & d_{1,n} \\ \vdots & \vdots & \ddots & \vdots \\ d_{n,0} & d_{n,1} & \cdots & d_{n,n} \end{bmatrix} \begin{bmatrix} t^0 \\ t^1 \\ \vdots \\ t^n \end{bmatrix}, \quad (8)$$

where D^n is a Dejdumrong monomial matrix of the element $d_{i,j}$ and

$$d_{i,j} = \begin{cases} (-1)^{j-i}3^i\binom{i+3}{j-i}, & \text{for } 0 \le i < \lceil\frac{n}{2}\rceil - 1, \\ (-1)^{j-i}3^i\binom{n-i}{j-i}, & \text{for } i = \lceil\frac{n}{2}\rceil - 1, \\ (-1)^{j-i}2(3^{i-1})\binom{i}{j-i}, & \text{for } i = \frac{n}{2} \text{ and } n \text{ is even}, \\ (-1)^{j-i}3^{n-i}\binom{n-i}{j-i}, & \text{for } i = \lceil\frac{n}{2}\rceil + 1, \\ (-1)^{j-n+i-1}3^{n-i}\binom{n-i}{j-n+i-3}, & \text{for } \lceil\frac{n}{2}\rceil + 1 < i \le n. \end{cases} \quad (9)$$

Proposition 6 (NB1 Monomial Form) An NB1 curve of degree n, $N^n(t)$, with $n + 1$ control points, denoted by $G = \{P_i\}_{i=0}^n$, can be formed by the power basis form as follows:

$$N^n(t) = G \cdot N^n \cdot T =$$

$$[P_0 P_1 \ \cdots \ P_n] \begin{bmatrix} g_{0,0} & g_{0,1} & \cdots & g_{0,n} \\ g_{1,0} & g_{1,1} & \cdots & g_{1,n} \\ \vdots & \vdots & \ddots & \vdots \\ g_{n,0} & g_{n,1} & \cdots & g_{n,n} \end{bmatrix} \begin{bmatrix} t^0 \\ t^1 \\ \vdots \\ t^n \end{bmatrix}, \qquad (10)$$

where N^n is a NB1 monomial matrix of the element and $g_{i,j}$

$$g_{i,j} = \begin{cases} (-1)^{j-i} \binom{\lfloor \frac{n}{2} \rfloor - 1 + i}{i} \binom{\lfloor \frac{n}{2} \rfloor}{j-i}, & \text{for } 0 \leq i \leq \lfloor \frac{n}{2} \rfloor - 2, \\ (-1)^{j-i} \binom{2i}{i} \binom{i+2}{j-i}, & \text{for } i = \lfloor \frac{n}{2} \rfloor - 1, \\ (-1)^{j-i} 2 \binom{2i-2}{i-1} \binom{n-i}{j-i}, & \text{for } i = \lfloor \frac{n}{2} \rfloor, \\ (-1)^{j-i} 2 \binom{2(n-i-1)}{n-i-1} \binom{n-i}{j-i}, & \text{for } i = \lceil \frac{n}{2} \rceil, \\ (-1)^{j-n+i} \binom{2(n-i)}{n-i} \binom{n-i}{j-n+i-2}, & \text{for } i = \lceil \frac{n}{2} \rceil + 1, \\ (-1)^{j - \lceil \frac{n}{2} \rceil} \binom{\lfloor \frac{n}{2} \rfloor - 1 + n - i}{n-i} \binom{n-i}{j - \lceil \frac{n}{2} \rceil}, & \text{for } \lceil \frac{n}{2} \rceil + 2 \leq i \leq n. \end{cases} \qquad (11)$$

3 THE CONVERSIONS AMONG CURVES USING MONOMIAL MATRICES

In commercial applications, it is sometimes beneficial to transform one curve into another, e.g., to increase computational speed and to prove the formulae of degree elevation and degree reduction, etc. Consequently, by taking advantage of the techniques and the mathematical properties of matrix multiplication, we can turn the basis of a monomial (A) into another basis of monomial (B) written as $B.A^{-1}$. Conversely, a curve defined by a monomial matrix (A) can be transformed into another typewith a monomial matrix (B) by $A.B^{-1}$

4 SUMMARY

In this work, monomial matrix functions for curves in CAGD (Said-Ball, Wang-Ball, DP, Dejdumrongand NB1 curves) are investigated to facilitate the use of their monomial matrix forms. A novel methodof curve representations, i.e., the monomial matrix form, provides a more flexible computation forrepresenting curves in CAGD. The monomial methods can be applied for computing the conversionsamong these CAGD curves since writing a program that utilizes matrix operations is far more efficientthan performing symbolic calculations. This strategy, therefore, produces simpler functions which arebetter alternatives for computer programming languages.

REFERENCES

[1] I. Faux and M. Pratt: *Computational Geometry for Design and Manufacture* (Ellis Horwood, 1979).

[2] G. Chang: Matrix formulation of Bézier technique, Computer Aided Design, Vol. 14 No. 6 (1982), p. 345–350.

[3] M. Mortenson: *Geometric Modeling* (Wiley, 1985).

[4] A. A. Ball: CONSURF, Part one: Introduction to conic lifting title, Computer Aided Design, Vol.6 (1974), p. 243–249.

[5] A. A. Ball: CONSURF, Part two: Description of the algorithms, Computer Aided Design, Vol. 7 (1975), p. 237–242.

[6] A. A. Ball: CONSURF, Part three: How the program is used, Computer Aided Design, Vol. 9 (1977), p. 9–12.

[7] H. B. Said: Generalized Ball Curve and Its Recursive Algorithm, ACM Transactions on Graphics, Vol. 8 (1989), p. 360–371.

[8] G. J. Wang: Ball Curve of High Degree and Its Geometric Properties, Appl. Math.: A Journal of Chinese Universities, Vol. 2 (1987), p. 126–140.

[9] Delgado and J. M. Peña : A Shape Preserving Representation with an EvaluationAlgorithm of Linear Complexity, Computer Aided Geometric Design, Vol. 20 No. 1 (2003), p. 1–20.

[10] S.M. Hu, G.Z. Wang and T.G. Jin: Properties of Two Types of Generalized Ball Curves, ComputerAided Design, Vol. 28 No. 2 (1996), p. 125–133.

[11] H.Y. Wu: Unifying Representation of Bézier Curve AndGenaralized Ball Curves, Appl. Math.J. Chinese Univ. Ser. B, Vol. 5 No. 1 (2000), p. 109–121.

[12] D. Yu and X. Chen: Another Type Of Generalized Ball Curves And Surfaces, ActaMathematicaScientia, Vol. 27 No. 4 (2007), p. 897–907.

[13] H.N.Phien and N.Dejdumrong: Efficient Algorithms for Bézier curves, Computer Aided GeometricDesign, Vol. 17 (2000), p. 247–250.

[14] N. Dejdumrong: Efficient Algorithms for Non-Rational and Rational Bézier Curves, The 5th International Conference on Computer Graphics, Imaging and Visualization, (2008), p. 109–114.

[15] N. Dejdumrong: A New Bivariate Basis Representation for Bézier -based Triangular Patches with Quadratic Complexity, Computers and Mathematics with Applications, Vol. 61 No. 8(2011), p. 2292–2295.

[16] G. Farin: *Curves and Surfaces for Computer Aided Geometric Design* (Edition 5, Academic Press Inc, London, 2002).

Information Technology and Computer Application Engineering – Liu, Sung & Yao (Eds)
© 2014 Taylor & Francis Group, London, ISBN 978-1-138-00079-7

Isolated zero-voltage-switching buck converter based on full-bridge topology

Hyun Lark Do
Department of Electronic & Information Engineering, Seoul National University of Science and Technology, Seoul, South Korea

ABSTRACT: An isolated Zero-Voltage-Switching (ZVS) buck converter based on full-bridge topology is proposed in this paper. In the proposed converter, a clamping capacitor is across the second bridge switches to reset the isolation transformer and provide ZVS features to power switches. By interleaving operation, current ripples at input and output stages are reduced. Due to the ZVS characteristic of power switches, the switching loss is significantly reduced and the efficiency is improved compared with the conventional full-bridge buck converter. Steady-state analysis for the proposed converter is presented. Experimental results based on a prototype are provided to verify the effectiveness and feasibility of the proposed converter.

Keywords: Buck converter, full-bridge converter, zero-voltage-switching, zero-current-switching

1 INTRODUCTION

Pulse-width-modulation (PWM) DC-DC converters have been the most common switching DC-DC converters. However, their switching frequency is limited due to large switching losses. To reduce the weight and size of a DC-DC converter, its switching frequency should be raised. To reduce switching losses and raise switching frequency, many soft-switching techniques have been adopted for DC–DC converters [1, 2]. Phase-shift full-bridge converters are widely used for step-down applications due to their advantages such as low voltage stresses of the switching devices, a fixed switching frequency, and ZVS operation of power switches. However, large circulating current causes large conduction losses [3, 4]. The asymmetrical full-bridge converter was proposed in [5]. It has various advantages such as zero switching loss, no conduction loss penalty, and fixed switching frequency. However, voltage stresses across power switches are fixed to input voltage even when steep conversion ratio is required. In order to overcome these problems, an isolated ZVS buck converter based on full-bridge topology is proposed in this paper. The proposed converter witch is shown in Fig. 1 uses active clamp techniques. Instead of DC-blocking capacitor in the conventional asymmetric full-bridge buck converter, a clamping capacitor is connected across the 2nd bridge to clamp the 2nd bridge switch voltages and provide ZVS operation. The proposed converter features clamped switch voltages, fixed switching frequency, soft-switching operations of all power switches, and low voltage stresses in small duty cycle operation. Therefore, the proposed converter shows high efficiency and it is suitable to a steep conversion application.

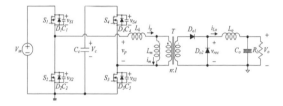

Figure 1. Circuit diagram of the proposed converter.

2 PRINCIPLE OF OPERATION

The voltage across the clamping capacitor C_c is assumed constant during a switching period. The transformer T is modeled as magnetizing inductance L_m, leakage inductance L_k, and an ideal transformer which has a turn ratio of n: 1. The theoretical waveforms are shown in Fig. 2. Before t_0, S_1, S_3, and D_{o1} are conducting.

At t_0, the switches S_1 and S_3 are simultaneously turned off. Then, the primary current is flowing through the output capacitances of power switches. As a result, voltages v_1 and v_3 across C_1 and C_3 increase linearly, whereas voltages v_2 and v_4 across C_2 and C_4 decrease at the same rate. This mode ends at t_1, when voltages v_1 and v_3 increase to $V_{in}/2$. At t_1, the primary voltage becomes zero and the output diode D_{o2} also begins to conduct. Since both output diodes are conducting, the transformer's secondary side is shorted. Then, the magnetizing current i_m remains constant and the output capacitances of power switches and L_k form a series-resonant circuit. The voltage v_2 continues to decrease with a resonant manner, whereas the voltage v_3 continues to increase. This mode ends when v_2 becomes zero and v_3 becomes V_c, respectively. At t_2, v_2

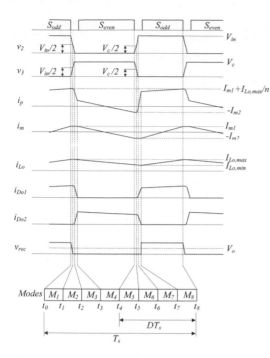

Figure 2.   Theoretical waveforms.

of this mode, i_{Lo} and i_m arrive their maximum values $I_{Lo,max}$ and I_{m1}.

The clamping capacitor voltage V_c can be easily obtained by applying the volt-second balance law to the primary voltage v_p. The voltage V_c is given by

$$V_o = \frac{V_{in}}{1-D}. \tag{1}$$

From Fig. 2, since the diode current i_{Do1} decreases from $I_{Lo,max}$ to zero during $d_1 T_s$, d_1 is given by

$$d_1 = \frac{I_{Lo,max} L_k}{n V_c T_s}. \tag{2}$$

Similarly, since i_{Do1} increases from zero to $I_{Lo,min}$ during $d_2 T_s$, d_2 is given by

$$d_2 = \frac{I_{Lo,min} L_k}{n V_{in} T_s}. \tag{3}$$

Since the average current flowing through Cc should be zero under steady state, the average value of i_p during $(1-D)T_s$ is zero as follows:

$$(1-D)T_s I_{m1} + \frac{I_{Lo,max} d_1 T_s}{n} - \frac{V_c (d_1 T_s)^2}{2 L_k} - \frac{V_c [(1-D)T_s - d_1 T_s]^2}{2(L_m + L_k)} = 0 \tag{4}$$

From (4), I_{m1} can be obtained by

$$I_{m1} = \frac{-\dfrac{I_{Lo,max} d_1}{n} + \dfrac{V_c d_1^2 T_s}{2 L_k} + \dfrac{V_c (1-D-d_1)^2 T_s}{2(L_m + L_k)}}{1-D}. \tag{5}$$

Since im decreases from I_{m1} to $-I_{m2}$ during $(1-D-d_1)T_s$, I_{m2} is given by

$$I_{m2} = -I_{m1} - \frac{V_c (1-D-d_1)T_s}{L_m + L_k}. \tag{6}$$

If d_1 and d_2 are sufficiently small, I_{m1} and I_{m2} are simplified as follows:

$$I_{m1} = I_{m2} \approx \frac{D V_{in} T_s}{2(L_m + L_k)}. \tag{7}$$

Since the average inductor voltage should be zero, the output voltage V_o is the average value of v_{rec} as follows:

$$\frac{V_o}{V_{in}} = \frac{D - d_2}{n} \cdot \frac{L_m}{L_m + L_k} \approx \frac{D}{n}. \tag{8}$$

If d_2 is very small and L_m is sufficiently larger than L_k, the voltage gain is simplified as D/n. The simplified voltage gain of the proposed converter is the same as the conventional continuous-conduction-mode buck converter.

becomes zero, the body diode D_2 of S_2 begins to conduct. Then, gate signal is applied to S_2. Therefore, S_2 is turned on with ZVS. Similarly, S_4 is also turned on with ZVS. Since $-V_c$ is applied to L_k, the primary current i_p decreases linearly. Consequently, the output diode current i_{Do1} decreases linearly, whereas i_{Do2} increases linearly. The current i_{Lo} continues to decrease with the same slope as in mode 2. At the end of this mode, the diode current i_{Do1} becomes zero. At t_3, i_{Do1} becomes zero and D_{o1} is turned off. The output inductor current i_{Lo} flows through D_{o2}. Since the primary voltage v_p is $-V_c$, the primary current i_p decreases linearly. At t_4, the switches S_2 and S_4 are simultaneously turned off. Then, voltages v_2 and v_4 across C_2 and C_4 increase linearly, whereas voltages v_1 and v_3 across C_1 and C_3 decrease at the same rate. This mode ends at t_5, when voltages v_2 and v_3 increase to $V_c/2$. At t_5, the primary voltage becomes zero and both output diodes are on. Since both output diodes are conducting, the transformer's secondary side is shorted and C_1 through C_4 and L_k form a series-resonant circuit. The voltage v_2 continues to increase with a resonant manner, whereas the voltage v_3 continue to decrease. This mode ends when v_2 becomes V_{in} and v_3 becomes zero, respectively. When v_2 becomes V_{in}, v_1 becomes zero and the body diode D_1 begins to conduct. Then, the switch S_1 is turned on. Consequently, the ZVS turn-on of S_1 is achieved. Similarly, S_3 is turned on with ZVS. Since input voltage V_{in} is applied to L_k in this mode, the primary current i_p increases. At the end of this mode, the diode current i_{Do2} becomes zero. At t_7, i_{Do2} becomes zero and D_{o2} is turned off. Since the primary voltage v_p is V_{in}, the current i_p increases linearly. At the end

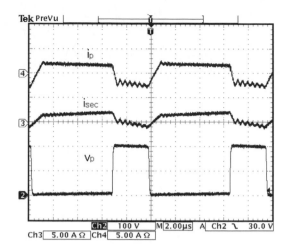

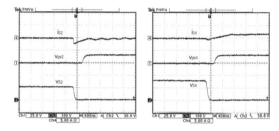

Figure 3. Measured key waveforms.

Figure 4. Soft-switching waveforms of S_2 and S_3.

3 EXPERIMENTAL RESULTS

In order to verify the performance of the proposed isolated ZVS buck converter based on full-bridge topology, a prototype has been built and tested. The output voltage V_o is 48 V. Maximum output power $P_{o,max}$ is 100 W. Input voltage range is from 80 V to 120 V. The operating frequency is 100 kHz. The transformer turn ratio n is selected as 1. The total serial inductance L_s is selected as 33uH. The magnetizing inductance is selected as 520uH and the capacitance of C_c is selected as 6.6uF. The output filter consists of $L_o = 100$ uH and $C_o = 940$ uF. Fig. 3 shows the measured key waveforms of the prototype under full load. These measured waveforms agree with the theoretical analysis. The ZVS operations of the power switches are shown in Fig. 4. Before the gate pulses are applied to the switches, the voltages across the switches go to zero and the currents flowing through the switches don't change their direction. Due to the ZVS operation, the switching losses are significantly reduced and the system efficiency is improved. The measured efficiency of the prototype is around 92.5%.

4 CONCLUSION

An isolated ZVS buck converter based on full-bridge topology has been proposed. Due to soft-switching characteristic, the switching losses are significantly reduced and the system efficiency is improved. Limited duty cycle problem of the conventional asymmetric full-bridge buck converter is overcome. A prototype was built to verify the performance of the proposed converter and the experimental results were provided to verify the feasibility of the proposed converter.

REFERENCES

[1] L. Qin, S. Xie, and H. Zhou. A Novel Family of PWM Converters Based on Improved ZCS Switch Cell, Proc. IEEE PESC (2007), pp. 2725–2730.
[2] H. –J. Chiu, Y. –K. Lo, H. –C. Lee, S. –J. Cheng, Y. –C. Yan, C. –Y. Lin, T. –H. Wang, and S. –C. Mou, A Single-stage soft-switching flyback converter for power-factor-correction applications, IEEE Trans. Ind. Electron. (2010), pp. 2187–2190.
[3] E. Adib and H. Farzanehfard, Zero-voltage transition current-fed full-bridge PWM converter, IEEE Trans. Power Electron. (2009), pp. 1041–1047.
[4] Y. Jang and M. M. Jovanovic, A new family of full-bridge ZVS converters," IEEE Trans. Power Electron. (2004), pp. 701–708.
[5] P. Imbertson and N. Mohan, Asymmetrical duty cycle permits zero switching loss in PWM circuits with no conduction loss penalty, IEEE Trans. Ind. Appl. (1993), pp. 121–125.

Information Technology and Computer Application Engineering – Liu, Sung & Yao (Eds)
© 2014 Taylor & Francis Group, London, ISBN 978-1-138-00079-7

Simplified power factor correction converter based on SEPIC topology

Hyun Lark Do

Department of Electronic & Information Engineering, Seoul National University of Science and Technology, Seoul, South Korea

ABSTRACT: In this paper, a simplified Power Factor Correction (PFC) converter based on Single-Ended Primary Inductor Converter (SEPIC) topology is proposed. In the proposed converter, a bridge diode at the input stage is deleted to reduce the conduction loss and improve the efficiency. Since the proposed converter operates under discontinuous conduction mode, the input current naturally follows the input voltage waveforms and provides high power factor without complex control. The presented theoretical analysis is verified by a prototype of 100 kHz and 150 W. Also, the measured efficiency of the proposed converter has reached a value of 90.8% at the maximum output power.

Keywords: power factor correction, SEPIC, bridge diode, bridgeless

1 INTRODUCTION

Conventional boost converters are usually adopted as a PFC stage. Especially, a boost converter in discontinuous-conduction-mode (DCM) and its modified topologies are widely used because high power factor can be achieved without complex control scheme. However, the output voltage of the DCM boost converter stage should be much larger than the peak value of the input voltage in order to obtain high power factor. Similarly, buck converters can be used as a PFC stage. Its output voltage is lower than the peak value of the input voltage. However, its input current has dead angles. Also, a SEPIC converter is adopted for PFC applications [1]–[5]. Since it has a step up/down capability, the output voltage can be lower than the peak value of the input voltage. All of these PFC circuits have a full-bridge diode. Therefore, conduction losses caused by this full-bridge diode lower down power conversion efficiency especially at low ac line. In order to overcome these problems, a PFC converter based on SEPIC topology is proposed in this paper. The circuit diagram of the proposed PFC converter is shown in Fig. 1. Since there is no bridge diode, the conduction loss can be reduced and the efficiency can be raised.

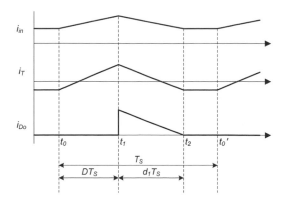

Figure 1. Circuit diagram of the proposed PFC converter.

Figure 2. Theoretical waveforms.

2 PRINCIPLE OF OPERATION

Fig. 2 shows the theoretical waveforms of the proposed converter whin a switching period. It is assumed that the input voltage v_{ac}, the capacitor voltage V_{C1}, and the output voltage V_o are constant within a switching period. To simplify the analysis, the operation anlaysis is carried out only for the case of $v_{ac} > 0$. Gate pulses according to the polarity of v_{ac} are shown in Fig. 3. In this case, the diode D_1 and the switch S_2 is always conducting. Before t_0, the switch S_1 and the output diode D_o are turned off. The input current iin flows through L_1, L_2, D_1, C_1, L_m, and S_2.

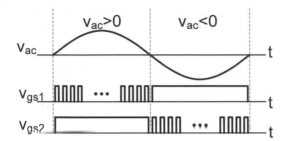

Figure 3. Gate pulses according to the polarity of v_{ac}.

Mode 1 $[t_0, t_1]$: At t_0, the switch S_1 is turned on. Since the input voltage v_{ac} is applied to the inductors L_1 and L_2, the input current i_{ac} increases linearly as follows:

$$i_{ac}(t) = i_{ac}(t_0) + \frac{V_{ac}}{L_1 + L_2}(t - t_0).\tag{1}$$

Similarly, the capacitor voltage V_{C1} is applied to the primary inductance L_p of T and the primary current i_T increases linearly as follows:

$$i_T(t) = i_T(t_0) + \frac{V_{c1}}{L_p}(t - t_0).\tag{2}$$

Mode 2 $[t_1, t_2]$: At t_1, the switch S_1 is turned off and the output diode D_o is turned on. Since $-(nV_o + V_{c1} - v_{ac})$ is applied to the input inductors L_1 and L_2, the input current decreases linearly as follows:

$$i_{ac}(t) = i_{ac}(t_1) - \frac{V_{c1} + nV_o - v_{ac}}{L_1 + L_2}(t - t_1).\tag{3}$$

Since the reflected output voltage is applied to the magnetizing inductance of T, the magnetizing current decreases linearly. Also, the output current decreases linearly.

Mode 3 $[t_2, t_3]$: At t_2, the output current arrives at zero and the output diode D_o is turned off. Then, the constant input current flows through the primary side of the transformer T, C_1, D_1, and S_2.

Since the average voltage across the primary side of the transformer should be zero at steady state, the voltage gain is given by

$$\frac{V_o}{V_{in}} = \frac{D}{nd_1}.\tag{4}$$

3 EXPERIMENTAL RESULTS

In order to verify the performance of the proposed PFC converter based on SEPIC converter, a prototype has been built and tested. The input voltage v_{ac} is 180Vac. The switching frequency is 100kHz. The maximum output power is 150 W. The input inductors L_1 and L_2 are selected as 2mH. The capacitor C_1 is selected

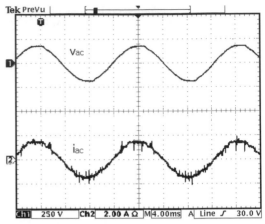

Figure 4. Experimental waveforms: v_{ac} and i_{ac}.

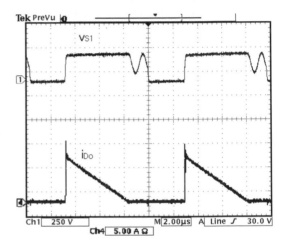

Figure 5. Experimental waveforms: v_{S1} and i_{Do}.

as 0.47uF. The transformer turn ratio is selected as 1 and the primary inductance is selected as 89uH. Fig. 4 shows the experimental waveforms of the proposed PFC converter at full load. The input current i_{ac} follows the line voltage. The measured power factor is above 0.99. Fig. 5 shows the switch voltage v_{S1} and the output diode current i_{Do}. The measured efficiency is 90.8% which is higher than that of the conventional SEPIC PFC converter by 1%. Due to the absence of a bridge diode, the conduction loss is reduced and the efficiency is improved.

4 CONCLUSION

A simplified power factor correction (PFC) converter based on single-ended primary inductor converter (SEPIC) topology has been proposed. To verify the performances and feasibility of the proposed PFC converter, a prototype was built and tested. The proposed PFC converter showed higher efficiency than the conventional SEPIC PFC converter. Due to the absence of a bridge diode, the conduction loss is reduced and the efficiency is improved by 1%.

REFERENCES

[1] P. F. Melo, R. Gules, E. F. R. Romaneli, and R. C. Annunziato, A modified SEPIC converter for high-power-factor rectifier and universal input voltage applications, IEEE Trans. Power Electron. (2010), pp. 310–321.

[2] E.H. Ismail, Bridgeless SEPIC rectifier with unity power factor and reduced conduction losses, IEEE Trans. Ind. Electron. (2009), pp. 1147–1157.

[3] D.S.L. Simonetti, J. Sebastian, and J. Uceda, The discontinuous conduction mode sepic and cuk power factor preregulators: analysis and design, IEEE Trans. Ind. Electron. (1997), pp. 630–637.

[4] J.–M. Kwon, et al, Continuous-conduction-mode SEPIC converter with low reverse-recovery loss for power factor correction, IET Proc. Electr. Power Appl. (2006), pp. 673–681.

[5] J.C.W. Lam and P.K. Jain, A high-power-factor single-stage single-switch electronic ballast for compact fluorescent lamps, IEEE Trans. Power Electron. (2010), pp. 2045–2058.

Analysis of distributed network management model based on independent self-organizing domain

Ma Kun Guo
Department of Information Construction, Academy of National Defense, Wuhan, Hubei, China

Yi Min Yu
Department of Command and Control, Academy of National Defense, Wuhan, Hubei, China

ABSTRACT: Traditional main and spare frame technology cannot meet what modern warfare demands of the military communication network management system in terms of invulnerability. Based on the logic center, this paper proposes an independent self-organizing domain strategy, which can extend the ability responding to a single point of fault to several segments and improve the invulnerability and survivability of the whole network without any addition of network components and hardware.

Keywords: network management, distributed, main and spare frame, invulnerability

1 INTRODUCTION

The Wide Area-covered hierarchical communication network needs a reliable and effective network management system for maintenance to make sure of the normal operation of the business system. The network management system with high invulnerability and survivability is very important for complex communication network. Because of the ever- changing environment of war and the uncertainty of the possible destruction of network facilities, the traditional work mode of the main frame and spare frame can't meet the requirements in terms of survivability from network, so in this case, based on the distributed location back-up mechanism, the logic definition of the concentrated network management center into each disperse node can ensure its survivability through good distributed self-organizing environment protocol regulations.

The distributed self-organizing environment mainly consists of several independent self-organizing domains, each domain consisting of all logical entities in the independent logic network management center; the orderly interconnection of main intra-domain entities between hierarchical self-organizing domains thus constitutes a complete distributed self-organizing environment. Therefore, the in-depth analysis of self-organizing domains will play a key role on the design of self-organizing environment protocol regulations. Based on the analyses of the physical fault models and the changing scenes of network topologies, this paper will emphatically discuss the establishment and maintenance of logic topologies in the self-organizing domains and put forward a kind of independent

self-organizing domain protocol regulations with logic network management center to increase the invulnerability of the network management.

2 TOPOLOGY STRUCTURE AND SELF-ORGANIZING DOMAIN

The so-called network invulnerability is that when in the network happen certainty or random failures (such as link or node failures), the network can maintains or restores its performance to an acceptable degree. The measurement of random invulnerability mainly depends on the network topology structure [1].

Two kinds of topology structures. In the distributed self-organizing environment, the topology structure can be divided into the physical topology and the logical topology. In the physical topology, each entity is a really existing physical device, and has a unique physical address PA, as shown in Figure 1. Among these physical entities there exist a certain logical relation, which does not mean that the informational interaction lies between these entities, but focuses on geographical and administrative management establishment, forming an fully connective interworking environment between physical entities.

The base on which the distributed self-organizing environment works is the logical topology structure made up of a series of logical entities with a tree structure. The tree structure of the logical topology is similar to that of the physical system, the difference between the two of which is that the link line represents the

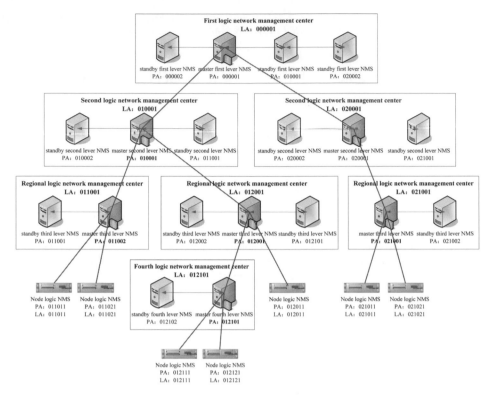

Figure 1.   Network Topology Structure.

generation and maintenance of the information inter-active path in a business system. Only one logical entity is added to the information tree in every network management center while the other logical entities and the main entities are coordinated by self-organizing relations [2].

The components of domain.   Entities, channels and rules are the three elements of the self-organizing domain. A self-organizing domain contains a certain number of logical network management entities, and its physical device may be the server of the physical network management center at the same or lower level. The parameters that need to be planned and allocated in advance are the maximum number of logical enti-ties in a domain and which physical entities should be added to a domain. The channel is the informa-tion transmission path connecting each logical entity within the domain, and the requirement of the self-organizing basic logic is that all entities can exchange the necessary information between each other.

Rules are the soul of the self-organizing domain, which determines the need and the way how to exchange signaling and data information between log-ical entities. In the self-organizing environment, the equivalence between logical entities requires that the unity of time be the base on which logic is dis-tributed, that the selection and replacement of the dynamic center be the core of rules between entities, and that the consistency of data between entities be

the prerequisite to ensuring the normal operation of the business system.

3   TOPOLOGY STRUCTURES AND ITS CHANGES

The topology structure solves the problems of the channel establishment and maintenance of the self-organizing domain, and it serves as the base on which the self-organizing rules are run [3]. Different from the simple carrying relationship in the general network model, the equivalent and no transcendental self-organizing logic will bring the greater signaling scatter overhead. Therefore, rules should be designed with a close topological relationship, making full use of the implicit information from the underlying channel.

Fault properties of the physical connectivity. Fully connected physical networks display diverse fault states, Figure 2 shows the basic physical con-nectivity fault scenes in the self-organizing domain, and the states of all physical connectivity faults can be available by these kinds of basic scenes [4].

The figure displays the fully connected self-organizing domain containing five entities, which shows that there exist three basic scenes: the Single Point of Failure happens when a certain entity stops running or when an entity fails to be linked to the net-work, as the result of which the said entity breaks away

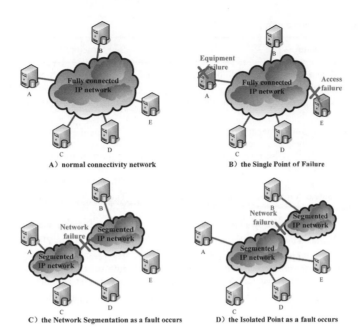

A）normal connectivity network

B）the Single Point of Failure

C）the Network Segmentation as a fault occurs

D）the Isolated Point as a fault occurs

Figure 2.　Physical Fault Models.

from the fully connected environment, completely losing serving ability; the Network Segmentation as a fault occurs in the fully connected IP network itself, the result of which is that the network is cut into two separate fragments, within which entities can still keep the fully physical connectivity between each other; the Isolated Point is a special network segment, within which only one entity is contained in a fully connected segmentation.

The Ways of Topology Construction. In the equivalently distributed IP network, the self-organizing domain entity at the network edge and linked to the network is absolutely impossible to perceive the changes of the physical connectivity. Therefore in the self-organizing domain, it is necessary to construct a logic topology structure to link every entity and perceive the changes of the physical connectivity by way of the signaling interaction between all logical entities [5].

The topology structure is basically required to truly reflect the connective state of the physical network. In a fully connected physical network, both the bus and star structure can realize this purpose because they use the connectivity transfer law while the ring structure actually budgets transcendental assumptions on the physical topology, which can not be adopted in the self-organizing domain. Table 1 makes a simple comparison on applicability and the ability between the bus and the star topology structure.

Carefully compare the construction modes of the three topologies, and the basic conclusion can be drawn that in the independent self-organizing domain the star structure should be adopted to establish each intra-domain entity logical topology.

4　THE TOPOLOGY RULE ANALYSES OF IN THE SELF-ORGANIZING DOMAIN

The topology analyses in the self-organizing domain. The star structure is actually a form in which the connective relationship between intra-domain entities tends to be stable, and the connective state between intra-domain entities is constantly changing, so the topology structure should go through a course of constant establishment and maintenance [6]. Figure 3 shows the changing scenes of topology.

The analyses of the changing scenes of topology in Figure 3 are as follows:

1) The generation of the initial star topology is an ideal scene. All the entities within the same domain get started meanwhile, form the Entity 3-centered star topology after a period of time's competition, and then monitor the changes in connectivity by way of this topology;

2) After the basic star topology is formed, subsequently Entity 7 gets booted, which joins the original star topology as the result of its consultation with the center of Entity. This scene is actually the basic situation in which the system loads and runs;

3) When the original network connectivity changes, a complete self-organizing domain is split into two or more fragments, and the fragment containing the original center remains to maintain the original star topology (the number of leaves decreases), the entities in other fragments forming another independent self-organizing domain, a new center through competition and consultation;

Table 1. The Comparison between the Two.

Factors	Bus Structure	Star Structure
Transfer Mode	Work in a broadcast way; all the entities receive the signaling from any other	Signaling interaction of all entities can be achieved through the temporary center
Physical Demands	Underlying network has the multicast ability	Underlying network has the unicast ability
Perceiving Way	By monitoring the signaling being broadcast, each entity independently perceives the connectivity of other entities and forms the knowledge of the topological state	Through the temporary center, the changes in connectivity are perceived and judged to form the knowledge of the topological state; Inform other entities of the topological state through the temporary center
Perceiving Ability	The Single Point of Failure, Segmentation and the Isolated Point	The Single Point of Failure, Segmentation and the Isolated Point
Perceiving Time	Perceive rapidly and synchronously	Perceive rapidly and synchronously
Processing Load	An balanced amount between entities	The center takes a heavier load
Rule Complexity	The perceiving rules are simpler but the selecting rules are complex and it's more difficult to consistently and synchronously finish the data	The perceiving rules are simpler but the selecting rules are complex. It's simple to settle the problem of consistency of the data
Environmental Adaptability	It has a stricter standard on both the underlying multicast business capacity and transmission quality. The allocation is more complex and it's difficult for a new entity to enter	It only requests the underlying level to have the unicast ability and has no too many demands on the allocations of other devices
Conclusion	Deny this program	Adopt this program

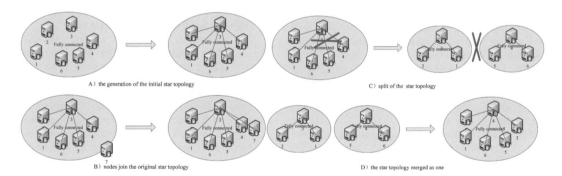

A) the generation of the initial star topology

C) split of the star topology

B) nodes join the original star topology

D) the star topology merged as one

Figure 3. The Changing Scenes of Topology.

4) Each sub-domain of the same self-organizing domain needs to continuously probe the changes in connectivity with other sub-domains while maintaining its sub-domain topology, and when the network regains connectivity, the two sub-domains will be merged as one, removing the center of one certain sub-domain and forming another star topology based on the center of the other sub-domain.

From the analyses above we can see that when a network fault happens, the topology in the self-organization domain always changes in one or more star structures. The star structure is just a network state when a self-organizing domain is relatively steady. Between the stars there exists no mutual perception ability, which requires that the center of each sub-domain to search for all possible sub-domains at any time to form and reflect the topology with true connectivity of the self-organizing domain.

The analyses of the establishment and maintenance of the logic topology. The establishment and maintenance of the logic topology is the most basic problem for the self-organizing domain work, which needs a set of strict and comprehensive protocols and rules for support. The establishment and maintenance of the logic topology can be summarized as three types of entities and three kinds of operation. Three types of entities refer to the temporary central node (the main logic entity), the leaf node (the spare logical entity), and the idle node (the still uncertain main or spare logical entity). The definition of the three types of entities not only shows the role positioning within the self-organizing domain at a specific point in time but also shows the possible state transition of the same

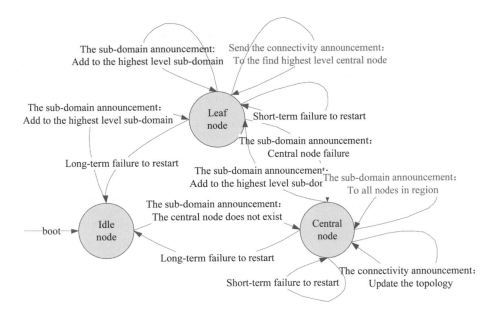

Figure 4. The Entity State Transition.

logical entity. Figure 4 shows the way of the entity state transition.

The entity state transition is based on the fact that it has accomplished the analyses of the sub-domain receiving information and the machine loading state. The sub-domain announcement or the connectivity announcement is the most basic signaling method in the self-organizing domain.

The rule planning analyses of the establishment and maintenance of the logic topology. The rules of the establishment and maintenance of the logic topology is the state transition logic that in the self-organizing domain each logic entity accomplishes according to the external signaling and the loading mode. The description of the basic functions and planning demands of the rules is given below:

1) The state transition of the idle node

After being powered up, the entity enters the idle state, and in a period of quiet time monitors the sub-domain notice from the central node; when it still receives no sub-domain notice at the end of the period of silence, the entity then transforms itself into the temporary central node; when it receives the notices from one or more central nodes, according to the selection rules of the central node, the entity automatically joins the central node sub-domain with the highest level, transforming itself into the temporary leaf node of the said sub-domain and transmitting the connectivity notice to the corresponding central node.

2) The state transition of the leaf node

The leaf node surely belongs to a certain sub-domain, and according to a defined time interval periodically sends the connectivity notice signaling to the central node of the sub-domain, showing that it is still active; the active leaf node is always monitoring the sub-domain notices from each sub-domain central node. After receiving the sub-domain notice, the temporary leaf node turns into a formal sub-domain leaf node.

When receiving the notices from multiple sub-domains at the same time with the original central node still active, the original central node comparing the priority of each sub-domain and automatically joining the central node with the highest level, will transform itself into the temporary leaf node of this very sub-domain and transmit the connectivity notice to the corresponding central node.

After the fault restarts, the leaf node maintains the original state if the time for the automatic judgment of the fault period is a short-time restart, and it turns into the idle node if the restart takes long.

3) The state transition of the central node

After it continuously monitors the sub-domain notice for a period of time, and if it finds no presence of other sub-domains, the temporary central node then converts into a formal bus node; at certain time interval the central node transmits the sub-domain notice signaling to all the nodes within the respective organizing domain; the central node continues to monitor the connectivity notice from the leaf node, maintains the star topology of this domain, and updates the sub-domain notice; the central node continues to monitor continuously monitors the notices from other sub-domains, and if it finds the presence of other sub-domains with a higher priority than itself, the central node automatically joins the central node sub-domains with the highest level, transforming itself into a temporary leaf node of the said sub-domain and sending the connectivity notice to the corresponding central node. After the fault restarts, the leaf node maintains the original state if the time for the automatic judgment of the fault period is a short-time restart, and it turns into the idle node if the restart takes long.

5 CONCLUSION

Based on the logical network management center, this paper puts forward an independent self-organizing domain strategy, which can not only further abstract mathematical models to simulate and verify the feasibility, but also can be used to guide the establishment of the self-organizing environment with the multi-level network management system distribution, which thus enhances the NMS survivability.

The next step is to design a distributed self-organizing agreement prototype and adopt the network emulator OPNET for further simulation tests with the hope that through these efforts the protocol standards more suitable for the actual project concerning the network invulnerability can be found, and that some references can be provided for the research on the NMS invulnerability in complex network environment.

REFERENCES

Koroma J, Li W. A Generalized Model for Network survivablity[c]. The 2003 Conference on Diversity in Computing (TAPIA'O3), Atlanta, Georgia, USA. 2003.

Ellison R, Linger R. Survivable Network System Analysis: a Case Study [J]. Software, 1999, 16(4): 70–77.

Jha S, Wing M. Survivability Analysis of Networked Systems[c]. ICSE 2001, Toronto, 2001.

Ye Yousun. Analysis of military communication network and system integration [M]. National Defence Indystry Press, 2005: 46–49.

Krings A W, Azadmanesh A. A Graph Based Model for Survivability Applications[J]. European Journal of Operational Research, 2005, 164(3): 680–689.

Traldi L. Commentary on: Reliability polynomials and link importance in Networks[J]. IEEE Trans Reliability, 2000, 49(3): 322.

Information Technology and Computer Application Engineering – Liu, Sung & Yao (Eds)
© *2014 Taylor & Francis Group, London, ISBN 978-1-138-00079-7*

Research on defensive posture for distributed simulation

M.L. Liu, J. Zhang & Bo Yun Liu
Scientific Institution, Naval University of Engineering Wuhan, China

ABSTRACT: Aim at the vulnerability of simulation networks, a concept of dynamic defensive posture in distribute simulation networks was examined. A concept of defensive posture has been proposed that it is provided by knowledge of whether and how simulation networks resources are vulnerable to attack. The constituent elements of defensive posture were discussed. Finally, a variety of research problems for which the solutions would contribute significantly to our ability to identify a network's defensive posture was proposed.

Keywords: Defence; network; Distributed Simulation

1 INTRODUCTION

The distributed simulation networks are complex and dynamic environments. With the rapidly development of the relationship and interdependence of those applications, and other factors make network difficult to understand, even if they have been carefully designed. This is particularly true with respect to the security of the network, for it is often not clear whether security is strengthened or weakened by changes in the network configuration and the security often be neglected when operations come to be the most attention point.

This paper introduces the concept of network defensive posture of distributed simulation. Roughly speaking, to know the defensive posture of the network is to know which elements of the network are exposed to potential attack by a malicious agent, and the extent to which such an attack would affect the network's operation. A defensive posture analysis system is an application that can determine the network's defensive posture.

1.1 *Background*

A simulation administrator needs to understand network security, and how network management decisions affect that security at the phase of system design. The complexity of large networks is a typical model of large-scale of simulation, however, in which user activities, security policies, local configurations, and software vulnerabilities interact, makes it difficult for the administrator to know what is happening on the network, much less understand the significance of the activity. Generally, the operation security problem is often result of the corresponding low-level detail problems.

1.2 *Security view*

Security view is a concept that has found application in many different areas In the context of network security management, the security view consists in a valid interpretation of the meaning of network activity, an understanding of its likely consequences for the provision of network services and enforcement of security policies, and the capacity to make informed network management decisions .This requires a tightly coupled knowledge of network management, network security information and simulation security operations, including knowledge of the network infrastructure and vulnerabilities. The primary aim is to be aware not just of the low-level details of network events, but of their high-level impact on the operations and services which the network supports. To have security view is to have a clear and correct picture of the network's state as it evolves in time and an understanding of how that state affects network services.

1.3 *Critical resources*

An essential element of security view is knowledge of the critical resources on the simulation network. Critical resources are defined relative to the missions or operations which the simulation is supporting which include database, RTI, computer network etc. They are the resources without which one or more important services or capabilities would be compromised. The assessment of critical resources would likely realize as assigning a value to each of the network elements, where the value reflects its importance in supporting the simulation mission.

1.4 *Vulnerability of simulation network*

The set of network resources vulnerable to an attacker starting from a particular point in the network defines the partial network exposure relative to that point. The step-by-step description of how the attacker can carry out the attack is called an attack path. The total exposure is specified by giving all vulnerable targets and the possible attack paths to those targets. It is important to understand that attacks can consist of more than one stage: the attacker may move through multiple hosts and exploit multiple vulnerabilities or configuration problems before finally reaching the target. Also, such multi-stage attacks may in general originate from any point inside or outside the network.

1.5 *Defensive posture*

The defensive posture of a network is the set of exposed, critical resources on the network. It combines knowledge of the critical resources with the exposed resources. From an operational point of view, the defensive posture is a current, prioritized catalogue of existing security weaknesses requiring attention.

It is important to understand the relationship of defensive posture to risk, for they are not identical. Defensive posture states which of the organization's most critical assets are vulnerable to attack, but it does not say how probable an attack would be. Risk, on the other hand, folds into defensive posture an estimate of the probability of a particular attack being launched against the network. Risk is often used to prioritize vulnerability instances. An attack could be possible against a critical asset, but if the probability of the attack is low the actual risk to the network is low, and the network operator may decide to focus his efforts on other problems. A problem with using risk to prioritize is that often the probability of an attack is difficult to specify precisely. There are cases where the probability of a particular attack is known to be high, such as, the use of risk introduces an intrinsic uncertainty to the prioritized list of threats against the network, possibly distorting that list if the probabilities are estimated wrongly. By contrast, defensive posture is assessed by combining knowledge of which assets are most important with an analysis of whether and how those assets could be attacked.

2 EXPLOREA DEFENSIVE POSTURE

2.1 *Define the most critical resources*

The simulation is explored to provide services and information in support of some set of goals or missions. For a particular set of goals, it is reasonable to expect that certain assets whether services, information, or devices will be critical to success. The criticality of each asset is always defined relative to the goals and priorities of the mission. For instance, a equipment simulation has the different critical resources of a battle simulation. A significant challenge for a defensive posture analysis system is to accept a high-level, prioritized description of the mission's requirements, and map that description onto lower-level network services, information, and devices.

2.2 *Find out the expoure resources*

The exposure is the set of assets that are vulnerable to attack, and is always defined relative to a particular starting point. After all, an attacker who starts with administrator privileges on a central network server canlikelyattackmoretargetsthananunprivilegedattackerontheInternet. We should be able to specify which assets are accessible to an attacker who begins at any particular location, whether inside or outside the network. A full specification of location may also include the attacker's privileges, since privileges can affect the attacker's reach. The exposure includes not only those assets directly accessible to the attacker, but also those which can be reached by multiple steps or stages. There three issues should be considerate:

* From which network locations is asset exposed to attack?
* How could asset be compromised by an attacker at a particular location?
* Which safeguards are protecting asset?

2.3 *Pitch the most critical and expoure resources*

Defensive posture identifies the exposed, critical resources on the network. When the two issues are brought together, the most critical assets vulnerable to attack must be confirmed. Given a list of network mission priorities, the steps will be as follows:

(a) Determining the most critical assets
(b) Whether and how they are vulnerable to attack
(c) Produce an ordered list of exposed, critical assets.

The ranking of critical assets may be ordered simply based on the criticality of the vulnerable asset, or according to some more elaborate scheme that also considers other factors, such as attack complexity, directness, probability of success, and so forth. The precise mixture of criteria by which a list of attacks should be ranked has yet to be determined, but certainly the criticality of the targeted asset will be the central, if not sole, consideration.

Not only the problem discussed upon should be clarified, but also the consequences of a given attack must be research. If a particular asset is vulnerable to attack, the impact of a successful attack would be answered. The impact is one important criterion relevant to assessing the severity of the attack. An attack that completely compromises an essential service, or even affects other services running from the same hardware, for instance, is more severe than one which partially compromises the service. An outstanding problem, however, is how to precisely specify both the type and extent of impact. One way to characterize the type is in terms of confidentiality, integrity, and availability. If a asset is a sensitive file, the attacker would be able to read it , change it, or even delete it, It is often less clear, however, what compromising confidentiality or integrity means for a service or a device. As for the extent of impact, it is again not clear how this can be described precisely enough to be meaningful, but generically enough to be implemented in an automated system. This is a matter requiring further clarification.

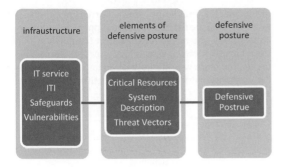

Figure 1. Flow diagram of the defensive postrue.

3 ELEMENTS OF DEFENSIVE POSTURE

As the figure1 shown, the output of this diagram is defensive posture. It is easy to see that to confirm a defensive posture, the input knowledge of the operational requirements, IT infrastructure (ITI), network safeguards, known software vulnerabilities are needed.

3.1 *Operations requirements*

The operations requirements are high-level, prioritized descriptions of the services and information the network must support or provide. They are specified by a force commander, and are derived from the mission of the organization or group using the network. They may be dynamic. For instance, the commander may require VoIP services and encrypted email services with high priority and access to an image server with medium priority, for the duration of a mission, and also require video conferencing services with high priority on the morning of one particular day. In practice, the operations requirements would probably be defined by providing a template of available services to the force commander, and having him indicate which are needed, when they are needed, and with what priority. He, or the network commander who decides which network resources can best meet the requirements, should also indicate which services require redundancy in case of failure. The commander should been couraged to be soberly realistic about prioritization of network services to avoid having everything marked as high priority.

3.2 *IT infrastructure (ITI)*

The IT infrastructure (ITI) is a crucial element needed to determine the defensive posture. To accurately identify the network's exposed resources; a thorough model of the network's structure and state is needed. Knowing the ITI involves knowing how many hosts are on the network, their connectivity, what operating systems and software each host has installed, and how that software is configured; it involves knowing what servers are present on the network, the services they

provide, and the interdependencies among these services; it includes knowing the access controls present on files and applications, and the access permissions granted to users; it involves knowing the configuration of network firewalls and routers, so as to determine the network connectivity and allowed traffic flow. This information must be collected and used to construct a network model on which analyses, such as searches for attack paths, can be performed.

3.3 *Software vulnerabilities*

The software include two aspects: one is the platform of the host in a simulation networks, the other is the operation system of a simulation. Some attacks against a network proceed by exploiting one or more vulnerabilities present in software on the network. These vulnerabilities are typically the result of defects in the design of the software, and, if exploited by a sufficiently capable attacker, may permit the attacker to obtain unauthorized access to hosts, adversely affect network performance, or otherwise compromise the confidentiality, integrity, and availability of network resources. Ideally, no such vulnerabilities would be present on the network, but realistically this ideal is unattainable. New vulnerabilities in deployed software are always being discovered, and there will often be a gap between the announcement of vulnerability and the availability of a patch. Even after a patch become available, administrators are sometimes prevented from applying it immediately. If the organization's policy permits users to install their own software, it would expect that users will frequently fail to keep their applications patched.

4 CONCLUSION

4.1 *Network modeling*

The model of the network is the foundation for the analysis of exposed and critical resources and is, of course, a critical part of a defensive posture model. Several important questions must be addressed in the design of the model: What elements of the network must be modeled, and what elements can be ignored? At what level of granularity should the network be modeled? The answers to these and related questions will depend on the range of attacks that the model attempts to identify. A minimal model would likely include privilege escalation attacks; a more comprehensive model might also include denial of service attacks, eavesdropping and sniffing attacks on data in transit, or data tampering; more complex still would be models of social engineering attacks, or attacks on the physical network infrastructure. A good model would also permit the network administrator to evaluate the effects of changes to the network configuration before the changes are actually made; alterations could be made in the model, and the security consequences evaluated prior to deployment.

4.2 *Asset valuation schema*

Essential to a quantitative assessment of asset criticality, and potentially also to assessment of a successful attack's impact on the network, is some means of assigning value to the network assets. The value in question, whether expressed numerically or categorically (for example, on a High-Medium-Low scale), should indicate the level of support the asset provides to the meeting of the network's current priorities. It may also indicate the nature of the support provided, or, equivalently, the negative impact on the network that would result from the asset's loss or compromise. This impact measure would likely be specified in terms of the conventional metrics of confidentiality, integrity, and availability.

A consideration on infrastructure of distributed simulation

M.L. Liu, J. Zhang & Bo Yun Liu
Scientific Institution, Naval University of Engineering, Wuhan, China

ABSTRACT: In a distributed simulation ,there are a lot of simulations which are produced by pure-visual, semi-practicality etc. The infrastructure of the system would be considerate for the running, reuse and integration. The High Level Architecture (HLA) is a distributed simulation architecture designed to facilitate interoperability and promote software reuse within the Modeling & Simulation (M&S) community. In HLA, Federates communicate via a distributed middleware called the Run-Time Infrastructure (RTI). The HLA specifies the interface between each federate and the RTI but does not specify how the RTI is implemented. This paper discusses the technical, political and economic considerations one must weigh when selecting a HLA RTI implementation, which play an important role in domain of management of distributed simulation.

Keywords: HLA; infrastructrue; Distributed Simulation; RTI

1 INTRODUCTION

This paper discusses the technical, political and economic considerations one must weigh when selecting an RTI implementation for a HLA compliant distributed simulation. Researching into these issues was the high-level initiated work in a large project, which play a general role in other projects: Visual battle Environments, 3D modeling, and effecting etc. What must be considerate for achieving data transport to realize the function upon has been presently engaged in a joint HLA/RTI selection process.

1.1 Background

The HLA is a distributed simulation architecture designed to facilitate interoperability and promote software reuse within the M&S community. The HLA was originally defined by the U.S. DMSO and is now an IEEE standard (IEEE 1516). In the HLA, the unit of software reuse is the federate. A group of communicating federates at run-time is called a federation. Federates communicate with each other via a distributed middleware called the RTI. The RTI provides the services that allow federates to manage the federation's global state. Contrary to popular belief, the RTI is not a centralized software entity that federates plug into. Rather, each federate communicates with its own local copy of the RTI software library (called the Local Run-time Component, or LRC). The LRCs then communicate amongst themselves to coordinate the execution of the federation. Thus, the RTI

library encapsulates the complexities of inter-federate communication.

1.2 Related works

The HLA specifies the interface between a federate's simulation logic and the RTI library. More precisely, HLA defines two interfaces: the RTI ambassador interface and the federate ambassador interface. The RTI ambassador interface is implemented by the LRC and defines the services that the federate can invoke on the RTI. The federate ambassador interface is implemented by the simulation logic and defines the events that the RTI can signal to the federate. The HLA does not, however, specify how the LRC is implemented. The RTI vendor is free to choose how the LRC is organized internally as well as the inter-LRC communication protocol. Thus, all federates within a federation must agree to use the same RTI library. To complicate matters, there are two interface specifications currently in use. The first is the DMSO specification we referred to as 1.3-NG. The second is the newer 1516 specification, defined by the IEEE 1516 standard and based on the 1.3-NG interface. Federates can conform to either or both standards, and likewise, RTI libraries can conform to either or both. In order for federates to communicate with an RTI, however, all must support the same interface specification. Several RTI implementations exist that conform to either the 1.3-NG or 1516 interfaces (or both). It is therefore natural to wonder which RTI implementation best suites the needs of a particular community.

The rest of the paper is organized as follows: Section 2 discussed the technical factors. In section3 and section4, the political and economic factors, respectively, Conclusion is mentioned in section 5.

2 TECHNOLOGICAL CONSIDERATIONS

2.1 Challenges by RTIs

There is no single RTI implementation suitable for all federations. Different RTI vendors build their RTI with different design goals. For example, some may provide a more efficient time management implementation and are thus better suited to constructive rather than virtual simulations. In general, one should approach the decision of which RTI to use on a per federation basis. It is important to understand that, in general, the choice of RTI library used by a particular federate is a compile-time decision. That is, switching to another RTI library will require recompilation of the federate. As the practical experience, the factors that must be weighted when choosing the RTI is follows:

* The reusing of a existed simulation.
* The need to develop new federates
* The required level of federation performance
* Choice of Federation Object Model (FOM)
* Choice of computing platform
* Choice of development tools

The degree to which the above factors are relevant varies considerably. In fact, some of these factors have little or no bearing on the RTI selection process but are discussed nonetheless to dispel potential misconceptions. Each such issue is discussed further in the following sections.

2.2 Reusing of existing simulation components

The following distinguishes between three types of simulation components that one may wish to reuse within a federation

* Distributed Interactive Simulation (DIS) compliant simulations
* Federates that comply with the IEEE 1516 interface
* Simulations that are neither DIS nor HLA compliant but do provide a programmable interface

DIS simulations require what is commonly referred to as a DIS-to-HLA bridge (or gateway). This is a software component that translates between the DIS Protocol Data Units (PDUs) and HLA objects/interactions.

The second situation is a better/easier solution that requires no source code changes is to choose an RTI implementation that supports both interfaces. Some RTI vendors choose to package this as a separate product while others provide this capability as part of the RTI itself.

The final category of simulation components that may be required to participate in a HLA federation are those that may be neither DIS nor HLA compliant, but those that provide some programmable way to access and/or alter the simulation's state. This type of simulation component must be encapsulated in a federate that exposes the simulation's state as one or more HLA objects/interactions. Some of these HLA-wrappers already exist for simulation frameworks such as .The choice of RTI is inconsequential since the HLA-wrapper is just another federate that needs no knowledge of the RTI's implementation.

2.3 Federate development

The choice of RTI can have a large impact on the development speed of new federates. The original RTI implementations were largely considered as a black boxes – the federate developer could not examine the RTI's state at runtime. Later implementations addressed this problem with various debugging/monitoring tools that report the RTI's internal activities (publications, subscriptions, object updates/reflections, interaction sends/receives, lookahead values, etc.). Such tools can be either open or closed. Closed tools provide a user interface that presents the relevant information. Open tools provide the ability to plug custom code into the RTI to both monitor and affect the behaviour of the RTI. Open tools enable more advanced diagnostics to be performed on a federation at run-time.

2.4 Performance

As previously discussed, no two RTI implementations are equal. It is therefore natural to wonder which RTI performs best in a given simulation. For example:

* Which RTI is most efficient when executing over a wide area network?
* Which RTI most efficiently implements the Data Distribution Management (DDM) service group?
* Which RTI provides the lowest latency communication?

It is important to distinguish between RTI performance and federation performance. In general, the former is a subset of the latter. That is, there are several factors that can affect a federation's performance that are not directly linked to the selected RTI. For example, a time regulating federate performing an expensive computation will slow the execution of all time constrained federates. In this case, improving the performance of the federation requires faster hardware for the computationally expensive federate or a redesign of the federate itself (perhaps replacing the expensive computation with an acceptable approximation).

2.5 FOM

The Federation Object Model (FOM) defines the vocabulary of data exchange within a federation. All federates participating within the same federation must support the same FOM.

There is a misconception amongst many researchers entering the world of HLA simulations that the choice of FOM is somehow linked to the choice of RTI implementation. While its true that a federation must agree on a single FOM, all RTIs can process all FOMs.

Unfortunately, the standard of fom has great different among the various standards of RTI/HLA version, That is, the file format of each differs so an RTI implementation may only recognize the file format that corresponds to the interface it supports. However, the differences between the formats are largely syntactic and easily convertible. The issue of FOM-agility is often brought up when discussing RTIs. FOM-agility is defined as the ability of a federate designed for one FOM to be reused in a federation using another semantically equivalent yet syntactically distinct FOM. That is, relevant objects/interactions in the new federation must somehow be mapped to objects/interactions understood by the reused federate. FOM-agility can be supported at different levels – one of which is at the RTI itself. Some RTI implementations support a plug-in architecture that allows the federate to supply a translation layer to accomplish the necessary FOM mappings. However, FOM-agility is not a direct factor in choosing an RTI.

2.6 *Computing platform*

A computing platform refers to the mix of hardware, operating system and programming language used to implement an application. This paper distinguishes between three platform related issues:

* The implementation language of the RTI
* The RTI's supported hardware and operating systems
* The RTI's supported language bindings

The most important factor in choosing an RTI is whether there is an implementation for your chosen hardware and operating system(s). In this case, the cross-platform nature of a Java-based RTI is advantageous since it can run without modification on any Java-enabled platform. In a heterogeneous computing environment, it is important that the chosen RTI vendor support all your platforms since, as previously stated, federates within the same federation must use the same RTI implementation.

2.7 *Development tools*

The services provided by the RTI are both complex and low-level. Third-party tools exist that both isolate and abstract the RTI's functionality and provide a simpler application programming interface (API) to the federate developer. While some RTI vendors also provide such tools, they are not (and should not) be dependent on that vendor's RTI.

3 POLITICAL CONSIDERATIONS

As part of the RTI selection process, consideration of the political issues is also necessary. Such issues are often implicit in the various software components used within an M&S environment along with issues such as availability, timeliness, interoperability, costing and standards compliance. They are also heavily influenced by less objective matters, such as anecdotal experience and technological, personal and organizational biases. While not necessarily "scientific", it is important for decision makers to realize that they do exist and that they can colour the perception of other factors in the decision process.

4 FINANCIAL CONSIDERATIONS

The final, yet extremely important issue affecting RTI selection is that of economics. When purchasing RTIs for federation development, one needs to consider that the pricing schemes adopted by vendors show significant variation. The price of a RTI is mainly based on the federate that it can support, it is generally exist a basic price ,the higher the price, the more federate that the RTI can support. So, before purchase a RIT, the following questions must be solved:

* How many federates are needed in a simulation
* What about redundancy
* Computing and compareing the various RTI's totla price for the practical simulation, getting the optimiza purchase solusion.

Thus when purchasing commercial RTIs, careful consideration should be given to a variety of factors that contribute to the cost of the package.

5 CONCLUSION

This paper has outlined the three broad areas of concern that need to be considered in selection of technologies associated with the use of the High Level Architecture, or HLA. In each of these areas, different specific issues were highlighted and briefly addressed. It is again stressed that this paper is intended to serve as an initial overview and introduction to these various issues and is not a final compendium.

As obvious from the breadth and depth of the technical issues associated with distributed simulation technologies, further work in understanding and addressing the performance as well as technical flexibility and constraints associated with any specific HLA technologies, including RTIs, is warranted.

Analysis on warpage of support structure of computer hard disk for optimum processing by precision injection molding

Yung Ning Wang & Her Shing Wang
Department of Industrial Engineering & Management, National Taipei University of Technology, Taipei, Taiwan

Wen Tsung Ho
Department of Business Administration, Takming University of Science and Technology, Taipei, Taiwan

Yi Lin & Yung Kang Shen
Department of Business Administration, Takming University of Science and Technology, Taipei, Taiwan
Graduate Institute of Industrial and Business Management, National Taipei University of Technology, Taipei, Taiwan
School of Dental Technology, College of Oral Medicine, Taipei Medical University, Taipei, Taiwan

ABSTRACT: This study emphasizes on warpage of support structure of computer hard disk for optimum processing by precision injection molding. Firstly, the authors use Moldflow software to analyze the runner's balance on multi-cavities for the support structure of computer hard disk. Then this study accords to these data to manufacture the real mold. This study uses different processing parameters (injection speed, injection pressure, mold temperature, packing time and melt temperature) to fabricate support structure of computer hard disk using precision injection molding. The authors use Taguchi method to find out which factor is more important for warpage of support structure of computer hard disk. The results show that the most important factor for warpage of support structure is followed by packing time, injection speed, melt temperature, packing time and the last one is the injection pressure.

Keywords: Warpage, Precision injection molding, Numerical simulation, Optimum processing

1 INTRODUCTION

Precision injection molding is used to manufacture micro structures and is among the most common and versatile methods of mass-producing complex plastic parts. Sul *et al*. [1] demonstrated that the mold temperature is the key processing parameter in injection molding. Shen *et al*. [2] applied micro-injection molding and micro-injection compression molding to form light-guiding plate microstructures. Their test results showed that mold temperature was the main factor in both processes. Wan *et al*. [3] studied of the window frame manufacturing by injection molding was carried out with aid of Moldflow analysis. The results showed that natural fiber composite was suitable to fabricate window frame.

This study fabricated support structure of computer hard disk from plastic. The ABS+PC material that performed best was then used to fabricate previous product by precision injection molding. The processing windows for support structure of computer hard disk are discussed. The aim of this study was to determine the minimum warpage of support structure by precision injection molding using Taguchi method. Finally, the optimal processing method of support

structure of computer hard disk for mass producing was determined.

2 EXPERIMENTAL METHOD

The original support structure of computer hard disk shows on Figure 1. A two-plate mold is utilized during precision injection molding. The mold cavities of previous part are fabricated by computer numerical control (CNC) process. The mold material is NAK-80. The inlet gate of the two-plate mold has a sidewall pin gate. All experiments use a injection molding machine (220S; ARBURG, Germany). The machine controlling mold temperature is a 300S (REGLOPLAS, Switzerland). This study uses the ABS+PC-C120 HF (GE, USA) material to fabricate the support structure of computer hard disk via precision injection molding. In this research, the authors want to find the optimal processing window for the support structure of computer hard disk. The processing parameters are injection speed (A), injection pressure (B), mold temperature (C), packing time (D), and melt temperature (E). Table 1 shows the process parameter values for the ABS+PC material using precision injection molding.

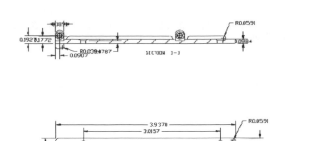
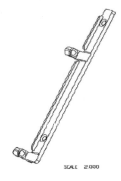

Figure 1. The dimensions of support structure of computer hard disk.

Table 1. Processing parameters for ABS+PC.

Level Parameter	Level 1	Level 2	Level 3	Level 4
A. Injection speed (mm/sec)	97	98	99	100
B. Injection press. (MPa)	55	60	65	70
C. Mold temp. (°C)	60	70	80	90
D. Packing time (Sec.)	3	4	5	6
E. Melt temp. (°C)	215	220	225	230

Because the types and values of processing parameters are too much, this research uses the Taguchi method to get the optimal processing conditions for minimum warpage of the support structure of computer hard disk (L16 orthogonal array).

Analysis of warpage of support structure is the primary task on this study. This research uses the optimal processing to find the relationship between processing parameters and warpage of support structure of computer hard disk. A three-dimension (3D) laser scanner LSH-II-150 (Hawk, Nextc, England) is used to determine the warpage of support structure of computer hard disk.

3 NUMERICAL METHOD

This study uses Moldflow software to simulate the warpage of support structure of computer hard disk. The governing equations for mass, momentum, and energy conservation for a non-isothermal, generalized Newtonian fluid are continuity equation, momentum equation and energy equation. Boundary and initial conditions:

$$\vec{u} = 0; \quad T = T_w \; ; \; \frac{\partial P}{\partial n} = 0 \text{ at } z = \pm h \text{ (on mold wall)} \quad (1)$$

$$\frac{\partial \vec{u}}{\partial z} = \frac{\partial T}{\partial z} = 0 \text{ at } z = 0 \text{ (on centerline)} \quad (2)$$

$$P = 0 \text{ on flow front} \quad (3)$$

$$\vec{u} = u(x, y, z, t) \text{ on inlet} \quad (4)$$

In this study, the governing equations are solved using the control volume finite element method. For details of numerical simulations, see Shen *et al.* [4]. The 3D mesh in Moldflow analysis is utilized to examine the support structure of computer hard disk. The mesh is four-node tetrahedral element. The model has 10500 meshes and 9512 nodes for the support structure. Calculation time for this case is about 357 seconds. A personal computer (PC) that has a Pentium 4 2.8 G CPU, 1 G memory, and 120 G hard drive is used. The Moldflow software is the 5.0 version.

4 RESULTS AND DISCUSSION

The warpage analysis for L9 case on numerical simulation reveals on Figure 2 The maximum warpage demonstrates on the ends of support structure. The value is 0.4849 mm.

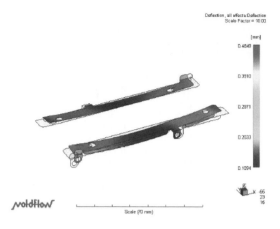

Figure 2. Warpage analysis for L9 case on numerical simulation.

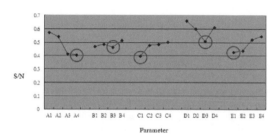

Figure 3. Variation of the S/N ratio with factor level for warpage of support structure.

Figure 3 indicates that the S/N ratio of support structure by precision injection molding. The optimal factor levels that can statistically result in the minimum warpage for precision injection molding are A4B3C1D3E1. These optimized factor levels represent an injection speed of 100 mm/sec, an injection pressure of 65 MPa, a mold temperature of 60°C, a packing time of 5 sec. and a melt temperature of 215°C. The results also appear that the packing time is the most important factor of processing parameter, then is injection speed, then is melt temperature, then is mold temperature and injection pressure is the unimportant factor. The higher injection speed and higher injection pressure can short the filling time of support structure on precision injection molding. This situation also causes the melt temperature of plastic to increase and decrease the viscosity of melt plastic. So the flow situation can be easy on cavity and this causes the smaller warpage of support structure. Suitable difference between melt temperature and mold temperature let the smaller difference of temperature distribution of plastic on filling stage of precision injection molding. This situation may lead the smaller warpage. The suitable higher packing time can add up the more melt plastic going into the cavity of mold

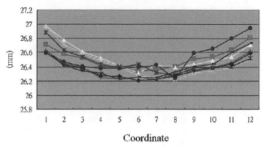

Figure 4. Measurement values (L1~L8).

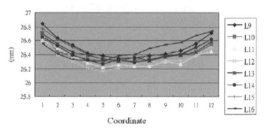

Figure 5. Measurement values (L9~L16).

and supplement the shrinkage due to temperature difference of product. This situation also can reduce the warpage of support structure of computer hard disk. The measurement values of warpage of support structure (group 1, group 2) list on Figure 4 and 5. The results appear that the experiment of L11 has the minimum value and these indicate that the L11 has the good processing. The average warpages of support structure are listed on Table 2 for measurement and numerical simulation. The results appear that the L11 condition has the minimum warpage whether experiment or numerical simulation. The L11 processing conditions include the injection speed of 99 mm/sec, injection pressure of 65 MPa, mold temperature of 60°C, packing time of 4 sec. and melt temperature of 230°C. The values of processing parameter are very close to the optimal processing. L4 condition shows the maximum value of warpage of support structure for experiment and numerical simulation. The L4 processing conditions include the injection speed of 97 mm/sec, injection pressure of 70 MPa, mold temperature of 90°C, packing time of 6 sec. and melt temperature of 230°C. The values of processing parameter are far away the optimal processing. Figure 6 reveals that the warpage difference of support structure between experiment and numerical simulation. The results reveal that the L14 has the maximum difference for experiment and numerical simulation and L11 has the minimum difference between experiment and numerical simulation. This study appears the same results on warpage analysis whether experiment and numerical simulation.

Table 2. Warpage for measurement and numerical simulation.

Measurement	Highest point	Lowest point	Difference	Numerical simulation	Highest point	Lowest point	Difference
L1	0.603	0.21	0.393	L1	0.465	0.065	0.399
L2	0.815	0.338	0.477	L2	0.617	0.155	0.461
L3	0.972	0.323	0.649	L3	0.796	0.177	0.618
L4	0.966	0.212	0.754	L4	1.017	0.204	0.812
L5	0.886	0.286	0.6	L5	0.795	0.183	0.611
L6	0.951	0.244	0.707	L6	0.9	0.204	0.695
L7	0.634	0.223	0.411	L7	0.555	0.146	0.408
L8	0.727	0.248	0.479	L8	0.606	0.15	0.455
L9	0.84	0.348	0.492	L9	0.545	0.069	0.484
L10	0.713	0.302	0.411	L10	0.484	0.079	0.405
L11	0.601	0.224	0.377	L11	0.484	0.109	0.375
L12	0.577	0.167	0.41	L12	0.469	0.084	0.388
L13	0.702	0.312	0.39	L13	0.479	0.097	0.381
L14	0.652	0.075	0.577	L14	0.472	0.09	0.382
L15	0.772	0.308	0.464	L15	0.522	0.07	0.451
L16	0.738	0.309	0.429	L16	0.493	0.092	0.401

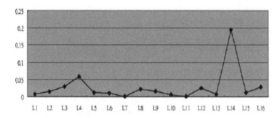

Figure 6. Difference of warpage between experiment and numerical simulation.

5 CONCLUSIONS

The goal of this research is to fabricate support structure of computer hard disk by precision injection molding. The packing time is the most important factor of processing parameters for minimum warpage by precision injection molding. The optimal factor levels in the minimum warpage by precision injection molding are predicted to be A4B3C1D3E1.

REFERENCES

[1] Y.C. Sul, J. Shah and L. Lin: *Implementation and analysis of polymeric microstructure replication by micro injection molding*, J. Micromech. Microeng. Vol. 14 (2004), p. 415–422.
[2] Y.K. Shen: *Comparison of height replication properties of micro-injection molding and micro-injection compression molding for production of microstructures on lightguiding plate*, Plast. Rubber. Compos. Vol. 36 (2007), p. 77–84.
[3] A.W.A.R Wan, T.S. Lee and A.R. Rahmat: *Injection molding simulation analysis of natural fiber composite window frame*, J. Mater. Process. Tech. Vol. 197 (2008), p. 22–30.
[4] Y.C. Chiang, H.C. Cheng, C.F. Huang, J.J. Lee, Y. Lin and Y.K. Shen, *Warpage phenomenon of time-wall injection molding*, Int. J. Adv. Manuf. Tech. Vol. 55 (2011) p. 517–526.

Author index